Lecture Notes in Computer Science   9666

Commenced Publication in 1973
Founding and Former Series Editors:
Gerhard Goos, Juris Hartmanis, and Jan van Leeuwen

More information about this series at http://www.springer.com/series/7410

Marc Fischlin · Jean-Sébastien Coron (Eds.)

Advances in Cryptology – EUROCRYPT 2016

35th Annual International Conference
on the Theory and Applications of Cryptographic Techniques
Vienna, Austria, May 8–12, 2016
Proceedings, Part II

 Springer

Editors
Marc Fischlin
Technische Universität Darmstadt
Darmstadt
Germany

Jean-Sébastien Coron
University of Luxembourg
Luxembourg
Luxembourg

ISSN 0302-9743   ISSN 1611-3349 (electronic)
Lecture Notes in Computer Science
ISBN 978-3-662-49895-8   ISBN 978-3-662-49896-5 (eBook)
DOI 10.1007/978-3-662-49896-5

Library of Congress Control Number: 2016935585

LNCS Sublibrary: SL4 – Security and Cryptology

Printed on acid-free paper

This Springer imprint is published by Springer Nature
The registered company is Springer-Verlag GmbH Berlin Heidelberg

Preface

Eurocrypt 2016, the 35th annual International Conference on the Theory and Applications of Cryptographic Techniques, was held in Vienna, Austria, during May 8–12, 2016. The conference was sponsored by the International Association for Cryptologic Research (IACR). Krzysztof Pietrzak (IST Austria), together with Joël Alwen, Georg Fuchsbauer, Peter Gaži (all IST Austria), and Eike Kiltz (Ruhr-Universität Bochum), were responsible for the local organization. They were supported by a local organizing team consisting of Hamza Abusalah, Chethan Kamath, and Michal Rybár (all IST Austria). We are indebted to them for their support and smooth collaboration.

The conference program followed the now established parallel track system where the works of the authors were presented in two concurrently running tracks. As in the previous edition of Eurocrypt, one track was labeled $\mathcal{R}$ (for real) and the other one was labeled $\mathcal{I}$ (for ideal). Only the invited talks, the tutorial, the best paper, papers with honorable mentions, and the final session of the conference spanned over both tracks.

The proceedings of Eurocrypt contain 62 papers selected from 274 submissions, which corresponds to a record number of submissions in the history of Eurocrypt. Each submission was anonymized for the reviewing process and was assigned to at least three of the 55 Program Committee members. Submissions co-authored by committee members were assigned to at least four members. Committee members were allowed to submit at most one paper, or two if both were co-authored. The reviewing process included a first-round notification followed by a rebuttal for papers that made it to the second round. After extensive deliberations the Program Committee accepted 62 papers. The revised versions of these papers are included in these two-volume proceedings.

The committee decided to give the Best Paper Award to "Tightly Secure CCA-Secure Encryption Without Pairings" by Romain Gay, Dennis Hofheinz, Eike Kiltz, and Hoeteck Wee. The two runners-up to the award, "Indistinguishability Obfuscation from Constant-Degree Graded Encoding Schemes" by Huijia Lin and "Essentially Optimal Robust Secret Sharing with Maximal Corruptions" by Allison Bishop, Valerio Pastro, Rajmohan Rajaraman, Daniel and Wichs, received honorable mentions. All three papers received invitations for the *Journal of Cryptology*.

The program also included invited talks by Karthikeyan Bhargavan, entitled "Protecting Transport Layer Security from Legacy Vulnerabilities", Bart Preneel, entitled "The Future of Cryptography" (IACR distinguished lecture), and Christian Collberg, entitled "Engineering Code Obfuscation." In addition, Emmanuel Prouff gave a tutorial about "Securing Cryptography Implementations in Embedded Systems." All the speakers were so kind as to also provide a short abstract for the proceedings.

We would like to thank all the authors who submitted papers. We know that the Program Committee's decisions, especially rejections of very good papers that did not find a slot among the sparse number of accepted papers, can be very disappointing. We sincerely hope that the rejected works eventually get the attention they deserve.

We are also indebted to the Program Committee members and all external reviewers for their voluntary work, especially since the newly established and unified page limits and the increasing number of submissions induce quite a workload. It has been an honor to work with everyone. The committee's work was tremendously simplified by Shai Halevi's submission software and his support, including running the service on IACR servers.

Finally, we thank everyone else—speakers, session chairs, and rump session chairs —for their contribution to the program of Eurocrypt 2016.

May 2016   Marc Fischlin
                                          Jean-Sébastien Coron

Eurocrypt 2016

The 35th Annual International Conference on the Theory and Applications of Cryptographic Techniques

Vienna, Austria
May 8–12, 2016

Track $\mathcal{I}$

General Chair

Krzysztof Pietrzak   IST Austria

Program Chairs

Marc Fischlin   Technische Universität Darmstadt, Germany
Jean-Sébastien Coron   University of Luxembourg, Luxembourg

Program Committee

Michel Abdalla   Ecole Normale Superieure and CNRS, France
Shweta Agrawal   IIT Delhi, India
Elette Boyle   IDC Herzliya, Israel
Christina Brzuska   TU Hamburg-Harburg, Germany
Ran Canetti   Tel Aviv University, Israel, and Boston University, USA
David Cash   Rutgers University, USA
Dario Catalano   University of Catania, Italy
Jean-Sébastien Coron   University of Luxembourg, Luxembourg
Cas Cremers   University of Oxford, UK
Yevgeniy Dodis   New York University, USA
Nico Döttling   Aarhus University, Denmark
Pooya Farshim   Queen's University Belfast, UK
Jean-Charles Faugère   Inria Paris-Rocquencourt, France
Sebastian Faust   Ruhr University Bochum, Germany
Dario Fiore   IMDEA Software Institute, Spain
Marc Fischlin   TU Darmstadt, Germany
Georg Fuchsbauer   IST, Austria
Juan A. Garay   Yahoo Labs, USA
Vipul Goyal   Microsoft Research, India
Tim Güneysu   University of Bremen, Germany
Shai Halevi   IBM, USA

Goichiro Hanaoka	AIST, Japan
Martin Hirt	ETH Zurich, Switzerland
Dennis Hofheinz	Karlsruhe KIT, Germany
Tibor Jager	Ruhr University Bochum, Germany
Abhishek Jain	Johns Hopkins University, USA
Aniket Kate	Purdue University, USA
Dmitry Khovratovich	University of Luxembourg, Luxembourg
Vadim Lyubashevsky	Ecole Normale Superieure, France
Sarah Meiklejohn	University College London, UK
Mridul Nandi	Indian Statistical Institute, Kolkata, India
María Naya-Plasencia	Inria, France
Svetla Nikova	KU Leuven, Belgium
Adam O'Neill	Georgetown University, USA
Claudio Orlandi	Aarhus University, Denmark
Josef Pieprzyk	Queensland University of Technology, Australia
Mariana Raykova	Yale University, USA
Thomas Ristenpart	Cornell Tech, USA
Matthieu Rivain	CryptoExperts, France
Arnab Roy	Fujitsu Laboratories of America, USA
Benedikt Schmidt	IMDEA Software Institute, Spain
Thomas Schneider	TU Darmstadt, Germany
Berry Schoenmakers	TU Eindhoven, The Netherlands
Peter Schwabe	Radboud University, The Netherlands
Yannick Seurin	ANSSI, France
Thomas Shrimpton	University of Florida, USA
Nigel P. Smart	University of Bristol, UK
John P. Steinberger	Tsinghua University, China
Ron Steinfeld	Monash University, Australia
Emmanuel Thomé	Inria Nancy, France
Yosuke Todo	NTT, Japan
Dominique Unruh	University of Tartu, Estonia
Daniele Venturi	Sapienza University of Rome, Italy
Ivan Visconti	University of Salerno, Italy
Stefan Wolf	USI Lugano, Switzerland

External Reviewers

Divesh Aggarwal	Kazumaro Aoki	Subhadeep Banik
Shashank Agrawal	Afonso Arriaga	Harry Bartlett
Adi Akavia	Gilad Asharov	Lejla Batina
Martin Albrecht	Gilles Van Assche	Carsten Baum
Joël Alwen	Nuttapong Attrapadung	Aemin Baumeler
Prabhanjan Ananth	Christian Badertscher	Christof Beierle
Ewerton Rodrigues	Thomas Baignères	Sonia Belaïd
Andrade	Josep Balasch	Fabrice Benhamouda
Elena Andreeva	Foteini Baldimtsi	David Bernhard

Ritam Bhaumik
Begül Bilgin
Nir Bitansky
Matthieu Bloch
Andrey Bodgnanov
Cecilia Boschini
Vitor Bosshard
Christina Boura
Florian Bourse
Cerys Bradley
Zvika Brakerski
Anne Broadbent
Dan Brown
Seyit Camtepe
Anne Canteaut
Angelo De Caro
Avik Chakraborti
Nishanth Chandran
Melissa Chase
Rahul Chatterjee
Yilei Chen
Jung Hee Cheon
Céline Chevalier
Alessandro Chiesa
Seung Geol Choi
Tom Chothia
Arka Rai Choudhuri
Kai-Min Chung
Yu-Chi Chen
Michele Ciampi
Michael Clear
Aloni Cohen
Ran Cohen
Katriel Cohn-Gordon
Sandro Coretti
Cas Cremers
Dana Dachman-Soled
Yuanxi Dai
Nilanjan Datta
Bernardo Machado David
Gareth T. Davies
Ed Dawson
Jean Paul Degabriele
Martin Dehnel-Wild
Jeroen Delvaux
Grégory Demay

Daniel Demmler
David Derler
Vasil Dimitrov
Yarkin Doroz
Léo Ducas
François Dupressoir
Frederic Dupuis
Avijit Dutta
Stefan Dziembowski
Keita Emura
Antonio Faonio
Serge Fehr
Claus Fieker
Matthieu Finiasz
Viktor Fischer
Jean-Pierre Flori
Pierre-Alain Fouque
Tore Kasper Frederiksen
Tommaso Gagliardoni
Steven Galbraith
David Galindo
Chaya Ganesh
Luke Garratt
Romain Gay
Peter Gaži
Daniel Genkin
Craig Gentry
Hossein Ghodosi
Satrajit Ghosh
Benedikt Gierlichs
Kristian Gjøsteen
Aleksandr Golovnev
Alonso Gonzalez
Dov Gordon
Louis Goubin
Jens Groth
Aurore Guillevic
Sylvain Guilley
Siyao Guo
Divya Gupta
Sourav Sen Gupta
Helene Flyvholm Haagh
Tzipora Halevi
Michael Hamburg
Carmit Hazay
Gottfried Herold

Susan Hohenberger
Justin Holmgren
Pavel Hubacek
Tsung-Hsuan Hung
Christopher Huth
Michael Hutter
Andreas Hülsing
Vincenzo Iovino
Håkon Jacobsen
Aayush Jain
Jérémy Jean
Claude-Pierre Jeannerod
Evan Jeffrey
Ashwin Jha
Daniel Jost
Charanjit Jutla
Ali El Kaafarani
Liang Kaitai
Saqib A. Kakvi
Chethan Kamath
Bhavana Kanukurthi
Pierre Karpman
Elham Kashefi
Tomasz Kazana
Marcel Keller
Dakshita Khurana
Aggelos Kiayias
Paul Kirchner
Elena Kirshanova
Ágnes Kiss
Fuyuki Kitagawa
Ilya Kizhvatov
Thorsten Kleinjung
Vlad Kolesnikov
Venkata Koppala
Luke Kowalczyk
Ranjit Kumaresan
Kaoru Kurosawa
Felipe Lacerda
Virginie Lallemand
Adeline Langlois
Enrique Larraia
Sebastian Lauer
Gregor Leander
Chin Ho Lee
Tancrède Lepoint

Gaëtan Leurent
Benoît Libert
Huijia (Rachel) Lin
Wei-Kai Lin
Bin Liu
Dongxi Liu
Yunwen Liu
Steve Lu
Atul Luykx
Bernardo Magri
Mohammad Mahmoody
Subhamoy Maitra
Hemanta Maji
Giulio Malavolta
Avradip Mandal
Daniel Masny
Takahiro Matsuda
Christian Matt
Willi Meier
Sebastian Meiser
Florian Mendel
Bart Mennink
Eric Miles
Kevin Milner
Ilya Mironov
Arno Mittelbach
Ameer Mohammad
Payman Mohassel
Hart Montgomery
Amir Moradi
François Morain
Paweł Morawiecki
Pedro Moreno-Sanchez
Nicky Mouha
Pratyay Mukherjee
Elke De Mulder
Anderson Nascimento
Muhammad Naveed
Phong Nguyen
Ivica Nikolic
Tobias Nilges
Peter Sebastian Nordholt
Koji Nuida
Maciej Obremski
Frederique Elise Oggier
Emmanuela Orsini

Mohammad Ali
    Orumiehchi
Elisabeth Oswald
Ekin Ozman
Jiaxin Pan
Giorgos Panagiotakos
Omkant Pandey
Omer Paneth
Dimitris Papadopoulos
Kostas Papagiannopoulos
Bryan Parno
Valerio Pastro
Chris Peikert
Ludovic Perret
Leo Paul Perrin
Christophe Petit
Krzysztof Pietrzak
Benny Pinkas
Oxana Poburinnaya
Bertram Poettering
Joop van de Pol
Antigoni Polychroniadou
Manoj Prabhakaran
Thomas Prest
Emmanuel Prouff
Jörn Müller-Quade
Tal Rabin
Kenneth Radke
Carla Rafols
Mario Di Raimondo
Samuel Ranellucci
Pavel Raykov
Francesco Regazzoni
Omer Reingold
Michał Ren
Guénaël Renault
Oscar Reparaz
Vincent Rijmen
Ben Riva
Tim Ruffing
Ulrich Rührmair
Yusuke Sakai
Amin Sakzad
Benno Salwey
Kai Samelin
Yu Sasaki

Alessandra Scafuro
Christian Schaffner
Tobias Schneider
Peter Scholl
Jacob Schuldt
Gil Segev
Nicolas Sendrier
Abhi Shelat
Leonie Simpson
Shashank Singh
Luisa Siniscalchi
Boris Skoric
Ben Smith
Juraj Somorovsky
John Steinberger
Noah
    Stephens-Dawidovitz
Björn Tackmann
Vanessa Teague
Sidharth Telang
R. Seth Terashima
Stefano Tessaro
Adrian Thillard
Susan Thomson
Mehdi Tibouchi
Jacques Traoré
Daniel Tschudi
Hoang Viet Tung
Aleksei Udovenko
Margarita Vald
Maria Isabel Gonzalez
    Vasco
Meilof Veeningen
Vesselin Velichkov
Alexandre Venelli
Muthuramakrishnan
    Venkitasubramaniam
Frederik Vercauteren
Marion Videau
Vinod Vikuntanathan
Gilles Villard
Damian Vizar
Emmanuel Volte
Christine van Vredendaal
Niels de Vreede
Qingju Wang

Bogdan Warinschi
Hoeteck Wee
Carolyn Whitnall
Daniel Wichs
Alexander Wild
David Wu

Jürg Wullschleger
Masahiro Yagisawa
Shota Yamada
Kan Yasuda
Scott Yilek
Kazuki Yoneyama

Ching-Hua Yu
Samee Zahur
Mark Zhandry
Zongyang Zhang
Vassilis Zikas
Michael Zohner

Contents – Part II

Lattices

Leakage

Indifferentiability

Multi-Party Computation II

Obfuscation

Automated Analysis, Functional Encryption, and Non-malleable Codes

Contents – Part I

Masking

Fully Homomorphic Encryption

Cryptanalysis II

Number Theory

Hash Functions

Multilinear Maps

Message Authentication Codes

Attacks on SSL/TLS

Real-World Protocols

Robust Designs

Lattice Reduction

Zero-Knowledge Arguments for Lattice-Based Accumulators: Logarithmic-Size Ring Signatures and Group Signatures Without Trapdoors

Benoît Libert[1](✉), San Ling[2], Khoa Nguyen[2](✉), and Huaxiong Wang[2]

[1] Ecole Normale Supérieure de Lyon, Laboratoire LIP, Lyon, France
benoit.libert@ens-lyon.fr
[2] School of Physical and Mathematical Sciences,
Nanyang Technological University, Singapore
{lingsan,khoantt,hxwang}@ntu.edu.sg

Abstract. An accumulator is a function that hashes a set of inputs into a short, constant-size string while preserving the ability to efficiently prove the inclusion of a specific input element in the hashed set. It has proved useful in the design of numerous privacy-enhancing protocols, in order to handle revocation or simply prove set membership. In the lattice setting, currently known instantiations of the primitive are based on Merkle trees, which do not interact well with zero-knowledge proofs. In order to efficiently prove the membership of some element in a zero-knowledge manner, the prover has to demonstrate knowledge of a hash chain without revealing it, which is not known to be efficiently possible under well-studied hardness assumptions. In this paper, we provide an efficient method of proving such statements using involved extensions of Stern's protocol. Under the Small Integer Solution assumption, we provide zero-knowledge arguments showing possession of a hash chain. As an application, we describe new lattice-based group and ring signatures in the random oracle model. In particular, we obtain: (i) The first lattice-based ring signatures with logarithmic size in the cardinality of the ring; (ii) The first lattice-based group signature that does not require any GPV trapdoor and thus allows for a much more efficient choice of parameters.

1 Introduction

Cryptographic accumulators were introduced by Benaloh and de Mare [10] as alternative to digital signatures in the design of distributed protocols. While initially used in time-stamping and membership testing mechanisms [10], they found numerous applications in the context of fail-stop signatures [7], anonymous credentials [1,19,20,44], group signatures [68], anonymous *ad hoc* authentication [28], digital cash [6,22,54], set membership proofs [63,69] or authenticated data structures [59,60] (see [27] for further examples).

In a nutshell, an accumulator is a sort of algebraic hash function that maps a large set R of inputs into a short, constant-size accumulator value u such that an efficiently computable short witness w provides evidence that a given input was

© International Association for Cryptologic Research 2016
M. Fischlin and J.-S. Coron (Eds.): EUROCRYPT 2016, Part II, LNCS 9666, pp. 1–31, 2016.
DOI: 10.1007/978-3-662-49896-5_1

indeed incorporated into the hashed set. In order to be useful, the size of the witness should be much smaller than the cardinality of the input set. An extension, suggested by Camenisch and Lysyanskaya [20], allows the accumulator value to be updated over time, by adding or deleting elements of the hashed set while preserving the ability to efficiently update witnesses. For most applications, the usual security requirement mandates the infeasibility of computing an accumulator value u and a valid witness w for an element x outside the set of hashed inputs. This is made possible by public-key techniques like the existence of a trapdoor (e.g., the factorization of an RSA modulus or the discrete logarithm of some public group element) hidden behind public parameters.

So far, number theoretic realizations have been divided into two main families. The first one relies on groups of hidden order [7,10,15,47] and includes proposals based on the Strong RSA assumption [7,43]. The second main family [19,57] was first explored by Nguyen [57] and appeals to bilinear maps (a.k.a. pairings) and assumptions of variabe size like the Strong Diffie-Hellman assumption [14]. Strong-RSA-based candidates enjoy the advantage of short public parameters and they easily extend into universal accumulators [43] (where non-membership witnesses can show that a given input was not accumulated). While pairing-based schemes [19,57] usually require linear-size public parameters in the number of elements to be hashed, they are useful in applications [6,22] where we want to limit the number of elements to be hashed. A third family (e.g., [59]) of constructions relies on Merkle trees [50] rather than number theoretic assumptions. Its main disadvantage is that the use of hash trees makes it hardly compatible with efficient zero-knowledge proofs, which are inevitable ingredients of privacy-preserving protocols [1,19,20,68]. In fact, currently known methods [9,15] for reconciling Merkle trees and zero-knowledge proofs require non-standard assumptions in groups of hidden order [15] or the machinery of SNARKs, which inherently rely on non-falsifiable [55] knowledge assumptions [35].

Despite its wide range of applications, the accumulator primitive still has a relatively small number of efficient realizations. For the time being, most known solutions require non-standard *ad hoc* assumptions like Strong RSA or Strong Diffie-Hellman. To our knowledge, the only exception is a generic construction from vector commitments [24], which leaves open the problem of candidates based on the standard Computational Diffie-Hellman assumption (in groups without a bilinear map) or zero-knowledge-friendly lattice-based schemes. In this paper, we describe a new construction based on standard lattice assumptions which interacts nicely with zero-knowledge proofs despite the use of Merkle trees. We show that this new construction enables new, unexpected applications to the design of lattice-based ring signatures and group signatures.

OUR CONTRIBUTIONS. We describe a lattice-based accumulator[1] that enables short zero-knowledge arguments of membership. Our construction relies on a

---

[1] A lattice-based accumulator was previously claimed in [38]. However, the generation of witnesses can only be performed using the secret key of the system. Moreover, their scheme is seemingly not compact due to the required choice of parameters.

Merkle hash tree which is computed in a special way that makes it compatible with efficient protocols for proving possession of a secret value (i.e., a leaf of the tree) that is properly accumulated in the root of the tree. More specifically, our system allows demonstrating the knowledge of a hash chain from the considered secret leaf to the root in a zero-knowledge manner. This building block enables many interesting applications. In particular, we use it to design lattice-based ring and group signatures with dramatic improvements over the existing constructions. In the random oracle model, we obtain:

- The first lattice-based ring signature with logarithmic signature size in the cardinality of the ring. So far, all suggested proposals have linear size in the number of ring members.
- A lattice-based group signature with much shorter public key, signature length, and weaker hardness assumptions than all earlier realizations.

Our ring signature does not require any other setup assumption than having all users agree on a modulus q, a lattice dimension n and a random matrix $\mathbf{A} \in \mathbb{Z}_q^{n \times m}$ (which can be derived from a random oracle). It provably satisfies the strong security definitions put forth by Bender, Katz and Morselli [11].

Our group signature is analyzed in the setting of static groups using the definitions of Bellare, Micciancio and Warinschi [8]. Its salient feature (which it shares with our ring signature) is that, unlike all earlier candidates [33,41,42,46, 58], it does not require the use of a trapdoor (as defined by Gentry, Peikert and Vaikuntanathan [31]) consisting of a short basis of some lattice. It thus eliminates one of the frequently cited reasons [49] for which lattice-based signatures tend to be impractical. In fact, our group signature departs from previously used design principles – which are all inspired in some way by the general construction of [8] – in that, surprisingly, it does not even require an ordinary digital signature to begin with. All we need is a lattice-based accumulator with a compatible zero-knowledge argument system for arguing knowledge of a hash chain.

OUR TECHNIQUES. Our accumulator proceeds by computing a Merkle tree using a hash function based on the Small Integer Solution (SIS) problem, which is a variant of the hash functions considered in [4,32,53] previously considered by Papamanthou et al. [59]. Instead of hashing a vector $\mathbf{x} \in \{0,1\}^m$ by computing its syndrome $\mathbf{A} \cdot \mathbf{x} \in \mathbb{Z}_q^n$ via a random matrix $\mathbf{A} \in \mathbb{Z}_q^{n \times m}$, it outputs the coordinate-wise binary decomposition $\mathsf{bin}(\mathbf{A} \cdot \mathbf{x} \mod q) \in \{0,1\}^{m/2}$ of the syndrome to obtain the two-fold compression factor that is needed for iteratively applying the function in a Merkle tree. However, Papamanthou et al. [59] did not consider the problem of proving knowledge of a hash chain in a zero-knowledge fashion. The main technical novelty that we introduce is thus a method for demonstrating knowledge of a Merkle-tree hash chain using the framework of Stern's protocol [67].

Using this method, we build ring and group signatures with logarithmic size in the number of ring or group members involved. Our constructions are conceptually simple. Each user's private key is a random m-bit vector $\mathbf{x} \in \{0,1\}^m$ and the matching public key is the binary expansion $\mathbf{d} = \mathsf{bin}(\mathbf{A} \cdot \mathbf{x} \mod q) \in \{0,1\}^{m/2}$

of the corresponding syndrome. In order to sign a message, the user considers an accumulation $\mathbf{u} \in \{0,1\}^{m/2}$ of all users' public keys $R = (\mathbf{d}_0, \ldots, \mathbf{d}_{N-1})$ – which is obtained by dynamically forming the ring R in the ring signature and simply consists of the group public key in the group signature – and generates a Stern-type argument that: (i) His public key $\mathbf{d}_j$ belongs to the hashed set R; (ii) He knows the underlying secret $\mathbf{d}_j = \mathsf{bin}(\mathbf{A} \cdot \mathbf{x}_j \bmod q)$; (iii) – for the group signature) He has honestly encrypted the binary representation of the integer j determining his position in the tree to a ciphertext attached in the signature. In order to acquire anonymity in the strongest sense (i.e., where the adversary is granted access to a signature opening oracle), we apply the Naor-Yung paradigm [56] to Regev's cryptosystem [64], as was previously considered in [12]. As pointed out earlier, the advantage of not relying on an ordinary digital signature[2] lies in that it does not require any party (i.e., neither the group manager nor the group members in the case of group signatures) to have a GPV trapdoor [31] consisting of a short lattice basis. As emphasized by Lyubashevsky [49], explicitly avoiding the use of such trapdoors allows for drastically more efficient choices of parameters. As by-products, our scheme features much smaller group public key and users' secret keys, produces shorter signatures, and relies on weaker hardness assumptions than all of the existing lattice-based group signature schemes [21,33,41,46,58] in the BMW model [8].

In the following, we give an estimated efficiency comparison among our group signature and the previous 2 most efficient schemes with CCA-anonymity, by Ling et al. [46] and Nguyen et al. [58]. The estimations are done with parameter $n = 2^8$, group size $N = 1024$, and soundness error 2^{-80} for the NIZKs.

- Ling et al.'s scheme requires $q = \mathcal{O}(\log N \cdot n^2)$, $m \geq 2n \log q$, so we set $q = 2^{18}$ and $m = 2^9 \cdot 18$. The infinity norm bound for discrete Gaussian samples is 2^6. The scheme produces group public key size 65.8 MB; user's secret key size 13.5 KB (a Boyen signature [17]); and signature size 1.20 GB.
- Nguyen et al.'s scheme requires $q > m^{8.5}$, $m \geq 2n \log q$, so we set $q = 2^{142}$ and $m = 2^9 \cdot 142$. The scheme produces group public key size 2.15 GB; user's secret key size 90 GB (a trapdoor in $\mathbb{Z}^{3m \times 3m}$ with $(\log m)$-bit entries); and signature size 500 MB.
- Our scheme works with $q = 2^8$, $m = 2^9 \cdot 8$, and parameters $p = 32719$, $m_E = 7980$ for the encryption layer. The scheme features public key size 4.9 MB; user's secret key size 3.25 KB; and it produces signatures of size 61.5 MB.

RELATED WORK. While originally suggested as a 3-move code-based identification scheme, Stern's protocol was adapted to the lattice setting by Kawachi et al. [40] and extended by Ling et al. [45] into an argument system for the Inhomogeneous Small Integer Solution (ISIS) problem. In particular, Ling et al. gave a method, called *decomposition-extension* framework, which allows arguing knowledge of an integer vector $\mathbf{x} \in \mathbb{Z}^m$ of norm $\|\mathbf{x}\|_\infty \leq \beta$ such that $\mathbf{A} \cdot \mathbf{x} = \mathbf{u} \in \mathbb{Z}_q^n$

---

[2] Recall that all $\mathcal{O}(\log N)$-size group signatures employ a signature scheme in the standard model (for which all known constructions use trapdoors) in order to smoothly interact with zero-knowledge proofs.

without leaving any gap between the vector computed by the knowledge extractor and the actual witness $\mathbf{x}$. As shown in [46], the technique of Ling et al. [45] can be used to prove more involved statements such as the possession of a Boyen signature [17] on a message encrypted by a dual Regev ciphertext [31]. Here, we take one step further and develop a zero-knowledge argument of knowledge (ZKAoK) that a specific element of some universe belongs to a hashed set.

Ring signatures were introduced by Rivest, Shamir and Tauman-Kalai [65] with the motivation of hiding the identity of a source (e.g., a whistleblower in a political scandal) while providing guarantees of trustworthiness. Bender, Katz and Morselli [11] gave stringent security definitions while constructions with sub-linear signature size were given by Chandran, Groth and Sahai [25]. The celebrated results of Gentry, Peikert and Vaikuntanathan [31] inspired a number of lattice-based ring signatures. The state-of-the-art construction probably stems from the framework of Brakerski and Tauman-Kalai [18], which results in linear-size in the number of ring members. The same holds for all known Fiat-Shamir-like lattice-based ring signatures (e.g., [2,40]), although some of them do not require a trapdoor. Thus far, the only logarithmic-size ring signatures [16,36] arise from the results of Groth and Kohlweiss [36] and it is not clear how to extend them to the lattice setting.

The notion of group signatures dates back to Chaum and Van Heyst [26]. While viable constructions were given in the seminal paper by Ateniese, Camenisch, Joye and Tsudik [5], their security notions remained poorly understood until the work of Bellare, Micciancio and Warinschi [8]. The first lattice-based proposal came out with the results of Gordon, Katz and Vaikuntanathan [33], which inspired a number of follow-up works describing new systems with a better asymptotic efficiency [41,46,58] or additional properties [21,42]. For the time being, the most efficient candidates are the recent concurrent proposals of Nguyen et al. and Ling et al. [46,58]. As it turns out, except for one scheme [12] that mixes lattice-based and discrete-logarithm-related assumptions, all currently available candidates [21,41,42,46,58] utilize a GPV trapdoor, either to perform the setup of the system or to trace signatures (or both). Our results thus provide the first system that completely eliminates GPV trapdoors.

At a high level, our ZKAoK system is partially inspired by the way Langlois et al. [42] made use of the Bonsai tree technique [23] since it proves knowledge of a solution to a SIS problem determined by the user's position in a tree. However, there are fundamental differences since our tree is built in a bottom-up (rather than top-down) manner and we do not perform any trapdoor delegation.

2 Preliminaries

NOTATIONS. We assume that all vectors are column vectors. The concatenation of matrices $\mathbf{A} \in \mathbb{Z}^{k \times i}$, $\mathbf{B} \in \mathbb{Z}^{k \times j}$ is denoted by $[\mathbf{A}|\mathbf{B}] \in \mathbb{Z}^{k \times (i+j)}$. For $b \in \{0, 1\}$, we denote the bit $1 - b \in \{0, 1\}$ by $\bar{b}$. For a positive integer i, we let $[i]$ be the set $\{1, \ldots, i\}$. If S is a finite set, $x \xleftarrow{\$} S$ means that x is chosen uniformly at random from S. All logarithms are of base 2. The addition in $\mathbb{Z}_2$ is denoted by $\oplus$.

In this section, we first recall the average-case lattice problems SIS and LWE, together with their hardness results; and the notion of statistical zero-knowledge arguments of knowledge. The definitions and security requirements of cryptographic accumulators, ring signatures, and group signatures are deferred to their respective Sects. 3, 4, and 5.

2.1   Average-Case Lattice Problems

Definition 1 ([3,31]). *The* $\mathsf{SIS}^\infty_{n,m,q,\beta}$ *problem is as follows: Given uniformly random matrix* $\mathbf{A} \in \mathbb{Z}_q^{n \times m}$, *find a non-zero vector* $\mathbf{x} \in \mathbb{Z}^m$ *such that* $\|\mathbf{x}\|_\infty \leq \beta$ *and* $\mathbf{A} \cdot \mathbf{x} = \mathbf{0} \bmod q$.

If $m, \beta = \mathsf{poly}(n)$, and $q > \beta \cdot \widetilde{\mathcal{O}}(\sqrt{n})$, then the $\mathsf{SIS}^\infty_{n,m,q,\beta}$ problem is at least as hard as the worst-case lattice problem SIVP_γ for some $\gamma = \beta \cdot \widetilde{\mathcal{O}}(\sqrt{nm})$ (see [31,52]). Specifically, when $\beta = 1$, $q = \widetilde{\mathcal{O}}(n)$, $m = 2n\lceil \log q \rceil$, the $\mathsf{SIS}^\infty_{n,m,q,1}$ problem is at least as hard as $\mathsf{SIVP}_{\widetilde{\mathcal{O}}(n)}$.

In the last decade, numerous SIS-based cryptographic primitives have been proposed. In this work, we will extensively employ 2 such constructions:

- Our Merkle tree accumulator is built upon a specific family of collision-resistant hash functions, which is a syntactic modification (*i.e.*, it takes two inputs, instead of one) of the one presented in [3,53]. A similar scheme that works with larger SIS norm bound β was proposed in [59].
- Our zero-knowledge argument systems use the statistically hiding and computationally binding string commitment scheme from [40].

For appropriate setting of parameters, the security of the above two constructions can be based on the worst-case hardness of $\mathsf{SIVP}_{\widetilde{\mathcal{O}}(n)}$.

In the group signature in Sect. 5, we will employ the multi-bit version of Regev's encryption scheme [64], presented in [39,62]. The scheme is based on the hardness of the LWE problem.

Definition 2 ([64]). *Let* $n, m_E \geq 1$, $p \geq 2$, *and let* χ *be a probability distribution on* $\mathbb{Z}$. *For* $\mathbf{s} \in \mathbb{Z}_p^n$, *let* $A_{\mathbf{s},\chi}$ *be the distribution obtained by sampling* $\mathbf{a} \xleftarrow{\$} \mathbb{Z}_q^n$ *and* $e \hookleftarrow \chi$, *and outputting* $(\mathbf{a}, \mathbf{s}^\top \cdot \mathbf{a} + e) \in \mathbb{Z}_p^n \times \mathbb{Z}_p$. *The* $\mathsf{LWE}_{n,p,\chi}$ *problem asks to distinguish* m_E *samples chosen according to* $A_{\mathbf{s},\chi}$ (*for* $\mathbf{s} \xleftarrow{\$} \mathbb{Z}_p^n$) *and* m_E *samples chosen according to the uniform distribution over* $\mathbb{Z}_p^n \times \mathbb{Z}_p$.

If p is a prime power, χ is the discrete Gaussian distribution $D_{\mathbb{Z},\alpha p}$, where $\alpha p \geq 2\sqrt{n}$, then $\mathsf{LWE}_{n,p,\chi}$ is as least as hard as $\mathsf{SIVP}_{\widetilde{\mathcal{O}}(n/\alpha)}$ (see [51,52,61,64]).

2.2   Zero-Knowledge Arguments of Knowledge

We will work with statistical zero-knowledge argument systems, namely, interactive protocols where the zero-knowledge property holds against *any* cheating verifier, while the soundness property only holds against *computationally bounded* cheating provers. More formally, let the set of statements-witnesses $R = \{(y, w)\} \in \{0, 1\}^* \times \{0, 1\}^*$ be an NP relation. A two-party game $\langle \mathcal{P}, \mathcal{V} \rangle$ is called an interactive argument system for the relation R with soundness error e if the following two conditions hold:

- Completeness. If $(y, w) \in R$ then $\Pr\left[\langle \mathcal{P}(y, w), \mathcal{V}(y) \rangle = 1\right] = 1$.
- Soundness. If $(y, w) \notin R$, then $\forall$ PPT $\widehat{\mathcal{P}}$: $\Pr[\langle \widehat{\mathcal{P}}(y, w), \mathcal{V}(y) \rangle = 1] \leq e$.

An argument system is called statistical zero-knowledge if for any $\widehat{\mathcal{V}}(y)$, there exists a PPT simulator $\mathcal{S}(y)$ producing a simulated transcript that is statistically close to the one of the real interaction between $\mathcal{P}(y, w)$ and $\widehat{\mathcal{V}}(y)$. A related notion is argument of knowledge, which requires the witness-extended emulation property. For protocols consisting of 3 moves (*i.e.*, commitment-challenge-response), witness-extended emulation is implied by *special soundness* [34], where the latter assumes that there exists a PPT extractor which takes as input a set of valid transcripts with respect to all possible values of the 'challenge' to the same 'commitment', and outputs w' such that $(y, w') \in R$.

The statistical zero-knowledge arguments of knowledge (sZKAoK) presented in this work are Stern-type [67]. In particular, they are Σ-protocols in the generalized sense defined in [12,37] (where 3 valid transcripts are needed for extraction, instead of just 2). Several recent works rely on Stern-type protocols to design lattice-based [42,45,46] and code-based [29,37] constructions.

3   A Lattice-Based Accumulator with Supporting Zero-Knowledge Argument of Knowledge

Throughout the paper, we will work with positive integers n, q, k, m, where: n is the security parameter; $q = \widetilde{\mathcal{O}}(n)$; $k = \lceil \log q \rceil$; and $m = 2nk$. We identify $\mathbb{Z}_q$ by the set $\{0, \ldots, q - 1\}$. We define the "powers-of-2" matrix

$$
\mathbf{G} = \begin{bmatrix} 1\ 2\ 4\ \ldots\ 2^{k-1} & & & \\ & 1\ 2\ 4\ \ldots\ 2^{k-1} & & \\ & & \ldots & \\ & & & 1\ 2\ 4\ \ldots\ 2^{k-1} \end{bmatrix} \in \mathbb{Z}_q^{n \times nk}.
$$

Note that for every $\mathbf{v} \in \mathbb{Z}_q^n$, we have $\mathbf{v} = \mathbf{G} \cdot \mathsf{bin}(\mathbf{v})$, where $\mathsf{bin}(\mathbf{v}) \in \{0, 1\}^{nk}$ denotes the binary representation of $\mathbf{v}$.

3.1   Cryptographic Accumulators

An *accumulator scheme* is a tuple of algorithms (TSetup, TAcc, TWitness, TVerify) defined as follows:

TSetup(n) On input security parameter n, output the public parameter pp.
TAcc$_{pp}$ On input a set $R = \{\mathbf{d}_0, \ldots, \mathbf{d}_{N-1}\}$ of N data values, output an accumulator value $\mathbf{u}$.

TWitness$_{pp}$ On input a data set R and a value $\mathbf{d}$, output $\perp$ if $\mathbf{d} \notin R$; otherwise output a witness w for the fact that $\mathbf{d}$ is accumulated in TAcc(R). (Typically, the size of w should be short (e.g., constant or logarithmic in N) to be useful.)

TVerify$_{pp}$ On input accumulator value $\mathbf{u}$ and a value-witness pair $(\mathbf{d}, w)$, output 1 (which indicates that $(\mathbf{d}, w)$ is valid for the accumulator $\mathbf{u}$) or 0.

An accumulator scheme is called correct if for all $pp \leftarrow$ TSetup(n), we have TVerify$_{pp}$(TAcc$_{pp}$(R), $\mathbf{d}$, TWitness$_{pp}$(R, $\mathbf{d}$)) = 1 for all $\mathbf{d} \in R$.

The security of an accumulator scheme, as defined in [7,20], says that it is infeasible to prove that a value $\mathbf{d}^*$ was accumulated in a value $\mathbf{u}$ if it was not. This property is formalized as follows.

Definition 3. *An accumulator scheme* (TSetup, TAcc, TWitness, TVerify) *is called secure if for all PPT adversaries $\mathcal{A}$:*

$$\Pr\big[pp \leftarrow \mathsf{TSetup}(n); (R, \mathbf{d}^*, w^*) \leftarrow \mathcal{A}(pp) :$$

$$\mathbf{d}^* \notin R \wedge \mathsf{TVerify}_{pp}(\mathsf{TAcc}_{pp}(R), \mathbf{d}^*, w^*) = 1\big] = \mathsf{negl}(n).$$

3.2   A Family of Lattice-Based Collision-Resistant Hash Functions

We now describe the specific family of lattice-based collision-resistant hash functions, upon which our Merkle hash tree will be built.

Definition 4. *The function family $\mathcal{H}$ mapping $\{0,1\}^{nk} \times \{0,1\}^{nk}$ to $\{0,1\}^{nk}$ is defined as $\mathcal{H} = \{h_{\mathbf{A}} \mid \mathbf{A} \in \mathbb{Z}_q^{n \times m}\}$, where for $\mathbf{A} = [\mathbf{A}_0 | \mathbf{A}_1]$ with $\mathbf{A}_0, \mathbf{A}_1 \in \mathbb{Z}_q^{n \times nk}$, and for any $(\mathbf{u}_0, \mathbf{u}_1) \in \{0,1\}^{nk} \times \{0,1\}^{nk}$, we have:*

$$h_{\mathbf{A}}(\mathbf{u}_0, \mathbf{u}_1) = \mathsf{bin}\big(\mathbf{A}_0 \cdot \mathbf{u}_0 + \mathbf{A}_1 \cdot \mathbf{u}_1 \bmod q\big) \in \{0,1\}^{nk}.$$

Note that $h_{\mathbf{A}}(\mathbf{u}_0, \mathbf{u}_1) = \mathbf{u} \Leftrightarrow \mathbf{A}_0 \cdot \mathbf{u}_0 + \mathbf{A}_1 \cdot \mathbf{u}_1 = \mathbf{G} \cdot \mathbf{u} \bmod q.$

Lemma 1. *The function family $\mathcal{H}$, defined in 4 is collision-resistant, assuming the hardness of the $\mathsf{SIVP}_{\widetilde{\mathcal{O}}(n)}$ problem.*

Proof. Given $\mathbf{A} = [\mathbf{A}_0 | \mathbf{A}_1] \xleftarrow{\$} \mathbb{Z}_q^{n \times m}$, if one can find two *distinct* pairs $(\mathbf{u}_0, \mathbf{u}_1) \in \big(\{0,1\}^{nk}\big)^2$ and $(\mathbf{v}_0, \mathbf{v}_1) \in \big(\{0,1\}^{nk}\big)^2$ such that $h_{\mathbf{A}}(\mathbf{u}_0, \mathbf{u}_1) = h_{\mathbf{A}}(\mathbf{v}_0, \mathbf{v}_1) \bmod q$, then one can obtain a *non-zero* vector $\mathbf{z} = \begin{pmatrix} \mathbf{u}_0 - \mathbf{v}_0 \\ \mathbf{u}_1 - \mathbf{v}_1 \end{pmatrix} \in \{-1, 0, 1\}^m$ such that $\mathbf{A} \cdot \mathbf{z} = \mathbf{A}_0 \cdot (\mathbf{u}_0 - \mathbf{v}_0) + \mathbf{A}_1 \cdot (\mathbf{u}_1 - \mathbf{v}_1) = \mathbf{G} \cdot h_{\mathbf{A}}(\mathbf{u}_0, \mathbf{u}_1) - \mathbf{G} \cdot h_{\mathbf{A}}(\mathbf{v}_0, \mathbf{v}_1) = \mathbf{0} \bmod q.$

In other words, $\mathbf{z}$ is a valid solution to the $\mathsf{SIS}_{n,m,q,1}^\infty$ problem associated with matrix $\mathbf{A}$. The lemma then follows from the worst-case to average-case reduction from $\mathsf{SIVP}_{\widetilde{\mathcal{O}}(n)}$. $\square$

3.3 Our Merkle-Tree Accumulator

We now give the construction of a Merkle tree with $N = 2^\ell$ leaves, where ℓ is a positive integer, based on the family of lattice-based hash function $\mathcal{H}$ defined above.

TSetup(n). Sample $\mathbf{A} \xleftarrow{\$} \mathbb{Z}_q^{n \times m}$, and output $pp = \mathbf{A}$.

TAcc$_\mathbf{A}(R = \{\mathbf{d}_0 \in \{0,1\}^{nk}, \ldots, \mathbf{d}_{N-1} \in \{0,1\}^{nk}\})$. For every $j \in [0, N-1]$, let $(j_1, \ldots, j_\ell) \in \{0,1\}^\ell$ be the binary representation of j, and let $\mathbf{d}_j = \mathbf{u}_{j_1, \ldots, j_\ell}$. Form the tree of depth $\ell = \log N$ based on the N leaves $\mathbf{u}_{0,0,\ldots,0}, \ldots, \mathbf{u}_{1,1,\ldots,1}$ as follows:

1. At depth $i \in [\ell]$, the node $\mathbf{u}_{b_1,\ldots,b_i} \in \{0,1\}^{nk}$, for all $(b_1, \ldots, b_i) \in \{0,1\}^i$, is defined as $h_\mathbf{A}(\mathbf{u}_{b_1,\ldots,b_i,0}, \mathbf{u}_{b_1,\ldots,b_i,1})$.

2. At depth 0: The root $\mathbf{u} \in \{0,1\}^{nk}$ is defined as $h_\mathbf{A}(\mathbf{u}_0, \mathbf{u}_1)$.

The algorithm outputs the accumulator value $\mathbf{u}$.

TWitness$_\mathbf{A}(R, \mathbf{d})$. If $\mathbf{d} \notin R$, return $\perp$. Otherwise, $\mathbf{d} = \mathbf{d}_j$ for some $j \in [0, N-1]$ with binary representation $(j_1, \ldots, j_\ell)$. Output the witness w defined as:

$$w = ((j_1, \ldots, j_\ell), (\mathbf{u}_{j_1,\ldots,j_{\ell-1},\bar{j}_\ell}, \ldots, \mathbf{u}_{j_1,\bar{j}_2}, \mathbf{u}_{\bar{j}_1})) \in \{0,1\}^\ell \times (\{0,1\}^{nk})^\ell,$$

for $\mathbf{u}_{j_1,\ldots,j_{\ell-1},\bar{j}_\ell}, \ldots, \mathbf{u}_{j_1,\bar{j}_2}, \mathbf{u}_{\bar{j}_1}$ computed by algorithm TAcc$_\mathbf{A}(R)$.

TVerify$_\mathbf{A}(\mathbf{u}, \mathbf{d}, w)$. Let the given witness w be of the form:

$$w = ((j_1, \ldots, j_\ell), (\mathbf{w}_\ell, \ldots, \mathbf{w}_1)) \in \{0,1\}^\ell \times (\{0,1\}^{nk})^\ell.$$

The algorithm recursively computes the path $\mathbf{v}_\ell, \mathbf{v}_{\ell-1}, \ldots, \mathbf{v}_1, \mathbf{v}_0 \in \{0,1\}^{nk}$ as follows: $\mathbf{v}_\ell = \mathbf{d}$ and

$$\forall i \in \{\ell-1, \ldots, 1, 0\} : \mathbf{v}_i = \begin{cases} h_\mathbf{A}(\mathbf{v}_{i+1}, \mathbf{w}_{i+1}), & \text{if } j_{i+1} = 0; \\ h_\mathbf{A}(\mathbf{w}_{i+1}, \mathbf{v}_{i+1}), & \text{if } j_{i+1} = 1. \end{cases}$$

Then it returns 1 if $\mathbf{v}_0 = \mathbf{u}$. Otherwise, it returns 0.

In Fig. 1, we give an illustrative example of a tree with $2^3 = 8$ leaves.

One can check that the above Merkle-tree accumulator scheme is correct. Furthermore, its security is based on the collision-resistance of the hash function family $\mathcal{H}$, which in turn is based on the hardness of SIVP$_{\tilde{\mathcal{O}}(n)}$.

Theorem 1. *The given accumulator scheme is secure in the sense of Definition 3, assuming the hardness of the* SIVP$_{\tilde{\mathcal{O}}(n)}$ *problem.*

Proof. Assuming that there exists a PPT adversary $\mathcal{B}$ who has non-negligible success probability in the security experiment of Definition 3. It receives a uniformly random matrix $\mathbf{A} \in \mathbb{Z}_q^{n \times m}$ generated by TSetup(n), and returns $(R = (\mathbf{d}_0, \ldots, \mathbf{d}_{N-1}), \mathbf{d}^*, w^*)$ such that $\mathbf{d}^* \notin R$ and TVerify$_\mathbf{A}(\mathbf{u}^*, \mathbf{d}^*, w^*) = 1$, where $\mathbf{u}^* = $ TAcc$_\mathbf{A}(R)$.

Parse $w^* = ((j_1^*, \ldots, j_\ell^*), (\mathbf{w}_\ell^*, \ldots, \mathbf{w}_1^*))$. Let $j^* \in [0, N-1]$ be the integer having binary representation $(j_1^*, \ldots, j_\ell^*)$ and let $\mathbf{u}_{j_1^*,\ldots,j_\ell^*} = \mathbf{d}_{j^*}, \mathbf{u}_{j_1^*,\ldots,j_{\ell-1}^*}, \ldots, \mathbf{u}_{j_1^*}, \mathbf{u}^*$

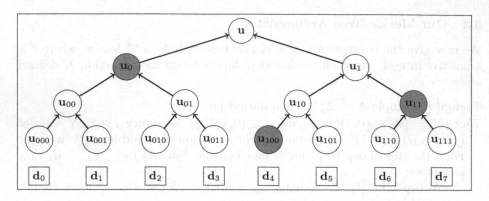

Fig. 1. A Merkle tree with $2^3 = 8$ leaves, which accumulates the data blocks $\mathbf{d}_0, \ldots, \mathbf{d}_7$ into the value $\mathbf{u}$ at the root. The bit string (101) and the gray nodes form a witness to the fact that $\mathbf{d}_5$ is accumulated in $\mathbf{u}$.

be the path from the leave $\mathbf{d}_{j^*}$ to the root of the tree generated by $\mathsf{TAcc}_{\mathbf{A}}(R)$. On the other hand, let $\mathbf{v}_\ell^* = \mathbf{d}^*, \mathbf{v}_{\ell-1}^*, \ldots, \mathbf{v}_1^*, \mathbf{v}_0^* = \mathbf{u}^*$ be the path computed by algorithm $\mathsf{TVerify}_{\mathbf{A}}(\mathbf{u}^*, \mathbf{d}^*, w^*)$. Note that $\mathbf{d}^* \neq \mathbf{d}_{j^*}$ since $\mathbf{d}^* \notin R$. Thus, comparing the two paths, we can find the smallest integer $k \in [\ell]$, such that $\mathbf{v}_k^* \neq \mathbf{u}_{j_1^*, \ldots, j_k^*}$. We then obtain a collision for $h_{\mathbf{A}}$ at the parent node of $\mathbf{u}_{j_1^*, \ldots, j_k^*}$. The theorem then follows from Lemma 1.   □

3.4   Zero-Knowledge AoK of an Accumulated Value

Our goal in this section is to construct a zero-knowledge argument system that allows prover $\mathcal{P}$ to convince verifier $\mathcal{V}$ that $\mathcal{P}$ knows a secret value that is properly accumulated into the root of the lattice-based Merkle tree described above. More formally, in our protocol, $\mathcal{P}$ convinces $\mathcal{V}$ on input $(\mathbf{A}, \mathbf{u})$ that $\mathcal{P}$ possesses a value-witness pair $(\mathbf{d}, w)$ such that $\mathsf{TVerify}_{\mathbf{A}}(\mathbf{u}, \mathbf{d}, w) = 1$. The associated relation $\mathrm{R}_{\mathrm{acc}}$ is defined as follows.

Definition 5

$$\mathrm{R}_{\mathrm{acc}} = \Big\{ \big((\mathbf{A}, \mathbf{u}) \in \mathbb{Z}_q^{n \times m} \times \{0, 1\}^{nk}; \mathbf{d} \in \{0, 1\}^{nk}, w \in \{0, 1\}^\ell \times (\{0, 1\}^{nk})^\ell \big) :$$

$$\mathsf{TVerify}_{\mathbf{A}}(\mathbf{u}, \mathbf{d}, w) = 1 \Big\}.$$

Before going into the details, we first introduce several supporting notations and techniques.

– We denote by B_m^{nk} the set of all vectors in $\{0, 1\}^m$ that have Hamming weight nk; and by $\mathcal{S}_m$ the set of all permutations of m elements.
– For $i \in \{nk, m\}$, for $b \in \{0, 1\}$ and for $\mathbf{v} \in \{0, 1\}^i$, we let $\mathsf{ext}(b, \mathbf{v})$ denote the vector $\mathbf{z} \in \{0, 1\}^{2i}$ of the form $\mathbf{z} = \begin{pmatrix} \bar{b} \cdot \mathbf{v} \\ b \cdot \mathbf{v} \end{pmatrix}$.

- For $b \in \{0,1\}$, for $\pi \in \mathcal{S}_m$, we define the permutation $F_{b,\pi}$ that transforms $\mathbf{z} = \begin{pmatrix} \mathbf{z}_0 \\ \mathbf{z}_1 \end{pmatrix} \in \mathbb{Z}_q^{2m}$ consisting of 2 blocks of size m into $F_{b,\pi}(\mathbf{z}) = \begin{pmatrix} \pi(\mathbf{z}_b) \\ \pi(\mathbf{z}_{\bar{b}}) \end{pmatrix}$. Namely, $F_{b,\pi}$ first rearranges the blocks of $\mathbf{z}$ according to b (it keeps the arrangement of blocks if $b = 0$, or swaps them if $b = 1$), then it permutes each block according to π.

Our strategy to achieve zero-knowledgeness will crucially rely on the following observation: For all $c, b \in \{0,1\}$, all $\pi, \phi \in \mathcal{S}_m$, and all $\mathbf{v}, \mathbf{w} \in \{0,1\}^m$, we have the equivalences

$$
\begin{cases}
\mathbf{z} = \mathsf{ext}(c, \mathbf{v}) \land \mathbf{v} \in \mathsf{B}_m^{nk} \iff F_{b,\pi}(\mathbf{z}) = \mathsf{Ext}(c \oplus b, \pi(\mathbf{v})) \land \pi(\mathbf{v}) \in \mathsf{B}_m^{nk}; \\
\mathbf{y} = \mathsf{ext}(\bar{c}, \mathbf{w}) \land \mathbf{w} \in \mathsf{B}_m^{nk} \iff F_{\bar{b},\pi}(\mathbf{y}) = \mathsf{Ext}(c \oplus b, \pi(\mathbf{w})) \land \pi(\mathbf{w}) \in \mathsf{B}_m^{nk}.
\end{cases}
\tag{1}
$$

Warm-up Step. Now, let $(\mathbf{d}, w)$ be such that $((\mathbf{A}, \mathbf{u}), \mathbf{d}, w) \in \mathsf{R}_{\mathrm{acc}}$, where w is of the form $w = ((j_1, \ldots, j_\ell), (\mathbf{w}_\ell, \ldots, \mathbf{w}_1))$, and let $\mathbf{v}_\ell = \mathbf{d}, \mathbf{v}_{\ell-1}, \ldots, \mathbf{v}_1, \mathbf{v}_0$ be the path computed by $\mathsf{TVerify}_{\mathbf{A}}(\mathbf{u}, \mathbf{d}, w)$. Note that $\mathbf{v}_0 = \mathbf{u}$ and:

$$
\forall i \in \{\ell-1, \ldots, 1, 0\} : \mathbf{v}_i = \begin{cases} h_{\mathbf{A}}(\mathbf{v}_{i+1}, \mathbf{w}_{i+1}), & \text{if } j_{i+1} = 0; \\ h_{\mathbf{A}}(\mathbf{w}_{i+1}, \mathbf{v}_{i+1}), & \text{if } j_{i+1} = 1. \end{cases}
\tag{2}
$$

We observe that relation (2) can be equivalently rewritten in a more compact form: $\forall i \in \{\ell-1, \ldots, 1, 0\}$,

$$
\mathbf{v}_i = \bar{j}_{i+1} \cdot h_{\mathbf{A}}(\mathbf{v}_{i+1}, \mathbf{w}_{i+1}) + j_{i+1} \cdot h_{\mathbf{A}}(\mathbf{w}_{i+1}, \mathbf{v}_{i+1}).
\tag{3}
$$

Equation (3) then can be interpreted as:

$$
\bar{j}_{i+1} \cdot (\mathbf{A}_0 \cdot \mathbf{v}_{i+1} + \mathbf{A}_1 \cdot \mathbf{w}_{i+1}) + j_{i+1} \cdot (\mathbf{A}_0 \cdot \mathbf{w}_{i+1} + \mathbf{A}_1 \cdot \mathbf{v}_{i+1}) = \mathbf{G} \cdot \mathbf{v}_i \mod q
$$

$$
\Leftrightarrow \mathbf{A} \cdot \begin{pmatrix} \bar{j}_{i+1} \cdot \mathbf{v}_{i+1} \\ j_{i+1} \cdot \mathbf{v}_{i+1} \end{pmatrix} + \mathbf{A} \cdot \begin{pmatrix} j_{i+1} \cdot \mathbf{w}_{i+1} \\ \bar{j}_{i+1} \cdot \mathbf{w}_{i+1} \end{pmatrix} = \mathbf{G} \cdot \mathbf{v}_i \mod q
$$

$$
\Leftrightarrow \mathbf{A} \cdot \mathsf{ext}(j_{i+1}, \mathbf{v}_{i+1}) + \mathbf{A} \cdot \mathsf{ext}(\bar{j}_{i+1}, \mathbf{w}_{i+1}) = \mathbf{G} \cdot \mathbf{v}_i \mod q.
$$

Therefore, to achieve our goal, it is necessary and sufficient to construct an argument system in which $\mathcal{P}$ convinces $\mathcal{V}$ in ZK that $\mathcal{P}$ knows $j_1, \ldots, j_\ell \in \{0,1\}^\ell$ and $\mathbf{v}_1, \ldots, \mathbf{v}_\ell, \mathbf{w}_1, \ldots, \mathbf{w}_\ell \in \{0,1\}^{nk}$ satisfying

$$
\begin{cases}
\mathbf{A} \cdot \mathsf{ext}(j_1, \mathbf{v}_1) + \mathbf{A} \cdot \mathsf{ext}(\bar{j}_1, \mathbf{w}_1) = \mathbf{G} \cdot \mathbf{u} \mod q; \\
\forall i \in [\ell-1] : \mathbf{A} \cdot \mathsf{ext}(j_{i+1}, \mathbf{v}_{i+1}) + \mathbf{A} \cdot \mathsf{ext}(\bar{j}_{i+1}, \mathbf{w}_{i+1}) = \mathbf{G} \cdot \mathbf{v}_i \mod q.
\end{cases}
\tag{4}
$$

To this end, we develop a Stern-type protocol [67], in which we adapt the extension technique from [45]. Specifically, we perform the following extensions:

- Extend matrix $\mathbf{A} = [\mathbf{A}_0 | \mathbf{A}_1]$ to matrix $\mathbf{A}^* = [\mathbf{A}_0 | \mathbf{0}^{n \times nk} | \mathbf{A}_1 | \mathbf{0}^{n \times nk}] \in \mathbb{Z}_q^{n \times 2m}$.
- Extend matrix $\mathbf{G}$ to matrix $\mathbf{G}^* = [\mathbf{G} | \mathbf{0}^{n \times nk}] \in \mathbb{Z}_q^{n \times m}$.

– Extend $\mathbf{v}_1, \ldots, \mathbf{v}_\ell, \mathbf{w}_1, \ldots, \mathbf{w}_\ell$ into $\mathbf{v}_1^*, \ldots, \mathbf{v}_\ell^*, \mathbf{w}_1^*, \ldots, \mathbf{w}_\ell^* \in \mathsf{B}_m^{nk}$, respectively. This is done by appending a length-nk vector of suitable Hamming weight to each of these vectors.

Let $\mathbf{z}_i = \mathsf{ext}(j_i, \mathbf{v}_i^*)$ and $\mathbf{y}_i = \mathsf{ext}(\bar{j}_i, \mathbf{w}_i^*)$ for each $i \in [\ell]$. Note that now the conditions in (4) can be equivalently rewritten as:

$$\begin{cases} \mathbf{A}^* \cdot \mathbf{z}_1 + \mathbf{A}^* \cdot \mathbf{y}_1 = \mathbf{G} \cdot \mathbf{u} \bmod q; \\ \forall i \in [\ell - 1]: \mathbf{A}^* \cdot \mathbf{z}_{i+1} + \mathbf{A}^* \cdot \mathbf{y}_{i+1} = \mathbf{G}^* \cdot \mathbf{v}_i^* \bmod q. \end{cases} \tag{5}$$

The Interactive Protocol. Having performed the above preparation and transformation steps, we now give a summary and sketch the main ideas of our interactive protocol, before formally describing it. The public parameters are n, q, k, m, ℓ, the "powers-of-2" matrix $\mathbf{G}$ and its extension $\mathbf{G}^*$.

Common inputs: $(\mathbf{A}, \mathbf{u})$. Both parties extend $\mathbf{A}$ to $\mathbf{A}^*$.

$\mathcal{P}$'s **inputs:** $\big((j_1, \ldots, j_\ell), (\mathbf{v}_1^*, \ldots, \mathbf{v}_\ell^*), (\mathbf{w}_1^*, \ldots, \mathbf{w}_\ell^*), (\mathbf{z}_1, \ldots, \mathbf{z}_\ell), (\mathbf{y}_1, \ldots, \mathbf{y}_\ell)\big)$.

$\mathcal{P}$'s **goal:** Prove in ZK that $\mathbf{v}_i^*, \mathbf{w}_i^* \in \mathsf{B}_m^{nk}$, $\mathbf{z}_i = \mathsf{ext}(j_i, \mathbf{v}_i^*)$, $\mathbf{y}_i = \mathsf{ext}(\bar{j}_i, \mathbf{w}_i^*)$ for all $i \in [\ell]$, and that (5) holds.

To achieve its goal, $\mathcal{P}$ employs the following strategies:

1. To prove in ZK that $\mathbf{v}_i^*, \mathbf{w}_i^* \in \mathsf{B}_m^{nk}$ and $\mathbf{z}_i = \mathsf{ext}(j_i, \mathbf{v}_i^*)$ and $\mathbf{y}_i = \mathsf{ext}(\bar{j}_i, \mathbf{w}_i^*)$ for all $i \in [\ell]$, the equivalences observed in (1) are exploited. Specifically, for each $i \in [\ell]$, $\mathcal{P}$ samples $\pi_i, \phi_i \xleftarrow{\$} \mathcal{S}_m$ and $b_i \xleftarrow{\$} \{0, 1\}$, then it demonstrates to $\mathcal{V}$ that:

$$\begin{cases} \pi_i(\mathbf{v}_i^*) \in \mathsf{B}_m^{nk} \;\wedge\; F_{b_i, \pi_i}(\mathbf{z}_i) = \mathsf{ext}(j_i \oplus b_i, \pi_i(\mathbf{v}_i^*)); \\ \phi_i(\mathbf{w}_i^*) \in \mathsf{B}_m^{nk} \;\wedge\; F_{\bar{b}_i, \pi_i}(\mathbf{y}_i) = \mathsf{ext}(j_i \oplus b_i, \phi_i(\mathbf{w}_i^*)). \end{cases} \tag{6}$$

Seeing (6), $\mathcal{V}$ should be convinced of the facts $\mathcal{P}$ wants to prove, while learning no additional information, thanks to the randomness of π_i, ϕ_i and b_i.

2. To prove in ZK that the ℓ equations in (5) hold, $\mathcal{P}$ samples uniformly random masking vectors $\mathbf{r}_\mathbf{v}^{(1)}, \ldots, \mathbf{r}_\mathbf{v}^{(\ell-1)} \xleftarrow{\$} \mathbb{Z}_q^m$; $\mathbf{r}_\mathbf{z}^{(1)}, \ldots, \mathbf{r}_\mathbf{z}^{(\ell)}, \mathbf{r}_\mathbf{y}^{(1)}, \ldots, \mathbf{r}_\mathbf{y}^{(\ell)} \xleftarrow{\$} \mathbb{Z}_q^{2m}$, and then it shows $\mathcal{V}$ that

$$\begin{cases} \mathbf{A}^* \cdot (\mathbf{z}_1 + \mathbf{r}_\mathbf{z}^{(1)}) + \mathbf{A}^* \cdot (\mathbf{y}_1 + \mathbf{r}_\mathbf{y}^{(1)}) - \mathbf{G} \cdot \mathbf{u} = \mathbf{A}^* \cdot \mathbf{r}_\mathbf{z}^{(1)} + \mathbf{A}^* \cdot \mathbf{r}_\mathbf{y}^{(1)} \bmod q; \\ \forall i \in [\ell - 1]: \mathbf{A}^* \cdot (\mathbf{z}_{i+1} + \mathbf{r}_\mathbf{z}^{(i+1)}) + \mathbf{A}^* \cdot (\mathbf{y}_{i+1} + \mathbf{r}_\mathbf{y}^{(i+1)}) - \mathbf{G}^* \cdot (\mathbf{v}_i^* + \mathbf{r}_\mathbf{v}^{(i)}) \\ \qquad\qquad = \mathbf{A}^* \cdot \mathbf{r}_\mathbf{z}^{(i+1)} + \mathbf{A}^* \cdot \mathbf{r}_\mathbf{y}^{(i+1)} - \mathbf{G}^* \cdot \mathbf{r}_\mathbf{v}^{(i)} \bmod q. \end{cases}$$

Let $\mathsf{COM} : \{0, 1\}^* \times \{0, 1\}^m \to \mathbb{Z}_q^n$ be the string commitment scheme from [40], which is statistically hiding and computationally binding if the $\mathsf{SIVP}_{\tilde{\mathcal{O}}(n)}$ problem is hard. The interaction between prover $\mathcal{P}$ and verifier $\mathcal{V}$ is described in Fig. 2.

1. Commitment. $\mathcal{P}$ samples randomness ρ_1, ρ_2, ρ_3 for COM and

$$\begin{cases} b_1, \ldots, b_\ell \xleftarrow{\$} \{0,1\}; \ \pi_1, \ldots, \pi_\ell, \phi_1, \ldots, \phi_\ell \xleftarrow{\$} \mathcal{S}_m; \\ \mathbf{r}_\mathbf{v}^{(1)}, \ldots, \mathbf{r}_\mathbf{v}^{(\ell-1)} \xleftarrow{\$} \mathbb{Z}_q^m; \ \mathbf{r}_\mathbf{z}^{(1)}, \ldots, \mathbf{r}_\mathbf{z}^{(\ell)}, \mathbf{r}_\mathbf{y}^{(1)}, \ldots, \mathbf{r}_\mathbf{y}^{(\ell)} \xleftarrow{\$} \mathbb{Z}_q^{2m}. \end{cases}$$

It then sends $\mathcal{V}$ commitment $\mathrm{CMT} = (C_1, C_2, C_3)$, where

$$\begin{cases} C_1 = \mathrm{COM}\big(\{b_i; \pi_i; \phi_i\}_{i=1}^\ell; \ \mathbf{A}^* \cdot \mathbf{r}_\mathbf{z}^{(1)} + \mathbf{A}^* \cdot \mathbf{r}_\mathbf{y}^{(1)}; \\ \qquad\qquad \{\mathbf{A}^* \cdot \mathbf{r}_\mathbf{z}^{(i+1)} + \mathbf{A}^* \cdot \mathbf{r}_\mathbf{y}^{(i+1)} - \mathbf{G}^* \cdot \mathbf{r}_\mathbf{v}^{(i)}\}_{i=1}^{\ell-1}; \ \rho_1) \\ C_2 = \mathrm{COM}\big(\{\pi_i(\mathbf{r}_\mathbf{v}^{(i)})\}_{i=1}^{\ell-1}; \ \{F_{b_i,\pi_i}(\mathbf{r}_\mathbf{z}^{(i)}); \ F_{b_i,\phi_i}(\mathbf{r}_\mathbf{y}^{(i)})\}_{i=1}^\ell; \ \rho_2) \\ C_3 = \mathrm{COM}\big(\{\pi_i(\mathbf{v}_i^* + \mathbf{r}_\mathbf{v}^{(i)})\}_{i=1}^{\ell-1}; \ \{F_{b_i,\pi_i}(\mathbf{z}_i + \mathbf{r}_\mathbf{z}^{(i)}); \ F_{b_i,\phi_i}(\mathbf{y}_i + \mathbf{r}_\mathbf{y}^{(i)})\}_{i=1}^\ell; \ \rho_3). \end{cases}$$

2. Challenge. Receiving CMT, $\mathcal{V}$ sends a challenge $Ch \xleftarrow{\$} \{1,2,3\}$ to $\mathcal{P}$.

3. Response. Depending on Ch, $\mathcal{P}$ sends the response RSP computed as follows:

- Case $Ch = 1$: For each $i \in [\ell-1]$, let $\mathbf{t}_\mathbf{v}^{(i)} = \pi_i(\mathbf{r}_\mathbf{v}^{(i)})$. For each $i \in [\ell]$, let:

$$a_i = j_i \oplus b_i; \ \mathbf{s}_\mathbf{v}^{(i)} = \pi_i(\mathbf{v}_i^*); \ \mathbf{s}_\mathbf{w}^{(i)} = \phi_i(\mathbf{w}_i^*); \ \mathbf{t}_\mathbf{z}^{(i)} = F_{b_i,\pi_i}(\mathbf{r}_\mathbf{z}^{(i)}); \ \mathbf{t}_\mathbf{y}^{(i)} = F_{b_i,\phi_i}(\mathbf{r}_\mathbf{y}^{(i)}).$$

Then let $\mathrm{RSP} = \big(\{\mathbf{t}_\mathbf{v}^{(i)}\}_{i=1}^{\ell-1}; \ \{a_i; \ \mathbf{s}_\mathbf{v}^{(i)}; \ \mathbf{t}_\mathbf{z}^{(i)}; \ \mathbf{s}_\mathbf{w}^{(i)}; \ \mathbf{t}_\mathbf{y}^{(i)}\}_{i=1}^\ell; \ \rho_2; \rho_3). \qquad (7)$

- Case $Ch = 2$: For each $i \in [\ell-1]$, let $\mathbf{e}_\mathbf{v}^{(i)} = \mathbf{v}_i^* + \mathbf{r}_\mathbf{v}^{(i)}$. For each $i \in [\ell]$, let:

$$c_i = b_i; \ \widehat{\pi}_i = \pi_i; \ \widehat{\phi}_i = \phi_i; \ \mathbf{e}_\mathbf{z}^{(i)} = \mathbf{z}_i + \mathbf{r}_\mathbf{z}^{(i)}; \ \mathbf{e}_\mathbf{y}^{(i)} = \mathbf{y}_i + \mathbf{r}_\mathbf{y}^{(i)}.$$

Then let $\mathrm{RSP} = \big(\{\mathbf{e}_\mathbf{v}^{(i)}\}_{i=1}^{\ell-1}; \ \{c_i; \ \widehat{\pi}_i; \ \widehat{\phi}_i; \ \mathbf{e}_\mathbf{z}^{(i)}; \ \mathbf{e}_\mathbf{y}^{(i)}\}_{i=1}^\ell; \ \rho_1; \rho_3). \qquad (8)$

- Case $Ch = 3$: For each $i \in [\ell-1]$, let $\mathbf{p}_\mathbf{v}^{(i)} = \mathbf{r}_\mathbf{v}^{(i)}$. For each $i \in [\ell]$, let:

$$d_i = b_i; \ \widetilde{\pi}_i = \pi_i; \ \widetilde{\phi}_i = \phi_i; \ \mathbf{p}_\mathbf{z}^{(i)} = \mathbf{r}_\mathbf{z}^{(i)}; \ \mathbf{p}_\mathbf{y}^{(i)} = \mathbf{r}_\mathbf{y}^{(i)}.$$

Then let $\mathrm{RSP} = \big(\{\mathbf{p}_\mathbf{v}^{(i)}\}_{i=1}^{\ell-1}; \ \{d_i; \ \widetilde{\pi}_i; \ \widetilde{\phi}_i; \ \mathbf{p}_\mathbf{z}^{(i)}; \ \mathbf{p}_\mathbf{y}^{(i)}\}_{i=1}^\ell; \ \rho_1; \rho_2). \qquad (9)$

Verification. Receiving RSP, $\mathcal{V}$ proceeds as follows.

- Case $Ch = 1$: Parse RSP as in (7). Check that $\mathbf{s}_\mathbf{v}^{(i)}, \mathbf{s}_\mathbf{w}^{(i)} \in \mathsf{B}_m^{nk}$ for all $i \in [\ell]$. Next, for each $i \in [\ell]$, let $\mathbf{s}_\mathbf{z}^{(i)} = \mathrm{ext}(a_i, \mathbf{s}_\mathbf{v}^{(i)})$ and let $\mathbf{s}_\mathbf{y}^{(i)} = \mathrm{ext}(a_i, \mathbf{s}_\mathbf{w}^{(i)})$. Then check that:

$$\begin{cases} C_2 = \mathrm{COM}\big(\{\mathbf{t}_\mathbf{v}^{(i)}\}_{i=1}^{\ell-1}; \ \{\mathbf{t}_\mathbf{z}^{(i)}; \ \mathbf{t}_\mathbf{y}^{(i)}\}_{i=1}^\ell; \ \rho_2), \\ C_3 = \mathrm{COM}\big(\{\mathbf{s}_\mathbf{v}^{(i)} + \mathbf{t}_\mathbf{v}^{(i)}\}_{i=1}^{\ell-1}; \ \{\mathbf{s}_\mathbf{z}^{(i)} + \mathbf{t}_\mathbf{z}^{(i)}; \ \mathbf{s}_\mathbf{y}^{(i)} + \mathbf{t}_\mathbf{y}^{(i)}\}_{i=1}^\ell; \ \rho_3). \end{cases} \qquad (10)$$

- Case $Ch = 2$: Parse RSP as in (8) and check that:

$$\begin{cases} C_1 = \mathrm{COM}\big(\{c_i; \widehat{\pi}_i; \widehat{\phi}_i\}_{i=1}^\ell; \ \mathbf{A}^* \cdot \mathbf{e}_\mathbf{z}^{(1)} + \mathbf{A}^* \cdot \mathbf{e}_\mathbf{y}^{(1)} - \mathbf{G} \cdot \mathbf{u}; \\ \qquad\qquad \{\mathbf{A}^* \cdot \mathbf{e}_\mathbf{z}^{(i+1)} + \mathbf{A}^* \cdot \mathbf{e}_\mathbf{y}^{(i+1)} - \mathbf{G}^* \cdot \mathbf{e}_\mathbf{v}^{(i)}\}_{i=1}^{\ell-1}; \ \rho_1) \\ C_3 = \mathrm{COM}\big(\{\widehat{\pi}_i(\mathbf{e}_\mathbf{v}^{(i)})\}_{i=1}^{\ell-1}; \ \{F_{c_i,\widehat{\pi}_i}(\mathbf{e}_\mathbf{z}^{(i)}); \ F_{\widehat{c}_i,\widehat{\phi}_i}(\mathbf{e}_\mathbf{y}^{(i)})\}_{i=1}^\ell; \ \rho_3). \end{cases} \qquad (11)$$

- Case $Ch = 3$: Parse RSP as in (9) and check that:

$$\begin{cases} C_1 = \mathrm{COM}\big(\{d_i; \widetilde{\pi}_i; \widetilde{\phi}_i\}_{i=1}^\ell; \ \mathbf{A}^* \cdot \mathbf{p}_\mathbf{z}^{(1)} + \mathbf{A}^* \cdot \mathbf{p}_\mathbf{y}^{(1)}; \\ \qquad\qquad \{\mathbf{A}^* \cdot \mathbf{p}_\mathbf{z}^{(i+1)} + \mathbf{A}^* \cdot \mathbf{p}_\mathbf{y}^{(i+1)} - \mathbf{G}^* \cdot \mathbf{p}_\mathbf{v}^{(i)}\}_{i=1}^{\ell-1}; \ \rho_1) \\ C_2 = \mathrm{COM}\big(\{\widetilde{\pi}_i(\mathbf{p}_\mathbf{v}^{(i)})\}_{i=1}^{\ell-1}; \ \{F_{d_i,\widetilde{\pi}_i}(\mathbf{p}_\mathbf{z}^{(i)}); \ F_{\widetilde{d}_i,\widetilde{\phi}_i}(\mathbf{p}_\mathbf{y}^{(i)})\}_{i=1}^\ell; \ \rho_2). \end{cases} \qquad (12)$$

In each case, $\mathcal{V}$ outputs 1 if all the conditions hold. Otherwise, it outputs 0.

Fig. 2. A zero-knowledge argument of knowledge for the relation $\mathrm{R}_{\mathrm{acc}}$.

3.5   Analysis of the Interactive Protocol

The properties of the given protocol are summarized in the following theorem.

Theorem 2. *The given interactive protocol has perfect completeness and communication cost $\widetilde{\mathcal{O}}(\ell \cdot n)$. If* COM *is a statistically hiding and computationally binding string commitment scheme, then it is a statistical zero-knowledge argument of knowledge for the relation* $\mathrm{R_{acc}}$.

Completeness and Communication Cost. Based on the discussion given in the previous section, it can be checked that the described protocol has perfect completeness, *i.e.*, if $\mathcal{P}$ is honest and follows the protocol, then $\mathcal{V}$ always outputs 1. It can also be seen that the communication cost of the protocol is $\widetilde{\mathcal{O}}(\ell \cdot m \cdot \log q) = \widetilde{\mathcal{O}}(\ell \cdot n)$ bits.

In order to prove that the protocol is a ZKAoK for the relation $\mathrm{R_{acc}}$, we will employ the standard simulation and extraction techniques for Stern-type protocols (see, *e.g.*, [40,45,46]).

Lemma 2 (Zero-Knowledge Property). *If* COM *is statistically hiding, then the interactive protocol in Fig. 2 is a statistical zero-knowledge argument.*

Proof. We construct a PPT simulator $\mathcal{S}$ interacting with a (possibly dishonest) verifier $\widehat{\mathcal{V}}$, such that, given only the public input, $\mathcal{S}$ outputs with probability negligibly close to $2/3$ a simulated transcript that is statistically close to the one produced by the honest prover in the real interaction. The simulator $\mathcal{S}$ begins by selecting a random $\overline{Ch} \in \{1, 2, 3\}$. This is a prediction of the challenge value that $\widehat{\mathcal{V}}$ will *not* choose.

Case $\overline{Ch} = 1$: Using linear algebra, $\mathcal{S}$ computes $\mathbf{z}_1', \ldots, \mathbf{z}_\ell', \mathbf{y}_1', \ldots, \mathbf{y}_\ell' \in \mathbb{Z}_q^{2m}$ and $\mathbf{v}_1', \ldots, \mathbf{v}_{\ell-1}' \in \mathbb{Z}_q^m$ such that

$$\begin{cases} \mathbf{A}^* \cdot \mathbf{z}_1' + \mathbf{A}^* \cdot \mathbf{y}_1' = \mathbf{G} \cdot \mathbf{u} \bmod q; \\ \forall i \in [1, \ell-1] : \ \mathbf{A}^* \cdot \mathbf{z}_{i+1}' + \mathbf{A}^* \cdot \mathbf{y}_{i+1}' = \mathbf{G}^* \cdot \mathbf{v}_i' \bmod q. \end{cases}$$

Then it samples randomness ρ_1, ρ_2, ρ_3 for COM and

$$\begin{cases} b_1, \ldots, b_\ell \xleftarrow{\$} \{0,1\}; \ \pi_1, \ldots, \pi_\ell, \phi_1, \ldots, \phi_\ell \xleftarrow{\$} \mathcal{S}_m; \\ \mathbf{r}_{\mathbf{v}}^{(1)}, \ldots, \mathbf{r}_{\mathbf{v}}^{(\ell-1)} \xleftarrow{\$} \mathbb{Z}_q^m; \ \mathbf{r}_{\mathbf{z}}^{(1)}, \ldots, \mathbf{r}_{\mathbf{z}}^{(\ell)}, \mathbf{r}_{\mathbf{y}}^{(1)}, \ldots, \mathbf{r}_{\mathbf{y}}^{(\ell)} \xleftarrow{\$} \mathbb{Z}_q^{2m}. \end{cases}$$

It then sends $\widehat{\mathcal{V}}$ commitment $\mathrm{CMT} = (C_1', C_2', C_3')$, where

$$\begin{cases} C_1' = \mathsf{COM}\big(\{b_i; \pi_i; \phi_i\}_{i=1}^\ell; \ \mathbf{A}^* \cdot \mathbf{r}_{\mathbf{z}}^{(1)} + \mathbf{A}^* \cdot \mathbf{r}_{\mathbf{y}}^{(1)}; \\ \qquad\qquad \{\mathbf{A}^* \cdot \mathbf{r}_{\mathbf{z}}^{(i+1)} + \mathbf{A}^* \cdot \mathbf{r}_{\mathbf{y}}^{(i+1)} - \mathbf{G}^* \cdot \mathbf{r}_{\mathbf{v}}^{(i)}\}_{i=1}^{\ell-1}; \ \rho_1 \big) \\ C_2' = \mathsf{COM}\big(\{\pi_i(\mathbf{r}_{\mathbf{v}}^{(i)})\}_{i=1}^{\ell-1}; \ \{F_{b_i, \pi_i}(\mathbf{r}_{\mathbf{z}}^{(i)}); \ F_{\overline{b}_i, \phi_i}(\mathbf{r}_{\mathbf{y}}^{(i)})\}_{i=1}^\ell; \ \rho_2 \big) \\ C_3' = \mathsf{COM}\big(\{\pi_i(\mathbf{v}_i' + \mathbf{r}_{\mathbf{v}}^{(i)})\}_{i=1}^{\ell-1}; \ \{F_{b_i, \pi_i}(\mathbf{z}_i' + \mathbf{r}_{\mathbf{z}}^{(i)}); \ F_{\overline{b}_i, \phi_i}(\mathbf{y}_i' + \mathbf{r}_{\mathbf{y}}^{(i)})\}_{i=1}^\ell; \ \rho_3 \big). \end{cases} \tag{13}$$

Receiving a challenge Ch from $\widehat{\mathcal{V}}$, the simulator responds as follows:

- If $Ch = 1$: Output $\perp$ and abort.
- If $Ch = 2$: Send RSP $= \left(\{v_i' + r_v^{(i)}\}_{i=1}^{\ell-1}; \{b_i; \pi_i; \phi_i; z_i' + r_z^{(i)}; y_i' + r_y^{(i)}\}_{i=1}^{\ell}; \rho_1; \rho_3 \right)$.
- If $Ch = 3$: Send RSP $= \left(\{r_v^{(i)}\}_{i=1}^{\ell-1}; \{b_i; \pi_i; \phi_i; r_z^{(i)}; r_y^{(i)}\}_{i=1}^{\ell}; \rho_1; \rho_2 \right)$.

Case $\overline{Ch} = 2$: $\mathcal{S}$ samples

$$
\begin{cases}
j_1', \ldots, j_\ell' \xleftarrow{\$} \{0,1\}; \ v_1', \ldots, v_\ell', w_1', \ldots, w_\ell' \xleftarrow{\$} B_m^{nk}; \\
b_1, \ldots, b_\ell \xleftarrow{\$} \{0,1\}; \ \pi_1, \ldots, \pi_\ell, \phi_1, \ldots, \phi_\ell \xleftarrow{\$} \mathcal{S}_m; \\
r_v^{(1)}, \ldots, r_v^{(\ell-1)} \xleftarrow{\$} \mathbb{Z}_q^m; \ r_z^{(1)}, \ldots, r_z^{(\ell)}, r_y^{(1)}, \ldots, r_y^{(\ell)} \xleftarrow{\$} \mathbb{Z}_q^{2m}.
\end{cases}
$$

It then computes $z_i' = \mathrm{ext}(j_i', v_i')$, $y_i' = \mathrm{ext}(\bar{j}_i', w_i')$ for each $i \in [\ell]$, and sends the commitment CMT computed in the same manner as in (13).

Receiving a challenge Ch from $\widehat{\mathcal{V}}$, it responds as follows:

- If $Ch = 1$: Send

$$
\text{RSP} = \left(\{\pi_i(r_v^{(i)})\}_{i=1}^{\ell-1}; \ \{j_i' \oplus b_i; \ \pi_i(v_i'); \ F_{b_i, \pi_i}(r_z^{(i)}); \ \phi_i(w_i'); \ F_{\bar{b}_i, \phi_i}(r_y^{(i)})\}_{i=1}^{\ell}; \ \rho_2; \rho_3 \right).
$$

- If $Ch = 2$: Output $\perp$ and abort.
- If $Ch = 3$: Send RSP computed as in the case $(\overline{Ch} = 1, Ch = 3)$.

Case $\overline{Ch} = 3$: The simulator proceeds with the preparation as in the case $\overline{Ch} = 2$ above. Then it sends the commitment CMT $:= (C_1', C_2', C_3')$, where C_2', C_3' are computed as in (13), while

$$
C_1' = \text{COM}\big(\{b_i; \pi_i; \phi_i\}_{i=1}^{\ell}; \ \mathbf{A}^* \cdot (z_1' + r_z^{(1)}) + \mathbf{A}^* \cdot (y_1' + r_y^{(1)}) - \mathbf{G} \cdot \mathbf{u};
$$
$$
\{\mathbf{A}^* \cdot (z_{i+1}' + r_z^{(i+1)}) + \mathbf{A}^* \cdot (y_{i+1}' + r_y^{(i+1)}) - \mathbf{G}^* \cdot (v_i' + r_v^{(i)})\}_{i=1}^{\ell-1}; \ \rho_1 \big).
$$

Receiving a challenge Ch from $\widehat{\mathcal{V}}$, it responds as follows:

- If $Ch = 1$: Send RSP computed as in the case $(\overline{Ch} = 2, Ch = 1)$.
- If $Ch = 2$: Send RSP computed as in the case $(\overline{Ch} = 1, Ch = 2)$.
- If $Ch = 3$: Output $\perp$ and abort.

We observe that, in every case we have considered above, since COM is statistically hiding, the distribution of the commitment CMT and the distribution of the challenge Ch from $\widehat{\mathcal{V}}$ are statistically close to those in the real interaction. Hence, the probability that the simulator outputs $\perp$ is negligibly close to $1/3$. Moreover, one can check that whenever the simulator does not halt, it will provide an accepted transcript, the distribution of which is statistically close to that of the prover in the real interaction. In other words, we have constructed a simulator that can successfully impersonate the honest prover with probability negligibly close to $2/3$. $\qquad\square$

To prove that our protocol is an argument of knowledge for the relation R_{acc}, it suffices to demonstrate that the protocol has the special soundness property [34].

Lemma 3 (Argument of Knowledge Property). *If* COM *is computationally binding, then there exists an efficient knowledge extractor K that, on input 3 valid responses* (RSP_1, RSP_2, RSP_3) *to the same commitment* CMT, *outputs a pair* $(\mathbf{d}' \in \{0,1\}^{nk}, w' \in \{0,1\}^{\ell} \times (\{0,1\}^{nk})^{\ell})$ *such that*

$$((\mathbf{A}, \mathbf{u}); \mathbf{d}', w') \in R_{acc}.$$

Proof. Let the 3 valid responses to $\text{CMT} = (C_1, C_2, C_3)$ be

$$\begin{cases} RSP_1 = (\ \{\mathbf{t}_{\mathbf{v}}^{(i)}\}_{i=1}^{\ell-1}; \ \{a_i; \ \mathbf{s}_{\mathbf{v}}^{(i)}; \ \mathbf{t}_{\mathbf{z}}^{(i)}; \ \mathbf{s}_{\mathbf{w}}^{(i)}; \ \mathbf{t}_{\mathbf{y}}^{(i)}\}_{i=1}^{\ell}; \ \rho_2; \rho_3), \\ RSP_2 = (\ \{\mathbf{e}_{\mathbf{v}}^{(i)}\}_{i=1}^{\ell-1}; \ \{c_i; \ \widehat{\pi}_i; \ \widehat{\phi}_i; \ \mathbf{e}_{\mathbf{z}}^{(i)}; \ \mathbf{e}_{\mathbf{y}}^{(i)}\}_{i=1}^{\ell}; \ \rho_1; \rho_3), \\ RSP_3 = (\ \{\mathbf{p}_{\mathbf{v}}^{(i)}\}_{i=1}^{\ell-1}; \ \{d_i; \ \widetilde{\pi}_i; \ \widetilde{\phi}_i; \ \mathbf{p}_{\mathbf{z}}^{(i)}; \ \mathbf{p}_{\mathbf{y}}^{(i)}\}_{i=1}^{\ell}; \ \rho_1; \rho_2). \end{cases}$$

The validity of RSP_1 implies that $\forall i \in [\ell] : \mathbf{s}_{\mathbf{v}}^{(i)}, \mathbf{s}_{\mathbf{w}}^{(i)} \in \mathsf{B}_m^{nk}$. Furthermore, it follows from the verification conditions given in (10), (11), (12), and from the computational binding property of COM that:

$$\mathbf{A}^* \cdot \mathbf{e}_{\mathbf{z}}^{(1)} + \mathbf{A}^* \cdot \mathbf{e}_{\mathbf{y}}^{(1)} - \mathbf{G} \cdot \mathbf{u} = \mathbf{A}^* \cdot \mathbf{p}_{\mathbf{z}}^{(1)} + \mathbf{A}^* \cdot \mathbf{p}_{\mathbf{y}}^{(1)} \bmod q,$$

and for all $i \in [1, \ell - 1]$: $\mathbf{t}_{\mathbf{v}}^{(i)} = \widetilde{\pi}_i(\mathbf{p}_{\mathbf{v}}^{(i)}); \ \mathbf{s}_{\mathbf{v}}^{(i)} + \mathbf{t}_{\mathbf{v}}^{(i)} = \widehat{\pi}_i(\mathbf{e}_{\mathbf{v}}^{(i)});$ and

$$\mathbf{A}^* \cdot \mathbf{e}_{\mathbf{z}}^{(i+1)} + \mathbf{A}^* \cdot \mathbf{e}_{\mathbf{y}}^{(i+1)} - \mathbf{G}^* \cdot \mathbf{e}_{\mathbf{v}}^{(i)} = \mathbf{A}^* \cdot \mathbf{p}_{\mathbf{z}}^{(i+1)} + \mathbf{A}^* \cdot \mathbf{p}_{\mathbf{y}}^{(i+1)} - \mathbf{G}^* \cdot \mathbf{p}_{\mathbf{v}}^{(i)} \bmod q,$$

and for all $i \in [\ell]$:

$$\begin{cases} c_i = d_i; \ \widehat{\pi}_i = \widetilde{\pi}_i; \ \widehat{\phi}_i = \widetilde{\phi}_i; \\ \mathbf{t}_{\mathbf{z}}^{(i)} = F_{d_i, \widetilde{\pi}_i}(\mathbf{p}_{\mathbf{z}}^{(i)}); \ \text{ext}(a_i, \mathbf{s}_{\mathbf{v}}^{(i)}) + \mathbf{t}_{\mathbf{z}}^{(i)} = F_{c_i, \widehat{\pi}_i}(\mathbf{e}_{\mathbf{z}}^{(i)}); \\ \mathbf{t}_{\mathbf{y}}^{(i)} = F_{\bar{d}_i, \widetilde{\phi}_i}(\mathbf{p}_{\mathbf{y}}^{(i)}); \ \text{ext}(a_i, \mathbf{s}_{\mathbf{w}}^{(i)}) + \mathbf{t}_{\mathbf{y}}^{(i)} = F_{\bar{c}_i, \widehat{\phi}_i}(\mathbf{e}_{\mathbf{y}}^{(i)}). \end{cases}$$

The knowledge extractor K now proceeds as follows. For each $i \in [\ell]$, let:

$$j_i = a_i \oplus c_i; \ \mathbf{v}_i^* = \widehat{\pi}_i^{-1}(\mathbf{s}_{\mathbf{v}}^{(i)}); \ \mathbf{w}_i^* = \widehat{\phi}_i^{-1}(\mathbf{s}_{\mathbf{w}}^{(i)}); \ \mathbf{z}_i = \mathbf{e}_{\mathbf{z}}^{(i)} - \mathbf{p}_{\mathbf{z}}^{(i)}; \ \mathbf{y}_i = \mathbf{e}_{\mathbf{y}}^{(i)} - \mathbf{p}_{\mathbf{y}}^{(i)}.$$

Note that $\widehat{\pi}_i(\mathbf{v}_i^*) = \mathbf{s}_{\mathbf{v}}^{(i)} \in \mathsf{B}_m^{nk}$, and thus $\mathbf{v}_i^* \in \mathsf{B}_m^{nk}$ (by (1)). Similarly, $\mathbf{w}_i^* \in \mathsf{B}_m^{nk}$. Furthermore, one has that:

- $F_{c_i, \widehat{\pi}_i}(\mathbf{z}_i) = \text{ext}(a_i, \mathbf{s}_{\mathbf{v}}^{(i)}) = \text{ext}(j_i \oplus c_i, \widehat{\pi}_i(\mathbf{v}_i^*))$. By (1), this implies $\mathbf{z}_i = \text{ext}(j_i, \mathbf{v}_i^*)$.
- $F_{\bar{c}_i, \widehat{\phi}_i}(\mathbf{y}_i) = \text{ext}(a_i, \mathbf{s}_{\mathbf{w}}^{(i)}) = \text{ext}(\bar{j}_i \oplus \bar{c}_i, \widehat{\phi}_i(\mathbf{w}_i^*))$. By (1), this implies $\mathbf{y}_i = \text{ext}(\bar{j}_i, \mathbf{w}_i^*)$.

Moreover, the following relations hold:

$$\begin{cases} \mathbf{A}^* \cdot \mathbf{z}_1 + \mathbf{A}^* \cdot \mathbf{y}_1 = \mathbf{G} \cdot \mathbf{u} \bmod q \\ \forall i \in [1, \ell - 1]: \ \mathbf{A}^* \cdot \mathbf{z}_{i+1} + \mathbf{A}^* \cdot \mathbf{y}_{i+1} = \mathbf{G}^* \cdot \mathbf{v}_i^* \bmod q \end{cases}$$

$$\Leftrightarrow \begin{cases} \mathbf{A}^* \cdot \mathsf{ext}(j_1, \mathbf{v}_1^*) + \mathbf{A}^* \cdot \mathsf{ext}(\bar{j}_i, \mathbf{w}_i^*) = \mathbf{G} \cdot \mathbf{u} \bmod q \\ \forall i \in [1, \ell - 1]: \ \mathbf{A}^* \cdot \mathsf{ext}(j_{i+1}, \mathbf{v}_{i+1}^*) + \mathbf{A}^* \cdot \mathsf{ext}(\bar{j}_{i+1}, \mathbf{w}_{i+1}^*) = \mathbf{G}^* \cdot \mathbf{v}_i^* \bmod q. \end{cases}$$

Now, by dropping the last nk coordinates from $\mathbf{v}_1^*, \ldots, \mathbf{v}_\ell^*, \mathbf{w}_1^*, \ldots, \mathbf{w}_\ell^*$, the knowledge extractor $\mathcal{K}$ obtains $\mathbf{v}_1', \ldots, \mathbf{v}_\ell', \mathbf{w}_1', \ldots, \mathbf{w}_\ell' \in \{0, 1\}^{nk}$, respectively. These vectors satisfy:

$$\begin{cases} \mathbf{A} \cdot \mathsf{ext}(j_1, \mathbf{v}_1') + \mathbf{A} \cdot \mathsf{ext}(\bar{j}_1, \mathbf{w}_1') = \mathbf{G} \cdot \mathbf{u} \bmod q \\ \forall i \in [1, \ell - 1]: \ \mathbf{A} \cdot \mathsf{ext}(j_{i+1}, \mathbf{v}_{i+1}') + \mathbf{A} \cdot \mathsf{ext}(\bar{j}_{i+1}, \mathbf{w}_{i+1}') = \mathbf{G} \cdot \mathbf{v}_i' \bmod q \end{cases}$$

$$\Leftrightarrow \begin{cases} \mathbf{v}_0' = \mathbf{u} \\ \forall i \in [0, \ell - 1]: \ \mathbf{v}_i' = \bar{j}_{i+1} \cdot h_{\mathbf{A}}(\mathbf{v}_{i+1}', \mathbf{w}_{i+1}') + j_{i+1} \cdot h_{\mathbf{A}}(\mathbf{w}_{i+1}', \mathbf{v}_{i+1}'). \end{cases}$$

Let $\mathbf{d}' = \mathbf{v}_\ell'$ and $w' = ((j_1, \ldots, j_\ell), (\mathbf{w}_\ell', \ldots, \mathbf{w}_1'))$, then $\mathsf{TVerify}_{\mathbf{A}}(\mathbf{u}, \mathbf{d}', w') = 1$. In other words, $(\mathbf{d}', w')$ satisfies $((\mathbf{A}, \mathbf{u}); \mathbf{d}', w') \in R_{\mathrm{acc}}$. This concludes the proof. $\qquad \square$

4   A Logarithmic-Size Ring Signature from Lattices

In this section, we construct a ring signature scheme [65] with signature size $\widetilde{\mathcal{O}}(\log N \cdot n)$, where N is the size of the ring, based on the hardness of lattice problem $\mathsf{SIVP}_{\widetilde{\mathcal{O}}(n)}$. We use the ZKAoK given in Sect. 3 as the building block.

4.1   Definitions

We recall the standard definitions and security requirements for ring signatures [11,36]. A ring signature scheme consists of a tuple of efficient algorithms $(\mathsf{RSetup}, \mathsf{RKgen}, \mathsf{RSign}, \mathsf{RVerify})$ for generating a public parameter, generating keys for users, signing messages, and verifying ring signatures, respectively.

$\mathsf{RSetup}(n)$: Generates public parameters pp which are made available to all users.
$\mathsf{RKgen}(pp)$: Generates a public key pk and the corresponding secret key sk.
$\mathsf{RSign}_{pp}(sk, M, R)$: Outputs a signature Σ on the message $M \in \{0, 1\}^*$ with respect to the ring $R = (pk_0, \ldots, pk_{N-1})$. It is required that (pk, sk) be a valid key pair produced by $\mathsf{RKgen}(pp)$ and that $pk \in R$.
$\mathsf{RVerify}_{pp}(M, R, \Sigma)$: Given a candidate signature Σ on a message M with respect to the ring of public keys R, this algorithm outputs 1 if Σ is deemed valid or 0 otherwise.

We next describe the following requirements for ring signatures: correctness, unforgeability with respect to insider corruption, and statistical anonymity.

The correctness requirement says that a user can always sign any message on behalf of a ring he belongs to. This is formalized as follows.

Definition 6 (Correctness). A ring signature (RSetup, RKgen, RSign, RVerify) is correct if for any $pp \leftarrow$ RSetup(n), any $(pk, sk) \leftarrow$ RKgen(pp), any R such that $pk \in R$, any $M \in \{0, 1\}^*$, we have RVerify$_{pp}(M, R, \text{RSign}_{pp}(sk, M, R)) = 1$.

A ring signature is unforgeable with respect to insider corruption if it is infeasible to forge a ring signature without controlling one of the ring members.

Definition 7 (Unforgeability w.r.t. insider corruption). *A ring signature scheme* (RSetup, RKgen, RSign, RVerify) *is unforgeable w.r.t. insider corruption if for all PPT adversaries $\mathcal{A}$,*

$$\Pr[pp \leftarrow RSetup(1^n); (M^\star, R^\star, \Sigma^\star) \leftarrow \mathcal{A}^{\text{PKGen,Sign,Corrupt}}(pp):$$
$$\text{RVerify}_{pp}(M^\star, R^\star, \Sigma^\star) = 1] \in \text{negl}(n),$$

where:

- PKGen *on the j-th query runs $(pk_j, sk_j) \leftarrow$ RKgen(pp) and returns pk_j.*
- Sign(j, M, R) *returns the output of* RSign$_{pp}(sk_j, M, R)$ *provided: (i) (pk_j, sk_j) has been generated by* PKGen*; (ii) $pk_j \in R$. Otherwise, it returns $\bot$.*
- Corrupt(j) *returns sk_j, provided that (pk_j, sk_j) has been generated by* PKGen.
- $\mathcal{A}$ *outputs $(M^\star, R^\star, \Sigma^\star)$ such that* Sign($\cdot, M^\star, R^\star$) *has not been queried. Moreover, $R^\star$ is non-empty and only contains public keys pk_j generated by* PKGen *for which j has not been corrupted.*

Definition 8. A ring signature scheme (RSetup, RKgen, RSign, RVerify) provides statistical anonymity if, for any (possibly unbounded) adversary $\mathcal{A}$,

$$\Pr\left[\begin{array}{c} pp \leftarrow \text{RSetup}(1^n); (M^\star, j_0, j_1, R^\star) \leftarrow \mathcal{A}^{\text{RKgen}(pp)}(pp) \\ b \xleftarrow{\$} \{0, 1\}; \Sigma^\star \leftarrow \text{RSign}_{pp}(sk_{j_b}, M^\star, R^\star) \end{array} : \mathcal{A}(\Sigma^\star) = b\right]$$
$$= 1/2 + \text{negl}(n),$$

where $pk_{j_0}, pk_{j_1} \in R^\star$.

Remark: Anonymity under full key exposure [11] requires that the randomness used by KeyGen be revealed to the adversary. In our construction, it does not make a difference since we assume computationally unbounded adversaries. A c-user ring signature scheme is a variant of ring signatures, that only supports rings of fixed size c. Here, we do not assume any upper bound on the size of a ring. Similarly to [36], we only assume that all users agree on pre-existing public parameters pp. In our scheme, these public parameters consist of a modulus q and a random matrix $\mathbf{A} \in \mathbb{Z}_q^{n \times 2nk}$ which can be derived from a random oracle. In this case, we only need all users to agree on the parameters q and n.

4.2 The Underlying Zero-Knowledge Protocol

The ring signature scheme that we will present next relies on a simple extension of the ZKAoK in Sect. 3. Specifically, one more layer is added: apart from proving that it has a secret value $\mathbf{d}$ that was properly accumulated to the root of the tree, $\mathcal{P}$ has to convince $\mathcal{V}$ that it knows a vector $\mathbf{x} \in \{0,1\}^m$ such that $\mathrm{bin}(\mathbf{A} \cdot \mathbf{x} \bmod q) = \mathbf{d}$, or equivalently, $\mathbf{A} \cdot \mathbf{x} = \mathbf{G} \cdot \mathbf{d} \bmod q$. The associated relation R_{ring} is defined as follows.

Definition 9. *Define the relation*

$$R_{\mathrm{ring}} = \Big\{ \big((\mathbf{A}, \mathbf{u}) \in \mathbb{Z}_q^{n \times m} \times \{0,1\}^{nk}; \mathbf{d} \in \{0,1\}^{nk}, w \in \{0,1\}^\ell \times (\{0,1\}^{nk})^\ell, $$

$$\mathbf{x} \in \{0,1\}^m \big) : \mathsf{TVerify}_{\mathbf{A}}(\mathbf{u}, \mathbf{d}, w) = 1 \wedge \mathbf{A} \cdot \mathbf{x} = \mathbf{G} \cdot \mathbf{d} \bmod q \Big\}.$$

A ZKAoK for R_{ring} can be obtained from the one in Sect. 3, where the new layer is handled by the same "extend-then-permute" technique. As before, the protocol relies on the string commitment scheme from [40], which is statistically hiding and computationally binding if the $\mathsf{SIVP}_{\widetilde{\mathcal{O}}(n)}$ problem is hard.

Lemma 4. *Let us assume that the $\mathsf{SIVP}_{\widetilde{\mathcal{O}}(n)}$ problem is hard. Then, there exists a statistical ZKAoK for the relation R_{ring} with perfect completeness and communication cost $\widetilde{\mathcal{O}}(\ell \cdot n)$. In particular:*

- *There exists an efficient simulator that, on input $(\mathbf{A}, \mathbf{u})$, outputs an accepted transcript which is statistically close to that produced by the real prover.*
- *There exists an efficient knowledge extractor that, on input 3 valid responses $(\mathrm{RSP}_1, \mathrm{RSP}_2, \mathrm{RSP}_3)$ to the same commitment CMT, outputs $(\mathbf{d}', w', \mathbf{x}')$ such that*

$$((\mathbf{A}, \mathbf{u}), \mathbf{d}', w', \mathbf{x}') \in R_{\mathrm{ring}}.$$

The full description and analysis of the argument system are given in the full version of the paper.

4.3   Description of the Ring Signature Scheme

We now will construct a ring signature scheme for rings of $N = 2^\ell$ users based on the Merkle-tree accumulator presented in Sect. 3. Our ring signature can be easily adapted for the case when the size of the ring is not a power of 2 (see Remark 1). The scheme uses parameters n, m, q defined as in Sect. 3, parameter $\kappa = \omega(\log n)$ that determines the number of protocol repetitions, and a random oracle $\mathcal{H}_{\mathsf{FS}} : \{0,1\}^* \to \{1,2,3\}^\kappa$.

$\mathsf{RSetup}(n)$: Sample $\mathbf{A} \xleftarrow{\$} \mathbb{Z}_q^{n \times m}$, and output $pp = \mathbf{A}$.

$\mathsf{RKgen}(pp = \mathbf{A})$: Pick $\mathbf{x} \xleftarrow{\$} \{0,1\}^m$, compute $\mathbf{d} = \mathrm{bin}(\mathbf{A} \cdot \mathbf{x} \bmod q) \in \{0,1\}^{nk}$, and output $(sk, pk) = (\mathbf{x}, \mathbf{d})$.

$\mathsf{RSign}_{pp}(sk, M, R)$: Given a ring $R = (\mathbf{d}_0, \ldots, \mathbf{d}_{N-1})$, where $\mathbf{d}_i \in \{0,1\}^{nk}$ for every $i \in [0, N-1]$, and $sk = \mathbf{x} \in \{0,1\}^m$ such that $\mathbf{d} = \mathrm{bin}(\mathbf{Ax} \bmod q) \in R$, this algorithm generates a ring signature Σ on $M \in \{0,1\}^*$ as follows:

1. Run algorithm $\mathsf{TAcc_A}(R)$ to build the Merkle tree based on R and the hash function $h_{\mathbf{A}}$, and obtain the root $\mathbf{u} \in \{0,1\}^{nk}$.

2. Run algorithm $\mathsf{TWitness_A}(R, \mathbf{d})$ to get a witness

$$w = \left((j_1, \ldots, j_\ell) \in \{0,1\}^\ell, (\mathbf{w}_\ell, \ldots, \mathbf{w}_1) \in (\{0,1\}^{nk})^\ell\right)$$

to the fact that $\mathbf{d}$ was properly accumulated in $\mathbf{u}$.

3. Generate a NIZKAoK Π_{ring} to demonstrate the possession of a valid pair $(sk, pk) = (\mathbf{x}, \mathbf{d})$ such that $\mathbf{d}$ is properly accumulated in $\mathbf{u}$. This is done by running the protocol in Sect. 4.2 with public input $(\mathbf{A}, \mathbf{u})$ and prover's witness $(\mathbf{x}, \mathbf{d}, w)$. The protocol is repeated $\kappa = \omega(\log n)$ times to achieve negligible soundness error and made non-interactive via the Fiat-Shamir heuristic as a triple $\Pi_{\mathsf{ring}} = (\{\mathrm{CMT}_i\}_{i=1}^\kappa, \mathrm{CH}, \{\mathrm{RSP}\}_{i=1}^\kappa)$, where

$$\mathrm{CH} = \mathcal{H}_{\mathsf{FS}}\left(M, (\{\mathrm{CMT}_i\}_{i=1}^\kappa, \mathbf{A}, \mathbf{u}, R\right) \in \{1, 2, 3\}^\kappa.$$

4. Let $\Sigma = \Pi_{\mathsf{ring}}$.

$\mathsf{RVerify}_{pp}(M, R, \Sigma)$: Given $pp = \mathbf{A}$, a message M, a ring $R = (\mathbf{d}_0, \ldots, \mathbf{d}_{N-1})$, and a signature Σ, this algorithm proceeds as follows:

1. Run algorithm $\mathsf{TAcc_A}(R)$ to compute the root $\mathbf{u}$ of the tree.

2. Parse Σ as $\Sigma = (\{\mathrm{CMT}_i\}_{i=1}^\kappa, (Ch_1, \ldots, Ch_\kappa), \{\mathrm{RSP}\}_{i=1}^\kappa)$. Return 0 if $(Ch_1, \ldots, Ch_\kappa) \neq \mathcal{H}_{\mathsf{FS}}(M, (\{\mathrm{CMT}_i\}_{i=1}^\kappa, \mathbf{A}, \mathbf{u}, R)$.

3. For each $i = 1$ to κ, run the verification phase of the protocol from Sect. 4.2 with public input $(\mathbf{A}, \mathbf{u})$ to check the validity of RSP_i with respect to CMT_i and Ch_i. If any of the conditions does not hold, then return 0. Otherwise, return 1.

4.4 Analysis of the Ring Signature Scheme

We first summarize the properties of the given ring signature scheme in the following theorem.

Theorem 3. *The ring signature scheme described in Sect. 4.3 is correct, and produces signatures of bit-size $\widetilde{\mathcal{O}}(n \cdot \log N)$. In the random oracle model, the scheme is unforgeable w.r.t. insider corruption based on the worst-case hardness of the $\mathsf{SIVP}_{\widetilde{\mathcal{O}}(n)}$ problem, and it is statistically anonymous.*

Correctness. The correctness of the ring signature scheme directly follows from the correctness of the accumulator scheme in Sect. 3 and the perfect completeness of the argument system in Sect. 4.2: A member of a ring can always obtain a tuple $(\mathbf{x}, \mathbf{d}, w)$ such that $((\mathbf{A}, \mathbf{u}), \mathbf{d}, w, \mathbf{x}) \in \mathrm{R}_{\mathsf{ring}}$, and thus, his signature on any message always get accepted by the verification algorithm.

Efficiency. Since the underlying protocol has communication cost $\widetilde{\mathcal{O}}(\ell \cdot n)$, the signatures produced by the scheme has bit-size $\widetilde{\mathcal{O}}(\kappa \cdot \ell \cdot n) = \widetilde{\mathcal{O}}(\log N \cdot n)$.

Unforgeability with Respect to Insider Corruption. For simplicity, the proof of unforgeability assumes that the cardinality of each ring $R^\star$ is a power of 2. However, this restriction can be easily eliminated, as we will see later on.

The proof of unforgeability relies on the following Lemma from [48].

Lemma 5 ([48], **Lemma 8**). *For any matrix $\mathbf{A} \in \mathbb{Z}_q^{n \times m}$ and a uniformly random $\mathbf{x} \in \{0,1\}^m$, the probability that there exists another $\mathbf{x}' \in \{0,1\}^m \backslash \{\mathbf{x}\}$ such that $\mathbf{A} \cdot \mathbf{x} = \mathbf{A} \cdot \mathbf{x}' \bmod q$ is at least $1 - 2^{n \cdot \log q - m}$.*

With $m = 2nk$ and $\mathbf{x} \overset{\$}{\leftarrow} \{0,1\}^m$, there exists $\mathbf{x}' \in \{0,1\}^m \backslash \{\mathbf{x}\}$ such that $\mathbf{A} \cdot \mathbf{x} = \mathbf{A} \cdot \mathbf{x}' \bmod q$ with overwhelming probability $1 - 2^{-nk}$.

Theorem 4. *The scheme provides unforgeability w.r.t. insider corruption in the random oracle model if the* $\mathsf{SIVP}_{\widetilde{\mathcal{O}}(n)}$ *problem is hard.* (The proof is available in the full version of the paper).

Statistical Anonymity. The proof of the following theorem relies on the statistical witness indistinguishability of the argument system of Lemma 4. The proof is straightforward and omitted.

Theorem 5. *The scheme provides statistical anonymity in the random oracle model.*

Remark 1. As already mentioned, we can handle arbitrary ring sizes. To this end, one option is to add dummy ring members $\mathbf{d}_{\mathsf{fake},1}, \ldots, \mathbf{d}_{\mathsf{fake},r_0}$ whose public keys are sampled obliviously of their private keys, by deriving them as $\mathbf{d}_{\mathsf{fake},j} = \mathsf{bin}(\mathcal{G}_0(j)) \in \{0,1\}^{nk}$ for each $j \in \{1, \ldots, r_0\}$, where $\mathcal{G}_0 : \mathbb{N} \to \mathbb{Z}_q^n$ is an additional random oracle. A simpler solution is to duplicate one of the actual ring members until reaching a multi-set whose cardinality is a power of two.

5   A Lattice-Based Group Signature Without Trapdoors

This section shows how to use our accumulator and argument systems to build a lattice-based group signature which is dramatically more efficient than previous proposals as it does not use any trapdoor. Indeed, surprisingly, the scheme does not rely on a standard digital signature to generate group members' private keys.

5.1   Definitions

We recall the standard definitions and security requirements for static group signatures [8]. A group signature scheme is a tuple of 4 polynomial-time algorithms (GKeygen, GSign, GVerify, GOpen) defined as follows:

- GKeygen: This is a probabilistic algorithm that takes as input $1^n, 1^N$, where $n \in \mathbb{N}$ is the security parameter and $N \in \mathbb{N}$ is the number of group users, and outputs a triple (gpk, gmsk, **gsk**), where gpk is the group public key; gmsk is the group manager's secret key; and **gsk** = (gsk[0], ..., gsk[$N-1$]), where for $j \in \{0, \ldots, N-1\}$, gsk[j] is the secret key for the group user of index j.
- GSign: is a randomized algorithm that inputs gpk, a secret key gsk[j] for some $j \in \{0, \ldots, N-1\}$, and a message M. It returns a group signature Σ on M.
- GVerify: This deterministic algorithm takes as input the group public key gpk, a message M, a purported signature Σ on M, and returns either 1 or 0.
- GOpen: This deterministic algorithm takes as input the group public key gpk, the group manager's secret key gmsk, a message M, a signature Σ on M, and returns an index $j \in \{0, \ldots, N-1\}$, or $\perp$ (to indicate failure).

$$
\begin{array}{|ll|}
\hline
 & \mathbf{Exp}_{\mathcal{GS},\mathcal{A}}^{\mathrm{trace}}(n, N) \\
 & (\mathsf{gpk}, \mathsf{gmsk}, \mathbf{gsk}) \leftarrow \mathsf{GKeygen}(1^n, 1^N) \\
 & \mathsf{st} \leftarrow (\mathsf{gmsk}, \mathsf{gpk}) \\
 & \mathcal{C} \leftarrow \emptyset \; ; \; K \leftarrow \varepsilon \; ; \; Cont \leftarrow \mathsf{true} \\
 & \text{while } (Cont = \mathsf{true}) \text{ do} \\
 & \quad (Cont, \mathsf{st}, j) \leftarrow \\
\mathbf{Exp}_{\mathcal{GS},\mathcal{A}}^{\mathrm{anon}\text{-}b}(n, N) & \mathcal{A}_1^{\mathcal{GS}.\mathsf{GSign}(\mathsf{gpk}, \mathbf{gsk}[\cdot], \cdot)}(\mathsf{st}, K) \\
(\mathsf{gpk}, \mathsf{gmsk}, \mathbf{gsk}) & \quad \text{if } Cont = \mathsf{true} \text{ then } \mathcal{C} \leftarrow \mathcal{C} \cup \{j\}; \\
\quad \leftarrow \mathsf{GKeyGen}(1^n, 1^N) & \quad K \leftarrow \mathbf{gsk}[j] \\
(\mathsf{st}, j_0, j_1, M^\star) & \quad \text{end if} \\
\quad \leftarrow \mathcal{A}_1^{\mathcal{GS}.\mathsf{GOpen}(\mathsf{gpk}, \mathsf{msk}, \cdot, \cdot)}(\mathsf{gpk}, \mathbf{gsk}) & \text{end while}; \\
\Sigma^\star \leftarrow \mathsf{GSign}(\mathsf{gpk}, \mathbf{gsk}[j_b], M^\star) & (M^\star, \Sigma^\star) \leftarrow \mathcal{A}_2^{\mathcal{GS}.\mathsf{GSign}(\mathsf{gpk}, \mathbf{gsk}[\cdot], \cdot)}(\mathsf{st}) \\
b' \leftarrow \mathcal{A}_2^{\mathcal{GS}.\mathsf{GOpen}(\mathsf{gpk}, \mathsf{msk}, \cdot, \cdot), \neg(M^\star, \Sigma^\star)}(\mathsf{st}, \Sigma^\star) & \text{if } \mathsf{GVerify}(\mathsf{gpk}, M^\star, \Sigma^\star) = 0, \; \text{Return } 0 \\
\text{Return } b' & \text{if } \mathsf{GOpen}(\mathsf{gpk}, \mathsf{gmsk}, M^\star, \Sigma^\star) = \perp, \\
 & \quad \text{Return } 1 \\
 & \text{if } \mathsf{GOpen}(\mathsf{gpk}, \mathsf{gmsk}, M^\star, \Sigma^\star) = j^\star \\
 & \quad \wedge (j^\star \in \{0, \ldots, N-1\} \setminus \mathcal{C}) \\
 & \quad \wedge (\text{no signing query involved}(j^\star, M^\star)) \\
 & \text{then Return } 1 \text{ else Return } 0 \\
\hline
\end{array}
$$

Fig. 3. Experiments for the definitions of anonymity and full traceability

Correctness. The correctness requirement is stated as follows. For all $n, N \in \mathbb{N}$, all $(\mathsf{gpk}, \mathsf{gmsk}, \mathbf{gsk})$ produced by $\mathsf{GKeygen}(1^n, 1^N)$, all $j \in \{0, \ldots, N-1\}$, and any message $M \in \{0, 1\}^*$, we have $\mathsf{GVerify}\big(\mathsf{gpk}, M, \mathsf{GSign}(\mathsf{gpk}, \mathbf{gsk}[j], M)\big) = 1$ and $\mathsf{GOpen}\big(\mathsf{gpk}, \mathsf{gmsk}, M, \mathsf{GSign}(\mathbf{gsk}[j], M)\big) = j$.

In static groups, the security model of Bellare, Micciancio and Warinschi subsumes the desirable security properties of group signatures using two security notions called *full anonymity* and *full traceability*.

Full Anonymity. Full anonymity requires that, without the group manager's secret key, no efficient adversary can infer the identity of a user from its signatures. The adversary should even be unable to distinguish signatures from two distinct users j_0, j_1, even knowing their private keys $\mathbf{gsk}[j_0], \mathbf{gsk}[j_1]$. Moreover, this should remain true even when the adversary is granted access to an oracle that opens arbitrary message-signature pairs $(M, \Sigma) \neq (M^\star, \Sigma^\star)$, where $(M^\star, \Sigma^\star)$ is the challenge pair generated by the challenger on behalf of user j_b, for some $b \in \{0, 1\}$. Formally, the attacker, modeled as a two-stage adversary $\mathcal{A} = (\mathcal{A}_1, \mathcal{A}_2)$, is run in the first experiment depicted in Fig. 3. The adversary's advantage is defined as

$$
\mathbf{Adv}_{\mathcal{GS},\mathcal{A}}^{\mathrm{anon}}(n, N) = \big| \Pr[\mathbf{Exp}_{\mathcal{GS},\mathcal{A}}^{\mathrm{anon}\text{-}1}(n, N) = 1] - \Pr[\mathbf{Exp}_{\mathcal{GS},\mathcal{A}}^{\mathrm{anon}\text{-}0}(n, N) = 1] \big| .
$$

Definition 10 (Full anonymity, [8]). *A group signature is fully anonymous if, for any polynomial N and any PPT adversary $\mathcal{A}$, $\mathbf{Adv}_{\mathcal{GS},\mathcal{A}}^{\mathrm{anon}}(n, N)$ is a negligible function in the security parameter n.*

Full Traceability. Full traceability mandates that all signatures, even those created by colluding users *and* the group manager who pool their secrets together, be traceable to a member of the coalition. The attacker is modeled as a two-stage adversary $\mathcal{A} = (\mathcal{A}_1, \mathcal{A}_2)$ which is run in the second experiment of Fig. 3, where it is further granted access to an oracle $\mathcal{GS}.\mathsf{GSign}(\mathsf{gpk}, \mathsf{gsk}[\cdot], \cdot)$ that returns signatures on behalf of any honest group member. Its success probability against $\mathcal{GS}$ is measured as

$$\mathbf{Succ}_{\mathcal{GS},\mathcal{A}}^{\mathrm{trace}}(n, N) = \Pr[\mathbf{Exp}_{\mathcal{GS},\mathcal{A}}^{\mathrm{trace}}(n, N) = 1].$$

Definition 11 (Full traceability, [8]). *A group signature scheme $\mathcal{GS}$ is fully traceable if for any polynomial N and any PPT adversariy $\mathcal{A}$, the probability $\mathbf{Succ}_{\mathcal{GS},\mathcal{A}}^{\mathrm{trace}}(n, N)$ is negligible in the security parameter n.*

5.2 The Underlying Zero-Knowledge Protocol

The group signature scheme that we will present in Sect. 5.3 relies on an extension of the ZKAoK in Sect. 4.2. An encryption layer is added, and the prover additionally has to prove that the given 2 Regev ciphertexts both encrypt the *same* $(j_1, \ldots, j_\ell)^\top$ that was included in w. The associated relation is defined as follows.

Definition 12. *Define* $\mathrm{R}_{\mathrm{group}} = \left\{ (\mathbf{A}, \mathbf{u}, \mathbf{B}, \mathbf{P}_1, \mathbf{P}_2, \mathbf{c}_1, \mathbf{c}_2), \mathbf{d}, w, \mathbf{x}, \mathbf{r}_1, \mathbf{r}_2 \right\}$ *as a relation where*

$$\begin{cases} \mathbf{A} \in \mathbb{Z}_q^{n \times m}; \ \mathbf{u} \in \{0,1\}^{nk}; \ \mathbf{B} \in \mathbb{Z}_p^{n \times m_E}; \\ \forall i \in \{1, 2\} : \mathbf{P}_i \in \mathbb{Z}_p^{\ell \times m_E}; \ \mathbf{c}_i = (\mathbf{c}_{i,1}, \mathbf{c}_{i,2}) \in \mathbb{Z}_p^n \times \mathbb{Z}_p^\ell; \\ \mathbf{d} \in \{0,1\}^{nk}; \ w = ((j_1, \ldots, j_\ell), (\mathbf{w}_\ell, \ldots, \mathbf{w}_1)) \in \{0,1\}^\ell \times (\{0,1\}^{nk})^\ell; \\ \mathbf{x} \in \{0,1\}^m; \ \mathbf{r}_1, \mathbf{r}_2 \in \{0,1\}^{m_E} \end{cases}$$

satisfy

$$\begin{cases} \mathsf{TVerify}_\mathbf{A}(\mathbf{u}, \mathbf{d}, w) = 1 \ \wedge \ \mathbf{A} \cdot \mathbf{x} = \mathbf{G} \cdot \mathbf{d} \bmod q \\ \forall i \in \{1, 2\} : \mathbf{c}_{i,1} = \mathbf{B} \cdot \mathbf{r}_i \bmod p \ \wedge \ \mathbf{c}_{i,2} = \mathbf{P}_i \cdot \mathbf{r}_i + \lfloor \tfrac{p}{2} \rfloor \cdot (j_1, \ldots, j_\ell)^\top \bmod p. \end{cases}$$

To prove in ZK that the vector $(j_1, \ldots, j_\ell)^T$ involved in the new layer is the *same* $(j_1, \ldots, j_\ell)^T$ that was included in w, we introduce the following technique.

- For each $c \in \{0, 1\}$, let $\mathsf{extbit}(c) = \begin{pmatrix} \bar{c} \\ c \end{pmatrix} \in \{0, 1\}^2$.
- For each $b \in \{0, 1\}$, we define the permutation T_b that transforms vector $\mathbf{z} = \begin{pmatrix} z_0 \\ z_1 \end{pmatrix} \in \mathbb{Z}_p^2$ into vector $T_b(\mathbf{z}) = \begin{pmatrix} z_b \\ z_{\bar{b}} \end{pmatrix}$.

Observe that the following equivalence holds: For all $b \in \{0, 1\}$ and all $\mathbf{z} \in \mathbb{Z}_p^2$,

$$\mathbf{z} = \mathsf{extbit}(j_i) \quad \Leftrightarrow \quad T_b(\mathbf{z}) = \mathsf{extbit}(j_i \oplus b). \tag{14}$$

In Stern's framework, this equivalence allows us to prove in ZK the possession of the bit j_i, for every $i \in [\ell]$, by extending j_i to $\mathsf{extbit}(j_i)$ and then, by permuting it with a one-time pad b_i. Furthermore, to prove that the same j_i is involved in both layers, we will use the same one-time pad in both layers of the protocol.

Embedding this new technique into the protocol in Sect. 4.2, we obtain an argument system for the relation R_{group}. As for the previous two protocols, they also rely on the string commitment scheme from [40], which is statistically hiding and computationally binding if the $\mathsf{SIVP}_{\tilde{\mathcal{O}}(n)}$ problem is hard.

Lemma 6. *Assume that the $\mathsf{SIVP}_{\tilde{\mathcal{O}}(n)}$ problem is hard. Then, there exists a statistical ZKAoK for the relation R_{group} with perfect completeness and communication cost $\tilde{\mathcal{O}}(\ell \cdot n) + \mathcal{O}((m_E + \ell) \cdot \log p)$. In particular:*

- *There exists an efficient simulator that, on input $(\mathbf{A}, \mathbf{u}, \mathbf{B}, \mathbf{P}_1, \mathbf{P}_2, \mathbf{c}_1, \mathbf{c}_2)$, outputs an accepted transcript which is statistically close to that produced by the real prover.*
- *There exists an efficient knowledge extractor that, on input 3 valid responses $(\mathrm{RSP}_1, \mathrm{RSP}_2, \mathrm{RSP}_3)$ to the same commitment CMT, outputs $(\mathbf{d}', w', \mathbf{x}', \mathbf{r}_1', \mathbf{r}_2')$ such that*

$$((\mathbf{A}, \mathbf{u}, \mathbf{B}, \mathbf{P}_1, \mathbf{P}_2, \mathbf{c}_1, \mathbf{c}_2), \mathbf{d}', w', \mathbf{x}', \mathbf{r}_1', \mathbf{r}_2') \in R_{\text{group}}.$$

The full description and analysis of the argument system are given in the full version of the paper.

5.3   Our Construction

Let n be the security parameter, and $N = 2^\ell = \mathsf{poly}(n)$ be the maximum expected number of group users. Parameters m, q, k, κ and the random oracle $\mathcal{H}_{\mathsf{FS}}$ are defined as in the ring signature scheme in Sect. 4.3. To employ the ℓ-bit version of Regev's encryption scheme, we will also need prime modulus $p = \tilde{\mathcal{O}}(n^{1.5})$, parameter $m_E = 2(n + \ell)\lceil \log p \rceil$, and an LWE error distribution $\chi = D_{\mathbb{Z}, 2\sqrt{n}}$.

GKeygen$(1^n, 1^N)$: This algorithm begins by sampling a uniformly random matrix $\mathbf{A} \xleftarrow{\$} \mathbb{Z}_q^{n \times m}$. Then, it performs the following steps:

1. For each $j \in [0, N - 1]$, sample a random binary vector $\mathbf{x}_j \xleftarrow{\$} \{0, 1\}^m$ and compute $\mathbf{d}_j = \mathsf{bin}(\mathbf{A} \cdot \mathbf{x}_j \bmod q) \in \{0, 1\}^{nk}$. In the unlikely event that $\{\mathbf{d}_j\}_{j=0}^{N-1}$ are not pairwise distinct, restart the process. Otherwise, define the set $R = (\mathbf{d}_0, \ldots, \mathbf{d}_{N-1})$.
2. Run algorithm $\mathsf{TAcc}_\mathbf{A}(R)$ to build the Merkle tree based on R and the hash function $h_\mathbf{A}$, and obtain the root $\mathbf{u} \in \{0, 1\}^{nk}$.

3. For each $j \in [0, N - 1]$, run algorithm $\mathsf{TWitness}_\mathbf{A}(R, \mathbf{d}_j)$ to output a witness

$$w^{(j)} = \left((j_1, \ldots, j_\ell) \in \{0, 1\}^\ell, (\mathbf{w}_\ell^{(j)}, \ldots, \mathbf{w}_1^{(j)}) \in (\{0, 1\}^{nk})^\ell \right)$$

to the fact that $\mathbf{d}_j$ was accumulated in $\mathbf{u}$. (Note that $(j_1, \ldots, j_\ell)$ is the binary representation of j.) Then define $\mathsf{gsk}[j] = (\mathbf{x}_j, \mathbf{d}_j, w^{(j)})$.
4. Sample $\mathbf{B} \xleftarrow{\$} \mathbb{Z}_p^{n \times m_E}$. For $i \in \{1, 2\}$, sample $\mathbf{S}_i \xleftarrow{\$} \mathbb{Z}_p^{n \times \ell}$, $\mathbf{E}_i \hookleftarrow \chi^{\ell \times m_E}$, and compute $\mathbf{P}_i = \mathbf{S}_i^\top \cdot \mathbf{B} + \mathbf{E}_i \in \mathbb{Z}_p^{\ell \times m_E}$.
5. Output

$$\mathsf{gpk} := \{\mathbf{A}, \mathbf{u}, \mathbf{B}, \mathbf{P}_1, \mathbf{P}_2\}; \quad \mathsf{gmsk} := \mathbf{S}_1; \quad \mathsf{gsk} := (\mathsf{gsk}[0], \ldots, \mathsf{gsk}[N - 1]).$$

GSign($\mathsf{gpk}, \mathsf{gsk}[j], M$): To sign $M \in \{0, 1\}^*$ using $\mathsf{gsk}[j] = (\mathbf{x}_j, \mathbf{d}_j, w^{(j)})$, where $w^{(j)} = ((j_1, \ldots, j_\ell), (\mathbf{w}_\ell^{(j)}, \ldots, \mathbf{w}_1^{(j)}))$, the user conducts the following steps:

1. Encrypt $(j_1, \ldots, j_\ell) \in \{0, 1\}^\ell$ twice using Regev's encryption scheme. Namely, for each $i \in \{1, 2\}$, sample $\mathbf{r}_i \xleftarrow{\$} \{0, 1\}^{m_E}$ and compute

$$\mathbf{c}_i = (\mathbf{c}_{i,1}, \mathbf{c}_{i,2})$$
$$= \left(\mathbf{B} \cdot \mathbf{r}_i \bmod p, \ \mathbf{P}_i \cdot \mathbf{r}_i + \lceil \tfrac{p}{2} \rceil \cdot (j_1, \ldots, j_\ell)^\top \bmod p \right) \in \mathbb{Z}_p^n \times \mathbb{Z}_p^\ell.$$

2. Generate a NIZKAoK Π_{group} in order to demonstrate the possession of a valid tuple $\tau = (\mathbf{x}_j, \mathbf{d}_j, w^{(j)}, \mathbf{r}_1, \mathbf{r}_2)$, where $w^{(j)} = ((j_1, \ldots, j_\ell), (\mathbf{w}_\ell^{(j)}, \ldots, \mathbf{w}_1^{(j)}))$, such that:
   (a) $\mathbf{A} \cdot \mathbf{x}_j = \mathbf{G} \cdot \mathbf{d}_j \bmod q$ and $\mathsf{TVerify}_\mathbf{A}(\mathbf{u}, \mathbf{d}_j, w^{(j)}) = 1$.
   (b) $\mathbf{c}_1$ and $\mathbf{c}_2$ are both correct encryptions of $(j_1, \ldots, j_\ell)$ with randomness $\mathbf{r}_1$ and $\mathbf{r}_2$, respectively.
   This is done by running the protocol in Sect. 5.2 with public input $(\mathbf{A}, \mathbf{u}, \mathbf{B}, \mathbf{P}_1, \mathbf{P}_2, \mathbf{c}_1, \mathbf{c}_2)$ and prover's witness τ defined above. The protocol is repeated $\kappa = \omega(\log n)$ times to achieve negligible soundness error and made non-interactive via the Fiat-Shamir heuristic as a triple $\Pi_{\mathsf{group}} = (\{\mathrm{CMT}_i\}_{i=1}^\kappa, \mathrm{CH}, \{\mathrm{RSP}\}_{i=1}^\kappa)$, where

$$\mathrm{CH} = \mathcal{H}_{\mathsf{FS}}\left(M, (\{\mathrm{CMT}_i\}_{i=1}^\kappa, \mathbf{A}, \mathbf{u}, \mathbf{B}, \mathbf{P}_1, \mathbf{P}_2, \mathbf{c}_1, \mathbf{c}_2) \right) \in \{1, 2, 3\}^\kappa.$$

3. Output the group signature $\Sigma = (\Pi_{\mathsf{group}}, \mathbf{c}_1, \mathbf{c}_2)$.

GVerify(gpk, M, Σ): This algorithm proceeds as follows:

1. Parse Σ as $\Sigma = (\{\mathrm{CMT}_i\}_{i=1}^\kappa, (Ch_1, \ldots, Ch_\kappa), \{\mathrm{RSP}\}_{i=1}^\kappa, \mathbf{c}_1, \mathbf{c}_2)$.
   If $(Ch_1, \ldots, Ch_\kappa) \neq \mathcal{H}_{\mathsf{FS}}\left(M, (\{\mathrm{CMT}_i\}_{i=1}^\kappa, \mathbf{A}, \mathbf{u}, \mathbf{B}, \mathbf{P}_1, \mathbf{P}_2, \mathbf{c}_1, \mathbf{c}_2) \right)$, then return 0.
2. For each $i = 1$ to κ, run the verification phase of the protocol in Sect. 5.2 with public input $(\mathbf{A}, \mathbf{u}, \mathbf{B}, \mathbf{P}_1, \mathbf{P}_2, \mathbf{c}_1, \mathbf{c}_2)$ to check the validity of RSP_i w.r.t. CMT_i and Ch_i. If any of the conditions does not hold, then return 0.
3. Return 1.

GOpen(gpk, gmsk, Σ, M): On input gmsk $= \mathbf{S}_1$ and a group signature $\Sigma = (\Pi_{\text{group}}, \mathbf{c}_1, \mathbf{c}_2)$ on message M, this algorithm decrypts $\mathbf{c}_1 = (\mathbf{c}_{1,1}, \mathbf{c}_{1,2})$ and returns an index $j \in [0, N-1]$, as follows:

1. Compute $(j'_1, \ldots, j'_\ell) = \mathbf{c}_{1,2} - \mathbf{S}_1^\top \cdot \mathbf{c}_{1,1} \in \mathbb{Z}_p^\ell$.
2. For each $i \in [\ell]$, if j'_i is closer to 0 than to $\lceil \frac{p}{2} \rfloor$ modulo p, then let $j_i = 0$; otherwise, let $j_i = 1$.
3. Output index $j \in [0, N-1]$ that has binary representation $(j_1, \ldots, j_\ell)$.

Efficiency. The public key consists of a constant number of matrices over $\mathbb{Z}_q$ and $\mathbb{Z}_p$, where q and p are small moduli. The group signature has bit-size $\kappa \cdot \left(\widetilde{\mathcal{O}}(\ell \cdot n) + \mathcal{O}((m_E + \ell) \cdot \log p) \right) = \widetilde{\mathcal{O}}(\log N \cdot n)$. The scheme is dramatically more efficient than previous lattice-based realizations of group signatures. Indeed, its most important advantage is that it does not require any party to hold a GPV trapdoor. As observed by Lyubashevsky [49], lattice-based signatures without trapdoor can be made significantly more efficient.

Correctness. The correctness of algorithm GVerify follows directly from the correctness of the accumulator scheme in Sect. 3, and the completeness of the argument system in Sect. 5.2. As for the correctness of algorithm GOpen, it suffices to note that

$$\mathbf{c}_{1,2} - \mathbf{S}_1^\top \cdot \mathbf{c}_{1,1} = (\mathbf{S}_1^\top \cdot \mathbf{B} + \mathbf{E}_1) \cdot \mathbf{r}_1 + \left\lceil \frac{p}{2} \right\rfloor \cdot (j_1, \ldots, j_\ell)^\top - \mathbf{S}_1^\top \cdot \mathbf{B} \cdot \mathbf{r}_1$$

$$= \mathbf{E}_1 \cdot \mathbf{r}_1 + \left\lceil \frac{p}{2} \right\rfloor \cdot (j_1, \ldots, j_\ell)^\top \bmod p,$$

and $\|\mathbf{E}_1 \cdot \mathbf{r}_1\|_\infty < p/4$ with overwhelming probability, for the given setting of parameters, and the decryption algorithm should return $(j_1, \ldots, j_\ell)^\top$.

Security. The full traceability property of our scheme is stated in Theorem 6. In the proof, which is given in the full version of the paper we prove that any adversary with noticeable probability of evading traceability implies an algorithm for either breaking the security of the underlying accumulator of Sect. 3, breaking the computational soundness of the argument system in Sect. 5.2, or solving an instance of the $\mathsf{SIS}_{n,m,q,1}^\infty$ problem.

Theorem 6. *The scheme provides full traceability in the random oracle model if the* $\mathsf{SIVP}_{\widetilde{\mathcal{O}}(n)}$ *problem is hard.*

The proof of full anonymity relies on the fact that applying the Naor-Yung paradigm [56] to Regev's cryptosystem yields an IND-CCA2 secure cryptosystem. (A similar argument was used by Benhamouda *et al.* [12] for an NTRU-like encryption scheme.) Indeed, the argument system of Definition 12 implies that $\mathbf{c}_1$ and $\mathbf{c}_2$ encrypt the same message. In the random oracle model, it was already observed by Fouque and Pointcheval [30] (see [13] for a more general treatment) that applying the Fiat-Shamir heuristic to Σ-protocols gives simulation-sound proofs [66]. Similarly to [13,30], the proof of Theorem 7 relies on the fact that applying Fiat-Shamir to the argument system of Definition 12 yields a simulation-sound NIZK argument in the random oracle model if the underlying commitment is computationally binding. This holds even though this argument system does not have the standard special soundness property (*i.e.*, three accepting conversations for distinct challenges are necessary to extract a witness). Simulation-soundness is actually implied by Lemma 6: suppose that $\mathbf{c}_1$ and $\mathbf{c}_2$ encrypt distinct ℓ-bit strings. This means that there exists no vector $(\mathbf{r}_1^T \mid \mathbf{r}_2^T)^T$ such that

$$\begin{bmatrix} \mathbf{B} & -\mathbf{B} \\ \mathbf{P}_1 & -\mathbf{P}_2 \end{bmatrix} \cdot \begin{bmatrix} \mathbf{r}_1 \\ \mathbf{r}_2 \end{bmatrix} = \begin{bmatrix} \mathbf{c}_{1,1} - \mathbf{c}_{2,1} \\ \mathbf{c}_{2,1} - \mathbf{c}_{2,2} \end{bmatrix}.$$

Now, recall that the computational soundness of all Stern-type protocols is proved by showing that the knowledge extractor obtains either a set of valid witnesses or breaks the binding property of the underlying commitment. Given that the witnesses do not exist if the statement is false, by rewinding a simulation-soundness adversary sufficiently many times, the knowledge extractor necessarily extracts two openings of a given commitment.

The proof of Theorem 7 is similar to [66] and given in the full version of the paper.

Theorem 7. *The scheme provides full anonymity if the* $\mathsf{LWE}_{n,p,\chi}$ *problem is hard, and if the argument system is simulation-sound.*

Acknowledgements. We thank Damien Stehlé for useful discussions and the anonymous reviewers of EUROCRYPT 2016 for helpful comments. The first author was funded by the "Programme Avenir Lyon Saint-Etienne de l'Université de Lyon" in the framework of the programme "Investissements d'Avenir" (ANR-11-IDEX-0007). San Ling, Khoa Nguyen and Huaxiong Wang were supported by the "Singapore Ministry of Education under Research Grant MOE2013-T2-1-041".

References

1. Acar, T., Nguyen, L.: Revocation for delegatable anonymous credentials. In: Catalano, D., Fazio, N., Gennaro, R., Nicolosi, A. (eds.) PKC 2011. LNCS, vol. 6571, pp. 423–440. Springer, Heidelberg (2011)
2. Aguilar Melchor, C., Bettaieb, S., Boyen, X., Fousse, L., Gaborit, P.: Adapting Lyubashevsky's signature schemes to the ring signature setting. In: Youssef, A., Nitaj, A., Hassanien, A.E. (eds.) AFRICACRYPT 2013. LNCS, vol. 7918, pp. 1–25. Springer, Heidelberg (2013)
3. Ajtai, M.: Generating hard instances of lattice problems (extended abstract). In: STOC, pp. 99–108. ACM (1996)
4. Ajtai, M.: Generating hard instances of the short basis problem. In: Wiedermann, J., Emde Boas, P., Nielsen, M. (eds.) ICALP 1999. LNCS, vol. 1644, p. 1. Springer, Heidelberg (1999)
5. Ateniese, G., Camenisch, J.L., Joye, M., Tsudik, G.: A practical and provably secure coalition-resistant group signature scheme. In: Bellare, M. (ed.) CRYPTO 2000. LNCS, vol. 1880, p. 255. Springer, Heidelberg (2000)

6. Au, M.H., Wu, Q., Susilo, W., Mu, Y.: Compact e-cash from bounded accumulator. In: Abe, M. (ed.) CT-RSA 2007. LNCS, vol. 4377, pp. 178–195. Springer, Heidelberg (2006)
7. Barić, N., Pfitzmann, B.: Collision-free accumulators and fail-stop signature schemes without trees. In: Fumy, W. (ed.) EUROCRYPT 1997. LNCS, vol. 1233, pp. 480–494. Springer, Heidelberg (1997)
8. Bellare, M., Micciancio, D., Warinschi, B.: Foundations of group signatures: formal definitions, simplified requirements, and a construction based on general. In: Biham, E. (ed.) EUROCRYPT 2003. LNCS, vol. 2656, pp. 614–629. Springer, Heidelberg (2003)
9. Ben-Sasson, E., Chiesa, A., Garman, C., Green, M., Miers, I., Tromer, E., Virza, M.: Zerocash: decentralized anonymous payments from bitcoin. In: IEEE S&P, pp. 459–474. IEEE (2014)
10. Benaloh, J.C., de Mare, M.: One-way accumulators: a decentralized alternative to digital signatures. In: Helleseth, T. (ed.) EUROCRYPT 1993. LNCS, vol. 765, pp. 274–285. Springer, Heidelberg (1994)
11. Bender, A., Katz, J., Morselli, R.: Ring signatures: stronger definitions, and constructions without random Oracles. J. Cryptol. 22(1), 114–138 (2009)
12. Benhamouda, F., Camenisch, J., Krenn, S., Lyubashevsky, V., Neven, G.: Better zero-knowledge proofs for lattice encryption and their application to group signatures. In: Sarkar, P., Iwata, T. (eds.) ASIACRYPT 2014. LNCS, vol. 8873, pp. 551–572. Springer, Heidelberg (2014)
13. Bernhard, D., Fischlin, M., Warinschi, B.: Adaptive proofs of knowledge in the random Oracle model. In: Katz, J. (ed.) PKC 2015. LNCS, vol. 9020, pp. 629–649. Springer, Heidelberg (2015)
14. Boneh, D., Boyen, X.: Efficient selective-ID secure identity-based encryption without random Oracles. In: Cachin, C., Camenisch, J.L. (eds.) EUROCRYPT 2004. LNCS, vol. 3027, pp. 223–238. Springer, Heidelberg (2004)
15. Boneh, D., Corrigan-Gibbs, H.: Bivariate polynomials modulo composites and their applications. In: Sarkar, P., Iwata, T. (eds.) ASIACRYPT 2014. LNCS, vol. 8873, pp. 42–62. Springer, Heidelberg (2014)
16. Bootle, J., Cerulli, A., Chaidos, P., Ghadafi, E., Groth, J., Petit, C.: Short accountable ring signatures based on DDH. In: Pernul, G., et al. (eds.) ESORICS. LNCS, vol. 9326, pp. 243–265. Springer, Heidelberg (2015). doi:10.1007/978-3-319-24174-6_13
17. Boyen, X.: Lattice mixing and vanishing trapdoors: a framework for fully secure short signatures and more. In: Nguyen, P.Q., Pointcheval, D. (eds.) PKC 2010. LNCS, vol. 6056, pp. 499–517. Springer, Heidelberg (2010)
18. Brakerski, Z., Kalai, Y.T.: A framework for efficient signatures, ring signatures and identity based encryption in the standard model. IACR Cryptol. ePrint Archive 2010:86 (2010)
19. Camenisch, J., Kohlweiss, M., Soriente, C.: An accumulator based on bilinear maps and efficient revocation for anonymous credentials. In: Jarecki, S., Tsudik, G. (eds.) PKC 2009. LNCS, vol. 5443, pp. 481–500. Springer, Heidelberg (2009)
20. Camenisch, J.L., Lysyanskaya, A.: Dynamic accumulators and application to efficient revocation of anonymous credentials. In: Yung, M. (ed.) CRYPTO 2002. LNCS, vol. 2442, p. 61. Springer, Heidelberg (2002)
21. Camenisch, J., Neven, G., Rückert, M.: Fully anonymous attribute tokens from lattices. In: Visconti, I., Prisco, R. (eds.) SCN 2012. LNCS, vol. 7485, pp. 57–75. Springer, Heidelberg (2012)

22. Canard, S., Gouget, A.: Multiple denominations in E-cash with compact transaction data. In: Sion, R. (ed.) FC 2010. LNCS, vol. 6052, pp. 82–97. Springer, Heidelberg (2010)
23. Cash, D., Hofheinz, D., Kiltz, E., Peikert, C.: Bonsai trees, or how to delegate a lattice basis. In: Gilbert, H. (ed.) EUROCRYPT 2010. LNCS, vol. 6110, pp. 523–552. Springer, Heidelberg (2010)
24. Catalano, D., Fiore, D.: Vector commitments and their applications. In: Kurosawa, K., Hanaoka, G. (eds.) PKC 2013. LNCS, vol. 7778, pp. 55–72. Springer, Heidelberg (2013)
25. Chandran, N., Groth, J., Sahai, A.: Ring signatures of sub-linear size without random Oracles. In: Arge, L., Cachin, C., Jurdziński, T., Tarlecki, A. (eds.) ICALP 2007. LNCS, vol. 4596, pp. 423–434. Springer, Heidelberg (2007)
26. Chaum, D., van Heyst, E.: Group signatures. In: Davies, D.W. (ed.) EUROCRYPT 1991. LNCS, vol. 547, pp. 257–265. Springer, Heidelberg (1991)
27. Derler, D., Hanser, C., Slamanig, D.: Revisiting cryptographic accumulators, additional properties and relations to other primitives. In: Nyberg, K. (ed.) CT-RSA 2015. LNCS, vol. 9048, pp. 127–144. Springer, Heidelberg (2015)
28. Dodis, Y., Kiayias, A., Nicolosi, A., Shoup, V.: Anonymous identification in *Ad Hoc* groups. In: Cachin, C., Camenisch, J.L. (eds.) EUROCRYPT 2004. LNCS, vol. 3027, pp. 609–626. Springer, Heidelberg (2004)
29. Ezerman, M.F., Lee, H.T., Ling, S., Nguyen, K., Wang, H.: A provably secure group signature scheme from code-based assumptions. In: Iwata, T., et al. (eds.) ASIACRYPT 2015. LNCS, vol. 9452, pp. 260–285. Springer, Heidelberg (2015). doi:10.1007/978-3-662-48797-6_12
30. Fouque, P.-A., Pointcheval, D.: Threshold cryptosystems secure against chosen-ciphertext attacks. In: Boyd, C. (ed.) ASIACRYPT 2001. LNCS, vol. 2248, p. 351. Springer, Heidelberg (2001)
31. Gentry, C., Peikert, C., Vaikuntanathan, V.: Trapdoors for hard lattices and new cryptographic constructions. In: STOC, pp. 197–206. ACM (2008)
32. Goldreich, O., Goldwasser, S., Halevi, S.: Collision-free hashing from lattice problems. ECCC **3**(42) (1996)
33. Gordon, S.D., Katz, J., Vaikuntanathan, V.: A group signature scheme from lattice assumptions. In: Abe, M. (ed.) ASIACRYPT 2010. LNCS, vol. 6477, pp. 395–412. Springer, Heidelberg (2010)
34. Groth, J.: Evaluating security of voting schemes in the universal composability framework. In: Jakobsson, M., Yung, M., Zhou, J. (eds.) ACNS 2004. LNCS, vol. 3089, pp. 46–60. Springer, Heidelberg (2004)
35. Groth, J.: Short pairing-based non-interactive zero-knowledge arguments. In: Abe, M. (ed.) ASIACRYPT 2010. LNCS, vol. 6477, pp. 321–340. Springer, Heidelberg (2010)
36. Groth, J., Kohlweiss, M.: One-out-of-many proofs: or how to leak a secret and spend a coin. In: Oswald, E., Fischlin, M. (eds.) EUROCRYPT 2015. LNCS, vol. 9057, pp. 253–280. Springer, Heidelberg (2015)
37. Jain, A., Krenn, S., Pietrzak, K., Tentes, A.: Commitments and efficient zero-knowledge proofs from learning parity with noise. In: Wang, X., Sako, K. (eds.) ASIACRYPT 2012. LNCS, vol. 7658, pp. 663–680. Springer, Heidelberg (2012)
38. Jhanwar, M.P., Safavi-Naini, R.: Compact accumulator using lattices. IACR Cryptology ePrint Archive: Report 2014/1015, February 2015
39. Kawachi, A., Tanaka, K., Xagawa, K.: Multi-bit cryptosystems based on lattice problems. In: Okamoto, T., Wang, X. (eds.) PKC 2007. LNCS, vol. 4450, pp. 315–329. Springer, Heidelberg (2007)

40. Kawachi, A., Tanaka, K., Xagawa, K.: Concurrently secure identification schemes based on the worst-case hardness of lattice problems. In: Pieprzyk, J. (ed.) ASIACRYPT 2008. LNCS, vol. 5350, pp. 372–389. Springer, Heidelberg (2008)
41. Laguillaumie, F., Langlois, A., Libert, B., Stehlé, D.: Lattice-based group signatures with logarithmic signature size. In: Sako, K., Sarkar, P. (eds.) ASIACRYPT 2013, Part II. LNCS, vol. 8270, pp. 41–61. Springer, Heidelberg (2013)
42. Langlois, A., Ling, S., Nguyen, K., Wang, H.: Lattice-based group signature scheme with verifier-local revocation. In: Krawczyk, H. (ed.) PKC 2014. LNCS, vol. 8383, pp. 345–361. Springer, Heidelberg (2014)
43. Li, J., Li, N., Xue, R.: Universal accumulators with efficient nonmembership proofs. In: Katz, J., Yung, M. (eds.) ACNS 2007. LNCS, vol. 4521, pp. 253–269. Springer, Heidelberg (2007)
44. Lin, Z., Hopper, N.: Jack: scalable accumulator-based nymble system. In: WPES, pp. 53–62. ACM (2010)
45. Ling, S., Nguyen, K., Stehlé, D., Wang, H.: Improved zero-knowledge proofs of knowledge for the ISIS problem, and applications. In: Kurosawa, K., Hanaoka, G. (eds.) PKC 2013. LNCS, vol. 7778, pp. 107–124. Springer, Heidelberg (2013)
46. Ling, S., Nguyen, K., Wang, H.: Group signatures from lattices: simpler, tighter, shorter, ring-based. In: Katz, J. (ed.) PKC 2015. LNCS, vol. 9020, pp. 427–449. Springer, Heidelberg (2015)
47. Lipmaa, H.: Secure accumulators from Euclidean rings without trusted setup. In: Bao, F., Samarati, P., Zhou, J. (eds.) ACNS 2012. LNCS, vol. 7341, pp. 224–240. Springer, Heidelberg (2012)
48. Lyubashevsky, V.: Lattice-based identification schemes secure under active attacks. In: Cramer, R. (ed.) PKC 2008. LNCS, vol. 4939, pp. 162–179. Springer, Heidelberg (2008)
49. Lyubashevsky, V.: Lattice signatures without trapdoors. In: Pointcheval, D., Johansson, T. (eds.) EUROCRYPT 2012. LNCS, vol. 7237, pp. 738–755. Springer, Heidelberg (2012)
50. Merkle, R.C.: A certified digital signature. In: Brassard, G. (ed.) CRYPTO 1989. LNCS, vol. 435, pp. 218–238. Springer, Heidelberg (1990)
51. Micciancio, D., Mol, P.: Pseudorandom Knapsacks and the sample complexity of LWE search-to-decision reductions. In: Rogaway, P. (ed.) CRYPTO 2011. LNCS, vol. 6841, pp. 465–484. Springer, Heidelberg (2011)
52. Micciancio, D., Peikert, C.: Hardness of SIS and LWE with small parameters. In: Canetti, R., Garay, J.A. (eds.) CRYPTO 2013, Part I. LNCS, vol. 8042, pp. 21–39. Springer, Heidelberg (2013)
53. Micciancio, D., Regev, O.: Worst-case to average-case reductions based on Gaussian measures. SIAM J. Comput. $37(1)$, 267–302 (2007)
54. Miers, I., Garman, C., Green, M., Rubin, A.D.: Zerocoin: anonymous distributed e-cash from bitcoin. In: IEEE S&P, pp. 397–411. IEEE (2013)
55. Naor, M.: On cryptographic assumptions and challenges. In: Boneh, D. (ed.) CRYPTO 2003. LNCS, vol. 2729, pp. 96–109. Springer, Heidelberg (2003)
56. Naor, M., Yung, M.: Public-key cryptosystems provably secure against chosen ciphertextattacks. In: STOC, pp. 427–437. ACM (1990)
57. Nguyen, L.: Accumulators from bilinear pairings and applications. In: Menezes, A. (ed.) CT-RSA 2005. LNCS, vol. 3376, pp. 275–292. Springer, Heidelberg (2005)
58. Nguyen, P.Q., Zhang, J., Zhang, Z.: Simpler efficient group signatures from lattices. In: Katz, J. (ed.) PKC 2015. LNCS, vol. 9020, pp. 401–426. Springer, Heidelberg (2015)

59. Papamanthou, C., Shi, E., Tamassia, R., Yi, K.: Streaming authenticated data structures. In: Johansson, T., Nguyen, P.Q. (eds.) EUROCRYPT 2013. LNCS, vol. 7881, pp. 353–370. Springer, Heidelberg (2013)
60. Papamanthou, C., Tamassia, R., Triandopoulos, N.: Authenticated hash tables. In: ACM-CCS, pp. 437–448. ACM (2008)
61. Peikert, C.: Public-key cryptosystems from the worst-case shortest vectorproblem: extended abstract. In: STOC, pp. 333–342. ACM (2009)
62. Peikert, C., Vaikuntanathan, V., Waters, B.: A framework for efficient and composable oblivious transfer. In: Wagner, D. (ed.) CRYPTO 2008. LNCS, vol. 5157, pp. 554–571. Springer, Heidelberg (2008)
63. Prabhakaran, M., Xue, R.: Statistically hiding sets. In: Fischlin, M. (ed.) CT-RSA 2009. LNCS, vol. 5473, pp. 100–116. Springer, Heidelberg (2009)
64. Regev, O.: On lattices, learning with errors, random linear codes, and cryptography. In: STOC, pp. 84–93. ACM (2005)
65. Rivest, R.L., Shamir, A., Tauman, Y.: How to leak a secret. In: Boyd, C. (ed.) ASIACRYPT 2001. LNCS, vol. 2248, p. 552. Springer, Heidelberg (2001)
66. Sahai, A.: Non-malleable non-interactive zero knowledge and adaptive chosen-ciphertext security. FOCS 1999, 543–553 (1999)
67. Stern, J.: A new paradigm for public key identification. IEEE Trans. Inf. Theor. 42(6), 1757–1768 (1996)
68. Tsudik, G., Xu, S.: Accumulating composites and improved group signing. In: Laih, C.-S. (ed.) ASIACRYPT 2003. LNCS, vol. 2894, pp. 269–286. Springer, Heidelberg (2003)
69. Xue, R., Li, N., Li, J.: Algebraic construction for zero-knowledge sets. J. Comput. Sci. Technol. 23(2), 166–175 (2008)

Adaptively Secure Identity-Based Encryption from Lattices with Asymptotically Shorter Public Parameters

Shota Yamada[✉]

National Institute of Advanced Industrial Science
and Technology (AIST), Tokyo, Japan
yamada-shota@aist.go.jp

Abstract. In this paper, we present two new adaptively secure identity-based encryption (IBE) schemes from lattices. The size of the public parameters, ciphertexts, and private keys are $\tilde{O}(n^2 \kappa^{1/d})$, $\tilde{O}(n)$, and $\tilde{O}(n)$ respectively. Here, n is the security parameter, κ is the length of the identity, and $d \in \mathbb{N}$ is a flexible constant that can be set arbitrary (but will affect the reduction cost). Ignoring the poly-logarithmic factors hidden in the asymptotic notation, our schemes achieve the best efficiency among existing adaptively secure IBE schemes from lattices. In more detail, our first scheme is anonymous, but proven secure under the LWE assumption with approximation factor $n^{\omega(1)}$. Our second scheme is not anonymous, but proven adaptively secure assuming the LWE assumption for all polynomial approximation factors.

As a side result, based on a similar idea, we construct an attribute-based encryption scheme for branching programs that simultaneously satisfies the following properties for the first time: Our scheme achieves compact secret keys, the security is proven under the LWE assumption with polynomial approximation factors, and the scheme can deal with unbounded length branching programs.

1 Introduction

Background. Identity-based encryption (IBE) is an advanced form of public key encryption (PKE) where any string such as an email address can be used as a public key. The notion of IBE was proposed by Shamir in 1984 [42]. Since then, it took nearly 20 years for the first realizations of IBE [10,18,41] to appear. Boneh and Franklin [10] and Sakai, Ohgishi, and Kasahara [41] used groups equipped with efficiently computable bilinear maps to construct the first IBE. On the other hand, Cocks [18] used quadratic residue for a composite modulus. These constructions are only proven secure in the random oracle model. In subsequent works, pairing-based schemes in the standard model appeared [8,9,15,47,48]. While earlier works [8,15] focus on the constructions that are only selectively secure, later works [9,47,48] focus on a much more realistic security, i.e., adaptive security.

Another important line of research is construction of IBE from lattices. The first lattice-based IBE was proposed in the seminal work by Gentry, Peikert,

© International Association for Cryptologic Research 2016
M. Fischlin and J.-S. Coron (Eds.): EUROCRYPT 2016, Part II, LNCS 9666, pp. 32–62, 2016.
DOI: 10.1007/978-3-662-49896-5_2

and Vaikuntanathan [25] in the random oracle model. Later, constructions in the standard model were proposed [1,12,16]. To achieve adaptive security in the lattice-based settings, we have to either rely on an analogue of Waters' hash [47] or an admissible hash [9,16]. In any case, we require $O(\kappa)$ number of basic matrices in the public parameters (master public key), where κ is the bit length of the identities. This results in very large public parameters with size $\tilde{O}(n^2\kappa)$. Here, n is the security parameter (dimension of the lattices). On the other hand, in the selectively secure variant of lattice IBE in [1], we only require small constant number of basic matrices in the public parameters. This stands in sharp contrast to pairing-based settings, in which we have adaptively secure IBE schemes [17,31] that are as efficient as selectively secure ones [8], up to only small constant factors. A natural important question is:

> *Can we construct adaptively secure IBE schemes from lattices, which is as efficient as selectively secure ones? In particular, can we reduce the size of the public parameters?*

Difficulties. A natural approach to achieve short public parameters in lattice based IBE schemes would be to mimic the technique for pairing based IBE schemes. However, all IBE schemes with short public parameters based on pairings are constructed using dual system encryption methodology [48], for which there is still no lattice analogue. The realization of the dual system encryption methodology in the lattice settings is an important open problem [38]. Another possible approach would be to use a technique from Naccache's IBE scheme [36], as is done in [44]. Using this approach, we can obtain a scheme with the public parameters shorter by a factor of u, at the cost of 2^u-loss in security. Therefore, using this approach, we are only allowed to reduce the size of public parameters up to logarithmic factor.

Our Contribution. Instead of taking the above approaches, we use a technique unique to the lattice setting. Namely, we use the fully homomorphic computation of trapdoors, which is recently devised in [11] to reduce the size of the public parameters. We obtain the following two different IBE schemes with trade-off between the security, efficiency, and underlying hardness assumptions. See Table 1 in Sect. 6 for the overview.

- We propose an adaptively secure and anonymous IBE with asymptotically short parameters. In particular, the size of the public parameters, ciphertexts, and private keys are $\tilde{O}(n^2\kappa^{1/d})$, $\tilde{O}(n)$, and $\tilde{O}(n)$ respectively. Here, $d \in \mathbb{N}$ is a flexible constant which can be set arbitrary. Ignoring poly-logarithmic factors hidden in the asymptotic notation, our scheme achieves the best efficiency among all previous adaptively secure IBE schemes from lattices. The security of the scheme is proven under the LWE assumption with super-polynomial approximation factors.
- We propose an adaptively secure IBE (without anonymity) that achieves asymptotically the same efficiency as the above scheme. The difference from the above scheme is that our scheme can be proven secure assuming the

LWE assumption with all polynomial approximation factor. The assumption is weaker than the one used in the above scheme, but the sizes of the public parameters, ciphertexts, and private keys are larger than the above scheme by a super-constant factor.

In the second construction, different from lattice IBE schemes in the literature [1,2,12,16], we have to rely on the LWE assumption for *all* polynomial approximation factors, rather than some *fixed* polynomial approximation factor (e.g., $O(n^3)$). The interesting feature of the reduction is that the problem we reduce the security to varies according to the power of the adversary. More specifically, as the number of key extraction queries grows or as the advantage of the adversary drops, we would need the LWE assumption with larger approximation factor. This is somewhat similar to the security proof based on the q-type assumptions (e.g., [24]), in which the problem that the reduction algorithm solves depends on the number of key extraction queries made by the adversary. However, unlike the q-type assumptions, our assumptions enjoy reduction to the worst case lattice problems [13,37,40].

To present our schemes in a unified manner, we define the new notion of parametrized IBE (PIBE). The syntax of PIBE is the same as that of ordinary IBE except that it is parametrized by a variable c. As for the security, roughly speaking, we require the advantage of any adversary to be at most $1/n^c$ if the number of key extraction queries is bounded by n^c. In the case of c is a super-constant function, the notion of PIBE corresponds to that of (ordinary) IBE. We then construct a specific PIBE scheme from the LWE assumption. By setting c to be a super-constant function, we obtain our first IBE scheme. Our second IBE scheme is obtained by running several instances of the PIBE scheme in parallel with different values of c. This is captured as a generic conversion from PIBE to (ordinary) IBE.

We note that our IBE schemes might not be as efficient as previous adaptively secure lattice IBE schemes [1,12] for a practical choice of parameters, due to the super-constant factors hidden in the asymptotic notation. However, we believe that our technique would be of theoretical interest. In particular, the security proof of our PIBE scheme is based on the traditional partitioning technique [47] with some novel ideas. In addition, our technique used in the generic construction of IBE from PIBE, inspired by [7], would be useful for other settings.

Other Application of Our Technique. As a side result, we show an application of our technique to attribute-based encryption (ABE). In particular, we obtain the first ABE scheme that simultaneously satisfies the following properties: an unbounded length branching program is usable as an attribute, the sizes of the private keys are compact, the security is proven under the LWE problem for all polynomial approximation factors. We obtain such a scheme by applying a simple conversion to the recent ABE scheme for branching programs by Gorbunov and Vinayagamurthy [28]. The idea for the conversion is similar in spirit to our PIBE-to-IBE conversion. We note that the original ABE scheme of [28] is either based on the super-polynomial LWE while dealing with unbounded length branching programs or based on the polynomial LWE while only dealing with bounded length branching programs. The details appear in the full version [50].

Related Works. We can obtain efficient PKE as well as IBE schemes over ideal lattices [22, 45]. By switching to the ring setting, we can generally reduce the size of the public parameters by an factor of $O(n)$. However, we have to rely on the ring LWE (RLWE) assumption [33, 34], which is a stronger assumption than the LWE assumption.

The techniques for constructing IBE and signatures are somewhat similar and related. Indeed, we can obtain secure signature from (adaptively) secure IBE, via the Naor transformation [10]. A construction of short signature with short public parameters from weak assumptions has been an important research topic. This problem has been addressed by several previous works [4,7,23,30,32]. However, their techniques heavily depend on the fact that we can convert a non-adaptively secure signature scheme into adaptively secure (or equivalently, EUF-CMA secure) one by using chameleon hash functions [43]. There is no known analogue of the conversion in the setting of IBE. We also note that our technique of converting PIBE into IBE is similar to the "on the fly adaptation technique" in [21], which was used to improve the efficiency and the reduction cost of the Naor-Reingold PRF.

2 Overview of Our Technique

2.1 Overview of the Construction

We follow the general framework for constructing lattice-based IBE schemes, which is an abstraction of many existing schemes [1,2,16]. In the template, we associate each identity ID with the following matrix:

$$(\mathbf{A}|\mathsf{H}(\mathsf{ID})) \in \mathbb{Z}_q^{n \times (m+m')}$$

where $\mathbf{A} \in \mathbb{Z}_q^{n \times m}$ and $\mathsf{H}(\cdot)$ is a function that maps an identity to a matrix in $\mathbb{Z}_q^{n \times m'}$ for some $n, m, m' \in \mathbb{N}$ and some prime number q. A ciphertext for an identity ID includes a vector of the following form:

$$\mathbf{s}^\top(\mathbf{A}|\mathsf{H}(\mathsf{ID})) + (\mathbf{x}_1^\top|\mathbf{x}_2^\top)$$

where $\mathbf{s}$ is a random vector in $\mathbb{Z}_q^n$ and $\mathbf{x}_1 \in \mathbb{Z}_q^m$ and $\mathbf{x}_2 \in \mathbb{Z}_q^{m'}$ are small error terms. A private key is a short vector $\mathbf{e} \in \mathbb{Z}^{m+m'}$ that satisfies

$$(\mathbf{A}|\mathsf{H}(\mathsf{ID}))\mathbf{e} = \mathbf{u} \mod q$$

for some fixed $\mathbf{u} \in \mathbb{Z}_q^n$. In the adaptively secure variant of the IBE scheme in [1], the function $\mathsf{H}(\mathsf{ID})$ is defined as

$$\mathsf{H}(\mathsf{ID}) = \mathbf{B}_0 + \sum_{\{i \in [1,\kappa] \mid \mathsf{ID}_i = 1\}} \mathbf{B}_i$$

where $\mathbf{B}_0, \mathbf{B}_1, \ldots, \mathbf{B}_\kappa \in \mathbb{Z}_q^{n \times m}$ are matrices that are included in the public parameters and ID_i is the i-th bit of the bit string $\mathsf{ID} \in \{0,1\}^\kappa$. We typically set $\kappa = O(n)$ and require rather long public parameters $\mathbf{B}_0, \mathbf{B}_1, \ldots, \mathbf{B}_\kappa$.

Our first idea is to use the technique called fully homomorphic trapdoor computation, which is introduced in [11], to reduce the size of the public parameters. Namely, we set $\ell = \lceil \sqrt{\kappa} \rceil$ and the public parameters as matrices $\mathbf{B}_{1,1} \ldots, \mathbf{B}_{1,\ell}$, $\mathbf{B}_{2,1} \ldots, \mathbf{B}_{2,\ell} \in \mathbb{Z}_q^{n \times m}$. We also introduce an injective map $S : \{0,1\}^\kappa \to 2^{[\ell] \times [\ell]}$ that maps an identity to a subset of the set $[\ell] \times [\ell]$. Then, we change the definition of the function as

$$H(\mathsf{ID}) = \mathbf{B}_0 + \sum_{(i,j) \in S(\mathsf{ID})} \mathbf{B}_{1,i} \cdot \mathbf{G}^{-1}(\mathbf{B}_{2,j}),$$

where $\mathbf{G}$ is a gadget matrix whose trapdoor is publicly known [35] and $\mathbf{G}^{-1}$ is a deterministic function[1] that maps a matrix in $\mathbf{U} = \mathbb{Z}_q^{n \times m}$ to a matrix in $\mathbf{V} = \{0,1\}^{m \times m}$ such that $\mathbf{G}\mathbf{V} = \mathbf{U}$. By this change, we are able to reduce the number of basic matrices from $O(\kappa)$ to $O(\sqrt{\kappa})$.[2]

2.2   Overview of the Security Proof

We prove the security of the scheme under the LWE assumption. Let the input to the reduction algorithm be $\mathbf{A} \in \mathbb{Z}_q^{n \times m}$ and $\mathbf{v} \in \mathbb{Z}_q^m$. The task of the algorithm is to distinguish whether $\mathbf{v}^\top = \mathbf{s}^\top \mathbf{A} + \mathbf{x}^\top \mod q$ for some $\mathbf{s} \in \mathbb{Z}_q^n$ and small $\mathbf{x} \in \mathbb{Z}^m$, or, $\mathbf{v}$ is a random vector. In the security proof, we pick random $y_0, y_{1,1}, \ldots, y_{1,\ell}, y_{2,1}, \ldots, y_{2,\ell} \in \mathbb{Z}_q$ from certain domains, whose sizes grow proportion to the number of key extraction queries Q that the adversary makes (similarly to in [47]). Since we assume that Q is much smaller than q, these random values are bounded by some "small" polynomial. Then, the reduction algorithm picks $\mathbf{R}_0, \mathbf{R}_{i,j} \xleftarrow{\$} \{-1,1\}^{m \times m}$ and embeds these values into the public parameters as

$$\mathbf{B}_0 = \mathbf{A}\mathbf{R}_0 + y_0 \mathbf{G}, \qquad \mathbf{B}_{i,j} = \mathbf{A}\mathbf{R}_{i,j} + y_{i,j} \mathbf{G}$$

for $(i,j) \in \{1,2\} \times [1,\ell]$. Then, we have

$$H(\mathsf{ID}) = (\mathbf{A}\mathbf{R}_0 + y_0 \mathbf{G}) + \sum_{(i,j) \in S(\mathsf{ID})} (\mathbf{A}\mathbf{R}_{1,i} + y_{1,i} \mathbf{G}) \cdot \mathbf{G}^{-1}(\mathbf{B}_{2,j})$$

$$= (\mathbf{A}\mathbf{R}_0 + y_0 \mathbf{G}) + \sum_{(i,j) \in S(\mathsf{ID})} (\mathbf{A}\mathbf{R}_{1,i} \mathbf{G}^{-1}(\mathbf{B}_{2,j}) + y_{1,i} \mathbf{B}_{2,j})$$

---

[1] Note that we are abusing the notation here. $\mathbf{G}^{-1}$ is not an inverse matrix of $\mathbf{G}$, but a function.

[2] For the sake of simplicity, we present a scheme that is a special case of our scheme in Sect. 5. More generally, we can further reduce the number of basic matrices from $O(\sqrt{\kappa})$ to be $O(\kappa^{1/d})$ for any constant $d \in \mathbb{N}$.

$$= \mathbf{A} \underbrace{\left(\mathbf{R}_0 + \sum_{(i,j) \in S(\mathsf{ID})} (\mathbf{R}_{1,i} \mathbf{G}^{-1}(\mathbf{B}_{2,j}) + y_{1,i} \mathbf{R}_{2,j}) \right)}_{:= \mathbf{R}_{\mathsf{ID}}, \text{ which is "small"}}$$

$$+ \underbrace{\left(y_0 + \sum_{(i,j) \in S(\mathsf{ID})} y_{1,i} y_{2,j} \right)}_{:= \mathsf{F}_{\mathbf{y}}(\mathsf{ID})} \cdot \mathbf{G}.$$

$$= \mathbf{A} \mathbf{R}_{\mathsf{ID}} + \mathsf{F}_{\mathbf{y}}(\mathsf{ID}) \mathbf{G}.$$

The reduction algorithm has a trapdoor for the matrix $(\mathbf{A} \| \mathsf{H}(\mathsf{ID}))$ if $\mathsf{F}_{\mathbf{y}}(\mathsf{ID}) \neq 0 \mod q$ and thus can simulate a private key for such an identity ID. ($\mathbf{R}_{\mathsf{ID}}$ corresponds to the $\mathbf{G}$-trapdoor [35] of $(\mathbf{A} \| \mathsf{H}(\mathsf{ID}))$). On the other hand, the reduction algorithm expects the challenge identity $\mathsf{ID}^\star$ to satisfy $\mathsf{F}_{\mathbf{y}}(\mathsf{ID}^\star) = 0$, for which it does not know the trapdoor. If these conditions are not satisfied, the reduction fails. We have to estimate the probability that it does not abort. In particular, we have to show that

$$\Pr[\mathsf{F}_{\mathbf{y}}(\mathsf{ID}^\star) = 0 \wedge \mathsf{F}_{\mathbf{y}}(\mathsf{ID}_1) \neq 0 \ldots \wedge \mathsf{F}_{\mathbf{y}}(\mathsf{ID}_Q) \neq 0] \tag{1}$$

is noticeable. Here, $\mathsf{ID}_1, \ldots, \mathsf{ID}_Q$ are identities for which key extraction queries are made. By a similar analysis to [6,47], to show a lower bound for the probability of (1), it suffices to show an upper bound for the following probability

$$\Pr[\mathsf{F}_{\mathbf{y}}(\mathsf{ID}^\star) = 0 \wedge \mathsf{F}_{\mathbf{y}}(\mathsf{ID}_i) = 0] \tag{2}$$

for identities $\mathsf{ID}^\star$ and ID_i where $\mathsf{ID}^\star \neq \mathsf{ID}_i$. To show an upper bound for (2), we first observe that

$$\mathsf{F}_{\mathbf{y}}(\mathsf{ID}^\star) = 0 \wedge \mathsf{F}_{\mathbf{y}}(\mathsf{ID}_i) = 0$$
$$\Leftrightarrow \mathsf{F}_{\mathbf{y}}(\mathsf{ID}^\star) = 0 \wedge \mathsf{F}_{\mathbf{y}}(\mathsf{ID}_i) - \mathsf{F}_{\mathbf{y}}(\mathsf{ID}^\star) = 0$$

$$\Leftrightarrow \underbrace{\left(y_0 + \sum_{(j,k) \in S(\mathsf{ID}^\star)} y_{1,j} y_{2,k} = 0 \right)}_{\text{Event (A)}}$$

$$\wedge \underbrace{\left(\sum_{(j,k) \in S(\mathsf{ID}_i)} y_{1,j} y_{2,k} - \sum_{(j,k) \in S(\mathsf{ID}^\star)} y_{1,j} y_{2,k} = 0 \right)}_{\text{Event (B)}}.$$

The value of y_0 is clearly independent of the Event (B). Therefore, we can easily estimate the probability of Event (A) occurring, conditioned on that Event (B) occurs. Thus, it suffices to show an upper bound on the probability of Event (B) occurring. This can be accomplished by using the Schwartz-Zippel lemma.

Proof Continued. Based on the idea we have explained above, we can simulate key extraction queries with sufficiently high success probability. However, two problems remain in order to complete the security proof.

(C) In the above discussion, we assumed that q is much larger than Q. Therefore, if q is bounded by some polynomial, so is Q. In such a setting, we can only prove "bounded" security, where the number of key extraction queries is bounded by a predetermined polynomial.

(D) Furthermore, we are not able to generate a properly distributed challenge ciphertext, as we explain below.

Let us explain the problem (D). Assume that for the challenge identity $\mathsf{ID}^\star$, we have $F_{\mathbf{y}}(\mathsf{ID}^\star) = 0$ and thus $H(\mathsf{ID}^\star) = \mathbf{A}\mathbf{R}_{\mathsf{ID}^\star}$. To prove security, we have to embed the LWE problem instance $\mathbf{A}$ and $\mathbf{v}$ into the challenge ciphertext, where $\mathbf{v}^\top = \mathbf{s}^\top \mathbf{A} + \mathbf{x}^\top$ or $\mathbf{v}$ a random vector. A natural way to do this is to implicitly set $\mathbf{x}_1 = \mathbf{x}$ and $\mathbf{x}_2 = \mathbf{R}_{\mathsf{ID}^\star}^\top \mathbf{x}$ and compute the challenge ciphertext as

$$\mathbf{s}^\top (\mathbf{A}|H(\mathsf{ID})) + (\mathbf{x}_1|\mathbf{x}_2) = (\mathbf{v}^\top |\mathbf{v}^\top \mathbf{R}_{\mathsf{ID}^\star}).$$

The problem with this approach is that the vector $\mathbf{x}_2$ is highly correlated to the value of $\mathbf{R}_{\mathsf{ID}^\star}$, which includes the information of $\mathbf{y} = (y_0, \{y_{i,j}\}_{(i,j) \in [1,2] \times [1,\ell]})$ and additionally $\mathbf{R}_0, \mathbf{R}_{1,1} \ldots, \mathbf{R}_{1,\ell}, \mathbf{R}_{2,1} \ldots, \mathbf{R}_{2,\ell}$. While a similar (but simpler) problem is resolved in a previous work [1] using a generalized form of the leftover hash lemma [20], we are not able to do the same argument due to the additional correlation to $\mathbf{y}$.

We can resolve the problem by a standard technique. Namely, we "smudge out" or "eat" the problematic term $\mathbf{R}_{\mathsf{ID}^\star}^\top \mathbf{x}$ by adding a large enough term $\mathbf{x}' \in \mathbb{Z}_q^m$ to it. This makes the error terms essentially statistically independent from $\mathbf{R}_{\mathsf{ID}^\star}$. The size of the term $\mathbf{x}'$ should be super-polynomially larger than the size of $\mathbf{R}_{\mathsf{ID}^\star}^\top \mathbf{x}$, but it should be polynomially smaller than q. Therefore, the size of q should be super-polynomially large, which also resolves the problem (C) at the same time. Appropriately setting the parameters, we obtain our new adaptively secure and anonymous IBE scheme.

2.3 An Additional Idea

However, making q super-polynomially large is not quite desirable because of the following two reasons. Firstly, this would negatively impact the performance of the system. Secondly, since the error term (in our case $\mathbf{x}$) is super-polynomially smaller compared to q, the corresponding LWE problem becomes easier. While we are not able to resolve the first problem, we present an idea to avoid the second problem.

Our first observation is that for any constant $c \in \mathbb{N}$, by making q and $\mathbf{x}'$ sufficiently large (but polynomial size), we can show that any PPT adversary whose number of key extraction queries is bounded by n^c cannot break the security of IBE with advantage non-negligibly larger than $1/n^c$. Of course, this is not sufficient because we need the adversary to have only negligible (rather than inverse of polynomial) advantage, even if the number of key extraction queries is unbounded.

In order to accomplish this, we prepare several instances of IBE scheme with different size of q. We call each instance of the IBE scheme as a sub-scheme. The number of sub-schemes is *super-constant* (rather than super-polynomial) and therefore the resulting scheme is still efficient. The size of q varies from very small polynomial to super-polynomial. Furthermore, we "glue" them so that an adversary must break the security of all of the sub-schemes, in order to break the resulting IBE scheme. This can easily be accomplished by splitting the message by k-out-of-k secret sharing scheme, and then encrypt them by each of the sub-schemes.

In the security proof, we assume an PPT adversary $\mathcal{A}$ that breaks the resulting IBE scheme. Since $\mathcal{A}$ is polynomial time and has non-negligible advantage, there exists some constant $c \in \mathbb{N}$ such that the number of the key extraction queries that $\mathcal{A}$ makes is smaller than n^c and $\mathcal{A}$'s advantage is non-negligibly larger than $1/n^c$. Thus, there exists at least one sub-scheme whose size of q fits for $\mathcal{A}$, and q is polynomial size. We transform the adversary A into another adversary $\mathcal{B}$ that breaks the sub-scheme. Since q is polynomial size, we can reduce the security to the LWE assumption with *polynomial* approximation factor. Note that similar technique is used in [21] to improve the efficiency and the reduction cost of the Naor-Reingold PRF. There, the reduction algorithm chooses the target sub-scheme based on the number of queries that the adversary makes. In our reduction, we choose the target depending on the advantage of the adversary in addition to the number of key extraction queries.

To present our results in a unified and modular manner, we introduce the notion of PIBE. Roughly speaking, PIBE is an IBE scheme that is parametrized by a variable c. Our technique to avoid super-polynomial factor we discussed above can be generalized to be a generic conversion from PIBE to IBE. Furthermore, our scheme we discussed in the previous subsection also can be captured as a special case of PIBE, in that c is set to be a super-constant.

3 Preliminaries

Notation. We denote by $[n]$ a set $\{1, 2, \ldots, n\}$ for any integer $n \in \mathbb{N}$. We treat a vector as a column vector. If $\mathbf{A}_1$ is $n \times m$ and $\mathbf{A}_2$ is $n \times m'$ matrix, then $(\mathbf{A}_1 | \mathbf{A}_2)$ denotes the $n \times (m + m')$ matrix formed by concatenating $\mathbf{A}_1$ and $\mathbf{A}_2$. We use similar notation for vectors. A function $f : \mathbb{N} \to \mathbb{R}_{\geq 0}$ is said to be negligible, if for all c, there exists N such that $f(n) < 1/n^c$ for all $n > N$. We denote by $\mathsf{negl}(n)$ a negligible function. We denote by $x \xleftarrow{\$} X$ the process of sampling a value x according to the distribution X. Similarly, for a finite set S, we denote by $x \xleftarrow{\$} S$ the process of sampling a value x according to the uniform distribution over S. Statistical distance between two random variables X and Y with support Ω is defined as $\Delta(X; Y) = \frac{1}{2} \sum_{s \in \Omega} |\Pr[X = s] - \Pr[Y = s]|$. For ensembles of random variable $\{X(n)\}_{n \in \mathbb{N}}$ and $\{Y(n)\}_{n \in \mathbb{N}}$, we say that they are $\mathsf{negl}(n)$-close if $\Delta(X(n); Y(n)) = \mathsf{negl}(n)$.

3.1 Identity-Based Encryption

Syntax. Let $\mathcal{ID}$ be the ID space of the scheme. If a collision resistant hash function $CRH : \{0,1\}^* \to \mathcal{ID}$ is available, one can use an arbitrary string as an identity. An IBE scheme is defined by the following four algorithms.

$\mathsf{Setup}(1^n) \to (\mathsf{mpk}, \mathsf{msk})$: The setup algorithm takes as input a security parameter 1^n and outputs a master public key mpk and a master secret key msk.

$\mathsf{KeyGen}(\mathsf{mpk}, \mathsf{msk}, \mathsf{ID}) \to \mathsf{sk_{ID}}$: The key generation algorithm takes as input the master public key mpk, the master secret key msk, and an identity $\mathsf{ID} \in \mathcal{ID}$. It outputs a private key $\mathsf{sk_{ID}}$. We assume that ID is implicitly included in $\mathsf{sk_{ID}}$.

$\mathsf{Encrypt}(\mathsf{mpk}, \mathsf{ID}, \mathsf{M}) \to C$: The encryption algorithm takes as input a master public key mpk, an identity $\mathsf{ID} \in \mathcal{ID}$, and a message M, It outputs a ciphertext C.

$\mathsf{Decrypt}(\mathsf{mpk}, \mathsf{sk_{ID}}, C) \to \mathsf{M}$ or $\bot$: The decryption algorithm takes as input the master public key mpk, a private key $\mathsf{sk_{ID}}$, and a ciphertext C. It outputs the message M or $\bot$, which means that the ciphertext is not in a valid form.

Correctness. We require correctness of decryption: that is, for all n, all $\mathsf{ID} \in \mathcal{ID}$, and all M in the specified message space, $\Pr[\mathsf{Decrypt}(\mathsf{mpk}, \mathsf{sk_{ID}}, \mathsf{Encrypt}(\mathsf{mpk}, \mathsf{ID}, \mathsf{M})) = \mathsf{M}] = 1 - \mathsf{negl}(n)$ holds, where the probability is taken over the randomness used in $(\mathsf{mpk}, \mathsf{msk}) \xleftarrow{\$} \mathsf{Setup}(1^n)$, $\mathsf{sk_{ID}} \xleftarrow{\$} \mathsf{KeyGen}(\mathsf{mpk}, \mathsf{msk}, \mathsf{ID})$, and $\mathsf{Encrypt}(\mathsf{mpk}, \mathsf{ID}, \mathsf{M})$.

Security. We now define the security for an IBE scheme Π. This security notion is defined by the following game between a challenger and an adversary $\mathcal{A}$.

- **Setup.** At the outset of the game, the challenger runs $\mathsf{Setup}(1^n) \to (\mathsf{mpk}, \mathsf{msk})$ and gives mpk to $\mathcal{A}$.

- **Phase 1.** $\mathcal{A}$ may adaptively make key-extraction queries. If $\mathcal{A}$ submits $\mathsf{ID} \in \mathcal{ID}$ to the challenger, the challenger returns $\mathsf{sk_{ID}} \leftarrow \mathsf{KeyGen}(\mathsf{mpk}, \mathsf{msk}, \mathsf{ID})$.

- **Challenge Phase.** At some point, $\mathcal{A}$ outputs a message M and an identity $\mathsf{ID}^* \in \mathcal{ID}$, on which it wishes to be challenged. Then, the challenger picks a random coin $\mathsf{coin} \xleftarrow{\$} \{0,1\}$ and a random ciphertext $C \xleftarrow{\$} \mathcal{C}$ from the ciphertext space. If $\mathsf{coin} = 0$, it runs $\mathsf{Encrypt}(\mathsf{mpk}, \mathsf{ID}^*, \mathsf{M}) \to C^*$ and gives the challenge ciphertext C^* to $\mathcal{A}$. If $\mathsf{coin} = 1$, it sets the challenge ciphertext as $C^* = C$ and gives it to $\mathcal{A}$.

- **Phase 2.** After the challenge query, $\mathcal{A}$ may continue to make key-extraction queries, with the added restriction that $\mathsf{ID} \neq \mathsf{ID}^*$.

- **Guess.** Finally, $\mathcal{A}$ outputs guess a $\widehat{\mathsf{coin}}$ for coin. The advantage of $\mathcal{A}$ is defined as $\mathsf{Adv}^{\mathsf{IBE}}_{\mathcal{A}, \Pi} = \left| \Pr[\widehat{\mathsf{coin}} = \mathsf{coin}] - \frac{1}{2} \right|$. We say that Π is adaptively anonymous, if the advantage of any PPT $\mathcal{A}$ is negligible.

We also define adaptive security (without anonymity) for Π via a similar game to the above. To define adaptive security, we change the challenge phase as follows.

- **Challenge Phase.** $\mathcal{A}$ outputs two messages M_0, M_1 and an identity $\mathsf{ID}^\star \in \mathcal{ID}$, on which it wishes to be challenged. Then, the challenger picks a random coin coin $\overset{\$}{\leftarrow} \{0,1\}$, runs $\mathsf{Encrypt}(\mathsf{mpk}, \mathsf{ID}^\star, \mathsf{M}_{\mathsf{coin}}) \to C^\star$, and gives the challenge ciphertext $C^\star$ to $\mathcal{A}$.

We also say that Π is adaptively secure, if the advantage of any PPT $\mathcal{A}$ is negligible. We note that the adaptive anonymity implies the adaptive security. Namely, the former is a stronger security notion.

3.2 Lattice Preliminaries

For positive integers q, m, n, a matrix $\mathbf{A} \in \mathbb{Z}_q^{n \times m}$, and a vector $\mathbf{u} \in \mathbb{Z}_q^m$, the m-dimensional integer lattice $\Lambda_q^{\mathbf{u}}(\mathbf{A})$ is defined as $\Lambda_q^{\mathbf{u}}(\mathbf{A}) = \{\mathbf{e} \in \mathbb{Z}^m : \mathbf{A}\mathbf{e} = \mathbf{u} \mod q\}$. $\Lambda_q^\perp(\mathbf{A})$ denotes $\Lambda_q^{\mathbf{0}}(\mathbf{A})$. Let $D_{\Lambda,\mathbf{c},\sigma}$ denote the discrete Gaussian distribution over Λ with center $\mathbf{c}$ and parameter γ. When $\mathbf{c}$ is omitted, we set $\mathbf{c} = \mathbf{0}$.

Matrix Norms. For a vector $\mathbf{u}$, we let $\|\mathbf{u}\|$ and $\|\mathbf{u}\|_\infty$ denote its ℓ_2 and ℓ_∞ norm respectively. For a matrix $\mathbf{R} \leq \mathbb{Z}^{k \times m}$ we denote three matrix norms:

$\|\mathbf{R}\|$ denotes the ℓ_2 length of the longest column of $\mathbf{R}$.
$\|\mathbf{R}\|_{\mathrm{GS}}$ denotes $\|\tilde{\mathbf{R}}\|$ where $\tilde{\mathbf{R}}$ is the result of applying Gram-Schmidt to the columns of $\mathbf{R}$.
$\|\mathbf{R}\|_2$ is the operator norm of $\mathbf{R}$ defined as $\|\mathbf{R}\|_2 = \sup_{\|\mathbf{x}\|=1} \|\mathbf{R}\mathbf{x}\|$.

We have that the following lemma holds [1].

Lemma 1. *Let m, n, q be positive integers with $m > n$, $\mathbf{A} \in \mathbb{Z}_q^{n \times m}$ be a matrix, $\mathbf{u} \in \mathbb{Z}_q^n$ be a vector, $\mathbf{T}_\mathbf{A}$ be a basis for $\Lambda_q^\perp(\mathbf{A})$, and $\sigma > \|\mathbf{T}_\mathbf{A}\|_{\mathrm{GS}} \cdot \omega(\sqrt{\log m})$. Then we have $\Pr[\mathbf{x} \overset{\$}{\leftarrow} D_{\Lambda_q^{\mathbf{u}}(\mathbf{A}),\sigma} : \|\mathbf{x}\| > \sqrt{m}\sigma] < \mathsf{negl}(n)$.*

Trapdoor Generators and Related Operations

Lemma 2. *Let $n, m, q > 0$ be integers with q prime. There are polynomial time algorithms such that*

1. *([3,5]):* $\mathsf{TrapGen}(1^n, 1^m, q) \to (\mathbf{A}, \mathbf{T}_\mathbf{A})$
   a randomized algorithm that, when $m \geq 6n\lceil \log q \rceil$, outputs a full rank matrix $\mathbf{A} \in \mathbb{Z}_q^{n \times m}$ and a basis $\mathbf{T}_\mathbf{A} \in \mathbb{Z}^{m \times m}$ for $\Lambda_q^\perp(\mathbf{A})$ such that $\mathbf{A}$ is $\mathsf{negl}(n)$-close to uniform and $\|\mathbf{T}_\mathbf{A}\|_{\mathrm{GS}} = O(\sqrt{n \log q})$ with all but negligible probability in n.
2. *([16]):* $\mathsf{SampleLeft}(\mathbf{A}, \mathbf{F}, \mathbf{u}, \mathbf{T}_\mathbf{A}, \sigma) \to \mathbf{e}$
   a randomized algorithm that, given a full rank matrix $\mathbf{A} \in \mathbb{Z}_q^{n \times m}$, a matrix $\mathbf{F} \in \mathbb{Z}_q^{n \times m}$, a vector $\mathbf{u} \in \mathbb{Z}_q^n$, a basis $\mathbf{T}_\mathbf{A}$ for $\Lambda_q^\perp(\mathbf{A})$, and a Gaussian parameter $\sigma > \|\mathbf{T}_\mathbf{A}\|_{\mathrm{GS}} \cdot \omega(\sqrt{\log m})$, outputs a vector $\mathbf{e} \in \mathbb{Z}^{2m}$ sampled from a distribution which is $\mathsf{negl}(n)$-close to $D_{\Lambda_q^{\mathbf{u}}(\mathbf{A}|\mathbf{F}),\sigma}$.
3. *([1]):* $\mathsf{SampleRight}(\mathbf{A}, \mathbf{G}, \mathbf{R}, y, \mathbf{u}, \mathbf{T}_\mathbf{G}, \sigma) \to \mathbf{e}$ *where $\mathbf{F} = \mathbf{A}\mathbf{R} + y\mathbf{G}$*
   a randomized algorithm that, given a full rank matrix $\mathbf{A}, \mathbf{G} \in \mathbb{Z}_q^{n \times m}$, $y \in \mathbb{Z}_q \backslash \{0\}$, a matrix $\mathbf{R} \in \mathbb{Z}^{m \times m}$, a vector $\mathbf{u} \in \mathbb{Z}_q^n$, a basis $\mathbf{T}_\mathbf{G}$ for $\Lambda_q^\perp(\mathbf{G})$, and a Gaussian parameter $\sigma > \|\mathbf{T}_\mathbf{G}\|_{\mathrm{GS}} \cdot \|\mathbf{R}\|_2 \cdot \omega(\sqrt{\log m})$ outputs a vector $\mathbf{e} \in \mathbb{Z}^{2m}$ sampled from a distribution which is $\mathsf{negl}(n)$-close to $D_{\Lambda_q^{\mathbf{u}}(\mathbf{A}|\mathbf{F}),\sigma}$.

4. *([35]): Let $m > n\lceil \log q \rceil$. Then there is a fixed full-rank matrix $\mathbf{G} \in \mathbb{Z}_q^{n \times m}$ such that the lattice $\Lambda_q^\perp(\mathbf{G})$ has a publicly known basis $\mathbf{T_G} \in \mathbb{Z}^{m \times m}$ with $\|\mathbf{T_G}\|_{GS} \leq \sqrt{5}$. Furthermore, there exists a deterministic polynomial-time algorithm $\mathbf{G}^{-1}$ which takes the input $\mathbf{U} \in \mathbb{Z}_q^{n \times m}$ and outputs $\mathbf{R} = \mathbf{G}^{-1}(\mathbf{U})$ such that $\mathbf{R} \in \{0, 1\}^{m \times m}$ and $\mathbf{GR} = \mathbf{U}$.*

Note that in the above, we are abusing notation and $\mathbf{G}^{-1}$ is not a matrix but rather a function. Namely, for any $\mathbf{U}$ there are many choices of $\mathbf{R}$ such that $\mathbf{GR} = \mathbf{U}$, and $\mathbf{G}^{-1}(\mathbf{U})$ deterministically outputs a particular short matrix from this set. Since we have $\|\mathbf{R}\|_2 \leq m$ for any $\mathbf{R} \in \{-1, 0, 1\}^{m \times m}$, $\|\mathbf{G}^{-1}(\mathbf{U})\|_2 \leq m$ holds for any $\mathbf{U} \in \mathbb{Z}_q^{n \times m}$.

Learning with Errors. The learning with errors (LWE) problem was introduced by Regev who showed that solving it on the average is as hard as (quantumly) solving several standard lattice problems in the worst case.

Definition 1 (LWE). *For an integers n, $m = m(n)$, a prime integer $q = q(n) > 2$, an error distribution $\chi = \chi(n)$ over $\mathbb{Z}_q$, and an PPT algorithm $\mathcal{A}$, an advantage for the learning with errors problem $\mathsf{dLWE}_{n,m,q,\chi}$ of $\mathcal{A}$ is defined as follows:*

$$\mathsf{Adv}_{\mathcal{A}}^{\mathsf{dLWE}_{n,m,q,\chi}} = |\Pr[\mathcal{A}(\mathbf{A}, \mathbf{s}^\top \mathbf{A} + \mathbf{x}^\top) \to 1] - \Pr[\mathcal{A}(\mathbf{A}, \mathbf{v}^\top) \to 1]|$$

where $\mathbf{A} \xleftarrow{\$} \mathbb{Z}_q^{n \times m}$, $\mathbf{s} \xleftarrow{\$} \mathbb{Z}_q^n$, $\mathbf{x} \xleftarrow{\$} \chi^m$, $\mathbf{v} \xleftarrow{\$} \mathbb{Z}_q^m$. We say that $\mathsf{dLWE}_{n,m,q,\chi}$ assumption holds if $\mathsf{Adv}_{\mathcal{A}}^{\mathsf{dLWE}_{n,m,q,\chi}}$ is negligible for all PPT $\mathcal{A}$.

Let $B = B(n) \in \mathbb{N}$. A family of distributions $\chi = \{\chi_n\}$ is called B-bounded if $\Pr[\chi \in [-B, B]] = 1$. For any constant $d > 0$ and sufficiently large q, Regev [40] through a quantum reduction showed that taking χ as a q/n^d-bounded (truncated) discretized Gaussian distribution, the $\mathsf{dLWE}_{n,m,q,\chi}$ problem is as hard as approximating the worst-case GapSVP to $n^{O(d)}$ factors, which is believed to be hard. In subsequent works, (partial) dequantization of the Regev's reduction were achieved [13,37]. More generally, let $\chi_{\max} < q$ be the bound on the noise distribution. The difficulty of the problem is measured by the ratio $q/\chi_{\max}$. This ratio is always bigger than 1 and the smaller it is the harder the problem. The problem appears to remain hard even when $q/\chi_{\max} < 2^{n^\epsilon}$ for some fixed ϵ that is $0 < \epsilon < 1/2$.

3.3 Basic Facts

Injective Map. Let d and κ be some integers. Furthermore, let ℓ be $\ell = \lceil \kappa^{1/d} \rceil$. Then, an element of $[1, \kappa]$ can be written as an element of $[1, \ell]^d$ using some canonical map. Furthermore, it is also possible to write a subset of $[1, \kappa]$ as a subset of $[1, \ell]^d$, by naturally extending the canonical map. By identifying a bit string in $\{0, 1\}^\kappa$ with a subset of $[1, \kappa]$ (for example, by regarding the former as the indicator vector of a subset of $[1, \kappa]$), we can define an efficiently computable injective map S that maps a bit string $\mathsf{ID} \in \{0, 1\}^\kappa$ to a subset $S(\mathsf{ID})$ of $[1, \ell]^d$.

The following lemma can be shown by a simple calculation.

Lemma 3 (Smudging out Lemma). Let $\mathbf{x}_0 \in \mathbb{Z}^m$ be a (fixed) vector such that $\|\mathbf{x}_0\|_\infty \leq \delta$ and let $\mathbf{x} \in \mathbb{Z}^m$ be a random vector that is chosen as $\mathbf{x} \xleftarrow{\$} [-B', B']^m$. Then, two distributions $\mathbf{x}_0 + \mathbf{x}$ and $\mathbf{x}$ are within statistical distance $m\delta/B'$.

As observed in [1,40], the following lemma is obtained as a corollary to the (general) leftover hash lemma.

Lemma 4 (Leftover Hash Lemma). Let $q \in \mathbb{N}$ be an odd prime and let $m > (n+1)\log q + \omega(\log n)$. Let $\mathbf{R} \xleftarrow{\$} \{-1, 1\}^{m \times m}$ and $\mathbf{A}, \mathbf{A}' \xleftarrow{\$} \mathbb{Z}_q^{n \times m}$ be uniformly random matrices. Then the distribution of $(\mathbf{A}, \mathbf{AR})$ is $\mathsf{negl}(n)$-close to the distribution of $(\mathbf{A}, \mathbf{A}')$.

The following lemma is implicitly shown in [6].

Lemma 5. Let $a_1, \ldots, a_n \in \mathbb{R}$ be real numbers such that $|\sum_{i=1}^n a_i| = \epsilon$ and $\sum_{i=1}^n |a_i| \leq 1/2$. Furthermore, let $\gamma_1, \ldots, \gamma_n \in \mathbb{R}$ be real numbers such that $0 < \gamma_{\min} \leq \gamma_i \leq \gamma_{\max}$ for $i \in [n]$. Then, we have $|\sum_{i=1}^n \gamma_i a_i| \geq \gamma_{\min}\epsilon - (\gamma_{\max} - \gamma_{\min})/2$.

4 Parametrized IBE

In this section, we introduce the notion of parametrized IBE (PIBE), which is an slight extension of the ordinary notion of IBE. The syntax and the security notion for PIBE is almost the same, except that it is parametrized by an integer c. Roughly speaking, the larger c becomes, the more secure PIBE becomes. In particular, when c is super-constant in n, the security notion for PIBE corresponds to that for ordinary IBE. However, in our construction of PIBE in Sect. 5, in order to prove the security of the scheme for super-constant c, we need to assume super-polynomial LWE, which is a stronger assumption than the assumption that is needed for constant c. In this section, to base the scheme on a weaker assumption, we provide generic construction of adaptively secure IBE scheme from PIBE scheme that is secure *only for constant c*.

4.1 Definition of Parametrized IBE

Here, we define PIBE. The syntax of PIBE is the same as ordinary IBE except that the Setup algorithm is parametrized by an integer $c = c(n)$. Namely, Setup takes as inputs 1^n and 1^c and outputs a master public key mpk and a master secret key msk. Other algorithms, KeyGen, Encrypt, and Decrypt are defined as in ordinary IBE. We require that these algorithms work within a time that is polynomial in n and c.

As for the security, we define advantage $\mathsf{Adv}_{\mathcal{A},\Pi}^{\mathsf{PIBE}}$ of an adversary $\mathcal{A}$ for a PIBE scheme Π via a game that is almost the same as that of an ordinary IBE scheme. The only difference is that mpk and msk are generated by $\mathsf{Setup}(1^n, 1^c)$ at the beginning of the game. The rest of the game is the same. We say that the scheme is c-adaptively anonymous, if for any PPT adversary $\mathcal{A}$ such that $Q(n) \leq n^c/2 - 1$,

$$\frac{\mathsf{Adv}^{\mathsf{PIBE}}_{\mathcal{A},\Pi}}{Q+1} < \frac{1}{n^c} + \mathsf{negl}(n) \qquad (3)$$

holds for some negligible function $\mathsf{negl}(n)$. Here $Q = Q(n)$ is the upper bound for the number of key extraction queries made by $\mathcal{A}$ during the game.

When $c(n)$ is a constant, the c-adaptive anonymity is an weaker security notion than the adaptive anonymity for IBE, since it allows an adversary to have non-negligible advantage. Furthermore, there is a bound on the number of key extraction queries. On the other hand, when $c(n)$ is super-constant, the security definition of c-adaptive anonymity corresponds to that of adaptive anonymity for (ordinary) IBE. More precisely, we have the following theorem.

Theorem 1. *If Π = (Setup, KeyGen, Encrypt, Decrypt) is c'-adaptively anonymous for some super constant function $c'(n) = \omega(1)$ such that $c'(n) < \mathsf{poly}(n)$, Π' = (Setup', KeyGen, Encrypt, Decrypt) is adaptively anonymous (as an ordinary IBE) if we set $\mathsf{Setup}'(1^n) = \mathsf{Setup}(1^n, 1^{c'(n)})$.*

Proof. Since $c'(n) < \mathsf{poly}(n)$, Setup', KeyGen, Encrypt, and Decrypt run in polynomial time. In addition, since $c'(n) = \omega(1)$ and thus $n^{c'}$ is super-polynomial, there is no bound on the number of key extraction queries for the adversary in the c'-adaptive anonymity game. Furthermore, since $1/n^{c'}$ is a negligible function, by Eq. (3), we have

$$\mathsf{Adv}^{\mathsf{PIBE}}_{\mathcal{A},\Pi} < (Q+1)\left(\frac{1}{n^{c'}} + \mathsf{negl}(n)\right) = \mathsf{negl}(n)$$

for any adversary $\mathcal{A}$. Thus, Π' defined as above is adaptively anonymous.

Comparison with Bounded Collusion IBE. Our notion of PIBE is similar to the notion of bounded collusion IBE [19] (also called k-resilient IBE [29]), in that adversaries only learn private keys of an a-priori bounded number of identities. The security requirement for the former is weaker than that for the latter, because we allow adversaries to have non-negligible advantages (in the case of c is a constant). On the other hand, we pose more severe requirement on the efficiency for the former. We require the algorithms of PIBE to work in polynomial time in c, rather than in n^c. Because of this, existing bounded collusion IBE schemes [19, 26, 29, 46, 49] do not satisfy the requirement of PIBE.

4.2 IBE from PIBE

In this section, we show a conversion from a PIBE scheme Π = (PIBE.Setup, PIBE.KeyGen, PIBE.Encrypt, PIBE.Decrypt) to an (ordinary) IBE scheme Π' = (IBE.Setup, IBE.KeyGen, IBE.Encrypt, IBE.Decrypt). In the following, let $\eta(n)$ be any function such that $\eta(n) = \omega(1)$ (e.g., $\eta(n) = \log\log(n)$). We also let the message space of Π and Π' be $\{0,1\}^{\ell_M}$ for some $\ell_M \in \mathbb{N}$.

$\mathsf{IBE.Setup}(1^n)$: It runs $\mathsf{PIBE.Setup}(1^n, 1^i) \to (\mathsf{mpk}^{(i)}, \mathsf{msk}^{(i)})$ for $i = 1, \ldots, \eta$. It outputs

$$\mathsf{mpk} = (\mathsf{mpk}^{(1)}, \mathsf{mpk}^{(2)}, \ldots, \mathsf{mpk}^{(\eta)}) \quad \text{and} \quad \mathsf{msk} = (\mathsf{msk}^{(1)}, \mathsf{msk}^{(2)}, \ldots, \mathsf{msk}^{(\eta)}).$$

IBE.KeyGen(mpk, msk, ID): It runs PIBE.KeyGen($\mathsf{mpk}^{(i)}, \mathsf{msk}^{(i)}, \mathsf{ID}) \to \mathsf{sk}^{(i)}_{\mathsf{ID}}$ for $i = 1, \ldots, \eta$. It outputs

$$\mathsf{sk}_{\mathsf{ID}} = (\mathsf{sk}^{(1)}_{\mathsf{ID}}, \mathsf{sk}^{(2)}_{\mathsf{ID}}, \ldots, \mathsf{sk}^{(\eta)}_{\mathsf{ID}}).$$

Encrypt(mpk, ID, M): To encrypt $\mathsf{M} = \{0,1\}^{\ell_M}$, it picks random $\mathsf{M}^{(i)} \in \{0,1\}^{\ell_M}$ for $i \in [\eta]$ subject to constraint that $\mathsf{M} = \bigoplus_{i=1}^{\eta} \mathsf{M}^{(i)}$, where $\bigoplus$ denotes bitwise exclusive or. Then it runs

$$\mathsf{PIBE.Encrypt}(\mathsf{mpk}^{(i)}, \mathsf{ID}, \mathsf{M}^{(i)}) \to C^{(i)} \qquad \text{for} \quad i = 1, \ldots, \eta.$$

Finally, it outputs the ciphertext $C = (C^{(1)}, \ldots, C^{(\eta)})$.

Decrypt(mpk, $\mathsf{sk}_{\mathsf{ID}}$, C): It first parses the ciphertext and the private key as $C \to (C^{(1)}, \ldots, C^{(\eta)})$ and $\mathsf{sk}_{\mathsf{ID}} \to (\mathsf{sk}^{(1)}_{\mathsf{ID}}, \ldots, \mathsf{sk}^{(\eta)}_{\mathsf{ID}})$. Then, it runs

$$\mathsf{PIBE.Decrypt}(\mathsf{mpk}^{(i)}, \mathsf{sk}^{(i)}_{\mathsf{ID}}, C^{(i)}) \to \mathsf{M}^{(i)} \qquad \text{for} \quad i = 1, \ldots, \eta.$$

Finally, it outputs $\mathsf{M} = \bigoplus_{i=1}^{\eta} \mathsf{M}^{(i)}$.

Correctness of the scheme can be shown very easily. The following theorem addresses the security of the scheme. Note that the resulting IBE scheme is not anonymous even if the original PIBE scheme is anonymous.

Theorem 2. *Assume that PIBE Π is secure for all (constant) $c \in \mathbb{N}$. Then, Π' is adaptively secure as an (ordinary, not parametrized) IBE scheme.*

Proof. Assume an adversary $\mathcal{A}$ that breaks Π' with non-negligible probability. Since $\mathcal{A}$ is a PPT algorithm, there exist constants $c' \in \mathbb{N}$ and $c'' \in \mathbb{N}$ such that

- The advantage $\epsilon(n)$ of $\mathcal{A}$ is greater than $2/n^{c'}$ for infinitely many n.
- The number $Q(n)$ of key extraction queries that $\mathcal{A}$ makes is bounded by $n^{c''}/2 - 1$.

Let i^* be $i^* = c' + c''$. Then, we have

$$\frac{\epsilon(n)}{2(Q(n) + 1)} - \frac{1}{n^{i^*}} \geq \frac{2}{n^{c' + c''}} - \frac{1}{n^{i^*}} = \frac{1}{n^{i^*}} \qquad (4)$$

for infinitely many n. In particular, $\epsilon/2(Q+1) - 1/n^{i^*}$ cannot be bounded by any negligible function. To show the theorem, we construct an adversary $\mathcal{B}$ against i^*-adaptive anonymity of PIBE Π from $\mathcal{A}$. In the following, we assume $\eta \geq i^*$. Since $\eta(n) = \omega(1)$, this holds for sufficiently large n.

Setup. First, PIBE.Setup($1^n, 1^{i^*}) \to (\mathsf{mpk}^{(i^*)}, \mathsf{msk}^{(i^*)})$ is run and $\mathsf{mpk}^{(i^*)}$ is given to $\mathcal{B}$. Then, $\mathcal{A}$ runs PIBE.Setup($1^n, 1^i) \to (\mathsf{mpk}^{(i)}, \mathsf{msk}^{(i)})$ for $i = [1, \eta] \backslash \{i^*\}$ and sets $\mathsf{mpk} = (\mathsf{mpk}^{(1)}, \mathsf{mpk}^{(2)}, \ldots, \mathsf{mpk}^{(\eta)})$. $\mathcal{B}$ keeps $\mathsf{msk}^{(i)}$ for $i \in [1, \eta] \backslash \{i^*\}$ secret, and returns mpk to $\mathcal{A}$.

Phases 1 and 2. When $\mathcal{A}$ makes a key extraction query for an identity ID, $\mathcal{B}$ queries a private key for the same ID to its challenger. Then,

PIBE.KeyGen($\mathsf{mpk}^{(i^*)}, \mathsf{msk}^{(i^*)}, \mathsf{ID}) \to \mathsf{sk}_{\mathsf{ID}}^{(i^*)}$ is run and $\mathsf{sk}_{\mathsf{ID}}^{(i^*)}$ is given to $\mathcal{B}$. Then $\mathcal{B}$ runs PIBE.KeyGen($\mathsf{mpk}^{(i)}, \mathsf{msk}^{(i^*)}, \mathsf{ID}) \to \mathsf{sk}_{\mathsf{ID}}^{(i)}$ for $i \in [1, \eta] \backslash \{i^*\}$ and returns $\mathsf{sk}_{\mathsf{ID}} = (\mathsf{sk}_{\mathsf{ID}}^{(1)}, \ldots, \mathsf{sk}_{\mathsf{ID}}^{(\eta)})$ to $\mathcal{A}$.

Challenge. When $\mathcal{A}$ makes a challenge query for $(\mathsf{ID}^*, \mathsf{M}_0, \mathsf{M}_1)$, $\mathcal{B}$ first picks random $\mathsf{M}^{(i)} \xleftarrow{\$} \{0,1\}^{\ell_M}$ for $i \in [1, \eta] \backslash \{i^*\}$. Then, it sets

$$\mathsf{M}_b^{(i^*)} = \mathsf{M}_b \oplus \left(\bigoplus_{i \in [1, \eta] \backslash \{i^*\}} \mathsf{M}^{(i)} \right) \qquad \text{for} \qquad b \in \{0, 1\}$$

and runs PIBE.Encrypt($\mathsf{mpk}^{(i)}, \mathsf{ID}, \mathsf{M}^{(i)}) \to C^{(i)}$ for $i \in [1, \eta] \backslash \{i^*\}$. Then, it picks random coin $\mathsf{coin}' \xleftarrow{\$} \{0, 1\}$ and makes the challenge query for $(\mathsf{ID}^*, \mathsf{M}_{\mathsf{coin}'}^{(i^*)})$ to its challenger. Then, the challenger picks a coin $\mathsf{coin} \xleftarrow{\$} \{0, 1\}$ and returns C^* to $\mathcal{B}$. If $\mathsf{coin} = 0$, we have PIBE.Encrypt($\mathsf{mpk}^{(i^*)}, \mathsf{ID}^*, \mathsf{M}_{\mathsf{coin}'}^{(i^*)}) \to C^*$. Otherwise, C^* is a random element of the ciphertext space. Given C^*, $\mathcal{B}$ returns the challenge ciphertext

$$(C^{(1)}, \ldots, C^{(i^*-1)}, C^*, C^{(i^*+1)}, \ldots, C^{(\eta)})$$

to $\mathcal{A}$.

Guess. Finally, $\mathcal{A}$ outputs a guess $\widehat{\mathsf{coin}}$ for coin'. If $\widehat{\mathsf{coin}} = \mathsf{coin}'$, $\mathcal{B}$ outputs 0 as its guess for coin and outputs 1 otherwise.

Analysis. We can see that $\mathcal{B}$ is a valid adversary for the parametrized IBE Π since $\mathcal{A}$ does not make a key extraction query for ID^*. Furthermore, $\mathcal{B}$ makes the same number of key extraction queries as $\mathcal{A}$ and in particular, we have $Q(n) < n^{i^*}/2 - 1$. It is easy to see that the view of the adversary $\mathcal{A}$ corresponds to that in adaptive security game for IBE Π' when $\mathsf{coin} = 0$. It can also be seen that the view of the adversary is independent of coin' when $\mathsf{coin} = 1$. Therefore, we have

$$\mathsf{Adv}_{\mathcal{B}, \Pi}^{\mathsf{PIBE}} = \left| \frac{1}{2} \Pr[\widehat{\mathsf{coin}} = \mathsf{coin}' | \mathsf{coin} = 0] + \frac{1}{2} \Pr[\widehat{\mathsf{coin}} \neq \mathsf{coin}' | \mathsf{coin} = 1] - \frac{1}{2} \right|$$

$$= \frac{1}{2} \left| \Pr[\widehat{\mathsf{coin}} = \mathsf{coin}' | \mathsf{coin} = 0] - \frac{1}{2} \right| = \frac{1}{2} \epsilon(n).$$

Thus, by Eq. (4), $\mathcal{B}$ is a successful attacker against the i^*-adaptive anonymity of Π.

More Efficient Conversion. In the above conversion, we run η instances of PIBE scheme in parallel. The number of instances can be reduced to $O(\log \eta)$. We briefly sketch the construction and the security proof for it. Let us assume that η is a power of 2. In the setup algorithm of the variant, we run PIBE.Setup($1^n, 1^i) \to (\mathsf{mpk}^{(i)}, \mathsf{msk}^{(i)})$ for $i = 1, 2, 4, \ldots, 2^i, \ldots, 2^{\log \eta}(= \eta)$, instead of $i = 1, 2, \ldots, \eta$. Other algorithms are defined similarly to the above. In the security proof, the target of the reduction algorithm is set to be i^* such that $2^{i^*-1} \leq c' + c'' < 2^{i^*}$.

5 Our Construction of PIBE from Lattices

Here, we show our constructions of PIBE from lattices. By setting the parameter c super-constant or applying the conversions in Sect. 4.2, we obtain IBE schemes that provide trade-off between the efficiency, security, and the underlying assumptions. (See Sect. 6 for the overview). In this section, we first introduce some functions that will be needed to describe our construction. Then, we show our construction of PIBE scheme for single-bit message space. We then prove the security of the scheme. Finally, we discuss extension of the scheme to the multi-bit variant.

5.1 Homomorphic Computation

Let d be a natural number. We introduce a function $\mathsf{PubEval}_d : (\mathbb{Z}_q^{n \times m})^d \to \mathbb{Z}_q^{n \times m}$ which takes a set of matrices $\mathbf{B}_1, \mathbf{B}_2, \ldots, \mathbf{B}_d \in \mathbb{Z}_q^{n \times m}$ as inputs and outputs a matrix in $\mathbb{Z}_q^{n \times m}$. The function is defined recursively as follows:

$$\mathsf{PubEval}_d(\mathbf{B}_1, \ldots, \mathbf{B}_d) = \begin{cases} \mathbf{B}_1 & \text{if } d = 1 \\ \mathbf{B}_1 \cdot \mathbf{G}^{-1}(\mathsf{PubEval}_{d-1}(\mathbf{B}_2, \ldots, \mathbf{B}_d)) & \text{if } d \geq 2. \end{cases}$$

We have that the following lemma holds. The proof appears in the full version.

Lemma 6. *Let $\mathbf{A}, \mathbf{B}_1, \ldots, \mathbf{B}_d$ be matrices in $\mathbb{Z}_q^{n \times m}$ and $\mathbf{R}_1, \ldots, \mathbf{R}_d$ be matrices in $\mathbb{Z}^{m \times m}$ such that $\mathbf{B}_i = \mathbf{A}\mathbf{R}_i + y_i \mathbf{G}$ for $i \in [d]$. Furthermore, we assume that $\|\mathbf{R}_i\|_2 \leq m$, $|y_i| \leq \delta$ for $i \in [d]$, and $\delta > m$. Then, there exists an efficient algorithm $\mathsf{TrapEval}_d$ that takes $\mathbf{R}_1, \ldots, \mathbf{R}_d, y_1, \ldots, y_d$ as inputs and outputs $\mathbf{R}'$ such that*

$$\mathsf{PubEval}_d(\mathbf{B}_1, \ldots, \mathbf{B}_d) = \mathbf{A}\mathbf{R}' + y_1 \cdots y_d \cdot \mathbf{G} \tag{5}$$

and $\|\mathbf{R}'\|_2 \leq md\delta^{d-1}$.

5.2 Our Construction

In the following, we present our PIBE scheme. Let d be a (flexible) constant. In addition, let the identity space of the scheme be $\mathcal{ID} = \{0,1\}^\kappa$ for some $\kappa \in \mathbb{N}$ and the message space be $\{0,1\}$. For our construction, we consider an efficiently computable injective map S that maps an identity $\mathsf{ID} \in \{0,1\}^\kappa$ to a subset $S(\mathsf{ID})$ of $[1, \ell]^d$, where $\ell = \lceil \kappa^{1/d} \rceil$. Such a map can be constructed easily as we explained in Sect. 3.3. We would typically set $\kappa = O(n)$, and thus $\ell = O(n^{1/d})$ in such a case.

$\mathsf{Setup}(1^n, 1^c)$: On input 1^n and 1^c, it sets the parameters q, m, σ, B, B', and a distribution χ as specified in Sect. 5.3, where q is a prime number. Then, it picks random matrices $\mathbf{B}_0 \xleftarrow{\$} \mathbb{Z}_q^{n \times m}$, $\mathbf{B}_{i,j} \xleftarrow{\$} \mathbb{Z}_q^{n \times m}$ for $(i,j) \in [d, \ell]$ and a vector $\mathbf{u} \xleftarrow{\$} \mathbb{Z}_q^n$. It also picks $\mathsf{TrapGen}(1^n, 1^m, q) \to (\mathbf{A}, \mathbf{T_A}) \in \mathbb{Z}_q^{n \times m} \times \mathbb{Z}^{m \times m}$ such that $\|\mathbf{T_A}\|_{\mathsf{GS}} = O(\sqrt{n \log q})$. It finally outputs

$$\mathsf{mpk} = (\mathbf{A}, \mathbf{B}_0, \{\mathbf{B}_{i,j}\}_{(i,j) \in [d,\ell]}, \mathbf{u}) \quad \text{and} \quad \mathsf{msk} = \mathbf{T_A}.$$

In the following, we use a deterministic function $H: \mathcal{ID} \rightarrow \mathbb{Z}_q^{n \times m}$ that is defined as follows.

$$H(ID) = \mathbf{B}_0 + \sum_{(j_1, \ldots, j_d) \in S(ID)} \mathsf{PubEval}_d(\mathbf{B}_{1,j_1}, \mathbf{B}_{2,j_2}, \ldots, \mathbf{B}_{d,j_d}) \in \mathbb{Z}_q^{n \times m}.$$

KeyGen(mpk, msk, ID): It first computes $H(ID)$ and picks $\mathbf{e} \in \mathbb{Z}^{2m}$ such that

$$(\mathbf{A} | H(ID)) \cdot \mathbf{e} = \mathbf{u}$$

by running $\mathsf{SampleLeft}(\mathbf{A}, H(ID), \mathbf{u}, \mathbf{T_A}, \sigma) \rightarrow \mathbf{e}$. It returns $\mathsf{sk}_{ID} = \mathbf{e}$.

Encrypt(mpk, ID, b): To encrypt a message $b \in \{0, 1\}$, it picks $\mathbf{s} \xleftarrow{\$} \mathbb{Z}_q^n$, $x_0 \xleftarrow{\$} \chi$, $\mathbf{x}_1 \xleftarrow{\$} \chi^m$, $\mathbf{x}_2 \xleftarrow{\$} [-B', B']^m$ and computes

$$c_0 = \mathbf{s}^\top \mathbf{u} + x_0 + b \cdot \lceil q/2 \rceil, \qquad \mathbf{c}_1^\top = \mathbf{s}^\top (\mathbf{A} | H(ID)) + (\mathbf{x}_1^\top | \mathbf{x}_2^\top).$$

Finally, it returns the ciphertext $C = (c_0, \mathbf{c}_1)$.

Decrypt(mpk, sk_{ID}, C): To decrypt a ciphertext $C = (c_0, \mathbf{c}_1)$ using a private key $\mathsf{sk}_{ID} := \mathbf{e}$, it first computes

$$w = c_0 - \mathbf{c}_1^\top \cdot \mathbf{e} \in \mathbb{Z}_q.$$

Then it returns 1 if $|w - \lceil q/2 \rceil| < \lceil q/4 \rceil$ and 0 otherwise.

5.3   Correctness and Parameter Selection

When the cryptosystem is operated as specified, we have during decryption,

$$w = c_0 - \mathbf{c}_1^\top \cdot \mathbf{e} = b \cdot \lceil q/2 \rceil + \underbrace{x_0 - (\mathbf{x}_1^\top | \mathbf{x}_2^\top) \cdot \mathbf{e}}_{\text{error term}}.$$

Lemma 7. *Assuming $B' > B$, the error term is bounded by $O(B'\sigma m)$ with overwhelming probability.*

Proof. Since χ is B-bounded distribution, with overwhelming probability, we have

$$|x_0 - (\mathbf{x}_1^\top | \mathbf{x}_2^\top) \cdot \mathbf{e}| \leq |x_0| + |(\mathbf{x}_1^\top | \mathbf{x}_2^\top) \cdot \mathbf{e}| \leq |x_0| + \|(\mathbf{x}_1^\top | \mathbf{x}_2^\top)\| \cdot \|\mathbf{e}\|$$

$$\leq B + \max\{B, B'\} \cdot \sqrt{2m} \cdot \sigma\sqrt{2m} = O(B'\sigma m).$$

The second inequality above follows from Cauchy-Schwartz and the third inequality follows from Lemma 1.

Parameter Selection. Now, to satisfy the correctness requirement and make the security proof work, we need that

- the error term is less than $q/5$ with overwhelming probability (i.e., $\Omega(B'\sigma m) < q$),
- that q is sufficiently large so that the simulation works (i.e., $q > \Theta(\kappa(dn^c)^d))$,
- that TrapGen can operate (i.e., $m \geq 6n\lceil \log q \rceil$),
- that the leftover hash lemma (Lemma 4) can be applied in the security proof (i.e., $m = (n+1)\log q + \omega(\log n)$),
- that σ is sufficiently large so that SampleLeft and SampleRight work, (i.e., $\sigma > O(\sqrt{n \log q}) \cdot \omega(\sqrt{\log m})$ and $\sigma > m(1 + \kappa d^d n^{c(d-1)}) \cdot \omega(\sqrt{\log m})$, where the latter condition turns out to be more restrictive),
- that the "noise smudging step" in the security proof works (i.e., $m^{5/2}(1 + \kappa d^d n^{c(d-1)})B/B' \leq d/(\kappa+1)(dn^c)^{d+1}$. See Eq. (11)).

To satisfy the above requirements, we set the parameters as follows:

$$m = O(n \log q), \qquad q = O(n^{3c(d-1)+3c'+6}), \qquad \chi = D_{\mathbb{Z}, \sqrt{n}},$$

$$\sigma = m\kappa n^{c(d-1)} \cdot \omega(\sqrt{\log m}), \quad B = O(n), \qquad B' = O(m^{5/2}\kappa^2 n^{2cd+1}),$$

where c' is a constant such that $\kappa = O(n^{c'})$. Typically, we would set $c' = 1$.

5.4   Security Proof

The following theorem addresses the security of the scheme. The proof is based on the partitioning technique, similarly to [1,6,12,47]. For simplicity, we opt to use the framework of [6] in our analysis, which does not require the artificial abort step [47]. The analysis with the artificial abort step is also possible, and it might lead to a scheme with slightly better efficiency (up to constant factors).

Theorem 3. *The above scheme is c-adaptive anonymous assuming* $\mathsf{dLWE}_{n,m+1,q,\chi}$ *is hard, where the ciphertext space is* $\mathcal{C} = \mathbb{Z}_q \times \mathbb{Z}_q^{2m}$.

Proof. Let $\mathcal{A}$ be a PPT adversary that breaks c-adaptive anonymity of the scheme. In addition, let $\epsilon = \epsilon(n)$ and $Q = Q(n)$ be its advantage and the upper bound of the number of key extraction queries, respectively. Without loss of generality, we assume that $\mathcal{A}$ always makes exactly Q key extraction queries. Let us define $\tilde{c}$ as a constant that satisfies

$$Q \leq \frac{n^{\tilde{c}}}{2} - 1 \quad \text{and} \quad \frac{\epsilon}{Q+1} - \frac{1}{n^{\tilde{c}}} = \mathsf{nonneg}(n) \tag{6}$$

where $\mathsf{nonneg}(n)$ is some non-negligible function. We explain such $\tilde{c}$ always exist. In the case of $c = c(n)$ is a constant, we simply let $\tilde{c} = c$. Let us consider the case of $c(n) = \omega(1)$. Since $\mathcal{A}$ is a PPT algorithm, there exists a constant c' such that $Q(n) \leq n^{c'}/2 - 1$. Furthermore, since $\mathcal{A}$ breaks c-adaptive anonymity of the scheme and $1/n^c$ is negligible, $\epsilon/(Q+1)$ is non-negligible. Therefore, there exists a constant c'' such that $\epsilon/(Q+1) > 2/n^{c''}$ holds for infinitely many n. By setting $\tilde{c} = \max\{c', c''\}$, we are done. We note that in any case, $\tilde{c}(n) \leq c(n)$ holds for sufficiently large n.

We show the security of the scheme via the following games. In each game, a value coin$'$ $\in \{0,1\}$ is defined. While it is set coin$'$ = coin in the first game, these values might be different in the later games. In the following, we define X_i be the event that coin$'$ = coin.

Game$_0$: This is the real security game. Recall that since the ciphertext space is $\mathcal{C} = \mathbb{Z}_q \times \mathbb{Z}_q^{2m}$, in the challenge phase, the challenge ciphertext is set as $C^\star = (c_0, \mathbf{c}_1) \xleftarrow{\$} \mathbb{Z}_q \times \mathbb{Z}_q^{2m}$ if coin = 1. At the end of the game, $\mathcal{A}$ outputs a guess $\widehat{\text{coin}}$ for coin. Finally, the challenger sets coin$'$ = $\widehat{\text{coin}}$. By the definition, we have

$$\left| \Pr[X_0] - \frac{1}{2} \right| = \left| \Pr[\text{coin}' = \text{coin}] - \frac{1}{2} \right| = \left| \Pr[\widehat{\text{coin}} = \text{coin}] - \frac{1}{2} \right| = \epsilon.$$

Game$_1$: In this game, we change Game$_0$ so that the challenger performs the following additional step at the end of the game. First, the challenger picks $\mathbf{y} = (y_0, \{y_{i,j}\}_{(i,j)\in[d,\ell]})$ as

$$y_0 \xleftarrow{\$} [-(\kappa+1)(dn^{\tilde{c}})^d + 1, 0] \quad \text{and} \quad y_{i,j} \xleftarrow{\$} [1, dn^{\tilde{c}}] \quad \text{for} \quad (i,j) \in [d] \times [\ell].$$

We define a function $\mathsf{F}_\mathbf{y} : \mathcal{ID} \to \mathbb{Z}_q$ as follows:

$$\mathsf{F}_\mathbf{y}(\mathsf{ID}) = y_0 + \sum_{(j_1,\ldots,j_d)\in S(\mathsf{ID})} y_{1,j_1} \cdots y_{d,j_d}.$$

Then the challenger checks whether the following condition holds:

$$\mathsf{F}_\mathbf{y}(\mathsf{ID}^\star) = 0 \ \wedge \ \mathsf{F}_\mathbf{y}(\mathsf{ID}_1) \neq 0 \ \wedge \ \mathsf{F}_\mathbf{y}(\mathsf{ID}_2) \neq 0 \ \wedge \ \cdots \ \wedge \ \mathsf{F}_\mathbf{y}(\mathsf{ID}_Q) \neq 0 \quad (7)$$

where $\mathsf{ID}^\star$ is the challenge identity, and $\mathsf{ID}_1, \ldots, \mathsf{ID}_Q$ are identities for which $\mathcal{A}$ has made key extraction queries. If it does not hold, the challenger ignores the output $\widehat{\text{coin}}$ of $\mathcal{A}$, and sets coin$'$ $\xleftarrow{\$} \{0,1\}$. In this case, we say that the challenger aborts. If condition (7) holds, the challenger sets coin$'$ = $\widehat{\text{coin}}$. As we will show in Lemma 8, we have

$$\left| \Pr[X_1] - \frac{1}{2} \right| \geq \frac{1}{\kappa+1} \cdot \left(\frac{1}{dn^{\tilde{c}}}\right)^d \cdot \left(\epsilon - \frac{Q}{n^{\tilde{c}}}\right).$$

So as not to interrupt the proof of Theorem 3, we intentionally skip the proof for the time being.

Game$_2$: In this game, we change the way $\mathbf{B}_0$ and $\mathbf{B}_{i,j}$ are chosen. At the beginning of the game, the challenger picks $\mathbf{R}_0, \mathbf{R}_{i,j} \xleftarrow{\$} \{-1,1\}^{m\times m}$ for $(i,j) \in [d] \times [\ell]$. It also picks $\mathbf{y}$ as in Game$_1$. Then, $\mathbf{A}$, $\mathbf{B}_0$, and $\mathbf{B}_{i,j}$ are defined as

$$\mathbf{B}_0 = \mathbf{A}\mathbf{R}_0 + y_0\mathbf{G}, \qquad \mathbf{B}_{i,j} = \mathbf{A}\mathbf{R}_{i,j} + y_{i,j}\mathbf{G} \qquad (8)$$

for $(i,j) \in [d] \times [\ell]$. The rest of the game is the same as in Game$_1$.

Then, we bound $|\Pr[X_2] - \Pr[X_1]|$. By Lemma 4, the distributions

$$(\mathbf{A}, \ \mathbf{AR}_0 + y_0\mathbf{G}, \ \{\mathbf{AR}_{i,j} + y_{i,j}\mathbf{G}\}) \quad \text{and} \quad (\mathbf{A}, \ \mathbf{B}_0, \ \{\mathbf{B}_{i,j}\})$$

are negl(n)-close, where $\mathbf{B}_0, \mathbf{B}_{i,j} \xleftarrow{\$} \mathbb{Z}_q^{n \times m}$. Therefore, we have $|\Pr[X_1] - \Pr[X_2]| = \mathsf{negl}(n)$.

Before describing the next game, we define $\mathbf{R}_{\mathsf{ID}}$ for an identity $\mathsf{ID} \in \mathcal{ID}$ as

$$\mathbf{R}_{\mathsf{ID}} = \mathbf{R}_0 + \sum_{(j_1,\ldots,j_d) \in S(\mathsf{ID})} \mathsf{TrapEval}(\mathbf{R}_{1,j_1}, \ldots, \mathbf{R}_{d,j_d}, y_{1,j_1}, \ldots, y_{d,j_d}). \quad (9)$$

Note that by Lemma 6, we have

$$\|\mathbf{R}_{\mathsf{ID}}^{\top}\|_2 = \|\mathbf{R}_{\mathsf{ID}}\|_2$$

$$\leq \|\mathbf{R}_0\|_2 + \sum_{(j_1,\ldots,j_d) \in S(\mathsf{ID})} \|\mathsf{TrapEval}(\mathbf{R}_{1,j_1}, \ldots, \mathbf{R}_{d,j_d}, y_{1,j_1}, \ldots, y_{d,j_d})\|_2$$

$$\leq \left(m + \kappa(md \cdot (dn^{\tilde{c}})^{d-1})\right) \leq m(1 + \kappa d^d n^{c(d-1)}) \quad (10)$$

for any $\mathsf{ID} \in \mathcal{ID}$. The last inequality above follows from $\tilde{c} \leq c$.

Game$_3$: In this game, we change the way the challenge ciphertext is created when coin $= 0$. If coin $= 0$, to create the challenge ciphertext Game$_3$ challenger first picks $\mathbf{s} \xleftarrow{\$} \mathbb{Z}_q^n$, $x_0 \xleftarrow{\$} \chi$, $\mathbf{x}_1 \xleftarrow{\$} \chi^m$, $\mathbf{x}_2 \xleftarrow{\$} [-B', B']^m$ and computes $\mathbf{R}_{\mathsf{ID}^\star}$. Then, the challenge ciphertext $C^\star = (c_0, \mathbf{c}_1)$ is computed as

$$c_0 = \mathbf{s}^{\top}\mathbf{u} + x_0 + b \cdot \lceil q/2 \rceil, \qquad \mathbf{c}_1^{\top} = \mathbf{s}^{\top}(\mathbf{A}|\mathsf{H}(\mathsf{ID}^\star)) + (\mathbf{x}_1^{\top}|\mathbf{x}_1^{\top}\mathbf{R}_{\mathsf{ID}^\star} + \mathbf{x}_2^{\top})$$

where $b \in \{0, 1\}$ is the message chosen by $\mathcal{A}$.
We then proceed to bound $|\Pr[X_3] - \Pr[X_2]|$. Since $\mathbf{x}_1$ is chosen from a B-bounded distribution, we have

$$\|\mathbf{R}_{\mathsf{ID}^\star}^{\top}\mathbf{x}_1\|_\infty \leq \|\mathbf{R}_{\mathsf{ID}^\star}^{\top}\mathbf{x}_1\|_2 \leq \|\mathbf{R}_{\mathsf{ID}^\star}^{\top}\|_2 \cdot \|\mathbf{x}_1\| \leq m^{3/2}(1 + \kappa d^d n^{c(d-1)})B.$$

When all randomness other than $\mathbf{x}_2$ in this game is fixed, the distributions $\mathbf{x}_2$ and $\mathbf{R}_{\mathsf{ID}^\star}^{\top} \cdot \mathbf{x}_1 + \mathbf{x}_2$ are within statistical distance

$$m\|\mathbf{R}_{\mathsf{ID}^\star}^{\top}\mathbf{x}_1\|_\infty/B' = m^{5/2}(1 + \kappa d^d n^{c(d-1)})B/B' \leq \frac{d}{\kappa + 1} \cdot \left(\frac{1}{dn^c}\right)^{d+1} \quad (11)$$

by Lemma 3. Averaging over all other randomness, we have that the distribution of the challenge ciphertext is within statistical distance $d/(\kappa+1)(dn^c)^{d+1}$ from the previous game, when coin $= 0$. In the case of coin $= 1$, the view of $\mathcal{A}$ is unchanged. Therefore, we conclude that the view of $\mathcal{A}$ in this game is within statistical distance $d/(\kappa + 1)(dn^c)^{d+1}$ from the previous game. Thus, we have

$$|\Pr[X_2] - \Pr[X_3]| \leq \frac{d}{\kappa + 1} \cdot \left(\frac{1}{dn^c}\right)^{d+1}.$$

Game$_4$: Recall that in the previous game, the challenger aborts at the end of the game, if the condition (7) is not satisfied. In this game, we change the game so that the challenger aborts as soon as the abort condition becomes true. Since this is only a conceptual change, we have $\Pr[X_3] = \Pr[X_4]$.

Game$_5$: In this game, we change the way the matrix $\mathbf{A}$ is sampled. Namely, Game$_5$ challenger picks $\mathbf{A} \xleftarrow{\$} \mathbb{Z}_q^{n \times m}$ instead of generating it with a trapdoor. By Lemma 2, this makes only negligible difference. Furthermore, we also change the way the key extraction queries are answered. When $\mathcal{A}$ makes a key extraction query for an identity ID, the challenger first computes $\mathbf{R}_{\mathsf{ID}}$ as in Eq. (9). By the definition of $\mathbf{R}_{\mathsf{ID}}$, it holds that

$$\mathsf{H}(\mathsf{ID}) = \mathbf{A} \cdot (\mathbf{R}_{\mathsf{ID}} + \mathsf{F}_{\mathbf{y}}(\mathsf{ID})\mathbf{G}).$$

If $\mathsf{F}_{\mathbf{y}}(\mathsf{ID}) = 0$, it aborts, as the previous game. Otherwise, it runs

$$\mathsf{SampleRight}(\mathbf{A}, \mathbf{G}, \mathbf{R}_{\mathsf{ID}}, \mathsf{F}_{\mathbf{y}}(\mathsf{ID}), \mathbf{u}, \mathbf{T}_{\mathbf{G}}, \sigma) \to \mathbf{e},$$

and returns $\mathbf{e}$ to $\mathcal{A}$. Note that the private key was sampled as

$$\mathsf{SampleLeft}(\mathbf{A}, \mathsf{H}(\mathsf{ID}), \mathbf{u}, \mathbf{T}_{\mathbf{A}}, \sigma) \to \mathbf{e}$$

in the previous game. By Eq. (10) and the choice of σ, the output distribution of SampleRight is $\mathsf{negl}(n)$-close to $D_{\Lambda_q^{\mathbf{u}}(\mathbf{A}|\mathsf{H}(\mathsf{ID})),\sigma}$. Similarly, by the choice of σ, the output distribution of SampleLeft is also $\mathsf{negl}(n)$-close to $D_{\Lambda_q^{\mathbf{u}}(\mathbf{A}|\mathsf{H}(\mathsf{ID})),\sigma}$. Therefore, the above change alters the view of the adversary only negligibly. Thus, we have $|\Pr[X_4] - \Pr[X_5]| = \mathsf{negl}(n)$.

Game$_6$: In this game, we change the way the challenge ciphertext is created when coin $= 0$. If coin $= 0$, to create the challenge ciphertext for the identity ID* and the message b, Game$_6$ challenger first picks $v_0 \xleftarrow{\$} \mathbb{Z}_q$, $\mathbf{v}_1 \xleftarrow{\$} \mathbb{Z}_q^m$, $\mathbf{x}_2 \xleftarrow{\$} [-B', B']^m$ and computes $\mathbf{R}_{\mathsf{ID}^\star}$. Then, it sets the challenge ciphertext $C^\star = (c_0, \mathbf{c}_1)$ as

$$c_0 = v_0 + b \cdot \lceil q/2 \rceil, \qquad \mathbf{c}_1^\top = (\mathbf{v}_1^\top | \mathbf{v}_1^\top \mathbf{R}_{\mathsf{ID}^\star}) + (\mathbf{0}_m^\top | \mathbf{x}_2^\top).$$

As we will show in Lemma 9, assuming $\mathsf{dLWE}_{n,m+1,q,\chi}$ is hard, we have $|\Pr[X_5] - \Pr[X_6]| = \mathsf{negl}(n)$.

Game$_7$: In this game, we change the challenge ciphertext to be a random vector, regardless of whether coin $= 0$ or coin $= 1$. Namely, Game$_7$ challenger generates the challenge ciphertext $(c_0, \mathbf{c}_1)$ as $c_0 \xleftarrow{\$} \mathbb{Z}_q$ and $\mathbf{c}_1 \xleftarrow{\$} \mathbb{Z}_q^m$.

We now proceed to bound $|\Pr[X_7] - \Pr[X_6]|$. Since Game$_6$ and Game$_7$ differ only in the creation of the challenge ciphertext when coin $= 0$, we focus on this case. First, it is easy to see that c_0 is uniformly random over $\mathbb{Z}_q$ in both of Game$_6$ and Game$_7$. We also have to show that the distribution of $\mathbf{c}_1$ is $\mathsf{negl}(n)$-close to the uniform distribution over $\mathbb{Z}_q^{2m}$. To see this, it suffices to show that $(\mathbf{v}_1^\top | \mathbf{v}_1^\top \mathbf{R}_{\mathsf{ID}^\star})$ is distributed statistically close to uniform distribution over $\mathbb{Z}_q^{2m}$. Observe that the following distributions are $\mathsf{negl}(n)$-close:

$$(\mathbf{A}, \mathbf{A}\mathbf{R}_0, \mathbf{v}_1^\top, \mathbf{v}_1^\top \mathbf{R}_0) \approx (\mathbf{A}, \mathbf{A}', \mathbf{v}_1^\top, {\mathbf{v}_1'}^\top) \approx (\mathbf{A}, \mathbf{A}\mathbf{R}_0, \mathbf{v}_1^\top, {\mathbf{v}_1'}^\top), \qquad (12)$$

where $\mathbf{A}, \mathbf{A}' \xleftarrow{\$} \mathbb{Z}_q^{n \times m}$, $\mathbf{R}_0 \xleftarrow{\$} \{-1,1\}^{m \times m}$, $\mathbf{v}_1, \mathbf{v}_1' \xleftarrow{\$} \mathbb{Z}_q^m$. It can be seen that the first and the second distributions are $\mathsf{negl}(n)$-close, by applying Lemma 4 for $(\mathbf{A}^\top | \mathbf{v})^\top \in \mathbb{Z}_{(n+1) \times m}$ and $\mathbf{R}_0$. It can also be seen that the second and the third distributions are $\mathsf{negl}(n)$-close, by applying the same lemma for $\mathbf{A}$ and $\mathbf{R}_0$. From the above, we have that the following distributions are statistically close:

$$(\mathbf{A}, \mathbf{A}\mathbf{R}_0, \mathbf{v}_1, \mathbf{v}_1^\top \mathbf{R}_{\mathsf{ID}}^*)$$

$$= \left(\mathbf{A}, \mathbf{A}\mathbf{R}_0, \mathbf{v}_1, \mathbf{v}_1^\top \left(\mathbf{R}_0 + \sum_{\substack{(j_1,\ldots,j_d) \\ \in S(\mathsf{ID})}} \mathsf{TrapEval}(\mathbf{R}_{1,j_1}, \ldots, \mathbf{R}_{d,j_d}, y_{1,j_1}, \ldots, y_{d,j_d}) \right) \right)$$

$$\approx \left(\mathbf{A}, \mathbf{A}\mathbf{R}_0, \mathbf{v}_1, \mathbf{v}_1'^\top + \mathbf{v}_1^\top \left(\sum_{\substack{(j_1,\ldots,j_d) \\ \in S(\mathsf{ID})}} \mathsf{TrapEval}(\mathbf{R}_{1,j_1}, \ldots, \mathbf{R}_{d,j_d}, y_{1,j_1}, \ldots, y_{d,j_d}) \right) \right)$$

$$\approx (\mathbf{A}, \mathbf{A}\mathbf{R}_0, \mathbf{v}_1, \mathbf{v}_1'^\top)$$

where $\mathbf{A}, \mathbf{A}' \xleftarrow{\$} \mathbb{Z}_q^{n \times m}$, $\mathbf{R}_0 \xleftarrow{\$} \{-1,1\}^{m \times m}$, $\mathbf{v}_1, \mathbf{v}_1' \xleftarrow{\$} \mathbb{Z}_q^m$. The second and the third distributions above are $\mathsf{negl}(n)$-close by Eq. (12). Therefore, we may conclude that $|\Pr[X_6] - \Pr[X_7]| = \mathsf{negl}(n)$.

Analysis. From the above, we have

$$\left| \Pr[X_7] - \frac{1}{2} \right| = \left| \Pr[X_1] - \frac{1}{2} + \sum_{i=1}^{6} \Pr[X_{i+1}] - \Pr[X_i] \right|$$

$$\geq \left| \Pr[X_1] - \frac{1}{2} \right| - \sum_{i=1}^{6} |\Pr[X_{i+1}] - \Pr[X_i]|$$

$$\geq \frac{1}{\kappa + 1} \cdot \left(\frac{1}{dn^{\tilde{c}}} \right)^d \cdot \left(\epsilon - \frac{Q}{n^{\tilde{c}}} \right) - \frac{d}{\kappa + 1} \cdot \left(\frac{1}{dn^c} \right)^{d+1} - \mathsf{negl}(n)$$

$$\geq \frac{1}{\kappa + 1} \cdot \left(\frac{1}{dn^{\tilde{c}}} \right)^d \cdot \left(\epsilon - \frac{Q}{n^{\tilde{c}}} \right) - \frac{d}{\kappa + 1} \cdot \left(\frac{1}{dn^{\tilde{c}}} \right)^{d+1} - \mathsf{negl}(n)$$

$$= \frac{1}{\kappa + 1} \cdot \left(\frac{1}{dn^{\tilde{c}}} \right)^d \cdot (Q+1) \cdot \left(\frac{\epsilon}{Q+1} - \frac{1}{n^{\tilde{c}}} \right) - \mathsf{negl}(n)$$

$$= \frac{1}{\mathsf{poly}(n)} \cdot \left(\frac{\epsilon}{Q+1} - \frac{1}{n^{\tilde{c}}} \right) - \mathsf{negl}(n). \tag{13}$$

The third inequality above follows from $c \geq \tilde{c}$. Since the challenge ciphertext is independent from the value of coin in Game_7, we have $\Pr[X_7] = 1/2$ and thus $|\Pr[X_7] - 1/2| = 0$. Therefore, from inequality (13), $\epsilon/(Q+1) < 1/n^{\tilde{c}} + \mathsf{negl}(n)$ follows. However, this contradicts to Eq. (6).

To complete the proof of Theorem 3, it remains to show Lemmas 8 and 9.

Lemma 8. *For any PPT adversary* $\mathcal{A}$, *we have*

$$\left| \Pr[X_1] - \frac{1}{2} \right| \geq \frac{1}{\kappa+1} \cdot \left(\frac{1}{dn^{\tilde{c}}} \right)^d \cdot \left(\epsilon - \frac{Q}{n^{\tilde{c}}} \right).$$

Proof. For a sequence of identities $\mathbb{ID} = (\mathsf{ID}^{\star}, \mathsf{ID}_1, \ldots, \mathsf{ID}_Q) \in \mathcal{ID}^{Q+1}$, we define $\gamma(\mathbb{ID})$ as

$$\gamma(\mathbb{ID}) = \Pr_{\mathbf{y}}[\mathsf{F}_{\mathbf{y}}(\mathsf{ID}^{\star}) = 0 \wedge \mathsf{F}_{\mathbf{y}}(\mathsf{ID}_1) \neq 0 \wedge \mathsf{F}_{\mathbf{y}}(\mathsf{ID}_2) \neq 0 \wedge \cdots \wedge \mathsf{F}_{\mathbf{y}}(\mathsf{ID}_Q) \neq 0]$$

where the probability is taken over $\mathbf{y} = (y_0, \{y_{i,j}\}_{(i,j) \in [d,\ell]})$, which is chosen as specified in Game_1. To show the lemma, we first show the following claim, which gives an upper and lower bounds for $\gamma(\mathbb{ID})$.

Claim. For any $\mathbb{ID} = (\mathsf{ID}^{\star}, \mathsf{ID}_1, \ldots, \mathsf{ID}_Q)$ such that $\mathsf{ID}^{\star} \neq \mathsf{ID}_i$ for all $i \in [Q]$,

$$\frac{1}{\kappa+1} \cdot \left(\frac{1}{dn^{\tilde{c}}} \right)^d \cdot \left(1 - \frac{Q}{n^{\tilde{c}}} \right) \leq \gamma(\mathbb{ID}) \leq \frac{1}{\kappa+1} \cdot \left(\frac{1}{dn^{\tilde{c}}} \right)^d.$$

Proof. Showing the upper bound of the probability is very easy. For any $\{y_{i,j}\}$, there exists exactly one $y_0 \in [-(\kappa+1)(dn^{\tilde{c}})^d + 1, 0]$ such that $\mathsf{F}_{\mathbf{y}}(\mathsf{ID}^{\star}) = 0$, since for any $\{y_{i,j}\}_{(i,j) \in [d] \times [\ell]}$ and ID, we have

$$0 \leq \sum_{(j_1, \ldots, j_d) \in S(\mathsf{ID})} y_{1,j_1} \cdots y_{d,j_d} \leq \sum_{(j_1, \ldots, j_d) \in S(\mathsf{ID})} (dn^{\tilde{c}})^d < (\kappa+1)(dn^{\tilde{c}})^d$$

Therefore, we have

$$\gamma(\mathbb{ID}) \leq \Pr_{\mathbf{y}}[\mathsf{F}_{\mathbf{y}}(\mathsf{ID}^{\star}) = 0] = \frac{1}{\kappa+1} \cdot \left(\frac{1}{dn^{\tilde{c}}} \right)^d.$$

We then proceed to show the lower bound.

$$\gamma(\mathbb{ID}) = \Pr_{\mathbf{y}}[\mathsf{F}_{\mathbf{y}}(\mathsf{ID}^{\star}) = 0 \wedge \mathsf{F}_{\mathbf{y}}(\mathsf{ID}_1) \neq 0 \wedge \mathsf{F}_{\mathbf{y}}(\mathsf{ID}_2) \neq 0 \wedge \cdots \wedge \mathsf{F}_{\mathbf{y}}(\mathsf{ID}_Q) \neq 0]$$

$$\geq \Pr_{\mathbf{y}}[\mathsf{F}_{\mathbf{y}}(\mathsf{ID}^{\star}) = 0] - \sum_{i \in [Q]} \Pr_{\mathbf{y}}[\mathsf{F}_{\mathbf{y}}(\mathsf{ID}^{\star}) = 0 \wedge \mathsf{F}_{\mathbf{y}}(\mathsf{ID}_i) = 0]$$

$$= \frac{1}{\kappa+1} \cdot \left(\frac{1}{dn^{\tilde{c}}} \right)^d - \sum_{i \in [Q]} \Pr_{\mathbf{y}}[\mathsf{F}_{\mathbf{y}}(\mathsf{ID}^{\star}) = 0 \wedge \mathsf{F}_{\mathbf{y}}(\mathsf{ID}_i) = 0]. \tag{14}$$

It suffices to show an upper bound for $\Pr[\mathsf{F}_{\mathbf{y}}(\mathsf{ID}^{\star}) = 0 \wedge \mathsf{F}_{\mathbf{y}}(\mathsf{ID}_i) = 0]$. For $i \in [Q]$, we have

$$\Pr_{\mathbf{y}}[\mathsf{F}_{\mathbf{y}}(\mathsf{ID}^{\star}) = 0 \wedge \mathsf{F}_{\mathbf{y}}(\mathsf{ID}_i) = 0]$$

$$= \Pr_{\mathbf{y}}[\mathsf{F}_{\mathbf{y}}(\mathsf{ID}^{\star}) = 0 \wedge \mathsf{F}_{\mathbf{y}}(\mathsf{ID}^{\star}) - \mathsf{F}_{\mathbf{y}}(\mathsf{ID}_i) = 0]$$

$$= \Pr_{\mathbf{y}}[F_{\mathbf{y}}(ID^\star) = 0 \mid F_{\mathbf{y}}'(ID^\star, ID_i) = 0] \cdot \Pr_{\mathbf{y}}[F_{\mathbf{y}}'(ID^\star, ID_i) = 0]$$

$$= \Pr_{\mathbf{y}}\left[y_0 = - \sum_{\substack{(j_1,\ldots,j_d) \\ \in S(ID^\star)}} y_{1,j_1} \cdots y_{d,j_d} \,\middle|\, F_{\mathbf{y}}'(ID^\star, ID_i) = 0 \right] \cdot \Pr_{\mathbf{y}}[F_{\mathbf{y}}'(ID^\star, ID_i) = 0]$$

$$= \frac{1}{\kappa + 1} \cdot \left(\frac{1}{dn^{\tilde{c}}} \right)^d \cdot \Pr_{\mathbf{y}}[F_{\mathbf{y}}'(ID^\star, ID_i) = 0]. \tag{15}$$

In the above, we defined $F_{\mathbf{y}}'(ID^\star, ID_i)$ as

$$F_{\mathbf{y}}'(ID^\star, ID_i) := F_{\mathbf{y}}(ID^\star) - F_{\mathbf{y}}(ID_i)$$

$$= \sum_{(j_1,\ldots,j_d) \in S(ID^\star)} y_{1,j_1} \cdots y_{d,j_d} - \sum_{(j_1,\ldots,j_d) \in S(ID_i)} y_{1,j_1} \cdots y_{d,j_d}.$$

The last equation in Eq. (15) follows since y_0 is independent from $F_{\mathbf{y}}'(ID^\star, ID_i)$. (Observe that y_0 does not appear in the definition of $F_{\mathbf{y}}'(ID^\star, ID_i)$.)

We then finally bound $\Pr_{\mathbf{y}}[F_{\mathbf{y}}'(ID^\star, ID_i) = 0]$. Since $ID^\star \neq ID_i$ and S is an injective map, we have $S(ID^\star) \neq S(ID_i)$. Therefore, there exists $(j_1^\star, \ldots, j_d^\star) \in [\ell]^d$ such that $(j_1^\star, \ldots, j_d^\star) \in S(ID^\star) \triangle S(ID_i)$, where $S(ID^\star) \triangle S(ID_i)$ denotes the symmetric difference of $S(ID^\star)$ and $S(ID_i)$. Thus, $F_{\mathbf{y}}'(ID^\star, ID_i)$ is not a zero-polynomial when we regard it as a polynomial in indeterminates $\{y_{j,k}\}_{(j,k) \in [d] \times [\ell]}$. Since each $y_{j,k}$ is uniformly random over $[1, dn^{\tilde{c}}]$ and $F_{\mathbf{y}}'(ID^\star, ID_i)$ is a polynomial with degree d, by the Schwartz-Zippel lemma, it follows that

$$\Pr_{\mathbf{y}}[F_{\mathbf{y}}'(ID^\star, ID_i) = 0] \leq \frac{d}{dn^{\tilde{c}}} \leq \frac{1}{n^{\tilde{c}}}.$$

By combining this with Eqs. (14) and (15), the claim follows.

We then proceed to show a lower bound for $|\Pr[X_1] - 1/2|$. For $ID = (ID^\star, ID_1, \ldots, ID_Q)$ such that $ID^\star \neq ID_i$ for all $i \in [Q]$, we define $\gamma_{\max}$ and $\gamma_{\min}$ as the largest and the smallest value of $\gamma(ID)$ taken over all such ID, respectively. We define $Q(ID)$ as the event that $\mathcal{A}$ chooses $ID^\star$ as its challenge identity and it makes key extraction queries for $ID_1, \ldots, ID_Q$. We also define Abort as the event that the challenger aborts. Then, we have

$$\left| \Pr[X_1] - \frac{1}{2} \right| = \left| \Pr[\mathsf{coin}' = \mathsf{coin}] - \frac{1}{2} \right|$$

$$= \left| \sum_{ID} \Pr[Q(ID)] \cdot \Pr[\mathsf{coin}' = \mathsf{coin} | Q(ID)] - \frac{1}{2} \right|$$

$$= \left| \sum_{ID} \Pr[Q(ID)] \cdot \Big(\Pr[\mathsf{coin}' = \mathsf{coin} \wedge \neg\mathsf{Abort} | Q(ID)] \right.$$

$$\left. + \Pr[\mathsf{coin}' = \mathsf{coin} \wedge \mathsf{Abort} | Q(ID)] - \frac{1}{2} \Big) \right|$$

$$= \left| \sum_{\mathbb{ID}} \Pr[Q(\mathbb{ID})] \cdot \left(\Pr[\widehat{\text{coin}} = \text{coin}|Q(\mathbb{ID})] \cdot \gamma(\mathbb{ID}) + \frac{1}{2} \cdot (1 - \gamma(\mathbb{ID})) - \frac{1}{2} \right) \right|$$

$$= \left| \sum_{\mathbb{ID}} \gamma(\mathbb{ID}) \cdot \Pr[Q(\mathbb{ID})] \cdot \left(\Pr[\widehat{\text{coin}} = \text{coin}|Q(\mathbb{ID})] - \frac{1}{2} \right) \right|$$

$$\geq \gamma_{\min} \cdot \epsilon - \frac{\gamma_{\max} - \gamma_{\min}}{2}.$$

In the third equation above, we used the fact $\sum_{\mathbb{ID}} \Pr[Q(\mathbb{ID})] = 1$. The fourth equation above follows from the fact that the probability of the abort is $\gamma(\mathbb{ID})$, when conditioned on $Q(\mathbb{ID})$ (regardless of the value of $\widehat{\text{coin}}$). The last inequality above follows by Lemma 5, since we have

$$\left| \sum_{\mathbb{ID}} \Pr[Q(\mathbb{ID})] \left(\Pr[\widehat{\text{coin}} = \text{coin}|Q(\mathbb{ID})] - \frac{1}{2} \right) \right|$$

$$= \left| \sum_{\mathbb{ID}} \Pr[\widehat{\text{coin}} = \text{coin} \wedge Q(\mathbb{ID})] - \frac{1}{2} \right| = \left| \Pr[\widehat{\text{coin}} = \text{coin}] - \frac{1}{2} \right| = \epsilon$$

and

$$\sum_{\mathbb{ID}} \left| \Pr[Q(\mathbb{ID})] \cdot \left(\Pr[\widehat{\text{coin}} = \text{coin}|Q(\mathbb{ID})] - \frac{1}{2} \right) \right| \leq \sum_{\mathbb{ID}} \Pr[Q(\mathbb{ID})] \cdot \frac{1}{2} = \frac{1}{2}.$$

We complete the proof of Lemma 8 by observing

$$\gamma_{\min} \cdot \epsilon - \frac{\gamma_{\max} - \gamma_{\min}}{2}$$

$$\geq \frac{1}{\kappa + 1} \cdot \left(\frac{1}{dn^{\tilde{c}}} \right)^d \cdot \left(1 - \frac{Q}{n^{\tilde{c}}} \right) \cdot \epsilon - \frac{1}{2(\kappa + 1)} \cdot \left(\frac{1}{dn^{\tilde{c}}} \right)^d \cdot \left(1 - \left(1 - \frac{Q}{n^{\tilde{c}}} \right) \right)$$

$$\geq \frac{1}{\kappa + 1} \cdot \left(\frac{1}{dn^{\tilde{c}}} \right)^d \cdot \left(\epsilon - \frac{Q}{n^{\tilde{c}}} \right).$$

The last inequality follows from $\epsilon \leq 1/2$.

Lemma 9. *For any PPT adversary $\mathcal{A}$, there exists another PPT adversary $\mathcal{B}$ such that*

$$|\Pr[X_5] - \Pr[X_6]| \leq \mathsf{Adv}_{\mathcal{B}}^{\mathsf{dLWE}_{n,m+1,q,\chi}}.$$

In particular, under the $\mathsf{dLWE}_{n,m+1,q,\chi}$ assumption, we have $|\Pr[X_5] - \Pr[X_6]| = \mathsf{negl}(n)$.

Proof. Suppose an adversary $\mathcal{A}$ that has non-negligible advantage in distinguishing Game_5 and Game_6. We use $\mathcal{A}$ to construct an LWE algorithm denoted $\mathcal{B}$, which proceeds as follows.

Instance. $\mathcal{B}$ is given the problem instance of LWE $(\mathbf{A}', \mathbf{v}') \in \mathbb{Z}_q^{n \times (m+1)} \times \mathbb{Z}_q^{m+1}$. Let the first column of $\mathbf{A}'$ be $\mathbf{u} \in \mathbb{Z}_q^n$ and the last m column be $\mathbf{A} \in \mathbb{Z}_q^{n \times m}$. It also sets the first coefficient of $\mathbf{v}'$ be v_0 and the last m coefficients be $\mathbf{v}_1$.

Setup. To construct master public key mpk, $\mathcal{B}$ first picks $\mathbf{y}$ as in Game$_1$. It also picks $\mathbf{R}_0, \mathbf{R}_{i,j} \xleftarrow{\$} \{-1, 1\}^{m \times m}$ and sets $\mathbf{B}_0$ and $\mathbf{B}_{i,j}$ as Eq. (8). Finally, it returns mpk $= (\mathbf{A}, \mathbf{B}_0, \{\mathbf{B}_{i,j}\}_{(i,j) \in [d,\ell]}, \mathbf{u})$ to $\mathcal{A}$. $\mathcal{B}$ also picks a random bit coin $\xleftarrow{\$} \{0, 1\}$ and keeps it secret.

Phases 1 and 2. When $\mathcal{A}$ makes a key extraction query for ID, $\mathcal{B}$ first computes $\mathsf{F_y}(\mathsf{ID})$. It aborts and sets coin$' \xleftarrow{\$} \{0, 1\}$ if $\mathsf{F_y}(\mathsf{ID}) = 0$. Otherwise, $\mathcal{B}$ generates the private key as in Game$_5$.

Challenge Query. When $\mathcal{A}$ makes the challenge query for the challenge identity $\mathsf{ID}^\star$ and the message b, $\mathcal{B}$ first computes $\mathsf{F_y}(\mathsf{ID}^\star)$. Then, it aborts and sets coin$' \xleftarrow{\$} \{0, 1\}$ if $\mathsf{F_y}(\mathsf{ID}^\star) \neq 0$. Otherwise, it proceeds as follows. If coin $= 0$, it computes $\mathbf{R}_{\mathsf{ID}^\star}$ and picks $\mathbf{x}_2 \xleftarrow{\$} [-B', B']^m$. Then, it sets the challenge ciphertext as

$$c_0 = v_0 + b \cdot \lceil q/2 \rceil, \qquad \mathbf{c}_1^\top = (\mathbf{v}_1^\top | \mathbf{v}_1^\top \mathbf{R}_{\mathsf{ID}^\star}) + (\mathbf{0}_m^\top | \mathbf{x}_2^\top)$$

and returns $C^\star = (c_0, \mathbf{c}_1)$ to $\mathcal{A}$. In the case of coin $= 1$, $\mathcal{B}$ picks $c_0 \xleftarrow{\$} \mathbb{Z}_q$, $\mathbf{c}_1 \xleftarrow{\$} \mathbb{Z}_q^{2m}$ and returns the challenge ciphertext $C^\star = (c_0, \mathbf{c}_1)$ to $\mathcal{A}$.

Guess. At last, $\mathcal{A}$ outputs its guess $\widehat{\text{coin}}$ (if the abort condition has not been satisfied). Then, $\mathcal{B}$ sets coin$' = \widehat{\text{coin}}$. Finally, $\mathcal{B}$ outputs 1 if coin$' =$ coin and 0 otherwise.

Analysis. We now show that $\mathcal{B}$ perfectly simulates the view of $\mathcal{A}$ in Game$_5$ if $(\mathbf{A}', \mathbf{v}')$ is a valid LWE sample (i.e., $\mathbf{v}'^\top = \mathbf{s}^\top \mathbf{A}' + \mathbf{x}^\top$ for $\mathbf{s} \xleftarrow{\$} \mathbb{Z}_q^n$ and $\mathbf{x} \xleftarrow{\$} \chi^{m+1}$), and Game$_6$ if $\mathbf{v}' \xleftarrow{\$} \mathbb{Z}_q^{m+1}$. Note that these games differ only in the generation of the challenge ciphertext in the case of coin $= 0$. Furthermore, it is easy to see that the simulation of the master public key, **Phases 1 and 2**, and the challenge ciphertext for the case of coin $= 1$ are perfect. Therefore, in the following, we focus on the generation of the challenge ciphertext in the case of coin $= 0$.

We first show that if $(\mathbf{A}', \mathbf{v}')$ is a valid LWE sample, i.e., $\mathbf{v}'^\top = \mathbf{s}^\top \mathbf{A}' + \mathbf{x}^\top$ for $\mathbf{s} \xleftarrow{\$} \mathbb{Z}_q^n$ and $\mathbf{x} \xleftarrow{\$} \chi^{m+1}$, the distribution of the challenge ciphertext corresponds to that of Game$_5$. Let us denote $\mathbf{x}^\top = (x_0, \mathbf{x}_1^\top)$ and assume that $\mathsf{F_y}(\mathsf{ID}^\star) = 0$ holds. Then, we have

$$c_0 = v_0 + b \cdot \lceil q/2 \rceil = (\mathbf{u}^\top \mathbf{s} + x_0) + b \cdot \lceil q/2 \rceil \qquad \text{and}$$
$$\begin{aligned}
\mathbf{c}_1 &= (\mathbf{v}_1^\top | \mathbf{v}_1^\top \mathbf{R}_{\mathsf{ID}^\star}) + (\mathbf{0}_m^\top | \mathbf{x}_2^\top) \\
&= \left(\mathbf{s}^\top \mathbf{A} + \mathbf{x}_1^\top | (\mathbf{s}^\top \mathbf{A} + \mathbf{x}_1^\top) \mathbf{R}_{\mathsf{ID}^\star} \right) + (\mathbf{0}_m^\top | \mathbf{x}_2^\top) \\
&= \mathbf{s}^\top (\mathbf{A} | \mathbf{A} \mathbf{R}_{\mathsf{ID}^\star}) + (\mathbf{x}_1^\top | \mathbf{x}_1^\top \mathbf{R}_{\mathsf{ID}^\star} + \mathbf{x}_2^\top) \\
&= \mathbf{s}^\top (\mathbf{A} | \mathsf{H}(\mathsf{ID}^\star)) + (\mathbf{x}_1^\top | \mathbf{x}_1^\top \mathbf{R}_{\mathsf{ID}^\star} + \mathbf{x}_2^\top).
\end{aligned}$$

The last equation follows because $\mathsf{F_y}(\mathsf{ID}^\star) = 0$. Therefore, the challenge ciphertext is distributed as in Game$_5$ in this case. It is easy to see that the challenge ciphertext is distributed as in Game$_6$, if $\mathbf{v}' \xleftarrow{\$} \mathbb{Z}_q^{m+1}$.

Therefore, we have $\mathsf{Adv}_{\mathcal{B}}^{\mathsf{dLWE}_{n,m+1,q,\chi}} = |\Pr[X_5] - \Pr[X_6]|$ as desired.

5.5   Multi-bit Encryption

Here, we explain that our scheme can be extended to deal with multi-bit messages without much increasing the sizes of public parameters and ciphertexts, similarly to [1,39]. To modify the scheme so that it can encrypt messages with N-bit, we replace $\mathbf{u} \in \mathbb{Z}_q^n$ in mpk with $\mathbf{u}_1, \ldots, \mathbf{u}_N \in \mathbb{Z}_q^n$. The component $c_0 = \langle \mathbf{u}, \mathbf{s} \rangle + x_0 + b\lceil \frac{q}{2} \rceil$ in the ciphertext is replaced with $c_0 = \{\langle \mathbf{u}_i, \mathbf{s} \rangle + x_{0,i} + b_i \lceil \frac{q}{2} \rceil\}_{i=1}^N$ where $x_{0,i} \xleftarrow{\$} \chi$ and $b_i \in \{0, 1\}$ is the i-th bit of the message. Furthermore, the private key is changed to be short vectors $\mathbf{e}_1, \ldots, \mathbf{e}_N \in \mathbb{Z}^m$ such that $(\mathbf{A}|\mathsf{H}(\mathsf{ID}))\mathbf{e}_i = \mathbf{u}_i$ for $i = 1, \ldots, N$. We can prove the security for the variant from $\mathsf{dLWE}_{n,m+N,q,\chi}$ by naturally extending the proof of Theorem 3.

As for the efficiency, the size of the master public key and the ciphertexts become $O((\ell m + N)n \log q)$ and $O((m + N) \log q)$ respectively, and these are asymptotically the same as the case of single-bit encryption when $N < O(m)$. The case of $N > O(m)$ can also be handled without increasing the size of parameters, by employing the KEM-DEM approach. Namely, we encrypt a random ephemeral key of sufficient length (e.g., $O(n)$) by IBE and then encrypt the message by the ephemeral key using a symmetric cipher.

6   Comparisons and Discussions

From the PIBE scheme in Sect. 5, we can obtain the following new IBE schemes:

- By setting $c = \omega(1)$, we obtain adaptively anonymous IBE by Theorem 1. However, we have to rely on super-polynomial LWE assumption, namely, $\mathsf{dLWE}_{n,m,q,\chi}$ with $q/\chi_{\max} = n^{\omega(1)}$.
- By applying PIBE-to-IBE conversion in Sect. 4.2 to our PIBE in Sect. 5, we obtain (non-anonymous) adaptively secure IBE from polynomial LWE. More precisely, the security of the scheme can be proven under the assumption that $\mathsf{dLWE}_{n,m,q,\chi}$ is hard for all $q/\chi_{\max} = \mathsf{poly}(n)$.

For concreteness, we would set $c(n) = O(\log \log n)$ in the first construction, and $c(n) = \log \log n$ and $\eta(n) = \log \log n$ for the second construction. Ignoring poly-logarithmic factors hidden in the asymptotic notation $\tilde{O}(\cdot)$, both of our schemes achieve the best efficiency among existing adaptively secure IBE schemes. See Table 1 for the comparison. Comparing in more details, ciphertexts and private keys of both of our schemes are longer than [1,12] by a super-constant factor. This is because we need to use super polynomially large q. On the other hand, in both of our schemes, the sizes of master public keys are asymptotically smaller than [1,12], even though we have to use larger q. This is because we require smaller number of basic matrices in the master public keys. Our first scheme is more efficient than our second scheme by super-constant factors, because the conversion in Sect. 4.2 incurs super-constant efficiency loss. We

Table 1. Comparison of IBE from the LWE assumption in the Standard Model.

Schemes	$|\mathsf{mpk}|$	$|C|$	$|\mathsf{sk_{ID}}|$	Anon?	Selective or adaptive	$q/\chi_{\max}$ for LWE assumption
[1]	$\tilde{O}(n^2)$	$\tilde{O}(n)$	$\tilde{O}(n)$	Yes	Selective	Fixed poly(n)
[16]	$\tilde{O}(n^2\kappa)$	$\tilde{O}(n\kappa)$	$\tilde{O}(n^2)$	Yes	Adaptive	Fixed poly(n)
[1,12][a]	$\tilde{O}(n^2\kappa)$	$\tilde{O}(n)$	$\tilde{O}(n)$	Yes	Adaptive	Fixed poly(n)
Ours: Section 5 + Theorem 1	$\tilde{O}(n^2\kappa^{1/d})$	$\tilde{O}(n)$	$\tilde{O}(n)$	Yes	Adaptive	$n^{\omega(1)}$
Ours: Section 5 + Theorem 2	$\tilde{O}(n^2\kappa^{1/d})$	$\tilde{O}(n)$	$\tilde{O}(n)$	No	Adaptive	All poly(n)

[a]In the security proof for the adaptively secure variant of IBE in [1], we have a restriction that $q > Q$. Namely, only bounded form of the security is proven. This restriction is removed in the refined analysis due to Boyen [12].

also note that our security reduction is very loose even compared to non-tight reduction of [1,12]. The security degrades exponentially as d grows. Therefore, in order to have polynomial reduction, we have to set d to be a (possibly small) constant.

In the table, we compare IBE schemes from the LWE assumption in the standard model. $|\mathsf{mpk}|$, $|C|$, and $|\mathsf{sk_{ID}}|$ show the size of the master public keys, ciphertexts, and private keys, respectively. κ denotes the length of the identity (which corresponds to the output length of the collision resistant hash if we first hash the bit string representing identity in the scheme). $d \in \mathbb{N}$ is a flexible constant, which can be set to be any value. "Anon?" shows whether the scheme is anonymous. "Selective/Adaptive" shows whether the scheme is selectively secure or adaptively secure. "$q/\chi_{\max}$" for LWE assumption refers to the ratio of the modulus to the error size of the underlying LWE assumption used in the security reduction. "Fixed poly(n)" means that the corresponding scheme is proven secure under the LWE assumption with $q/\chi_{\max}$ being some fixed polynomial (e.g., n^3). "All poly(n)" mean that we have to assume the LWE assumption for all polynomial $q/\chi_{\max}$.

Acknowledgement. The author would like to thank all members of the study group "Shin-Akarui-Angou-Benkyou-Kai" for fruitful discussion. In particular, the author thanks Shuichi Katsumata for his comments on improving the presentation, Goichiro Hanaoka and Jacob. C.N. Schuldt for their helpful advice in the rebuttal phase. The author also thanks the anonymous reviewers of Eurocrypt 2016 for their insightful comments.

References

1. Agrawal, S., Boneh, D., Boyen, X.: Efficient lattice (H)IBE in the standard model. In: Gilbert, H. (ed.) EUROCRYPT 2010. LNCS, vol. 6110, pp. 553–572. Springer, Heidelberg (2010)
2. Agrawal, S., Boneh, D., Boyen, X.: Lattice basis delegation in fixed dimension and shorter-ciphertext hierarchical IBE. In: Rabin, T. (ed.) CRYPTO 2010. LNCS, vol. 6223, pp. 98–115. Springer, Heidelberg (2010)
3. Ajtai, M.: Generating hard instances of the short basis problem. In: Wiedermann, J., Van Emde Boas, P., Nielsen, M. (eds.) ICALP 1999. LNCS, vol. 1644, pp. 1–9. Springer, Heidelberg (1999)

4. Alperin-Sheriff, J.: Short signatures with short public keys from homomorphic trapdoor functions. In: Katz, J. (ed.) PKC 2015. LNCS, vol. 9020, pp. 236–255. Springer, Heidelberg (2015)
5. Alwen, J., Peikert, C.: Generating shorter bases for hard random lattices. In: STACS, pp. 75–86 (2009)
6. Bellare, M., Ristenpart, T.: Simulation without the artificial abort: simplified proof and improved concrete security for waters' IBE scheme. In: Joux, A. (ed.) EURO-CRYPT 2009. LNCS, vol. 5479, pp. 407–424. Springer, Heidelberg (2009)
7. Böhl, F., Hofheinz, D., Jager, T., Koch, J., Seo, J.H., Striecks, C.: Practical signatures from standard assumptions. In: Johansson, T., Nguyen, P.Q. (eds.) EURO-CRYPT 2013. LNCS, vol. 7881, pp. 461–485. Springer, Heidelberg (2013)
8. Boneh, D., Boyen, X.: Efficient selective-ID secure identity-based encryption without random oracles. In: Cachin, C., Camenisch, J.L. (eds.) EUROCRYPT 2004. LNCS, vol. 3027, pp. 223–238. Springer, Heidelberg (2004)
9. Boneh, D., Boyen, X.: Secure identity based encryption without random oracles. In: Franklin, M. (ed.) CRYPTO 2004. LNCS, vol. 3152, pp. 443–459. Springer, Heidelberg (2004)
10. Boneh, D., Franklin, M.: Identity-based encryption from the weil pairing. In: Kilian, J. (ed.) CRYPTO 2001. LNCS, vol. 2139, pp. 213–229. Springer, Heidelberg (2001)
11. Boneh, D., Gentry, C., Gorbunov, S., Halevi, S., Nikolaenko, V., Segev, G., Vaikuntanathan, V., Vinayagamurthy, D.: Fully key-homomorphic encryption, arithmetic circuit ABE and compact garbled circuits. In: Nguyen, P.Q., Oswald, E. (eds.) EUROCRYPT 2014. LNCS, vol. 8441, pp. 533–556. Springer, Heidelberg (2014)
12. Boyen, X.: Lattice mixing and vanishing trapdoors: a framework for fully secure short signatures and more. In: Nguyen, P.Q., Pointcheval, D. (eds.) PKC 2010. LNCS, vol. 6056, pp. 499–517. Springer, Heidelberg (2010)
13. Brakerski, Z., Langlois, A., Peikert, C., Regev, O., Stehlé, D.: Classical hardness of learning with errors. In: STOC, pp. 575–584 (2013)
14. Brakerski, Z., Vaikuntanathan, V.: Lattice-based FHE as secure as PKE. In: ITCS, pp. 1–12 (2014)
15. Canetti, R., Halevi, S., Katz, J.: A forward-secure public-key encryption scheme. In: EUROCRYPT, pp. 255–271 (2003)
16. Cash, D., Hofheinz, D., Kiltz, E., Peikert, C.: Bonsai trees, or how to delegate a lattice basis. In: Gilbert, H. (ed.) EUROCRYPT 2010. LNCS, vol. 6110, pp. 523–552. Springer, Heidelberg (2010)
17. Chen, J., Wee, H.: Fully, (almost) tightly secure IBE and dual system groups. In: Canetti, R., Garay, J.A. (eds.) CRYPTO 2013, Part II. LNCS, vol. 8043, pp. 435–460. Springer, Heidelberg (2013)
18. Cocks, C.: An identity based encryption scheme based on quadratic residues. In: Honary, B. (ed.) IMA 2001. LNCS, vol. 2260, pp. 360–363. Springer, Heidelberg (2001)
19. Dodis, Y., Katz, J., Xu, S., Yung, M.: Key-insulated public key cryptosystems. In: Knudsen, L.R. (ed.) EUROCRYPT 2002. LNCS, vol. 2332, pp. 65–82. Springer, Heidelberg (2002)
20. Dodis, Y., Ostrovsky, R., Reyzin, L., Smith, A.: Fuzzy extractors: how to generate strong keys from biometrics and other noisy data. SIAM J. Comput. $38(1)$, 97–139 (2008)
21. Döttling, N., Schröder, D.: Efficient pseudorandom functions via on-the-fly adaptation. In: Gennaro, R., Robshaw, M. (eds.) CRYPTO 2015. LNCS, vol. 9215, pp. 329–350. Springer, Heidelberg (2015)

22. Ducas, L., Lyubashevsky, V., Prest, T.: Efficient identity-based encryption over NTRU lattices. In: Sarkar, P., Iwata, T. (eds.) ASIACRYPT 2014, Part II. LNCS, vol. 8874, pp. 22–41. Springer, Heidelberg (2014)
23. Ducas, L., Micciancio, D.: Improved short lattice signatures in the standard model. In: Garay, J.A., Gennaro, R. (eds.) CRYPTO 2014, Part I. LNCS, vol. 8616, pp. 335–352. Springer, Heidelberg (2014)
24. Gentry, C.: Practical identity-based encryption without random oracles. In: Vaudenay, S. (ed.) EUROCRYPT 2006. LNCS, vol. 4004, pp. 445–464. Springer, Heidelberg (2006)
25. Gentry, C., Peikert, C., Vaikuntanathan, V.: Trapdoors for hard lattices and new cryptographic constructions. In: STOC, pp. 197–206 (2008)
26. Goldwasser, S., Lewko, A., Wilson, D.A.: Bounded-collusion IBE from key homomorphism. In: Cramer, R. (ed.) TCC 2012. LNCS, vol. 7194, pp. 564–581. Springer, Heidelberg (2012)
27. Gorbunov, S., Vaikuntanathan, V., Wee, H.: Attribute-based encryption for circuits. In: STOC, pp. 545–554 (2013)
28. Gorbunov, S., Vinayagamurthy, D.: Riding on asymmetry: efficient ABE for branching programs. In: Iwata, T., Cheon, J.H. (eds.) ASIACRYPT 2015. LNCS, vol. 9452, pp. 549–573. Springer, Heidelberg (2015). doi:10.1007/978-3-662-48797-6_23
29. Heng, S.-H., Kurosawa, K.: k-resilient identity-based encryption in the standard model. In: Okamoto, T. (ed.) CT-RSA 2004. LNCS, vol. 2964, pp. 67–80. Springer, Heidelberg (2004)
30. Hohenberger, S., Waters, B.: Short and stateless signatures from the RSA assumption. In: Halevi, S. (ed.) CRYPTO 2009. LNCS, vol. 5677, pp. 654–670. Springer, Heidelberg (2009)
31. Jutla, C.S., Roy, A.: Shorter quasi-adaptive NIZK proofs for linear subspaces. In: Sako, K., Sarkar, P. (eds.) ASIACRYPT 2013, Part I. LNCS, vol. 8269, pp. 1–20. Springer, Heidelberg (2013)
32. Lyubashevsky, V., Micciancio, D.: Asymptotically efficient lattice-based digital signatures. In: Canetti, R. (ed.) TCC 2008. LNCS, vol. 4948, pp. 37–54. Springer, Heidelberg (2008)
33. Lyubashevsky, V., Peikert, C., Regev, O.: On ideal lattices and learning with errors over rings. In: Gilbert, H. (ed.) EUROCRYPT 2010. LNCS, vol. 6110, pp. 1–23. Springer, Heidelberg (2010)
34. Lyubashevsky, V., Peikert, C., Regev, O.: A toolkit for ring-LWE cryptography. In: Johansson, T., Nguyen, P.Q. (eds.) EUROCRYPT 2013. LNCS, vol. 7881, pp. 35–54. Springer, Heidelberg (2013)
35. Micciancio, D., Peikert, C.: Trapdoors for lattices: simpler, tighter, faster, smaller. In: Pointcheval, D., Johansson, T. (eds.) EUROCRYPT 2012. LNCS, vol. 7237, pp. 700–718. Springer, Heidelberg (2012)
36. Naccache, D.: Secure and practical identity-based encryption. IET Inf. Secur. 1(2), 59–64 (2007)
37. Peikert, C.: Public-key cryptosystems from the worst-case shortest vector problem: extended abstract. In: STOC, pp. 333–342 (2009)
38. Peikert, C.: A decade of lattice cryptography. IACR Cryptology ePrint Archive, Report 2015/939
39. Peikert, C., Vaikuntanathan, V., Waters, B.: A framework for efficient and composable oblivious transfer. In: Wagner, D. (ed.) CRYPTO 2008. LNCS, vol. 5157, pp. 554–571. Springer, Heidelberg (2008)

40. Regev, O.: On lattices, learning with errors, random linear codes, and cryptography. In: STOC, pp. 843–873 (2005)
41. Sakai, R., Ohgishi, K., Kasahara, M.: Cryptosystems based on pairing over elliptic curve. In: The 2000 Symposium on Cryptography and Information Security (2000). (in Japanese)
42. Shamir, A.: Identity-based cryptosystems and signature schemes. In: Blakely, G.R., Chaum, D. (eds.) CRYPTO 1984. LNCS, vol. 196, pp. 47–53. Springer, Heidelberg (1985)
43. Shamir, A., Tauman, Y.: Improved online/offline signature schemes. In: Kilian, J. (ed.) CRYPTO 2001. LNCS, vol. 2139, pp. 355–367. Springer, Heidelberg (2001)
44. Singh, K., Pandurangan, C., Banerjee, A.K.: Adaptively secure efficient lattice (H)IBE in standard model with short public parameters. In: Bogdanov, A., Sanadhya, S. (eds.) SPACE 2012. LNCS, vol. 7644, pp. 153–172. Springer, Heidelberg (2012)
45. Stehlé, D., Steinfeld, R., Tanaka, K., Xagawa, K.: Efficient public key encryption based on ideal lattices. In: Matsui, M. (ed.) ASIACRYPT 2009. LNCS, vol. 5912, pp. 617–635. Springer, Heidelberg (2009)
46. Tessaro, S., Wilson, D.A.: Bounded-collusion identity-based encryption from semantically-secure public-key encryption: generic constructions with short ciphertexts. In: Krawczyk, H. (ed.) PKC 2014. LNCS, vol. 8383, pp. 257–274. Springer, Heidelberg (2014)
47. Waters, B.: Efficient identity-based encryption without random oracles. In: Cramer, R. (ed.) EUROCRYPT 2005. LNCS, vol. 3494, pp. 114–127. Springer, Heidelberg (2005)
48. Waters, B.: Dual system encryption: realizing fully secure IBE and HIBE under simple assumptions. In: Halevi, S. (ed.) CRYPTO 2009. LNCS, vol. 5677, pp. 619–636. Springer, Heidelberg (2009)
49. Yamada, S., Hanaoka, G., Kunihiro, N.: Two-dimensional representation of cover free families and its applications: short signatures and more. In: Dunkelman, O. (ed.) CT-RSA 2012. LNCS, vol. 7178, pp. 260–277. Springer, Heidelberg (2012)
50. Yamada, S.: Adaptively Secure Identity-Based Encryption from Lattices with Asymptotically Shorter Public Parameters. Cryptology ePrint Archive, Report/140 (2016). http://eprint.iacr.org/2016/140

Online/Offline OR Composition
of Sigma Protocols

Michele Ciampi[1]([✉]), Giuseppe Persiano[2], Alessandra Scafuro[3],
Luisa Siniscalchi[1], and Ivan Visconti[1]

[1] DIEM, University of Salerno, Salerno, Italy
{mciampi,lsiniscalchi,visconti}@unisa.it
[2] DISA-MIS, University of Salerno, Salerno, Italy
giuper@gmail.com
[3] Boston University and Northeastern University, Boston, USA
scafuro@bu.edu

Abstract. Proofs of partial knowledge allow a prover to prove knowledge of witnesses for k out of n instances of NP languages. Cramer, Schoenmakers and Damgård [10] provided an efficient construction of a 3-round public-coin witness-indistinguishable (k, n)-proof of partial knowledge for any NP language, by cleverly combining n executions of Σ-protocols for that language. This transform assumes that all n instances are fully specified before the proof starts, and thus directly rules out the possibility of choosing some of the instances after the first round.

Very recently, Ciampi et al. [6] provided an improved transform where one of the instances can be specified in the last round. They focus on $(1, 2)$-proofs of partial knowledge with the additional feature that one instance is defined in the last round, and could be *adaptively* chosen by the verifier. They left as an open question the existence of an efficient $(1, 2)$-proof of partial knowledge where no instance is known in the first round. More in general, they left open the question of constructing an efficient (k, n)-proof of partial knowledge where knowledge of *all* n instances can be postponed. Indeed, this property is achieved only by inefficient constructions requiring NP reductions [19].

In this paper we focus on the question of achieving *adaptive-input* proofs of partial knowledge. We provide through a transform the first efficient construction of a 3-round public-coin witness-indistinguishable (k, n)-proof of partial knowledge where *all* instances can be decided in the third round. Our construction enjoys *adaptive-input* witness indistinguishability. Additionally, the proof of knowledge property remains also if the adversarial prover selects instances adaptively at last round as long as our transform is applied to a proof of knowledge belonging to the widely used class of proofs of knowledge described in [9,21]. Since knowledge of instances and witnesses is not needed before the last round, we have that the first round can be precomputed and in the online/offline setting our performance is similar to the one of [10].

Our new transform relies on the DDH assumption (in contrast to the transforms of [6,10] that are unconditional).

Keywords: Σ-protocols · WI · PoKs · Delayed and adaptive input

© International Association for Cryptologic Research 2016
M. Fischlin and J.-S. Coron (Eds.): EUROCRYPT 2016, Part II, LNCS 9666, pp. 63–92, 2016.
DOI: 10.1007/978-3-662-49896-5_3

1   Introduction

Proofs of knowledge (PoKs) are ubiquitous in cryptographic protocols. When enjoying additional features such as honest-verifier zero knowledge (HVZK), witness indistinguishability (WI) or zero knowledge (ZK), they are used as building blocks in basically any protocol for secure computation. As such, the degree of security and efficiency achieved by the underlying PoKs directly (and dramatically) impacts on the security and efficiency of the larger protocol. For instance, very efficient WI PoKs for specific languages, such as Discrete Log and DDH, have been instrumental for constructing efficient maliciously secure two-party computation (see [17] and references within). Furthermore, stronger security notions of PoKs, such as soundness, WI and ZK in presence of *adaptive-input* selection, are useful for constructing round-efficient protocols [18,25].

Proofs of Partial Knowledge. In [10], Cramer et al. showed how to construct *efficient* PoKs for compound statements starting from Σ-protocols. More precisely, the compound statement consists of n instances, and the goal is to prove knowledge of witnesses for at least k of the n instances. As such, these proofs are named "*proofs of partial knowledge*" in [10]. The transform of [10] cleverly combines n parallel executions of PoKs that are Σ-protocols in an efficient 3-round public-coin perfect WI (k,n)-proof of partial knowledge. A similar result was given in [26].

Note that, if efficiency is not a concern, proofs of partial knowledge were already possible (with *computational* WI, though) thanks to the general construction of Lapidot and Shamir (LaSh) [19]. Proving compound statements via LaSh constructions however requires expensive NP reductions. On the other hand, LaSh PoKs provide a stronger security guarantee: honest players use the instances specified in the statements only in the last round, and security holds even if the adversarial verifier (resp., prover) chooses the instances adaptively after having seen the first (resp., second) round. LaSh's construction is therefore an *adaptive-input* WI proof of partial knowledge for all NP. As mentioned above, this property can be instrumental to save at least one round of communication, when the proof of partial knowledge is used in a larger protocol. The construction shown in [10], instead, although efficient, does not provide any form of adaptivity, as all the n instances must be fully specified before the protocols starts. As a consequence, the better efficiency of [10] can be paid in additional rounds compared to [19] when the construction is used in larger applications.

The Proof of Partial Knowledge of [6]. A very recent work by Ciampi et al. [6] makes a first preliminary step towards closing the gap between [10,19]. [6] proposes a different transform for WI proofs of partial knowledge that gives some adaptiveness at the price of generality. Namely, their technique yields to a $(1,2)$-proof of partial knowledge where the knowledge of one of the two instances can be postponed to the last round. In more details, they show a PoK for a statement "$x_0 \in L_0 \lor x_1 \in L_1$" such that x_0 and x_1 are not immediately needed (in contrast to [10]). The honest prover needs x_0 to run the 1st round while x_1 is needed only in the 3rd round along with a witness for either x_0 or x_1. The verifier needs to see x_0 and x_1 only at the end, in order to accept/reject the proof. Ciampi et al. [6] defined the property of delayed input requiring that the *honest* prover does not need to know the instance to start the protocol. In other words, the need of the input is delayed to the very last round. For clarity, we stress that a delayed-input protocol is not necessarily secure against inputs that have been adaptively chosen. Indeed, their technique yields a proof of partial knowledge that is delayed input for one of the instances but is not adaptively secure against malicious provers (although it is adaptive-input WI). The security achieved by their transform is sufficient for their target applications.

The Open Question and its Importance. The above preliminary progress leaves open the following fascinating question: can we design an efficient transform that yields an adaptive-input WI (k, n)-proof of partial knowledge where *all* n instances are known only in the last round?

Previous efficient transforms require the a-priori knowledge of all instances or of one out of two instances, even if the corresponding languages admit efficient delayed-input Σ-protocols. For the sake of concreteness, assume one wants to prove knowledge of the discrete logarithm of at least one of g^{x_0} or g^{x_1}. There exists a very efficient Σ-protocol Σ^{dl} due to Schnorr [27], for proving knowledge of one discrete log and that also enjoys the delayed-input property, i.e., the prover can compute the first round without knowing the instance g^x. However, when we apply known transforms to combine Σ^{dl}, the resulting protocol loses the delayed-input property, as it will still need either both instances g^{x_0} and g^{x_1}, if using [10], or at least one g^{x_0}, to be specified in advance if using [6].

1.1 Our Results

In this work we study the above open question and give various positive answers.

Σ-Protocols and Adaptive-input Selection. We shed light on the relation between delayed-input Σ-protocols and adaptive-input Σ-protocols. Recall that a Σ-protocols enjoys a special soundness[1] property, which means that given two accepting transcripts for the same statement having the same first round, one can efficiently extract a witness for that statement.

We show that in general Σ-protocols are delayed-input but are not adaptive-input sound; that is, they are not sound if the malicious prover can choose the statements adaptively. Indeed, in Sect. 4.1 we show how a malicious prover, based on the second round played by the verifier, can craft a false statement that will make the verifier accept (and the extractor of special soundness fail even when the statement is true). The attack applies to very popular Σ-protocols like Schnorr's protocol for discrete logarithm (DLog), the protocol for proving equality of DLogs for Diffie-Hellman (DH) tuples and the protocol of [22] for

---

[1] In literature special soundness is often generalized to $\ell > 2$ accepting transcripts with the bound of ℓ being polynomial in the security parameter.

proving knowledge of committed messages. These protocols all fall into a well known class of protocols studied by Cramer in [9] and Maurer in [21].

The above issue was already noticed in [1] for the case of non-interactive zero-knowledge arguments obtained from Σ-protocols by applying the Fiat-Shamir transform [14]. Indeed there are in literature some incorrect use of the Fiat-Shamir transform where the instance is not given in input to the random oracle. As a consequence an adversarial prover can first create a transcript and then can try to find an instance not in the language such that the transcript is accepting. Of course in the random-oracle model the above issue has the trivial fix consisting of giving the instance as input to the random oracle to generate the challenge. This fix is meaningless in the standard model that is the focus of our work.

We then analyze the transform of [6], that is delayed-input with respect to one instance only. We observe that when [6] combines protocols belonging to the class of [9, 21], it also suffers from the same attack, when the malicious prover is allowed to adaptively choose his input. Therefore the transform of [6] is not adaptive-input sound. We stress however, that in the applications targeted in [6] the input that is specified only in the last round is chosen by the verifier. As such, for their applications they do not need any form of adaptive-input soundness, but only adaptive-input witness-indistinguishability (which they achieve). Moreover, the special soundness of their transform preserves security w.r.t. adaptive-input selection. Summing up, [6] correctly defines and achieves delayed-input Σ-protocols and adaptive-input WI and uses it in the applications. However adaptive-input special soundness is not defined and not achieved in their work.

Adaptive-input Special-sound Σ-protocols. In light of the above discussion, a natural question is whether we can upgrade the security of the class of Σ-protocols that are delayed input, but not adaptive-input sound. Towards this, we first clarify the conceptual gap between adaptive-input selection and the adaptiveness considered in [6] by defining formally adaptive-input special soundness. Then we show a compiler that takes as input any delayed-input Σ-protocol belonging to the class specified in [9, 21], and outputs a Σ-protocol, that is adaptive-input sound, i.e., it is sound even when the malicious prover adaptively chooses his input in the last round. The main idea behind this compiler is to force the prover to send correctly the first round of the Σ-protocol through another parallel run of the Σ-protocol. This allows for the extraction of any witness in the proof of knowledge. The compiler is shown in Sect. 4.2. We also show (in Sect. 5) that nevertheless, [6]'s transform preserves the adaptivity of the Σ-protocols that are combined. Namely, on input Σ-protocols that are already adaptive-input special sound and WI, the [6]'s transform outputs a $(1, 2)$-proof of partial knowledge that is an adaptive-input proof of knowledge as well.

Adaptive-input (k, n)-proofs of Partial Knowledge. The main contribution of this paper is a new transform that yields the first efficient (k, n)-proofs of partial knowledge where *all* n instances can be specified in the last round.

Our new transform takes as input a delayed-input Σ-protocol for a relation $\mathcal{R}$, and outputs a 3-round public-coin WI special-sound (k, n)-proof of partial knowledge for the relation $(\mathcal{R} \vee \cdots \vee \mathcal{R})$ where no instance is known at the beginning. The security of our transform is based on the DDH assumption. The WI property of the resulting protocol holds also with respect to adaptive-input selection, while the PoK property holds also in case of adaptive-input selection only if the underlying Σ-protocol is adaptive-input special sound.

We also show a transform that admits instances taken from different relations. Interestingly, this construction makes use as subprotocol of the first construction where instances are taken from the same relation.

1.2 Our Technique

We provide a technique for composing a delayed-input Σ-protocol for a relation $\mathcal{R}$ into a delayed-input Σ-protocol for the (k, n)-proof of partial knowledge for relation $(\mathcal{R} \vee \ldots \vee \mathcal{R})$. For a better understanding of our technique, it is instructive to see why the transform of [10] (resp., [6]) requires that all n (resp., 1 out of 2) instances are specified before the protocol starts.

Limitations of Previous Transforms. Let $\Sigma_{\mathcal{R}}$ be a delayed-input Σ-protocol, and let $(\mathcal{R} \vee \ldots \vee \mathcal{R})$ be the relation for which we would like to have a (k, n)-proof of partial knowledge. The technique of [10] works as follows. The prover P, on input the instances $(x_1 \in \mathcal{R} \vee \ldots \vee x_n \in \mathcal{R})$, runs protocols $\Sigma_{\mathcal{R}}, \ldots, \Sigma_{\mathcal{R}}$ in parallel. P gets only k witnesses for k different instances but it needs to somehow generate an accepting transcript for *all* instances. How to prove the remaining $n - k$ instances without having the witness? The idea of [10] consists simply in letting the prover generate the $n - k$ transcripts (corresponding to the instances for which he did not get the witnesses) using the HVZK simulator S associated to the Σ-protocol. Additionally [10] introduces a mechanism that allows the prover to control the value of exactly $(n - k)$ of the challenges played by V, so that the prover can force the transcripts computed by the simulator in $(n - k)$ positions.

So, why does the transform of [10] need *all* instances to be known already in the 1st round? The answer is that P needs to run S already in the 1st round, and S expects the instance as input. Similar arguments apply for [6] as it requires that 1 instance out of 2 is known already in the 1st round.

The Core Idea of Our Technique. Previous transforms fail because the prover runs the HVZK simulator to compute the 1st round of some of the transcripts of $\Sigma_{\mathcal{R}}$. Our core idea is to provide mechanisms allowing P to postpone the use of the simulator to the 3rd round. The main challenge is to implement mechanisms that are very efficient and preserve soundness and WI of the composed Σ-protocol. We stress that we want to solve the open problems in full, and thus none of the instances are known at the beginning of the protocol. To be more explicit, in the 1st round, the prover starts with the following statement $(? \in L_{\mathcal{R}} \vee \ldots \vee ? \in L_{\mathcal{R}})$.

Assume we have a (k, n)-equivocal commitment scheme that allows the prover to compute n commitments such that k of them are binding and the remaining $n - k$ are equivocal, and the verifier cannot distinguish between the two types of commitment, where the k positions that are binding must be chosen already in the commitment phase (a similar tool is constructed in [24]). With this gadget in hand, we can construct a delayed-input (k, n)-proof of partial knowledge $\Sigma_{k,n}^{OR}$

as follows. Let (a, c, z) denote generically the 3 messages exchanged during the execution of a Σ-protocol $\Sigma_{\mathcal{R}}$.

In the 1st round, P honestly computes a_i for the i-th execution of $\Sigma_{\mathcal{R}}$. Here we are using the fact that $\Sigma_{\mathcal{R}}$ is delayed-input, and thus a_i can be computed without using the instance. Then he commits to $a_1, \ldots, a_n$ using the (k, n)-equivocal commitment scheme discussed above, where the k binding positions are randomly chosen. Thus, the 1st round of protocol $\Sigma_{k,n}^{\mathsf{OR}}$ consists of n commitments. In the 2nd round V simply sends a single challenge c according to $\Sigma_{\mathcal{R}}$. In the 3rd round, P obtains the n instances $x_1, \ldots, x_n$ and k witnesses. At this point, for the instances x_i for which he did not receive the witness, he will use the HVZK simulator to compute an accepting transcript $(\tilde{a}_i, c, \tilde{z}_i)$ and then equivocate the $(n - k)$ equivocal commitments so that they decommit to the new generated $\tilde{a}_i$. For the k remaining instances he will honestly compute the 3rd round using the committed input a_i. Intuitively, soundness follows from the fact that k commitments are binding, and from the soundness of $\Sigma_{\mathcal{R}}$. WI follows from the hiding of the equivocal commitment scheme and the HVZK property of $\Sigma_{\mathcal{R}}$. Note that in this solution we are crucially using the fact that we are composing the *same* Σ-protocol so that P can use any of the a_i committed in the 1st round to compute an honest transcript. This technique thus falls short as soon as we want to compose arbitrary Σ-protocols together. Nevertheless, this transformation turns to be useful for the case of different Σ-protocols.

(k, n)-*equivocal Commitment Scheme.* A (k, n)-equivocal commitment scheme allows a sender to compute n commitments $\mathsf{com}_1, \ldots, \mathsf{com}_n$ such that k of them are binding and $n - k$ are equivocal. We will use the language DH of DH tuples and we will call non-DH a tuple that is not a DH tuple. We will implement a (k, n)-equivocal commitment scheme very efficiently under the DDH assumption as follows. In the commitment phase, the sender computes n tuples $T_1 = (g_1, A_1, B_1, X_1), \ldots, T_n = (g_n, A_n, B_n, X_n)$ and proves that k out of n tuples are *not* in DH (i.e., they are non-DH tuples). We show that this can be done using the classical [10] (k, n)-proof of partial knowledge that can be obtained starting with a Σ-protocol Σ^{ddh} for DH. We then use the well known [4,5,12,17] fact that Σ-protocols can be used to construct an instance-dependent trapdoor commitment scheme, where the sender can equivocate if he knows the witness for the instance. Thus, each tuple T_i can be used to compute an instance-dependent trapdoor commitment com_i using Σ^{ddh}. com_i will be equivocal if T_i was indeed a DH tuple, it will be binding otherwise. Because the sender proves that k tuples are not in DH, it holds that there are at least k binding commitment. Hiding follows from the WI property of [10] and the HVZK of Σ^{ddh}. Commitment and decommitment can be completed in 3 rounds.

The Case of Different Σ-protocols. We now consider the case where we want to compose $\Sigma_1, \ldots, \Sigma_n$ for possibly different relations. Our (k, n)-equivocal commitment does not help here because each a_i is specific to protocol Σ_i, and cannot be arbitrarily mixed and matched once the k witnesses are known.

For this case we thus use a different trick. We ask the prover to commit to each a_i twice, once using a binding commitment and once using an equivocal commitment. This again can be very efficiently implemented from the DDH assumption as follows. For each i, P generates tuples T_i^0 and T_i^1, that are such that at most one can be a DH tuple. It then commits to a_i twice using the instance-dependent trapdoor commitment associated to tuple T_i^0 and tuple T_i^1. Because at most one of the two tuples is a DH tuple, at most one of the commitments of a_i can be later equivocated. Thus the 1st round of our transformation consists of 2 commitments of a_i for $1 \leq i \leq n$. In the 3rd round, when P receives instances $x_1, \ldots, x_n$ and k witnesses, he proceeds at follows. For each i, if P knows the witness for x_i, he will open the binding commitment for position i, and compute z_i using the honest prover procedure of Σ_i. Instead, if P does not have a witness for x_i, he will compute a new $\tilde{a}_i, z_i$ using the simulator on input x_i, c and open the equivocal commitment in position i. At the end, for each position i, one commitment has remained unopened.

This mechanism allows an honest prover to complete the proof with the knowledge of only k witnesses. However, what stops a malicious prover to always open the equivocal commitments and thus complete the proof without knowing any of the witnesses? We avoid this problem by requiring P to prove that, among the n tuples corresponding to the unopened commitments, at least k out of n tuples are DH tuples. This directly means that k of the opened commitments were constructed over non-DH tuples, and therefore are binding.

Now note that proving this theorem requires an (k, n)-proof of partial knowledge in order to implement Σ^{ddh}, where the instance to prove, i.e., the tuple that will be unopened, is known only in the 3rd round when P knows for which instances he is able to open a binding commitment. Here we crucially use the (k, n)-proof of partial knowledge for the same Σ-protocol developed above making sure to first run our compiler that strengthen Σ^{ddh} with respect to statements adaptively selected by a malicious prover.

1.3   Comparison with the State of the Art

In Table 1 we compare our results with the relevant related work. We consider [19], a 3-round public-coin WIPoK that is *fully adaptive-input* and that works for any NP language. We also consider [10] that proposed efficient 3-round public-coin WI proofs of partial knowledge (though, without supporting any adaptivity). Finally, we consider [6] since it was the only work that faced the problem of combining together efficiency and some form of delayed-input instances. The last row refers to our main result that allows to postpone knowledge of all the instances to the last round. The 2nd column refers to the computational assumptions needed by [19] (i.e., one-way permutations) and our main result (i.e., DDH assumption). The 3rd column specifies the type of WI depending on the adaptive selection of the instances from the adversarial verifier. The 4th column specifies the soundness depending on the adaptive selection of the instances from the adversarial prover.

Table 1. Comparison with previous work.

	Assumption	Adaptive WI	Adaptive PoK	NP reduction
LaSh90 [19]	OWP	k out of n (all adaptive)	k out of n (all adaptive)	Yes
CDS94 [10]	/	/	/	No
CPSSV16 [6]	/	1 out of 2 (1 adaptive)	/	No
This work (main result)	DDH	k out of n (all adaptive)	k out of n (all adaptive)	No

1.4   Online/Offline Computations

Our result has the advantage that the prover can compute the first round without knowing instances and witnesses. The first round is therefore an *offline phase*. When the prover interacts with the verifier (*online phase*) he sends the first round precomputed and computes only the third round of the protocol. We stress that [10] requires to know the instances already to compute the first round. Furthermore the work of [19] allows the prover to compute the first round offline but in the online phase the prover must perform an NP reduction.

In Table 2[2] we compare the effort of the prover in the online phase in our work and in [10,19]. We consider a prover that proves knowledge of discrete logarithms for 1 instance out of 2 instances (1st column) and a prover that proves knowledge of discrete logarithms for k instances out of n instances (2nd column). As we noted, above in the online phase of [19] the prover computes an NP reduction (2nd row). For our construction and the one of [10] we count the number of modular[3] exponentiations that are computed in the online phase (3rd and 4th rows). Below we briefly describe how we have computed the above costs. In [10] the number of exponentiations is $2n - k$. This comes from the fact that the first round of Schnorr's Σ-protocol requires one exponentiation while the simulator requires two exponentiations. In [10] the simulator is executed $2(n-k)$ times and moreover k exponentiations are needed to run the prover of Schnorr's protocol.

Table 2. Comparison with previous work proving knowledge of discrete logarithms. The table illustrates the computations of the prover in the online phase.

	$(1,2)$ DLogs	(k,n) DLogs
LaSh90 [19]	NP-reduction	NP-reduction
CDS94 [10]	3 exps	$2n - k$ exps
CPSSV16 [6]	4 exps	/
This work (main result)	2 exps [4 exps]	$2(n - k)$ exps [$4(n - k)$ exps]

---

[2] The actual amount of computations significantly depends on the precise versions of the subprotocols used in the construction. The adaptive-input special-sound versions of the subprotocols are more expensive than their non-adaptive counterparts.

[3] We will omit the word *modular* from now on.

In the (1,2)-proof of partial knowledge of [6], the 1st round requires 3 exponentiations. Indeed, in the 1st round of [6] the prover runs Schnorr's simulator and computes the 1st round of Schnorr's Σ-protocol. The 3rd round of [6] has a different analysis depending on which witness is used. When the prover of [6] uses the witness for an adaptively chosen instance, then there is no addition exponentiation. Otherwise, another execution of Schnorr's simulator is required. For this reason, in the worst case the 3rd round of [6] costs two exponentiations. Note that in the execution of the construction of [6] 4 exponentiations are performed in the online phase, since only the 1st round of Schnorr's Σ-protocol can be precomputed.

The final row corresponds to our main result and shows the general case of k instances out of n. Our construction involves $10n - k$ exponentiations. Indeed a commitment computed according to the commitment scheme described previously based on DH tuples costs 4 exponentiations. In our construction in the 1st round we sample $n - k$ DH tuples and k non-DH, sampling a DH/non-DH tuple costs 3 exponentiations, so this operation costs $3n$. Also in the 1st round we compute $n - k$ equivocal commitments and k binding commitments, and this sums up to $2n + 2k$ modular exponentiations. Furthermore the prover computes the 1st round of Schnorr's Σ-protocol n times and this costs n exponentiations. Moreover it has to run [10] to prove knowledge of witnesses for k instances out of n instances, and this costs $2n - k$ exponentiations. The only operations that involve exponentiations at the third round are the $n - k$ executions of the simulator of Schnorr's Σ-protocol. Therefore the online phase costs $2(n - k)$.

The adaptive-input special-sound version of our construction costs $13n - 3k$ exponentiations. Consider that in the adaptive-input special-sound version of Schnorr's Σ-protocol an execution of the simulator costs 4 exponentiations. Moreover computing the 1st round involves 2 exponentiations. Hence the first round of our adaptive-input special-sound construction involves $6n + k$ exponentiations and the online phase costs $4(n - k)$ exponentiations.

The exponentiations in square brackets specify the cost of our main result when Schnorr's Σ-protocol is transformed into an adaptive-input special-sound Σ-protocol. The analysis for the case of 1 out of 2 is similar with $k = 1$ and $n = 2$ but in this case, in the offline phase, we do not consider the cost of [10] since the correctness of the pair of tuples can be self-verified.

2   Preliminaries

We use λ as security parameter. $A(x)$ denotes the probability distribution of the output of a probabilistic algorithm A when running with x as input. We will use $A(x; r)$ to denote the randomness r used by A. PPT stands for probabilistic polynomial time.

If $\mathcal{R}$ is a subset of $\{0,1\}^\star \times \{0,1\}^\star$ for which membership of (x, w) to $\mathcal{R}$ can be decided in time polynomial in $|x|$ then we say that $\mathcal{R}$ is a polynomial-time relation and w is a witness for the instance x. Given a polynomial-time relation $\mathcal{R}$, $L_{\mathcal{R}}$ defined as $L_{\mathcal{R}} = \{x | \exists w : (x, w) \in \mathcal{R}\}$ is an NP language. For generality, we define $\hat{L}_\mathcal{R}$ to be the *input language* that includes both $L_\mathcal{R}$ and all well formed instances that do not have a witness, as already done in [15]. It follows that $L_\mathcal{R} \subseteq \hat{L}_\mathcal{R}$ and membership in $\hat{L}_\mathcal{R}$ can be tested in polynomial time. In proof systems for relation $\mathcal{R}$, the verifier runs the protocol only if the common input x belongs to $\hat{L}_\mathcal{R}$, while it rejects immediately common inputs not in $\hat{L}_\mathcal{R}$.

Given two interactive machines M_0 and M_1, we denote by $\langle M_0(x_0), M_1(x_1)\rangle(x_2)$ the output of M_1 when running on input x_1 with M_0 running on input x_0, both running on common input x_2.

Definition 1. *A pair $(\mathcal{P}, \mathcal{V})$ of PPT interactive machines is a complete protocol for an* NP*-language L with relation $\mathcal{R}$ if the following property holds:*

– *Completeness. For every* common input $x \in L$ *and witness w such that $(x, w) \in \mathcal{R}$, it holds that* $\mathrm{Prob}\,[\,\langle \mathcal{P}(w), \mathcal{V}\rangle(x) = 1\,] = 1$.

Definition 2. *A complete protocol $(\mathcal{P}, \mathcal{V})$ is a proof system for an* NP*-language L with relation $\mathcal{R}$ if the following property holds:*

– *Soundness. For every interactive machine $\mathcal{P}^\star$ there exists a negligible function ν such that for every $x \notin L$:* $\mathrm{Prob}\,[\,\langle \mathcal{P}^\star, \mathcal{V}\rangle(x) = 1\,] \leq \nu(|x|)$.

A proof system $(\mathcal{P}, \mathcal{V})$ is public coin *if $\mathcal{V}$ sends only random bits.*

Definition 3 ([11]). *Let $k : \{0,1\}^* \to [0,1]$ be a function. A protocol $(\mathcal{P}, \mathcal{V})$ is a proof of knowledge for the relation $\mathcal{R}$ with knowledge error k if the following properties are satisfied:*

– *Completeness: if $\mathcal{P}$ and $\mathcal{V}$ follow the protocol on input x and private input w to $\mathcal{P}$ where $(x, w) \in \mathcal{R}$, then $\mathcal{V}$ always accepts.*
– *Knowledge Soundness: there exists a constant $c > 0$ and a probabilistic oracle machine* Extract, *called the* extractor, *such that for every interactive prover $\mathcal{P}^\star$ and every input x, the machine* Extract *satisfies the following condition. Let $\epsilon(x)$ be the probability that $\mathcal{V}$ accepts on input x after interacting with $\mathcal{P}^\star$. If $\epsilon(x) > k(x)$, then upon input x and oracle access to $\mathcal{P}^\star$, the machine* Extract *outputs a string w such that $(x, w) \in \mathcal{R}$ within an expected number of steps bounded by $|x|^c/(\epsilon(x) - k(x))$.*

A *transcript* τ of an execution of a public-coin protocol $\Pi = (\mathcal{P}, \mathcal{V})$ for statement x consists of the sequence of messages exchanged $\mathcal{P}$ and $\mathcal{V}$. We say that τ is *accepting* if $\mathcal{V}$ outputs 1. Two accepting transcripts (a, c, z) and (a', c', z') for a 3-round public coin proof system with the same common input constitute a *collision* iff $a = a'$ and $c \neq c'$. The one message sent by the verifier $\mathcal{V}$ in a 3-round public coin proof system is called the *challenge*.

Σ-protocols. The most common form of proof system used in practice consists of 3-round protocols referred to as Σ-protocols. For several useful languages there exist efficient Σ-protocols, and they are easy to work with as already shown in many transforms [2,3,8,12,20,22,23,28,29].

Definition 4. *A 3-round public-coin protocol $\Pi = (\mathcal{P}, \mathcal{V})$ is a Σ-protocol for an NP-language L with polynomial-time relation $\mathcal{R}$ iff the following additional properties are satisfied:*

- *Completeness. When $\mathcal{P}, \mathcal{V}$ execute the protocol on input x and private input w to $\mathcal{P}$ where $(x, w) \in \mathcal{R}$, the verifier $\mathcal{V}$ always accepts.*
- *Special Soundness. There exists an efficient algorithm Extract that, on input x and a collision for x, outputs a witness w such that $(x, w) \in \mathcal{R}$.*
- *Special Honest Verifier Zero Knowledge (special HVZK, SHVZK). There exists a PPT simulator algorithm S that, on input an instance $x \in L$ and challenge c, outputs (a, z) such that (a, c, z) is an accepting w.r.t. x. Moreover, the distribution of the output of S on input (x, c) is perfectly[4] indistinguishable from the distribution of the transcript obtained when $\mathcal{V}$ sends c as challenge and $\mathcal{P}$ runs on common input x and any private input w such that $(x, w) \in \mathcal{R}$.*

A security parameter 1^{λ} for a Σ-protocol represents challenge length. Therefore we have that a Σ-protocol with a sufficiently large security parameter 1^{λ} is also a proof system.

Theorem 1 ([10]). *Every Σ-protocol is Perfect WI.*

Theorem 2 ([11]). *Let Π be a Σ-protocol for a relation $\mathcal{R}$ with security parameter λ. Then Π is a proof of knowledge with knowledge error $2^{-\lambda}$.*

From the above theorem we have that every Σ-protocol with a sufficiently long challenge is a proof of knowledge with negligible knowledge error. We observe that in the proof of the above theorem only completeness and special soundness of the Σ-protocol are used. Therefore the theorem regardless of HVZK. Furthermore, using the same proof approach used in the security proof of this theorem we can consider a relaxed notion of special soundness, t-special soundness, requiring $t \geq 2$ transcripts to extract the witness, with $t = \text{poly}(\lambda)$. This is still sufficient to obtain a proof of knowledge with negligible soundness error when the challenge is sufficiently long.

Therefore in this work when interested in proving the proof of knowledge property we will without loss of generality just prove t-special soundness for a polynomially bounded t and completeness.

Definition 5 (Delayed-Input Σ-protocol [6]). *A Σ-protocol $\Pi = (\mathcal{P}, \mathcal{V})$ for a relation $\mathcal{R}$ is delayed-input if $\mathcal{P}$ computes the first round having as input only the security parameter 1^{λ} and $\ell = |x|$.[5]*

---

[4] In this work we stick with the requirement of perfect SHVZK for Σ-protocols. Various other papers in literature considered also special *computational* HVZK.

[5] For simplicity in the rest of the paper we do not specify anymore that the algorithms $\mathcal{P}, \mathcal{V}$ take as input ℓ when the instance x is not known.

2.1   Adaptive-Input Special Soundness and Proof of Knowledge

The special soundness of a Σ-protocol strictly requires the statement $x \in L$ to be unchanged in the 2 accepting transcripts. We introduce a stronger notion referred to as *adaptive-input special soundness*. Roughly speaking, we require that it is possible to extract witnesses from a collision even if the two accepting 3-round transcripts are for two different instances. It is easy to see that adaptive-input special soundness implies extraction against provers that choose the theorem to be proved after seeing the challenge.

Definition 6. *A Σ-protocol Π for relation $\mathcal{R}$ enjoys* adaptive-input special soundness *if there exists an efficient algorithm* AExtract *that, on input accepting 3-round transcripts (a, c_1, z_1) for input x_1 and (a, c_2, z_2) for input x_2, outputs witnesses w_1 and w_2 such that $(x_1, w_1) \in \mathcal{R}$ and $(x_2, w_2) \in \mathcal{R}$.*

In this work we also define a protocol $\Pi = (\mathcal{P}, \mathcal{V})$ that is adaptive-input proof of knowledge. The adaptive-input proof of knowledge property is the same as the proof of knowledge property, with the difference that the adversarial prover $\mathcal{P}^*$ can choose the statement when the last round is played. We require that the instance x given in output by AExtract must be perfect indistinguishable from an instance x' given in output by $\mathcal{P}^*$ in an execution of Π with $\mathcal{V}$. The previous discussion about proving the proof of knowledge property from ℓ-special soundness also applies when proving adaptive-input proof of knowledge from adaptive-input ℓ-special soundness.

2.2   Adaptive-Input Witness Indistinguishability

The notion of *adaptive-input WI* formalizes security of the prover with respect to an adversarial verifier $\mathcal{A}$ that adaptively chooses the input instance to the protocol; that is, after seeing the first message of the prover. More specifically, for a delayed-input 3-round complete protocol Π, we consider game $\mathsf{ExpAWI}_{\Pi, \mathcal{A}}$ between a challenger $\mathcal{C}$ and an adversary $\mathcal{A}$ in which the instance x and two witnesses w_0 and w_1 for x are chosen by $\mathcal{A}$ *after* seeing the first message of the protocol played by the challenger. The challenger then continues the game by randomly selecting one of the two witnesses, w_b, and by computing the third message by running the prover's algorithm on input the instance x, the selected witness w_b and the challenge received from the adversary. The adversary wins the game if she can guess which of the two witnesses was used by the challenger.

We now define the adaptive-input WI experiment $\mathsf{ExpAWI}_{\Pi, \mathcal{A}}(\lambda, \mathsf{aux})$. This experiment is parameterized by a delayed-input 3-round complete protocol $\Pi = (\mathcal{P}, \mathcal{V})$ for a relation $\mathcal{R}$ and by PPT adversary $\mathcal{A}$. The experiment has as input the security parameter λ and auxiliary information aux for $\mathcal{A}$.

$\mathsf{ExpAWI}_{\Pi, \mathcal{A}}(\lambda, \mathsf{aux})$:

1. $\mathcal{C}$ randomly selects coin tosses r and runs $\mathcal{P}$ on input $(1^\lambda; r)$ to obtain a;
2. $\mathcal{A}$, on input a and aux, outputs instance x, witnesses w_0 and w_1 such that $(x, w_0), (x, w_1) \in \mathcal{R}$, challenge c and internal state state;

3. $\mathcal{C}$ randomly selects $b \leftarrow \{0,1\}$ and runs $\mathcal{P}$ on input (x, w_b, c) to obtain z;

4. $b' \leftarrow \mathcal{A}((a, c, z), \mathsf{aux}, \mathsf{state})$;

5. if $b = b'$ then output 1 else output 0.

We set $\mathsf{AdvAWI}_{\Pi,\mathcal{A}}(\lambda, \mathsf{aux}) = \left| \mathrm{Prob}\left[\mathsf{ExpAWI}_{\Pi,\mathcal{A}}(\lambda, \mathsf{aux}) = 1 \right] - \frac{1}{2} \right|$.

Definition 7 (Adaptive-Input Witness Indistinguishability). *A delayed-input 3-round complete protocol Π is adaptive-input WI if for any PPT adversary $\mathcal{A}$ there exists a negligible function ν such that for any $\mathsf{aux} \in \{0,1\}^*$ it holds that $\mathsf{AdvAWI}_{\Pi,\mathcal{A}}(\lambda, \mathsf{aux}) \leq \nu(\lambda)$.*

About DDH. The DDH assumption posits the hardness of distinguishing a randomly selected DH tuple from a randomly selected non-DH tuple with respect to a *group generator* algorithm IG. For sake of concreteness, we consider a specific group generator that, on input 1^λ, randomly selects a λ-bit prime p such that $q = (p-1)/2$ is also prime and outputs the (description of the) order q group $\mathcal{G}$ of the quadratic residues modulo p along with a random generator g of $\mathcal{G}$.

2.3   A Σ-Protocol for Partial Knowledge of DH/Non-DH Tuples

Let $\mathcal{G}$ be a cyclic group of order p. We say that $T = (g, A, B, X) \in \mathcal{G}^4$ is *oneNDH* if there exits $\alpha, \beta \in Z_p$ such that $A = g^\alpha, B = g^\beta, X = g^{\alpha\beta+1}$. In this section we describe a Σ-protocol for proving that at least k out of n tuples are oneNDH. The Σ-protocol is based on the one of [10] and we stress that, just as in [10], the Σ-protocol is perfect WI.

Formally, for $1 \leq k \leq n-1$, we construct Σ-protocol $\Pi_{k,n}^{nddh} = (\mathcal{P}_{k,n}, \mathcal{V}_{k,n})$ for the polynomial-time relation

$$\mathsf{NDH}_{k,n} = \{ (((g_1, A_1, B_1, X_1), \ldots, (g_n, A_n, B_n, X_n)), (\alpha_{i_1}, \ldots, \alpha_{i_k}, \beta_{i_1}, \ldots, \beta_{i_k})) :$$

$$1 \leq i_1 < \cdots < i_k \leq n \wedge A_{i_j} = g_{i_j}^{\alpha_{i_j}} \wedge B_{i_j} = g_{i_j}^{\beta_{i_j}} \wedge X_{i_j} = g_{i_j}^{\alpha_{i_j}\beta_{i_j}+1}, \text{ for } j = 1, \ldots, k\}$$

of the sequences of the n-tuples such that at least k of them are oneNDH. The prover $\mathcal{P}_{k,n}$ and the verifier $\mathcal{V}_{k,n}$ of $\Pi_{k,n}^{nddh}$, on input n tuples $(g_1, A_1, B_1, X_1), \ldots$ $\ldots, (g_n, A_n, B_n, X_n)$ constructs tuples (g_i, A_i, B_i, Y_i) setting $Y_i = X_i/g_i$, for $i = 1, \ldots, n$.

Then prover and verifier start Σ-protocol Σ^{ddh} of [10] for proving that at least k of n constructed tuples are DH.

Theorem 3. *For every n and $1 \leq k \leq n-1$, $\Pi_{k,n}^{ddh}$ is a Σ-protocol for the polynomial-time relation $\mathsf{NDH}_{k,n}$ with perfect WI.*

Proof. The perfect WI property follows from the perfect WI of [10]. The proof is then completed by the following two simple observations. If at least k of the input tuples are oneNDH then at least k of the *constructed* tuples (g_i, A_i, B_i, Y_i) are DH and the prover has a witness of this fact. On the other hand, if fewer than k of the input tuples are oneNDH then the transformed tuples contain fewer than k DH tuples.

2.4   Commitments from Σ-protocols

We define the notion of an Instance-Dependent Trapdoor Commitment scheme associated with a polynomial-time relation $\mathcal{R}$ and show a construction that uses Σ-protocols and fits this definition.

Definition 8 (Instance-Dependent Trapdoor Commitment Scheme). *Let $\mathcal{R}$ be a polynomial-time relation. An* Instance-Dependent Trapdoor Commitment *(a IDTC, in short) scheme for $\mathcal{R}$ with message space M is a quadruple of PPT algorithms* $(\mathsf{Com}, \mathsf{Dec}, (\mathsf{Fake}_1, \mathsf{Fake}_2))$ *where* Com *is the randomized commitment algorithm that takes as input an instance $x \in \hat{L}_{\mathcal{R}}$ (with $|x| = \mathrm{poly}(\lambda)$) and a message $m \in M$ and outputs* commitment com *and decommitment* dec. Dec *is the* verification *algorithm that takes as input $(x, \mathsf{com}, \mathsf{dec}, m)$ and decides whether m is the decommitment of* com.

$(\mathsf{Fake}_1, \mathsf{Fake}_2)$ are randomized algorithms. Fake_1 takes as input an instance x, a witness w s.t. $(x, w) \in \mathcal{R}$ ($|x| = \mathrm{poly}(\lambda)$) and outputs commitment com, *and* equivocation information rand. *Fake_2 takes as input x, w, m, and rand, and outputs* dec *s.t.* Dec, *on input $(x, \mathsf{com}, \mathsf{dec}, m)$, accepts m as decommitment of* com.

An Instance-Dependent Trapdoor Commitment scheme has the following properties:

- **Correctness:** *for all $x \in \hat{L}_{\mathcal{R}}$, all $m \in M$, it holds that*

$$\mathrm{Prob}\left[\, (\mathsf{com}, \mathsf{dec}) \leftarrow \mathsf{Com}(x, m) : \mathsf{Dec}(x, \mathsf{com}, \mathsf{dec}, m) = 1 \,\right] = 1.$$

- **Binding:** *if $x \notin L$ then for every commitment* com *there exists at most one message m s.t. $\mathsf{Dec}(x, \mathsf{com}, \mathsf{dec}, m) = 1$ for any value* dec.
- **Hiding:** *for every receiver $\mathcal{A}$, for every auxiliary information* aux, *for all $x \in L_{\mathcal{R}}$ and all for $m_0, m_1 \in M$, it holds that*

$$\mathrm{Prob}\left[\, b \leftarrow \{0,1\}; (\mathsf{com}, \mathsf{dec}) \leftarrow \mathsf{Com}(1^\lambda, x, m_b) : b = \mathcal{A}(\mathsf{aux}, x, \mathsf{com}, m_0, m_1) \,\right] \leq \frac{1}{2}.$$

- **Trapdoorness:** *the following two families of probability distributions are perfect indistinguishable (namely the two probability distributions coincide for all (x, w, m) such that $(x, w) \in \mathcal{R}$ and $m \in M$):*

$$\{(\mathsf{com}, \mathsf{rand}) \leftarrow \mathsf{Fake}_1(x, w); \mathsf{dec} \leftarrow \mathsf{Fake}_2(x, w, m, \mathsf{rand}) : (\mathsf{com}, \mathsf{dec})\}$$

$$\{(\mathsf{com}, \mathsf{dec}) \leftarrow \mathsf{Com}(x, m) : (\mathsf{com}, \mathsf{dec})\}.$$

IDTC from Σ-protocol. Our construction follows similar constructions of [11, 13, 17]. Let $\Pi = (\mathcal{P}, \mathcal{V})$ be a Σ-protocol for polynomial-time relation $\mathcal{R}$ with the associated NP-language $L_{\mathcal{R}}$ and challenge length λ. Let S be the special HVZK simulator for Π and let (x, w) be s.t. $(x, w) \in \mathcal{R}$. Now we show an IDTC $\mathsf{CS}^\Pi = (\mathsf{Com}^\Pi, \mathsf{Dec}^\Pi, (\mathsf{Fake}_1^\Pi, \mathsf{Fake}_2^\Pi))$.

- Com^Π takes as input instance x and message $m \in \{0,1\}^\lambda$, sets $(\mathsf{com}, \mathsf{dec}) \leftarrow S(x, m)$ and outputs $(\mathsf{com}, \mathsf{dec})$.

- Dec^{Π} takes as input instance x and transcript $(\text{com}, m, \text{dec})$, runs $\mathcal{V}$ on input the instance and the transcript and returns $\mathcal{V}$'s output.
- Fake_1^{Π} takes as input instance x and witness w, samples random string ρ and runs $\mathcal{P}$ on input $(1^{\lambda}, x, w; \rho)$ to get the 1st message a of Π. Fake_1^{Π} sets $\text{rand} = \rho$, $\text{com} = \text{a}$ and outputs $(\text{com}, \text{rand})$.
- Fake_2^{Π} takes as input x, w, m, rand and runs $\mathcal{P}$ on input $(1^{\lambda}, x, w, m, \text{rand})$ to get the 3rd message z of Π. Fake_1^{Π} sets $\text{dec} = \text{z}$ and outputs dec.

Theorem 4. CS^{Π} *is an* $IDTC$.

Proof. The security proof relies only on the properties of Π. Correctness follows from the completeness of Π. Binding follows from the special soundness of Π. Hiding and Trapdoorness follow from the SHVZK and the completeness of Π.

3 Adaptive-Input (k,n)-Proof of Partial Knowledge

In this section we describe in details our new transform for compound statements. For the high-level overview the reader is referred to Sect. 1.2.

Let $\mathcal{R}$ be a polynomial-time relation admitting a delayed-input Σ-protocol $\Pi = (\mathcal{P}, \mathcal{V})$. Recall that delayed-input means that the prover does not need the instances of the statement to play the 1st round.

We describe a compiler that on input Π for $\mathcal{R}$ outputs a delayed-input WIPoK $\Pi^k = (\mathcal{P}^k, \mathcal{V}^k)$ for the (k, n)-threshold relation $\mathcal{R}_k$ defined as follows

$$\mathcal{R}_k = \big\{((x_1, \ldots, x_n), (w_{i_1}, \ldots w_{i_k})) : 1 \le i_1 < \cdots < i_k \le n$$

$$\text{and } (x_{i_j}, w_{i_j}) \in \mathcal{R}, \text{ for } j = 1, \ldots, k \text{ and } x_j \in \hat{L}_{\mathcal{R}}, \text{ for } j = 1, \ldots, n\big\}.$$

The main tools involved in our construction are the protocol $\Pi_{k,n}^{nddh}$ described in Sect. 2.3, and an IDTC scheme described in Sect. 2.4. More precisely the IDTC scheme is constructed using a Σ-protocol for DDH. Therefore given a tuple $T = (g, A, B, X)$ (either DH or non-DH), a message m and a randomness r, we can compute (com, dec) using the scheme described in Sect. 2.4. If T is a DH tuple, with $A = g^{\alpha}$, then α represents the trapdoor for the commitment com and dec is equal to $\perp$. In this case given a randomness r, com, the tuple T and α for every message m it is possible to compute dec such that a receiver accepts com as a commitment of the message m.

1st Round. $\mathcal{P}^k \Rightarrow \mathcal{V}^k$:
1. Set $(\mathcal{G}, p, g) \leftarrow \text{IG}(1^{\lambda})$.
2. Randomly choose tuples $T_1 = (g_1, A_1, B_1, X_1), \ldots, T_n = (g_n, A_n, B_n, X_n)$ of elements of $\mathcal{G}$ under the constraint that exactly k are oneNDH and $n - k$ are DH, along with $\alpha_1, \ldots, \alpha_n$ such that $A_i = g_i^{\alpha_i}$, for $i = 1, \ldots, n$.
3. Let $b_1, \ldots, b_k$ denote the indices of the k oneNDH tuples and $\tilde{b}_1, \ldots, \tilde{b}_{n-k}$ denote the indices of the $n - k$ DH tuples.
4. Run the prover of $\Pi_{k,n}^{nddh}$ on input $T = (T_1, \ldots, T_n)$, witnesses $(\alpha_{b_1}, \ldots, \alpha_{b_k})$ and randomness $r_{k,n}$ thus obtaining message $a_{k,n}$. Send $a_{k,n}$ to $\mathcal{V}^k$.

5. For $i = 1, \ldots, n$:

   Compute the first round a_i of Π by running $\mathcal{P}$ with randomness r_i.

   Compute pair $(\mathsf{com}_i, \mathsf{dec}_i)$ of commitment and decommitment of a_i using T_i.

   Send (T_i, com_i) to $\mathcal{V}^k$.

2nd Round. $\mathcal{V}^k \Rightarrow \mathcal{P}^k$: randomly select a challenge c and send it to $\mathcal{P}^k$.

3rd Round. $\mathcal{P}^k \Rightarrow \mathcal{V}^k$:

1. Receive inputs $(x_1, \ldots, x_n)$ and witnesses $(w_{d_1}, \ldots, w_{d_k})$ for inputs $x_{d_1}, \ldots, x_{d_k}$ (we denote by $\tilde{d}_1, \ldots, \tilde{d}_{n-k}$ the indices of the inputs for which no witness has been provided).

2. Compute the third round of $\Pi_{k,n}^{nddh}$ using c as challenge to get $z_{n,k}$ and send it to $\mathcal{V}^k$.

3. Pick a random permutation σ of $\{1, \ldots, k\}$ to associate each of the k oneNDH tuples $T_{b_1}, \ldots, T_{b_k}$ with one of the k inputs $x_{d_1}, \ldots, x_{d_k}$ for which a witness is available.

4. For $i = 1, \ldots, k$:

   Set $j = d_{\sigma(i)}$ and $t_j = b_i$.

   Compute z_j by running $\mathcal{P}$ on input (x_j, w_j), a_{t_j}, randomness r_{t_j} and challenge c.

   Set $M_j = (j, t_j, \mathsf{dec}_{t_j}, a_{t_j}, z_j)$.

5. Pick a random permutation τ of $\{1, \ldots, n - k\}$ to associate each of the $n - k$ DH tuples $T_{\tilde{b}_1}, \ldots, T_{\tilde{b}_{n-k}}$ to one of the k inputs $x_{\tilde{d}_1}, \ldots, x_{\tilde{d}_{n-k}}$ for which no witness is available.

6. For $i = 1, \ldots, n - k$:

   Set $j = \tilde{d}_{\tau(i)}$ and $t_j = \tilde{b}_i$.

   Run simulator S on input x_j and c obtaining (a_j, z_j).

   Use trapdoor α_{t_j} to compute decommitment dec_{t_j} of com_{t_j} as a_j.

   Set $M_j = (j, t_j, \mathsf{dec}_{t_j}, a_j, z_j)$.

7. For $j = 1, \ldots, n$: send M_j to $\mathcal{V}^k$.

$\mathcal{V}^k$ accepts if and only if all the following conditions are satisfied:

1. $(a_{n,k}, c, z_{n,k})$ is an accepting transcript for $\mathcal{V}_{k,n}^{nddh}$ with input T.

2. All t_j's are distinct.

3. For $j = 1, \ldots, n$: dec_{t_j} is a valid decommitment of com_{t_j} with respect to T_{t_j}.

4. For $j = 1, \ldots, n$: (a_j, c, z_j) is accepting for $\mathcal{V}$ with input x_j.

We will show now that Π^k is a (adaptive-Input) PoK and is adaptive-input WI for the relation $\mathcal{R}_k$.

3.1   (Adaptive-Input) Proof of Knowledge

Theorem 5. *Protocol Π^k is a proof of knowledge for $\mathcal{R}_k$.*

Proof. The completeness property follows from the completeness of protocols $\Pi_{k,n}^{nddh}$ and Π, and from the correctness and trapdorness property of the Instance-Dependent Trapdoor Commitment scheme used.

Now we proceed by proving that our protocol is $((n-1) \cdot k + 2)$-special sound and then, using the arguments of Sect. 2 about the proof of knowledge property of protocols that enjoy t-special soundness, we can conclude the proof claiming that Π^k is a proof of knowledge. There exists an efficient extractor that, for any sequence $(x_1, \ldots, x_n)$ of n inputs and for any set of $N = (n-1) \cdot k + 2$ accepting transcripts of Π^k that share the same first message and have different challenges, outputs the witnesses of k of the n inputs. The extractor is based on the following observations.

First of all, observe that, by the special soundness of $\Pi_{n,k}$, it is possible to extract the witness that k of the tuple $T_1, \ldots, T_n$ appearing in the first message are oneNDH. Let us denote by $b_1, \ldots, b_k$ the indices of the oneNDH tuples. This implies that commitments $\mathsf{com}_{b_1}, \ldots, \mathsf{com}_{b_k}$ that appear in the shared first round of the N transcripts will be opened to the same strings $a_{b_1}, \ldots, a_{b_k}$. We also observe that if two transcripts use the same input x_i with the same oneNDH tuple T_{b_i} then we can extract two transcripts of the Σ-protocol Π that share the same first message and have two different challenges. By the special soundness of Π there exists an extractor that efficiently extracts a witness. In other words, in order to be able to extract a witness for x_i, x_i has to be associated with the same oneNDH tuple in two distinct transcripts.

The extractor willing to get k witnesses considers the N transcripts one at the time and stops as soon as it reaches a *special* transcript C^l in which, for $j = 1, \ldots, k$, tuple T_{b_j} is associated with input x_{d_j} in C^l and in at least a transcript C^{l_j} with $l_j < l$. Clearly, once such a transcript is reached the extractor has obtained k witnesses. Now observe that a pair (oneNDH tuple, input x_i) can be used to eliminate at most one transcript. Moreover, there are $n \cdot k$ such pairs and the first transcript exhibits exactly k of these pairs. Therefore the set of N input transcripts contains at least one special transcript.

Theorem 6. *If Π is adaptive-input special sound then Π^k is an adaptive-input proof of knowledge for $\mathcal{R}_k$.*

Proof. We prove the following stronger statement. There exists an efficient algorithm that on input 2 accepting transcripts (a, c_1, z_1) (a, c_2, z_2) for Π^k, where

- the first one is accepting with respect to a sequence of n theorems $(x_1^1, \ldots, x_n^1)$,
- the second one is accepting with respect to a sequence of n (potentially different from the previous one) theorems $(x_1^2, \ldots, x_n^2)$,
- share the same first round and
- have different challenges, outputs, for each of the two sequence, k witnesses (for a total of $2 \cdot k$ witnesses).

The extractor is based on the following observations.

First of all, observe that, by the special soundness of protocol $\Pi_{n,k}$, it is possible to extract the witness certifying that k of the tuple $T_1, \ldots, T_n$ appearing in the first message are oneNDH let us denote by $b_1, \ldots, b_k$ the indices of the oneNDH tuples. This implies that commitments $\mathsf{com}_{b_1}, \ldots, \mathsf{com}_{b_k}$ that appear in the common first round of the N transcripts will be opened to the same strings $a_{b_1}, \ldots, a_{b_k}$.

To conclude the proof we observe that if two transcripts use the same oneNDH tuple T_{b_i} then we can obtain two transcripts of Σ-protocol Π that share the same first message and have two different challenges. By the adaptive-input special-soundness property of Π there exists an extractor that outputs a witness.

3.2 Adaptive-Input Witness Indistinguishability

Here we prove that Π^k is WI even when $\mathcal{A}$ can select instances and witnesses adaptively after receiving the first round. We have the following theorem.

Theorem 7. *Under the DDH assumption, if Π is SHVZK for $\mathcal{R}$ then Π^k is adaptive-input WI for relation $\mathcal{R}_k$.*

Proof. Let us fix a PPT adversary $\mathcal{A}$ and let us denote by X and W^0 and W^1 the instance and the witnesses of Π^k output by $\mathcal{A}$ at Step 2 of $\mathsf{ExpAWI}_{\Pi^k,\mathcal{A}}$. More precisely, we let $X = (x_1, \ldots, x_n)$ be the sequence of n instances output by $\mathcal{A}$ and $W^0 = ((w_1^0, d_1^0), \ldots (w_k^0, d_k^0))$ and $W^1 = ((w_1^1, d_1^1), \ldots (w_k^1, d_k^1))$ the two sequences of witnesses. We remark that $(x_{d_i^b}, w_i^b) \in \mathcal{R}$ for $i = 1, \ldots, k$ and $b = 0, 1$ and that $i \neq j$ implies that $d_i^0 \neq d_j^0$ and $d_i^1 \neq d_j^1$.

Let $m \leq k$ be the number of instances of Π in X for which W^1 contains a witness but W^0 does not. Obviously, since W^0 and W^1 contain witnesses for the same number k of instances of Π in X, it must be the case that m is also the number of instances of Π in X for which W^0 contains a witness and W^1 does not. We can rename the instances of X, so that W^0 and W^1 can be written as

$$W^0 = ((w_1^0, m+1), \ldots, (w_m^0, 2m), (w_{m+1}^0, 2m+1), (w_k^0, m+k)) \text{ and}$$

$$W^1 = ((w_1^1, 1), \ldots, (w_m^1, m), (w_{m+1}^1, 2m+1), \ldots, (w_k^1, m+k)).$$

For our proof we now consider the case in which $m = 0$ and $m \neq 0$. When $m = 0$ we have that W^1 and W^0 contains witnesses for the same theorems (for Π). Therefore by the perfect-WI property of Π^6 we can claim that if $m = 0$ then $\mathsf{AdvAWI}_{\Pi^k,\mathcal{A}}(\lambda, \mathsf{aux}) = 0$. Now we consider the more interesting case where $m \neq 0$.

We define the intermediate sequences of witnesses $W_1, \ldots, W_k$ in the following way.

1. For $i = 0, \ldots, m$: W_i consists of witnesses

$$W_i = ((w_1^1, 1), \ldots, (w_i^1, i), (w_{i+1}^0, m+i+1), \ldots$$
$$\ldots, (w_m^0, 2m), (w_{m+1}^0, 2m+1), \ldots, (w_{m+k}^0, m+k)).$$

Note that W_i contains witnesses for $(x_1, \ldots, x_i, x_{m+1+i}, \ldots, x_{2m})$. Moreover, W_0 coincides with W^0 and in W_m the first m witnesses are from W^1 and the remaining are from W^0.

---

[6] We observe that Σ-protocols enjoy SHVZK and therefore by Theorem 1 we can claim that every Σ-protocol is also perfect WI.

2. For $i = m + 1, \ldots, k$: W_i consists of witnesses

$$W_i = ((w_1^1, 1), \ldots, (w_m^1, m), (w_{m+1}^1, 2m + 1), \ldots$$
$$\ldots, (w_{m+i}^1, m + i), (w_{m+i+1}^0, m + i + 1), \ldots, (w_{m+k}^0, m + k)).$$

It is easy to see that W_k coincides with W^1.

For $i = 0, \ldots, k$, we define hybrid experiment $\mathcal{H}_i$ as the experiment in which the challenger $\mathcal{C}$ uses sequence of witnesses W_i to complete the third step of the experiment $\mathsf{ExpAWI}_{\Pi^k, \mathcal{A}}$. Clearly, $\mathcal{H}_0$ is the experiment $\mathsf{ExpAWI}_{\Pi^k, \mathcal{A}}$ when $\mathcal{C}$ picks $b = 0$ and $\mathcal{H}_k$ is the same experiment when $\mathcal{C}$ picks $b = 1$. We conclude the proof by showing that, for $i = 0, \ldots, k - 1$, $\mathcal{H}_i$ and $\mathcal{H}_{i+1}$ are indistinguishable.

We start by proving indistinguishability of $\mathcal{H}_i$ and $\mathcal{H}_{i+1}$ for $i = 0, \ldots, m - 1$. We remind the reader that, in $\mathcal{H}_i$ and $\mathcal{H}_{i+1}$, the challenger $\mathcal{C}$ uses witnesses for the following k inputs:

$\mathcal{H}_i$	$x_1 \cdots x_i$		x_{m+i+1}	$x_{m+i+2} \cdots x_{2m}$	$x_{2m+1} \cdots x_{m+k}$
$\mathcal{H}_{i+1}$	$x_1 \cdots x_i$	x_{i+1}		$x_{m+i+2} \cdots x_{2m}$	$x_{2m+1} \cdots x_{m+k}$

To prove indistinguishability of $\mathcal{H}_i$ and $\mathcal{H}_{i+1}$ we consider six intermediate hybrids: $\mathcal{H}_i^1, \ldots, \mathcal{H}_i^6$.

1. $\mathcal{H}_i^1(\lambda, \mathsf{aux})$ differs from $\mathcal{H}_i(\lambda, \mathsf{aux})$ in the way that the accepting transcript for the theorem x_{i+1} is computed. More precisely in $\mathcal{H}_i(\lambda, \mathsf{aux})$ the SHVZK simulator of Π was used to compute the transcript for x_{i+1} while in $\mathcal{H}_i^1(\lambda, \mathsf{aux})$ the transcript for x_{i+1} is computed using the honest-prover procedure that has also w_{i+1} as input. To prove the indistinguishability of the hybrids we can easily invoke the SHVZK property of Π. We remark that this is possible only because the commitment of the first round of Π with respect to the theorem x_{i+1} is hiding.
2. $\mathcal{H}_i^2(\lambda, \mathsf{aux})$ differs from $\mathcal{H}_i^1(\lambda, \mathsf{aux})$ in the way the tuples used to compute the commitments are chosen. More precisely k tuples oneNDH are chosen, $n - k - 1$ tuple DH are chosen and the last tuple is chosen non-DH. The additional non-DH tuple is used to compute the commitment of the first message of Π that will be associated to the theorem x_{i+1} in the third round. Even in this case is possible to compute an accepting transcript for Π^k because $k + 1$ witnesses are used instead of k, therefore there is no problem if $k + 1$ commitments are binding. The indistinguishability between the two hybrids is ensured by the DDH-assumption.
3. $\mathcal{H}_i^3(\lambda, \mathsf{aux})$ the only difference between this hybrid experiment and $\mathcal{H}_i^2(\lambda, \mathsf{aux})$ is that instead of a non-DH tuple, a oneNDH tuple is chosen. As in the previous hybrid experiment the considered tuple is used to compute the commitment of the first message of Π that will be associated to the theorem x_{i+1} in the third round. The indistinguishability between the two hybrids is ensured by the DDH-assumption.
4. $\mathcal{H}_i^4(\lambda, \mathsf{aux})$. The differences between this hybrid and $\mathcal{H}_i^3(\lambda, \mathsf{aux})$ are that we use k tuples oneNDH $n - k - 1$ tuples DH and one tuple non-DH. In this case

the additional non-DH tuple is used to commit the first round of Π that will be use as the first round of the accepting transcript with respect to x_{m+i+1}. By the DDH-assumption and perfect WI property of $\Pi_{n,k}$ we can claim that this hybrid is indistinguishable from the previous one.

5. $\mathcal{H}_i^5(\lambda, \textbf{aux})$. The differences between this hybrid and $\mathcal{H}_i^4(\lambda, \text{aux})$ are that we again use k oneNDH and $n-k$ DH tuples. In this case the additional DH tuple is used to commit to the first round of Π that will be used as the first round of the accepting transcript with respect to x_{m+i+1}. By the DDH-assumption we can claim that this hybrid is indistinguishable from the previous one.

6. $\mathcal{H}_i^6(\lambda, \textbf{aux})$ differs from $\mathcal{H}_i^5(\lambda, \text{aux})$ in the way that the accepting transcript for the theorem x_{m+i+1} is computed. More precisely in $\mathcal{H}_i^6(\lambda, \text{aux})$ the honest-prover procedure of Π was used to compute the accepting transcript for x_{m+i+1}. In $\mathcal{H}_i^5(\lambda, \text{aux})$ the transcript for x_{m+i+1} is computed using the SHVZK simulator of Π. To prove the indistinguishability of this hybrid we invoke the SHVZK property of Π. We remark that this is possible only because the commitment of the first round of Π with respect to the theorem x_{m+i+1} is hiding. We observe that this hybrid is equal to $\mathcal{H}_{i+1}(\lambda, \text{aux})$.

Now we are able to complete the first part of the proof observing that[7].

$$\mathcal{H}_i(\lambda, \text{aux}) \approx \mathcal{H}_i^1(\lambda, \text{aux}) \approx \cdots \approx \mathcal{H}_i^6(\lambda, \text{aux}) = \mathcal{H}_{i+1}(\lambda, \text{aux}).$$

We have thus proved that $\mathcal{H}_0$ and $\mathcal{H}_m$ are indistinguishable. To complete the proof, we need to prove that $\mathcal{H}_{m+i}$ and $\mathcal{H}_{m+i+1}$ are indistinguishable for $i = 0, \ldots, k-1$. This follows directly from the observation that $\mathcal{H}_{m+i}$ and $\mathcal{H}_{m+i+1}$ only differ in the witness used for x_{2m+i+1} as in $\mathcal{H}_{m+i}$ the witness from W^0 is used by $\mathcal{C}$ whereas in $\mathcal{H}_{m+i+1}$ $\mathcal{C}$ uses the witness from W^1. Indistinguishability follows directly from the Perfect WI of Π.

4   On Adaptive-Input Special-Soundness of Σ-Protocols

In this section we show that Σ-Protocols are not secure when the adversarial prover can choose the statement adaptively, when playing the 3rd round. These issues for the case of the Fiat-Shamir transform were noted in [1].

We then show an efficient compiler that on input a Σ-protocol belonging to the general class considered in [9, 21], outputs a Σ-protocol that is secure against adaptively chosen statements.

4.1   Soundness Issues in Delayed-Input Σ-Protocols

We start by showing that the notion of adaptive-input special soundness is non-trivial in the sense that there are Σ-protocols that are not special sound when the statement is chosen adaptively at the 3rd round.

---

[7] See the full version of this work [7] for a formal description of the hybrid experiments.

Issues with Soundness. Let us consider the following well-known Σ-protocol Π_{DH} for relation DH. On common input $T = (g, A, B, X)$ and private input α such that $A = g^\alpha$ and $X = B^\alpha$ for the prover, the following steps are executed. We denote by q the size of the group $\mathcal{G}$.

1. $\mathcal{P}$ picks $r \in \mathbb{Z}_q$ at random and computes and sends $a = g^r$, $x = B^r$ to $\mathcal{V}$;
2. $\mathcal{V}$ chooses a random challenge $c \in \mathbb{Z}_q$ and sends it to $\mathcal{P}$;
3. $\mathcal{P}$ computes and sends $z = r + c\alpha$ to $\mathcal{V}$;
4. $\mathcal{V}$ accepts if and only if: $g^z = a \cdot A^c$ and $B^z = x \cdot X^c$.

We now show that the above Σ-protocol is not special sound when an adversarial prover selects X adaptively.

Consider the following two conversations $((a = g^r, x = B^s), c_1, z_1 = r + \alpha \cdot c_1)$ and $((a = g^r, x = B^s), c_2, z_2 = r + \alpha \cdot c_2)$ respectively for tuples (g, A, B, X_1) and (g, A, B, X_2) where $A = g^\alpha$, $X_1 = g^{\gamma_1}$ and $X_2 = g^{\gamma_2}$ and $\gamma_i = \frac{z_i - s}{c_i} = \alpha + \frac{r - s}{c_i}$, for $i = 1, 2$. It is easy to see that both conversations are accepting (for their respective inputs) and that, if $r \neq s$, neither tuple is a DH tuple and therefore no witness can be extracted. Notice that this is a very strong soundness attack since the adversarial prover can succeed in convincing the verifier even though the statement is false. A similar argument can be used to prove that the Σ-protocol of [22] for relation $\mathsf{Com} = \{((g, h, G, H, m), r) : G = g^r \text{ and } H = h^{r+m}\}$ does not enjoy adaptive-input special soundness.

Issues with Special Soundness. Let us now consider the case of Schnorr's Σ-protocol [27] for relation $\mathsf{DLog} = \{((\mathcal{G}, g, Y), y) : g^y = Y\}$. Clearly, this is a different case since there is no false theorem to prove, but the attack can only consist in proving a statement violating special soundness (i.e., even though there are two accepting transcripts with the same first message no witness can be extracted).

In Schnorr's protocol, the prover on input $(Y, y) \in \mathsf{DLog}$ starts by sending $a = g^r$, for a randomly chosen $r \in Z_q$. Upon receiving challenge c, $\mathcal{P}$ replies by computing $z = r + yc$. $\mathcal{V}$ accepts (a, c, z) if $g^z = a \cdot Y^c$.

Consider now accepting transcripts (a, c_i, z_i) with respect to inputs Y_i, $i = 1, 2$. In this case, to extract witnesses y_i s.t. $((\mathcal{G}, g, Y_i), y_i) \in \mathsf{DLog}$ one has to solve the following system with unknowns r, y_1, and y_2.

$$\begin{cases} z_1 = r + c_1 \cdot y_1 \\ z_2 = r + c_2 \cdot y_2 \end{cases}$$

Clearly the system above has q solutions and thus it gives no information on any of the two witnesses.

4.2   A Compiler for Adaptive-Input Special Soundness

In this section we show how to upgrade special soundness to adaptive-input special-soundness in all Σ-protocols belonging to the interesting class of Σ-protocols proposed in [9,21].

We show a compiler that obtains a Σ-protocol Π_f^a for proving knowledge of the pre-image of a homomorphic function. Our compiler takes as input a Σ-protocol $\Pi_f = (\mathcal{P}_f, \mathcal{V}_f)$ for the same generic relation that includes Schnorr's [27], Guillou-Quisquater [16] and the Σ-protocol for DH tuples as special cases [9,21].

Let $(\mathcal{G}, \star)$ and $(\mathcal{H}, \otimes)$ be two groups with efficient operations and let $f : \mathcal{G} \to \mathcal{H}$ be a one-way homomorphism from $\mathcal{G}$ to $\mathcal{H}$. That is, for all $x, y \in \mathcal{G}$, we have that $f(x \star y) = f(x) \otimes f(y)$ and it is infeasible to compute w from $f(w)$ for a randomly chosen w. In protocol Π_f for relation $\mathcal{R}_f = \{(x, w) : x = f(w)\}$, prover and verifier receive as input a description of the groups $\mathcal{G}$ and $\mathcal{H}$ and $x \in \mathcal{H}$. The prover receives w such that $x = f(w)$ as a private input. The prover and verifier execute the following steps:

1. $\mathcal{P}_f$ picks $r \leftarrow \mathcal{G}$, sets $a \leftarrow f(r)$ and sends a to $\mathcal{V}_f$;
2. $\mathcal{V}_f$ randomly selects a challenge c and sends it to $\mathcal{P}_f$;
3. $\mathcal{P}_f$ on input r, x, w and c computes $z = r \star w^c$ and sends it to $\mathcal{V}_f$;
4. $\mathcal{V}_f$ accepts if and only if $f(z) = a \otimes x^c$.

It is easy to see that this protocol can be instantiated to give Schnorr's [27] and Guillou-Quisquater [16] Σ-protocols as special cases. Theorem 3 of [21] describes necessary conditions for Π_f to be special sound. Specifically, given a collision (a, c_1, z_1) and (a, c_2, z_2) for common input x, it is possible to extract w such that $x = f(w)$ if integer y and element $u \in \mathcal{G}$ are known and it holds that

1. $\gcd(c_1 - c_2, y) = 1$;
2. $f(u) = x^y$.

It is not difficult to see that this is the case when the protocol is instantiated for all the relations described above. We also observe that, since Schnorr's protocol is a special case of this protocol, protocol Π_f does not enjoy adaptive-input special soundness.

From Π_f to Π_f^a. We next show how to efficiently transform this Σ-protocol into one that enjoys adaptive-input special soundness. The underlying idea is that an adaptive attack against such protocol consists in misbehaving when playing the first round. For instance, in the case of the Σ-protocol for DH, $\mathcal{P}^*$ has to send a non-DH tuple in the 1st round while instead the protocol asks for a DH tuple. We therefore can convert Π_f into Π_f^a by asking the prover to also give an auxiliary proof where it proves knowledge of the randomness used to correctly compute the first round of Π_f. Notice that on this auxiliary proof an adaptive-input selection attack can not take place since the adversarial prover is stuck with the content of the 1st round of Π_f that therefore specifies already the statement to prove. We now show that special soundness allows to get the randomness used to compute the first round of Π_f. Then the same argument shown in Theorem 3 of [21] allows to extract the witness from a single transcript.

We now discuss the compiler and why it works more formally.

Let us start with the following observation. Consider an accepting transcript (a, c, z) of Π_f for input x. If r such that $f(r) = a$ is available, then it is possible to compute a witness w for x. Indeed, from Theorem 3 of [21] it follows that we can compute w as $w = u^\alpha \star (z \star r^{-1})^\beta$, where α and β are such that $y \cdot \alpha + c \cdot \beta = 1$[8]. We use an argument already used in Theorem 3 of [21] in order to prove that $f(w) = x$, where $w = u^\alpha \star (z \star r^{-1})^\beta$. First we observe that $f(z) = a \otimes x^c$ and this implies that $f(z \star r^{-1}) = x^c$. Then we observe (like in [21]) that $f(w) = f(u^\alpha \star (z \star r^{-1})^\beta) = f(u)^\alpha \otimes f(z \star r^{-1})^\beta = x^{y\alpha} + x^{\beta c} = x$ that proves that $f(w) = x$.

Consider protocol Π_f^a consisting of the parallel execution of two instances of Π_f. For common input x, the first instance of Π_f is executed on common input x, whereas in the second instance the common input is the first message a of the first instance. The verifier of Π_f^a sends the same challenge to both instances and accepts if and only if it accepts in both instances. Since in a collision the first message is fixed, both transcripts have the same first message a and therefore we can invoke special soundness to extract r such that $f(r) = a$. Once r is available, we apply the observation above and extract witnesses for x_1 and x_2 (the two inputs of the two 3-round transcripts constituting the collision). We have thus the following theorem.

Theorem 8. *If there exists a Σ-protocol Π_f for $\mathcal{R}_f$, then there exists a Σ-protocol Π_f^a for $\mathcal{R}_f$ that enjoys adaptive-input special soundness.*

5   On the Adaptive-Input Soundness of [6]'s Transform

Ciampi et al. in [6] show a compiler that takes as input two Σ-protocols, Π_0 and Π_1 for languages L_0 and L_1, and outputs a new Σ-protocol Π^{OR} for $L_0 \vee L_1$ in which the instance for the language L_1 is required by the prover only in the 3rd round. The compiler requires that Π_1 be delayed input and they show that the output of the compiler is a Σ-protocol, therefore it enjoys special soundness. In this section we assume that Π_1 is adaptive-input special sound, and we will show that Π^{OR} enjoys also adaptive-input special soundness.

5.1   Overview of the Construction of [6]

We start with a succinct description of the main building block used in of [6].

t-Instance-Dependent Trapdoor Commitment. Ciampi et al. in [6] define the notion of a t-Instance-Dependent Trapdoor Commitment (t-IDTC) scheme. Such a scheme works with respect to an polynomial-time relation $\mathcal{R}$. More formally, given the pair x, w s.t. $(x, w) \in \mathcal{R}$, it is possible to compute a commitment (with respect to some massage space M) using only the instance x, and the message. After that it is possible to open the commitment if one knows the randomness used in the commitment phase, or it is possible to equivocate the commitment using the witness w.

---

[8] By the first condition of Theorem 3 of [21] we have that $\gcd(c, y) = 1$ and thus α and β can be computed by using the extended Euclidean gcd algorithm.

The t-IDTC scheme is defined by a triple of PPT algorithm (TCom, TDec, TFake) where TCom, TDec are the honest commitment and decommitment procedures and TFake is the equivocation procedure that, given a witness for an instance x, equivocates any commitment computed using x as input of TCom. The properties of a t-IDTC scheme are: correctness, hiding, trapdoor and t-Special Extractability. The property of t-Special Extractability informally says that if the sender opens the same commitment in t different ways, then it is possible to efficiently extract the witness w. For more details see [6].

The authors of [6] show how to construct a 2-IDTC schemes are perfect hiding, perfect trapdoor and 2-Special Extractable from Σ-protocols.

In the rest of this section when a player runs the algorithm TCom on input x, m, obtains the pair (com, dec) where com is the commitment of the message m, and dec is the decommitment value. To check if dec is a valid decommitment of com with respect to the message m, we use the algorithm TDec. To compute a fake opening of the commitment com with respect to a message $m' \neq m$ a player can use the algorithm TFake using as input (com, dec).

The Construction of [6]. Let $\mathcal{R}_0$ be a relation admitting a t-IDTC scheme, with $t = 2$ or $t = 3$. Let $\mathcal{R}_1$ be a relation admitting an delayed-input Σ-protocol Π_1 with associated simulator S^1.

We show a Σ-protocol $\Pi^{OR} = (\mathcal{P}^{OR}, \mathcal{V}^{OR})$ for the OR relation:

$$\mathcal{R}^{OR} = \left\{ ((x_0, x_1), w) : ((x_0, w) \in \mathcal{R}_0 \wedge x_1 \in \hat{L}_{\mathcal{R}_1}) \text{ OR } ((x_1, w) \in \mathcal{R}_1 \wedge x_0 \in \hat{L}_{\mathcal{R}_0}) \right\}.$$

The initial common input is x_0 and the other input x_1 and the witness w for (x_0, x_1) are available to the prover only at the 3rd round. We let $b \in \{0, 1\}$ be such that $(x_b, w) \in \mathcal{R}_b$. The construction of [6] is described below.

Common input: $(x_0, 1^\lambda)$, where λ is the length of the instance of $\hat{L}_{\mathcal{R}_1}$.

1. $\mathcal{P}^{OR}$ executes the following steps:
   1.1. pick random r_1 and compute the 1st round a_1 of the delayed-input Σ-protocol Π_1;
   1.2. compute a pair (com, dec_1) of commitment and decommitment of a_1;
   1.3. send com to $\mathcal{V}^{OR}$.
2. $\mathcal{V}^{OR}$ sends a random challenge c.
3. $\mathcal{P}^{OR}$ on input $((x_0, x_1), c, (w, b))$ s.t. $(x_b, w) \in \mathcal{R}_b$ executes the following steps:
   3.1. If $b = 1$, compute the 3rd round of Π_1, z_1, using as input (x_1, w, c);
   3.2. Send (dec_1, a_1, z_1) to $\mathcal{V}^{OR}$r;
   3.3. If $b = 0$, run simulator S^1 on input x_1 and c obtaining (a_2, z_2); use trapdoor to compute decommitment dec_2 of com as a_2;
   3.4. Send (dec_2, a_2, z_2) to $\mathcal{V}^{OR}$.
4. V^{OR} accepts if and only if the following conditions are satisfied:
   4.1. (a, c, z) is an accepting conversation for x_1;
   4.2. dec is a valid decommitment of com for a message a.

5.2 Adaptive-Input Security of Π^{OR}

We now show that Π^{OR} preserves the adaptive-input special soundness of the underlying Σ-protocol.

Theorem 9. *If $\mathcal{R}_0$ admits a 2-IDTC and $\mathcal{R}_1$ admits a delayed-input adaptive-input special-sound Σ-protocol, then Π^{OR} is an adaptive-input special-sound Σ-protocol.*

Proof. The claim follows from the adaptive-input special soundness of the underlying Σ-protocol Π_1 and from the 2-Special Extractability property of the 2-IDTC scheme. More formally, consider an accepting transcript $(\mathsf{com}, c, (z, a, \mathsf{dec}))$ for input (x_0, x_1) and an accepting transcript $(\mathsf{com}, c', (z', a', \mathsf{dec}'))$ for input (x_0, x_1'), where $c' \neq c$ and x_1 is potentially different from x_1'. We observe that:

- if $a = a'$ then by the property of adaptive-input special soundness of Π_1 there exists an efficient extractor AExtract that, given as input $((a, c, z), x_1)$ and $((a', c', z'), x_1')$, outputs w_1 and w_1' s.t. $(x_1, w_1) \in \mathcal{R}_1$ and $(x_1', w_1') \in \mathcal{R}_1$;
- if $a \neq a'$, then dec and dec' are two openings of com with respect to x_0 for messages $a \neq a'$; then we can obtain a witness w_0 by the 2-Special Extractability of the 2-IDTC scheme.

A similar arguments can be used to show that if $\mathcal{R}_0$ admits a 3-IDTC and $\mathcal{R}_1$ admits a delayed-input Σ-protocol with adaptive-input special soundness, then Π^{OR} enjoys the adaptive-input proof of knowledge property.

6 Extension to Multiple Relations

In this section, we generalize the result of Sect. 3 to the case of different relations. More specifically, given delayed-input Σ-protocols $\Pi_1, \ldots, \Pi_n$ for polynomial-time relations $\mathcal{R}_1, \ldots, \mathcal{R}_n$, we construct, for some positive constant k, Adaptive-Input Proof of Partial Knowledge Γ for the *threshold* polynomial-time relation

$$\mathcal{R}^{\mathsf{thres}} = \left\{ \left((x_1, \ldots, x_n, k), ((w_1, d_1) \ldots (w_k, d_k)) \right) : 1 \leq d_1 < \cdots < d_k \leq n \right.$$

$$\left. \text{and } (x_{d_i}, w_i) \in \mathcal{R}_i \text{ for } i = 1, \ldots, k \text{ and } x_1 \in \hat{L}_1, \ldots, x_n \in \hat{L}_n \right\}.$$

We remind the reader that $\hat{L}_1, \ldots, \hat{L}_n$ are the input languages associated with the polynomial-time relations $\mathcal{R}_1, \ldots, \mathcal{R}_n$.

Protocol Γ uses delayed-input protocol Π^k, in the adaptive-input special-soundness version, presented in Sect. 3 for relation $\mathsf{NDH}_{k,n}$. We remark that protocol $\Pi_{k,n}^{\mathsf{ddh}}$ of Sect. 2.3 would not work here since the prover of Γ learns the actual statement to be proved just before the third round.

1st Round. Γ.Prover $\Rightarrow$ Γ.Verifier:

Γ.Prover receives as unary inputs the security parameter λ, the number n of theorems that will be given as input at the beginning of the third round, and the number k of witnesses that will be provided.

1. Set $(\mathcal{G}, p, g) \leftarrow \mathsf{IG}(1^\lambda)$.
2. For $j = 1, \ldots, n$
   2.1. Randomly sample a non-DH tuple $T_j^0 = (g_j, A_j, B_j, X_j)$ over $\mathcal{G}$, along with α_j such that $A_j = g_j^{\alpha_j}$.
   2.2. Set $Y_j = B_j^{\alpha_j}$ and $T_j^1 = (g_j, A_j, B_j, Y_j)$ (note that the quadruple T_j^1 is by construction a DH tuple).
3. Select a random string $R_{k,n}$ and use it to compute the first round message $a_{k,n}$ of Π^k by running prover $\mathcal{P}^k$.
   Send $a_{k,n}$ to $\Gamma.\mathsf{Verifier}$.
4. For $j = 1, \ldots, n$
   4.1. Select random strings R_j^0 and R_j^1 and use them to compute the first rounds a_j^0 and a_j^1 of Π_j by running prover $\mathcal{P}_j$.
   4.2. Compute the pair $(\mathsf{com}_j^0, \mathsf{dec}_j^0)$ of commitment and decommitment of the message a_j^0 using non-DH tuple T_j^0.
   4.3. Compute the commitment com_j^1 (of the message a_j^1) using the DH tuple T_j^1.
   4.4. Send pairs $(T_j^0, \mathsf{com}_j^0)$ and $(T_j^1, \mathsf{com}_j^1)$ in random order to $\Gamma.\mathsf{Verifier}$.

2nd Round. $\Gamma.\mathsf{Verifier} \Rightarrow \Gamma.\mathsf{Prover}$: $\Gamma.\mathsf{Verifier}$ randomly selects a challenge c and sends it to $\Gamma.\mathsf{Prover}$.

3rd Round. $\Gamma.\mathsf{Prover} \Rightarrow \Gamma.\mathsf{Verifier}$:

$\Gamma.\mathsf{Prover}$ receives theorems $x_1, \ldots, x_n$ and, for $d_1 < \ldots < d_k$, witnesses $w_1, \ldots, w_k$ for theorems $x_{d_1}, \ldots, x_{d_k}$, respectively. We let $\tilde{d}_1 < \ldots < \tilde{d}_{n-k}$ denote the indices of the theorems for which no witness has been provided.

1. For $l = 1, \ldots, k$
   1.1. Use j as a shorthand for d_l.
   1.2. Set $U_j = T_j^1$ and $\hat{U}_j = T_j^0$.
   1.3. Compute third round z_j of Π_j by running prover $\mathcal{P}_j$ on input (x_j, w_i), randomness R_j^0 used to compute the first round a_j^0, and a challenge c.
   1.4. Set $M_j = (a_j^0, z_j, \mathsf{dec}_j^0, \hat{U}_j)$.
2. For $l = 1, \ldots, n - k$
   2.1. Set $j = \tilde{d}_l$.
   2.2. Set $U_j = T_j^0$ and $\hat{U}_j = T_j^1$.
   2.3. Run the simulator S_j of Π_j on input x_j and c therefore obtaining $(\tilde{a}_j^1, z_j)$.
   2.4. Use the trapdoor α_j to compute the decommitment dec_j^1 of com_j^1 as $\tilde{a}_j^1$.
   2.5. Set $M_j = (\tilde{a}_j^1, z_j, \mathsf{dec}_j^1, \hat{U}_j)$.
3. For $l = 1, \ldots, n$ send M_l to $\Gamma.\mathsf{Verifier}$.
4. Compute the third round $z_{k,n}$ of Π^k by running prover $\mathcal{P}^k$ of Π^k on input tuples $(U_1, \ldots, U_n)$, witnesses $\alpha_{d_1}, \ldots, \alpha_{d_k}$ and randomness $R_{k,n}$ used to compute the first round $a_{k,n}$.

$\Gamma.\mathsf{Verifier}$ accepts if and only if the following conditions are satisfied.
1. Check that $(a_{n,k}, c, z_{n,k})$ is an accepting conversation for $\mathcal{V}^k$ for input $U_1, \ldots, U_n$.

2. For $i = 1 \ldots n$

    Check that tuples T_i^0 and T_i^1 differ only in the last component.

    Check that $\{U_i, \hat{U}_i\} = \{T_i^0, T_i^1\}$.

    Write M_i as $M_i = (a_i, z_i, \mathsf{dec}_i, \hat{U}_i)$.

    Check that dec_i is a decommitment of one of com_i^0 and com_i^1 as a_i with respect to tuple $\hat{U}_i$.

    Check that (a_i, c, z_i) is an accepting conversation for Π_i on input x_i.

Theorem 10. *Γ is a proof of knowledge.*

Proof. The completeness property follows from the completeness of protocols Π^k and Π_i, for $i \in \{1, \ldots, n\}$, and from the correctness and trapdoorness property of the Instance-Dependent Trapdoor Commitment scheme used.

Now we proceed by proving that our protocol is $(2n + k)$-special sound and then, using the arguments of Sect. 2 about the proof of knowledge property of protocols that enjoy t-special soundness, we can conclude the proof claiming that Γ is a proof of knowledge. In more details, we prove that there exists an efficient extractor which, for any sequence $(x_1, \ldots, x_n)$ of n inputs and for any set of $2n + k$ accepting conversations of Γ that share the same first message and have different challenges, outputs the witness of w_i s.t. $(x_i, w_i) \in \mathcal{R}_i$ for some $i \in \{1, \ldots, n\}$. The extractor considers a set of $2n + k$ accepting conversations a, c^j, z^j (with $j = 1, \ldots, 2n + k$) such that they share the same first message and have different challenges. For each a, c^j, z^j (with $j = 1, \ldots, 2n + k$) processed by the extractor one of the following two cases is possible.

1. There are two conversations of Σ-protocol Π^i for theorem x_i that share the same first message a_i and have two different challenges. Then by the special soundness property of Π^i one can efficiently get a witness w_i for theorem x_i.
2. If the new accepting transcript a, c^j, z^j does not allow the extractor to obtain the witness then a new non-DH tuple is used for the first time in the accepting conversation a, c^j, z^j.

The proof ends with the observation that the algorithm stops after k times that the first case occurs, while the second case occurs at most $2n$ times.

Theorem 11. *Under the DDH assumption, if Π_i is SHVZK for $\mathcal{R}_i$, for $i \in \{1, \ldots, n\}$, then Γ is adaptive-input WI for $\mathcal{R}^{\mathsf{thres}}$, for a constant k.*

Proof Sketch. The definition of adaptive-input WI gives to the adversary $\mathcal{A}$ the power to choose both theorems and witnesses upon receiving the first message from the challenger. This implies that in Γ the first round should be computed without knowing which witnesses will be chosen by $\mathcal{A}$, and without knowing for what instances the witnesses will be available in the third round. It is easy to see that the first round of Γ is independent from the $\mathcal{A}$ could have. Unfortunately if we follow the same proof of Theorem 7, considering a similar sequence of hybrids experiments, we will have to define hybrid experiments in which the first round depends on which witnesses will be received from $\mathcal{A}$ at the second round. This implies that the only way for the challenger to complete these hybrid experiments consists in guessing the instances that correspond to the witnesses that will be received. This explains why k is a constant.

We now explain with more details the differences between the security proof of Theorem 7 and the one needed for protocol Γ. The security proof of Theorem 7 works for every k because the n instances $x_1, \ldots, x_n$ that will be sent by $\mathcal{A}$ in the protocol Π^k belong to the same NP-language L. Furthermore the Σ-protocol Π used in Π^k is delayed input. Hence for a first round a_i of Π it is possible to create an accepting transcript (a_i, c, z) for a theorem x_j, for $i, j \in \{1, \ldots, n\}$ (if one has the witness w_j clearly). Therefore the assignment of the values $a_1, \ldots, a_n$ committed in the first round with the theorems $x_1, \ldots, x_n$ is made only at the third round. This property holds in all hybrid experiments. Now we consider the protocol Γ. The arguments described above are clearly not applicable to prove that Γ is adaptive-input WI.

More in details, during the first round of Γ, for each language L_i we compute the first message of protocol Π_i and commit to it twice using the instance-dependent trapdoor commitment associated to a DH tuple and to a non-DH tuple, for $i \in \{1, \ldots, n\}$. Hence for each a_i we compute an equivocal commitment and a binding commitment. First note that these two commitments are linked to a fixed language L_i (in contrast to the first round of Π^k). When in the security proof we need to consider the hybrid experiment in which $n + 1$ non-DH tuples (one non-DH tuple per pair except one pair where both tuples are non-DH) are used (as in the proof of Theorem 7), we have to commit the first round of a_i, for some $i \in \{1, \ldots, n\}$, using two commitments that are perfectly binding. Therefore the only way that we have to compute an accepting transcript with respect to the language L_i consists in using the witness for the instance x_i that will be sent by $\mathcal{A}$. Unfortunately we have no guarantee that $\mathcal{A}$ will send w_i, and thus the experiment will have to try again. For lack of space, further details can be found in the full version of this work.

Acknowledgments. We thank the anonymous reviewers of Eurocrypt 2016 for many insightful comments and suggestions. This work has been supported in part by "GNCS - INdAM" and in part by the EU COST Action IC1306.

References

1. Bernhard, D., Pereira, O., Warinschi, B.: How not to prove yourself: pitfalls of the fiat-shamir heuristic and applications to helios. In: Proceedings of the Advances in Cryptology - ASIACRYPT 2012–18th International Conference on the Theory and Application of Cryptology and Information Security, Beijing, China, 2–6 December 2012, pp. 626–643 (2012)
2. Blundo, C., Persiano, G., Sadeghi, A.-R., Visconti, I.: Improved security notions and protocols for non-transferable identification. In: Jajodia, S., Lopez, J. (eds.) ESORICS 2008. LNCS, vol. 5283, pp. 364–378. Springer, Heidelberg (2008)
3. Catalano, D., Dodis, Y., Visconti, I.: Mercurial commitments: minimal assumptions and efficient constructions. In: Halevi, S., Rabin, T. (eds.) TCC 2006. LNCS, vol. 3876, pp. 120–144. Springer, Heidelberg (2006)

4. Catalano, D., Visconti, I.: Hybrid trapdoor commitments and their applications. In: Caires, L., Italiano, G.F., Monteiro, L., Palamidessi, C., Yung, M. (eds.) ICALP 2005. LNCS, vol. 3580, pp. 298–310. Springer, Heidelberg (2005)
5. Catalano, D., Visconti, I.: Hybrid commitments and their applications to zero-knowledge proof systems. Theor. Comput. Sci. **374**(1–3), 229–260 (2007)
6. Ciampi, M., Persiano, G., Scafuro, A., Siniscalchi, L., Visconti, I.: Improved OR-composition of sigma-protocols. In: Kushilevitz, E., Malkin, T. (eds.) TCC 2016-A. LNCS, vol. 9563, pp. 112–141. Springer, Heidelberg (2016)
7. Ciampi, M., Persiano, G., Scafuro, A., Siniscalchi, L., Visconti, I.: Online/offline OR composition of sigma protocols. Cryptology ePrint Archive, Report 2016/175 (2016). http://eprint.iacr.org/
8. Ciampi, M., Persiano, G., Siniscalchi, L., Visconti, I.: A transform for NIZK almost as efficient and general as the fiat-shamir transform without programmable random oracles. In: Kushilevitz, E., Malkin, T. (eds.) TCC 2016-A. LNCS, vol. 9563, pp. 83–111. Springer, Heidelberg (2016)
9. Cramer, R., Damgård, I.B.: Zero-knowledge proofs for finite field arithmetic or: can zero-knowledge be for free? In: Krawczyk, H. (ed.) CRYPTO 1998. LNCS, vol. 1462, pp. 424–441. Springer, Heidelberg (1998)
10. Cramer, R., Damgård, I.B., Schoenmakers, B.: Proof of partial knowledge and simplified design of witness hiding protocols. In: Desmedt, Y.G. (ed.) CRYPTO 1994. LNCS, vol. 839, pp. 174–187. Springer, Heidelberg (1994)
11. Damgård, I.: On Σ-protocol (2010). http://www.cs.au.dk/~ivan/Sigma.pdf
12. Damgård, I., Groth, J.: Non-interactive and reusable non-malleable commitment schemes. In: Proceedings of the 35th Annual ACM Symposium on Theory of Computing, June 9–11, 2003, San Diego, CA, USA, pp. 426–437 (2003)
13. Damgård, I.B., Nielsen, J.B.: Perfect hiding and perfect binding universally composable commitment schemes with constant expansion factor. In: Yung, M. (ed.) CRYPTO 2002. LNCS, vol. 2442, pp. 581–596. Springer, Heidelberg (2002)
14. Fiat, A., Shamir, A.: How to prove yourself: practical solutions to identification and signature problems. In: Odlyzko, A.M. (ed.) CRYPTO 1986. LNCS, vol. 263, pp. 186–194. Springer, Heidelberg (1987)
15. Garay, J.A., MacKenzie, P., Yang, K.: Strengthening zero-knowledge protocols using signatures. J. Cryptology **19**(2), 169–209 (2006)
16. Guillou, L.C., Quisquater, J.-J.: A practical zero-knowledge protocol fitted to security microprocessor minimizing both transmission and memory. In: Günther, C.G. (ed.) EUROCRYPT 1988. LNCS, vol. 330, pp. 123–128. Springer, Heidelberg (1988)
17. Hazay, C., Lindell, Y.: Efficient Secure Two-Party Protocols - Techniques and Constructions. Information Security and Cryptography. Springer, Heidelberg (2010)
18. Katz, J., Ostrovsky, R.: Round-optimal secure two-party computation. In: Franklin, M. (ed.) CRYPTO 2004. LNCS, vol. 3152, pp. 335–354. Springer, Heidelberg (2004)
19. Lapidot, D., Shamir, A.: Publicly verifiable non-interactive zero-knowledge proofs. In: Menezes, A., Vanstone, S.A. (eds.) CRYPTO 1990. LNCS, vol. 537, pp. 353–365. Springer, Heidelberg (1991)
20. Lindell, Y.: An efficient transform from sigma protocols to NIZK with a CRS and non-programmable random oracle. In: Dodis, Y., Nielsen, J.B. (eds.) TCC 2015, Part I. LNCS, vol. 9014, pp. 93–109. Springer, Heidelberg (2015)
21. Maurer, U.: Zero-knowledge proofs of knowledge for group homomorphisms. Des. Codes Crypt. **77**(2), 1–14 (2015)

22. Micciancio, D., Petrank, E.: Simulatable commitments and efficient concurrent zero-knowledge. In: Proceedings of the Advances in Cryptology - EUROCRYPT 2003, International Conference on the Theory and Applications of Cryptographic Techniques, Warsaw, Poland, 4–8 May 2003, pp. 140–159 (2003)
23. Ostrovsky, R., Pandey, O., Visconti, I.: Efficiency preserving transformations for concurrent non-malleable zero knowledge. In: Micciancio, D. (ed.) TCC 2010. LNCS, vol. 5978, pp. 535–552. Springer, Heidelberg (2010)
24. Ostrovsky, R., Richelson, S., Scafuro, A.: Round-optimal black-box two-party computation. In: Gennaro, R., Robshaw, M. (eds.) CRYPTO 2015. LNCS, vol. 9216, pp. 339–358. Springer, Heidelberg (2015)
25. Pass, R.: Simulation in quasi-polynomial time, and its application to protocol composition. In: Biham, E. (ed.) EUROCRYPT 2003. LNCS, vol. 2656, pp. 160–176. Springer, Heidelberg (2003)
26. Santis, A.D., Crescenzo, G.D., Persiano, G., Yung, M.: On monotone formula closure of SZK. In: 35th Annual Symposium on Foundations of Computer Science, Santa Fe, New Mexico, USA, 20–22 November 1994, pp. 454–465 (1994)
27. Schnorr, C.-P.: Efficient identification and signatures for smart cards. In: Brassard, G. (ed.) CRYPTO 1989. LNCS, vol. 435, pp. 239–252. Springer, Heidelberg (1990)
28. Visconti, I.: Efficient zero knowledge on the internet. In: Bugliesi, M., Preneel, B., Sassone, V., Wegener, I. (eds.) ICALP 2006. LNCS, vol. 4052, pp. 22–33. Springer, Heidelberg (2006)
29. Yung, M., Zhao, Y.: Generic and practical resettable zero-knowledge in the bare public-key model. In: Proceedings of the Advances in Cryptology - EUROCRYPT 2007, 26th Annual International Conference on the Theory and Applications of Cryptographic Techniques, Barcelona, Spain, 20–24 May 2007, pp. 129–147 (2007)

Constant-Round Leakage-Resilient
Zero-Knowledge from Collision Resistance

Susumu Kiyoshima[✉]

NTT Secure Platform Laboratories, Tokyo, Japan
kiyoshima.susumu@lab.ntt.co.jp

Abstract. We construct a constant-round leakage-resilient zero-knowledge argument system under the existence of collision-resistant hash function family. That is, using collision-resistant hash functions, we construct a constant-round zero-knowledge argument system such that for any cheating verifier that can obtain arbitrary amount of leakage of the prover's state, there exists a simulator that can simulate the adversary's view by obtaining at most the same amount of leakage of the witness. Previously, leakage-resilient zero-knowledge protocols were constructed only under a relaxed security definition (Garg-Jain-Sahai, CRYPTO'11) or under the DDH assumption (Pandey, TCC'14).

Our leakage-resilient zero-knowledge argument system satisfies an additional property that it is simultaneously leakage-resilient zero-knowledge, meaning that both zero-knowledgeness and soundness hold in the presence of leakage.

1 Introduction

Zero-knowledge (ZK) *proofs* and *arguments* [14] are interactive proof/argument systems with which the prover can convince the verifier of the correctness of a mathematical statement while providing *zero additional knowledge*. This "zero additional knowledge" property is formalized thorough the *simulation paradigm*. Specifically, an interactive proof or argument is said to be zero-knowledge if for any adversarial verifier there exists a *simulator* that can output a simulated view of the adversary.

Recently, Garg et al. [12] introduced a new notion of zero-knowledgeness called *leakage-resilient zero-knowledge* (LRZK). Roughly speaking, LRZK is a notion of zero-knowledgeness in the setting where adversarial verifiers can obtain arbitrary leakage on the entire state of the honest prover (including the witness and the randomness) during the entire protocol execution. LRZK is motivated by the studies of *side-channel attacks* (e.g., [2,18,27]), which demonstrated that adversaries might be able to obtain leakage of honest parties' secret states by attacking physical implementations of cryptographic algorithms.

Informally speaking, LRZK requires that the protocol does not reveal anything beyond the validity of the statement *and the leakage that the adversary obtained*. More formally, LRZK is defined as follows. In the definition of LRZK,

© International Association for Cryptologic Research 2016
M. Fischlin and J.-S. Coron (Eds.): EUROCRYPT 2016, Part II, LNCS 9666, pp. 93–123, 2016.
DOI: 10.1007/978-3-662-49896-5_4

the cheating verifier is allowed to make arbitrary number of *leakage queries* during the interaction with a honest prover, where each leakage query f is answered by $f(w, \text{tape})$ for the witness w and the randomness tape that the honest prover generated thus far. On the other hand, the simulator is allowed to make queries to the *leakage oracle* $\mathcal{L}_w$, which is parametrized by the witness w of the honest prover and outputs $f(w)$ on input any function f. LRZK is then defined by requiring that for any cheating verifier V^* there exists a simulator $\mathcal{S}$ such that for any $\ell \in \mathbb{N}$, when V^* obtains ℓ bits of leakage of the prover's state via leakage queries, $\mathcal{S}$ can simulate the view of V^* by obtaining ℓ bits of leakage of the witness via queries to the leakage oracle $\mathcal{L}_w$.[1]

In [12], Garg et al. showed a proof system that satisfies a weaker notion of LRZK called $(1 + \epsilon)$-LRZK. Specifically, they showed that for any $\epsilon > 0$, there exists a proof system such that when V^* obtains ℓ bits of leakage from the prover, a simulator can simulate the verifier's view by obtaining at most $(1 + \epsilon) \cdot \ell$ bits of leakage from $\mathcal{L}_w$. The round complexity of this protocol is at least $\omega(\log n)/\epsilon$, and its security is proven under a standard general assumption (the existence of statistically hiding commitment scheme that is public-coin w.r.t. the receiver).

A natural question left open by [12] is whether we can construct a LRZK protocol without weakening the security requirement. That is, the question is whether we can reduce ϵ to 0 in the protocol of [12]. This question is particularly of theoretical interest because reducing ϵ to 0 is optimal in the sense that λ-LRZK for $\lambda < 0$ is impossible to achieve in the plain model [12].

Recently, this question was solved affirmatively by Pandey [23], who constructed the first LRZK argument system by using the DDH assumption and collision-resistant hash functions. Pandey's protocol has only constant number of rounds; therefore, it follows that asymptotically optimal round complexity can be achievable even in the presence of leakage.

A question that is explicitly left open by Pandey [23, Section 1] is whether we can construct LRZK protocols under a standard *general* assumption. In fact, although the protocol of Pandey [23] is superior to the protocol of Garg et al. [12] in terms of both leakage resilience (LRZK v.s. $(1 + \epsilon)$-LRZK) and round complexity (constant v.s. $\omega(\log n)/\epsilon$), the assumption of the former is seemingly much stronger than that of the latter (the DDH assumption v.s. the existence of statistically hiding commitment scheme that is public-coin w.r.t. the receiver, which is implied by, say, the existence of collision-resistant hash function family or even the existence of one-way functions[2]).

Question. *Can we construct a (constant-round) leakage-resilient zero-knowledge protocol under standard general assumptions?*

---

[1] In [22], it is pointed out that nowadays *leakage tolerance* is the commonly accepted term for this security notion. In this paper, however, we use the term "leakage resilience" for this security notion for consistency with previous works [12,23].

[2] A constant-round one can be constructed from collision-resistant hash functions [10,21] and a polynomial-round one can be constructed from one-way functions [15].

1.1 Our Results

In this paper, we answer the above question affirmatively by constructing a LRZK protocol from collision-resistant hash functions (CRHFs). Like the protocol of [23], our protocol has only constant number of rounds. Also, our protocol has an additional property that it is public coin (w.r.t. the verifier).

Theorem. *Assume the existence of collision-resistant hash function family. Then, there exists a constant-round public-coin leakage-resilient zero-knowledge argument for* $\mathcal{NP}$.

Simultaneously Leakage-Resilient Zero-Knowledge. Our protocol has an additional property that it is *simultaneously leakage-resilient zero-knowledge* [12], meaning that not only zero-knowledgeness but also soundness holds in the presence of leakage. The *leakage-resilient (LR) soundness* (i.e., soundness in the presence of leakage) of our protocol follows immediately from its public-coin property. In fact, any public-coin interactive proof/argument system is LR sound for arbitrary amount of leakage of the verifier because the verifier has no secret state in public-coin protocols.

To the best of our knowledge, our protocol is the first simultaneously LRZK protocol. The $(1 + \epsilon)$-LRZK protocol of Garg et al. [12] is LR sound in a weak sense—it is LR sound when there is an a-priori upper bound on the amount of leakage—but is not LR sound when the amount of leakage is unbounded,[3] and similarly, the LRZK protocol of Pandey [23] is also not LR sound with unbounded amount of leakage. In contrast, our protocol is sound even when cheating verifiers obtain arbitrary amount of leakage.

The summary of the previous results and ours is given in Table 1. In the table, "bounded-LR sound" means that the soundness holds when there is an a-priori upper bound on the amount of leakage from the verifier.

Table 1. Summary of the results on LRZK protocols. The round complexity of the protocol of [12] depends on the assumption that is used to instantiate the underlying statistically-hiding commitment scheme; in particular, when only one-way functions (OWFs) are used, there is a polynomial additive overhead because statistically hiding commitment schemes currently require polynomial number of rounds in this case [15].

	LR ZKness	LR soundness	#(round)	Assumptions
[12]	$(1 + \epsilon)$-LRZK	Bounded-LR sound	$\mathsf{poly}(n) + \omega(\log n)/\epsilon$	OWFs
			$\omega(\log n)/\epsilon$	CRHFs
[23]	LRZK	-	$O(1)$	DDH + CRHFs
This work	LRZK	LR sound	$O(1)$	CRHFs

---

[3] This is because in the protocol of [12], the verifier commits to the challenge bits of Blum's Hamiltonicity protocol in advance and hence an cheating prover can easily break the soundness by obtaining the challenge bits via leakage.

1.2   Related Works

Several works study interactive protocols in the presence of arbitrary leakage in the models other than the plain model, e.g., the work about leakage-tolerant UC-secure protocols in the CRS model [5], the work about non-transferable interactive proof systems in the CRS model with leak-free input encoding/updating phase [1], and the works about secure computation protocols in the CRS model with leak-free preprocessing/input-encoding phase and constant fraction of honest parties [6–8]. We remind the readers that, like [12,23], this work considers LRZK protocols in the plain model without any leak-free phase.

In [22], Ostrovsky et al. showed an impossibility result about black-box LRZK in the model with only leak-free input-encoding phase (i.e., without CRS and preprocessing). We notice that this impossibility result does not contradict our result since the definition of LRZK in [22] is different from the one we use (i.e., the definition given by [12]). Specifically, in the definition of [22], the simulator is not allowed to obtain any leakage, whereas in the definition that we use, the simulator can obtain the same amount of leakage as the cheating verifier. (In other words, Ostrovsky et al. [22] considers leakage resilience whereas we consider leakage tolerance; see Footnote 1.)

2   Overview of Our Techniques

2.1   Previous Techniques

Since our techniques rely on the techniques that are used in the previous LRZK protocols of [12,23], we start by recalling these protocols.

Protocol of [12]. In [12], Garg et al. constructed a $(1 + \epsilon)$-leakage-resilient zero-knowledge proof system, i.e., a proof system such that when V^* obtains ℓ bits of leakage from the prover, its view can be simulated by obtaining at most $(1 + \epsilon) \cdot \ell$ bits of leakage from $\mathcal{L}_w$.

A key idea behind the protocol of [12] is to give the simulator two independent ways of cheating—one for simulating prover's messages and the other for simulating leakages. Concretely, Garg et al. constructed their protocol by combining two well-known techniques of constant-round zero-knowledge protocols—the technique by Goldreich and Kahan [13] that requires the verifier to commit to its challenges in advance and the technique by Feige and Shamir [11] that uses equivocal commitment schemes. They then proved the security by considering a simulator that simulates the prover's messages by extracting the challenges and simulates the leakages by using the equivocality of the commitment scheme.

In more details, the protocol of [12] consists of the following two phases. In the first phase, the verifier uses an extractable commitment scheme to commit to a challenge string ch of Blum's Hamiltonicity protocol and trapdoor information td of an equivocal commitment scheme.[4] In the second phase, the prover and

---

4 Actually, there is a coin-tossing protocol that determines the parameter of the equivocal commitment, and td is the trapdoor for biasing the outcome of the coin-tossing.

the verifier execute Blum's Hamiltonicity protocol that is instantiated with the equivocal commitment scheme. In simulation, the simulator extracts ch and td in the first phase and then simulates the prover's messages and the leakages in the second phase by using the knowledge of ch and td in the following way. (For simplicity, we assume that Blum's protocol is executed only once instead of many times in parallel.)

When the extracted challenge ch is 0, the simulator commits to a randomly permuted graph of statement G, and after V^* decommits the challenge ch (which must be 0), the simulator decommits the commitment to the permuted graph of G.

    Notice that the simulator does exactly the same things as a honest prover. Hence, the simulator can simulate prover's randomness tape easily and therefore can answer any leakage query f from V^* by querying $f(\cdot, \mathsf{tape})$ to $\mathcal{L}_w$.

When the extracted challenge ch is 1, the simulator commits to a randomly chosen cycle graph H at the beginning and then partially decommits it in the last step so that only the edges on the cycle are revealed.

    When V^* makes a leakage query, the simulator answers it by using w and td to compute randomness that "explains" the commitment to H as a commitment to a permuted graph of G. (Recall that the prover is supposed to commit to a permuted graph of G.) Specifically, the simulator answers a leakage query f from V^* by querying the following function $\tilde{f}(\cdot)$ to $\mathcal{L}_w$.

1. On input w, function $\tilde{f}$ first computes a permutation π that maps the Hamiltonian cycle w in G to the cycle in H (i.e., computes π such that $\pi(G)$ has the same cycle as H).

2. Then, by using equivocality[5] with trapdoor td, it computes randomness tape that explains the commitment to H as a commitment to $\pi(G)$ (i.e., it computes tape such that committing to $\pi(G)$ with randomness tape will generate the same commitment as the one that the simulator has sent to V^* by committing to H).

3. Finally, it outputs $f(w, \mathsf{tape})$.

Notice that since $\pi(G)$ has the same cycle as H, the simulated leakages (from which V^* may be able to compute $\pi(G)$) are consistent with the cycle of H that is decommitted by the simulator in the last step.

    We remark that the reason why the protocol of [12] satisfies only $(1 + \epsilon)$-LRZK (rather than standard LRZK) is that the extraction of ch and td involves the rewinding of V^*. In fact, since V^* can make new leakage queries after being rewound, the simulator need to obtain new leakages from $\mathcal{L}_w$ in each rewinding and hence the simulator need to obtain more bits of leakage than V^*.

Protocol of [23]. In [23], Pandey constructed a constant-round LRZK argument system under the DDH assumption. Roughly speaking, Pandey's idea is

---

[5] What is actually used here is *adaptive security*, which guarantees that for each underlying commitment, it is possible to compute randomness tape_0 and tape_1 such that tape_b explains the commitment as a commitment to b for each $b \in \{0, 1\}$.

to replace the rewinding simulation technique in the protocol of [12] with the "straight-line" simulation technique of Barak [3]. In particular, Pandey replaced the first phase of the protocol of [12] with the following one.

1. First, the prover and the verifier execute an encrypted version of so called *Barak's preamble* [3,24,25], which determines a "fake statement" that is false except with negligible probability.
2. Next, the prover and the verifier execute Yao's garbled circuit protocol [28] in which the prover can obtain ch and td only when it has a valid witness for the fake statement.

From the security of the encrypted Barak's preamble, no cheating prover can make the fake statement true; hence, ch and td are hidden from the cheating prover. In contrast, a non-black-box simulator can make the fake statement true by using the knowledge of the code of the verifier; hence, the simulator can obtain ch and td without rewinding V^*. An issue is that, to guarantee leakage resilience, it is required that Yao's protocol is executed in a way that all prover's messages are pseudorandom (since otherwise it is hard to simulate randomness that explains the simulated prover's messages as honest prover's messages during the simulation of the leakages). Since Yao's protocol involves executions of an oblivious transfer protocol (in which the prover behaves as a receiver), this property is hard to satisfy. Pandey solved this problem by using the DDH assumption, under which there exists an oblivious transfer protocol such that all receiver's messages are indistinguishable from random group elements.

2.2 Our Techniques

The reason why the protocols of [12,23] either guarantee only weaker security or rely on a stronger assumption is that the simulation involves extraction from V^*. In fact, in [12], the simulator need to obtain more amount of leakage than V^* because it rewinds V^* during extraction, and in [23], the DDH assumption is required because Yao's protocol is used for extraction.

Based on this observation, our strategy is to modify the protocols of [12,23] so that no extraction is required in simulation. We first remove the extraction of trapdoor td and next remove the extraction of challenge ch.

Removing Extraction of Trapdoor td. We first modify the protocols of [12,23] so that leakages can be simulated without extracting the trapdoor td of an equivocal commitment scheme.

Our main tool is Hamiltonicity commitment scheme H-Com [9,11], which is a well-known instance-dependent equivocal commitment scheme based on Blum's Hamiltonicity protocol. H-Com is parametrized by a graph G with $q = \mathsf{poly}(n)$ vertices. To commit to 0, the committer chooses a random permutation π and commits to the adjacent matrix of $\pi(G)$ using any commitment scheme Com; to decommit, the committer reveals π and decommits all the entries of the matrix. To commit to 1, the committer commits to the adjacent matrix of a random

q-cycle graph; to decommit, the committer decommits only the entries that corresponds to the edges on the cycle. H-Com satisfies equivocality when G has a Hamiltonian cycle; this is because after committing to 0, the committer can decommit it to both 0 and 1 given a Hamiltonian cycle w in G.

Given H-Com, we remove the extraction of td by combining H-Com with an encrypted variant of Barak's preamble. Specifically, we replace the equivocal commitment scheme in the protocols of [12,23] with H-Com that depends on the fake statement G' that is obtained by the encrypted Barak's preamble. From the security of Barak's preamble, any cheating prover cannot make G' true and hence it cannot use the equivocality of H-Com, whereas the simulator can make G' true and hence it can use the equivocality of H-Com as desired.

Remark 1. As observed in [23], it is not straightforward to use the encrypted Barak's preamble in the presence of leakage. Roughly speaking, in the encrypted Barak's preamble, the prover commits to its messages instead of sending them in clear, and in the proof of soundness, it is required that the prover's messages are extractable from the commitments. The problem is that it is not easy to guarantee this extractability in the presence of leakage (this is because the prover's messages are typically not pseudorandom in the techniques of extractability). Pandey [23] solved this problem by having the prover use a specific extractable commitment scheme based on the DDH assumption. In this paper, we instead have the prover use a commitment scheme that satisfies only very weak extractability but the prover's messages of which are pseudorandom and the security of which is based on the existence of CRHFs.[6] For details, see Sect. 4.1.

Removing Extraction of Challenge *ch.* Next, we modify the protocols of [12,23] so that prover's messages can be simulated without extracting the challenge *ch* of Hamiltonicity protocol.

We first notice that although the simulator can use equivocality without extraction as shown above, it is not easy for the simulator to use equivocality for simulating prover's messages. This is because when the leakages to V^* includes the randomness that is used for commitments, V^* may be able to determine the committed values from the leakages and therefore equivocation may be detected by V^*.

As our main technical tool, then, we introduce a specific instance-dependent equivocal commitment scheme GJS-Com that we obtain by considering the technique of [12] on Hamiltonicity protocol in the context of H-Com. Recall that, as explained in Sect. 2.1, in [12] Garg et al. use Blum's Hamiltonicity protocol that is instantiated with an equivocal commitment scheme. Here, we use H-Com that is instantiated with an equivocal commitment scheme (i.e., we use H-Com in which the adjacent matrix is committed to by an equivocal commitment scheme). The equivocal commitment scheme that we use here is, as above, H-Com that depends on the fake statement generated by the encrypted Barak's preamble.[7]

---

[6] This extractability is used only in the proof of soundness. Hence, the proof of zero-knowledgeness works even in the presence of this extractable commitment scheme.

[7] Actually, we use an adaptively secure H-Com [9,19]. See Footnote 5.

Hence, the commitment scheme GJS-Com is a version of H-Com that is instantiated by using H-Com itself as the underling commitment scheme.[8] GJS-Com depends on two statements of the Hamiltonicity problem: The "outer" H-Com (the H-Com that is implemented with H-Com) depends on the real statement G, and the "inner" H-Com (the H-Com that is used to implement H-Com) depends on the fake statement G'. GJS-Com inherits equivocality from the outer H-Com, i.e., given a witness for the real statement G, a GJS-Com commitment to 0 can be decommitted to both 0 and 1.

Since GJS-Com is obtained by considering the technique of [12] in the context of H-Com, it satisfies a property that is useful for proving LRZK property. First, observe that given GJS-Com, the second phase of the LRZK protocol of [12] (i.e., Hamiltonicity protocol phase) can be viewed as follows.

1. The prover commits to 0 by using GJS-Com.
2. The verifier reveals the challenge $ch \in \{0, 1\}$ that is committed to in the first phase.
3. When $ch = 0$, the prover decommits the GJS-Com commitment to 0 honestly, and when $ch = 1$, the prover decommits it to 1 by using the equivocality with the knowledge of Hamiltonian cycle w in G.

When the second phase of the protocol of [12] is viewed in this way, the key property that is used in the simulation of the leakages in [12] is the following.

– Given a Hamiltonian cycles in G and G', a GJS-Com commitment to 1 (in which a random cycle graph is committed) can be "explained" as a commitment to 0 (in which a permutation of G is committed) by using the equivocality of the inner H-Com.

   Furthermore, even after being explained as a commitment to 0, the commitment can later be decommitted to 1 in a consistent way with the explained randomness (cf. function $\tilde{f}$ in Sect. 2.1).

Because of this property, even when the simulator commits to 1 instead of 0 using GJS-Com to simulate the messages, the simulator can answer any leakage query f from V^* by querying $\mathcal{L}_w$ a function $\tilde{f}$ that, on input w, computes randomness tape that explains the commitment to 1 as a commitment to 0 and then outputs $f(w, \text{tape})$.

   A problem of this property is that it can be used only in a very limited situation. Specifically, this property can be used only when the simulator knows which GJS-Com commitment will be decommitted to 1, and this is the reason why the extraction of ch is required in the simulation strategy of [12,23]. Hence, to remove the extraction of ch, we need to use GJS-Com in a way that, given a witness for the fake statement, the simulator can predict which value each GJS-Com commitment will be decommitted to.

   Our key observation is that we can use this property if we use GJS-Com to implement the Hamiltonicity protocol *in which the fake statement is proven*. Concretely, we consider the following protocol.

---

[8] In the "inner" H-Com, the underlying commitment scheme is Com as before.

1. The prover and the verifier execute an encrypted variant of Barak's preamble. Let G' be the fake statement and let q' be the number of the nodes of G'.
2. (a) The prover commits to a $q' \times q'$ zero matrix by using GJS-Com.
   (b) The verifier sends a challenge $ch \in \{0, 1\}$.
   (c) When $ch = 0$, the prover sends a random permutation π over G' to the verifier and then decommit the GJS-Com commitments to the adjacent matrix of $\pi(G')$ by using the equivocality of GJS-Com with the knowledge of a witness for the real statement.

   When $ch = 1$, the prover chooses a random q'-cycle graph H and decommits some of the GJS-Com commitments to 1 by using the equivocality of GJS-Com so that the decommitted entries of the matrix correspond to the cycle in H.
   (d) When $ch = 0$, the verifier verifies whether the decommitted graph is $\pi(G')$. When $ch = 1$, the verifier verifies whether the decommitted entries corresponds to a q'-cycle in a graph.

Since any charting prover cannot make the fake statement G' true, GJS-Com is statistically binding when the real statement G is false, and hence soundness follows. In contrast, the simulator can cheat in Barak's preamble so that it knows a Hamiltonian cycle w' in the fake statement G', and therefore it can simulate the prover's messages by "honestly" proving the fake statement, i.e., by committing to $\pi(G')$ in step 2(a) for a randomly chosen π and then revealing the entire graph $\pi(G')$ or only the cycle $\pi(w')$ depending on the value of ch. Furthermore, since in step 2(a) the simulator do know which value each GJS-Com commitment will be decommitted to (the commitments to the edges on $\pi(w')$ will be always decommitted to 1 and others will be decommitted honestly or will not be decommitted), the simulator can simulate the leakage in the same way as in the protocol of [12] by using the property of GJS-Com described above.

This completes the overview of our techniques. The details are given in what follows.

3 Preliminaries

3.1 Notations

We use n to denote the security parameter. For any $k \in \mathbb{N}$, we use $[k]$ to denote the set $\{1, \ldots, k\}$. For any randomized algorithm Algo, we use $\mathsf{Algo}(x; r)$ to denote the execution of Algo with input x and randomness r, and we use $\mathsf{Algo}(x)$ to denote the execution of Algo with input x and uniform randomness.

We use $\mathbf{L}_{\mathrm{HC}}$ to denote the languages of the Hamiltonian graphs. For any $G \in \mathbf{L}_{\mathrm{HC}}$, we use $\mathbf{R}_{\mathrm{HC}}(G)$ to denote the set of the Hamiltonian cycles in G. Generally, for any language $\mathbf{L}$ and any instance $x \in \mathbf{L}$, we use $\mathbf{R}_{\mathbf{L}}(x)$ to denote the set of the witnesses for $x \in \mathbf{L}$.

For any two-party protocol $\langle A, B \rangle$, we use $\mathsf{trans}\,[A(x) \leftrightarrow B(y)]$ to denote a random variable representing the transcript of the interaction between A and B with input x and y respectively, and use $\mathsf{output}_A\,[A(x) \leftrightarrow B(y)]$ (resp.,

$\text{output}_B [A(x) \leftrightarrow B(y)])$ to denote a random variable representing the output of A (resp., B) in the interaction between A and B with input x and y respectively.

3.2 Leakage-Resilient Zero-Knowledge

We recall the definition of leakage-resilient zero-knowledgeness [12]. For convenience, we use a slightly different formulation of the definition.

For any interactive proof system $\langle P, V \rangle$, any PPT cheating receiver V^*, any statement $x \in \mathbf{L}$, any witness $w \in \mathbf{R_L}(x)$, and any oracle machine $\mathcal{S}$ called *simulator*, consider the following two experiments.

$\underline{\text{REAL}_{V^*}(x, w, z)}$

1. Execute $V^*(x, z)$ with a honest prover $P(x, w)$ of $\langle P, V \rangle$.
   During the interaction, V^* can make arbitrary number of adaptive leakage queries on the state of P. A leakage query consists of an efficiently compatible function f_i (described as a circuit) and it is answered with $f_i(w, \mathsf{tape})$, where tape is the randomness used by P so far.
2. Output the view of V^*.

$\underline{\text{IDEAL}_{\mathcal{S}}(x, w, z)}$

1. Execute $\mathcal{S}(x, z)$ with access to a leakage oracle $\mathcal{L}_w$. A query to $\mathcal{L}_w$ consists of an efficiently computable function f and answered with $f(w)$. Let τ be the output of $\mathcal{S}$.
2. If τ is not a valid view of V^*, the output of the experiment is $\bot$. Otherwise, let ℓ be the total length of the leakage that V^* obtains in τ. If the total length of the answers that $\mathcal{S}$ obtained from $\mathcal{L}_w$ is larger than ℓ, the output of the experiment is $\bot$. Otherwise, the output is τ.

Let $\text{REAL}_{V^*}(x, w, z)$ be the random variable representing the output of $\text{REAL}_{V^*}(x, w, z)$ and $\text{IDEAL}_{\mathcal{S}}(x, w, z)$ be the random variable representing the output of $\text{IDEAL}_{\mathcal{S}}(x, w, z)$.

Definition 1. *An interactive argument system $\langle P, V \rangle$ for a language $\mathbf{L}$ with witness relation $\mathbf{R}$ is **leakage-resilient zero knowledge** if for every PPT machine V^* and every sequence $\{w_x\}_{x \in L}$ such that $(x, w_x) \in \mathbf{R_L}$, there exists a PPT oracle machine $\mathcal{S}$ such that the following hold.*

Indistinguishability Condition

$$\{\text{REAL}_{V^*}(x, w_x, z)\}_{x \in L, z \in \{0,1\}^*} \approx \{\text{IDEAL}_{\mathcal{S}}(x, w_x, z)\}_{x \in L, z \in \{0,1\}^*} .$$

Leakage-length condition. *For every $x \in \mathbf{L}$ and $z \in \{0, 1\}^*$,*

$$\Pr[\text{IDEAL}_{\mathcal{S}}(x, w_x, z) = \bot] = 0.$$

3.3 Commitment Scheme

Recall that commitment schemes are two-party protocols between a committer C and a receiver R. We say that a commitment is *valid* if there exists a value to which it can be decommitted. We denote by value($\cdot$) a function that, on input a commitment (i.e., a transcript in the commit phase), outputs its committed value if it is uniquely determined and outputs $\perp$ otherwise.

3.4 Naor's Commitment

We recall Naor's statistically binding commitment scheme Com, which can be constructed from one-way functions [16,20].

Commit Phase. The commit phase consists of two rounds. In the first round, the receiver sends a random $3n$-bit string $r \in \{0,1\}^{3n}$. In the second round, the committer chooses a random seed $s \in \{0,1\}^n$ for a pseudorandom generator PRG : $\{0,1\}^n \rightarrow \{0,1\}^{3n}$ and then sends PRG(s) if it wants to commit to 0 and sends PRG(s) $\oplus r$ if it wants to commit to 1.

We use Com$_r(\cdot)$ to denote an algorithm that, on input $b \in \{0,1\}$, computes a commitment to b as above by using r as the first-round message.

Decommit Phase. In the decommit phase, the committer reveals the seed s.

Security. Com is statistically binding and computational hiding. Furthermore, the binding and hiding property hold even when the same first-round message r is used in multiple commitments.

Committing to Strings. For any $\ell \in \mathbb{N}$, we can commit to an ℓ-bit string by simply committing to each bit using Com. We notice that the same first-round message r can be used in all the commitments.

We abuse the notation and use Com$_r(\cdot)$ to denote an algorithm that, on input $m \in \{0,1\}^*$, computes a commitment to m as above by using r as the first-round message. Notice that Com$_r(\cdot)$ has pseudorandom range. Thus, by using an algorithm Com$_{pub}$ that outputs a random $3n\ell$-bit string on input 1^ℓ, we can obtain a "fake commitment" that is indistinguishable from a real commitment.

3.5 Hamiltonicity Commitment

We recall a well-known instance-dependent commitment scheme H-Com [9,11] that is based on Blum's zero-knowledge proof for Hamiltonicity.

Commit Phase. H-Com is parametrized by a graph G. Let q be the number of its vertices. To commit to 0, the committer chooses a random permutation π over the vertices of G and then commits to the adjacent matrix of $\pi(G)$ by using Com. To commit to 1, the committer chooses a random q-cycle graph and then commits to its adjacent matrix by using Com.

We use H-Com$_{G,r}(\cdot)$ to denote an algorithm that, on input $b \in \{0,1\}$, computes a commitment to b as above by using r as the first-round message of all the Com commitments.

Decommit Phase. When the committer committed to 0, it reveals π, and also reveals all the entries of the adjacent matrix by decommitting all the Com commitments. When the committer committed to 1, it reveals only the entries corresponding to the edges on the q-cycle by decommitting the Com commitments in which these entries are committed.

Security. H-Com is computationally hiding, and it is statistically binding when $G \notin \mathbf{L}_{\mathrm{HC}}$.

Equivocality. When $G \in \mathbf{L}_{\mathrm{HC}}$, a commitment to 0 can be decommitted to 1 given a Hamiltonian cycle $w \in \mathbf{R}_{\mathrm{HC}}(G)$ in G. Specifically, a commitment to 0 can be decommitted to 1 by decommitting the entries that corresponds to the edges on $\pi(w)$ (i.e., the cycle that is obtained by applying π on w).

3.6   Adaptive Hamiltonicity Commitment

We recall the adaptively secure Hamiltonicity commitment scheme AH-Com, which was used in, e.g., [9,19].

Commit Phase. AH-Com is parametrized by a graph G. Let q be the number of its vertices. To commit to 0, the committer does the same things as in H-Com; i.e., it chooses a random permutation π over the vertices of G and then commits to the adjacent matrix of $\pi(G)$ by using Com. To commit to 1, the committer chooses a random q-cycle graph and then commits to its adjacent matrix in the following way: For all the entries corresponding to the edges on the q-cycle, it commits to 1 by using Com, and for all the other entries, it simply sends random $3n$-bit strings instead of committing to 0. (Since Com has pseudorandom range, random $3n$-bit strings are indistinguishable from Com commitments.)

We use $\mathsf{AH\text{-}Com}_{G,r}(\cdot)$ to denote an algorithm that, on input $b \in \{0,1\}$, computes a commitment to b as above by using r as the first-round message of all the Com commitments.

Decommit Phase. To decommit, the committer reveals all the randomness used in the commit phase. We use $\mathsf{AH\text{-}Dec}_r(\cdot, \cdot, \cdot)$ to denote an algorithm that, on input c, b, ρ such that $\mathsf{AH\text{-}Com}_r(b; \rho) = c$, outputs a decommitment d as above.

Security. Like H-Com, AH-Com is computationally hiding both when $G \in \mathbf{L}_{\mathrm{HC}}$ and when $G \notin \mathbf{L}_{\mathrm{HC}}$, and it is statistically binding when $G \notin \mathbf{L}_{\mathrm{HC}}$.

Adaptive Security. When $G \in \mathbf{L}_{\mathrm{HC}}$, a commitment to 0 can be "explained" as a valid commitment to 1 given a witness $w \in \mathbf{R}_{\mathrm{HC}}(G)$. Specifically, for a commitment c to 0, we can compute ρ such that $\mathsf{AH\text{-}Com}(1; \rho) = c$. This is because commitments to the entries that do not correspond to the edges on $\pi(w)$ are indistinguishable from random strings.

Formally, there exists an algorithm AH-ExplainAsOne such that for security parameter $n \in \mathbb{N}$, graphs $G \in \mathbf{L}_{\mathrm{HC}}$, witness $w \in \mathbf{R}_{\mathrm{HC}}(G)$, and string $r \in \{0,1\}^{3n}$, the following hold.

Correctness. Given witness $w \in \mathbf{R}_{HC}(G)$ and c, ρ such that $\text{AH-Com}_{G,r}(0; \rho) = c$, $\text{AH-ExplainAsOne}_{G,r}$ outputs ρ' such that $\text{AH-Com}_{G,r}(1; \rho') = c$.

Indistinguishability. Consider the following two probabilistic experiments.

$\underline{\text{EXP}_0^{AH}(n, G, w, r)}$

/* commit to 1 and reveal randomness */

    1. Computes $c \leftarrow \text{AH-Com}_{G,r}(1)$.

        Let ρ_1 be the randomness used in AH-Com.

    2. Output (c, ρ_1).

$\underline{\text{EXP}_1^{AH}(n, G, w, r)}$

/* commit to 0 and explain it as commitment to 1 */

    1. Computes $c \leftarrow \text{AH-Com}_{G,r}(0)$.

        Let ρ_0 be the randomness used in AH-Com.

        Compute $\rho_1 := \text{AH-ExplainAsOne}_{G,r}(w, c, \rho_0)$.

    2. Output (c, ρ_1).

Let $\text{EXP}_b^{AH}(n, G, w, r)$ be the random variable representing the output of $\text{EXP}_b^{AH}(n, G, w, r)$ for each $b \in \{0, 1\}$. Then, the following two ensembles are computationally indistinguishable.

$$\left\{ \text{EXP}_0^{AH}(n, G, w, r) \right\}_{n \in \mathbb{N}, G \in \mathbf{L}_{HC}, w \in \mathbf{R}_{HC}(G), r \in \{0,1\}^{3n}}$$

$$\left\{ \text{EXP}_1^{AH}(n, G, w, r) \right\}_{n \in \mathbb{N}, G \in \mathbf{L}_{HC}, w \in \mathbf{R}_{HC}(G), r \in \{0,1\}^{3n}}$$

3.7 Barak's Non-black-box Zero-Knowledge Protocols

As explained in Sect. 2, in our LRZK protocol, we use a variant of so called "encrypted" Barak's preamble [24,25], which is based on the preamble stage of Barak's non-black-box zero-knowledge protocol [3]. In this section, we recall Barak's non-black-box zero-knowledge protocol. Our variant of encrypted Barak's preamble is described in Sect. 4.1.

Barak's non-black-box zero-knowledge protocol is constructed from any collision-resilient hash function family $\mathcal{H}$. Informally speaking, Barak's protocol BarakZK proceeds as follows.

Protocol BarakZK

1. The verifier V sends a random hash function $h \in \mathcal{H}$ and the first-round message $r_1 \in \{0, 1\}^{3n}$ of Com to the prover P.
2. P sends $c \leftarrow \text{Com}_{r_1}(0^n)$ to V. Then, V sends random string r_2 to P.
3. P proves the following statement by a witness-indistinguishable argument.
   - $x \in L$, or
   - $(h, c, r_2) \in \Lambda$, where $(h, c, r_2) \in \Lambda$ holds if and only if there exists a machine Π such that c is a commitment to $h(\Pi)$ and Π outputs r_2 in $n^{\log \log n}$ steps.

Note that the statement proven in the last step is not in $\mathcal{NP}$. Thus, P proves this statement by a witness-indistinguishable *universal argument* (WIUA), with

---

Stage 1:
  The verifier V_B sends a random hash function $h \in \mathcal{H}$ to the prover P_B, where the domain of h is $\{0,1\}^*$ and the range of h is $\{0,1\}^n$. V_B also sends $r_1 \in \{0,1\}^{3n}$ (the first-round message of Com) to P_B.

Stage 2:
  1. P_B computes $c \leftarrow \mathsf{Com}_{r_1}(0^n)$ and send c to V_B.
  2. V_B sends random $r_2 \in \{0,1\}^{n+n^2}$ to P_B.

Stage 3: P_B proves statement $(h, r_1, c, r_2) \in \Lambda$ by using UA.
  1. V_B sends the first-round message α.
  2. P_B sends the second-round message β.
  3. V_B sends the third-round message γ.
  4. P_B sends the fourth-round message δ.

....

Language Λ:
  $(h, r_1, c, r_2) \in \Lambda$ if and only if there exist
    − a machine Π
    − randomness rand for Com
    − a string y such that $|y| \leq n^2$
  such that
    − $c = \mathsf{Com}_{r_1}(h(\Pi); \mathsf{rand})$, and
    − $\Pi(c, y)$ outputs r_2 within $n^{\log \log n}$ steps.

Fig. 1. Encrypted Barak's preamble $\langle P_B, V_B \rangle$.

which P can prove any statement in $\mathcal{NEXP}$. Intuitively, BarakZK is sound since $\Pi(c) \neq r$ holds with overwhelming probability even when a cheating prover P^* commits to $h(\Pi)$ for a machine Π. On the other hand, the zero-knowledge property can be proven by using a simulator that commits to $h(\Pi)$ such that Π is a machine that emulates the cheating verifier V^*; since $\Pi(c) = V^*(c) = r$ holds from the definition, the simulator can give a valid proof in the last step.

For our purpose, it is convenient to consider a variant of BarakZK that we denote by $\langle P_B, V_B \rangle$. $\langle P_B, V_B \rangle$ is the same as BarakZK except that in the last step, instead of proving $x \in L \lor (h, c, r_2) \in \Lambda$ by using WIUA, P proves $(h, c, r_2) \in \Lambda$ by using four-round public-coin universal argument system UA [4]. (Hence, $\langle P_B, V_B \rangle$ is no longer zero-knowledge protocol.) The formal description of $\langle P_B, V_B \rangle$ is shown in Fig. 1. We remark that in $\langle P_B, V_B \rangle$, the language proven in the last step is replaced with a slightly more complex language as in, e.g., [3,23–25]. This replacement is important for using $\langle P_B, V_B \rangle$ in the setting of leakage-resilient zero-knowledge, because the cheating verifier can obtain arbitrary information (i.e., leakage) before sending r_2.

In essentially the same way as the soundness of BarakZK, we can prove the following lemma on $\langle P_B, V_B \rangle$, which roughly states that there exists a "hard" language $\mathbf{L}_B$ on the transcript of $\langle P_B, V_B \rangle$ such that no cheating prover can generate a transcript that is included in $\mathbf{L}_B$.

Language L_B:
$\tau = (h, r_1, c, r_2, \alpha, \beta, \gamma, \delta) \in L_B$ if and only if $(\alpha, \beta, \gamma, \delta)$ is an accepting transcript of UA for statement $(h, r_1, c, r_2) \in \Lambda$.

Fig. 2. A "hard" language L_B.

Lemma 1 (Soundness). *Let L_B be the language defined in Fig. 2. Then, for any cheating prover P^* against $\langle P_B, V_B \rangle$, any $n \in \mathbb{N}$, and any $z \in \{0,1\}^*$,*

$$\Pr\left[\tau \leftarrow \mathsf{trans}\left[P^*(1^n, z) \leftrightarrow V_B(1^n)\right] : \tau \in L_B\right] \leq \mathsf{negl}(n).$$

A proof sketch of this lemma is given in the full version of this paper [17].

3.8 Somewhat Extractable Commitment Scheme

As we mentioned in Remark 1 in Sect. 2.2, in our variant of encrypted Barak's preamble, we use a commitment scheme that satisfies only very weak extractability, which we call *somewhat extractability*. An important point is that since only very weak extractability is required, we can construct a somewhat extractable commitment scheme such that the committer sends only pseudorandom messages. Furthermore, we can construct such a scheme from one-way functions.

Concretely, we consider the commitment scheme SWExtCom in Fig. 3. SWExtCom is the same as the extractable commitment scheme of [26] except that in the last step, the committer simply reveals the values that it committed to in the first step (instead of decommitting the commitments). Because of this simplification, SWExtCom does not satisfy extractability in the standard sense. Still, it is not hard to see that SWExtCom satisfies extractability in the sense that, given two valid commitments c and c' such that the transcripts of the commit stage are identical but those of the challenge stage are different, the committed value of c can be extracted. Formally, SWExtCom satisfies the following extractability.

Lemma 2 (Somewhat Extractability). *Let us say that two commitments $c = (\{c_{i,b}\}_{i \in [n], b \in \{0,1\}}, \{e_i\}_{i \in [n]}, \{a_{i,e_i}\}_{i \in [n]})$ and $c' = (\{c'_{i,b}\}_{i \in [n], b \in \{0,1\}}, \{e'_i\}_{i \in [n]}, \{a'_{i,e_i}\}_{i \in [n]})$ are **admissible** if*

- *$c_{i,b} = c'_{i,b}$ for every $i \in [n]$ and $b \in \{0,1\}$,*
- *there exists $i^* \in [n]$ such that $e_{i^*} \neq e'_{i^*}$, and*
- *the committed value of $c_{i,b}$ is uniquely determined for every $i \in [n]$ and $b \in \{0,1\}$.*

Let $\mathsf{Extract}(\cdot, \cdot)$ be the algorithm shown in Fig. 3. Then, for any two admissible commitments c and c', if both c and c' are valid, $\widetilde{v} \overset{\text{def}}{=} \mathsf{Extract}(c, c')$ is equal to $\mathsf{value}(c)$ (i.e., $\widetilde{v}$ is the committed value of c).

Commit phase. The committer C and the receiver R receive common inputs 1^n. To commit to $v \in \{0,1\}^n$, the committer C does the following with the receiver R.

Commit stage. For each $i \in [n]$, the committer C chooses a pair of random n-bit strings $(a_{i,0}, a_{i,1})$ such that $a_{i,0} \oplus a_{i,1} = v$. Then, for each $i \in [n]$ in parallel, C commits to $a_{i,0}$ and $a_{i,1}$ by using Com. For each $i \in [n]$ and $b \in \{0,1\}$, let $c_{i,b}$ be the commitment to $a_{i,b}$.

Challenge stage. R sends random n-bit string $e = (e_1, \ldots, e_n)$ to C.

Reply stage. For each $i \in [n]$, C sends a_{i,e_i} to R.

Decommit phase. C sends v to R and decommits $c_{i,b}$ to $a_{i,b}$ for all $i \in [n]$ and $b \in \{0,1\}$. R checks whether $a_{1,0} \oplus a_{1,1} = \cdots = a_{n,0} \oplus a_{n,1} = v$ holds and whether $a_{1,e_1}, \ldots, a_{n,e_n}$ are equal to the values that were revealed in the commit phase.

....

Extracting algorithm Extract.

On input two commitments $c = (\{c_{i,b}\}_{i \in [n], b \in \{0,1\}}, \{e_i\}_{i \in [n]}, \{a_{i,e_i}\}_{i \in [n]})$ and $c' = (\{c'_{i,b}\}_{i \in [n], b \in \{0,1\}}, \{e'_i\}_{i \in [n]}, \{a'_{i,e_i}\}_{i \in [n]})$ such that $c_{i,b} = c'_{i,b}$ for every $i \in [n]$ and $b \in \{0,1\}$, do the following.

1. Find any $i \in [n]$ such that $e_i \neq e'_i$. If no such i exist, output fail.
2. Output $\widetilde{v} \overset{\text{def}}{=} a_{i,e_i} \oplus a'_{i,e'_i}$.

Fig. 3. A somewhat extractable commitment scheme SWExtCom.

Proof. First, when c and c' are valid, $a_{i^*,e_{i^*}}$ and $a'_{i^*,e'_{i^*}}$ are the committed values of $c_{i^*,e_{i^*}}$ and $c_{i^*,e'_{i^*}}$ (since otherwise, any decommitments of c and c' would be rejected because the decommitted values of $c_{i^*,e_{i^*}}$ and $c_{i^*,e'_{i^*}}$ are not consistent with $a_{i^*,e_{i^*}}$ and $a'_{i^*,e'_{i^*}}$). Second, when c and c' are valid, the committed value of c can be computed by XORing the committed values of $c_{i^*,e_{i^*}}$ and $c_{i^*,e'_{i^*}}$ (since otherwise, any decommitments of c and c' would be rejected). From these, the lemma follows. $\square$

A nice property of SWExtCom is that all the messages that the committer sends in the commit phase are pseudorandom. Formally, we have the following lemma.

Lemma 3 (Existence of Public-Coin Fake Committing Algorithm).
Let C be a honest committer algorithm of SWExtCom. There exists a PPT public-coin algorithm C_{pub} such that for any PPT cheating receiver R^* that interacts with C in the commit phase of SWExtCom, the following ensembles are computationally indistinguishable.

- $\{\text{output}_{R^*}[C(v) \leftrightarrow R^*(1^n, z)]\}_{n \in \mathbb{N}, v \in \{0,1\}^n, z \in \{0,1\}^*}$
- $\{\text{output}_{R^*}[C_{\text{pub}}(1^n) \leftrightarrow R^*(1^n, z)]\}_{n \in \mathbb{N}, v \in \{0,1\}^n, z \in \{0,1\}^*}$

Proof (sketch). C_{pub} is an algorithm that is the same as C except that, instead of sending commitments of Com, it sends fake commitments of Com using Com_{pub} (i.e., sends random strings with the same length as the Com commitments). Since Com has pseudorandom range, the indistinguishability can be proven by using a standard hybrid argument (in which the commitments of Com are replaced with random strings one by one). The formal proof is omitted. □

4 Building Blocks

4.1 Special-Purpose Encrypted Barak's Preamble

In our LRZK protocol, we use a variant of so called "encrypted" Barak's preamble [24,25]. The encrypted Barak's preamble is the same as (a variant of) Barak's non-black-box zero-knowledge protocol $\langle P_{\text{B}}, V_{\text{B}} \rangle$ in Sect. 3.7 except that P_{B} commits to its UA messages β and δ instead of sending them in clear. In this paper, we use a variant in which, instead of giving valid commitments, P_{B} gives fake commitments of Com and SWExtCom by using Com_{pub} and C_{pub}. A nice property of this variant is that the prover sends only random strings; as will become clear later, this property is useful for constructing leakage-resilient protocols. The formal description of this variant, which we denote by $\langle \mathbb{P}_{\text{B}}, \mathbb{V}_{\text{B}} \rangle$, is shown in Fig. 4.

We first show that, as in the case of $\langle P_{\text{B}}, V_{\text{B}} \rangle$, there exists a "hard" language on the transcript of $\langle \mathbb{P}_{\text{B}}, \mathbb{V}_{\text{B}} \rangle$.

Lemma 4 (Soundness). *Let $\mathbb{L}_{\text{B}}$ be the language defined in Fig. 5. Then, for any cheating prover $\mathbb{P}^*$ against $\langle \mathbb{P}_{\text{B}}, \mathbb{V}_{\text{B}} \rangle$, any $n \in \mathbb{N}$, and any $z \in \{0,1\}^*$,*

$$\Pr\left[\tau \leftarrow \text{trans}\left[\mathbb{P}^*(1^n, z) \leftrightarrow \mathbb{V}_{\text{B}}(1^n)\right] : \tau \in \mathbb{L}_{\text{B}}\right] \leq \text{negl}(n).$$

Proof. Assume for contradiction that there exists $\mathbb{P}^*$ such that for infinitely many n's, there exists $z \in \{0,1\}^*$ such that

$$\Pr\left[\tau \leftarrow \text{trans}\left[\mathbb{P}^*(1^n, z) \leftrightarrow \mathbb{V}_{\text{B}}(1^n)\right] : \tau \in \mathbb{L}_{\text{B}}\right] \geq \frac{1}{p(n)}$$

for a polynomial $p(\cdot)$. We use $\mathbb{P}^*$ to construct a cheating prover P^* against $\langle P_{\text{B}}, V_{\text{B}} \rangle$ and show that it contradicts the soundness of $\langle P_{\text{B}}, V_{\text{B}} \rangle$ (i.e., Lemma 1).

Consider the following cheating prover P^* against $\langle P_{\text{B}}, V_{\text{B}} \rangle$. First, P^* internally invokes $\mathbb{P}^*$. Then, while externally interacting with a honest V_{B} of $\langle P_{\text{B}}, V_{\text{B}} \rangle$, P^* interacts with internal $\mathbb{P}^*$ as a verifier of $\langle \mathbb{P}_{\text{B}}, \mathbb{V}_{\text{B}} \rangle$ in the following way.

- In Stage 1 and 2 (of $\langle \mathbb{P}_{\text{B}}, \mathbb{V}_{\text{B}} \rangle$), P^* forwards all messages from external V_{B} to internal $\mathbb{P}^*$ and forwards all messages from internal $\mathbb{P}^*$ to external V_{B}. (Notice that the verifier of $\langle P_{\text{B}}, V_{\text{B}} \rangle$ and that of $\langle \mathbb{P}_{\text{B}}, \mathbb{V}_{\text{B}} \rangle$ are identical.) Let (h, r_1, c, r_2) be the transcript of these stages.
- In Stage 3-1, P^* forwards α from external V_{B} to internal $\mathbb{P}^*$.

Stage 1:
    The verifier $\mathbb{V}_B$ sends a random hash function $h \in \mathcal{H}$ to the prover $\mathbb{P}_B$. $\mathbb{V}_B$ also sends $r_1 \in \{0,1\}^{3n}$ (the first-round message of Com) to $\mathbb{P}_B$.

Stage 2:
    1. $\mathbb{P}_B$ gives a fake commitment c of Com to $\mathbb{V}_B$ by running $c \leftarrow \mathsf{Com}_{\mathrm{pub}}(1^n)$.
    2. $\mathbb{V}_B$ sends random $r_2 \in \{0,1\}^{n+n^2}$ to $\mathbb{P}_B$.

Stage 3 (Encrypted UA):
    1. $\mathbb{V}_B$ sends the first-round message α of UA for statement $(h, r_1, c, r_2) \in \Lambda$.
    2. $\mathbb{P}_B$ gives a fake commitment of SWExtCom to $\mathbb{V}_B$ by running $C_{\mathrm{pub}}(1^n)$. Let $\widehat{\beta}$ be the fake commitment (i.e., the transcript of this step).
    3. $\mathbb{V}_B$ sends the third-round message γ of UA for statement $(h, r_1, c, r_2) \in \Lambda$.
    4. $\mathbb{P}_B$ gives a fake commitment of SWExtCom to $\mathbb{V}_B$ by running $C_{\mathrm{pub}}(1^n)$. Let $\widehat{\delta}$ be the fake commitment.

....

Language Λ (same as the one in Fig. 1):
    $(h, r_1, c, r_2) \in \Lambda$ if and only if there exist
    – a machine Π
    – randomness rand for Com
    – a string y such that $|y| \leq n^2$
such that
    – $c = \mathsf{Com}_{r_1}(h(\Pi); \mathsf{rand})$, and
    – $\Pi(c, y)$ outputs r_2 within $n^{\log \log n}$ steps.

Fig. 4. Special-purpose encrypted Barak's preamble $\langle \mathbb{P}_B, \mathbb{V}_B \rangle$.

– In Stage 3-2, P^* interacts with internal $\mathbb{P}^*$ as a honest receiver of SWExtCom and obtains $\widehat{\beta}_1$. Let st be the current state of $\mathbb{P}^*$. Then, P^* rewinds $\mathbb{P}^*$ to the point just before the challenge stage of SWExtCom, interacts with $\mathbb{P}^*$ again, and obtains $\widehat{\beta}_2$. Then, P^* computes a potential committed value $\widetilde{\beta} \overset{\text{def}}{=} \mathsf{Extract}(\widehat{\beta}_1, \widehat{\beta}_2)$ of $\widehat{\beta}_1$ (recall that Extract is the extracting algorithm of SWExtCom shown in Fig. 3) and sends $\widetilde{\beta}$ to external V_B.
– In Stage 3-3, P^* receives γ from V_B and sends it to internal $\mathbb{P}^*$ (which is restarted from state st).
– In Stage 3-4, P^* interacts with internal $\mathbb{P}^*$ as a honest receiver of SWExtCom and obtains $\widehat{\delta}_1$. Then, P^* rewinds $\mathbb{P}^*$ to the point just before the challenge stage of SWExtCom, interacts with $\mathbb{P}^*$ again, and obtains $\widehat{\delta}_2$. Then, P^* computes $\widetilde{\delta} := \mathsf{Extract}(\widehat{\delta}_1, \widehat{\delta}_2)$ and sends $\widetilde{\delta}$ to external V_B.

Whenever internal $\mathbb{P}^*$ aborts, P^* also aborts.

Before analyzing the success probability of P^*, we first introduce some terminologies regarding the internally emulated interaction between $\mathbb{P}^*$ and $\mathbb{V}_B$. Let $\tau = (h, r_1, c, r_2, \alpha, \widehat{\beta}_1, \gamma, \widehat{\delta}_1)$ be its transcript. Notice that since P^* emulates $\mathbb{V}_B$ for internal $\mathbb{P}^*$ perfectly, we have $\tau \in \mathbb{L}_B$ with probability at least $1/p(n)$.

Language $\mathbb{L}_B$:

$(h, r_1, c, r_2, \alpha, \widehat{\beta}, \gamma, \widehat{\delta}) \in \mathbb{L}_B$ if and only if there exist
 - decommitments $d_1, d_2 \in \{0, 1\}^{\mathsf{poly}(n)}$ for SWExtCom
 - the second-round and the fourth-round messages $\beta, \delta \in \{0, 1\}^n$ of UA

such that
 - d_1 is a valid decommitment of $\widehat{\beta}$ to β, and
 - d_2 is a valid decommitment of $\widehat{\delta}$ to δ, and
 - $(\alpha, \beta, \gamma, \delta)$ is an accepting transcript of UA for statement $(h, r_1, c, r_2) \in \Lambda$.

Fig. 5. Language $\mathbb{L}_B$.

- We say that a transcript τ_1 up until the commit stage of SWExtCom in Stage 3-2 is *good* if under the condition that τ_1 is a prefix of τ, the probability that $\tau \in \mathbb{L}_B$ holds is at least $1/2p(n)$.
- We say that a transcript τ_2 up until the commit stage of SWExtCom in Stage 3-4 is *good* if (1) a prefix of τ_2 up until the commit stage of SWExtCom in Stage 3-2 is good and (2) under the condition that τ_2 is a prefix of τ, the probability that $\tau \in \mathbb{L}_B$ holds is at least $1/4p(n)$.

We then analyze the success probability of P^* as follows. Let GOOD_1 be the event that a prefix of τ up until the commit stage of SWExtCom in Stage 3-2 is good, and let GOOD_2 be the event that a prefix of τ up until the commit stage of SWExtCom in Stage 3-4 is good. From an average argument, we have

$$\Pr\left[\mathrm{GOOD}_1\right] \geq \frac{1}{2p(n)} \quad \text{and} \quad \Pr\left[\mathrm{GOOD}_2 \mid \mathrm{GOOD}_1\right] \geq \frac{1}{4p(n)}.$$

Hence, we have

$$\Pr\left[\mathrm{GOOD}_2\right] = \Pr\left[\mathrm{GOOD}_1 \wedge \mathrm{GOOD}_2\right] \geq \frac{1}{8\left(p(n)\right)^2}. \tag{1}$$

Also, from the definition of GOOD_2, we have

$$\Pr\left[\tau \in \mathbb{L}_B \mid \mathrm{GOOD}_2\right] \geq \frac{1}{4p(n)}. \tag{2}$$

Hence, from Eqs. (1) and (2), we have

$$\Pr\left[\mathrm{GOOD}_1 \wedge \mathrm{GOOD}_2 \wedge \tau \in \mathbb{L}_B\right] = \Pr\left[\mathrm{GOOD}_2 \wedge \tau \in \mathbb{L}_B\right] \geq \frac{1}{32\left(p(n)\right)^3}. \tag{3}$$

Next, we observe that when the transcript up until the commit stage of SWExtCom in Stage 3-2 is good, $\mathbb{P}^*$ gives a valid commitment of SWExtCom in Stage 3-2 with probability at least $1/2p(n)$, and similarly, when the transcript up until the commit stage of SWExtCom in Stage 3-4 is good, $\mathbb{P}^*$ gives a valid commitment of SWExtCom in Stage 3-4 with probability at least $1/4p(n)$.

(This is because when the transcript is in $\mathbb{L}_B$, the SWExtCom commitments in Stage 3-2 and 3-4 are valid.) Hence, under the condition that $\mathsf{GOOD}_1 \wedge \mathsf{GOOD}_2 \wedge \tau \in \mathbb{L}_B$, the probability that both of $\widehat{\beta}_2$ and $\widehat{\delta}_2$ are valid is at least $1/8(p(n))^2$. Also, from the definition of $\mathbb{L}_B$, both of $\widehat{\beta}_1$ and $\widehat{\delta}_1$ are valid when $\tau \in \mathbb{L}_B$, and furthermore, $\widehat{\beta}_1$ and $\widehat{\beta}_2$ (resp, $\widehat{\delta}_1$ and $\widehat{\delta}_2$) are admissible except with negligible probability. Hence, from Lemma 2, for $\widetilde{\beta} = \mathsf{Extract}(\widehat{\beta}_1, \widehat{\beta}_2)$ and $\widetilde{\delta} = \mathsf{Extract}(\widehat{\delta}_1, \widehat{\delta}_2)$ we have

$$\Pr\left[\widetilde{\beta} = \mathsf{value}(\widehat{\beta}_1) \wedge \widetilde{\delta} = \mathsf{value}(\widehat{\delta}_1) \mid \mathsf{GOOD}_1 \wedge \mathsf{GOOD}_2 \wedge \tau \in \mathbb{L}_B\right]$$
$$\geq \frac{1}{8(p(n))^2} - \mathsf{negl}(n). \tag{4}$$

Hence, from Eqs. (3) and (4), we have

$$\Pr\left[\mathsf{GOOD}_1 \wedge \mathsf{GOOD}_2 \wedge \tau \in \mathbb{L}_B \wedge \widetilde{\beta} = \mathsf{value}(\widehat{\beta}_1) \wedge \widetilde{\delta} = \mathsf{value}(\widehat{\delta}_1)\right]$$
$$\geq \frac{1}{256(p(n))^5} - \mathsf{negl}(n).$$

Notice that from the definition of $\mathbb{L}_B$, when $\tau \in \mathbb{L}_B \wedge \widetilde{\beta} = \mathsf{value}(\widehat{\beta}_1) \wedge \widetilde{\delta} = \mathsf{value}(\widehat{\delta}_1)$, it holds that $(\alpha, \widetilde{\beta}, \gamma, \widetilde{\delta})$ is an accepting UA proof for $(h, r_1, c, r_2) \in \Lambda$. Hence, we have

$$\Pr\left[(h, r_1, c, r_2, \alpha, \widetilde{\beta}, \gamma, \widetilde{\delta}) \in \mathbf{L}_B\right] \geq \frac{1}{256(p(n))^5} - \mathsf{negl}(n),$$

which contradicts Lemma 1.   □

We next note that a non-black-box simulator can simulate the transcript τ in such a way that $\tau \in \mathbb{L}_B$ holds, and the simulator can additionally output a witness for $\tau \in \mathbb{L}_B$.

Lemma 5 (Simulatability). *Let $\mathbb{L}_B$ be the language defined in Fig. 5. Then, for any PPT cheating verifier V^* against $\langle \mathbb{P}_B, V_B \rangle$, there exists a PPT simulator S such that the following hold.*

- *Let $S_1(x, z)$ be the random variable representing the first output of $S(x, z)$. Then, the following indistinguishability holds.*

$$\{\mathsf{view}_{V^*}\left[\mathbb{P}_B(1^n) \leftrightarrow V^*(1^n, z)\right]\}_{n \in \mathbb{N}, z \in \{0,1\}^*} \approx \{S_1(1^n, z)\}_{n \in \mathbb{N}, z \in \{0,1\}^*}$$

- *For any $n \in \mathbb{N}$ and $z \in \{0,1\}^*$, the following holds.*

$$\Pr\left[\begin{matrix} (v, w) \leftarrow S(1^n, z); \\ \text{reconstruct transcript } \tau \text{ from view } v \text{ of } V^* \end{matrix} : w \in \mathbf{R}_{\mathbb{L}_B}(\tau)\right] \geq 1 - \mathsf{negl}(n)$$

This lemma can be proven in essentially the same way as the zero-knowledge property of Barak's non-black-box zero-knowledge protocol. A proof sketch is given in the full version [17].

4.2 Special-Purpose Instance-Dependent Commitment

In our LRZK protocol, we use a special-purpose instance-dependent commitment scheme GJS-Com, which is shown in Fig. 6. GJS-Com is parametrized by two graphs, G and G', and obtained by modifying Hamiltonicity commitment scheme H-Com$_{G,r}$ in such a way that the adjacent matrix is committed to by using AH-Com$_{G',r}$ instead of Com$_r$. GJS-Com inherits many properties from H-Com—hiding, binding, and equivocality—and additionally, thanks to the adaptive security of AH-Com, it provides adaptive security in the following sense: When $G \in \mathbf{L}_{HC}$ and $G' \in \mathbf{L}_{HC}$, a commitment to 1 can be explained as a valid commitment to 0, and furthermore, even after being explained as a commitment to 0, it can be decommitted to 1 in a consistent way. Details follow.

Parameters:
- Security parameter n.
- Two graphs G and G', where the number of vertices in G is $q = \mathsf{poly}(n)$ and that in G' is $q' = \mathsf{poly}'(n)$.

Inputs:
- C has secret input $b \in \{0,1\}$, which is the value to be committed to.

Commit phase:
1. R sends the first-round message $r \in \{0,1\}^{3n}$ of Com.
2. **To commit to 0,** C chooses a random permutation π over the vertices of G, computes $H_0 := \pi(G)$, and commits to its adjacent matrix $A_0 = \{a_{0,i,j}\}_{i,j\in[q]}$ by using AH-Com$_{G',r}$, i.e., sends $c_{i,j} \leftarrow$ AH-Com$_{G',r}(a_{0,i,j})$ for every $i,j \in [q]$.

   To commit to 1, C chooses a random q-cycle graph H_1 and commits to its adjacent matrix $A_1 = \{a_{1,i,j}\}_{i,j\in[q]}$ by using AH-Com$_{G',r}$, i.e., sends $c_{i,j} \leftarrow$ AH-Com$_{G',r}(a_{1,i,j})$ for every $i,j \in [q]$.

   Let GJS-Com$_{G,G',r}(\cdot)$ be a function that, on input $b \in \{0,1\}$, computes a commitment to b as above by considering r as the first-round message from the receiver.

Decommit phase:
- **When C committed to 0,** it reveals π and decommits $c_{i,j}$ to $a_{0,i,j}$ for every $i,j \in [q]$. R verifies whether the decommitted matrix is the adjacent matrix of $\pi(G)$.
- **When C committed to 1,** it decommits $c_{i,j}$ to 1 for every i,j such that edge $(i.j)$ is on the q-cycle in H_1 (i.e., every i,j such that $a_{1,i,j} = 1$). R verifies whether the decommitted entries correspond to the edges on a Hamilton cycle.

   Let GJS-Dec$_r(\cdot)$ be a function that, on input (c,b,ρ) such that GJS-Com$_{G,G',r}(b;\rho) = c$, outputs a decommitment to b as above.

Fig. 6. Special-purpose instance-dependent commitment GJS-Com.

Lemma 6 (Hiding and Binding). GJS-Com *is computationally hiding. Furthermore, it is statistically binding when* $G \notin \mathbf{L}_{HC}$ *and* $G' \notin \mathbf{L}_{HC}$.

Lemma 7 (Equivocality). *There exists an algorithm* GJS-EquivToOne *that is parametrized by graphs* G, G' *and a string* $r \in \{0,1\}^{3n}$ *and satisfies the following: When* $G \in L_{\mathrm{HC}}$, *on input any* $w \in R_{\mathrm{HC}}(G)$ *and any* c *and* ρ *such that* GJS-Com$_{G,G',r}(0; \rho) = c$, GJS-EquivToOne$_{G,G',r}$ *outputs a valid decommitment of* c *to* 1.

Proofs of these two lemmas are straightforward. We give the proofs in the full version [17].

Lemma 8 (Adaptive Security). *There exists an algorithm* GJS-ExplainAsZero *that is parametrized by graphs* G, G' *and a string* $r \in \{0,1\}^{3n}$ *and satisfies the following.*

Correctness. *When* $G, G' \in L_{\mathrm{HC}}$, *on input any* $w \in R_{\mathrm{HC}}(G)$ *and* $w' \in R_{\mathrm{HC}}(G')$ *and any* c *and* ρ_1 *such that* GJS-Com$_{G,G',r}(1; \rho_1) = c$, GJS-ExplainAsZero$_{G,G',r}$ *outputs* ρ_0 *such that* GJS-Com$_{G,G',r}(0; \rho_0) = c$.

Indistinguishability. *For security parameter* $n \in \mathbb{N}$, *graphs* $G, G' \in L_{\mathrm{HC}}$, *witnesses* $w \in R_{\mathrm{HC}}(G)$ *and* $w' \in R_{\mathrm{HC}}(G')$, *and string* $r \in \{0,1\}^{3n}$, *consider the following two probabilistic experiments.*

$\underline{\mathrm{EXP}_0^{\mathrm{GJS}}(n, G, G', w, w', r)}$
   /* commit to 0 and decommit it to 1 using equivocality */
   1. *Compute* $c \leftarrow$ GJS-Com$_{G,G',r}(0)$.
      Let ρ_0 *be the randomness used in* GJS-Com.
   2. *Compute* $d_1 :=$ GJS-EquivToOne$_{G,G',r}(c, w, \rho_0)$.
   3. *Output* (c, ρ_0, d_1).

$\underline{\mathrm{EXP}_1^{\mathrm{GJS}}(n, G, G', w, w', r)}$
   /* commit & decommit to 1 and explain it as commitment to 0 */
   1. *Compute* $c \leftarrow$ GJS-Com$_{G,G',r}(1)$.
      Let ρ_1 *be the randomness used in* GJS-Com.
      Compute $d_1 :=$ GJS-Dec$_{G,G',r}(c, 1, \rho)$.
   2. *Compute* $\rho_0 :=$ GJS-ExplainAsZero$_{G,G',r}(c, w, w', \rho_1)$.
   3. *Output* (c, ρ_0, d_1).

Let $\mathrm{EXP}_b^{\mathrm{GJS}}(n, G, G', w, w', r)$ *be the random variable representing the output of* $\mathrm{EXP}_b^{\mathrm{GJS}}(n, G, G', w, w', r)$ *for each* $b \in \{0,1\}$. *Then, the following two ensembles are computationally indistinguishable.*

$$\left\{ \mathrm{EXP}_0^{\mathrm{GJS}}(n, G, G', w, w', r) \right\}_{n \in \mathbb{N}, G, G' \in L_{\mathrm{HC}}, w \in R_{\mathrm{HC}}(G), w' \in R_{\mathrm{HC}}(G'), r \in \{0,1\}^{3n}}$$

$$\left\{ \mathrm{EXP}_1^{\mathrm{GJS}}(n, G, G', w, w', r) \right\}_{n \in \mathbb{N}, G, G' \in L_{\mathrm{HC}}, w \in R_{\mathrm{HC}}(G), w' \in R_{\mathrm{HC}}(G'), r \in \{0,1\}^{3n}}$$

Proof (sketch). GJS-ExplainAsZero is shown in Fig. 7. A key idea is that given the ability to explain AH-Com commitments to 0 as AH-Com commitments to 1, we can explain a GJS-Com commitment to 1 (which is AH-Com commitments to the adjacent matrix of a cycle graph) as a GJS-Com commitment to 0 (which is AH-Com commitments to the adjacent matrix of a Hamiltonian graph G). Intuitively, this is because a cycle graph can be transformed to any Hamiltonian graph by appropriately adding edges (which corresponds to changing some entries of the adjacent matrix from 0 to 1). A formal proof is given in the full version [17].   □

Parameter:
 - Graphs $G, G' \in \mathbf{L}_{HC}$
 - String $r \in \{0,1\}^{3n}$

Input:
 - Witnesses $w \in \mathbf{R}_{HC}(G)$ and $w' \in \mathbf{R}_{HC}(G')$
 - Commitment c and randomness ρ_1 s.t. $\mathsf{GJS\text{-}Com}_{G,G',r}(1; \rho_1) = c$

Output:
  1. Parse c as $\{c_{i,j}\}_{i,j \in [q]}$, where each $c_{i,j}$ is a AH-Com commitment. Also, from ρ_1, reconstruct $A_1 = \{a_{1,i,j}\}_{i,j \in [q]}$ and $\{\sigma_{1,i,j}\}_{i,j \in [q]}$ such that A_1 is the adjacent matrix of a q-cycle graph H_1 and $\mathsf{AH\text{-}Com}_{G',r}(a_{1,i,j}; \sigma_{1,i,j}) = c_{i,j}$ for every $i, j \in [q]$.
  2. Choose a random permutation π under the condition that a q-cycle in $H_0 \stackrel{\text{def}}{=} \pi(G)$ coincides with the q-cycle in H_1 (i.e., H_0 has the same cycle as H_1).[a] Let $A_0 = \{a_{0,i,j}\}_{i,j \in [q]}$ be the adjacent matrix of H_0.
  3. For every $i, j \in [q]$, define $\sigma_{0,i,j}$ by $\sigma_{0,i,j} \stackrel{\text{def}}{=} \sigma_{1,i,j}$ when $a_{0,i,j} = a_{1,i,j}$ and by $\sigma_{0,i,j} \stackrel{\text{def}}{=} \mathsf{AH\text{-}ExplainAsOne}_{G',r}(w', c_{i,j}, \sigma_{1,i,j})$ when $a_{0,i,j} \neq a_{1,i,j}$.[b]
  4. Outputs $\rho_0 \stackrel{\text{def}}{=} (\pi, \{\sigma_{0,i,j}\}_{i,j \in [q]})$.

---

[a] Given w, this can be done efficiently.
[b] When $a_{0,i,j} \neq a_{i,j}$, it holds that $a_{0,i,j} = 1$ and $a_{1,i,j} = 0$.

Fig. 7. GJS-ExplainAsZero.

5 Our Leakage-Resilient Zero-Knowledge Argument

Theorem 1. *Assume the existence of collision-resistant hash function family. Then, there exists a constant-round public-coin leakage-resilient zero-knowledge argument system* LR-ZK.

Proof. LR-ZK is shown in Fig. 8. Since $\langle \mathbb{P}_B, \mathbb{V}_B \rangle$ can be constructed from any collision-resistant hash function family, and SWExtCom can be constructed from any one-way function (which can be obtained from any collision-resistant hash function family), LR-ZK can be constructed from any collision-resistant hash function family. Also, by inspection, it can be seen that LR-ZK is public-coin and has constant number of rounds.

Roughly speaking, the soundness of LR-ZK can be proven as follows. From the soundness of $\langle \mathbb{P}_B, \mathbb{V}_B \rangle$, we have $\tau \notin \mathbb{L}_B$ (and hence $G' \notin \mathbf{L}_{HC}$) in Stage 1 except with negligible probability. Hence, $\mathsf{GJS\text{-}Com}_{G,G'}$ is statistically binding except with negligible probability, and thus we can use essentially the same argument as in the proof of the soundness of Blum's Hamiltonicity protocol to show that any cheating prover can give valid response in Stage 2-3 of all n iterations only with negligible probability. The formal proof is given in the full version [17].

In the following, we prove leakage-resilient zero-knowledgeness.

Lemma 9. LR-ZK *is leakage-resilient zero-knowledge.*

Input.
   – Common input is graph $G \in \mathbf{L}_{HC}$.
   Let $n \stackrel{\text{def}}{=} |G|$, and q be the number of vertices in G.
   – Private input to the prover P is witness $w \in \mathbf{R}_{HC}(G)$.

Stage 1.
   – P and V execute special-purpose encrypted Barak's preamble $\langle \mathbb{P}_B, \mathbb{V}_B \rangle$.
   Let τ be the transcript.
   – P and V reduce statement "$\tau \in \mathbb{L}_B$" to Hamiltonicity problem via general $\mathcal{NP}$ reduction. Let G' be the graph that P and V obtained. Let q' be the number of vertices in G'.

Stage 2.
   – V sends the first-round message $r \in \{0, 1\}^{3n}$ of Com to P.
   – P and V do the following for n times in parallel.

   1. P commits to a $q' \times q'$ zero matrix in a bit-by-bit manner by using $\mathsf{GJS\text{-}Com}_{G,G',r}$. That is, P sends $c_{i,j} \leftarrow \mathsf{GJS\text{-}Com}_{G,G',r}(0)$ to V for every $i, j \in [q']$. Let $\rho_{i,j}$ be the randomness that was used to compute $c_{i,j}$.

   2. V sends a random bit $ch \in \{0, 1\}$ to P.

   3. **When $ch = 0$:**
      • P chooses a random permutation π and computes $H_0 := \pi(G')$. Let $A_0 = \{a_{0,i,j}\}_{i,j \in [q']}$ be the adjacent matrix of H_0.
      • P sends π to V and decommits the $\mathsf{GJS\text{-}Com}$ commitments in Stage 2-1 to A_0 by using the equivocality of $\mathsf{GJS\text{-}Com}$. That is, for every $i, j \in [q]$, P sends a honest decommitment $d_{i,j} := \mathsf{GJS\text{-}Dec}_{G,G',r}(c_{i,j}, 0, \rho_{i,j})$ to V when $a_{0,i,j} = 0$ and sends a fake decommitment $d_{i,j} := \mathsf{GJS\text{-}EquivToOne}_{G,G',r}(c_{i,j}, w_0, \rho_{i,j})$ to V when $a_{0,i,j} = 1$.
      • V computes $H_0 = \pi(G')$ and verifies whether the decommitted matrix is equal to the adjacent matrix of H_0.

   When $ch = 1$:
      • P chooses a random q'-cycle graph H_1. Let $A_1 = \{a_{1,i,j}\}_{i,j \in [q']}$ be the adjacent matrix of H_1.
      • P decommits $c_{i,j}$ to $a_{1,i,j}$ for every i, j such that $a_{1,i,j} = 1$ (i.e., for every i, j such that edge (i, j) is on the q'-cycle of H_1). That is, for every such i and j, P sends a fake decommitment $d_{i,j} := \mathsf{GJS\text{-}EquivToOne}_{G,G',r}(c_{i,j}, w_0, \rho_{i,j})$ to V.
      • V checks whether the decommitted entries of the matrix correspond to the edges on a q'-cycle.

Fig. 8. Constant-round leakage-resilient zero-knowledge argument LR-ZK.

In the following, we prove this lemma only w.r.t. a simplified version of LR-ZK in which Stage 2-1, 2-2, and 2-3 are executed only once (instead of executed n times in parallel). The proof w.r.t. LR-ZK can be obtained by modifying the following proof in a straight-forward way.

Proof. Without loss of generality, we assume that after receiving each message from the prover, the cheating verifier makes exactly a single leakage query.

To see that we indeed do not lose generality, observe that instead of making two queries f_1 and f_2, the cheating verifier can always query a single query f such that, on input witness w and prover's randomness tape, it computes the first leakage $L_1 := f_1(w, \mathsf{tape})$, chooses the second query f_2 adaptively, computes the second leakage $L_2 := f_2(w, \mathsf{tape})$, and outputs (L_1, L_2).

Description of the Simulator. Given access to leakage oracle $\mathcal{L}_w$ and input (G, z), our simulator $\mathcal{S}$ simulates the view of cheating verifier V^* by internally invoking $V^*(G, z)$ and interacting with it as follows.

Simulating Messages and Leakages in Stage 1. Roughly speaking, $\mathcal{S}$ simulates the messages in Stage 1 by interacting with V^* in the same way as the simulator of $\langle \mathbb{P}_B, \mathbb{V}_B \rangle$ (cf. Lemma 5). To simulate the leakages in Stage 1, $\mathcal{S}$ uses the fact that Stage 1 of LR-ZK is public coin w.r.t. the prover and therefore all the randomness that a honest prover generates during Stage 1 is the messages themselves. Specifically, $\mathcal{S}$ simulates the leakages by considering the messages msgs that it has sent to V^* thus far as the randomness of the prover. An issue is that due to the existence of leakage queries, $\mathcal{S}$ cannot use the simulator of $\langle \mathbb{P}_B, \mathbb{V}_B \rangle$ in a modular way. Nonetheless, $\mathcal{S}$ can still use the technique used in the simulator of $\langle \mathbb{P}_B, \mathbb{V}_B \rangle$ as long as the length of the leakages is bounded by n^2. (Notice that when the length of leakage exceeds n^2, $\mathcal{S}$ can simply obtain a Hamiltonian cycle w of G from $\mathcal{L}_w$.)

Formally, $\mathcal{S}$ interacts with V^* as follows.

1. After receiving h and r_1 from V^*, $\mathcal{S}$ sends $c \leftarrow \mathsf{Com}_{r_1}(h(V^*))$ to V^*. Let rand be the randomness that was used in this step.
   Leakage query: When V^* makes a leakage query f, $\mathcal{S}$ does the following.
   – Let $\mathsf{tape} := c$.
   – If the output length of f is more than n^2, $\mathcal{S}$ obtains w from $\mathcal{L}_w$ and returns $f(w \,\|\, \mathsf{tape})$ to V^*.
   – Otherwise, $\mathcal{S}$ queries $f(\cdot, \mathsf{tape})$ to $\mathcal{L}_w$, obtains reply L from $\mathcal{L}_w$, and forwards L to V^*.
   If $\mathcal{S}$ obtained w, from now on $\mathcal{S}$ interacts with V^* in exactly the same way as a honest prover. Otherwise, do the following.
2. After receiving r_2 and α from V^*, $\mathcal{S}$ computes the second-round UA message β by using witness (V^*, rand, L) and then honestly commits to β by using SWExtCom. Let $\widehat{\beta}$ be the commitment and d_1 be the decommitment.
   Leakage query: When V^* makes a leakage query f, $\mathcal{S}$ sets $\mathsf{tape} := \mathsf{msgs}$, queries $f(\cdot, \mathsf{tape})$ to $\mathcal{L}_w$, and forwards the reply from $\mathcal{L}_w$ to V^*, where msgs are the messages that $\mathcal{S}$ has sent to V^* thus far.
3. After receiving γ from V^*, $\mathcal{S}$ computes the fourth-round UA message δ and then honestly commits to δ by using SWExtCom. Let $\widehat{\delta}$ be the commitment and d_2 be the decommitment.
   Leakage query: When V^* makes a leakage query f, $\mathcal{S}$ answers it in exactly the same way as above.

Let $\tau \stackrel{\text{def}}{=} (h, r_1, c, r_2, \alpha, \widehat{\beta}, \gamma, \widehat{\delta})$ and $\bar{w} \stackrel{\text{def}}{=} (d_1, d_2, \beta, \delta)$. Since (V^*, rand, L) is a valid witness for $(h, r_1, c, r_2) \in \Lambda$, we have $\tau \in \mathbb{L}_{\mathsf{B}}$ and $\bar{w} \in \mathbf{R}_{\mathbb{L}_{\mathsf{B}}}(\tau)$. Let G' and w' be the graph and its Hamiltonian cycle that are obtained by reducing statement "$\tau \in \mathbb{L}_{\mathsf{B}}$" to Hamiltonicity problem through the $\mathcal{NP}$ reduction.

Simulating Messages Stage 2. If S obtained w during Stage 1, it interacts with V^* in the same way as a honest prover. Otherwise, S interacts with V^* as follows. The idea is that, since S know a witness w' for $G' \in \mathbf{L}_{\mathsf{HC}}$, S can correctly respond to the challenge for both $ch = 0$ and $ch = 1$ by committing to a random permutation of G' in the first step.

1. S chooses a random permutation π and computes $H := \pi(G')$. Then, S commits to the adjacent matrix $A = \{a_{i,j}\}_{i,j \in [q']}$ of H by using $\mathsf{GJS\text{-}Com}_{G,G',r}$. That is, S sends $c_{i,j} \leftarrow \mathsf{GJS\text{-}Com}_{G,G',r}(a_{i,j})$ to V^* for every $i, j \in [q']$. Let $\{\rho_{i,j}\}_{i,j \in [q']}$ be the randomness used in the $\mathsf{GJS\text{-}Com}$ commitments and $\pi(w')$ be the Hamiltonian cycle in H that is obtained by applying π on Hamiltonian cycle w' in G'.
2. S receives a random bit $ch \in \{0, 1\}$ from V^*.
3. **When $ch = 0$,** S sends π to V and decommits $c_{i,j}$ to $a_{i,j}$ honestly for every $i, j \in [q']$. That is, S sends $d_{i,j} := \mathsf{GJS\text{-}Dec}_{G,G',r}(c_{i,j}, a_{i,j}, \rho_{i,j})$ to V for every $i, j \in [q']$.
   When $ch = 1$, S decommits $c_{i,j}$ to 1 honestly for every i, j such that edge (i, j) is on the Hamiltonian cycle $\pi(w')$ in H. That is, for every such i and j, S sends $d_{i,j} := \mathsf{GJS\text{-}Dec}_{G,G',r}(c_{i,j}, a_{i,j}, \rho_{i,j})$ to V^*.

Simulating Leakage Queries in Stage 2. When V^* makes a leakage query f, S simulates the leakage as follows. Recall that in Stage 2-1, a honest prover commits to a $q' \times q'$ zero matrix whereas S commits to the adjacent matrix of H. Hence, S simulates the leakage by "explaining" commitments $\{c_{i,j}\}_{i,j \in [q']}$ to $\{a_{i,j}\}_{i,j \in [q']}$ as commitments to $\{0\}$ by using the adaptive security of $\mathsf{GJS\text{-}Com}$ and the knowledge of w'. Concretely, S does the following.

- First, for each $i, j \in [q']$, S constructs a function $F_{i,j}(\cdot)$ such that on input w, it outputs $\widetilde{\rho}_{i,j}$ such that $\mathsf{GJS\text{-}Com}_{G,G',r}(0; \widetilde{\rho}_{i,j}) = c_{i,j}$. Concretely, when $a_{i,j} = 0$, $F_{i,j}(\cdot)$ is a function that always outputs $\rho_{i,j}$, and when $a_{i,j} = 1$, $F_{i,j}(\cdot) \stackrel{\text{def}}{=} \mathsf{GJS\text{-}ExplainAsZero}_{G,G',r}(c_{i,j}, \cdot, w', \rho_{i,j})$.
- Next, S constructs a function f such that on input w, it computes $\mathsf{tape} := \mathsf{msgs} \| \{F_{i,j}(w)\}_{i,j \in [q']}$ and outputs $f(w, \mathsf{tape})$.
- Finally, S queries f to $\mathcal{L}_w$ and forwards the reply from $\mathcal{L}_w$ to V^*.

Amount of Total Leakage. From the construction of S, it always obtains at most the same amount of leakages as V^*.

Indistinguishability of Views. For any cheating verifier V^* and any sequence $\{w_G\}_{G \in \mathbf{L}_{\mathsf{HC}}}$ such that $w_G \in \mathbf{R}_{\mathsf{HC}}(G)$, we show the following indistinguishability.

$$\{\mathsf{REAL}_{V^*}(G, w_G, z)\}_{G \in \mathbf{L}_{\mathsf{HC}}, z \in \{0,1\}^*} \approx \{\mathsf{IDEAL}_S(G, w_G, z)\}_{G \in \mathbf{L}_{\mathsf{HC}}, z \in \{0,1\}^*}. \quad (5)$$

Toward this end, we consider the following hybrid experiments.

Hybrid $\mathrm{HYB}_0(G, z)$ is identical with experiment $\mathrm{REAL}_{V^*}(G, w, z)$. That is, V^* interacts with honest $P(G, w)$ and obtains leakage that is computed honestly based on witness w and the prover's randomness. The outputs of this hybrid is the view of V^*.

Hybrid $\mathrm{HYB}_1(G, z)$ is the same as HYB_0 except for the following.

- In Stage 1, a honest prover is replaced with the simulator. That is, c is computed by committing to $h(V^*)$, $\widehat{\beta}$ is computed by committing to β, and $\widehat{\delta}$ is computed by committing to δ.

  Let τ and $\tilde{w}$ be the statement and the witness generated in it. Let G' and w' be the graph and its Hamiltonian cycle that are obtained by reducing statement "$\tau \in \mathbb{L}_B$" to Hamiltonicity problem through the $\mathcal{NP}$ reduction.

- The leakage queries are answered by considering that the randomness generated by the prover during Stage 1 is equal to the messages sent to V^* during Stage 1.

Hybrid $\mathrm{HYB}_2(G, z)$ is the same as HYB_1 except for the following.

- As in $\mathcal{S}$, a random permutation π is chosen randomly at the beginning of Stage 2-1. Let $H \overset{\text{def}}{=} \pi(G')$, and $A = \{a_{i,j}\}_{i,j \in [q']}$ be the adjacent matrix of H. Let $\pi(w')$ be the Hamiltonian cycle in H that is obtained by applying π on Hamiltonian cycle w' in G'.

  We remark that in this hybrid, the prover still commits to a $q' \times q'$ zero matrix as in HYB_1. Also, the leakage query immediately after Stage 2-1 is answered in exactly the same way as in HYB_1. In particular, when the leakage query is answered, π is not included in the randomness generated by the prover in Stage 2-1.

- In Stage 2-3, graph H_0 or H_1 is chosen as follows.

  When $ch = 0$, $H_0 := H$.

  When $ch = 1$, H_1 is the graph that is obtained by removing every edge in H except for the ones on Hamiltonian cycle $\pi(w')$.

  The leakage query immediately after Stage 2-3 is answered in the same way as in HYB_1 by considering that H_0 or H_1 was chosen during Stage 2-3 as in HYB_1.

Hybrid $\mathrm{HYB}_3(G, z)$ is the same as HYB_2 except for the following.

- In Stage 2-1, for every $i, j \in [q']$, commitment $c_{i,j}$ is computed by committing to $a_{i,j}$ (instead of 0), i.e., $c_{i,j} \leftarrow \mathsf{GJS\text{-}Com}_{G,G',r}(a_{i,j})$.

- In Stage 2-3, for every $i, j \in [q']$, if commitment $c_{i,j}$ need to be decommitted, it is decommitted to $a_{i,j}$ honestly.

- When the leakage queries are answered during Stage 2, the randomness $\rho_{i,j}$ used for computing $c_{i,j}$ is simulated by $\widetilde{\rho}_{i,j}$ that is computed by function $F_{i,j}$ as in $\mathcal{S}$ for every $i, j \in [q']$.

Hybrid $\mathrm{HYB}_4(G, z)$ is identical with $\mathrm{IDEAL}_{\mathcal{S}}(x, w, z)$. That is, $\mathcal{S}(G, z)$ is executed given access to $\mathcal{L}_w$. The outputs of this hybrid is that of $\mathcal{S}$.

Claim 1. *The output of $\mathrm{HYB}_0(G, z)$ and that of $\mathrm{HYB}_1(G, z)$ are computationally indistinguishable.*

Proof. HYB$_1$ differs from HYB$_0$ only in that fake commitments of Com and SWExtCom are replaced with real commitments. Hence, the indistinguishability follows from the security of Com$_{\text{pub}}$ and C_{pub} (see Sects. 3.4 and 3.8).   □

Claim 2. *The output of HYB$_1$(G, z) and that of HYB$_2$(G, z) are computationally indistinguishable.*

Proof. This claim can be proven by inspection. Observe that HYB$_2$ differs from HYB$_1$ only in the way graph H_0 or H_1 is chosen in Stage 2. When $ch = 0$, the distribution of H_0 in HYB$_2$ is the same as that in HYB$_1$ since H_0 is obtained both in HYB$_2$ and HYB$_1$ by applying a random permutation on G'. When $ch = 1$, the distribution of H_1 in HYB$_2$ is the same as that in HYB$_1$ since the Hamiltonian cycle w' in G' is mapped to a random q-cycle by π. Hence, the output of HYB$_2$ is identically distributed with that of HYB$_1$.   □

Claim 3. *The output of HYB$_2$(G, z) and that of HYB$_3$(G, z) are computationally indistinguishable.*

Proof. Assume for contradiction that for infinitely many $G \in \mathbf{L}_{\text{HC}}$, there exists $z \in \{0, 1\}^*$ such that a distinguisher $\mathcal{D}$ distinguishes the output of HYB$_2$(G, z) and that of HYB$_3$(G, z) with advantage $1/p(n)$ for a polynomial $p(\cdot)$. Fix any such G and z. To derive a contradiction, we consider the following intermediate hybrids.

Hybrid hyb$_{2:0}$(G, z) is identical with HYB$_2$(G, z).
Hybrid hyb$_{2:k}$(G, z), where $k \in [q'^2]$, is the same as HYB$_{2:k-1}$ except for the
 following. Let $u \overset{\text{def}}{=} \lfloor (k-1)/q' \rfloor + 1$ and $v \overset{\text{def}}{=} k - \lfloor (k-1)/q' \rfloor \cdot q'$.
 – In Stage 2-1, commitment $c_{u,v}$ is computed by committing to $a_{u,v}$ (instead of 0), i.e., $c_{u,v} \leftarrow \text{GJS-Com}_{G,G',r}(a_{u,v})$.
 – In Stage 2-3, if commitment $c_{u,v}$ need to be decommitted, it is decommitted to $a_{u,v}$ honestly.
 – When the leakage queries are answered during Stage 2, the randomness $\rho_{u,v}$ used for computing $c_{u,v}$ is simulated by $\widetilde{\rho}_{u,v}$ that is computed by function $F_{u,v}$ as in $\mathcal{S}$.

Clearly, HYB$_{2:q'^2}$ is identical with HYB$_3$. Hence, there exists $k^* \in [q'^2]$ such that the output of HYB$_{2:k^*-1}$ and that of HYB$_{2:k^*}$ can be distinguished with advantage $1/q'^2 p(n)$. Furthermore, from an average argument, there exists a prefix σ of the execution of HYB$_{k^*-1}$ up until permutation π is chosen in Stage 2-1 (i.e., just before $\{c_{i,j}\}_{i,j \in [q']}$ is sent to V^*) such that under the condition that a prefix of the execution is σ, the output of HYB$_{2:k^*-1}$ and that of HYB$_{2:k^*}$ can be distinguished with advantage $1/q'^2 p(n)$. Notice that σ determines G', w', r, $\{a_{i,j}\}_{i,j \in [q']}$.

We derive a contradiction by showing that we can break the adaptive security of GJS-Com (Lemma 8). Specifically, we show that $\text{EXP}_0^{\text{GJS}}(n, G, G', w, w', r)$ and $\text{EXP}_1^{\text{GJS}}(n, G, G', w, w', r)$ can be distinguished with advantage $1/q'^2 p(n)$. Toward this end, consider the following distinguisher $\mathcal{D}'$.

 – Externally, $\mathcal{D}'$ takes (c, ρ_0, d_1) as well as (n, G, G', w, w', r) as input. $\mathcal{D}'$ also takes (σ, z) as non-uniform input.

– Internally, $\mathcal{D}'$ invokes V^* and simulates $\text{HYB}_{2:k^*-1}(G, z)$ for V^* from σ honestly except for the following. Let $u^* \overset{\text{def}}{=} \lfloor (k^* - 1)/q' \rfloor + 1$ and $v^* \overset{\text{def}}{=} k^* - \lfloor (k^* - 1)/q' \rfloor \cdot q'$. Notice that it must hold that $a_{u^*,v^*} = 1$ since $\text{HYB}_{2:k^*}$ is identical with $\text{HYB}_{2:k^*-1}$ when $a_{u^*,v^*} = 0$.
  - In Stage 2-1, commitment c_{u^*,v^*} is defined by setting $c_{u^*,v^*} := c$.
  - In Stage 2-3, when commitment c_{u^*,v^*} is decommitted, it is decommitted to $a_{u^*,v^*} = 1$ by sending d_1.
  - When the leakage queries are answered during Stage 2, the randomness ρ_{u^*,v^*} used for computing c_{u^*,v^*} is simulated by setting $\widetilde{\rho}_{u^*,v^*} := \rho_0$.
  Let view be the view of V^*. Then, $\mathcal{D}'$ outputs $\mathcal{D}(\text{view})$.

When $(c, \rho_0, d_1) \leftarrow \text{EXP}_0^{\text{GJS}}(n, G, G', w, w', r)$ (i.e., when c is a commitment to 0, ρ_0 is the randomness that is used to generate c, and d_1 is a decommitment to 1 that is computed by GJS-EquivToOne), $\mathcal{D}'$ emulates $\text{HYB}_{2:k^*-1}$ for V^* perfectly. On the other hand, when $(c, \rho_0, d_1) \leftarrow \text{EXP}_1^{\text{GJS}}(n, G, G', w, w', r)$ (i.e., when c is a commitment to 1, ρ_0 is randomness that is computed by GJS-ExplainAsZero, and d_1 is a decommitment to 1 that is computed honestly), $\mathcal{D}'$ emulates $\text{HYB}_{2:k^*}$ for V^* perfectly. Hence, from our assumption, $\mathcal{D}'$ distinguishes $\text{EXP}_0^{\text{GJS}}(n, G, G', w, w', r)$ and $\text{EXP}_1^{\text{GJS}}(n, G, G', w, w', r)$ with advantage $1/q'^2 p(n)$, and therefore we reach a contradiction. $\square$

Claim 4. *The output of* $\text{HYB}_3(G, z)$ *and that of* $\text{HYB}_4(G, z)$ *are computationally indistinguishable.*

Proof. In HYB_3, the prover interacts with V^* in exactly the same way as $\mathcal{S}$. Hence, the claim follows. $\square$

Equation (5) follows from these claims. This concludes the proof of Lemma 9. $\square$

This concludes the proof of Theorem 1. $\square$

Acknowledgments. The author would like to thank the anonymous reviewers for their helpful comments.

References

1. Ananth, P., Goyal, V., Pandey, O.: Interactive proofs under continual memory leakage. In: Garay, J.A., Gennaro, R. (eds.) CRYPTO 2014, Part II. LNCS, vol. 8617, pp. 164–182. Springer, Heidelberg (2014)
2. Anderson, R., Kuhn, M.: Tamper resistance: a cautionary note. In: WOEC, pp. 1–11 (1996)
3. Barak, B.: How to go beyond the black-box simulation barrier. In: FOCS, pp. 106–115 (2001)
4. Barak, B., Goldreich, O.: Universal arguments and their applications. SIAM J. Comput. **38**(5), 1661–1694 (2008)
5. Bitansky, N., Canetti, R., Halevi, S.: Leakage-tolerant interactive protocols. In: Cramer, R. (ed.) TCC 2012. LNCS, vol. 7194, pp. 266–284. Springer, Heidelberg (2012)

6. Bitansky, N., Dachman-Soled, D., Lin, H.: Leakage-tolerant computation with input-independent preprocessing. In: Garay, J.A., Gennaro, R. (eds.) CRYPTO 2014, Part II. LNCS, vol. 8617, pp. 146–163. Springer, Heidelberg (2014)
7. Boyle, E., Garg, S., Jain, A., Kalai, Y.T., Sahai, A.: Secure computation against adaptive auxiliary information. In: Canetti, R., Garay, J.A. (eds.) CRYPTO 2013, Part I. LNCS, vol. 8042, pp. 316–334. Springer, Heidelberg (2013)
8. Boyle, E., Goldwasser, S., Jain, A., Kalai, Y.T.: Multiparty computation secure against continual memory leakage. In: STOC, pp. 1235–1254 (2012)
9. Canetti, R., Lindell, Y., Ostrovsky, R., Sahai, A.: Universally composable two-party and multi-party secure computation. In: STOC, pp. 494–503 (2002)
10. Damgård, I., Pedersen, T.P., Pfitzmann, B.: Statistical secrecy and multibit commitments. IEEE Trans. Inf. Theor. 44(3), 1143–1151 (1998)
11. Feige, U., Shamir, A.: Zero knowledge proofs of knowledge in two rounds. In: Brassard, G. (ed.) CRYPTO 1989. LNCS, vol. 435, pp. 526–544. Springer, Heidelberg (1990)
12. Garg, S., Jain, A., Sahai, A.: Leakage-resilient zero knowledge. In: Rogaway, P. (ed.) CRYPTO 2011. LNCS, vol. 6841, pp. 297–315. Springer, Heidelberg (2011)
13. Goldreich, O., Kahan, A.: How to construct constant-round zero-knowledge proof systems for NP. J. Cryptol. 9(3), 167–190 (1996)
14. Goldwasser, S., Micali, S., Rackoff, C.: The knowledge complexity of interactive proof systems. SIAM J. Comput. 18(1), 186–208 (1989)
15. Haitner, I., Nguyen, M., Ong, S.J., Reingold, O., Vadhan, S.P.: Statistically hiding commitments and statistical zero-knowledge arguments from any one-way function. SIAM J. Comput. 39(3), 1153–1218 (2009)
16. Håstad, J., Impagliazzo, R., Levin, L.A., Luby, M.: A pseudorandom generator from any one-way function. SIAM J. Comput. 28(4), 1364–1396 (1999)
17. Kiyoshima, S.: Constant-round leakage-resilient zero-knowledge from collision resistance. Cryptology ePrint Archive, Report 2015/1235 (2015). http://eprint.iacr.org/
18. Kocher, P.C.: Timing attacks on implementations of Diffie-Hellman, RSA, DSS, and other systems. In: Koblitz, N. (ed.) CRYPTO 1996. LNCS, vol. 1109, pp. 104–113. Springer, Heidelberg (1996)
19. Lindell, Y., Zarosim, H.: Adaptive zero-knowledge proofs and adaptively secure oblivious transfer. J. Cryptol. 24(4), 761–799 (2011)
20. Naor, M.: Bit commitment using pseudorandomness. J. Cryptol. 4(2), 151–158 (1991)
21. Naor, M., Yung, M.: Universal one-way hash functions and their cryptographic applications. In: STOC, pp. 33–43 (1989)
22. Ostrovsky, R., Persiano, G., Visconti, I.: Impossibility of black-box simulation against leakage attacks. In: Gennaro, R., Robshaw, M. (eds.) CRYPTO 2015. LNCS, vol. 9216, pp. 130–149. Springer, Heidelberg (2015)
23. Pandey, O.: Achieving constant round leakage-resilient zero-knowledge. In: Lindell, Y. (ed.) TCC 2014. LNCS, vol. 8349, pp. 146–166. Springer, Heidelberg (2014)
24. Pass, R., Rosen, A.: Concurrent non-malleable commitments. In: FOCS, pp. 563–572 (2005)
25. Pass, R., Rosen, A.: New and improved constructions of non-malleable cryptographic protocols. In: STOC, pp. 533–542 (2005)
26. Pass, R., Wee, H.: Black-box constructions of two-party protocols from one-way functions. In: Reingold, O. (ed.) TCC 2009. LNCS, vol. 5444, pp. 403–418. Springer, Heidelberg (2009)

27. Quisquater, J., Samyde, D.: ElectroMagnetic Analysis (EMA): measures and counter-measures for smart cards. In: Attali, I., Jensen, T. (eds.) E-smart 2001. LNCS, vol. 2140, pp. 200–210. Springer, Heidelberg (2001)

28. Yao, A.C.C.: How to generate and exchange secrets (extended abstract). In: FOCS, pp. 162–167 (1986)

Constrained Pseudorandom Functions
for Unconstrained Inputs

Apoorvaa Deshpande[1]([⊠]), Venkata Koppula[2]([⊠]), and Brent Waters[2]

[1] Brown University, Providence, USA
acdeshpa@cs.brown.edu
[2] University of Texas at Austin, Austin, USA
{kvenkata,bwaters}@cs.utexas.edu

Abstract. A constrained pseudo random function (PRF) behaves like a standard PRF, but with the added feature that the (master) secret key holder, having secret key K, can produce a constrained key, $K\{f\}$, that allows for the evaluation of the PRF on all inputs satisfied by the constraint f. Most existing constrained PRF constructions can handle only bounded length inputs. In a recent work, Abusalah et al. [1] constructed a constrained PRF scheme where constraints can be represented as Turing machines with unbounded inputs. Their proof of security, however, requires risky "knowledge type" assumptions such as differing inputs obfuscation for circuits and SNARKs.

In this work, we construct a constrained PRF scheme for Turing machines with unbounded inputs under weaker assumptions, namely, the existence of indistinguishability obfuscation for circuits (and injective pseudorandom generators).

1 Introduction

Constrained pseudorandom functions (PRFs), as introduced by [7,9,23], are a useful extension of standard PRFs [18]. A constrained PRF system is defined with respect to a *family of constraint functions*, and has an additional algorithm Constrain. This algorithm allows a (master) PRF key holder, having PRF key K, to produce a *constrained* PRF key $K\{f\}$ corresponding to a constraint f. This constrained key $K\{f\}$ can be used to evaluate the PRF at all points x accepted by f (that is, $f(x) = 1$). The security notion ensures that even when given multiple constrained keys $K\{f_1\}, \ldots, K\{f_Q\}$, PRF evaluation at a point

A. Deshpande—This work was done while the author was visiting the Simons Institute for the Theory of Computing, supported by the Simons Foundation and by the DIMACS/Simons Collaboration in Cryptography through NSF grant #CNS-1523467.
B. Waters—Supported by NSF CNS-0952692, CNS-1228599 and CNS-1414082. DARPA through the U.S. Office of Naval Research under Contract N00014-11-1-0382, Google Faculty Research award, the Alfred P. Sloan Fellowship, Microsoft Faculty Fellowship, and Packard Foundation Fellowship.

M. Fischlin and J.-S. Coron (Eds.): EUROCRYPT 2016, Part II, LNCS 9666, pp. 124–153, 2016.
DOI: 10.1007/978-3-662-49896-5_5

not accepted by any of the functions f_i 'looks' uniformly random to a computationally bounded adversary. Since their inception, constrained PRFs have found several applications such as broadcast encryption, identity-based key exchange, policy-based key distribution [7] and multi-party key exchange [8]. In particular, even the most basic class of constrained PRFs called puncturable PRFs has found immense application in the area of program obfuscation through the 'punctured programming' technique introduced by [25]. The initial works of [7,9,23] showed that the [18] PRF construction can be modified to construct a basic class of constrained PRFs called prefix-constrained PRFs (which also includes puncturable PRFs). Boneh and Waters [7] also showed a construction for the richer class of circuit-constrained PRFs[1] using multilinear maps [14]. Since then, we have seen great progress in this area, leading to constructions from different cryptographic assumptions [4,8,10] and constructions with additional properties [1,4,10,12]. However, all the above mentioned works have a common limitation: the corresponding PRF can handle only bounded length inputs.

The problem of constructing constrained PRFs with unbounded length was studied in a recent work by Abusalah, Fuchsbauer and Pietrzak [1], who also showed motivating applications such as broadcast encryption with unbounded recipients and multi-party identity based non-interactive key exchange with no apriori bound on number of parties. Abusalah et al. construct a constrained PRF scheme where the constraint functions are represented as Turing machines with unbounded inputs. The scheme is proven secure under the assumption that differing input obfuscation ($di\mathcal{O}$) for circuits exists. Informally, this assumption states that there exists an 'obfuscation' program $\mathcal{O}$ that takes as input a circuit C, and outputs another circuit $\mathcal{O}(C)$ with the following security guarantee: if an efficient adversary can distinguish between $\mathcal{O}(C_1)$ and $\mathcal{O}(C_2)$, then there exists an efficient extraction algorithm that can find an input x such that $C_1(x) \neq C_2(x)$. However, the $di\mathcal{O}$ assumption is believed to be a risky one due to its 'extractability nature'. Furthermore, the work of [16] conjectures that there exist certain function classes for which $di\mathcal{O}$ is impossible to achieve.

A natural direction then is to try to base the security on the relatively weaker assumption of indistinguishability obfuscation ($i\mathcal{O}$) for circuits. An obfuscator $\mathcal{O}$ is an indistinguishability obfuscator for circuits if for any two circuits C_1 and C_2 that have identical functionality, their obfuscations $\mathcal{O}(C_1)$ and $\mathcal{O}(C_2)$ are computationally indistinguishable. Unlike $di\mathcal{O}$, there are no known impossibility results for $i\mathcal{O}$, and moreover, there has been recent progress [2,6,17] towards the goal of constructing $i\mathcal{O}$ from standard assumptions. This brings us to the central question of our work:

Can we construct constrained PRFs for Turing machines under the assumptions that indistinguishability obfuscation and one-way functions exist?

Our starting point is three recent works that build indistinguishability obfuscation for Turing Machines with bounded length inputs using $i\mathcal{O}$ for circuits [5,11,24]. The works of [5,11] show how to do this where the encoding time

---
[1] Where the constraints can be any boolean circuit.

and size of the obfuscated program grows with the maximum space used by the underlying program, whereas the work of [24] achieves this with no such restriction. An immediate question is whether we can use a Turing machine obfuscator for constructing constrained PRFs for Turing machines, similar to the circuit-constrained PRF construction of [8]. However, as mentioned above the Turing machine obfuscator constructions are restricted to Turing Machines with bounded size inputs[2]. Thus, we are unable to use the Turning Machine obfuscation scheme in a black box manner and have to introduce new techniques to construct constrained PRFs for unbounded sized inputs.

Our Results: The main result of our work is as follows.

Theorem 1 (informal). *Assuming the existence of secure indistinguishability obfuscators and injective pseudorandom generators, there exists a constrained PRF scheme that is selectively secure.*

Selective Security vs. Adaptive Security: Selective security is a security notion where the adversary must specify the 'challenge input' before receiving constrained keys. A stronger notion, called adaptive security, allows the adversary to query for constrained keys before choosing the challenge input. While adaptive security should be the ideal target, achieving adaptive security with only polynomial factor security loss (i.e. without 'complexity leveraging') has been challenging, even for circuit based constrained PRFs. Currently, the best known results for adaptive security either require superpolynomial security loss [13], or work for very restricted functionalities [20], or achieve non-collusion based security [10] or achieve it in the random oracle mode [19].

Moreover, for many applications, it turns out that selective security is sufficient. For example, the widely used punctured programming technique of [25] only requires selectively secure puncturable PRFs. Similarly, as discussed in [1], selectively secure constrained PRFs with unbounded inputs can be used to construct broadcast encryption schemes with unbounded recipients and identity based non-interactive key exchange (ID-NIKE) protocol with no apriori bound on number of parties. Therefore, as a corollary of Theorem 1, we get both these applications using only indistinguishability obfuscation and injective pseudorandom generators. Interestingly, two recent works have shown direct constructions for both these problems using $i\mathcal{O}$. Zhandry [26] showed a broadcast encryption scheme with unbounded recipients, while Khurana et al. [22] showed an ID-NIKE scheme with unbounded number of parties.

We also show how our construction above can be easily adapted to get selectively secure attribute based encryption for Turing machines with unbounded inputs, which illustrates the versatility of our techniques above.

---

[2] The restriction to bounded length inputs is due to the fact that their iO analysis requires a hybrid over all possible inputs. They absorb this loss by growing the size of the obfuscated program polynomially in the input size using complexity leveraging and a sub-exponential hardness assumption on the underlying circuit iO. Currently, there is no known way to avoid this.

Theorem 2 (informal). *Assuming the existence of secure indistinguishability obfuscators and injective pseudorandom generators, there exists an ABE scheme for Turing machines that is selectively secure.*

Recently, Ananth and Sahai [3] had an exciting result where they show adaptively secure functional encryption for Turing machines with unbounded inputs. While our adaptation is limited to ABE, we believe that the relative simplicity of our construction is an interesting feature. In addition, we were able to apply our tail-hybrid approach to get an end-to-end polynomial time reduction.

1.1   Overview of Our Constrained PRF Construction

To begin, let us consider the simple case of standard PRFs with unbounded inputs. Any PRF (with sufficient input size) can be extended to handle unbounded inputs by first compressing the input using a collision-resistant hash function (CRHF), and then computing the PRF on this hash value. Abusalah et al. [1] showed that by using $di\mathcal{O}$, this approach can be extended to work for constrained PRFs. However, the proof of security relies on the extractability property of $di\mathcal{O}$ in a fundamental way. In particular, this approach will not work if $i\mathcal{O}$ is used instead of $di\mathcal{O}$ because general CRHFs are not '$i\mathcal{O}$-compatible'[3] (see Sect. 2 for a more detailed discussion on $i\mathcal{O}$-compatibility).

Challenges of a similar nature were addressed in [24] by introducing new tools and techniques that guarantee program functional equivalence at different stages of the proof. Let us review one such tool called *positional accumulators*, and see why it is $i\mathcal{O}$-compatible. A positional accumulator scheme is a cryptographic primitive used to provide a short commitment to a much larger storage. This commitment (also referred to as an accumulation of the storage) has two main features: succinct verifiability (there exists a short proof to prove that an element is present at a particular position) and succinct updatability (using short auxiliary information, the accumulation can be updated to reflect an update to the underlying storage). The scheme also has a setup algorithm which generates the parameters, and can operate in two computationally indistinguishable modes. It can either generate parameters 'normally', or it can be *enforcing* at a particular position p. When parameters are generated in the enforcing mode, the accumulator is *information-theoretically binding* to position p of the underlying storage. This information theoretic enforcing property is what makes it compatible for proofs involving $i\mathcal{O}$.

Returning to our constrained PRF problem, we need a special hash function that can be used with $i\mathcal{O}$. That brings us to the main insight of our work: the KLW positional accumulator can be repurposed to be an $i\mathcal{O}$-friendly hash function.[4] Besides giving us an $i\mathcal{O}$-friendly

---

[3] Consider the following toy example. Let C_0, C_1 be circuits such that $C_0(x, y) = 0 \; \forall (x, y)$ and $C_1(x, y) = 1$ iff $\mathsf{CRHF}(x) = \mathsf{CRHF}(y)$ for $x \neq y$. Now, under the $di\mathcal{O}$ assumption, the obfuscations of C_0 and C_1 are computationally indistinguishable. However, we cannot get the same guarantee by using $i\mathcal{O}$, since the circuits are not functionally identical.

[4] More formally, it gives us an $i\mathcal{O}$ friendly universal one way hash function.

hash function, this also puts the input in a data structure that is already suitable for the KLW framework.[5]

Our Construction: We will now sketch out our construction. Our constrained PRF scheme uses a puncturable PRF F with key k. Let $\mathsf{Hash\text{-}Acc}(x)$ represent the accumulation of storage initialized with input $x = x_1 \ldots x_n$. The PRF evaluation (in our scheme) is simply $F(k, \mathsf{Hash\text{-}Acc}(x))$.

The interesting part is the description of our constrained keys, and how they can be used to evaluate at an input x. The constrained key for machine M consists of two programs. The first one is an obfuscated circuit which takes an input, and outputs a signature on that input. The second one is an obfuscated circuit which essentially computes the next-step of the Turing machine, and eventually, if it reaches the 'accepting state', it outputs $F(k, \mathsf{Hash\text{-}Acc}(x))$. This circuit also performs additional authenticity checks to prevent illegal inputs - it takes a signature and accumulator as input, verifies the signature and accumulator before computing the next step, and finally updates the accumulator and outputs a signature on the new state and accumulator.

Evaluating the PRF at input x using the constrained key consists of two steps. The first one is the initialization step, where the evaluator first computes $\mathsf{Hash\text{-}Acc}(x)$ and then computes a signature on $\mathsf{Hash\text{-}Acc}(x)$ using the signing program. Then, it iteratively runs the obfuscated next-step circuit (also including $\mathsf{Hash\text{-}Acc}(x)$ as input at each time step) until the circuit either outputs the PRF evaluation, or outputs $\perp$. While this is similar to the KLW message hiding encoding scheme, there are some major differences. One such difference is with regard to accumulation of the input. In KLW, the input is accumulated by the 'honest' encoding party, while in our case, the (possibly corrupt) evaluator generates the accumulation and feeds it at each step of the iteration. As a result, the KLW proof for message-hiding encoding scheme needs to be tailored to fit our setting.

Proof of Security: Recall we are interested in proving selective security, where the adversary sends the challenge input x^* before requesting for constrained keys. Our goal is to replace the (master) PRF key k in all constrained keys with one that is punctured at $\mathsf{acc\text{-}inp}^* = \mathsf{Hash\text{-}Acc}(x^*)$. Once this is done, the security of puncturable PRFs guarantees that the adversary cannot distinguish between $F(k, \mathsf{acc\text{-}inp}^*)$ and a truly random string. Let us focus our attention on one constrained key query corresponding to machine M, and suppose M runs for t^* steps on input x^* and finally outputs 'reject'.

To replace k with a punctured key, we need to ensure that the obfuscated program for M does not reach the 'accepting state' on inputs with

---

[5] We note that the *somewhat statistically binding* hash of [21] has a similar spirit to positional accumulators in that they have statistical binding at a selected position. However, they are not sufficient for our purposes as positional accumulators provide richer semantics such as interleaved reads, writes, and overwrites that are necessary here.

acc-inp = acc-inp*. This is done via two main hybrid steps. First, we alter the program so that it does not reach the accepting state within t^* steps on inputs with acc-inp = acc-inp*. Then, we have the *tail hybrid*, where we ensure that on inputs with acc-inp = acc-inp*, the program does not reach accepting state even at time steps $t > t^*$. For the first step, we follow the KLW approach, and define a sequence of t^* sub-hybrids, where in the ith hybrid, the obfuscated circuit does not reach accepting state at time steps $t \leq i$ for inputs with acc-inp = acc-inp*. We use the KLW selective enforcement techniques to show that consecutive hybrids are computationally indistinguishable.

We have a novel approach for handling the tail hybrids Let $T(= 2^\lambda)$ denote the upper bound on the running time of any machine M on any input. In KLW, the tail hybrid step was handled by defining $T - t^*$ intermediate hybrids. If we adopt a similar approach for our construction, it results in an exponential factor security loss, which is undesirable for our application[6]. Our goal would be to overcome this to get an end to end polynomial reduction to $i\mathcal{O}$. Therefore, we propose a modification to our scheme which will allow us to handle the tail hybrid with only a polynomial factor security loss. First, let us call the time step 2^i as the ith *landmark*, while the interval $[2^i, 2^{i+1} - 1]$ is the ith interval. The obfuscated program now takes a PRG seed as input at each time step, and performs some additional checks on the input PRG seed. At time steps just before a landmark, it outputs a new (pseudorandomly generated) PRG seed, which is then used in the next interval. Using standard $i\mathcal{O}$ techniques, we can show that if the program outputs $\bot$ just before a landmark, then we can alter the program indistinguishably so that it outputs $\bot$ at all time steps in the next interval. Since we know that the program outputs $\bot$ at (acc-inp*, $t^* - 1$), we can ensure that the program outputs $\bot$ for all (acc-inp*, t) such that $t^* \leq t \leq 2t^*$. Proceeding inductively, we can ensure that the program never reaches accepting state if acc-inp = acc-inp*.

1.2 Attribute Based Encryption for Turing Machines with Unbounded Inputs

We will now describe our ABE scheme for Turing machines with unbounded inputs. Let $\mathcal{PKE}$ be a public key encryption scheme. Our ABE scheme's master secret key is a puncturable PRF key k and the public key is an obfuscated program Prog-PK and accumulator parameters. The program Prog-PK takes as input a string acc-inp, computes $r = F(k, \text{acc-inp})$ and uses r as randomness for PKE.setup. It finally outputs the $\mathcal{PKE}$ public key. To encrypt a message m for attribute x, one must first accumulate the input x, then feed the accumulated input to Prog-PK to get a $\mathcal{PKE}$ public key pk, and finally encrypts m using public key pk. The secret keys corresponding to Turing machine M is simply the constrained PRF key for M. This key can be used to compute $F(k, \text{Hash-Acc}(x))$ if $M(x) = 1$, and therefore can decrypt messages encrypted for x.

---

[6] An exponential loss in the security proof of randomized encodings in KLW was acceptable because the end goal was indistinguishability obfuscation, which already requires an exponential number of hybrids.

1.3 Paper Organization

We present the required preliminaries in Sect. 2 and the notions of constrained PRFs for Turing machines in Sect. 3. The construction of our constrained PRF scheme can be found in Sect. 4, while our ABE scheme can be found in Sect. 5. Due to space constraints, part of our constrained PRF security proof is deferred to the full version of the paper.

2 Preliminaries

2.1 Notations

In this work, we will use the following notations for Turing machines.

Turing Machines: A Turing machine is a 7-tuple $M = \langle Q, \Sigma_{\text{tape}}, \Sigma_{\text{inp}}, \delta, q_0, q_{\text{ac}}, q_{\text{rej}} \rangle$ with the following semantics:

- Q is the set of states with start state q_0, accept state q_{ac} and reject state q_{rej}.
- Σ_{inp} is the set of inputs symbols.
- Σ_{tape} is the set of tape symbols. We will assume $\Sigma_{\text{inp}} \subset \Sigma_{\text{tape}}$ and there is a special blank symbol '␣' $\in \Sigma_{\text{tape}} \setminus \Sigma_{\text{inp}}$.
- $\delta : Q \times \Sigma_{\text{tape}} \to Q \times \Sigma_{\text{tape}} \times \{+1, -1\}$ is the transition function.

2.2 Obfuscation

We recall the definition of indistinguishability obfuscation from [15,25].

Definition 1 *(Indistinguishability Obfuscation). Let $\mathcal{C} = \{\mathcal{C}_\lambda\}_{\lambda \in \mathbb{N}}$ be a family of polynomial-size circuits. Let $i\mathcal{O}$ be a uniform PPT algorithm that takes as input the security parameter λ, a circuit $C \in \mathcal{C}_\lambda$ and outputs a circuit C'. $i\mathcal{O}$ is called an indistinguishability obfuscator for a circuit class $\{\mathcal{C}_\lambda\}$ if it satisfies the following conditions:*

- *(Preserving Functionality) For all security parameters $\lambda \in \mathbb{N}$, for all $C \in \mathcal{C}_\lambda$, for all inputs x, we have that $C'(x) = C(x)$ where $C' \leftarrow i\mathcal{O}(1^\lambda, C)$.*
- *(Indistinguishability of Obfuscation) For any (not necessarily uniform) PPT distinguisher $\mathcal{B} = (Samp, \mathcal{D})$, there exists a negligible function $negl(\cdot)$ such that the following holds: if for all security parameters $\lambda \in \mathbb{N}, \Pr[\forall x, C_0(x) = C_1(x) : (C_0; C_1; \sigma) \leftarrow Samp(1^\lambda)] > 1 - negl(\lambda)$, then*

$$| \Pr[\mathcal{D}(\sigma, i\mathcal{O}(1^\lambda, C_0)) = 1 : (C_0; C_1; \sigma) \leftarrow Samp(1^\lambda)]$$
$$- \Pr[\mathcal{D}(\sigma, i\mathcal{O}(1^\lambda, C_1)) = 1 : (C_0; C_1; \sigma) \leftarrow Samp(1^\lambda)]| \leq negl(\lambda).$$

In a recent work, [15] showed how indistinguishability obfuscators can be constructed for the circuit class *P/poly*. We remark that $(Samp, \mathcal{D})$ are two algorithms that pass state, which can be viewed equivalently as a single stateful algorithm $\mathcal{B}$. In our proofs we employ the latter approach, although here we state the definition as it appears in prior work.

2.3   iO-Compatible Primitives

In this section, we define extensions of some cryptographic primitives that makes them 'compatible' with indistinguishability obfuscation[7]. All of the primitives described here can be constructed from $i\mathcal{O}$ and one way functions. Their constructions can be found in [24].

Splittable Signatures. A splittable signature scheme is a normal deterministic signature scheme, augmented by some additional algorithms and properties that we require for our application. Such a signature scheme has four different kinds of signing/verification key pairs. First, we have the standard signing/verification key pairs, where the signing key can compute signatures on any message, and the verification key can verify signatures corresponding to any message. Next, we have 'all-but-one' signing/verification keys. These keys, which correspond to a special message m^*, work for all messages except m^*; that is, the signing key can sign all messages except m^*, and the verification key can verify signatures for all messages except m^* (it does not accept any signature corresponding to m^*). Third, we have 'one' signing/verification keys. These keys correspond to a special message m', and can only be used to sign/verify signatures for m'. For all other messages, the verification algorithm does not accept any signatures. Finally, we have the rejection verification key which does not accept any signatures. The setup algorithm outputs a standard signing/verification key together with a rejection verification key, while a 'splitting' algorithm uses a standard signing key to generate 'all-but-one' and 'one' signing/verification keys.

At a high level, we require the following security properties. First, the standard verification key and the rejection verification key must be computationally indistinguishable. Intuitively, this is possible because an adversary does not have any secret key or signatures. Next, we require that if an adversary is given an 'all-but-one' secret key for message m^*, then he/she cannot distinguish between a standard verification key and an 'all-but-one' verification key corresponding to m^*. We also have a similar property for the 'one' keys. No PPT adversary, given a 'one' signing key, can distinguish between a standard verification key and a 'one' verification key. Finally, we have the 'splittability' property, which states that the keys generated by splitting one signing key are indistinguishable from the case where the 'all-but-one' key pair and the 'one' key pair are generated from different signing keys.

We will now formally describe the syntax and correctness/security properties of splittable signatures.

Syntax: A splittable signature scheme $\mathcal{S}$ for message space $\mathcal{M}$ consists of the following algorithms:

Setup-Spl(1^λ). The setup algorithm is a randomized algorithm that takes as input the security parameter λ and outputs a signing key SK, a verification key VK and *reject-verification key* $\mathrm{VK}_{\mathrm{rej}}$.

---

[7] In the full version of our paper, we describe a toy example to illustrate why we need to extend/modify certain primitives in order to use them with $i\mathcal{O}$.

Sign-Spl(SK, m). The signing algorithm is a deterministic algorithm that takes as input a signing key SK and a message $m \in \mathcal{M}$. It outputs a signature σ.

Verify-Spl(VK, m, σ). The verification algorithm is a deterministic algorithm that takes as input a verification key VK, signature σ and a message m. It outputs either 0 or 1.

Split(SK, m^*). The splitting algorithm is randomized. It takes as input a secret key SK and a message $m^* \in \mathcal{M}$. It outputs a signature $\sigma_{one} =$ Sign-Spl(SK, m^*), a one-message verification key VK_{one}, an all-but-one signing key SK_{abo} and an all-but-one verification key VK_{abo}.

Sign-Spl-abo(SK_{abo}, m). The all-but-one signing algorithm is deterministic. It takes as input an all-but-one signing key SK_{abo} and a message m, and outputs a signature σ.

Correctness: Let $m^* \in \mathcal{M}$ be any message. Let (SK, VK, VK_{rej}) $\leftarrow$ Setup-Spl(1^λ) and ($\sigma_{one}, VK_{one}, SK_{abo}, VK_{abo}$) $\leftarrow$ Split(SK, m^*). Then, we require the following correctness properties:

1. For all $m \in \mathcal{M}$, Verify-Spl(VK, m, Sign-Spl(SK, m)) $= 1$.
2. For all $m \in \mathcal{M}, m \neq m^*$, Sign-Spl(SK, m) = Sign-Spl-abo(SK_{abo}, m).
3. For all σ, Verify-Spl(VK_{one}, m^*, σ) = Verify-Spl(VK, m^*, σ).
4. For all $m \neq m^*$ and σ, Verify-Spl(VK, m, σ) = Verify-Spl(VK_{abo}, m, σ).
5. For all $m \neq m^*$ and σ, Verify-Spl(VK_{one}, m, σ) = 0.
6. For all σ, Verify-Spl(VK_{abo}, m^*, σ) = 0.
7. For all σ and all $m \in \mathcal{M}$, Verify-Spl(VK_{rej}, m, σ) = 0.

Security: We will now define the security notions for splittable signature schemes. Each security notion is defined in terms of a security game between a challenger and an adversary $\mathcal{A}$.

Definition 2 (VK_{rej} **indistinguishability**). *A splittable signature scheme $\mathcal{S}$ is said to be VK_{rej} indistinguishable if any PPT adversary $\mathcal{A}$ has negligible advantage in the following security game:*

Exp-$VK_{rej}(1^\lambda, \mathcal{S}, \mathcal{A})$:

1. *Challenger computes* (SK, VK, VK_{rej}) $\leftarrow$ Setup-Spl(1^λ). *Next, it chooses* $b \leftarrow \{0,1\}$. *If* $b = 0$, *it sends VK to* $\mathcal{A}$. *Else, it sends* VK_{rej}.
2. $\mathcal{A}$ *sends its guess* b'.

$\mathcal{A}$ *wins if* $b = b'$.

We note that in the game above, $\mathcal{A}$ never receives any signatures and has no ability to produce them. This is why the difference between VK and VK_{rej} cannot be tested.

Definition 3 (VK_{one} **indistinguishability**). *A splittable signature scheme $\mathcal{S}$ is said to be VK_{one} indistinguishable if any PPT adversary $\mathcal{A}$ has negligible advantage in the following security game:*

$\mathsf{Exp\text{-}VK_{one}}(1^\lambda, \mathcal{S}, \mathcal{A})$:

1. $\mathcal{A}$ sends a message $m^* \in \mathcal{M}$.
2. Challenger computes $(\mathrm{SK}, \mathrm{VK}, \mathrm{VK_{rej}}) \leftarrow \mathsf{Setup\text{-}Spl}(1^\lambda)$. Next, it computes $(\sigma_{one}, \mathrm{VK_{one}}, \mathrm{SK_{abo}}, \mathrm{VK_{abo}}) \leftarrow \mathsf{Split}(\mathrm{SK}, m^*)$. It chooses $b \leftarrow \{0, 1\}$. If $b = 0$, it sends $(\sigma_{one}, \mathrm{VK_{one}})$ to $\mathcal{A}$. Else, it sends $(\sigma_{one}, \mathrm{VK})$ to $\mathcal{A}$.
3. $\mathcal{A}$ sends its guess b'.

$\mathcal{A}$ wins if $b = b'$.

We note that in the game above, $\mathcal{A}$ only receives the signature σ_{one} on m^*, on which VK and $\mathrm{VK_{one}}$ behave identically.

Definition 4 ($\mathrm{VK_{abo}}$ indistinguishability). *A splittable signature scheme $\mathcal{S}$ is said to be $\mathrm{VK_{abo}}$ indistinguishable if any PPT adversary $\mathcal{A}$ has negligible advantage in the following security game:*

$\mathsf{Exp\text{-}VK_{abo}}(1^\lambda, \mathcal{S}, \mathcal{A})$:

1. $\mathcal{A}$ sends a message $m^* \in \mathcal{M}$.
2. Challenger computes $(\mathrm{SK}, \mathrm{VK}, \mathrm{VK_{rej}}) \leftarrow \mathsf{Setup\text{-}Spl}(1^\lambda)$. Next, it computes $(\sigma_{one}, \mathrm{VK_{one}}, \mathrm{SK_{abo}}, \mathrm{VK_{abo}}) \leftarrow \mathsf{Split}(\mathrm{SK}, m^*)$. It chooses $b \leftarrow \{0, 1\}$. If $b = 0$, it sends $(\mathrm{SK_{abo}}, \mathrm{VK_{abo}})$ to $\mathcal{A}$. Else, it sends $(\mathrm{SK_{abo}}, \mathrm{VK})$ to $\mathcal{A}$.
3. $\mathcal{A}$ sends its guess b'.

$\mathcal{A}$ wins if $b = b'$.

We note that in the game above, $\mathcal{A}$ does not receive or have the ability to create a signature on m^*. For all signatures $\mathcal{A}$ can create by signing with $\mathrm{SK_{abo}}$, $\mathrm{VK_{abo}}$ and VK will behave identically.

Definition 5 (Splitting indistinguishability). *A splittable signature scheme $\mathcal{S}$ is said to be splitting indistinguishable if any PPT adversary $\mathcal{A}$ has negligible advantage in the following security game:*

$\mathsf{Exp\text{-}Spl}(1^\lambda, \mathcal{S}, \mathcal{A})$:

1. $\mathcal{A}$ sends a message $m^* \in \mathcal{M}$.
2. Challenger computes $(\mathrm{SK}, \mathrm{VK}, \mathrm{VK_{rej}}) \leftarrow \mathsf{Setup\text{-}Spl}(1^\lambda)$, $(\mathrm{SK}', \mathrm{VK}', \mathrm{VK'_{rej}}) \leftarrow \mathsf{Setup\text{-}Spl}(1^\lambda)$. Next, it computes $(\sigma_{one}, \mathrm{VK_{one}}, \mathrm{SK_{abo}}, \mathrm{VK_{abo}}) \leftarrow \mathsf{Split}(\mathrm{SK}, m^*)$, $(\sigma'_{one}, \mathrm{VK'_{one}}, \mathrm{SK'_{abo}}, \mathrm{VK'_{abo}}) \leftarrow \mathsf{Split}(\mathrm{SK}', m^*)$. It chooses $b \leftarrow \{0, 1\}$. If $b = 0$, it sends $(\sigma_{one}, \mathrm{VK_{one}}, \mathrm{SK_{abo}}, \mathrm{VK_{abo}})$ to $\mathcal{A}$. Else, it sends $(\sigma'_{one}, \mathrm{VK'_{one}}, \mathrm{SK_{abo}}, \mathrm{VK_{abo}})$ to $\mathcal{A}$.
3. $\mathcal{A}$ sends its guess b'.

$\mathcal{A}$ wins if $b = b'$.

In the game above, $\mathcal{A}$ is either given a system of $\sigma_{one}, \mathrm{VK_{one}}, \mathrm{SK_{abo}}, \mathrm{VK_{abo}}$ generated together by one call of $\mathsf{Setup\text{-}Spl}$ or a "split" system of $(\sigma'_{one}, \mathrm{VK'_{one}}, \mathrm{SK_{abo}}, \mathrm{VK_{abo}})$ where the all but one keys are generated separately from the signature and key for the one message m^*. Since the correctness conditions do not link the behaviors for the all but one keys and the one message values, this split generation is not detectable by testing verification for the σ_{one} that $\mathcal{A}$ receives or for any signatures that $\mathcal{A}$ creates honestly by signing with SK_{abo}.

Positional Accumulators. An accumulator can be seen as a special hash function mapping unbounded[8] length strings to fixed length strings. It has two additional properties: succinct verifiability and succinct updatability. Let $\mathsf{Hash\text{-}Acc}(\cdot)$ be the hash function mapping $x = x_1 \ldots x_n$ to y. Then, succinct verifiability means that there exists a 'short' proof π to prove that bit x_i is present at the ith position of x. Note that this verification only requires the hash value y and the short proof π. Succinct updatability means that given y, a bit x'_i, position i and some 'short' auxiliary information, one can update y to obtain $y' = \mathsf{Hash\text{-}Acc}(x_1 \ldots x'_i \ldots x_n)$. We will refer to y as the tape, and x_i the symbol written at position i.

The notion of accumulators is not sufficient for using with $i\mathcal{O}$, and we need a stronger primitive called *positional accumulators* that is iO-compatible. In a positional accumulator, we have three different setup modes. The first one is the standard setup which outputs public parameters and the initial accumulation corresponding to the empty tape. Next, we have the read-enforced setup mode. In this mode, the algorithm takes as input a sequence of k pairs $(\mathsf{sym}_i, \mathsf{pos}_i)$ which represent the first k symbols written and their positions. It also takes as input the enforcing position pos, and outputs public parameters and an accumulation of the empty tape. As the name might suggest, this mode is read enforcing at position pos - if the first k symbols written are $(\mathsf{sym}_1, \ldots, \mathsf{sym}_k)$, and their write positions are $(\mathsf{pos}_1, \ldots, \mathsf{pos}_k)$, then there *exists* exactly one opening for position pos: the correct symbol written at pos. Similarly, we have a write-enforcing setup which takes as input k (symbol, position) pairs $\{(\mathsf{sym}_i, \mathsf{pos}_i)\}_{i \leq k}$ representing the first k writes, and outputs public parameters and an accumulation of the empty tape. The write-enforcing property states that if $(\mathsf{sym}_i, \mathsf{pos}_i)$ are the first k writes, and acc_{k-1} is the correct accumulation after the first $k-1$ writes, then there is a unique accumulation after the kth write (irrespective of the auxiliary string). Note that both the read and write enforcing properties are information theoretic. This is important when we are using these primitives with indistinguishability obfuscation.

For security, we require that the different setup modes are computationally indistinguishable. We will now give a formal description of the syntax and properties. A positional accumulator for message space $\mathcal{M}_\lambda$ consists of the following algorithms.

- $\mathsf{Setup\text{-}Acc}(1^\lambda, T) \rightarrow (PP, \mathsf{acc}_0, \mathsf{STORE}_0)$: The setup algorithm takes as input a security parameter λ in unary and an integer T in binary representing the maximum number of values that can stored. It outputs public parameters PP, an initial accumulator value acc_0, and an initial storage value STORE_0.

---

[8] Unbounded, but polynomial in the security parameter.

- Setup-Acc-Enf-Read($1^\lambda, T, (m_1, \text{INDEX}_1), \ldots, (m_k, \text{INDEX}_k), \text{INDEX}^*) \rightarrow$ (PP, $\text{acc}_0, \text{STORE}_0$): The setup enforce read algorithm takes as input a security parameter λ in unary, an integer T in binary representing the maximum number of values that can be stored, and a sequence of symbol, index pairs, where each index is between 0 and $T-1$, and an additional INDEX* also between 0 and $T-1$. It outputs public parameters PP, an initial accumulator value acc_0, and an initial storage value STORE_0.
- Setup-Acc-Enf-Write($1^\lambda, T, (m_1, \text{INDEX}_1), \ldots, (m_k, \text{INDEX}_k)) \rightarrow$ (PP, acc_0, STORE_0): The setup enforce write algorithm takes as input a security parameter λ in unary, an integer T in binary representing the maximum number of values that can be stored, and a sequence of symbol, index pairs, where each index is between 0 and $T-1$. It outputs public parameters PP, an initial accumulator value acc_0, and an initial storage value STORE_0.
- Prep-Read(PP, STORE_{in}, INDEX) $\rightarrow (m, \pi)$: The prep-read algorithm takes as input the public parameters PP, a storage value STORE_{in}, and an index between 0 and $T-1$. It outputs a symbol m (that can be ϵ) and a value π.
- Prep-Write(PP, STORE_{in}, INDEX) $\rightarrow aux$: The prep-write algorithm takes as input the public parameters PP, a storage value STORE_{in}, and an index between 0 and $T-1$. It outputs an auxiliary value aux.
- Verify-Read(PP, $\text{acc}_{in}, m_{read}, \text{INDEX}, \pi) \rightarrow \{True, False\}$: The verify-read algorithm takes as input the public parameters PP, an accumulator value acc_{in}, a symbol, m_{read}, an index between 0 and $T-1$, and a value π. It outputs $True$ or $False$.
- Write-Store(PP, STORE_{in}, INDEX, m) $\rightarrow \text{STORE}_{out}$: The write-store algorithm takes in the public parameters, a storage value STORE_{in}, an index between 0 and $T-1$, and a symbol m. It outputs a storage value STORE_{out}.
- Update(PP, $\text{acc}_{in}, m_{write}, \text{INDEX}, aux) \rightarrow \text{acc}_{out}$ or $Reject$: The update algorithm takes in the public parameters PP, an accumulator value acc_{in}, a symbol m_{write}, and index between 0 and $T-1$, and an auxiliary value aux. It outputs an accumulator value acc_{out} or $Reject$.

In general we will think of the Setup-Acc algorithm as being randomized and the other algorithms as being deterministic. However, one could consider non-deterministic variants.

Correctness: We consider any sequence $(m_1, \text{INDEX}_1), \ldots, (m_k, \text{INDEX}_k)$ of symbols $m_1, \ldots, m_k$ and indices $\text{INDEX}_1, \ldots, \text{INDEX}_k$ each between 0 and $T-1$. We fix any PP, $\text{acc}_0, \text{STORE}_0 \leftarrow$ Setup-Acc($1^\lambda, T$). For j from 1 to k, we define STORE_j iteratively as $\text{STORE}_j := $ Write-Store(PP, $\text{STORE}_{j-1}, \text{INDEX}_j, m_j$). We similarly define aux_j and acc_j iteratively as $aux_j := $ Prep-Write(PP, $\text{STORE}_{j-1}, \text{INDEX}_j$) and $\text{acc}_j := Update($PP, $\text{acc}_{j-1}, m_j, \text{INDEX}_j, aux_j)$. Note that the algorithms other than Setup-Acc are deterministic, so these definitions fix precise values, not random values (conditioned on the fixed starting values PP, $\text{acc}_0, \text{STORE}_0$).

We require the following correctness properties:

1. For every INDEX between 0 and $T - 1$, Prep-Read(PP, STORE$_k$, INDEX) returns m_i, π, where i is the largest value in $[k]$ such that INDEX$_i$ = INDEX. If no such value exists, then $m_i = \epsilon$.
2. For any INDEX, let $(m, \pi) \leftarrow$ Prep-Read(PP, STORE$_k$, INDEX). Then Verify-Read(PP, acc$_k$, m, INDEX, π) = $True$.

Remarks on Efficiency: In our construction, all algorithms will run in time polynomial in their input sizes. More precisely, Setup-Acc will be polynomial in λ and $\log(T)$. Also, accumulator and π values should have size polynomial in λ and $\log(T)$, so Verify-Read and Update will also run in time polynomial in λ and $\log(T)$. Storage values will have size polynomial in the number of values stored so far. Write-Store, Prep-Read, and Prep-Write will run in time polynomial in λ and T.

Security: Let Acc = (Setup-Acc, Setup-Acc-Enf-Read, Setup-Acc-Enf-Write, Prep-Read, Prep-Write, Verify-Read, Write-Store, Update) be a positional accumulator for symbol set $\mathcal{M}$. We require Acc to satisfy the following notions of security.

Definition 6 (Indistinguishability of Read Setup). *A positional accumulator Acc is said to satisfy indistinguishability of read setup if any PPT adversary $\mathcal{A}$'s advantage in the security game Exp-Setup-Acc(1^λ, Acc, $\mathcal{A}$) is at most negligible in λ, where Exp-Setup-Acc is defined as follows.*

Exp-Setup-Acc(1^λ, Acc, $\mathcal{A}$)

1. *Adversary chooses a bound $T \in \Theta(2^\lambda)$ and sends it to challenger.*
2. *$\mathcal{A}$ sends k messages $m_1, \ldots, m_k \in \mathcal{M}$ and k indices INDEX$_1$, ..., INDEX$_k \in \{0, \ldots, T - 1\}$ to the challenger.*
3. *The challenger chooses a bit b. If $b = 0$, the challenger outputs (PP, acc$_0$, STORE$_0$) $\leftarrow$ Setup-Acc(1^λ, T). Else, it outputs (PP, acc$_0$, STORE$_0$) $\leftarrow$ Setup-Acc-Enf-Read $(1^\lambda, T, (m_1, \text{INDEX}_1), \ldots, (m_k, \text{INDEX}_k))$.*
4. *$\mathcal{A}$ sends a bit b'.*

$\mathcal{A}$ wins the security game if $b = b'$.

Definition 7 (Indistinguishability of Write Setup). *A positional accumulator Acc is said to satisfy indistinguishability of write setup if any PPT adversary $\mathcal{A}$'s advantage in the security game Exp-Setup-Acc(1^λ, Acc, $\mathcal{A}$) is at most negligible in λ, where Exp-Setup-Acc is defined as follows.*

Exp-Setup-Acc(1^λ, Acc, $\mathcal{A}$)

1. *Adversary chooses a bound $T \in \Theta(2^\lambda)$ and sends it to challenger.*
2. *$\mathcal{A}$ sends k messages $m_1, \ldots, m_k \in \mathcal{M}$ and k indices INDEX$_1$, ..., INDEX$_k \in \{0, \ldots, T - 1\}$ to the challenger.*

3. *The challenger chooses a bit* b. *If* $b = 0$, *the challenger outputs*
   $(\mathrm{PP}, \mathrm{acc}_0, \mathrm{STORE}_0) \leftarrow$ Setup-Acc$(1^\lambda, T)$. *Else, it outputs*
   $(\mathrm{PP}, \mathrm{acc}_0, \mathrm{STORE}_0) \leftarrow$ Setup-Acc-Enf-Write
   $(1^\lambda, T, (m_1, \mathrm{INDEX}_1), \ldots, (m_k, \mathrm{INDEX}_k))$.
4. $\mathcal{A}$ *sends a bit* b'.

$\mathcal{A}$ *wins the security game if* $b = b'$.

Definition 8 (Read Enforcing). *Consider any* $\lambda \in \mathbb{N}$, $T \in \Theta(2^\lambda)$, $m_1, \ldots, m_k \in \mathcal{M}$, $\mathrm{INDEX}_1, \ldots, \mathrm{INDEX}_k \in \{0, \ldots, T-1\}$ *and any* $\mathrm{INDEX}^* \in \{0, \ldots, T-1\}$. *Let* $(\mathrm{PP}, \mathrm{acc}_0, \mathrm{STORE}_0) \leftarrow$ Setup-Acc-Enf-Read $(1^\lambda, T, (m_1, \mathrm{INDEX}_1), \ldots, (m_k, \mathrm{INDEX}_k), \mathrm{INDEX}^*)$. *For* j *from 1 to* k, *we define* STORE_j *iteratively as* $\mathrm{STORE}_j :=$ Write-Store$(\mathrm{PP}, \mathrm{STORE}_{j-1}, \mathrm{INDEX}_j, m_j)$. *We similarly* *define* aux_j *and* acc_j *iteratively as* $aux_j :=$ Prep-Write$(\mathrm{PP}, \mathrm{STORE}_{j-1}, \mathrm{INDEX}_j)$ *and* $\mathrm{acc}_j := Update(\mathrm{PP}, \mathrm{acc}_{J-1}, m_j, \mathrm{INDEX}_j, aux_j)$. *Acc is said to be* Read enforcing *if* Verify-Read$(\mathrm{PP}, \mathrm{acc}_k, m, \mathrm{INDEX}^*, \pi) = True$, *then either* $\mathrm{INDEX}_1, \ldots,$ $\mathrm{INDEX}_K \mathrm{INDEX}_1, \ldots, \mathrm{INDEX}_K \mathrm{INDEX}_1, \ldots, \mathrm{INDEX}_K \mathrm{INDEX}_1, \ldots, \mathrm{INDEX}_K \mathrm{INDEX}^* \notin$ $\{\mathrm{INDEX}_1, \ldots, \mathrm{INDEX}_K\}$ *and* $m = \epsilon$, *or* $m = m_i$ *for the largest* $i \in [k]$ *such that* $\mathrm{INDEX}_i = \mathrm{INDEX}^*$. *Note that this is an information-theoretic property: we are requiring that for all other symbols* m, *values of* π *that would cause* Verify-Read *to output* True *at* INDEX^* *do no exist.*

Definition 9 (Write Enforcing). *Consider any* $\lambda \in \mathbb{N}$, $T \in \Theta(2^\lambda)$, $m_1, \ldots,$ $m_k \in \mathcal{M}$, $\mathrm{INDEX}_1, \ldots, \mathrm{INDEX}_k \in \{0, \ldots, T-1\}$. *Let* $(\mathrm{PP}, \mathrm{acc}_0, \mathrm{STORE}_0) \leftarrow$ Setup-Acc-Enf-Write$(1^\lambda, T, (m_1, \mathrm{INDEX}_1), \ldots, (m_k, \mathrm{INDEX}_k))$. *For* j *from 1 to* k, *we define* STORE_j *iteratively as* $\mathrm{STORE}_j :=$ Write-Store$(\mathrm{PP}, \mathrm{STORE}_{j-1}, \mathrm{INDEX}_j, m_j)$. *We similarly define* aux_j *and* acc_j *iteratively as* $aux_j :=$ Prep-Write$(\mathrm{PP}, \mathrm{STORE}_{j-1}, \mathrm{INDEX}_j)$ *and* $\mathrm{acc}_j := Update(\mathrm{PP}, \mathrm{acc}_{J-1}, m_j, \mathrm{INDEX}_j, aux_j)$. *Acc is said to be* write enforcing *if* Update$(\mathrm{PP}, \mathrm{acc}_{k-1}, m_k, \mathrm{INDEX}_k, aux) = \mathrm{acc}_{out} \neq$ *Reject, for any* aux, *then* $\mathrm{acc}_{out} = \mathrm{acc}_k$. *Note that this is an information-theoretic property: we are requiring that an* aux *value producing an accumulated value other than* acc_k *or Reject does not exist.*

Iterators. In this section, we define the notion of *cryptographic iterators*. A cryptographic iterator essentially consists of a small state that is updated in an iterative fashion as messages are received. An update to apply a new message given current state is performed via some public parameters.

Since states will remain relatively small regardless of the number of messages that have been iteratively applied, there will in general be many sequences of messages that can lead to the same state. However, our security requirement will capture that the normal public parameters are computationally indistinguishable from specially constructed "enforcing" parameters that ensure that a particular *single* state can be only be obtained as an output as an update to precisely one other state, message pair. Note that this enforcement is a very localized property to a particular state, and hence can be achieved information-theoretically when we fix ahead of time where exactly we want this enforcement to be.

Syntax: Let ℓ be any polynomial. An iterator $\mathcal{I}$ with message space $\mathcal{M}_\lambda = \{0,1\}^{\ell(\lambda)}$ and state space $\mathcal{S}_\lambda$ consists of three algorithms - Setup-Itr, Setup-Itr-Enf and Iterate defined below.

Setup-Itr($1^\lambda, T$). The setup algorithm takes as input the security parameter λ (in unary), and an integer bound T (in binary) on the number of iterations. It outputs public parameters PP and an initial state $v_0 \in \mathcal{S}_\lambda$.

Setup-Itr-Enf($1^\lambda, T, \mathbf{m} = (m_1, \ldots, m_k)$). The enforced setup algorithm takes as input the security parameter λ (in unary), an integer bound T (in binary) and k messages $(m_1, \ldots, m_k)$, where each $m_i \in \{0,1\}^{\ell(\lambda)}$ and k is some polynomial in λ. It outputs public parameters PP and a state $v_0 \in \mathcal{S}$.

Iterate(PP, v_{in}, m). The iterate algorithm takes as input the public parameters PP, a state v_{in}, and a message $m \in \{0,1\}^{\ell(\lambda)}$. It outputs a state $v_{\text{out}} \in \mathcal{S}_\lambda$.

For simplicity of notation, we will drop the dependence of ℓ on λ. Also, for any integer $k \leq T$, we will use the notation Iterate^k(PP, $v_0, (m_1, \ldots, m_k)$) to denote Iterate(PP, v_{k-1}, m_k), where $v_j =$ Iterate(PP, v_{j-1}, m_j) for all $1 \leq j \leq k-1$.

Security: Let $\mathcal{I} = $ (Setup-Itr, Setup-Itr-Enf, Iterate) be an iterator with message space $\{0,1\}^\ell$ and state space $\mathcal{S}_\lambda$. We require the following notions of security.

Definition 10 (Indistinguishability of Setup). *An iterator $\mathcal{I}$ is said to satisfy indistinguishability of Setup phase if any PPT adversary $\mathcal{A}$'s advantage in the security game* Exp-Setup-Itr($1^\lambda, \mathcal{I}, \mathcal{A}$) *at most is negligible in λ, where* Exp-Setup-Itr *is defined as follows.*

Exp-Setup-Itr($1^\lambda, \mathcal{I}, \mathcal{A}$)

1. *The adversary $\mathcal{A}$ chooses a bound $T \in \Theta(2^\lambda)$ and sends it to challenger.*
2. *$\mathcal{A}$ sends k messages $m_1, \ldots, m_k \in \{0,1\}^\ell$ to the challenger.*
3. *The challenger chooses a bit b. If $b = 0$, the challenger outputs (PP, v_0) $\leftarrow$ Setup-Itr($1^\lambda, T$). Else, it outputs (PP, v_0) $\leftarrow$ Setup-Itr-Enf($1^\lambda, T, 1^k, \mathbf{m} = (m_1, \ldots, m_k)$).*
4. *$\mathcal{A}$ sends a bit b'.*

$\mathcal{A}$ wins the security game if $b = b'$.

Definition 11 (Enforcing). *Consider any $\lambda \in \mathbb{N}$, $T \in \Theta(2^\lambda)$, $k < T$ and $m_1, \ldots, m_k \in \{0,1\}^\ell$. Let (PP, v_0) $\leftarrow$ Setup-Itr-Enf($1^\lambda, T, \mathbf{m} = (m_1, \ldots, m_k)$) and $v_j =$ Iterate^j(PP, $v_0, (m_1, \ldots, m_j)$) for all $1 \leq j \leq k$. Then, $\mathcal{I} = $ (Setup-Itr, Setup-Itr-Enf, Iterate) is said to be enforcing if*

$$v_k = \text{Iterate}(\text{PP}, v', m') \implies (v', m') = (v_{k-1}, m_k).$$

Note that this is an information-theoretic property.

2.4 Attribute Based Encryption

An ABE scheme where policies are represented by Turing machines comprises of the following four algorithms (ABE.setup, ABE.enc, ABE.keygen, ABE.dec):

- ABE.setup(1^λ) $\to$ ($\mathrm{PK_{ABE}}, \mathrm{MSK_{ABE}}$): The setup algorithm takes as input the security parameter λ and outputs the public key $\mathrm{PK_{ABE}}$ and the master secret key $\mathrm{MSK_{ABE}}$
- ABE.enc($m, x, \mathrm{PK_{ABE}}$) $\to$ ct: The encryption algorithm takes as input the message m, the attribute string x of unbounded length and the public key $\mathrm{PK_{ABE}}$ and it outputs the corresponding ciphertext ct_x specific to the attribute string.
- ABE.keygen($\mathrm{MSK_{ABE}}, M$) $\to$ SK$\{M\}$: The key generation algorithm takes as input $\mathrm{MSK_{ABE}}$ and a Turing machine M and outputs the secret key SK$\{M\}$ specific to M
- ABE.dec(SK$\{M\}$, ct) $\to m$ or $\bot$: The decryption algorithm takes in SK$\{M\}$ and ciphertext ct and outputs either a message m or $\bot$.

The *correctness* of the scheme guarantees that if ABE.enc($m, x, \mathrm{PK_{ABE}}$) $\to \mathrm{ct}_x$ and ABE.keygen($\mathrm{MSK_{ABE}}, M$) $\to$ SK$\{M\}$ then ABE.dec(SK$\{M\}$, ct_x) $\to m$.

2.5 Selective Security

Consider the following experiment between a challenger $\mathcal{C}$ and a stateful adversary $\mathcal{A}$:

- **Setup Phase:** $\mathcal{A}$ sends the challenge attribute string x^* of his choice to $\mathcal{C}$. $\mathcal{C}$ runs the ABE.setup(1^λ) and sends across $\mathrm{PK_{ABE}}$ to $\mathcal{A}$.
- **Pre-Challenge Query Phase:** $\mathcal{A}$ gets to query for secret keys corresponding to Turing machines. For each query M such that $M(x^*) = 0$, the challenger computes SK$\{M\} \leftarrow$ ABE.keygen($\mathrm{MSK_{ABE}}, .$) and sends it to $\mathcal{A}$.
- **Challenge Phase:** $\mathcal{A}$ sends two messages m_0, m_1 with $|m_0| = |m_1|$, the challenger chooses bit b uniformly at random and outputs $\mathrm{ct}^* = $ ABE.enc($m_b, x^*, \mathrm{PK_{ABE}}$).
- **Post-Challenge Query Phase:** This is identical to the Pre-Challenge Phase.
- **Guess:** Finally, $\mathcal{A}$ sends its guess b' and wins if $b = b'$.

The advantage of $\mathcal{A}$, $\mathsf{Adv}_{\mathcal{A}}^{ABE}(\lambda)$ in the above experiment is defined to be $|\Pr[b' = b] - \frac{1}{2}|$.

Definition 12. *An ABE scheme is said to be* selectively secure *if for all PPT adversaries $\mathcal{A}$, the advantage $\mathsf{Adv}_{\mathcal{A}}^{ABE}(\lambda)$ is a negligible function in λ.*

3   Constrained Pseudorandom Functions for Turing Machines

The notion of constrained pseudorandom functions was introduced in the concurrent works of [7,9,23]. Informally, a constrained PRF extends the notion of standard PRFs, enabling the master PRF key holder to compute 'constrained keys' that allow PRF evaluations on certain inputs, while the PRF evaluation on remaining inputs 'looks' random. In the above mentioned works, these constraints could only handle bounded length inputs. In order to allow unbounded inputs, we need to ensure that the constrained keys correspond to polynomial time Turing Machines. A formal definition is as follows.

Let $\mathcal{M}_\lambda$ be a family of Turing machines with (worst case) running time bounded by 2^λ. Let $\mathcal{K}$ denote the key space, $\mathcal{X}$ the input domain and $\mathcal{Y}$ the range space. A pseudorandom PRF : $\mathcal{K} \times \mathcal{X} \to \mathcal{Y}$ is said to be *constrained* with respect to the Turing machine family $\mathcal{M}_\lambda$ if there is an additional key space $\mathcal{K}_c$, and three algorithms PRF.setup, PRF.constrain and PRF.eval as follows:

- PRF.setup(1^λ) is a PPT algorithm that takes the security parameter λ as input and outputs a key $K \in \mathcal{K}$.
- PRF.constrain(K, M) is a PPT algorithm that takes as input a PRF key $K \in \mathcal{K}$ and a Turing machine $M \in \mathcal{M}_\lambda$ and outputs a constrained key $K\{M\} \in \mathcal{K}_c$.
- PRF.eval($K\{M\}, x$) is a deterministic polynomial time algorithm that takes as input a constrained key $K\{M\} \in \mathcal{K}_c$ and $x \in \mathcal{X}$ and outputs an element $y \in \mathcal{Y}$. Let $K\{M\}$ be the output of PRF.constrain(K, M). For correctness, we require the following:

$$\text{PRF.eval}(K\{M\}, x) = F(K, x) \text{ if } M(x) = 1.$$

For simplicity of notation, we will use PRF($K\{M\}, x$) to denote PRF.eval($K\{M\}, x$).

3.1   Security of Constrained Pseudorandom Functions

Intuitively, we require that even after obtaining several constrained keys, no polynomial time adversary can distinguish a truly random string from the PRF evaluation at a point not accepted by the queried Turing machines. In this work, we achieve a weaker notion of security called *selective security*, which is formalized by the following security game between a challenger and an adversary Att.

Let PRF : $\mathcal{K} \times \mathcal{X} \to \mathcal{Y}$ be a constrained PRF with respect to a Turing machine family $\mathcal{M}$. The security game consists of three phases.

Setup Phase: The adversary sends the challenge input x^*. The challenger chooses a random key $K \leftarrow \mathcal{K}$ and a random bit $b \leftarrow \{0,1\}$. If $b = 0$, the challenger outputs PRF(K, x^*). Else, the challenger outputs a random element $y \leftarrow \mathcal{Y}$.

Query Phase: In this phase, Att is allowed to ask for the following queries:

- **Evaluation Query.** Att sends $x \in \mathcal{X}$, and receives $\mathsf{PRF}(K, x)$.
- **Key Query.** Att sends a Turing machine $M \in \mathcal{M}$ such that $M(x^*) = 0$, and receives $\mathsf{PRF.constrain}(K, M)$.

Guess: Finally, A outputs a guess b' of b.

A wins if $b = b'$ and the advantage of Att is defined to be $\mathsf{Adv}_{\mathsf{Att}}(\lambda) = \left| \Pr[\mathsf{Att\ wins}] - 1/2 \right|$.

Definition 13. *The PRF* PRF *is a secure constrained PRF with respect to* $\mathcal{M}$ *if for all PPT adversaries* A $\mathsf{Adv}_{\mathsf{Att}}(\lambda)$ *is negligible in* λ.

3.2 Puncturable Pseudorandom Functions

A special class of constrained PRFs, called *puncturable PRFs*, was introduced in the work of [25]. In a puncturable PRF, the constrained key queries correspond to points in the input domain, and the constrained key is one that allows PRF evaluations at all points except the punctured point.

Formally, a PRF $\mathsf{F} : \mathcal{K} \times \mathcal{X} \to \mathcal{Y}$ is a puncturable pseudorandom function if there is an additional key space $\mathcal{K}_p$ and three polynomial time algorithms $\mathsf{F.setup}$, $\mathsf{F.eval}$ and $\mathsf{F.puncture}$ as follows:

- $\mathsf{F.setup}(1^\lambda)$ is a randomized algorithm that takes the security parameter λ as input and outputs a description of the key space $\mathcal{K}$, the punctured key space $\mathcal{K}_p$ and the PRF F.
- $\mathsf{F.puncture}(K, x)$ is a randomized algorithm that takes as input a PRF key $K \in \mathcal{K}$ and $x \in \mathcal{X}$, and outputs a key $K_x \in \mathcal{K}_p$.
- $\mathsf{F.eval}(K_x, x')$ is a deterministic algorithm that takes as input a punctured key $K_x \in \mathcal{K}_p$ and $x' \in \mathcal{X}$. Let $K \in \mathcal{K}$, $x \in \mathcal{X}$ and $K_x \leftarrow \mathsf{F.puncture}(K, x)$. For correctness, we need the following property:

$$\mathsf{F.eval}(K_x, x') = \begin{cases} F(K, x') & \text{if } x \neq x' \\ \bot & \text{otherwise} \end{cases}$$

The selective security notion is analogous to the security notion of constrained PRFs.

4   Construction

A High Level Description of Our Construction: Our constrained PRF construction uses a puncturable PRF F as the base pseudorandom function. The setup algorithm chooses a puncturable PRF key K together with the public parameters of the accumulator and an accumulation of the empty tape (it also outputs additional parameters for the authenticity checks described in the next paragraph). To evaluate the constrained PRF on input x, one first accumulates the input x. Let y denote this accumulation. The PRF evaluation is $F(K, y)$.

Next, let us consider the constrained key for Turing machine M. The major component of this key is an obfuscated program Prog. At a very high level, this program evaluates the next-step circuit of M. Its main inputs are the time step t, hash y of the input and the symbol, state, position of TM used at step t. Using the state and symbol, it computes the next state and the symbol to be written. If the state is accepting, it outputs $F(K, y)$, else it outputs the next state and symbol. However, this is clearly not enough, since the adversary could pass illegal states and symbols as inputs. So the program first performs some additional authenticity checks then evaluates the next state, symbol, and finally outputs authentication required for the next step evaluation. These authenticity checks are imposed via the accumulator, signature scheme and iterator. For these checks, Prog takes additional inputs: accumulation of the current tape acc, proof π that the input symbol is the correct symbol at the tape-head position, auxiliary string aux to update the accumulation, iterated value and signature σ. The iterated value and the signature together ensure that the correct state and accumulated value is input at each step, while the accumulation ensures that the adversary cannot send a wrong symbol. Finally, to perform the 'tail-cutting', the program requires an additional input seed. The first and last step of the program are for checking the validity of seed, and to output the new seed if required. The constrained key also has another program Init-Sign which is used to sign the accumulation of the input. In the end, if all the checks go through, the final output will be the PRF evaluation using the constrained key.

Formal Description: Let Acc = (Setup-Acc, Setup-Acc-Enf-Read, Setup-Acc-Enf-Write, Prep-Read, Prep-Write, Verify-Read, Write-Store, Update) be a positional accumulator, Itr = (Setup-Itr, Setup-Itr-Enf, Iterate) an iterator, $\mathcal{S}$ = (Setup-Spl, Sign-Spl, Verify-Spl, Split, Sign-Spl-abo) a splittable signature scheme and PRG : $\{0,1\}^\lambda \rightarrow \{0,1\}^{2\lambda}$ a length doubling injective pseudorandom generator.

Let F be a puncturable pseudorandom function whose domain and range are chosen appropriately, depending on the accumulator, iterator and splittable signature scheme. For simplicity, we assume that F takes inputs of bounded length, instead of fixed length inputs. This assumption can be easily removed by using different PRFs for different input lengths (in our case, we will require three different fixed-input-length PRFs). Also, to avoid confusion, the puncturable PRF keys (both master and punctured) are represented using lower case letters (e.g. k, $k\{z\}$), while the constrained PRF keys are represented using upper case letters (e.g. K, $K\{M\}$).

- PRF.setup(1^λ): The setup algorithm takes the security parameter λ as input. It first chooses a puncturable PRF keys $k \leftarrow$ F.setup(1^λ). Next, it runs the accumulator setup to obtain $(\text{PP}_{\text{Acc}}, \text{acc}_0, \text{STORE}_0) \leftarrow$ Setup-Acc(1^λ). The master PRF key is $K = (k, \text{PP}_{\text{Acc}}, \text{acc}_0, \text{STORE}_0)$.
- PRF Evaluation: To evaluate the PRF with key $K = (k, \text{PP}_{\text{Acc}}, \text{acc}_0, \text{STORE}_0)$ on input $x = x_1 \ldots x_n$, first 'hash' the input using the accumulator. More formally, let Hash-Acc(x) = acc_n, where for all $j \leq n$, acc_j is defined as follows:

- $\text{STORE}_j = \text{Write-Store}(\text{PP}_\text{Acc}, \text{STORE}_{j-1}, j-1, x_j)$
- $aux_j = \text{Prep-Write}(\text{PP}_\text{Acc}, \text{STORE}_{j-1}, j-1)$
- $\text{acc}_j = \text{Update}(\text{PP}_\text{Acc}, \text{acc}_{j-1}, x_j, j-1, aux_j)$

The PRF evaluation is defined to be $F(k, \text{Hash-Acc}(x))$.

- PRF.constrain$(K = (k, \text{PP}_\text{Acc}, \text{acc}_0, \text{STORE}_0), M)$: The constrain algorithm first chooses puncturable PRF keys $k_1, \ldots, k_\lambda$ and $k_{\text{sig},A}$ and runs the iterator setup to obtain $(\text{PP}_\text{Itr}, \text{it}_0) \leftarrow \text{Setup-Itr}(1^\lambda, T)$. Next, it computes an obfuscation of program Prog (defined in Fig. 1) and Init-Sign (defined in Fig. 2). The constrained key $K\{M\} = (\text{PP}_\text{Acc}, \text{acc}_0, \text{STORE}_0, \text{PP}_\text{Itr}, \text{it}_0, i\mathcal{O}(\text{Prog}), i\mathcal{O}(\text{Init-Sign}))$.

- PRF Evaluation using Constrained Key: Let $K\{M\} = (\text{PP}_\text{Acc}, \text{acc}_0, \text{STORE}_0, \text{PP}_\text{Itr}, \text{it}_0, P_1, P_2)$ be a constrained key corresponding to machine M, and $x = x_1, \ldots, x_n$ the input. As in the evaluation using master PRF key, first compute $\text{acc-inp} = \text{Hash-Acc}(x)$.

To begin the evaluation, compute a signature on the initial values using the program P_2. Let $\sigma_0 = P_2(\text{acc-inp})$.

Suppose M runs for t^* steps on input x. Run the program P_1 iteratively for t^* steps. Set $\text{pos}_0 = 0$, $\text{seed}_0 = ``"$, and for $i = 1$ to t^*, compute

1. Let $(\text{sym}_{i-1}, \pi_{i-1}) = \text{Prep-Read}(\text{PP}_\text{Acc}, \text{STORE}_{i-1}, \text{pos}_{i-1})$.
2. Compute $aux_{i-1} \leftarrow \text{Prep-Write}(\text{PP}_\text{Acc}, \text{STORE}_{i-1}, \text{pos}_{i-1})$.
3. Let $\text{out} = P_1(i, \text{seed}_{i-1}, \text{pos}_{i-1}, \text{sym}_{i-1}, \text{st}_{i-1}, \text{acc}_{i-1}, \pi_{i-1}, aux_{i-1},$
   $\text{acc-inp}, \text{it}_{i-1}, \sigma_{i-1})$.

   If $j = t^*$, output out. Else, parse out as $(\text{sym}_{w,i}, \text{pos}_i, \text{st}_i, \text{acc}_i, \text{it}_i, \sigma_i, \text{seed}_i)$.
4. Compute $\text{STORE}_i = \text{Write-Store}(\text{PP}_\text{Acc}, \text{STORE}_{i-1}, \text{pos}_{i-1}, \text{sym}_{w,i})$.

The output at step t^* is the PRF evaluation using the constrained key.

4.1   Proof of Selective Security

Theorem 1. *Assuming $i\mathcal{O}$ is a secure indistinguishability obfuscator, F is a selectively secure puncturable pseudorandom function, Acc is a secure positional accumulator, Itr is a secure positional iterator and S is a secure splittable signature scheme, the constrained PRF construction described in Sect. 4 is selectively secure as defined in Definition 13.*

Our security proof will consist of a sequence of computationally indistinguishable hybrid experiments. Recall that we are proving selective security, where the adversary sends the challenge input x^* before receiving any constrained keys.

Sequence of Hybrid Experiments: We will first set up some notation for the hybrid experiments. Let q denote the number of constrained key queries made by the adversary. Let x^* denote the challenge input chosen by the adversary, $(k, \text{PP}_\text{Acc}, \text{acc}_0, \text{STORE}_0)$ the master key chosen by challenger, $\text{acc-inp}^* = \text{Hash-Acc}(x^*)$ as defined in the construction. Let M_j denote the jth constrained key query, and t_j^* be the running time of machine M_j on input x^*, and τ_j be the

---

Program Prog

Constants : Turing machine $M = \langle Q, \Sigma_{\text{tape}}, \delta, q_0, q_{\text{ac}}, q_{\text{rej}} \rangle$,
time bound T
Public parameters $PP_{\text{Acc}}, PP_{\text{ltr}}$
Puncturable PRF keys $k, k_1, \ldots, k_\lambda, k_{\text{sig},A}$

Inputs : Time t, String seed, position pos_{in}, symbol sym_{in},
TM state st_{in} Accumulator value acc_{in}, proof π,
auxiliary value aux, accumulation of input acc-inp
Iterator value it_{in}, signature σ_{in}.

1. Let μ be an integer such that $2^\mu \le t < 2^{\mu+1}$.
   If $\text{PRG}(\text{seed}) \ne \text{PRG}(\text{F}(k_\mu, \text{acc-inp}))$ and $t > 1$, output $\bot$.

2. If $\text{Verify-Read}(PP_{\text{Acc}}, \text{acc}_{\text{in}}, \text{sym}_{\text{in}}, \text{pos}_{\text{in}}, \pi) = 0$ output $\bot$.

3. (a) Let $r_A = \text{F}(k_{\text{sig},A}, (\text{acc-inp}, t-1))$.
       Compute $(\text{SK}_A, \text{VK}_A, \text{VK}_{A,\text{rej}}) = \text{Setup-Spl}(1^\lambda; r_A)$.
   (b) Let $m_{\text{in}} = (\text{it}_{\text{in}}, \text{st}_{\text{in}}, \text{acc}_{\text{in}}, \text{pos}_{\text{in}})$.
       If $\text{Verify-Spl}(\text{VK}_A, m_{\text{in}}, \sigma_{\text{in}}) = 0$ output $\bot$.

4. (a) Let $(\text{st}_{\text{out}}, \text{sym}_{\text{out}}, \beta) = \delta(\text{st}_{\text{in}}, \text{sym}_{\text{in}})$ and $\text{pos}_{\text{out}} = \text{pos}_{\text{in}} + \beta$.
   (b) If $\text{st}_{\text{out}} = q_{\text{rej}}$ output $\bot$.
       Else if $\text{st}_{\text{out}} = q_{\text{ac}}$ output $\text{F}(k, \text{acc-inp})$.

5. (a) Compute $\text{acc}_{\text{out}} = \text{Update}(PP_{\text{Acc}}, \text{acc}_{\text{in}}, \text{sym}_{\text{out}}, \text{pos}_{\text{in}}, aux)$.
       If $\text{acc}_{\text{out}} = Reject$, output $\bot$.
   (b) Compute $\text{it}_{\text{out}} = \text{Iterate}(PP_{\text{ltr}}, \text{it}_{\text{in}}, (\text{st}_{\text{in}}, \text{acc}_{\text{in}}, \text{pos}_{\text{in}}))$.

6. (a) Let $r'_A = \text{F}(k_{\text{sig},A}, (\text{acc-inp}, t))$.
       Compute $(\text{SK}'_A, \text{VK}'_A, \text{VK}'_{A,\text{rej}}) = \text{Setup-Spl}(1^\lambda; r'_A)$.
   (b) Let $m_{\text{out}} = (\text{it}_{\text{out}}, \text{st}_{\text{out}}, \text{acc}_{\text{out}}, \text{pos}_{\text{out}})$ and $\sigma_{\text{out}} = \text{Sign-Spl}(\text{SK}'_\alpha, m_{\text{out}})$.

7. If $t+1 = 2^{\mu+1}$, set $\text{seed}' = \text{F}(k_{\mu+1}, \text{acc-inp})$.
   Else, set $\text{seed}' = \text{\_}$.

8. Output $\text{pos}_{\text{out}}, \text{sym}_{\text{out}}, \text{st}_{\text{out}}, \text{acc}_{\text{out}}, \text{it}_{\text{out}}, \sigma_{\text{out}}, \text{seed}'$.

Fig. 1. Program Prog

Program Init-Sign

Constants: Puncturable PRF key $k_{sig,A}$, Initial TM state q_0, Iterator value it_0

Input: Accumulation of input acc-inp

1. Let $F(k_{sig,A}, (\text{acc-inp}, 0)) = r_{sig}$. Compute $(SK, VK, VK_{rej}) = $ Setup-Spl$(1^\lambda; r_{sig})$.
2. Output $\sigma = $ Sign-Spl$(SK, (it_0, q_0, \text{acc-inp}, 0))$.

Fig. 2. Program Init-Sign

smallest power of two greater than t_j. The program Prog_j denotes the program Prog with machine M_j hardwired.

Hybrid_0: This corresponds to the real experiment.

Next, we define q hybrid experiments $\text{Hybrid}_{0,j}$ for $1 \leq j \leq q$.

$\text{Hybrid}_{0,j}$: Let Prog-1 denote the program defined in Fig. 3. In this experiment, the challenger sends an obfuscation of the program Prog-1$_i$ (Prog-1 with machine M_i hardwired) for the ith query if $i \leq j$. For the remaining queries, the challenger outputs an obfuscation of Prog_i.

Hybrid_1: This experiment is identical to hybrid $\text{Hybrid}_{0,q}$. In this experiment, the challenger sends an obfuscation of Prog-1$_i$ for all constrained key queries.

Hybrid_2: In this experiment, the challenger punctures the PRF key k at input acc-inp* and uses the punctured key for all key queries. More formally, after receiving the challenge input x^*, it chooses $(PP_{Acc}, acc_0, STORE_0) \leftarrow$ Setup-Acc(1^λ) and computes acc-inp$^* = $ Hash-Acc(x^*). It then chooses a PRF key k and computes $k\{\text{acc-inp}^*\} \leftarrow$ F.puncture$(k, \text{acc-inp}^*)$. Next, it receives constrained key queries for machines $M_1, \ldots, M_q$. For each query, it chooses $(PP_{ltr}, it_0) \leftarrow$ Setup-Itr(1^λ) and PRF keys $k_1, \ldots, k_\lambda, k_{sig,A}$. It computes an obfuscation of Prog-1$\{M_i, PP_{Acc}, PP_{ltr}, k\{\text{acc-inp}^*\}, k_{sig,A}\}$.

$$\text{Hybrid}_0 \equiv \text{Hybrid}_{0,0} \xrightarrow{1} \text{Hybrid}_{0,1} \xrightarrow{1} \cdots \xrightarrow{1} \text{Hybrid}_{0,q} \equiv \text{Hybrid}_1$$

$$\Big\downarrow i\mathcal{O}$$

$$0 \overset{\text{PPRF}}{\approx} \text{Hybrid}_2$$

Analysis. Let $\text{Adv}_i^{\mathcal{A}}$ denote the advantage of any PPT adversary $\mathcal{A}$ in the hybrid experiment Hybrid_i (similarly, let $\text{Adv}_{0,j}^{\mathcal{A}}$ denote the advantage of $\mathcal{A}$ in the intermediate hybrid experiment $\text{Hybrid}_{0,j}$).

Program Prog-1

Constants : Turing machine $M = \langle Q, \Sigma_{\text{tape}}, \delta, q_0, q_{\text{ac}}, q_{\text{rej}} \rangle$, time $t^* \in [T]$

Public parameters for accumulator PP_{Acc},

Public parameters for Iterator PP_{Itr}

Puncturable PRF keys $k, k_1, \ldots, k_\lambda, k_{\text{sig},A} \in \mathcal{K}$

Hardwired accumulated value acc-inp^*

Inputs: Time t, String seed, position pos_{in}, symbol sym_{in}, TM state st_{in}

Accumulator value acc_{in}, proof π, auxiliary value aux,

accumulation of input acc-inp, Iterator value it_{in}, signature σ_{in}.

1. Let μ be an integer such that $2^\mu \leq t < 2^{\mu+1}$.

   If $\text{PRG}(\text{seed}) \neq \text{PRG}(\text{F}(k_\mu, \text{acc-inp}))$ and $t > 1$, output $\perp$.

2. If $\text{Verify-Read}(\text{PP}_{\text{Acc}}, \text{acc}_{\text{in}}, \text{sym}_{\text{in}}, \text{pos}_{\text{in}}, \pi) = 0$ output $\perp$.

3. (a) Let $r_{\text{sig}} = \text{F}(k_{\text{sig},A}, t - 1)$. Compute $(\text{SK}, \text{VK}, \text{VK}_{\text{rej}}) = \text{Setup-Spl}(1^\lambda; r_{\text{sig}})$.

   (b) Let $m_{\text{in}} = (\text{it}_{\text{in}}, \text{st}_{\text{in}}, \text{acc}_{\text{in}}, \text{pos}_{\text{in}}, \text{acc-inp})$. If $\text{Verify-Spl}(\text{VK}, m_{\text{in}}, \sigma_{\text{in}}) = 0$ output $\perp$.

4. (a) Let $(\text{st}_{\text{out}}, \text{sym}_{\text{out}}, \beta) = \delta(\text{st}_{\text{in}}, \text{sym}_{\text{in}})$ and $\text{pos}_{\text{out}} = \text{pos}_{\text{in}} + \beta$.

   (b) If $\text{st}_{\text{out}} = q_{\text{rej}}$ output $\perp$.

   (c) If $\text{st}_{\text{out}} = q_{\text{ac}}$ and $\text{acc-inp} \neq \text{acc-inp}^*$, output $\text{F}(k, \text{acc-inp})$.

   Else If $\text{st}_{\text{out}} = q_{\text{ac}}$ output $\perp$.

5. (a) Compute $\text{acc}_{\text{out}} = \text{Update}(\text{PP}_{\text{Acc}}, \text{acc}_{\text{in}}, \text{sym}_{\text{out}}, \text{pos}_{\text{in}}, aux)$. If $w_{\text{out}} = Reject$, output $\perp$.

   (b) Compute $\text{it}_{\text{out}} = \text{Iterate}(\text{PP}_{\text{Itr}}, \text{it}_{\text{in}}, (\text{st}_{\text{in}}, \text{acc}_{\text{in}}, \text{pos}_{\text{in}}))$.

6. (a) Let $r'_{\text{sig}} = \text{F}(k_{\text{sig},A}, (\text{acc-inp}, t))$. Compute $(\text{SK}', \text{VK}', \text{VK}'_{\text{rej}}) \leftarrow \text{Setup-Spl}(1^\lambda; r'_{\text{sig}})$.

   (b) Let $m_{\text{out}} = (\text{it}_{\text{out}}, \text{st}_{\text{out}}, \text{acc}_{\text{out}}, \text{pos}_{\text{out}}, \text{acc-inp})$ and $\sigma_{\text{out}} = \text{Sign-Spl}(\text{SK}', m_{\text{out}})$.

7. If $t + 1 = 2^{\mu+1}$, set $\text{seed}' = \text{F}(k_{\mu+1}, \text{acc-inp})$.

   Else, set $\text{seed}' = \lrcorner$.

8. Output $\text{pos}_{\text{out}}, \text{sym}_{\text{out}}, \text{st}_{\text{out}}, \text{acc}_{\text{out}}, \text{it}_{\text{out}}, \sigma_{\text{out}}, \text{seed}'$.

Fig. 3. Program Prog-1

Recall $\mathsf{Hybrid}_{0,0}$ corresponds to the experiment Hybrid_0, and $\mathsf{Hybrid}_{0,q}$ corresponds to the experiment Hybrid_1. Using the following lemma, we can show that $|\mathsf{Adv}_0^{\mathcal{A}} - \mathsf{Adv}_1^{\mathcal{A}}| \leq \mathrm{negl}(\lambda)$.

Lemma 1. *Assuming F is a puncturable PRF, Acc is a secure positional accumulator, Itr is a secure positional iterator, $\mathcal{S}$ is a secure splittable signature scheme and $i\mathcal{O}$ is a secure indistinguishability obfuscator, for any PPT adversary $\mathcal{A}$, $|\mathsf{Adv}_{0,j}^{\mathcal{A}} - \mathsf{Adv}_{0,j+1}^{\mathcal{A}}| \leq \mathrm{negl}(\lambda)$.*

The proof of this lemma involves multiple hybrids. We include a high level outline of the proof in Appendix A, while the complete proof can be found in the full version of our paper.

Lemma 2. *Assuming $i\mathcal{O}$ is a secure indistinguishability obfuscator, for any PPT adversary $\mathcal{A}$, $|\mathsf{Adv}_1^{\mathcal{A}} - \mathsf{Adv}_2^{\mathcal{A}}| \leq \mathrm{negl}(\lambda)$.*

Proof. Let us assume for now that the adversary makes exactly one constrained key query corresponding to machine M_1. This can be naturally extended to the general case via a hybrid argument.

Note that the only difference between the two hybrids is the PRF key hardwired in Prog-1. In one case, the challenger sends an obfuscation of $P_1 = \mathsf{Prog}$-$1\{M_1, \mathsf{PP}_{\mathsf{Acc}}, \mathsf{PP}_{\mathsf{Itr}}, k, k_1, \ldots, k_\lambda, k_{\mathrm{sig},A}\}$, while in the other, it sends an obfuscation of $P_2 = \mathsf{Prog}$-$1\{M_1, \mathsf{PP}_{\mathsf{Acc}}, \mathsf{PP}_{\mathsf{Itr}}, k\{\mathsf{acc\text{-}inp}^*\}, k_1, \ldots, k_\lambda, k_{\mathrm{sig},A}\}$. To prove that these two hybrids are computationally indistinguishable, it suffices to show that the P_1 and P_2 are functionally identical. Note that program P_1 computes $F(k, \mathsf{acc\text{-}inp})$ only if $\mathsf{acc\text{-}inp} \neq \mathsf{acc\text{-}inp}^*$. As a result, using the correctness property of puncturable PRFs, the programs have identical functionality.

Lemma 3. *Assuming F is a selectively secure puncturable PRF, for any PPT adversary $\mathcal{A}$, $|\mathsf{Adv}_2^{\mathcal{A}}| \leq \mathrm{negl}(\lambda)$.*

Proof. Suppose there exists a PPT adversary $\mathcal{A}$ such that $|\mathsf{Adv}_2^{\mathcal{A}}| = \epsilon$. We will use $\mathcal{A}$ to construct a PPT algorithm $\mathcal{B}$ that breaks the security of the puncturable PRF F.

To begin with, $\mathcal{B}$ receives the challenge input x^* from $\mathcal{A}$. It chooses $(\mathsf{PP}_{\mathsf{Acc}}, \mathsf{acc}_0, \mathsf{STORE}_0) \leftarrow \mathsf{Setup\text{-}Acc}(1^\lambda)$. It then computes $\mathsf{acc\text{-}inp}^* = \mathsf{Hash\text{-}Acc}(x^*)$, and sends $\mathsf{acc\text{-}inp}^*$ to the PRF challenger as the challenge input. It receives a punctured key k' and an element y (which is either the pseudorandom evaluation at $\mathsf{acc\text{-}inp}^*$ or a truly random string in the range space). $\mathcal{B}$ sends y to $\mathcal{A}$ as the challenge response.

Next, it receives multiple constrained key requests. For the ith query corresponding to machine M_i, $\mathcal{B}$ chooses PRF keys $k_1, \ldots, k_\lambda, k_{\mathrm{sig},A} \leftarrow \mathsf{F.setup}(1^\lambda)$, $(\mathsf{PP}_{\mathsf{Itr}}, \mathsf{it}_0) \leftarrow \mathsf{Setup\text{-}Itr}(1^\lambda)$ and computes an obfuscation of Prog-$1\{M_i, \mathsf{PP}_{\mathsf{Acc}}, \mathsf{PP}_{\mathsf{Itr}}, k', k_1, \ldots, k_\lambda, k_{\mathrm{sig},A}\}$. It sends this obfuscated program to $\mathcal{A}$ as the constrained key.

Finally, after all constrained key queries, $\mathcal{A}$ sends its guess b', which $\mathcal{B}$ forwards to the challenger. Note that if $\mathcal{A}$ wins the security game against PRF, then $\mathcal{B}$ wins the security game against F. This concludes our proof.

5   Attribute Based Encryption for Turing Machines

In this section, we describe an ABE scheme where policies are associated with Turing machines, and as a result, attributes can be strings of unbounded length. Our ABE scheme is very similar to the constrained PRF construction described in Sect. 4.

Let $\mathcal{PKE}$ = (PKE.setup, PKE.enc, PKE.dec) be a public key encryption scheme and F a puncturable PRF for Turing machines, with algorithms PRF.setup and PRF.constrain. Consider the following ABE scheme:

- ABE.setup(1^λ). The setup algorithm chooses a puncturable PRF key $k \leftarrow$ F.setup(1^λ) and $(\text{PP}_{\text{Acc}}, \text{acc}_0, \text{STORE}_0) \leftarrow$ Setup-Acc($1^\lambda, T$). Next, it computes an obfuscation of Prog-PK$\{k\}$ (defined in Fig. 4). The public key $\text{PK}_{\text{ABE}} = (\text{PP}_{\text{Acc}}, \text{acc}_0, \text{STORE}_0, i\mathcal{O}(\text{Prog-PK}\{k\}))$, while the master secret key is $\text{MSK}_{\text{ABE}} = k$.
- ABE.enc($m, x, \text{PK}_{\text{ABE}}$). Let $\text{PK}_{\text{ABE}} = (\text{PP}_{\text{Acc}}, \text{acc}_0, \text{STORE}_0, \text{Program}_{pk})$ and $x = x_1 \ldots x_n$. As in Sect. 4, the encryption algorithm first 'accumulates' the attribute x using the accumulator public parameters. Let acc-inp $= \text{acc}_n$, where for all $j \leq n$, acc_j is defined as follows:
  - $\text{STORE}_j = \text{Write-Store}(\text{PP}_{\text{Acc}}, \text{STORE}_{j-1}, j-1, x_j)$
  - $aux_j = \text{Prep-Write}(\text{PP}_{\text{Acc}}, \text{STORE}_{j-1}, j-1)$
  - $\text{acc}_j = \text{Update}(\text{PP}_{\text{Acc}}, \text{acc}_{j-1}, x_j, j-1, aux_j)$.

  Next, the accumulated value is used to compute a PKE public key. Let pk $=$ $\text{Program}_{pk}(\text{acc-inp})$. Finally, the algorithm outputs ct $= \text{PKE.enc}(m, \text{pk})$.
- ABE.keygen($\text{MSK}_{\text{ABE}}, M$). Let $\text{MSK}_{\text{ABE}} = k$ and $M = $ a Turing machine. The ABE key corresponding to M is exactly the constrained key corresponding to M, as defined in Sect. 4. In particular, the key generation algorithm chooses $(\text{PP}_{\text{ltr}}, \text{it}_0) \leftarrow$ Setup-Itr($1^\lambda, T$) and a puncturable PRF key $k_{\text{sig}, A}$, and computes an obfuscation of $\text{Prog}\{M, k, k_{\text{sig}}, \text{PP}_{\text{Acc}}, \text{PP}_{\text{ltr}}\}$ (defined in Fig. 1) and $\text{Init-Sign}\{k_{\text{sig}, A}$ (defined in Fig. 2). The secret key $\text{SK}\{M\} = (\text{PP}_{\text{ltr}}, \text{it}_0, i\mathcal{O}(\text{Prog}), i\mathcal{O}(\text{Init-Sign}))$.
- ABE.dec($\text{SK}\{M\}, \text{ct}, x$). Let $\text{SK}\{M\} = (\text{PP}_{\text{ltr}}, \text{it}_0, \text{Program}_1, \text{Program}_2)$, and suppose M accepts x in t^* steps. As in the constrained key PRF evaluation, the decryption algorithm first obtains a signature using Program_2 and then runs Program_1 for t^* steps, until it outputs the pseudorandom string r. Using this PRF output r, the decryption algorithm computes $(\text{pk}, \text{sk}) = \text{PKE.setup}(1^\lambda; r)$ and then decrypts ct using sk. The algorithm outputs $\text{PKE.dec}(\text{sk}, \text{ct})$.

5.1   Proof of Security

We will first define a sequence of hybrid experiments, and then show that any two consecutive hybrid experiments are computationally indistinguishable.

Program Prog-PK

Constants: Puncturable PRF key k
Input: Accumulation of input acc-inp

1. Let $F(k, \text{acc-inp}) = r$. Compute $(\text{pk}, \text{sk}) = \text{PKE.setup}(1^\lambda; r)$.
2. Output pk.

Fig. 4. Program Prog-PK

Sequence of Hybrid Experiments

Hybrid H_0: This corresponds to the selective security game. Let x^* denote the challenge input, and $\text{acc-inp}^* = \text{Hash-Acc}(x^*)$.

Hybrid H_1: In this hybrid, the challenger sends an obfuscation of Prog-1 instead of Prog. Prog-1, on inputs corresponding to acc-inp^*, never reaches the accepting state q_{ac}. This is similar to Hybrid$_1$ of the constrained PRF security proof in Sect. 4.1.

Hybrid H_2: In this hybrid, the challenger first punctures the PRF key k at acc-inp^*. It computes $k' \leftarrow \text{F.puncture}(k, \text{acc-inp}^*)$ and $(\text{pk}^*, \text{sk}^*) = \text{PKE.setup}(1^\lambda; F(k, \text{acc-inp}^*))$. Next, it uses k' and pk^* to define Prog-PK$'\{k', \text{pk}^*\}$ (see Fig. 5). It sends an obfuscation of Prog-PK$'$ as the public key. Next, for each of the secret key queries, it sends an obfuscation of Prog-1. However unlike the previous hybrid, Prog-1 has k' hardwired instead of k.

Program Prog-PK$'$

Constants: Punctured PRF key k', Hardwired accumulation acc-inp^* and public key pk^*.
Input: Accumulation of input acc-inp

1. If $\text{acc-inp} = \text{acc-inp}^*$, set $\text{pk} = \text{pk}^*$.
   Else let $F(k', \text{acc-inp}) = r$. Compute $(\text{pk}, \text{sk}) = \text{PKE.setup}(1^\lambda; r)$.
2. Output pk.

Fig. 5. Program Prog-PK$'$

Hybrid H_3: In this hybrid, the challenger chooses $(\text{pk}^*, \text{sk}^*) \leftarrow \text{PKE.setup}(1^\lambda)$; that is, the public key is computed using true randomness. It then hardwires pk^* in Prog-PK. The secret key queries are same as in previous hybrids.

Analysis. Let $\mathsf{Adv}_i^{\mathcal{A}}$ denote the advantage of $\mathcal{A}$ in hybrid H_i.

Lemma 4. *Assuming $i\mathcal{O}$ is a secure indistinguishability obfuscator, Acc is a secure positional accumulator, Itr is a secure iterator, $\mathcal{S}$ is a secure splittable signature scheme and F is a secure puncturable PRF, for any adversary $\mathcal{A}$, $|\mathsf{Adv}_0^{\mathcal{A}} - \mathsf{Adv}_1^{\mathcal{A}}| \leq negl(\lambda)$.*

The proof of this lemma is identical to the proof of Lemma 1.

Lemma 5. *Assuming $i\mathcal{O}$ is a secure indistinguishability obfuscator, for any PPT adversary $\mathcal{A}$, $|\mathsf{Adv}_1^{\mathcal{A}} - \mathsf{Adv}_2^{\mathcal{A}}| \leq negl(\lambda)$.*

Proof. Similar to the proof of Lemma 2, k can be replaced with k' in all the secret key queries, since $F(k, \mathsf{acc\text{-}inp}^*)$ is never executed. As far as Prog-PK and Prog-PK$'$ are concerned, $(\mathsf{pk}^*, \mathsf{sk}^*)$ is set to be $\mathsf{PKE.setup}(1^{\lambda}; F(k, \mathsf{acc\text{-}inp}^*))$, and therefore, the programs are functionally identical.

Lemma 6. *Assuming F is a selectively secure puncturable PRF, for any PPT adversary $\mathcal{A}$, $|\mathsf{Adv}_2^{\mathcal{A}} - \mathsf{Adv}_3^{\mathcal{A}}| \leq negl(\lambda)$.*

Proof. The proof of this follows immediately from the security definition of puncturable PRFs. Suppose there exists an adversary that can distinguish between H_2 and H_3 with advantage ϵ. Then, there exists a PPT algorithm $\mathcal{B}$ that can break the selective security of F. $\mathcal{B}$ first receives x^* from the adversary. It computes $\mathsf{acc\text{-}inp}^*$, sends $\mathsf{acc\text{-}inp}^*$ to the PRF challenger and receives k', y, where y is either the PRF evaluation at $\mathsf{acc\text{-}inp}^*$, or a truly random string. Using y, it computes $(\mathsf{pk}^*, \mathsf{sk}^*) = \mathsf{PKE.setup}(1^{\lambda}; y)$, and uses k', pk^* to define the public key $i\mathcal{O}(\mathsf{Prog\text{-}PK}'\{k', \mathsf{pk}^*\})$. The secret key queries are same in both hybrids, and can be answered using k' only. As a result, $\mathcal{B}$ simulates either H_2 or H_3 perfectly. This concludes our proof.

Lemma 7. *Assuming $\mathcal{PKE}$ is a secure public key encryption scheme, for any PPT adversary $\mathcal{A}$, $\mathsf{Adv}_3^{\mathcal{A}} \leq negl(\lambda)$.*

Proof. Suppose there exists a PPT adversary $\mathcal{A}$ such that $\mathsf{Adv}_3^{\mathcal{A}} = \epsilon$. Then there exists a PPT adversary $\mathcal{B}$ that breaks IND-CPA security of $\mathcal{PKE}$. $\mathcal{B}$ receives a public key pk^* from the challenger. It chooses PRF key k, punctures it at $\mathsf{acc\text{-}inp}^*$ and sends the public key $i\mathcal{O}\{\mathsf{Prog\text{-}PK}'\}$. Next, it responds to the secret key queries, and finally, on receiving challenge messages m_0, m_1, it forwards them to the challenger, and receives ct^*, which it forwards to the adversary. The post challenge key query phase is also simulated perfectly, since it has all the required components.

A   Proof Outline of Lemma 1

In this section, we provide an outline of the proof of Lemma 1. The detailed proof is included in the full version of our paper. Let us assume the key query is for TM M, and M does not accept the challenge input x^*, and let $\mathsf{acc\text{-}inp}^*$ denote the accumulation of x^*. Our goal in this hybrid is to ensure that the program will never output $F(K, \text{acc-inp}^*)$. This is done via a sequence of hybrids, where we use the security properties of splittable signatures, accumulators and iterators together with $i\mathcal{O}$ security.

Preprocessing Hybrid: The first step is to modify the program Prog to allow additional valid signatures without being detected. In particular, we have an additional PRF key in the program, and this generates 'bad' signing/verification keys. The program first checks if the input signature is accepted by the usual 'good' verification key. If not, it checks if it is accepted by the 'bad' verification key. If the incoming signature is bad, then the output signature is also computed using the bad signing key. Let us call this hybrid Hyb-1. This switch is indistinguishable because the Init-Sign program only outputs a good signature, and we use the rejection-verification key indistinguishability property to show that this change is indistinguishable.

Intermediate Hybrids Hyb-$(1, i)$: Next, we gradually ensure that the program does not output the PRF evaluation on acc-inp* in the first i steps. If $i = T$, then we are done. Here, we need to define our intermediate hybrid carefully. In the ith intermediate hybrid, the program does not output PRF evaluation if $t \leq i$. Moreover, if acc-inp = acc-inp*, it only accepts good signatures for the first $i - 1$ steps. For the ith step, if acc-inp = acc-inp*, it accepts only good signatures, but outputs a bad signature if the input iterated value, accumulated value or state are not the correct ones for time step i (here, the program has the correct values for step i hardwired). We now need to go from step Hyb-$(1, i)$ to step Hyb-$(1, i + 1)$.

For this, we will first ensure that if acc-inp = acc-inp*, the only signature accepted at step $i + 1$ is the one corresponding to the correct (iterated value, accumulated value, state) input tuple at step $i+1$. Intuitively, this is true because the program, at step i, outputs a bad signature for all other tuples. To enforce this, we use the properties of the splittable signature schemes. Next, we make the accumulator read-enforcing. This would mean that both the state and symbol input at step $i + 1$ are the correct ones. As a result, the program cannot output the PRF evaluation at step $i + 1$ if acc-inp = acc-inp*. So now, the state and symbol output at step $i + 1$ also have to be the correct ones. To ensure that the accumulated value and iterated value output are also correct, we make the accumulator write-enforcing and iterator enforcing respectively. Together, these will ensure that the transition from Hyb-$(1, i)$ and Hyb-$(1, i + 1)$ are computationally indistinguishable.

Continuing this way, we can ensure, step by step, that the program does not output the PRF evaluation on acc-inp*. However, the approach described above will require exponential hybrids. To make the number of intermediate hybrids polynomial, we use the 'tail-cutting' technique described in Sect. 1. Note that the program, after t^* steps, only outputs $\perp$. Suppose t^* is a power of two. Using a PRG trick, we can wipe out steps t^* to $2t^*$ in one shot. At every step where t is a power of two, the program outputs a new PRG seed, and this PRG seed's validity is checked till t reaches the next power of two. Now, if no PRG seed is output at step t^*, then using the PRG security, one can ensure that the PRG seed validity check fails. As a result, for all $t \in (t^*, 2t^*)$, the program outputs $\bot$.

References

1. Abusalah, H., Fuchsbauer, G., Pietrzak, K.: Constrained prfs for unbounded inputs. IACR Cryptology ePrint Archive 2014, 840 (2014). http://eprint.iacr.org/2014/840
2. Ananth, P., Jain, A.: Indistinguishability obfuscation from compact functional encryption. IACR Cryptology ePrint Archive 2015, 173 (2015). http://eprint.iacr.org/2015/173
3. Ananth, P., Sahai, A.: Functional encryption for turing machines. Cryptology ePrint Archive, Report 2015/776 (2015). http://eprint.iacr.org/
4. Banerjee, A., Fuchsbauer, G., Peikert, C., Pietrzak, K., Stevens, S.: Key-homomorphic constrained pseudorandom functions. In: Dodis, Y., Nielsen, J.B. (eds.) TCC 2015, Part II. LNCS, vol. 9015, pp. 31–60. Springer, Heidelberg (2015). http://dx.doi.org/10.1007/978-3-662-46497-7_2
5. Bitansky, N., Garg, S., Lin, H., Pass, R., Telang, S.: Succinct randomized encodings and their applications. In: Proceedings of the Forty-Seventh Annual ACM Symposium on Theory of Computing, STOC 2015, Portland, OR, USA, 14–17 June 2015, pp. 439–448 (2015). http://doi.acm.org/10.1145/2746539.2746574
6. Bitansky, N., Vaikuntanathan, V.: Indistinguishability obfuscation from functional encryption. IACR Cryptology ePrint Archive 2015, 163 (2015)
7. Boneh, D., Waters, B.: Constrained pseudorandom functions and their applications. In: Sako, K., Sarkar, P. (eds.) ASIACRYPT 2013, Part II. LNCS, vol. 8270, pp. 280–300. Springer, Heidelberg (2013)
8. Boneh, D., Zhandry, M.: Multiparty key exchange, efficient traitor tracing, and more from indistinguishability obfuscation. In: Garay, J.A., Gennaro, R. (eds.) CRYPTO 2014, Part I. LNCS, vol. 8616, pp. 480–499. Springer, Heidelberg (2014)
9. Boyle, E., Goldwasser, S., Ivan, I.: Functional signatures and pseudorandom functions. In: Krawczyk, H. (ed.) PKC 2014. LNCS, vol. 8383, pp. 501–519. Springer, Heidelberg (2014)
10. Brakerski, Z., Vaikuntanathan, V.: Constrained key-homomorphic PRFs from standard lattice assumptions. In: Dodis, Y., Nielsen, J.B. (eds.) TCC 2015, Part II. LNCS, vol. 9015, pp. 1–30. Springer, Heidelberg (2015). http://dx.doi.org/10.1007/978-3-662-46497-7_1
11. Canetti, R., Holmgren, J., Jain, A., Vaikuntanathan, V.: Succinct garbling and indistinguishability obfuscation for RAM programs. In: Proceedings of the Forty-Seventh Annual ACM on Symposium on Theory of Computing, STOC 2015, Portland, OR, USA, 14–17 June 2015, pp. 429–437 (2015). http://doi.acm.org/10.1145/2746539.2746621
12. Chandran, N., Raghuraman, S., Vinayagamurthy, D.: Constrained pseudorandom functions: Verifiable and delegatable. Cryptology ePrint Archive, Report 2014/522 (2014). http://eprint.iacr.org/
13. Fuchsbauer, G., Konstantinov, M., Pietrzak, K., Rao, V.: Adaptive security of constrained prfs. IACR Cryptology ePrint Archive 2014, 416 (2014)
14. Garg, S., Gentry, C., Halevi, S.: Candidate multilinear maps from ideal lattices. In: Johansson, T., Nguyen, P.Q. (eds.) EUROCRYPT 2013. LNCS, vol. 7881, pp. 1–17. Springer, Heidelberg (2013)

15. Garg, S., Gentry, C., Halevi, S., Raykova, M., Sahai, A., Waters, B.: Candidate indistinguishability obfuscation and functional encryption for all circuits. In: FOCS (2013)
16. Garg, S., Gentry, C., Halevi, S., Wichs, D.: On the implausibility of differing-inputs obfuscation and extractable witness encryption with auxiliary input. In: Garay, J.A., Gennaro, R. (eds.) CRYPTO 2014, Part I. LNCS, vol. 8616, pp. 518–535. Springer, Heidelberg (2014)
17. Gentry, C., Lewko, A., Sahai, A., Waters, B.: Indistinguishability obfuscation from the multilinear subgroup elimination assumption. Cryptology ePrint Archive, Report 2014/309 (2014). http://eprint.iacr.org/
18. Goldreich, O., Goldwasser, S., Micali, S.: How to construct random functions (extended abstract). In: FOCS, pp. 464–479 (1984)
19. Hofheinz, D., Kamath, A., Koppula, V., Waters, B.: Adaptively secure constrained pseudorandom functions. IACR Cryptology ePrint Archive 2014, 720 (2014). http://eprint.iacr.org/2014/720
20. Hohenberger, S., Koppula, V., Waters, B.: Adaptively secure puncturable pseudorandom functions in the standard model. IACR Cryptology ePrint Archive 2014, 521 (2014). http://eprint.iacr.org/2014/521
21. Hubacek, P., Wichs, D.: On the communication complexity of secure function evaluation with long output. In: Proceedings of the 2015 Conference on Innovations in Theoretical Computer Science, ITCS 2015, Rehovot, Israel, 11–13 January 2015, pp. 163–172 (2015)
22. Khurana, D., Rao, V., Sahai, A.: Multi-party key exchange for unbounded parties from indistinguishability obfuscation. In: Iwata, T., Cheon, J.H. (eds.) ASIACRYPT 2015. LNCS, vol. 9452, pp. 52–75. Springer, Heidelberg (2015). doi:10. 1007/978-3-662-48797-6_3
23. Kiayias, A., Papadopoulos, S., Triandopoulos, N., Zacharias, T.: Delegatable pseudorandom functions and applications. In: ACM Conference on Computer and Communications Security, pp. 669–684 (2013)
24. Koppula, V., Lewko, A.B., Waters, B.: Indistinguishability obfuscation for turing machines with unbounded memory. In: Proceedings of the Forty-Seventh Annual ACM on Symposium on Theory of Computing, STOC 2015, NY, USA, pp. 419–428 (2015). http://doi.acm.org/10.1145/2746539.2746614
25. Sahai, A., Waters, B.: How to use indistinguishability obfuscation: deniable encryption, and more. In: STOC, pp. 475–484 (2014)
26. Zhandry, M.: Adaptively secure broadcast encryption with small system parameters (2014)

Pseudorandom Functions in Almost Constant Depth from Low-Noise LPN

Yu Yu[1,2,3]($\boxtimes$) and John Steinberger[4]

[1] Department of Computer Science and Engineering,
Shanghai Jiao Tong University, Shanghai, China
yyuu@sjtu.edu.cn
[2] State Key Laboratory of Information Security,
Institute of Information Engineering, Chinese Academy of Sciences,
Beijing 100093, China
[3] State Key Laboratory of Cryptology, P.O. Box 5159, Beijing 100878, China
[4] Institute for Interdisciplinary Information Sciences,
Tsinghua University, Beijing, China
jpsteinb@gmail.com

Abstract. Pseudorandom functions (PRFs) play a central role in symmetric cryptography. While in principle they can be built from any one-way functions by going through the generic HILL (SICOMP 1999) and GGM (JACM 1986) transforms, some of these steps are inherently sequential and far from practical. Naor, Reingold (FOCS 1997) and Rosen (SICOMP 2002) gave parallelizable constructions of PRFs in NC^2 and TC^0 based on concrete number-theoretic assumptions such as DDH, RSA, and factoring. Banerjee, Peikert, and Rosen (Eurocrypt 2012) constructed relatively more efficient PRFs in NC^1 and TC^0 based on "learning with errors" (LWE) for certain range of parameters. It remains an open problem whether parallelizable PRFs can be based on the "learning parity with noise" (LPN) problem for both theoretical interests and efficiency reasons (as the many modular multiplications and additions in LWE would then be simplified to AND and XOR operations under LPN).

In this paper, we give more efficient and parallelizable constructions of randomized PRFs from LPN under noise rate n^{-c} (for any constant $0 < c < 1$) and they can be implemented with a family of polynomial-size circuits with unbounded fan-in AND, OR and XOR gates of depth $\omega(1)$, where $\omega(1)$ can be any small super-constant (e.g., $\log\log\log n$ or even less). Our work complements the lower bound results by Razborov and Rudich (STOC 1994) that PRFs of beyond quasi-polynomial security are not contained in $AC^0(MOD_2)$, i.e., the class of polynomial-size, constant-depth circuit families with unbounded fan-in AND, OR, and XOR gates.

Furthermore, our constructions are security-lifting by exploiting the redundancy of low-noise LPN. We show that in addition to parallelizability (in almost constant depth) the PRF enjoys either of (or any tradeoff between) the following:

© International Association for Cryptologic Research 2016
M. Fischlin and J.-S. Coron (Eds.): EUROCRYPT 2016, Part II, LNCS 9666, pp. 154–183, 2016.
DOI: 10.1007/978-3-662-49896-5_6

- A PRF on a weak key of sublinear entropy (or equivalently, a uniform key that leaks any $(1 - o(1))$-fraction) has comparable security to the underlying LPN on a linear size secret.
- A PRF with key length λ can have security up to $2^{O(\lambda/\log \lambda)}$, which goes much beyond the security level of the underlying low-noise LPN.

where adversary makes up to certain super-polynomial amount of queries.

1 Introduction

LEARNING PARITY WITH NOISE. The computational version of learning parity with noise (LPN) assumption with parameters $n \in \mathbb{N}$ (length of secret), $q \in \mathbb{N}$ (number of queries) and $0 < \mu < 1/2$ (noise rate) postulates that it is computationally infeasible to recover the n-bit secret $s \in \{0,1\}^n$ given $(a \cdot s \oplus e, a)$, where a is a random $q \times n$ matrix, e follows Ber_μ^q, Ber_μ denotes the Bernoulli distribution with parameter μ (i.e., $\Pr[\mathrm{Ber}_\mu = 1] = \mu$ and $\Pr[\mathrm{Ber}_\mu = 0] = 1 - \mu$), '$\cdot$' denotes matrix vector multiplication over $\mathrm{GF}(2)$ and '$\oplus$' denotes bitwise XOR. The decisional version of LPN simply assumes that $a \cdot s \oplus e$ is pseudorandom (i.e., computationally indistinguishable from uniform randomness) given a. The two versions are polynomially equivalent [5,12,36].

HARDNESS OF LPN. The computational LPN problem represents a well-known NP-complete problem "decoding random linear codes" [9] and thus its worst-case hardness is well studied. LPN was also extensively studied in learning theory, and it was shown in [24] that an efficient algorithm for LPN would allow to learn several important function classes such as 2-DNF formulas, juntas, and any function with a sparse Fourier spectrum. Under a constant noise rate (i.e., $\mu = \Theta(1)$), the best known LPN solvers [13,40] require time and query complexity both $2^{O(n/\log n)}$. The time complexity goes up to $2^{O(n/\log\log n)}$ when restricted to $q = \mathrm{poly}(n)$ queries [42], or even $2^{O(n)}$ given only $q = O(n)$ queries [45]. Under low noise rate $\mu = n^{-c}$ ($0 < c < 1$), the security of LPN is less well understood: on the one hand, for $q = n + O(1)$ we can already do efficient distinguishing attacks with advantage $2^{-O(n^{1-c})}$ that match the statistical distance between the LPN samples and uniform randomness (see Remark 2); on the other hand, for (even super-)polynomial q the best known attacks [7,11,15,39,54] are not asymptotically better, i.e., still at the order of $2^{\Theta(n^{1-c})}$. We mention that LPN does not succumb to known quantum algorithms, which makes it a promising candidate for "post-quantum cryptography". Furthermore, LPN also enjoys simplicity and is more suited for weak-power devices (e.g., RFID tags) than other quantum-secure candidates such as LWE [52][1].

LPN-BASED CRYPTOGRAPHIC APPLICATIONS. LPN was used as a basis for building lightweight authentication schemes against passive [31] and even active

---

[1] The inner product of LWE requires many multiplications modulo a large prime p (polynomial in the security parameter), and in contrast the same operation for LPN is simply an XOR sum of a few AND products.

adversaries [35,36] (see [1] for a more complete literature). Recently, Kiltz et al. [38] and Dodis et al. [20] constructed randomized MACs based on the hardness of LPN, which implies a two-round authentication scheme with man-in-the-middle security. Lyubashevsky and Masny [43] gave an more efficient three-round authentication scheme from LPN (without going through the MAC transformation) and recently Cash, Kiltz, and Tessaro [16] reduced the round complexity to 2 rounds. Applebaum et al. [4] showed how to constructed a linear-stretch[2] pseudorandom generator (PRG) from LPN. We mention other not-so-relevant applications such as public-key encryption schemes [3,22,37], oblivious transfer [19], commitment schemes and zero-knowledge proofs [33], and refer to a recent survey [49] on the current state-of-the-art about LPN.

DOES LPN IMPLY LOW-DEPTH PRFS? Pseudorandom functions (PRFs) play a central role in symmetric cryptography. While in principle PRFs can be obtained via a generic transform from any one-way function [26,29], these constructions are inherently sequential and too inefficient to compete with practical instantiations (e.g., the AES block cipher) built from scratch. Motivated by this, Naor, Reingold [46] and Rosen [47] gave direct constructions of PRFs from concrete number-theoretic assumptions (such as decision Diffie-Hellman, RSA, and factoring), which can be computed by low-depth circuits in NC^2 or even TC^0. However, these constructions mainly established the feasibility result and are far from practical as they require extensive preprocessing and many exponentiations in large multiplicative groups. Banerjee, Peikert, and Rosen [6] constructed relatively more efficient PRFs in NC^1 and TC^0 based on the "learning with errors" (LWE) assumption. More specifically, they observed that LWE for certain range of parameters implies a deterministic variant which they call "learning with rounding" (LWR), and that LWR in turn gives rise to pseudorandom synthesizers [46], a useful tool for building low-depth PRFs. Despite that LWE is generalized from LPN, the derandomization technique used for LWE [6] does not seemingly apply to LPN, and thus it is an interesting open problem if low-depth PRFs can be based on (even a low-noise variant of) LPN (see a discussion in [49, Footnote 18]). In fact, we don't even know how to build low-depth weak PRFs from LPN. Applebaum [4] observed that LPN implies "weak randomized pseudorandom functions", which require independent secret coins on every function evaluation, and Akavia et al. [2] obtained weak PRFs in "$AC^0 \circ MOD_2$" from a relevant non-standard hard learning assumption.

OUR CONTRIBUTIONS. In this paper, we give constructions of low-depth PRFs from low-noise LPN (see Theorem 1 below), where the noise rate n^{-c} (for any constant $0 < c < 1$) encompasses the noise level of Alekhnovich [3] (i.e., $c = 1/2$) and higher noise regime. Strictly speaking, the PRFs we obtain are not contained in $AC^0(MOD_2)^3$, but the circuit depth $\omega(1)$ can be arbitrarily small (e.g.,

---

[2] A PRG $G: \{0,1\}^{\ell_1} \rightarrow \{0,1\}^{\ell_2}$ has linear stretch if the stretch factor ℓ_2/ℓ_1 equals some constant greater than 1.

[3] Recall that $AC^0(MOD_2)$ refers to the class of polynomial-size, constant-depth circuit families with unbounded fan-in AND, OR, and XOR gates.

$\log \log \log n$ or even less). This complements the negative result of Razborov and Rudich [51] (which is based on the works of Razborov and Smolensky [50,53]) that PRFs with more than quasi-polynomial security do not exist in $AC^0(MOD_2)$.

Theorem 1 (main results, informal). *Assume that the LPN problem with secret length n and noise rate $\mu = n^{-c}$ (for any constant $0 < c < 1$) is ($q = 1.001n$, $t = 2^{O(n^{1-c})}$, $\epsilon = 2^{-O(n^{1-c})}$)-hard[4]. Then,*

1. *for any $d = \omega(1)$, there exists a ($q' = n^{d/3}$, $t - q'$poly(n), $O(nq'\epsilon)$)-randomized-PRF on any weak key of Rényi entropy no less than $O(n^{1-c} \cdot \log n)$, or on an $n^{1-\frac{c}{2}}$-bit uniform random key with any $(1 - \frac{O(\log n)}{n^{c/2}})$-fraction of leakage (independent of the public coins of the PRF);*
2. *let $\lambda = \Theta(n^{1-c}\log n)$, for any $d = \omega(1)$, there exists a ($q' = \lambda^{\Theta(d)}$, $t' = 2^{O(\lambda/\log \lambda)}$, $\epsilon' = 2^{-O(\lambda/\log \lambda)}$)-randomized PRF with key length λ;*

where both PRFs are computable by polynomial-size depth-$O(d)$ circuits with unbounded-fan-in AND, OR and XOR gates.

ON LIFTED SECURITY. Note that there is nothing special with the factor 1.001, which can be replaced with any constant greater than 1. The first parallelizable PRF has security[5] comparable to the underlying LPN (with linear secret length) yet it uses a key of only sublinear entropy, or in the language of leakage resilient cryptography, a sublinear-size secret key with any $(1 - o(1))$-fraction of leakage (independent of the public coins). From a different perspective, let the security parameter λ be the key length of the PRF, then the second PRF can have security up to $2^{O(\lambda/\log \lambda)}$ given any $n^{\Theta(d)}$ number of queries. We use security-preserving PRF constructions without relying on k-wise independent hash functions. This is crucial for low-depth constructions as recent works [17,34] use (almost) $\omega(\log n)$-wise independent hash functions, which are not known to be computable in (almost) constant-depth even with unbounded fan-in gates. We remark that circuit depth $d = \omega(1)$ is independent of the time/advantage security of PRF, and is reflected only in the query complexity $q' = n^{\Theta(d)}$. This is reasonable in many scenarios as in practice the number of queries may depend not only on adversary's computing power but also on the amount of data available for cryptanalysis. It remains open whether the dependency of query complexity on circuit depth can be fully eliminated.

BERNOULLI-LIKE RANDOMNESS EXTRACTOR/SAMPLER. Of independent interests, we propose the following randomness extractor/sampler in constant depth and they are used in the first/second PRF constructions respectively.

---

[4] t and $1/\epsilon$ are upper bounded by $2^{O(n^{1-c})}$ due to known attacks.

[5] Informally, we say that a PRF has security T if it is $1/T$-indistinguishable from a random function for all oracle-aid distinguishers running in time T and making up to certain superpolynomial number of queries.

- A Bernoulli randomness extractor in $\text{AC}^0(\text{MOD}_2)$ that converts almost all entropy of a weak Rényi entropy source into Bernoulli noise distributions.
- A sampler in AC^0 that uses a short uniform seed and outputs a Bernoulli-like distribution of length m and noise rate μ, denoted as ψ_μ^m (see Algorithm 1).

Alekhnovich's cryptosystem [3] considers a random distribution of length m that has exactly μm 1's, which we denote as $\chi_{\mu m}^m$. The problem of sampling $\chi_{\mu m}^m$ dates back to [12], but the authors only mention that it can be done efficiently, and it is not known whether $\chi_{\mu m}^m$ can be sampled in $\text{AC}^0(\text{MOD}_2)$. Instead, Applebaum et al. [4] propose the following sampler for Bernoulli distribution Ber_μ^q using uniform randomness. Let $w = w_1 \cdots w_n$ be an n-bit uniform random string, and for convenience assume that μ is a negative power of 2 (i.e., $\mu = 2^{-v}$ for integer v). Let sample : $\{0,1\}^v \to \{0,1\}$ output the AND of its input bits, and let

$$e = (\mathsf{sample}(w_1 \cdots w_v), \cdots, \mathsf{sample}(w_{(q-1)v+1} \cdots w_{(q-1)v+v}))$$

so that $e \sim \text{Ber}_\mu^q$ for any $q \leq \lfloor n/\log(1/\mu) \rfloor$. Note that Ber_μ has Shannon entropy $\mathbf{H}_1(\text{Ber}_\mu) = \Theta(\mu \log(1/\mu))$ (see Fact A1), and thus the above converts a $(q\mathbf{H}_1(\text{Ber}_\mu)/n) = O(\mu)$-fraction of the entropy into Bernoulli randomness. It was observed in [4] that conditioned on e source w remains of $(1 - O(\mu))n$ bits of average min-entropy, which can be recycled into uniform randomness with a universal hash function h. That is, the two distributions are statistically close

$$(e, h(w), h) \overset{s}{\sim} (\text{Ber}_\mu^q, U_{(1-O(\mu))n}, h),$$

where U_q denotes a uniform distribution over $\{0,1\}^q$. The work of [4] then proceeded to a construction of PRG under noise rate $\mu = \Theta(1)$. However, for $\mu = n^{-c}$ the above only samples an $O(n^{-c})$-fraction of entropy. To convert more entropy into Bernoulli distributions, one may need to apply the above sample-then-recycle process to the uniform randomness recycled from a previous round (e.g., $h(w)$ of the first round) and repeat the process many times. However, this method is sequential and requires a circuit of depth $\Omega(n^c)$ to convert any constant fraction of entropy. We propose a more efficient and parallelizable extractor in $\text{AC}^0(\text{MOD}_2)$. As shown in Fig. 1, given any weak source of Rényi entropy $\Theta(n)$, we apply i.i.d. pairwise independent hash functions $h_1, \cdots, h_q$ (each of output length v) to w and then use sample on the bits extracted to get the Bernoulli distributions. We prove a lemma showing that this method can transform almost all entropy into Bernoulli distribution Ber_μ^q, namely, the number of extracted Bernoulli bits q can be up to $\Theta(n/\mathbf{H}_1(\text{Ber}_\mu))$. This immediately gives an equivalent formulation of the standard LPN by reusing matrix a to randomize the hash functions. For example, for each $1 \leq i \leq q$ denote by a_i the i-th row of a, let h_i be described by a_i, and let i-th LPN sample be $\langle a_i, s \rangle \oplus \mathsf{sample}(h_i(w))$. Note that the algorithm is non-trivial as $(h_1(w), \cdots, h_q(w))$ can be of length $\Theta(n^{1+c})$, which is much greater than the entropy of w.

The Bernoulli randomness extractor is used in the first PRF construction. For our second construction, we introduce a Bernoulli-like distribution ψ_μ^m that can be more efficiently sampled in AC^0 (i.e., without using XOR gates), and show that it can be used in place of Ber_μ^m with provable security.

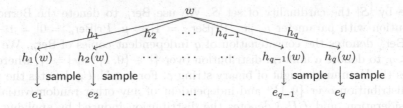

Fig. 1. An illustration of the proposed Bernoulli randomness extractor in $AC^0(MOD_2)$.

PRGs AND PRFs FROM LPN. It can be shown that standard LPN implies a variant where the secret s and noise vector e are sampled from Ber_μ^{n+q} or even ψ_μ^{n+q}. This allows us to obtain a randomized PRG G_a with short seed and polynomial stretch, where a denotes the public coin. We then use the technique of Goldreich, Goldwasser and Micali [26] with a $n^{\Theta(1)}$-ary tree of depth $\omega(1)$ (reusing public coin a at every invocation of G_a) and construct a randomized PRF (see Definition 4) $F_{k,a}$ with input length $\omega(\log n)$, secret key k and public coin a. This already implies PRFs of arbitrary input length by Levin's trick [41], i.e., $\bar{F}_{(k,h),a}(x) \stackrel{\text{def}}{=} F_{k,a}(h(x))$ where h is a universal hash function from any fixed-length input to $\omega(\log n)$ bits. Note that $\bar{F}_{(k,h),a}$ is computable in depth $\omega(1)$ (i.e., the depth of the GGM tree) for any small $\omega(1)$. However, the security of the above does not go beyond $n^{\omega(1)}$ due to a birthday attack. To overcome this, we use a simple and parallel method [8,44] by running a sub-linear number of independent[6] copies of $\bar{F}_{(k,h),a}$ and XORing their outputs, and we avoid key expansions by using pseudorandom keys (expanded using G_a or $F_{k,a}$) for all copies of $\bar{F}_{(k,h),a}$. We obtain our final security-preserving construction of PRFs by putting together all the above ingredients.

The rest of the paper is organized as follows: Sect. 2 gives background information about relevant notions and definitions. Section 3 presents the Bernoulli randomness extractor. Sections 4 and 5 give the two constructions of PRFs respectively. We include in Appendix A well-known lemmas and inequalities used, and refer to Appendix B for all the proofs omitted in the main text.

2   Preliminaries

NOTATIONS AND DEFINITIONS. We use $[n]$ to denote set $\{1, \ldots, n\}$. We use capital letters[7] (e.g., X, Y) for random variables and distributions, standard letters (e.g., x, y) for values, and calligraphic letters (e.g. $\mathcal{X}$, $\mathcal{E}$) for sets and events. The support of a random variable X, denoted by $\mathsf{Supp}(X)$, refers to the set of values on which X takes with non-zero probability, i.e., $\{x : \Pr[X = x] > 0\}$.

---

[6] By "independent" we mean that $\bar{F}_{(k,h),a}$ is evaluated on independent keys but still reusing the same public coin a.

[7] The two exceptions are G and F, which are reserved for PRGs and PRFs respectively.

Denote by $|\mathcal{S}|$ the cardinality of set $\mathcal{S}$. We use Ber_μ to denote the Bernoulli distribution with parameter μ, i.e., $\Pr[\mathsf{Ber}_\mu = 1] = \mu$, $\Pr[\mathsf{Ber}_\mu = 0] = 1 - \mu$, while Ber_μ^q denotes the concatenation of q independent copies of Ber_μ. We use χ_i^q, $i \leq q$, to denote a uniform distribution over $\{e \in \{0,1\}^q : |e| = i\}$, where $|e|$ denotes the Hamming weight of binary string e. For $n \in \mathbb{N}$, U_n denotes the uniform distribution over $\{0,1\}^n$ and independent of any other random variables in consideration, and $f(U_n)$ denotes the distribution induced by applying the function f to U_n. $X \sim D$ denotes that random variable X follows distribution D. We use $s \leftarrow S$ to denote sampling an element s according to distribution S, and let $s \xleftarrow{\$} \mathcal{S}$ denote sampling s uniformly from set $\mathcal{S}$.

ENTROPY DEFINITIONS. For a random variable X and any $x \in \mathsf{Supp}(X)$, the sample-entropy of x with respect to X is defined as

$$\mathbf{H}_X(x) \stackrel{\text{def}}{=} \log(1/\Pr[X = x])$$

from which we define the Shannon entropy, Rényi entropy and min-entropy of X respectively, i.e.,

$$\mathbf{H}_1(X) \stackrel{\text{def}}{=} \mathbb{E}_{x \leftarrow X}[\mathbf{H}_X(x)], \quad \mathbf{H}_2 \stackrel{\text{def}}{=} -\log \sum_{x \in \mathsf{Supp}(X)} 2^{-2\mathbf{H}_X(x)}, \quad \mathbf{H}_\infty(X) \stackrel{\text{def}}{=} \min_{x \subset \mathsf{Supp}(X)} \mathbf{H}_X(x).$$

For $0 < \mu < 1/2$, let $\mathbf{H}(\mu) \stackrel{\text{def}}{=} \mu \log(1/\mu) + (1 - \mu)\log(1/(1-\mu))$ be the binary entropy function so that $\mathbf{H}(\mu) = \mathbf{H}_1(\mathsf{Ber}_\mu)$. We know that $\mathbf{H}_1(X) \geq \mathbf{H}_2(X) \geq \mathbf{H}_\infty(X)$ with equality when X is uniformly distributed. A random variable X of length n is called an (n, λ)-Rényi entropy (resp., min-entropy) source if $\mathbf{H}_2(X) \geq \lambda$ (resp., $\mathbf{H}_\infty(X) \geq \lambda$). The *statistical distance* between X and Y, denoted by $\mathsf{SD}(X, Y)$, is defined by

$$\mathsf{SD}(X, Y) \stackrel{\text{def}}{=} \frac{1}{2} \sum_x |\Pr[X = x] - \Pr[Y = x]|$$

We use $\mathsf{SD}(X, Y | Z)$ as a shorthand for $\mathsf{SD}((X, Z), (Y, Z))$.

SIMPLIFYING NOTATIONS. To simplify the presentation, we use the following simplified notations. Throughout, n is the security parameter and most other parameters are functions of n, and we often omit n when clear from the context. For example, $\mu = \mu(n) \in (0, 1/2)$, $q = q(n) \in \mathbb{N}$, $t = t(n) > 0$, $\epsilon = \epsilon(n) \in (0, 1)$, and $m = m(n) = \mathsf{poly}(n)$, where poly refers to some polynomial.

Definition 1 (Computational/decisional LPN). *Let n be a security parameter, and let μ, q, t and ϵ all be functions of n. The **decisional** $\mathsf{LPN}_{\mu,n}$ problem (with secret length n and noise rate μ) is (q, t, ϵ)-hard if for every probabilistic distinguisher D running in time t we have*

$$\left| \Pr_{A,S,E}[\mathsf{D}(A, \ A \cdot S \oplus E) = 1] - \Pr_{A,U_q}[\mathsf{D}(A, U_q) = 1] \right| \leq \epsilon \tag{1}$$

where $A \sim U_{qn}$ is a $q \times n$ matrix, $S \sim U_n$ and $E \sim \mathsf{Ber}_\mu^q$. The **computational** *$\mathsf{LPN}_{\mu,n}$ problem is (q,t,ϵ)-hard if for every probabilistic algorithm D running in time t we have*

$$\Pr_{A,S,E}[\mathsf{D}(A,\ A{\cdot}S \oplus E) = (S,E)] \leq \epsilon,$$

where $A \sim U_{qn}$, $S \sim U_n$ and $E \sim \mathsf{Ber}_\mu^q$.

Definition 2 (LPN variants). *The decisional/computational* **X-$\mathsf{LPN}_{\mu,n}$** *is defined as per Definition 1 accordingly except that (S,E) follows distribution X. Note that standard $\mathsf{LPN}_{\mu,n}$ is a special case of X-$\mathsf{LPN}_{\mu,n}$ for $X \sim (U_n, \mathsf{Ber}_\mu^q)$.*

In respect of the randomized feature of LPN, we generalize standard PRGs/PRFs to equivalent randomized variants, where the generator/function additionally uses some public coins for randomization, and that seed/key can be sampled from a weak source (independent of the public coins).

Definition 3 (Randomized PRGs on weak seeds). *Let $\lambda \leq \ell_1 < \ell_2, \ell_3, t, \epsilon$ be functions of security parameter n. An efficient function family ensemble $\mathcal{G} = \{G_a : \{0,1\}^{\ell_1} \to \{0,1\}^{\ell_2}, a \in \{0,1\}^{\ell_3}\}_{n \in \mathbb{N}}$ is a (t,ϵ) randomized PRG on (ℓ_1, λ)-weak seed if for every probabilistic distinguisher D of running time t and every (ℓ_1, λ)-Rényi entropy source K it holds that*

$$\left| \Pr_{K, A \sim U_{\ell_3}}[\mathsf{D}(G_A(K), A) = 1] - \Pr_{U_{\ell_2}, A \sim U_{\ell_3}}[\mathsf{D}(U_{\ell_2}, A) = 1] \right| \leq \epsilon.$$

The stretch factor of $\mathcal{G}$ is ℓ_2/ℓ_1. Standard (deterministic) PRGs are implied by defining $G'(k,a) \stackrel{\text{def}}{=} (G_a(k), a)$ for a uniform random k.

Definition 4 (Randomized PRFs on weak keys). *Let $\lambda \leq \ell_1, \ell_2, \ell_3, \ell, t, \epsilon$ be functions of security parameter n. An efficient function family ensemble $\mathcal{F} = \{F_{k,a} : \{0,1\}^\ell \to \{0,1\}^{\ell_2}, k \in \{0,1\}^{\ell_1}, a \in \{0,1\}^{\ell_3}\}_{n \in \mathbb{N}}$ is a (q,t,ϵ) randomized PRF on (ℓ_1, λ)-weak key if for every oracle-aided probabilistic distinguisher D of running time t and bounded by q queries and for every (ℓ_1, λ)-Rényi entropy source K we have*

$$\left| \Pr_{K, A \sim U_{\ell_3}}[\mathsf{D}^{F_{K,A}}(A) = 1] - \Pr_{R, A \sim U_{\ell_3}}[\mathsf{D}^R(A) = 1] \right| \leq \epsilon(n),$$

where R denotes a random function distribution ensemble mapping from ℓ bits to ℓ_2 bits. Standard PRFs are a special case for empty a (or keeping $k' = (k,a)$ secret) on uniformly random key.

Definition 5 (Universal hashing). *A function family $\mathcal{H} = \{h_a : \{0,1\}^n \to \{0,1\}^m, a \in \{0,1\}^l\}$ is* **universal** *if for any $x_1 \neq x_2 \in \{0,1\}^n$ it holds that*

$$\Pr_{a \xleftarrow{\$} \{0,1\}^l} [h_a(x_1) = h_a(x_2)] \leq 2^{-m}.$$

Definition 6 (Pairwise independent hashing). *A function family* $\mathcal{H} = \{h_a: \{0,1\}^n \to \{0,1\}^m, a \in \{0,1\}^l\}$ *is **pairwise independent** if for any* $x_1 \neq x_2 \in \{0,1\}^n$ *and any* $v \in \{0,1\}^{2m}$ *it holds that*

$$\Pr_{a \xleftarrow{\$} \{0,1\}^l} [(h_a(x_1), h_a(x_2)) = v] = 2^{-2m}.$$

CONCRETE CONSTRUCTIONS. We know that for every $m \leq n$ there exists a pairwise independent (and universal) $\mathcal{H}$ with description length $l = \Theta(n)$, where every $h \in \mathcal{H}$ can be computed in $\text{AC}^0(\text{MOD}_2)$. For example, $\mathcal{H}_1$ and $\mathcal{H}_2$ defined below are universal and pairwise independent respectively:

$$\mathcal{H}_1 = \left\{ h_a : \{0,1\}^n \to \{0,1\}^m \mid h_a(x) \overset{\text{def}}{=} a \cdot x, a \in \{0,1\}^{n+m-1} \right\}$$

$$\mathcal{H}_2 = \left\{ h_{a,b} : \{0,1\}^n \to \{0,1\}^m \mid h_{a,b}(x) \overset{\text{def}}{=} a \cdot x \oplus b, a \in \{0,1\}^{n+m-1}, b \in \{0,1\}^m \right\}$$

where $a \in \{0,1\}^{n+m-1}$ is interpreted as an $m \times n$ Toeplitz matrix and '$\cdot$' and '$\oplus$' denote matrix-vector multiplication and addition over $\text{GF}(2)$ respectively.

3   Bernoulli Randomness Extraction in $\text{AC}^0(\text{MOD}_2)$

First, we state below a variant of the lemma (e.g., [28]) that taking sufficiently many samples of i.i.d. random variables yields an "almost flat" joint random variable, i.e., the sample-entropy of most values is close to the Shannon entropy of the joint random variable. The proof is included in Appendix B for completeness.

Lemma 1 (Flattening Shannon entropy). *For any* $n \in \mathbb{N}$, $0 < \mu < 1/2$ *and for any* $\Delta > 0$ *define*

$$\mathcal{E} \overset{\text{def}}{=} \left\{ e \in \{0,1\}^q : \mathbf{H}_{\text{Ber}_\mu^q}(e) \leq (1+\Delta)q\mathbf{H}(\mu) \right\}. \tag{2}$$

Then, we have $\Pr[\text{Ber}_\mu^q \in \mathcal{E}] \geq 1 - \exp^{-\frac{\min(\Delta, \Delta^2)\mu q}{3}}$.

Lemma 2 states that the proposed Bernoulli randomness extractor (see Fig. 1) extracts almost all entropy from a Rényi entropy (or min-entropy) source. We mention that the extractor can be considered as a parallelized version of the random bits recycler of Impagliazzo and Zuckerman [32] and the proof technique is also closely relevant to the crooked leftover hash lemma [14, 21].

Lemma 2 (Bernoulli randomness extraction). *For any* $m, v \in \mathbb{N}$ *and* $0 < \mu \leq 1/2$, *let* $W \in \mathcal{W}$ *be any* $(\lceil \log |\mathcal{W}| \rceil, m)$-*Rényi entropy source, let* $\mathcal{H}$ *be a family of pairwise independent hash functions mapping from* $\mathcal{W}$ *to* $\{0,1\}^v$, *let* $\mathbf{H} = (H_1, \ldots, H_q)$ *be a vector of i.i.d. random variables such that each* H_i *is uniformly distributed over* $\mathcal{H}$, *let* sample : $\{0,1\}^v \to \{0,1\}$ *be any Boolean function such that* sample$(U_v) \sim \text{Ber}_\mu$. *Then, for any constant* $0 < \Delta \leq 1$ *it holds that*

$$\text{SD}(\text{Ber}_\mu^q, \text{sample}(\mathbf{H}(W)) \mid \mathbf{H}) \leq 2^{\left((1+\Delta)q\mathbf{H}(\mu) - m\right)/2} + \exp^{-\frac{\Delta^2 \mu q}{3}},$$

where
$$\mathsf{sample}(\boldsymbol{H}(W)) \stackrel{\text{def}}{=} (\mathsf{sample}(H_1(W)), \ldots, \mathsf{sample}(H_q(W))).$$

Remark 1 (On entropy loss). The amount of entropy extracted (i.e., $q\mathbf{H}(\mu)$) can be almost as large as entropy of the source (i.e., m) by setting $m = (1+2\Delta)q\mathbf{H}(\mu)$ for any arbitrarily small constant Δ. Further, the leftover hash lemma falls into a special case for $v = 1$ (sample being an identity function) and $\mu = 1/2$.

Proof. Let set $\mathcal{E}$ be defined as in (2). For any $e \in \{0,1\}^q$ and $\boldsymbol{h} \in \mathcal{H}^q$, use shorthands $p_{\boldsymbol{h}} \stackrel{\text{def}}{=} \Pr[\boldsymbol{H} = \boldsymbol{h}]$, $p_{e|\boldsymbol{h}} \stackrel{\text{def}}{=} \Pr[\mathsf{sample}(\boldsymbol{h}(W)) = e]$ and $p_e \stackrel{\text{def}}{=} \Pr[\mathsf{Ber}_\mu^q = e]$. We have

$$\mathsf{SD}\big((\mathsf{Ber}_\mu^q, \boldsymbol{H}), \ (\mathsf{sample}(\boldsymbol{H}(W)), \boldsymbol{H})\big)$$

$$= \frac{1}{2} \sum_{\boldsymbol{h}\in\mathcal{H}^q, e\in\mathcal{E}} p_{\boldsymbol{h}} | \, p_{e|\boldsymbol{h}} - p_e \, | + \frac{1}{2} \sum_{\boldsymbol{h}\in\mathcal{H}^q, e\notin\mathcal{E}} p_{\boldsymbol{h}} | \, p_{e|\boldsymbol{h}} - p_e \, |$$

$$\leq \frac{1}{2} \sum_{\boldsymbol{h}\in\mathcal{H}^q, e\in\mathcal{E}} (\sqrt{p_{\boldsymbol{h}} \cdot p_e}) \cdot \left(\sqrt{\frac{p_{\boldsymbol{h}}}{p_e}} \, | \, p_{e|\boldsymbol{h}} - p_e \, | \, \right)$$

$$+ \frac{1}{2}\left(\sum_{\boldsymbol{h}\in\mathcal{H}^q, e\notin\mathcal{E}} p_{\boldsymbol{h}} p_{e|\boldsymbol{h}} + \sum_{\boldsymbol{h}\in\mathcal{H}^q, e\notin\mathcal{E}} p_{\boldsymbol{h}} p_e \right)$$

$$\leq \frac{1}{2}\sqrt{ \left(\sum_{\boldsymbol{h}\in\mathcal{H}^q, e\in\mathcal{E}} p_{\boldsymbol{h}} \cdot p_e \right) \cdot \left(\sum_{\boldsymbol{h}\in\mathcal{H}^q, e\in\mathcal{E}} \frac{p_{\boldsymbol{h}}}{p_e} \cdot (p_{e|\boldsymbol{h}} - p_e)^2 \right) + \Pr[\mathsf{Ber}_\mu^q \notin \mathcal{E}]}$$

$$\leq \frac{1}{2}\sqrt{ 1 \cdot \sum_{e\in\mathcal{E}} \left(\sum_{\boldsymbol{h}\in\mathcal{H}^q} \frac{p_{\boldsymbol{h}} p_{e|\boldsymbol{h}}^2}{p_e} - 2\sum_{\boldsymbol{h}\in\mathcal{H}^q} p_{\boldsymbol{h}} p_{e|\boldsymbol{h}} + \sum_{\boldsymbol{h}\in\mathcal{H}^q} p_{\boldsymbol{h}} p_e \right) + \exp^{-\frac{\Delta^2\mu q}{3}} }$$

$$\leq \frac{1}{2}\sqrt{|\mathcal{E}| \cdot 2^{-m}} + \exp^{-\frac{\Delta^2\mu q}{3}}$$

$$\leq 2^{\frac{(1+\Delta)q\mathbf{H}(\mu)-m}{2}} + \exp^{-\frac{\Delta^2\mu q}{3}},$$

where the second inequality is Cauchy-Schwarz, i.e., $|\sum a_i b_i| \leq \sqrt{(\sum a_i^2) \cdot (\sum b_i)^2}$ and (3) below, the third inequality follows from Lemma 1, and the fourth inequality is due to (4) and (5), i.e., fix any e (and thus fix p_e as well) we can substitute $p_e \cdot (2^{-m} + p_e)$ for $\sum_{\boldsymbol{h}\in\mathcal{H}^q} p_{\boldsymbol{h}} p_{e|\boldsymbol{h}}^2$, and p_e for both $\sum_{\boldsymbol{h}\in\mathcal{H}^q} p_{\boldsymbol{h}} p_{e|\boldsymbol{h}}$ and $\sum_{\boldsymbol{h}\in\mathcal{H}^q} p_{\boldsymbol{h}} p_e$, and the last inequality follows from the definition of $\mathcal{E}$ (see (2))

$$|\mathcal{E}| \ \leq \ 1/\min_{e\in\mathcal{E}} \Pr[\mathsf{Ber}_\mu^q = e] \ \leq \ 2^{(1+\Delta)q\mathbf{H}(\mu)}$$

which completes the proof.

Claim 1
$$\sum_{\boldsymbol{h}\in\mathcal{H}^q, e\notin\mathcal{E}} p_{\boldsymbol{h}} p_{e|\boldsymbol{h}} = \sum_{\boldsymbol{h}\in\mathcal{H}^q, e\notin\mathcal{E}} p_{\boldsymbol{h}} p_e = \Pr[\mathsf{Ber}_\mu^q \notin \mathcal{E}] \tag{3}$$

$$\forall e \in \{0,1\}^q : \sum_{h \in \mathcal{H}^q} p_h p_{e|h}^2 \leq p_e \cdot (2^{-m} + p_e) \tag{4}$$

$$\forall e \in \{0,1\}^q : \sum_{h \in \mathcal{H}^q} p_h p_{e|h} = \sum_{h \in \mathcal{H}^q} p_h p_e = p_e \tag{5}$$

Proof. Let $\boldsymbol{H}(W) \overset{\text{def}}{=} (H_1(W), \ldots, H_q(W))$. The pairwise independence of $\mathcal{H}$ implies that

$$\boldsymbol{H}(W) \sim (U_v^1, \ldots, U_v^q)$$

holds even conditioned on any fixing of $W = w$, and thus $\mathsf{sample}(\boldsymbol{H}(W)) \sim \mathsf{Ber}_\mu^q$. We have

$$\sum_{h \in \mathcal{H}^q, e \notin \mathcal{E}} p_h p_{e|h} = \Pr[\mathsf{sample}(\boldsymbol{H}(W)) \notin \mathcal{E}] = \Pr[\mathsf{Ber}_\mu^q \notin \mathcal{E}],$$

$$\forall e \in \{0,1\}^q : \sum_{h \in \mathcal{H}^q} p_h p_{e|h} = \Pr[\mathsf{sample}(\boldsymbol{H}(W)) = e] = \Pr[\mathsf{Ber}_\mu^q = e] = p_e,$$

$$\sum_{h \in \mathcal{H}^q, e \notin \mathcal{E}} p_h p_e = \sum_{h \in \mathcal{H}^q} p_h \cdot \sum_{e \notin \mathcal{E}} p_e = \Pr[\mathsf{Ber}_\mu^q \notin \mathcal{E}],$$

$$\forall e \in \{0,1\}^q : \sum_{h \in \mathcal{H}^q} p_h p_e = p_e \cdot \sum_{h \in \mathcal{H}^q} p_h = p_e.$$

Now fix any $e \in \{0,1\}^q$, and let W_1 and W_2 be random variables that are i.i.d. to W, we have

$$\sum_{h \in \mathcal{H}^q} p_h p_{e|h}^2$$

$$= \Pr_{W_1, W_2, \boldsymbol{H}}[\mathsf{sample}(\boldsymbol{H}(W_1)) = \mathsf{sample}(\boldsymbol{H}(W_2)) = e]$$

$$\leq \Pr_{W_1, W_2}[W_1 = W_2] \cdot \Pr_{W_1, \boldsymbol{H}}[\mathsf{sample}(\boldsymbol{H}(W_1)) = e]$$

$$\quad + \Pr_{\boldsymbol{H}}[\mathsf{sample}(\boldsymbol{H}(w_1)) = \mathsf{sample}(\boldsymbol{H}(w_2)) = e \mid w_1 \neq w_2]$$

$$\leq 2^{-m} \cdot p_e + \Pr[\mathsf{Ber}_\mu^q = e]^2 = 2^{-m} \cdot p_e + p_e^2,$$

where the second inequality is again due to the pairwise independence of $\mathcal{H}$, i.e., for any $w_1 \neq w_2$, $\boldsymbol{H}(w_1)$ and $\boldsymbol{H}(w_2)$ are i.i.d. to $(U_v^1, \ldots, U_v^q)$ and thus the two distributions $\mathsf{sample}(\boldsymbol{H}(w_1))$ and $\mathsf{sample}(\boldsymbol{H}(w_2))$ are i.i.d. to Ber_μ^q.

4   Parallelizable PRFs on Weak Keys

4.1   A Succinct Formulation of LPN

The authors of [22] observed that the secret of LPN is not necessary to be uniformly random and can be replaced with a Bernoulli distribution. We state a more quantitative version (than [22, Problem 2]) in Lemma 3 that Ber_μ^{n+q}-$\mathsf{LPN}_{\mu,n}$ (see Definition 2) is implied by standard LPN for nearly the same parameters except that standard LPN needs n more samples. The proof follows by a simple reduction and is included in Appendix B.

Lemma 3. *Assume that the decisional (resp., computational)* $\mathsf{LPN}_{\mu,n}$ *problem is* (q, t, ϵ)*-hard, then the decisional (resp., computational)* Ber_{μ}^{n+q}*-$\mathsf{LPN}_{\mu,n}$ problem is at least* $(q - (n + 2),\ t - \mathsf{poly}(n + q),\ 2\epsilon)$*-hard.*

Remark 2 (On the security of low-noise LPN). For $\mu = n^{-c}$, a trivial statistical test suggests (by the piling-up lemma) that any single sample of decisional Ber_{μ}^{n+q}-$\mathsf{LPN}_{\mu,n}$ is $(1/2 + 2^{-O(n^{1-c})})$-biased to 0. In other words, decisional Ber_{μ}^{n+q}-$\mathsf{LPN}_{\mu,n}$ is no more than $(q = 1, t = O(1), \epsilon = 2^{-O(n^{1-c})})$-hard and thus it follows (via the reduction of Lemma 3) that decisional $\mathsf{LPN}_{\mu,n}$ cannot have indistinguishability beyond $(q = n + 3, t = \mathsf{poly}(n), \epsilon = 2^{-O(n^{1-c})})$. Asymptotically, this is also the current state-of-the-art attack on low-noise LPN using $q = \mathsf{poly}(n)$ or even more samples.

4.2 A Direct Construction in Almost Constant Depth

To build a randomized PRG (on weak source w) from the succinct LPN, we first sample Bernoulli vector (s, e) from w (using random coins a), and then output $a{\cdot}s \oplus e$. Theorem 2 states that the above yields a randomized PRG on weak seed w and public coin a.

Theorem 2 (randomized PRGs from LPN). *Let n be a security parameter, let $\delta > 0$ be any constant, and let $\mu = n^{-c}$ for any $0 < c < 1$. Assume that decisional $\mathsf{LPN}_{\mu,n}$ problem is $((1 + 2\delta)n, t, \epsilon)$-hard, then $\mathcal{G} = \{G_a : \{0,1\}^{n^{1-\frac{c}{2}}} \to \{0,1\}^{\delta n}, a \in \{0,1\}^{\delta n \times n}\}_{n \in \mathbb{N}}$, where*

$$G_a(w) = a \cdot s \oplus e, s \in \{0,1\}^n, e \in \{0,1\}^{\delta n}$$

and $(s, e) = \mathsf{sample}(h_a(w))$, is a $(t - \mathsf{poly}(n), O(\epsilon))$-randomized PRG on $(n^{1-\frac{c}{2}}, 4c(1 + \delta^2)n^{1-c} \cdot \log n)$-weak seed with stretch factor $\delta{\cdot}n^{\frac{c}{2}}$.

Proof. We have by Lemma 3 that $((1 + 2\delta)n, t, \epsilon)$-hard decisional $\mathsf{LPN}_{\mu,n}$ implies $(\delta n, t - \mathsf{poly}(n), 2\epsilon)$-hard decisional $\mathsf{Ber}_{\mu}^{n+\delta n}$-$\mathsf{LPN}_{\mu,n}$, so the conclusion follows if we could sample $(s, e) \overset{\$}{\leftarrow} \mathsf{Ber}_{\mu}^{n+\delta n}$ from w. This follows from Lemma 2 by choosing $q = n + \delta n$, $\Delta = \delta$, and $m = 4c(1 + \delta)^2 n^{1-c} \cdot \log n$ such that the sampled noise vector is statistically close to $\mathsf{Ber}_{\mu}^{n+\delta n}$ except for an error bounded by

$$2^{\left((1+\Delta)q\mathbf{H}(\mu) - m\right)/2} + \exp^{-\frac{\Delta^2 \mu q}{3}}$$

$$\leq 2^{\left((1+\delta)^2 n\mathbf{H}(\mu) - 2(1+\delta)^2 n\mathbf{H}(\mu)\right)/2} + 2^{-\Omega(n^{1-c})}$$

$$= 2^{-\Omega(n^{1-c}\cdot\log n)} + 2^{-\Omega(n^{1-c})}$$

$$= 2^{-\Omega(n^{1-c})}$$

where recall by Fact A1 that $\mu\log(1/\mu) < \mathbf{H}(\mu) < \mu(\log(1/\mu) + 2)$ and thus $m > 2(1 + \delta^2)n^{1-c}(c\log n + 2) > 2(1 + \delta^2)n\mathbf{H}(\mu)$. We omit the above term since $\epsilon = 2^{-O(n^{1-c})}$ (see Remark 2).

We state a variant of the theorem by Goldreich, Goldwasser and Micali [26] on building PRFs from PRGs, where we consider PRGs with stretch factor 2^v for $v = O(\log n)$ (i.e., a balanced 2^v-ary tree) and use randomized (instead of deterministic) PRG G_a, reusing public coin a at every invocation of G_a.

Theorem 3 (PRFs from PRGs [26]). *Let n be a security parameter, let $v = O(\log n)$, $\lambda \leq m = n^{O(1)}$, $\lambda = \mathsf{poly}(n)$, $t = t(n)$ and $\epsilon = \epsilon(n)$. Let $\mathcal{G} = \{G_a : \{0,1\}^m \to \{0,1\}^{2^v \cdot m}, a \in \mathcal{A}\}_{n \in \mathbb{N}}$ be a (t, ϵ) randomized PRG (with stretch factor 2^v) on (m, λ)-weak seed. Parse $G_a(k)$ as 2^v blocks of m-bit strings:*

$$G_a(k) \stackrel{\text{def}}{=} G_a^{0\cdots00}(k) \| G_a^{0\cdots01}(k) \| \cdots \| G_a^{1\cdots11}(k)$$

where $G_a^{i_1\cdots i_v}(k)$ denotes the $(i_1 \cdots i_v)$-th m-bit block of $G_a(k)$. Then, for any $d \leq \mathsf{poly}(n)$ and $q = q(n)$, the function family ensemble $\mathcal{F} = \{F_{k,a} : \{0,1\}^{dv} \to \{0,1\}^{2^v \cdot m}, k \in \{0,1\}^m, a \in \mathcal{A}\}_{n \in \mathbb{N}}$, where

$$F_{k,a}(x_1 \cdots x_{dv}) \stackrel{\text{def}}{=} G_a(G_a^{x_{(d-1)v+1}\cdots x_{dv}}(\cdots G_a^{x_{v+1}\cdots x_{2v}}(G_a^{x_1\cdots x_v}(k))\cdots)),$$

is a $(q, t - q \cdot \mathsf{poly}(n), dq\epsilon)$ randomized PRF on (m, λ)-weak key.

ON POLYNOMIAL-SIZE CIRCUITS. The above GGM tree has $\Theta(2^{dv})$ nodes and thus it may seem that for $dv = \omega(\log n)$ we need a circuit of super-polynomial size to evaluate $F_{k,p}$. This is not necessary since we can represent the PRF in the following alternative form:

$$F_{k,a} = G_a \circ \underbrace{\mathsf{mux}_{x_{(d-1)v+1}\cdots x_{dv}} \circ G_a}_{G_a^{x_{(d-1)v+1}\cdots x_{dv}}} \circ \cdots \circ \underbrace{\mathsf{mux}_{x_{v+1}\cdots x_{2v}} \circ G_a}_{G_a^{x_{v+1}\cdots x_{2v}}} \circ \underbrace{\mathsf{mux}_{x_1\cdots x_v} \circ G_a}_{G_a^{x_1\cdots x_v}}$$

where '$\circ$' denotes function composition, each multiplexer $\mathsf{mux}_{i_1\cdots i_v} : \{0,1\}^{2^v m} \to \{0,1\}^m$ simply selects as output the $(i_1 \cdots i_v)$-th m-bit block of its input, and it can be implemented with $O(2^v \cdot m) = \mathsf{poly}(n)$ NOT and (unbounded fan-in) AND/OR gates of constant depth. Thus, for $v = O(\log n)$ function $F_{k,p}$ can be evaluated with a polynomial-size circuit of depth $O(d)$.

Lemma 4 (Levin's trick [41]). *For any $\ell \leq n \in \mathbb{N}$, let R_1 be a random function distribution over $\{0,1\}^\ell \to \{0,1\}^n$, let $\mathcal{H}$ be a family of universal hash functions from n bits to ℓ bits, and let H_1 be a function distribution uniform over $\mathcal{H}$. Let $R_1 \circ H_1(x) \stackrel{\text{def}}{=} R_1(H_1(x))$ be a function distribution over $\{0,1\}^n \to \{0,1\}^n$. Then, for any $q \in \mathbb{N}$ and any oracle aided D bounded by q queries, we have*

$$\left| \Pr_{R_1, H_1}[\mathsf{D}^{R_1 \circ H_1} = 1] - \Pr_R[\mathsf{D}^R = 1] \right| \leq \frac{q^2}{2^{\ell+1}},$$

where R is a random function distribution from n bits to n bits.

Theorem 4 (A direct PRF). *Let n be a security parameter, and let $\mu = n^{-c}$ for constant $0 < c < 1$. Assume that decisional $\mathsf{LPN}_{\mu,n}$ problem is $(\alpha n, t, \epsilon)$-hard*

*for any constant $\alpha > 1$, then for any (efficiently computable) $d = \omega(1) \leq O(n)$
and any $q \leq n^{d/3}$ there exists a $(q, t - q\,\mathrm{poly}(n), O(dq\epsilon) + q^2 n^{-d})$-randomized
PRF on $(n^{1-\frac{c}{2}}, O(n^{1-c}\log n))^8$-weak key*

$$\bar{\mathcal{F}} = \{\bar{F}_{k,a} : \{0,1\}^n \to \{0,1\}^n, k \in \{0,1\}^{n^{1-\frac{c}{2}}}, a \in \{0,1\}^{O(n^2)}\}_{n \in \mathbb{N}} \qquad (6)$$

*which is computable by a uniform family of polynomial-size depth-$O(d)$ circuits
with unbounded-fan-in AND, OR and XOR gates.*

Proof. For $\mu = n^{-c}$, we have by Theorem 2 that the decisional $(\alpha n, t, \epsilon)$-hard
$\mathrm{LPN}_{\mu,n}$ implies a $(t - \mathrm{poly}(n), O(\epsilon))$ randomized PRG in $\mathrm{AC}^0(\mathrm{MOD}_2)$ on $(n^{1-\frac{c}{2}},$
$O\,(n^{1-c}\log n)$)-weak seed k and public coin $a \in \{0,1\}^{O(n^2)}$ with stretch factor
$2^v = n^{\frac{c}{2}}$. We plug it into the GGM construction (see Theorem 3) with tree
depth $d' = 2d/c$ to get a $(q, t - q\,\mathrm{poly}(n), O(dq\epsilon))$ randomized PRF on weak
keys (of same parameters) with input length $d'v = d\log n$ and output length
$2^v \cdot n^{1-\frac{c}{2}} = n$ as below:

$$\mathcal{F} = \{F_{k,a} : \{0,1\}^{d\log n} \to \{0,1\}^n, k \in \{0,1\}^{n^{1-\frac{c}{2}}}, a \in \{0,1\}^{O(n^2)}\}_{n \in \mathbb{N}}. \qquad (7)$$

Now we expand k (e.g., by evaluating $F_{k,a}$ on a few fixed points) into a pseudo-
random $(\bar{k}, \bar{h}_1)$, where $\bar{k} \in \{0,1\}^{n^{1-\frac{c}{2}}}$ and $\bar{h}_1$ describes a universal hash function from n bits to $\ell = d\log n$ bits. Motivated by Levin's trick, we define a
domain-extended PRF $\bar{F}_{k,a}(x) \stackrel{\mathrm{def}}{=} F_{k,a} \circ \bar{h}_1(x)$. For any oracle-aided distinguisher D running in time $t - q\mathrm{poly}(n)$ and making q queries, denote with
$\delta_\mathsf{D}(F_1, F_2) \stackrel{\mathrm{def}}{=} |\mathrm{Pr}[\mathsf{D}^{F_1}(A) = 1] - \mathrm{Pr}[\mathsf{D}^{F_2}(A) = 1]|$ the advantage of D (who gets
public coin A as additional input) in distinguishing between function oracles F_1
and F_2. Therefore, we have by a triangle inequality

$$\delta_\mathsf{D}(F_{\bar{K},A} \circ \bar{H}_1, R) \leq \delta_\mathsf{D}(F_{\bar{K},A} \circ \bar{H}_1, F_{K,A} \circ H_1) + \delta_\mathsf{D}(F_{K,A} \circ H_1, R_1 \circ H_1)$$
$$+ \delta_\mathsf{D}(R_1 \circ H_1, R)$$
$$\leq O(dq\epsilon) + q^2 n^{-d},$$

where advantage is upper bounded by three terms, namely, the indistinguishability between $(\bar{K}, \bar{H}_1)$ and truly random (K, H_1), that between $F_{K,A}$ and random function R_1 (of the same input/output lengths as $F_{K,A}$), and that due to
Lemma 4. Note that A is independent of R_1, H_1 and R.

4.3 Going Beyond the Birthday Barrier

Unfortunately, for small $d = \omega(1)$ the security of the above PRF does not go
beyond super-polynomial (cf. term $q^2 n^{-d}$) due to a birthday attack. This situation can be handled using security-preserving constructions. Note the techniques
from [17,34] need (almost) $\Omega(d\log n)$-wise independent hash functions which we
don't know how to compute with unbounded fan-in gates of depth $O(d)$. Thus,

---

[8] Here the big-Oh omits a constant dependent on c and α.

we use a more intuitive and depth-preserving approach below by simply running a few independent copies and XORing their outputs. The essential idea dates backs to [8,44] and the technique receives renewed interest recently in some different contexts [23,25]. We mention that an alternative (and possibly more efficient) approach is to use the second security-preserving domain extension technique from [10] that requires a few pairwise independent hash functions and makes only a constant number of calls to the underlying small-domain PRFs. This yields the PRF stated in Theorem 5.

Lemma 5 (Generalized Levin's Trick [8,44]). *For any $\kappa, \ell \leq n \in \mathbb{N}$, let R_1, ..., R_κ be independent random function distributions over $\{0,1\}^\ell \to \{0,1\}^n$, let $\mathcal{H}$ be a family of universal hash functions from n bits to ℓ bits, and let H_1, $\cdots, H_\kappa$ be independent function distributions all uniform over $\mathcal{H}$. Let $F_{\boldsymbol{R},\boldsymbol{H}}$ be a function distribution (induced by $\boldsymbol{R} = (R_1, \ldots, R_\kappa)$ and $\boldsymbol{H} = (H_1, \ldots, H_\kappa)$) over $\{0,1\}^n \to \{0,1\}^n$ defined as*

$$F_{\boldsymbol{R},\boldsymbol{H}}(x) \stackrel{\text{def}}{=} \bigoplus_{i=1}^{\kappa} R_i(H_i(x)). \tag{8}$$

Then, for any $q \in \mathbb{N}$ and any oracle aided D bounded by q queries, we have

$$\left| \Pr[\mathsf{D}^{F_{\boldsymbol{R},\boldsymbol{H}}} = 1] - \Pr[\mathsf{D}^R = 1] \right| \leq \frac{q^{\kappa+1}}{2^{\kappa\ell}}$$

where R is a random function distribution over $\{0,1\}^n \to \{0,1\}^n$.

Finally, we get the first security-preserving construction below. To have comparable security to LPN with secret size n, it suffices to use a key of entropy $O(n^{1-c} \cdot \log n)$, or a uniform key of size $n^{1-\frac{c}{2}}$ with any $(1 - O(n^{-\frac{c}{2}} \log n))$-fraction of leakage (see Fact A7), provided that leakage is independent of public coin a.

Theorem 5 (A security-preserving PRF on weak key). *Let n be a security parameter, and let $\mu = n^{-c}$ for constant $0 < c < 1$. Assume that the decisional $\mathsf{LPN}_{\mu,n}$ problem is $(\alpha n, t, \epsilon)$-hard for any constant $\alpha > 1$, then for any (efficiently computable) $d = \omega(1) \leq O(n)$ and any $q \leq n^{d/3}$ there exists a $(q, t - q\mathrm{poly}(n), O(dq\epsilon))$-randomized PRF on $(n^{1-\frac{c}{2}}, O(n^{1-c} \cdot \log n))$-weak key*

$$\hat{\mathcal{F}} = \{\hat{F}_{k,a} : \{0,1\}^n \to \{0,1\}^n, k \in \{0,1\}^{n^{1-\frac{c}{2}}}, a \in \{0,1\}^{O(n^2)}\}_{n \in \mathbb{N}}$$

which are computable by a uniform family of polynomial-size depth-$O(d)$ circuits with unbounded-fan-in AND, OR and XOR gates.

Proof sketch. Following the proof of Theorem 4, we get a $(q, t - q\mathrm{poly}(n), O(dq\epsilon))$-randomized PRF $\mathcal{F} = \{F_{k,a}\}_{n \in \mathbb{N}}$ on weak keys (see (7)) with input length $d \log n$ and of depth $O(d)$. We define $\mathcal{F}' = \{F'_{(k,h),a} : \{0,1\}^n \to \{0,1\}^n, k \in \{0,1\}^{O(\kappa n^{1-\frac{c}{2}})}, h \in \mathcal{H}^\kappa, a \in \{0,1\}^{O(n^2)}\}_{n \in \mathbb{N}}$ where

$$F'_{(\boldsymbol{k},\boldsymbol{h}),a}(x) \stackrel{\text{def}}{=} \bigoplus_{i=1}^{\kappa} F_{k_i,a}(h_i(x)), \quad \boldsymbol{k} = (k_1, \cdots, k_\kappa), \boldsymbol{h} = (h_1, \cdots, h_\kappa).$$

Let $\delta_{\mathsf{D}}(F_1, F_2) \overset{\text{def}}{=} \big| \Pr[\mathsf{D}^{F_1}(A) = 1] - \Pr[\mathsf{D}^{F_2}(A) = 1] \big|$. We have that for any oracle-aided distinguisher running in time $t - q\mathsf{poly}(n)$ and making up to q queries, we have by a triangle inequality that

$$\delta_{\mathsf{D}}(F'_{(K,H),A},\, R) \leq \delta_{\mathsf{D}}(F'_{(K,H),A},\, F_{R,H}) + \delta_{\mathsf{D}}(F_{R,H},\, R)$$
$$\leq O(\kappa d q \epsilon) + n^{d(1-2\kappa)/3}$$
$$= O(\kappa d q \epsilon) + 2^{-\omega(n^{1-c})} = O(\kappa d q \epsilon),$$

where $F_{R,H}$ is defined as per (8), the first term of the second inequality is due to a hybrid argument (replacing every $F_{K_i,A}$ with R_i one at a time), the second term of the second inequality follows from Lemma 5 with $\ell = d\log n$ and $q \leq n^{d/3}$, and the equalities follow by setting $\kappa = n^{1-c}$ to make the first term dominant. Therefore, $F'_{(k,h),a}$ is almost the PRF as desired except that it uses a long key (k,h), which can be replaced with a pseudorandom one. That is, let $\hat{F}_{k,a}(x) \overset{\text{def}}{=} F'_{(k,h),a}(x)$ and $(k,h) \overset{\text{def}}{=} F_{k,a}(1)\|F_{k,a}(2)\| \cdots \|F_{k,a}(O(\kappa))$, which adds only a layer of gates of depth $O(d)$. □

5 An Alternative PRF with a Short Uniform Key

In this section, we introduce an alternative construction based on a variant of LPN (reducible from standard LPN) whose noise vector can be sampled in AC^0 (i.e., without using XOR gates). We state the end results in Theorem 6 that standard LPN with n-bit secret implies a low-depth PRF with key size $\Theta(n^{1-c}\log n)$. Concretely (and ideally), assume that computational LPN is $(q = 1.001n, t = 2^{n^{1-c}/3}, \epsilon = 2^{-n^{1-c}/12})$-hard, and let $\lambda = \Theta(n^{1-c}\log n)$, then for any $\omega(1) = d = O(\lambda/\log^2\lambda)$ there exists a parallelizable $(q' = \lambda^{\Theta(d)}, t' = 2^{\Theta(\lambda/\log\lambda)}, \epsilon' = 2^{-\Theta(\lambda/\log\lambda)})$-randomized PRF computable in depth $O(d)$ with secret key length λ and public coin length $O(\lambda^{\frac{1+c}{1-c}})$.

5.1 Main Results and Roadmap

Theorem 6 (A PRF with a compact uniform key). *Let n be a security parameter, and let $\mu = n^{-c}$ for constant $0 < c < 1$. Assume that the computational $\mathsf{LPN}_{\mu,n}$ problem is $(\alpha n, t, \epsilon)$-hard for any constant $\alpha > 1$ and efficiently computable ϵ, then for any (efficiently computable) $d = \omega(1) \leq O(n)$ and any $q' \leq n^{d/3}$ there exists a $(q', \Theta(t \cdot \epsilon^2 n^{1-2c}), O(dq'n^2\epsilon))$-randomized PRF on uniform key*

$$\tilde{\mathcal{F}} = \{\tilde{F}_{k,a} : \{0,1\}^n \to \{0,1\}^n, k \in \{0,1\}^{\Theta(n^{1-c}\cdot\log n)}, a \in \{0,1\}^{O(n^2)}\}_{n\in\mathbb{N}}$$

which are computable by a uniform family of polynomial-size depth-$O(d)$ circuits with unbounded-fan-in AND, OR and XOR gates.

We sketch the steps below to prove Theorem 6, where 'C-' and 'D-' stand for 'computational' and 'decisional' respectively.

1. Introduce distribution ψ_μ^m that can be sampled in AC^0.
2. $((1+\Theta(1))n,t,\epsilon)$-hard C- $\mathsf{LPN}_{\mu,n}$ $\implies$ $(\Theta(n),t-\mathsf{poly}(n),2\epsilon)$-hard C- Ber_μ^{n+q}-$\mathsf{LPN}_{\mu,n}$ (by Lemma 3).
3. $(\Theta(n),t,\epsilon)$-hard C- Ber_μ^{n+q}-$\mathsf{LPN}_{\mu,n}$ $\implies$ $(\Theta(n),t-\mathsf{poly}(n),O(n^{3/2-c}\epsilon))$-hard C- ψ_μ^{n+q}-$\mathsf{LPN}_{\mu,n}$ (by Lemma 9).
4. $(\Theta(n),t,\epsilon)$-hard C- ψ_μ^{n+q}-$\mathsf{LPN}_{\mu,n}$ $\implies$ $(\Theta(n),\Omega(t(\epsilon/n)^2),2\epsilon)$-hard D- ψ_μ^{n+q}-$\mathsf{LPN}_{\mu,n}$ (by Theorem 7).
5. $(\Theta(n),t,\epsilon)$-hard D- ψ_μ^{n+q}-$\mathsf{LPN}_{\mu,n}$ $\implies$ $(q,t-q\,\mathsf{poly}(n),O(dq'\epsilon))$-randomized PRF for any $d=\omega(1)$ and $q' \le n^{d/3}$, where the PRF has key length $\Theta(n^{1-c}\log n)$ and can be computed by polynomial-size depth-$O(d)$ circuits with unbounded-fan-in AND, OR and XOR gates. This is stated as Theorem 8.

5.2 Distribution ψ_μ^m and the ψ_μ^{n+q}-$\mathsf{LPN}_{\mu,n}$ Problem

We introduce a distribution ψ_μ^m that can be sampled in AC^0 and show that ψ_μ^{n+q}-$\mathsf{LPN}_{\mu,n}$ is implied by Ber_μ^{n+q}-$\mathsf{LPN}_{\mu,n}$ (and thus by standard LPN). Further, for $\mu = n^{-c}$ sampling ψ_μ^m needs $\Theta(mn^{-c}\log n)$ random bits, which asymptotically match the Shannon entropy of Ber_μ^m.

Algorithm 1. Sampling distribution ψ_μ^m in AC^0

Require: $2\mu m \log m$ random bits (assume WLOG that m is a power of 2)
Ensure: ψ_μ^m satisfies Lemma 6

1: Sample random $z_1,\ldots,z_{2\mu m}$ of Hamming weight 1, i.e., for every $i \in [m]$ $z_i \xleftarrow{\$} \{z \in \{0,1\}^m : |z| = 1\}$.
   {E.g., to sample z_1 with randomness $r_1 \ldots r_{\log m}$, simply let each $(b_1 \ldots b_{\log m})$-th bit of z_1 to be $r_1^{b_1} \wedge \cdots \wedge r_{\log m}^{b_{\log m}}$, where $r_j^{b_j} \overset{\text{def}}{=} r_j$ for $b_j = 0$ and $r_j^{b_j} \overset{\text{def}}{=} \neg r_j$ otherwise. Note that AC^0 allows NOT gates at the input level.}
2: Output the bitwise-OR of the vectors $z_1,\ldots,z_{2\mu m}$.
   {Note: we take a bitwise-OR (***not bitwise-XOR***) of the vectors.}

Lemma 6. *The distribution ψ_μ^m (sampled as per Algorithm 1) is $2^{-\Omega(\mu m \log(1/\mu))}$-close to a convex combination of $\chi_{\mu m}^m, \chi_{\mu m+1}^m, \ldots, \chi_{2\mu m}^m$.*

Proof. It is easy to see that ψ_μ^m is a convex combination of $\chi_1^m, \chi_2^m, \ldots, \chi_{2\mu m}^m$ as conditioned on $|\psi_\mu^m| = i$ (for any i) ψ_μ^m hits every $y \in \{0,1\}^m$ of Hamming weight $|y| = i$ with equal probability. Hence, it remains to show that those χ_j^m's with Hamming weight $j < \mu m$ sum to a fraction less than $2^{-\mu m(\log(1/\mu)-2)}$, i.e.,

$$\Pr[|\psi_\mu^m| < \mu m] = \sum_{y \in \{0,1\}^m : |y| < \mu m} \Pr[\psi_\mu^m = y]$$
$$< \mu^{2\mu m} \cdot 2^{m\mathbf{H}(\mu) - \frac{\log m}{2} + O(1)}$$
$$< \mu^{2\mu m} \cdot 2^{\mu m(\log(1/\mu)+2) + O(1)} = 2^{\mu m(-\log(1/\mu)+2) + O(1)}$$

where the first inequality is due to the partial sum of binomial coefficients (see Fact A5) and that for any fixed y with $|y| < \mu m$ $\psi_\mu^m = y$ happens only if the bit 1 of every z_i (see Algorithm 1) hits the 1's of y (each with probability less than μ independently) and the second inequality is Fact A1.

By definition of ψ_μ^{n+q} the sampled (s, e) has Hamming weight no greater than $2\mu(n + q)$ and the following lemma states that ψ_μ^{n+q}-$\mathsf{LPN}_{\mu,n}$ is almost injective.

Lemma 7 (ψ_μ^{n+q}-$\mathsf{LPN}_{\mu,n}$ is almost injective). *For $q = \Omega(n)$, define set*
$$\mathcal{Y} \overset{\text{def}}{=} \{(s, e) \in \{0,1\}^{n+q} : |(s, e)| \le (n + q)/\log n\}. \text{ Then, for every } (s, e) \in \mathcal{Y},$$
$$\Pr_{a \leftarrow U_{qn}} \left[\exists (s', e') \in \mathcal{Y} : (s', e') \ne (s, e) \wedge as \oplus e = as' \oplus e' \right] = 2^{-\Omega(q)}.$$

Proof. Let $\mathcal{H} \overset{\text{def}}{=} \{h_a : \{0,1\}^{n+q} \to \{0,1\}^q, a \in \{0,1\}^{qn}, h_a(s, e) \overset{\text{def}}{=} as \oplus e\}$ and it is not hard to see that $\mathcal{H}$ is a family of universal hash functions. We have
$$\log |\mathcal{Y}| = \log \sum_{i=0}^{(n+q)/\log n} \binom{n+q}{i} = O\big((n + q)\log\log n/\log n\big) = o(q),$$

where the approximation is due to Fact A5 and the conclusion immediately follows from Lemma 8.

Lemma 8 (The injective hash lemma (e.g. [55])). *For any integers $l_1 \le l_2, m$, let $\mathcal{Y}$ be any set of size $|\mathcal{Y}| \le 2^{l_1}$, and let $\mathcal{H} \overset{\text{def}}{=} \{h_a : \{0,1\}^m \to \{0,1\}^{l_2}, a \in \mathcal{A}, \mathcal{Y} \subseteq \{0,1\}^m\}$ be a family of universal hash functions. Then, for every $y \in \mathcal{Y}$ we have*
$$\Pr_{a \overset{\$}{\leftarrow} \mathcal{A}} \left[\exists y' \in \mathcal{Y} : y' \ne y \wedge h_a(y') = h_a(y) \right] \le 2^{l_1 - l_2}.$$

5.3   Computational Ber_μ^{n+q}-$\mathsf{LPN}_{\mu,n} \to$ Computational ψ_μ^{n+q}-$\mathsf{LPN}_{\mu,n}$

Lemma 9 non-trivially extends the well-known fact that the computational LPN implies the computational exact LPN, i.e., $(U_n, \chi_{\mu q}^q)$-$\mathsf{LPN}_{\mu,n}$.

Lemma 9. *Let $q = \Omega(n)$, $\mu = n^{-c}$ $(0 < c < 1)$ and $\epsilon = 2^{-O(n^{1-c})}$. Assume that the computational Ber_μ^{n+q}-$\mathsf{LPN}_{\mu,n}$ problem is (q, t, ϵ)-hard, then the computational ψ_μ^{n+q}-$\mathsf{LPN}_{\mu,n}$ problem is $(q, t - \mathrm{poly}(n + q), O(\mu(n + q)^{3/2}\epsilon))$-hard.*

Proof. Let $m = n + q$ and write $\mathsf{Adv}_\mathsf{D}(X) \overset{\text{def}}{=} \Pr_{a \overset{\$}{\leftarrow} U_{qn}, (s,e) \leftarrow X} [\mathsf{D}(a, a{\cdot}s \oplus e) = (s, e)]$. Towards a contradiction we assume that there exists D such that $\mathsf{Adv}_\mathsf{D}(\psi_\mu^m) > \epsilon'$, and we assume WLOG that on input (a, z) D always outputs (s', e') with $|(s', e')| \le 2\mu m$. That is, even if it fails to find any (s', e') satisfying $as' \oplus e' = z$ and $|(s', e')| \le 2\mu m$ it just outputs a zero vector. Lemma 6 states that ψ_μ^m is $2^{-\Omega(\mu n(\log(1/\mu))}$-close to a convex combination of $\chi_{\mu m}^m, \chi_{\mu m+1}^m, \ldots, \chi_{2\mu m}^m$, and thus there exists $j \in \{\mu m, \mu m + 1, \ldots, 2\mu m\}$

such that $\mathsf{Adv}_\mathsf{D}(\chi_j^m) > \epsilon' - 2^{-\Omega(n^{1-c}\log n)} > \epsilon'/2$, which further implies that $\mathsf{Adv}_\mathsf{D}(\mathsf{Ber}_{j/m}^m) = \Omega(\epsilon'/\sqrt{m})$ as $\mathsf{Ber}_{j/m}^m$ is a convex combination of $\chi_0^m, \ldots, \chi_m^m$, of which it hits χ_j^m with probability $\Omega(1/\sqrt{m})$ by Lemma 10. Next, we define D' as in Algorithm 2.

Algorithm 2. A Ber_μ^m-$\mathsf{LPN}_{\mu,n}$ solver D'

Require: a random Ber_μ^m-$\mathsf{LPN}_{\mu,n}$ instance $(a, z = a \cdot s \oplus e)$ as input
Ensure: a good chance to find out (s, e)

1: Sample $j^* \xleftarrow{\$} \{\mu m, \mu m + 1, \ldots, 2\mu m\}$ as a guess about j.
2: Compute $\mu' = j^*/m$.
3: $(s_1, e_1) \leftarrow \mathsf{Ber}_{\frac{\mu'-\mu}{1-2\mu}}^m$. {This makes $(a, z \oplus (as_1 \oplus e_1))$ a random $\mathsf{Ber}_{\mu'}^m$-$\mathsf{LPN}_{\mu',n}$ sample by the piling-up lemma (see Fact A6)}
4: $(s', e') \leftarrow \mathsf{D}(\ a,\ z \oplus (as_1 \oplus e_1)\ )$.
5: Output $(s' \oplus s_1, e' \oplus e_1)$. {$\mathsf{D}'$ succeeds iff $(s' \oplus s_1, e' \oplus e_1) = (s, e)$}

We denote $\mathcal{E}_{suc}$ the event that D succeeds in finding (s', e') such that $as' \oplus e' = z \oplus (as_1 \oplus e_1)$ and thus we have $a(s' \oplus s_1) \oplus (e' \oplus e_1) = z = as \oplus e$, where values are sampled as defined above. This however does not immediately imply $(s, e) = (s' \oplus s_1, e' \oplus e_1)$ unless conditioned on the event $\mathcal{E}_{inj}$ that $h_a(s, e) \overset{\text{def}}{=} a \cdot s \oplus e$ is injective on input (s, e).

$$\Pr_{a \leftarrow U_{qn},\ (s,e) \leftarrow \mathsf{Ber}_\mu^m,\ (s_1,e_1) \leftarrow \mathsf{Ber}_{\frac{\mu'-\mu}{1-2\mu}}^m,\ s' \leftarrow \mathsf{D}(a, y \oplus (as_1 \oplus e_1))} [(s' \oplus s_1, e' \oplus e_1) = (s, e)]$$

$$\geq \Pr[\mathcal{E}_{suc} \wedge \mathcal{E}_{inj}]$$

$$\geq \Pr[\mathcal{E}_{suc}] - \Pr[\neg\mathcal{E}_{inj}]$$

$$\geq \Pr[j^* = j] \cdot \mathsf{Adv}_\mathsf{D}(\mathsf{Ber}_{j/m}^m) - 2^{-\Omega(m/\log^2 n)}$$

$$= \Omega(\epsilon'/\mu m^{3/2}),$$

where the bound on event $\neg\mathcal{E}_{inj}$ is given below. We reach a contradiction by setting $\varepsilon' = \Omega(1) \cdot \mu m^{3/2}\epsilon$ for a large enough $\Omega(1)$ so that D' solves Ber_μ^m-$\mathsf{LPN}_{\mu,n}$ with probability greater than ϵ.

$$\Pr[\neg\mathcal{E}_{inj}]$$

$$\leq \Pr[\neg\mathcal{E}_{inj} \wedge (s, e) \in \mathcal{Y} \wedge (s' \oplus s_1, e' \oplus e_1) \in \mathcal{Y}]$$

$$\quad + \Pr[(s, e) \notin \mathcal{Y} \vee (s' \oplus s_1, e' \oplus e_1) \notin \mathcal{Y}]$$

$$\leq 2^{-\Omega(m)} + \Pr[(s, e) \notin \mathcal{Y}] + \Pr[(s' \oplus s_1, e' \oplus e_1) \notin \mathcal{Y}]$$

$$\leq 2^{-\Omega(m)} + \Pr_{(s,e) \leftarrow \mathsf{Ber}_\mu^m}[|(s, e)| \geq m/\log n] + \Pr_{(s_1,e_1) \leftarrow \mathsf{Ber}_{\frac{\mu'-\mu}{1-2\mu}}^m}[|(s_1, e_1)| \geq (\frac{1}{\log n} - 2\mu)m]$$

$$= 2^{-\Omega(m/\log^2 n)},$$

where $\mathcal{Y} \stackrel{\text{def}}{=} \{(s, e) \in \{0, 1\}^m : |(s, e)| < m/\log n\}$, the second inequality is from Lemma 7, the third inequality is that $|(u \oplus w)| \geq \kappa$ implies $|w| \geq \kappa - |u|$ and by definition of D string (s', e') has Hamming weight no greater than $2\mu m$, and the last inequality is a typical Chernoff-Hoeffding bound.

Lemma 10. *For $0 < \mu' < 1/2$ and $m \in \mathbb{N}$, we have that*

$$\Pr\left[|\mathsf{Ber}^m_{\mu'}| = \lceil \mu'm \rceil\right] = \Omega(1/\sqrt{m}).$$

5.4 C- ψ^{n+q}_μ-LPN$_{\mu,n}$ → D- ψ^{n+q}_μ-LPN$_{\mu,n}$ → $\omega(1)$-Depth PRFs

Next we show that the computational ψ^{n+q}_μ-LPN$_{\mu,n}$ problem implies its decisional counterpart. The theorem below is implicit in [5][9] and the case for ψ^{n+q}_μ-LPN$_{\mu,n}$ falls into a special case. Note that ψ^{n+q}_μ-LPN$_{\mu,n}$ is almost injective by Lemma 7, and thus its computational and decisional versions are equivalent in a sample-preserving manner. In fact, Theorem 7 holds even without the injective condition, albeit with looser bounds.

Theorem 7 (Sample preserving reduction [5]). *If the computational X-LPN$_{\mu,n}$ is (q, t, ϵ)-hard for any efficiently computable ϵ, and it satisfies the injective condition, i.e., for any $(s, e) \in \mathsf{Supp}(X)$ it holds that*

$$\Pr_{a \leftarrow U_{qn}}\left[\exists (s', e') \in \mathsf{Supp}(X) : (s', e') \neq (s, e) \wedge a \cdot s \oplus e = a \cdot s' \oplus e'\right] \leq 2^{-\Omega(n)}.$$

Then, the decisional X-LPN$_{\mu,n}$ is $(q, \Omega(t(\epsilon/n)^2), 2\epsilon)$-hard.

Theorem 8 (Decisional ψ^{n+q}_μ-LPN$_{\mu,n}$ → PRF). *Let n be a security parameter, and let $\mu = n^{-c}$ for any constant $0 < c < 1$. Assume that the decisional ψ^{n+q}_μ-LPN$_{\mu,n}$ problem is $(\delta n, t, \epsilon)$-hard for any constant $\delta > 0$, then for any (efficiently computable) $d = \omega(1) \leq O(n)$ and any $q' \leq n^{d/3}$ there exists a $(q', t - q'\mathsf{poly}(n), O(dq'\epsilon))$-randomized PRF (on uniform key) with key length $\Theta(n^{1-c}\log n)$ and public coin size $O(n^2)$, which are computable by a uniform family of polynomial-size depth-$O(d)$ circuits with unbounded-fan-in AND, OR and XOR gates.*

Proof sketch. The proof is essentially the same as that of Theorem 5, replacing the Bernoulli randomness extractor with the ψ^{n+q}_μ sampler. That is, decisional ψ^{n+q}_μ-LPN$_{\mu,n}$ for $q = \Theta(n)$ implies a constant-depth polynomial-stretch randomized PRG on seed length $2\mu(n + q)\log(n + q) = \Theta(n^{1-c}\log n)$ and output length $\Theta(n)$, which in turn implies a nearly constant-depth randomized PRF, where the technique in Lemma 5 is also used to make the construction security preserving. □

---

[9] Lemma 4.4 from the full version of [5] states a variant of Theorem 7 for uniformly random a and s, and arbitrary e. However, by checking its proof it actually only requires the matrix a to be uniform and independent of (s, e).

Acknowledgments. Yu Yu is more than grateful to Alon Rosen for motivating this work and many helpful suggestions, and he also thanks Siyao Guo for useful comments. The authors thank Ilan Komargodski for pointing out that the domain extension technique from [10] can also be applied to our constructions with improved efficiency. Yu Yu was supported by the National Basic Research Program of China Grant number 2013CB338004, the National Natural Science Foundation of China Grant (Nos. 61472249, 61572192). John Steinberger was funded by National Basic Research Program of China Grant 2011CBA00300, 2011CBA00301, the National Natural Science Foundation of China Grant 61361136003, and by the China Ministry of Education grant number 20121088050.

A   Well-Known Facts, Lemmas and Inequalities

Fact A1. Let $\mathbf{H}(\mu) \overset{\text{def}}{=} \mu \log(1/\mu) + (1 - \mu) \log(1/(1 - \mu))$ be the binary entropy function. Then, for any $0 < \mu < 1/2$ it holds that

$$\mu \log(1/\mu) < \mathbf{H}(\mu) < \mu(\log(1/\mu) + 2).$$

Proof.

$$\mu \log(1/\mu)$$
$$< \left(\mathbf{H}(\mu) = \mu \log(1/\mu) + (1 - \mu) \log(1/(1 - \mu)) \right)$$
$$= \mu \log(1/\mu) + (1 - \mu) \log(1 + \frac{\mu}{1 - \mu})$$
$$= \mu \log(1/\mu) + (1 - \mu) \frac{\ln(1 + \frac{\mu}{1-\mu})}{\ln 2}$$
$$\leq \mu \log(1/\mu) + \frac{\mu}{\ln 2} < \mu(\log(1/\mu) + 2),$$

where the first inequality is due to $(1 - \mu) \log(1/(1 - \mu)) > 0$, the second one follows from the elementary inequality $\ln(1 + x) \leq x$ for any $x > 0$, and the last inequality is simply $1 < 2 \ln 2$.

Lemma 11 (Chernoff bound). *For any $n \in \mathbb{N}$, let $X_1, \ldots, X_n$ be independent random variables and let $\bar{X} = \sum_{i=1}^{n} X_i$, where $\Pr[0 \leq X_i \leq 1] = 1$ holds for every $1 \leq i \leq n$. Then, for any $\Delta_1 > 0$ and $0 < \Delta_2 < 1$,*

$$\Pr[\bar{X} > (1 + \Delta_1) \cdot \mathbb{E}[\bar{X}]] < \exp^{-\frac{\min(\Delta_1, \Delta_1^2)}{3} \mathbb{E}[\bar{X}]},$$
$$\Pr[\bar{X} < (1 - \Delta_2) \cdot \mathbb{E}[\bar{X}]] < \exp^{-\frac{\Delta_2^2}{2} \mathbb{E}[\bar{X}]}.$$

Theorem 9 (The Hoeffding bound [30]). *Let $q \in \mathbb{N}$, and let $\xi_1, \xi_2, \ldots, \xi_q$ be independent random variables such that for each $1 \leq i \leq q$ it holds that $\Pr[a_i \leq \xi_i \leq b_i] = 1$. Then, for any $t > 0$ we have*

$$\Pr\left[\left| \sum_{i=1}^{q} \xi_i - \mathbb{E}[\sum_{i=1}^{q} \xi_i] \right| \geq t \right] \leq 2 \exp^{-\frac{2t^2}{\sum_{i=1}^{q}(b_i - a_i)^2}}.$$

Fact A2. *For any* $\sigma \in \mathbb{N}^+$, *the probability that a random* $(n + \sigma) \times n$ *Boolean matrix* $M \sim U_{(n+\sigma) \times n}$ *has full rank (i.e., rank* n*) is at least* $1 - 2^{-\sigma+1}$.

Proof. Consider matrix M being sampled column by column, and denote $\mathcal{E}_i$ to be the event that "column i is non-zero and neither is it any linear combination of the preceding columns (i.e., columns 1 to $i - 1$)".

$$
\begin{aligned}
\Pr[M \text{ has full rank}] &= \Pr[\mathcal{E}_1] \cdot \Pr[\mathcal{E}_2|\mathcal{E}_1] \cdots \cdot \Pr[\mathcal{E}_n|\mathcal{E}_{n-1}] \\
&= (1 - 2^{-(n+\sigma)}) \cdot (1 - 2^{-(n+\sigma)+1}) \cdots \cdot (1 - 2^{-(n+\sigma)+n-1}) \\
&> 2^{-\left(2^{-(n+\sigma)+1} + 2^{-(n+\sigma)+2} + \cdots + 2^{-(n+\sigma)+n}\right)} \\
&> 2^{-2^{-\sigma+1}} \\
&> \exp^{-2^{-\sigma+1}} \\
&> 1 - 2^{-\sigma+1}
\end{aligned}
$$

where the first inequality is due to Fact A4 and the last follows from Fact A3.

Fact A3. *For any* $x > 0$ *it holds that* $\exp^{-x} > 1 - x$.

Fact A4. *For any* $0 < x < \frac{2-\sqrt{2}}{2}$ *it holds that* $1 - x > 2^{-(\frac{2+\sqrt{2}}{2})x} > 2^{-2x}$.

Fact A5 (A partial sum of binomial coefficients ([27], p. 492)). *For any* $0 < \mu < 1/2$, *and any* $m \in \mathbb{N}$

$$
\sum_{i=0}^{m\mu} \binom{m}{i} = 2^{m\mathbf{H}(\mu) - \frac{\log m}{2} + O(1)}
$$

where $\mathbf{H}(\mu) \overset{\text{def}}{=} \mu \log(1/\mu) + (1 - \mu) \log(1/(1 - \mu))$ *is the binary entropy function.*

Fact A6 (Piling-up Lemma). *For any* $0 < \mu \leq \mu' < 1/2$, $(\text{Ber}_\mu \oplus \text{Ber}_{\frac{\mu'-\mu}{1-2\mu}}) \sim \text{Ber}_{\mu'}$.

Fact A7 (Min-entropy source conditioned on leakage). *Let* X *be any random variable over support* $\mathcal{X}$ *with* $\mathbf{H}_\infty(X) \geq l_1$, *let* $f : \mathcal{X} \to \{0, 1\}^{l_2}$ *be any function. Then, for any* $0 < \varepsilon < 1$, *there exists a set* $\mathcal{X}_1 \times \mathcal{Y}_1 \subseteq \mathcal{X} \times \{0, 1\}^{l_2}$ *such that* $\Pr[(X, f(X)) \in (\mathcal{X}_1 \times \mathcal{Y}_1)] \geq 1 - \varepsilon$ *and for every* $(x, y) \in (\mathcal{X}_1 \times \mathcal{Y}_1)$

$$
\Pr[X = x \mid f(X) = y] \leq 2^{-(l_1 - l_2 - \log(1/\varepsilon))}.
$$

B Lemmas and Proofs Omitted

Proof of Lemma 1. Recall that $\mathbf{H}(\mu) \overset{\text{def}}{=} \mu \log(1/\mu) + (1 - \mu) \log(1/(1 - \mu))$ equals to $\mathbf{H}_1(\text{Ber}_\mu)$. Parse Ber_μ^q as Boolean variables $E_1, \ldots, E_q$, and for each $1 \leq i \leq q$ define

$$
\xi_i \overset{\text{def}}{=}
\begin{cases}
1, & \text{if } E_i = 1 \\
\frac{\log(\frac{1}{1-\mu})}{\log(\frac{1}{\mu})}, & \text{if } E_i = 0
\end{cases}
$$

and thus we have that $\xi_1, \ldots, \xi_q$ are i.i.d. over $\{\frac{\log(1/(1-\mu))}{\log(1/\mu)}, 1\}$, each of expectation $\mathbf{H}(\mu)/\log(1/\mu)$.

$$\Pr\left[\mathrm{Ber}_\mu^q \in \mathcal{E}\right]$$
$$= 1 - \Pr\left[\sum_{i=1}^q \xi_i > (1+\Delta) \cdot \frac{q\mathbf{H}(\mu)}{\log(1/\mu)}\right]$$
$$> 1 - \exp^{-\frac{\min(\Delta, \Delta^2)q\mathbf{H}(\mu)}{3\log(1/\mu)}} > 1 - \exp^{-\frac{\min(\Delta, \Delta^2)\mu q}{3}},$$

where the inequality follows from the Chernoff bound (see Lemma 11) and we recall $\mathbf{H}(\mu) > \mu \log(1/\mu)$ by Fact A1.

Proof of Lemma 3.
DECISIONAL $\mathrm{LPN}_{\mu,n} \to$ DECISIONAL $\mathrm{Ber}_\mu^{n+q}\text{-}\mathrm{LPN}_{\mu,n}$
Assume for contradiction there exists a distinguisher D that

$$\Pr_{A,S,E}[\mathsf{D}(A,\ A{\cdot}S \oplus E) = 1] - \Pr_{A,U_{q-(n+2)}}[\mathsf{D}(A, U_{q-(n+2)}) = 1] > 2\epsilon,$$

where $A \sim U_{(q-(n+2))n}$, $S \sim \mathrm{Ber}_\mu^n$ and $E \sim \mathrm{Ber}_\mu^{q-(n+2)}$. To complete the proof, we show that there exists another D' (of nearly the same complexity as D) that on input $(a',b) \in \{0,1\}^{qn} \times \{0,1\}^q$ that distinguishes $(A', A' \cdot X \oplus \mathrm{Ber}_\mu^q)$ from (A', U_q) for $A' \sim U_{qn}$ and $X \sim U_n$ with advantage more than ϵ. We parse the $q \times n$ matrix a' and q-bit b as

$$a' = \begin{bmatrix} m \\ a \end{bmatrix}, \ b = (b_m, b_a) \tag{9}$$

where m and a are $(n+2) \times n$ and $(q-(n+2)) \times n$ matrices respectively, $b_m \in \{0,1\}^{n+2}$ and $b_a \in \{0,1\}^{q-(n+2)}$. Algorithm D' does the following: it first checks whether m has full rank or not, and if not it outputs a random bit. Otherwise (i.e., m has full rank), D' outputs $\mathsf{D}(a\bar{m}^{-1}, (a\bar{m}^{-1}){\cdot}b_{\bar{m}} \oplus b_a)$, where $\bar{m}$ is an $n \times n$ invertible submatrix of m and $b_{\bar{m}}$ is the corresponding[10] substring of b_m. Now we give the lower bound of the advantage in distinguishing the two distributions. On the one hand, when $(a',b) \leftarrow (A', (A' \cdot X) \oplus \mathrm{Ber}_\mu^q)$ and conditioned on that $\bar{m}$ is invertible, we have that

$$\bar{m} \cdot x \oplus s = b_{\bar{m}}$$
$$a \cdot x \oplus e = b_a \tag{10}$$

where $a \leftarrow U_{(q-(n+2))n}$, $x \leftarrow U_n$, $s \leftarrow \mathrm{Ber}_\mu^n$, and $e \leftarrow \mathrm{Ber}_\mu^{q-(n+2)}$, and it follows (by elimination of x) that $b_a = (a\bar{m}^{-1})s \oplus (a\bar{m}^{-1})b_{\bar{m}} \oplus e$, and thus $(a\bar{m}^{-1})b_{\bar{m}} \oplus b_a = (a\bar{m}^{-1})s \oplus e$. On the other hand, when $(a',b) \leftarrow (U_{qn}, U_q)$

---

[10] E.g., if $\bar{m}$ is the submatrix of m by keeping only the first n rows, then $b_{\bar{m}}$ is the n-bit prefix of b_m.

and conditioned on an invertible m it holds that $(a\bar{m}^{-1}, (a\bar{m}^{-1}) \cdot b_{\bar{m}} \oplus b_a)$ follows $(U_{(q-(n+2))n}, U_{q-(n+2)})$. Therefore, for $A \sim U_{(q-(n+2))n}$, $S \sim \mathsf{Ber}_\mu^n$ and $E \sim \mathsf{Ber}_\mu^{q-(n+2)}$ we have

$$\Pr[\mathsf{D}'(U_{qn}, U_{qn} \cdot U_n \oplus \mathsf{Ber}_\mu^q) = 1] - \Pr[\mathsf{D}'(U_{qn}, U_q) = 1]$$

$$\geq \Pr[\mathcal{E}_f] \cdot \left(\Pr_{A,S,E}[\mathsf{D}(A, \ A \cdot S \oplus E) = 1] - \Pr_{A,U_{q-(1+\delta)n}}[\mathsf{D}(A, U_{q-(1+\delta)n}) = 1] \right)$$

$$> (1 - 2^{-1})2\epsilon = \epsilon$$

where $\mathcal{E}_f$ denotes the event that $m \leftarrow U_{(n+2)\times n}$ has full rank whose lower bound probability is given in Fact A2.

COMPUTATIONAL $\mathsf{LPN}_{\mu,n} \rightarrow$ COMPUTATIONAL $\mathsf{Ber}_\mu^{n+q}\text{-}\mathsf{LPN}_{\mu,n}$

The reduction follows steps similar to that of the decisional version. Assume for contradiction there exists a distinguisher D that

$$\Pr_{A,S,E}[\mathsf{D}(A, \ A \cdot S \oplus E) = (S,E)] > 2\epsilon,$$

where $A \sim U_{(q-(n+2))n}$, $S \sim \mathsf{Ber}_\mu^n$ and $E \sim \mathsf{Ber}_\mu^{q-(n+2)}$, then there exists another D' that on input $(a', b = a'x \oplus e') \in \{0,1\}^{qn} \times \{0,1\}^q$ recovers (x, e') with probability more than ϵ. Similarly, D' parses (a', b) as in (9), checks if m has full rank and we define $\bar{m}$, $b_{\bar{m}}$ and $\mathcal{E}_f$ same as the above reduction. Let $(s^*, e^*) \leftarrow \mathsf{D}(a\bar{m}^{-1}, (a\bar{m}^{-1}) \cdot b_{\bar{m}} \oplus b_a)$. As analyzed above, conditioned on $\mathcal{E}_f$ we have $(a\bar{m}^{-1}) \cdot b_{\bar{m}} \oplus b_a = (a\bar{m}^{-1})s \oplus e$ where $(a\bar{m}^{-1}, s, e)$ follows distribution (A, S, E) defined above, and hence $(s^*, e^*) = (s, e)$ with probability more than 2ϵ. Once D' got s^*, it computes $x^* = \bar{m}^{-1} \cdot (b_{\bar{m}} \oplus s^*)$ (see (10)), $e'^* = a'x^* \oplus b$ and outputs (x^*, e'^*).

$$\Pr[\mathsf{D}'(A', A' \cdot X \oplus E') = (X, E')]$$

$$\geq \Pr[\mathcal{E}_f] \cdot \Pr_{A,S,E}[\mathsf{D}(A, \ A \cdot S \oplus E) = (S,E)]$$

$$> (1 - 2^{-1})2\epsilon = \epsilon$$

where $A' \sim U_{qn}$, $X \sim U_n$ and $E' \sim \mathsf{Ber}_\mu^q$. □

Proof of Lemma 5. To prove this indistinguishability result we use Patarin's H-coefficient technique in its modern transcript-based incarnation [18,48].

Without loss of generality the distinguisher D is deterministic and does not repeat queries. We refer to the case when the D's oracle is $F_{R,H}$ as the *real world* and to the case where the D's oracle is R as the *ideal world*.

D *transcript* consists of a sequence $(X_1, Y_1), \ldots, (X_q, Y_q)$ of query-answer pairs to its oracle, plus (and following the "transcript stuffing" technique of [18]) the vector $\boldsymbol{H} = H_1, \ldots, H_\kappa$ of hash functions, appended to the transcript after the distinguisher has made its last query; in the ideal world, $\boldsymbol{H}$ consists of a "dummy" κ-tuple $H_1, \ldots, H_\kappa$ that can be sampled after the distinguisher's last query, and is similarly appended to the transcript.

The probability space underlying the real world is $\Omega_{\mathsf{real}} \stackrel{\mathsf{def}}{=} \mathcal{H}^\kappa \times \mathcal{F}_{\ell \rightarrow n}^\kappa$ where $\mathcal{F}_{\ell \rightarrow n}$ is the set of all functions from ℓ bits to n bits, with uniform measure. The probability space underlying the ideal world is $\Omega_{\text{ideal}} \stackrel{\text{def}}{=} \mathcal{H}^\kappa \times \mathcal{F}_{n \to n}$ where $\mathcal{F}_{n \to n}$ is the set of all functions from n bits to n bits, also with uniform measure.

We can identify elements of Ω_{real} and/or Ω_{ideal} as "oracles" for D to interact with. We write D^ω for the transcript obtained when D interacts with oracle ω, where $\omega \in \Omega_{\text{real}}$ in the real world and $\omega \in \Omega_{\text{ideal}}$ in the ideal world. Thus, the real-world transcripts are distributed according to $D^{W_{\text{real}}}$ where W_{real} is uniformly distributed over Ω_{real}, while the ideal-world transcripts are distributed according to $D^{W_{\text{ideal}}}$ where W_{ideal} is uniformly distributed over Ω_{ideal}.

A transcript τ is *attainable* if there exists some $\omega \in \Omega_{\text{ideal}}$ such that $D^\omega = \tau$. (Which transcripts are attainable depends on D, but we assume a fixed D). A transcript $\tau = ((X_1, Y_1), \ldots, (X_q, Y_q), H_1, \ldots, H_\kappa)$ is *bad* if there exists some $i \in [q]$ such that

$$H_j(X_i) \in \{H_j(X_1), \ldots, H_j(X_{i-1})\}$$

for all $j \in \kappa$. We let T_{bad} be the set of bad attainable transcripts, T_{good} the set of non-bad attainable transcripts.

We will show that $\Pr[D^{W_{\text{real}}} = \tau] = \Pr[D^{W_{\text{ideal}}} = \tau]$ for all $\tau \in T_{\text{good}}$. In this case, by Patarin's H-coefficient technique [18], D's distinguishing advantage is upper bounded by $\Pr[D^{W_{\text{ideal}}} \in T_{\text{bad}}]$. We commence by upper bounding the later quantity, and then move to the former claim.

Let $\mathcal{E}_{i,j}$, $(i,j) \in [q] \times [\kappa]$, be the event that

$$H_j(X_i) \in \{H_j(X_1), \ldots, H_j(X_{i-1})\}$$

and let

$$\mathcal{E}_i = \mathcal{E}_{i,1} \wedge \cdots \wedge \mathcal{E}_{i,\kappa}.$$

Since the values $X_1, \ldots, X_q$ and the hash functions $H_1, \ldots, H_\kappa$ are uniquely determined by any $\omega \in \Omega_{\text{ideal}}$ or $\omega \in \Omega_{\text{real}}$, we can write $\mathcal{E}_i(W_{\text{ideal}})$ (in the ideal world) or $\mathcal{E}_i(W_{\text{real}})$ (in the real world) to emphasize that $\mathcal{E}_i$ is a deterministic predicate of the uniformly distributed oracle, in either world. Then

$$(D^{W_{\text{ideal}}} \in T_{\text{bad}}) \iff (\mathcal{E}_1(W_{\text{ideal}}) \vee \cdots \vee \mathcal{E}_q(W_{\text{ideal}})). \tag{11}$$

Moreover,

$$\Pr[\mathcal{E}_{i,j}(W_{\text{ideal}})] \leq (i-1)\frac{1}{2^\ell} \leq \frac{q}{2^\ell}$$

since the hash functions $H_1, \ldots, H_\kappa$ are chosen independently of everything in the ideal world, and by the universality of $\mathcal{H}$, and

$$\Pr[\mathcal{E}_i(W_{\text{ideal}})] \leq \left(\frac{q}{2^\ell}\right)^\kappa$$

since the events $\mathcal{E}_{i,1}, \ldots, \mathcal{E}_{i,\kappa}$ are independent in the ideal world; finally

$$\Pr[D^{W_{\text{ideal}}} \in T_{\text{bad}}] \leq q \left(\frac{q}{2^\ell}\right)^\kappa = \frac{q^{\kappa+1}}{2^{\ell\kappa}}$$

by (11) and by a union bound.

To complete the proof, we must show that $\Pr[D^{W_{\text{real}}} = \tau] = \Pr[D^{W_{\text{ideal}}} = \tau]$ for all $\tau \in T_{\text{good}}$. Clearly,

$$\Pr[D^{W_{\text{ideal}}} = \tau] = \frac{1}{2^{nq}} \cdot \frac{1}{|\mathcal{H}|^\kappa}$$

for all attainable τ. Moreover, if

$$\tau = ((x_1, y_1), \ldots, (x_q, y_q), h_1, \ldots, h_\kappa)$$

then it is easy to see that

$$\Pr[D^{W_{\text{real}}} = \tau \mid \boldsymbol{H}(W_{\text{real}}) = (h_1, \ldots, h_\kappa)] = \frac{1}{2^{nq}}$$

by induction on the number of distinguisher queries, using $\tau \in T_{\text{good}}$. (We write $\boldsymbol{H}(W_{\text{real}})$ for the $\boldsymbol{H}$-coordinate of W_{real}.) Since

$$\Pr[\boldsymbol{H}(W_{\text{real}}) = (h_1, \ldots, h_\kappa)] = \frac{1}{|\mathcal{H}|^\kappa}$$

this completes the proof. $\square$

Proof of Lemma 8.

$$\Pr_{a \xleftarrow{\$} \mathcal{A}} [\exists y \in \mathcal{Y} : y' \neq y \wedge h_a(y') = h_a(y)]$$

$$\leq \sum_{y' \in \mathcal{Y} \backslash \{y\}} \Pr_{a \xleftarrow{\$} \mathcal{A}} [h_a(y') = h_a(y)]$$

$$\leq |\mathcal{Y}| \cdot 2^{-l_2} \leq 2^{-(l_2 - l_1)},$$

where the first inequality is a union bound and the second inequality follows by the universality of $\mathcal{H}$. $\square$

Proof of Lemma 10. Assume WLOG that $\mu'm$ is integer and use shorthand $p_l \overset{\text{def}}{=} \Pr[|\text{Ber}_{\mu'}^m| = l]$ and thus

$$p_{\mu'm} = \binom{m}{\mu'm} \mu'^{\mu'm} (1 - \mu')^{m - \mu'm}$$

For $1 \leq i \leq \mu'm$, we have

$$p_{\mu'm-i} = \binom{m}{\mu'm - i} \mu'^{\mu'm - i} (1 - \mu')^{m - \mu'm + i}$$

$$= \frac{m! \cdot \mu'^{\mu'm} (1 - \mu')^{m - \mu'm}}{(\mu'm - i)!(m - \mu'm + i)!}$$

$$= p_{\mu'm} \frac{(\mu'm - i + 1)(\mu'm - i + 2) \ldots (\mu'm - i + i)}{(m - \mu'm + 1)(m - \mu'm + 2) \ldots (m - \mu'm + i)} \cdot \left(\frac{1 - \mu'}{\mu'}\right)^i$$

$$= p_{\mu'm} \frac{(1 - \frac{i-1}{\mu'm})(1 - \frac{i-2}{\mu'm}) \ldots (1 - \frac{0}{\mu'm})}{(1 + \frac{1}{m(1-\mu')})(1 + \frac{2}{m(1-\mu')}) \ldots (1 + \frac{i}{m(1-\mu')})}.$$

Similarly, for $1 \leq i \leq (1 - \mu')m$ we can show that

$$p_{\mu'm+i} = p_{\mu'm} \frac{(1 - \frac{0}{m(1-\mu')})(1 - \frac{1}{m(1-\mu')}) \ldots (1 - \frac{i-1}{m(1-\mu')})}{(1 + \frac{1}{\mu'm})(1 + \frac{2}{\mu'm}) \ldots (1 + \frac{i}{\mu'm})}.$$

Therefore, we have $p_{\mu'm} = \max\{p_i \mid 0 \leq i \leq m \}$ and thus complete the proof with the following

$$(1 + 2\sqrt{m}) \cdot p_{\mu'm} \geq \sum_{j=\mu'm-\min\{\sqrt{m},\mu'm\}}^{\mu'm+\sqrt{m}} p_j$$
$$\geq 1 - \Pr[\mid |\mathrm{Ber}_{\mu'}^m| - \mu'm| \geq \sqrt{m}]$$
$$\geq 1 - 2\exp^{-2} = \Omega(1)$$

where the last inequality is a Hoeffding bound. $\square$

References

1. Related work on LPN-based authentication schemes. http://www.ecrypt.eu.org/lightweight/index.php/HB
2. Akavia, A., Bogdanov, A., Guo, S., Kamath, A., Rosen, A.: Candidate weak pseudorandom functions in $\mathrm{AC}^0 \circ \mathrm{MOD}_2$. In: Innovations in Theoretical Computer Science, ITCS 2014, pp. 251–260 (2014)
3. Alekhnovich, M.: More on average case vs. approximation complexity. In: 44th Annual Symposium on Foundations of Computer Science (FOCS 2003), Cambridge, Massachusetts, pp. 298–307. IEEE (2003)
4. Applebaum, B., Cash, D., Peikert, C., Sahai, A.: Fast cryptographic primitives and circular-secure encryption based on hard learning problems. In: Halevi, S. (ed.) CRYPTO 2009. LNCS, vol. 5677, pp. 595–618. Springer, Heidelberg (2009)
5. Applebaum, B., Ishai, Y., Kushilevitz, E.: Cryptography with constant input locality. In: Menezes, A. (ed.) CRYPTO 2007. LNCS, vol. 4622, pp. 92–110. Springer, Heidelberg (2007). http://www.eng.tau.ac.il/bennyap/pubs/input-locality-full-revised-1.pdf
6. Banerjee, A., Peikert, C., Rosen, A.: Pseudorandom functions and lattices. In: Pointcheval, D., Johansson, T. (eds.) EUROCRYPT 2012. LNCS, vol. 7237, pp. 719–737. Springer, Heidelberg (2012)
7. Becker, A., Joux, A., May, A., Meurer, A.: Decoding random binary linear codes in $2^{n/20}$: how $1 + 1 = 0$ improves information set decoding. In: Pointcheval, D., Johansson, T. (eds.) EUROCRYPT 2012. LNCS, vol. 7237, pp. 520–536. Springer, Heidelberg (2012)
8. Bellare, M., Goldreich, O., Krawczyk, H.: Stateless evaluation of pseudorandom functions: security beyond the birthday barrier. In: Wiener, M. (ed.) CRYPTO 1999. LNCS, vol. 1666, pp. 270–287. Springer, Heidelberg (1999)
9. Berlekamp, E., McEliece, R.J., van Tilborg, H.: On the inherent intractability of certain coding problems. IEEE Trans. Inf. Theor. 24(3), 384–386 (1978)
10. Berman, I., Haitner, I., Komargodski, I., Naor, M.: Hardness preserving reductions via cuckoo hashing. In: Sahai, A. (ed.) TCC 2013. LNCS, vol. 7785, pp. 40–59. Springer, Heidelberg (2013)

11. Bernstein, D.J., Lange, T., Peters, C.: Smaller decoding exponents: ball-collision decoding. In: Rogaway, P. (ed.) CRYPTO 2011. LNCS, vol. 6841, pp. 743–760. Springer, Heidelberg (2011)
12. Blum, A., Furst, M.L., Kearns, M., Lipton, R.J.: Cryptographic primitives based on hard learning problems. In: Stinson, D.R. (ed.) CRYPTO 1993. LNCS, vol. 773, pp. 278–291. Springer, Heidelberg (1994)
13. Blum, A., Kalai, A., Wasserman, H.: Noise-tolerant learning, the parity problem, and the statistical query model. J. ACM **50**(4), 506–519 (2003)
14. Boldyreva, A., Fehr, S., O'Neill, A.: On notions of security for deterministic encryption, and efficient constructions without random oracles. In: Wagner, D. (ed.) CRYPTO 2008. LNCS, vol. 5157, pp. 335–359. Springer, Heidelberg (2008)
15. Canteaut, A., Chabaud, F.: A new algorithm for finding minimum-weight words in a linear code: application to McEliece's cryptosystem and to narrow-sense BCH codes of length 511. IEEE Trans. Inf. Theor. **44**(1), 367–378 (1998)
16. Cash, D., Kiltz, E., Tessaro, S.: Two-round man-in-the-middle security from LPN. In: Kushilevitz, E., et al. (eds.) TCC 2016-A. LNCS, vol. 9562, pp. 225–248. Springer, Heidelberg (2016)
17. Chandran, N., Garg, S.: Balancing output length and query bound in hardness preserving constructions of pseudorandom functions. In: Meier, W., Mukhopadhyay, D. (eds.) INDOCRYPT 2014. LNCS, vol. 8885, pp. 89–103. Springer, Cham (2014)
18. Chen, S., Steinberger, J.: Tight security bounds for key-alternating ciphers. In: Nguyen, P.Q., Oswald, E. (eds.) EUROCRYPT 2014. LNCS, vol. 8441, pp. 327–350. Springer, Heidelberg (2014)
19. David, B., Dowsley, R., Nascimento, A.C.A.: Universally composable oblivious transfer based on a variant of LPN. In: Gritzalis, D., Kiayias, A., Askoxylakis, I. (eds.) CANS 2014. LNCS, vol. 8813, pp. 143–158. Springer, Heidelberg (2014)
20. Dodis, Y., Kiltz, E., Pietrzak, K., Wichs, D.: Message authentication, revisited. In: Pointcheval, D., Johansson, T. (eds.) EUROCRYPT 2012. LNCS, vol. 7237, pp. 355–374. Springer, Heidelberg (2012)
21. Dodis, Y., Smith, A.: Entropic security and the encryption of high entropy messages. In: Kilian, J. (ed.) TCC 2005. LNCS, vol. 3378, pp. 556–577. Springer, Heidelberg (2005)
22. Döttling, N., Müller-Quade, J., Nascimento, A.C.A.: IND-CCA secure cryptography based on a variant of the LPN problem. In: Wang, X., Sako, K. (eds.) ASIACRYPT 2012. LNCS, vol. 7658, pp. 485–503. Springer, Heidelberg (2012)
23. Döttling, N., Schröder, D.: Efficient pseudorandom functions via on-the-fly adaptation. In: Gennaro, R., Robshaw, M. (eds.) CRYPTO 2015. LNCS, vol. 9215, pp. 329–350. Springer, Heidelberg (2015)
24. Feldman, V., Gopalan, P., Khot, S., Ponnuswami, A.K.: New results for learning noisy parities and halfspaces. In: 47th Symposium on Foundations of Computer Science, Berkeley, CA, USA, 21–24 October 2006, pp. 563–574. IEEE (2006)
25. Gazi, P., Tessaro, S.: Secret-key cryptography from ideal primitives: a systematic overview. In: 2015 IEEE Information Theory Workshop (ITW 2015), pp. 1–5 (2015)
26. Goldreich, O., Goldwasser, S., Micali, S.: How to construct random functions. J. ACM **33**(4), 792–807 (1986)
27. Graham, R.L., Knuth, D.E., Patashnik, O.: Concrete Mathematics: A Foundation for Computer Science, 2nd edn. Addison-Wesley Longman Publishing Co. Inc., Boston (1994)

28. Haitner, I., Reingold, O., Vadhan, S.P.: Efficiency improvements in constructing pseudorandom generators from one-way functions. In: Proceedings of the 42nd ACM Symposium on the Theory of Computing, pp. 437–446 (2010)
29. Håstad, J., Impagliazzo, R., Levin, L., Luby, M.: Construction of pseudorandom generator from any one-way function. SIAM J. Comput. **28**(4), 1364–1396 (1999)
30. Hoeffding, W.: Probability inequalities for sums of bounded random variables. J. Am. Stat. Assoc. **58**(301), 13–30 (1963)
31. Hopper, N.J., Blum, M.: Secure human identification protocols. In: Boyd, C. (ed.) ASIACRYPT 2001. LNCS, vol. 2248, pp. 52–66. Springer, Heidelberg (2001)
32. Impagliazzo, R., Zuckerman, D.: How to recycle random bits. In: 30th Annual Symposium on Foundations of Computer Science, Research Triangle Park, North Carolina, 30 October–1 November 1989, pp. 248–253. IEEE (1989)
33. Jain, A., Krenn, S., Pietrzak, K., Tentes, A.: Commitments and efficient zero-knowledge proofs from learning parity with noise. In: Wang, X., Sako, K. (eds.) ASIACRYPT 2012. LNCS, vol. 7658, pp. 663–680. Springer, Heidelberg (2012)
34. Jain, A., Pietrzak, K., Tentes, A.: Hardness preserving constructions of pseudorandom functions. In: Cramer, R. (ed.) TCC 2012. LNCS, vol. 7194, pp. 369–382. Springer, Heidelberg (2012)
35. Juels, A., Weis, S.A.: Authenticating pervasive devices with human protocols. In: Shoup, V. (ed.) CRYPTO 2005. LNCS, vol. 3621, pp. 293–308. Springer, Heidelberg (2005)
36. Katz, J., Shin, J.S.: Parallel and concurrent security of the HB and HB$^+$ protocols. In: Vaudenay, S. (ed.) EUROCRYPT 2006. LNCS, vol. 4004, pp. 73–87. Springer, Heidelberg (2006)
37. Kiltz, E., Masny, D., Pietrzak, K.: Simple chosen-ciphertext security from low-noise LPN. In: Krawczyk, H. (ed.) PKC 2014. LNCS, vol. 8383, pp. 1–18. Springer, Heidelberg (2014)
38. Kiltz, E., Pietrzak, K., Cash, D., Jain, A., Venturi, D.: Efficient authentication from hard learning problems. In: Paterson, K.G. (ed.) EUROCRYPT 2011. LNCS, vol. 6632, pp. 7–26. Springer, Heidelberg (2011)
39. Kirchner, P.: Improved generalized birthday attack. Cryptology ePrint Archive, Report 2011/377 (2011). http://eprint.iacr.org/2011/377
40. Levieil, É., Fouque, P.-A.: An improved LPN algorithm. In: De Prisco, R., Yung, M. (eds.) SCN 2006. LNCS, vol. 4116, pp. 348–359. Springer, Heidelberg (2006)
41. Levin, L.A.: One-way functions and pseudorandom generators. Combinatorica **7**(4), 357–363 (1987)
42. Lyubashevsky, V.: The parity problem in the presence of noise, decoding random linear codes, and the subset sum problem. In: Chekuri, C., Jansen, K., Rolim, J.D.P., Trevisan, L. (eds.) APPROX 2005 and RANDOM 2005. LNCS, vol. 3624, pp. 378–389. Springer, Heidelberg (2005)
43. Lyubashevsky, V., Masny, D.: Man-in-the-middle secure authentication schemes from LPN and weak PRFs. In: Canetti, R., Garay, J.A. (eds.) CRYPTO 2013, Part II. LNCS, vol. 8043, pp. 308–325. Springer, Heidelberg (2013)
44. Maurer, U.M.: Indistinguishability of random systems. In: Knudsen, L.R. (ed.) EUROCRYPT 2002. LNCS, vol. 2332, pp. 110–132. Springer, Heidelberg (2002)
45. May, A., Meurer, A., Thomae, E.: Decoding random linear codes in $\tilde{\mathcal{O}}(2^{0.054n})$. In: Wang, X., Lee, D.H. (eds.) ASIACRYPT 2011. LNCS, vol. 7073, pp. 107–124. Springer, Heidelberg (2011)
46. Naor, M., Reingold, O.: Number-theoretic constructions of efficient pseudo-random functions. In: 38th Annual Symposium on Foundations of Computer Science, Miami Beach, Florida, 20–22 October 1997, pp. 458–467. IEEE (1997)

47. Naor, M., Reingold, O., Rosen, A.: Pseudo-random functions and factoring. Electronic Colloquium on Computational Complexity (ECCC) TR01-064 (2001)
48. Patarin, J.: The "Coefficients H" technique. In: Avanzi, R.M., Keliher, L., Sica, F. (eds.) SAC 2008. LNCS, vol. 5381, pp. 328–345. Springer, Heidelberg (2009)
49. Pietrzak, K.: Cryptography from learning parity with noise. In: Bieliková, M., Friedrich, G., Gottlob, G., Katzenbeisser, S., Turán, G. (eds.) SOFSEM 2012. LNCS, vol. 7147, pp. 99–114. Springer, Heidelberg (2012)
50. Razborov, A.A.: Lower bounds on the size of bounded depth networks over a complete basis with logical addition. Mathematische Zametki **41**, 598–607 (1986). English Translation in Mathematical Notes of the Academy of Sciences of the USSR
51. Razborov, A.A., Rudich, S.: Natural proofs. In: Proceedings of the Twenty-Sixth Annual ACM Symposium on the Theory of Computing, Montréal, Québec, Canada, 23–25 May 1994, pp. 204–213 (1994)
52. Regev, O.: On lattices, learning with errors, random linear codes, and cryptography. In: Proceedings of the 37th Annual ACM Symposium on Theory of Computing (STOC 2005)
53. Smolensky, R.: Algebraic methods in the theory of lower bounds for Boolean circuit complexity. In: Proceedings of the 19th Annual ACM Symposium on Theory of Computing (STOC 1987), pp. 77–82 (1987)
54. Dong, T., Stern, J.: A method for finding codewords of small weight. In: Cohen, G., Wolfmann, J. (eds.) Coding Theory and Applications. LNCS, vol. 388, pp. 106–113. Springer, Heidelberg (2005)
55. Yu, Y., Gu, D., Li, X., Weng, J.: (Almost) optimal constructions of UOWHFs from 1-to-1, regular one-way functions and beyond. In: Gennaro, R., Robshaw, M. (eds.) CRYPTO 2015. LNCS, vol. 9216, pp. 209–229. Springer, Heidelberg (2015)

Secure Computation from Elastic Noisy Channels

Dakshita Khurana[1](✉), Hemanta K. Maji[2], and Amit Sahai[1]

[1] Department of Computer Science, Center for Encrypted Functionalities,
UCLA, Los Angeles, USA
{dakshita,sahai}@cs.ucla.edu
[2] Department of Computer Science, Purdue University, West Lafayette, USA
hmaji@purdue.edu

Abstract. Noisy channels enable unconditionally secure multi-party computation even against parties with unbounded computational power. But inaccurate noise estimation and adversarially determined channel characteristics render known protocols insecure. Such channels are known as unreliable noisy channels. A large body of work in the last three decades has attempted to construct secure multi-party computation from unreliable noisy channels, but this previous work has not been able to deal with most parameter settings.

In this work, we study a form of unreliable noisy channels where the unreliability is one-sided, that we name *elastic* noisy channels: thus, in one form of elastic noisy channel, an adversarial receiver can increase the reception reliability unbeknown to the sender, but the sender cannot change the channel characteristic.

Our work shows feasibility results for a large set of parameters for the elastic binary symmetric channel, significantly improving upon the best results obtainable using prior techniques. In a key departure from existing approaches, we use a more elemental correlated private randomness as an intermediate cryptographic primitive that exhibits only a rudimentary essence of oblivious transfer. Toward this direction, we introduce new information-theoretic techniques that are potentially applicable to other cryptographic settings involving unreliable noisy channels.

Keywords: Noisy channel · Unfair noisy channel · Elastic noisy channel · Oblivious transfer · Information-theoretic security · Secure computation

D. Khurana and A. Sahai—Research supported in part from a DARPA/ARL SAFE-WARE award, NSF Frontier Award 1413955, NSF grants 1228984, 1136174, 1118096, and 1065276, a Xerox Faculty Research Award, a Google Faculty Research Award, an equipment grant from Intel, and an Okawa Foundation Research Grant. This material is based upon work supported by the Defense Advanced Research Projects Agency through the ARL under Contract W911NF-15-C-0205. The views expressed are those of the author and do not reflect the official policy or position of the Department of Defense, the National Science Foundation, or the U.S. Government.
H.K. Maji—Work done while at UCLA.

M. Fischlin and J.-S. Coron (Eds.): EUROCRYPT 2016, Part II, LNCS 9666, pp. 184–212, 2016.
DOI: 10.1007/978-3-662-49896-5_7

1   Introduction

Secure multi-party computation [27,57] helps mutually distrusting parties to securely compute a function of their private data. General secure computation is impossible in the information-theoretic plain model for most cryptographically interesting functionalities even when parties are semi-honest [3,31,36,41–43]. This necessitates restrictions on the power of the adversaries, for example, honest majority [6,12,21,50], computational hardness assumptions [27,33] or physical cryptographic resources, like, noisy channels [4,17,19,37,38], correlated private randomness [19,38,44,54], trusted resources [10,34] or tamper-proof hardware [11,23,28,35,46].

Using cryptographic resources like noisy channels, it is possible to securely compute arbitrary functionalities with unconditional security guarantees against malicious computationally unbounded adversaries as well [4,17,19,37,38]. Aside from unconditional security, this line of work also offers advantages in efficiency [5,45,48]. Additionally, all invocations of the noisy channel can be performed in an offline phase that is independent of the target functionality to be securely computed [54]. But, the security analysis of these protocols crucially hinges on accurate knowledge of the channel characteristic. Inaccurately estimated or, even worse, adversarially determined channel characteristic can violate the security guarantees of known secure computation protocols that rely on noisy channels. We broadly call such channels unreliable noisy channels.

Over the last three decades, a lot of effort has been focussed towards performing information-theoretic secure multi-party computation using unreliable noisy channels, but with limited success. Weak forms of oblivious transfer[1] (OT) [7,8,17,22,55] and noisy channels [16,19,20,22,47,55,56] have been leveraged to perform secure computation with strong security guarantees, but only for limited settings of parameters. For example, the notion of an *unfair* noisy channel allows both the adversarial sender and the receiver to increase their knowledge of the other party's outputs or inputs to the channel. This model captures extremely general physical systems. Unfortunately, strong impossibility results exist for unfair channels [22], thus, significantly limiting the potential set of feasible parameters (Ref. Fig. 1).

Faced with these daunting impossibility results, in this work we ask whether security is possible in meaningful relaxations of the unfair noisy channel model. In particular, we study an unreliable noisy channel model, namely *elastic noisy channels*, where only one party, either the receiver or sender, but not both, can increase their knowledge of the other party's inputs and outputs to the channel. We show that an elastic noisy channel with sender advantage is equivalent to an elastic noisy channel with receiver advantage (see Sect. 5), and thus in the sequel, we focus on the case where the receiver can increase its knowledge of the sender's

---

[1] Oblivious Transfer [25,49,53] is a two-party functionality which takes $(x_0, x_1) \in \{0, 1\}^2$ as input from the sender and $c \in \{0, 1\}$ from the receiver and provides x_c as output to the receiver. Information-theoretic secure general multi-party computation can be constructed in the OT-hybrid [10,34].

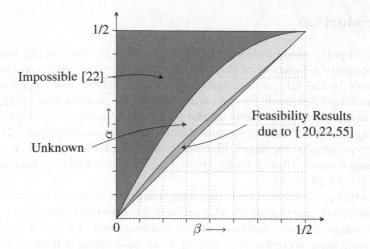

Fig. 1. Unfair binary symmetric channel parameters for binary symmetric channels. Honest channel flips the input symbol with probability α, where $0 < \alpha < 1/2$. Both the sender and the receiver can make the channel more reliable with flip probability β, where $0 < \beta \leqslant \alpha$.

inputs to the channel. Such a study is motivated, for example, by transmission and reception of information over physical wireless channels between physically separated parties. This is because in physical wireless systems, thermal noise is always present at the receiver's end and cannot be observed by a physically distant sender. Thus, the sender, even if malicious, cannot anticipate the entire error introduced at the receiver antenna. However, an adversarial receiver, on the other hand, can install a large super-cooled antenna to make its reception more reliable than the reception available to an honest receiver that uses an inexpensive antenna.

While this scenario is *one* example, our study is primarily motivated from a theoretical standpoint, in the face of severe impossibility results for the full unfair channel setting, where very little progress has been made despite decades of research. Interestingly, our elastic channel model avoids the impossibility results of [22] and, hence, holds the promise to yield secure multi-party computation protocols based on a wide range of parameters. Nevertheless, previous work achieve only quite weak results in the elastic noisy channel setting.

Our main result pertains to realization of information-theoretic secure multi-party computation using (α, β)-BSC, a binary symmetric channel where, informally,[2] an honest receiver obtains the sender's input bit flipped with probability α, while the adversarial receiver obtains an the sender's input bit flipped only with probability β, where $0 < \beta \leqslant \alpha < 1/2$. Figure 2 shows the set of feasible

---

[2] The actual definition of (α, β)-BSC uses a *degradation channel* model. The channel output is a degradation of the leakage. But for intuitive purposes the description presented here suffices. Section 2 provides a more detailed and accurate description.

parameters that can be achieved using the best previous techniques of [22,55]. The figure also illustrates the much larger set of possible (α, β) pairs for which it is possible to achieve secure multi-party computation on (α, β)-BSC using the techniques we develop in this paper. As a concrete example, if the best antenna in the market incurs only 5 % error, then prior techniques need to assume that the honest receiver uses a receiver with at most 14 % error. Our protocols, on the other hand, work even when the honest reception error is as high as 30 %.

New Ideas. The crux of this significant gain in feasibility parameters is a new perspective on how to securely realize OT from unreliable noisy channels. Over the last several decades, a common underlying theme of previous constructions is a reduction from unreliable noisy channels to weak OT using two-repetition of the underlying channel and the rejection sampling technique of [17] and, subsequently, amplifying the weak OT to a full-fledged OT [17,22,55]. The first reduction in this approach, we find, leads to a significant loss in parameters. We, instead, reduce from unreliable noisy channels to a correlated private randomness that provides extremely weak guarantees and ensures only a rudimentary essence of OT. In this respect, as a departure from prior techniques, our target correlated private randomness is closer to the notion of universal OT as proposed by Cachin [8]. Then, we morph this elemental correlated private randomness into a weak variant of OT using the weak converse of Shannon's Channel Coding Theorem [26,52] as utilized by [40] and fuzzy extractors [24]. Next, this weak variant of OT is amplified to (full-fledged) OT using techniques similar to those proposed in [55]. Section 1.2 provides a summary of our technical contributions and intuition of the protocol designs.

Looking ahead, we believe that the techniques introduced in this paper are of independent interest and are likely to find use in other areas of cryptography where noisy channels are analyzed.

1.1   Our Contributions

Our main contribution is to design protocols that securely realize oblivious transfer and therefore secure multi-party computation, from *elastic* binary symmetric channels. Before summarizing our results, we explain the notion of elastic channels.

Elastic Channels. We will model *elastic* variants of noisy channels as consisting of a pair of noisy channels where the channel for the honest receiver is a degradation of the channel for the adversarial receiver. In general, we view an (α, β)-BSC as a pair of channels, such the honest receiver has reception over a BSC with flip probability α, and an adversarial receiver has reception over a BSC with flip probability $\beta \leqslant \alpha$.

General Secure Computation. We prove that general secure computation is possible for a large range of parameters of elastic binary symmetric channels. In

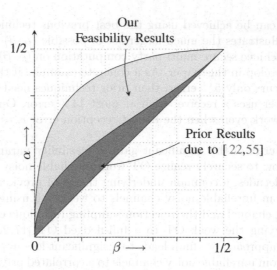

Fig. 2. Space of parameters (β, α), where $0 < \beta \leqslant \alpha < 1/2$, for which we construct secure computation protocol from (α, β)-BSC. The smaller dark region is the space for which such protocols can be obtained using prior techniques from [22,55] combined.

particular, we obtain oblivious transfer (OT) using elastic noisy channels, and then the OT functionality can be used to obtain general secure computation [10,27,36,57]. Our main theorem is as follows:

Theorem 1 (Elastic BSC Completeness). *There exists a universal constant* $c \in (0,1)$, *such that for all* $0 < \beta \leqslant \alpha < 1/2$, *if* $\alpha < \left(1 + (4\beta(1-\beta))^{-1/2}\right)^{-1}$ *then there exists a protocol* $\Pi_{\alpha,\beta}$ *such that,* $\Pi_{\alpha,\beta}$ *securely realizes the OT functionality* $\mathcal{F}_{OT}$ *when given access to* $((\alpha, \beta)\text{-BSC})^{\otimes\kappa}$ *channels with at most* $2^{-\kappa^c}$ *simulation error, where* κ *is the security parameter, with information-theoretic unconditional security against malicious adversaries.*

Refer to Fig. 2 for a summary of the parameter space in Theorem 1 and a comparison of our results with results from previous work[3]. Henceforth, we will use $\ell(\beta) := \left(1 + (4\beta(1-\beta))^{-1/2}\right)^{-1}$.

In addition to elastic noisy channels, both parties also communicate over reliable communication channels in our protocols. These reliable channels can be constructed from the (elastic) noisy channels themselves via standard techniques in error correcting codes (e.g. using polar codes [1,2,29]).

---

[3] When comparing to previous work, note that no previous work considered the setting of elastic channels. Instead, to provide some context, we plot parameters that would be obtained by combining techniques from [22,55] and adapting these to the setting of elastic channels. We do not attempt to combine also the results from [20], because of definitional differences.

Furthermore, we can strengthen our completeness theorems using techniques from [32, 34, 40] to achieve *constant rate*: that is, our protocols can produce $\Theta(\kappa)$ OTs with only $O(\kappa)$ total communication and only $O(\kappa)$ calls to the underlying elastic binary symmetric channels.

Corollary 1 (Constant Rate Elastic BSC Completeness). *For all* $0 < \beta \leqslant \alpha < 1/2$, *if* $\alpha < \left(1 + (4\beta(1-\beta))^{-1/2}\right)^{-1}$ *then, there exists a protocol* $\Pi_{\alpha,\beta}$ *and constants* $c_{\alpha,\beta}, d_{\alpha,\beta}$ *such that,* $\Pi_{\alpha,\beta}$ *securely realizes* $\mathcal{F}_{\mathrm{OT}}^{\otimes m}$ *when given access to* $((\alpha,\beta)\text{-BSC})^{\otimes \kappa}$ *channels with at most* $2^{-\kappa^{c_{\alpha,\beta}}}$ *simulation error and* $m = d_{\alpha,\beta}\kappa$.

1.2 Technical Overview

While our protocols have many ingredients and require a careful analysis, in this section we try to explain the core ideas in our scheme.

A New Take on Previous Approaches. We begin by re-interpreting previous approaches to realize oblivious transfer from noisy channels. Our new understanding of these methods helps abstract out their essence and better illustrate the bottlenecks in our setting. Then, we develop key ideas to achieve oblivious transfer even from channels with adversarial receiver-controlled characteristic, for a large range of parameters of such channels.

To obtain OT from a perfect BSC, a natural starting point is to have the sender pick appropriate codewords (typically simple repetition codes) and send them over the BSC to the receiver. The receiver must then partition the received outputs into two sets establishing two "virtual" channels with the following property: There exists a threshold R, such that one of the virtual channels has capacity $C^* > R$, while the other channel has capacity $\tilde{C} < R$. Moreover, the sender will be unable to tell which virtual channel is which.

In the protocol, the sender pushes information across the virtual channels at rate equal to R. The receiver recovers the information that is transmitted over the virtual channel with capacity $C^* > R$. But, he incurs errors decoding the information transmitted over the virtual channel with capacity $\tilde{C} < R$ because the weak converse of Shannon's Channel Coding Theorem [26, 52] kicks in. This decoding error can be amplified using fuzzy extractors [24], to completely erase the other message and guarantee statistical hiding.

But, we would like to design protocols that remain secure even given an (α, β)-BSC. In the following, we will use α-BSC to denote the channel used by the honest receiver; and β-BSC to denote the channel used by the adversarial receiver. Intuitively, the correctness of our protocol needs to be ensured even for an honest receiver who uses a channel prescribed as the "minimum system requirement" of the protocol description (the α-BSC). We also require that the same protocol be secure even against an adversarial receiver who can reduce the noise level significantly (using the β-BSC). Again, we will think of the problem as forcing the receiver to establish two virtual channels of noticeably different capacities. We require the capacity C^* of the better virtual channel established by the receiver using α-BSC, to be higher than the capacity $\tilde{C}$ of the worse virtual channel established by any adversarial receiver using the β-BSC. The sender will code at a suitable rate intermediate to C^* and $\tilde{C}$. Then, more information will be received over the C^* capacity channel in the honest scenario, than the information received over one of the two virtual channels (of capacity at most $\tilde{C}$) created by the adversarial receiver. This will give oblivious transfer.

Challenges in Our Setting. Let us re-examine our quantitative goal: Suppose the error of the best (adversarial) receiver in the market is 2 %, but honest receivers have 20 % error. The adversarial receiver can obtain much more information than the honest receiver, without the sender's knowledge. Yet, we want to establish two virtual channels such that the capacity of the better virtual channel established using the α-BSC, is higher than the capacity of the worse virtual channel established by any adversarial receiver using the β-BSC. Such an adversarial receiver is allowed to behave arbitrarily, in particular, it could distribute its total capacity equally between the two channels. Ensuring a capacity gap between the better honest and the worse adversarial capacities in this situation, seems to be a tall order. Indeed, previously the results of Wullschleger [55] could achieve this gap only if the honest adversarial receiver had an error at most 9 %.

Towards a Solution. Our first step is to try and relax this goal. Instead of directly shooting for 2-choose-1 oblivious transfer, we try to obtain a weaker form of oblivious transfer, namely $(n, 1, n-1)$ OT, where a sender has n messages, an honest receiver gets to choose 1 message, but a dishonest receiver gets $n-1$ messages of his choice. The sender gets no output. Using the 'virtual channel' intuition presented above, we want the receiver to set up n virtual channels (for some constant n), with a threshold R such that at least one of the n virtual channels set up by the honest receiver has capacity $C^* > R$, while at least one of the n virtual channels set up by the adversarial receiver has capacity less $\tilde{C} < R$. At this point, we have divided our objective into the following two sub-problems:

1. Reduce $(n, 1, n-1)$ OT to (α, β)-BSC
2. Reduce 2-choose-1 OT to $(n, 1, n-1)$ OT

The second result has been considered in the works of [18,51] and can also be demonstrated using techniques presented in [20,22,55] for the setting of weak erasure channels. While this reduction is not the focus of our work, for completeness we provide a protocol securely realizing OT from $(n, 1, n-1)$ OT in the full version, achieving security against malicious adversaries.

Now our main goal is to demonstrate the first reduction. Our next question is, what could be some reasonable ways to take an (α, β)-BSC and build several virtual channels outs of it with varying reliabilities?

A New Kind of Channel Decomposition. A logical starting point is to have the sender send λ repetitions of his bit over fresh instantiations of the (α, β)-BSC, and list all possible outputs obtained by the receiver. Each possible output could be used by the receiver to define a "virtual channel". On sending λ repetitions of a bit b, if the receiver obtains λ identical bits, then his confidence about the original bit b is extremely high. This is the most reliable channel, and will be set to be the choice channel (with capacity C^*) by the honest receiver.

Since errors are independently added at each invocation of the (α, β)-BSC, all receiver outputs with the same number of zeroes, irrespective of the positions of these zeroes, convey the same amount of information to the receiver. Thus, such outputs can be classified into the same equivalence class/virtual channel. Furthermore, for $\eta \in [0, \lfloor \lambda/2 \rfloor + 1]$, let $\mathbb{S}_\eta$ denote all output strings with either η zeroes, or η ones. That is, $\mathbb{S}_\eta$ includes all pairs of output strings of the form $\{0^\eta 1^{\lambda-\eta}, 0^{\lambda-\eta} 1^\eta\}$ and their permutations. This results in the creation of $\lfloor \frac{\lambda}{2} \rfloor + 1$ binary symmetric channels[4] of noticeably different capacities, such that the 'best' virtual channel of an honest receiver consists of outputs solely from $\mathbb{S}_0$. It is easy to see that the sender, who gets no output from the BSC, cannot distinguish between various virtual channels created by the receiver.

For security against an adversarial receiver, it suffices to ensure that the capacity of the virtual channel created using values in $\mathbb{S}_0$ corresponding to the α-BSC, is higher than the average capacity (over all possible channels) over all the outputs assembled by an adversarial receiver when he uses the β-BSC. We note that the receiver is *never* allowed to discard any of the outputs he received; he must necessarily divide and distribute them all into his virtual channels.

On analyzing this approach, we find that in fact as we increase λ, the situation improves for many parameters α, β. While both average adversarial and best honest capacities increase as λ increases, in fact the best honest capacity increases faster. Eventually, then, the best honest capacity becomes better than the average adversarial capacity and we obtain the following results (Ref. Fig. 4 for an example illustration of this phenomenon.). For any constants $0 < \beta \leqslant \alpha < \left(1 + \left(4\beta(1-\beta)\right)^{-1}\right)^{-1}$, there exists an efficiently computable constant $\lambda \in \mathbb{N}$ for which the above property holds. Figure 3 plots the space of these parameters for various values of λ and the limiting curve $\ell(\beta)$.

Although this completes our high-level overview, making these ideas work requires a careful use of the weak converse of Shannon's Channel Coding Theorem, Fuzzy Extractors and other protocol tools, as well as a careful setting of parameters. Refer Sect. 3 for more details about our construction.

Commitments. Enroute proving Theorem 1, we show that it is possible to obtain string commitments from any (α, β)-BSC, where $0 < \beta \leqslant \alpha < 1$[5]. Using techniques from [32,34,40], we can also obtain string commitments at a constant rate. We stress that we can obtain commitments from any (α, β) elastic BSC for all parameters $0 < \beta \leqslant \alpha < 1$, unlike our completeness result. Our result is formally stated in the following theorem:

---

[4] We observe that each set $\mathbb{S}_\eta$ can then be analyzed as a new BSC.

[5] This is in contrast to the setting of unfair noisy channels, which become trivial for a wide range of parameters.

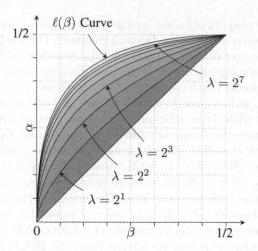

Fig. 3. For $\lambda \in \{2^1, \ldots, 2^7\}$, the space of points (β, α) for which the capacity of the virtual channel created using values in $\mathbb{S}_0$ corresponding to the α-BSC is higher than the average capacity (over all possible channels) over all the outputs assembled by an adversarial receiver when he uses the β-BSC. Finally the limiting $\ell(\beta)$ curve is plotted.

Theorem 2. *There exists a universal constant $c \in (0, 1)$, such that for all $0 < \beta \leqslant \alpha < 1/2$, there exists a protocol $\Pi_{\alpha,\beta}$, constant $d \in (0, 1)$ such that, $\Pi_{\alpha,\beta}$ securely realizes the string commitment functionality for strings of length $d\kappa$, $\mathcal{F}_{\mathsf{com}}(d\kappa)$, when given access to $((\alpha, \beta)\text{-BSC})^{\otimes \kappa}$ channels, with at most $2^{-\kappa^c}$ simulation error, where κ is the security parameter, with information-theoretic unconditional security against malicious adversaries.*

On Adversarial Senders. Finally, we note that noisy channels where only the sender can make the transmission more reliable (that is, sender-elastic binary symmetric channels) reduces to the case of elastic noisy channels with an adversarial receiver (receiver-elastic channels), using a tight reduction presented in Sect. 5. Our one-to-one transformation is optimal and tight.

1.3 Prior Work

There is a lot of literature on constructing secure computation based on noisy channels [16,17,19,32,38–40]. An elastic noisy channel, whose characteristic can be altered by adversarial parties, cannot be modeled as a functionality considered by the completeness theorems of [38,40,44]. However, the following channels in the literature, are related to the notion of elastic channels.

– Unfair Noisy Channels. Unfair noisy channels were formally defined by Damgård et al. [22]: in an unfair noisy channel, *both* the sender and the receiver can change the channel characteristic. Furthermore, the work of [22] showed strong impossibility results in this model. Several works

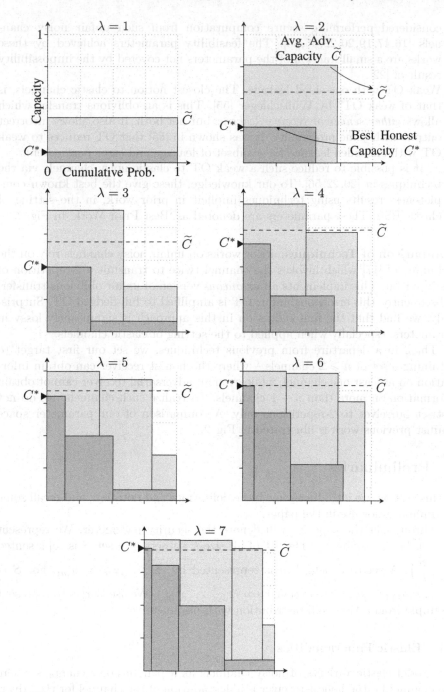

Fig. 4. Obtaining best honest capacity C^* higher than average adversarial capacity $\widetilde{C}$ for (α, β)-BSC, where $(\alpha, \beta) = (1/3, 1/6)$. Each graph represents the capacity profile of sub-channels in the decomposition of $(V, \widehat{V})$, where $\lambda \in \{1, \ldots, 7\}$. The lighter bars denote the adversarial receiver case and the darker bars represent the honest receiver case. When $\lambda = 7$, $C^* > \widetilde{C}$.

considered performing secure computation from such unfair noisy channels [16,17,19,20,22,55,56]. The feasibility parameters achieved by these works are a small fraction of the parameters not covered by the impossibility result of [22].

- Weak OT with one-sided leakage. The closest notion to elastic channels, is that of weak OT[6] by Wüllschleger [55]. This is an oblivious transfer which allows *either sender or receiver leakage*, but not both. It also allows incorrect output with some probability. It was shown in [55] that OT reduces to weak OT with one-sided leakage for a subset of leakage and error parameters.

It is possible to reduce such a weak OT to elastic noisy channels via the techniques in [20,22,56]. To our knowledge, these give the best known completeness results using techniques implicit in prior work, in the setting of elastic BSC. These parameters are denoted as 'Best Prior Work' in Fig. 2.

Comparison of Techniques. Prior works on unfair noisy channels rely on the technique of [17] which invokes the channel twice to transmit a 2-repetition of the input bit. This implements an *erroneous* version of unfair oblivious transfer. Subsequently, this erroneous unfair OT is amplified to full-fledged OT. Surprisingly, we find that the first reduction in this approach is significantly lossy in parameters, especially when applied to the setting of elastic channels.

Thus, in a departure from previous techniques, we set our first target to obtaining a set of $n \geqslant 2$ channels – where the honest receiver can obtain information on at least one channel, while even an adversarial receiver cannot obtain information on more than $n - 1$ channels. To realize such channels, we do not restrict ourselves to 2-repetitions only. A comparison of our parameter space against previous work is illustrated in Fig. 2.

2   Preliminaries

In this section, we introduce some basic definitions and notation, and recall some preliminaries for use in the paper.

Throughout the paper, κ will denote the security parameter. We represent the set $\{1, \ldots, n\}$ by $[n]$. The set of all size-k subsets of a set S is represented by $\binom{S}{k}$. A vector of length n is represented by $(x_1, \ldots, x_n) = x_{[n]}$. For $S = \{i_1, \ldots, i_{|S|}\} \subseteq [n]$, we represent $x_S = (x_{i_1}, \ldots, x_{i_{|S|}})$. We use $\mathrm{Ber}(p)$ to represent a sample from a Bernoulli distribution with parameter p.

2.1   Elastic Functionalities

We model elastic variants of noisy channels as a pair of noisy channels where the channel for the honest receiver is a degradation of the channel for the adversarial receiver. The input (say, bit b) is first transmitted over a more reliable

---

[6] Not to be confused with our notion of (n, k, ℓ)- OT which is complete for all constants $n, (1 < k, \ell < n)$.

(adversarial) channel to obtain leakage z. Then, z is transmitted over a second channel (z is further degraded) to obtain honest receiver output $\tilde{b}$, such that $\tilde{b}$ is effectively, the result of transmitting b over a less reliable channel. The honest receiver obtains output $\tilde{b}$ and the adversarial receiver obtains output leakage z as well as $\tilde{b}$. Note that in our modeling, the leakage z is strictly more informative than honest receiver output $\tilde{b}$. This is exactly why we chose to model elastic channels as degradation channels, as it allows more intuitive analysis. We formalize this notion, as follows, for specific instances of elastic noisy channels.

Definition 1 (Elastic Binary Symmetric Channel). *Let Ber(p) be a sample of Bernoulli distribution with parameter p. For any $0 < \beta \leqslant \alpha < 1/2$, an (α, β)-BSC channel is defined as follows.*

---

1. *Emulate β-BSC on input b: Obtain input b from the sender and sample $e_\ell \sim Ber(\beta)$, the compute $z = b \oplus e_\ell$.*
2. *Emulate γ-BSC on input leakage z: Sample $e' \sim Ber(\gamma)$ and compute $\tilde{b} = z \oplus e'$, where $\beta(1 - \gamma) + (1 - \beta)\gamma = \alpha$. Intuitively, γ is chosen such that $Ber(\alpha) \equiv Ber(\gamma) \oplus Ber(\beta)$.*
3. *Receiver output: Output $\tilde{b}$ to the receiver and, if the receiver is adversarial, then additionally output z to the receiver.*

Let B, Z and $\tilde{B}$ be the random variables corresponding to b, z and $\tilde{b}$, respectively. We have $\tilde{B} = B \oplus Ber(\alpha)$ and $Z = B \oplus Ber(\beta)$, such that $B \to Z \to \tilde{B}$.

Definition 2 $((n, k, \ell)$-OT). *For $0 < k \leqslant \ell < n$, (n, k, ℓ)-OT is defined as:*

---

1. *Sender inputs bits $x_{[n]}$ and receiver inputs set $T \in \binom{[n]}{k}$.*
2. *Output $\{x_{i:i \in T}\}$ to the receiver.*
3. *If the receiver is corrupted by the adversary, then obtain $S \in \binom{[n]}{\ell}$ such that $T \subseteq S$ from the adversary, and output $\{x_{i:i \in S}\}$ to the adversary.*

2-choose-1 bit OT is equivalent to $(2, 1, 1)$-OT.

2.2   Basic Information Theory

Entropy. The entropy of a distribution X is defined as: $\mathbb{E}_{x \sim X}[-\lg \mathsf{P}_{x' \sim X}[x' = x]]$. Given a joint distribution (X, Y), the mutual information is: $I(X; Y) = H(X) + H(Y) - H(X, Y)$.

Channel Capacity. The capacity of a channel W is defined to be $I(W) = \max_X I(X; W(X))$, where X is any probability distribution over the input space. If W is output symmetric, then $I(W) = I(U; W(U))$, where U is the uniform distribution over the input space.

For $0 \leqslant \varepsilon \leqslant 1$, the capacity of ε-BEC is $I(\varepsilon$-BEC$) = 1 - \varepsilon$; and the capacity of ε-BSC is $I(\varepsilon$-BSC$) = 1 - h(\varepsilon)$, where $h(x) := -x \lg(x) - (1 - x) \lg(1 - x)$ is the binary entropy.

$(\mathbf{A}, \mathbf{B}) \to (\mathbf{A}, \mathbf{C})$. For a joint distribution (A, B) and (A, C), if there exists f such that the distributions $(A, f(B))$ and (A, C) are identical, then we say $(A, B) \to (A, C)$. We say that $(A, B) \equiv (A, C)$, if $(A, B) \to (A, C)$ and $(A, C) \to (A, B)$.

$(\mathbf{J}, \mathbf{W_J})$. A channel (J, W_J) is defined as follows:

On input x, sample $j \sim J(x)$ and sample $z \sim W_j(x)$. Output (j, z). We say that a channel $W \equiv (J, W_J)$, if the distributions $(X, W(X)) \equiv (X, J(X), W_{J(X)}(X))$, for all input distributions X.

A binary-input memoryless channel with transition probabilities $(W|0)$ and $(W|1)$ for input symbols 0 and 1, respectively, is called output-symmetric if the probabilities of these two distributions are permutations of each other.

If $I(X; J(X)) = 0$ and all W_j channels are output symmetric, then the capacity of the channel W is $I(W) = \mathbb{E}_{j \sim J}[I(W_j)]$, where J is a fixed distribution over indices (say $J(0)$).[7]

Polar Codes. There are explicit rate achieving Polar Codes with efficient encoding and decoding parameters for ε-BEC and ε-BSC, for $0 \leqslant \varepsilon \leqslant 1$ [1,2,29].

Definition 3 *(Discrete Memoryless Channel). A discrete channel is defined to be a system $W : \mathcal{X} \to \mathcal{Y}$ between a sender and a receiver with sender (input) alphabet $\mathcal{X}$, receiver (output) alphabet $\mathcal{Y}$ and a probability transition matrix $W(y|x)$ specifying the probability that of obtaining output $y \in \mathcal{Y}$ conditioned on input $x \in \mathcal{X}$. The channel is said to be memoryless if the output distribution depends only on the input distribution and is conditionally independent of previous channel inputs and outputs.*

Imported Theorem 1 (Efficient Polar Codes [29]). *There is an absolute constant $\mu < \infty$ such that the following holds. Let W be a binary-input memoryless output-symmetric channel with capacity $I(W)$. Then there exists $a_W < \infty$ such that for all $\varepsilon > 0$ and all powers of two $N \geqslant a_W / \varepsilon^\mu$, there exists a deterministic* poly(N) *time construction of a binary linear code of block length N and rate at least $I(W) - \varepsilon$ and a deterministic $N \cdot$ poly$(\log N)$ decoding algorithm for the code with block error probability at most $2^{-N^{0.49}}$ for communication over W.*

Leftover Hash Lemma. The *min-entropy* of a discrete random variable X is defined to be $H_\infty(X) = -\log \max_{x \in \mathsf{Supp}(X)} \mathsf{P}[X = x]$. For a joint distribution (A, B), the *average min-entropy* of A w.r.t. B is defined as $\tilde{H}_\infty(A|B) = -\log(\mathbb{E}_{b \sim B}[2^{-H_\infty(A|B=b)}])$.

Imported Lemma 1 (Generalized Leftover Hash Lemma(LHL) [24]). *Let $\{H_x : \{0,1\}^n \to \{0,1\}^\ell\}_{x \in X}$ be a family of universal hash functions. Then, for any joint distribution (W, I):* $\mathsf{SD}\left((H_X(W), X, I), (\mathcal{U}_\ell, X, I)\right) \leqslant \frac{1}{2}\sqrt{2^{-\tilde{H}_\infty(W|I)} 2^\ell}$.

---

[7] Because W is also output symmetric.

Weak Converse of Shannon's Channel Coding Theorem. Let $W^{\otimes N}$ denote N independent instances of channel W, which takes as input alphabets from set $\{0,1\}$. Let the capacity of the channel W be C, for a constant $C > 0$. Let $\mathcal{C} \in \{0,1\}^N$ be a rate $R \in \{0,1\}$ code. Then, if the sender transmits a random codeword $\mathbf{c} \xleftarrow{\$} \mathcal{C}$ over $W^{\otimes N}$, the probability of error of the receiver in predicting is $P_e \geqslant 1 - \frac{1}{NR} - \frac{C}{R}$.

2.3 Chernoff-Hoeffding Bound for Hypergeometric Distribution

Imported Theorem 2 (Multiplicative Chernoff Bound for Binomial Random Variables [13, 30]). *Let $X_1, X_2, \ldots X_n$ be independent random variables taking values in $[0,1]$. Let $X = \sum_{i \in [n]} X_i$, and let $\mu = \mathbb{E}[X]$ denote the expected value of the X. Then, for any $\delta > 0$, the following hold.*

- $\Pr[X > (1+\delta)\mu] < \exp\left(-n D_{\mathsf{KL}}\left(\mu(1+\delta)\|\mu\right)\right)$.
- $\Pr[X > (1-\delta)\mu] < \exp\left(-n D_{\mathsf{KL}}\left(\mu(1-\delta)\|\mu\right)\right)$.

Imported Theorem 3 (Multiplicative Chernoff Bound for Hypergeometric Random Variables [14, 30]). *If X is a random variable with hypergeometric distribution, then it satisfies the Chernoff bounds given in Imported Theorem 2.*

2.4 Constant Rate OT Generation

Imported Theorem 4 ([32]). *Let π be a protocol which UC-securely realizes $\mathcal{F}_{\mathsf{OT}}$ in the f-hybrid with simulation error $1 - o(1)$. Then there exists a protocol ρ which UC-securely realizes $\mathcal{F}_{\mathsf{OT}}^{\otimes m}$ in the $f^{\otimes n}$-hybrid with simulation error $1 - \mathsf{negl}(\kappa)$, such that $n = \mathsf{poly}(\kappa)$ and $m = \Theta(n)$.*

3   Binary Symmetric Channels

3.1   Channel Decomposition

In an (α, β)-BSC, the capacity of each channel invocation in the adversarial receiver case is higher than the capacity when the receiver is honest. Despite this bottleneck, our aim is to (non-interactively) synthesize n new noisy channels such that the highest capacity of these channels when interacting with an honest receiver surpasses the capacity of at least one channel obtained by any adversarial receiver. Intuitively, this is achieved by decomposing the original elastic noisy channel into sub-channels such that the sub-channels are "receiver identifiable." Details are provided in the following paragraphs.

It is not evident how to directly decompose an elastic BSC into receiver identifiable sub-channels with the above property. So, we construct a different channel from BSC channels and, in turn, we decompose that channel.

Consider the channel C_ε (parameterized by $\lambda \in \mathbb{N}$) defined below. Given input bit b from the sender, pass b^λ through $(\varepsilon\text{-BSC})^{\otimes \lambda}$, i.e. λ independent copies of ε-BSC, and provide the output string to the receiver. The receiver receives an output string $\tilde{b}_{[\lambda]} \in \{0,1\}^{\lambda}$.

Let id(s) represent the number of minority bits in $s \in \{0,1\}^{\lambda}$.[8] So, we have id: $\{0,1\}^{\lambda} \to \{0,\ldots,\lfloor\lambda/2\rfloor\}$. Define $S_i \subseteq \{0,1\}^n$, as the set of all strings $s \in \{0,1\}^{\lambda}$ such that id$(s) = i$. Given an output string $\tilde{b}_{[\lambda]}$ of the channel $\widetilde{C}$, we interpret it output from the id$(\tilde{b}_{[\lambda]})$-th sub-channel.

Now, note that the sub-channel which takes as input $\{0^{\lambda}, 1^{\lambda}\}$ and outputs a string in S_i is (isomorphic to) an ε_i-BSC channel, for $i \in \{0,\ldots,\lfloor\lambda/2\rfloor\}$, where:

$$\varepsilon_i := \frac{\varepsilon^{\lambda-i} \cdot (1-\varepsilon)^i}{\varepsilon^{\lambda-i} \cdot (1-\varepsilon)^i + (1-\varepsilon)^{\lambda-i} \cdot \varepsilon^i} = \frac{\varepsilon^{\lambda-2i}}{\varepsilon^{\lambda-2i} + (1-\varepsilon)^{\lambda-2i}}$$

Note that ε_i is an increasing function of i. The probability that the i-th sub-channel is stochastically obtained by C_{ε} is:

$$p_i(\varepsilon) := \binom{\lambda}{i}\left(\varepsilon^{\lambda-i}(1-\varepsilon)^i + \varepsilon^i(1-\varepsilon)^{\lambda-i}\right)$$

Now, intuitively, we have decomposed C_{ε}, a channel synthesized from ε-BSC, into a convex linear combination of receiver identifiable sub-channels. More concretely, we have shown that: $C_{\varepsilon} \equiv \sum_{i=0}^{\lfloor\lambda/2\rfloor} p_i(\varepsilon) \cdot (\varepsilon_i\text{-BSC})$.

Now, for any $0 < \beta \leqslant \alpha < 1/2$, we consider the (α, β)-BSC channel. Analogous to the channel C_{ε}, we consider the channel $C_{\alpha,\beta}$. This is identical to the channel C_{ε} and $\varepsilon = \alpha$ when the receiver is honest, and $\varepsilon = \beta$ when the receiver is adversarial. The maximum capacity of sub-channels in the honest receiver case is: $C^* = 1 - h(\alpha_0)$, where $h(x) = -x\lg(x) - (1-x)\lg(1-x)$ is the binary entropy function. The average capacity of sub-channels in the adversarial receiver case is:

$$\widetilde{C} = 1 - \sum_{i=0}^{\lfloor\lambda/2\rfloor+1} p_i(\beta) \cdot h(\beta_i)$$

If we have $C^* > \widetilde{C}$, then we know that best capacity from α-BSC exceeds the average malicious capacity from β-BSC. We set $n = 1/p_0(\alpha)$ and create n-instantiations of the channel C_{ε}. Then one of the sub-channels in the honest receiver case has capacity C^*, while the average capacity of sub-channels in the adversarial receiver case is $\widetilde{C}$. So, out of the n sub-channels, there is one sub-channel in the honest receiver case which has capacity higher than some sub-channel in the adversarial receiver case.

The next question is: for what (α, β) does there exist a λ such that $C^* > \widetilde{C}$? In the following lemma, we show that, if $\alpha < \ell(\beta) := \left(1 + (4\beta(1-\beta))^{-1/2}\right)^{-1}$, then such a λ exists.

For $\alpha = 1/3$ and $\beta = 1/6$, Fig. 4 explains the receiver identifiable decomposition of $C_{\alpha,\beta}$ for increasing values of λ until $C^* > \widetilde{C}$.

---

[8] If s has equal number of 0s and 1s, then we define id$(s) := |s|/2$.

Lemma 1. *For constants* $0 < \alpha < \ell(\beta) := \left(1 + (4\beta(1-\beta))^{-1/2}\right)^{-1}$, *given an* (α, β)-*BSC, there exists a constant* $\lambda \in \mathbb{N}$ *such that it is possible for the receiver to sender-obliviously construct channels where the maximum capacity* C^* *of one sub-channel in the honest receiver case, over* α-*BSC, is greater than the average capacity* $\widetilde{C}$ *of all sub-channels in the adversarial receiver case, over* β-*BSC.*

Consider an elastic binary symmetric channel (α, β)-BSC. For a given a value of $\lambda \in \mathbb{N}$, define $\pi \colon \{0,1\} \rightarrow \{0,1\}^{\lambda}$ as $\pi(b) = b^{\lambda}$ (i.e. λ repetitions of the bit b). Corresponding to this, we obtain channels $(V, \widehat{V})$ corresponding to the honest and adversarial receiver respectively. We have $C^* = 1 - h(\alpha_0^{(\lambda)})$ and $\widetilde{C} = 1 - \sum_{i \in [\lfloor \lambda/2 \rfloor + 1]} p_i^{(\lambda)}(\beta) h(\beta_i^{(\lambda)})$. Define two functions: $h^*(x^{(\lambda)}) := h(x_0^{(\lambda)})$ and $\tilde{h}(x^{(\lambda)}) := \sum_{i \in [\lfloor \lambda/2 \rfloor + 1]} p_i^{(\lambda)}(x) h(x_i^{(\lambda)})$. Note that $C^* = 1 - h^*(\alpha^{(\lambda)})$ and $\widetilde{C} = 1 - \tilde{h}(\beta^{(\lambda)})$. Consider the following manipulation:

$$
\tilde{h}(x^{(\lambda)}) = \sum_{i \in S} p_i^{(\lambda)}(x) h(x_i^{(\lambda)}) > 2 \sum_{i \in S} p_i^{(\lambda)}(x) \cdot x_i^{(\lambda)}
$$

$$
= 2 \sum_{i \in S} \binom{\lambda}{i} x^i (1-x)^i \cdot x^{\lambda - 2i} = \sum_{i \in S} \binom{\lambda}{i} x^{\lambda - i} (1-x)^i
$$

This is a binomial distribution with mean $(1 - x)\lambda$. By using anti-concentration bound from [15]:

$$
\tilde{h}(x^{(\lambda)}) > \frac{1}{\lambda^2} \exp\left(-\lambda \mathsf{D}_{\mathsf{KL}}\left(1/2 \| x\right)\right)
$$

$$
= h\left(h^{-1}\left(\frac{1}{\lambda^2 \exp\left(\lambda \mathsf{D}_{\mathsf{KL}}\left(1/2 \| x\right)\right)}\right)\right)
$$

Next, we use the inequality $h^{-1}(x) \geqslant x/(2\log(6/x))$ from [9]. Set $t(x) = x/(2\log(6/x))$. This gives $\tilde{h}(x^{(\lambda)}) > h\left(t\left(\frac{1}{\lambda^2 \exp(\lambda \mathsf{D}_{\mathsf{KL}}(1/2\|x))}\right)\right)$. For any $x \in (0, 1/2)$, consider $\lambda \rightarrow \infty$. We analyze the behavior of $t\left(\frac{1}{\lambda^2 \exp(\lambda \mathsf{D}_{\mathsf{KL}}(1/2\|x))}\right)$.

Define a such that: $\frac{1}{\lambda^3 \exp(\lambda \mathsf{D}_{\mathsf{KL}}(1/2\|x))\mathsf{polylog}(\lambda)} \leqslant t\left(\frac{1}{\lambda^2 \exp(\lambda \mathsf{D}_{\mathsf{KL}}(1/2\|x))}\right) =:$ $\frac{1}{1+\left(\frac{1}{a}-1\right)^{\lambda}} = h^*(a^{(\lambda)})$ Observe that under these conditions $a \rightarrow a^* :=$ $\frac{1}{1+\exp(\mathsf{D}_{\mathsf{KL}}(1/2\|x))} = \frac{1}{1+\frac{1}{\sqrt{4x(1-x)}}}$. Now for any fixed x and $y < a^*$ (as defined above), for all sufficiently large $\lambda \in \mathbb{N}$ we have $\tilde{h}(x^{(\lambda)}) > h^*(y^{(\lambda)})$.

This shows that for $0 < \beta \leqslant \alpha < \left(1 + (4\beta(1-\beta))^{-1/2}\right)^{-1}$, there exists a constant $\lambda_{\alpha,\beta}$ such that for $\lambda \geqslant \lambda_{\alpha,\beta}$ we have $\tilde{h}(\beta^{(\lambda)}) > h^*(\alpha^{(\lambda)})$, i.e. $C^* > \widetilde{C}$. Furthermore, this bound is tight.

3.2 Semi-honest Completeness of (α, β)-BSC for $0 < \beta \leqslant \alpha < \ell(\beta)$

Consider the channel V_ϵ (parameterized by $\lambda \in \mathbb{N}$) which on input a bit b, passes b^{λ} through $(\epsilon$-BSC$)^{\otimes \lambda}$. Then, for the channels $(V, \widehat{V})$ constructed by sending a λ-repetition code via an (α, β)-BSC, let $C^* := \max_{j \in \mathsf{Supp}(J)} I(V_j)$ and $\widetilde{C} := I(\widehat{V})$.

We use Lemma 1 to compute $\lambda_{\alpha,\beta}$ corresponding to α, β where $0 < \beta \leqslant \alpha < \ell(\beta)$, such that $C^* > \widetilde{C}$, and use the capacity-inverting encoding $\pi_{\alpha,\beta}(b) = b^{\lambda_{\alpha,\beta}}$. For ease of notation, we will use λ to represent $\lambda_{\alpha,\beta}$.

Let n be an integer, such that $n = \frac{1}{\alpha^\lambda + (1-\alpha)^\lambda - \epsilon}$, where $\epsilon \in (0, \alpha^\lambda + (1-\alpha)^\lambda/2)$. Let $\delta = \frac{c^*_h}{\tilde{c}_m} - 1$. Pick a polar code of rational rate r where $\tilde{c}_m(1 + \delta/3) < r < \tilde{c}_m(1 + 2\delta/3)$, and block-length κ/n. Let enc, dec denote the encoding and decoding algorithms of this polar code. Then, Fig. 5 gives a protocol to UC-securely realize n-choose-1 OT using an (α, β)-BSC, in the semi-honest setting.

Inputs: $\mathcal{S}$ has inputs $(x_1, x_2, \ldots x_n) \in \{0,1\}^n$, $\mathcal{R}$ has input choice $c \in [n]$.
Hybrid: (α, β)-BSC for $0 < \beta \leqslant \alpha < \ell(\beta)$.
The protocol is parameterized by κ, a multiple of n.

1. **Correlation Generation:**
   For all $i \in [\kappa^2]$, $\mathcal{S}$ picks bit $b_i \in \{0,1\}$ and sends $b_{i,[\lambda]} = b_i^\lambda$ over the $\left((\alpha, \beta)\text{-BSC}^{\otimes\lambda}\right)$ to $\mathcal{R}$. Let $\mathcal{R}$ obtain output $\tilde{b}_{i,[\lambda]}$.

2. **Receiver Message:**
   Let $I = \{i : i \in [\kappa^2] \text{ and } \tilde{b}_{i,[\lambda]} \in \{0^\lambda, 1^\lambda\}\}$. Set $\tilde{b}_i = \tilde{b}_{i,1}$ for all $i \in I$.

   If $|I| < \kappa^2/n$, abort. Else, let $S_c \overset{\$}{\leftarrow} \begin{pmatrix} I \\ \kappa^2/n \end{pmatrix}$ and for all $\ell \in [n] \setminus \{c\}$,

   set $S_\ell \overset{\$}{\leftarrow} [\kappa^2] \setminus (S_c \cup (S_1 \cup S_2 \cup \cdots S_{\ell-1}))$. For all $\ell \in [n]$, let $S_\ell = \{\text{ind}_{\frac{(\ell-1)\kappa^2}{n}+1}, \text{ind}_{\frac{(\ell-1)\kappa^2}{n}+2}, \ldots \text{ind}_{\frac{\ell\kappa^2}{n}}\}$. Send $(S_1, S_2, \ldots S_n)$ to $\mathcal{S}$.

3. **Sender Message:**
   For $j \in [\kappa]$, $\ell \in [n]$, pick $m_{j,\ell,[r\kappa/n]} \overset{\$}{\leftarrow} \{0,1\}^{r\kappa/n}$, compute $m'_{j,\ell,[\kappa/n]} = \text{enc}(m_{j,\ell,[r\kappa/n]})$. For all $j \in [\kappa]$, $\ell \in [n]$, $i \in [\kappa/n]$, compute and send $y_{j,\ell,i} = m'_{j,\ell,i} \oplus \tilde{b}_{\text{ind}_{\frac{(\ell-1)\kappa^2}{n} + \frac{(j-1)\kappa}{n} + i}}$.

   For all $\ell \in [n]$, pick $h_\ell \overset{\$}{\leftarrow} \mathcal{H}$, a hash function from $\{0,1\}^{\kappa^2/n} \to \{0,1\}$. Compute $r_\ell = h_\ell(m_{1,\ell,[\kappa/n]}, m_{2,\ell,[\kappa/n]}, \ldots m_{\kappa,\ell,[\kappa/n]}) \oplus x_\ell$. For $\ell \in [n]$, send h_ℓ, r_ℓ to $\mathcal{R}$.

4. **Receiver Output:**
   For all $j \in [\kappa]$ and $i \in [\kappa/n]$, compute $m'_{j,c,i} = y_{j,c,i} \oplus \tilde{b}_{\text{ind}_{\frac{(c-1)\kappa^2}{n} + \frac{(j-1)\kappa}{n} + i}}$. Compute $m_{j,c,[r\kappa/n]} = \text{dec}(m'_{j,c,[\kappa/n]})$. Output $x_c = h_c(m_{1,c,[\kappa/n]}, m_{2,c,[\kappa/n]}, \ldots m_{\kappa,c,[\kappa/n]}) \oplus r_c$.

Fig. 5. n-choose-1 bit OT from (α, β)-BSC for $0 < \beta \leqslant \alpha < \ell(\beta)$.

Correctness. It is easy to see that the protocol correctly implements 2-choose-1 oblivious transfer.

Lemma 2. *For all $0 < \beta \leqslant \alpha < \ell(\beta)$, for all $(x_1, x_2, \ldots x_n) \in \{0,1\}^n$ and $c \in [n]$, the output of $\mathcal{R}$ equals x_c with probability at least $(1 - 2^{-\kappa^{0.4}})$.*

Proof. When the sender and the receiver are both honest, the expected fraction of receiver outputs in $\{0^\lambda, 1^\lambda\}$ is $\alpha^\lambda + (1 - \alpha)^\lambda - \epsilon$. Then, the probability that the receiver obtains less than $1/n = \alpha^\lambda + (1 - \alpha)^\lambda - \epsilon$ outputs in $\{0^\lambda, 1^\lambda\}$ is at most $2^{-\frac{\epsilon^2 \kappa}{\alpha^\lambda + (1-\alpha)^\lambda}}$, by the Chernoff bound. Moreover, by Imported Theorem 1, the decoding error when a code of block length κ/n is sent over κ channels at a rate constant lower than capacity, is at most $\kappa \cdot 2^{-\frac{\kappa^{0.49}}{n}}$.

It is easy to see that, conditioned on the receiver obtaining at least $1/n = \alpha^\lambda + (1 - \alpha)^\lambda - \epsilon$ outputs in $\{0^\lambda, 1^\lambda\}$ and no decoding error, the protocol is always correct. Thus, the output of $\mathcal{R}$ equals x_c with probability at least $(1 - 2^{-\kappa^{0.4}})$.

Receiver Security. The semi-honest simulation strategy Sim_S is given in Fig. 6.

The simulator Sim_S does the following.

1. Obtain inputs $(x_1, x_2, \ldots x_n)$ from S.
2. Follow honest strategy: pick $b_{[\kappa^2]} \xleftarrow{\$} \{0,1\}^{\kappa^2}$. Pass $b_{[\kappa^2]}^\lambda$ through an honest emulation of $((\alpha, \beta)\text{-BSC})^{\otimes \lambda \kappa^2}$ to generate $z_{[\kappa^2], [\lambda]}, \tilde{b}_{[\kappa^2], [\lambda]}$.
3. Generate $I = \left\{ i : i \in [\kappa^2], \tilde{b}_{i, [\lambda]} \in \{0^\lambda, 1^\lambda\} \right\}$. Set $\tilde{b}_i = \tilde{b}_{i,1}$ for all $i \in I$. If $|I| < \kappa^2/n$, then aborts$_{\mathsf{Sim}}$. Else send a random partition, $S_1, S_2, \ldots S_n$ of $[\kappa^2]$ to S.
4. For $j \in [\kappa]$ and $\ell \in [n]$, pick $m_{j,\ell,[r\kappa/n]} \xleftarrow{\$} \{0,1\}^{r\kappa/n}$, compute $m'_{j,\ell,[\kappa/n]} = \mathsf{enc}(m_{j,\ell,[r\kappa/n]})$. For all $j \in \kappa$, $\ell \in [n]$ and $i \in [\kappa/n]$, compute and send $y_{j,\ell,i} = m'_{j,\ell,i} \oplus \tilde{b}_{\mathsf{ind}_{\frac{(\ell-1)\kappa^2}{n} + \frac{(j-1)\kappa}{n} + i}}$.

   For all $\ell \in [n]$, pick $h \xleftarrow{\$} \mathcal{H}$, a family of universal hash functions. Compute $r_\ell = \left(h_\ell(m_{1,\ell,[\kappa/n]}, m_{2,\ell,[\kappa/n]}, \ldots m_{\kappa, \ell, [\kappa/n]}) \right) \oplus x_\ell$.

Fig. 6. Sender simulation strategy for n-choose-1 bit OT.

Lemma 3. *The simulation error for the semi-honest sender is at most* $1 - 2^{-\frac{\epsilon^2 \kappa}{\alpha^\lambda + (1-\alpha)^\lambda}}$.

Proof. The view of the sender is, $V_S := \{(x_1, x_2, \ldots x_n), b_{[\kappa^2]}, S_1, S_2, \ldots S_n\}$.

First, the probability of abort in the real view is at most $2^{-\frac{\epsilon^2 \kappa}{\alpha^\lambda + (1-\alpha)^\lambda}}$. Note that the simulator never aborts. But, conditioned on the receiver not aborting, we argue that the simulated sender view is identical to the real view.

For all $i \in [\kappa^2]$, the probability that $\tilde{b}_{i, [\lambda]} \in \{0^\lambda, 1^\lambda\}$, is an i.i.d. random variable, over the randomness of the (α, β)-BSC as well as the receiver. For some fixed size s such that $\kappa^2/n \leq s \leq \kappa^2$, in the view of the sender, $I : |I| = s$ is a random subset of $[\kappa]$ of size s, and S_c is a random partition of I of size $\kappa/2$. The other sets are a random partition of $[\kappa^2] \setminus S_c$, and thus all the sets are a random equal partition of $[\kappa^2]$. Thus, in this case the simulation is perfect.

Thus, the simulation error is exactly equal to the probability of abort, which is at most $2^{-\frac{\epsilon^2 \kappa}{\alpha^\lambda + (1-\alpha)^\lambda}}$.

Sender Security. The semi-honest simulation strategy $\mathsf{Sim}_{\mathcal{R}}$ is given in Fig. 7.

The simulator $\mathsf{Sim}_{\mathcal{R}}$ does the following.

1. Obtain input choice bit c and output θ from $\mathcal{R}$.
2. Pick $b_{[\kappa^2]} \xleftarrow{\$} \{0,1\}^{\kappa^2}$.

   Pass $b_{[\kappa^2]}^\lambda$ through an honest emulation of $((\alpha, \beta)\text{-BSC})^{\otimes \lambda \cdot \kappa^2}$ and generate $z_{[\kappa^2],[\lambda]}, \tilde{b}_{[\kappa^2],[\lambda]}$.
3. Generate $I = \{i : i \in [\kappa^2], \tilde{b}_i \in \{0^\lambda, 1^\lambda\}\}$. Set $\tilde{b}_i = \tilde{b}_{i,1}$ for all $i \in I$.

   Repeat until $|I| \geq \kappa^2/n$. Set $S_c \xleftarrow{\$} \binom{I}{\kappa^2/n}$. For all $\ell \in [n] \setminus \{c\}$,

   set $S_\ell \xleftarrow{\$} \binom{[\kappa^2] \setminus (S_c \cup S_1 \cup S_2 \cup \dots S_{\ell-1})}{\kappa^2/n}$. For all $\ell \in [n]$, let $S_\ell = \{\mathsf{ind}_{\frac{(\ell-1)\kappa^2}{n}+1}, \mathsf{ind}_{\frac{(\ell-1)\kappa^2}{n}+2}, \dots \mathsf{ind}_{\frac{\ell\kappa^2}{n}}\}$.
4. Set $x_c = \theta$, and set $x_\ell \xleftarrow{\$} \{0,1\}$ for all $\ell \in [n] \setminus \{c\}$.

   For $j \in [\kappa]$ and $\ell \in [n]$, pick $m_{j,\ell,[r\kappa/n]} \xleftarrow{\$} \{0,1\}^{r\kappa/n}$, compute $m'_{j,\ell,[\kappa/n]} = \mathsf{enc}(m_{j,\ell,[r\kappa/n]})$. For all $j \in \kappa$, $\ell \in [n]$ and $i \in [\kappa/n]$, compute $y_{j,\ell,i} = m'_{j,\ell,i} \oplus \tilde{b}_{\mathsf{ind}_{\frac{(\ell-1)\kappa^2}{n} + \frac{(j-1)\kappa}{n} + i}}$.

   For all $\ell \in [n]$, pick $h \xleftarrow{\$} \mathcal{H}$, a family of universal hash functions.
   Compute $r_\ell = \left(h_\ell(m_{1,\ell,[\kappa/n]}, m_{2,\ell,[\kappa/n]}, \dots m_{\kappa,\ell,[\kappa/n]})\right) \oplus x_\ell$.

Fig. 7. Receiver simulation strategy for n-choose-1 bit OT.

Lemma 4. *The simulation error for the semi-honest receiver is at most $2^{-\kappa\delta/4}$.*

Proof. The view of the receiver $V_{\mathcal{R}} := \{c, \theta, \tilde{b}_{[\kappa^2],[\lambda]}, z_{[\kappa^2],[\lambda]}, r_0, r_1\}$. The values $\tilde{b}_{[\kappa^2],[\lambda]}, z_{[\kappa^2],[\lambda]}$ are generated using honest sender strategy. There is no abort from the sender side in the (α, β)-BEC hybrid or the simulated view.

Consider channel S_c, composed of κ sub-channels of block-length (κ/n), each of capacity $\tilde{c}_h$. Recall that $B \to Z \to \tilde{B}$, where $B, Z, \tilde{B}$ are random variables denoting the sender input, leakage and receiver output respectively. Thus, the capacity of any sub-channel of S_c, can only increase when the receiver obtains additional leakage. For a semi-honest receiver, the capacity of each sub-channel of S_c is at least $\tilde{c}_h = c_m^*(1 + \delta)$ even when the receiver is adversarial and can change channel characteristic. The channels S_ℓ for $\ell \in [n] \setminus \{c\}$ are constructed by sampling sets of κ sub-channels at random, without replacement from the remaining set. Since, the overall average capacity of the adversarial receiver (semi-honest, but changes channel characteristic) is at most c_m^*, the average capacity of any sub-channel in this remaining set is at most $c_m^*(n-1-\delta)/(n-1)$. Then, there are at least a constant fraction $(n-1-\delta)/(n-1)$ sub-channels in this remaining set, each with capacity at most $c_m^* < r$.

Now, consider the event that there exists a channel S_ℓ for $\ell \in [n] \setminus \{c\}$, such that for more than $(\kappa - \sqrt{\kappa})$ sub-channels in S_ℓ, the sub-channel capacity is greater than c_m^*. This event occurs with probability at most $2^{-\kappa/3}$. We argue that conditioned on this event not happening, the simulated view is $(n-1)2^{-\kappa/3}$-close to the receiver view in the (α, β)-BSC hybrid.

For a channel with capacity c and a code of rate $r > c$, a weak converse of Shannon's channel coding theorem proves the decoding error is at least $1 - \frac{c}{r}$, therefore the min-entropy is at least $h_2(1 - \frac{c}{r})$. Then, an application of the Leftover Hash Lemma gives us that for a randomly chosen universal hash function h, if $\sqrt{\kappa}$ sub-channels have constant min-entropy $> \delta/2$, the hash value is at least $2^{-\kappa\delta/3}$ close to uniform. Thus for all channels S_ℓ where $\ell \in [n] \setminus \{c\}$, the output r_ℓ is $2^{-\kappa\delta/3}$ close to uniform. Moreover, r_c is computed using honest sender strategy, so the random variable r_c is identical in the (α, β)-BSC hybrid and simulated views. Thus, the total simulation error is $(n-1)2^{-\kappa\delta/3} + 2^{-\kappa/3} = n2^{-\kappa\delta/3} < 2^{-\kappa\delta/4}$.

3.3   Special-Malicious Completeness of (α, β)-BSC for $0 < \beta \leqslant \alpha < \ell(\beta)$

In fact, it is not difficult to prove that the protocol in Fig. 5 yields $(n, 1, n-1)$ OT in a special-malicious setting. In this setting, the receiver is allowed to behave maliciously, whereas the sender must (semi-)honestly send a repetition code in the first step of the protocol, and after this step the sender is allowed to behave maliciously. Please refer to the full version for a formal proof.

4   Full Malicious Completeness of Binary Symmetric Channels

4.1 $\mathcal{F}_{\mathsf{com}}$ from (α, β)-BSC for $0 < \beta \leqslant \alpha < 1/2$

The protocol is presented in Fig. 8, in terms of a polar code $\mathcal{C}$ over the binary alphabet, with block-length κ, rate $1 - o(1)$ and minimum distance $\omega(\kappa^{4/5})$.

Intuitively, the sender sends picks a codeword from the appropriate code and sends a 2-repetition of the codeword over the BSC, to the receiver. The commitment is statistically hiding because the capacity of the receiver is less than the rate of the code, and therefore there is constant prediction error for each codeword c_i for $i \in [\kappa]$. The commitment is statistically binding because the sender cannot flip too many bits, or send too many 'bad' indices to the receiver.

Inputs: S has input bit $b \in \{0,1\}$ and R has no input.
Hybrid: (α, β)-BSC for $0 < \beta \leqslant \alpha < 1$.
The protocol is parameterized by κ.

1. Commit Phase:
   (a) For all $i \in [\kappa]$, S picks codeword $\mathbf{c}_i = (c_{i,1}, c_{i,2}, \dots c_{i,\kappa}) \xleftarrow{\$} C$, and sends
       $\mathbf{c}_{i,[2]} = (c_{i,1}, c_{i,2}, \dots c_{i,\kappa}, c_{i,1}, c_{i,2}, \dots c_{i,\kappa})$ over the (α, β)-BSC to R. Let
       R obtain $\tilde{\mathbf{c}}_{i,[2]}$.
   (b) S picks $h \xleftarrow{\$} \mathcal{H}$, a universal hash function family mapping $\{0,1\}^{\kappa^2} \to$
       $\{0,1\}$, and sends $h, y = b \oplus h(\mathbf{c}_1, \mathbf{c}_2, \dots \mathbf{c}_\kappa)$ to R.
2. Reveal Phase:
   (a) For all $i \in [\kappa]$, S sends $b, \mathbf{c}_i = (c_{i,1}, c_{i,2}, \dots c_{i,\kappa})$ to R.
   (b) R accepts if all the following conditions hold:
       – For all $i \in [\kappa]$, $\mathbf{c}_i$ is a valid codeword.
       – For all $i \in [\kappa]$, set $I_{i,1} = \{j : (\tilde{c}_{i,j}, \tilde{c}_{i,\kappa+j}) = (1 - c_{i,j}), (1 - c_{i,j})\}$.
         Then $|I_{i,1}| \leqslant (1 - \alpha)^2 (\kappa + \kappa^{2/3})$.
       – For all $i \in [\kappa]$, set $I_{i,2} = \{j : \tilde{c}_{i,j} \neq \tilde{c}_{i,\kappa+j}\}$. Then $|I_{i,2}| \leqslant 2\alpha(1 - \alpha)(\kappa + \kappa^{2/3})$.
       – $b = y \oplus h(\mathbf{c}_1, \mathbf{c}_2, \dots \mathbf{c}_\kappa)$.

Fig. 8. UC-secure $\mathcal{F}_{\mathsf{com}}$ from (α, β)-BSC for $0 < \beta \leqslant \alpha < 1$.

If he does, he will be caught with overwhelming probability. If he sends a few bad/flipped bits, the minimum distance of the code will still hash them down to the same value.

Correctness. For honest sender strategy, using a Chernoff bound, it is possible to show that the size of I_1 and I_2 is bounded by $(1 - \alpha)^2(\kappa + \kappa^{2/3})$ and $2\alpha(1 - \alpha)(\kappa + \kappa^{2/3})$ with probability at least $1 - 2 \cdot 2^{-\kappa/3}$. Thus, when S and R are both honest, then R accepts Reveal(Commit(b)) for any $b \in \{0,1\}$ with probability at least $1 - 2^{-\kappa/4}$.

Receiver Security (Statistical Binding/Extractability). It suffices to consider a dummy sender S and malicious environment $\mathcal{Z}_S$, such that the dummy sender forwards all messages from $\mathcal{Z}_S$ to the honest receiver/simulator, and vice-versa.

Without loss of generality, the semi-honest simulation strategy Sim_S can be viewed to interact directly with $\mathcal{Z}_S$. Sim_S is described in Fig. 9.

Lemma 5. *The simulation error for the malicious sender is at most $2^{-\kappa^{0.5}}$.*

Proof. First, note that both the real and ideal views reject with probability 1 when $\mathbf{c}'_i$ is not a valid codeword, for any $i \in [\kappa]$. Next, if $|I_{i,1}| > 2\kappa^{2/3}$ or

The simulator $\text{Sim}_{\mathcal{S}}$ does the following.

1. **Commit Phase:**
   (a) For all $i \in [\kappa]$, obtain $h, y, \mathbf{c}_{i,[2]}$ from $\mathcal{Z}_{\mathcal{S}}$.
   (b) For all $i \in [\kappa]$, compute the nearest codeword $\tilde{\mathbf{c}}_i$ to $\mathbf{c}_i = \{c_{i,1}, c_{i,2} \ldots c_{i,\kappa}\}$.
   (c) Extract bit $b' = y \oplus h(\tilde{\mathbf{c}}_i, \tilde{\mathbf{c}}_2, \ldots \tilde{\mathbf{c}}_\kappa)$ and send it to the ideal $\mathcal{F}_{\text{com}}$ functionality.
2. **Reveal Phase:**
   (a) For all $i \in [\kappa]$, obtain $\mathbf{c}'_i$ from $\mathcal{Z}_{\mathcal{S}}$.
   (b) Allow the ideal functionality to output the extracted bit b' if all the following conditions hold (and otherwise reject):
   - For all $i \in [\kappa]$, $\mathbf{c}'_i$ is a valid codeword.
   - For all $i \in [\kappa]$, set $I_{i,1} = \{j : c'_{i,j} \neq c_{i,j}\}$. Then $|I_{i,1}| \leq 2\kappa^{2/3}$.
   - For all $i \in [\kappa]$, set $I_{i,2} = \{j : c_{i,j} \neq c_{i,\kappa+j}\}$. Then $|I_{i,2}| \leq 2\kappa^{2/3}$.

Fig. 9. Sender simulation strategy for $\mathcal{F}_{\text{com}}$.

$|I_{i,2}| > 2\kappa^{2/3}$, then the real view rejects with probability at least $(1 - 2^{-\kappa^{2/3}})$, whereas the ideal view always rejects.

Conditioned on the receiver not rejecting, it remains to argue that the bit b' extracted by the simulator (and later output to the receiver) is distributed identically in the hybrid and ideal worlds. Conditioned on not rejecting, for each $i \in [\kappa]$, the distance between $\mathbf{c}'_i$ and $\mathbf{c}_i$ is at most $|I_{i,1}| + |I_{i,2}| = 4\kappa^{2/3}$. Then, because the code has minimum distance $\omega(\kappa^{4/5})$, the nearest codeword $\tilde{\mathbf{c}}_i$ to $\mathbf{c}_i$ is actually $\mathbf{c}'_i$ itself. Therefore, the bit $b' = y \oplus h(\tilde{\mathbf{c}}_i, \tilde{\mathbf{c}}_2, \ldots \tilde{\mathbf{c}}_\kappa) = y \oplus h(\mathbf{c}'_1, \mathbf{c}'_2, \ldots \mathbf{c}'_\kappa)$ is distributed identically in the hybrid and ideal worlds in this case.

Thus the simulation error is at most $2.2^{-\kappa^{2/3}} < 2^{-\kappa^{0.5}}$.

Sender Security (Statistical Hiding/Equivocability). It suffices to consider a dummy receiver $\mathcal{R}$ and malicious environment $\mathcal{Z}_{\mathcal{R}}$, such that the dummy receiver forwards all messages from $\mathcal{Z}_{\mathcal{R}}$ to the honest receiver/simulator, and vice-versa.

Without loss of generality, the semi-honest simulation strategy $\text{Sim}_{\mathcal{R}}$ can be viewed to interact directly with $\mathcal{Z}_{\mathcal{R}}$. $\text{Sim}_{\mathcal{R}}$ is described in Fig. 10.

Lemma 6. *The simulation error for the malicious receiver is at most $2.2^{-\kappa}$.*

Proof. For all $i \in [\kappa]$ and honestly generated $\mathbf{c}_i$, the channel $\tilde{\mathbf{c}}_{i,[2]}$ has a constant fraction $2\beta(1 - \beta)$ bits of the form 01 or 10, which count as erasures. Thus, the capacity of each such channel is at most $1 - 2\beta(1 - \beta)$. Since the rate of the code sent over channel $\tilde{\mathbf{c}}_{i,[2]}$ is $1 - o(1)$, the entropy in the received string is at least $1 - \frac{1 - 2\beta(1-\beta)}{1 - o(1)} \approx 2\beta(1 - \beta)$. Therefore, via the leftover hash lemma, $h(\mathbf{c}_1, \mathbf{c}_2, \ldots \mathbf{c}_\kappa)$ is at least $1 - 2^{-\kappa}$ close to uniform, and therefore, y is at least $1 - 2^{-\kappa}$ close to uniform.

The simulator $\mathsf{Sim}_\mathcal{R}$ does the following.

1. Commit Phase:
   (a) Wait for the honest sender to send bit b' to the ideal $\mathcal{F}_{\mathsf{com}}$ functionality.
   (b) For all $i \in [\kappa]$, pick codeword $\mathbf{c}_i = (c_{i,1}, c_{i,2}, \ldots c_{i,\kappa}) \xleftarrow{\$} C$, and send $\mathbf{c}_{i,[2]} = (c_{i,1}, c_{i,2}, \ldots c_{i,\kappa}, c_{i,1}, c_{i,2}, \ldots c_{i,\kappa})$ over the (α, β)-BSC to $\mathcal{R}$. Obtain output $\tilde{\mathbf{c}}_{i,[2]}$ and leakage $\tilde{\mathbf{z}}_{i,[2]}$ for $\mathcal{R}$.
   (c) Pick $h \xleftarrow{\$} \mathcal{H}$, a universal hash function family mapping $\{0,1\}^{\kappa^2} \to \{0,1\}$, and send $y = h(\mathbf{c}_1, \mathbf{c}_2, \ldots \mathbf{c}_\kappa)$ to $\mathcal{R}$.
2. Reveal Phase:
   (a) Allow the ideal functionality to output the extracted bit b'.
   (b) If $b' = 0$, then output $\mathbf{c}_i, \mathbf{c}_2, \ldots \mathbf{c}_n$ to $\mathcal{R}$.
   (c) Else for all $i \in [\kappa]$,
       – Set codeword $\mathbf{c}'_i = \mathbf{c}_i$.
       – Set $I_i = \{j : \tilde{z}_{i,j} \neq \tilde{z}_{i,\kappa+j}\}$ (these are the erased indices).
       – Flip $c'_{i,j}$ at random indices $\mathsf{ind} \in I_i$, ensuring that $\mathbf{c}'_i$ remains a valid codeword.
   (d) Check if $h(\mathbf{c}'_1, \mathbf{c}'_2, \ldots \mathbf{c}'_\kappa) \neq h(\mathbf{c}_1, \mathbf{c}_2, \ldots \mathbf{c}_\kappa)$. If not, repeat step (c).

Fig. 10. Receiver simulation strategy for $\mathcal{F}_{\mathsf{com}}$.

Moreover, with probability at least $1 - 2^{-\kappa}$, it is possible to efficiently find a different set of codewords $\mathbf{c}'_i$ which hash to a different bit, for the same output $\tilde{\mathbf{c}}_i$ and $\tilde{\mathbf{z}}_i$ of the receiver.

4.2 Malicious Completeness of (α, β)-BSC for $0 < \beta \leqslant \alpha < \ell(\beta)$

To make the protocol in Sect. 3.3 secure against a general malicious sender instead of only a special-malicious one, we must ensure correctness of the repetition code sent in Step 1 by the sender. To ensure this, we make use of the commitment protocol $\mathcal{F}_{\mathsf{com}}$.

The functionality $\mathcal{F}_{\mathsf{com}}$ can be constructed from any (α, β)-BSC as demonstrated in Sect. 4.1. The sender and receiver use $\mathcal{F}_{\mathsf{com}}$ to toss random coins, and then implement a cut-and-choose based protocol to implement Step 1 of the special-malicious protocol. The protocol is presented in Fig. 11 in the $\mathcal{F}_{\mathsf{com}}$ and (α, β)-BSC hybrids. The protocol (including commitments) always uses the (α, β)-BSC from the sender to the receiver. Since OT can be reversed, this demonstrates fixed-role completeness of (α, β)-BSC for $0 < \beta \leqslant \alpha < \ell(\beta)$. Step 1 of the protocol in Sect. 3.3 is modified as follows.

Analysis. The sender and receiver use $\mathcal{F}_{\mathsf{com}}$ to toss common random coins.

In step 1, the sender sends λ-repetitions of κ^6 bits over the (α, β)-BSC. Additionally, he sends a commitment to each of these bits. Then, the parties pick a

---

Inputs: $\mathcal{S}$ has inputs $(x_0, x_1) \in \{0,1\}^2$ and $\mathcal{R}$ has input choice bit $c \in \{0,1\}$.
Hybrid: (α, β)-BSC for $0 < \beta \leqslant \alpha < \ell(\beta)$.

1. Correlation Generation:
    (a) Sender Message: For all $i \in [\kappa^6]$, $\mathcal{S}$ picks bit $b_i \in \{0,1\}$ and sends
        $b_{i,[\lambda]} = b_i^\lambda$ over the (α, β)-BSC to $\mathcal{R}$. Let $\mathcal{R}$ obtain output $\tilde{b}_{i,[\lambda]}$. $\mathcal{S}$
        sends $d_i = \text{com}(b_i)$ to $\mathcal{R}$.
    (b) Coin tossing in the well: Parties $\mathcal{S}$ and $\mathcal{R}$ use $\mathcal{F}_{\text{com}}$ to generate random
        coins in the following manner. $\mathcal{S}$ picks random $r_S \overset{\$}{\leftarrow} \{0,1\}^{\kappa^6}$ and sends
        $\text{com}(r_S)$ to $\mathcal{R}$. Then, $\mathcal{R}$ picks random $r_R \overset{\$}{\leftarrow} \{0,1\}^{\kappa^6}$ and sends r_R to $\mathcal{S}$.
        Then, $\mathcal{S}$ decommits to r_S, and if accepted, both parties obtain shared
        randomness $r = r_S \oplus r_R$.
    (c) Cut and Choose: Parties use randomness r to pick $S \overset{\$}{\leftarrow} \binom{[\kappa^6]}{\kappa^6/2}$ and

    $$T \overset{\$}{\leftarrow} \binom{[\kappa^6] \setminus S}{\kappa^2}.$$ $\mathcal{S}$ reveals b_i for all $i \in S$. Let $\mathcal{I} = \{i : \tilde{b}_i \neq \alpha^\lambda\}$. $\mathcal{R}$
        aborts if $|\mathcal{I}| > (1 - \alpha^\lambda)(\kappa^6/2 + \kappa^{3.1})$.
        Else, $\mathcal{S}$ and $\mathcal{R}$ use this set T, to continue the rest of the protocol
        according to Fig. 5.

Fig. 11. 2-choose-1 bit OT from (α, β)-BSC for $0 < \beta \leqslant \alpha < \ell(\beta)$.

random subset, consisting of half of the values sent in step 1, and the sender is
required to reveal these values.

Next, out of the remaining $\kappa^6/2$ commitments, both parties pick a random
subset of size κ^5. Then, with probability at least $(1 - 1/\kappa)$, this subset is such that
at most $\kappa^{3.1}$ of the values committed to do not match the repetition code (that
is, the statistical check would have passed passed). If the sender and receiver
pick a random set of κ^2 random values out of this set of κ^5 values, then with
probability at least $(1 - 1/\kappa^{1.2})$, all of them are correct repetition codes.

Therefore, we obtain a statistical OT which fails with probability at most
$2/\kappa^{1.2}$, we call such a functionality that fails with vanishing probability, $\mathcal{F}_{\widetilde{\text{OT}}}^{(\delta)}$,
which is formally described in Fig. 12. This functionality $\mathcal{F}_{\widetilde{\text{OT}}}^{(\delta)}$, can then be
compiled using [32,34] to obtain constant-rate OT, following [40]. We provide
the details of this compiler in the full version.

This completes the proof of Theorem 1.

Functionality $\mathcal{F}_{\widetilde{\text{OT}}}^{(\delta)}$. Parameterized by a function $\delta(\kappa)$.

- Set $b = 1$ with probability $b = \delta(\kappa)$, otherwise set $b = 0$.
- Provide the parties access to a 2-choose-1 bit OT functionality. If $b = 1$,
  let the adversary control the functionality.

Fig. 12. $\mathcal{F}_{\widetilde{\text{OT}}}^{(\delta)}$ Functionality

5   Conclusion

It is an interesting open problem to explore whether our completeness results extend to parameters $\alpha > \ell(\beta)$, or if there are impossibility results for this setting.

Unfair channels [22] give a theoretical model, general enough to capture many realistic noisy channels. However, in light of strong impossibility results for the completeness of unfair channels, we weaken the adversarial model resulting in what we call *elastic* noisy channels.

We show that this model circumvents the impossibility results in the unfair channel setting, and show a wide range of parameters for which elastic channels can be used to securely realize OT. We believe our techniques are of independent interest and can be leveraged, along with other ideas, to close the gap between the known feasible and infeasible parameters in the unfair channel setting.

5.1   Sender-Elastic Channels Reduction to (Receiver-) Elastic Channels

We can reduce sender-elastic BSC to a (receiver-) elastic BSC in the following manner. Suppose Alice is the sender and sends a bit b through the sender-elastic BSC. She receives a leakage $b \oplus E_1$, where $E_1 = \text{Ber}(\beta)$. Bob, the receiver, obtains $C = b \oplus E_1 \oplus E_2$, where $E_2 = \text{Ber}(\gamma)$ such that $\text{Ber}(\alpha) \equiv \text{Ber}(\beta) + \text{Ber}(\gamma)$.

We reverse this channel using the following technique. Bob defines $T := C \oplus R$, where R is a uniform random bit, and sends T to Alice. Alice now defines $S := b \oplus T$. Now, interpret R as the bit sent and S as the received bit. It is clear that this is a (α, γ)-BSC channel. And, it can also be formally argued that this one-to-one transformation is tight.

References

1. Arikan, E.: Channel polarization: a method for constructing capacity-achieving codes. In: Kschischang, F.R., Yang, E. (eds.) 2008 IEEE International Symposium on Information Theory, ISIT 2008, Toronto, ON, Canada, 6–11 July 2008, pp. 1173–1177. IEEE (2008). http://dx.doi.org/10.1109/ISIT.2008.4595172
2. Arikan, E.: Channel polarization: a method for constructing capacity-achieving codes for symmetric binary-input memoryless channels. IEEE Trans. Inf. Theor. **55**(7), 3051–3073 (2009). http://dx.doi.org/10.1109/TIT.2009.2021379
3. Beaver, D.: Perfect privacy for two-party protocols. In: Feigenbaum, J., Merritt, M. (eds.) Proceedings of DIMACS Workshop on Distributed Computing and Cryptography, vol. 2, pp. 65–77. American Mathematical Society (1989)
4. Beimel, A., Malkin, T., Micali, S.: The all-or-nothing nature of two-party secure computation. In: Wiener, M.J. (ed.) Advances in Cryptology - CRYPTO 1999. LNCS, vol. 1666, pp. 80–97. Springer, Heidelberg (1999)
5. Ben-David, A., Nisan, N., Pinkas, B.: FairplayMP: a system for secure multi-party computation. In: Ning, P., Syverson, P.F., Jha, S. (eds.) ACM 15th Conference on Computer and Communications Security, CCS 2008, pp. 257–266. ACM Press, Alexandria (27–31 October 2008)

6. Ben-Or, M., Goldwasser, S., Wigderson, A.: Completeness theorems for non-cryptographic fault-tolerant distributed computation (extended abstract). In: 20th Annual ACM Symposium on Theory of Computing, pp. 1–10. ACM Press, Chicago (2–4 May 1988)
7. Brassard, G., Crépeau, C., Wolf, S.: Oblivious transfers and privacy amplification. J. Cryptol. 16(4), 219–237 (2003). http://dx.doi.org/10.1007/s00145-002-0146-4
8. Cachin, C.: On the foundations of oblivious transfer. In: Nyberg, K. (ed.) Advances in Cryptology - EUROCRYPT 1998. LNCS, vol. 1403, pp. 361–374. Springer, Heidelberg (1998)
9. Calabro, C.: The exponential complexity of satisfiability problems. Ph.D. thesis (2009). http://www.escholarship.org/uc/item/0pk5w64k
10. Canetti, R., Lindell, Y., Ostrovsky, R., Sahai, A.: Universally composable two-party and multi-party secure computation. In: 34th Annual ACM Symposium on Theory of Computing, pp. 494–503. ACM Press, Montréal (19–21 May 2002)
11. Chandran, N., Goyal, V., Sahai, A.: New constructions for UC secure computation using Tamper-proof hardware. In: Smart, N.P. (ed.) EUROCRYPPT 2008. LNCS, vol. 4965, pp. 545–562. Springer, Heidelberg (2008)
12. Chaum, D., Crépeau, C., Damgård, I.: Multiparty unconditionally secure protocols (extended abstract). In: 20th Annual ACM Symposium on Theory of Computing, pp. 11–19. ACM Press, Chicago (2–4 May 1988)
13. Chernoff, H.: A measure of asymptotic efficiency for tests of a hypothesis based on the sum of observations. Ann. Math. Stat. 23, 493–507 (1952)
14. Chvátal, V.: The tail of the hypergeometric distribution. Discrete Math. 25(3), 285–287 (1979). http://www.sciencedirect.com/science/article/pii/0012365X799 00840
15. Cover, T.M., Thomas, J.A.: Elements of Information Theory, 2nd edn. Wiley, New York (2006)
16. Crépeau, C.: Efficient cryptographic protocols based on noisy channels. In: Fumy, W. (ed.) Advances in Cryptology - EUROCRYPT 1997. LNCS, vol. 1233, pp. 306–317. Springer, Heidelberg (1997)
17. Crépeau, C., Kilian, J.: Achieving oblivious transfer using weakened security assumptions (extended abstract). In: 29th Annual Symposium on Foundations of Computer Science, pp. 42–52. IEEE Computer Society Press, White Plains, New York (24–26 October 1988)
18. Crépeau, C., Kilian, J., Savvides, G.: Interactive hashing: an information theoretic tool (invited talk). In: Safavi-Naini, R. (ed.) ICITS 08: 3rd International Conference on Information Theoretic Security. LNCS, vol. 5155, pp. 14–28. Springer, Heidelberg (2008)
19. Crépeau, C., Morozov, K., Wolf, S.: Efficient unconditional oblivious transfer from almost any noisy channel. In: Blundo, C., Cimato, S. (eds.) SCN 04: 4th International Conference on Security in Communication Networks. LNCS, vol. 3352, pp. 47–59. Springer, Heidelberg (2005)
20. Damgård, I., Fehr, S., Morozov, K., Salvail, L.: Unfair noisy channels and oblivious transfer. In: Naor, M. (ed.) TCC 2004: 1st Theory of Cryptography Conference. LNCS, vol. 2951, pp. 355–373. Springer, Heidelberg (2004)
21. Damgård, I., Ishai, Y.: Scalable secure multiparty computation. In: Dwork, C. (ed.) Advances in Cryptology - CRYPTO 2006. LNCS, vol. 4117, pp. 501–520. Springer, Heidelberg (2006)

22. Damgård, I., Kilian, J., Salvail, L.: On the (im)possibility of basing oblivious transfer and bit commitment on weakened security assumptions. In: Stern, J. (ed.) Advances in Cryptology - EUROCRYPT 1999. LNCS, vol. 1592, pp. 56–73. Springer, Heidelberg (1999)
23. Damgård, I., Nielsen, J.B., Wichs, D.: Isolated proofs of knowledge and isolated zero knowledge. In: Smart, N.P. (ed.) Advances in Cryptology - EUROCRYPT 2008. LNCS, vol. 4965, pp. 509–526. Springer, Heidelberg (2008)
24. Dodis, Y., Ostrovsky, R., Reyzin, L., Smith, A.: Fuzzy extractors: how to generate strong keys from biometrics and other noisy data. SIAM J. Comput. **38**(1), 97–139 (2008). http://dx.doi.org/10.1137/060651380
25. Even, S., Goldreich, O., Lempel, A.: A randomized protocol for signing contracts. In: Chaum, D., Rivest, R.L., Sherman, A.T. (eds.) Advances in Cryptology - CRYPTO 1982, pp. 205–210. Plenum Press, New York (1982)
26. Gallager, R.: Information Theory and Reliable Communication. Wiley, New York (1968)
27. Goldreich, O., Micali, S., Wigderson, A.: How to play any mental game or a completeness theorem for protocols with honest majority. In: Aho, A. (ed.) 19th Annual ACM Symposium on Theory of Computing, pp. 218–229. City, New York (25–27 May 1987)
28. Goyal, V., Ishai, Y., Sahai, A., Venkatesan, R., Wadia, A.: Founding cryptography on tamper-proof hardware tokens. In: Micciancio, D. (ed.) TCC 2010: 7th Theory of Cryptography Conference. LNCS, vol. 5978, pp. 308–326. Springer, Heidelberg (2010)
29. Guruswami, V., Xia, P.: Polar codes: speed of polarization and polynomial gap to capacity. In: 54th Annual Symposium on Foundations of Computer Science, pp. 310–319. IEEE Computer Society Press, Berkeley (26–29 October 2013)
30. Hoeffding, W.: Probability inequalities for sums of bounded random variables. J. Am. Stat. Assoc. **58**(301), 13–30 (1963). http://www.jstor.org/stable/2282952
31. Impagliazzo, R., Luby, M.: One-way functions are essential for complexity based cryptography (extended abstract). In: 30th Annual Symposium on Foundations of Computer Science, pp. 230–235. IEEE Computer Society Press, Research Triangle Park (30 October–1 November 1989)
32. Ishai, Y., Kushilevitz, E., Ostrovsky, R., Prabhakaran, M., Sahai, A., Wullschleger, J.: Constant-rate oblivious transfer from noisy channels. In: Rogaway, P. (ed.) CRYPTO 2011. LNCS, vol. 6841, pp. 667–684. Springer, Heidelberg (2011)
33. Ishai, Y., Kushilevitz, E., Ostrovsky, R., Sahai, A.: Extracting correlations. In: 50th Annual Symposium on Foundations of Computer Science, pp. 261–270. IEEE Computer Society Press, Atlanta (25–27 October 2009)
34. Ishai, Y., Prabhakaran, M., Sahai, A.: Founding cryptography on oblivious transfer - efficiently. In: Wagner, D. (ed.) Advances in Cryptology - CRYPTO 2008. LNCS, vol. 5157, pp. 572–591. Springer, Heidelberg (2008)
35. Katz, J.: Universally composable multi-party computation using tamper-proof hardware. In: Naor, M. (ed.) Advances in Cryptology - EUROCRYPT 2007. LNCS, vol. 4515, pp. 115–128. Springer, Heidelberg (2007)
36. Kilian, J.: Founding cryptography on oblivious transfer. In: 20th Annual ACM Symposium on Theory of Computing, pp. 20–31. ACM Press, Chicago (2–4 May 1988)
37. Kilian, J.: A general completeness theorem for two-party games. In: 23rd Annual ACM Symposium on Theory of Computing, pp. 553–560. ACM Press, New Orleans (6–8 May 1991)

38. Kilian, J.: More general completeness theorems for secure two-party computation. In: 32nd Annual ACM Symposium on Theory of Computing, pp. 316–324. ACM Press, Portland (21–23 May 2000)

39. Korjik, V., Morozov, K.: Generalized oblivious transfer protocols based on noisy channels. In: Gorodetski, V.I., Skormin, V.A., Popyack, L.J. (eds.) MMM-ACNS 2001. LNCS, vol. 2052, pp. 219–229. Springer, Heidelberg (2001). http://dx.doi.org/10.1007/3-540-45116-1_22

40. Kraschewski, D., Maji, H.K., Prabhakaran, M., Sahai, A.: A full characterization of completeness for two-party randomized function evaluation. In: Nguyen, P.Q., Oswald, E. (eds.) EUROCRYPT 2014. LNCS, vol. 8441, pp. 659–676. Springer, Heidelberg (2014)

41. Künzler, R., Müller-Quade, J., Raub, D.: Secure computability of functions in the IT setting with dishonest majority and applications to long-term security. In: Reingold, O. (ed.) TCC 2009: 6th Theory of Cryptography Conference. LNCS, vol. 5444, pp. 238–255. Springer, Heidelberg (2009)

42. Kushilevitz, E.: Privacy and communication complexity. In: 30th Annual Symposium on Foundations of Computer Science. pp. 416–421. IEEE Computer Society Press, Research Triangle Park, North Carolina (30 October–1 November 1989)

43. Maji, H.K., Prabhakaran, M., Rosulek, M.: Complexity of multi-party computation problems: the case of 2-party symmetric secure function evaluation. In: Reingold, O. (ed.) TCC 2009: 6th Theory of Cryptography Conference. LNCS, vol. 5444, pp. 256–273. Springer, Heidelberg (2009)

44. Maji, H.K., Prabhakaran, M., Rosulek, M.: A unified characterization of completeness and triviality for secure function evaluation. In: Galbraith, S.D., Nandi, M. (eds.) Progress in Cryptology INDOCRYPT 2012. LNCS, vol. 7668, pp. 40–59. Springer, Heidelberg (2012)

45. Malkhi, D., Nisan, N., Pinkas, B., Sella, Y.: Fairplay - secure two-party computation system. In: Blaze, M. (ed.) Proceedings of the 13th USENIX Security Symposium, 9–13 August 2004, San Diego, CA, USA, pp. 287–302. USENIX (2004). http://www.usenix.org/publications/library/proceedings/sec04/tech/malkhi.html

46. Moran, T., Segev, G.: David and Goliath commitments: UC computation for asymmetric parties using tamper-proof hardware. In: Smart, N.P. (ed.) Advances in Cryptology - EUROCRYPT 2008, 27th Annual International Conference on the Theory and Applications of Cryptographic Techniques, Proceedings. LNCS, vol. 4965, pp. 527–544. Springer, Heidelberg (2008). http://dx.doi.org/10.1007/978-3-540-78967-3_30

47. Nascimento, A.C.A., Winter, A.J.: On the oblivious-transfer capacity of noisy resources. IEEE Trans. Inf. Theor. **54**(6), 2572–2581 (2008). http://dx.doi.org/10.1109/TIT.2008.921856

48. Nielsen, J.B., Nordholt, P.S., Orlandi, C., Burra, S.S.: A new approach to practical active-secure two-party computation. In: Safavi-Naini, R., Canetti, R. (eds.) Advances in Cryptology - CRYPTO 2012. LNCS, vol. 7417, pp. 681–700. Springer, Heidelberg (2012)

49. Rabin, M.: How to exchange secrets by oblivious transfer. Technical Report TR-81, Harvard Aiken Computation Laboratory (1981)

50. Rabin, T., Ben-Or, M.: Verifiable secret sharing and multiparty protocols with honest majority (extended abstract). In: 21st Annual ACM Symposium on Theory of Computing, pp. 73–85. ACM Press, Seattle (15–17 May 1989)

51. Savvides, G.: Interactive Hashing and Reductions Between Oblivious Transfer Variants. Ph.D. thesis, Montreal, Que., Canada, Canada, aAINR32237 (2007)

52. Shannon, C.E.: Communication theory of secrecy systems. Bell Syst. Tech. J. **28**(4), 656–715 (1949)
53. Wiesner, S.: Conjugate coding. SIGACT News **15**, 78–88 (1983). http://doi.acm.org/10.1145/1008908.1008920
54. Wolf, S., Wullschleger, J.: Oblivious transfer is symmetric. In: Vaudenay, S. (ed.) Advances in Cryptology - EUROCRYPT 2006. LNCS, vol. 4004, pp. 222–232. Springer, Heidelberg (2006)
55. Wullschleger, J.: Oblivious-transfer amplification. In: Naor, M. (ed.) Advances in Cryptology - EUROCRYPT 2007. LNCS, vol. 4515, pp. 555–572. Springer, Heidelberg (2007)
56. Wullschleger, J.: Oblivious transfer from weak noisy channels. In: Reingold, O. (ed.) TCC 2009: 6th Theory of Cryptography Conference. LNCS, vol. 5444, pp. 332–349. Springer, Heidelberg (2009)
57. Yao, A.C.C.: Theory and applications of trapdoor functions (extended abstract). In: 23rd Annual Symposium on Foundations of Computer Science. pp. 80–91. IEEE Computer Society Press, Chicago (3–5 November 1982)

All Complete Functionalities are Reversible

Dakshita Khurana[1]([✉]), Daniel Kraschewski[2], Hemanta K. Maji[3],
Manoj Prabhakaran[4], and Amit Sahai[1]

[1] Department of Computer Science, Center for Encrypted Functionalities,
UCLA, Los Angeles, USA
{dakshita,sahai}@cs.ucla.edu
[2] TNG Technology Consulting GmbH, Munich, Germany
daniel.kraschewski@tngtech.com
[3] Department of Computer Science, Purdue University, West Lafayette, USA
hmaji@purdue.edu
[4] Department of Computer Science, University of Illinois,
Urbana-Champaign, USA
mmp@uiuc.edu

Abstract. Crépeau and Santha, in 1991, posed the question of reversibility of functionalities, that is, which functionalities when used in one direction, could securely implement the identical functionality in the reverse direction. Wolf and Wullschleger, in 2006, showed that oblivious transfer is reversible. We study the problem of reversibility among 2-party SFE functionalities, which also enable general multi-party computation, in the information-theoretic setting.

We show that any functionality that enables general multi-party computation, when used in both directions, is reversible. In fact, we show that any such functionality can securely realize oblivious transfer when used in an a priori fixed direction. This result enables secure computation using physical setups that parties can only use in a particular direction due to inherent asymmetries in them.

D. Khurana and A. Sahai—Research supported in part from a DARPA/ARL SAFE-WARE award, NSF Frontier Award 1413955, NSF grants 1228984, 1136174, 1118096, and 1065276, a Xerox Faculty Research Award, a Google Faculty Research Award, an equipment grant from Intel, and an Okawa Foundation Research Grant. This material is based upon work supported by the Defense Advanced Research Projects Agency through the ARL under Contract W911NF-15-C-0205. The views expressed are those of the author and do not reflect the official policy or position of the Department of Defense, the National Science Foundation, or the U.S. Government.
D. Khurana, H.K. Maji, M. Prabhakaran and A. Sahai—Work done in part while visiting the Simons Institute for Theoretical Computer Science, supported by the Simons Foundation and by the DIMACS/Simons Collaboration in Cryptography through NSF grant #CNS-1523467.
D. Kraschewski—Part of the research leading to these results was done while the author was at KIT and Technion. Supported by the European Union's Tenth Framework Programme (FP10/2010-2016) under grant agreement no. 259426 – ERC Cryptography and Complexity.
M. Prabhakaran—Research supported by NSF grant 1228856.

© International Association for Cryptologic Research 2016
M. Fischlin and J.-S. Coron (Eds.): EUROCRYPT 2016, Part II, LNCS 9666, pp. 213–242, 2016.
DOI: 10.1007/978-3-662-49896-5_8

Keywords: Secure function evaluation · Information-theoretic security · UC-security · Reversibility of functionalities · Fixed-role reduction

1 Introduction

In 1991, Crépeau and Santha [7] posed the following question. Given oblivious transfers in one direction can we implement oblivious transfer in the opposite direction? That is, given oblivious transfers where Alice is the sender and Bob is the receiver, can we securely realize an oblivious transfer where Bob is the sender and Alice is the receiver? Wolf and Wullschleger [22] resolved this question in the affirmative. This result inspired several interesting results in cryptography, like offline generation of correlated private randomness independent of the target functionality being computed in secure computation [4,12] and (comparatively) easily introducing adaptive-security to secure computation protocols [18]. The proof of reversibility for oblivious transfer of [22] appears to be intimately tied to the specifics of the oblivious transfer functionality. Could reversibility, however, be a more general phenomenon?

Some functionalities, like simultaneous exchange, are inherently reversible. But we are most interested in functionalities which provide us general secure [3] multi-party computation [9,23], i.e. the *complete* functionalities. Restricted to the class of complete functionalities, the line of inquiry initiated in 1991 naturally leads to the following fundamental question.

Which Complete Functionalities can be Reversed?

We study this problem in the two-party setting for secure function evaluation (SFE) functionalities. Our work provides a full characterization of SFE functionalities that are reversible as well as sufficient for information-theoretic general secure multi-party computation. In fact, we show that *every* complete SFE functionality is reversible. In other words, we show that if using a functionality in both directions is powerful enough to enable general secure function evaluation, then in fact using the functionality in just one direction is enough.

Aside from its inherent theoretical appeal, the question of reversibility is also motivated by asymmetries that may be present in different systems. For example, if some physical phenomenon between two parties Alice and Bob is being utilized in order to carry out secure computations, it may be that only a powerful entity can play the role of Alice, but a weak device can play the role of Bob. In such an scenario, it would be critical to ensure that the cryptographic advantage offered by the physical phenomenon is sufficient for secure computation even if roles cannot be reversed.

We obtain our characterization of reversibility, in fact, by studying the more general problem of characterizing all 2-party complete functionalities that can be used in fixed roles to enable secure information-theoretic two-party computation, i.e. the characterization of fixed-role completeness.

1.1   Our Contributions

In this work, we study 2-party secure function evaluation (SFE) functionalities in the information-theoretic UC-setting [3]. Our first result shows that any complete 2-party SFE functionality is reversible.

Informal Theorem 1 (Reversibility Characterization). *Any complete 2-party SFE functionality $\mathcal{F}$ is reversible.*

Our construction is also constant rate. That is, n instances of the functionality in one direction is used to implement $\Theta(n)$ instances of the functionality in the reverse direction.

A functionality $\mathcal{F}$ is complete if it can be used (in both directions) to securely realize the oblivious transfer functionality. For the stronger security notion of fixed-role completeness, we show that any complete functionality, when used in fixed-role, is also complete.

Informal Theorem 2 (Fixed-Role Completeness Characterization). *Any complete 2-party SFE functionality $\mathcal{F}$ is also fixed-role complete.*

Similar to the previous result, this result is also constant rate. That is, using n instances of the $\mathcal{F}$ functionality in a fixed direction, we implement $\Theta(n)$ instances of the oblivious transfer functionality.

Additionally, we also show that the commitment functionality can be securely realized in the $\mathcal{F}$-hybrid if and only if $\mathcal{F}$ is complete (see Corollary 1). The proof is sketched in Sect. 1.4. This rules out the possibility of a functionality $\mathcal{F}$ which is of an *intermediate complexity* in the following sense: it enables the computation of the commitment functionality (a non-trivial functionality) but not the (all powerful) oblivious transfer functionality.

1.2   Prior Works

The problem of reversibility was initially posed by Crépeau and Santha [7] and the reversibility of oblivious transfer (and oblivious linear function evaluation) was exhibited by Wolf and Wullschleger [22].

There are several results characterizing completeness of functionalities in different settings. The oblivious transfer functionality was identified by Wiesner and Rabin [20,21]. Brassard et al. [2] showed the equivalence between various flavors of OT. In a seminal work, Kilian showed the active-completeness of OT [13]. Prior to this, the passive-completeness of OT was shown in [10,11]. Crépeau and Kilian showed that noisy channels are active-complete [5].

The first characterization of completeness appeared in the seminal work of Kilian [14]. In the asymmetric SFE setting, Beimel et al. [1] provided a characterization. Kilian, in another seminal work in 2000, vastly generalized these results [15]. Subsequent works extended Kilian's result for active-completeness in two different directions: [6] considered "channel functions;" [17] considered deterministic functions.

Recently, the full characterization Θ of 2-party complete functionalities in the semi-honest [19] and malicious [16] settings were obtained.

1.3   Technical Overview: Reversibility of Functionalities

Let $\mathcal{F}$ be a randomized two-party functionality between parties A and B, and let $\mathcal{F}_{\text{core}}$ denote the redundancy-free core of $\mathcal{F}$ (obtained after removing redundancies from $\mathcal{F}$, as described in Sect. 3.2 of our paper). Kraschewski et al. [16] showed that $\mathcal{F}$ is complete $\iff$ $\mathcal{F}_{\text{core}}$ is not simple.

To develop intuition for 'simple' functions, consider the following example of a 'simple' two-party functionality $\mathcal{F}_{\text{coin}}$. $\mathcal{F}_{\text{coin}}$ ignores the inputs of both parties and just outputs a common uniform independent random bit to both parties. The formal notion of a simple function generalizes this to arbitrary randomized functions, by ensuring that if the parties start with independent inputs, then conditioned on the "common information" present after evaluating $\mathcal{F}_{\text{core}}$, the views of the two players remain independent of each other. Naturally then, a *non-simple function* is one where the views of the two players are *not independent* conditioned on the "common information" present after evaluating $\mathcal{F}_{\text{core}}$ on independent inputs. For the rest of this exposition, we will assume that $\mathcal{F}$ is redundancy-free, and thus $\mathcal{F} = \mathcal{F}_{\text{core}}$.

Kraschewski et al. [16] also showed how to obtain UC commitments from either $A \to B$ or $B \to A$, but not necessarily in both directions, using any non-simple $\mathcal{F}$. W.l.o.g. for our case analysis and the examples below, we assume that $\mathcal{F}$ already gives commitments from $A \to B$.

The main technical challenge in our paper, is to obtain commitments from $B \to A$ using any complete (equivalently, non-simple) $\mathcal{F}$. This is done by partitioning all complete functionalities into three exhaustive cases: 1(a), 1(b) and 2. We will illustrate how we achieve this with the help of representative examples for each case (Figs. 1, 2 and 3). We define the notion of 'extreme views' and 'intersection' below, after which we describe our partition and explain the main ideas that allow us to obtain commitments in each case.

Extreme Views: Consider the example function matrices in Figs. 1, 2 and 3. For simplicity, these examples have no redundancies, and are therefore equivalent to their core. Alice views are rows, and each row is a tuple (x, w): where x is her input and w is the output she received. Bob views are columns and each column is a tuple (y, z), where y is his input and z is his output. $\perp$ denotes no input. Double-lines separate sets of columns that correspond to the same input of Bob. The entry in row (x, w) and column (y, z) denotes $\Pr_{\mathcal{F}}[(w, z) \mid (x, y)]$.

A view of Bob corresponds to a column in the matrix, labelled by the (input, output) for that view. An extreme view of Bob is a column that cannot be written as a convex linear combination of other columns in the matrix. Note that for any non-simple $\mathcal{F}$, both parties will have at least one extreme view.

Warmup: Extreme views guarantee binding. Looking ahead, extreme views will form an important part of our analysis. Consider the following illustrative situation: Suppose Alice and Bob invoke the functionality in Fig. 2 many times on uniformly random inputs (assume they picked their inputs honestly). After this, Bob is supposed to send Alice the indices of all executions where he received

$(1,0)$. Suppose malicious Bob instead decides to send to Alice some indices where his view was $(0,1)$ or $(0,0)$.

Note that corresponding to Bob's view $(1,0)$, Alice always obtains view $(\bot,1)$. On the other hand corresponding to Bob's view $(0,1)$, Alice obtains view $(\bot,0)$ with constant probability. Corresponding to Bob's view $(0,0)$, Alice always obtains view $(\bot,0)$. Since Bob cannot guess what view Alice obtained, if Bob tries to cheat by claiming that his view was $(1,0)$ when actually his view was $(0,1)$ or $(0,0)$, Alice will sometimes end up with a view of $(\bot,0)$ and thus immediately detect Bob's cheating with constant probability. This weakly binds Bob to his views. We use repetition techniques (error-correcting codes) to amplify this weak binding property.

More generally, since extreme views cannot be expressed as a convex linear combination of other views, it impossible for any party to obtain other views and claim that he obtained a specific extreme view without getting caught. In the example situation above, no convex linear combination of other views $(0,1)$ and $(0,0)$ can be claimed to be the extreme view $(1,0)$. The same thing is true for *all extreme views* in any functionality $\mathcal{F}$.

Intersecting Views: A view of Alice, V_A, intersects with a view of Bob, V_B, if the joint view (V_A, V_B) occurs with non-zero probability on invoking $\mathcal{F}$ with uniform distribution over both inputs.

Case Analysis. Given this terminology, we partition the set of all complete functionalities into three sets, corresponding to Cases 1(a), 1(b) and 2. [16] already show how to obtain commitments from any functionality in what we call Case 1(a). The major technical contribution of our paper is to obtain commitments from functionalities that lie in Cases 1(b) and 2.

We will now walk through these cases using example functionalities from Figs. 1, 2 and 3. We will first define Case 1(a), and then describe how we partition the remaining possibilities for complete functionalities into Cases 1(b) and 2. At this level, the fact that they are exhaustive will be trivial to see. For Cases 1(b) and 2, we will then explain the main ideas behind obtaining commitments from $B \to A$, with Y with the help of examples.

Alice \ Bob	$(\bot,0)$	$(\bot,1)$
$(\bot,0)$	1/2	1/6
$(\bot,1)$	0	1/3

Fig. 1. Case 1(a). Both columns are extreme.

Alice \ Bob	$(0,0)$	$(0,1)$	$(1,0)$
$(\bot,0)$	1/4	1/12	0
$(\bot,1)$	0	2/3	1

Fig. 2. Case 1(b). $(0,0)$ and $(1,0)$ are extreme. $\mathsf{col}(0,1) \equiv 1/3 \times \mathsf{col}(0,0) + 2/3 \times \mathsf{col}(1,0)$

Alice \ Bob	$(0,0)$	$(0,1)$	$(0,2)$	$(1,0)$	$(1,1)$
$(\bot,0)$	1/5	0	0	1/20	0
$(\bot,1)$	0	3/5	0	9/20	9/20
$(\bot,2)$	0	0	1/5	0	1/20

Fig. 3. Case 2. (0,0), (0,1) and (0,2) are extreme. $\mathsf{col}(1,0) \equiv 1/4 \times \mathsf{col}(0,0) + 3/4 \times \mathsf{col}(1,0)$. $\mathsf{col}(1,1) \equiv 1/4 \times \mathsf{col}(0,2) + 3/4 \times \mathsf{col}(1,0)$.

- **Case 1(a):** Kraschewski et al. [16] obtained commitments from $P_1 \to P_2$ using any functionality between parties P_1 and P_2 which has the following property: There exist at least 2 extreme views $(\mathcal{V}^1_{P_1}, \mathcal{V}^2_{P_1})$ of P_1 which intersect with the same view V_{P_2} of P_2, i.e. both joint views $(\mathcal{V}^1_{P_1}, V_{P_2})$ and $(\mathcal{V}^2_{P_1}, V_{P_2})$ occur with non-zero probability. They also show that any complete functionality must satisfy this property in at least one direction, either $P_1 \to P_2$ or $P_2 \to P_1$.

  Recall that we require commitments from $B \to A$. We define Case 1(a) as the set of all $\mathcal{F}$ which satisfy the above property in the $B \to A$ direction. That is, Case 1(a) consists of all $\mathcal{F}$ for which there exist at least 2 extreme views $(\mathcal{V}^1_B, \mathcal{V}^2_B)$ of Bob that intersect with the same view V_A of Alice, i.e. both joint views $(\mathcal{V}^1_B, V_A)$ and $(\mathcal{V}^2_B, V_A)$ occur with non-zero probability.

  Observe that in the example in Fig. 1, both Bob views $(\bot, 0)$ and $(\bot, 1)$ are extreme, and they intersect with common Alice view $(\bot, 0)$. Figure 1 satisfies the above property from $B \to A$ and lies in Case 1(a). Thus, [16] give $B \to A$ commitments for this case.

  At a very intuitive level, Bob is committed to the views he obtained. He reveals these views in the decommitment phase. The common intersecting view of Alice occurs sometimes, and in these instances, she does not know what view Bob obtained. This property is amplified to obtain hiding. As illustrated above, Bob cannot equivocate extreme views, and [16] used this property of the extreme views to obtain binding as illustrated above.

Remaining Cases are Exhaustive. Let $\mathcal{V}_B$ denote the set of all extreme views of Bob. Let $\widehat{Y}_B := \{y : \exists z, \text{ such that } (y, z) \in \mathcal{V}_B\}$, that is $\widehat{Y}_B$ denotes the set of Bob inputs, which have at least one corresponding view in $\mathcal{V}_B$. Let $\widehat{V}_B$ denote the set of all views of Bob that have some $y \in \widehat{Y}_B$ as input, i.e., $\widehat{V}_B = \{(y, z) : y \in \widehat{Y}_B, (y, z) \text{ occurs with non-zero probability}\}$. Note: $\widehat{V}_B$ contains all extreme Bob views, and may also contain some non-extreme Bob views.

- Case 1, i.e. Case 1(a) ∪ Case 1(b), consists of all complete functionalities for which two views in $\widehat{V}_B$ intersect with a common Alice view.
- Case 2 consists of all complete functionalities for which no two views in $\widehat{V}_B$ intersect with a common Alice view.

It is easy to see that Cases 1 and 2 are an exhaustive partition of all complete $\mathcal{F}$. Next,

- Case 1(a) consists of all functionalities $\mathcal{F}$ in Case 1, where there are at least two extreme views in $\widehat{V}_B$ that intersect with a common Alice view.

- Case 1(b) consists of all functionalities in Case 1 that are not in Case 1(a). In particular, the fact that $\mathcal{F}$ is in Case 1(b) requires that no two extreme views in $\widehat{V_B}$ intersect with a common Alice view. This means that either an extreme and non-extreme view of Bob in $\widehat{V_B}$ intersect with a common Alice view, or two non-extreme views of Bob in $\widehat{V_B}$ intersect with a common Alice view. Note that if two non-extreme views intersect, then an extreme and non-extreme view also intersect (by the definition of extreme views).

- **Case 1(b):** Recall that this case consists of complete functionalities for which an extreme and a non-extreme view of Bob in $\widehat{V_B}$ intersect with a common Alice view, for $\widehat{V_B}$ defined above. An illustrative example for this case is in Fig. 2 above. The views $(0, 0)$ and $(1, 0)$ of Bob are extreme, $\widehat{Y_B} = \{0, 1\}, \widehat{V_B} = \{(0,0), (0,1), (1,0)\}$. Moreover, views $(0, 0)$ and $(0, 1)$ in $\widehat{V_B}$ intersect with a common Alice view. Also, views $(1, 0)$ and $(0, 1)$ in $\widehat{V_B}$ intersect with a common Alice view. But no two extreme Bob views intersect with a common Alice view.

  To obtain $B \rightarrow A$ commitments, Alice and Bob invoke $\mathcal{F}$, with Alice using a uniform distribution over her inputs and Bob using a uniform distribution over inputs in $\widehat{Y_B}$. Assume for simplicity that Alice and Bob can be forced to use the correct distribution over their inputs. (This can be ensured using cut-and-choose techniques and extreme views of Bob.)

Binding. We split Bob's views into two categories: extreme and non-extreme. The main idea behind building commitments will be to ensure that he cannot obtain views in one category and later claim that they belong in another category. To understand this, consider the following example scenario w.r.t. the functionality in Fig. 2: Bob obtains view $(0, 0)$, which is an extreme view, and claims later that he obtained $(0, 1)$, which is a non-extreme view. We would like to prevent this situation. We would also like to prevent Bob from obtaining view $(0, 1)$, which is a non-extreme view, and later claiming that he obtained $(0, 0)$, which is an extreme view. In both these situations, we would like Alice to catch such a cheating Bob with high probability. Ensuring that she catches such a cheating Bob will (weakly) bind Bob to the category of views he obtained. Here is how we ensure this.

- Suppose Bob obtains $(0, 1)$ and later claims it was $(0, 0)$. By a similar argument as the warmup, Alice will catch him with constant probability: Note that Alice obtains view $(\perp, 1)$ with constant probability corresponding to Bob's view $(0, 1)$, but she never obtains view $(\perp, 1)$ corresponding to Bob's view $(0, 0)$. Since Bob doesn't know what view Alice obtained, if he actually obtained the view $(0, 1)$ and tried to claim that he obtained $(0, 0)$, Alice will sometimes end up with view $(\perp, 1)$ and detect Bob's cheating with constant probability. This can be amplified to full-fledged binding using error correction.

- Suppose Bob obtains $(0, 0)$ and claims that it was $(0, 1)$. In this case, the previous argument no longer works since $(0, 1)$ is not an extreme view. However, because both parties used uniform inputs, Bob will obtain some 'correct' distribution over his outputs. Also by the previous item, Bob cannot have obtained $(0, 1)$ and claim that it is $(0, 0)$. Thus, if he obtains $(0, 0)$ and claims that he obtained $(0, 1)$, then $(0, 1)$ will appear too often in his claimed views and Alice will detect this. In general, to equivocate extreme views to non-extreme views, Bob will have to "steal" probability mass from the extreme views and add more mass to the non-extreme views – which Alice will detect.

Hiding. For a uniform distribution over her inputs, with constant probability Alice obtains a common view that intersects both an extreme and a non-extreme view of Bob. Thus she cannot tell which category Bob's view was in, at the end of the commit stage. This gives a weak form of hiding which can then be amplified. For example in the functionality in Fig. 2, Alice's view $(\bot, 0)$ intersects with the extreme view $(0, 0)$ and non-extreme view $(0, 1)$ of Bob. Only one such intersection suffices to obtain hiding. For a complete analysis of this case, please refer to Sect. 5.

– **Case 2:** Recall that this case consists of complete functionalities for which no two views of Bob in $\widehat{V}_B$ intersect with a common Alice view, for $\widehat{V}_B$ defined above. Nevertheless, note that at least 2 views of Bob must intersect with a common Alice view, because otherwise $\mathcal{F}$ is trivial. Moreover, if two views outside $\widehat{V}_B$ intersect with a common Alice view, then both views must be non-extreme (by the definition of $\widehat{V}_B$). This means that at least one extreme and non-extreme view pair intersect with a common Alice view, which means that in this case necessarily, one Bob view inside $\widehat{V}_B$ and one outside $\widehat{V}_B$ intersect with a common Alice view.

In the illustrative example in Fig. 3, since the first three columns can be convex-linearly combined to obtain the fourth and fifth columns, only the first three views $(0, 0), (0, 1), (0, 2)$ of Bob are extreme. Moreover, all extreme views of Bob correspond to input 0, thus $\widehat{Y}_B = \{0\}, \widehat{V}_B = \{(0, 0), (0, 1), (0, 2)\}$ and views in $\widehat{V}_B$ do not intersect with any common Alice view. Note also that Bob's input 1 is not redundant, because the distribution over Alice's views induced by Bob's input 1 is different from the distribution induced by Bob's input 0.

To obtain $B \to A$ commitments in this case, Alice and Bob invoke $\mathcal{F}$ with Alice using a uniform distribution over her inputs and Bob using a uniform distribution over all his inputs.

Binding. We partition Bob's views into two categories: views inside $\widehat{V}_B$ and views outside $\widehat{V}_B$, then argue that he cannot equivocate between these categories. Again, here we only argue that a cheating Bob will be caught with constant probability – this can be amplified using error-correcting codes to obtain full-fledged binding.

In this case, it is not straightforward to argue that Bob can be forced to use a uniform (or some requisite) distribution over his inputs – in fact arguing this forms the crux of our binding argument. Consider the example in Fig. 3. Here are two representative strategies of a malicious Bob:

- Bob actually obtains view $(1, 0)$, and later claims that it was $(0, 1)$. However, unbeknownst to Bob, Alice may obtain view $(\bot, 0)$ and therefore detects Bob's cheating with constant probability. More generally, if Bob uses input 1 and claims that it is a 0, Alice will catch him with constant probability.
- Bob actually uses input 0 all the time, and later claims that in some invocations he used input 1. Here, we note that the distributions over Alice's views corresponding to Bob's inputs 0 and 1 in the example functionality are different. If this were not the case, then Bob's input 1 would be redundant. This means that Alice, by simply checking her output distribution, will catch Bob whenever he launches such an attack.

We generalize this argument (refer to Lemma 3) to show that in any redundancy-free core of a complete functionality, in Case 2, there exists at least one Bob input outside of $\widehat{Y}_B$ (this input is 1 in the representative example) which cannot be mimicked using any input in $\widehat{Y}_B$ (this input is 0 in this example).

Hiding. We show that there exists a common Alice view which intersects at least one Bob view in $\widehat{Y}_B$ (which is 0 in the representative example in Fig. 3) and one Bob view corresponding to the un-mimickable input outside $\widehat{Y}_B$ (which is a 1 in the example). In the example functionality, Alice's view $(\bot, 0)$ intersects with the views $(0, 0)$ in $\widehat{V}_B$ and $(1, 0)$ corresponding to input 1 outside $\widehat{Y}_B$. When using a uniform distribution over her inputs (this can be easily ensured), with constant probability Alice obtains this intersecting view. This gives a weak form of hiding which can then be amplified. A complete analysis of this case is in Sect. 6.

1.4 Technical Overview: Commitment Reducible Only to Complete SFE Functionalities

We have already shown what if f is a 2-party SFE which is malicious-complete then $\mathcal{F}_{\text{com}}$ fixed-role reduces to it. So, it suffices to show that if $\mathcal{F}$ has a simple core, then $\mathcal{F}_{\text{com}}$ does not reduce to $\mathcal{F}$. Suppose a protocol Π securely realizes $\mathcal{F}_{\text{com}}$ in the $\mathcal{F}$-hybrid, where $\mathcal{F}$ has a simple core. Note that, given a public transcript, since $\mathcal{F}$ has a simple core, a party can always sample joint-views consistent with it. Therefore, either each transcript can be equivocated or it is not hiding. Hence, we have the following result:

Corollary 1. *For every 2-party SFE $\mathcal{F}$, we have: $\mathcal{F}_{\text{com}} \sqsubseteq_{\text{UC}} \mathcal{F}$ iff $\mathcal{F}_{\text{OT}} \sqsubseteq_{\text{UC}} \mathcal{F}$.*

2 Preliminaries

In this section, we recall some primitives useful in stating unified completeness results for 2-party SFE in various security notions.

2.1 Secure Function Evaluation

A Functionality. Consider a two-party finite randomized functionality $\mathcal{F}$ between Alice and Bob, where Alice has input $x \in \mathcal{X}$ and Bob has input $y \in \mathcal{Y}$. They invoke the functionality with their respective inputs and obtain outputs $w \in \mathcal{W}$ and $z \in \mathcal{Z}$. We recall that such a functionality can be denoted by a matrix. The rows of this matrix are indexed by Alice views $(x, w) \in \mathcal{X} \times \mathcal{W}$ and columns are indexed by Bob views $(y, z) \in \mathcal{Y} \times \mathcal{Z}$. The entry in the cell in row (x, w) and column (y, z) equals $\Pr[w, z|x, y]$.

This matrix can also be viewed as a collection of stochastic sub-matrices, where each sub-matrix corresponds to some input $x \in \mathcal{X}$ of Alice and $y \in \mathcal{Y}$ of Bob. Each cell in this sub-matrix, with row indexed by Alice output w and column indexed by Bob output z equals $\Pr[w, z|x, y]$.

Graph of an SFE Functionality. Given a 2-party SFE $\mathcal{F}(f_A, f_B)$ we define a bipartite graph $G(\mathcal{F})$ as follows.

Definition 1. *Graph of a 2-party SFE. Given a SFE functionality $\mathcal{F}(f_A, f_B)$, its corresponding graph $G(\mathcal{F})$ is a weighted bipartite graph constructed as follows. Its partite sets are $X \times Z_A$ and $Y \times Z_B$. For every $(x, a) \in X \times Z_A$ and $(y, b) \in Y \times Z_B$, the edge joining these two vertices is assigned weight*

$$\mathrm{wt}\left((x, a), (y, b)\right) := \frac{\Pr_{r \xleftarrow{\$} R}[f_A(x, y, r) = a \ \wedge \ f_B(x, y, r) = b]}{|X \times Y|}$$

The choice of the normalizing constant $1/|X \times Y|$ is arbitrary. For this particular choice of constant, we can view the weight of an edge as representing the joint-distribution probability of input-output pairs seen by the two parties when $(x, y, r) \xleftarrow{\$} X \times Y \times R$.

The *kernel* of a 2-party function f is a function which outputs to the two parties only the "common information" that f makes available to them. To formalize this, we define a weighted bipartite graph $G(f)$ with partite sets $X \times W$ and $Y \times Z$, and for every $(x, w) \in X \times W$ and $(y, z) \in Y \times Z$, the edge joining these two vertices is assigned weight $\frac{p_f[w,z|x,y]}{|X \times Y|}$. The kernel of $\mathcal{F}$ is a randomized function which takes inputs $x \in X$ and $y \in Y$ from the parties, samples $(w, z) \xleftarrow{\$} f(x, y)$, and outputs to both parties the connected component of $G(\mathcal{F})$ which contains the edge $(x, w), (y, z)$.

2-Party Secure Function Evaluation. A two-party randomized function (also called a secure function evaluation (SFE) functionality) is specified by a single randomized function denoted as $f : \mathcal{X} \times \mathcal{Y} \to \mathcal{W} \times \mathcal{Z}$. Despite the notation, the range of f is, more accurately, the space of probability distributions over $\mathcal{W} \times \mathcal{Z}$. The functionality takes an input $x \in \mathcal{X}$ from Alice and an input $y \in \mathcal{Y}$ from Bob, and samples $(w, z) \in \mathcal{W} \times \mathcal{Z}$ according to the distribution $f(x, y)$; then it delivers w to Alice and z to Bob. Throughout, we shall denote the probability of outputs being (w, z) when Alice and Bob use inputs x and y respectively is represented by $\beta^{\mathcal{F}}[w, z|x, y]$. We use the following variables for the sizes of the sets $\mathcal{W}, \mathcal{X}, \mathcal{Y}, \mathcal{Z}$: $|\mathcal{X}| = m, |\mathcal{Y}| = n, |\mathcal{W}| = q, |\mathcal{Z}| = r$.

As is conventional in this field, in this paper, we shall restrict to function evaluations where m, n, q and r are constants, that is, as the security parameter increases the domains do not expand. (But the efficiency and security of our reductions are only polynomially dependent on m, n, q, r, so one could let them grow polynomially with the security parameter. We have made no attempt to optimize this dependency.) W.l.o.g., we shall assume that $\mathcal{X} = [m]$ (that is, the set of first m positive integers), $\mathcal{Y} = [n], \mathcal{W} = [q]$ and $\mathcal{Z} = [r]$.

We consider standard security notions in the information-theoretic setting: UC-security, standalone security and passive-security against computationally unbounded adversaries (and with computationally unbounded simulators). Using UC-security allows to compose our sub-protocols securely [3]. Error in security (simulation error) is always required to be negligible in the security parameter of the protocol, and the communication complexity of all protocols are required to be polynomial in the same parameter. However, we note that a protocol may invoke a sub-protocol with a security parameter other than its own (in particular, with a constant independent of its own security parameter).

Complete Functionalities. A two-party randomized function evaluation $\mathcal{F}$ is standalone-complete (respectively, UC-complete) against information theoretic adversaries if any functionality $\mathcal{G}$ can be standalone securely (respectively, UC securely) computed in the $\mathcal{F}$ hybrid. We shall also consider passive-complete functions where we consider security against passive (semi-honest) adversaries.

Redundancy-free core of a functionality. The core of a functionality is computed by removing redundant parts of the functionality f. A redundancy may be of two forms. It could consist of inputs which are useless for the adversary, that is, using another input gives the adversary strictly more *information* about the view of the (other) honest party, while the honest party cannot distinguish the cases in which the adversary used the less informative or the more informative input. In this case, the less informative input is called redundant and is removed to obtain the core of the functionality.

Another kind of redundancy is an output redundancy, where two or more outputs can be compressed into a single output if they convey identical information to the adversary about the honest party's view. As an example, consider a functionality in which when Bob's input is 0, if Alice's input is 0 then he receives 0, but if her input is 1, he receives the output symbol α with probability 3/4 and β with probability 1/4. Here, the two outcomes α and β give Bob the same information about Alice's input, and could be merged into a single output. We recall the formal linear algebraic definition of redundancies from Kraschewski et al. [16] in Sect. 3.2.

Simple core of functionalities. The core of a functionality f is *simple* if for parties starting with independent inputs, the views of the parties remain independent of each other conditioned on the common information after the function evaluation.

Recall that Kraschewski et al. [16] showed that a finite randomized functionality is complete if and only if the redundancy-free core of $\mathcal{F}$ is not simple.

Extreme views and mimicking inputs. Consider the matrix $\beta^{\mathcal{F}}$ obtained after removing the above-mentioned redundancies from the matrix $\mathcal{F}$. The entry in the cell in row (x, w) and column (y, z) is denoted by $\beta^{\mathcal{F}}_{x,w,y,z}$ and equals $\Pr[w, z|x, y]$.

Then a view (y, z) of Bob is an *extreme* view if the column indexed by (y, z) in $\beta^{\mathcal{F}}$ cannot be written as a convex linear combination of other columns in $\beta^{\mathcal{F}}$. Note that there necessarily exist at least two extreme views for each party in any non-trivial functionality. We say that a view (y, z) of Bob intersects with a view (x, w) of Alice if the entry $\beta^{\mathcal{F}}_{x,w,y,z} \neq 0$.

Let $Y_0 \subset \mathcal{Y}$ be a set of Bob inputs. We say that an input $y^* \in Y \backslash Y_0$ of Bob, is *mimicked* by Y_0, if there exists a probability distribution η over Y_0 such that Alice's view when Bob is choosing inputs from this distribution is indistinguishable from her view when Bob uses y^*.

2.2   Leftover Hash Lemma

The *min-entropy* of a discrete random variable X is defined to be $H_\infty(X) = -\log \max_{x \in \mathsf{Supp}(X)} \mathsf{p}^f[X = x]$. For a joint distribution (A, B), the *average min-entropy* of A w.r.t. B is defined as $\widetilde{H}_\infty(A|B) = -\log \left(\mathbb{E}_{b \sim B} \left[2^{-H_\infty(A|B=b)} \right] \right)$.

Imported Lemma 1 (Generalized Leftover Hash Lemma(LHL) [8]). *Let* $\{H_x : \{0,1\}^n \rightarrow \{0,1\}^\ell\}\}_{x \in X}$ *be a family of universal hash functions. Then, for any joint distribution* (W, I):$\mathsf{SD}\left((H_X(W), X, I), (\mathcal{U}_\ell, X, I) \right) \leq \frac{1}{2}\sqrt{2^{-\widetilde{H}_\infty(W|I)} 2^\ell}$.

3   Technical Tools

This section is mainly based on concepts introduced in [16].

3.1   Notation and Definitions

Consider the matrix $\beta^{\mathcal{F}}$ of the redundancy-free core of $\mathcal{F}$, whose columns are indexed by Bob views $(y, z) \in \mathcal{Y} \times \mathcal{Z}$ and rows are indexed by Alice views $(x, w) \in \mathcal{X} \times \mathcal{W}$. The entry in the cell in row (x, w) and column (y, z) is denoted by $\beta^{\mathcal{F}}_{x,w,y,z}$ and equals $\Pr[w, z|x, y]$.

We will also consider the compressed matrix $\beta^{\mathcal{F}}_B$ whose rows are indexed by Bob inputs y and rows are indexed by Alice views $(x, w) \in \mathcal{X} \times \mathcal{W}$. The entry in the cell in row (x, w) and column y is denoted by $\beta^{\mathcal{F}}_{x,w,y}$ and equals $\Pr[w|x, y]$.

The maps ϕ_A and ϕ_B. These maps define equivalence classes of views. Roughly, two rows (or columns) in $\beta^{\mathcal{F}}$ lie in the same equivalence class if they are scalar multiples of each other. Formally, for each $(x, w) \in \mathcal{X} \times \mathcal{W}$, let the vector $\beta^{\mathcal{F}}|(x, w) \in \mathbb{R}^{nr}$ be the row indexed by (x, w) in the matrix $\beta^{\mathcal{F}}$. Let $\phi_A : [m] \times [q] \rightarrow [\ell]$ (for a sufficiently large $\ell \leq mq$) be such that $\phi_A(x, w) = \phi_A(x', w')$ iff $\beta^{\mathcal{F}}|_{(x,w)} = c \cdot \beta^{\mathcal{F}}|_{(x',w')}$ for some positive scalar c. ϕ_B is defined similarly for column vectors indexed by Bob views (y, z).

3.2 Characterizing Irredundancy

Redundancy in a function allows at least one party to deviate in its behavior in the ideal world and not be detected (with significant probability) by an environment. In our protocols, which are designed to detect deviation, it is important to use a function in a form in which redundancy has been removed. We use definitions of irredundancy from [16], and give a brief overview here for completeness. There also exists an efficient algorithm to remove redundancies following [16].

Irredundancy of a 2-Party Secure Function Evaluation Function. Recall that a 2-party SFE function f with input domains, $X \times Y$ and output domain $W \times Z$ is defined by probabilities $\mathsf{p}^f[w, z | x, y]$. Output redundancies identify if the output can be compressed to remove aspects of the output that are useless for the adversary's goal of gaining information about the honest party's inputs. For input redundancy, we define left and right redundancy of f as follows. Below, $|X| = m, |Y| = n, |W| = q, |Z| = r$. To define left-redundancy, consider representing f by the matrices $\{P^x\}_{x \in X}$ where each P^x is an $nr \times q$ matrix with $P^x_{(y,z),w} = \mathsf{p}^f[w, y, z | x]$. Here, $\mathsf{p}^f[w, y, z | x] \triangleq \frac{1}{n} \mathsf{p}^f[w, z | x, y]$ (where we pick y independent of x, with uniform probability $\mathsf{p}^f[y|x] = \frac{1}{n}$).

Definition 2. *For an SFE function $f : X \times Y \to W \times Z$, represented by matrices $\{P^x\}_{x \in X}$, with $P^x_{(y,z),w} = \Pr[w, y, z | x]$, we say that an input $\hat{x} \in X$ is left-redundant if there is a set $\{(\alpha_x, M_x) | x \in X\}$, where $0 \leq \alpha_x \leq 1$ with $\sum_x \alpha_x = 1$, and each M_x is a $q \times q$ stochastic matrix such that if $\alpha_{\hat{x}} = 1$ then $M_{\hat{x}} \neq I$, and $P^{\hat{x}} = \sum_{x \in X} \alpha_x P^x M_x$. We say $\hat{x}$ is strictly left-redundant if it is left-redundant as above, but $\alpha_{\hat{x}} = 0$. We say $\hat{x}$ is self left-redundant if it is left-redundant as above, but $\alpha_{\hat{x}} = 1$ (and hence $M_{\hat{x}} \neq I$). We say that f is left-redundancy free if there is no $x \in X$ that is left-redundant.*

Right-redundancy notions for inputs $\hat{y} \in Y$ are defined analogously. f is said to be *redundancy-free* if it is left-redundancy free and right-redundancy free.

3.3 Statistically Testable Function Evaluation

Statistical tests [16] help ensure that a cut-and-choose technique can be used to verify an adversary's claims about what inputs it sent to a 2-party function and what outputs it received, when the verifier has access to only the other end of the function. It is important to note that such statistical tests can only be applied when an adversary declares (or commits to) his claimed inputs beforehand and is not allowed to adaptively choose his input claims adaptively based on function output. Kraschewski et al. [16] show that *evaluation of a 2-party function is statistically testable iff the function is redundancy free*. We repeat the statistical test game and the proof of the above statement in the full version of the paper.

3.4 Weak Converse of the Channel Coding Theorem, Generalization

A converse of the channel coding theorem states that message transmission is not possible over a noisy channel at a rate above its capacity, except with a non-vanishing rate of errors. We use a generalization of the (weak) converse of channel coding theorem due to [16] where the receiver can adaptively choose the channel based on its current view. Then if in at least a μ fraction of the transmissions, the receiver chooses channels which are noisy (i.e., has capacity less than that of a noiseless channel over the same input alphabet), it is possible to lower bound its probability of error in predicting the input codeword as a function of μ, an upper bound on the noisy channel capacities, and the rate of the code. We import the following lemma from [16].

Imported Lemma 2. *Let $\mathcal{F} = \{\mathcal{F}_1, \ldots, \mathcal{F}_K\}$ be a set of K channels which take as input alphabets from a set Λ, with $|\Lambda| = 2^\lambda$. Let $\mathcal{G} \subseteq [K]$ be such that for all $i \in \mathcal{G}$, the capacity of the channel $\mathcal{F}_i$ is at most $\lambda - c$, for a constant $c > 0$.*

Let $\mathcal{C} \subseteq \Lambda^N$ be a rate $R \in [0,1]$ code. Consider the following experiment: a random codeword $c_1 \ldots c_N \equiv \boldsymbol{c} \overset{s}{\leftarrow} \mathcal{C}$ is drawn and each symbol $c_1 \ldots c_N$ is transmitted sequentially; the channel used for transmitting each symbol is chosen (possibly adaptively) from the set $\mathcal{F}$ by the receiver.

Conditioned on the receiver choosing a channel in $\mathcal{G}$ for μ or more transmissions, the probability of error of the receiver in predicting $\boldsymbol{c}$ is

$$P_e \geq 1 - \frac{1}{NR\lambda} - \frac{1 - c\mu/\lambda}{R}.$$

4   Summary and Exhaustive Case Analysis

4.1 Summary

Given a 2-party SFE $\mathcal{F}$, we represent by $\mathcal{F}_{A \to B}$ the functionality which takes its first input from Alice and its second input from Bob. Similarly, we define the functionality $\mathcal{F}_{B \to A}$. We say $\mathcal{F}$ reduces to $\mathcal{G}$, represented by $\mathcal{F} \sqsubseteq_{\mathsf{UC}} \mathcal{G}$, if there exists a information-theoretic UC-secure protocol for $\mathcal{F}$ in the $\mathcal{G}$-hybrid. The functionality $\mathcal{F}^{\otimes n}$ represents n independent copies of the functionality $\mathcal{F}$.

We observe that Kraschewski et al. [16] obtain oblivious transfer using any finite randomized functionality $\mathcal{F}$ with a non-simple core, in a fixed direction, if there exist commitments in both directions. Furthermore, they already show that for any finite randomized functionality $\mathcal{F}$ with a non-simple core, commitments can be obtained from either Alice to Bob or from Bob to Alice.

Our main technical contribution will be to show that, in fact, for any finite randomized functionality $\mathcal{F}$ with a non-simple core, commitments can be obtained *both* from Alice to Bob *and* from Bob to Alice, by using $\mathcal{F}$ in a fixed direction.

Analogous to the above statement, we also have a statement where $\mathcal{F}_{A \to B}$ is replaced by $\mathcal{F}_{B \to A}$. Next, once we get $\mathcal{F}_{\mathsf{OT}}$ at constant rate, we can implement $\mathcal{F}_{B \to A}$ at constant rate using [12]. This gives our main result.

Theorem 1 (Reversible Characterization). *For every 2-party SFE $\mathcal{F}$: if $\mathcal{F}_{\mathsf{OT}} \sqsubseteq_{\mathrm{UC}} \mathcal{F}$ in the malicious setting (possibly using $\mathcal{F}$ in both directions), then there exists $c > 0$ such that $\mathcal{F}_{A \to B}^{\otimes \sigma} \sqsubseteq_{\mathrm{UC}} \mathcal{F}_{B \to A}^{\otimes \kappa}$ in the malicious setting and $\sigma \geq c\kappa$.*

Again, once we have commitments in both directions, by using the SFE functionality in only one direction, we can use the compiler of [16] to directly obtain the following theorem.

Theorem 2 (Fixed-Role Completeness Characterization). *For every 2-party SFE $\mathcal{F}$: $\mathcal{F}_{\mathsf{OT}} \sqsubseteq_{\mathrm{UC}} \mathcal{F}$ in the malicious setting (possibly using $\mathcal{F}$ in both directions) if and only if there exists $c > 0$ such that $\mathcal{F}_{\mathsf{OT}}^{\otimes \sigma} \sqsubseteq_{\mathrm{UC}} \mathcal{F}_{A \to B}^{\otimes \kappa}$ in the malicious setting and $\sigma \geq c\kappa$.*

4.2 Exhaustive Case Analysis

First, we will classify any functionality $\mathcal{F}$ with a non-simple redundancy-free core, into a set of exhaustive cases. In each case, we demonstrate that it is possible to obtain commitments using $\mathcal{F}$, from Bob to Alice. Let $\mathcal{V}_B$ denote the set of extreme Bob views, and $\widehat{Y}$ be the set of inputs of Bob that admit at least one extreme view, that is, $\widehat{Y} := \{y \colon \exists z, \text{ such that } (y, z) \in \mathcal{V}_B\}$. Let $\widehat{V}_B$ denote the set of all Bob views corresponding to inputs in $\widehat{Y}$, that is $\widehat{V}_B = \{(y, z) \colon y \in \widehat{Y}\}$. Our cases are listed in Table 1.

Table 1. Exhaustive summary of cases

1	There exists an Alice view with which ≥ 2 Bob views in $\widehat{V}_B$ intersect.
(a)	There exists an Alice view with which ≥ 2 *extreme* Bob views in $\widehat{V}_B$ intersect. In this case, it is possible to obtain commitments from Bob to Alice [16].
(b)	There exists an Alice view with which one extreme and ≥ 1 non-extreme Bob view in $\widehat{V}_B$ intersect.
2	No two Bob views in $\widehat{V}_B$ intersect with the same Alice view.

Claim. In a non-simple functionality $\mathcal{F}$, if no two extreme Bob views intersect with the same Alice view, then there exists an Alice view which intersects with one extreme and one non-extreme Bob view.

Proof. In a non-simple functionality $\mathcal{F}$, if no two extreme Bob views intersect with the same Alice view, then we have the following possibilities:

1. There is an Alice view intersecting an extreme and non-extreme Bob view,
2. Or, there is an Alice view which intersects 2 non-extreme Bob views,
3. Or, no Alice view intersects any two Bob views.

We show that $2 \implies 1$, and 3 contradicts the fact that $\mathcal{F}$ is non-simple.

Let the number of extreme views of Bob be γ. Denote the extreme views of Bob by (y_i^*, z_i^*), for $i \in [\gamma]$. Suppose Alice view $V_A = (x, z)$ intersects with two non-extreme Bob views $V_B^1 = (y_1, z_1)$ and $V_B^2 = (y_2, z_2)$. Then, the columns $\beta_{|(y_1,z_1)}^{\mathcal{F}}$ and $\beta_{|(y_2,z_2)}^{\mathcal{F}}$ of $\beta^{\mathcal{F}}$ have non-zero entries in the row corresponding to (x, z). Since both views (V_B^1, V_B^2) are non-extreme, the columns $\beta_{|(y_1,z_1)}^{\mathcal{F}}$ and $\beta_{|(y_2,z_2)}^{\mathcal{F}}$ of $\beta^{\mathcal{F}}$ can be expressed as a linear combination of extreme columns (y_i^*, z_i^*), for $i \in [\gamma]$. This means that there necessarily exists at least one extreme view $(y^*, z^*) \in \{(y_1^*, z_1^*), (y_2^*, z_2^*), \ldots (y_\gamma^*, z_\gamma^*)\}$ such that the column $\beta_{|(y^*,z^*)}^{\mathcal{F}}$ of $\beta^{\mathcal{F}}$ has a non-zero entry in the row corresponding to (x, z). This proves $2 \implies 1$.

Suppose that in a non-simple functionality $\mathcal{F}$, no view of Alice intersects with any two views of Bob. That is, every view of Alice intersects with at most one view of Bob. In this case, the common information/kernel obtained after function evaluation is the view of Bob. It is straightforward to see that both parties can independently sample their views, conditioned on any view of Bob. This completes the proof of this claim.

In the following sections, we construct commitments $\mathcal{F}_{com,B \to A}$, for any functionality $\mathcal{F}$ depending on which of the two cases it falls in.

We observe that in case there exists an Alice view with which at least two *extreme* Bob views in $\widehat{V}_B$ intersect, the protocol of [16] can be used to obtain commitments from Bob to Alice. We re-state their result in the following lemma. In the following lemma, we will recall appropriate notions of confusability from [16]. Any functionality $\mathcal{F}$ in which at least two *extreme* Bob views in $\widehat{V}_B$ intersect with a common Alice view, will be said to have a confusable $\mathfrak{b}^{\mathcal{F}}$.

Imported Lemma 3. *Denote the set of extreme views of Bob by* $\mathfrak{b}^{\mathcal{F}}$. *For each Alice view* (x, w) *denote by* $\mathfrak{b}^{\mathcal{F}}|_{(x,w)}$ *all the extreme views of Bob which intersect with the specific Alice view* (x, w). *That is,* $\mathfrak{b}^{\mathcal{F}}|_{(x,w)}$ *is the set of extreme views* (y, z) *of Bob such that the row in* $\beta^{\mathcal{F}}$ *indexed by* (y, z) *has a positive entry in the column indexed by* (x, w). $\mathfrak{b}^{\mathcal{F}}$ *is said to be confusable if there exists* $(x, w) \in \mathcal{X} \times \mathcal{W}$ *and two elements* $(y_1, z_1), (y_2, z_2) \in \mathfrak{b}^{\mathcal{F}}|_{(x,w)}$ *such that* $\phi_B(y_1, z_1) \neq \phi_B(y_2, z_2)$. $\mathfrak{a}^{\mathcal{F}}$ *is defined similarly for extreme views of Alice. Then,*

1. *If the redundancy-free core of $\mathcal{F}$ is simple, either $\mathfrak{a}^{\mathcal{F}}$ or $\mathfrak{b}^{\mathcal{F}}$ is confusable.*
2. *If $\mathfrak{a}^{\mathcal{F}}$ is confusable, it is possible to obtain commitments from Alice to Bob. If $\mathfrak{b}^{\mathcal{F}}$ is confusable, it is possible to obtain commitments from Bob to Alice.*

5   Case 1(b): Commitments

5.1   Construction

Let $\mathcal{V}_B$ denote the set of all extreme views of Bob and let $\widehat{Y}$ denote the set of all inputs of Bob that contain at least one extreme view, that is $\widehat{Y} := \{y \colon \exists z, \text{ such that } (y, z) \in \mathcal{V}_B\}$. Further, let $\widehat{V}_B$ denote the set of all Bob views corresponding to inputs in $\widehat{Y}$, that is $\widehat{V}_B = \{(y, z) \colon y \in \widehat{Y}\}$.

In this section, we demonstrate how to obtain commitments from any functionality $\mathcal{F}$ for which the following is true: $\widehat{V}_B$ "is confusable", that is, there exists an Alice view (x, w) and two distinct Bob views $(\widehat{Y}_1, \hat{z}_1)$ and $(\widehat{Y}_2, \hat{z}_2) \in \widehat{V}_b$ (where possibly $\widehat{Y}_1 = \widehat{Y}_2$) such that $\beta^{\mathcal{F}}_{x, \widehat{Y}_1, w, \hat{z}_1} \neq 0$ and $\beta^{\mathcal{F}}_{x, \widehat{Y}_2, w, \hat{z}_2} \neq 0$. The protocol is described in Fig. 4.

5.2 Proof of Security

Receiver Security (Statistical Binding/Extractability). In the UC setting, it suffices to consider a dummy sender $\mathcal{S}$ and malicious environment $\mathcal{Z}_{\mathcal{S}}$, such that the dummy sender forwards all messages from $\mathcal{Z}_{\mathcal{S}}$ to the honest receiver/simulator, and vice-versa. Without loss of generality, the malicious simulation strategy $\mathsf{Sim}_{\mathcal{S}}$ can be viewed to interact directly with $\mathcal{Z}_{\mathcal{S}}$. $\mathsf{Sim}_{\mathcal{S}}$ is described in Fig. 5.

Lemma 1. *There exists a constant c such that the simulation error for the malicious sender is at most $2^{-c\kappa}$.*

Proof. The simulator performs Steps $1(a), (b)$ and (c) as per the honest receiver strategy, and also emulates the functionality $\mathcal{F}$ honestly for the sender. It remains to show that the unique bit b' extracted by the simulator equals the bit b committed by the sender Bob. The crux of this proof relies on the fact that the protocol requires the sender to use one extreme view and on the minimum distance of the code used.

Bob cannot claim non-extreme views to be extreme. In the opening made by Bob, consider the positions where Bob claimed his view to be extreme, that is, $(y_i, z_i) = (y^*, z^*) \in \mathcal{V}_B$, such that the equivalence class of this view $\phi_B(y^*, z^*) = \Phi$. Consider the fraction of these positions where the actual view of Bob (y', z') such that $\phi_B(y', z') \neq \Phi$. In these positions, the expected view of Alice is given by a linear combination of the columns $\beta^{\mathcal{F}}|_{(y', z')}$ (with coordinates scaled appropriately). If this linear combination is not close to the vector $\beta^{\mathcal{F}}|_{(y^*, z^*)}$ (scaled appropriately) then with all but negligible probability, the opening will not be accepted by the receiver. On the other hand, if the linear combination is close to $\beta^{\mathcal{F}}|_{(y^*, z^*)}$, since $\beta^{\mathcal{F}}|_{(y^*, z^*)}$ is outside the linear span of other $\beta^{\mathcal{F}}|_{(y', z')}$ with $\phi_B(y'z') \neq \phi_B(y^*, z^*)$, only at a small number (sub-linear fraction, say $\kappa^{2/3}$) of places can Bob open to (y^*, z^*) but have had an actual view (y', z'). This is because, an extreme view can't be expressed as a linear combination of other views of Bob, without being detected by Alice with constant probability.

Bob uses close to uniform distribution over inputs in $\widehat{Y}_B$. Consider an input $y^* \in \widehat{Y}_B$ and let (y^*, z^*) denote its corresponding extreme view. Alice will not accept the extreme view (y^*, z^*) in the opening of Bob (except with probability $2^{-c\kappa^{2/3}}$) unless Bob actually obtained the particular view in all but $\kappa^{2/3}$ of these

Inputs: Sender $\mathcal{S}$ has input bit bit $\in \{0,1\}$ and receiver $\mathcal{R}$ has no input.

Hybrid: $\mathcal{F}$ for non-simple function $\mathcal{F}$, and $\widehat{Y}$ as defined above is confusable. $\mathcal{F}$ provides commitments (Com) from Alice to Bob.

The protocol is presented in terms of a $(\kappa, \kappa - \kappa^{15/16}, \Omega(\kappa^{15/16}))$-linear code $\mathcal{C}$ over the binary alphabet. (An explicit code is not necessary: the receiver can pick random $\Omega(\kappa^{15/16})$ "parity checks" to construct the code and announce it to the sender.) The protocol is parameterized by κ.

1. Commit Phase:
    (a) $\mathcal{R}$ (Alice) picks inputs $(X_1, X_2, \ldots, X_{2\kappa^2})$ uniformly from $\mathcal{X}^{2\kappa^2}$. She commits to each of them using fresh randomness and sends $\mathsf{Com}(X_1), \mathsf{Com}(X_2), \ldots \mathsf{Com}(X_{2\kappa^2})$ to $\mathcal{S}$.
    (b) $\mathcal{S}$ (Bob) picks inputs $(Y_1, Y_2, \ldots, Y_{2\kappa^2})$ from a uniform distribution over $\widehat{Y}^{2\kappa^2}$. $\mathcal{R}$ and $\mathcal{S}$ invoke $\mathcal{F}$, 2κ times, with inputs $(X_1, X_2, \ldots, X_{2\kappa^2})$ and $(Y_1, Y_2, \ldots, Y_{2\kappa^2})$ respectively.
    (c) Cut-and-Choose: $\mathcal{R}$ picks $r_1 \xleftarrow{\$} \{0,1\}^*$ and sends $\mathsf{Com}(r_1)$ to $\mathcal{S}$. $\mathcal{S}$ sends $r_2 \xleftarrow{\$} \{0,1\}^*$ to $\mathcal{R}$. $\mathcal{R}$ uses randomness $(r_1 \oplus r_2)$ to pick a subset $I \leftarrow \binom{2\kappa^2}{[\kappa^2]}$ of the κ^2 indices. $\mathcal{R}$ decommits to r_1. Furthermore, for all $i \in I$, $\mathcal{R}$ decommits to input X_i and also opens her view (X_i, W_i).
    $\mathcal{S}$ aborts if the decommitments are not correct, or the inputs of $\mathcal{R}$ are not close to a uniform distribution, or if (X_i, W_i) for $i \in I$ satisfy the consistency checks in the Left-Statistical-Tests.
    Else, $\mathcal{S}$ and $\mathcal{R}$ set $S = [2\kappa^2] \setminus I$ and reorder the indices in S to $[\kappa^2]$.
    (d) $\mathcal{S}$ does the following for all $i \in [\kappa]$.
        – Construct the jth characteristic vector $\mathbf{u}_j$ such that for all $i \in [\kappa]$, $u_{j,i} = 0$ if and only if $(Y_{j\kappa+i}, Z_{j\kappa+i}) \in \mathcal{V}$, else $u_{j,i} = 1$.
        – Pick κ random codewords $\mathbf{c}_1, \mathbf{c}_2, \ldots \mathbf{c}_\kappa \in \mathcal{C}^\kappa$. Pick $h \xleftarrow{\$} \mathcal{H}$, a universal hash function mapping $\{0,1\}^{\kappa^2} \to \{0,1\}$, and for $j \in [\kappa]$, compute $y = h(\mathbf{c}_1, \mathbf{c}_2, \ldots \mathbf{c}_\kappa) \oplus \mathsf{bit}, \mathsf{offset}_j = (\mathbf{c}_j \oplus \mathbf{u}_j)$. Send $(h, y, \mathsf{offset}_1, \mathsf{offset}_2, \ldots \mathsf{offset}_\kappa)$ to $\mathcal{R}$.
2. Reveal Phase:
    (a) $\mathcal{S}$ sets $b' = \mathsf{bit}, \mathbf{u}'_j = \mathbf{u}_j$ for $j \in [\kappa]$ and sends $b', \mathbf{u}'_1, \mathbf{u}'_2, \ldots \mathbf{u}'_\kappa$ to $\mathcal{R}$ as his opening. $\mathcal{S}$ also sends (Y_i, Z_i) for all $i \in [\kappa^2]$, to $\mathcal{R}$.
    (b) $\mathcal{R}$ accepts if all the following conditions hold:
        – For $j \in [\kappa]$, $\mathbf{c}_j = \mathbf{u}'_j \oplus \mathsf{offset}_j$, is a valid codeword.
        – $b' = h(\mathbf{c}_1, \mathbf{c}_2, \ldots \mathbf{c}_\kappa) \oplus y$.
        – For all $i \in [\kappa^2]$, (Y_i, Z_i) satisfy input-output frequency tests.

Fig. 4. $\mathcal{F}_{\mathsf{com}}$ in Case 1(b).

The simulator $\mathsf{Sim}_\mathcal{S}$ does the following.

1. Commit Phase:
   (a) $\mathsf{Sim}_\mathcal{S}$ picks inputs $(X_1, X_2, \ldots, X_{2\kappa^2})$ uniformly from $\mathcal{X}^{2\kappa^2}$. $\mathsf{Sim}_\mathcal{S}$ then commits to each of them using fresh randomness and sends $\mathsf{Com}(X_1), \mathsf{Com}(X_2), \ldots \mathsf{Com}(X_{2\kappa^2})$ to $\mathcal{S}$. Note that $\mathsf{Sim}_\mathcal{S}$ has the capability to equivocate these commitments.
   (b) $\mathsf{Sim}_\mathcal{S}$ obtains inputs $(Y_1, Y_2, \ldots Y_{2\kappa^2})$ from $\mathcal{S}$ and emulates the functionality $\mathcal{F}$ honestly for $\mathcal{S}$ with inputs $(X_1, X_2, \ldots X_{2\kappa^2})$ and $(Y_1, Y_2, \ldots Y_{2\kappa^2})$.
   (c) Cut-and-Choose: $\mathsf{Sim}_\mathcal{S}$ picks $r_1 \xleftarrow{\$} \{0,1\}^*$ and sends $\mathsf{com}_1 = \mathsf{Com}(r_1)$ to $\mathcal{S}$. $\mathcal{S}$ sends $r_2 \xleftarrow{\$} \{0,1\}^*$ to $\mathsf{Sim}_\mathcal{S}$. $\mathsf{Sim}_\mathcal{S}$ uses $(r_1' \oplus r_2)$ to pick subset $I \xleftarrow{\$} \binom{2\kappa^2}{[\kappa^2]}$ of the κ^2 indices. $\mathsf{Sim}_\mathcal{S}$ decommits com_1 to r_1 and, for all $i \in I$, $\mathsf{Sim}_\mathcal{S}$ decommits to input X_i and also opens the view (X_i, W_i). Set $S = [2\kappa^2] \setminus I$ and reorder the indices in S to $[\kappa^2]$.
   (d) $\mathsf{Sim}_\mathcal{S}$ obtains $(h, y, \mathsf{offset}_j)$ for $j \in [\kappa]$ from $\mathcal{S}$. It constructs characteristic vectors $\mathbf{u}_j$ such that for all $i \in S$, $u_i = 0$ if and only if $(Y_i, Z_i) \in \mathcal{V}$, else $u_i = 1$. It then computes $\tilde{c}_j = \mathbf{u}_j \oplus \mathsf{offset}_j$, sets c_j' to be the nearest codeword[a] to $\tilde{c}_j$, and sets bit $b' = y \oplus h(c_1', c_2', \ldots c_\kappa')$.
2. Reveal Phase:
   (a) Obtain $b', u_1', u_2', \ldots u_\kappa', (Y_i, Z_i)$ for all $i \in [\kappa^2]$ from $\mathcal{S}$ as his opening.
   (b) Allow the ideal functionality to output the extracted bit b' if all the following conditions hold (and otherwise reject):
   - $(u_j' \oplus \mathsf{offset}_j)$ is a valid codeword for $j \in [\kappa]$.
   - (Y_i, Z_i) for all $i \in [\kappa^2]$ satisfy input-output frequency tests.

---

[a] If the nearest codeword is not unique, then $\mathsf{Sim}_\mathcal{S}$ commits to an arbitrary bit.

Fig. 5. Sender simulation strategy in Case 1(b).

indices. In order to obtain the view (y^*, z^*) in $1/\hat{\mathcal{Y}}_B \times \beta^\mathcal{F}_{z^*|y^*}$ fraction of indices, Bob should have used the input y^* to the functionality with probability at least $1/|\hat{\mathcal{Y}}_B|$.

Bob cannot equivocate outputs. Since Bob uses all inputs in $\hat{Y}_B$ with nearly the correct probability (except on $O(\kappa^{2/3})$ indices, then in the real and simulated worlds, he also obtains views in $\hat{V}_B$ with nearly the expected probability. Furthermore, he cannot obtain views not in $\mathcal{V}_B$ and pretend that they were in $\mathcal{V}_B$ except for $O(\kappa^{7/8})$ indices. Therefore, he cannot obtain views in $\mathcal{V}_B$ and pretend that they were not in $\mathcal{V}_B$ except for $O(\kappa^{7/8})$ indices, otherwise he will fail the frequency tests on the outputs.

To summarize,

The simulator $\mathsf{Sim}_{\mathcal{R}}$ does the following.

1. Commit Phase:
   (a) $\mathsf{Sim}_{\mathcal{R}}$ obtains commitments $c_1, c_2, \ldots c_{2\kappa^2}$ from $\mathcal{R}$.
   (b) $\mathsf{Sim}_{\mathcal{R}}$ obtains inputs $(X_1, X_2, \ldots X_{2\kappa^2})$ from $\mathcal{R}$ and emulates the functionality $\mathcal{F}$ honestly for $\mathcal{R}$ with inputs $(X_1, X_2, \ldots X_{2\kappa^2})$ and $(Y_1, Y_2, \ldots Y_{2\kappa^2})$.
   (c) Cut-and-Choose: $\mathsf{Sim}_{\mathcal{R}}$ obtains com_1 from $\mathcal{R}$. $\mathsf{Sim}_{\mathcal{R}}$ sends $r_2 \xleftarrow{\$} \{0,1\}^*$ to $\mathcal{R}$. $\mathcal{R}$ decommits to r_1 and sends subset $I \xleftarrow{\$} \binom{2\kappa^2}{[\kappa^2]}$ of the κ^2 indices. For all $i \in I$, $\mathsf{Sim}_{\mathcal{R}}$ obtains decommitments X_i and also the openings (X_i, W_i). $\mathsf{Sim}_{\mathcal{R}}$ aborts if the decommitments are not correct, or the inputs of $\mathcal{R}$ are not from a uniform distribution, or if (X_i, W_i) for $i \in I$ do not satisfy the consistency checks in Left-Statistical-Tests.
   (d) $\mathsf{Sim}_{\mathcal{R}}$ follows honest strategy to commit to a uniformly random bit $bit' \xleftarrow{\$} \{0,1\}$.

Fig. 6. Receiver simulation strategy in Case 1(b).

- For any input $y^* \in \hat{\mathcal{Y}}_B$, if Alice accepts the decommitment, Bob should have actually used the input to the functionality $\mathcal{F}$ in exactly $1/|\hat{\mathcal{Y}}_B|$ fraction of the places, except cheating in at most $\kappa^{2/3}$ indices.
- For any (extreme) view $(y^*, z^*) \in \widehat{V}_B$, Bob cannot have claimed to obtain (y^*, z^*) at specific indices unless he obtained the view in (y^*, z^*) at all but $O(\kappa^{7/8})$ of these indices.
- For any non-extreme view $(y^*, z^*) \in \widehat{V}_B$, Bob cannot have claimed to obtain (y^*, z^*) at specific indices unless he actually obtained *some* non-extreme view at all but $O(\kappa^{7/8})$ of these indices.

By using a code such that the minimum distance of the code $(\Omega(\kappa^{15/16}))$ is much larger than the number of positions where the sender can cheat as above $(O(\kappa^{7/8}))$, we guarantee that the sender is bound to his committed bit.

Specifically, the simulator computes the nearest codeword to the codeword extracted from the sender, and uses this to extract his committed bit. The sender cannot equivocate this codeword without cheating in $\Omega(\kappa^{15/16})$ views, and if he does so, his decommitment is not accepted except with probability at least $(1 - 2^{-c\kappa})$. This completes the proof of this lemma. x;w N2=3 0. If not, it aborts the protocol.

Sender Security (Statistical Hiding/Equivocability). It suffices to consider a dummy receiver $\mathcal{R}$ and malicious environment $\mathcal{Z}_{\mathcal{R}}$, such that the dummy receiver forwards all messages from $\mathcal{Z}_{\mathcal{R}}$ to the honest sender/simulator, and vice-versa. Without loss of generality, the malicious simulation strategy $\mathsf{Sim}_{\mathcal{R}}$ can be viewed to interact directly with $\mathcal{Z}_{\mathcal{R}}$. $\mathsf{Sim}_{\mathcal{R}}$ is described in Fig. 6.

Lemma 2. *There exists a constant c such that the simulation error for the malicious receiver is at most $2^{-c\kappa}$.*

Proof. Consider the use of the function f as a "channel", which accepts $x_{i,j}$ from Alice, $c_{i,j}$ from Bob, samples $(y_{i,j}, w_{i,j}, z_{i,j})$ and outputs $z_{i,j}$ to Bob, and $a_{i,j} \oplus c_{i,j}$ to Alice where $a_{i,j} = \phi_B(y_{i,j}, z_{i,j})$.

The cut-and-choose verification in Step 1(c) ensures that Alice uses (close to) a uniform distribution over her inputs. This is done by invoking Left-Statistical-Tests on committed inputs $X_1, x_2 \ldots X_{2\kappa^2}$ of Alice, and her claimed outputs $W_1, W_2, \ldots W_{2\kappa^2}$.

This test ensures that she obtains the view (x, w) that intersects with an extreme and a non-extreme view in $\widehat{V}_B$ in at least $\beta^{\mathcal{F}}_{|x,z} \kappa^2 - O(\kappa)$ invocations. At all these invocations, given her view, Alice has confusion about whether the corresponding view of Bob was extreme or non-extreme. Therefore, the views obtained by Alice act as a channel transmitting information about the corresponding views of Bob. It is that the capacity of this channel is a constant, that is less than 1.

Then we appeal to an extension of the weak converse of Shannon's Channel Coding Theorem (Imported Lemma 2) to argue that since the code has rate $1 - o(1)$, Alice errs in decoding each codeword with at least a constant probability. We need this extension of the (weak) converse of the channel coding theorem to handle that the facts that:

1. The receiver can adaptively choose the channel characteristic, by picking $y_{i,j}$ adaptively, and
2. Some of the channel characteristics that can be chosen include a noiseless channel, but the number of times such a characteristic can be used cannot be large (except with negligible probability). The reason this restriction can be enforced is because Alice's view intersects with views of Bob corresponding to characteristic index 0 and 1.

Then, applying the Leftover Hash Lemma, we get that for a universal hash function h, if Bob sends κ codewords over such a channel, the output of the hash function is at least $1 - 2^{-c\kappa}$ close to uniform. Thus, the simulation error is at most $2^{-c\kappa}$.

6 Case 2: Commitments

As before, let $\mathcal{V}_B$ denote the set of all extreme views of Bob and let $\widehat{Y}$ denote the set of all inputs of Bob that contain at least one extreme view, that is $\widehat{Y} := \{y \colon \exists z, \text{ such that } (y, z) \in \mathcal{V}_B\}$. Further, let $\widehat{V}_B$ denote the set of all Bob views corresponding to inputs in $\widehat{Y}$, that is $\widehat{V}_B = \{(y, z) \colon y \in \widehat{Y}\}$.

In this section, we demonstrate how to construct commitments from any function $\mathcal{F}$ for which the following is true: $\widehat{V}_B$ has no confusion, that is no two Bob views in $\widehat{V}_B$ intersect with the same Alice view. In other words, all views corresponding to all inputs $y \in \widehat{Y}$ are extreme and also disjoint.

First, we make the following basic observation about disjoint extreme views. Let $\mathcal{V}_B$ denote the set of extreme views of Bob. If there is no Alice view V_A which intersects two or more Bob views in $\mathcal{V}_B$, then each Bob view in $\mathcal{V}_B$ is in one-to-one correspondence with the equivalence class ϕ of Alice views. In particular, each Bob view (y, z) in $\mathcal{V}_B$ reveals $\phi(V_A)$ for any view V_A which the Bob view (y, z) intersects. Then, we note that for all inputs $\hat{y}$ in $\hat{Y}$, each output view $(\hat{y}, \hat{z})$ completely reveals the equivalence class ϕ of Alice views. The following lemma is imported from [16].

Imported Lemma 4 [16]. *Suppose $\hat{Y} \subseteq Y$ is a set of inputs, where each view $(\hat{y}, z)$ for each input $\hat{y} \in \hat{Y}$ is completely revealing about the equivalence class ϕ of Alice views. If some input $y^* \in Y \backslash \hat{Y}$ can be fully-mimicked by $\hat{Y}$ then y^* is a strictly redundant input.*

Note that if $y \notin Y_0$ can be mimicked by Y_0, it does not necessarily mean that y^* is redundant, because for redundancy there must exist a probabilistic mapping from $Y_0 \times Z$ to $y^* \times Z$. However, if Y_0 are all completely revealing about the equivalence class ϕ of Alice views, it can be shown that y^* is indeed redundant. For completeness, we repeat the formal proof from [16] in the full version.

Lemma 3. *Suppose $\hat{Y} \subseteq Y$ is a set of inputs, where each view $(\hat{y}, z)$ for each input $\hat{y} \in \hat{Y}$ is completely revealing about an equivalence class of Alice views. Let $Y' = Y \backslash \hat{Y}$. If every input in Y' can be mimicked using a probability distribution over other inputs that assigns constant non-zero weight to $\hat{Y}$, then every input in Y' is strictly redundant.*

Proof. Our proof follows along the lines of Gaussian elimination, removing one variable dependency at a time. As is the case with Gaussian elimination, the invariant we maintain is that the ith variable does not influence anything beyond the ith constraint. Our proof uses an inductive argument where the above invariant is iteratively maintained in each iteration.

Consider inputs $y^* \in Y'$ that can be mimicked using non-zero constant weight in $\hat{Y}$. We prove that if all inputs $y^* \in Y'$ can be mimicked using non-zero constant weight in $\hat{Y}$, then they can in fact be *fully* mimicked only by $\hat{Y}$. Once we prove this, we can invoke Imported Lemma 4 to prove that all such inputs y^* must be strictly redundant. We first set up some notation for the proof.

Notation. Let $Y' = \{y_1^*, y_2^*, \ldots y_\ell^*\}$ and $\hat{Y} = \{\hat{y}_1, \hat{y}_2, \ldots \hat{y}_{|\mathcal{Y}|-\ell}\}$, where $\ell < |\mathcal{Y}|$. Let M be an $\ell \times (\ell + 1)$ matrix whose entries are set such that for all $i \in [\ell]$, $y_i^* = \sum_{j \in [\ell]} (M_{i,j}) y_j^* + \sum_{j \in [|\mathcal{Y}|-\ell]} \mathsf{p}_{i,j} \hat{y}_j$. Then $M_{i,(\ell+1)} = \sum_{j \in [|\mathcal{Y}|-\ell]} \mathsf{p}_{i,j}$.

That is, for $(i, j) \in [\ell] \times [\ell]$, the row M_i denotes the probability distribution over inputs y_j^* used to mimic the input y_i^*. The entry $M_{i,\ell+1}$ denotes the total weight of inputs in $\hat{Y}$ assigned by the probability distribution, for mimicking the input y_i^*.

Transformation. Assume, contrary to the statement of the lemma, that every entry $M_{i,\ell+1}$ for all $i \in [1, \ell]$ is a non-zero constant, denote the ith such entry by c_i. We give a series of transformations on M, such that the resulting matrix M' has non-zero entries only in the $(\ell+1)^{th}$ column. This suffices to prove that all inputs can be fully mimicked using some distribution over inputs *only* in $\hat{Y}$, therefore proving the lemma.

We inductively set $M_{i,j} = 0$ for all $(i,j) \in [1,k] \times [1,k]$.

Base Case. In the base case, if $M_{1,1} = 0$, we are done.

Else we can rewrite the first row equations as:

$$y_1^* = \sum_{j \in [\ell]} (M_{i,j}) y_j^* + \sum_{j \in [|\mathcal{Y}|-\ell]} \mathfrak{p}_{i,j} \hat{y}_j \tag{1}$$

$$= M_{1,1} y_1^* + \sum_{j \in [2,\ell]} (M_{i,j}) y_j^* + \sum_{j \in [|\mathcal{Y}|-\ell]} \mathfrak{p}_{i,j} \hat{y}_j \tag{2}$$

$$y_1^* - M_{1,1} y_1^* = \sum_{j \in [2,\ell]} (M_{1,j}) y_j^* + \sum_{j \in [|\mathcal{Y}|-\ell]} \mathfrak{p}_{1,j} \hat{y}_j \tag{3}$$

$$y_1^*(1 - M_{1,1}) = \sum_{j \in [2,\ell]} (M_{1,j}) y_j^* + \sum_{j \in [|\mathcal{Y}|-\ell]} \mathfrak{p}_{1,j} \hat{y}_j \tag{4}$$

If $M_{1,1} \neq 0$, we rewrite this as:

$$y_1^* = \sum_{j \in [2,\ell]} \frac{M_{1,j}}{(1 - M_{1,1})} y_j^* + \sum_{j \in [|\mathcal{Y}|-\ell]} \frac{\mathfrak{p}_{1,j}}{(1 - M_{1,1})} \hat{y}_j \tag{5}$$

At the end of this manipulation, we have an equivalent system of equations represented by matrix M', such that $M'_{1,1} = 0$ and for all $j \in [\ell]$, $M'_{1,j} = \frac{M_{1,j}}{(1-M_{1,1})}$. In shorthand, we denote this by $M_{1,1} \to 0, M_{1,j} \to \frac{M_{1,j}}{(1-M_{1,1})}$ for $j \in [2,\ell]$.

Inductive Hypothesis. Assume that after the kth transformation, all entries $M_{i,j} = 0$ for $(i,j) \in [1,k] \times [1,k]$. This gives us, that for $i' \in [1,k]$, the probability distribution over other inputs for mimicking inputs $y_{i'}^*$ are of the form:

$$y_{i'}^* = \sum_{j \in [k+1,\ell]} (M_{k+1,j}) y_j^* + \sum_{j \in [|\mathcal{Y}|-\ell]} \mathfrak{p}_{k+1,j} \hat{y}_j \tag{6}$$

Induction Step. This consists of the following two transformations:

1. The probability distribution over other inputs for mimicking the input y_{k+1}^* can be written as:

$$y_{k+1}^* = \sum_{j \in [k]} (M_{k+1,j}) y_j^* + \sum_{j \in [k+1,\ell]} (M_{k+1,j}) y_j^* + \sum_{j \in [|\mathcal{Y}|-\ell]} \mathfrak{p}_{k+1,j} \hat{y}_j \tag{7}$$

Then, it is possible to substitute the first k terms in this equation using Eq. 6 to obtain another equation of the form:

$$y_{k+1}^* = \sum_{j \in [k+1,\ell]} (M'_{k+1,j}) y_j^* + \sum_{j \in [|\mathcal{Y}|-\ell]} \mathfrak{p}'_{k+1,j} \hat{y}_j, \tag{8}$$

for suitably modified values $(M'_{k+1,j})$ and $\mathfrak{p}'_{k+1,j}$.
At the end of this set of transformations, for all $j \in [k], M_{k+1,j} \to 0$ and for $j \in [k+1,\ell], M_{k+1,j} \to M'_{k+1,j}$.

2. Now, we can write the $(k+1)^{th}$ row Eq. 8 as:

$$y^*_{k+1} = (M'_{k+1,k+1})y^*_{k+1} + \sum_{j \in [k+2,\ell]} (M'_{k+1,j})y^*_j + \sum_{j \in [|\mathcal{Y}|-\ell]} \mathfrak{p}'_{k+1,j}\hat{y}_j \quad (9)$$

If $M_{k+1,k+1} \neq 0$, this can be rewritten as:

$$y^*_{k+1} = \sum_{j \in [k+2,\ell]} \frac{M'_{k+1,j}}{(1 - M'_{k+1,k+1})}y^*_j + \sum_{j \in [|\mathcal{Y}|-\ell]} \frac{\mathfrak{p}_{k+1,j}}{(1 - M'_{k+1,k+1})}\hat{y}_j \quad (10)$$

At the end of this transformation, the matrix entry $M'_{k+1,k+1} \to 0$.

3. Substituting Eq. 10 into the first k rows, we get that for $i' \in [1, \kappa + 1]$, the probability distribution over other inputs for mimicking inputs $y^*_{i'}$ are of the form:

$$y^*_{i'} = \sum_{j \in [k+1,\ell]} (M''_{k+1,j})y^*_j + \sum_{j \in [|\mathcal{Y}|-\ell]} \mathfrak{p}''_{k+1,j}\hat{y}_j \quad (11)$$

At the end of these transformations, we obtain an matrix $\bar{M}$ representing an equivalent system of equations, such that for all $(i,j) \in [\ell] \times [\ell]$, $\bar{M}_{i,j} = 0$ and $\bar{M}_{i,\ell+1} \neq 0$. This completes the proof of this lemma.

Now, suppose that for all inputs $y^* \in Y \backslash \hat{Y}$, Bob can mimic y^* using non-zero weight in $\hat{Y}$. Then, since Lemma 3 proves that all inputs $y^* \in Y \backslash \hat{Y}$ can be written as a convex linear combination of inputs entirely in $\hat{Y}$. This contradicts Imported Lemma 4. Since the functionalities we study only have a constant-sized domain, it is always easy to find such an input y^*.

6.1 Construction

The protocol is described in Fig. 7. Without loss of generality, we can assume that there exists a commitment protocol from Alice to Bob. We construct a commitment protocol with Bob as sender, and Alice as receiver.

6.2 Proof of Security

Receiver Security (Statistical Binding/Extractability). In the UC setting, it suffices to consider a dummy sender $\mathcal{S}$ and malicious environment $\mathcal{Z}_\mathcal{S}$, such that the dummy sender forwards all messages from $\mathcal{Z}_\mathcal{S}$ to the honest receiver/simulator, and vice-versa. Without loss of generality, the malicious simulation strategy $\mathsf{Sim}_\mathcal{S}$ can be viewed to interact directly with $\mathcal{Z}_\mathcal{S}$. $\mathsf{Sim}_\mathcal{S}$ is described in Fig. 8.

Lemma 4. *There exists a constant c such that the simulation error for the malicious sender is at most $2^{-c\kappa}$.*

Inputs: Sender S has input bit bit $\in \{0, 1\}$ and receiver R has no input.
Hybrid: $\mathcal{F}$ for non-simple function $\mathcal{F}$ where $\hat{Y}$ is not confusable, and y^* as defined above, exists. $\mathcal{F}$ provides commitments (Com) from Alice to Bob.
The protocol is presented in terms of a $(\kappa', \kappa' - \kappa'^{8/9}, \omega(\kappa^{15/16}))$-linear code C over the binary alphabet. (An explicit code is not necessary: the receiver can pick random $\omega(\kappa^{7/8})$ "parity checks" to construct the code and announce it to the sender.) The protocol is parameterized by κ.

1. Commit Phase:
   (a) For all $i \in [\kappa]$, R (Alice) picks inputs $(X_1, X_2, \ldots, X_{2\kappa^2})$ from a uniform distribution over $\mathcal{X}^{2\kappa^2}$. Alice also sends $\mathsf{Com}(X_1), \mathsf{Com}(X_2), \ldots \mathsf{Com}(X_{2\kappa^2})$ to Bob.
   (b) S (Bob) picks inputs $(Y_1, Y_2, \ldots, Y_{2\kappa^2})$ from a uniform distribution over $(\hat{Y} \cup y^*)^{2\kappa^2}$. R and S invoke $\mathcal{F}$, 2κ times, with inputs $(X_1, X_2, \ldots, X_{2\kappa^2})$ and $(Y_1, Y_2, \ldots, Y_{2\kappa^2})$ respectively.
   (c) Cut-and-Choose: R picks $r_1 \xleftarrow{\$} \{0, 1\}^*$ and sends $\mathsf{Com}(r)$ to S. S sends $r_2 \xleftarrow{\$} \{0, 1\}^*$ to R. R uses $r_1 \oplus r_2$ to pick a subset $I \xleftarrow{\$} \binom{2\kappa^2}{[\kappa^2]}$ of the κ^2 indices. R decommits to r_1. Furthermore, for all $i \in I$, R decommits to input X_i and also opens her view (X_i, W_i).
   S aborts if the decommitments are not correct, or the inputs of R are not close to a uniform distribution, or if (X_i, W_i) for $i \in I$ satisfy the consistency checks in the Left-Statistical-Tests.
   (d) S does the following for all $i \in [\kappa]$.
   - Construct the jth characteristic vector $\mathbf{u}_j$ such that for all $i \in [\kappa]$, $u_{j,i} = 0$ if and only if $Y_{j\kappa+i} \in \hat{Y}$, else $u_{j,i} = 1$.
   - Pick κ random codewords $\mathbf{c}_1, \mathbf{c}_2, \ldots \mathbf{c}_\kappa \in C^\kappa$. Pick $h \xleftarrow{\$} \mathcal{H}$, a universal hash function mapping $\{0, 1\}^{\kappa^2} \to \{0, 1\}$, and for $j \in [\kappa]$, compute $y = h(\mathbf{c}_1, \mathbf{c}_2, \ldots \mathbf{c}_\kappa) \oplus \text{bit}$, $\text{offset}_j = (\mathbf{c}_j \oplus \mathbf{u}_j)$. Send $(h, y, \text{offset}_1, \text{offset}_2, \ldots \text{offset}_\kappa)$ to R.
2. Reveal Phase:
   (a) S sets $b' = b$, $\mathbf{u}' = \mathbf{u}$ and sends b', u' to R as his opening. S also sends (Y_i, Z_i) for all $i \in [\kappa^2]$, to R.
   (b) R accepts if all the following conditions hold:
   - For $j \in [\kappa]$, $\mathbf{c}_j = \mathbf{u}'_j \oplus \text{offset}_j$, is a valid codeword.
   - $b' = h(\mathbf{c}_1, \mathbf{c}_2, \ldots \mathbf{c}_\kappa) \oplus y$.
   - Input-output frequency tests on (Y_i, Z_i) pass for all $i \in [\kappa^2]$.

Fig. 7. $\mathcal{F}_{\mathsf{com}}$ in Case 2.

Proof. The simulator performs Steps $1(a), (b)$ and (c) as per the honest receiver strategy, and also emulates the functionality $\mathcal{F}$ honestly for the sender. It remains to show that the unique bit b' extracted by the simulator equals the bit b committed by the sender Bob. The crux of this proof relies on the fact that the

The simulator Sim_S does the following.

1. Commit Phase:
   (a) Sim_S picks inputs $(X_1, X_2, \ldots, X_{2\kappa^2})$ uniformly from $\mathcal{X}^{2\kappa^2}$. Sim_S then commits to each of them using fresh randomness and sends $\mathsf{Com}(X_1), \mathsf{Com}(X_2), \ldots \mathsf{Com}(X_{2\kappa^2})$ to S. Note that Sim_S has the capability to equivocate these commitments.
   (b) Sim_S obtains inputs $(Y_1, Y_2, \ldots Y_{2\kappa^2})$ from S and emulates the functionality $\mathcal{F}$ honestly for S with inputs $(X_1, X_2, \ldots X_{2\kappa^2})$ and $(Y_1, Y_2, \ldots Y_{2\kappa^2})$.
   (c) Cut-and-Choose: Sim_S picks $r_1 \xleftarrow{\$} \{0,1\}^*$ and sends $\mathsf{com}_1 = \mathsf{Com}(r_1)$ to S. S sends $r_2 \xleftarrow{\$} \{0,1\}^*$ to Sim_S. Sim_S uses $(r_1' \oplus r_2)$ to pick subset $I \xleftarrow{\$} \binom{2\kappa^2}{[\kappa^2]}$ of the κ^2 indices. Sim_S decommits com_1 to r_1 and, for all $i \in I$, Sim_S decommits to input X_i and also opens the view (X_i, W_i). Set $S = [2\kappa^2] \setminus I$ and reorder the indices in S to $[\kappa^2]$.
   (d) Sim_S obtains $(h, y, \mathsf{offset}_j)$ for $j \in [\kappa]$ from S. It constructs characteristic vectors $\mathbf{u}_j$ such that for all $i \in S$, $u_{j,i} = 0$ if and only if $Y_{j\kappa+i} \in \hat{Y}$, else $u_i = 1$. It then computes $\tilde{c}_j = \mathbf{u}_j \oplus \mathsf{offset}_j$, sets c_j' to be the nearest codeword[a] to $\tilde{c}_j$, and sets bit $b' = y \oplus h(c_1', c_2', \ldots c_\kappa')$.
2. Reveal Phase:
   (a) Obtain $b', u_1', u_2', \ldots u_\kappa', (Y_i, Z_i)$ for all $i \in [\kappa^2]$ from S as his opening.
   (b) Allow the ideal functionality to output the extracted bit b' if all the following conditions hold (and otherwise reject):
     − $(u_j' \oplus \mathsf{offset}_j)$ is a valid codeword for $j \in [\kappa]$.
     − (Y_i, Z_i) for all $i \in [\kappa^2]$ satisfy the input-output frequency tests in the Right-Statistical-Tests.

---
[a] If the nearest codeword is not unique, then Sim_S commits to an arbitrary bit.

Fig. 8. Sender simulation strategy in Case 2.

protocol requires the sender to use all his extreme views, and some non-extreme views; and on the minimum distance of the code used.

Bob cannot claim non-extreme views to be extreme. Equivalently, Bob cannot claim an input outside $\hat{Y}_B$ to be an input inside $\hat{Y}_B$. In the opening made by Bob, consider the positions where Bob claimed his view to be $(y_i, z_i) = (y^*, z^*) \in \mathcal{V}_B$, such that the equivalence class of this view $\phi_B(y^*, z^*) = \Phi$. Consider the fraction of these positions where the actual view of Bob (x', w') such that $\phi_B(y', z') \neq \Phi$.

In these positions, the expected view of Alice is given by a linear combination of the columns $\beta^{\mathcal{F}}|_{(y', z')}$ (with coordinates scaled appropriately). If this linear combination is not close to the vector $\beta^{\mathcal{F}}|_{(y^*, z^*)}$ (scaled appropriately) then with all but negligible probability, the opening will not be accepted by the receiver. On the other hand, if the linear combination is close to $\beta^{\mathcal{F}}|_{(y^*, z^*)}$, since $\beta^{\mathcal{F}}|_{(y^*, z^*)}$

is outside the linear span of other $\beta^{\mathcal{F}}|_{(y',z')}$ with $\phi_B(y'z') \neq \phi_B(y^*, z^*)$, only at a small number (sub-linear fraction, say $\kappa^{2/3}$) of places can Bob open to (y^*, z^*) but have had an actual view (y', z'). Thus, extreme views cannot be claimed to be obtained as a result of using inputs which exclusively yield non-extreme views.

Bob cannot claim an input inside $\widehat{Y}_B$ to be outside $\widehat{Y}_B$. By Lemma 3, we also know that y^* cannot be mimicked with any non-zero weight in $\hat{Y}$, without getting caught by the receiver in the Right-Statistical-Tests. Thus, it is not possible to use inputs in $\hat{Y}$ and equivocate them to y^*. This gives that the sender cannot equivocate at more that $O(\kappa^{2/3})$ indices.

Bob cannot equivocate. By using a code such that the minimum distance of the code $(\Omega(\kappa^{3/4}))$ is much larger than the number of positions where the sender can cheat in one of the two situations above $(O(\kappa^{2/3})$, we guarantee that the sender is bound to his committed bit.

Specifically, the simulator computes the nearest codeword to the codeword extracted from the sender, and uses this to extract his committed bit. The sender cannot equivocate this codeword without cheating in $\Omega(\kappa^{3/4})$ views, and if he does so, his decommitment is not accepted except with probability at least $(1 - 2^{-c\kappa})$. This completes the proof of this lemma.

Sender Security (Statistical Hiding/Equivocability). It suffices to consider a dummy receiver $\mathcal{R}$ and malicious environment $\mathcal{Z}_{\mathcal{R}}$, such that the dummy receiver forwards all messages from $\mathcal{Z}_{\mathcal{R}}$ to the honest sender/simulator, and vice-versa. Without loss of generality, the malicious simulation strategy $\mathsf{Sim}_{\mathcal{R}}$ can be viewed to interact directly with $\mathcal{Z}_{\mathcal{R}}$. $\mathsf{Sim}_{\mathcal{R}}$ is described in Fig. 9.

Lemma 5. *There exists a constant c such that the simulation error for the malicious receiver is at most $2^{-c\kappa}$.*

Proof. Consider the use of the function f as a "channel", which accepts $x_{i,j}$ from Alice, $c_{i,j}$ from Bob, samples $(y_{i,j}, w_{i,j}, z_{i,j})$ and outputs $z_{i,j}$ to Bob, and $a_{i,j} \oplus c_{i,j}$ to Alice where $a_{i,j} = \phi_B(y_{i,j}, z_{i,j})$.

The cut-and-choose verification in Step 1(c) ensures that Alice uses (close to) a uniform distribution over her inputs. Then, she obtains the view (x, w) that intersects with an extreme and a non-extreme view in $\widehat{V}_B$ in at least a constant fraction of the invocations. At all these invocations, given her view, Alice has confusion about whether the corresponding view of Bob was extreme of non-extreme. Formally, we can show that the capacity of the above channel is a constant, that is less than 1.

Then we appeal to an extension of the weak converse of Shannon's Channel Coding Theorem (Imported Lemma 2) to argue that since the code has rate 1, Alice errs in decoding each codeword with at least a constant probability. We need this extension of the (weak) converse of the channel coding theorem to handle that the facts that:

The simulator $\mathsf{Sim}_\mathcal{R}$ does the following.

1. Commit Phase:
   (a) $\mathsf{Sim}_\mathcal{R}$ obtains commitments $c_1, c_2, \ldots c_{2\kappa^2}$ from $\mathcal{R}$.
   (b) $\mathsf{Sim}_\mathcal{R}$ obtains inputs $(X_1, X_2, \ldots X_{2\kappa^2})$ from $\mathcal{R}$ and emulates the functionality $\mathcal{F}$ honestly for $\mathcal{R}$ with inputs $(X_1, X_2, \ldots X_{2\kappa^2})$ and $(Y_1, Y_2, \ldots Y_{2\kappa^2})$.
   (c) Cut-and-Choose: $\mathsf{Sim}_\mathcal{R}$ obtains com_1 from $\mathcal{R}$. $\mathsf{Sim}_\mathcal{R}$ sends $r_2 \xleftarrow{\$} \{0,1\}^*$ to $\mathcal{R}$. $\mathcal{R}$ decommits to r_1 and sends subset $I \xleftarrow{\$} \binom{2\kappa^2}{[\kappa^2]}$ of the κ^2 indices. For all $i \in I$, $\mathsf{Sim}_\mathcal{R}$ obtains decommitments X_i and also the openings (X_i, W_i). $\mathsf{Sim}_\mathcal{R}$ aborts if the decommitments are not correct, or the inputs of $\mathcal{R}$ are not from a uniform distribution, or if (X_i, W_i) for $i \in I$ do not satisfy the consistency checks in Left-Statistical-Tests.
   (d) $\mathsf{Sim}_\mathcal{R}$ follows honest sender strategy to commit to a uniformly random bit $\mathsf{bit}' \xleftarrow{\$} \{0,1\}^*$.

Fig. 9. Receiver simulation strategy in Case 2.

1. The receiver can adaptively choose the channel characteristic, by picking $y_{i,j}$ adaptively, and
2. Some of the channel characteristics that can be chosen include a noiseless channel, but the number of times such a characteristic can be used cannot be large (except with negligible probability). The reason this restriction can be enforced is because Alice's view intersects with views of Bob corresponding to characteristic index 0 and 1.

Then, applying the Leftover Hash Lemma, we get that for a universal hash function h, if Bob sends κ codewords over such a channel, the output of the hash function is at least $1 - 2^{-c\kappa}$ close to uniform. Thus, the simulation error is at most $2^{-c\kappa}$.

References

1. Beimel, A., Malkin, T., Micali, S.: The all-or-nothing nature of two-party secure computation. In: Wiener, M.J. (ed.) CRYPTO 1999. LNCS, vol. 1666, pp. 80–97. Springer, Heidelberg (1999)
2. Brassard, G., Crépeau, C., Robert, J.M.: Information theoretic reductions among disclosure problems. In: 27th Annual Symposium on Foundations of Computer Science, Toronto, Ontario, Canada, 27–29 October 1986, pp. 168–173. IEEE Computer Society Press (1986)
3. Canetti, R.: Security and composition of multiparty cryptographic protocols. J. Cryptol. 13(1), 143–202 (2000)

4. Canetti, R., Lindell, Y., Ostrovsky, R., Sahai, A.: Universally composable two-party and multi-party secure computation. In: 34th Annual ACM Symposium on Theory of Computing, Montréal, Québec, Canada, 19–21 May 2002, pp. 494–503. ACM Press (2002)
5. Crépeau, C., Kilian, J.: Achieving oblivious transfer using weakened security assumptions (extended abstract). In: 29th Annual Symposium on Foundations of Computer Science, White Plains, New York, 24–26 October 1988, pp. 42–52. IEEE Computer Society Press (1988)
6. Crépeau, C., Morozov, K., Wolf, S.: Efficient unconditional oblivious transfer from almost any noisy channel. In: Blundo, C., Cimato, S. (eds.) SCN 2004. LNCS, vol. 3352, pp. 47–59. Springer, Heidelberg (2005)
7. Crépeau, C., Sántha, M.: On the reversibility of oblivious transfer. In: Davies, D.W. (ed.) EUROCRYPT 1991. LNCS, vol. 547, pp. 106–113. Springer, Heidelberg (1991)
8. Dodis, Y., Ostrovsky, R., Reyzin, L., Smith, A.: Fuzzy extractors: how to generate strong keys from biometrics and other noisy data. SIAM J. Comput. **38**(1), 97–139 (2008). http://dx.org/10.1137/060651380
9. Goldreich, O., Micali, S., Wigderson, A.: How to play any mental game or a completeness theorem for protocols with honest majority. In: Aho, A. (ed.) 19th Annual ACM Symposium on Theory of Computing, New York City, New York, USA, 25–27 May 1987, pp. 218–229 (1987)
10. Goldreich, O., Vainish, R.: How to solve any protocol problem - an efficiency improvement. In: Pomerance, C. (ed.) CRYPTO 1987. LNCS, vol. 293, pp. 73–86. Springer, Heidelberg (1988)
11. Haber, S., Micali, S.: Unpublished manuscript (1986)
12. Ishai, Y., Prabhakaran, M., Sahai, A.: Founding cryptography on oblivious transfer - efficiently. In: Wagner, D. (ed.) CRYPTO 2008. LNCS, vol. 5157, pp. 572–591. Springer, Heidelberg (2008)
13. Kilian, J.: Founding cryptography on oblivious transfer. In: 20th Annual ACM Symposium on Theory of Computing, Chicago, Illinois, USA, 2–4 May 1988, pp. 20–31. ACM Press (1988)
14. Kilian, J.: A general completeness theorem for two-party games. In: 23rd Annual ACM Symposium on Theory of Computing, New Orleans, Louisiana, USA, 6–8 May 1991, pp. 553–560. ACM Press (1991)
15. Kilian, J.: More general completeness theorems for secure two-party computation. In: 32nd Annual ACM Symposium on Theory of Computing, Portland, Oregon, USA, 21–23 May 2000, pp. 316–324. ACM Press (2000)
16. Kraschewski, D., Maji, H.K., Prabhakaran, M., Sahai, A.: A full characterization of completeness for two-party randomized function evaluation. In: Nguyen, P.Q., Oswald, E. (eds.) EUROCRYPT 2014. LNCS, vol. 8441, pp. 659–676. Springer, Heidelberg (2014)
17. Kraschewski, D., Müller-Quade, J.: Completeness theorems with constructive proofs for finite deterministic 2-party functions. In: Ishai, Y. (ed.) TCC 2011. LNCS, vol. 6597, pp. 364–381. Springer, Heidelberg (2011)
18. Lindell, Y.: Adaptively secure two-party computation with erasures. Cryptology ePrint Archive, Report 2009/031 (2009). http://eprint.iacr.org/2009/031
19. Maji, H.K., Prabhakaran, M., Rosulek, M.: A unified characterization of completeness and triviality for secure function evaluation. In: Galbraith, S.D., Nandi, M. (eds.) INDOCRYPT 2012. LNCS, vol. 7668, pp. 40–59. Springer, Heidelberg (2012)
20. Rabin, M.: How to exchange secrets by oblivious transfer. Technical Report TR-81, Harvard Aiken Computation Laboratory (1981)

242   D. Khurana et al.

21. Wiesner, S.: Conjugate coding. SIGACT News **15**, 78–88. http://doi.acm.org/10. 1145/1008908.1008920
22. Wolf, S., Wullschleger, J.: Oblivious transfer is symmetric. In: Vaudenay, S. (ed.) EUROCRYPT 2006. LNCS, vol. 4004, pp. 222–232. Springer, Heidelberg (2006)
23. Yao, A.C.C.: Protocols for secure computations (extended abstract). In: 23rd Annual Symposium on Foundations of Computer Science, Chicago, Illinois, 3–5 November 1982, pp. 160–164. IEEE Computer Society Press (1982)

On the Power of Hierarchical Identity-Based Encryption

Mohammad Mahmoody[✉] and Ameer Mohammed

University of Virginia, Charlottesville, USA
mohammad@cs.virginia.edu, am8zv@virginia.edu

Abstract. We prove that there is no fully black-box construction of collision-resistant hash functions (CRH) from hierarchical identity-based encryption (HIBE) with arbitrary polynomial number of identity levels. To the best of our knowledge this is the first limitation proved for HIBE. As a corollary, we obtain a series of separations that are not directly about HIBE or CRH but are interesting on their own right. Namely, we show that primitives such as IBE and CCA-secure public-key encryption cannot be used in a black-box way to construct fully homomorphic encryption or any primitive that implies CRH in a black-box way.

Our proof relies on the reconstruction paradigm of Gennaro and Trevisan (FOCS 2000) and Haitner et al. (FOCS 2007) and extends their techniques for one-way and trapdoor permutations to the setting of HIBE. A main technical challenge in the proof of our separation stems from the *adaptivity* of the HIBE adversary who is allowed to obtain keys for different identities *before* she selects the attacked identity. Our main technical contribution is to develop compression/reconstruction techniques that can be achieved relative to such adaptive attackers.

Keywords: Hierarchical identity-based encryption · Collision resistant hashing · Homomorphic encryption · Black-box separations

1 Introduction

Modern cryptography is based on well-defined hardness assumptions and formal proofs of security. For example, a sequence [19, 21, 24, 27, 31, 34, 35, 41, 48] of fundamental work has led to constructions of private key encryption, pseudorandom generators, pseudorandom functions and permutations, bit commitment, and digital signatures solely based on the assumption that one-way function exists. On the other hand, cryptographic primitives such as public key encryption, oblivious transfer, and key agreement that are perhaps more "structured" are not known to be implied by one-way functions alone. The goal of founding cryptography on *minimal* assumptions has led to an extensive study of the power and limitation of cryptographic primitives. As a result, for every (newly

M. Mahmoody — Supported by NSF CAREER award CCF-1350939.

A. Mohammed — Supported by University of Kuwait.

M. Fischlin and J.-S. Coron (Eds.): EUROCRYPT 2016, Part II, LNCS 9666, pp. 243–272, 2016.
DOI: 10.1007/978-3-662-49896-5_9

introduced) primitive $\mathcal{P}$, researchers aim to answer two questions: (1) What are the minimal computational assumptions necessary for constructing $\mathcal{P}$? (2) What are the power and limitation of $\mathcal{P}$ as a computational assumption? In particular, what other cryptographic primitives could be constructed from $\mathcal{P}$?

Hierarchical Identity-Based Encryption. In this work, we study the limitations of the power of *identity based encryption* and its *hierarchical* variant as strong forms of encryption. A traditional public-key encryption scheme allows Alice to send messages to Bob privately over a public channel knowing only Bob's public key. An identity-based encryption scheme does not require Alice to know a specific individual's public-key and allows Alice to encrypt messages for Bob by *only* knowing Bob's identity and a single master public key that is the same for all identities. A decryption key for Bob can also be generated using the (single) master secret key and Bob's (public) identity. The notion of IBE was first proposed by Shamir [44]. Later on many papers did try to construct IBE schemes (e.g., [33] presented an scheme with a rather slightly inefficient key generation) but the first fully functional IBE was first constructed by Boneh and Franklin [7] based on assumptions about bilinear maps.

A *hierarchical* identity based encryption scheme (see Definition 8) takes the versatility of IBE to the next level: each identity's decryption key can be considered as a master secret key on its own to generate decryption keys for "sub-identities". So as the name suggests it allows delegating encrypting power in a hierarchy of identities. HIBE was first defined and constructed in [17,25] where the security was based on the hardness of the Bilinear Diffie-Hellman problem in the random oracle model. Later, Boneh and Boyen [4] proposed a more efficient HIBE scheme in the standard (plain) model but only achieved selective-ID security. This construction was further improved in [5], and in [1] Agrawal et al. showed how to construct fully-secure efficient HIBE based on the learning with errors (LWE) assumption [39].

It was shown by [6] that IBE can be used to obtain CCA secure public-key encryption in a black-box way, and Gentry et al. [16] showed a perhaps surprising application of IBE to garbling RAM programs. Canetti, Halevi, and Katz [10] showed how to achieve forward-secure encryption scheme from IBE and HIBE. More recently, Naor and Ziv [37] used HIBE in their construction of Primary-Secondary-Resolver Membership Proof Systems. In this work, we study the following question about HIBE as a cryptographic primitive/assumption:

What are the limitations of the power of IBE/HIBE? Namely, what crypto primitives can or cannot be constructed from IBE/HIBE?

The Black-Box Framework. We study our main question in the black-box framework of [28]. Impagliazzo and Rudich [28] were the first to develop such framework which enabled the possibility of ruling out the existence of an important and powerful class of reductions between primitives. The work of Reingold, Trevisan, and Vadhan [40] formalized this framework further and established a taxonomy for the field. Intuitively, many cryptographic constructions are black-box in the sense that (1) the algorithm implementing the construction of $\mathcal{Q}$ uses another cryptographic primitive $\mathcal{P}$ (e.g., one-way functions) only as an oracle, and (2) the *security reduction* takes any adversary **Adv** who breaks $\mathcal{Q}^{\mathcal{P}}$ as an oracle and turns it into an attack against $\mathcal{P}$. Black-box constructions are also considered important due to their (typical) efficiency advantage over their non-black-box counterparts. Following the work of [28] a sequence of results known as "black-box separations" emerged in which limitation of the power of cryptographic primitives are proved with respect to black-box constructions/reductions. In this work we study the power of *fully* black-box reductions as defined in [40] which is the most common form of black-box constructions used in cryptography.

1.1   Our Results

In this work we prove a black box separation result for hierarchical IBE which, to the best of our knowledge, is the first such result. Namely, we show that there is no fully black box construction of collision-resistant hash functions (CRH) from HIBE schemes.

Theorem 1 (Main Theorem). *There is no black-box construction of collision-resistant hash functions from hierarchical identity-based encryption with an arbitrary polynomial number of levels.*

Separating Homomorphic Encryption from HIBE. A primary corollary of our main theorem above is that HIBE does not imply fully homomorphic encryption (FHE) in a black-box way. This follows from the result of Ishai, Kushilevitz, and Ostrovsky [29], where they show that FHE implies (private-coin) CRH in a black-box way. Note that while their result achieves only private-coin CRH, which does not in general imply public-coin CRH [26], our separating oracle is oblivious to whether the CRH is public-coin or not so the proof works either way. Furthermore, Theorem 1 together with the result of [6] implies that CCA secure public-key encryption does not imply CRH or FHE in a black-box way. Since CCA-secure public key encryption can be constructed from trapdoor permutations one might think that this separation follows form the work of Haitner et al. [23] who ruled out black-box constructions of CRH from trapdoor-permutations. However the construction of CCA secure encryption from TDP is *non-black-box* [30,36,42] due to its use of non-interactive zero knowledge proofs (Fig. 1).[1]

The work of [29] provides several other primitives (than FHE) whose existence also implies CRH in a black-box way. These primitives include: one-round private information retrieval (PIR) protocols as well as homomorphic one-way commitments. As a direct corollary, our main theorem above extends to all these

---

[1] We shall also note that even the techniques behind the proof of [23] do not seem to extend to a separation of CCA secure encryption from TDPs, even though CCA secure encryption could be constructed from TDP and random oracles [3]. The reason is that the "collision finding oracle" (see Definition 9) of [23,45] prevents the random TDP oracle from being independent of the other subroutines of the oracle to be used as a random oracle.

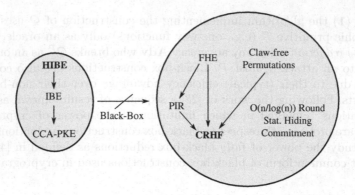

Fig. 1. Our works shows that primitives on the left side do not imply any of the primitives on the right side in a black-box way.

primitives as well. Our separations also holds when the goal of the construction is to achieve statistically hiding commitment schemes with $o(n/\log n)$ round complexity (where n is the security parameter).[2] However, for simplicity of the presentation here we focus on the simpler primitive of CRH.

Previous Separations Regarding IBE. In this work we present the first black-box separation for *hierarchical* IBE. However previous separations about IBE are known. The work of Boneh et al. [8] proved that there is no black-box construction of IBE from trapdoor permutations. Furthermore, Goyal et al. [22] proved the first *limitation* for IBE and separated it from fuzzy IBE schemes [43]. Our techniques are quite different from those of [22]. At a high level, [22] uses a random IBE oracle and aims at breaking any fuzzy IBE construction relative to this oracle with only a polynomial number of queries to this oracle, while our approach uses a random HIBE oracle together with the collision-finding oracle of [23,45] (see Sect. 2.4). The challenging part of our proof is to show that the random HIBE oracle remains secure in the presence of the collision-finding oracle rather than finding a poly-query attack to break CRH of the oracle.

Comparison with [2]. A corollary of our Theorem 1 for the case of IBE could be also derived from the concurrent and independent beautiful work of [2]. Asharov and Segev [2] showed that there is no black-box construction of CRH from indistinguishability obfuscation (iO) and one-way functions, even if the iO primitive could be applied to OWF in a non-black-box way. Technically, they prove this result using an oracle O with OWF and iO subroutines while the iO sub-oracle could accept circuits with OWF gates in them. On the other hand Waters [47] showed how to construct general (adaptively secure) functional encryption [9] from iO and OWFs. Note that IBE (and not HIBE) is a special case of func-

---

[2] Statistically hiding commitment is known to be implied by CRH [12,35] and so proving separation for statistically hiding commitment is stronger than a similar result for CRH.

tional encryption. The construction of [47] also uses OWF in a non-black-box way but only for applying the iO to circuits with OWF gates, therefore it can be implemented using the separating oracle of [2]. In other words, [2] shows how to construct an oracle relative to which the functional encryption (and thus IBE) construction of [47] exists, but relative to the same oracle no CRH is secure.

As we will describe below, our proof is quite different from that of [2] and our compression techniques do not have a counterpart in [2]. While the result of [2] handles richer class of functional encryption schemes beyond IBE, our result extends to arbitrary levels of hierarchies.

1.2   Technical Overview

In this subsection we give an overview of our techniques in the proof of our main theorem. Here we focus on the case of IBE (i.e., HIBE with one hierarchy) since some of the main challenges already show up for the (adaptively secure) IBE, and after resolving them we can scale the proof up easily with the number of hierarchy levels.

We first need to recall tools and techniques from Haitner et al. [15,23] which we use as building blocks in our proof. Our starting point for the proof of Theorem 1 is the result of Haitner et al. [23] that separates CRH from trapdoor permutations. In their separation, they employed an oracle $O = (T, \mathrm{Sam}^T)$ where T is a random trapdoor permutation and Sam^T is (a variant of) the collision finding oracle of Simon [45] (see Definition 9) which returns a collision (x, x') for any given input circuit C with T gates. It is easy to see that relative to O there is no secure construction of CRH that only uses T gates. To derive the separation it is enough to show that relative to O, T remains a secure TDP (see Lemma 10 for a proof). The main technical argument in [23] was to show that a random trapdoor permutation T remains secure in the presence of Sam^T. We will sketch some components of this proof first, and then we will discuss how to use these tools in addition to our new compression lemmas to prove our result.

Hardness of Random Permutations in the Presense of Sam. The seminal work of [14] showed that a random permutation $\pi \colon \{0,1\}^n \mapsto \{0,1\}^n$ is hard to invert by introducing the compresson/reconstruction technique. For any supposed adversary A^π who inverts an $1/\mathrm{poly}(n)$ fraction of $\{0,1\}^n$, [14] showed how to represent the permutation π with $\Omega(n)$ bits *fewer* than it is takes to represent a random permutation. That would be sufficient to show that the probability that π is "easy" for such A is at most exponentially small $2^{-\Omega(n)}$.[3]

The work of [23] extended the result of [14] by showing that the hardness of random permutations π holds, even if we allow the adversary access the collision finding oracle Sam^π that takes as input any circuit C with π gates (that shrinks its input) and returns a collision for it (chosen from a specific distribution). This immediately implies a (fully) black-box separation for CRH from TDP since

---

[3] In fact [15] achieves exponential compression and *doubly* exponentially small probability of success for a fixed A. This lets them do a union bound over all poly-sized circuits A when π is chosen at random.

for every black-box construction $\mathcal{H}$ it gives an oracle (π, Sam^π) relative to which secure TDP exists (i.e., π) but the implementation $\mathcal{H}$ of CRH using π is insecure. This is indeed sufficient to derive the fully black-box separation [18,40].

Extension to TDPs. [23] extended their result to trapdoor permutations using a similar argument to that of [13] who also proved the hardness of random trapdoor permutation as follows. Let $T = (G, F, F^{-1})$ be a random trapdoor permutation in which G is a random permutation over $\{0,1\}^n$, $F[pk](\cdot)$ is a random permutation for every $pk \in \{0,1\}^n$ and $F^{-1}[sk](\cdot)$ be the inverse of $F[pk](\cdot)$ whenever $G(sk) = pk$. Let $A^{(T,\mathrm{Sam}^T)}$ be an adversary who inverts a random y under a random public key pk with probability $1/\mathrm{poly}(n)$. Then A is implicitly doing either of the following. **Case 1:** A finds $sk = G^{-1}(pk)$ for the given random pk. In this case A has inverted the random permutation G. **Case 2:** A succeeds at finding $x = F^{-1}[sk](y)$ *without* asking $G(sk) = pk$. Since A does not ask $G(sk) = pk$ the permutation $F[pk](\cdot)$ would be a random permutation inverted by A. In both cases the existence of A leads to an efficient-query algorithm inverting random permutations in the presence of Sam. However, as we saw this is impossible.

Beyond TDP: The Case of IBE and HIBE. First, let us recall IBE informally (see Definition 8 for a formal definition). An *identity-based encryption* (IBE) scheme for security parameter n and messages $\{0,1\}^n$ is defined using PPTs (Setup, KeyGen, Enc, Dec):

- Setup(MSK) = MPK takes as input a random master secret key and generates the master public key MPK.
- KeyGen[MSK]$(id) = td$ generates a trapdoor td for a given identity id.
- Enc[MPK, id]$(x) = y$ encrypts x under identity id and outputs ciphertext y.
- Dec[MPK, id, td]$(y) = x$ decrypts y using the trapdoor td and gets back x.

Security Variants. The basic CPA security of an IBE requires that no adversary can distinguish between the encryptions of any two messages of his choice even if he is allowed to *choose* the challenge identity id^* of the encryption after getting trapdoors for identities $id \neq id^*$ of his choice. Here we first focus on the basic CPA security for IBE, but our full proof handles the stronger notion of CCA secure IBE/HIBE. For simplicity in this exposition we assume that the IBE's adversary aims at decrypting a randomly selected message x (encrypted under challenge identity id^*).[4]

As in the case of TDPs, here our goal is to design an oracle $O = (U, \mathrm{Sam}^U)$ such that U implements IBE/HIBE in way that it remains secure even in the presence of the Sam^U oracle. We first start by a direct generalization of the above arguments to oracles with more than one level of trapdoor, and then will see what aspects of the security of IBE will require us to change our oracle and the proof of hardness accordingly. Our first try is to use hierarchical random trapdoor

---

[4] By the Goldreich Levin lemma [20] these two notions of security are equivalent in a black-box way.

permutations to implement, but to prove the full fledged adaptive security of IBE/HIBE we will change this oracle to use injective random functions.

Hierarchical Random Permutations. We first use a direct adaptation from random TDPs to implement our IBE oracle. Let $U = (S, G, F, F^{-1})$ be an oracle that we call a "random IBE oracle" defined and used as follows. S implements the setup and is a random permutation over $\{0, 1\}^n$ that maps master secret key MSK to master public key MPK. $G[\text{MSK}](\cdot)$ implements the trapdoor generation and is a random permutation over $\{0, 1\}^n$ that maps identities to their trapdoor. Finally $F[\text{MPK}, id](\cdot)$ and $F^{-1}[\text{MPK}, td])$ are random permutations over $\{0, 1\}^n$ that are used for encryption and decryption.

Our goal here would be to show that any adversary A who breaks the IBE implemented by U by accessing $O = (U, \text{Sam}^U)$ will be forced to invert some random permutation (to derive the contradiction). Let us list the possible cases through which an adversary can win the security game:

- **Case 1:** A finds MSK $= S^{-1}(\text{MPK})$.
- **Case 2:** A does not find MSK, but it finds the trapdoor td^* for id^*.
- **Case 3:** A does not find the trapdoor for identity id^*, but it still manages to invert the challenge ciphertext y.

Problem with **Case 2**: *Adaptivity of Adversary.* Similarly to the case of random TDP, **Case 1** and **Case 3** will imply that A is indeed inverting a random permutation, which can only happen with negligible probability. However, **Case 2** does not imply so for two reasons:

1. While inverting id^* to find its trapdoor, the adversary A is allowed to obtain trapdoors $td \neq td^*$ for other identities $id \neq id^*$.
2. The adversary gets to *choose* id^* as opposed to being given a random one.

In the following we will show how to resolve the two issues above. First we will ignore the second issue above by working with a weaker security definition for IBE in which the adversary does not choose id^* but rather is given a random one (but still gets to ask the trapdoor for other identities). And then we will describe our new oracle to handle both issues above.

Attacks Against Random *Challenge Identity.* Note that we are focusing on the scenario in which the adversary A^{U, Sam^U} breaks the IBE by causing Case 2 to occur with non-negligible probability. We are also assuming, for now, that the adversary does *not* choose an identity of his choice, bur rather finds the trapdoor of a *random* identity. Our goal is to reduce this case to when (a variant of) A is essentially inverting a random permutation using Sam oracle. Think of the permutation $G^{-1}[\text{MPK}](\cdot)$ as the inverse of $G[\text{MSK}]$. It can be verified that when A succeeds in Case 2, it is in fact inverting G^{-1} on a random point while he is able to ask some "inversion queries" to G^{-1} before "inverting" the random challenge id^* on its own.

To rule out successful attackers against random challenge identity we generalize the results of [14] for hardness of random permutations π to the setting in which the adversary is allowed to ask inversion queries to $\pi^{-1}(\cdot)$ in addition to direct queries to $\pi(\cdot)$. Namely, we show that a random permutation, is *adaptively* one-way [38] even in the presence of the Sam oracle.

Lemma 2 (Informal: Adaptive/CCA hardness of random permutations in the presence of Sam**).** *For any permutation π over $\{0,1\}^n$, define $O = (\pi, \pi^{-1})$ to be an oracle giving access to π in both directions. Let A be a poly(n)-query circuit that accepts a challenge $y^* \in \{0,1\}^n$ and whose goal is to find $x = \pi^{-1}(y^*)$ using O while all queries to O are allowed except for $\pi^{-1}(y^*)$. Then, with overwhelming probability over the choice of random permutation π it holds that: the probability of success for A is negligible, even if it is allowed to call the Sam^O oracle.*

We show that the argument of [15] for the case of "basic" (i.e., non-CCA) attackers does indeed extend to the CCA setting described in Lemma 2. We also show that the $\mathrm{Sam}^{(\pi, \pi^{-1})}$ oracle will not help the attacker much, using the techniques of [23] (see Sect. 3 for a proof of this more general setting and the full version of this paper [32] for a simpler variant restricted to the setting where Sam oracle does not exist).

Fully Adaptive Attacks. The most challenging aspect of our proof for handling general IBE attacks stems from the fact that here the adversary is allowed to *choose* the challenge identity. Note that such an adversary does not necessarily invert a "randomly sampled" identity id^* to win according to Case 2, and it has full control over id^*. Unfortunately, we are not able to prove a variant of Lemma 2 in which the adversary chooses y^* itself since as soon as we sample and fix the permutation π, there is always a *trivial* adversary whose description depends on π and "knows" a pair $\pi(x) = y$ (through non-uniformity) and that proposes y as the challenge and inverts it!

Compression Amplification. The way we can rule out such attacks in the context of IBE/HIBE (for Case 2) is by relying on the fact that the adversary needs to succeed for a randomly given master public-key out of his choice. The idea is that even though one cannot prove a strong hardness of inverting π when the adversary chooses the inversion point y, we can still compress the description of π by $\approx \Omega(n)$ bits. Even though the above compression for the oracle achieved in Case 2 as described above is quite small (i.e., $\Omega(n)$ as opposed to the needed $2^{\Omega(n)}$ bits of compression) we show how to *amplify* this compression for our random IBE oracle. The key point is that the adversary A who has $1/\mathrm{poly}(n)$ chance of winning in Case 2 is still winning in Case 2 for $2^n/\mathrm{poly}(n)$ number of the master public keys given to him. As a result, we achieve $\Omega(n) \cdot 2^n/\mathrm{poly}(n)$ bits of compression on the *total* representation of the random IBE oracle, which is sufficient to derive its hardness against poly-size (oracle-aided) circuits.

Using Random Injective Functions for Trapdoor Generation. A seemingly minor technical obstacle against our compression amplification argument sketched in the previous paragraph is that we need to represent the key (MPK or MSK) for each of the sub-oracles $G[\text{MSK}](\cdot)$ for which we achieve $\Omega(n)$ bits of compression, so that we can reconstruct the full oracle (S, G, F, F^{-1}). Unfortunately, if we do so, we will lose all the (little) compression that we could achieve for representation of $G[\text{MSK}](\cdot)$ (for many MSKs). To resolve this issue, we use random *injective* functions with large image length to generate the identity trapdoors. This enables us to gain compression for representation of each $G[\text{MSK}](\cdot)$ over which the attacker succeeds in finding matching $G[\text{MSK}](id^*) = td^*$ even if we explicitly represent the corresponding MSK. We formally state and prove this simple, yet useful building block of our compression argument (when there is no Sam oracle) in the full version [32], which we will invoke when we switch to using injective functions for generating identity trapdoors.

Extension to HIBE. Our proof sketched above scales well with the number of identity hierarchies. We will do so by expanding Case 2 of the analysis into many more subcases corresponding to each level. However, the fundamental difference between the first (master key general) and last (encryption/decryption) levels compared to other levels (generating identity trapdoors) remain different and is handled as described above using injective functions with long output.

Comparison to [11]. At an abstract level, our compression amplification technique described above allows us to achieve exponential compression for primitives that are of the "family" form (here we are interpreting the MPK as an index over which the (sub) primitive and the attack are launched) and can potentially be applied to more applications. In particular, we conjecture that our technique could give an alternative approach to that of Chung et al. [11] whose goal was to show how to achieve hardness against non-uniform attackers (circuits) when the primitive is of the "family" form. [11] achieved this by employing an information theoretic lemma by Unruh [46], while our approach uses the compression technique of [15].

2 Preliminaries

For any $n \in \mathbb{N}$, let Π_n be a family of permutations where $\pi \leftarrow \Pi_n$ is a random permutation mapping $\{0,1\}^n$ to $\{0,1\}^n$. Furthermore, let $\mathcal{F}_{n,m}$ be a family of *injective* functions where $f \leftarrow \mathcal{F}_{n,m}$ is random injective function mapping $\{0,1\}^n$ to $\{0,1\}^m$. For a set S by $x \leftarrow S$ we refer to the process of sampling x uniformly at random from S. We use $[i..j]$ for $\{i, \ldots, j\}$ and $[N]$ for $[1..N]$.

2.1 Black-Box Constructions

We use the following definition of black-box constructions due to Reingold et al. [40]. Unless specified otherwise, in this work we use the terms black-box and fully black-box equivalently.

Definition 3 *[Fully black-box constructions [40]]. A fully black-box construction of a primitive $\mathcal{Q}$ from a primitive $\mathcal{P}$ consists of two PPT algorithms (Q, S):*

1. *Implementation: For any oracle P that implements $\mathcal{P}$, Q^P implements $\mathcal{Q}$.*
2. *For any oracle P implementing $\mathcal{P}$ and for any oracle adversary A successfully breaking the security of Q^P, it holds that $S^{P,A}$ breaks the security of P.*

Even though the notion above was formalized in [40], the original work of Impagliazzo and Rudich were the first to note that black-box constructions "relativize"; namely they hold relative to any oracle. Thus to rule out black-box constructions it is sufficient to rule out relativizing constructions. The following argument has its roots in the work of Gertner et al. [18] and is a strengthening of this argument. Informally speaking it asserts that it is enough to choose the separating oracle after (and based on) a candidate construction. Another interpretation of this technique is known as the two-oracle technique of Hsiao and Reyzin [26]. Here we describe this lemma and prove it for sake of completeness. This lemma is implicitly used in the work of Haitner et al. [23]. Here we abstract out this lemma and use it in our proof of Sect. 3.

Lemma 4. *For any primitives $\mathcal{P}$ and $\mathcal{Q}$ let $O = (O_1, O_2)$ be a randomized oracle with two subroutines such that:*

1. *Primitive $\mathcal{P}$ could be implemented using (any sample for) the first part O_1.*
2. *Any fixed (computationally unbounded) poly(n)-query adversary B who could call both of O_1 or O_2 could break the implementation of $\mathcal{P}$ relative to O_1 with probability negl(n) where n is the security parameter. This probability is over the selection of O as well as attacker B.*
3. *For any implementation Q of $\mathcal{Q}$ that only calls O_1 there is a poly(n)-query attacker A who breaks Q^{O_1} with probability $\geq 1 - 1/n^2$ where the probability is over O and the attacker A.*

Then there is no fully black-box construction of $\mathcal{Q}$ from $\mathcal{P}$.

Breaking a primitive could mean different thing for different primitives, but in this paper we deal with $\mathcal{Q}$ being CRH whose attackers have to find collisions with non-negligible probability.

Proof. For sake of contradiction, suppose (Q, S) is a fully black-box construction of $\mathcal{Q}$ from $\mathcal{P}$. Sample $O = (O_1, O_2)$ and use the implementation of $\mathcal{P}$ that exists relative to O_1 to get implementation Q^{O_1} of $\mathcal{Q}$, and let A be the attacker who breaks this scheme (whose existence is guaranteed by the property 3 of O). Since A succeeds with probability $1 - 1/n^2$ and $\sum_n 1/n^2 = O(1)$ by Borel-Cantelli lemma, with measure one over the choice of O, A succeeds for an infinite set of security parameters. We call such A a good adversary relative to O.

Now, consider $S^{P,A}$ where P is the implementation of $\mathcal{P}$ using O_1 and A is the above adversary. By the definition of fully black-box constructions, for any sampled O such that A is a good adversary relative to O, $S^{P,A}$ will break P^{O_1} also for an infinite sequence of security parameters. Therefore, with measure one over the choice of O, $S^{P,A}$ will break P^{O_1} for an infinite sequence of security parameters. But we will show below that this cannot happen.

Let us "merge" the algorithm A into S and consider $B^O = (S^A)^O$ as a new poly(n)-query attacker who calls O and tries to break P^{O_1} directly. By property 2 of O, this attacker would have negl(n) chance of doing so. By an averaging argument, for each fixed security parameter, with probability $1 - \text{negl}(n) \geq 1 - 1/n^2$ over the choice of O it holds that B^O breaks P^{O_1} with probability at most negl$(n) = \alpha(n)$. By another application of Borel-Cantelli lemma, it follows that with measure one over the choice of O it holds that: the number of security parameters for which $B^O = (S^A)^O$ breaks P^{O_1} with probability more than $\alpha(n)$ is finite, which is a contradiction.

2.2 Collision-Resistant Hash Functions

In this work we define collision-resistant hash functions only for those that shrink their input by a factor of two. It is well known that any CRH with only one bit of shrinkage could be turned into one defined below. We use this definition as it simplifies some of the arguments needed for the separation.

Definition 5. *For $m = poly(n)$, a collision resistant hash function $H = \{h \mid h \colon \{0,1\}^m \times \{0,1\}^n \mapsto \{0,1\}^{n/2}\}$ is a family of functions such that for any PPT adversary A there is a negligible function $\epsilon(n)$ such that:*

$$\Pr_{d \leftarrow \{0,1\}^m} [A(d) = (x_1, x_2) \in \{0,1\}^{2n} \wedge x_1 \neq x_2 \wedge h_d(x_1) = h_d(x_2)] \leq \epsilon(n).$$

where $h_d(x) = h(d, x)$.

2.3 Hierarchical Identity-Based Encryption

Definition 6 (Identity vector). *For $i \geq 0$, an i-level identity vector $\text{ID}_i = \langle id_0, ..., id_i \rangle$ is a tuple of i identities, where $id_j \in \{0,1\}^* \ \forall \ j \in [0, i]$. Furthermore, the corresponding private-key for identity vector ID_i is given as td_{ID_i}. If $i < 0$, we let $\text{ID}_i = \epsilon$, the empty vector.*

Definition 7 (Prefix vector). *We define a prefix vector for identity vector $\text{ID}_i = \langle id_0, ..., id_i \rangle$ as any tuple $\langle s_0, ..., s_j \rangle$ such that $j \leq i$ and $s_k = id_k$ for $0 \leq k \leq j$. We denote the set of all prefix vectors of ID_i as $pre(\text{ID}_i)$.*

Definition 8 (Hierarchical identity-based encryption [25]). *Given security parameter n, an l-depth hierarchical identity-based encryption (l-HIBE) scheme for messages in $\mathcal{M}$ and ciphertext space $\mathcal{C}$ consists of $l + 3$ PPT algorithms $(\text{Setup}, \{\text{KeyGen}_i\}_{i=1}^l, \text{Enc}, \text{Dec})$ defined as follows. (For simplicity and without loss of generality we assume that n is the security parameter as well as the length of the master secret and public keys.)*

- Setup(1^n) *takes as input security parameter n and outputs a pair of keys $(\text{MSK}, \text{MPK}) \in \{0,1\}^n \times \{0,1\}^n$. We let $\text{ID}_0 = \langle id_0 \rangle = \langle \text{MPK} \rangle$ and $td_{\text{ID}_0} = \langle td_0 \rangle = \text{MSK}$.*

- For $i \in [1, l]$, $\mathrm{KeyGen}_i(\mathrm{ID}_{i-1}, td_{\mathrm{ID}_{i-1}}, id_i)$ takes as input the parent identity
  vector ID_{i-1}, the corresponding private-key $td_{\mathrm{ID}_{i-1}}$ and identity id_i then outputs the corresponding i-level private key vector td_{ID_i}.[5]
- $\mathrm{Enc}(\mathrm{ID}_l, x)$ takes as input the full public identity vector ID_l, and a message
  $x \in \mathcal{M}$, and outputs ciphertext $y \in \mathcal{C}$.[6]
- $\mathrm{Dec}(\mathrm{ID}_l, td_{\mathrm{ID}_l}, y)$ takes as input the identity vector ID_l, a corresponding
  private-key vector td_{ID_l}, and ciphertext $y \in \mathcal{C}$, and it returns the message
  $x \in \mathcal{M}$.

Correctness. Given any $(\mathrm{MSK}, \mathrm{MPK}) \leftarrow \mathrm{Setup}(1^n)$, an HIBE scheme must sat-
isfy $\mathrm{Dec}(\mathrm{ID}_l, td_{\mathrm{ID}_l}, \mathrm{Enc}(\mathrm{ID}_l, x)) = x$ for all $x \in \mathcal{M}$ and all $(\mathrm{ID}_l, td_{\mathrm{ID}_l})$ where
td_{ID_l} is the corresponding private-key of $\mathrm{ID}_l = \langle \mathrm{MPK}, id_1, ..., id_l \rangle$, the identity
vector obtained through an iterative call to KeyGen_i.

Security. An HIBE scheme is said to be CCA secure if for all adaptive PPT
adversaries A:

$$\Pr[CCA_A^{HIBE}(1^n) = 1] \leq \frac{1}{2} + \mathrm{negl}(n)$$

where CCA_A^{HIBE} is shown in Fig. 2. In Step 2, A can adaptively ask key genera-
tion queries for ID_i to oracle H_{MSK} which returns td_{ID_i}, the private-key associated with this identity vector, by recursively applying the key generation procedure
$\mathrm{KeyGen}_i(\mathrm{ID}_{i-1}, td_{\mathrm{ID}_{i-1}}, id_i) = td_{\mathrm{ID}_i}$ given that $td_{\mathrm{ID}_0} = \mathrm{MSK}$. Its chosen identity
ID_l^A must not be asked as a query to H_{MSK}. Furthermore, A can adaptively ask
decryption queries $D_{\mathrm{MSK}}(\mathrm{ID}_l, c)$ to decrypt ciphertext $c \in \mathcal{C}$ with respect to any
identity ID_l. In Step 4, A can still issue queries to H_{MSK} but only for identities
ID_i that are not in $\mathrm{pre}(\mathrm{ID}_l^A)$, and it can still issue queries to D_{MSK} but not for
inputs (ID_l^A, c), where c is the challenge ciphertext.

     Note that, for $l = 0$, this reduces to a standard CCA secure public-key encryp-
tion system, and for $l = 1$ this reduces to a CCA-secure IBE scheme.

Definitional Variations. The standard CCA security of HIBE as given in the
previous definition can be weakened in multiple ways. We present here some
variations of the security definition that we might refer to later, noting only the
differences from the original definition.

- CPA (resp. CCA1): The adversary's capabilities are limited to chosen plaintext (resp. non-adaptive chosen ciphertext) attacks.
- rID-CCA/rID-CPA: Instead of having the adversary choose ID_l^A, the target
  identity will be chosen uniformly at random by the challenger and provided
  to the adversary along with MPK.

---

[5] Note that we define a key generation algorithm for each level (as opposed to a single
algorithm) in order to simplify our HIBE construction using our ideal oracle.

[6] Some of the subsequent definitions of HIBE use a more general definition in which
one can encrypt messages under partial identity vectors $\mathrm{ID}_i = \langle id_0, ..., id_i \rangle$ of depth
$i < \ell$. Our impossibility result directly extends to this more general setting as well.
However, for sake of simplicity here we focus on the original definition of [25].

Experiment $CCA_A^{HIBE}(1^n)$:
1. $(\text{MSK}, \text{MPK}) \leftarrow \text{Setup}(1^n)$
2. $(x_0, x_1, \text{ID}_l^A) \leftarrow A^{H_{\text{MSK}}(.), D_{\text{MSK}}(.,.)}(\text{MPK})$
3. $c \leftarrow \text{Enc}(\text{ID}_l^A, x_b)$ *where* $b \overset{\$}{\leftarrow} \{0,1\}$
4. $b' \leftarrow A^{H_{\text{MSK}}(.), D_{\text{MSK}}(.,.)}(c)$
5. *Output* $(b = b')$

Fig. 2. The CCA_A^{HIBE} experiment

– OW-CCA/OW-CPA: Instead of distinguishing between two ciphertexts, the goals of the adversary here is to "invert" the given challenge ciphertext and find the corresponding (randomly selected) message. These notions could be combined into notions like OW-rID-CPA, OW-CCA, etc.

Black-box Construction of HIBE. The definition of a black-box construction of CRH from HIBE could be derived from Definitions 3 and 8.

2.4 Collision Finding (Sam) Oracle

In this section, we define the collision finding oracle Sam of [23,45]. Roughly speaking and in its simplest form, Sam is a (possibly inefficient) algorithm that accepts some description of a circuit C and outputs (x, x'), both uniformly distributed, such that $C(x) = C(x')$. This oracle was originally introduced by Simon [45] and then was extended into the nested Sam oracle of Haitner et al. [23].[7]

Definition 9 (Collision-finding oracle [23,45]). *For any arbitrary oracle* O, *the algorithm* $\text{Sam}_r^O(C)$ *for* C *with input length* n *samples a uniformly random* $x \in \{0,1\}^n$ *and then (after sampling* x*) samples another uniformly random point* x' *conditioned on* $C(x) = C(x')$.[8] *It then returns* (x, x'). *Note that this oracle is* randomized *but the randomness* r_C *is independent for each circuit* C *(and is sampled only once). The randomness of* Sam *for each query is provided by the randomized function* $r(C) = r_C$ *that for each circuit query* C *provides a random point in the inputs of* C *as well as a random permutation over the input space of* C*. The first is used to sample* x *and then the random permutation is enumerated till we get a collision* x'.

It is easy to see that using Sam one can efficiently find collisions for any circuit C whose output length m is smaller than its input length n. Specifically, if $n > m$

---

[7] In this work we focus on using the simpler collision finding oracle that is *not* interactive. However, all of our separation results hold with respect to the stronger Sam oracle of "low" (i.e., $o(n/\log n)$) as well. We refer the proof for the more general case to the full version [32] of the paper.

[8] Note that the returned "collision" (x, x') is *not* necessarily distributed like a uniformly sampled collision from all possible collisions.

then one guarantees the existence of some pair (x, x') such that $C(x) = C(x')$ (i.e., a collision), which results in Sam successfully returning such a pair. The following lemma provides a general tool to use Sam for separating primitives from CRH.

Lemma 10. *Let $\mathcal{P}$ be any primitive and $\mathcal{Q}$ represent the collision-resistant hash function primitive. Let $U = (O, \mathrm{Sam}^O)$ be a randomized oracle with two subroutines such that: (1) Primitive $\mathcal{P}$ could be implemented using O. (2) Any fixed (even computationally unbounded) $\mathrm{poly}(n)$-query adversary B who could call both of O and Sam^O could break the implementation of $\mathcal{P}$ relative to O with probability $\mathrm{negl}(n)$ where n is the security parameter. Then there is no fully black-box construction of collision-resistant hash functions from $\mathcal{P}$.*

Proof. The lemma almost directly follows from Lemma 4; we just have to prove the third property needed by Lemma 4. In fact, for any implementation Q of CRH that only calls O there is a 1-query attacker A who breaks Q^O with probability $1 - \mathrm{negl}(n)$. All A does is to take d as the index of hash function, turn $h_d(\cdot)$ into a $\mathrm{poly}(n)$-size circuit C with input length n, and call Sam^O over C and outputs the result. It is easy to see that since h is shrinking its input by a factor of two, with $1 - \mathrm{negl}(n)$ probability over the first sampled point x_1, the number of "siblings" of x_1 relative to the function $h_d(\cdot)$ are exponentially large, and therefore the two colliding points (x_1, x_2) returned by Sam^O will be different points with probability $1 - \mathrm{negl}(n)$.

3   Separating Hierarchical IBE from Collision Resistant Hashing

In this section we will formally prove our main Theorem 1. Namely, we will prove that there exists no fully black-box construction of collision-resistant hash functions from l-level hierarchical identity-based encryption for any polynomial $l = \mathrm{poly}(n)$.

Theorem 11. *For any security parameter n and an arbitrary polynomial number of levels $\ell = \mathrm{poly}(n)$ there is no fully black-box construction of collision-resistant hash functions from ℓ-level OW-CCA secure hierarchical identity-based encryption scheme.*

Corollary of Theorem 11. The above theorem, together with the result of Boneh et al. [6] shows the separation of (standard) CCA-secure HIBE from CRH. In particular, we apply Theorem 11 for $\ell + 1$ levels of identity with one-way CCA security (in fact one-way CPA also suffices). Then using Goldreich-Levin lemma and bit-by-bit encryption, we can achieve CPA security as a black-box, and the result of [6] gives us an CCA secure HIBE for ℓ levels of identity relative to the same oracle.

To Prove Theorem 11, we will state and use the following claim, which shows the existence of a separating oracle for which a secure implementation of a hierarchical IBE exists.

Claim 12. *There exists a randomized oracle $U = (O, \text{Sam}^O)$ such that the following holds:*

1. *An implementation P of OW-CCA-secure l-HIBE exists relative to O.*
2. *Any* $\text{poly}(n)$-*query adversary A with access to U can break P only with negligible probability.*

Proof (of Theorem 11). Using Claim 12, we can assume the existence of oracle $U = (O, \text{Sam}^O)$ for which OW-CCA-secure l-HIBE exists. Since such an oracle exists, an immediate application of Lemma 10 yields that there is no black-box construction of collision-resistant hash functions from OW-CCA-secure l-HIBE.

We now dedicate the rest of this section to proving Claim 12. We first start in Sect. 3.1 by giving a formal description of the first subroutine O of our separating oracle, which represents an (idealized) random hierarchical trapdoor injective function, so that we will later use it to implement HIBE in Sect. 3.2. The proof proper starts in Sect. 3.3 where we will use the randomized oracle $U = (O, \text{Sam}^O)$ such that **(1)** a OW-CCA-secure l-level HIBE implementation exists relative to O, and **(2)** the HIBE implementation remains secure against any $\text{poly}(n)$-query computationally unbounded adversary even after adding Sam^O.

3.1 Description of Oracle O

In this section, we describe the distribution of our oracle O, which will be used to show that with overwhelming probability over the choice of this oracle, ℓ-level OW-CCA HIBE exists relative to it. In the following definition we will use our notation of identity vectors as defined in Definition 6 but, for simplicity of presentation, all of our identities will be strings of length n. Our proof can be directly extended to handle the case of unbounded-length identities as well, but for sake of the simplicity of the presentation we focus on the case of bounded-length identities; see the full version of this paper [32] for a sketch of the modifications needed for the case of unbounded identity lengths.

Note About Notation. In definition below, we use several functions as part of the oracle. The *inverse* of some of these functions are also involved in the definition of these functions (in a recursive way). So we will define (injective) functions like $f^{-1}(\cdot)$ and then subsequently use $f(\cdot)$ for $f^{-1^{-1}}(\cdot)$. Also, for the sake of clarity and separating actual inputs from indices, we use the notation of $f[x](y)$ to denote $f(x, y)$ if we envision x as the index (or key) and y as the actual input.

Definition 13 (Random Hierarchical Trapdoor Injective Functions).
For any security parameter integer $n \in \mathbb{N}$ and number of hierarchies $\ell = \ell(n)$, let $m = 10n \cdot \ell$. Our random hierarchical injective function oracle O_n consists of $2\ell+3$ subroutines: $\{h_0^{-1}, g_1, h_1^{-1}, \ldots, g_{\ell+1}, h_{\ell+1}^{-1}\}$ distributed as follows. (Although functions $h_i(\cdot)$ are not publicly available as subroutines of O we still use them to describe the subroutines of O more easily.)

- *The key generation oracle $h_0^{-1}(\cdot) := S(\cdot)$, the encryption oracle $h_{\ell+1}^{-1}[\mathrm{ID}_\ell](\cdot)$ $:= F[\mathrm{ID}_\ell](\cdot)$ and the decryption oracle $g_{\ell+1}[\mathrm{ID}_{\ell-1}, td_\ell](\cdot) := F^{-1}[\mathrm{ID}_\ell](\cdot)$ are random permutations over $\{0,1\}^n$.*
- *For $i \in [1..\ell]$, and index $\mathrm{ID}_{i-1} = (\mathrm{MPK}, id_1, \ldots, id_{i-1}) \in \{0,1\}^{i \cdot n}$, let $h_i[\mathrm{ID}_{i-1}](\cdot) : \{0,1\}^n \mapsto \{0,1\}^m$ be a random injective function. We define $h_i^{-1}[\mathrm{ID}_{i-1}](td_i) = id_i$ if $h_i[\mathrm{ID}_{i-1}](id_i) = td_i$ for some $id_i \in \{0,1\}^n$ and we define $h_i^{-1}[\mathrm{ID}_{i-1}](td_i) = \perp$ if no such id_i exists.*
- *For $i \in [1..\ell]$ and $(\mathrm{ID}_{i-2}, td_{i-1}) \in \{0,1\}^{(i-1)n+m}$ function $g_i[\mathrm{ID}_{i-2}, td_{i-1}](\cdot) : \{0,1\}^n \mapsto \{0,1\}^m$ is defined as follow: For given input id_i, if we have $h_{i-1}^{-1}[\mathrm{ID}_{i-2}](td_{i-1}) = id_{i-1}$ for some $id_{i-1} \neq \perp$, then $g_i[\mathrm{ID}_{i-2}, td_{i-1}](id_i) = h_i[\mathrm{ID}_{i-1}](id_i)$. If no such id_{i-1} exists, then $g_i[\mathrm{ID}_{i-2}, td_{i-1}](id_i) = \perp$.*

Our actual oracle (for applying Lemma 10) will be (O, Sam^O) where O is sampled from the distribution of the oracles of Definition 13, and Sam^O is the Sam oracle as defined in Sect. 2.4 where the input circuits to Sam^O are allowed to have O-gates.

3.2 Implementing ℓ-level HIBE Using Oracle O

Here we show how to use the oracle O of Definition 13 to implement ℓ-level HIBE. Then we will turn into proving the security which is the main part of the proof. Although the *security* of the sampled O is intuitive, due to the *fully random* nature of the permutations used in the implementation, since our actual oracle also has Sam^O attached to it, the proof of security becomes nontrivial.

Intuition. We use the subroutine $h_0^{-1}(\cdot)$ to generate the master public key. We use $g_i(\cdot)$ to generate the ID's of the i'th layer, and we use $h_{\ell+1}^{-1}(\cdot)$ to encrypt and $g_{\ell+1}(\cdot)$ to decrypt. Therefore, for sake of clarity of the presentation, we will use the following alternative names for the subroutines of the first and last layers: $h_0^{-1}(\cdot) = S(\cdot), h_{\ell+1}^{-1}[\cdot](\cdot) = F[\cdot](\cdot), g_{\ell+1}[\cdot](\cdot) = F^{-1}[\cdot](\cdot)$. We will also treat the master public key as the identity of level zero, the ciphertexts as identity of the level $\ell + 1$, and the plaintexts as the trapdoors of the level $\ell + 1$.

Note that we elected to use random trapdoor *injective functions* to represent $h_i^{-1}[.](.) $ in O as opposed to using random trapdoor permutations as one would first naturally assume. This is to prevent the adversary from trivially breaking the scheme using a call to the $h_i^{-1}[.](.) $ subroutines. Specifically, if we used a trapdoor permutation, since the adversary can *choose* the challenge identity, it can first call $h_i^{-1}[.](.) $ for any $i \in [1..l]$ to find an identity for his own (say randomly selected) trapdoor of level i and announce that specific identity as the one he will use in the attack! Therefore, it was crucial that either we remove the subroutines $h_i^{-1}[.](.) $ from O or change the oracle in way that mitigates such an attack. We thus chose to use random injective functions with a sparse range to make it hard for an adversary to discover a valid trapdoor td_i such that $h_i^{-1}[\mathrm{ID}_{i-1}](td_i) \neq \perp$ for any ID_{i-1}.

Construction 14 (Implementing ℓ-level HIBE Using Oracle O). *For any security parameter n, and oracle O_n sampled according to Definition 13,* *we will implement an l-HIBE scheme as follows. Our message space and the identities of each level are all $\{0,1\}^n$. To get unbounded message length, we will use bit-by-bit encryption after the scheme is turned into an CPA secure scheme. For larger $\mathrm{poly}(n)$ identity lengths, we will change the security parameter n into $\mathrm{poly}(n)$ in the first step of the construction. As described below, for any identity vector ID_i, we will represent its trapdoor td_{ID_i} as $(\mathrm{ID}_{i-1}, td_i)$ for the "correct" td_i. The algorithms for the constructed scheme work as follows:*

- Setup(1^n) : *Choose* MSK $\in \{0,1\}^n$ *uniformly at random then get* MPK $=$ $S(\mathrm{MSK})$. *We let* $\mathrm{ID}_0 = \langle id_0 \rangle = \langle \mathrm{MPK} \rangle$ *and* $td_{\mathrm{ID}_0} = td_0 = \mathrm{MSK}$. *Output* (MSK, MPK).
- *For* $i \in [1..l]$, $\mathrm{KeyGen}_i(\mathrm{ID}^*_{i-1}, td_{\mathrm{ID}_{i-1}}, id_i)$ *where* $\mathrm{ID}^*_{i-1} = \langle id^*_0, ..., id^*_{i-1} \rangle$: *Parse* $td_{\mathrm{ID}_{i-1}} = (\mathrm{ID}_{i-2}, td_{i-1})$. *If* $\mathrm{ID}^*_{i-2} = \mathrm{ID}_{i-2}$ *then set* $td_i = g_i[\mathrm{ID}_{i-2}, td_{i-1}](id_i)$ *and output* $td_{\mathrm{ID}_i} = (\mathrm{ID}_{i-1}, td_i)$. *Otherwise, output* $\perp$.
- $\mathrm{Enc}(\mathrm{ID}_l, m)$: *Output* $F(\mathrm{ID}_{l-1}, id_l, m)$.
- $\mathrm{Dec}(\mathrm{ID}^*_l, td_{\mathrm{ID}_l}, c)$: *Parse* $td_{\mathrm{ID}_l} = (\mathrm{ID}_{l-1}, td_l)$. *If* $\mathrm{ID}^*_{l-1} = \mathrm{ID}_{l-1}$ *then output* $F^{-1}(\mathrm{ID}_{l-1}, td_l, c)$. *Otherwise, output* $\perp$.

3.3   Security of Implemented HIBE Relative to O

We prove that the constructed HIBE of Construction 14 (using the oracle O of Definition 13) is OW-CCA secure relative to (O, Sam^O). The proof has the following two steps:

1. **Compression and Reconstruction.** Assuming the existence of any (deterministic) adversary A who can break the OW-CCA security of the constructed HIBE (using O of Definition 13) with probability $\epsilon \geq 1/\mathrm{poly}(n)$ and $q = \mathrm{poly}(n)$ queries, we show how to **(1)** compress O to represent it using a "fewer" bits than is necessary to represent a general sampled O, and **(2)** show how to reconstruct O. This compression is relative to the fixed adversary A and both compression and reconstruction heavily depend on A. The number of bits saved in the representation of O will directly imply a bound on the number of such oracles that A can successfully attack. This would imply that with overwhelming probability over the choice of O it will not be a good oracle for A's attack. As usual with reconstruction-type arguments, the bound obtained with this argument allows us to even do a union bound over all possible adversaries that are implemented as circuits of polynomial size. Thus, with overwhelming probability no efficient attacker would exist.
2. **Adding** Sam^O. The second step of the proof shows that adding Sam^O (and allowing A to call it) does not interfere with the compression and reconstruction. The argument of this step is identical to that of [23] but we provide our sketch of the proof in later in this section.

Formalizing the Adversary A. Without loss of generality, we assume that A is a *deterministic* adversary who asks q queries (to O and the challenger combined) and wins against a fixed oracle O with probability $\geq \epsilon$:

$$\Pr_{(\mathrm{MPK}, y)} [A(\mathrm{MPK}, y) = (id^*_1, \ldots, id^*_\ell, x) \mid F(\mathrm{MPK}, id^*_1, \ldots, id^*_\ell, x) = y] \geq \epsilon$$

where A is participating in the OW-CCA security game, i.e. no vector of identities $(id_1^*, \ldots, id_\ell^*)$ is given to the adversary and he is the one who chooses them, but A is given a random master public key MPK and a random ciphertext y and he wants to invert y in a CCA attack. We can assume A is deterministic since we are working with non-uniform adversaries and we will prove that our oracle is secure against all circuits, and a non-uniform attacker can always fix its randomness to the "best" value.

Notation. Throughout this section, we might occasionally use the simplifying notation in which $\text{MPK} = id_0^*$, $y = id_{\ell+1}^*$, and so the full identity vector of the attack is simply $\text{ID}_{\ell+1}^* = (\text{MPK}, id_1^*, \ldots, id_\ell^*, y)$, but the first and last components of this vector are not chosen by the adversary but are selected at random. In addition we use td_i^* to denote the corresponding trapdoor for id_i^* with respect to the prefix ID_{i-1}^*. So we will also have the selected $\text{MSK} = td_0^*$ and $x = td_{\ell+1}^*$ for x the inverse of y.

Putting A in Canonical Form. We will modify A as follows.

1. Whenever A wants to output x as its final answer, it queries the encryption on x by calling $F[\text{ID}_\ell^*](x)$.
2. Whenever A is about to ask the query $g_i[\text{ID}_{i-2}, td_{i-1}](id_i)$ (which returns td_i) A will first ask the query $h_{i-1}^{-1}[\text{ID}_{i-3}, id_{i-2}](td_{i-1})$ from O (that returns id_{i-1} corresponding to td_{i-1} for prefix ID_{i-2}). This modification potentially increases the number of queries asked by A by a factor of two which we can safely ignore (and assume that A is in canonical form to start with).

Now we define the following events, which somehow capture the "level" in which the adversary finds a relevant trapdoor. This "trapdoor" could be finding the message x itself (which as described before, is interpreted as the "trapdoor" for its corresponding challenge ciphertext y), or it could be finding the relevant master secret key MSK (which we interpret as the trapdoor of the "identity" MPK), or it could be finding a trapdoor somewhere in between for id_i for $i \in [\ell]$. Note that finding trapdoor for smaller i is a "stronger" attack that lets A find the relevant trapdoors for bigger i and eventually invert y.

Definition 15 (Events E_i). *For $i \in [0..\ell+1]$ we say that the event E_i has happened during the execution of A in the (OW-CCA) security game, if A calls the query $h_i^{-1}[\text{ID}_{i-1}^*](td_i^*)$ and receives an answer other than $\perp$. We also let E_{-1} be an empty event (and so $\neg E_{-1}$ holds with probability one).*

The first canonical modification of A implies that the success probability of A (and the notation we use to denote x as $td_{\ell+1}^*$ and $\text{MPK} = id_0^*$) simply means that what we are assuming about A's success attack is equivalent to saying:

$$\Pr_{\text{ID}_{\ell+1}^*} [E_{\ell+1}] \geq \epsilon.$$

In the following we assume this is the case.

Lemma 16. *For events E_i's defined according to Definition 15, there exists $i \in [0..\ell+1]$ such that $\Pr_{\text{ID}^*_{\ell+1}}[E_i \wedge \neg E_{i-1}] \geq \epsilon/(\ell+2)$.*

Proof. It holds that:

$$E_{\ell+1} \subseteq \bigcup_{i \in [0..\ell+1]} (E_i \wedge \neg E_{i-1} \wedge \cdots \wedge \neg E_0) \subseteq \bigcup_{i \in [0..\ell+1]} (E_i \wedge \neg E_{i-1}).$$

Also note that $\Pr[E_{\ell+1}] \geq \epsilon$. Therefore, there should exists an index i for which $\Pr[E_i \wedge \neg E_{i-1}] \geq \epsilon/(\ell+2)$.

Fixing Parameters. In the rest of the proof, we fix i to the value that satisfies Lemma 16, and we let $\epsilon' = \epsilon/(\ell+2)$. However, we will *not* always fix the master public-key MPK nor the challenge ciphertext y using an averaging argument. Whether or not we fix either of them will depend on i.

The Sub-Oracle $h(\cdot) = \{h_0(\cdot), \ldots, h_{\ell+1}(\cdot)\}$. So far we only defined $h^{-1}(\cdot)$ without referring to $h(\cdot)$ (which was not a subroutine provided as part of the oracle O). Here we introduce this subroutine module and allow A to call it in a "restricted" way. To elaborate, note that in the CCA security game, the adversary can call the oracles $H_{\text{MSK}}(.)$ and $D_{\text{MSK}}(.,.)$, which allow him get the trapdoors for any identity as long as it is not a prefix of the challenge identity, and get decryption of any message other than challenge ciphertext y. Both of these queries are special cases of queries that are the inverse of $h^{-1}(\cdot)$. Namely, for $i \in [1..\ell+1]$ let $h_i(\text{ID}_i) = h_i[\text{ID}_{i-1}](id_i)$ be defined to be equal to td_i whenever $h_i^{-1}[\text{ID}_{i-2}, id_{i-1}](td_i) = h_i^{-1}[\text{ID}_{i-1}](td_i) = id_i$. Then any query of the adversary A to *both* of the oracles $H_{\text{MSK}}(.), D_{\text{MSK}}(.,.)$ is a special case of a query to $h_i(\cdot)$ for some $i \in [1..\ell+1]$. For simplicity, we will even define h_0 even though the adversary is not calling such queries directly (since the MPK $= id_0^*$ is given by the challenger and is fixed). The restriction on adversary's queries to $H_{\text{MSK}}(.), D_{\text{MSK}}(.,.)$ translates into a natural restriction on how he accesses the oracle $h(\cdot)$: none of these queries are allowed to be a prefix of $\text{ID}^*_{\ell+1}$.

Step 1: Compression and Reconstruction of O Without the Presence of Sam^O. We now begin the first step of the proof by showing how we can use a fixed adversary A (with the behaviour and capabilities that were described earlier in this section) to compress the description of oracle O.

Full Representation of O With No Compression. To represent a general oracle O fully (while there is no attacker) without redundant information, it suffices to represent only the injective oracles $h_i[\text{ID}_{i-1}](\cdot)$ for all $i \in [0..\ell+1]$ and all $\text{ID}_{i-1} \in \{0,1\}^{in}$. Note that for $i = \{0, \ell+1\}$ these injective functions happen to be a permutation as well! Now any query to $h_i^{-1}[\text{ID}_{i-1}](\cdot)$ can be answered using its corresponding explicitly represented inverse function $h_i[\text{ID}_{i-1}](\cdot)$. To answer the $g_i[\text{ID}_{i-2}, td_{i-1}](id_i)$ queries, we employ induction on i. Recall that the master public key generation $S(\cdot)$ sub-routine of O is the same as $h_0^{-1}(\cdot)$. Now, for $i \in [1..\ell+1]$ and a given query $g_i[\text{ID}_{i-2}, td_{i-1}](id_i)$, we can first find the relevant identity of td_{i-1} by looking up the value of $h_{i-1}^{-1}[\mathrm{ID}_{i-2}](td_{i-1})$, whose answer is represented in the description of the permutation $h_{i-1}[\mathrm{ID}_{i-2}](\cdot)$, and get id_{i-1}. This will enable us to find $h_i(\mathrm{ID}_i) = h_i[\mathrm{ID}_{i-1}](id_i)$, which is also the answer to $g_i[\mathrm{ID}_{i-2}, td_{i-1}](id_i)$.

Intuition Behind the Compression of O Relative to A. Here we describe the high level idea behind how to compress O relative to A, using the ideas described in Lemma 2. At a very high level we will compress O as follows.

1. If $i = l + 1$ (i.e., adversary wins by inverting $y = id_{l+1}$ without finding any trapdoor for any of the identities he proposes): In this case, we apply a similar idea to Lemma 2 and compress O by representing it using three pieces of information: the description of the fixed master public key $id_0^* = \mathrm{MPK}^*$ that maximizes the adversary's success probability, the full description of $h_i^{-1}[\mathrm{ID}_{i-1}](\cdot)$ for all $i \in [0..\ell]$ and all $\mathrm{ID}_{i-1} \in \{0,1\}^{i \cdot n}$, and the "compressed" description of $h_{l+1}^{-1}[\mathrm{ID}_l^*] = F[\mathrm{ID}_l^*]$ where $\mathrm{ID}_l^* = \langle id_0^*, id_1^*, ..., id_l^* \rangle$ for some adversarially chosen identities $(id_1^*, ..., id_l^*)$.

   The idea behind compressing $h_{l+1}^{-1}[\mathrm{ID}_l^*](\cdot)$ is as follows. Note that the identity vector ID_l^* (and its corresponding trapdoor) determines a single permutation that is described in *different directions* by $h_{l+1}^{-1}[\mathrm{ID}_l^*](\cdot)$, $h_{l+1}[\mathrm{ID}_l^*](\cdot)$, and $g_{l+1}[\mathrm{ID}_{l-1}^*, td_l^*](\cdot)$. The main difference between these three permutations is that $h_{l+1}^{-1}[\mathrm{ID}_l^*](\cdot)$ provides the access from trapdoors to identities, while the other two provide the access in the opposite direction. The algorithm A is "inverting" a random ciphertext id_{l+1} with respect to the identity vector ID_l^*, and it finds its related td_{l+1} with probability ϵ'. Now, if we can show that A's access to the three permutations $h_{l+1}^{-1}[\mathrm{ID}_l^*](\cdot)$, $h_{l+1}[\mathrm{ID}_l^*](\cdot)$, and $g_{l+1}[\mathrm{ID}_{l-1}^*, td_l^*](\cdot)$ does not let him find td_{l+1} "trivially" we can apply the compression algorithm of Lemma 2. The queries that let A find td_{l+1} trivially are the two queries $h_{l+1}[\mathrm{ID}_l^*](id_{l+1})$ and $g_{l+1}[\mathrm{ID}_{l-1}^*, td_l^*](id_{l+1})$. However, we already know that none of these queries are asked by A (before he asks $h_{l+1}^{-1}[\mathrm{ID}_l^*](td_{l+1})$). The reason that A is not asking $h_{l+1}[\mathrm{ID}_l^*](\cdot)$ is that $h(\cdot)$ is not a subroutine publicly available as part of oracle O, and is only provided due to the CCA nature of the attack, yet this particular query $h_{l+1}[\mathrm{ID}_l^*](id_{l+1})$ is prohibited to be asked by A since it violates the CCA attack's requirements (asking this query is akin to allowing the adversary to query the oracle with the challenge!). In addition, the reason that A is not asking $g_{l+1}[\mathrm{ID}_{l-1}^*, td_l^*](id_{l+1})$ is that if he does so, he would be asking the query $h_l^{-1}[\mathrm{ID}_{l-1}^*](td_l^*)$ right before that, which means the event E_l is happening (which we already assumed is not happening). Therefore, the behaviour of A lets us apply the compression algorithm of Lemma 2 to compress the description of $h_{l+1}^{-1}[\mathrm{ID}_l^*](\cdot)$.

2. If $i = 0$ (i.e., adversary wins by finding the master secret key): In this case, we apply a similar idea to Lemma 2 and compress O by representing it using three pieces of information: the description of the fixed challenge ciphertext $id_{l+1}^* = y^*$ that maximizes the adversary's success probability, the full description of $h_i^{-1}[\mathrm{ID}_{i-1}](\cdot)$ for all $i \in [1..\ell + 1]$ and all $\mathrm{ID}_{i-1} \in \{0,1\}^{i \cdot n}$, and the

"compressed" description of h_0^{-1} (which corresponds to S). The same idea that was described for $i = (l+1)$ applies here as well except that we need not care about queries that could be used to trivially find td_0.

3. If $i \in [1..\ell]$: This is the part of the proof for the OW-CCA security game that differs from the other two cases significantly. First we will fix y to whatever that maximizes the winning probability of the adversary. Now, the only remaining randomness (over which the adversary wins with some probability $\geq \epsilon'$) is the randomly selected master public key. We call a MPK *good* (for adversary) if the adversary manages to make $(E_i \wedge \neg E_{i-1})$ hold given this MPK (and the fixed challenge ciphertext y).

We compress O as follows. We represent each "subtree" of the oracle O that correspond to different MPKs differently. For MPKs that are not good, we will give a full representation. However, for each good MPK, we will represent the part of O that corresponds to this MPK in a compressed manner using the basic compression algorithm of Lemma 2 (for the case of inverting random injective functions) applied to the *single* injective function $h_i[\text{ID}_{i-1}^*](\cdot)$. We will also represent (ID_{i-1}^*) (for this particular MPK) but the number of bits that we save by compressing $h_i[\text{ID}_{i-1}^*](\cdot)$ is more than $|\text{ID}_{i-1}^*|$ because $m \gg n \cdot \ell$. Finally, since the adversary is succeeding for all good MPK's and there are super-polynomially many of them, our compression algorithm compresses the description of O by a super-polynomial number of bits.

Now we proceed with stating the formal claims that we will focus on from now on for proving that there exists a secure HIBE with respect to O. In particular, we will have two claims: one for handling the two identical cases of $i = 0$ and $i = l+1$ (where an adversary would find the corresponding $\text{MSK}^* = td_0^*$ or $x^* = td_{l+1}^*$), and another for treating the case of $i \in [1..l]$ (where an adversary finds an intermediate trapdoor) as mentioned during the intuition.

Claim 17. *Let O be an l-level random hierarchical trapdoor injective function oracle and n be the security parameter. Let $A = (A_1, A_2)$ be a q-query circuit that accepts a master public key $\text{MPK} \in \{0,1\}^n$, chooses an identity vector $\text{ID}_l^* = \langle \text{MPK}, id_1^*, ..., id_l^* \rangle$, then receives a challenge ciphertext $y^* \leftarrow F[\text{ID}_l^*](x^*) = h_{l+1}^{-1}[\text{ID}_l^*](x^*)$ for a random $x^* \in \{0,1\}^n$. Then for $q \leq 2^{n/5}$, $\epsilon' \geq 2^{-n/5}$, and $i \in \{0, l+1\}$ (and large enough n) we have:*

$$\Pr_{O}\left[\Pr_{\substack{\text{MPK}\leftarrow\{0,1\}^n \\ x^*\leftarrow\{0,1\}^n}} \left[A_1^O(\text{MPK}) = (\text{ID}_l^*, \sigma), A_2^O(\sigma, \text{MPK}, h_{l+1}^{-1}[\text{ID}_l^*](x^*)) = td_i^* \right] \geq \epsilon' \right] \leq 2^{-2^{n/2}}$$

Therefore, the oracle O can be represented using $\alpha - 2^{n/2}$ bits where α is the number of bits required to represent a random O.

Claim 18. *Let O be an l-level random hierarchical trapdoor injective function oracle and n be the security parameter. Let A be a q-query circuit that accepts a master public key $\text{MPK}^* \in \{0,1\}^n$, chooses an identity vector $\text{ID}_l^* =*

$\langle \text{MPK}^*, id_1^*, ..., id_l^* \rangle$, then receives a challenge ciphertext $y^* \leftarrow F[\text{ID}_l^*](x^*) = h_{l+1}^{-1}[\text{ID}_l^*](x^*)$ for a random $x^* \in \{0,1\}^n$. Then for $q \leq 2^{n/5}$, $\epsilon' \geq 2^{-n/5}$, $m = 10nl$, and $i \in [1, l]$ (and large enough n) we have:

$$
\Pr_O \left[\Pr_{\substack{\text{MPK} \leftarrow \{0,1\}^n \\ x^* \leftarrow \{0,1\}^n}} \left[\text{ID}_l^* \leftarrow A^O(\text{MPK}) : A^O(\text{MPK}, h_{l+1}^{-1}[\text{ID}_l^*](x^*)) = td_i^* \right] \geq \epsilon' \right] \leq 2^{-2^{3n/5}}
$$

Therefore, the oracle O can be represented using $\alpha - 2^{3n/5}$ bits where α is the number of bits required to represent a random O.

Proof (of Claim 17). We show here the compression of the oracle in case $i = 0$ or $i = (l+1)$ since these two cases are identical in nature. As described before, the compressed representation of O will contain the fixed i, the fixed master public key or challenge ciphertext, as well as full representation of permutations $h_j^{-1}[\text{ID}_{j-1}](\cdot)$ for all $j \in [0..\ell + 1]$ and all $\text{ID}_{j-1} \neq \text{ID}_{i-1}^*$. In the following we describe how to represent the description of $h_i^{-1}[\text{ID}_{i-1}^*](\cdot)$ by describing the encoding and decoding algorithms.

Encoder: Fix i and let $c = |i - (l+1)|$. Fix id_c^* such that Lemma 16 is satisfied. Note that id_c^* represents the fixed master public key if $i = (l+1)$, and represents the fixed ciphertext challenge when $i = 0$. Let $N = 2^n$ and let $I \subseteq \{0,1\}^n$ be the set of i-level identities $id_i^* \in \{0,1\}^n$ for which A can successfully find their corresponding trapdoor td_i^* (so $|I| \geq \epsilon' N$), and let $Y = \emptyset$. The encoder works as follows:

1. Remove the lexicographically first element $\widetilde{id_i^*}$ from I and put it in Y
2. Run $A^O(id_c^*, \widetilde{id_i^*})$. If A asks query:
   - $h_i^{-1}[\text{ID}_{i-1}^*](td_i^*) = id_i^*$ and $id_i^* \in I$ then remove id_i^* from I
   - $h_i[\text{ID}_{i-1}^*](id_i^*)$ and $id_i^* \in I$ then remove id_i^* from I
   Note that since event E_{i-1} does not happen here, A will not call function $g_i[\text{ID}_{i-2}^*, td_{i-1}^*](\cdot)$
3. If $|I| \neq \emptyset$, go back to Step 1. Otherwise go to the next step.
4. Output the following:
   - Description of i
   - Description of id_c^*
   - The compressed representation of $h_i^{-1}[\text{ID}_{i-1}^*]$ which consists of:
     - Description of $Y \subseteq I$
     - Description of $X = h_i[\text{ID}_{i-1}^*](Y)$
     - Description of $Z = \{(id_i^*, h_i[\text{ID}_{i-1}^*](id_i^*) \mid id_i^* \in \{0,1\}^n \backslash Y\}$, which describes the remaining part of the permutation necessary to reconstruct it.
   - The full representation of all the other permutations $\mathcal{H} = \{h_j^{-1}[\text{ID}_{j-1}] \mid j \in [0, .., l+1], \text{ID}_{j-1} \neq \text{ID}_{i-1}^*\}$

Decoder: Given A, the descriptions of X, Y, Z, and $\mathcal{H}$, and the description of i and id_c^*, the decoder can reconstruct O as follows:

1. Remove first lexicographically ordered id_i^* from Y and call it $\widetilde{id_i^*}$
2. Run $A^O(id_c^*, \widetilde{id_i^*})$. If A asks query:
   - $h_i^{-1}[\mathrm{ID}_{i-1}^*](td_i^*)$ for any $td_i^* \in \{0,1\}^n$:
     - If $td_i^* \notin X$: value of $h_i^{-1}[\mathrm{ID}_{i-1}^*](td_i^*)$ is given by Z.
     - If $td_i^* \in X$ and $h_i^{-1}[\mathrm{ID}_{i-1}^*](td_i^*) <_{lex} \widetilde{id_i^*}$: value of $h_i^{-1}[\mathrm{ID}_{i-1}^*](td_i^*)$ would have been precomputed before this call.
     - If $td_i^* \in X$ and $h_i^{-1}[\mathrm{ID}_{i-1}^*](td_i^*) = \widetilde{id_i^*}$: A has *hit* $\widetilde{id_i^*}$ and found its corresponding trapdoor, so we set $h_i^{-1}[\mathrm{ID}_{i-1}^*](td_i^*) = \widetilde{id_i^*}$.
   - $h_i[\mathrm{ID}_{i-1}^*](id_i^*) = td_i^*$ for any $id_i^* \in \{0,1\}^n \setminus \widetilde{id_i^*}$
     - If $td_i^* \notin X$: value of $h_i[\mathrm{ID}_{i-1}^*](id_i^*)$ is given by Z.
     - If $td_i^* \in X$ and $id_i^* <_{lex} \widetilde{id_i^*}$: value of $h_i[\mathrm{ID}_{i-1}^*](id_i^*)$ would have been precomputed before this call and can be inferred from the description of $h_i^{-1}[\mathrm{ID}_{i-1}^*]$
   - $h_j^{-1}[\mathrm{ID}_{j-1}](td_j)$ for any $td_j \in \{0,1\}^n$, and either $j \neq i$ or $\mathrm{ID}_{j-1} \neq \mathrm{ID}_{i-1}^*$: the result id_j can be obtained using the given full representation of $h_j^{-1}[\mathrm{ID}_{j-1}]$
   - $h_j[\mathrm{ID}_{j-1}](id_j)$ for any $id_j \in \{0,1\}^n$, and either $j \neq i$ or $\mathrm{ID}_{j-1} \neq \mathrm{ID}_{i-1}^*$: the result td_j can be obtained using the given full representation of $h_j^{-1}[\mathrm{ID}_{j-1}]$
   - $g_j[\mathrm{ID}_{j-2}, td_{j-1}](id_j)$ for any $td_j \in \{0,1\}^n$, and either $j \neq i$ or $\mathrm{ID}_{j-1} \neq \mathrm{ID}_{i-1}^*$: due to the canonical behaviour of A, $h_{j-1}^{-1}[\mathrm{ID}_{j-2}](td_{j-1})$ will be called first to get id_{j-1}. Then we can find the desired trapdoor td_j using $h_j[\mathrm{ID}_{j-1}](id_j)$ whose answer is represented in $h_j^{-1}[\mathrm{ID}_{j-1}]$.
3. If $|Y| = \emptyset$ then stop. Otherwise go to step 1.

Since for each id_i^* that is inserted into Y we remove at most q from I, the size of Y is at least $a := |I|/(q+1) \geq \epsilon' N/(q+1)$. Let $Enc(O)$ represent the size (in bits) of the compressed oracle. The only difference in the description of the permutations between $Enc(O)$ and O is that in the compressed oracle we are saving on the representation of $h_i^{-1}[\mathrm{ID}_{i-1}^*]$. Specifically, while $h_i^{-1}[\mathrm{ID}_{i-1}^*]$ requires $\log N!$ bits to be fully represented in O, we only need $2\log\binom{N}{a} + \log((N-a)!)$ to represent the compressed $h_i^{-1}[\mathrm{ID}_{i-1}^*]$, which consists of X, Y and Z. Furthermore, we need $n + \log(l+2) = O(n+l)$ to represent id_c^* and i. Thus, the amount of bits we save in our compression is:

$$\log N! - 2\log\binom{N}{a} - \log((N-a)!) - O(n+l)$$

and since $l = \mathrm{poly}(n)$, the overhead we incur due to representing the index i and identity id_c^* in the compressed oracle is relatively insignificant. In particular, the fraction of oracles O on which A can do ϵ'-well is at most:

$$\frac{2^{|Enc(O)|}}{2^{|O|}} = 2^{2\log\binom{N}{a}+\log((N-a)!)+O(n+l)-\log N!}$$

$$= \frac{\binom{N}{a}^2 (N-a)!}{N!} \cdot 2^{O(n+l)}$$

$$= \frac{\binom{N}{a}}{a!} \cdot 2^{O(n+l)}$$

$$\leq \left(\frac{Ne^2}{a^2}\right)^a \cdot 2^{O(n+l)}$$

If we let $q \leq 2^{n/5}$ and $\epsilon' \geq 2^{-n/5}$ we get that $a \geq \dfrac{2^{-n/5}2^n}{2^{n/5}+1} \geq 2^{3n/5}/2$.

So the upper bound reduces to $\left(\dfrac{(4)2^n e^2}{2^{6n/5}}\right)^a 2^{O(n+l)} = \left(\dfrac{4e^2}{2^{n/5}}\right)^a 2^{O(n+l)} \leq$

$2^{-a+O(n+l)} \leq 2^{-2^{3n/5-1}+O(n+l)} \leq 2^{-2^{n/2}}$ for sufficiently large n.

Proof (of Claim 18). We show here the compression of the oracle in case $i \in [1..l]$. In the following we describe how to represent the description of the injective function $h_i[\text{ID}_{i-1}^*](\cdot)$ by describing the encoding and decoding algorithms.

Encoder: Fix i and y such that Lemma 16 is satisfied. Let $N = 2^n, M = 2^m$ and let $I \subseteq \{0,1\}^n$ be the set of master public keys MPK $\in \{0,1\}^n$ for which A can successfully find *some* ID_i^* such that id_i^* was obtained by calling $h_i^{-1}[\text{ID}_{i-1}^*](td_i^*)$ without any prior invocation to $h_i[\text{ID}_{i-1}^*](id_i^*)$. Thus, $|I| \geq \epsilon'N$. Initialize the set $Y = \emptyset$. The encoder works as follows:

1. Remove the lexicographically first element $\widetilde{\text{MPK}}$ from I and put it in Y
2. Run $A^O(\widetilde{\text{MPK}}, y)$. If A asks query:
   – $h_i[\text{ID}_{i-1}^*](id_i^*)$ and $id_0^* \in I$ then remove id_0^* from I
   Note that since event E_{i-1} does not happen here, A will not call function $g_i[\text{ID}_{i-2}^*, td_{i-1}^*](\cdot)$
3. If $|I| \neq \emptyset$, go back to Step 1. Otherwise go to the next step.
4. Output the following:
   – Description of i and y
   – Description of $Y \subseteq I$
   – For each MPK $= id_0^* \in Y$:
     • Description of ID_{i-1}^* on which A was successful
     • The compressed representation of $h_i[\text{ID}_{i-1}^*]$ which consists of:
       * Description of point id_i^*
       * Description of the injective function $h_i'[\text{ID}_{i-1}^*] : [N-1] \to [M]$ on all points except id^*
       * The query index $k \in [q]$ during which $h_i^{-1}[\text{ID}_{i-1}^*](td_i^*) = id_i^*$ is called
     • The full representation of all the other injective functions $\mathcal{H} = \{h_j[\text{ID}_{j-1}] \mid \text{ID}_{j-1} \neq \text{ID}_{i-1}^*\}$
   – The full representation of all injective functions for "bad" MPK: $\mathcal{R} = \{h_j[\text{ID}_{j-1}] \mid id_0 \notin Y\}$

Decoder: Given A, the descriptions of $i, y, Y, \mathcal{H}, \mathcal{R}$, and the $|Y|$ compressed representations of $h_i[\mathrm{ID}^*_{i-1}]$ (including ID^*_{i-1} the query index k for each representation), the decoder can reconstruct O as follows:

1. Remove first lexicographically ordered MPK from Y and call it $\widetilde{\mathrm{MPK}}$. Let ID^*_{i-1} be the target identity that specifies which function has been compressed.
2. Reconstruct all the answers of $h_i(\mathrm{ID}^*_{i-1})$ using $h'_i[\mathrm{ID}^*_{i-1}]$ except for the value of $h_i(\mathrm{ID}^*_{i-1})(id^*_i)$ which is yet to be determined
3. Run $A^O(\widetilde{\mathrm{MPK}}, y)$. If A asks query:
   - $h_i^{-1}[\mathrm{ID}^*_{i-1}](td^*_i)$ for any $td^*_i \in \{0,1\}^m$:
     - If this is the kth query then we have found the corresponding id^*_i so set $h_i[\mathrm{ID}^*_{i-1}](id^*_i) = td^*_i$
     - Otherwise answer using $h'_i[\mathrm{ID}^*_{i-1}]$
   - $h_i[\mathrm{ID}^*_{i-1}](id^*_i) = td^*_i$ for any $id^*_i \in \{0,1\}^n$: answer using $h'_i[\mathrm{ID}^*_{i-1}]$
   - $h_j^{-1}[\mathrm{ID}_{j-1}](td_j)$ for any $td_j \in \{0,1\}^m$, and $id_0 \in Y$ and $\mathrm{ID}_{j-1} \neq \mathrm{ID}^*_{i-1}$: answer using the given full representation from $\mathcal{H}$. The same applies for $h_j[\mathrm{ID}_{j-1}]$ queries.
   - $h_j^{-1}[\mathrm{ID}_{j-1}](td_j)$ for any $td_j \in \{0,1\}^m$, and $id_0 \notin Y$: answer using the given full representation from $\mathcal{R}$. The same applies for $h_j[\mathrm{ID}_{j-1}]$ queries.
   - $g_j[\mathrm{ID}_{j-2}, td_{j-1}](id_j)$ for any $td_j \in \{0,1\}^m$: due to the canonical behaviour of A, $h_{j-1}^{-1}[\mathrm{ID}_{j-2}](td_{j-1})$ will be called first to get id_{j-1}. Then we can find the desired trapdoor td_j using $h_j[\mathrm{ID}_{j-1}](id_j)$ whose answer is represented in the description of $h_j^{-1}[\mathrm{ID}_{j-1}]$.
4. If $|Y| = \emptyset$ then stop. Otherwise go to step 1.

Since for each MPK that is inserted into Y we remove at most q from I, the size of Y is at least $a := |I|/(q+1) \geq \epsilon' N/(q+1)$. Let $Enc(O)$ represent the size (in bits) of the compressed oracle. The only difference in the description of the injective functions between $Enc(O)$ and O is that in the compressed oracle, we are saving on the representation of $h_i[\mathrm{ID}^*_{i-1}]$ for the master public keys represented in Y. Specifically, for each $\mathrm{MPK} \in Y$, we are compressing a single injective function from requiring $\alpha_{N,M} = \Pi_{i \in [N]}(M-i-1)$ bits to $\alpha_{N-1,M}$ bits whilst incurring an overhead of at most $n(l+1) + \log(q)$ to represent ID^*_i and the query index k. Thus if $q < 2^{n/5}$ and $m = 10nl$ and $l = \mathrm{poly}(n)$, we have that, for each good MPK, the net savings (in bits) of:

$$\log(\alpha_{N,M}) - \log(\alpha_{N-1,M}) - nl - n - \log(q) = \log\left(\frac{\alpha_{N,M}}{\alpha_{N-1,M}}\right) - nl - n - \log(q)$$
$$= \log(M - N - 1) - nl - n - \log(q)$$
$$\geq \log(M/2) - nl - n - \log(2^{n/5})$$
$$\geq m - 1 - nl - n - (n-1)$$
$$= 10nl - nl - 2n \geq 6nl$$

Since we have $a \geq \epsilon' N/(q+1)$ "good" master public keys and given that $q < 2^{n/5}$ and $\epsilon' > 2^{-n/5}$, the total number of savings we get is at least $a \times 6nl \geq (6nl)2^{3n/5}/2 \geq 2^{3n/5}$ for sufficiently large n.

Step 2: Adding the Sam^O Oracle. The second step of the proof shows that giving access to the oracle Sam^O to A does not interfere with the compression and reconstruction procedures. The argument of this step is identical to that of [23]. However, we sketch the steps of this argument for sake of completeness. Our goal here is to show that the same compression level of the oracle O (relative to which the adversary "succeeds" with non-negligible probability) could be obtained even when we add the Sam^O oracle (with an *arbitrary* fixed randomness) and allow A to call it. This would show that, with high probability over the choice of the oracle O and the randomness of the oracle Sam^O the implementation of HIBE using O remains secure.

Looking ahead, the only change would be that this time we need to use the *augmented* query complexity of the attacker instead, and we lose a factor of 3 in the success probability of A. Therefore, the hardness of the constructed HIBE using the sampled oracle O would be almost the same as before (as a parameter of ϵ and the augmented query complexity q). The augmented query complexity of an attacker A is equal to its standard query complexity to the oracle O plus the total number of *indirect* O queries in the form of O gates that are planted in circuits that are queried to from Sam^O by the adversary. The new modified proof goes through the following two steps.

1. First note that the job of the adversary is essentially to "hit" the preimage of the challenge ciphertext y or an identity id_i^*, or the master public key. This event could be either as a result of a direct query to O or as a result of an *indirect* query to O through a circuit C asked to Sam. The exact hitting event that the adversary is looking for depends on which case $E_i \wedge \neg E_{i-1}$ we are focusing on, but let us assume we are dealing with a fixed i and the adversary is able to make the event $E_i \wedge \neg E_{i-1}$ happen with a decent chance by "hitting" the trapdoor td_i^* of id_i^*. An *indirect hitting* of td_i^* would happen if and only if the adversary sends a circuit C to Sam$^O(C)$ with O gates in it and returns a collision (x, x') and either of $C(x)$ or $C(x')$ hits td_i^*.

   A crucial argument due to [23] shows that one can always modify the attacker to ask a few more queries so that it hits its goal td_i^* *directly* (before it happens indirectly) with a probability that is at most a factor of 3 less than the total probability of hitting it (directly or indirectly). The intuition behind this argument is that the distributions of the two points x and x' are both uniform over the inputs of C (even though they are distributed in a correlated way). So, if the adversary chooses a random point x'' and evaluates $C(x'')$ before asking C from the Sam^O oracle, it keeps the chance of hitting td_i^* directly at least half of hitting it indirectly![9]

   In the second part of the argument below we will safely assume that the event $E_i \wedge \neg E_{i-1} \wedge \neg$SamHit is happening with non-negligible probability, while SamHit refers to an the event that td_i^* is being hit first indirectly through a Sam query by the adversary.

---

[9] The actual argument is more subtle, but the main idea is the linearity of expectation over different probabilities.

2. The second part of the proof shows that if we start with the guarantee that $E_i \wedge \neg E_{i-1} \wedge \neg\mathsf{SamHit}$ is happening with a noticeable probability, we can achieve the *same* compression of the oracle O *even if we fix the randomness of Sam*. This argument indeed holds because of the way our compression and decompression algorithms work. Note that at the heart of our compression and decompression algorithms we basically run the adversary over different inputs till it hits a special point. What is crucial in these arguments is that while we have not hit the final point of interest we can still continue the execution of the adversary and hope that the answer to the current queries are already reconstructed. Now if we have the guarantee that no $\mathsf{Sam}^O(C)$ query by adversary is hitting td_i^* indirectly, and if we have already fixed the randomness of the Sam oracle, we can still run the encoding and decoding algorithms with almost no change. Namely, suppose C is a circuit query to Sam. The first thing Sam does is to run C on a random (but not fixed) input x. We have the guarantee that the execution of $C(x)$ does not encounter any query whose answer is not already reconstructed. Moreover, the second point x' is the lexicographically first input to C such that $C(x) = C(x')$ where x' is being chosen from a random permutation (that is also fixed!) over the inputs of C. To find the same x' while doing the reconstruction, all we have to do is to run C over all inputs one by one using the same permutation order (that is now fixed) till we manage to finish an execution $C(x')$ that happens to output the same $C(x)$. This means that we can run the same encoding and decoding algorithms even in the presence of Sam oracle.

Proof (of Claim 12). Given the implementation of the HIBE scheme using O in Construction 14, we prove the first part of the claim, by referring to Claims 17 and 18. In particular, the combined claims show that for any given adversary of the HIBE scheme whose goal is to invert its challenge ciphertext for an identity vector of its choice, the probability of doing so is negligible in the security parameter when it is trying to invert an identity at level i. Thus, a union bound over all possible $i \in [l]$, where $l = \mathrm{poly}(n)$ still results in negligible probability of success. The second part of the claim (that is, that the HIBE is secure even in the presence of Sam^O) follows by extension from the discussion in Sect. 3.3, and in particular from the techniques of [23].

Acknowledgement. We thank Vinod Vaikuntanathan for pointing out to us the connection between our results and the work of [2].

References

1. Agrawal, S., Boneh, D., Boyen, X.: Efficient lattice (H)IBE in the standard model. In: Gilbert, H. (ed.) EUROCRYPT 2010. LNCS, vol. 6110, pp. 553–572. Springer, Heidelberg (2010)
2. Asharov, G., Segev, G.: Limits on the power of indistinguishability obfuscation and functional encryption. Cryptology ePrint Archive, Report 2015/341 (2015). http://eprint.iacr.org/

3. Bellare, M., Rogaway, P.: Random oracles are practical: a paradigm for designing efficient protocols. In: ACM Conference on Computer and Communications Security, pp. 62–73 (1993)
4. Boneh, D., Boyen, X.: Efficient selective-ID secure identity-based encryption without random oracles. In: Cachin, C., Camenisch, J.L. (eds.) EUROCRYPT 2004. LNCS, vol. 3027, pp. 223–238. Springer, Heidelberg (2004)
5. Boneh, D., Boyen, X., Goh, E.-J.: Hierarchical identity based encryption with constant size ciphertext. In: Cramer, R. (ed.) EUROCRYPT 2005. LNCS, vol. 3494, pp. 440–456. Springer, Heidelberg (2005)
6. Boneh, D., Canetti, R., Halevi, S., Katz, J.: Chosen-ciphertext security from identity-based encryption. SIAM J. Comput. 36(5), 1301–1328 (2006)
7. Boneh, D., Franklin, M.K.: Identity-based encryption from the Weil pairing. SIAM J. Comput. 32(3), 586–615 (2003)
8. Boneh, D., Papakonstantinou, P.A., Rackoff, C., Vahlis, Y., Waters, B.: On the impossibility of basing identity based encryption on trapdoor permutations. In: FOCS, pp. 283–292. IEEE Computer Society (2008)
9. Boneh, D., Sahai, A., Waters, B.: Functional encryption: definitions and challenges. In: Ishai, Y. (ed.) TCC 2011. LNCS, vol. 6597, pp. 253–273. Springer, Heidelberg (2011)
10. Canetti, R., Halevi, S., Katz, J.: A forward-secure public-key encryption scheme. In: Biham, E. (ed.) EUROCRYPT 2003. LNCS, vol. 2656, pp. 255–271. Springer, Heidelberg (2003)
11. Chung, K.-M., Lin, H., Mahmoody, M., Pass, R.: On the power of nonuniformity in proofs of security. In: Proceedings of the 4th Conference on Innovations in Theoretical Computer Science, pp. 389–400. ACM (2013)
12. Damgård, I.B., Pedersen, T.P., Pfitzmann, B.: On the existence of statistically hiding bit commitment schemes and fail-stop signatures. J. Cryptol. 10(3), 163–194 (1997)
13. Gennaro, G., Katz.: Lower bounds on the efficiency of encryption and digital signature schemes. In: STOC: ACM Symposium on Theory of Computing (STOC) (2003)
14. Gennaro, R., Gertner, Y., Katz, J., Trevisan, L.: Bounds on the efficiency of generic cryptographic constructions. SIAM J. Comput. 35(1), 217–246 (2005)
15. Gennaro, R., Trevisan, L.: Lower bounds on the efficiency of generic cryptographic constructions. In: FOCS, pp. 305–313 (2000)
16. Gentry, C., Halevi, S., Raykova, M., Wichs, D.: Garbled ram revisited, part i. Cryptology ePrint Archive, Report 2014/082 (2014). http://eprint.iacr.org/
17. Gentry, C., Silverberg, A.: Hierarchical ID-based cryptography. In: Zheng, Y. (ed.) ASIACRYPT 2002. LNCS, vol. 2501, pp. 548–566. Springer, Heidelberg (2002)
18. Gertner, Y., Kannan, S., Malkin, T., Reingold, O., Viswanathan, M.: The relationship between public key encryption and oblivious transfer. In: FOCS, pp. 325–335 (2000)
19. Goldreich, O., Goldwasser, S., Micali, S.: How to construct random functions. J. ACM 33(4), 792–807 (1986)
20. Goldreich, O., Levin, L.A.: A hard-core predicate for all one-way functions. In: Proceedings of 21st STOC, pp. 25–32. ACM (1989)
21. Goldwasser, S., Micali, S.: Probabilistic encryption. J. Comput. Syst. Sci. 28(2), 270–299 (1984)
22. Goyal, V., Kumar, V., Lokam, S., Mahmoody, M.: On black-box reductions between predicate encryption schemes. In: Cramer, R. (ed.) TCC 2012. LNCS, vol. 7194, pp. 440–457. Springer, Heidelberg (2012)

23. Haitner, I., Hoch, J.J., Reingold, O., Segev, G.: Finding collisions in interactive protocols - a tight lower bound on the round complexity of statistically-hiding commitments. In: 48th Annual IEEE Symposium on Foundations of Computer Science (FOCS), 20–23 October 2007, Providence, RI, USA, pp. 669–679. IEEE Computer Society (2007)
24. Håstad, J., Impagliazzo, R., Levin, L.A., Luby, M.: A pseudorandom generator from any one-way function. SIAM J. Comput. **28**(4), 1364–1396 (1999)
25. Horwitz, J., Lynn, B.: Toward hierarchical identity-based encryption. In: Knudsen, L.R. (ed.) EUROCRYPT 2002. LNCS, vol. 2332, pp. 466–481. Springer, Heidelberg (2002)
26. Hsiao, C.-Y., Reyzin, L.: Finding collisions on a public road, or do secure hash functions need secret coins? In: Franklin, M. (ed.) CRYPTO 2004. LNCS, vol. 3152, pp. 92–105. Springer, Heidelberg (2004)
27. Impagliazzo, R., Luby, M.: One-way functions are essential for complexity based cryptography (extended abstract). In: FOCS, pp. 230–235 (1989)
28. Impagliazzo, R., Rudich, S.: Limits on the provable consequences of one-way permutations. In: Proceedings of the 21st Annual ACM Symposium on Theory of Computing (STOC), pp. 44–61. ACM Press (1989)
29. Ishai, Y., Kushilevitz, E., Ostrovsky, R.: Sufficient conditions for collision-resistant hashing. In: Kilian, J. (ed.) TCC 2005. LNCS, vol. 3378, pp. 445–456. Springer, Heidelberg (2005)
30. Lindell, Y.: A simpler construction of CCA2-secure public-key encryption under general assumptions. J. Cryptol. **19**(3), 359–377 (2006)
31. Luby, M., Rackoff, C.: How to construct pseudorandom permutations from pseudorandom functions. SIAM J. Comput. **17**(2), 373–386 (1988)
32. Mahmoody, M., Mohammed, A.: On the power of hierarchical identity-based encryption. Cryptology ePrint Archive, Report 2015/815 (2015). http://eprint.iacr.org/
33. Maurer, U.M., Yacobi, Y.: Non-interactive public-key cryptography. In: Davies, D.W. (ed.) Advances in Cryptology EUROCRYPT 1991. LNCS, vol. 547, pp. 498–507. Springer, Heidelberg (1991)
34. Naor, M.: Bit commitment using pseudorandomness. J. Cryptol. **4**(2), 151–158 (1991)
35. Naor, M., Yung, M.: Universal one-way hash functions and their cryptographic applications. In: Proceedings of the 21st Annual ACM Symposium on Theory of Computing (STOC), pp. 33–43. ACM Press (1989)
36. Naor, M., Yung, M.: Public-key cryptosystems provably secure against chosen ciphertext attacks. In: Proceedings of the 22nd STOC, pp. 427–437. ACM Press (1990)
37. Naor, M., Ziv, A.: Primary-secondary-resolver membership proof systems. In: Dodis, Y., Nielsen, J.B. (eds.) TCC 2015, Part II. LNCS, vol. 9015, pp. 199–228. Springer, Heidelberg (2015)
38. Pandey, O., Pass, R., Vaikuntanathan, V.: Adaptive one-way functions and applications. In: Wagner, D. (ed.) CRYPTO 2008. LNCS, vol. 5157, pp. 57–74. Springer, Heidelberg (2008)
39. Regev, O.: On lattices, learning with errors, random linear codes, and cryptography. In: STOC: ACM Symposium on Theory of Computing (STOC) (2005)
40. Reingold, O., Trevisan, L., Vadhan, S.P.: Notions of reducibility between cryptographic primitives. In: Naor, M. (ed.) TCC 2004. LNCS, vol. 2951, pp. 1–20. Springer, Heidelberg (2004)

41. Rompel, J.: One-way functions are necessary and sufficient for secure signatures. In: STOC, pp. 387–394 (1990)
42. Sahai, A.: Non-malleable non-interactive zero knowledge and adaptive chosen-ciphertext security. In: Proceedings of the 40th Annual Symposium on Foundations of Computer Science (FOCS), pp. 543–553 (1999)
43. Sahai, A., Waters, B.: Fuzzy identity-based encryption. In: Cramer, R. (ed.) EUROCRYPT 2005. LNCS, vol. 3494, pp. 457–473. Springer, Heidelberg (2005)
44. Shamir, A.: Identity-based cryptosystems and signature schemes. In: Blakely, G.R., Chaum, D. (eds.) CRYPTO 1984. LNCS, vol. 196, pp. 47–53. Springer, Heidelberg (1985)
45. Simon, D.R.: Findings collisions on a one-way street: can secure hash functions be based on general assumptions? In: Nyberg, K. (ed.) EUROCRYPT 1998. LNCS, vol. 1403, pp. 334–345. Springer, Heidelberg (1998)
46. Unruh, D.: Random oracles and auxiliary input. In: Menezes, A. (ed.) CRYPTO 2007. LNCS, vol. 4622, pp. 205–223. Springer, Heidelberg (2007)
47. Brent Waters.: A punctured programming approach to adaptively secure functional-encryption. Cryptology ePrint Archive, Report 2014/588 (2014). http://eprint.iacr.org/
48. Yao, A.C.: Theory and applications of trapdoor functions. In: Proceedings of the 23rd FOCS, pp. 80–91. IEEE (1982)

On the Impossibility of Tight Cryptographic Reductions

Christoph Bader, Tibor Jager$^{(\boxtimes)}$, Yong Li, and Sven Schäge$^{(\boxtimes)}$

Horst Görtz Institute for IT Security, Ruhr-University Bochum,
Bochum, Germany
sschaege@gmail.com

Abstract. The existence of *tight* reductions in cryptographic security proofs is an important question, motivated by the theoretical search for cryptosystems whose security guarantees are truly independent of adversarial behavior and the practical necessity of concrete security bounds for the theoretically-sound selection of cryptographic parameters. At Eurocrypt 2002, Coron described a *meta-reduction* technique that allows to prove the *impossibility* of tight reductions for certain digital signature schemes. This seminal result has found many further interesting applications. However, due to a technical subtlety in the argument, the applicability of this technique beyond digital signatures in the *single-user* setting has turned out to be rather limited. We describe a new meta-reduction technique for proving such impossibility results, which improves on known ones in several ways. It enables interesting novel applications, including a formal proof that for certain cryptographic primitives (including public-key encryption/key encapsulation mechanisms and digital signatures), the security loss incurred when the primitive is transferred from an idealized single-user setting to the more realistic multi-user setting is *impossible* to avoid, and a lower tightness bound for non-interactive key exchange protocols. Moreover, the technique allows to rule out tight reductions from a very general class of non-interactive complexity assumptions. Furthermore, the proofs and bounds are simpler than in Coron's technique and its extensions.

1 Introduction

Provable Security. In modern cryptography, new cryptosystems are usually constructed together with a *proof of security*. Usually this security proof consists of a reduction Λ (in a complexity-theoretic sense), which turns an efficient adversary $\mathcal{A}$ into a machine $\Lambda^{\mathcal{A}}$ solving a well-studied, assumed-to-be-hard computational problem. Under the assumption that this computational problem is not efficiently solvable, this implies that the cryptosystem is secure. This approach is usually called "provable security", it is inspired by the analysis of relations between computational problems in complexity theory, and allows to show that

T. Jager—Supported by DFG grant JA 2445/1-1.
S. Schäge—Supported by UbiCrypt, DFG grant GRK 1817/1.

M. Fischlin and J.-S. Coron (Eds.): EUROCRYPT 2016, Part II, LNCS 9666, pp. 273–304, 2016.
DOI: 10.1007/978-3-662-49896-5_10

breaking the security of a cryptosystem is at least as hard as solving a certain well-defined hard computational problem.

The Security Loss in Reduction-Based Security Proofs. The "quality" of a reduction can be measured by comparing the running time and success probability of $\Lambda^\mathcal{A}$ to the running time and success probability of attacker $\mathcal{A}$. Ideally, $\Lambda^\mathcal{A}$ has about the same running time and success probability as $\mathcal{A}$. However, most security proofs describe reductions where $\Lambda^\mathcal{A}$ has either a significantly larger running time or a significantly smaller success probability than $\mathcal{A}$ (or both). Thus, the reduction "loses" efficiency and/or efficacy.

Since provable security is inspired by classical complexity theory, security proofs have traditionally been formulated asymptotically. The running time and success probability of Turing machines are modeled as functions in a *security parameter* $k \in \mathbb{N}$. Let $t_{\Lambda^\mathcal{A}}(k)$ denote the running time and $\epsilon_{\Lambda^\mathcal{A}}(k)$ denote the success probability of $\Lambda^\mathcal{A}$. Likewise, let $t_\mathcal{A}(k)$ and $\epsilon_\mathcal{A}(k)$ denote the running time and success probability of $\mathcal{A}$. Then it holds that

$$t_{\Lambda^\mathcal{A}}(k)/\epsilon_{\Lambda^\mathcal{A}}(k) = \ell(k) \cdot t_\mathcal{A}(k)/\epsilon_\mathcal{A}(k)$$

for some "loss" $\ell(k)$. A reduction Λ is considered *efficient*, if its loss $\ell(k)$ is bounded by a polynomial. Note that in this approach the concrete size of polynomial ℓ (i.e., its degree and the size of its coefficients) does not matter. As common in classical complexity theory, it was considered sufficient to show that ℓ is polynomially-bounded.

Concrete Security Proofs, the Notion of Tightness, and Its Relevance. In order to deploy a cryptosystem in practice, the size of cryptographic parameters (like for instance the length of moduli or the size of underlying algebraic groups) has to be selected. However, the asymptotic approach described above does not allow to derive concrete recommendations for such parameters, as it only shows that sufficiently large parameters *exist*. This is because the size of parameters depends on the concrete value of ℓ, the loss of the reduction. A larger loss requires larger parameters.

The more recent approach, termed *concrete security*, makes the concrete security loss of a reduction explicit. This allows to derive concrete recommendations for parameters in a theoretically sound way (see e.g. [7] for a detailed treatment). Ideally, $\ell(k)$ is constant. In this case the reduction is said to be *tight*.[1] The existence of cryptosystems whose security is independent of deployment parameters is of course an interesting theoretical question in its own right. Moreover, it has a strong practical motivation, because the tightness of a reduction directly influences the selection of the size of cryptographic parameters, and thus has a direct impact to the efficiency of cryptosystems.

---

[1] When speaking of tight reductions in this paper, we mean tight reductions from non-interactive computational problems, like integer factorization, the discrete logarithm problem, etc., rather than (often trivial) tight reductions from interactive or contrived non-standard computational problems, which sometimes are very similar to the assumption that the cryptosystem is secure.

Coron's Result and Its Refinements. Coron [18] considered the existence of tight reductions for *unique*[2] signature schemes in the single user setting, and described a "rewinding argument" (cf. Goldwasser *et al.* [27]), which allowed to prove lower tightness bounds for such signature schemes. In particular, Coron considered "simple"[3] reductions, which convert a forger F breaking the security[4] of a unique signature scheme into a machine solving a computationally hard problem Π. He showed that any such reduction yields an algorithm $\mathcal{B}$ solving Π *directly* with probability $\epsilon_{\mathcal{B}}$, where

$$\epsilon_{\mathcal{B}} \geq \epsilon_{\Lambda} - \frac{\epsilon_F}{\exp(1) \cdot n} \cdot \left(1 - \frac{n}{|\mathcal{M}|}\right)^{-1}. \tag{1}$$

Here ϵ_{Λ} is the success probability of Λ, ϵ_F is the success probability of the signature forger F used by Λ, n is the number of signatures queried by F in the EUF-CMA security experiment, and $|\mathcal{M}|$ is the size of the message space. Note that if $|\mathcal{M}| \gg n$, which is a reasonable for signature schemes, then the bound in (1) essentially implies that the success probability of ϵ_{Λ} of the reduction can not substantially exceed $\epsilon_F/(\exp(1) \cdot n)$, unless there exists an algorithm $\mathcal{B}$ solving Π efficiently. The latter, however, contradicts the hardness assumption on Π. This result was later revisited by Kakvi and Kiltz [31], and generalized by Hofheinz *et al.* [30] to (non-unique) signature schemes with efficiently rerandomizable signatures, see also Appendix A.

Limitations of Known Meta-Reductions. Unfortunately, Coron's result has found only limited applications beyond digital signatures in the single-user setting. Most previous works [18,30,31] consider this setting, the (to our best knowledge) only exception is due to Lewko and Waters [33], which considers hierarchical identity-based encryption. Why isn't it possible to apply it to other primitives? One reason is that the bound in Eq. (1) ceases to be useful for reasonable values of ϵ_{Λ} and ϵ_F if $n \approx |\mathcal{M}|$. This can be easily seen by setting $n = |\mathcal{M}| - 1$. The assumption that $|\mathcal{M}| \gg n$ is a prerequisite for the arguments in [18,30,31] to work, thus, it is not possible to apply this technique to settings, where the assumption $|\mathcal{M}| \gg n$ is *not* reasonable.

Therefore Coron's technique is not applicable when $|\mathcal{M}|$ is polynomially-bounded. However, such a situation appears often when considering cryptographic primitives beyond digital signatures in the single-user setting. Consider, for instance, a security model where the adversary is provided with

---

[2] For a unique signature scheme there exists exactly one unique valid signature for each message. For instance, important instantiations of the famous Full-Domain Hash construction are unique signature schemes, see [31].

[3] Intuitively, a "simple" reduction is a reduction which has black-box access to the adversary, and runs the adversary only *sequentially*. Most reductions in cryptographic security proofs are of this type. A more precise definition is given in the body of the paper.

[4] In the sense of existential unforgeability under chosen-message attacks (EUF-CMA, cf. Definition 18).

$\mathcal{M} = \{pk_1, \ldots, pk_n\}$, where $pk_1, \ldots, pk_n$ is a list of public keys. The adversary may learn all but one of the corresponding secret keys, and is considered successful if it "breaks security" with respect to an uncorrupted key. This is a quite common setting, which occurs for instance in security models for signatures or public-key encryption in the multi-user setting with corruptions [3,4], all common security models for authenticated key exchange [4,9,15], and non-interactive key exchange [25] protocols. *How can we analyze the existence of inherent tightness bounds in these settings?*

Our Contributions. We develop a new meta-reduction technique, which is also applicable in settings where $|\mathcal{M}|$ is polynomially bounded. In comparison to [18,30,31], we achieve the simpler bound

$$\epsilon_\mathcal{B} \geq \epsilon_\mathcal{A} - 1/n.$$

which is independent of $|\mathcal{M}|$.

Our new technique allows to rule out tight reductions from any non-interactive complexity assumption (cf. Definition 5). This includes also "decisional" assumptions (like decisional Diffie-Hellman). It avoids the combinatorial lemma of Coron [18, Lemma 1], which has a relatively technical proof. Our approach does not require such a combinatorial argument, but is more "direct".

This simplicity allows us to describe a generalized experiment with an abstract computable relation that captures the necessary properties for our tightness bounds. Then we explain that the standard security experiments for many cryptographic primitives are specific instances of this abstract experiment.

Technical Idea. To describe our technical idea, let us consider the example of digital signatures in the single-user settings, as considered in [18,30,31], for this introduction. As sketched above, the result will later be generalized and applied to other settings as well. We consider a weakened signature security definition, where the security experiment proceeds as follows.

1. The adversary receives as input a verification key vk along with n random but pairwise distinct messages $m_1, \ldots, m_n$.
2. The adversary selects an index j^*, and receives in response $n - 1$ signatures σ_i for all messages m_i with $i \neq j^*$.
3. Finally, the adversary wins the experiment if it outputs σ^* that is a valid signature for m_{j^*} with respect to j^*.

Note that this is a very weak security definition, because the adversary is only able to observe signatures of random messages. However, note also that any lower tightness bound for such a weaker security definition implies a corresponding bound for any stronger definition. In particular, the above definition is weaker than the standard security definition *existential unforgeability under chosen message attacks* considered in [18,30,31], where messages may be adaptively chosen by the adversary.

Essentially, we argue that once a reduction has started the adversary in Step 1 of the above experiment, and thus has "committed" to a verification key

vk and messages $m_1, \ldots, m_n$, there can only be a single choice of j^* for which this reduction is able to output valid signatures σ_i for all $i \neq j^*$. Thus, for any adversary which chooses j^* uniformly at random the reduction has probability at most $1/n$ to succeed. We prove this by contradiction, by showing essentially that any reduction which is successful for two distinct choices of j^*, say j_0, j_1, can be used to construct a machine that breaks the underlying security assumption directly.

Technically, we proceed in two steps: first we describe an *inefficient* adversary against the reduction which chooses j^* uniformly random, and computes the signature σ^* for m_{j^*} by exhaustive search. Next, we show that this adversary can efficiently be simulated by our meta-reduction, if the reduction could succeed for two different choices j_0 and j_1 after committing to $(vk, m_1, \ldots, m_n)$. The meta-reduction simulates the inefficient adversary by rewinding the reduction. Essentially, if the reduction could succeed for two different values j_0, j_1, then it must also be able output the signatures for *all* n messages. Therefore we start the reduction and let it run until it reaches a "break point" where it outputs $(vk, m_1, \ldots, m_n)$. Next, we run the reduction n-times, each time starting from the break point and using a different index j, to search for two values j_0, j_1 such that $j_0 \neq j_1$ such that the reduction outputs valid signatures for all-but-one messages. If indeed there exist two such indices j_0, j_1, then we now have learned signatures for all messages $(m_1, \ldots, m_n)$ which are valid w.r.t. vk. Thus, we can run the reduction one last time from the break point, this time to the end, using index j_0 (or equivalently j_1), and we simulate the inefficient adversary using the fact that we know a valid signature for m_{j_0} (or m_{j_1}). Importantly, in the last execution of the reduction we are able to simulate the inefficient adversary perfectly, so the reduction will help us to break the non-interactive complexity assumption.

We caution that the rigorous proof of the above is more complex than the intuition provided in this introduction, and we have to put restrictions on the signature scheme, which depend on the considered application. For instance, when considering signatures in the single-user setting as above, we have to require that signatures are efficiently re-randomizable. In the generalized setting we will consider other applications, which require different but usually simple-to-check properties, like for instance that for each public key vk there exists a *unique* secret key. In this way, our result provides simple criteria to check whether a cryptographic construction can have a tight proof at all. At the same time it implicitly provides guidelines for the construction of tightly secure cryptographic schemes, since all tightly secure constructions must circumvent our result in one way or the other.

The fact that we consider a weakened security experiment has several nice features. We think that the approach and its analysis described above are much simpler than previous works, which enables more involved impossibility results. We will show that it achieves a simpler bound and yields a qualitatively stronger result, as it even rules out tight reductions for such weak security experiments. Like previous works, we only consider reductions that execute the adversary sequentially and in a black-box fashion. We stress that most reductions in cryptography have this property.

We generalize the above idea from signature schemes in a single-user setting to abstract relations, which capture the relevant properties required for our impossibility argument to go through. We show that this abstraction allows to apply the result relatively easily to other cryptographic primitives, by describing applications to public-key encryption and signatures in the multi-user setting, and non-interactive key exchange.

Overview of Applications. A first, immediate application of our new technique are strengthened versions of the results of [18,30,31], but with significantly simpler proofs and tightness bounds even for weaker security notions (which is a stronger result). In contrast to previous works [18,30,31], the impossibility results hold also for "decisional" complexity assumptions.

Additionally, the fact that our meta-reduction does not require the combinatorial lemma of Coron enables further, novel applications in settings with polynomially-bounded spaces (where Coron's result worked only for exponential-sized spaces). As a first novel application of our generalized theorem, we analyze the tightness loss that occurs when security proofs in idealized single-user settings are transferred to the more realistic multi-user setting. Classical security models for standard cryptographic primitives often consider an idealized setting. For instance, the standard IND-CPA and IND-CCA security experiments for public-key encryption consider a setting with only one challenge public key and only a single challenge ciphertext. This is of course unrealistic for many practical applications. Public-key encryption is typically used in settings where an attacker sees many public keys and ciphertexts, and is (potentially) able to corrupt secret keys adaptively. Even though there is a reduction from breaking security in the multi-user setting to breaking security in the idealized setting, this reduction comes with a security loss which is linear in the number of users and ciphertexts. We show that under certain conditions (e.g., for schemes where there exists a *unique* secret key for each public key) this loss is impossible to avoid. This gives an insight into which properties a cryptosystem must or must not meet in order to allow a tight reduction in the multi-user setting.

Another novel application is the analysis of the existence of *non-interactive* key exchange (NIKE). In non-interactive key exchange (NIKE) two parties are able to derive a common shared secret. However, in contrast to traditional key exchange protocols, they do not need to exchange any messages. Besides the secret key of one party the key derivation algorithm only requires the availability of the public key of the communication partner. Security is defined solely by requiring indistinguishability of the derived shared secret from a random value. We show how to apply our main result to rule out tight reductions for a large class of NIKE protocols from a standard assumption in any sufficiently strong security model (such as the CKS-heavy model from [25]).

On Certified Public Keys and the Results of Kakvi and Kiltz. Several years after the publication of the paper of Coron [18] it has turned out that this paper contains a subtle technical flaw. Essentially, it is implicitly assumed that the value output by the reduction to the adversary is a *correct* signature public key (recall that Coron considered only digital signature schemes in the single-user setting). This misses the fact that a reduction may possibly output *incorrect* keys which are computationally indistinguishable from correct ones. Indeed, such keys lead to the technical problem that a meta-reduction may *not* be able to simulate the adversary constructed in the meta-reduction of Coron correctly.

This flaw was identified and corrected by Kakvi and Kiltz [31]. Essentially, Kakvi and Kiltz enforce that the reduction outputs only public keys which can be efficiently recognized as correct, by introducing the notion of *certified* public keys. A different (but similar in spirit), slightly more general approach is due to Hofheinz *et al.* [30], who require that signatures are *efficiently re-randomizable* with respect to the public key output by from the reduction (regardless of whether this key is correct or not). Both these approaches [30,31] essentially overcome the subtle issue from Coron's paper by ensuring that the adversaries simulated by the meta-reductions are always able to output *correctly distributed* signatures.

In this paper, we introduce the notion of *efficiently re-randomizable relations* to overcome the subtle issue pointed out by Kakvi and Kiltz [31]. This notion further generalizes the approach of [30] in a way that suits our more general setting.

Relation to Tightly-Secure Constructions. There exist various constructions of tightly-secure cryptosystems, which have to avoid our impossibility results in one way or another. The signature schemes constructed in [1,10,19,29,32,36], for example, are tightly-secure in a single-user setting. They avoid our impossibility result because they do not have unique signatures or no efficient re-randomization algorithm is known. The same holds for the signature schemes derived from the IBE schemes of [11,17]. Bader *et al.* [4] constructed signature schemes with tight security even in the multi-user setting with adaptive secret-key corruptions. Again, our impossibility results are avoided here because signatures are not efficiently re-randomizable. The encryption schemes of Bellare, Boldyreva and Micali [6] are tightly-secure in a multi-user setting, but only without corruptions. We consider impossibility results for the multi-user setting with corruptions. The key encapsulation mechanism presented in [4] is tightly-secure even in a multi-user setting with corruptions. It avoids our impossibility result because it does not have unique secret keys.

More Related Work. Since their introduction by Boneh and Venkatesan in 1998 [12] meta-reductions have proven to be a versatile tool in many areas of provably security. Previous works have mainly used meta-reductions to derive impossibility results and efficiency/security bounds on signatures schemes [5,20–22,24,26,34,37], blind-signature schemes [23] and encryption systems [35]. In particular, among these results there exist several works that consider the existence of (tight) security proofs for the Schnorr signature scheme [5,24,26,34,37].

The results in [13,14] use meta-reductions to derive relationships among cryptographic one-more type problems. Lewko and Waters [33], building on [30], showed that under certain conditions it is impossible to prove security of hierarchical IBE (HIBE) schemes. To this end, Lewko and Waters extend the approach of [30] from signatures to hierarchical IBE to show that for certain HIBE schemes an *exponential* tightness loss is impossible to avoid. Finally, the inexistence of certain meta-reductions was considered in [22].

Outline. We begin with considering essentially the same setting as Coron and follow-up works [18,30,31], namely digital signatures in the single-user setting, as an instructive example. We prove a strengthened variant of the results of [18,30,31]. This allows us to explain how our new technique works in a known setting, which may be helpful for readers already familiar with these works. A generalized, much more abstract version will be presented in Sects. 4 and 5 gives many further interesting applications, which seem not achievable using the previous approach of [18,30,31].

2 The New Meta-reduction Technique

2.1 Preliminaries

Notation. We write $[n]$ to denote the set $[n] := \{1, 2, \ldots, n\}$, and for $j \in [n]$ we write $[n \backslash j]$ to denote the set $[n] \backslash \{j\}$. If A is a set then $a \leftarrow^{\$} A$ denotes the action of sampling a uniformly from A. Given a set A we denote by U_A the uniform distribution on A. If A is a Turing machine (TM) then $a \leftarrow A(x; r)$ denotes that A outputs a when run with input x and random coins r. By $A(x)$ we denote the distribution of $a \leftarrow A(x; r)$ over the uniform choice of r. If x is a binary string, then $|x|$ denotes its length. If M is a Turing machine, we denote by $\widehat{M}$ its description as a bitstring.

If $t : \mathbb{N} \rightarrow \mathbb{N}$ and there exists a constant c such that $t(k) \leq k^c$ for all but finitely many $k \in \mathbb{N}$, then we say that $t \in \mathsf{poly}(k)$. We denote by $\mathsf{poly}^{-1}(k)$ the set $\mathsf{poly}^{-1}(k) := \{\delta : \frac{1}{\delta} \in \mathsf{poly}(k)\}$. We say that $\epsilon : \mathbb{N} \rightarrow [0, 1]$ is *negligible* if for all $c \in \mathbb{N}$ it holds that $\epsilon(k) > k^{-c}$ is true only for at most finitely many $k \in \mathbb{N}$. We write $\epsilon \in \mathsf{negl}(k)$ to denote that ϵ is negligible.

Digital Signatures. A digital signature scheme $\mathsf{SIG} = (\mathsf{Setup}, \mathsf{Gen}, \mathsf{Sign}, \mathsf{Vfy})$ is a four-tuple of PPT-TMs:

Public Parameters. The public parameter generation machine $\Pi \leftarrow^{\$}$ $\mathsf{Setup}(1^k)$ takes the security parameter k as input and returns public parameters Π.

Key Generation. The key generation machine takes as input public parameters Π and outputs a key pair, $(vk, sk) \leftarrow^{\$} \mathsf{Gen}(\Pi)$.

Signing. The signing machine takes as input a secret key sk and a message m and returns a signature $\sigma \leftarrow^{\$} \mathsf{Sign}(sk, m)$.

Verification. The verification machine, on input a public key vk, a signature σ and a message m, outputs 0 or 1, $\mathsf{Vfy}(vk, m, \sigma) \in \{0, 1\}$.

Game UF-SMA$_{\text{SIG}}^{n,\mathcal{A}}\left(1^k\right)

Game UF-SMA$_{\text{SIG}}^{n,\mathcal{A}}\left(1^k\right)$
$\Pi \leftarrow^{\$} \text{SIG.Setup}(1^k); \rho_{\mathcal{A}} \leftarrow^{\$} \{0,1\}^k$
$(vk, sk) \leftarrow^{\$} \text{SIG.Gen}(\Pi)$
$m_1, \dots, m_n \leftarrow^{\$} \mathcal{M}$ s.t. $m_i \neq m_j$ for all $i \neq j$
$\sigma_i \leftarrow^{\$} \text{SIG.Sign}(sk, m_i)$ for all $i \in [n]$
$(j, st) \leftarrow \mathcal{A}_1(vk, (m_i)_{i \in [n]}; \rho_{\mathcal{A}})$
$\sigma_j \leftarrow \mathcal{A}_2\left(st, (\sigma_i)_{i \in [n \setminus j]}\right)$
return $\text{SIG.Vfy}(vk, m_j, \sigma_j)$

Fig. 1. The UF-SMA-security game with attacker $\mathcal{A} = (\mathcal{A}_1, \mathcal{A}_2)$.

Unique and Re-Randomizable Signatures. Let $\Sigma(vk, m) := \{\sigma : \text{Vfy}(vk, m, \sigma) = 1\}$ denote the set of all valid signatures σ w.r.t. a given message m and verification key vk.

Definition 1 (Unique signatures). *We say that* SIG *is a* unique *signature scheme, if* $|\Sigma(vk, m)| = 1$ *for all* vk *and* m.

Definition 2 (Re-randomizable signatures). *We say that* SIG *is* t_{ReRand}-*re-randomizable, if there exists a TM* SIG.ReRand *which takes as input* (vk, m, σ) *and outputs a signature* $\sigma' \leftarrow^{\$}$ SIG.ReRand(vk, m, σ) *with the following properties.*

1. SIG.ReRand *runs in time at most* t_{ReRand}
2. *If* $\text{Vfy}(vk, m, \sigma) = 1$, *then* σ' *is distributed uniformly over* $\Sigma(vk, m)$.

Remark 1. Note that we do not put any bounds on t_{ReRand}. Thus, any signature scheme is t_{ReRand}-re-randomizable for sufficiently large t_{ReRand}. However, there are many examples of signature schemes which are *efficiently* re-randomizable, like the class of schemes considered in [30]. In particular, all unique signature schemes are efficiently re-randomizable by the Turing machine $\sigma \leftarrow^{\$}$ SIG.ReRand(vk, m, σ) which simply outputs its input σ.

Unforgeability Under Static Message Attacks. The UF-SMA security experiment is depicted in Fig. 1.

Definition 3. *Let* UF-SMA$_{\text{SIG}}^{n,\mathcal{A}}\left(1^k\right)$ *denote the UF-SMA security experiment depicted in Fig. 1, executed with signature scheme* SIG *and attacker* $\mathcal{A} = (\mathcal{A}_1, \mathcal{A}_2)$. *We say that* $\mathcal{A}$ $(t_{\mathcal{A}}, n, \epsilon_{\mathcal{A}})$-*breaks the* UF-SMA-*security of* SIG, *if it runs in time* $t_{\mathcal{A}}$ *and*

$$\Pr\left[\text{UF-SMA}_{\text{SIG}}^{n,\mathcal{A}}\left(1^k\right) \Rightarrow 1\right] \geq \epsilon_{\mathcal{A}}.$$

Remark 2. Observe that the messages in the UF-SMA security experiment from Fig. 1 are chosen at random (but pairwise distinct). We do this for simplicity, but stress that for our tightness bound we actually do not have to make any assumption about the distribution of messages, apart from being pairwise distinct. For instance, the messages could alternatively be the lexicographically first n messages of the message space, for instance.

Non-interactive Complexity Assumptions. The following very general definition of non-interactive complexity assumptions is due to Abe et al. [2].

Definition 4. *A non-interactive complexity assumption $N = (\mathsf{T}, \mathsf{V}, \mathsf{U})$ consists of three TMs. The instance generation machine $(c, w) \leftarrow^{\$} \mathsf{T}(1^k)$ takes the security parameter as input, and outputs a problem instance c and a witness w. U is a probabilistic polynomial-time machine, which takes as input c and outputs a candidate solution s. The verification TM V takes as input (c, w) and a candidate solution s. If $\mathsf{V}(c, w, s) = 1$, then we say that s is a correct solution to the challenge c.*

Intuitively, U is a probabilistic polynomial-time machine which implements a suitable "trivial" attack strategy for N. This algorithm is used to define what "breaking" N with non-trivial success probability means, cf. Definition 5 below and [2].

Consider the following experiment $\mathsf{NICA}_N^B(1^k)$.

1. The experiment runs the instance generator of N to generate a problem instance $(c, w) \leftarrow^{\$} \mathsf{T}(1^k)$. Then it samples uniformly random coins $\rho_B \leftarrow^{\$} \{0, 1\}^k$ for B.
2. B is executed on input (c, ρ_B), it outputs a candidate solution s.
3. The experiment returns whatever $\mathsf{V}(c, w, s)$ returns.

Definition 5. *We say that B (t, ϵ)-breaks assumption N, if Λ runs in time $t(k)$ and it holds that*

$$\left| \Pr\left[\mathsf{NICA}_N^B\left(1^k\right) \Rightarrow 1 \right] - \Pr\left[\mathsf{NICA}_N^\mathsf{U}\left(1^k\right) \Rightarrow 1 \right] \right| \geq \epsilon(k)$$

where the probability is taken over the random coins consumed by T and the uniformly random choices of ρ_B and ρ_N respectively.

Simple Reductions From Non-interactive Complexity Assumptions to Breaking UF-SMA-*Security.* A reduction from breaking the UF-SMA-security of a signature scheme SIG to breaking the security of a non-interactive complexity assumption $N = (\mathsf{T}, \mathsf{V}, \mathsf{U})$ is a TM, which turns an attacker $\mathcal{A} = (\mathcal{A}_1, \mathcal{A}_2)$ according to Definition 3 into a TM $\Lambda^{\mathcal{A}}$ according to Definition 5.

Following [18, 30, 31, 33], we will consider a specific class of reductions in the sequel. We consider reductions having *black-box access* to the attacker, and which execute the attacker *only once* and *without rewinding*. We will generalize this later to reductions that may execute the attacker several times sequentially. Following [33], we call such reductions *simple*. At first sight we heavily constrain the class of reductions to that our result applies. However, as explained in [33], we include reductions that perform hybrid steps. Moreover, most reductions in cryptography are simple.

For preciseness and clarity, we define such a reduction as a triplet of Turing machines $\Lambda = (\Lambda_1, \Lambda_2, \Lambda_3)$. From these TMs and an attacker $\mathcal{A} = (\mathcal{A}_1, \mathcal{A}_2)$, we construct a Turing machine $\Lambda^{\mathcal{A}}$ for a non-interactive complexity assumption as follows.

1. Machine $\Lambda^{\mathcal{A}}$ receives as input a challenge c of the considered non-interactive complexity assumption, as well as random coins $\rho_\Lambda \leftarrow^\$ \{0,1\}^k$. It first runs $\Lambda_1(c, \rho_\Lambda)$, which returns the input to $\mathcal{A}_1$, consisting of a verification key vk, a sequence of messages $(m_i)_{i \in [n]}$, and random coins $\rho_{\mathcal{A}}$, as well as some state st_{Λ_2}.
2. Then $\Lambda^{\mathcal{A}}$ executes the attacker $\mathcal{A}_1$ on input $(vk, (m_i)_{i \in [n]}, \rho_{\mathcal{A}})$, which returns an index $j^* \in [n]$ and some state $st_{\mathcal{A}}$.
3. TM Λ_2 receives as input j^* and state st_{Λ_2}, and returns a list of signatures $(\sigma_i)_{i \in [n \setminus j^*]}$ and an updated state st_{Λ_3}.
4. The attacker $\mathcal{A}_2$ is executed on $(\sigma_i)_{i \in [n \setminus j^*]}$ and state $st_{\mathcal{A}}$, it returns a signature σ^*.
5. Finally, $\Lambda^{\mathcal{A}}$ runs $\Lambda_3(\sigma^*, j^*, st_{\Lambda_3})$, which produces a candidate solution s, and outputs s.

Definition 6. *We say that a Turing machine $\Lambda = (\Lambda_1, \Lambda_2, \Lambda_3)$ is a simple $(t_\Lambda, n, \epsilon_\Lambda, \epsilon_{\mathcal{A}})$-reduction from breaking $N = (T, V, U)$ to breaking the* UF-SMA-*security of* SIG, *if for any TM $\mathcal{A}$ that $(t_{\mathcal{A}}, n, \epsilon_{\mathcal{A}})$-breaks the* UF-SMA *security of* SIG, *TM $\Lambda^{\mathcal{A}}$ $(t_\Lambda + t_{\mathcal{A}}, \epsilon_\Lambda)$-breaks N.*

Definition 7. *Let $\ell : \mathbb{N} \to \mathbb{N}$. We say that reduction Λ loses ℓ, if there exists an adversary $\mathcal{A}$ that $(t_{\mathcal{A}}, n, \epsilon_{\mathcal{A}})$-breaks the* UF-SMA *security of* SIG, *such that $\Lambda^{\mathcal{A}}$ $(t_\Lambda + t_{\mathcal{A}}, \epsilon_\Lambda)$-breaks N with*

$$\frac{t_\Lambda(k) + t_{\mathcal{A}}(k)}{\epsilon_\Lambda(k)} \geq \ell(k) \cdot \frac{t_{\mathcal{A}}(k)}{\epsilon_{\mathcal{A}}(k)}.$$

Remark 3. The quotient $t_{\mathcal{A}}(k)/\epsilon_{\mathcal{A}}(k)$ of the running time $t_{\mathcal{A}}(k)$ and the success probability $\epsilon_{\mathcal{A}}(k)$ of a Turing machine $\mathcal{A}$ is called the *work factor* of $\mathcal{A}$ [8]. Thus, the factor ℓ in Definition 6 relates the work factor of attacker $\mathcal{A}$ to the work factor of TM $\Lambda^{\mathcal{A}}$, which allows us to measure the *tightness* of a cryptographic reduction. The smaller ℓ, the tighter is the reduction.

2.2   Bound for Simple Reductions Without Rewinding

For simplicity, we will consider reductions that have access to a "perfect" adversary $\mathcal{A}$, which $(t_{\mathcal{A}}, \epsilon_{\mathcal{A}})$-breaks the signature scheme with $\epsilon_{\mathcal{A}} = 1$. We explain in Sect. 2.4 why the extension to adversaries with $\epsilon_{\mathcal{A}} < 1$ is straightforward.

Theorem 1. *Let $N = (T, V, U)$ be a non-interactive complexity assumption, $n \in \mathsf{poly}(k)$ and let* SIG *be a signature scheme. For any simple $(t_\Lambda, n, \epsilon_\Lambda, 1)$-reduction from breaking N to breaking the* UF-SMA-*security of* SIG, *there exists a Turing machine $\mathcal{B}$ that $(t_\mathcal{B}, \epsilon_\mathcal{B})$-breaks N where*

$$t_\mathcal{B} \leq n \cdot t_\Lambda + n \cdot (n-1) \cdot t_{\mathsf{Vfy}} + t_{\mathsf{ReRand}} \quad \text{and} \quad \epsilon_\mathcal{B} \geq \epsilon_\Lambda - 1/n.$$

Here, t_{ReRand} *is the time required to re-randomize a signature, and* t_{Vfy} *is the running time of the verification machine of* SIG.

Proof. Our proof structure follows the structure of [30] (also used in [33]). That is, we first describe a hypothetical, inefficient adversary, then we show how to simulate it efficiently for certain reductions.

The Hypothetical Adversary. The hypothetical adversary $\mathcal{A} = (\mathcal{A}_1, \mathcal{A}_2)$ consists of two procedures that work as follows.

$\mathcal{A}_1 (vk, (m_i)_{i \in [n]}; \rho_{\mathcal{A}})$. On input a public key vk and messages $m_1, \ldots, m_n$, $\mathcal{A}_1$ samples $j \leftarrow^{\$} [n]$ uniformly random and outputs (j, st), where $st = (vk, (m_i)_{i \in [n]}, j)$.

$\mathcal{A}_2((\sigma_i)_{i \in [n] \backslash j}, st)$. $\mathcal{A}_2$ checks whether SIG.Vfy$(vk, m_i, \sigma_i) = 1$ for all $i \in [n] \backslash j$. If this holds, then it samples a uniformly random signature $\sigma_j \leftarrow^{\$} \Sigma(vk, m_j)$ for m_j. Finally, it outputs σ_j.

Note that $\mathcal{A}$ $(t_{\mathcal{A}}, 1)$-breaks the UF-SMA-security of SIG. Note also that the second step of this adversary may not be efficiently computable, which is why we call this adversary *hypothetical.*

Simulating $\mathcal{A}$. Consider the following TM $\mathcal{B}$, which runs reduction $\Lambda = (\Lambda_1, \Lambda_2, \Lambda_3)$ as a subroutine and attempts to break N. $\mathcal{B}$ receives as input $c \leftarrow^{\$} \mathsf{T}(1^k)$. It maintains an array A with n entries, which are all initialized to $\emptyset$, and proceeds as follows.

1. $\mathcal{B}$ first runs $(vk, (m_i)_{i \in [n]}, \rho_{\mathcal{A}}, st_{\Lambda_2}) \leftarrow^{\$} \Lambda_1(c; \rho_{\Lambda})$ for uniformly random $\rho_{\Lambda} \leftarrow^{\$} \{0,1\}^k$.
2. Next, $\mathcal{B}$ runs $\Lambda_2(j, st_{\Lambda_2})$ for each $j \in [n]$. Let $((\sigma_{i,j})_{i \in [n] \backslash j}, st_{\Lambda_3, j})$ denote the output of the j-th execution of Λ_2. Whenever Λ_2 outputs $(\sigma_{i,j})_{i \in [n] \backslash j}$ such that
$$\mathsf{SIG.Vfy}(vk, m_i, \sigma_{i,j}) = 1 \text{ for all } i \in [n] \backslash j$$
then it sets $A[i] \leftarrow \sigma_{i,j}$ for all $i \in [n] \backslash j$.
3. $\mathcal{B}$ samples $j^* \leftarrow^{\$} [n]$. Then it proceeds as follows.
   - If there exists an index $i \in [n] \backslash j^*$ such that $\mathsf{SIG.Vfy}(vk, \dot{m}_i, \sigma_{i,j^*}) \neq 1$, then $\mathcal{B}$ sets $\sigma^* := \bot$.
   - Otherwise, if $\mathsf{SIG.Vfy}(vk, m_i, \sigma_{i,j^*}) = 1$ for all $i \in [n] \backslash j^*]$, then $\mathcal{B}$ computes
$$\sigma^* \leftarrow^{\$} \mathsf{SIG.ReRand}(vk, m_{j^*}, A[j^*]).$$
4. Finally, $\mathcal{B}$ runs $s \leftarrow \Lambda_3(\sigma^*, j^*, st_{\Lambda_3, j^*})$ and outputs s. Note that the state st_{Λ_3, j^*} used to execute Λ_3 corresponds to the state returned by Λ_2 on its j^*-th execution.

Running Time of $\mathcal{B}$. $\mathcal{B}$ essentially runs each part of Turing machine $\Lambda = (\Lambda_1, \Lambda_2, \Lambda_3)$ once, plus $n - 1$ additional executions of Λ_2. Moreover, it executes

SIG.Vfy $n(n-1)$ times, and the re-randomization TM SIG.ReRand once. Thus, the total running time of $\mathcal{B}$ is at most

$$t_{\mathcal{B}} \leq n \cdot t_\Lambda + n \cdot (n-1) \cdot t_{\mathsf{Vfy}} + t_{\mathsf{ReRand}}.$$

Success Probability of $\mathcal{B}$. To analyze the success probability of $\mathcal{B}$, let us define an event bad. Intuitively, this event occurs, if j^* is the *only* (with respect to state st_{Λ_2}) value such that $\Lambda_2(st_{\Lambda_2}, j)$ outputs signatures which are all valid. More formally, for both experiments $\mathsf{NICA}_N^{\mathcal{B}}(1^k)$ and $\mathsf{NICA}_N^{\Lambda^A}(1^k)$, let st_{Λ_2} denote the (in both experiments unique) value computed by $\Lambda_1(c; \rho_\Lambda)$, and let j^* denote the (in both experiments unique) value given as input to $\Lambda_3(\sigma^*, j^*, st_{\Lambda_3, j^*})$. We say that bad occurs (in either $\mathsf{NICA}_N^{\mathcal{B}}(1^k)$ or $\mathsf{NICA}_N^{\Lambda^A}(1^k)$), if $\mathsf{pred}(st_{\Lambda_2}, j^*) = 1 \wedge \mathsf{pred}(st_{\Lambda_2}, j) = 0 \ \forall \ j \in [n\backslash j^*]$, where predicate pred is defined as

$$\mathsf{pred}(st_{\Lambda_2}, j) = 1$$
$$\iff \bigwedge_{i \in [n\backslash j]} \mathsf{SIG.Vfy}(vk, m_i, \sigma_i) = 1, \text{ where } ((\sigma_i)_{i\in[n\backslash j]}, st_{\Lambda_3}) \leftarrow \Lambda_2(st_{\Lambda_2}, j).$$

Note that pred is well-defined, because Λ_2 is a deterministic TM.

Let us write $S(\mathcal{F})$ shorthand for the event $\mathsf{NICA}_N^{\mathcal{F}}(1^k) \Rightarrow 1$ to abbreviate our notation. Then, it holds that

$$\left| \Pr[S(\mathcal{B})] - \Pr[S(\Lambda^A)] \right| \leq \left| \Pr[S(\mathcal{B}) \cap \neg\mathsf{bad}] - \Pr[S(\Lambda^A) \cap \neg\mathsf{bad}] \right| + \Pr[\mathsf{bad}]. \tag{2}$$

Bounding $\Pr[\mathsf{bad}]$. Recall that event bad occurs only if

$$\mathsf{pred}(st_{\Lambda_2}, j^*) = 1 \wedge \mathsf{pred}(st_{\Lambda_2}, j) = 0 \ \forall \ j \in [n\backslash j^*] \tag{3}$$

where st_{Λ_2} is the value computed by $\Lambda_1(c; \rho_\Lambda)$, and j^* is the value given as input to $\Lambda_3(\sigma^*, j^*, st_{\Lambda_3, j^*})$. Suppose that indeed st_{Λ_2} is such that there exist at least one $j^* \in [n]$ such that (3) holds. We claim that even then we have

$$\Pr[\mathsf{bad}] \leq 1/n. \tag{4}$$

To see this, note first that for each st_{Λ_2} there can be at most one value j^* that satisfies (3). Moreover, both the hypothetical adversary $\mathcal{A}$ and the adversary simulated by $\mathcal{B}$ choose $j^* \leftarrow^\$ [n]$ independently and uniformly random, which yields (4).

Proving $\Pr[S(\mathcal{B}) \cap \neg\mathsf{bad}] = \Pr[S(\Lambda^A) \cap \neg\mathsf{bad}]$. Note that $\mathcal{B}$ executes in particular

1. $(vk, (m_i)_{i\in[n]}, st_{\Lambda_2}) \leftarrow^\$ \Lambda_1(c; \rho_\Lambda)$
2. $((\sigma_{i,j^*})_{i\in[n\backslash j^*]}, st_{\Lambda_3}) \leftarrow^\$ \Lambda_2(j^*, st_{\Lambda_2})$
3. $s \leftarrow \Lambda_3(\sigma^*, j^*, st_{\Lambda_3})$.

We show that if $\neg\mathsf{bad}$ occurs, then $\mathcal{B}$ simulates the hypothetical adversary $\mathcal{A}$ perfectly. To this end, consider the distribution of σ^* computed by $\mathcal{B}$ in following two cases.

1. Machine $\Lambda_2(j^*, st_{\Lambda_2})$ outputs $((\sigma_{i,j^*})_{i\in[n\setminus j^*]}, st_{\Lambda_3,j^*})$ such that there exists an index $i \in [n\setminus j^*]$ with $\mathsf{SIG.Vfy}(vk, m_i, \sigma_{i,j^*}) \neq 1$.
   In this case, $\mathcal{A}$ would compute $\sigma^* := \bot$. $\mathcal{B}$ also sets $\sigma^* := \bot$ in this case.

2. TM $\Lambda_2(j^*, st_{\Lambda_2})$ outputs $((\sigma_{i,j^*})_{i\in[n\setminus j^*]}, st_{\Lambda_3,j^*})$ such that for all $i \in [n\setminus j^*]$ it holds that

$$\mathsf{SIG.Vfy}(vk, m_i, \sigma_{i,j^*}) = 1.$$

In this case, $\mathcal{A}$ would output a uniformly random signature $\sigma^* \xleftarrow{\$} \Sigma(vk, m_{j^*})$. Note that in this case $\mathcal{B}$ outputs a re-randomized signature $\sigma^* \xleftarrow{\$} \mathsf{SIG.ReRand}(vk, m_{j^*}, A[j^*])$, which is a uniformly distributed valid signature for m_{j^*} provided that $A[j^*] \neq \emptyset$. The latter happens whenever bad does not occur.

Thus, $\mathcal{B}$ simulates $\mathcal{A}$ perfectly in either case, provided that $\neg$bad. This implies $S(\mathcal{B}) \cap \neg$bad $\iff S(\Lambda^{\mathcal{A}}) \cap \neg$bad, which yields

$$\Pr[S(\mathcal{B}) \cap \neg\mathsf{bad}] = \Pr[S(\Lambda^{\mathcal{A}}) \cap \neg\mathsf{bad}]. \tag{5}$$

Finishing the Proof of Theorem 1. By plugging (4) and (5) into Inequality (2), we obtain

$$\left| \Pr[S(\mathcal{B})] - \Pr[S(\Lambda^{\mathcal{A}})] \right| \leq 1/n$$

which implies

$$\epsilon_{\mathcal{B}} = |\Pr[S(\mathcal{B})] - \Pr[S(U)]| \geq |\Pr[S(\Lambda)] - \Pr[S(U)]| - 1/n = \epsilon_\Lambda - 1/n.$$

2.3   Interpretation

Assuming that no adversary $\mathcal{B}$ is able to (t_N, ϵ_N)-break the security of NICA with $t_N = t_{\mathcal{B}} = n \cdot t_\Lambda + n \cdot (n-1) \cdot t_{\mathsf{Vfy}} + t_{\mathsf{ReRand}}$, we must have $\epsilon_{\mathcal{B}} \leq \epsilon_N$. By Theorem 1, we thus must have

$$\epsilon_\Lambda \leq \epsilon_{\mathcal{B}} + 1/n \leq \epsilon_N + 1/n$$

for all reductions Λ. In particular, the hypothetical adversary $\mathcal{A}$ constructed in the proof of Theorem 1 is an example of an adversary such that

$$\frac{t_\Lambda + t_{\mathcal{A}}}{\epsilon_\Lambda} \geq \frac{t_{\mathcal{A}}}{\epsilon_N + 1/n} = (\epsilon_N + 1/n)^{-1} \cdot \frac{t_{\mathcal{A}}}{1} = (\epsilon_N + 1/n)^{-1} \cdot \frac{t_{\mathcal{A}}}{\epsilon_{\mathcal{A}}}.$$

Thus, any reduction Λ from breaking the security of NICA N to breaking the UF-SMA-security of signature scheme SIG loses (in the sense of Definition 7) at least a factor of $\ell \geq 1/(\epsilon_N + 1/n)$. In particular, note that $\ell \approx n$ if ϵ_N is very small. This yields the following informal theorem.

Theorem 2 (Informal). *Any simple reduction from breaking the security of NICA N to breaking the UF-SMA-security (or any stronger security notion, like EUF-CMA-security, cf. Definition 19) of signature scheme SIG that provides efficient signature re-randomization loses a factor that is at least linear in the number n of sign queries issued by the attacker, or N is easy to solve.*

Remark 4. Since a unique signature scheme is trivially efficiently re-randomizable, Theorem 2 applies also to unique signature schemes.

$$
\begin{array}{|l|}
\hline
\text{TM } r\text{-}\varLambda^{\mathcal{A}}(c;\rho_\varLambda) \\
st_{\varLambda_{1,1}} \leftarrow \varLambda_0\,(c,\rho_\varLambda) \\
\textbf{for } 1 \le l \le r \textbf{ do:} \\
\quad (vk^l,(m_i^l)_{i\in[n]},\rho_\mathcal{A},st_{\varLambda_{l,2}}) \leftarrow \varLambda_{l,1}(st_{\varLambda_{l,1}}) \\
\quad (j^{l*},st_\mathcal{A}) \leftarrow \mathcal{A}_1(vk^l,(m_i^l)_{i\in[n]};\rho_\mathcal{A}) \\
\quad ((\sigma_i^l)_{i\in[n\setminus j^{l*}]},st_{\varLambda_{l,3}}) \leftarrow \varLambda_{l,2}(j^{l*},st_{\varLambda_{l,2}}) \\
\quad \sigma_{j^{l*}}^l \leftarrow \mathcal{A}_2((\sigma_i^l)_{i\in[n\setminus j^{l*}]},st_\mathcal{A}) \\
\quad st_{\varLambda_{l+1,1}} \leftarrow \varLambda_{l,3}\left(\sigma_{j^{l*}}^l,j^{l*},st_{\varLambda_{l,3}}\right) \\
s \leftarrow \varLambda_3\left(st_{\varLambda_{r+1,1}}\right) \\
\textbf{return } s \\
\hline
\end{array}
$$

Fig. 2. TM $r\text{-}\varLambda^{\mathcal{A}}$ that solves a non-interactive complexity assumption according to Definition 5, constructed from a r-simple reduction $r\text{-}\varLambda = \left(\varLambda_0, (\varLambda_{l,1},\varLambda_{l,2},\varLambda_{l,3})_{l\in[r]}, \varLambda_3\right)$ and an attacker $\mathcal{A} = (\mathcal{A}_1, \mathcal{A}_2)$.

2.4 Extension to "Non-perfect" Adversaries

Note that the proof of Theorem 1 trivially generalizes to $(t_\varLambda, n, \epsilon_\varLambda, \epsilon_\mathcal{A})$-reductions with $\epsilon_\mathcal{A} < 1$, that is, reductions that have access to an adversary which has success probability $\epsilon_\mathcal{A} < 1$. To this end, we first would have to describe a hypothetical adversary, which has success probability $\epsilon_\mathcal{A}$. This is simple, because we can simply let the hypothetical adversary constructed above toss a biased coin χ with $\Pr[\chi = 1] = \epsilon_\mathcal{A}$, such that $\mathcal{A}$ outputs σ^* only if $\chi = 1$. Note that in the proof of Theorem 1 we are even able to simulate a perfect adversary $\mathcal{A}$. Therefore we would also be able to simulate the non-perfect adversary sketched above, by tossing a biased coin χ and outputting σ^* only if $\chi = 1$. This yields the following theorem.

Theorem 3. *Let $N = (\mathsf{T},\mathsf{V},\mathsf{U})$ be a non-interactive complexity assumption, $n \in \mathsf{poly}(k)$ and let SIG be a signature scheme. For any simple $(t_\varLambda, n, \epsilon_\varLambda, \epsilon_\mathcal{A})$-reduction from breaking the UF-SMA-security of SIG to breaking N, there exists a Turing machine $\mathcal{B}$ that $(t_\mathcal{B}, \epsilon_\mathcal{B})$-breaks N where*

$$
t_\mathcal{B} \le n \cdot t_\varLambda + n \cdot (n-1) \cdot t_{\mathsf{Vfy}} + t_{\mathsf{ReRand}} \qquad \text{and} \qquad \epsilon_\mathcal{B} \ge \epsilon_\varLambda - 1/n.
$$

Here, t_{ReRand} is the time to re-randomize a given valid signature over a message and t_{Vfy} is the time needed to execute the verification machine of SIG.

3 Bound for Reductions with Sequential Rewinding

Theorem 1 applies only to reductions that run the forger only once. Here we show that under assumptions similar to that in Theorem 1 the work factor of any reduction that is allowed to run or rewind the adversary r times *sequentially* cannot decrease significantly below $\frac{n}{r}$ if N is hard.

Let r be an upper bound on the number of times that the adversary can be rewound by the reduction. We then consider a reduction $r\text{-}\Lambda$ as a $3 \cdot r + 2$-tuple of Turing machines $r\text{-}\Lambda = \left(\Lambda_0, (\Lambda_{l,1}, \Lambda_{l,2}, \Lambda_{l,3})_{l \in [r]}, \Lambda_3 \right)$. Let now $\mathcal{A} = (\mathcal{A}_1, \mathcal{A}_2)$ be an attacker against the UF-SMA-security of SIG. From these TMs we construct a Turing machine $r\text{-}\Lambda^{\mathcal{A}}$ that solves a NICA N as depicted in Fig. 2. We shortly explain Fig. 2 here.

Λ_0. $r\text{-}\Lambda$ inputs a challenge c of the considered non-interactive complexity assumption and random coins ρ_Λ. It processes these inputs by running Λ_0 which outputs a state st_Λ.

$\Lambda_l = (\Lambda_{l,1}, \Lambda_{l,2}, \Lambda_{l,3})$. Now, for each $l \in [r]$, we have a triplet of TMs $\Lambda_l = (\Lambda_{l,1}, \Lambda_{l,2}, \Lambda_{l,3})$ that has black box access to attacker $\mathcal{A} = (\mathcal{A}_1, \mathcal{A}_2)$. Note that the state st_Λ may be passed over from $\Lambda_{l,3}$ to $\Lambda_{l+1,1}$ (and Λ_3) while the state $st_\mathcal{A}$ of $\mathcal{A}_2$ may not be passed over to the next execution of $\mathcal{A}_1$.

$\Lambda_{l,1}$. $\Lambda_{l,1}$ inputs the current state $st_{\Lambda_{l,1}}$ and outputs a public key vk^l, distinct messages $m_i^l, i \in [n]$, a random tape $\rho_\mathcal{A}$ for $\mathcal{A}_1$ and a state $st_{\Lambda_{l,2}}$. Next, $\mathcal{A}_1$ is run on input $\left(vk^l, (m_i)_{i \in [n]} ; \rho_\mathcal{A} \right)$ and returns a state $st_\mathcal{A}$ and an index j^l.

$\Lambda_{l,2}$. On input index j^l and state $st_{\Lambda_{l,2}}$, $\Lambda_{l,2}$ returns signatures $\left(\sigma_i^l \right)_{i \in [n \setminus j]}$ and state $st_{\Lambda_{l,2}}$. Now, $\mathcal{A}_2$ is run on $\left((\sigma_i^l)_{i \in [n \setminus j^l]}, st_\mathcal{A} \right)$ and returns $\sigma_{j^l}^l$.

$\Lambda_{l,3}$. $\Lambda_{l,3}$ inputs the signature output by $\mathcal{A}_{l,2}$ and the current state $st_{\Lambda_{l,2}}$. It returns the state $st_{\Lambda_{l+1,1}}$.

Λ_3. Finally, Λ_3 inputs the current state of $r\text{-}\Lambda$ and returns s. $r\text{-}\Lambda$ is considered successful if $V(c, w, s) = 1$.

Definition 8. *We say that a Turing machine* $r\text{-}\Lambda = \left(\Lambda_0, (\Lambda_{l,1}, \Lambda_{l,2}, \Lambda_{l,3})_{l \in [r]}, \Lambda_3 \right)$ *is an* r-*simple* $(t_\Lambda, n, \epsilon_\Lambda, \epsilon_\mathcal{A})$-*reduction from breaking* $N = (T, V, U)$ *to breaking the* UF-SMA-*security of* SIG, *if for any TM* $\mathcal{A}$ *that* $(t_\mathcal{A}, n, \epsilon_\mathcal{A})$-*breaks the* UF-SMA *security of* SIG, *TM* $r\text{-}\Lambda^{\mathcal{A}}$ *(as constructed above)* $(t_\Lambda + r \cdot t_\mathcal{A}, \epsilon_\Lambda)$-*breaks* N.

Definition 9. *Let* $\ell : \mathbb{N} \to \mathbb{N}$. *We say that an* r-*simple reduction* Λ *from breaking a non-interactive complexity assumption* N *to breaking the* UF-SMA *security of a signature scheme* SIG *loses* ℓ *if there exists an adversary* $\mathcal{A}$ *that* $(t_\mathcal{A}, n, \epsilon_\mathcal{A})$-*breaks such that* $\Lambda^{\mathcal{A}}$ $(t_\Lambda + r \cdot t_\mathcal{A}, \epsilon_\Lambda)$-*breaks* N *where*

$$\frac{t_\Lambda(k) + r \cdot t_\mathcal{A}(k)}{\epsilon_\Lambda} \geq \ell(k) \cdot \frac{t_\mathcal{A}(k)}{\epsilon_\mathcal{A}(k)}.$$

Theorem 4. *Let* $N = (T, V, U)$ *be a non-interactive complexity assumption,* $n, r \in \mathsf{poly}(k)$ *and let* SIG *be a signature scheme. Then for any* r-*simple* $(t_\Lambda, n, \epsilon_\Lambda, 1)$-*reduction* Λ *from breaking* N *to breaking the* UF-SMA-*security of* SIG *there exists a TM* $\mathcal{B}$ *that* $(t_\mathcal{B}, \epsilon_\mathcal{B})$-*breaks* N *where*

$$t_\mathcal{B} \leq r \cdot n \cdot t_\Lambda + r \cdot n \cdot (n-1) \cdot t_{\mathsf{Vfy}} + r \cdot t_{\mathsf{ReRand}}$$
$$\epsilon_\mathcal{B} \geq \epsilon_\Lambda - \frac{r}{n}.$$

Here, t_{ReRand} *is the time to re-randomize a given valid signature over a message and* t_{Vfy} *is the time needed to run the verification machine of* SIG.

The proof of this theorem is structured as the proof of Theorem 1. We again first consider a hypothetical attacker $\mathcal{A}$ (cf. Page 11) that breaks the UF-SMA-security of SIG. Next, when we show how to simulate $\mathcal{A}$, we basically apply the technique from the proof of Theorem 1 r times. A detailed proof can be found in the full version of this paper.

3.1   Interpretation

Assuming that no adversary $\mathcal{B}$ is able to (t_N, ϵ_N)-break the security of NICA with $t_N = t_{\mathcal{B}} = r \cdot n \cdot t_{\mathcal{A}} + r \cdot n \cdot (n-1) \cdot t_{Vfy} + r \cdot t_{ReRand}$, we must have $\epsilon_{\mathcal{B}} \leq \epsilon_N$. By Theorem 4, we thus must have

$$\epsilon_{\mathcal{A}} \leq \epsilon_{\mathcal{B}} + r/n \leq \epsilon_N + r/n$$

for all reductions Λ. In particular, the hypothetical adversary $\mathcal{A}$ constructed in the proof of Theorem 1 is an example of an adversary such that

$$\frac{t_{\Lambda} + r \cdot t_{\mathcal{A}}}{\epsilon_{\Lambda}} \geq \frac{r \cdot t_{\mathcal{A}}}{\epsilon_N + r/n} = (\epsilon_N + r/n)^{-1} \cdot r \cdot \frac{t_{\mathcal{A}}}{1} = (\epsilon_N + r/n)^{-1} \cdot r \cdot \frac{t_{\mathcal{A}}}{\epsilon_{\mathcal{A}}}.$$

Thus, any reduction Λ from breaking the security of NICA N to breaking the UF-SMA-security of signature scheme SIG loses (in the sense of Definition 7) at least a factor of $\ell \geq r/(\epsilon_N + r/n)$. In particular, note that $\ell \approx n$ if ϵ_N is very small.

4   A Generalized Meta-reduction

In this section we state and prove our main result, which generalizes the results from Sect. 2. Essentially, we observe that for the proof to work we do not need all structural elements a signature scheme possesses. In particular we do not require dedicated parameter generation-, key generation- and sign-algorithms. Instead, we consider an abstract security experiment with the following properties:

1. The values that are publicly available "induce a relation" $R(x, y)$ that is efficiently verifiable for the adversary during the security experiment.
2. The adversary is provided with statements $y_1, \ldots, y_n$ at the beginning of the security experiment and has access to an oracle that when queried y_i returns x_i such that $R(x_i, y_i), i \in [n]$.
3. If the adversary is able to output x_j such that $R(x_j, y_j)$ and it did not query its oracle on y_j, this is sufficient to win the security game.

Remark 5. To show the usefulness of such an abstract experiment, we note that for instance the security experiments for public key encryption or key encapsulation mechanisms in the multi-user setting with corruptions [4], or digital signature schemes in the multi-user (MU) setting with corruptions [3,4], naturally satisfy these properties as follows. Essentially, we define a relation $R(sk, pk)$ over pairs of public keys and secret keys such that $R(sk, pk) = 1$ whenever sk "matches" pk. The adversary is provided with public keys at the beginning of the experiment, and is able to obtain secret keys corresponding to public keys of its choice. Finally, if the adversary is able to output an uncorrupted secret key, it is clearly able to compute a signature over a message that was not signed before (i.e., winning the signature security game) or decrypt the challenge ciphertext (i.e., winning the PKE/KEM security game). Thus, all three requirements are satisfied. Details on how to apply the result to, e.g., digital signatures and PKE/KEMs in the multi user setting with corruptions we refer to Sect. 5.

4.1 Definitions

Re-randomizable Relations. Let $R \subseteq X \times Y$ be a relation. For (x, y) with $R(x, y) = 1$ we call x the *witness* and y the *statement.* We use $X(R, y)$ to denote the set

$$X(R, y) := \{x : R(x, y) = 1\}$$

of all witnesses x for statement y with respect to R. We denote by $L(R) := \{y : \exists\, x \,\text{s.t.}\, R(x, y) = 1\} \subseteq Y$ the language consisting of statements in R.

In the sequel we will consider *computable* relations. We will therefore identify a relation R with a machine $\widehat{R}$ that computes R. We say that a relation R is t_{Vfy}-computable, if there is a deterministic Turing machine $\widehat{R}$ that runs in time at most $t_{\mathsf{Vfy}}(|x| + |y|)$ such that $\widehat{R}(x, y) = R(x, y)$.

Definition 10. *Let $\mathcal{R} := \{R_i\}_{i \in I}$ be a family of computable relations. We say that $\mathcal{R}$ is t_{ReRand}-re-randomizable if there is a probabilistic Turing machine $\mathcal{R}.\mathsf{ReRand}$ that inputs $(\widehat{R}_i, y, x)$, runs in time at most t_{ReRand}, and outputs x' which is uniformly distributed over $X(R, y_i)$ whenever $R_i(x, y) = 1$, with probability 1.*

Example 1. Digital signatures in the single user setting, as considered in Sect. 2, may be described in terms of families of relations. We set $R_{\Pi, vk}$ to the relation over signatures and messages that is defined by a verification key vk. In this case, we have that $X(R, y) = \Sigma(vk, y)$ is the set of all valid signatures over message y with respect to public key vk. Note that the family of relations $(R_{\Pi, vk})_{\Pi, vk}$ is t_{ReRand}-re-randomizable, if the signature scheme is t_{ReRand}-re-randomizable (cf. Definition 2).

Witness Unforgeability Under Static Statement Attacks. We will consider a weak security experiment for computable relations, which is inspired by the UF-SMA-security experiment considered in Sect. 2, but abstract and general enough to be applicable in other useful settings. Jumping slightly ahead, we will show in Sect. 5 that this includes applications to signatures, public-key encryption, key encapsulation mechanisms in the multi-user setting, and non-interactive key exchange.

$$
\begin{array}{|l|}
\hline
\text{Game UF-SSA}_{\mathcal{R}}^{n,\mathcal{A}}\,(1^k) \\
\hline
R = R_i \leftarrow^{\$} \mathcal{R} \\
y_1, \ldots, y_n \leftarrow^{\$} L(R) \text{ s.t. } y_i \neq y_j \text{ for all} \\
i \neq j \\
x_i \leftarrow^{\$} X(R, y_i) \text{ for all } i \in [n] \\
(j, st) \leftarrow \mathcal{A}_1\big(\widehat{R}, (y_i)_{i\in[n]}; \rho_{\mathcal{A}}\big) \\
x_j \leftarrow \mathcal{A}_2\big(st, (x_i)_{i\in[n\setminus j]}\big) \\
\textbf{return } R(x_j, y_j) \\
\hline
\end{array}
$$

Fig. 3. The UF-SSA-security game with attacker $\mathcal{A} = (\mathcal{A}_1, \mathcal{A}_2)$.

$$
\begin{array}{|l|}
\hline
\text{TM } r\text{-}\Gamma^{\mathcal{A}}(c; \rho_{\mathcal{A}}) \\
\hline
st_\Gamma \leftarrow \Gamma_0\,(c, \rho_\Gamma) \\
\textbf{for } 1 \leq l \leq r \textbf{ do:} \\
\quad \Big(\widehat{R^l}, (y_i^l)_{i\in[n]}, \rho_{\mathcal{A}}, st_\Gamma\Big) \leftarrow \Gamma_{l,1}(st_\Gamma) \\
\quad (j^l, st_{\mathcal{A}}) \leftarrow \mathcal{A}_1\Big(\widehat{R^l}, (y_i^l)_{i\in[n]}; \rho_{\mathcal{A}}\Big) \\
\quad \Big((x_i^l)_{i\in[n\setminus j^l]}, st_\Gamma\Big) \leftarrow \Gamma_{l,2}\,(j^l, st_\Gamma) \\
\quad x_j^l \leftarrow \mathcal{A}_2\Big((x_i^l)_{i\in[n\setminus j^l]}, st_{\mathcal{A}}\Big) \\
\quad st_\Gamma \leftarrow \Gamma_{l,3}\,(x_j^l, st_\Gamma) \\
s \leftarrow \Gamma_3\,(st_\Gamma) \\
\textbf{return } s \\
\hline
\end{array}
$$

Fig. 4. TM $r\text{-}\Gamma^{\mathcal{A}}$ that solves a non-interactive complexity assumption according to Definition 5, constructed from a r-simple reduction $r\text{-}\Gamma = \Big(\Gamma_0, (\Gamma_{l,1}, \Gamma_{l,2}, \Gamma_{l,3})_{l\in[r]}, \Gamma_3\Big)$ and an attacker $\mathcal{A} = (\mathcal{A}_1, \mathcal{A}_2)$.

The security experiment is described in Fig. 3. It is parametrized by a family $\mathcal{R}$ of computable relations, $\mathcal{R} = \{R_i\}_{i\in I}$, and the number n of statements the adversary $\mathcal{A} = (\mathcal{A}_1, \mathcal{A}_2)$ is provided with. These statements need to be pairwise distinct. $\mathcal{A}$ may *non-adaptively* ask for witnesses for *all but one* statement, and is considered successful if it manages to output a "valid" witness for the remaining statement.

Definition 11. *Let $\mathcal{R} = \{R_i\}_{i\in I}$ be a family of computable relations. We say that an adversary $\mathcal{A} = (\mathcal{A}_1, \mathcal{A}_2)$ (t, n, ϵ)-breaks the witness unforgeability under static statement attacks of $\mathcal{R}$ if it runs in time t and*

$$
\Pr\left[\text{UF-SSA}_{\mathcal{R}}^{n}(\mathcal{A}) \Rightarrow 1\right] \geq \epsilon
$$

where $\text{UF-SSA}_{\mathcal{R}}^{n}(\mathcal{A})$ is the security game depicted in Fig. 3.

Simple Reductions From Non-interactive Complexity Assumptions to Breaking UF-SSA-Security. Informally, a reduction from breaking the UF-SSA-security of a family of relations $\mathcal{R}$ to breaking the security of a non-interactive complexity assumption $N = (\mathsf{T}, \mathsf{U}, \mathsf{V})$ is a Turing machine Γ, which turns an attacker $\mathcal{A} = (\mathcal{A}_1, \mathcal{A}_2)$ against $\mathcal{R}$ according to Definition 11 into a TM $\Gamma^{\mathcal{A}}$ that breaks N according to Definition 5. As in Sect. 2, we will only consider simple reductions, i.e., reductions that have *black-box* access to the attacker and that may run the attacker at most r times *sequentially*.

We define a reduction from breaking the security of $\mathcal{R}$ to breaking N as an $(3r + 2)$-tuple of TMs $\Gamma = \left(\Gamma_0, (\Gamma_{l,1}, \Gamma_{l,2}, \Gamma_{l,3})_{l \in [r]}, \Gamma_3 \right)$, which turn a TM $\mathcal{A}$ breaking the security of $\mathcal{R}$ into a TM $\Gamma^{\mathcal{A}}$ breaking N, as described in Fig. 4. Note that this Turing machine works almost identical to that considered in Sect. 3, except that we consider a more general class of relations.

Definition 12. *We say that a TM* r-$\Gamma = \left(\Gamma_0, (\Gamma_{l,1}, \Gamma_{l,2}, \Gamma_{l,3})_{l \in [r]}, \Gamma_3 \right)$ *is an* r-*simple* $(t_\Gamma, n, \epsilon_\Gamma, \epsilon_{\mathcal{A}})$-*reduction from breaking* $N = (\mathsf{T}, \mathsf{V}, \mathsf{U})$ *to breaking the* UF-SSA-*security of a family of relations* $\mathcal{R}$, *if for any TM* $\mathcal{A}$ *that* $(t_{\mathcal{A}}, n, \epsilon_{\mathcal{A}})$-*breaks the* UF-SSA *security of* $\mathcal{R}$, *TM* r-$\Gamma^{\mathcal{A}}$ *(cf. Fig. 4)* $(t_\Lambda + r \cdot t_{\mathcal{A}}, \epsilon_\Lambda)$-*breaks* N.

We define the loss of an r-simple reduction r-Γ from breaking N to breaking the UF-SSA-security of a family of computable relations $\mathcal{R}$ similar to Definition 9.

4.2 Main Result

In this Section we establish the following result that generalizes Theorem 4.

Theorem 5. *Let* $N = (\mathsf{T}, \mathsf{V}, \mathsf{U})$ *be a non-interactive complexity assumption,* $n, r \in \mathsf{poly}(k)$ *and let* $\mathcal{R}$ *be a family of computable relations. Then for any* r-*simple* $(t_\Gamma, n, \epsilon_\Gamma, 1)$-*reduction* Γ *from breaking* N *to breaking the* UF-SSA-*security of* $\mathcal{R}$ *there exists a TM* $\mathcal{B}$ *that* $(t_{\mathcal{B}}, \epsilon_{\mathcal{B}})$-*breaks* N *where*

$$t_{\mathcal{B}} \leq r \cdot n \cdot t_\Gamma + r \cdot n \cdot (n - 1) \cdot t_{\mathsf{Vfy}} + r \cdot t_{\mathsf{ReRand}}$$

$$\epsilon_{\mathcal{B}} \geq \epsilon_\Gamma - \frac{r}{n}.$$

Here, t_{ReRand} *is the time to re-randomize a given valid witness and* t_{Vfy} *is the maximum time needed to compute* $R \in \mathcal{R}$.

The proof of Theorem 5 is nearly identical to the proof of Theorem 4, and therefore omitted. Also the interpretation of Theorem 5 is nearly identical to the interpretation described in Sect. 2.3. Assuming that no adversary $\mathcal{B}$ is able to $(t_{\mathsf{N}}, \epsilon_{\mathsf{N}})$-break the security of NICA with $t_{\mathsf{N}} = t_{\mathcal{B}} = r \cdot n \cdot t_\Lambda + r \cdot n \cdot (n - 1) \cdot t_{\mathsf{Vfy}} + r \cdot t_{\mathsf{ReRand}}$, we must have $\epsilon_{\mathcal{B}} \leq \epsilon_{\mathsf{N}}$. Thus, if $\mathcal{R}$ is efficiently computable and re-randomizable, the loss of any simple reduction from breaking N to breaking the UF-SSA-security of $\mathcal{R}$ is at least linear in n.

5   New Applications

5.1   Signatures in the Multi-user Setting

Definitions. The syntax of digital signature schemes is defined in Sect. 2. Here, we define additional properties of signature schemes that are required to establish our result. Let $\mathsf{SIG} = (\mathsf{Setup}, \mathsf{Gen}, \mathsf{Sign}, \mathsf{Vfy})$ be a signature scheme. In the sequel we require *perfect correctness*, i.e., that for all $k \in \mathbb{N}$, all $\Pi \leftarrow^{\$} \mathsf{Setup}(1^k)$, all $(vk, sk) \leftarrow^{\$} \mathsf{Gen}(\Pi)$ and all m it holds that:

$$\Pr \Big[\mathsf{SIG}.\mathsf{Vfy}(vk, m, \sigma) = 1 : \sigma \leftarrow^{\$} \mathsf{SIG}.\mathsf{Sign}(sk, m) \Big] = 1.$$

Moreover, let $\Pi \leftarrow^{\$} \mathsf{Setup}(1^k)$ and let us recall that Π is contained in vk. We require an additional deterministic TM $\mathsf{SKCheck}_\Pi$ that takes as input strings sk and pk and outputs 0 or 1 such that:

$$\mathsf{SKCheck}_\Pi(pk, sk) = 1$$
$$\Longleftrightarrow$$
$$\Pr \big[\mathsf{Vfy}(pk, m, \sigma) = 1 : m \leftarrow^{\$} |\mathcal{M}| \wedge \sigma \leftarrow^{\$} \mathsf{Sign}(sk, m) \big] = 1.$$

That is, $\mathsf{SKCheck}$ takes inputs sk and pk and returns 1 if and only if pk is a valid public key and sk is a corresponding secret key. Since we require perfect correctness for signature schemes, we have $\mathsf{SKCheck}(vk, sk) = 1$ whenever $(vk, sk) \leftarrow^{\$} \mathsf{Gen}(\Pi)$.

Definition 13. (Key re-randomization). *We say that a signature encryption scheme* SIG *is* t_{ReRand}-*key re-randomizable if there exists a Turing machine* $\mathsf{SIG}.\mathsf{ReRand}$ *that runs in time at most* t_{ReRand}, *takes as input* $\Pi(vk, sk)$ *and returns* sk *uniformly distributed over* $\{sk : \mathsf{SKCheck}_\Pi(vk, sk) = 1\}$ *whenever* $\mathsf{SKCheck}_\Pi(vk, sk) = 1$.

Example 2. If we consider, for example, the Waters signature scheme [38], a public key consists among others of elements $g, g_1, g_2 \in \mathcal{G}$ where $g_1 = g^\alpha$. The key generation algorithm outputs a corresponding secret key as $sk = g_2^\alpha$. However, there may be other secret keys that might be accepted by $\mathsf{SKCheck}$.

To investigate this issue we shortly recall the signing and verification algorithms of [38]. The signing algorithm, when given as input a secret key and a message returns $\sigma = (\sigma_1, \sigma_2) = (g^r, sk \cdot (H(m))^r)$ where r is uniformly random chosen from $\mathbb{Z}_p$. Verification returns $e(g_1, g_2) =^? e(g, \sigma_2) \cdot e(\sigma_1, H(m))^{-1} = e(g, sk) \cdot e(g, H(m))^r \cdot e(g, H(m))^{-r}$.

We observe that by definition of $\mathsf{SKCheck}$ we must have $\mathsf{SKCheck}(vk, sk) = 1 \Leftrightarrow e(g_1, g_2) = e(g, sk)$. Thus there is an efficient $\mathsf{SKCheck}$ procedure. Moreover, since there is only one value that satisfies this equation in prime order groups we have an efficient secret key re-randomization algorithm, namely, the identity map. This is all that is to verify before applying our result.

Game MU-EUF-CMA-C$_{\text{SIG}}^{n,\mu}(\mathcal{A})$

$\Pi \leftarrow^{\$} \text{SIG.Setup}(1^k)$	$\mathcal{O}.\text{Sign}(m, i)$
$(vk_i, sk_i) \leftarrow^{\$} \text{SIG.Gen}(\Pi)$	**if** $\|Q_i\| \geq \mu$ **return** $\perp$
$\rho_{\mathcal{A}} \leftarrow^{\$} \{0,1\}^k$	$Q_i \leftarrow Q_i \cup \{m\}$
$Q^{\text{Corrupt}} = Q_1 = \ldots = Q_n \leftarrow \emptyset$	**return** $\sigma \leftarrow^{\$} \text{SIG.Sign}(sk_i, m)$
$({}^*i, m^*, \sigma^*) \leftarrow \mathcal{A}^{\mathcal{O}.\text{Sign}(\cdot,\cdot), \mathcal{O}.\text{Corrupt}(\cdot)} \left((vk_i)_{i \in [n]} ; \rho_{\mathcal{A}} \right)$	$\mathcal{O}.\text{Corrupt}(i)$
return $vk_{i^*} \notin Q^{\text{Corrupt}} \wedge m^* \notin Q_{i^*} \wedge \text{SIG.Vfy}(vk_{i^*}, m^*, \sigma^*)$	$Q^{\text{Corrupt}} \leftarrow Q^{\text{Corrupt}} \cup \{vk_i\}$
	return sk_i

Fig. 5. MU-EUF-CMA-C-security game. The attacker has access to a signing oracle $\mathcal{O}.\text{Sign}$ and a corrupt oracle $\mathcal{O}.\text{Corrupt}$.

Security Definition. The MU-EUF-CMA-C-security game is depicted in Fig. 5. Here the adversary $\mathcal{A}$ is provided with public keys $vk_1, \ldots, vk_n$ of the signature scheme. It may now adaptively issue *sign* and *corrupt*-queries. To issue a sign query it specifies a message m and a public key $vk_i, i \in [n]$ and obtains a valid signature σ over m that is valid with respect to vk_i. In order to issue a corrupt query, $\mathcal{A}$ specifies an index $i \in [n]$ and obtains a secret key sk_i that "matches" vk_i. Finally, $\mathcal{A}$ outputs a triplet (i, m, σ) and is considered successful if it did neither issue a corrupt query for i nor a sign query for (m, vk_i) and at the same time σ is valid over m with respect to vk_i.

Definition 14 (MU-EUF-CMA-C-security). *We say that an adversary* (t, n, μ, ϵ)-*breaks the* MU-EUF-CMA-C-*security of a signature scheme* SIG *if it runs in time* t *and*

$$\Pr\left[\text{MU-EUF-CMA-C}_{\text{SIG}}^{n,\mu}(\mathcal{A}) \Rightarrow 1\right] \geq \epsilon.$$

Definition 15. *We say that a Turing machine* r-Γ *is an* r-*simple* $(t_\Lambda, n, \mu, \epsilon_\Lambda, \epsilon_{\mathcal{A}})$-*reduction from breaking* $N = (\text{T}, \text{V}, \text{U})$ *to breaking the* MU-EUF-CMA-C-*security of* SIG, *if for any TM* $\mathcal{A}$ *that* $(t_{\mathcal{A}}, n, \mu, \epsilon_{\mathcal{A}})$-*breaks the* MU-EUF-CMA-C *security of* SIG, *TM* $\Lambda^{\mathcal{A}}$ $(t_\Lambda + r \cdot t_{\mathcal{A}}, \epsilon_\Lambda)$-*breaks* N.

The loss of an r-simple reduction Γ from breaking N to breaking the MU-EUF-CMA-C-security of SIG is defined similar to Definition 7.

Defining a Suitable Relation. Let SIG $= (\text{Setup}, \text{Gen}, \text{Sign}, \text{Vfy})$ be a signature scheme and let I be the range of Setup. We set $\mathcal{R}_{\text{SIG}} = \{R_\Pi\}_{\Pi \in I}$ where $R_\Pi(x, y) := \text{SKCheck}_\Pi(y, x)$. Now, if SIG is t_{ReRand}-key re-randomizable then $\mathcal{R}_{\text{SIG}}$ is t_{ReRand} re-randomizable.

UF-SSA Security for $\mathcal{R}_{\text{SIG}}$ is Weaker Than MU-EUF-CMA-C-*Security for* SIG. Let now SIG be a perfectly correct signature scheme and let $\mathcal{R}_{\text{SIG}}$ be derived from SIG as described in Sect. 5.1.

Claim. If there is an attacker $\mathcal{A}$ that (t, n, e)-breaks the UF-SSA-security for $\mathcal{R}_{\text{SIG}}$ then there is an attacker $\mathcal{B}$ that $(t', n, 0, \epsilon')$-breaks the MU-EUF-CMA-C-security of SIG with $t' = \mathcal{O}(t)$ and $\epsilon' \geq \epsilon$.

Proof. We construct $\mathcal{B}$ that $(t', n, 0, \epsilon')$-breaks the MU-EUF-CMA-C-security of SIG, given black box access to $\mathcal{A}$ as follows:

1. $\mathcal{B}$ is called on input a set of public key $(vk)_{i \in [n]}$ and random tape ρ. Recall that Π are contained in vk. First, $\mathcal{B}$ samples and $\rho_{\mathcal{A}}$, the random coins of $\mathcal{A}$. After that, it runs $(j, st_{\mathcal{A}}) \leftarrow \mathcal{A}_1 \left(\Pi, (vk)_{i \in [n]}, \rho_{\mathcal{A}} \right)$.
2. $\mathcal{B}$ will issue a corrupt-query to oracle $\mathcal{O}$.Corrupt for all $i \in [n \backslash j]$. It will obtain sk_i such that $\mathsf{SKCheck}_\Pi(vk_i, sk_i)$. Next, $\mathcal{B}$ runs $sk_j \leftarrow^{\$} \mathcal{A}_2 \left((sk_i)_{i \in [n \backslash j]}, st_{\mathcal{A}} \right)$. Note that $\mathsf{SKCheck}_\Pi(vk_j, sk_j) = 1$ with probability ϵ.
3. $\mathcal{B}$ samples $m \leftarrow^{\$} \mathcal{M}$ and computes $\sigma \leftarrow^{\$} \mathsf{SIG.Sign}(sk_j, m)$ and outputs (j, m, σ). Note that $vk_j \notin Q^{\mathsf{Corrupt}}$ and $m \notin Q_j$. Moreover, by the property of SKCheck we have $\mathsf{SIG.Vfy}(vk_j, m, \sigma) = 1$.

Tightness Bound

Theorem 6 (informal). *Any simple reduction from breaking the security of a NICA N to breaking the MU-EUF-CMA-C-security of a perfectly correct signature scheme SIG (cf. Definition 15) that provides* efficient *key re-randomization and that supports an efficient SKCheck loses a factor that is linear in the number of public keys the attacker is provided with and that it may corrupt, or N is easy to solve.*

We prove the Theorem via the following technical Theorem, which follows immediately from Theorem 5.

Theorem 7. *Let $N = (\mathsf{T}, \mathsf{V}, \mathsf{U})$ be a non-interactive complexity assumption, $n, r \in \mathsf{poly}(k)$ and let $\mathcal{R}_{\mathsf{SIG}}$ be a family of computable relations as described above. Then for any r-simple $(t_\Gamma, n, \epsilon_\Gamma, 1)$-reduction Γ from breaking N to breaking the UF-SSA-security of $\mathcal{R}_{\mathsf{SIG}}$ there exists a TM $\mathcal{B}$ that $(t_\mathcal{B}, \epsilon_\mathcal{B})$-breaks N where*

$$t_\mathcal{B} \leq r \cdot n \cdot t_\Gamma + r \cdot n \cdot (n-1) \cdot t_{\mathsf{Vfy}} + r \cdot t_{\mathsf{ReRand}}$$

$$\epsilon_\mathcal{B} \geq \epsilon_\Gamma - \frac{r}{n}.$$

Here, t_{ReRand} is the time to re-randomize a given valid witness and t_{Vfy} is the maximum time needed to compute $R \in \mathcal{R}_{\mathsf{SIG}}$.

5.2   Public-Key Encryption in the Multi-user Setting

Our main result also applies to public key encryption in the multi-user setting with corruptions (and a similar result for key encapsulation mechanisms is straightforward). In the following, we only sketch the main steps to establishing our result. The full version contains a detailed, formal treatment. We start off by first defining MU-IND-CPA-C-security (Fig. 6), a security definition for public key encryption schemes PKE = (Setup, Gen, Enc, Dec) in the multi-user setting with corruptions. To apply our main result, we again have to formally define a

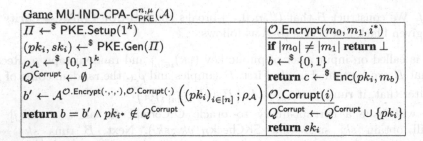

Fig. 6. MU-IND-CPA-C-security game. The attacker has access to an encryption oracle $\mathcal{O}.\mathsf{Encrypt}$ which may be queried only once and a corrupt oracle $\mathcal{O}.\mathsf{Corrupt}$.

family $\mathcal{R}_{\mathsf{PKE}}$ of suitable computable relations. To this end (and similar to the case of digital signatures in the multi user setting), we require the existence of an additional TM $\mathsf{SKCheck}_\Pi$ for $\Pi \leftarrow^{\$} \mathsf{Setup}(1^k)$ such that

$$\mathsf{SKCheck}_\Pi(pk, sk) = 1 \iff \Pr\left[\mathsf{Dec}(sk, \mathsf{Enc}(pk, m)) = m : m \leftarrow^{\$} \mathcal{M}\right] = 1.$$

That is, $\mathsf{SKCheck}$ takes inputs sk and pk and returns 1 if and only if pk is a PKE public key and sk is a secret key corresponding to public key pk. To define our suitable relation, we set $\mathcal{R}_{\mathsf{PKE}} = \{R_\Pi\}_{\Pi \in I}$ where $R_\Pi(x, y) := \mathsf{SKCheck}_\Pi(y, x)$ and I is the set of all public parameters that can be output by Setup. Finally, we show that MU-IND-CPA-C-security for PKE is stronger than UF-SSA-security for $\mathcal{R}_{\mathsf{PKE}}$. Via our main result, this immediately proves that any security reduction must have a security loss that is (at least) linear in the number of public keys considered in the MU-IND-CPA-C-security experiment.

5.3 Non-interactive Key Exchange

In this section we will show how to apply our main result to non-interactive key exchange (NIKE) [25]. This case differs from the cases considered before in that we will have to define a relation $R(x, y)$, which is not efficiently verifiable, given just x and y. Instead, we will need additional information, which will be available in the NIKE security experiment. Formally, we consider again UF-SSA-security for some relation R but model $\mathcal{A}_2$ as an oracle machine. The responses of the oracle may depend on the output of $\mathcal{A}_1$. We explain that this makes it possible to extend the range of covered cryptographic primitives to NIKE.

Definitions. Following [16, 25], a NIKE protocol consists of three PPT-TMs with the following syntax:

Public Parameters. On input 1^k, the public parameter generation machine $\Pi \leftarrow^{\$} \mathsf{NIKE.Setup}(1^k)$ outputs a set Π of system parameters.

Key Generation. The key generation machine takes as input Π and outputs a random key pair (sk_i, pk_i) for party i, i.e. $(sk_i, pk_i) \leftarrow^{\$} \mathsf{NIKE.Gen}(\Pi)$. We assume that pk contains Π and 1^k.

Shared Key Generation. The deterministic shared key machine SharedKey takes as input (sk_i, pk_j) and outputs a shared key $K_{i,j}$ in time t_{Vfy}, where $K_{i,j} = \bot$ if $i = j$.

We require perfect correctness, that is,

$$\Pr\left[\mathsf{SharedKey}(sk_i, pk_j) = \mathsf{SharedKey}(sk_j, pk_i)\right] = 1$$

for all $\Pi \leftarrow^{\$} \mathsf{NIKE.Setup}(1^k)$ and $(pk_i, sk_i), (pk_j, sk_j) \leftarrow^{\$} \mathsf{NIKE.Gen}(\Pi)$.

We require an additional Turing machine PKCheck that inputs strings Π and pk and evaluates to true if pk is in the range of $\mathsf{NIKE.Gen}(\Pi)$. Moreover, whenever two public keys pk and pk' are accepted by PKCheck, we require that the respective shared key is uniquely determined, given only pk and pk'. In the sequel we will denote this key by $K(pk, pk')$ and call NIKE *unique*. The pairing-based NIKE scheme from [25] satisfies uniqueness.

NIKE Security. There exists several different, but polynomial-time equivalent [25] security models for NIKE. Of course the tightness of a reduction depends on the choice of the security model. Indeed, the weakest security model considered in [25] is the *CKS-light* model. However, this model is strongly idealized. The reduction from breaking security in a stronger and more realistic security model (called the CKS model in [25]) to breaking security in this idealized model loses a factor of n^2, where n is the number of users. We show that this loss is inherent for NIKE schemes with the properties defined above.

CKS-Security for NIKE. The *CKS-security* experiment is depicted in Fig. 7.

Game $\mathsf{CKS}^{n,\mathcal{A}}_{\mathsf{NIKE}}(1^k)$	
$\Pi \leftarrow^{\$} \mathsf{Setup}(1^k)$	$\mathcal{O}.\mathsf{Corrupt}(i)$
$(pk_i, sk_i) \leftarrow^{\$} \mathsf{NIKE.Gen}(\Pi)$	$Q^{\mathsf{Corrupt}} \leftarrow Q^{\mathsf{Corrupt}} \cup \{pk_i\}$
$\rho_{\mathcal{A}} \leftarrow^{\$} \{0,1\}^k$	**return** sk_i
$Q^{\mathsf{Corrupt}} = Q^{\mathsf{Reveal}} \leftarrow \emptyset$	
$b' \leftarrow^{\$} \mathcal{A}^{\mathcal{O}.\mathsf{Corrupt}(\cdot),\mathcal{O}.\mathsf{Reveal}(\cdot,\cdot),\mathcal{O}.\mathsf{Test}(\cdot,\cdot)}\left(\Pi, (pk_i)_{i \in [n]}; \rho_{\mathcal{A}}\right)$	$\mathcal{O}.\mathsf{Reveal}(i,j)$
return $b' = b \wedge pk_{i^*}, pk_{j^*} \notin Q^{\mathsf{Corrupt}} \wedge (i^*, j^*) \notin Q^{\mathsf{Reveal}}$	$Q^{\mathsf{Reveal}} \leftarrow Q^{\mathsf{Reveal}} \cup \{(i,j)\}$
	return $\mathsf{SharedKey}(sk_i, pk_j)$
$\mathcal{O}.\mathsf{Test}(i^*, j^*)$	
$K_0 \leftarrow \mathsf{SharedKey}(sk_{i^*}, pk_{j^*}); K_1 \leftarrow^{\$} \mathsf{SharedKey}(\cdot, \cdot)$	
$b \leftarrow^{\$} \{0,1\}$	
return K_b	

Fig. 7. CKS-Security game for NIKE. Oracle $\mathcal{O}.\mathsf{Test}$ may be queried only once. K_1 is sampled uniform from the range of SharedKey.

Definition 16. *We say that an adversary* $\mathcal{A}$ (t, n, ϵ)*-breaks the* CKS*-security of a non-interactive key exchange protocol* NIKE *if it runs in time at most t and*

$$\Pr\left[\mathsf{CKS}^{n,\mathcal{A}}_{\mathsf{NIKE}}(1^k) \Rightarrow 1\right] \geq \epsilon.$$

Definition 17. *We say that a Turing machine* r-Γ *is an* r-simple $(t_\Lambda, n, \epsilon_\Lambda, \epsilon_{\mathcal{A}})$-*reduction from breaking* $N = (T, V, U)$ *to breaking the* CKS-*security of* NIKE, *if for any TM* $\mathcal{A}$ *that* $(t_{\mathcal{A}}, n, \epsilon_{\mathcal{A}})$-*breaks the* CKS *security of* NIKE, *TM* $\Lambda^{\mathcal{A}}$ $(t_\Lambda + r \cdot t_{\mathcal{A}}, \epsilon_\Lambda)$-*breaks* N.

The loss of an r-simple reduction Γ from breaking the security of N to breaking the CKS-security of NIKE is defined similar to Definition 7.

Defining a Suitable Relation. Let NIKE $=$ (Setup, Gen, SharedKey) be a unique NIKE scheme and let I be the range of Setup. We set $\mathcal{R}_{\mathsf{NIKE}} = \{R_\Pi\}_{\Pi \in I}$ where

$$R_\Pi(x, (y_1, y_2)) = 1 \Leftrightarrow x = K(y_1, y_2).$$

Let us fix Π for the moment. Note that the attacker is provided with $\tilde{n} = (n-1)\cdot n$ R_Π statements if it is provided with n NIKE-public keys.

Let now $\mathcal{A} = (\mathcal{A}_1, \mathcal{A}_2)$ denote an attacker against the UF-SSA-security of $\mathcal{R}_{\mathsf{NIKE}}$. Because R may not be efficiently verifiable, we let $\mathcal{A}_2$ have oracle access to Oracle Corrupt$_{i^*, j^*}$ that returns secret key sk_i when queried on input $i \in [n \backslash \{i^*, j^*\}]$. Here $K(pk_{i^*}, pk_{j^*})$ is the shared key that $\mathcal{A}$ needs to compute to break the UF-SSA security of $\mathcal{R}$ and n is the number of public keys that $\mathcal{A}$ is provided with (note that this leads to $\tilde{n}$ NIKE shared keys).

UF-SSA-*Security for* $\mathcal{R}_{\mathsf{NIKE}}$ *is Weaker Than* CKS-*Security for* NIKE. Next, we show that any adversary that breaks the UF-SSA-security of $\mathcal{R}_{\mathsf{NIKE}}$ then there is an attacker that breaks the CKS-security of NIKE.

Claim. If there is an attacker $\mathcal{A}$ that $(t, \tilde{n}, \epsilon)$-breaks the UF-SSA-security of $\mathcal{R}_{\mathsf{NIKE}}$ then there is an attacker $\mathcal{B}$ that (t', n, ϵ')-breaks the CKS-security of NIKE with $t' = \mathcal{O}(t)$ and $\epsilon' \geq \epsilon$.

Proof. We construct $\mathcal{B}$ that (t', n, ϵ')-breaks the CKS-security of NIKE, given black box access to $\mathcal{A}$ as follows:

1. $\mathcal{B}$ is called on input a set of public keys $(pk)_{i \in [n]}$ and random tape ρ. Recall that Π is contained in pk. First, $\mathcal{B}$ samples and $\rho_{\mathcal{A}}$, the random coins of $\mathcal{A}$. Next, it runs $((i^*, j^*), st_{\mathcal{A}}) \leftarrow \mathcal{A}_1 \left(\Pi, (pk)_{i \in [n]}, \rho_{\mathcal{A}} \right)$. Note that n public keys define $n \cdot (n-1)$ statements for R_Π. The one that $\mathcal{A}$ will compute is determined by i^* and j^*.
2. $\mathcal{B}$ will issue a reveal-query to oracle $\mathcal{O}$.Reveal for all $(i, j) \in [n]^2 \backslash \{(i^*, j^*)\}, i \neq j$. It will obtain $K_{i,j} = \mathsf{SharedKey}(sk_i, pk_j)$. Next, $\mathcal{B}$ runs

$$K^* \xleftarrow{\$} \mathcal{A}_2^{\mathcal{O}.\mathsf{Corrupt}_{i^*, j^*}(\cdot)} \left((K_{i,j})_{(i,j) \in [n]^2 \backslash \{i^*, j^*\}, i \neq j}, st_{\mathcal{A}} \right).$$

$\mathcal{B}$ provides $\mathcal{A}$ with oracle Corrupt$_{i^*, j^*}$ by forwarding all queries to oracle $\mathcal{O}$.Corrupt() and forwarding the response back to $\mathcal{A}$. Note that, using sk_i, $\mathcal{A}$ may efficiently check whether $K_{i,j} = \mathsf{SharedKey}(sk_i, pk_j)$ for all $j \in [n]$. By assumption it holds that $K^* = \mathsf{SharedKey}(sk_{i^*}, pk^{j^*})$ with probability at least ϵ.

3. Next, $\mathcal{B}$ issues (i^*, j^*) to oracle $\mathcal{O}.\mathsf{Test}()$ which will respond with K. $\mathcal{B}$ returns 0 if $K = K^*$ and 1 otherwise. Note that by construction of oracle $\mathsf{Corrupt}_{i^*,j^*}$ it holds that $i^*, j^* \notin Q^{\mathsf{Corrupt}}$. Moreover, by the perfect correctness of NIKE and the uniqueness of shared keys $\mathcal{B}$ is successful whenever $\mathcal{A}$ is successful.

Tightness Bounds

Theorem 8 (informal). *Any simple reduction from breaking the security of a NICA N to breaking the CKS-security of a perfectly correct, unique NIKE scheme NIKE (cf. Definition 16) that supports an efficient PKCheck loses a factor that is quadratic in the number of public keys the attacker is provided with and that it may corrupt, or N is easy to solve.*

We prove the Theorem via the following technical Theorem.

Theorem 9. *Let $N = (\mathsf{T}, \mathsf{V}, \mathsf{U})$ be a non-interactive complexity assumption, $\tilde{n}, r \in \mathsf{poly}(k)$ and let $\mathcal{R}_{\mathsf{NIKE}}$ be a family of computable relations as described above. Then for any r-simple $(t_\Gamma, \tilde{n}, \epsilon_\Gamma, 1)$-reduction Γ from breaking N to breaking the UF-SSA-security of $\mathcal{R}_{\mathsf{NIKE}}$ there exists a TM $\mathcal{B}$ that $(t_\mathcal{B}, \epsilon_\mathcal{B})$-breaks N where*

$$t_\mathcal{B} \leq r \cdot \tilde{n} \cdot t_\Gamma + r \cdot \tilde{n} \cdot (\tilde{n} - 1) \cdot t_{\mathsf{Vfy}} \text{ and } \epsilon_\mathcal{B} \geq \epsilon_\Gamma - \frac{r}{\tilde{n}}.$$

Here, t_{Vfy} is the maximum time needed to compute $R \in \mathcal{R}_{\mathsf{NIKE}}$ with access to $\mathsf{Corrupt}_{i^*,j^*}$.

Interpretation. As mentioned before, if the attacker is provided with $\tilde{n}$ statements, it is provided only with $\approx \sqrt{\tilde{n}}$ public keys. Thus, the loss of any r-simple reduction is *quadratic* in the number of public keys if the underlying problem is assumed to be hard.

Our lower bound for NIKE can easily be generalized to systems where keys are derived from $\ell = O(\log(k))$ parties for security parameter k. Syntactically, the difference is that SharedKey now takes as input $\ell - 1$ public keys and a single secret key. Now, the attacker obtains $\tilde{n}$ statements and $\approx \tilde{n}^{1/\ell}$ public keys. Thus, the loss of any r-simple reduction grows with an exponent of ℓ in the number of public keys.

Extending the Result to Interactive Key Exchange. On the one hand, our NIKE bounds do not carry over directly to arbitrary interactive key exchange protocols, because these do not necessarily meet the properties of NIKE schemes that we need to put up. In particular, we have to require that any pair of NIKE public keys uniquely determines the corresponding shared key (which limits the generality of the result, but appears very reasonable for natural (and possibly all) NIKE constructions, in particular it holds for the NIKE schemes of [25]). This requirement does not hold for interactive AKE protocols, where the shared key may additionally depend on ephemeral random values (nonces or Diffie-Hellman shares, for example) exchanged between parties.

On the other hand, our tightness bounds for signatures and public-key encryption (with unique/re-randomizable secret keys, in the multi-user setting with corruptions) directly imply tightness bounds for AKE protocols that use these primitives, and where the attacker is able to adaptively corrupt the secret keys of these signature/PKE schemes. Note that this includes the vast majority of all known AKE constructions. The tightly-secure key exchange protocol of [4] overcomes this hurdle by using a signature scheme that does not have unique/re-randomizable secret keys, and this is used in a crucial way (cf. the "Naor-Yung trick for signatures" in [4]).

A   Summary of Coron's Meta-reduction and Its Generalizations

EUF-CMA-security is commonly considered the standard security definition for digital signature schemes [28]. The security game is depicted in Fig. 8.

Game EUF-CMA$_{\mathsf{SIG}}^{n,\mathcal{A}}(1^k)$	
$\Pi \leftarrow^{\$} \mathsf{Setup}(1^k)$	$\mathcal{O}.\mathsf{Sign}(m)$
$(vk, sk) \leftarrow^{\$} \mathsf{Gen}(\Pi)$	**if** $\|Q\| \geq n$ **return** $\bot$
$\rho_{\mathcal{A}} \leftarrow^{\$} P_{\mathcal{A}}$	$Q \leftarrow Q \cup \{m\}$
$Q \leftarrow \emptyset$	**return** $\sigma \leftarrow^{\$} \mathsf{SIG}.\mathsf{Sign}(sk, m)$
$(m^*, \sigma^*) \leftarrow \mathcal{A}^{\mathcal{O}.\mathsf{Sign}(\cdot)}(vk; \rho_{\mathcal{A}})$	
return $m^* \notin Q \wedge \mathsf{SIG}.\mathsf{Vfy}(vk, m^*, \sigma^*)$	

Fig. 8. EUF-CMA-Security game. When called, the attacker has access to a signing oracle $\mathcal{O}.\mathsf{Sign}$.

Definition 18. (EUF-CMA-security). *We say that an attacker* (t, n, ϵ)*-breaks the* EUF-CMA*-security of a signature scheme* SIG *if it runs in time* t *and*

$$\Pr\left[\mathsf{EUF\text{-}CMA}_{\mathsf{SIG}}^{n,\mathcal{A}}(1^k) \Rightarrow 1\right] \geq \epsilon.$$

Definition 19. *We say that a Turing machine* $r\text{-}\Gamma$ *is an* r*-simple* $(t_\Lambda, n, \epsilon_\Lambda, \epsilon_{\mathcal{A}})$*-reduction from breaking* $N = (\mathsf{T}, \mathsf{V}, \mathsf{U})$ *to breaking the* EUF-CMA*-security of* SIG, *if for any TM* $\mathcal{A}$ *that* $(t_{\mathcal{A}}, n, \epsilon_{\mathcal{A}})$*-breaks the* EUF-CMA *security of* SIG, *TM* $\Lambda^{\mathcal{A}}$ $(t_\Lambda + r \cdot t_{\mathcal{A}}, \epsilon_\Lambda)$*-breaks* N.

Definition 20. *Let* $\ell : \mathbb{N} \to \mathbb{N}$. *We say that an* r*-simple reduction* Γ *from breaking* N *to breaking the* EUF-CMA*-security of* SIG *loses* ℓ, *if there exists an adversary* $\mathcal{A}$ *that* $(t_{\mathcal{A}}, n, \epsilon_{\mathcal{A}})$*-breaks the* EUF-CMA *security of* SIG, *such that* $\Lambda^{\mathcal{A}}$ $(t_\Lambda + t_{\mathcal{A}}, \epsilon_\Lambda)$*-breaks* N *with*

$$\frac{t_\Lambda(k) + t_{\mathcal{A}}(k)}{\epsilon_\Lambda(k)} \geq \ell(k) \cdot \frac{t_{\mathcal{A}}(k)}{\epsilon_{\mathcal{A}}(k)}.$$

The following lemma is due to Hofheinz *et al.* [30] and generalizes a result from Coron [18].

Lemma 1 ([18,30]). *Let N be a (t_N, ϵ_N)-secure non-interactive complexity assumption where $\epsilon_N \in \mathsf{negl}(k)$ and let SIG be a* unique *signature scheme with message space of size 2^l. If Γ is a $(t_\Gamma, n, \epsilon_\Gamma)$-reduction from breaking N to breaking the EUF-CMA-security of SIG and $t_N \geq 2 \cdot t_\Gamma + t_{\mathsf{ReRand}}$ then*

$$\epsilon_\Gamma \leq \epsilon_{\mathcal{A}} \cdot \frac{\exp(-1)}{n} \cdot \left(1 - \frac{n}{2^l}\right)^{-1} + \mathsf{negl}(k). \qquad \square$$

Coron [18] and Hofheinz *et al.* [30] conclude that we have $\epsilon_\Lambda = \mathcal{O}\left(\frac{\epsilon_{\mathcal{A}}}{n}\right)$. The conclusion builds on the fact that $2^l \gg n$. This is reasonable for most digital signatures schemes.

B   UF-SMA-Security Is Strictly Weaker Than EUF-CMA-Security

We show that any attacker $\mathcal{A}$ that breaks the UF-SMA-security of a signature scheme SIG implies an attacker $\mathcal{A}'$ that breaks the EUF-CMA-security (depicted in Fig. 8) of SIG in roughly the same running time and with the same probability of success. Moreover UF-SMA-security and EUF-CMA-security are *not* polynomially equivalent.

Claim. Let SIG be a signature scheme. If there is an attacker $\mathcal{A}$ that (t, n, ϵ)-breaks the UF-SMA-security of a signature scheme SIG then there is an attacker $\mathcal{B}$ that (t', n, ϵ')-breaks the EUF-CMA-security of SIG where $t' = \mathcal{O}(t)$ and $\epsilon' \geq \epsilon$.

Proof. We construct $\mathcal{B}$ that (t', n, ϵ')-breaks the EUF-CMA-security of SIG, given black box access to $\mathcal{A}$ as follows:

1. $\mathcal{B}$ is called on input a public key vk and random tape ρ. First, $\mathcal{B}$ samples n *distinct* messages $m_1, \ldots, m_n$ from the message space and $\rho_{\mathcal{A}}$, the random coins of $\mathcal{A}$. After that, it runs $(j, st_{\mathcal{A}}) \leftarrow \mathcal{A}_1\left(vk, (m_i)_{i \in [n]}, \rho_{\mathcal{A}}\right)$.
2. $\mathcal{B}$ will issue a sign-query to oracle Sign for all messages $m_i, i \in [n \backslash j]$. It will obtain $\sigma_i \leftarrow^\$ \mathsf{SIG.Sign}(sk, m_i)$. Note that σ_i is a valid signature over m_i with respect to vk. Next, $\mathcal{B}$ runs $\sigma_j \leftarrow^\$ \mathcal{A}_2\left((\sigma_i)_{i \in [n \backslash j]}, st_{\mathcal{A}}\right)$ which is valid with probability ϵ.
3. $\mathcal{B}$ outputs (m_j, σ_j). Note that due to the fact that $m_i \neq m_j$ for all $i \neq j$, this is a valid forgery which is valid with probability at least ϵ.

Let SIG be a signature scheme with exponential message space $\mathcal{M}$. Let $m \leftarrow^\$ \mathcal{M}$. Then we define a signature scheme $\mathsf{SIG}'(m)$ that works exactly like SIG except the $\mathsf{SIG}'(m)$-verification machine will accept 0 as a valid signature over m.

Claim. Suppose that no adversary (t, n, ϵ)-breaks the EUF-CMA-security of SIG. Then the following holds: 1. There is no adversary that (t, n, ϵ')-breaks the UF-SMA-security of $\mathsf{SIG}'(m)$ with $\epsilon' \geq \epsilon + \frac{n}{|\mathcal{M}|}$. 2. There exists a trivial attack strategy that $(\mathcal{O}(1), 0, 1)$-breaks the EUF-CMA-security of $\mathsf{SIG}'(m)$.

Proof. 1. Recall that at the beginning of the UF-SMA security experiment, $\mathcal{A}$ is called on input a verification key and n distinct messages that are sampled uniformly from $\mathcal{M}$. Now, the probability that $m_i = m$ for $i \in [n]$ is upper bounded by $\frac{n}{|\mathcal{M}|}$. However, if for all $i \in [n]$ we have $m_i \neq m$ then we can apply the previous claim. When called on vk, $\mathcal{A}$ simply outputs $(m, 0)$ which is a valid forgery.

References

1. Abdalla, M., Fouque, P.-A., Lyubashevsky, V., Tibouchi, M.: Tightly-secure signatures from lossy identification schemes. In: Pointcheval, D., Johansson, T. (eds.) EUROCRYPT 2012. LNCS, vol. 7237, pp. 572–590. Springer, Heidelberg (2012)
2. Abe, M., Groth, J., Ohkubo, M.: Separating short structure-preserving signatures from non-interactive assumptions. In: Lee, D.H., Wang, X. (eds.) ASIACRYPT 2011. LNCS, vol. 7073, pp. 628–646. Springer, Heidelberg (2011)
3. Bader, C.: Efficient signatures with tight real world security in the random-oracle model. In: Gritzalis, D., Kiayias, A., Askoxylakis, I. (eds.) CANS 2014. LNCS, vol. 8813, pp. 370–383. Springer, Heidelberg (2014)
4. Bader, C., Hofheinz, D., Jager, T., Kiltz, E., Li, Y.: Tightly-secure authenticated key exchange. In: Dodis, Y., Nielsen, J.B. (eds.) TCC 2015, Part I. LNCS, vol. 9014, pp. 629–658. Springer, Heidelberg (2015)
5. Baldimtsi, F., Lysyanskaya, A.: On the security of one-witness blind signature schemes. In: Sako, K., Sarkar, P. (eds.) ASIACRYPT 2013, Part II. LNCS, vol. 8270, pp. 82–99. Springer, Heidelberg (2013)
6. Bellare, M., Boldyreva, A., Micali, S.: Public-key encryption in a multi-user setting: security proofs and improvements. In: Preneel, B. (ed.) EUROCRYPT 2000. LNCS, vol. 1807, pp. 259–274. Springer, Heidelberg (2000)
7. Bellare, M., Ristenpart, T.: Simulation without the artificial abort: simplified proof and improved concrete security for waters' IBE scheme. Cryptology ePrint Archive, Report 2009/084 (2009). http://eprint.iacr.org/2009/084
8. Bellare, M., Ristenpart, T.: Simulation without the artificial abort: simplified proof and improved concrete security for waters' ibe scheme. In: Joux, A. (ed.) EUROCRYPT 2009. LNCS, vol. 5479, pp. 407–424. Springer, Heidelberg (2009)
9. Bellare, M., Rogaway, P.: Entity authentication and key distribution. In: Stinson, D.R. (ed.) CRYPTO 1993. LNCS, vol. 773, pp. 232–249. Springer, Heidelberg (1994)
10. Blazy, O., Kakvi, S.A., Kiltz, E., Pan, J.: Tightly-secure signatures from chameleon hash functions. In: Katz, J. (ed.) PKC 2015. LNCS, vol. 9020, pp. 256–279. Springer, Heidelberg (2015)
11. Blazy, O., Kiltz, E., Pan, J.: (Hierarchical) identity-based encryption from affine message authentication. In: Garay, J.A., Gennaro, R. (eds.) CRYPTO 2014, Part I. LNCS, vol. 8616, pp. 408–425. Springer, Heidelberg (2014)
12. Boneh, D., Venkatesan, R.: Breaking RSA may not be equivalent to factoring. In: Nyberg, K. (ed.) EUROCRYPT 1998. LNCS, vol. 1403, pp. 59–71. Springer, Heidelberg (1998)
13. Bresson, E., Monnerat, J., Vergnaud, D.: Separation results on the "one-more" computational problems. In: Malkin, T. (ed.) CT-RSA 2008. LNCS, vol. 4964, pp. 71–87. Springer, Heidelberg (2008)

14. Brown, D.R.L.: Irreducibility to the one-more evaluation problems: more may be less. Cryptology ePrint Archive, Report 2007/435 (2007). http://eprint.iacr.org/
15. Canetti, R., Krawczyk, H.: Analysis of key-exchange protocols and their use for building secure channels. In: Pfitzmann, B. (ed.) EUROCRYPT 2001. LNCS, vol. 2045, pp. 453–474. Springer, Heidelberg (2001)
16. Cash, D.M., Kiltz, E., Shoup, V.: The twin Diffie-Hellman problem and applications. In: Smart, N.P. (ed.) EUROCRYPT 2008. LNCS, vol. 4965, pp. 127–145. Springer, Heidelberg (2008)
17. Chen, J., Wee, H.: Fully, (almost) tightly secure IBE and dual system groups. In: Canetti, R., Garay, J.A. (eds.) CRYPTO 2013, Part II. LNCS, vol. 8043, pp. 435–460. Springer, Heidelberg (2013)
18. Coron, J.-S.: Optimal security proofs for PSS and other signature schemes. In: Knudsen, L.R. (ed.) EUROCRYPT 2002. LNCS, vol. 2332, pp. 272–287. Springer, Heidelberg (2002)
19. Cramer, R., Damgård, I.B.: New generation of secure and practical RSA-based signatures. In: Koblitz, N. (ed.) CRYPTO 1996. LNCS, vol. 1109, pp. 173–185. Springer, Heidelberg (1996)
20. Dodis, Y., Oliveira, R., Pietrzak, K.: On the generic insecurity of the full domain hash. In: Shoup, V. (ed.) CRYPTO 2005. LNCS, vol. 3621, pp. 449–466. Springer, Heidelberg (2005)
21. Dodis, Y., Reyzin, L.: On the power of claw-free permutations. In: Cimato, S., Galdi, C., Persiano, G. (eds.) SCN 2002. LNCS, vol. 2576, pp. 55–73. Springer, Heidelberg (2003)
22. Fischlin, M., Fleischhacker, N.: Limitations of the meta-reduction technique: the case of Schnorr signatures. In: Johansson, T., Nguyen, P.Q. (eds.) EUROCRYPT 2013. LNCS, vol. 7881, pp. 444–460. Springer, Heidelberg (2013)
23. Fischlin, M., Schröder, D.: On the impossibility of three-move blind signature schemes. In: Gilbert, H. (ed.) EUROCRYPT 2010. LNCS, vol. 6110, pp. 197–215. Springer, Heidelberg (2010)
24. Fleischhacker, N., Jager, T., Schröder, D.: On tight security proofs for Schnorr signatures. In: Sarkar, P., Iwata, T. (eds.) ASIACRYPT 2014. LNCS, vol. 8873, pp. 512–531. Springer, Heidelberg (2014)
25. Freire, E.S.V., Hofheinz, D., Kiltz, E., Paterson, K.G.: Non-interactive key exchange. In: Kurosawa, K., Hanaoka, G. (eds.) PKC 2013. LNCS, vol. 7778, pp. 254–271. Springer, Heidelberg (2013)
26. Garg, S., Bhaskar, R., Lokam, S.V.: Improved bounds on security reductions for discrete log based signatures. In: Wagner, D. (ed.) CRYPTO 2008. LNCS, vol. 5157, pp. 93–107. Springer, Heidelberg (2008)
27. Goldwasser, S., Micali, S., Rackoff, C.: The knowledge complexity of interactive proof systems. SIAM J. Comput. 18(1), 186–208 (1989)
28. Goldwasser, S., Micali, S., Rivest, R.L.: A digital signature scheme secure against adaptive chosen-message attacks. SIAM J. Comput. 17(2), 281–308 (1988)
29. Hofheinz, D., Jager, T.: Tightly secure signatures and public-key encryption. In: Safavi-Naini, R., Canetti, R. (eds.) CRYPTO 2012. LNCS, vol. 7417, pp. 590–607. Springer, Heidelberg (2012)
30. Hofheinz, D., Jager, T., Knapp, E.: Waters signatures with optimal security reduction. In: Fischlin, M., Buchmann, J., Manulis, M. (eds.) PKC 2012. LNCS, vol. 7293, pp. 66–83. Springer, Heidelberg (2012)
31. Kakvi, S.A., Kiltz, E.: Optimal security proofs for full domain hash, revisited. In: Pointcheval, D., Johansson, T. (eds.) EUROCRYPT 2012. LNCS, vol. 7237, pp. 537–553. Springer, Heidelberg (2012)

304   C. Bader et al.

32. Katz, J., Wang, N.: Efficiency improvements for signature schemes with tight security reductions. In: Jajodia, S., Atluri, V., Jaeger, T. (eds.) ACM CCS 2003, pp. 155–164. ACM Press, October 2003

33. Lewko, A., Waters, B.: Why proving HIBE systems secure is difficult. In: Nguyen, P.Q., Oswald, E. (eds.) EUROCRYPT 2014. LNCS, vol. 8441, pp. 58–76. Springer, Heidelberg (2014)

34. Paillier, P., Vergnaud, D.: Discrete-log-based signatures may not be equivalent to discrete log. In: Roy, B. (ed.) ASIACRYPT 2005. LNCS, vol. 3788, pp. 1–20. Springer, Heidelberg (2005)

35. Paillier, P., Villar, J.L.: Trading one-wayness against chosen-ciphertext security in factoring-based encryption. In: Lai, X., Chen, K. (eds.) ASIACRYPT 2006. LNCS, vol. 4284, pp. 252–266. Springer, Heidelberg (2006)

36. Schäge, S.: Tight proofs for signature schemes without random oracles. In: Paterson, K.G. (ed.) EUROCRYPT 2011. LNCS, vol. 6632, pp. 189–206. Springer, Heidelberg (2011)

37. Seurin, Y.: On the exact security of schnorr-type signatures in the random oracle model. In: Pointcheval, D., Johansson, T. (eds.) EUROCRYPT 2012. LNCS, vol. 7237, pp. 554–571. Springer, Heidelberg (2012)

38. Waters, B.: Efficient identity-based encryption without random oracles. In: Cramer, R. (ed.) EUROCRYPT 2005. LNCS, vol. 3494, pp. 114–127. Springer, Heidelberg (2005)

On the Size of Pairing-Based
Non-interactive Arguments

Jens Groth[✉]

University College London, London, UK
j.groth@ucl.ac.uk

Abstract. Non-interactive arguments enable a prover to convince a verifier that a statement is true. Recently there has been a lot of progress both in theory and practice on constructing highly efficient non-interactive arguments with small size and low verification complexity, so-called succinct non-interactive arguments (SNARGs) and succinct non-interactive arguments of knowledge (SNARKs).

Many constructions of SNARGs rely on pairing-based cryptography. In these constructions a proof consists of a number of group elements and the verification consists of checking a number of pairing product equations. The question we address in this article is how efficient pairing-based SNARGs can be.

Our first contribution is a pairing-based (preprocessing) SNARK for arithmetic circuit satisfiability, which is an NP-complete language. In our SNARK we work with asymmetric pairings for higher efficiency, a proof is only 3 group elements, and verification consists of checking a single pairing product equations using 3 pairings in total. Our SNARK is zero-knowledge and does not reveal anything about the witness the prover uses to make the proof.

As our second contribution we answer an open question of Bitansky, Chiesa, Ishai, Ostrovsky and Paneth (TCC 2013) by showing that linear interactive proofs cannot have a linear decision procedure. It follows from this that SNARGs where the prover and verifier use generic asymmetric bilinear group operations cannot consist of a single group element. This gives the first lower bound for pairing-based SNARGs. It remains an intriguing open problem whether this lower bound can be extended to rule out 2 group element SNARGs, which would prove optimality of our 3 element construction.

Keywords: SNARKs · Non-interactive zero-knowledge arguments · Linear interactive proofs · Quadratic arithmetic programs · Bilinear groups

J. Groth—The research leading to these results has received funding from the European Research Council under the European Union's Seventh Framework Programme (FP/2007-2013)/ERC Grant Agreement n. 307937 and the Engineering and Physical Sciences Research Council grant EP/J009520/1. This work was done in part while the author was visiting the Simons Institute for the Theory of Computing, supported by the Simons Foundation and by the DIMACS/Simons Collaboration in Cryptography through NSF grant #CNS-1523467.

M. Fischlin and J.-S. Coron (Eds.): EUROCRYPT 2016, Part II, LNCS 9666, pp. 305–326, 2016.
DOI: 10.1007/978-3-662-49896-5_11

1   Introduction

Goldwasser et al. [GMR89] introduced zero-knowledge proofs that enable a prover to convince a verifier that a statement is true without revealing anything else. They have three core properties:

Completeness: Given a statement and a witness, the prover can convince the verifier.

Soundness: A malicious prover cannot convince the verifier of a false statement.

Zero-knowledge: The proof does not reveal anything but the truth of the statement, in particular it does not reveal the prover's witness.

Blum et al. [BFM88] extended the notion to *non-interactive* zero-knowledge (NIZK) proofs in the common reference string model. NIZK proofs are useful in the construction of non-interactive cryptographic schemes, e.g., digital signatures and CCA-secure public key encryption.

The amount of communication is an important performance parameter for zero-knowledge proofs. Kilian [Kil92] gave the first sublinear communication zero-knowledge argument that sends fewer bits than the size of the statement to be proved. Micali [Mic00] proposed sublinear size NIZK arguments by letting the prover in a communication efficient zero-knowledge argument compute the verifier's challenges using a cryptographic function.

Groth et al. [GOS12,GOS06,Gro06,GS12] introduced pairing-based NIZK proofs, yielding the first linear size proofs based on standard assumptions. Groth [Gro10] combined these techniques with ideas from interactive zero-knowledge arguments [Gro09] to give the first constant size NIZK arguments. Lipmaa [Lip12] used an alternative construction based on progression-free sets to reduce the size of the common reference string.

Groth's constant size NIZK argument is based on constructing a set of polynomial equations and using pairings to efficiently verify these equations. Gennaro et al. [GGPR13] found an insightful construction of polynomial equations based on Lagrange interpolation polynomials yielding a pairing-based NIZK argument with a common reference string size proportional to the size of the statement and witness. They gave two types of polynomial equations: quadratic span programs for proving boolean circuit satisfiability and quadratic arithmetic programs for proving arithmetic circuit satisfiability. Lipmaa [Lip13] suggested more efficient quadratic span programs using error correcting codes, and Danezis et al. [DFGK14] refined quadratic span programs to square span programs that give NIZK arguments consisting of 4 group elements for boolean circuit satisfiability.

Following these theoretical advances there has been exciting work on building concrete implementations. Most efficient implementations refine the quadratic arithmetic program approach of Gennaro et al. [GGPR13] and combine it with a compiler producing a suitable quadratic arithmetic program that is equivalent to the statement to be proven [PHGR13,BCG+13,BCTV14b,CTV15,CFH+15].

One powerful motivation for building efficient non-interactive arguments is verifiable computation. A client can outsource a complicated computational task to a server in the cloud and get back the results. To convince the client that the computation is correct the server may include a non-interactive argument of correctness with the result. However, since the verifier does not have many computational resources this only makes sense if the argument is compact and computationally light to verify, i.e., it is a succinct non-interactive argument (SNARG) or a succinct non-interactive argument of knowledge (SNARK). While pairing-based SNARGs are efficient for the verifier, the computational overhead for the prover is still orders of magnitude too high to warrant use in outsourced computation [Wal15] and further efficiency improvements are needed. In their current state, SNARKs that are zero-knowledge already have uses when proving statements about private data though. Zero-knowledge SNARKs are for instance key ingredients in the virtual currency proposals Pinnocchio coin [DFKP13] and Zerocash [BCG+14].

In parallel with developments in pairing-based NIZK arguments there has been interesting work on understanding SNARKs. Gentry and Wichs [GW11] showed that SNARGs must necessarily rely on non-falsifiable assumptions, and Bitansky et al. [BCCT12] proved designated verifier SNARKs exist if and only if extractable collision-resistant hash functions exist. Of particular interest in terms of efficiency is a series of works studying how SNARKs compose [Val08, BCCT13, BCTV14a]. They show among other things that a preprocessing SNARK with a long common reference string can be used to build a fully succinct SNARK with a short common reference string.

Bitansky et al. [BCI+13] give an abstract model of SNARKs that rely on linear encodings of field elements. Their information theoretic framework called linear interactive proofs (LIPs) capture proof systems where the prover is restricted to using linear operations in computing her messages. Given a LIP it can be converted to a publicly verifiable SNARK using pairing-based techniques or to a designated verifier using additively homomorphic encryption techniques.

1.1 Our Contribution

Succinct NIZK. We construct a NIZK argument for arithmetic circuit satisfiability where a proof consists of only 3 group elements. In addition to being small, the proof is also easy to verify. The verifier just needs to compute a number of exponentiations proportional to the statement size and check a single pairing product equation, which only has 3 pairings. Our construction can be instantiated with any type of pairings including Type III pairings, which are the most efficient pairings.

The argument has perfect completeness and perfect zero-knowledge. For soundness we take an aggressive stance and rely on a security proof in the generic bilinear group model in order to get optimal performance. This stance is partly justified by Gentry and Wichs [GW11] that rule out SNARGs based on standard falsifiable assumptions. However, following Abe et al. [AGOT14] we do provide a hedge against cryptanalysis by proving our construction secure in the symmetric pairing setting. For optimal efficiency it makes sense to use our NIZK argument in the asymmetric setting, however, by providing a security proof in the symmetric setting we get additional security: even if cryptanalytic advances yield a

Table 1. Comparison for boolean circuit satisfiability with ℓ-bit statement, m wires and n fan-in 2 logic gates. Notation: $\mathbb{G}$ means group elements, M means multiplications, E means exponentiations and P means pairings with subscripts indicating the relevant group. It is possible to get a CRS size of $m + 2n$ elements in $\mathbb{G}_1$ and n elements in $\mathbb{G}_2$ but we have chosen to include some precomputed values in the CRS to reduce the prover's computation, see Sect. 3.2.

	CRS size	Proof size	Prover comp.	Verifier comp.	PPE
[DFGK14]	$2m + n - 2\ell$ $\mathbb{G}_1$, $m + n - \ell$ $\mathbb{G}_2$	3 $\mathbb{G}_1$, 1 $\mathbb{G}_2$	$m + n - \ell$ E_1	ℓ M_1, 6 P	3
This work	$3m + n$ $\mathbb{G}_1$, m $\mathbb{G}_2$	2 $\mathbb{G}_1$, 1 $\mathbb{G}_2$	n E_1	ℓ M_1, 3 P	1

Table 2. Comparison for arithmetic circuit satisfiability with ℓ-element statement, m wires, n multiplication gates. Notation: $\mathbb{G}$ means group elements, E means exponentiations and P means pairings. We compare symmetric pairings in the first two rows and asymmetric pairings in the last two rows.

	CRS size	Proof size	Prover comp.	Verifier comp.	PPE
[PHGR13]	$7m + n - 2\ell$ $\mathbb{G}$	8 $\mathbb{G}$	$7m + n - 2\ell$ E	ℓ E, 11 P	5
This work	$m + 2n$ $\mathbb{G}$	3 $\mathbb{G}$	$m + 3n - \ell$ E	ℓ E, 3 P	1
[SVdV15]	$6m + n - 2\ell$ $\mathbb{G}_1$, m $\mathbb{G}_2$	7 $\mathbb{G}_1$, 1 $\mathbb{G}_2$	$6m + n - 6\ell$ E_1, $m - \ell$ E_2	2ℓ E_1, ℓE_2, 12 P	5
This work	$m + 2n$ $\mathbb{G}_1$, n $\mathbb{G}_2$	2 $\mathbb{G}_1$, 1 $\mathbb{G}_2$	$m + 3n - \ell$ E_1, n E_2	ℓ E_1, 3 P	1

hitherto unknown efficiently computable isomorphism between the source groups this does not necessarily lead to a break of our scheme. We therefore have a unified NIZK argument that can be instantiated with any type of pairing, yielding both optimal efficiency and optimal generic bilinear group resilience.

We give a performance comparison for boolean circuit satisfiability in Table 1 and for arithmetic circuit satisfiability in Table 2 of the size of the common reference string (CRS), the size of the proof, the prover's computation, the verifier's computation, and the number of pairing product equations used to verify a proof. We perform better than the state of the art on all efficiency parameters.

In both comparisons the number of wires exceeds the number of gates, $m \geq n$, since each gate has an output wire. We expect for typical cases that the statement size ℓ will be small compared to m and n. In both tables, we have excluded the size of representing the relation for which we give proofs. In the boolean circuit satisfiability case, we are considering arbitrary fan-in 2 logic gates. In the arithmetic circuit satisfiability case we work with fan-in 2 multiplication gates where each input factor can be a weigthed sum of other wires. We assume each multiplication gate input depends on a constant number of wires; otherwise the cost of evaluating the relation itself may exceed the cost of the subsequent proof generation.

We note that [PHGR13] uses symmetric bilinear groups where $\mathbb{G}_1 = \mathbb{G}_2$ and we are therefore comparing with a symmetric bilinear group instantiation of our scheme, which saves n elements in the common reference string. However, in the implementation of their system, called Pinocchio, asymmetric pairings are used for better efficiency. The switch to asymmetric pairings only requires minor modifications, see e.g. [SVdV15].

SIZE MATTERS. While the reduction in proof size to 3 group elements and the reduction in verification time is nice in itself, we would like to highlight that it is particularly important when composing SNARKs. [BCCT13,BCTV14a] show that preprocessing SNARKs with a long CRS can be composed to yield fully succinct SNARKs with a short CRS. The transformations split the statement into smaller pieces, prove each piece is correct by itself, and recursively construct proofs of knowledge of other proofs that jointly show the pieces are correct and fit together. In the recursive construction of proofs, it is extra beneficial when the proofs are small and easy to verify since the resulting statements "there exists a proof satisfying the verification equation. . . " become small themselves. So we gain both from the prover's lower computation and from the fact that the statements in the recursive composition are smaller since we have a more efficient verification procedure for our SNARK. We estimate that in the scalable and fully succinct zero-knowledge SNARKs by Ben-Sasson et al. [BCTV14a] that use two related elliptic curves to prove statements about each other, the prover's computation will be reduced by up to an order of magnitude.

TECHNIQUE. All pairing-based SNARKs in the literature follow a common paradigm where the prover computes a number of group elements using generic group operations and the verifier checks the proof using a number of pairing product equations. Bitansky et al. [BCI+13] formalize this paradigm through the definition of linear interactive proofs (LIPs). A linear interactive proof works over a finite field and the prover's and verifier's messages consist of vectors of field elements. It furthermore requires that the prover computes her messages using only linear operations. Once we have the LIP, it can then be compiled into a SNARK by executing the equations "in the exponent" using pairing-based cryptography. One source of our efficiency gain is that we design a LIP system for arithmetic circuits where the prover only sends 3 field elements. In comparison, the quadratic arithmetic programs by [GGPR13,PHGR13] correspond to LIPs where the prover sends 4 field elements.

A second source of efficiency gain compared to previous work is a more aggressive compilation of the LIP. Bitansky et al. [BCI+13] propose a transformation in the symmetric bilinear group setting, where each field element gets compiled into two group elements. They then use a knowledge of exponent assumption to argue that the prover knows the relevant field elements. A less conservative choice would be to compile each field element into a single group element. This improves efficiency but security requires stronger assumptions since we the scheme may be secure in the generic group model but we can no longer use the knowledge of exponent assumption. It is also possible to make a choice between these two extremes, Parno et al. [PHGR13] for instance have a LIP with 4 field elements, which gets compiled into 7 group elements. In this paper we have opted for maximal efficiency and compile each field element in the LIP into a single group element and argue security in the generic group model.

We prefer to work with asymmetric bilinear groups for their higher efficiency than symmetric bilinear groups. This means that there is more to the story than the number of field elements the prover sends in the LIP and the choice of how aggressive a compilation we use. When working with asymmetric bilinear groups, a field element can appear as an exponent in the first source group, the second source group, or both. Our LIP is carefully designed such that each field element gets compiled into a single source group element in order to minimize the proof size to 3 group elements in total.

Lower Bounds. Working towards ever more efficient non-interactive arguments, it is natural to ask what the minimal proof size is. We will show that pairing-based SNARGs with a single group element proof cannot exist. This result relates to an open question raised by Bitansky et al. [BCI+13], whether there are LIPs with a linear decision procedure for the verifier. Such a linear decision procedure would be quite useful; it could for instance enable the construction of SNARGs based on ElGamal encryption.

We answer this open problem negatively by proving that LIPs with a linear decision procedure do not exist. A consequence of this is that any pairing-based SNARG must pair group elements from the proof together to make the decision procedure quadratic instead of linear. Working over asymmetric bilinear groups we must therefore have elements in both source groups in order to do such a pairing. This rules out the existence of 1 group element SNARGs, regardless of whether it is zero-knowledge or not, and shows our NIZK argument has close to optimal proof size. It remains an intriguing open problem to completely close the gap by either constructing a SNARG with exactly one element from each source group $\mathbb{G}_1$ and $\mathbb{G}_2$, or alternatively rule out the existence of such a SNARG.

2   Preliminaries

Given two functions $f, g : \mathbb{N} \to [0,1]$ we write $f(\lambda) \approx g(\lambda)$ when $|f(\lambda) - g(\lambda)| = \lambda^{-\omega(1)}$. We say that f is *negligible* when $f(\lambda) \approx 0$ and that f is *overwhelming* when $f(\lambda) \approx 1$. We will use λ to denote a security parameter, with the intuition that as λ grows we would like to have stronger security.

We write $y = A(x; r)$ when algorithm A on input x and randomness r, outputs y. We write $y \leftarrow A(x)$ for the process of picking randomness r at random and setting $y = A(x; r)$. We also write $y \leftarrow S$ for sampling y uniformly at random from the set S. We will assume it is possible to sample uniformly at random from sets such as $\mathbb{Z}_p$.

Following Abe and Fehr [AF07] we write $(y; z) \leftarrow (\mathcal{A} \parallel \mathcal{X}_\mathcal{A})(x)$ when $\mathcal{A}$ on input x outputs y and $\mathcal{X}_\mathcal{A}$ on the same input (including random coins) outputs z.

2.1   Bilinear Groups

We work with bilinear groups $(p, \mathbb{G}_1, \mathbb{G}_2, \mathbb{G}_T, e)$ with the following properties:

- $\mathbb{G}_1, \mathbb{G}_2, \mathbb{G}_T$ are groups of prime order p
- $e : \mathbb{G}_1 \times \mathbb{G}_2 \to \mathbb{G}_T$ is a bilinear map, i.e., $e(U^a, V^b) = e(U, V)^{ab}$

- If G is a generator for $\mathbb{G}_1$ and H is a generator for $\mathbb{G}_2$ then $e(G,H)$ is a generator for $\mathbb{G}_T$
- There are efficient algorithms for computing group operations, evaluating the bilinear map, deciding membership of the groups, deciding equality of group elements and sampling generators of the groups. We refer to these as the generic bilinear group operations.

There are many ways to set up bilinear groups both as symmetric bilinear groups where $\mathbb{G}_1 = \mathbb{G}_2$ and as asymmetric bilinear groups where $\mathbb{G}_1 \neq \mathbb{G}_2$. Galbraith et al. [GPS08] classify bilinear groups as Type I where $\mathbb{G}_1 = \mathbb{G}_2$, Type II where there is an efficiently computable non-trivial homomorphism Ψ : $\mathbb{G}_2 \to \mathbb{G}_1$, and Type III where no such efficiently computable homomorphism exists in either direction between $\mathbb{G}_1$ and $\mathbb{G}_2$. Type III bilinear groups are the most efficient type of bilinear groups and hence the most relevant for practical applications. We give the lower bound for Type III bilinear groups and but our construction works without change for all 3 types of bilinear groups.

2.2 Non-interactive Zero-Knowledge Arguments of Knowledge

Let $\mathcal{R}$ be a relation generator that given a security parameter λ in unary returns a polynomial time decidable binary relation R. For pairs $(\phi, w) \in R$ we call ϕ the statement and w the witness. We define $\mathcal{R}_\lambda$ to be the set of possible relation $\mathcal{R}$ may output given 1^λ. The relation generator may also output some side information, an auxiliary input z, which will be given to the adversary. An efficient prover publicly verifiable non-interactive argument for $\mathcal{R}$ is a quadruple of probabilistic polynomial algorithms (Setup, Prove, Vfy, Sim) such that

$(\sigma, \tau) \leftarrow \mathsf{Setup}(R)$: The setup takes as input a security parameter λ and a relation $R \in \mathcal{R}_\lambda$ and returns a common reference string σ and a simulation trapdoor τ for the relation R.

$\pi \leftarrow \mathsf{Prove}(R, \sigma, \phi, w)$: The prover algorithm takes as input a common reference string σ and $(\phi, w) \in R$ and returns an argument π.

$0/1 \leftarrow \mathsf{Vfy}(R, \sigma, \phi, \pi)$: The verification algorithm takes as input a common reference string σ, a statement ϕ and an argument π and returns 0 (reject) or 1 (accept).

$\pi \leftarrow \mathsf{Sim}(R, \tau, \phi)$: The simulator takes as input a simulation trapdoor and statement ϕ and returns an argument π.

Definition 1. *We say* (Setup, Prove, Vfy) *is a non-interactive argument for $\mathcal{R}$ if it has perfect completeness and computational soundness as defined below.*

Definition 2. *We say* (Setup, Prove, Vfy, Sim) *is a perfect non-interactive zero-knowledge argument of knowledge for $\mathcal{R}$ if it has perfect completeness, perfect zero-knowledge and computational knowledge soundness as defined below.*

PERFECT COMPLETENESS. Completeness says that, given any true statement, an honest prover should be able to convince an honest verifier. For all $\lambda \in \mathbb{N}$, $R \in \mathcal{R}_\lambda$, $(\phi, w) \in R$

$$\Pr\Big[(\sigma,\tau)\leftarrow\mathsf{Setup}(R);\pi\leftarrow\mathsf{Prove}(R,\sigma,\phi,w):\mathsf{Vfy}(R,\sigma,\phi,\pi)=1\Big]=1.$$

PERFECT ZERO-KNOWLEDGE. An argument is zero-knowledge if it does not leak any information besides the truth of the statement. We say (Setup, Prove, Vfy, Sim) is perfect zero-knowledge if for all $\lambda\in\mathbb{N}$, $(R,z)\leftarrow\mathcal{R}(1^\lambda)$, $(\phi,w)\in R$ and all adversaries $\mathcal{A}$

$$\Pr\Big[(\sigma,\tau)\leftarrow\mathsf{Setup}(R);\pi\leftarrow\mathsf{Prove}(R,\sigma,\phi,w):\mathcal{A}(R,z,\sigma,\tau,\pi)=1\Big]$$
$$=\Pr\Big[(\sigma,\tau)\leftarrow\mathsf{Setup}(R);\pi\leftarrow\mathsf{Sim}(R,\tau,\phi):\mathcal{A}(R,z,\sigma,\tau,\pi)=1\Big].$$

COMPUTATIONAL SOUNDNESS. We say (Setup, Prove, Vfy, Sim) is sound if it is not possible to prove a false statement, i.e., convince the verifier if no witness exists. Let L_R be the language consisting of statements for which there exist matching witnesses in R. Formally, we require that for all non-uniform polynomial time adversaries $\mathcal{A}$

$$\Pr\left[\begin{array}{l}(R,z)\leftarrow\mathcal{R}(1^\lambda);(\sigma,\tau)\leftarrow\mathsf{Setup}(R);(\phi,\pi)\leftarrow\mathcal{A}(R,z,\sigma):\\ \phi\notin L_R\text{ and }\mathsf{Vfy}(R,\sigma,\phi,\pi)=1\end{array}\right]\approx 0.$$

COMPUTATIONAL KNOWLEDGE SOUNDNESS. Strengthening the notion of soundness, we call (Setup, Prove, Vfy, Sim) an argument of knowledge if there is an extractor that can compute a witness whenever the adversary produces a valid argument. The extractor gets full access to the adversary's state, including any random coins. Formally, we require that for all non-uniform polynomial time adversaries $\mathcal{A}$ there exists a non-uniform polynomial time extractor $\mathcal{X}_\mathcal{A}$ such that

$$\Pr\left[\begin{array}{l}(R,z)\leftarrow\mathcal{R}(1^\lambda);(\sigma,\tau)\leftarrow\mathsf{Setup}(R);((\phi,\pi);w)\leftarrow(\mathcal{A}\parallel\mathcal{X}_\mathcal{A})(R,z,\sigma):\\ (\phi,w)\notin R\text{ and }\mathsf{Vfy}(R,\sigma,\phi,\pi)=1\end{array}\right]\approx 0.$$

PUBLIC VERIFIABILITY AND DESIGNATED VERIFIER PROOFS. We can naturally generalize the definition of a non-interactive argument by splitting σ into two parts σ_P and σ_V used by the prover and verifier respectively. We say the non-interactive argument is publicly verifiable when σ_V can be deduced from σ_P. Otherwise we refer to it as a designated verifier argument. For designated verifier arguments it is possible to relax soundness and knowledge soundness such that the adversary only sees σ_P but not σ_V.

SNARGS AND SNARKS. A non-interactive argument where the verifier runs in polynomial time in $\lambda+|\phi|$ and the proof size is polynomial in λ is called a preprocessing succinct non-interactive argument (SNARG) if it sound, and a preprocessing succinct argument of knowledge (SNARK) if it is knowledge sound. If we also restrict the common reference string to be polynomial in λ we say the non-interactive argument is a fully succinct SNARG or SNARK. Bitansky et al. [BCCT13] show that preprocessing SNARKs can be composed to yield fully succinct SNARKs. The focus of this paper is on preprocessing SNARKs.

BENIGN RELATION GENERATORS. Bitansky et al. [BCPR14] show that indistinguishability obfuscation implies that for every candidate SNARK there are auxiliary output distributions that enable the adversary to create a valid proof without it being possible to extract the witness. Assuming also public coin differing input obfuscation and other cryptographic assumptions, Boyle and Pass [BP15] strengthen this impossibility to show that there is an auxiliary output distribution that defeats witness extraction for all candidate SNARKs. These counter examples, however, rely on specific auxiliary input distributions. We will therefore in the following assume the relationship generator is *benign* in the sense that the relation and the auxiliary input are distributed in such a way that SNARKs can exist.

2.3 Quadratic Arithmetic Programs

Consider an arithmetic circuit consisting of addition and multiplication gates over a finite field $\mathbb{F}$. We may designate some of the input/output wires as specifying a statement and use the rest of the wires in the circuit to define a witness. This gives us a binary relation R consisting of statement wires and witness wires that satisfy the arithmetic circuit, i.e., make it consistent with the designated input/output wires.

Generalizing arithmetic circuits, we may be interested in relations described by equations over a set of variables. Some of the variables correspond to the statement; the remaining variables correspond to the witness. The relation consists of statements and witnesses that satisfy all the equations. The equations will be over $a_0 = 1$ and variables $a_1, \ldots, a_m \in \mathbb{F}$ and be of the form

$$\sum a_i u_{i,q} \cdot \sum a_i v_{i,q} = \sum a_i w_{i,q},$$

where $u_{i,q}, v_{i,q}, w_{i,q}$ are constants in $\mathbb{F}$ specifying the qth equation.

We observe that addition and multiplication gates are special cases of such equations so such systems of arithmetic constraints do indeed generalize arithmetic circuits. A multiplication gate can for instance be described as $a_i \cdot a_j = a_k$ (using $u_i = 1, v_j = 1$ and $w_k = 1$ and setting the remaining constants for this gate to 0). Addition gates are handled for free in the sums defining the equations, i.e., if $a_i + a_j = a_k$ and a_k is multiplied by a_ℓ, we may simply write $(a_i + a_j) \cdot a_\ell$ and skip the calculation of a_k.

Following Gennaro et al. [GGPR13] we can reformulate the set of arithmetic constraints as a quadratic arithmetic program assuming $\mathbb{F}$ is large enough. Given n equations we pick arbitrary distinct $r_1, \ldots, r_n \in \mathbb{F}$ and define $t(x) = \prod_{q=1}^{n}(x - r_q)$. Furthermore, let $u_i(x), v_i(x), w_i(x)$ be degree $n - 1$ polynomials such that

$$u_i(r_q) = u_{i,q} \qquad v_i(r_q) = v_{i,q} \qquad w_i(r_q) = w_{i,q} \qquad \text{for} \quad i = 0, \ldots, m, q = 1, \ldots, n.$$

We now have that $a_0 = 1$ and the variables $a_1, \ldots, a_m \in \mathbb{F}$ satisfy the n equations if and only if in each point $r_1, \ldots, r_q$

$$\sum_{i=0}^{m} a_i u_i(r_q) \cdot \sum_{i=0}^{m} a_i v_i(r_q) = \sum_{i=0}^{m} a_i w_i(r_q).$$

Since $t(X)$ is the lowest degree monomial with $t(r_q) = 0$ in each point, we can reformulate this condition as

$$\sum_{i=0}^{m} a_i u_i(X) \cdot \sum_{i=0}^{m} a_i v_i(X) \equiv \sum_{i=0}^{m} a_i w_i(X) \bmod t(X).$$

Formally, we will be working with quadratic arithmetic programs R that have the following description

$$R = (\mathbb{F}, \text{aux}, \ell, \{u_i(X), v_i(X), w_i(X)\}_{i=0}^{m}, t(X)),$$

where $\mathbb{F}$ describes a finite field, aux is some auxiliary information, $1 \leq \ell \leq m$, $u_i(X), v_i(X), w_i(X), t(X) \in \mathbb{F}[X]$ and $u_i(X), v_i(X), w_i(X)$ have strictly lower degree than n, the degree of $t(X)$. A quadratic arithmetic program with such a description defines the following binary relation, where we define $a_0 = 1$,

$$R = \left\{ (\phi, w) \left| \begin{array}{l} \phi = (a_1, \ldots, a_\ell) \in \mathbb{F}^\ell \\ w = (a_{\ell+1}, \ldots, a_m) \in \mathbb{F}^{m-\ell} \\ \sum_{i=0}^{m} a_i u_i(X) \cdot \sum_{i=0}^{m} a_i v_i(X) \equiv \sum_{i=0}^{m} a_i w_i(X) \bmod t(X) \end{array} \right. \right\}.$$

We say $\mathcal{R}$ is a quadratic arithmetic program generator if it generates relations of the form given above with fields of size larger than $2^{\lambda-1}$.

Relations can arise in many different ways in practice. It may be that the relationship generator is deterministic or it may be that it is randomized. It may be that first the field $\mathbb{F}$ is generated and then the rest of the relation is built on top of the field. Or it may be that the polynomials are specified first and then a random field is chosen. To get maximal flexibility we have chosen our definitions to be agnostic with respect to the exact way the field and the relation is generated, the different options can all be modelled by appropriate choices of relation generators.

Looking ahead, we will in our pairing-based NIZK arguments let the auxiliary information aux specify a bilinear group. It may seem a bit surprising to make the choice of bilinear group part of the relation generator but this provides a better model of settings where the relation is built on top of an already existing bilinear group. Again, there is no loss of generality in this choice, one can think of a traditional setting where the relation is chosen first and then the bilinear group is chosen at random as the special case where the relation generator works in two steps, first choosing the relation and then picking a random bilinear group. Of course letting the relation generator pick the bilinear group is another good reason that we need to assume it is benign; an appropriate choice of bilinear group is essential for security.

2.4   Linear Interactive Proofs

Bitansky et al. [BCI+13] give a useful characterization of the information theoretic underpinning of recent SNARK constructions. A two-move algebraic linear interactive proof (LIP) of degree (d_Q, d_D) for a relation generator $\mathcal{R}$, where we assume the relations specify a finite field $\mathbb{F}$, is a non-interactive argument system where the algorithms work as follows:

$(\boldsymbol{\sigma}, \boldsymbol{\tau}) \leftarrow \mathsf{Setup}(R)$: It creates an arithmetic circuit of multiplicative depth d_Q that takes as input randomness $\boldsymbol{r} \in \mathbb{F}^\mu$ and returns vectors $\boldsymbol{\sigma} \in \mathbb{F}^m$ and $\boldsymbol{\tau} \in \mathbb{F}^n$. We will for notational simplicity assume that $\boldsymbol{\sigma}$ always contains 1 as an entry such that there is no distinction between affine and linear functions of $\boldsymbol{\sigma}$.

$\boldsymbol{\pi} \leftarrow \mathsf{Prove}(R, \boldsymbol{\sigma}, \phi, w)$: The prover operates in two stages:
  - First it runs $\Pi \leftarrow \mathsf{ProofMatrix}(R, \phi, w)$, where $\mathsf{ProofMatrix}$ is a probabilistic polynomial time algorithm that generates a matrix $\Pi \in \mathbb{F}^{k \times m}$.
  - Then it computes the proof as $\boldsymbol{\pi} = \Pi\boldsymbol{\sigma}$.

$0/1 \leftarrow \mathsf{Vfy}(R, \boldsymbol{\sigma}, \phi, \boldsymbol{\pi})$: The verifier runs in two stages:
  - First it runs a deterministic polynomial time algorithm $\boldsymbol{t} \leftarrow \mathsf{Test}(R, \phi)$ to get an arithmetic circuit $\boldsymbol{t} : \mathbb{F}^{m+k} \to \mathbb{F}^\eta$ of multiplicative depth d_D.
  - It then accepts the proof if and only if $\boldsymbol{t}(\boldsymbol{\sigma}, \boldsymbol{\pi}) = \mathbf{0}$.

The degrees and dimensions $d_Q, d_D, \mu, m, n, k, \eta$ may be constants or polynomials in the security parameter λ.

Definition 3 (Linear Interactive Proof). *The tuple* $(\mathsf{Setup}, \mathsf{Prove}, \mathsf{Vfy})$ *is a linear interactive proof for* $\mathcal{R}$ *if it has perfect completeness and statistical knowledge soundness against affine prover strategies as defined below.*

STATISTICAL KNOWLEDGE SOUNDNESS AGAINST AFFINE PROVER STRATEGIES. An LIP has knowledge soundness against affine prover strategies if a witness can be extracted from a successful proof matrix Π. More precisely, there is a polynomial time extractor $\mathcal{X}$ such that for all adversaries $\mathcal{A}$

$$\Pr\left[\begin{array}{c} (R, z) \leftarrow \mathcal{R}(1^\lambda); (\boldsymbol{\sigma}, \boldsymbol{\tau}) \leftarrow \mathsf{Setup}(R); (\phi, \Pi) \leftarrow \mathcal{A}(R, z); w \leftarrow \mathcal{X}(R, \phi, \Pi) : \\ \Pi \in \mathbb{F}^{m \times k} \wedge \mathsf{Vfy}(R, \boldsymbol{\sigma}, \phi, \Pi\boldsymbol{\sigma}) = \mathbf{0} \wedge (\phi, w) \notin R \end{array} \right] \approx 0.$$

NON-INTERACTIVE ARGUMENTS FROM LINEAR INTERACTIVE PROOFS. LIPs are useful concepts because they can be compiled into publicly verifiable non-interactive arguments using pairings and designated verifier non-interactive arguments using Paillier encryption [BCI+13]. If we work in the pairing setting, the intuition is that an algebraic LIP of degree $(d_Q, 2)$ can be executed "in the exponents": The common reference string contains exponentiations of the field elements in $\boldsymbol{\sigma}$. The prover computes the proof as multi-exponentiations of group elements, corresponding to linear operations on the field elements in $\boldsymbol{\sigma}$. The verifier checks the argument by verifying a number of pairing product equations (equations formed by multiplying together the results of pairings), which corresponds to checking quadratic equations in the exponents. We will see this methodology applied in the following section.

3   Constructions of Non-interactive Arguments

We will construct a pairing-based NIZK argument for quadratic arithmetic programs where proofs consist of only 3 group elements. We give the construction in two steps, first we construct a LIP, and then we convert the LIP into a pairing-based NIZK argument.

3.1   Linear Interactive Proofs for Quadratic Arithmetic Programs

We will now construct a LIP for quadratic arithmetic program generators that outputs relations of the form

$$R = (\mathbb{F}, \text{aux}, \ell, \{u_i(X), v_i(X), w_i(X)\}_{i=0}^m, t(X)).$$

The relation defines a language of statements $(a_1, \ldots, a_\ell) \in \mathbb{F}^\ell$ and witnesses $(a_{\ell+1}, \ldots, a_m) \in \mathbb{F}^{m-\ell}$ such that with $a_0 = 1$

$$\sum_{i=0}^m a_i u_i(X) \cdot \sum_{i=0}^m a_i v_i(X) = \sum_{i=0}^m a_i w_i(X) + h(X)t(X),$$

for some degree $n - 2$ quotient polynomial $h(X)$, where n is the degree of $t(X)$.

$(\sigma, \tau) \leftarrow \mathsf{Setup}(R)$: Pick $\alpha, \beta, \gamma, \delta, x \leftarrow \mathbb{F}^*$. Set $\tau = (\alpha, \beta, \gamma, \delta, x)$ and

$$\sigma = \left(\alpha, \beta, \gamma, \delta, \{x^i\}_{i=0}^{n-1}, \left\{ \frac{\beta u_i(x) + \alpha v_i(x) + w_i(x)}{\gamma} \right\}_{i=0}^\ell, \right.$$
$$\left. \left\{ \frac{\beta u_i(x) + \alpha v_i(x) + w_i(x)}{\delta} \right\}_{i=\ell+1}^m, \left\{ \frac{x^i t(x)}{\delta} \right\}_{i=0}^{n-2} \right).$$

$\pi \leftarrow \mathsf{Prove}(R, \sigma, a_1, \ldots, a_m)$: Pick $r, s \leftarrow \mathbb{F}$ and compute a $3 \times (m + 2n + 4)$ matrix Π such that $\pi = \Pi\sigma = (A, B, C)$ where

$$A = \alpha + \sum_{i=0}^m a_i u_i(x) + r\delta \qquad B = \beta + \sum_{i=0}^m a_i v_i(x) + s\delta$$

$$C = \frac{\sum_{i=\ell+1}^m a_i \left(\beta u_i(x) + \alpha v_i(x) + w_i(x)\right) + h(x)t(x)}{\delta} + As + rB - rs\delta.$$

$0/1 \leftarrow \mathsf{Vfy}(R, \sigma, a_1, \ldots, a_\ell)$: Compute a quadratic multi-variate polynomial t such that $t(\sigma, \pi) = 0$ corresponds to the test

$$A \cdot B = \alpha \cdot \beta + \frac{\sum_{i=0}^\ell a_i \left(\beta u_i(x) + \alpha v_i(x) + w_i(x)\right)}{\gamma} \cdot \gamma + C \cdot \delta.$$

Accept the proof if the test passes.

$\pi \quad \leftarrow \quad \mathsf{Sim}(R, \tau, a_1, \ldots, a_\ell)$: Pick $A, B \quad \leftarrow \quad \mathbb{F}$ and compute $C = \frac{AB - \alpha\beta - \sum_{i=0}^{\ell} a_i(\beta u_i(x) + \alpha v_i(x) + w_i(x))}{\delta}$. Return $\pi = (A, B, C)$.

Before formally proving this is a LIP, let us give a little intuition behind the different components. The role of α and β is to ensure A, B and C are consistent with each other in the choice of $a_0, \ldots, a_m$. The product $\alpha \cdot \beta$ in the verification equation guarantees that A and B involve non-trivial α and β components. This means the product $A \cdot B$ involves a linear dependence on α and β, and we will later prove that this linear dependence can only be balanced out by C with a consistent choice of $a_0, \ldots, a_m$ in all three of A, B and C. The role of γ and δ is to make the two latter products of the verification equation independent from the first product, by dividing the left factors with γ and δ respectively. This prevents mixing and matching of elements intended for different products in the verification equation. Finally, we use r and s to randomize the proof to get zero-knowledge.

Theorem 1. *The construction above yields a LIP with perfect completeness, perfect zero-knowledge and statistical knowledge soundness against affine prover strategies.*

Proof. Perfect completeness is straightforward to verify. Perfect zero-knowledge follows from both real proofs and simulated proofs having uniformly random field elements A, B. These elements uniquely determine C through the verification equation, so real proofs and simulated proofs have identical probability distributions.

What remains is to demonstrate that for any affine prover strategy with non-negligible success probability we can extract a witness. When using an affine prover strategy we have

$$A = A_\alpha \alpha + A_\beta \beta + A_\gamma \gamma + A_\delta \delta + A(x) + \sum_{i=0}^{\ell} A_i \frac{\beta u_i(x) + \alpha v_i(x) + w_i(x)}{\gamma}$$

$$+ \sum_{i=\ell+1}^{m} A_i \frac{\beta u_i(x) + \alpha v_i(x) + w_i(x)}{\delta} + A_h(x) \frac{t(x)}{\delta},$$

for known field elements $A_\alpha, A_\beta, A_\gamma, A_\delta, A_i$ and polynomials $A(x), A_h(x)$ of degrees $n - 1$ and $n - 2$, respectively that correspond to the first row of the matrix Π. We can write out B and C in a similar fashion from the second and third rows of Π.

We now view the verification equation as an equality of multi-variate Laurent polynomials. By the Schwartz-Zippel lemma the prover has negligible success probability unless the verification equation holds when viewing A, B and C as formal polynomials in indeterminates $\alpha, \beta, \gamma, \delta, x$.

The terms with indeterminate α^2 are $A_\alpha B_\alpha \alpha^2 = 0$, which means $A_\alpha = 0$ or $B_\alpha = 0$. Since $AB = BA$ we can without loss of generality assume $B_\alpha = 0$. The terms with indeterminate $\alpha\beta$ give us $A_\alpha B_\beta + A_\beta B_\alpha = A_\alpha B_\beta = 1$. This means

$AB = (AB_\beta)(A_\alpha B)$ so we can without loss of generality after rescaling assume $A_\alpha = B_\beta = 1$. The terms with indeterminate β^2 now give us $A_\beta B_\beta = A_\beta = 0$. We have now simplified A and B constructed by the adversary to be of the form

$$A = \alpha + A_\gamma \gamma + A_\delta \delta + A(x) + \cdots \qquad B = \beta + B_\gamma \gamma + B_\delta \delta + B(x) + \cdots.$$

Next, let us consider the terms involving $\frac{1}{\delta^2}$. We have

$$\left(\sum_{i=\ell+1}^{m} A_i \left(\beta u_i(x) + \alpha v_i(x) + w_i(x) \right) + A_h(x) t(x) \right) \cdot$$
$$\left(\sum_{i=\ell+1}^{m} B_i \left(\beta u_i(x) + \alpha v_i(x) + w_i(x) \right) + B_h(x) t(x) \right) = 0,$$

showing either the left factor is 0 or the right factor is 0. By symmetry, let us without loss of generality assume $\sum_{i=\ell+1}^{m} A_i \left(\beta u_i(x) + \alpha v_i(x) + w_i(x) \right) + t(x) A_t(x) = 0$. The terms in $\alpha \frac{\sum_{i=\ell+1}^{m} B_i (\beta u_i(x) + \alpha v_i(x) + w_i(x)) + B_h(x) t(x)}{\delta} = 0$ now show us that also $\sum_{i=\ell+1}^{m} B_i \left(\beta u_i(x) + \alpha v_i(x) + w_i(x) \right) + B_h(x) t(x) = 0$.

The terms involving $\frac{1}{\gamma^2}$ give us

$$\sum_{i=0}^{\ell} A_i \left(\beta u_i(x) + \alpha v_i(x) + w_i(x) \right) \cdot \sum_{i=0}^{\ell} B_i \left(\beta u_i(x) + \alpha v_i(x) + w_i(x) \right) = 0,$$

showing either the left factor is 0 or the right factor is 0. By symmetry, let us without loss of generality assume $\sum_{i=0}^{\ell} A_i \left(\beta u_i(x) + \alpha v_i(x) + w_i(x) \right) = 0$. The terms in $\alpha \frac{\sum_{i=0}^{m} B_i (\beta u_i(x) + \alpha v_i(x) + w_i(x))}{\gamma} = 0$ now show us $\sum_{i=0}^{\ell} B_i \left(\beta u_i(x) + \alpha v_i(x) + w_i(x) \right) = 0$ as well.

The terms $A_\gamma \beta \gamma = 0$ and $B_\gamma \alpha \gamma = 0$ show us that $A_\gamma = 0$ and $B_\gamma = 0$. We now have

$$A = \alpha + A(x) + A_\delta \delta \qquad B = \beta + B(x) + B_\delta \delta.$$

The remaining terms in the verification equation that involve α give us $\alpha B(x) = \sum_{i=0}^{\ell} a_i \alpha v_i(x) + \sum_{i=\ell+1}^{m} C_i \alpha v_i(x)$. The terms involving β give us $\beta A(x) = \sum_{i=0}^{\ell} a_i \beta u_i(x) + \sum_{i=\ell+1}^{m} C_i \beta u_i(x)$. Defining $a_i = C_i$ for $i = \ell+1, \ldots, m$ we now have

$$A(x) = \sum_{i=0}^{m} a_i u_i(x) \qquad B(x) = \sum_{i=0}^{m} a_i v_i(x).$$

Finally, we look at the terms involving powers of x to get

$$\sum_{i=0}^{m} a_i u_i(x) \cdot \sum_{i=0}^{m} a_i v_i(x) = \sum_{i=0}^{m} a_i w_i(x) + C_h(x) t(x).$$

This shows that $(a_{\ell+1}, \ldots, a_m) = (C_{\ell+1}, \ldots, C_m)$ is a witness for the statement $(a_1, \ldots, a_\ell)$.   □

2 FIELD ELEMENT LIPs. It is natural to ask whether the number of field elements the prover sends in the LIP can be reduced further. The square span programs of Danezis et al. [DFGK14] give rise to 2 field element LIPs for boolean circuit satisfiability. It is also possible to get a 2-element LIP for arithmetic circuit satisfiability by rewriting the circuit into one that only uses squaring gates, each multiplication gate $a \cdot b = c$ can be rewritten as a $(a+b)^2 - (a-b)^2 = 4c$. When an arithmetic circuit only has squaring gates we get $u_i(x) = v_i(x)$ for all i. By choosing $r = s$ in the LIP, we now have that $B = A + \beta - \alpha$, so the prover only needs to send two elements A and C to make a convincing proof. Rewriting the arithmetic circuit to only use squaring gates may double the number of gates and also requires some additional wires for the subtraction of the squares, so the reduction of the size of the LIP comes at a significant computational cost though.

3.2   NIZK Arguments for Quadratic Arithmetic Programs

We will now give a pairing-based NIZK argument for quadratic arithmetic programs. We consider relation generators $\mathcal{R}$ that return relations of the form

$$R = (p, \mathbb{G}_1, \mathbb{G}_2, \mathbb{G}_T, e, \ell, \{u_i(X), v_i(X), w_i(X)\}_{i=0}^m, t(X)),$$

with $|p| = \lambda$. The relation defines a field $\mathbb{Z}_p$ and a language of statements $(a_1, \ldots, a_\ell) \in \mathbb{Z}_p^\ell$ and witnesses $(a_{\ell+1}, \ldots, a_m) \in \mathbb{Z}_p^{m-\ell}$ such that with $a_0 = 1$

$$\sum_{i=0}^m a_i u_i(X) \cdot \sum_{i=0}^m a_i v_i(X) = \sum_{i=0}^m a_i w_i(X) + h(X)t(X),$$

for some degree $n - 2$ quotient polynomial $h(X)$.

We will construct the pairing-based argument by using the LIP from the previous section "in the exponents". An important design feature of the LIP is that the elements A, B and C are only used once in the verification equation and therefore it is easy to assign them to different source groups such that the verification equation can be carried out using a pairing product equation. Since pairing-friendly elliptic curves can be constructed such that the group element representations are smaller in $\mathbb{G}_1$ than in $\mathbb{G}_2$ [GPS08] we choose to assign A and C to the first source group and B to the second source group for maximal efficiency. This gives us the following NIZK argument.

$(\sigma, \tau) \leftarrow \mathsf{Setup}(R)$: Pick arbitrary generators G and H for $\mathbb{G}_1$ and $\mathbb{G}_2$. Pick $\alpha, \beta, \gamma, \delta, x \leftarrow \mathbb{Z}_p^*$. Define $\tau = (\alpha, \beta, \gamma, \delta, x)$ and compute

$$\sigma = \left(\begin{array}{c} G^\alpha, G^\beta, H^\beta, H^\gamma, G^\delta, H^\delta, \left\{ G^{x^i} \right\}_{i=0}^{n-1}, \left\{ H^{x^i} \right\}_{i=0}^{n-1}, \\ \left\{ G^{\frac{\beta u_i(x) + \alpha v_i(x) + w_i(x)}{\gamma}} \right\}_{i=0}^\ell, \left\{ G^{\frac{\beta u_i(x) + \alpha v_i(x) + w_i(x)}{\delta}} \right\}_{i=\ell+1}^m, \left\{ G^{\frac{x^i t(x)}{\delta}} \right\}_{i=0}^{n-2} \end{array} \right).$$

$\pi \leftarrow \mathsf{Prove}(R, \sigma, a_1, \ldots, a_m)$: Pick $r, s \leftarrow \mathbb{Z}_p$ and compute $\pi = (A, B, C)$, where

$$A = G^{\alpha + \sum_{i=0}^{m} a_i u_i(x) + r\delta} \qquad\qquad B = H^{\beta + \sum_{i=0}^{m} a_i v_i(x) + s\delta}$$

$$C = G^{\frac{\sum_{i=\ell+1}^{m} a_i(\beta u_i(x) + \alpha v_i(x) + w_i(x)) + h(x)t(x)}{\delta} + s\left(\alpha + \sum_{i=0}^{m} a_i u_i(x)\right) + r\left(\beta + \sum_{i=0}^{m} a_i v_i(x)\right) + rs\delta}.$$

$0/1 \leftarrow \mathsf{Vfy}(R, \sigma, a_1, \ldots, a_\ell, \pi)$: Parse $\pi = (A, B, C) \in \mathbb{G}_1 \times \mathbb{G}_2 \times \mathbb{G}_1$. Accept the proof if and only if

$$e(A, B) = e(G^\alpha, H^\beta) e(G^{\frac{\sum_{i=0}^{\ell} a_i(\beta u_i(x) + \alpha v_i(x) + w_i(x))}{\gamma}}, H^\gamma) e(C, H^\delta).$$

$\pi \leftarrow \mathsf{Sim}(R, \tau, a_1, \ldots, a_\ell)$: Pick $r, s \leftarrow \mathbb{Z}_p$ and compute a simulated proof $\pi = (A, B, C)$ as

$$A = G^r \qquad B = H^s \qquad C = G^{\frac{rs - \alpha\beta - \sum_{i=0}^{\ell} a_i(\beta u_i(x) + \alpha v_i(x) + w_i(x))}{\delta}}.$$

Theorem 2. *The protocol given above is a non-interactive zero-knowledge argument with perfect completeness and perfect zero-knowledge. It has statistical knowledge soundness against adversaries that only use a polynomial number of generic bilinear group operations.*

Proof. Perfect completeness follows by direct verification. Perfect zero-knowledge follows from the fact that both in real proofs and simulated proofs A, B are uniformly random group elements and through the verification equation uniquely determine C.

To see that we have statistical knowledge soundness against generic adversaries first note that any test the adversary can do on the common reference string corresponds to an equality test of Laurent polynomials. Either the polynomials match formally, or by the Schwartz-Zippel lemma there is negligible probability of them matching up over the random choices of $\alpha, \beta, \gamma, \delta, x$. The adversary therefore has negligible probability of learning anything it did not already know about the common reference string using only generic group operations. What remains is the possibility that the adversary computes A, B and C as exponentiations of group elements to known field elements. This corresponds exactly to an affine prover strategy on the LIP "in the exponents" and by the knowledge soundness of the LIP we can extract a witness from these known field elements. $\qquad\square$

Efficiency. The proof size is 2 elements in $\mathbb{G}_1$ and 1 element in $\mathbb{G}_2$. The common reference string contains a description of the relation R, n elements in $\mathbb{Z}_p$, $m + 2n + 3$ elements in $\mathbb{G}_1$, and $n + 3$ elements in $\mathbb{G}_2$.

The verifier does not need to know the entire common reference string, it suffices to know

$$\sigma_V = \left(p, \mathbb{G}_1, \mathbb{G}_2, \mathbb{G}_T, e, H^\gamma, H^\delta, \left\{ G^{\frac{\beta u_i(x) + \alpha v_i(x) + w_i(x)}{\gamma}} \right\}_{i=0}^{\ell}, e(G^\alpha, H^\beta) \right).$$

The verifier's reference string only contains a description of the bilinear group, $\ell + 1$ elements in $\mathbb{G}_1$, 2 elements in $\mathbb{G}_2$, and 1 element in $\mathbb{G}_T$.

The verification consists of checking that the proof consists of three appropriate group elements and checking a single pairing product equation. The verifier computes ℓ exponentiations in $\mathbb{G}_1$, a small number of group multiplications, and 3 pairings (assuming $e(G^\alpha, H^\beta)$ is precomputed in the verifier's reference string).

The prover has to compute the polynomial $h(X)$. The prover can compute the polynomial evaluations

$$\sum_{i=0}^{m} a_i u_i(r_q) = \sum_{i=0}^{m} a_i u_{i,q} \quad \sum_{i=0}^{m} a_i v_i(r_q) = \sum_{i=0}^{m} a_i v_{i,q} \quad \sum_{i=0}^{m} a_i w_i(r_q) = \sum_{i=0}^{m} a_i w_{i,q}$$

for $q = 1, \ldots, n$. It depends on the relation how long time this computation takes; if it arises from an arithmetic circuit where each multiplication gate connects to a constant number of wires, the relation will be sparse and the computation will be linear in n. Since the polynomials have degree $n - 1$ they are completely determined by these evaluation points. If $r_1, \ldots, r_n$ are roots of unity for a suitable prime p she can compute $h(X)$ using standard Fast Fourier Transform techniques in $O(n \log n)$ operations in $\mathbb{Z}_p$. The prover can also compute the coefficients of $\sum_{i=0}^{m} a_i u_i(X)$ and $\sum_{i=0}^{m} a_i v_i(X)$ using FFT techniques. Having all the coefficients, the prover does $m + 3n - \ell + 3$ exponentiations in $\mathbb{G}_1$ and $n + 1$ exponentiations in $\mathbb{G}_2$.

Asymptotically the exponentiations are the dominant cost as the security parameter grows. However, in practice the multiplications that go into the FFT computations may be more costly for moderate security parameters and large statements. In that case, it may be worth to use a larger common reference string that contains precomputed $G^{u_i(x)}, G^{v_i(x)}, H^{v_i(x)}$ elements for $i = 0, \ldots, m$ such that A and B can be constructed directly instead of the prover having to compute the coefficients of $\sum_{i=0}^{m} a_i u_i(X)$ and $\sum_{i=0}^{m} a_i v_i(X)$ and then do the exponentiations. In the case of boolean circuits we have $a_i \in \{0, 1\}$ and the prover can with such precomputed elements just do m group multiplications for each when computing A and B. We have for this reason let the CRS be longer in Table 1 to get a low computational cost for the prover.

4 Lower Bounds for Non-interactive Arguments

It is an intriguing question how efficient non-interactive arguments can be. We will now give a lower bound showing that pairing-based non-interactive arguments must have proofs with at least 2 group elements if one-way functions exist. More precisely, we look at pairing-based arguments where the common reference string contains a description of a bilinear group and a number of group elements, the proof consists of a number of group elements computed by the prover using generic group operations, and the verifier checks the proof using generic bilinear group operations. We will show that for such pairing-based argument systems, the proof needs to have elements from both $\mathbb{G}_1$ and $\mathbb{G}_2$ if the language includes hard decisional problems as defined below.

Let us consider sampleable decisional problems for a relation R, where there are two sampling algorithms Yes and No. Yes samples statements and witnesses in the relation. No samples statements outside the language L_R defined by the relation. We are interested in relations where it is hard to tell whether a statement ϕ has been sampled by Yes or No.

Definition 4. *We say the relation generator $\mathcal{R}$ has hard decisional problems if there are two efficient algorithms Yes and No such that for $(R, z) \leftarrow \mathcal{R}(1^\lambda)$ we have $\mathsf{Yes}(R) \to (\phi, w) \in R$ and $\mathsf{No}(R) \to \phi \notin L_R$ with overwhelming probability, and for all non-uniform polynomial time distinguishers $\mathcal{A}$*

$$\Pr\left[(R, z) \leftarrow \mathcal{R}(1^\lambda); \phi_0 \leftarrow \mathsf{No}(R); (\phi_1, w_1) \leftarrow \mathsf{Yes}(R); b \leftarrow \{0, 1\} : \mathcal{A}(R, z, \phi_b) = b\right] \approx \frac{1}{2}.$$

If one-way functions exist, we can construct pseudorandom generators. A pseudorandom generator can be used to generate a pseudorandom string, a Yes-instance, with the seed being the witness. To get a No-instance we sample a uniform random string, which with overwhelming probability is not pseudorandom. If the relation R is NP-complete, or just expressive enough to capture pseudorandom generators, then it has a hard decisional problem.

4.1   Linear Interactive Proofs Cannot Have Linear Decision Procedures

We will now prove that LIPs cannot have a linear decision procedure. This answers an open question raised by Bitansky et al. [BCI+13]. The result holds even if we consider designated verifier LIPs and instead of knowledge soundness only consider the weaker notion of soundness that we now define.

Definition 5 (Statistical Soundness Against Affine Prover Strategies). *We say a LIP is (adaptively) sound against affine prover strategies if for all adversaries $\mathcal{A}$*

$$\Pr\left[\begin{array}{l}(R, z) \leftarrow \mathcal{R}(1^\lambda); (\boldsymbol{\sigma}_P, \boldsymbol{\sigma}_V, \boldsymbol{\tau}) \leftarrow \mathsf{Setup}(R); (\phi, \Pi) \leftarrow \mathcal{A}(R, z) \\ \boldsymbol{\pi} = \Pi \boldsymbol{\sigma}_P; \boldsymbol{t} \leftarrow \mathsf{Test}(R, \phi) : \phi \notin L_R \wedge \boldsymbol{t}(\boldsymbol{\sigma}_V, \boldsymbol{\pi}) = \boldsymbol{0}\end{array}\right] \approx 0.$$

Theorem 3. *There are no 2-move algebraic linear interactive proofs with a linear decision procedure for relation generators with hard decisional problems.*

Proof. When the decision procedure is linear, the test $\boldsymbol{t}(\boldsymbol{\sigma}_V, \boldsymbol{\pi}) = \boldsymbol{0}$ can be rewritten as $T\Pi\boldsymbol{\sigma}_P = T'\boldsymbol{\sigma}_V$, where the matrices $T \in \mathbb{F}^{\eta \times k}$ and $T' \in \mathbb{F}^{\eta \times m_V}$ can be efficiently computed from $\boldsymbol{t}$.

Let us now construct an adversary $\mathcal{A}$ that given R and ϕ has a good chance of determining whether ϕ is sampled as a Yes-instance or a No-instance. First, $\mathcal{A}$ repeatedly runs $(\phi_i, w_i) \leftarrow \mathsf{Yes}(R)$ and computes the matching proof and test matrices Π_i and (T_i, T_i'). Let V be the vector space generated by the tuples $(T_i\Pi_i, T_i')$. The adversary keeps sampling tuples until there is more than 50% chance that a new tuple $(T_i\Pi_i, T_i')$ already belongs to V. We will in polynomial time with overwhelming probability sample such a vector space V since there are at most $\eta(m_P + m_V)$ linearly independent tuples.

Now the adversary looks at the statement ϕ that it is trying to classify as a Yes-instance or a No-instance. It computes the test matrices T and T' for ϕ and then tries to solve $(T\Pi, T') = \sum_i r_i(T_i\Pi_i, T_i')$ for $\Pi \in \mathbb{F}^{k \times m_P}$ and $r_i \in \mathbb{F}$. This is a system of linear equations and can therefore be solved efficiently. If a solution is found it guesses $\phi \in L_R$ and if no solution is found it guesses $\phi \notin L_R$.

Let us first analyze the case where $\phi \in L_R$. Since this is a Yes-instance there is more than 50% chance that there is a solution Π such that $(T\Pi, T')$ belongs to the vector space V, so the adversary has 50% chance of guessing $\phi \in L_R$.

Next, let us analyze the case where $\phi \notin L_R$. If we run the setup algorithm $(\boldsymbol{\sigma}_P, \boldsymbol{\sigma}_V, \boldsymbol{\tau}) \leftarrow \mathsf{Setup}(R)$ and $\phi \notin L_R$ we have negligible probability for $T\Pi\boldsymbol{\sigma}_P = T'\boldsymbol{\sigma}_V$. However, by completeness we have for all tuples in V that $T_i\Pi_i\boldsymbol{\sigma}_P = T_i'\boldsymbol{\sigma}_V$. If there were a matrix Π such that $(T\Pi, T') = \sum_i r_i(T_i\Pi, T_i')$ we would have $T\Pi\boldsymbol{\sigma}_P = \sum_i r_i T_i\Pi_i\boldsymbol{\sigma}_P = \sum_i r_i T_i'\boldsymbol{\sigma}_V = T'\boldsymbol{\sigma}_V$, so soundness implies this probability is negligible. The adversary guesses $\phi \notin L_R$ with overwhelming probability. $\square$

4.2   Lower Bound for the Size of Generic Pairing-Based Non-interactive Arguments

We will now show that a generic pairing-based non-interactive argument over Type III groups must have elements in both $\mathbb{G}_1$ and $\mathbb{G}_2$. The intuition behind this argument is that if we have a unilateral argument with only elements in $\mathbb{G}_1$ or only elements in $\mathbb{G}_2$, then the verification equations become linear and the impossibility result for LIPs apply.

Before we get started with the proof, let us define some useful notation. Define for a vector $v = (v_1, \ldots, v_n)$ that $G^v = (G^{v_1}, \ldots, G^{v_n})$. Define for a vector of group elements G^v and a matrix A that $(G^v)^A = G^{vA}$. Also, define for two vectors of group elements $e(G^v, H^w) = \prod_{i=1}^n e(G^{v_i}, H^{w_i})$.

We will consider pairing-based argument systems (Setup, Prove, Vfy) where the proofs consist of group elements and where the algorithms only use generic group operations. Let us be explicit about how such a system operates and the consequences of using generic group operations.

$(\sigma, \tau) \leftarrow \mathsf{Setup}(R)$: The relation contains a description of a bilinear group $(p, \mathbb{G}_1, \mathbb{G}_2, \mathbb{G}_T, e)$ and the common reference string contains group elements in $\mathbb{G}_1, \mathbb{G}_2, \mathbb{G}_T$. Let us fix generators G and H for $\mathbb{G}_1$ and $\mathbb{G}_2$ and write the vectors of group elements in $\mathbb{G}_1, \mathbb{G}_2$ and $\mathbb{G}_T$ as $\boldsymbol{\Sigma}_1 = G^{\sigma_1}$, $\boldsymbol{\Sigma}_2 = H^{\sigma_2}$ and $\boldsymbol{\Sigma}_T = e(G, H)^{\sigma_T}$. We want to avoid that the prover can learn non-trivial information about the discrete logarithms $\sigma_1, \sigma_2, \sigma_T$ using generic bilinear group operations. An example of such a pathological case is a common reference string with group elements G, G^b, where b is a bit. The prover can easily recover the bit b by guessing it and verifying the guess with generic group operations. We say the common reference string is *disclosure-free* if for any pairing product equation on the group elements in $\boldsymbol{\Sigma}_1, \boldsymbol{\Sigma}_2$ and $\boldsymbol{\Sigma}_T$

it is possible with overwhelming probability to predict whether the equation holds or not, when we know the distribution of the common reference string but where we do not know the actual group elements.

$\pi \leftarrow$ Prove(R, σ, ϕ, w): A prover using generic group operations and working on a disclosure-free common reference string has negligible chance of learning any non-trivial information about the common reference string group elements. This means her only viable mode of operation is to pick matrices Π_1, Π_2 and Π_T and compute the proof by setting $\pi = (\psi_1, \psi_2, \psi_T)$, where

$$\psi_1 = \Sigma_1^{\Pi_1} \qquad \psi_2 = \Sigma_2^{\Pi_2} \qquad \psi_T = \Sigma_T^{\Pi_T}.$$

$0/1 \leftarrow$ Vfy(R, σ, ϕ, π): A verifier using generic group operations can only verify a proof by mapping ϕ to matrices and vectors $\{A_q, B_q, C_q, D_q, e_q, f_q\}_{q=1}^Q$ of elements in $\mathbb{Z}_p$ and checking pairing product equations of the form

$$e(\Sigma_1^{A_q}, \Sigma_2)e(\psi_1^{B_q}, \Sigma_2)e(\Sigma_1^{C_q}, \psi_2)e(\psi_1^{D_q}, \psi_2) = \Sigma_T^{e_q} \cdot \psi_T^{f_q}.$$

We note that there is no loss of generality in excluding multi-exponentiation equations in $\mathbb{G}_1$ or $\mathbb{G}_2$; such equations can be translated to pairing product equations by pairing them with G or H.

We now get the following corollary to Theorem 3.

Corollary 1. *A pairing-based non-interactive argument with a disclosure-free common reference string and algorithms using generic group operations cannot exist for relation generators with hard decisional problems unless the proofs have elements both in $\mathbb{G}_1$ and $\mathbb{G}_2$.*

Proof. When the common reference string is disclosure free and the algorithms use generic operations they must work as outlined above. Taking discrete logarithms we get verification equations of the form

$$\sigma_1 A_q \sigma_2 + \pi_1 B_q \sigma_2 + \sigma_1 C_q \pi_2 + \pi_1 D_q \pi_2 = \sigma_T e_q + \pi_T f_q,$$

where $\psi_1 = G^{\pi_1}$ and $\psi_2 = H^{\pi_2}$ and $\psi_T = e(G, H)^{\pi_T}$. If either π_1 or π_2 are empty, there are no $\pi_1 D_q \pi_2$ parts in the verification equations. Observe also that without loss of generality we can assume all the entries in the outer product of σ_1 and σ_2 are given in σ_T (this does not affect disclosure-freeness) so we can set $A_q = 0$ in every equation. This means all the verification equations are linear. Since the verification equations correspond to verifying a LIP "in the exponents" it follows from the impossibility of having LIPs with a linear decision procedure that the proof must have that both π_1 and π_2 are non-trivial and therefore that the proof has elements both in $\mathbb{G}_1$ and $\mathbb{G}_2$. $\qquad\square$

Acknowledgments. We thank Eran Tromer for interesting discussions about the performance of SNARK implementations and the anonymous reviewers for their helpful reviews.

References

[AF07] Abe, M., Fehr, S.: Perfect NIZK with adaptive soundness. In: Vadhan, S.P. (ed.) TCC 2007. LNCS, vol. 4392, pp. 118–136. Springer, Heidelberg (2007)

[AGOT14] Abe, M., Groth, J., Ohkubo, M., Tibouchi, M.: Unified, minimal and selectively randomizable structure-preserving signatures. In: Lindell, Y. (ed.) TCC 2014. LNCS, vol. 8349, pp. 688–712. Springer, Heidelberg (2014)

[BCCT12] Bitansky, N., Canetti, R., Chiesa, A., Tromer, E.: From extractable collision resistance to succinct non-interactive arguments of knowledge, and back again. In: Innovations in Theoretical Computer Science, pp. 326–349 (2012)

[BCCT13] Bitansky, N., Canetti, R., Chiesa, A., Tromer, E.: Recursive composition and bootstrapping for SNARKS and proof-carrying data. In: STOC, pp. 111–120 (2013)

[BCG+13] Ben-Sasson, E., Chiesa, A., Genkin, D., Tromer, E., Virza, M.: SNARKs for C: verifying program executions succinctly and in zero knowledge. In: Canetti, R., Garay, J.A. (eds.) CRYPTO 2013, Part II. LNCS, vol. 8043, pp. 90–108. Springer, Heidelberg (2013)

[BCG+14] Ben-Sasson, E., Chiesa, A., Garman, C., Green, M., Miers, I., Tromer, E., Virza, M.: Zerocash: decentralized anonymous payments from bitcoin. In: IEEE Symposium on Security and Privacy, pp. 459–474 (2014)

[BCI+13] Bitansky, N., Chiesa, A., Ishai, Y., Ostrovsky, R., Paneth, O.: Succinct non-interactive arguments via linear interactive proofs. In: Sahai, A. (ed.) TCC 2013. LNCS, vol. 7785, pp. 315–333. Springer, Heidelberg (2013)

[BCPR14] Bitansky, N., Canetti, R., Paneth, O., Rosen, A.: On the existence of extractable one-way functions. In: STOC, pp. 505–514 (2014)

[BCTV14a] Ben-Sasson, E., Chiesa, A., Tromer, E., Virza, M.: Scalable zero knowledge via cycles of elliptic curves. In: Garay, J.A., Gennaro, R. (eds.) CRYPTO 2014, Part II. LNCS, vol. 8617, pp. 276–294. Springer, Heidelberg (2014)

[BCTV14b] Ben-Sasson, E., Chiesa, A., Tromer, E., Virza, M.: Succinct non-interactive zero knowledge for a von Neumann architecture. In: USENIX, pp. 781–796 (2014)

[BFM88] Blum, M., Feldman, P., Micali, S.: Non-interactive zero-knowledge and its applications. In: STOC, pp. 103–112 (1988)

[BP15] Boyle, E., Pass, R.: Limits of extractability assumptions with distributional auxiliary input. In: Iwata, T., Cheon, J.H. (eds.) ASIACRYPT 2015. LNCS, vol. 9453, pp. 236–261. Springer, Heidelberg (2015). doi:10.1007/978-3-662-48800-3_10

[CFH+15] Costello, C., Fournet, C., Howell, J., Kohlweiss, M., Kreuter, B., Naehrig, M., Parno, B., Zahur, S.: Geppetto: versatile verifiable computation. In: IEEE Symposium on Security and Privacy, pp. 253–270 (2015)

[CTV15] Chiesa, A., Tromer, E., Virza, M.: Cluster computing in zero knowledge. In: Oswald, E., Fischlin, M. (eds.) EUROCRYPT 2015. LNCS, vol. 9057, pp. 371–403. Springer, Heidelberg (2015)

[DFGK14] Danezis, G., Fournet, C., Groth, J., Kohlweiss, M.: Square span programs with applications to succinct NIZK arguments. In: Sarkar, P., Iwata, T. (eds.) ASIACRYPT 2014. LNCS, vol. 8873, pp. 532–550. Springer, Heidelberg (2014)

[DFKP13] Danezis, G., Fournet, C., Kohlweiss, M., Parno, B.: Pinocchio coin: building zerocoin from a succinct pairing-based proof system. In: PETShopCCS (2013)

[GGPR13] Gennaro, R., Gentry, C., Parno, B., Raykova, M.: Quadratic span programs and succinct NIZKs without PCPs. In: Johansson, T., Nguyen, P.Q. (eds.) EUROCRYPT 2013. LNCS, vol. 7881, pp. 626–645. Springer, Heidelberg (2013)

[GMR89] Goldwasser, S., Micali, S., Rackoff, C.: The knowledge complexity of interactive proofs. SIAM J. Comput. 18(1), 186–208 (1989)

[GOS06] Groth, J., Ostrovsky, R., Sahai, A.: Non-interactive zaps and new techniques for NIZK. In: Dwork, C. (ed.) CRYPTO 2006. LNCS, vol. 4117, pp. 97–111. Springer, Heidelberg (2006)

[GOS12] Groth, J., Ostrovsky, R., Sahai, A.: New techniques for noninteractive zero-knowledge. J. ACM 59(3), 11:1–11:35 (2012)

[GPS08] Galbraith, S.D., Paterson, K.G., Smart, N.P.: Pairings for cryptographers. Discrete Appl. Math. 156(16), 3113–3121 (2008)

[Gro06] Groth, J.: Simulation-sound NIZK proofs for a practical language and constant size group signatures. In: Lai, X., Chen, K. (eds.) ASIACRYPT 2006. LNCS, vol. 4284, pp. 444–459. Springer, Heidelberg (2006)

[Gro09] Groth, J.: Linear algebra with sub-linear zero-knowledge arguments. In: Halevi, S. (ed.) CRYPTO 2009. LNCS, vol. 5677, pp. 192–208. Springer, Heidelberg (2009)

[Gro10] Groth, J.: Short pairing-based non-interactive zero-knowledge arguments. In: Abe, M. (ed.) ASIACRYPT 2010. LNCS, vol. 6477, pp. 321–340. Springer, Heidelberg (2010)

[GS12] Groth, J., Sahai, A.: Efficient noninteractive proof systems for bilinear groups. SIAM J. Comput. 41(5), 1193–1232 (2012)

[GW11] Gentry, C., Wichs, D.: Separating succinct non-interactive arguments from all falsifiable assumptions. In: STOC, pp. 99–108 (2011)

[Kil92] Kilian, J.: A note on efficient zero-knowledge proofs and arguments. In: STOC, pp. 723–732 (1992)

[Lip12] Lipmaa, H.: Progression-free sets and sublinear pairing-based non-interactive zero-knowledge arguments. In: Cramer, R. (ed.) TCC 2012. LNCS, vol. 7194, pp. 169–189. Springer, Heidelberg (2012)

[Lip13] Lipmaa, H.: Succinct non-interactive zero knowledge arguments from span programs and linear error-correcting codes. In: Sako, K., Sarkar, P. (eds.) ASIACRYPT 2013, Part I. LNCS, vol. 8269, pp. 41–60. Springer, Heidelberg (2013)

[Mic00] Micali, S.: Computationally sound proofs. SIAM J. Comput. 30(4), 1253–1298 (2000)

[PHGR13] Parno, B., Howell, J., Gentry, C., Raykova, M.: Pinocchio: nearly practical verifiable computation. In: IEEE Symposium on Security and Privacy, pp. 238–252 (2013)

[SVdV15] Schoenmakers, B., Veeningen, M., de Vreede, N.: Trinocchio: privacy-friendly outsourcing by distributed verifiable computation. In: Cryptology ePrint Archive, Report 2015/480 (2015)

[Val08] Valiant, P.: Incrementally verifiable computation or proofs of knowledge imply time/space efficiency. In: Canetti, R. (ed.) TCC 2008. LNCS, vol. 4948, pp. 1–18. Springer, Heidelberg (2008)

[Wal15] Walfish, M.: A wishlist for verifiable computation: an applied CS perspective. Presentation at the Securing Computation Workshop at the Simons Institute for the Theory of Computing, UC Berkeley (2015)

Efficient Zero-Knowledge Arguments for Arithmetic Circuits in the Discrete Log Setting

Jonathan Bootle[1]($\boxtimes$), Andrea Cerulli[1], Pyrros Chaidos[1],
Jens Groth[1], and Christophe Petit[2]

[1] University College London, London, UK
{jonathan.bootle.14,andrea.cerulli.13,pyrros.chaidos.10,j.groth}@ucl.ac.uk
[2] University of Oxford, Oxford, UK
christophe.f.petit@gmail.com

Abstract. We provide a zero-knowledge argument for arithmetic circuit satisfiability with a communication complexity that grows logarithmically in the size of the circuit. The round complexity is also logarithmic and for an arithmetic circuit with fan-in 2 gates the computation of the prover and verifier is linear in the size of the circuit. The soundness of our argument relies solely on the well-established discrete logarithm assumption in prime order groups.

At the heart of our new argument system is an efficient zero-knowledge argument of knowledge of openings of two Pedersen multicommitments satisfying an inner product relation, which is of independent interest. The inner product argument requires logarithmic communication, logarithmic interaction and linear computation for both the prover and the verifier.

We also develop a scheme to commit to a polynomial and later reveal the evaluation at an arbitrary point, in a verifiable manner. This is used to build an optimized version of the constant round square root complexity argument of Groth (CRYPTO 2009), which reduces both communication and round complexity.

Keywords: Sigma-protocol · Zero-knowledge argument · Arithmetic circuit · Discrete logarithm assumption

1 Introduction

Zero-knowledge proofs and arguments are ubiquitous in cryptography today, with prominent applications in authentication protocols, multi-party computa-

The research leading to these results has received funding from the European Research Council under the European Union's Seventh Framework Programme (FP/2007-2013) / ERC Grant Agreement n. 307937 and EPSRC grant EP/J009520/1.

P. Chaidos—Was supported by an EPSRC scholarship (EP/G037264/1 – Security Science DTC).

M. Fischlin and J.-S. Coron (Eds.): EUROCRYPT 2016, Part II, LNCS 9666, pp. 327–357, 2016.
DOI: 10.1007/978-3-662-49896-5_12

tion, encryption primitives, electronic voting systems and verifiable computation protocols.

Informally, a zero-knowledge argument involves two parties, the prover and the verifier, and allows the prover to prove to the verifier that a particular statement is true, without revealing anything else about the statement itself. Statements are of the form $u \in L$, where L is a language in NP. We call w a witness for a statement u if $(u, w) \in R$, where R is a polynomial time decidable binary relation associated with L. We require the zero-knowledge argument to be complete, sound and zero-knowledge.

Completeness: A prover with a witness w for $u \in L$ can convince the verifier of this fact.

Soundness: A prover cannot convince a verifier when $u \notin L$.

Zero-knowledge: The interaction should not reveal anything to the verifier except that $u \in L$. In particular, it should not reveal the prover's witness w.

Our goal is to build an efficient argument system for the satisfiability of an arithmetic circuit, i.e., a circuit that consists of addition and multiplication gates over a finite field $\mathbb{Z}_p$. Moreover we want to base the security of this argument solely on the discrete logarithm assumption: this will provide both strong security guarantees and good efficiency since there exists no known attacks better than generic ones for well-chosen elliptic curve subgroups.

The most efficient zero-knowledge arguments solely based on the discrete logarithm assumption are Groth's protocol based on linear algebra [21] and its variant by Seo [36]. Both of these protocols have a communication complexity that is proportional to the square root of the circuit size. This square root complexity has since then appeared as a (perhaps fundamental) barrier for discrete logarithm-based arguments for circuit satisfiability.

1.1   Our Contributions

We provide an honest verifier zero-knowledge argument for arithmetic circuit satisfiability based on the discrete logarithm assumption that only requires a *logarithmic* communication complexity. Our argument has perfect completeness and perfect special honest verifier zero-knowledge. Soundness is computational and based on the discrete logarithm assumption. We require a logarithmic number of moves, and both the prover and verifier have linear computational complexity. The argument is therefore efficient on all parameters with the biggest improvement being in the communication complexity.

Improved Square Root Complexity Argument. We start from the circuit satisfiability argument of Groth [21], which requires 7 moves and has square root communication complexity in the *total* number of gates. In this argument the prover commits to all the wires using homomorphic multicommitments, verifies addition gates using the homomorphic properties, and uses a product argument to show that the multiplication gates are satisfied.

We first improve Groth's argument into a 5 moves argument with square root communication complexity in the number of *multiplication gates* only. We achieve fewer moves compared to [21] by avoiding generic reductions to linear algebra statements. We remove the communication cost of the addition gates in the argument by providing a technique that can directly handle a set of Hadamard products and linear relations together. Another efficiency improvement is a subroutine to commit to a polynomial and later reveal its evaluation at an arbitrary point in a verifiable manner. In Sect. 3 we provide a protocol to perform this task, which has a square root communication complexity with respect to the degree of the polynomial, and which may be of independent interest.

Logarithmic Complexity Argument. In spite of all these improvements, the above argument still requires a square root communication complexity with respect to multiplication gates. In the first move the prover commits to all circuit wires using $3m$ commitments to n elements each, where $mn = N$ is a bound on the number of multiplication gates, and in the last move after receiving a challenge he opens one commitment that can be constructed from the previous ones and the challenge. By setting $m \approx n$ we get a minimal communication complexity of $O(\sqrt{N})$.

Our key idea to break this square root communication complexity barrier is to replace the last opening step in this protocol by an argument of knowledge of the opening values. Using specific properties of Pedersen multicommitments, namely homomorphic properties with respect to the keys, we rewrite this argument as an argument of knowledge of openings of two homomorphic commitments, satisfying an inner product relation. In Sect. 4 we provide an argument system for this problem, which only requires a logarithmic communication with respect to the vector sizes. The argument is built in a recursive way, reducing the size and complexity of the statement further in each recursion step. Using this inner product argument as a subroutine we obtain an arithmetic circuit satisfiability argument with logarithmic communication.

Implementation. In Sect. 6 we report on an implementation of our arguments. To show the practicality of our results we compare the efficiency of our implementation to that of Pinocchio [34]. Pinocchio is a practical verifiable computation scheme allowing a constrained client to outsource computation of a function to a powerful worker and to efficiently verify the outcome of the function. It uses quadratic arithmetic programs, a generalisation of arithmetic circuits, and for some functions achieves verification that is faster than local computation. While we do not achieve comparably fast verification, we compare favourably in terms of prover computation, and do so under simpler assumptions.

1.2 Related Work

Zero-knowledge proofs were invented by Goldwasser et al. [18]. It is useful to distinguish between zero-knowledge *proofs*, with statistical soundness, and zero-knowledge *arguments* with computational soundness. In general proofs can only have computational zero-knowledge, while arguments may have perfect zero-knowledge. Goldreich et al. [16] showed that all languages in NP have zero-knowledge proofs while Brassard et al. [8] showed that all languages in NP have zero-knowledge arguments with perfect zero-knowledge.

Gentry et al. [14] used fully homomorphic encryption to construct zero-knowledge proofs where the communication complexity corresponds to the size of the witness. However, proofs cannot in general have communication that is smaller than the witness size unless surprising results about the complexity of solving SAT instances hold [15,17].

Kilian [27] showed that in contrast to zero-knowledge proofs, zero-knowledge arguments can have very low communication complexity. His construction relied on the PCP theorem though, and did not yield a practical scheme.

Schnorr [35] and Guillou and Quisquater [25] gave early examples of practical zero-knowledge arguments for concrete number theoretic problems. Extending Schnorr's protocols, there have been many constructions of zero-knowledge arguments based on the discrete logarithm assumption. Cramer and Damgård [10] gave a zero-knowledge argument for arithmetic circuit satisfiability, which has linear communication complexity.

Currently the most efficient discrete logarithm based zero-knowledge arguments for arithmetic circuits are the ones by Groth [21] and Seo [36], which are constant move arguments with a communication proportional to the square root of the circuit size. Using pairing-based cryptography instead of just relying on the discrete logarithm assumption, Groth [20] extended these techniques to give a zero-knowledge argument with a cubic root communication complexity.

There are recent works giving a logarithmic communication complexity for specific languages. Bayer and Groth [2] show that one can prove that a polynomial evaluated at a secret committed value gives a certain output with a logarithmic communication complexity and Groth and Kohlweiss [24] show that one can prove that one out of N commitments contain 0 with logarithmic communication complexity. These results are for very specific types of statements (with low circuit depth) and the techniques do not seem to generalize to arbitrary NP languages.

An exciting line of research [4–7,13,22,24,30,34] has developed many proposals for succinct non-interactive arguments (SNARGs) yielding pairing-based constructions where the arguments consist of a constant number of group elements. However, they all rely on a common reference string (with a special structure) and non-falsifiable knowledge extractor assumptions. In contrast, the arguments we develop here are based solely on the discrete logarithm assumption, and use a small common reference string which is independent of the circuit.

Table 1 compares the most efficient previous zero-knowledge arguments based on the discrete logarithm assumption with our scheme, when allowing for 5 moves or a logarithmic number of moves. Using 5 moves, our scheme requires significantly less computation than [36]. On the other hand when using a logarithmic number of moves and applying a reduction similar to [1], our scheme dramatically improves the communication costs with respect to all previous work without incurring any significant overhead. We note that [1] uses the reduction to reduce computation whereas we use it to reduce communication.

Table 1. Efficiency comparison between our arguments and the most efficient interactive zero-knowledge arguments relying on discrete logarithm. We express communication in number of group elements $\mathbb{G}$ and field elements $\mathbb{Z}_p$ and computation costs in number of exponentiations over $\mathbb{G}$ and multiplications over $\mathbb{Z}_p$. The efficiency displayed is for a circuit with N multiplication gates.

Reference	Moves	Communication		Prover complexity		Verifier complexity	
		$\mathbb{G}$	$\mathbb{Z}_p$	exp.	mult.	exp.	mult.
[10]	3	$6N$	$5N+2$	$6N$	$6N$	$6N$	0
[21]	7	$9\sqrt{N}+4$	$7\sqrt{N}+6$	$\frac{6N}{\log N}$	$O(N\log N)$	$\frac{39\sqrt{N}}{\log N}$	$O(N)$
[21]	$2\log N+5$	$2\sqrt{N}$	$7\sqrt{N}$	$\frac{6N}{\log N}$	$O(N)$	$\frac{18\sqrt{N}}{\log N}$	$O(N)$
[36]	5	$30\sqrt{N}$	$7\sqrt{N}$	$\frac{6N}{\log N}$	$O(N\log N)$	$\frac{77\sqrt{N}}{\log N}$	$O(N)$
This paper	5	$2\sqrt{N}$	$2\sqrt{N}$	$\frac{6N}{\log N}$	$3N\log N$	$\frac{8\sqrt{3N}}{\log N}$	$O(N)$
This paper	$2\log N+1$	$4\log N+7$	$2\log N+6$	$12N$	$O(N)$	$4N$	$O(N)$

As part of our construction we give a protocol for committing to a polynomial and later revealing an evaluation of the polynomial in a given point. Kate et al. [26] have also provided protocols to commit to polynomials and then evaluate them at a given point in a verifiable way. Their protocols only require a constant number of commitments but security relies on pairing assumptions. Our polynomial commitment protocol has square root communication complexity but relies solely on the discrete logarithm assumption.

2 Preliminaries

We write $y = A(x; r)$ when the algorithm A on input x and randomness r, outputs y. We write $y \leftarrow A(x)$ for the process of picking randomness r at random and setting $y = A(x; r)$. We also write $y \leftarrow S$ for sampling y uniformly at random from the set S. We will assume one can sample uniformly at random from sets such as $\mathbb{Z}_p$ and $\mathbb{Z}_p^*$.

Algorithms in our schemes receive a security parameter λ as input (sometimes implicitly written in unary). The intuition is that the higher the security parameter, the lower the risk of the scheme being broken. Given two functions $f, g : \mathbb{N} \to [0,1]$ we write $f(\lambda) \approx g(\lambda)$ when $|f(\lambda) - g(\lambda)| = \lambda^{-\omega(1)}$. We say that f is *negligible* when $f(\lambda) \approx 0$ and that f is *overwhelming* when $f(\lambda) \approx 1$.

Throughout the paper we let $\mathbb{G}$ be a group of prime order p. Let $\boldsymbol{g} = (g_1, \ldots, g_n) \in \mathbb{G}^n$ and $\boldsymbol{f} = (f_1, \ldots, f_n) \in \mathbb{Z}_p^n$. We write $\boldsymbol{g}^{\boldsymbol{f}}$ for the multi-exponentiation $\boldsymbol{g}^{\boldsymbol{f}} = \prod_{i=1}^n g_i^{f_i}$. A multi-exponentiation of size n can be computed at a cost of roughly $\frac{n}{\log n}$ single group exponentiations using the multi-exponentiation techniques of [28, 31, 32].

2.1 The Discrete Logarithm Assumption

Let GGen be an algorithm that on input 1^λ returns $(\mathbb{G}, p, g)$ such that $\mathbb{G}$ is the description of a finite cyclic group of prime order p, where $|p| = \lambda$, and g is a generator of $\mathbb{G}$.

Definition 1 (Discrete Logarithm Assumption). *The discrete logarithm assumption holds relative to* GGen *if for all non-uniform polynomial time adversaries* $\mathcal{A}$

$$\Pr\left[(\mathbb{G}, p, g) \leftarrow \mathrm{GGen}(1^\lambda); h \leftarrow \mathbb{G}; a \leftarrow \mathcal{A}(\mathbb{G}, p, g, h) : g^a = h\right] \approx 0$$

In this definition, the value a is called the discrete logarithm of h in the basis g. Note that the discrete logarithm assumption is defined with respect to a particular group generator algorithm GGen. According to current state-of-the-art cryptanalytic techniques, to get a security level of $2^{-\lambda}$ the group generator may for example return well-chosen elliptic curve groups where group elements can be represented with $O(\lambda)$ bits or multiplicative subgroups of finite fields with a large characteristic where group elements can be represented with $O(\lambda^3)$ bits. It is well-known that the discrete logarithm assumption is equivalent to the following assumption.

Definition 2 (Discrete Logarithm Relation Assumption). *For all* $n \geq 1$ *and all non-uniform polynomial time adversaries* $\mathcal{A}$

$$\Pr\left[\begin{array}{l}(\mathbb{G}, p, g) \leftarrow \mathrm{GGen}(1^\lambda); g_1, \ldots, g_n \leftarrow \mathbb{G}; \\ a_0, \ldots, a_n \leftarrow \mathcal{A}(\mathbb{G}, p, g, \{g_i\}_i)\end{array} : \exists a_i \neq 0 \ and \ g^{a_0} \prod_{i=1}^{n} g_i^{a_i} = 1\right] \approx 0$$

We call such a product $g^{a_0} \prod_{i=1}^{n} g_i^{a_i} = 1$ a non-trivial discrete logarithm relation.

2.2 Pedersen Commitments

A non-interactive commitment scheme allows a sender to create a commitment to a secret value. She may later open the commitment and reveal the value in a verifiable manner. A commitment should be hiding, i.e., not reveal the secret value, and binding in the sense that a commitment cannot be opened to two different values.

Formally, a non-interactive commitment scheme is a pair of probabilistic polynomial time algorithms (CGen, Com). The setup algorithm $ck \leftarrow \mathrm{CGen}(1^\lambda)$ generates a commitment key ck. The commitment key specifies a message space $\mathcal{M}_{ck}$, a randomness space $\mathcal{R}_{ck}$ and a commitment space $\mathcal{C}_{ck}$. The commitment algorithm combined with the commitment key specifies a function $\mathrm{Com}_{ck} : \mathcal{M}_{ck} \times \mathcal{R}_{ck} \rightarrow \mathcal{C}_{ck}$. Given a message $m \in \mathcal{M}_{ck}$ the sender picks uniformly at random $r \leftarrow \mathcal{R}_{ck}$ and computes the commitment $c = \mathrm{Com}_{ck}(m; r)$.

Definition 3 (Perfectly Hiding). *We say a non-interactive commitment scheme* (CGen, Com) *is perfectly hiding if a commitment does not reveal the* *committed value. For all non-uniform polynomial time stateful interactive adversaries $\mathcal{A}$*

$$\Pr\begin{bmatrix} ck \leftarrow \mathrm{CGen}(1^\lambda); (m_0, m_1) \leftarrow \mathcal{A}(ck); \\ b \leftarrow \{0,1\}; c \leftarrow \mathrm{Com}_{ck}(m_b) \end{bmatrix} : \mathcal{A}(c) = b \end{bmatrix} = \frac{1}{2}$$

where $\mathcal{A}$ outputs $m_0, m_1 \in \mathcal{M}_{ck}$.

Definition 4 (Binding). *A non-interactive commitment scheme $(\mathrm{CGen}, \mathrm{Com})$ is computationally binding if a commitment can only be opened to one value. For all non-uniform polynomial time adversaries $\mathcal{A}$*

$$\Pr\begin{bmatrix} ck \leftarrow \mathrm{CGen}(1^\lambda); & \mathrm{Com}_{ck}(m_0; r_0) = \mathrm{Com}_{ck}(m_1; r_1) \\ (m_0, r_0, m_1, r_1) \leftarrow \mathcal{A}(ck) & and \ m_0 \neq m_1 \end{bmatrix} \approx 0$$

where $\mathcal{A}$ outputs $m_0, m_1 \in \mathcal{M}_{ck}$ and $r_0, r_1 \in \mathcal{R}_{ck}$.

We say a commitment scheme is homomorphic if for all valid keys ck the message, randomness and commitment spaces are abelian groups and for all messages $m_0, m_1 \in \mathcal{M}_{ck}$ and randomness $r_0, r_1 \in \mathcal{R}_{ck}$ we have

$$\mathrm{Com}_{ck}(m_0; r_0) \cdot \mathrm{Com}_{ck}(m_1; r_1) = \mathrm{Com}_{ck}(m_0 + m_1; r_0 + r_1).$$

The most prominent example of a homomorphic perfectly hiding commitment scheme is the Pedersen commitment scheme. Pedersen commitments have the form $c = g^r h^m$ where g, h are group elements specified in the commitment key. The opening of a Pedersen commitment is $(m, r) \in \mathbb{Z}_p^2$, from which anybody can recompute the commitment c and verify it was a valid commitment. Since Pedersen commitments are random group elements, they are perfectly hiding. On the other hand, breaking the binding property of Pedersen commitments corresponds to breaking the discrete logarithm assumption.

We will be using a variant of Pedersen commitments that allow us to commit to multiple values at once. The commitment key is $ck = (\mathbb{G}, p, g, g_1, \ldots, g_n)$ and a commitment is of the form $c = g^r \prod_{i=1}^n g_i^{m_i}$. We write $c = \mathrm{Com}_{ck}(m_1, \ldots, m_n; r)$ for this operation.

With the Pedersen commitment scheme in mind, we will assume throughout the paper that the message space is $\mathbb{Z}_p^n$ and the randomness space is $\mathbb{Z}_p$. The constructions we have in Sects. 3 and 5.1 require a perfectly hiding, homomorphic commitment scheme so we are not limited to using the Pedersen commitment scheme. However, in Sects. 4 and 5.2, we will rely on specific properties of the Pedersen scheme and work directly on the group elements in the key.

2.3 Zero-Knowledge Arguments of Knowledge

Let R be a polynomial time decidable binary relation, i.e., a relation that defines a language in NP. We call w a witness for a statement u if $(u, w) \in R$.

In the arguments we consider a prover $\mathcal{P}$ and a verifier $\mathcal{V}$, both of which are probabilistic polynomial time interactive algorithms. The transcript produced by $\mathcal{P}$ and $\mathcal{V}$ when interacting on inputs s and t is denoted by $tr \leftarrow \langle \mathcal{P}(s), \mathcal{V}(t) \rangle$. We write $\langle \mathcal{P}(s), \mathcal{V}(t) \rangle = b$ depending on whether the verifier rejects, $b = 0$, or accepts, $b = 1$.

Definition 5 (Argument of Knowledge). *The pair $(\mathcal{P}, \mathcal{V})$ is called an argument of knowledge for the relation R if we have perfect completeness and statistical witness-extended emulation as defined below.*

Definition 6 (Perfect Completeness). *$(\mathcal{P}, \mathcal{V})$ has perfect completeness if for all non-uniform polynomial time adversaries $\mathcal{A}$*

$$\Pr \left[(u, w) \leftarrow \mathcal{A}(1^\lambda) : (u, w) \notin R \text{ or } \langle \mathcal{P}(u, w), \mathcal{V}(u) \rangle = 1 \right] = 1$$

To define an argument of knowledge we follow Groth and Ishai [23] that borrowed the term witness-extended emulation from Lindell [29]. Informally, their definition says that given an adversary that produces an acceptable argument with some probability, there exists an emulator that produces a similar argument with the same probability together with a witness w. Note that the emulator is allowed to rewind the prover and verifier's interaction to any previous move.

Definition 7 (Statistical Witness-Extended Emulation). *$(\mathcal{P}, \mathcal{V})$ has statistical witness-extended emulation if for all deterministic polynomial time $\mathcal{P}^*$ there exists an expected polynomial time emulator $\mathcal{E}$ such that for all interactive adversaries $\mathcal{A}$*

$$\Pr \left[(u, s) \leftarrow \mathcal{A}(1^\lambda); tr \leftarrow \langle \mathcal{P}^*(u, s), \mathcal{V}(u) \rangle : \mathcal{A}(tr) = 1 \right]$$
$$\approx \Pr \left[\begin{array}{l} (u, s) \leftarrow \mathcal{A}(1^\lambda); (tr, w) \leftarrow \mathcal{E}^{\langle \mathcal{P}^*(u,s), \mathcal{V}(u) \rangle}(u) : \\ \mathcal{A}(tr) = 1 \text{ and if } tr \text{ is accepting then } (u, w) \in R \end{array} \right]$$

where the oracle called by $\mathcal{E}^{\langle \mathcal{P}^*(u,s), \mathcal{V}(u) \rangle}$ permits rewinding to a specific point and resuming with fresh randomness for the verifier from this point onwards.

In the definition, s can be interpreted as the state of $\mathcal{P}^*$, including the randomness. So, whenever $\mathcal{P}^*$ is able to make a convincing argument when in state s, $\mathcal{E}$ can extract a witness. This is why we call it an argument of knowledge.

Definition 8 (Public Coin). *An argument $(\mathcal{P}, \mathcal{V})$ is called public coin if the verifier chooses his messages uniformly at random and independently of the messages sent by the prover, i.e., the challenges correspond to the verifier's randomness ρ.*

An argument is zero-knowledge if it does not leak information about the witness beyond what can be inferred from the truth of the statement. We will present arguments that have special honest verifier zero-knowledge in the sense that if the verifier's challenges are known in advance, then it is possible to simulate the entire argument without knowing the witness.

Definition 9 (Perfect Special Honest Verifier Zero-Knowledge). *A public coin argument* $(\mathcal{P}, \mathcal{V})$ *is called a* perfect special honest verifier zero knowledge *(SHVZK)* *argument for* R *if there exists a probabilistic polynomial time simulator* $\mathcal{S}$ *such that for all interactive non-uniform polynomial time adversaries* $\mathcal{A}$

$$\Pr\Big[(u, w, \rho) \leftarrow \mathcal{A}(1^\lambda); tr \leftarrow \langle \mathcal{P}(u, w), \mathcal{V}(u; \rho) \rangle : (u, w) \in R \text{ and } \mathcal{A}(tr) = 1\Big]$$

$$= \Pr\Big[(u, w, \rho) \leftarrow \mathcal{A}(1^\lambda); tr \leftarrow \mathcal{S}(u, \rho) : (u, w) \in R \text{ and } \mathcal{A}(tr) = 1\Big]$$

where ρ *is the public coin randomness used by the verifier.*

Full Zero-Knowledge. In real life applications special honest verifier zero-knowledge may not suffice since a malicious verifier may give non-random challenges. However, it is easy to convert an SHVZK argument into a full zero-knowledge argument secure against *arbitrary* verifiers in the common reference string model using standard techniques [12,19]. The conversion can be very efficient and only costs a small additive overhead.

The Fiat-Shamir Heuristic. The Fiat-Shamir transformation takes an interactive public coin argument and replaces the challenges with the output of a cryptographic hash function. The idea is that the hash function will produce random looking output and therefore be a suitable replacement for the verifier. The Fiat-Shamir heuristic yields a non-interactive zero-knowledge argument in the random oracle model [3].

The transformation can be applied to our arguments to make them non-interactive at the cost of using the random oracle model in the security proofs. From an efficiency point of view this is especially useful for the arguments in Sects. 4 and 5.2, reducing a logarithmic number of moves to a single one.

A General Forking Lemma. Suppose that we have a $(2\mu + 1)$-move public-coin argument with μ challenges, $x_1, \ldots, x_\mu$ in sequence. Let $n_i \geq 1$ for $1 \leq i \leq \mu$. Consider $\prod_{i=1}^{\mu} n_i$ accepting transcripts with challenges in the following tree format. The tree has depth μ and $\prod_{i=1}^{\mu} n_i$ leaves. The root of the tree is labelled with the statement. Each node of depth $i < \mu$ has exactly n_i children, each labelled with a distinct value for the ith challenge x_i.

This can be referred to as an $(n_1, \ldots, n_\mu)$-tree of accepting transcripts. All of our arguments allow a witness to be extracted efficiently from an appropriate tree of accepting transcripts. This is a natural generalisation of special-soundness for Sigma-protocols, where $\mu = 1$ and $n = 2$. For simplicity in the following lemma, we assume that the challenges are chosen uniformly from $\mathbb{Z}_p$ where $|p| = \lambda$, but any sufficiently large challenge space would suffice. We refer to the full version of the paper for a proof of the forking lemma.

Lemma 1 (Forking Lemma). *Let $(\mathcal{P}, \mathcal{V})$ be a $(2\mu+1)$-move, public coin inter-active protocol. Let χ be a witness extraction algorithm that always succeeds in extracting a witness from an $(n_1, \ldots, n_\mu)$-tree of accepting transcripts in proba-bilistic polynomial time. Assume that $\prod_{i=1}^{\mu} n_i$ is bounded above by a polynomial in the security parameter λ. Then $(\mathcal{P}, \mathcal{V})$ has witness-extended emulation.*

3   Commitments to Polynomials

In this section, we present a protocol to commit to a polynomial $t(X)$ and later reveal the evaluation of $t(X)$ at any point $x \in \mathbb{Z}_p^*$ together with a proof that enables a verifier to check that the evaluation is correct with respect to the committed $t(X)$. We will consider Laurent polynomials $t(X) \in \mathbb{Z}_p[X, X^{-1}]$ i.e. polynomials in which we allow terms of negative degree. This protocol will be used as a subroutine for the arguments described in Sects. 5.1 and 5.2.

A simple solution for this problem would be to send commitments to coeffi-cients of $t(X)$ individually, from which the evaluation of $t(X)$ at any particular point can be verified using the homomorphic properties. This solution requires d group elements to be sent, where d is the number of non-zero coefficients in $t(X)$. As we shall show it is possible to reduce the communication costs to $O(\sqrt{d})$ group elements, where $d = d_2 + d_1$ if $t(X) = \sum_{k=-d_1}^{d_2} t_k X^k$.

For clarity we first informally describe our protocol for a standard (not Lau-rent) polynomial $t(X) = \sum_{k=0}^{d} t_k X^k$. We then extend this informal description to Laurent polynomials with zero constant term. We finally provide a formal description of the protocol and analyze its security and efficiency.

Main Idea for Standard Polynomials. Let $t(X) = \sum_{k=0}^{d} t_k X^k$ be a polynomial with coefficients in $\mathbb{Z}_p$ and assume $d + 1 = mn$. We can write $t(X) = \sum_{i=0}^{m-1} \sum_{j=0}^{n-1} t_{i,j}(X) X^{in+j}$ and arrange the coefficients in a $m \times n$ matrix

$$\begin{pmatrix} t_{0,0} & t_{0,1} & \cdots t_{0,n-1} \\ t_{1,0} & t_{1,1} & \cdots t_{1,n-1} \\ \vdots & \vdots & \vdots \\ t_{m-1,0} & t_{n-1,1} & \cdots t_{m-1,n-1} \end{pmatrix}$$

Now, $t(X)$ can be evaluated by multiplying the matrix by row and column vectors.

$$t(X) = \begin{pmatrix} 1 \ X^n \cdots X^{(m-1)n} \end{pmatrix} \begin{pmatrix} t_{0,0} & t_{0,1} & \cdots t_{0,n-1} \\ t_{1,0} & t_{1,1} & \cdots t_{1,n-1} \\ \vdots & \vdots & \vdots \\ t_{m-1,0} & t_{n-1,1} & \cdots t_{m-1,n-1} \end{pmatrix} \begin{pmatrix} 1 \\ X \\ \vdots \\ X^{n-1} \end{pmatrix}$$

The idea behind the protocol is to commit to the rows of this matrix using commitments $T_0, \ldots, T_{m-1}$. Later, when given an evaluation point $x \in \mathbb{Z}_p$ we can use the homomorphic property of the commitment scheme to compute the commitment $\prod_{i=0}^{m-1} T_i^{x^{in}}$ to the vector

$$\bar{t} = \begin{pmatrix} 1 & x^n & \cdots & x^{(m-1)n} \end{pmatrix} \begin{pmatrix} t_{0,0} & t_{0,1} & \cdots & t_{0,n-1} \\ t_{1,0} & t_{1,1} & \cdots & t_{1,n-1} \\ \vdots & \vdots & & \vdots \\ t_{m-1,0} & t_{m-1,1} & \cdots & t_{m-1,n-1} \end{pmatrix}$$

The prover opens this latter commitment and now it is easy to compute $v = t(x)$ from $\bar{t}$ and x.

The problem with this straightforward solution is that it leaks partial information about the coefficients of $t(X)$. We remedy this by inserting some blinding values $u_1, \ldots, u_{n-1}$ to hide the weighted sum of the coefficients in each column. However, we make sure that the blinding values cancel each other out so that we still get the correct evaluation of the polynomial. More precisely, we commit to the rows of the following $(m+1) \times n$ matrix

$$T = \begin{pmatrix} t_{0,0} & t_{0,1} - u_1 & \cdots & t_{0,n-2} - u_{n-2} & t_{0,n-1} - u_{n-1} \\ t_{1,0} & t_{1,1} & \cdots & t_{1,n-2} & t_{1,n-1} \\ \vdots & \vdots & & & \vdots \\ t_{m-1,0} & t_{m-1,1} & \cdots & t_{m-1,n-2} & t_{m-1,n-1} \\ u_1 & u_2 & \cdots & u_{n-1} & 0 \end{pmatrix}$$

with U being a commitment to the last row. This time

$$t(X) = \begin{pmatrix} 1 & X^n & \cdots & X^{(m-1)n} & X \end{pmatrix} T \begin{pmatrix} 1 \\ X \\ X^2 \\ \vdots \\ X^{n-1} \end{pmatrix}$$

We now open $U^x \prod_{i=0}^{m-1} T_i^{x^{in}}$ by revealing the vector

$$\bar{t} = \begin{pmatrix} 1 & x^n & \cdots & x^{(m-1)n} & x \end{pmatrix} T$$

This still allows us to compute $t(x)$, but due to the blinders we no longer leak information about the coefficients of $t(X)$. In fact, each element of $\bar{t}$ is uniformly random, conditional on their weighted sum being equal to $t(x)$, which the prover intends for the verifier to learn anyway.

Extension to Laurent Polynomials. Let now $t(X)$ be a Laurent polynomial $t(X) = \sum_{i=-d_1}^{d_2} t_i X^i$ with constant term $t_0 = 0$. Let m_1, m_2, n be positive integers such that $d_1 = nm_1$ and $d_2 = nm_2$ and write $t(X) = X^{-m_1 n} t'(X) + X t''(X)$ for degree $d_1 - 1$ and $d_2 - 1$ polynomials $t'(X), t''(X) \in \mathbb{Z}_p[X]$. We can write $t'(X) = \sum_{i=0}^{m_1-1} \sum_{j=0}^{n-1} t'_{i,j} X^{in+j}$ and $t''(X) = \sum_{i=0}^{m_2-1} \sum_{j=0}^{n-1} t''_{i,j} X^{in+j}$.

We can arrange the coefficients of $t'(X)$ and $t''(X)$ in a $(m_1 + m_2) \times n$ matrix T. We commit to both $t'(X)$ and $t''(X)$ simultaneously by committing to the rows of the matrix using commitments T_i' and T_i''. As when committing to polynomials we add blinders $u_1, \ldots, u_{n-1}$ and make a commitment U to the additional last row arising from this.

$$T = \begin{pmatrix} t_{0,0}' & t_{0,1}' & \cdots & t_{0,n-1}' \\ t_{1,0}' & t_{1,1}' & \cdots & t_{1,n-1}' \\ \vdots & \vdots & & \vdots \\ t_{m_1-1,0}' & t_{m_1-1,1}' & \cdots & t_{m_1-1,n-1}' \\ t_{0,0}'' & t_{0,1}'' - u_1 & \cdots & t_{0,n-1}'' - u_{n-1} \\ t_{1,0}'' & t_{1,1}'' & \cdots & t_{1,n-1}'' \\ \vdots & \vdots & & \vdots \\ t_{m_2-1,0}'' & t_{m_2-1,1}'' & \cdots & t_{m_2-1,n-1}'' \\ u_1 & u_2 & \cdots & 0 \end{pmatrix} = \begin{pmatrix} t_0' \\ t_1' \\ \vdots \\ t_{m_1-1}' \\ t_0'' \\ t_1'' \\ \vdots \\ t_{m_2-1}'' \\ u \end{pmatrix}$$

Define vectors

$$Z = Z(X) = \left(X^{-m_1 n}, X^{-(m_1-1)n}, \ldots, X^{-n}, X, X^{n+1}, \ldots, X^{(m_2-1)n+1}, X^2 \right)$$

$$X = X(X) = \begin{pmatrix} 1 \\ X \\ \vdots \\ X^{n-1} \end{pmatrix}$$

and we have $t(X) = ZTX$.

To evaluate at $x \in \mathbb{Z}_p^*$ we open $\left(\prod_{i=0}^{m_1-1} (T_i')^{x^{(i-m_1)n}} \right) \left(\prod_{i=0}^{m_2-1} (T_i'')^{x^{in+1}} \right) U^{x^2}$ to the vector $\bar{t} = Z(x)T$. This allows us to compute $t(x)$ as $\bar{t}X(x)$. The blinders hide the weighted sums of each column as before, and now the verifier is able to compute $t(x)$ without gaining additional information about its coefficients.

Evaluation Protocol. Our protocol is made of the following three algorithms.

- PolyCommit$(ck, m_1, m_2, n, t(X)) \to (\mathsf{pc}, \mathsf{st})$: Take as input a commitment key ck and a Laurent polynomial $t(X) = \sum_{i=-m_1 n}^{nm_2} t_i X^i$ with constant coefficient $t_0 = 0$. Pick blinders $u_1, \ldots, u_{n-1} \leftarrow \mathbb{Z}_p$ and randomness $\tau_u, \tau_0', \ldots, \tau_{m_1-1}',$ $\tau_0'', \ldots, \tau_{m_2-1}'' \leftarrow \mathbb{Z}_p$. Set $\tau = (\tau_0', \ldots, \tau_{m_1-1}', \tau_0'', \ldots, \tau_{m_2-1}'', \tau_u)$. Compute

$$T_i' = \mathrm{Com}_{ck}(t_i'; \tau_i'), \qquad T_i'' = \mathrm{Com}_{ck}(t_i''; \tau_i''), \qquad U = \mathrm{Com}_{ck}(u; \tau_u)$$

  Return a polynomial commitment $\mathsf{pc} = \left(\{T_i'\}_{i=0}^{m_1-1}, \{T_i''\}_{i=0}^{m_2-1}, U \right)$ and private information $\mathsf{st} = (t(X), \tau)$.
- PolyEval$(\mathsf{st}, x) \to \mathsf{pe}$: Compute

$$\bar{t} = Z(x)T, \qquad \bar{\tau} = Z(x) \cdot \tau$$

  Return $\mathsf{pe} = (\bar{t}, \bar{\tau})$.

- PolyVerify$(ck, m_1, m_2, n, \mathsf{pc}, \mathsf{pe}, x) \to v$: The verifier checks whether

$$\mathrm{Com}_{ck}(\bar{t}; \bar{\tau}) = \left(\prod_{i=0}^{m_1-1} (T_i')^{x^{(i-m_1)n}} \right) \left(\prod_{i=0}^{m_2-1} (T_i'')^{x^{in+1}} \right) U^{x^2}$$

If the check is satisfied the verifier returns $v = t(x) = \bar{t}\boldsymbol{X}(x)$.
Otherwise, the verifier rejects pe as invalid with respect to pc and x and
returns $v = \bot$.

Security Properties. We define three security properties for our protocol: completeness, l-special soundness, and special-honest-verifier zero-knowledge. Later, the protocol is used as a sub-protocol inside our zero-knowledge arguments-of-knowledge. These properties will help us to prove the completeness, witness-extended emulation, and special honest verifier zero knowledge for the zero knowledge argument.

The definition of completeness simply guarantees that if PolyCommit, PolyVerify are carried out honestly, then PolyVerify will accept and return a commitment to the evaluation of the polynomial.

Definition 10 (Perfect Completeness). *(PolyCommit, PolyEval, PolyVerify) has perfect completeness if for all non-uniform polynomial time adversaries $\mathcal{A}$*

$$\Pr \left[\begin{array}{l} (ck, m_1, m_2, n, t(X), x) \leftarrow \mathcal{A}(1^\lambda) \\ (\mathsf{pc}, \mathsf{st}) \leftarrow \mathrm{PolyCommit}(ck, m_1, m_2, n, t(X)) \\ \mathsf{pe} \leftarrow \mathrm{PolyEval}(\mathsf{st}, x) \\ v \leftarrow \mathrm{PolyVerify}(ck, m_1, m_2, n, \mathsf{pc}, \mathsf{pe}, x) \end{array} : v = t(x) \right] = 1$$

where ck is a key for a homomorphic commitment scheme, $t(X)$ is a Laurent polynomial of degrees $d_1 = m_1 n, d_2 = m_2 n$ and $x \in \mathbb{Z}_p^*$.

The definition of l-Special Soundness says that given l accepting evaluations for different evaluation points, but from the same commitment pc, then it is possible to extract either a valid Laurent polynomial $t(X)$ with zero constant term that is consistent with the evaluations produced or a breach in the binding property of the commitment scheme. Furthermore, any other accepting evaluations for the same commitment will also be evaluations of $t(X)$.

Definition 11 Statistical l-Special Soundness). *(PolyCommit, PolyEval, PolyVerify) is statistically l-special sound if there exists a probabilistic polynomial time algorithm χ that, given l accepting transcripts with the same commitment pc, either extracts the committed polynomial $t(X)$, or extracts a break of the binding property of the underlying commitment scheme. For all adversaries $\mathcal{A}$ and all $L \geq l$*

$$\Pr \left[\begin{array}{l} ck \leftarrow \mathrm{CGen}(1^\lambda) \\ (m_1, m_2, n, \mathsf{pc}, x_1, \mathsf{pe}_1, \ldots, x_L, \mathsf{pe}_L) \leftarrow \mathcal{A}(ck) \\ \mathrm{Parse\ } \mathsf{pe}_i = (\bar{t}_i, \bar{\tau}_i) \\ (T, \boldsymbol{\tau}) \leftarrow \chi(ck, m_1, m_2, n, \mathsf{pc}, x_1, \mathsf{pe}_1, \ldots, x_l, \mathsf{pe}_l) \\ v_i \leftarrow \mathrm{PolyVerify}(ck, m_1, m_2, n, \mathsf{pc}, \mathsf{pe}_i, x_i) \end{array} : \begin{array}{c} \forall i: \ v_i = \boldsymbol{Z}(x_i)T\boldsymbol{X}(x_i) \\ \mathrm{or\ } \exists j \mathrm{\ s.t.} \\ \mathrm{Com}_{ck}(\bar{t}_j; \bar{\tau}_j) = \\ \mathrm{Com}_{ck}(\boldsymbol{Z}(x_j)T; \boldsymbol{Z}(x_j)\boldsymbol{\tau}), \\ \mathrm{where\ } \bar{t}_j \neq \boldsymbol{Z}(x_j)T \end{array} \right] \approx 1,$$

where $x_1, \ldots, x_l$ are distinct, $x_i \in Z_p^*$, $\mathrm{pe}_i \in \mathbb{Z}_p^n \times \mathbb{Z}_p$, $T \in \mathbb{Z}_p^{(m_1+m_2) \times n}$, and $\tau \in \mathbb{Z}_p^{m_1+m_2}$.

Perfect special honest verifier zero-knowledge means that given any value v and evaluation point x, it is possible to simulate pc and pe, distributed exactly as in a real execution of the protocol where v was the evaluation of $t(X)$ at x.

Definition 12 (Perfect Special Honest Verifier Zero Knowledge). (PolyCommit, PolyEval, PolyVerify) *has* perfect special honest verifier zero knowledge *(SHVZK) if there exists a probabilistic polynomial time simulator* S *such that for all interactive non-uniform polynomial time adversaries* $\mathcal{A}$

$$\Pr \begin{bmatrix} (ck, m_1, m_2, n, t(X), x) \leftarrow \mathcal{A}(1^\lambda) \\ (\mathrm{pc}, \mathrm{st}) \leftarrow \mathrm{PolyCommit}(ck, m_1, m_2, n, t(X)) : \mathcal{A}(\mathrm{pc}, \mathrm{pe}) = 1 \\ \mathrm{pe} \leftarrow \mathrm{PolyEval}(\mathrm{st}, x) \end{bmatrix}$$

$$= \Pr \begin{bmatrix} (ck, m_1, m_2, n, t(X), x) \leftarrow \mathcal{A}(1^\lambda) \\ (\mathrm{pc}, \mathrm{pe}) \leftarrow S(ck, m_1, m_2, n, x, t(x)) \end{bmatrix} : \mathcal{A}(\mathrm{pc}, \mathrm{pe}) = 1 \end{bmatrix}$$

where ck *is a key for a homomorphic commitment scheme,* $t(X)$ *is a Laurent polynomial of degrees* $d_1 = m_1 n, d_2 = m_2 n$ *and* $x \in \mathbb{Z}_p^*$.

Theorem 1. *The polynomial commitment protocol has perfect completeness, perfect special honest verifier zero-knowledge and* $(m_1 + m_2)n + 1$*-special soundness for extracting either a breach of the binding property of the commitment scheme or openings to the polynomial.*

We refer to the full version of the paper for the proof.

Efficiency. We now discuss the efficiency of the above protocol when instantiated with the Pedersen multicommitment scheme. The outputs pc, pe of the polynomial commitment protocol have sizes of $m_1 + m_2 + 1$ group elements and $n + 1$ field elements respectively. The computational cost of computing pc is dominated by computing commitments T_i' and T_i'', corresponding to $m_1 + m_2$ n-wide multi-exponentiations. Using multi-exponentiation techniques as in [28,31,32], the total cost is roughly $\frac{(m_1+m_2)n}{\log n}$ group exponentiations. The main cost for computing pe is dominated by the $n(m_1 + m_2)$ field multiplications required to compute ZT. The dominant cost in PolyVerify is to check the verification equation. This costs roughly $\frac{m_1+m_2+n}{\log(m_1+m_2+n)}$ group exponentiations.

4   Recursive Argument for Inner Product Evaluation

We will now give an inner product argument of knowledge of two vectors $\boldsymbol{a}, \boldsymbol{b} \in \mathbb{Z}_p^n$ such that $A = \boldsymbol{g}^{\boldsymbol{a}}$, $B = \boldsymbol{h}^{\boldsymbol{b}}$ and $\boldsymbol{a} \cdot \boldsymbol{b} = z$, given $z \in \mathbb{Z}_p$, $A, B \in \mathbb{G}$ and $\boldsymbol{g}, \boldsymbol{h} \in \mathbb{G}^n$. The argument will be used later as a subroutine where zero-knowledge is not required, so the prover could in principle just reveal the witness $\boldsymbol{a}, \boldsymbol{b}$

to the verifier. In the following we show how to use interaction to reduce the communication from linear to logarithmic in n, the length of the vectors.

The basic step in our inner product argument is a 2-move reduction to a smaller statement using techniques similar to [1]. It will suffice for the prover to reveal the witness for the smaller statement in order to convince the verifier about the validity of the original statement. In the full argument, prover and verifier recursively run the reduction to obtain increasingly smaller statements. The argument is then concluded with the prover revealing a witness for a very small statement. The outcome of this is a $O(\log n)$-move argument with an overall communication of $O(\log n)$ group and field elements. The inner product argument will be used in the next section to build a logarithmic size argument for circuit satisfiability.

Due to the obvious relationship with Pedersen commitments, we will think of multi-exponentiations $\boldsymbol{g}^{\boldsymbol{a}}$ and $\boldsymbol{h}^{\boldsymbol{b}}$ as commitments with randomness set equal to zero, and to $\boldsymbol{a}, \boldsymbol{b}$ as openings with respect to commitment keys $\boldsymbol{g}, \boldsymbol{h}$.

4.1   Main Idea

We now describe the basic step in our argument. Consider the common input for both prover and verifier to be of the form $(\mathbb{G}, p, \boldsymbol{g}, A, \boldsymbol{h}, B, z, m)$ where m divides n, the length of the vectors. For arbitrary n one can always reduce to the case where $m|n$ by appending at most $m - 1$ random group elements to $\boldsymbol{g}$ and $\boldsymbol{h}$.

We split the bases for the multi-exponentiations into m sets $\boldsymbol{g} = (\boldsymbol{g}_1, \ldots, \boldsymbol{g}_m)$ and $\boldsymbol{h} = (\boldsymbol{h}_1, \ldots, \boldsymbol{h}_m)$, where each set has size $\frac{n}{m}$. We want to prove knowledge of vectors $\boldsymbol{a} = (\boldsymbol{a}_1, \ldots, \boldsymbol{a}_m)$ and $\boldsymbol{b} = (\boldsymbol{b}_1, \ldots, \boldsymbol{b}_m)$ such that

$$A = \boldsymbol{g}^{\boldsymbol{a}} = \prod_{i=1}^{m} \boldsymbol{g}_i^{\boldsymbol{a}_i} \qquad B = \boldsymbol{h}^{\boldsymbol{b}} = \prod_{i=1}^{m} \boldsymbol{h}_i^{\boldsymbol{b}_i} \qquad \boldsymbol{a} \cdot \boldsymbol{b} = \sum_{i=1}^{m} \boldsymbol{a}_i \cdot \boldsymbol{b}_i = z$$

The key idea is for the prover to replace A with A', a commitment to a shorter vector $\boldsymbol{a}' = \sum_{i=1}^{m} \boldsymbol{a}_i x^i$, given a random challenge $x \leftarrow \mathbb{Z}_p^*$ provided by the verifier. In the argument, the prover first computes and sends

$$A_k = \prod_{i=\max(1,1-k)}^{\min(m,m-k)} \boldsymbol{g}_i^{\boldsymbol{a}_{i+k}} \quad \text{for } k = 1 - m, \ldots, m - 1$$

corresponding to the products over the diagonals of the following matrix

$$
\begin{array}{cccc}
 & a_1 & a_2 & \cdots & a_m \\
\end{array}
$$

$$
\begin{array}{c}
\boldsymbol{g}_1 \\
\vdots \\
\boldsymbol{g}_{m-1} \\
\boldsymbol{g}_m
\end{array}
\left(
\begin{array}{cccc}
\boldsymbol{g}_1^{\boldsymbol{a}_1} & \boldsymbol{g}_1^{\boldsymbol{a}_2} & \cdots & \boldsymbol{g}_1^{\boldsymbol{a}_m} \\
\ddots & \boldsymbol{g}_2^{\boldsymbol{a}_2} & \ddots & \vdots \\
\boldsymbol{g}_{m-1}^{\boldsymbol{a}_1} & \ddots & \ddots & \boldsymbol{g}_{m-1}^{\boldsymbol{a}_m} \\
\boldsymbol{g}_m^{\boldsymbol{a}_1} & \boldsymbol{g}_m^{\boldsymbol{a}_2} & \cdots & \boldsymbol{g}_m^{\boldsymbol{a}_m}
\end{array}
\right)
\begin{array}{c}
 \\
A_{m-1} \\
\vdots \\
A_{m-2}
\end{array}
$$

$$
\begin{array}{cccc}
A_{1-m} & A_{2-m} & \cdots & A_0 = A
\end{array}
$$

Notice that $A_0 = A$ is already known to the verifier since it is part of the statement. The verifier now sends a random challenge $x \leftarrow \mathbb{Z}_p^*$.

At this point, both the prover and the verifier can compute $g' := \prod_{i=1}^m g_i^{x^{-i}}$ and $A' := \prod_{k=1-m}^{m-1} A_k^{x^k}$. If the prover is honest then we have $A' = (g')^{a'}$, namely A' is a commitment to a' under the key g'. Furthermore, even if the prover is dishonest, we can show that if the prover can open A' with respect to the key g' for $2m - 1$ different challenges, then we can extract opening $(a_1, \ldots, a_m)$ corresponding to $A = \prod_{i=1}^m g_i^{a_i}$.

The same type of argument can be applied in parallel to B with the inverse challenge x^{-1} giving us a sum of the form $b' = \sum_{i=1}^m b_i x^{-i}$ and a new base $h' = \prod_{i=1}^m h_i^{x^i}$.

All that remains is to demonstrate that z is the constant term in the product $a' \cdot b' = \sum_{i=1}^m a_i x^i \cdot \sum_{j=1}^m b_j x^{-j}$. Similarly to A and B, the prover sends values

$$z_k = \sum_{i=\max(1,1-k)}^{\min(m,m-k)} a_i \cdot b_{i+k} \quad \text{for } k = 1 - m, \ldots, m - 1$$

where $z_0 = z = \sum_{i=1}^m a_i \cdot b_i$, and shows that $z' := a' \cdot b' = \sum_{k=1-m}^{m-1} z_k x^{-k}$.

To summarise, after the challenge x has been sent, both parties compute g', A', h', B', z' and then run an argument for the knowledge of a', b' of length $\frac{n}{m}$. Given $n = m_\mu m_{\mu-1} \cdots m_1$, we recursively apply this reduction over the factors of n to obtain, after $\mu - 1$ iterations, vectors of length m_1. The prover concludes the argument by revealing a short witness associated with the last statement.

4.2   Formal Description

We now give a formal description of the argument of knowledge introduced above.

Common input: $(\mathbb{G}, p, g, A, h, B, z, m_\mu = m, m_{\mu-1} = m', \ldots, m_1)$ such that $g, h \in \mathbb{G}^n$, $A, B \in \mathbb{G}$ and $n = \prod_{i=1}^\mu m_i$.
Prover's witness: $(a_1, \ldots, a_m, b_1, \ldots, b_m)$ satisfying

$$A = \prod_{i=1}^m g_i^{a_i} \qquad B = \prod_{i=1}^m h_i^{b_i} \qquad \sum_{i=1}^m a_i \cdot b_i = z$$

Argument if $\mu = 1$:
    P → V: Send $(a_1, \ldots, a_m, b_1, , \ldots, b_m)$.
    P ← V: Accept if and only if

$$A = \prod_{i=1}^m g_i^{a_i} \qquad B = \prod_{i=1}^m h_i^{b_i} \qquad \sum_{i=1}^m a_i b_i = z$$

Reduction if $\mu \neq 1$:
   P $\rightarrow$ V: Send $A_{1-m}, B_{1-m}, z_{1-m}, \ldots, A_{m-1}, B_{m-1}, z_{m-1}$ where

$$A_k = \prod_{i=\max(1,1-k)}^{\min(m,m-k)} g_i^{a_{i+k}} \qquad B_k = \prod_{i=\max(1,1-k)}^{\min(m,m-k)} h_i^{b_{i+k}} \qquad z_k = \sum_{i=\max(1,1-k)}^{\min(m,m-k)} a_i \cdot b_{i+k}$$

Observe $A_0 = A, B_0 = B, z_0 = z$ so they can be omitted from the message.
   P $\leftarrow$ V: $x \leftarrow \mathbb{Z}_p^*$.
Both prover and verifier compute a reduced statement of the form

$$(\mathbb{G}, p, \boldsymbol{g}', A', \boldsymbol{h}', B', z', m_{\mu-1}, \ldots, m_1)$$

where

$$\boldsymbol{g}' = (g_1', \ldots, g_{m'}') = \prod_{i=1}^m g_i^{x^{-i}} \qquad A' = \prod_{k=1-m}^{m-1} A_k^{x^k}$$

$$\boldsymbol{h}' = (h_1', \ldots, h_{m'}') = \prod_{i=1}^m h_i^{x^i} \qquad B' = \prod_{k=1-m}^{m-1} B_k^{x^{-k}} \qquad z' = \sum_{k=1-m}^{m-1} z_k x^{-k}$$

The prover computes a new witness as $(a_1', \ldots, a_{m'}') = \sum_{i=1}^m \boldsymbol{a}_i x^i$ and $(b_1', \ldots, b_{m'}') = \sum_{i=1}^m \boldsymbol{b}_i x^{-i}$.

Security Analysis.

Theorem 2. *The argument has perfect completeness and statistical witness extended emulation for either extracting a non-trivial discrete logarithm relation or a valid witness.*

Proof. Perfect completeness can be verified directly. To prove witness-extended emulation we start by giving an extractor that either extracts a witness for the original statement or a non-trivial discrete logarithm relation.

For $\mu = 1$ we have (perfect) witness-extended emulation since the prover reveals a witness and the verifier checks it.

Before discussing extraction in the recursive step, note that if we get a non-trivial discrete logarithm relation for $g_1', \ldots, g_{m'}'$ then we also get a non-trivial discrete logarithm relation for $g_1, \ldots, g_m$, since $x \neq 0$. A similar argument applies to $h_1', \ldots, h_{m'}'$ and $h_1, \ldots, h_m$.

Now, assume we get witness $\boldsymbol{a}', \boldsymbol{b}'$ such that

$$A' = \prod_{k=1-m}^{m-1} A_k^{x^k} = \left(\prod_{i=1}^m g_i^{x^{-i}} \right)^{\boldsymbol{a}'} \quad B' = \prod_{k=1-m}^{m-1} B_k^{x^{-k}} = \left(\prod_{i=1}^m h_i^{x^i} \right)^{\boldsymbol{b}'} \quad \boldsymbol{a}' \cdot \boldsymbol{b}' = \sum_{k=1-m}^{m-1} z_k x^{-k}$$

for $2m - 1$ different challenges $x \in \mathbb{Z}_p^*$. We will show that they yield either a witness for the original statement, or a non-trivial discrete logarithm relation for either $g_1, \ldots, g_m$ or $h_1, \ldots, h_m$.

Take $2m - 1$ different challenges $x \in \mathbb{Z}_p^*$. They form a shifted Vandermonde matrix with rows $(x^{1-m}, x^{2-m}, \ldots, x^{m-1})$. By taking appropriate linear combinations of the vectors we can obtain any unit vector $(0, \ldots, 0, 1, 0, \ldots, 0)$. Taking the same linear combinations of the $2m - 1$ equations

$$\prod_{k=1-m}^{m-1} A_k^{x^k} = \left(\prod_{i=1}^{m} g_i^{x^{-i}}\right)^{a'} \quad \text{we get vectors } a_{k,i} \text{ such that} \quad A_k = \prod_{i=1}^{m} g_i^{a_{k,i}}$$

For each of the $2m - 1$ challenges, we now have $\prod_{k=1-m}^{m-1} A_k^{x^k} = \left(\prod_{i=1}^{m} g_i^{x^{-i}}\right)^{a'}$, which means that *for all i* we have

$$x^{-i} a' = \sum_{k=1-m}^{m-1} a_{k,i} x^k$$

unless we encounter a non-trivial discrete logarithm relation for $g_1, \ldots, g_m$. This means that $a' = \sum_{k=1-m}^{m-1} a_{k,i} x^{k+i}$ for all i, and in particular $\sum_{k=1-m}^{m-1} a_{k,i} x^{k+i} = \sum_{k=1-m}^{m-1} a_{k,1} x^{k+1} = \sum_{k=1-m}^{m-1} a_{k,m} x^{k+m}$. Matching terms of degree outside $\{1, \ldots, m\}$ reveals $a_{k,i} = 0$ for $k + i \notin \{1, \ldots, m\}$. Defining $a_i = a_{0,i}$, and matching terms of similar degree we get

$$a_{k,i} = \begin{cases} a_{k+i} & \text{if } k+i \in \{1, \ldots, m\} \\ 0 & \text{otherwise} \end{cases}$$

This means

$$a' = \sum_{k=1-m}^{m-1} a_{k,1} x^{k+1} = \sum_{k=0}^{m-1} a_{k+1} x^{k+1} = \sum_{i=1}^{m} a_i x^i$$

A similar analysis of $B_{1-m}, \ldots, B_{m-1}$ and openings b' for $2m - 1$ different challenges $x^{-1} \in \mathbb{Z}_p^*$ gives us either a non-trivial discrete logarithm relation for $h_1, \ldots, h_m$ or vectors b_i such that $b' = \sum_{i=1}^{m} b_i x^{-i}$ and $B = \prod_{i=1}^{m} h_i^{b_i}$.

Finally, with $\sum_{i=1}^{m} a_i x^i \cdot \sum_{j=1}^{m} b_j x^{-j} = \sum_{k=1-m}^{m-1} z_k x^{-k}$ for $2m - 1$ different challenges we get $z = z_0 = \sum_{i=1}^{m} a_i \cdot b_i$.

We can now apply the forking lemma to a tree of size $(2m_\mu - 1)(2m_{\mu-1} - 1) \cdots (2m_2 - 1) \leq n^2$, which is polynomial in λ, to conclude that the argument has witness-extended emulation. $\qquad\square$

Efficiency. The recursive argument uses $2\mu - 1$ moves. The communication cost of all steps sums up to $4 \sum_{i=2}^{\mu}(m_i - 1)$ group elements and $2 \sum_{i=2}^{\mu}(m_i - 1) + 2m_1$ field elements.

At each iteration, the main cost for the prover is computing the A_k and B_k values, using less than $\frac{4(m_\mu^2 m_{\mu-1} \ldots m_1)}{\log(m_\mu \ldots m_1)}$ group exponentiations via multi-exponentiation techniques, and the z_k values using $m_\mu^2 m_{\mu-1} \cdots m_1$ field multiplications. The cost of computing the reduced statements is dominated by

$\frac{2(m_\mu m_{\mu-1}...m_1)}{\log m_\mu}$ group exponentiations for both the prover and the verifier. In the case where $m_\mu = \ldots = m_1 = m$, the verifier complexity is bounded above by $\frac{2m^\mu}{\log m}\frac{m}{m-1}$ group exponentiations. The prover complexity is bounded above by $\frac{6m^{\mu+1}}{\log m}\frac{m}{m-1}$ group exponentiations and $m^{\mu+1}\frac{m}{m-1}$ field multiplications.

Zero-Knowledge Version. The above argument can be modified to become zero-knowledge. We leave the details to the reader as zero-knowledge is not needed for our use of this argument in the next section.

5 Logarithmic Communication Argument for Arithmetic Circuit Satisfiability

In this section, we revisit zero knowledge arguments for the satisfiability of an arithmetic circuit under the discrete logarithm assumption. We will explain how to build an argument with square root communication complexity, and superior efficiency to the argument of [21]. We then observe that our new argument involves computing a large inner product, and can achieve as good as logarithmic communication complexity by using our recursive inner product argument.

At a high level, we transform an arithmetic circuit into two kinds of equations. Multiplication gates are directly represented as equations of the form $a \cdot b = c$, where a, b, c represent the left, right and output wires. We will arrange these values in matrix form producing a Hadamard matrix product. This process will lead to duplicate values, when a wire is the output of one multiplication gate and the input of another, or when it is used as input multiple times. We keep track of this by using a series of linear constraints. For example, if the output of the first multiplication gate is the right input of the second, we would write $c_1 - b_2 = 0$.

We also add linear constraints representing the addition and multiplication by constant gates of the circuit. We then rewrite those equations so that the only wires that are referenced in the equations are those linked to (non-constant) multiplication gates. We describe this process in Appendix A.

Finally, we fold both the Hadamard matrix product and the linear constraints into a single polynomial equation, where a Laurent polynomial has 0 as its constant term, and use the construction of Sect. 3 to prove this. We can optionally integrate the inner product argument of Sect. 4 to reduce communication.

Our technique improves on the efficiency of [21] by making three main changes, each resulting in efficiency improvements.

1. We do not need commitments to the input and output wires of addition gates. We handle addition gates with linear consistency equations thus yielding a significant performance improvement proportional to the number of addition gates. This parallels [13] who also manage to eliminate addition gates when constructing Quadratic Arithmetic Programs from circuits.

2. We avoid black-box reductions to zero-knowledge arguments for generic linear algebra statements and instead design an argument directly for arithmetic circuit satisfiability. As a result, our square-root argument has only 5 moves, while the argument from [21] requires 7 moves. We note that [36] reduced the complexity of [21] to 5 moves as well, but at a significant computational overhead whereas we also reduce the computational cost.
3. We use our protocol from Sect. 3 to reduce the communication costs of a polynomial commitment.

These improvements give us a square root communication complexity with respect to the number of multiplication gates in the circuit. This is because for a circuit with $N = mn$ multiplication gates, the prover makes $3m$ commitments to wire values in his first move, and later provides an opening consisting of n field elements to a homomorphic combination of these commitments. Optimising the parameters by choosing $m \approx n \approx \sqrt{N}$ leads to square root complexity.

In our square root complexity argument, the verifier uses the n field elements to check an inner product relation. Our key idea to reduce communication further is to use our inner product evaluation argument instead of sending these field elements. This allows for verification of the inner product, and also provides an argument of knowledge of the opening of the commitment. We no longer need to open a large commitment, leading to a drastic reduction in communication complexity depending on the settings for the inner product argument.

Below we give a first informal exposition of our arguments, and follow with a formal description.

Reduction of Circuit Satisfiability to a Hadamard Matrix Product and Linear Constraints. We consider an arithmetic circuit containing $N = mn$ multiplication gates over a field $\mathbb{Z}_p$. Without loss of generality, we assume that the circuit has been pre-processed (see the full version of the paper for a way to do this), so that the input and the output wires feed into and go out from multiplication gates only. We number the multiplication gates from 1 to N and we arrange the inputs and outputs of these gates into three $m \times n$ matrices A, B and C such that the (i, j) entries of the matrices correspond to the left input, right input and output of the same multiplication gate.

An arithmetic circuit can be described as a system of equations in the entries of the above matrices. The multiplication gates define a set of N equations

$$A \circ B = C \tag{1}$$

where $\circ$ is the Hadamard (entry-wise) product. The circuit description also contains constraints on the wires between multiplication gates. Denoting the rows of the matrices A, B, C as

$$\boldsymbol{a}_i = (a_{i,1}, \ldots, a_{i,n}) \quad \boldsymbol{b}_i = (b_{i,1}, \ldots, b_{i,n}) \quad \boldsymbol{c}_i = (c_{i,1}, \ldots, c_{i,n}) \quad \text{for } i \in \{1, \ldots, m\}$$

these constraints can be expressed as $Q < 2N$ linear equations of inputs and outputs of multiplication gates of the form

$$\sum_{i=1}^{m} \boldsymbol{a}_i \cdot \boldsymbol{w}_{q,a,i} + \sum_{i=1}^{m} \boldsymbol{b}_i \cdot \boldsymbol{w}_{q,b,i} + \sum_{i=1}^{m} \boldsymbol{c}_i \cdot \boldsymbol{w}_{q,c,i} = K_q \quad \text{for } q \in \{1, \ldots, Q\} \quad (2)$$

for constant vectors $\boldsymbol{w}_{q,a,i}, \boldsymbol{w}_{q,b,i}, \boldsymbol{w}_{q,c,i}$ and scalars K_q.

For example, suppose that the circuit contains a single addition gate, with $a_{1,1}$ and $a_{1,2}$ as inputs, and $b_{1,1}$ as output. In this case, $Q = 1$ and we would set $\boldsymbol{w}_{1,a,1} = (1, 1, 0, \ldots, 0), \boldsymbol{w}_{1,b,1} = (-1, 0, \ldots, 0)$, and all other $\boldsymbol{w}$ vectors would be set to $\boldsymbol{0}$. Then (2) would simply read

$$a_{1,1} + a_{1,2} - b_{1,1} = 0$$

to capture the constraint imposed by the addition gate.

In total, to capture all multiplications and linear constraints, we have $N + Q$ equations that the wires must satisfy in order for the circuit to be satisfiable.

Reduction to a Single Polynomial Equation. Let Y be a formal indeterminate. We will reduce the $N + Q$ equations above to a single polynomial equation in Y by embedding each equation into a distinct power of Y. In our argument we will then require the prover to prove that this single equation holds when replacing Y by a random challenge received from the verifier.

Let $\boldsymbol{Y}'$ denote the vector $(Y^m, \ldots, Y^{mn})$ and $\boldsymbol{Y}$ denote $(Y, Y^2, \ldots, Y^m)$. Then, we can multiply (1) by $\boldsymbol{Y}$ from the left and $\boldsymbol{Y}'^T$ on the right to obtain $\boldsymbol{Y}(A \circ B)\boldsymbol{Y}'^T = \boldsymbol{Y}CY'^T$, or equivalently

$$\sum_{i=1}^{m} Y^i(\boldsymbol{a}_i \circ \boldsymbol{b}_i) \cdot \boldsymbol{Y}' = \sum_{i=1}^{m} Y^i(\boldsymbol{c}_i \cdot \boldsymbol{Y}')$$

Since $(\boldsymbol{a} \circ \boldsymbol{b}) \cdot \boldsymbol{Y}' = \boldsymbol{a} \cdot (\boldsymbol{b} \circ \boldsymbol{Y}')$, we obtain the following expression

$$\sum_{i=1}^{m} \boldsymbol{a}_i \cdot (\boldsymbol{b}_i \circ \boldsymbol{Y}')Y^i = \left(\sum_{i=1}^{m} \boldsymbol{c}_i Y^i \cdot \boldsymbol{Y}'\right)$$

This is easily seen to be equivalent to (1), because $a_{i,j}b_{i,j} = c_{i,j}$ appears in the coefficients of Y^{i+jm}, and $i + jm$ takes every value from $m + 1$ to $M = N + m$ exactly once.

Moreover, the Q linear constraints on the wires in Eq. 2 are satisfied if and only if

$$\sum_{q=1}^{Q} \left(\sum_{i=1}^{m} \boldsymbol{a}_i \cdot \boldsymbol{w}_{q,a,i} + \sum_{i=1}^{m} \boldsymbol{b}_i \cdot \boldsymbol{w}_{q,b,i} + \sum_{i=1}^{m} \boldsymbol{c}_i \cdot \boldsymbol{w}_{q,c,i}\right) Y^q = \sum_{q=1}^{Q} K_q Y^q$$

since the qth constraint arises from comparing the coefficients of Y^q. Combining the two polynomial equations by adding them after multiplying the latter by Y^M, and swapping summations, we see that the circuit is satisfied if and only if

$$\left(\sum_{i=1}^{m} \boldsymbol{a}_i \cdot (\boldsymbol{b}_i \circ \boldsymbol{Y}') Y^i\right) + \sum_{i=1}^{m} \boldsymbol{a}_i \cdot \left(\sum_{q=1}^{Q} \boldsymbol{w}_{q,a,i} Y^{M+q}\right) + \sum_{i=1}^{m} \boldsymbol{b}_i \cdot \left(\sum_{q=1}^{Q} \boldsymbol{w}_{q,b,i} Y^{M+q}\right)$$

$$+ \sum_{i=1}^{m} \boldsymbol{c}_i \cdot \left(-Y^i \boldsymbol{Y}' + \sum_{q=1}^{Q} \boldsymbol{w}_{q,c,i} Y^{M+q}\right) \quad = \quad \left(\sum_{q=1}^{Q} K_q Y^{M+q}\right)$$

Let us define

$$\boldsymbol{w}_{a,i}(Y) = \sum_{q=1}^{Q} \boldsymbol{w}_{q,a,i} Y^{M+q} \qquad\qquad \boldsymbol{w}_{b,i}(Y) = \sum_{q=1}^{Q} \boldsymbol{w}_{q,b,i} Y^{M+q}$$

$$\boldsymbol{w}_{c,i}(Y) = -Y^i \boldsymbol{Y}' + \sum_{q=1}^{Q} \boldsymbol{w}_{q,c,i} Y^{M+q} \qquad\qquad K(Y) = \sum_{q=1}^{Q} K_q Y^{M+q}$$

Then the circuit is satisfied if and only if

$$\sum_{i=1}^{m} \boldsymbol{a}_i \cdot (\boldsymbol{b}_i \circ \boldsymbol{Y}') Y^i + \sum_{i=1}^{m} \boldsymbol{a}_i \cdot \boldsymbol{w}_{a,i}(Y) + \sum_{i=1}^{m} \boldsymbol{b}_i \cdot \boldsymbol{w}_{b,i}(Y) + \sum_{i=1}^{m} \boldsymbol{c}_i \cdot \boldsymbol{w}_{c,i}(Y) - K(Y) = 0$$

$$(3)$$

In the argument, the prover will commit to $\boldsymbol{a}_i, \boldsymbol{b}_i$ and $\boldsymbol{c}_i$. The verifier will then issue a random challenge $y \leftarrow \mathbb{Z}_p^*$ and the prover will convince the verifier that the committed values satisfy Eq. 3, evaluated on y. If the committed values do not satisfy the polynomial equation, the probability the equality holds for a random y is negligible, so the prover is unlikely to be able to convince the verifier.

5.1   Square Root Communication Argument

In order to show that (3) is satisfied, we craft a special Laurent polynomial $t(X)$ in a second formal indeterminate X, whose constant coefficient is exactly twice the left-hand side of (3). Therefore, this polynomial will have zero constant term if and only if (3) is satisfied. In our argument this is proved using the polynomial commitment protocol of Sect. 3. We define

$$r(X) := \sum_{i=1}^{m} \boldsymbol{a}_i y^i X^i + \sum_{i=1}^{m} \boldsymbol{b}_i X^{-i} + X^m \sum_{i=1}^{m} \boldsymbol{c}_i X^i + \boldsymbol{d} X^{2m+1}$$

$$s(X) := \sum_{i=1}^{m} \boldsymbol{w}_{a,i}(y) y^{-i} X^{-i} + \sum_{i=1}^{m} \boldsymbol{w}_{b,i}(y) X^i + X^{-m} \sum_{i=1}^{m} \boldsymbol{w}_{c,i}(y) X^{-i}$$

$$r'(X) := r(X) \circ \boldsymbol{y}' + 2s(X)$$

$$t(X) := r(X) \cdot r'(X) - 2K(y)$$

Here $\boldsymbol{y}'$ is the vector $\boldsymbol{Y}'$ evaluated at y, and $\boldsymbol{d}$ is a blinding vector consisting of random scalars that the prover commits to in the first round. In the square root argument the prover will reveal $r(x)$ for a randomly chosen challenge $x \in \mathbb{Z}_p^*$, and the blinding vector d ensures that we can reveal $r(x)$ without leaking information about a_i, b_i and c_i. We also observe that $s(x)$ is efficiently computable from public information about the circuit and the challenges.

We have designed these polynomials such that the constant term of $r \cdot (r \circ y')$ is equal to $2 \sum_{i=1}^{m} a_i \cdot (b_i \circ y')y^i$ and the constant term of $r \cdot s$ is equal to $\sum_{i=1}^{m} a_i \cdot w_{a,i}(y) + \sum_{i=1}^{m} b_i \cdot w_{b,i}(y) + \sum_{i=1}^{m} c_i \cdot w_{c,i}(y)$. We conclude that the constant term of $t(X)$ is exactly twice the left-hand side of (3), and is therefore zero if and only if the circuit is satisfied.

We are now in a position to describe an argument with square root communication complexity.

The prover first commits to vectors a_i, b_i, c_i and d and the verifier replies with a challenge $y \leftarrow \mathbb{Z}_p^*$. The prover computes $t(X)$ and commits to it by using the algorithm PolyCommit defined in Sect. 3. Then, the verifier sends a random challenge $x \leftarrow \mathbb{Z}_p^*$ and the prover responds by revealing $r(x)$ and blinded openings pe of $t(X)$ obtained by running algorithm PolyEval as described in Sect. 3.

The verifier first checks that $r(x)$ is consistent with the previously sent commitments of a_i, b_i, c_i and d using the homomorphic properties of the commitment scheme. She also computes $s(x), r'(x)$ and K. Then, she computes $v = t(x)$ using the PolyVerify algorithm of Sect. 3, and checks if $v = r(x) \cdot r'(x) - 2K$. The verifier accepts the argument if both checks are satisfied.

As described so far, the argument requires communicating $O(m)$ group elements and $O(n)$ field elements, so setting $m \approx n$ leads to square root communication. The argument improves on [21,36] by requiring only 5 moves without computational overhead and significantly reduces the computational complexity. However, breaking this ostensible square root communication barrier requires new ideas that we describe in the next section.

5.2 Breaking the Square Root Barrier

The square root complexity argument described above was designed so that the verifier uses $r = r(x)$ to check the inner product $v = r \cdot r' - 2K$, where v is the evaluation of a committed polynomial at x. Sending r has a cost of n field elements. In order to break the square root barrier we try to avoid sending r directly so that we can then let n be larger and m be smaller and thus globally lower the communication of the argument.

Rather than sending r to the verifier, the prover could instead send commitments to r and r', and use our inner product argument to show that $v + 2K$ was a correctly formed inner product. In fact, the prover does not even need to send commitments to r and r'! The verifier can compute a commitment to $r(x)$ directly from A_i, B_i, C_i and D, the commitments to a_i, b_i, c_i and d which were previously used to check that r is correctly formed

$$\mathrm{Com}_{ck}(r;0) = \mathrm{Com}_{ck}(0;-\rho) \left[\prod_{i=1}^{m} A_i^{x^i y^i}\right] \left[\prod_{i=1}^{m} B_i^{x^{-i}}\right] \left[\prod_{i=1}^{m} C_i^{x^{m+i}}\right] D^{x^{2m+1}} = g^r$$

where ρ is an appropriate randomness value, which is sent by the prover to the verifier, and the vector $\boldsymbol{g} = (g_1, \ldots, g_n)$ for a given commitment key $ck = (\mathbb{G}, p, g, g_1, \ldots, g_n)$.

As for a commitment to $\boldsymbol{r}'$, we observe that the Pedersen commitment, besides its well-known homomorphic properties with respect to the message and the randomness, also has the useful property that it is homomorphic *with respect to the commitment key*. Specifically, let $\boldsymbol{h} = (g_1^{y^{-m}}, \ldots, g_n^{y^{-mn}})$, so that $\boldsymbol{g}^{\boldsymbol{r}} = \boldsymbol{h}^{\boldsymbol{r} \circ \boldsymbol{y}'}$. Multiplying $\boldsymbol{g}^{\boldsymbol{r}}$ by $\boldsymbol{h}^{2s}$, the verifier obtains $\mathrm{Com}_{ck'}(\boldsymbol{r}'; 0) = \boldsymbol{h}^{\boldsymbol{r}'}$, with respect to the new commitment key ck' which uses $\boldsymbol{h}$ instead of $\boldsymbol{g}$. We note that $\boldsymbol{h}$ and $\boldsymbol{s} = \boldsymbol{s}(x)$ can be computed by the verifier.

Now the prover and verifier can run the inner product argument with statement

$$(\mathbb{G}, p, \boldsymbol{g}, r, \boldsymbol{h}, r', v + 2K, m_\mu, m_{\mu-1}, \ldots, m_1) \qquad \text{where}$$

$$ck = (\mathbb{G}, p, g, \boldsymbol{g}) \qquad\qquad n = m_\mu m_{\mu-1} \cdots m_1$$

$$\boldsymbol{g} = (g_1, g_2, \ldots, g_n) \qquad\qquad \boldsymbol{h} = (g_1^{y^{-m}}, g_2^{y^{-2m}}, \ldots, g_n^{y^{-mn}})$$

$$R = \mathrm{Com}_{ck}(0; -\rho) \left[\prod_{i=1}^m A_i^{x^i y^i}\right]\left[\prod_{i=1}^m B_i^{x^{-i}}\right]\left[\prod_{i=1}^m C_i^{x^{m+i}}\right] D^{x^{2m+1}} = \boldsymbol{g}^{\boldsymbol{r}}$$

$$R' = R \cdot \boldsymbol{h}^{2s} = \boldsymbol{h}^{\boldsymbol{r}'}$$

and the prover's witness is $\boldsymbol{r}, \boldsymbol{r}'$.

The values of $m_\mu, \ldots, m_1$ can be chosen according to the desired efficiency of the circuit satisfiability argument.

5.3 Formal Description

We now give the formal description of the above arguments of knowledge for the satisfiability of an arithmetic circuit C. Both prover and verifier take the move parameter μ as common input. For square root communication complexity, the inner product argument is not used and we set $\mu = 0$. For $\mu > 0$, the common input includes the values $(m_\mu, \ldots, m_1)$ used in the inner product argument. The description of the arithmetic circuit C is given as a number N of multiplication gates and the values $\boldsymbol{w}_{q,a,i}, \boldsymbol{w}_{q,b,i}, \boldsymbol{w}_{q,c,i}$, which specify linear consistency constraints between the input and output values of the multiplication gates.

Common Input: $(ck, C, N, m, n, m_1', m_2', n', m_\mu, \ldots, m_1, \mu)$ where ck is a commitment key, C is the description of an arithmetic circuit with $N = mn$ multiplication gates, μ is the move parameter and $n = m_\mu \cdots m_1$. Parameters (m_1', m_2', n') are set to satisfy both $3m \le m_1' n'$ and $4m + 2 \le m_2' n'$.

Prover's Witness: Satisfying assignments $\boldsymbol{a}_i, \boldsymbol{b}_i$ and $\boldsymbol{c}_i$ to the wires of C.

Argument:

   P $\rightarrow$ V: Pick randomness $\alpha_1, \beta_1, \gamma_1, \ldots, \alpha_m, \beta_m, \gamma_m, \delta \leftarrow \mathbb{Z}_p$ and blinding vector $\boldsymbol{d} \leftarrow \mathbb{Z}_p^n$. Compute for $i \in \{1, \ldots, m\}$

$$A_i = \mathrm{Com}(\boldsymbol{a}_i; \alpha_i) \quad B_i = \mathrm{Com}(\boldsymbol{b}_i; \beta_i) \quad C_i = \mathrm{Com}(\boldsymbol{c}_i; \gamma_i) \quad D = \mathrm{Com}(\boldsymbol{d}; \delta).$$

      Send to the verifier $A_1, B_1, C_1, \ldots, A_m, B_m, C_m, D$.

   P $\leftarrow$ V: $y \leftarrow \mathbb{Z}_p^*$.

As argued before, the circuit determines vectors of polynomials $\boldsymbol{w}_{a,i}(Y)$, $\boldsymbol{w}_{b,i}(Y)$, $\boldsymbol{w}_{c,i}(Y)$ and $K(Y)$ such that C is satisfiable if and only if

$$\sum_{i=1}^{m} \boldsymbol{a}_i \cdot (\boldsymbol{b}_i^T \circ Y')Y^i + \sum_{i=1}^{m} \boldsymbol{a}_i \cdot \boldsymbol{w}_{a,i}(Y) + \sum_{i=1}^{m} \boldsymbol{b}_i \cdot \boldsymbol{w}_{b,i}(Y) + \sum_{i=1}^{m} \boldsymbol{c}_i \cdot \boldsymbol{w}_{c,i}(Y) = K(Y)$$

where $Y' = (Y^m, \ldots, Y^{mn})$. Given y, both the prover and verifier can compute $K = K(y)$, $\boldsymbol{w}_{a,i} = \boldsymbol{w}_{a,i}(y)$, $\boldsymbol{w}_{b,i} = \boldsymbol{w}_{b,i}(y)$ and $\boldsymbol{w}_{c,i} = \boldsymbol{w}_{c,i}(y)$.

$\mathbf{P} \rightarrow \mathbf{V}$: Compute Laurent polynomials $\boldsymbol{r}, \boldsymbol{s}, \boldsymbol{r}'$, which have vector coefficients, and Laurent polynomial t, in the indeterminate X

$$\boldsymbol{r}(X) = \sum_{i=1}^{m} \boldsymbol{a}_i y^i X^i + \sum_{i=1}^{m} \boldsymbol{b}_i X^{-i} + X^m \sum_{i=1}^{m} \boldsymbol{c}_i X^i + \boldsymbol{d} X^{2m+1}$$

$$\boldsymbol{s}(X) = \sum_{i=1}^{m} \boldsymbol{w}_{a,i} y^{-i} X^{-i} + \sum_{i=1}^{m} \boldsymbol{w}_{b,i} X^i + X^{-m} \sum_{i=1}^{m} \boldsymbol{w}_{c,i} X^{-i}$$

$$\boldsymbol{r}'(X) = \boldsymbol{r}(X) \circ \boldsymbol{y}' + 2\boldsymbol{s}(X)$$

$$t(X) = \boldsymbol{r}(X) \cdot \boldsymbol{r}'(X) - 2K = \sum_{k=-3m}^{4m+2} t_k X^k$$

When the wires $\boldsymbol{a}_i, \boldsymbol{b}_i, \boldsymbol{c}_i$ correspond to a satisfying assignment, the Laurent polynomial $t(X)$ will have constant term $t_0 = 0$.

Commit to $t(X)$ by running

$$(\mathsf{pc}, \mathsf{st}) \leftarrow \mathsf{PolyCommit}(ck, m_1', m_2', n', t(X))$$

Send pc to the verifier.

$\mathbf{P} \leftarrow \mathbf{V}$: $x \leftarrow \mathbb{Z}_p^*$

$\mathbf{P} \rightarrow \mathbf{V}$: Compute $\mathsf{PolyEval}(\mathsf{st}, x) \rightarrow \mathsf{pe}$, and

$$\boldsymbol{r} = \sum_{i=1}^{m} \boldsymbol{a}_i x^i y^i + \sum_{i=1}^{m} \boldsymbol{b}_i x^{-i} + x^m \sum_{i=1}^{m} \boldsymbol{c}_i x^i + \boldsymbol{d} x^{2m+1}$$

$$\rho = \sum_{i=1}^{m} \alpha_i x^i y^i + \sum_{i=1}^{m} \beta_i x^{-i} + x^m \sum_{i=1}^{m} \gamma_i x^i + \delta x^{2m+1}$$

- If $\mu = 0$: the inner product argument is not used. The prover sends $(\mathsf{pe}, \boldsymbol{r}, \rho)$ to the verifier.
- If $\mu > 0$: the inner product argument is used. The prover computes $\boldsymbol{r}' = \boldsymbol{r}'(x)$ and sends (pe, ρ) to the verifier.

Verification: First, the verifier computes

$$\mathsf{PolyVerify}(ck, m_1', m_2', n', \mathsf{pc}, \mathsf{pe}, x) \rightarrow v$$

and rejects the argument if $v = \bot$.

- If $\mu = 0$: the inner product argument is not used. The verifier computes $r' = r \circ y' + 2s(x)$, and accepts only if

$$r \cdot r' - 2K = v$$
$$\text{Com}_{ck}(r; \rho) = \left[\prod_{i=1}^{m} A_i^{x^i y^i} \right] \left[\prod_{i=1}^{m} B_i^{x^{-i}} \right] \left[\prod_{i=1}^{m} C_i^{x^{m+i}} \right] D^{x^{2m+1}}$$

- If $\mu > 0$: prover and verifier run the inner product argument with common input

$$(\mathbb{G}, p, \boldsymbol{g}, R, \boldsymbol{h}, R', v + 2K, m_\mu, m_{\mu-1}, \ldots, m_1) \qquad \text{where}$$

$ck = (\mathbb{G}, p, g, \boldsymbol{g})$

$\boldsymbol{g} = (g_1, g_2, \ldots, g_n)$

$R = \text{Com}_{ck}(0; -\rho) \left[\prod_{i=1}^{m} A_i^{x^i y^i} \right] \left[\prod_{i=1}^{m} B_i^{x^{-i}} \right] \left[\prod_{i=1}^{m} C_i^{x^{m+i}} \right] D^{x^{2m+1}} = \boldsymbol{g}^r$

$n = m_\mu m_{\mu-1} \cdots m_1$

$\boldsymbol{h} = (g_1^{y^{-m}}, g_2^{y^{-2m}}, \ldots, g_n^{y^{-mn}})$

$R' = R \cdot \boldsymbol{h}^{2s(x)} = \boldsymbol{h}^{r'}$

and the prover's witness is r and r'.

The verifier accepts if the inner product argument is accepting.

Security Analysis. In the full version of the paper, we prove the following.

Theorem 3. *The argument for satisfiability of an arithmetic circuit has perfect completeness, perfect special honest verifier zero-knowledge and statistical witness-extended emulation for extracting either a breach of the binding property of the commitment scheme or a witness for the satisfiability of the circuit.*

Efficiency.

Square Root Communication. When we set $\mu = 0$, the argument above has a communication cost of $m_1' + m_2' + 2 + 1 + 3m$ commitments and $n + n' + 2$ field elements. Setting $m \approx \sqrt{\frac{N}{3}}$, $n \approx \sqrt{3N}$, $n' \approx \sqrt{7m}$, $m_1' \approx 3\sqrt{\frac{m}{7}}$ and $m_2' \approx 4\sqrt{\frac{m}{7}}$ we get a total communication complexity where the total number of group and field elements sent is as low as possible and approximately $2\sqrt{N}$ each. The main computational cost for the prover is computing the initial commitments, corresponding to $\frac{3mn}{\log n}$ group exponentiations. The prover can compute $t(X)$ using FFT-based techniques. Assuming that p is of a suitable form [9], the dominant number of multiplications for this process is $\frac{3}{2}mn \log m$. The main cost in the verification is computing $s(X)$ given the description of the circuit which requires in the worst case Qn multiplications in $\mathbb{Z}_p$, considering arbitrary fan-in addition gates. In case of $O(N)$-size circuits with fan-in 2 gates, computing $s(X)$ requires $O(N)$ multiplications. Evaluating $s(x)$ requires $3N$ multiplications. The last verification equation costs roughly $\frac{(n+3m)}{\log n + 3m}$ group exponentiations to the verifier.

$(\mu + 1)$-*Root Communication.* We can reduce communication by using $\mu = O(1)$ iterations of the inner product argument. Choosing $m = N^{\frac{1}{\mu+1}}$, $n = N^{\frac{\mu}{\mu+1}}$ and $m_i = (\frac{N}{m})^{\frac{1}{\mu}}$ will give us a communication complexity of $4\mu N^{\frac{1}{\mu+1}}$ group elements and $2\mu N^{\frac{1}{\mu+1}}$ field elements. The prover's complexity is dominated by $\frac{6\mu N}{\log N}$ group exponentiations and fewer than $\frac{3N}{2\mu} \log N$ field multiplications. The verifier's cost is dominated by $\frac{2\mu N}{\log N}$ group exponentiations and $O(N)$ field multiplications.

Logarithmic Communication. By increasing the number of iteration of the inner product argument we can further reduce the communication complexity.

To minimize the communication, we set $\mu = \log N - 1$, $n = \frac{N}{2}$, $m = m_i = 2$, $m'_1 = 2$, $m'_2 = 3$ and $n' = 4$ in the above argument gives us $2 \log N + 1$ moves. The total communication amounts to $4 \log N + 7$ group elements and $2 \log N + 6$ field elements. The prover computational cost is dominated by $12N$ group exponentiations, and $O(N)$ multiplications in $\mathbb{Z}_p$. The main verification cost is bounded by $4N$ group exponentiations and $O(N)$ multiplications in $\mathbb{Z}_p$.

Alternatively, we can optimize the computation while maintaining logarithmic communication by setting $\mu = \log N - \log \log 2N$, $m = \log N$, $n = \frac{N}{\log N}$, $n' \approx \sqrt{7 \log N}$, $m'_1 \approx 3\sqrt{\frac{\log N}{7}}$, $m'_2 \approx 4\sqrt{\frac{\log N}{7}}$, $m_i = 2$ for $1 \leq i \leq \mu$. In this way we obtain a $2 \log N - 2 \log \log N + 1$ moves argument. With respect to the previous settings, we now save $2 \log \log N$ moves by starting the inner product argument with a smaller statement. The resulting communication is at most $7 \log N + \sqrt{7 \log N}$ group elements and at most $2 \log N + \sqrt{7 \log N}$ field elements. Thus, the prover computation is dominated by $\frac{3N}{\log N}$ group exponentiations and $11N \log \log N$ field multiplications. For the verifier, it is bounded from above by $\frac{4N}{\log N \log \log N}$ group exponentiations and $O(N)$ field multiplications.

6   Implementation Using Python

To verify the practicality of our construction we produced a proof of concept implementation in Python using the NumPy [33] package. The more costly operations are executed natively: we use Petlib [11] to outsource elliptic curve operations to the OpenSSL library, and also use a small C++ program to calculate the polynomial multiplication producing $t(X)$ using NTL [37]. Our implementation is single-threaded, but the operations performed are easily parallelisable.

Our implementation accepts the circuit description format used by Pinocchio [34], which it preprocesses to remove addition and multiplication by constant gates, encoding them as a constraint table. Pinocchio also supports split gates, taking as input a single arithmetic wire and producing a fixed number of binary wires as outputs, so the binary wires correspond to the binary representation of the arithmetic wire. We handle split gates by adding appropriate multiplication gates and constraints to ensure binary wires can only carry zeroes or ones, and that their values scaled by the appropriate powers of 2 sum up to the gate's input.

Performance Comparison. We compared the performance of our implementation to that of Pinocchio [34] for a set of circuits produced by Pinocchio's toolchain. The circuits implement multiplication of a vector by a fixed matrix, multiplication of two matrices, evaluation of a multivariate polynomial, and other applications for which we refer to [34]. We used an i5-4690K running Pinocchio under Windows 10 and our software under Ubuntu 14.04 for the tests.

We note here that Pinocchio operates in a pairing-based setting, using knowledge of exponent assumptions, whereas we operate in the discrete log setting. Even so, we feel the comparison is meaningful, as we are not aware of previous implementations of circuit-evaluation arguments in our setting.

Table 2. Performance comparison between our implementation and Pinocchio. Pinocchio was set to use public verifiability and zero-knowledge.

Application	Mult. gates	This work										Pinocchio (Constant)				
		Square root					Logarithmic									
		Key		Proof			Key		Proof			Key		Proof		
		Gen	Size	Prove	Verify	Size	Gen	Size	Prove	Verify	Size	Gen	Size	Prove	Verify	Size
		s	B	s	s	B	s	B	s	s	B	s	B	s	s	B
Vector matrix	600	0.07	1120	0.38	0.25	6K	0.03	3872	0.55	0.31	3552	0.42	0.3M	0.23	.023	288
Product	1000	0.10	1440	0.76	0.61	8K	0.06	6464	1.05	0.67	3744	0.93	0.5M	0.53	.035	288
Matrix	347K	1.1	19K	14.7	3.4	76K	5.3	618K	49.9	22.9	5792	47.3	97.9M	167.4	.201	288
Product	1343K	2.7	37K	60.8	12.7	160K	18.6	2.2M	187.0	81.7	6496	170.4	374.8M	706.8	.503	288
Polynomial	203K	1.0	14K	30.0	2.1	88K	3.3	383K	53.1	14.0	5440	24.4	55.9M	146.8	.007	288
Evaluation	571K	1.7	24K	97.0	5.6	160K	8.3	962K	164.5	36.0	6272	60.2	156.8M	422.1	.007	288
Image	86K	0.7	9K	2.6	1.0	44K	1.5	171K	11.4	6.2	5120	15.2	23.6M	25.1	.007	288
Matching	278K	1.2	17K	7.4	2.9	72K	4.2	490K	34.3	18.1	5920	38.9	75.8M	88.8	.007	288
Shortest	366K	1.5	19K	9.3	3.7	52K	5.6	644K	45.6	23.9	5792	50.4	99.6M	130.7	.015	288
Paths	1400K	2.6	38K	35.1	12.6	72K	19.2	2.2M	169.8	84.0	6496	177.6	381.4M	523.3	.026	288
Gas	144K	0.8	12K	8.8	6.1	64K	2.3	271K	23.7	13.9	5440	22.6	39.6M	47.6	.007	288
Simulation	283K	1.2	17K	26.7	20.7	160K	4.3	503K	54.8	34.5	5920	45.9	77.7M	103.1	.007	288
SHA-1	24K	0.18	5K	3.7	3.3	24K	0.5	54K	6.5	4.3	4992	7.9	6.5M	9.0	.007	288

From the comparison in Table 2, it is clear that our implementation is extremely competitive in terms of prover computation, with the square root version outperforming Pinocchio by a factor larger than 10 for some applications. There is a significant amount of variance in terms of the speedups achieved. The worst cases are those where the number of constraints is high in comparison with the number of multiplication gates: the calculation of $s(X)$ is performed entirely in Python and thus becomes the dominant term in the computation. We expect that in a fully compiled implementation, optimisation would prevent this issue.

The logarithmic communication version is slower in comparison but still outperforms Pinocchio for most applications. The performance also becomes more even, as the constraints are irrelevant in the recursive part.

Our verification times are much higher than Pinocchio's, which can often verify circuit evaluation faster than native execution of an equivalent program. As with the prover, some speedups can be gained by moving to a compiled language, but we would still not expect to match Pinocchio's performance; our verification cost would still be linear. Our proofs are considerably larger as well, especially for the square root version.

Our key generation is simply a commitment key generation, and is not application-specific. Therefore, it can be easily amortised even across different circuits. For a circuit with N multiplication gates, the size of our commitment key is $\sqrt{N}$ elements for the square root version and $\frac{N}{\log N}$ for the log version. In comparison, Pinocchio's key generation is bound to specific circuits and produces keys of size $8N$. Thus, if the keys need to be communicated, our arguments are competitive in terms of total communication if the number of circuit evaluations is up to $\sqrt{N}$ for the square root version, and up to $\frac{N}{\log N}$ for the log version.

A   Arithmetic Circuits

Our satisfiability arguments consider arithmetic circuits described as a list of multiplication gates together with a set of linear consistency equations relating the inputs and outputs of the gates. In this section, we show how to reduce an arbitrary arithmetic circuit to this format.

An arithmetic circuit over a field $\mathbb{Z}_p$ and variables $(a_1, \ldots, a_m)$ is a directed acyclic graph whose vertices are called gates. Gates of in-degree 0 are inputs to the circuit and labelled with some a_i or a constant field element. All other gates are labelled $+$ or $\times$. We may consider fan-in 2 circuits, in which case all of the $+$ and $\times$ gates have in-degree 2, or arbitrary fan-in circuits.

We show how to remove addition and multiplication-by-constant gates from an arithmetic circuit A, and replace them with bilinear consistency equations on the inputs and outputs of the remaining gates, such that satisfiability of the equations is equivalent to satisfiability in the original circuit.

Let B be the sub-circuit of A containing all wires and gates before a multiplication gate, with m input wires and n output wires. Label the m inputs of B with the unit vectors $e_i = (0, \ldots, 1, \ldots, 0)$ of length m. For every addition gate with inputs labelled as x, y, label the output wire as $x + y$. For every multiplication-by-constant gate with inputs x and constant c label the output with cx. By proceeding inductively, the n outputs of B are now labelled with vectors of length m representing them as linear combinations of the inputs.

We can now remove the gates of B from A. We also remove any multiplication gates whose inputs are the inputs of the new circuit. Now we simply repeat the process of finding consistency equations until we have considered the whole of A. In Fig. 1 there is an example of a circuit together and the corresponding consistency equations.

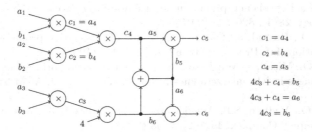

Fig. 1. A simple arithmetic circuit, and the corresponding consistency equations.

References

1. Bayer, S., Groth, J.: Efficient zero-knowledge argument for correctness of a shuffle. In: Pointcheval, D., Johansson, T. (eds.) EUROCRYPT 2012. LNCS, vol. 7237, pp. 263–280. Springer, Heidelberg (2012)
2. Bayer, S., Groth, J.: Zero-knowledge argument for polynomial evaluation with application to blacklists. In: Johansson, T., Nguyen, P.Q. (eds.) EUROCRYPT 2013. LNCS, vol. 7881, pp. 646–663. Springer, Heidelberg (2013)
3. Bellare, M., Rogaway, P.: Random oracles are practical: a paradigm for designing efficient protocols. In: ACM Conference on Computer and Communications Security – CCS 1993, pp. 62–73 (1993)
4. Ben-Sasson, E., Chiesa, A., Genkin, D., Tromer, E., Virza, M.: SNARKs for C: verifying program executions succinctly and in zero knowledge. In: Canetti, R., Garay, J.A. (eds.) CRYPTO 2013, Part II. LNCS, vol. 8043, pp. 90–108. Springer, Heidelberg (2013)
5. Ben-Sasson, E., Chiesa, A., Tromer, E., Virza, M.: Succinct non-interactive zero knowledge for a von Neumann architecture. In: USENIX Security Symposium 2014, pp. 781–796 (2014)
6. Bitansky, N., Canetti, R., Chiesa, A., Tromer, E.: From extractable collision resistance to succinct non-interactive arguments of knowledge, and back again. In: Innovations in Theoretical Computer Science – ITCS 2012, pp. 326–349 (2012)
7. Bitansky, N., Canetti, R., Chiesa, A., Tromer, E.: Recursive composition and bootstrapping for SNARKS and proof-carrying data. In: Symposium on Theory of Computing Conference – TCC 2013, pp. 111–120 (2013)
8. Brassard, G., Chaum, D., Crépeau, C.: Minimum disclosure proofs of knowledge. J. Comput. Syst. Sci. 37(2), 156–189 (1988)
9. Cantor, D.G.: On arithmetical algorithms over finite fields. J. Comb. Theor. Ser. A 50(2), 285–300 (1989)
10. Cramer, R., Damgård, I.B.: Zero-knowledge proofs for finite field arithmetic or: can zero-knowledge be for free? In: Krawczyk, H. (ed.) CRYPTO 1998. LNCS, vol. 1462, pp. 424–441. Springer, Heidelberg (1998)
11. Danezis, G.: Petlib: a Python library that implements a number of privacy enhancing technologies (PETs) (2015). https://github.com/gdanezis/petlib
12. Garay, J.A., MacKenzie, P., Yang, K.: Strengthening zero-knowledge protocols using signatures. J. Cryptology 19(2), 169–209 (2006)
13. Gennaro, R., Gentry, C., Parno, B., Raykova, M.: Quadratic span programs and succinct NIZKs without PCPs. In: Johansson, T., Nguyen, P.Q. (eds.) EUROCRYPT 2013. LNCS, vol. 7881, pp. 626–645. Springer, Heidelberg (2013)
14. Gentry, C., Groth, J., Ishai, Y., Peikert, C., Sahai, A., Smith, A.: Using fully homomorphic hybrid encryption to minimize non-interactive zero-knowledge proofs. J. Cryptology 28(4), 820–843 (2015)
15. Goldreich, O., Håstad, J.: On the complexity of interactive proofs with bounded communication. Inf. Process. Lett. 67(4), 205–214 (1998)
16. Goldreich, O., Micali, S., Wigderson, A.: Proofs that yield nothing but their validity or all languages in NP have zero-knowledge proof systems. J. ACM 38(3), 691–729 (1991)
17. Goldreich, O., Vadhan, S.P., Wigderson, A.: On interactive proofs with a laconic prover. Comput. Complex. 11(1–2), 1–53 (2002)
18. Goldwasser, S., Micali, S., Rackoff, C.: The knowledge complexity of interactive proofs. SIAM J. Comput. 18(1), 186–208 (1989)

19. Groth, J.: Honest verifier zero-knowledge arguments applied. Ph.D. thesis, University of Aarhus (2004)
20. Groth, J.: Efficient zero-knowledge arguments from two-tiered homomorphic commitments. In: Wang, X., Lee, D.H. (eds.) ASIACRYPT 2011. LNCS, vol. 7073, pp. 431–448. Springer, Heidelberg (2011)
21. Groth, J.: Linear algebra with sub-linear zero-knowledge arguments. In: Halevi, S. (ed.) CRYPTO 2009. LNCS, vol. 5677, pp. 192–208. Springer, Heidelberg (2009)
22. Groth, J.: Short pairing-based non-interactive zero-knowledge arguments. In: Abe, M. (ed.) ASIACRYPT 2010. LNCS, vol. 6477, pp. 321–340. Springer, Heidelberg (2010)
23. Groth, J., Ishai, Y.: Sub-linear zero-knowledge argument for correctness of a shuffle. In: Smart, N.P. (ed.) EUROCRYPT 2008. LNCS, vol. 4965, pp. 379–396. Springer, Heidelberg (2008)
24. Groth, J., Kohlweiss, M.: One-out-of-many proofs: or how to leak a secret and spend a coin. In: Oswald, E., Fischlin, M. (eds.) EUROCRYPT 2015. LNCS, vol. 9057, pp. 253–280. Springer, Heidelberg (2015)
25. Guillou, L.C., Quisquater, J.-J.: A practical zero-knowledge protocol fitted to security microprocessor minimizing both transmission and memory. In: Günther, C.G. (ed.) EUROCRYPT 1988. LNCS, vol. 330, pp. 123–128. Springer, Heidelberg (1988)
26. Kate, A., Zaverucha, G.M., Goldberg, I.: Constant-size commitments to polynomials and their applications. In: Abe, M. (ed.) ASIACRYPT 2010. LNCS, vol. 6477, pp. 177–194. Springer, Heidelberg (2010)
27. Kilian, J.: A note on efficient zero-knowledge proofs and arguments. In: Symposium on Theory of Computing Conference – TCC 1992, pp. 723–732 (1992)
28. Lim, C.H.: Efficient multi-exponentiation and application to batch verification of digital signatures, manuscript (2000). http://dasan.sejong.ac.kr/chlim/pub/multi_exp.ps
29. Lindell, Y.: Parallel coin-tossing and constant-round secure two-party computation. J. Cryptology 16(3), 143–184 (2003)
30. Lipmaa, H.: Progression-free sets and sublinear pairing-based non-interactive zero-knowledge arguments. In: Cramer, R. (ed.) TCC 2012. LNCS, vol. 7194, pp. 169–189. Springer, Heidelberg (2012)
31. Möller, B.: Algorithms for multi-exponentiation. In: Vaudenay, S., Youssef, A.M. (eds.) SAC 2001. LNCS, vol. 2259, pp. 165–180. Springer, Heidelberg (2001)
32. Möller, B., Rupp, A.: Faster multi-exponentiation through caching: accelerating (EC) DSA signature verification. In: Ostrovsky, R., De Prisco, R., Visconti, I. (eds.) SCN 2008. LNCS, vol. 5229, pp. 39–56. Springer, Heidelberg (2008)
33. Oliphant, T.E.: A guide to NumPy, vol. 1. Trelgol Publishing, USA (2006)
34. Parno, B., Howell, J., Gentry, C., Raykova, M.: Pinocchio: nearly practical verifiable computation. In: IEEE Symposium on Security and Privacy, pp. 238–252 (2013)
35. Schnorr, C.P.: Efficient signature generation by smart cards. J. Cryptology 4(3), 161–174 (1991)
36. Seo, J.H.: Round-efficient sub-linear zero-knowledge arguments for linear algebra. In: Catalano, D., Fazio, N., Gennaro, R., Nicolosi, A. (eds.) PKC 2011. LNCS, vol. 6571, pp. 387–402. Springer, Heidelberg (2011)
37. Shoup, V.: NTL: a library for doing number theory (2001). http://www.shoup.net/ntl/

On the Complexity of Scrypt and Proofs of Space in the Parallel Random Oracle Model

Joël Alwen[1], Binyi Chen[2], Chethan Kamath[1], Vladimir Kolmogorov[1], Krzysztof Pietrzak[1]([✉]), and Stefano Tessaro[2]

[1] IST Austria, Klosterneuburg, Austria
krzpie@gmail.com
[2] University of California, Santa Barbara, USA

Abstract. We study the time- and memory-complexities of the problem of computing labels of (multiple) randomly selected challenge-nodes in a directed acyclic graph. The w-bit label of a node is the hash of the labels of its parents, and the hash function is modeled as a random oracle. Specific instances of this problem underlie both proofs of space [Dziembowski et al. CRYPTO'15] as well as popular memory-hard functions like scrypt. As our main tool, we introduce the new notion of a *probabilistic parallel entangled pebbling game*, a new type of combinatorial pebbling game on a graph, which is closely related to the labeling game on the same graph.

As a first application of our framework, we prove that for scrypt, when the underlying hash function is invoked n times, the cumulative memory complexity (CMC) (a notion recently introduced by Alwen and Serbinenko (STOC'15) to capture amortized memory-hardness for parallel adversaries) is at least $\Omega(w \cdot (n/\log(n))^2)$. This bound holds for adversaries that can store many natural functions of the labels (e.g., linear combinations), but still not arbitrary functions thereof.

We then introduce and study a combinatorial quantity, and show how a sufficiently small upper bound on it (which we conjecture) extends our CMC bound for scrypt to hold against *arbitrary* adversaries.

We also show that such an upper bound solves the main open problem for proofs-of-space protocols: namely, establishing that the *time complexity* of computing the label of a random node in a graph on n nodes (given an initial kw-bit state) reduces tightly to the time complexity for black pebbling on the same graph (given an initial k-node pebbling).

1 Introduction

The common denominator of *password hashing* (e.g., as in PKCS#5 [13]) and *proofs of work* [7,12] is the requirement for a certain computation to be sufficiently expensive, while still remaining feasible. In this context, "expensive" has traditionally meant high *time complexity*, but recent hardware advances have shown this requirement to be too weak, with fairly inexpensive tailored-made ASIC devices for Bitcoin mining and password cracking gaining increasingly widespread usage.

© International Association for Cryptologic Research 2016
M. Fischlin and J.-S. Coron (Eds.): EUROCRYPT 2016, Part II, LNCS 9666, pp. 358–387, 2016.
DOI: 10.1007/978-3-662-49896-5_13

In view of this, a much better requirement is *memory-hardness*, i.e., the *product* of the memory (a.k.a. space) *and* the time required to solve the task at hand (this is known as the *space-time* (ST) complexity) should be large. The ST complexity is widely considered to be a good estimate of the product of the area and the time (AT) complexity of a circuit solving the task [3,5,16], and thus increasing ST complexity appears to incur a higher dollar cost for building custom circuits compared to simply increasing the required raw computing power alone. Motivated by this observation, Percival [16] developed scrypt, a candidate memory-hard function for password hashing and key derivation which has been well received in practice (e.g., it underlies the Proof of Work protocols of LiteCoin [14], one of the currently most prevalent cryptocurrencies in terms of market capitalization [1]). This has made memory-hardness one of the main desiderata in candidates for the recent password-hashing competition, including its winner, Argon2 [4]. Dziembowski *et al.* [9] introduce the concept of *proofs of space* (PoSpace), where the worker (or miner) can either dedicate a large amount of storage space, and then generate proofs extremely efficiently, or otherwise must pay a large time cost for every proof generated. The PoSpace protocol has also found its way into a recent proposal for digital currency [15].

Our contributions, in a nutshell. Cryptanalytic attacks [3,5,6,17] targeting candidate memory-hard functions [2,4,11,17] have motivated the need for developing constructions with *provable security guarantees*. With the exception of [3], most candidate memory-hard functions come without security proofs and those that do (e.g. [11,16,17]) only consider a severely restricted class of algorithms and complexity notions, as we discuss below. A primary goal of this paper is to advance the foundations of memory-hardness, and we make progress along several fronts.

We develop a new class of probabilistic pebbling games on graphs – called *entangled pebbling games* – which are used to prove results on the memory-hardness of tasks such as computing scrypt for large non-trivial classes of adversaries. Moreover, we show how to boost these results to hold against *arbitrary* adversaries in the parallel random oracle model (pROM) [3] under the conjecture that a new combinatorial quantity which we introduce is (sufficiently) bounded.

A second application of the techniques introduced in this paper considers Proofs of Space. We show that time lower bounds on the pebbling complexity of a graph imply time lower bounds in the pROM model agains *any* adversary. The quantitative bounds we get depend on the combinatorial value we introduce, and assuming our conjecture, are basically tight. This solves, modulo the conjecture, the main problem left open in the Proofs of Space paper [9].

Sequentially memory-hard functions. Recall that scrypt[1] uses a hash function $h : \{0,1\}^* \rightarrow \{0,1\}^w$ (e.g., SHA-256), and proceeds in two phases, given an

---

[1] In fact, what we describe here is only a subset of the whole scrypt function, called ROMix. ROMix is the actual core of the scrypt function, and we will use the generic name "scrypt" for in the following. ROMix (with some minor modification and extensions) also underlies one of the two variants of the winner Argon [4] of the recent password hashing competition https://password-hashing.net/, namely the data-dependent variant Argon2d.

input X. It first computes $X_i = h^i(X)$ for all $i \in [n]$, and with $S_0 = X_n$, it then computes $S_1, \ldots, S_n$ where

$$S_i = h(S_{i-1} \oplus X_{\text{int}(S_{i-1})})$$

where $\text{int}(S)$ reduces an w-bit string S to an integer in $[n]$. The final output is S_n. Note that is possible to evaluate scrypt on input X using $n \cdot w$ bits of memory and in time linear in n, by keeping the values $X_1, \ldots, X_n$ stored in memory once they are computed. However, the crucial point is that there is no apparent way to save memory – for example, to compute S_i, we need to know $X_{\text{int}(S_{i-1})}$, and under the assumption that $\text{int}(S_{i-1})$ is (roughly) uniformly random in $[n]$, an evaluator without memory needs to do linear work (in n) to recover this value before continuing with the execution. This gives a constant-memory, $O(n^2)$ time algorithm to evaluate scrypt. In fact, as stated by Percival [16], the actual hope is that no matter how much time $T(n)$ and how much memory $S(n)$ an adversarial evaluator invests, we always have $S(n) \cdot T(n) \geq n^{2-\epsilon}$ for all $\epsilon > 0$, *even if the evaluator can parallelize its computation arbitrarily.*

Percival's analysis of scrypt assumes that h is a random oracle. The analysis is limited in two ways: (1) It only considers adversaries which can only store random oracle outputs in their memory. (2) The bound measures memory complexity in terms of the *maximum* memory resources $S(n)$. The latter is undesirable, since the ultimate goal of an adversary performing a brute-force attack is to evaluate scrypt on *as many inputs as possible*, and if the large memory usage is limited to a small fraction of the computing time, a much higher *amortized complexity* can be achieved.

Alwen and Serbinenko (AS) [3] recently addressed these shortcomings, and delivered provably sequentially memory-hard functions in the so-called *parallel random oracle model* (pROM), developing new and better complexity metrics tailored to capturing amortized hardness. While their work falls short of delivering guarantees for scrypt-like functions, it serves as an important starting point for our work, and we give a brief overview.

From sequential memory-hardness to pebbling. AS consider adversaries attempting to evaluate a function $\mathcal{H}^h$ (which makes calls to some underlying hash function h, modeled as a random oracle). These adversaries proceed in rounds: in each round i, the adversary can make an *unbounded* number of *parallel* queries to h, and then pass on a state σ_i to the next round. The complexity of the adversary is captured by its *cumulative memory complexity* (CMC) given by $\sum_i |\sigma_i|$. One then denotes as $\text{cmc}^{\text{pROM}}(\mathcal{H})$ the expected CMC of the best adversary where the expectation is over the choice of RO h and coins of the adversary. We stress that CMC exhibits some very important features: *First,* a lower bound appears to yield a reasonable lower bound on the AT complexity metric. *Second,* In contrast to the ST complexity the CMC of a task also gives us a lower-bound on the electricity consumption of performing the task. This is because storing data in volatile memory for, say, the time it takes to evaluate h consumes a significant amount of electricity. Thus CMC tells us something not only about the dollar cost of building a custom circuit for computing a task but also about the dollar cost of actually running it. While the former can be amortized over the life of the device, the later represents a recurring fee.

AS study sequentially memory-hard functions naturally defined by a single-source and single-sink directed acyclic graph (DAG) $G = (V, E)$. The label of a vertex $i \in V$ with parents $\{p_1, \ldots, p_d\}$ (i.e., $(p_j, v) \in E$ for $i = 1, \ldots, d$) is defined as $\ell_i = h(i, \ell_{p_1}, \ldots, \ell_{p_d})$. Note that the labels of all vertices can be recursively computed starting with the sources. The function $\mathsf{label}(G, h)$ is now simply the label ℓ_v of the sink v. There is a natural connection between $\mathsf{cmc}^{\mathsf{pROM}}(\mathsf{label}(G, h))$ for a randomly chosen h and the *cumulative pebbling complexity* (CC) of the graph G.[2] CC is defined in a game where one can place pebbles on the vertices of V, according to the following rules: In every step of the game, new pebbles can be placed on any vertex for which all parents of v have pebbles on them (in particular, pebbles can always be placed on sources), and pebbles can always be removed. The game is won when a pebble has been placed on the sink. The CC of a strategy for pebbling G is defined as $\sum_i |S_i|$, where S_i is the set of vertices on which a pebble is placed at the end of the ith step, and the CC of G – denoted $\mathsf{cc}(G)$ – is the CC of the best strategy.

Indeed, $\mathsf{cc}(G)$ captures the CMC of *restricted* pROM adversaries computing $\mathsf{label}(G, h)$ for which every state σ_i *only consists of random oracle outputs*, i.e., of vertex labels. A pebble on v is equivalent to the fact that σ_i contains ℓ_v. However, a full-fledged pROM adversary *has no reason to be restricted to such a strategy* – it could for example store as part of its state σ_i a particular encoding of the information accumulated so far. Nonetheless, AS show that (up to a negligible extent) such additional freedom does *not* help in computing $\mathsf{label}(G, h)$. They complement this with an efficiently constructible class of constant-degree DAGs G_n on n vertices such that $\mathsf{cc}(G_n) = \Omega(n^2/\mathrm{polylog}(n))$.

Unfortunately however, the framework of [3] does not extend to functions like scrypt, as they are *data dependent*, i.e., the values which need to be input to h are determined *at run-time*. While this makes the design far more intuitive, AS's techniques crucially rely on the relationship between intermediate values in the computation being laid out a priori in a *data-independent* fashion.

Our contributions. This paper validates the security of scrypt-like functions with two types of results – results for *restricted* adversaries, as well as results for arbitrary adversaries under a combinatorial conjecture. Our results also have direct implications on proofs of space, but we postpone this discussion to ease presentation.

(1) PROBABILISTIC PEBBLING GAMES. We introduce a generalization pebble of pebbling games on a DAG $G = (V, E)$ with dynamic challenges uniformly sampled from a set $C \subseteq V$. With the same pebbling rules as before, we now proceed over n rounds, and at every round, a challenge c_i is drawn uniformly

---

[2] A similar connection, for a weaker pebbling game, was first exploited to construct functions for which evaluation requires many cache memory in [8] and more recently to build one-time computable functions [10] as well as in the security proofs the memory-hard functions in [11, 17].

at random from C. The player's goal is to place a pebble on c_i, before moving to the next round, and learning the next challenge c_{i+1}. The game terminates when the last challenge has been covered by a pebble. One can similarly associate with G a *labeling game* computeLabel in the pROM, where the goal is instead to compute the *label* ℓ_{c_i} of c_i, rather than placing a pebble on it. For instance, the computation of scrypt is tightly connected to the computeLabel played on the line graph L_n with vertices $[n] = \{1, 2, \ldots, n\}$, edges $\{(i, i+1) : i \in [n-1]\}$, and challenges $C = [n]$ (as detailed in Sect. 2.5). The labels to be computed in this game are those needed to advance the computation in the second half of the scrypt computation, and the challenges (in the actual scrypt function) are computed from hash-function outputs.

In fact, it is not hard to see that in computeLabel for some graph G a pROM adversary that only stores random-oracle generated outputs can easily be turned into a player for the pebble for graph G. This is particular true for $G = L_n$, and thus lower bounding the CC of an adversary playing pebble on L_n also yields a lower bound on the CMC of computing (the second half of) scrypt. Our first result provides such a lower bound.

Theorem 1. For any constant $\delta > 0$, the CC of an adversary playing pebble on the line graph L_n with challenges $[n]$ is $\Omega_\delta(n^2/\log^2(n))$ with probability $1 - \delta$ over the choice of all challenges.[3]

To appreciate this result, it should be noted that it inherently relies on the choice of the challenges being *independent* of the adversary playing the game – indeed, *if the challenges are known a priori*, techniques from [3] directly give a strategy with CC $O(n^{1.5})$ for the above game. Also this result already improves on Percival's analysis (which, implicitly, places similar restrictions on class of pROM algorithms considered), as Theorem 1 uses the CC of the (simple) pebbling of a graph, and thus it actually generalized to a lower bound on the *amortized* complexity of computing multiple scrypt instances in the pROM.[4]

(2) ENTANGLED PEBBLING. The above result is an important first step – to the best of our knowledge all known evaluation attacks against memory-hard functions indeed *only store hash labels directly or not at all* and thus fit into this model – but we ask the question whether the model can be strengthened. For example, an adversary could store the XOR $\ell_i \oplus \ell_j$ of two labels (which only takes w bits) and depending on possible futures of the game, recover both labels given any one of them. As we will see, this *can* help. As a middle ground between capturing pROM security for arbitrary adversaries and the above pebbling adversaries, we introduce a new class of pebbling games, called *entanglement pebbling games*, which constitutes a combinatorial abstraction for such adversaries.

In such games, an adversary can place on a set $\mathcal{Y} \subseteq V$ an "entangled pebble" $\langle \mathcal{Y} \rangle_t$ for some integer $0 \le t \le |\mathcal{Y}|$. The understanding here is that placing an individual pebble on any t vertices $v \in \mathcal{Y}$ – which we see as a special case of $\langle v \rangle_0$

---

[3] The subscript δ in Ω_δ denotes that the hidden constant depends on δ.

[4] This follows from a special case of the Lemma in [3] showing that CC of a graph is equal to the sum of the CCs the graphs disconnected components.

entangled pebble – is equivalent to having individual pebbles on all vertices in $\mathcal{Y}$. The key point is that keeping an entangled pebble $\langle \mathcal{Y} \rangle_t$ costs *only* $|\mathcal{Y}| - t$, and depending on challenges, we may take different choices as to which t pebbles we use to "disentangle" $\langle \mathcal{Y} \rangle_t$. Also, note that in order to *create* such an entangled pebble, on all elements of $\mathcal{Y}$ there must be either an individual pebble, or such pebble can easily be obtained by disentangling existing entangled pebbles.

In the pROM labeling game, an entangled pebble $\langle \mathcal{Y} \rangle_t$ corresponds to an encoding of length $w \cdot (|\mathcal{Y}| - t)$ of the w-bit labels $\{\ell_i \ : \ i \in \mathcal{Y}\}$ such that given any t of those labels, we can recover all the remaining ones. Such an encoding can be obtained as follows: Fix $2d - t$ elements $x_1, \ldots, x_{2d-t}$ in the finite field $\mathbb{F}_{2^w}$. Let $\mathcal{Y} = \{y_1, \ldots, y_d\}$, and consider the (unique) degree $d - 1$ polynomial $p(.)$ over the finite field $\mathbb{F}_{2^w}$ (whose element are represented as w-bit strings) such that

$$\forall i \in [d] \ : \ p(x_i) = \ell_{y_i}.$$

The encoding now simply contains $\{p(x_{d+1}), \ldots, p(x_{2d-t})\}$, i.e., the evaluation of this polynomial on $d - t$ points. Note that given this encoding and any t labels $\ell_i, i \in \mathcal{Y}$, we have the evaluation of $p(.)$ on d points, and thus can reconstruct $p(.)$. Once we know $p(.)$, we can compute all the labels $\ell_{y_i} = p(i)$ in $\mathcal{Y}$.

In general, we prove (in the full version) that entangled pebbling is *strictly more powerful* (in terms of minimizing the expected CC) than regular pebbling. Fortunately, we will also show that for the probabilistic pebbling game on the line graph L_n entangled pebbling cannot outperform regular ones.

Theorem 2. For any constant $\delta > 0$, the CC of an entangled pebbling adversary playing pebble on graph L_n is $\Omega_\delta(n^2/\log^2(n))$ with probability $1 - \delta$ over the choice of all challenges.

Interestingly, the proof is a simple adaptation of the proof of for the non-entangled case. This result can again be interpreted as providing a guarantee in the label game in the pROM for L_n for the class of adversaries that can be abstracted by entangled pebbling strategies.

(3) ARBITRARY ADVERSARIES. So far we have only discussed (entangled) pebbling lower bounds, which then imply lower bounds for restricted adversaries in the pROM model. In Sect. 4 we consider security against arbitrary adversaries. Our main results there show that there is a tight connection between the complexity of playing computeLabel and a combinatorial quantity γ_n that we introduce. We show two results. The first lower-bounds the *time* complexity of playing computeLabel for *any* graph G while the second lower-bounds the *CMC* of playing computeLabel for L_n (and thus scrypt).

1. For any DAG $G = (V, E)$ with $|V| = n$, with high probability over the choice of the random hash function h, the pROM *time* complexity to play computeLabel for graph G, for any number of challenges, using h and when starting with any state of size $k \cdot w$ is (roughly) at least the time complexity needed to play pebble on G with the same number of challenges and starting with an initial pebbling of size roughly $\gamma_n \cdot k$.
2. The pROM CMC for pebble for L_n is $\Omega(n^2/\log^2(n) \cdot \gamma_n)$.

At this point, we do not have any non-trivial upper bound on γ_n but we *conjecture* that γ_n grows very small (if at all) as a function of n. The best lower bound we have is $\gamma_5 > 3/2$. Note that γ does not need to be constant in n – we would get non-trivial statements even if γ_n were to *grow* moderately as a function of n, i.e. $\gamma_n = \mathrm{polylog}(n)$ or $\gamma_n = n^\epsilon$ for some small $\epsilon > 0$.

Therefore, assuming our conjecture on γ_n, the first result in fact solves the main open problem from the work of Dziembowski et al. [9] on proofs of space. The second result yields, in particular, a near-quadratic lower bound on the CMC of evaluating scrypt for arbitrary pROM adversaries.

2   Pebbling, Entanglement, and the pROM

In this section, we first present both a notion of parallel pebbling of graphs with probabilistic challenges, and then extend this to our new notion of entangled pebbling games. Next, we discuss some generic relations between entangled and regular pebbling, before finally turning to defining the parallel random-oracle model (pROM), and associated complexity metrics.

Throughout, we use the following notation for common sets $\mathbb{N} := \{0, 1, 2, \ldots\}$, $\mathbb{N}^+ := \mathbb{N} \setminus \{0\}$, $\mathbb{N}_{\leq c} := \{0, 1, \ldots, c\}$ and $[c] := \{1, 2, \ldots, c\}$. For a distribution $\mathcal{D}$ we write $x \in \mathcal{D}$ to denote sampling x according to $\mathcal{D}$ in a random experiment.

2.1   Probabilistic Graph Pebbling

Throughout, let $G = (V, E)$ denote a directed acyclic graph (DAG) with vertex set $V = [n]$. For a vertex $i \in V$, we denote by $\mathsf{parent}(i) = \{j \in V : (j, i) \in E\}$ the parents of i. The *m-round, probabilistic parallel pebbling game* between a

---

$\mathsf{pebble}(G, C, m, \mathbf{T}, P_{\mathsf{init}})$: The m-round parallel pebbling game for DAG $G = (V, E)$, challenge set $C \subseteq V$ and initial pebbling configuration $P_{\mathsf{init}} \subseteq V$ is played between a challenger and a pebbler $\mathbf{T}$.

1. Initialise $\mathsf{cnt} := 0$, $\mathsf{round} := 0$, $P_{\mathsf{cnt}} := P_{\mathsf{init}}$ and $\mathsf{cost} := 0$.
2. A challenge $c \leftarrow C$ is chosen uniformly from C and passed to $\mathbf{T}$.
3. $\mathsf{cost} := \mathsf{cost} + |P_{\mathsf{cnt}}|$.
4. $\mathbf{T}$ choses a new pebbling configuration $P_{\mathsf{cnt}+1}$ which must satisfy

$$\forall i \in P_{\mathsf{cnt}+1} \setminus P_{\mathsf{cnt}} \; : \; \mathsf{parent}(i) \in P_{\mathsf{cnt}} \qquad (1)$$

5. $\mathsf{cnt} := \mathsf{cnt} + 1$.
6. If $c \notin P_{\mathsf{cnt}}$ go to step 3.   *c not yet pebbled*
7. $\mathsf{round} := \mathsf{round} + 1$. If $\mathsf{round} < m$ go to step 2, otherwise if $\mathsf{round} = m$ the experiment is over, the output is the final count cnt and the cumulative cost cost.

Fig. 1. Description of the m-round, probabilistic parallel pebbling game

player T on a graph $G = (V, E)$ with challenge nodes $C \subseteq V$ is defined in Fig. 1. The *cumulative black pebbling complexity* is defined as

$$\mathsf{cc}(G, C, m, \mathsf{T}, P_{\mathsf{init}}) := \underset{\mathsf{pebble}(G,C,m,\mathsf{T},P_{\mathsf{init}})}{\mathbb{E}} [\mathsf{cost}]$$

$$\mathsf{cc}(G, C, m, k) := \min_{\substack{\mathsf{T}, P_{\mathsf{init}} \subseteq V \\ |P_{\mathsf{init}}| \leq k}} \{\mathsf{cc}(G, C, m, \mathsf{T}, P_{\mathsf{init}})\}$$

Similarly, the *time cost* is defined as

$$\mathsf{time}(G, C, m, \mathsf{T}, P_{\mathsf{init}}) := \underset{\mathsf{pebble}(G,C,m,\mathsf{T},P_{\mathsf{init}})}{\mathbb{E}} [\mathsf{cnt}]$$

$$\mathsf{time}(G, C, m, k) := \min_{\substack{\mathsf{T}, P_{\mathsf{init}} \subseteq V \\ |P_{\mathsf{init}}| \leq k}} \{\mathsf{time}(G, C, m, \mathsf{T}, P_{\mathsf{init}})\}$$

The above notions consider the expected cost of a pebbling, thus even if, say $\mathsf{cc}(G, C, m, k)$, is very large, this could be due to the fact that for a tiny fraction of challenge sequences the complexity is very high, while for all other sequences it is very low. To get more robust security notions, we will define a more fine-grained notion which will guarantee that the complexity is high on all but some ϵ fraction on the runs.

$$\mathsf{cc}_\epsilon(G, C, m, \mathsf{T}, P_{\mathsf{init}}) := \inf \left\{ \gamma \ \middle| \ \underset{\mathsf{pebble}(G,C,m,\mathsf{T},P_{\mathsf{init}})}{\mathbb{P}} [\mathsf{cost} \geq \gamma] \geq 1 - \epsilon \right\}$$

$$\mathsf{cc}_\epsilon(G, C, m, k) := \min_{\substack{\mathsf{T}, P_{\mathsf{init}} \subseteq V \\ |P_{\mathsf{init}}| \leq k}} \{\mathsf{cc}_\epsilon(G, C, m, \mathsf{T}, P_{\mathsf{init}}\}$$

$$\mathsf{time}_\epsilon(G, C, m, \mathsf{T}, P_{\mathsf{init}}) := \inf \left\{ \gamma \ \middle| \ \underset{\mathsf{pebble}(G,C,m,\mathsf{T},P_{\mathsf{init}})}{\mathbb{P}} [\mathsf{cnt} \geq \gamma] \geq 1 - \epsilon \right\}$$

$$\mathsf{time}_\epsilon(G, C, m, k) := \min_{\substack{\mathsf{T}, P_{\mathsf{init}} \subseteq V \\ |P_{\mathsf{init}}| \leq k}} \{\mathsf{time}_\epsilon(G, C, m, \mathsf{T}, P_{\mathsf{init}}\}$$

In general, we cannot upper bound cc in terms of cc_ϵ if $\epsilon > 0$ (same for time in terms of time_ϵ), but in the other direction it is easy to show that

$$\mathsf{cc}(G, C, m, \mathsf{T}, P_{\mathsf{init}}) \geq \mathsf{cc}_\epsilon(G, C, m, \mathsf{T}, P_{\mathsf{init}})(1 - \epsilon)$$

2.2 Entangled Graph Pebbling

In the above pebbling game, a node is always either pebbled or not and there is only one type of pebble which we will hence forth refer to as a "black" pebble. We will now introduce a more general game, where T can put "entangled" pebbles.

A *t-entangled pebble*, denoted $\langle \mathcal{Y} \rangle_t$, is defined by a subset of nodes $\mathcal{Y} \subseteq [n]$ together with an integer $t \in \mathbb{N}_{\leq |\mathcal{Y}|}$. Having black pebble on all nodes $\mathcal{Y}$ now corresponds to the special case $\langle \mathcal{Y} \rangle_0$. Entangled pebbles $\langle \mathcal{Y} \rangle_t$ now have the following behaviour. Once any subset of $\mathcal{Y}$ of size (at least) t contains black pebbles then all $v \in \mathcal{Y}$ immediately receive a black pebble (regardless of whether their parents already contained black pebbles or not). We define the *weight* of an entangled pebble as:

$$|\langle \mathcal{Y} \rangle_t|_{\updownarrow} := |\mathcal{Y}| - t.$$

More generally, an *(entangled) pebbling configuration* is defined as a set $P = \{\langle \mathcal{Y}_1 \rangle_{t_1}, \ldots, \langle \mathcal{Y}_z \rangle_{t_z}\}$ of entangled pebbles and its weight is

$$|P|_{\updownarrow} := \sum_{i \in [s]} |\langle \mathcal{Y}_i \rangle_{t_i}|_{\updownarrow}.$$

The rule governing how a pebbling configuration P_{cnt} can be updated to configuration $P_{\mathsf{cnt}+1}$ – which previously was the simple property eq.(1) – are now a bit more involved. To describe them formally we need the following definition.

Definition 1 (Closure). *The* closure *of an entangled pebbling configuration* $P = \{\langle \mathcal{Y}_1 \rangle_{t_1}, \ldots, \langle \mathcal{Y}_s \rangle_{t_s}\}$ – *denoted* closure(S) – *is defined recursively as follows: initialise* $\Lambda = \emptyset$ *and then*

$$\text{while } \exists j \in [s] : (\mathcal{Y}_j \nsubseteq \Lambda) \wedge (\Lambda \cap \mathcal{Y}_j \geq t_j) \text{ set } \Lambda := \Lambda \cup \mathcal{Y}_j$$

once Λ *cannot be further extended using the rule above we define* closure$(S) = \Lambda$.

Note that closure(S) is non-empty iff there's at least one set of t-entangled pebbles $\langle \mathcal{Y} \rangle_t$ in P with $t = 0$. Equipped with this notion we can now specify how a given pebbling configuration can be updated.

Definition 2 (Valid Update). *Let* $P = \{\langle \mathcal{Y}_1 \rangle_{t_1}, \ldots, \langle \mathcal{Y}_m \rangle_{t_s}\}$ *be an entangled pebbling configuration. Further,*

- *Let* $\mathcal{V}_1 :=$ closure(P).
- *Let* $\mathcal{V}_2 := \{i : \mathsf{parent}(i) \subseteq \mathcal{V}_1\}$. *These are the nodes that can be pebbled using the black pebbling rules (Eq. 1).*

Now $P' = \{\langle \mathcal{Y}'_1 \rangle_{t'_1}, \ldots, \langle \mathcal{Y}'_{s'} \rangle_{t'_s}\}$ *is a* valid update *of* P *if for every* $\langle \mathcal{Y}'_{j'} \rangle_{t'_{j'}}$, *one of the two conditions is satisfied*

1. $\mathcal{Y}'_{j'} \subseteq (\mathcal{V}_1 \cup \mathcal{V}_2)$.
2. $\exists i$ *with* $\mathcal{Y}'_{j'} = \mathcal{Y}_i$ *and* $t'_j \geq t_i$. *That is,* $\langle \mathcal{Y}'_{j'} \rangle_{t'_{j'}}$ *is an entangled pebble* $\langle \mathcal{Y}_i \rangle_{t_i}$ *that is already in* P, *but where we potentially have increased the threshold from* t_i *to* $t'_{j'}$.

The entangled pebbling game pebble$^{\updownarrow}(G, C, m, \mathsf{T})$ is now defined like the game pebble(G, C, m, T) above, except that T is allowed to choose entangled pebblings. We give it in Fig. 2. The *cumulative entangled pebbling complexity* and the *entangled time complexity* of this game are defined analogously to those of the simple pebbling game – we just replace cc with cc$^{\updownarrow}$ and time with time$^{\updownarrow}$ in our notation to account for entanglement being considered. In the full version, we show that entanglement can indeed improve the cumulative complexity with respect to unentangled pebbling. However, in the next section, we will show that this is not true with respect to *time* complexity.

$\mathsf{pebble}^{\ddagger}(G, C, m, \mathsf{T}, P_{\mathsf{init}})$: The m-round parallel, entangled pebbling game for DAG $G = (V, E)$, challenge set $C \subseteq V$ and initial entagled pebbling configuration P_{init}

1. Initialise $\mathsf{cnt} := 0$, $\mathsf{round} := 0$, $P_{\mathsf{cnt}} := P_{\mathsf{init}}$ and $\mathsf{cost} := 0$.
2. A challenge $c \leftarrow C$ is chosen uniformly from C and passed to T.
3. $\mathsf{cost} := \mathsf{cost} + |P_{\mathsf{cnt}}|_{\ddagger}$.
4. T choses a new pebbling configuration $P_{\mathsf{cnt}+1}$ which must be a valid update of P_{cnt}.
5. $\mathsf{cnt} := \mathsf{cnt} + 1$.
6. If $c \notin \mathsf{closure}(P_{\mathsf{cnt}})$ go to step 3.   *c not yet pebbled*
7. $\mathsf{round} := \mathsf{round} + 1$. If $\mathsf{round} < m$ go to step 2. Otherwise if $\mathsf{round} = m$ end the experiment and output the final count cnt and cumulative cost cost.

Fig. 2. The entangled pebbling game $\mathsf{pebble}^{\updownarrow}(G, C, m, \mathsf{T})$.

2.3 Entanglement Does Not Improve Time Complexity

We show that in terms of *time* complexity, entangled pebbling are no more efficient than normal pebbles.

Lemma 3 (Entangled Time = Simple Time). *For any* $G, C, m, \mathsf{T}^{\updownarrow}, P_{\mathsf{init}}^{\updownarrow}$ *and* $\epsilon \geq 0$ *there exist a* $\mathsf{T}, P_{\mathsf{init}}$ *such that* $|P_{\mathsf{init}}| \leq |P_{\mathsf{init}}^{\updownarrow}|_{\updownarrow}$ *and*

$$\mathsf{time}(G, C, m, \mathsf{T}, P_{\mathsf{init}}) \leq \mathsf{time}^{\updownarrow}(G, C, m, \mathsf{T}^{\updownarrow}, P_{\mathsf{init}}^{\updownarrow}) \tag{2}$$

$$\mathsf{time}_{\epsilon}(G, C, m, \mathsf{T}, P_{\mathsf{init}}) \leq \mathsf{time}_{\epsilon}^{\updownarrow}(G, C, m, \mathsf{T}^{\updownarrow}, P_{\mathsf{init}}^{\updownarrow}) \tag{3}$$

in particular

$$\mathsf{time}^{\updownarrow}(G, C, m, k) = \mathsf{time}(G, C, m, k) \qquad \mathsf{time}_{\epsilon}^{\updownarrow}(G, C, m, k) = \mathsf{time}_{\epsilon}(G, C, m, k) \tag{4}$$

Proof. The $\geq$ directions in Eq. (4) follows directly from the fact that a black pebbling is a special case of an entangled pebbling. The $\leq$ direction follows from Eqs. (2) and (3). Below we prove Eq. (2), the proof for Eq. (3) is almost analogous.

We say that a player $\mathsf{A}_{\mathsf{greedy}}$ for a normal or entangled pebbling is "greedy", if its strategy is simply to pebble everything possible in every round and never remove pebbles. Clearly, $\mathsf{A}_{\mathsf{greedy}}$ is optimal for time complexity, i.e.,

$$\forall G, C, m, P_{\mathsf{init}} : \min_{\mathsf{T}} \mathsf{time}(G, C, m, \mathsf{T}, P_{\mathsf{init}}) = \mathsf{time}(G, C, m, \mathsf{A}_{\mathsf{greedy}}, P_{\mathsf{init}}) \tag{5}$$

$$\forall G, C, m, P_{\mathsf{init}}^{\updownarrow} : \min_{\mathsf{T}} \mathsf{time}^{\updownarrow}(G, C, m, \mathsf{T}, P_{\mathsf{init}}^{\updownarrow}) = \mathsf{time}^{\updownarrow}(G, C, m, \mathsf{A}_{\mathsf{greedy}}, P_{\mathsf{init}}^{\updownarrow}) \tag{6}$$

We next describe how to derive an initial black pebbling P_{init}^{*} from an entangled pebbling $P_{\mathsf{init}}^{\updownarrow}$ of cost $|P_{\mathsf{init}}^{*}| \leq |P_{\mathsf{init}}^{\updownarrow}|_{\updownarrow}$ such that

$$\mathsf{time}(G, C, m, \mathsf{A}_{\mathsf{greedy}}, P_{\mathsf{init}}^{*}) \leq \mathsf{time}^{\updownarrow}(G, C, m, \mathsf{A}_{\mathsf{greedy}}, P_{\mathsf{init}}^{\updownarrow}) \tag{7}$$

Note that this then proves Eq. (2) (with $\mathsf{A}_{\mathsf{greedy}}, P_{\mathsf{init}}{}^*$ being $\mathsf{T}, P_{\mathsf{init}}$ in the statement of the lemma) as

$$\mathsf{time}^{\mathsf{\uparrow}}(G, C, m, \mathsf{T}^{\mathsf{\uparrow}}, P_{\mathsf{init}}{}^{\mathsf{\uparrow}}) \geq \mathsf{time}^{\mathsf{\uparrow}}(G, C, m, \mathsf{A}_{\mathsf{greedy}}, P_{\mathsf{init}}{}^{\mathsf{\uparrow}}) \tag{8}$$

$$\geq \mathsf{time}(G, C, m, \mathsf{A}_{\mathsf{greedy}}, P_{\mathsf{init}}{}^*) \tag{9}$$

It remains to prove Eq. (7). For every share $\langle \mathcal{Y} \rangle_t \in P_{\mathsf{init}}{}^{\mathsf{\uparrow}}$ we observe which $|\mathcal{Y}| - t$ pebbles are the last ones to become available[5] in the random experiment $\mathsf{pebble}^{\mathsf{\uparrow}}(G, C, m, \mathsf{T}^{\mathsf{\uparrow}}, P_{\mathsf{init}}{}^{\mathsf{\uparrow}})$, and we add these pebbles to P_{init} if they're not already in there.

Note that then $|P_{\mathsf{init}}| \leq |P_{\mathsf{init}}{}^{\mathsf{\uparrow}}|_{\mathsf{\updownarrow}}$ as required. Moreover Eq. (7) holds as at any timestep, the nodes available in $\mathsf{pebble}^{\mathsf{\uparrow}}(G, C, m, \mathsf{A}_{\mathsf{greedy}}, P_{\mathsf{init}}{}^{\mathsf{\uparrow}})$ are nodes already pebbled in $\mathsf{pebble}(G, C, m, \mathsf{A}_{\mathsf{greedy}}, P_{\mathsf{init}}{}^*)$ at the same timestep.   □

2.4   The Parallel Random Oracle Model (pROM)

We turn to an analogue of the above pebbling games n the parallel random oracle model (pROM) [3]. In particular, let $G = (V, E)$ be a DAG with a dedicated set $C \subseteq V$ of challenge edges, we identify the vertices with $V = [n]$. A labelling $\ell_1, \dots, \ell_n$ of G's verticies using a hash functiotn $\mathsf{h} : \{0,1\}^* \to \{0,1\}^w$ is defined as follows. Let $\mathsf{parent}(i) = \{j \in V : (j, i) \in E\}$ denote the parents of i, then

$$\ell_i = \mathsf{h}(i, \ell_{p_1}, \dots, \ell_{p_d}) \quad \text{where} \quad (p_1, \dots, p_d) = \mathsf{parent}(i) \tag{10}$$

Note that if i is a source, then its label is simply $\ell_i = \mathsf{h}(i)$.

$\mathsf{computeLabel}(G, C, m, \mathsf{A}, \sigma_{\mathsf{init}}, \mathsf{h} : \{0,1\}^* \to \{0,1\}^w)$:

1. Initialise $\mathsf{cnt} := 0$, $\mathsf{round} := 0$, $\sigma_{\mathsf{cnt}} := \sigma_{\mathsf{init}}$ and $\mathsf{cost} := 0$.
2. A challenge $c \leftarrow C$ is chosen uniformly from C.
3. $(q_1, \dots, q_s, \ell) \leftarrow \mathsf{A}(c, \sigma_{\mathsf{cnt}})$ A *choses parallel* h *queries and (optionally) a guess for* ℓ_c
4. $\mathsf{cost} := \mathsf{cost} + |\sigma_{\mathsf{cnt}}| + s \cdot w$.
5. $(\sigma_{\mathsf{cnt}+1}) \leftarrow \mathsf{A}(c, \sigma_{\mathsf{cnt}}, \mathsf{h}(q_1), \dots, \mathsf{h}(q_s))$ A *outputs next state*
6. $\mathsf{cnt} := \mathsf{cnt} + 1$
7. If $\ell = \bot$ (no guess in this round) go to step 3.
8. If $\ell \neq \ell_c$ (wrong guess) set $\mathsf{cost} = \infty$ and abort.
9. $\mathsf{round} := \mathsf{round} + 1$. If $\mathsf{round} = m$ end the experiment. Otherwise go to step 2.
10. $\mathsf{round} := \mathsf{round} + 1$. If $\mathsf{round} < m$ go to step 2. Otherwise if $\mathsf{round} = m$ end the experiment and output the final count cnt and cumulative cost cost.

Fig. 3. The labeling game $\mathsf{computeLabel}(G, C, m, \mathsf{A}, \sigma_{\mathsf{init}}, \mathsf{h})$.

---

[5] A pebble is available if it's in the closure of the current entangled pebbling configuration, also note that $\mathsf{A}_{\mathsf{greedy}}$'s strategy is deterministic and independent of the challenges it gets, so the "last nodes to become available" is well defined.

We consider a game $\mathsf{computeLabel}(G, C, m, \mathsf{A}, \sigma_{\mathsf{init}}, \mathsf{h})$ where an algorithm A must m times consecutively compute the label of a node chosen at random from C. A gets an initial state $\sigma_0 = \sigma_{\mathsf{init}}$. The *cumulative memory complexity* is defined as follows.

$$\mathsf{cmc}^{\mathsf{pROM}}(G, C, m, \mathsf{A}, \sigma_{\mathsf{init}}, \mathsf{h}) = \underset{\mathsf{computeLabel}(G,C,m,\mathsf{A},\sigma_{\mathsf{init}},\mathsf{h})}{\mathbb{E}} [\mathsf{cost}]$$

$$\mathsf{cmc}^{\mathsf{pROM}}(G, C, m, \sigma_{\mathsf{init}}) = \min_{\mathsf{A}} \underset{\mathsf{h} \leftarrow \mathcal{H}}{\mathbb{E}} \, \mathsf{cmc}^{\mathsf{pROM}}(G, C, m, \mathsf{A}, \sigma_{\mathsf{init}}, \mathsf{h})$$

The *time complexity* of a given adversary is

$$\mathsf{time}^{\mathsf{pROM}}(G, C, m, \mathsf{A}, \sigma_{\mathsf{init}}, \mathsf{h}) = \underset{\mathsf{computeLabel}(G,C,m,\mathsf{A},\sigma_{\mathsf{init}},\mathsf{h})}{\mathbb{E}} [\mathsf{cnt}]$$

We will also consider this notion against the best adversaries from some restricted class of adversaries, in this case we put the class as subscript, like

$$\mathsf{cmc}^{\mathsf{pROM}}_{\mathcal{A}}(G, C, m, \sigma_{\mathsf{init}}) = \min_{\mathsf{A} \in \mathcal{A}} \underset{\mathsf{h} \leftarrow \mathcal{H}}{\mathbb{E}} \, \mathsf{cmc}^{\mathsf{pROM}}(G, C, m, \mathsf{A}, \sigma_{\mathsf{init}}, \mathsf{h})$$

As for pebbling, also here we will consider the more meaningful ϵ variants of these notions

$$\mathsf{cmc}^{\mathsf{pROM}}_{\epsilon}(\mathsf{G}, C, m, \mathsf{A}, \sigma_{\mathsf{init}}, \mathsf{h}) = \inf\left\{ \gamma \,\middle|\, \underset{\mathsf{computeLabel}(G,C,m,\mathsf{A},\sigma_{\mathsf{init}},\mathsf{h})}{\mathbb{P}} [\mathsf{cost} \geq \gamma] \geq 1 - \epsilon \right\}$$

$$\mathsf{cmc}^{\mathsf{pROM}}_{\epsilon}(G, C, m, \sigma_{\mathsf{init}}) = \min_{\mathsf{A}} \underset{\mathsf{h} \leftarrow \mathcal{H}}{\mathbb{E}} \, \mathsf{cmc}^{\mathsf{pROM}}_{\epsilon}(G, C, m, \mathsf{A}, \sigma_{\mathsf{init}}, \mathsf{h})$$

$$\mathsf{time}^{\mathsf{pROM}}_{\epsilon}(\mathsf{G}, C, m, \mathsf{A}, \sigma_{\mathsf{init}}, \mathsf{h}) = \inf\left\{ \gamma \,\middle|\, \underset{\mathsf{computeLabel}(G,C,m,\mathsf{A},\sigma_{\mathsf{init}},\mathsf{h})}{\mathbb{P}} [\mathsf{cnt} \geq \gamma] \geq 1 - \epsilon \right\}$$

2.5   scrypt and the computeLabel Game

We informally discuss the relation between evaluating scrypt in the pROM and the $\mathsf{computeLabel}$ game for the line graph (described below) and, and explain why we will focus on the latter. A similar discussion can be made for Argon2d.

First, recall that scrypt uses a hash function $\mathsf{h} : \{0,1\}^* \to \{0,1\}^w$, and proceeds in two phases, given an input X. In the first phase it computes $X_i = \mathsf{h}^i(X)$ for all $i \in [n]$,[6] and in the second phase, setting $S_0 = X_n$, it computes $S_1, \ldots, S_n$ defined recursively to be

$$S_i = \mathsf{h}(S_{i-1} \oplus X_{\mathsf{int}(S_{i-1})})$$

where $\mathsf{int}(S)$ reduces a w-bit string S to an integer in $[n]$ such that if S is uniform random then $\mathsf{int}(S)$ is (close to) uniform over $[n]$. The final output of $\mathsf{scrypt}^h_n(X) = S_n$. To show that scrypt is memory-hard, we need to lower-bound the CMC required to compute it in the pROM.

---

[6] Here $\mathsf{h}^i(X)$ denotes iteratively applying h i times to the input X.

We argue that to obtain this bound it suffices to restrict our attention to the minimal final value of cost in $\mathsf{cmc}^{\mathsf{PROM}}(L_n, [n], n)$ where $L_n = (V, E)$ is the line graph where $V = [n]$ and $E = \{(i, i + 1) \; : \; i \in [n - 1]\}$. Intuitively this is rather easy to see. Clearly any algorithm which hopes to evaluate scrypt with more than negligble probability must, at some point, compute all X_i values and all S_j values since guessing them is almost impossible. Moreover until S_{i-1} has been computed the value of $\mathsf{int}(S_{i-1})$ – i.e. the challenge label needed to compute S_i – is uniform random and independent, just like the distribution of ith challenge $c \leftarrow C$ in the $\mathsf{computeLabel}$ game. In other words once an algorithm has computed the values $X_1, \ldots, X_n$ computing the values of $S_1, \ldots, S_n$ corresponds exactly to playing the $\mathsf{computeLabel}$ game on graph L_n with challenge set $[n]$ for n rounds. The initial state is exactly the state given to the algorithm as input in the step where it first computes X_n. It is immediate that, when restricted to strategies which don't simply guess relevant outputs of h, then any strategy for computing the values $S_1, \ldots, S_n$ corresponds to a strategy for playing $\mathsf{computeLabel}(L_n, [n], n)$.

In summary, once A has finished the first phase of evaluating scrypt, the second phase essentially corresponds to playing the $\mathsf{computeLabel}$ game on the graph L_n with challenge set $[n]$ for n rounds. The initial state σ_{init} in $\mathsf{computeLabel}$ is the state given to A as input in the first step of round 1 (i.e. in the step when A first computes X_n). It is now immediate that (when restricted to strategies which don't simply guess relevant outputs of h) then any strategy A for computing the second phase of scrypt is essentially a strategy for playing $\mathsf{computeLabel}(L_n, [n], n)$. Clearly the total CMC of A when computing both phases of scrypt is at least the CMC of computing just the second. Thus our lowerbound on $\mathsf{cmc}^{\mathsf{PROM}}(L_n, [n], n)$ in Theorem 15 also gives us a lower bound on the CMC of scrypt_n. (The proof is rather tedious, and omitted from this version of the paper).

Simple Algorithms. Theorem 15 below will make no restrictions on the algorithm playing $\mathsf{computeLabel}$, at the cost of relying on γ_n, for which we only conjecture an upper bound. We do not need such conjectures if we restrict our attention to *simple algorithms* from the class $\mathcal{A}_{SA}$: A simple algorithms $\mathsf{A} \in \mathcal{A}_{SA}$ is one which either stores a value X_i directly in its intermediary states[7] or stores nothing about the value of X_i at all. (They are however permitted to store arbitrary other information in their states.) For example a simple algorithm may not store, say, $X_i \oplus X_j$ or just the first 20 bits of X_i. We note that, to the best of our knowledge, all algorithms in the literature for computing scrypt (or any memory-hard function for that matter) are indeed of this form. For simple algorithms, then we obtain an unconditional lower-bound on the CMC of scrypt by using Theorem 4 below, which only consider pebbling games.

Much as in the more general case above, for the set of algorithms $\mathcal{A}_{SA}$ we can now draw a parallel between computing phase two of scrypt in the pROM and playing the game pebble on the graph L_n with challenge set $[n]$ for n rounds.

---

[7] or at least an equivalent encoding of X_i.

Therefore Theorem 4 immediatly gives us a lower-bound on the CMC of $\texttt{scrypt}_n$ for all algorithms in $\mathcal{A}_{SA}$.

Entangled Adversaries. In fact we can even relax our restrictions on algorithms computing $\texttt{scrypt}$ to the class $\mathcal{A}_{EA}$ of *entangled* algorithms while still obtaining an unconditional lower-bound on the CMC of $\texttt{scrypt}$. In addition to what is permitted for simple algorithms we also allow storing "entangled" information about the values of $X_1, \ldots, X_n$ of the following form. For any subset $L \subseteq [n]$ and integer $t \in [|L|]$ an algorithm can store an encoding of $X_L = \{X_i\}_{i \in L}$ such that if it obtains any t values in L then it can immediatly output all remaining $|L| - t$ values in L with no further information or queries to h. One such encoding uses polynomial interpolation as described in the introduction. Indeed, this motivates our definition of entangled pebbles above.

As shown in the full version, the class $\mathcal{A}_{EA}$ is (in general) strictly more powerful $\mathcal{A}_{SA}$ when it comes to minimizing CMC. Thus we obtain a more general unconditional lower-bound on the CMC of $\texttt{scrypt}$ using Theorem 9 which lower-bounds $\mathsf{cc}^{\downarrow}(L_n, [n], n, n)$, the entangled cumulative pebbling complexity of L_n.

3 Pebbling Lower Bounds for the Line Graph

In this section, we prove lower bounds for the cumulative complexity of the n-round probabilistic pebbling game on the line graph L_n with challenges from $[n]$. We will start with the case without entanglement (i.e., dealing only with black pebbles) which captures the essence of our proof, and then below, extend our proof approach to the entangled case.

Theorem 4 (Pebbling Complexity of the Line Graph). *For all $0 \le k \le n$, and constant $\delta > 0$,*

$$\mathsf{cc}[\delta](L_n, C = [n], n, k) = \Omega_\delta \left(\frac{n^2}{\log^2(n)} \right).$$

We note in passing that the above theorem can be extended to handle a different number of challenges $t \ne n$, as it will be clear in the proof. We dispense with the more general theorem, and stick with the simpler statement for the common case $t = n$ motivated by $\texttt{scrypt}$. The notation Ω_δ indicates that the constant hidden in the Ω depends on δ.

In fact, we also note that our proof allows for more concrete statements as a function of δ, which may be constant. However, not surprisingly, the bound becomes weaker the smaller δ is, but note that if we are only interested in the expectation $\mathsf{cc}(L_n, C = [n], n, k)$, then applying the result with $\delta = O(1)$ (e.g., $\frac{1}{2}$) is sufficient to obtain a lower bound of $\Omega \left(\frac{n^2}{\log^2 n} \right)$.

Proof intuition – the expectation game. Before we turn to the formal proof, we give some high-level intuition. It turns out that most of the proof is going to in fact lower bound the cc of a much simpler game, where the goal is far simpler than covering challenges from $[n]$ with a pebble. In fact, the game will be completely deterministic.

The key observation is that every time a new challenge c_i is drawn, and the player has reached a certain pebbling configuration P, then there is a well-defined expected number $\Phi(P)$ of steps the adversary needs to take *at least* in order to cover the random challenge. We refer to $\Phi(P)$ as the *potential* of P. In particular, the best strategy is the greedy one, which looks at the largest $j = j(c_i) \leq c_i$ on which a pebble is placed, i.e., $j \in P$, and then needs to output a valid sequence of *at least* $c_i - j$ further pebbling configurations, such that the last configuration contains c_i. Note if $j = c_i$, we still need to perform one step to output a valid configuration. Therefore, $\Phi(P)$ is the expected value of $\max(1, c_i - j(c_i))$. We will consider a new game – called the *expectation game* – which has the property that at the beginning of every stage, the challenger just computes $\Phi(P)$, and expects the player T to take $\Phi(P)$ legal steps until T can move to the next stage.

Note that these steps can be totally arbitrary – there is no actual challenge any more to cover. Still, we will be interested in lower bounding the *cumulative complexity* of such a strategy for the expectation game, and it is not obvious how T can keep the cc low. Indeed:

- If the potential is high, say $\Phi(P) = \Omega(n)$, then this means that linearly many steps must be taken to move to the next stage, and since every configuration contains at least one pebble, we pay a cumulative cost of $\Omega(n)$ for the present stage.
- Conversely, if the potential $\Phi(P)$ is *low* (e.g., $O(1)$), then we can expect to be faster. However we will show that this implies that there are *many* pebbles in P (at least $\Omega(n/\Phi(P))$), and thus one can expect high cumulative cost again, i.e.,, linear $\Omega(n)$.

However, there is a catch – the above statements refer to the *initial configurations*. The fact that we have many pebbles at the beginning of a stage and at its end, does not mean we have many pebbles *throughout the whole stage*. Even though the strategy T is forced to pay $\Phi(P)$ steps, the strategy may try to drop as many pebbles as possible for a while, and then adding them back again. *Excluding that this can happen is the crux of our proof.* We will indeed show that for the expectation game, any strategy incurs cumulative complexity $\Omega(n^2/\log^2(n))$ roughly. The core of the analysis will be understanding the behavior of the potential function throughout a stage.

Now, we can expect that a low-cc strategy T for the original parallel pebbling game on L_n gives us one for the expectation game too – after all, for every challenge, the strategy T needs to perform roughly $\Phi(P)$ steps from the initial pebbling configuration when learning the challenge. This is *almost* correct, but again, there is a small catch. The issue is that $\Phi(P)$ is only an expectation, yet we want to have the guarantee that we go for $\Phi(P)$ steps with sufficiently high probability (this is particularly crucial if we want to prove a statement which is parameterized by δ). However, this is fairly simple (if somewhat tedious) to overcome – the idea is that we partition the n challenges into n/λ groups of λ challenges. For every such group, we look at the initial configuration P when learning the first of the next λ challenges, and note that with sufficiently high probability (roughly $e^{-\Omega(\lambda^2)}$ by a Chernoff bound) there will be one challenge (among these λ ones) which is at least (say) $\Phi(P)/2$ away from the closest pebble. This allows us to reduce a strategy for the n-challenge pebbling game on L_n to a strategy for the (n/λ)-round expectation game. The value of λ can be chosen small enough not to affect the overall analysis.

Proof (Theorem 4). As the first step in the proof, we are going to reduce playing the game $\mathsf{pebble}(L_n, C = [n], n, \mathsf{T}, P_{\mathsf{init}})$, for an arbitrary player T and initial pebbling configuration P_{init} ($|P_{\mathsf{init}}| \leq k$), to a simpler (and somewhat different) pebbling game, which we refer to as the *expectation game*.

To this end, we introduce first the concept of a *potential function* $\Phi : 2^{[n]} \to \mathbb{N}$. The *potential* of a pebbling configuration $P = \{\ell_1, \ell_2, \ldots, \ell_m\} \subseteq [n]$ is

$$\Phi(P) := \frac{m}{n} + \frac{1}{n} \sum_{i=0}^{m} (1 + \ldots + (\ell_{i+1} - \ell_i - 1))$$

$$= \frac{m}{n} + \frac{1}{2n} \sum_{i=0}^{m} (\ell_{i+1} - \ell_i) \cdot (\ell_{i+1} - \ell_i - 1) = \frac{1}{2n} \sum_{i=0}^{m} (\ell_{i+1} - \ell_i)^2 - \frac{n+1-2m}{2n}$$

Here $m = |P|$ and we let $\ell_0 = 0$ and $\ell_{m+1} = n + 1$ as notational placeholders. Indeed, $\Phi(P)$ is the expected number of moves required (by an optimal strategy) to pebble a random challenge starting from the pebbling configuration P, where the expectation is over the choice of the random challenge. (Note in particular it is required to pay at least one move even if a pebble is already on the challenge node.) In other words, $\Phi(P)$ is exactly $\mathsf{time}(L_n, [n], 1, \mathsf{T}^*, P)$ for the optimal strategy T^*.

Now we are ready to introduce the *expectation game* which has no challenge. At the beginning of every stage, the challenger only computes $\Phi(P)$, and expects the player T to take $\Phi(P)$ steps until he can move to the next stage. The game $\mathsf{expect}(n, t, \mathsf{T}, P_{\mathsf{init}})$ is played by a pebbler T as depicted in Fig. 4.

In the following, for a (randomized) pebbler T and initial configuration P_{init}, we write $\mathsf{expect}_{n,t}(\mathsf{T}, P_{\mathsf{init}})$ for the output of the expectation game; note the output only depends on the randomness of pebbler T and configuration P_{init}. We similarly define the *cumulative complexity of the expectation game*

$$\mathsf{cc}[\delta](\mathsf{expect}_{n,t}(\mathsf{T}, P_{\mathsf{init}})) := \inf \left\{ \gamma \;\middle|\; \mathop{\mathbb{P}}_{\mathsf{expect}(n,t,\mathsf{T},P_{\mathsf{init}})} [\mathsf{cost} \geq \gamma] \geq 1 - \epsilon \right\}$$

$$\mathsf{cc}[\delta](\mathsf{expect}_{n,t,k}) := \min_{\substack{\mathsf{T}, P_{\mathsf{init}} \subseteq V \\ |P_{\mathsf{init}}| \leq k}} \left\{ \mathsf{cc}[\delta](\mathsf{expect}_{n,t}(\mathsf{T}, P_{\mathsf{init}})) \right\}$$

The expectation game $\mathsf{expect}_{n,t,k}$ has an important feature: because the randomness is only over the pebbler's coins, these coins can be fixed to their optimal

expect($n, t, \mathsf{T}, P_{\mathsf{init}}$): The t-round expectation game of parameter n and an initial pebbling configuration $P_{\mathsf{init}} \subseteq V$ is played by challenger and player T as follows.

1. Initialize $\mathsf{cnt} := 0$, $\mathsf{round} := 0$, $P_{\mathsf{cnt}} := P_{\mathsf{init}}$ and $\mathsf{cost} := |P_{\mathsf{init}}|$.
2. Player T submits a sequence of *non-empty* pebbling configurations $(P_{\mathsf{round},1}, \ldots, P_{\mathsf{round},t_{\mathsf{round}}}) \subset [n]^{\times t_{\mathsf{round}}}$,
3. Let $P_{\mathsf{round},0} := P_{\mathsf{cnt}}$. Check if $t_{\mathsf{round}} \geq \Phi(P_{\mathsf{cnt}})$ and $\forall i \in [t_{\mathsf{round}}]$

$$\forall v \in P_{\mathsf{round},i} \setminus P_{\mathsf{round},i-1} : \mathsf{parent}(v) \in P_{\mathsf{round},i-1} .$$

   If check fails, output $\mathsf{cnt} = \mathsf{cost} = \infty$ and halt.
4. $\mathsf{cnt} := \mathsf{cnt} + t_{\mathsf{round}}$.
5. $\mathsf{cost} := \mathsf{cost} + \sum_{j=1}^{t_{\mathsf{round}}} |P_{\mathsf{round},j}|$.
6. $P_{\mathsf{cnt}} := P_{\mathsf{round},t_{\mathsf{round}}}$.
7. $\mathsf{round} := \mathsf{round} + 1$. If $\mathsf{round} < t$ go to step 2, otherwise if $\mathsf{round} = t$ the experiment is over, the output is the final count cnt and the cumulative cost cost.

Fig. 4. The Expectation Game

choice without making the overall cc worse. This implies that $\mathsf{cc}_\delta(\mathsf{expect}_{n,t,k}) = \mathsf{cc}_0(\mathsf{expect}_{n,t,k})$ for all $\delta \geq 0$. In particular, we use the shorthand $\mathsf{cc}(\mathsf{expect}_{n,t,k})$ for the latter.

The remainder of the proof consists of the following two lemmas. Below, we combine these two lemmas in the final statement, before turning to their proofs. (The proof of Lemma 5 is deferred to the full version for lack of space, and relies on the intuition given above.)

Lemma 5 (Reduction to the Expectation Game). *For all n, t, k, λ, and any $\delta > 3\mu(t, \lambda)$, we have*

$$\mathsf{cc}(\mathsf{expect}_{n,t,k}) = \mathsf{cc}_{\delta-3\mu(t,\lambda)}(\mathsf{expect}_{n,t,k}) \leq 2 \cdot \mathsf{cc}_\delta(L_n, C = [n], t \cdot \lambda, k) ,$$

where $\mu(t, \lambda) = t \cdot e^{-\lambda^2/8}$.

To give some intuition about the bound, note that in general, for every $\delta' \leq \delta$, we have $\mathsf{cc}[\delta'](\mathsf{expect}_{n,t,k}) \leq \mathsf{cc}[\delta](\mathsf{expect}_{n,t,k})$. This is because if a c is such that for all T and P_{init} we have $\Pr \mathsf{expect}_{n,t}(\mathsf{T}, P_{\mathsf{init}}) \geq c \geq 1 - \delta'$, then also $\Pr \mathsf{expect}_{n,t}(\mathsf{T}, P_{\mathsf{init}}) \geq c \geq 1 - \delta$. Thus the set from which we are taking the supremum only grows bigger as δ increases. In the specific case of Lemma 5, the $3\mu(t, \lambda)$ offset captures the loss of our reduction.

Lemma 6 (CC Complexity of the Expectation Game). *For all $t, 0 \leq k \leq n$ and $\epsilon > 0$, we have*

$$\mathsf{cc}(\mathsf{expect}_{n,t,k}) \geq \left\lfloor \frac{\epsilon t}{2} \right\rfloor \cdot \frac{n^{1-\epsilon}}{6} .$$

To conclude the proof before turning to the proofs of the above two lemmas, we choose t, λ such that $t \cdot \lambda = n$, and $\mu(t, \lambda) = t \cdot e^{-\lambda^2/8} < \delta/3$. We also set $\epsilon = 0.5 \log\log(n)/\log(n)$, and note that in this case $n^{1-\epsilon} = n/\sqrt{\log(n)}$. In particular, we can set $\lambda = O(\sqrt{\log t})$, and can choose e.g. $t = n/\sqrt{\log n}$. Then, by Lemma 6,

$$\mathsf{cc}(\mathsf{expect}_{n,t,k}) \geq \left\lfloor \frac{\epsilon t}{2} \right\rfloor \cdot \frac{n^{1-\epsilon}}{6} = \Omega\left(\frac{n^2}{\log^2(n)}\right) .$$

This concludes the proof of Theorem 4.

Proof (Proof of Lemma 6). First we observe if a pebbling configuration P has potential Φ, the size $|P|$ of the pebbling configuration (i.e., the number of vertices on which a pebble is placed) will be at least $\frac{n}{6\cdot\Phi}$. We give a formal proof for completeness.[8]

Lemma 7. *For every* non-empty *pebbling configuration $P \subseteq [n]$, we have*

$$\Phi(P) \cdot |P| \geq \frac{n}{6} .$$

Proof. Let $m = |P| \geq 1$, by definition of *potential*:

$$\Phi(P) = \frac{1}{2n} \sum_{i=0}^{m} (\ell_{i+1} - \ell_i)^2 - \frac{n+1-2m}{2n} ,$$

where $\ell_0 = 0$ and $\ell_{m+1} = n+1$ are notational placeholders. Since $\Phi(P) \geq 1$ and $m \geq 1$, we have $\frac{n+1-2m}{2n} \leq \frac{1}{2} \leq \frac{1}{2} \cdot \Phi(P)$. Therefore

$$\Phi(P) \geq \frac{2}{3} \cdot \frac{1}{2n} \sum_{i=0}^{m} (\ell_{i+1} - \ell_i)^2 ,$$

since $m \geq \frac{m+1}{2}$, multiply the left side by m and the right side by $\frac{m+1}{2}$, we have

$$\Phi(P) \cdot m \geq \frac{2}{3} \left(\frac{1}{2n} \sum_{i=0}^{m} (\ell_{i+1} - \ell_i)^2 \right) \cdot \frac{m+1}{2} = \frac{1}{6n} \left(\sum_{i=0}^{m} (\ell_{i+1} - \ell_i)^2 \right) \cdot (m+1)$$

Therefore $\Phi(P) \cdot m \geq \frac{n}{6}$ follows, since by Cauchy-Schwarz Inequality we have

$$\left(\sum_{i=0}^{m} (\ell_{i+1} - \ell_i)^2 \right) \cdot (m+1) \geq \left(\sum_{i=0}^{m} (\ell_{i+1} - \ell_i) \right)^2 \geq n^2 .$$

$\square$

---

[8] Note that the contra-positive is not necessarily true. A simple counter-example is when pebbles are placed on vertices $[0, n/2]$ of $C_{1=n}$ (that is, $|P| = O(n)$). The expected number of moves in this case is still $\Omega(n)$.

Also, the following claim provides an important property of the potential function.

Lemma 8. *In one iteration, the potential can decrease by at most one.*

Proof. Consider an arbitrary configuration $P = \{\ell_1, \ell_2, \ldots, \ell_m\} \subseteq [n]$. The best that a pebbling algorithm can do to decrease the potential is to place new pebbles next to *all* the current pebbles – let's call the new configuration P'. That is,

$$P' = \{\ell_1, \ell_1 + 1, \ell_2, \ell_2 + 1, \ldots, \ell_m, \ell_m + 1\} \subseteq [n].$$

The potential of the new configuration is

$$\Phi(P') = \frac{1}{2n} \left(\ell_1^2 + \sum_{i=1}^{m} 1 + (\ell_{i+1} - (\ell_i + 1))^2 \right) - \frac{n+1-2|P'|}{2n} \tag{11}$$

$$= \frac{1}{2n} \left(m + \sum_{i=0}^{m} \left((\ell_{i+1} - \ell_i)^2 - 2(\ell_{i+1} - \ell_i) + 1 \right) \right) - \frac{n+1-2|P'|}{2n} \tag{12}$$

$$\geq \frac{1}{2n} \left(m + \sum_{i=0}^{m} \left((\ell_{i+1} - \ell_i)^2 - 2(\ell_{i+1} - \ell_i) + 1 \right) \right) - \frac{n+1-2m}{2n} \tag{13}$$

$$\geq \Phi(P) + \frac{m}{n} - \frac{1}{n} \sum_{i=0}^{m} (\ell_{i+1} - \ell_i) \geq \Phi(P) - 1 \tag{14}$$

where the first inequality holds because $|P'| \geq m$. $\qquad\square$

Assume without loss of generality the pebbler T is legal and deterministic. Consider a particular round $i \in [t]$ of the expectation game. Let P and P' denote the initial and final pebbling configurations in the i-th round, and let us denote by $\phi_i = \Phi(P)$ the potential of the initial configuration in round i. Depending on the value of $\Phi(P')$, we classify the pebbling sequence from P to P' into three different categories:

<u>Type 1:</u> $\Phi(P') > \phi_i \cdot n^{\epsilon/2}$; or
<u>Type 2:</u> $\Phi(P') \leq \phi_i \cdot n^{\epsilon/2}$ – we have two sub-cases:
    <u>Type 2a:</u> the potential was *always* less than $\phi_i \cdot n^{\epsilon}$ for all the intermediate pebbling configurations from P to P'; or
    <u>Type 2b:</u> the potential went above $\phi_i \cdot n^{\epsilon}$ for some intermediate configuration.

With each type, we associate a cost that the pebbling algorithm has to pay, which lower bounds the contribution to the cumulative complexity of the pebbling configurations generated during this stage. The pebbling algorithm can carry out pebbling of Type 1 for *free*[9] – however, the latter two have accompanying costs.

---

[9] The cost might be greater than zero, but setting it to zero doesn't affect the lower bound.

- For pebbling sequences of Type 2a, the corresponding cumulative cost is at least $\phi_i \cdot \frac{n}{6 \cdot \phi_i n^\epsilon} = \frac{1}{6} n^{1-\epsilon}$ since by Lemma 7, the size of the pebbling configuration is never less than $\frac{n}{6\phi_i n^\epsilon}$ during all intermediate iterations and in stage i valid pebbler must produce at least ϕ_i configurations.
- For sequences of Type 2b, by Lemma 8, it follows that in a Type 2b sequence it takes at least $\phi_i(n^\epsilon - n^{\epsilon/2})$ steps to decrease the potential from $\phi \cdot n^\epsilon$ to $\phi_i \cdot n^{\epsilon/2}$, and the size of the pebbling configuration is at least $\frac{n}{6\phi_i n^\epsilon}$ in every intermediate step by Lemma 7. Therefore, the cumulative cost is at least

$$\phi_i(n^\epsilon - n^{\epsilon/2}) \cdot \frac{n}{6\phi_i n^\epsilon} \geq \frac{n}{6} - \frac{n^{1-\epsilon/2}}{6} \geq \frac{1}{6} n^{1-\epsilon},$$

where the last inequality follows for sufficiently large n.

To conclude the proof, we partition the $t \geq \lceil 2/\epsilon \rceil$ rounds into groups of consecutive $\lceil 2/\epsilon \rceil$ phases. We observe that any group *must* contain at least one pebbling sequence of Type 2: otherwise, with ϕ being the potential at the beginning of the first of theses $2/\epsilon$ phases, the potential at the end would be strictly larger than

$$\phi n^{\frac{\epsilon}{2} \cdot \frac{2}{\epsilon}} \geq \phi \cdot n > n/2$$

which cannot be, as the potential can be at most $\frac{n}{2}$. By the above, however, the cumulative complexity of each group of phases is at least $\frac{n^{1-\epsilon}}{6}$, and thus we get

$$cc(\mathsf{expect}_{n,t,k}) \geq \left\lfloor \frac{\epsilon t}{2} \right\rfloor \cdot \frac{n^{1-\epsilon}}{6}, \tag{15}$$

which concludes the proof of Lemma 6. □

As the second result, we show that the above theorem also holds for the entangled case.

Theorem 9 (Entangled Pebbling Complexity of the Line Graph). *For all $0 \leq k \leq n$ and constant $\delta > 0$,*

$$cc_\delta^\updownarrow(L_n, C = [n], n, k) = \Omega\left(\frac{n^2}{\log^2 n}\right).$$

Luckily, it will not be necessary to repeat the whole proof. We will give now a proof sketch showing that in essence, the proof follows by repeating the same format and arguments as the one for Theorem 4, using Lemma 3 as a tool.

Proof (Sketch). One can prove the theorem following exactly the same framework of Theorem 4, with a few differences. First off, we define a natural entangled version of the expectation game where, in addition to allowing entanglement in a pebbling configuration, we define the potential as

$$\Phi^\updownarrow(P) = \mathsf{time}^\updownarrow(L_n, C = [n], 1, \mathsf{T}^{*,\updownarrow}, P),$$

i.e., the expected time complexity for *one* challenge of an optimal entangled strategy $\mathsf{T}^{*,\updownarrow}$ starting from the (entangled) pebbling configuration P.

First off, a proof similar to the one of Lemma 5, based on a Chernoff bound, can be used to show that if we separate challenges in t chunks of λ challenges each, and we look at the configuration P at the beginning of each of the t chunks, then there exists at least one challenge (out of λ) which requires spending time $\Phi^\updownarrow(P)$ to be covered, except with small probability.

A lower bound on the cumulative complexity of the (entangled) expectaton game follows exactly the same lines as the proof as Lemma 6. This is because the following two facts (which correspond to the two lemmas in the proof of Lemma 6) are true also in the setting with entanglement:

- First off, for every P and $\mathbf{T}^{*,\updownarrow}$ such that $\Phi^\updownarrow(P) = \mathsf{time}^\updownarrow(L_n, C = [n], 1, \mathbf{T}^{*,\updownarrow}, P)$, Lemma 3 guarantees that there exist a (regular) pebbling strategy $\mathbf{T}'$ and a (regular) pebbling configuration P' such that $|P|_\updownarrow \geq |P'|$ and

$$\Phi^\updownarrow(P) = \mathsf{time}^\updownarrow(L_n, C = [n], 1, \mathbf{T}^{*,\updownarrow}, P)$$
$$\geq \mathsf{time}(L_n, C = [n], 1, \mathbf{T}', P') \geq \Phi(P') .$$

Therefore, by Lemma 7,

$$|P|_\updownarrow \cdot \Phi^\updownarrow(P) \geq |P'| \cdot \Phi(P') \geq \frac{n}{6} . \tag{16}$$

- Second, the potential can decrease by at most one when making an arbitrary step from one configuration P to one configuration P'. This is by definition – assume it were not the case, and $\Phi^\updownarrow(P') < \Phi^\updownarrow(P) - 1$. Then, there exists a strategy to cover a random challenge starting from P which first moves to P' in one step, and then applies the optimal strategy achieving expected time $\Phi^\updownarrow(P')$. The expected number of steps taken by this strategy is smaller than $\Phi^\updownarrow(P)$, contradicting the fact that $\Phi^\updownarrow(P)$ is the optimal number of steps required by any strategy.   □

4 From Pebbling to pROM

4.1 Trancscipts and Traces

Below we define the notion of a trace and transcript, which will allow us to relate the computeLabel and pebble$^\updownarrow$ experiments. For any possible sequence of challenges $c \in C^m$, let cnt_c denote the number of steps (i.e., the variable cnt) made in the computeLabel$(G, C, m, \mathsf{A}, \sigma_{\mathsf{init}}, \mathsf{h})$ experiment conditioned on the m challenges being c (note that once c is fixed, the entire experiment is deterministic, so cnt_c is well defined). Let $\tau_c = q_1|q_2|\ldots|q_{\mathsf{cnt}_c}$ be the *trace* of the computation: here $q_1 \subset [n]$ means that the first batch of parallel queries are the queries required to output the labels $\{\ell_i, i \in q_1\}$, etc.

For example, for the Graph in Fig. 5, $\tau_7 = 2|4, 5|7$ corresponds to a first query $\ell_2 = \mathsf{h}(2)$, then two parallel queries $\ell_4 = \mathsf{h}(4, \ell_1), \ell_5 = \mathsf{h}(5, \ell_2)$, and then the final query computing the label of the challenge $\ell_7 = \mathsf{h}(7, \ell_4, \ell_5, \ell_6)$.

A trace as a pebbling. We can think of a trace as a parallel pebbling, e.g., $\tau_7 = 2|4,5|7$ means we pebble node 2 in the first step, nodes 4, 5 in the second, and 7 in the last step. We say that an initial (entangled) pebbling configuration P_{init} is consistent with a trace τ, if starting from P_{init}, τ is a valid pebbling sequence. E.g., consider again the traces $\tau_7 = 2|4,5|7, \tau_8 = 3|6|8$ for the graph in Fig. 5, then $P_{\text{init}} = \{1, 5, 6\}$ is consistent with τ_7 and τ_8, and it's the smallest initial pebbling having this property. In the entangled case, $P_{\text{init}}^{\updownarrow} = \{\langle 1 \rangle_0, \langle 5, 6 \rangle_1\}$ is consistent with τ_7, τ_8. Note that in the entangled case we only need a pebbling configuration of weight 2, whereas the smallest pebbling configuration for the standard pebbling game has weight 3. In fact, there are traces where the gap between the smallest normal and entangled pebbling configuration consistent with all the traces can differ by a factor $\Theta(n)$.

Turning a trace into a transcript. We define the implications T_c of a trace $\tau_c = q_1|q_2|\ldots|q_{\text{cnt}_c}$ as follows. For $i = 1, \ldots, \text{cnt}_c$, we add the implication $(v_i) \rightarrow (f_i)$, where $v_i \subset [n]$ denotes all the vertices whose labels have appeared either as inputs or outputs in the experiment so far, and f_i denotes the labels contained in the inputs from this round which have never appeared before (if the guess for the challenge label in this round is non-empty, i.e., $\ell \neq \perp$, then we include ℓ in f_i).

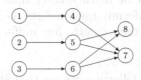

Fig. 5. Graph used in Example 10.

Example 10. Consider the graph from Fig. 5 with $m = 1$ and challenge set $C = \{7, 8\}$, and traces

$$\tau_7 = 2|4,5|7 \quad \text{and} \quad \tau_8 = 3|6|8$$

We have

$$T_7 = \{(2) \rightarrow 1, (1, 2, 4, 5) \rightarrow 6\} \quad T_8 = \{(3, 6) \rightarrow 5\} \tag{17}$$

where e.g. $(2) \rightarrow 1$ is in there as the first query is $\ell_2 = h(2)$, and the second query is $\ell_4 = h(4, \ell_1)$ and in parallel $\ell_5 = h(5, \ell_2)$. At this point we so far only observed the label $v_2 = \{\ell_2\}$, so the label $f_2 = \{\ell_1\}$ used as input in this query is fresh, which means we add the implication $(2) \rightarrow 1$.

Above we formalised how to extract a transcript T_c from $(G, C, m, \mathsf{A}, \sigma_{\text{init}}, \mathsf{h})$, with

$$T(G, C, m, \mathsf{A}, \sigma_{\text{init}}, \mathsf{h}) = \cup_{c \in C^m} T_c$$

we denote the union of all T_c's.

4.2 Extractability, Coverability and a Conjecture

In this section we introduce the notion of extractability and coverability of a transcript. Below we first give some intuition what these notions have to do with the computeLabel and pebble$^\downarrow$ experiments.

Extractability intuition. Consider the experiment computeLabel$(G, C, m, \mathsf{A}, \sigma_{\mathsf{init}}, \mathsf{h})$. We can invoke A on some particular challenge sequence $c \in C^m$, and if at some point A makes a query whose input contains a label ℓ_i which has not appeared before, we can "extract" this value from $(\mathsf{A}, \sigma_{\mathsf{init}})$ without actually querying h for it. More generally, we can run A on several challenge sequences scheduling queries in a way that will maximise the number of labels that can be extracted from $(\mathsf{A}, \sigma_{\mathsf{init}})$. To compute this number, we don't need to know the entire input/output behaviour of A for all possible challenge sequences, but the transcript $T = T(G, C, m, \mathsf{A}, \sigma_{\mathsf{init}}, \mathsf{h})$ is sufficient. Recall that T contains implication like $(1, 5, 6) \to 3$, which means that for some challenge sequence, there's some point in the experiment where A has already seen the labels ℓ_1, ℓ_5, ℓ_6, and at this point makes a query whose input contains a label ℓ_3 (that has not been observed before). Thus, given σ_{init} and ℓ_1, ℓ_5, ℓ_6 we can learn ℓ_3.

We denote with $ex(T)$ the maximum number of labels that can be extracted from T. If the labels are uniformly random values in $\{0, 1\}^w$, then it follows that σ_{init} will almost certainly not be much smaller than $ex(T) \cdot w$, as otherwise we could compress $w \cdot ex(T)$ uniformly random bits (i.e., the extracted labels) to a string which is shorter than their length, but uniformly random values are not compressible.

Coverability intuition. In the following, we say that an entangled pebbling experiment pebble$^\downarrow(G, C, m, \mathsf{P}, P_{\mathsf{init}}{}^\downarrow)$ mimics the computeLabel$(G, C, m, \mathsf{A}, \sigma_{\mathsf{init}}, \mathsf{h})$ experiment if for every challenge sequence the following is true: whenever A makes a query to compute some label $\ell_i = h(i, \ell_{p_1}, \ldots, \ell_{p_t})$, P puts a (normal) pebble on i. For this $P_{\mathsf{init}}{}^\downarrow$ must contain (entangled) pebbles that allow to cover every implication in T (as defined above), e.g., if $(1, 5, 6) \to 3 \in T$, then from the initial pebbling $P_{\mathsf{init}}{}^\downarrow$ together with the pebbles $\langle 1 \rangle_0, \langle 5 \rangle_0, \langle 6 \rangle_0$ seen so far it must be possible derive $\langle 3 \rangle_0$, i.e., $\langle 3 \rangle_0 \in \mathsf{closure}(P_{\mathsf{init}}{}^\downarrow \cup \langle 1 \rangle_0, \langle 5 \rangle_0, \langle 6 \rangle_0\})$. We say that such an initial state $P_{\mathsf{init}}{}^\downarrow$ covers T. We're interested in the maximum possible ratio of $\max_T$ over $[n]$ $\min_{P_{\mathsf{init}}{}^\downarrow, P_{\mathsf{init}}{}^\downarrow \text{ covers } T} |P_{\mathsf{init}}{}^\downarrow|_\updownarrow / ex(T)$, which we'll denote with γ_n, thus, if any T is k extractable, it can be covered by an initial pebbling $P_{\mathsf{init}}{}^\downarrow$ of weight $\gamma_n \cdot k$. The best current lower bound we have on γ_n is 1.5, we conjecture that γ_n is small, polylog(n) or even constant. We will prove in Sect. 4.3 that pebbling time complexity implies pROM time complexity for any graph, and in Sect. 4.4 that CC complexity implies cumulative complexity in the pROM model for the scrypt graph. The loss in our reductions will depend on γ_n. Assuming $\gamma_n = \Theta(1)$ we get the best bounds one can hope for, but already $\gamma_n \in o(n)$ would give the first non-trivial bounds on pROM complexity.

Definitions. Let $n \in \mathbb{N}$. An "implication" $(\mathcal{X}) \to z$ given by a value $z \in [n]$ and a subset $\mathcal{X} \subset [n] \setminus z$ means that "knowing $\mathcal{X}$ gives z for free". We use $(\mathcal{X}) \to \mathcal{Z}$ as a shortcut for the set of implications $\{(\mathcal{X}) \to z \ : \ z \in \mathcal{Z}\}$.

A transcript is a set of of implications. Consider a transcript $T = \{\alpha_1, \ldots, \alpha_\ell\}$, each α_i being an implication. We say that a transcript T is k $(0 \le k \le n)$ extractable if there exists an extractor E that makes at most $n - k$ queries in the following game:

- At any time E can query for a value in $[n]$.
- Assume E has values $\mathcal{L} \subset [n]$ and there exists an implication $(\mathcal{X}) \to z \in T$ where $\mathcal{X} \subset \mathcal{L}$, then E gets the value z "for free".
- The game is over when E has received all of $[n]$.

Every (even an empty) transcript T is 0 extractable as E can always simply ignore T and query for $1, 2, \ldots, n$. Let

$$ex(T) = \max_k (T \text{ is } k\text{-extractable})$$

Example 11. Let $n = 5$ and consider the transcript

$$T = \{(1,2) \to 3, (2,3) \to 1, (3,4) \to 2, (1) \to 4\} \tag{18}$$

This transcript is 2 but not 3 extractable. To see 2 extractability consider the E which first asks for 1, then gets 4 for free (due to $(1) \to 4$), next E asks for 2 and gets 3 for free (due to $(1,2) \to 3$).

A set S of entangled pebbles covers an implication $(\mathcal{X}) \to z$ if $z \in \mathsf{closure}(S \cup \langle \mathcal{X} \rangle_0)$, with $\mathsf{closure}$ as defined in Definition 1.

Definition 12 (k-coverable). *We say that a transcript T is k-coverable if there exists a set of entangled pebbles S of total weight k such that every implication in T is covered by S. With $cw(T)$ we denote the minimum weight of an S covering T:*

$$cw(T) = \min_{S \text{ that covers } T} |S|_\updownarrow$$

Note that every transcript is trivially n coverable by using the pebble $\langle 1, \ldots, n \rangle_0$ of weight n which covers every possible implication. For the 2 extractable transcript from Example 11, a set of pebbles of total weight 2 covering it is

$$S = \{\langle 1,2,3 \rangle_2, \langle 1,4 \rangle_1\} \tag{19}$$

For example $(3,4) \to 2$ is covered as $2 \in \mathsf{closure}(\langle 1,2,3 \rangle_2, \langle 1,4 \rangle_1, \langle 3,4 \rangle_0) = \{1,2,3,4\}$: we first can set $\Gamma = \{3,4\}$ (using $\langle 3,4 \rangle_0$), then $\Gamma = \{1,3,4\}$ using $\langle 1,4 \rangle_1$, and then $\Gamma = \{1,2,3,4\}$ using $\langle 1,2,3 \rangle_2$.

We will be interested in the size of the smallest cover for a transcript T. One could conjecture that every k-extractable transcript is k-coverable. Unfortunately this is not true, consider the transcript

$$T^* = \{(2,5) \to 1, (1,3) \to 2, (2,4) \to 3, (3,5) \to 4, (1,4) \to 5\} \tag{20}$$

We have $ex(T^*) = 2$ (e.g. via query $2, 4, 5$ and extract $1, 3$ using $(2, 5) \rightarrow 1, (2, 4) \rightarrow 3$), but it's not 2-coverable (a cover of weight 3 is e.g. $\{\langle 5, 1 \rangle_1\}, \langle 2, 3, 4 \rangle_1\}$). With γ_n we denote the highest coverability vs extractability ration that a transcript over $[n]$ can have:

Conjecture 13. Let

$$\gamma_n = \max_{T \text{ over } [n]} \min_{S \text{ that covers } T} \frac{|S|_{\updownarrow}}{ex(T)} = \max_{T \text{ over } [n]} \frac{cw(T)}{ex(T)}$$

then (weak conjecture) $\gamma_n \in \text{polylog}(n)$, or even (strong conjecture) $\gamma_n \in \Theta(1)$.

By the example Eq. (20) above, γ_n is at least $\gamma_n \geq \gamma_5 \geq 3/2$. We will update the full version of this paper as we get aware on progress on (dis)proving this conjecture. In the full version we also introduce another parameter $shannon(w)$, which can give better lower bounds on the size of a state required to realize a given transcript in terms of Shannon entropy.

4.3   Bounding pROM Time Using Pebbling Time

We are ultimately interested in proving lower bounds on time and cumulative complexity in the parallel ROM model. We first show that pebbling time complexity implies time complexity in the pROM model, the reduction is optimal up to a factor γ_n. Under conjecture 13, this basically answers the main open problem left in the Proofs of Space paper [9]. In the theorem below we need the label length w to in the order of $m \log(n)$ to get a lower bound on $|\sigma_{\text{init}}|$. For the proofs of space application, where $m = 1$, this is a very weak requirement, but for scrypt, where $m = n$, this means we require rather long labels (the number of queries q will be $\leq n^2$, so the $\log(q)$ term can be ignored).

Theorem 14. *Consider any* $G = (V, E), C \subseteq V, m \in \mathbb{N}, \epsilon \geq 0$ *and algorithm* A. *Let* $n = |V|$ *and* γ_n *be as in Conjecture 13. Let* $\mathcal{H}$ *contain all functions* $\{0, 1\}^* \rightarrow \{0, 1\}^w$, *then with probability* $1 - 2^{-\Delta}$ *over the choice of* h $\leftarrow \mathcal{H}$ *the following holds for every* $\sigma_{\text{init}} \in \{0, 1\}^*$. *Let* q *be an upper bound on the total number of* h *queries made by* A *and let*

$$k = \frac{|\sigma_{\text{init}}| + \Delta}{(w - m \log(n) - \log(q))}$$

(so $|\sigma_{\text{init}}| \approx k \cdot w$ *for sufficiently large* w*), then*

$$\text{time}^{pROM}(G, C, m, \text{A}, \sigma_{\text{init}}, \text{h}) \geq \text{time}(G, C, m, \lceil k \cdot \gamma_n \rceil)$$

and for every $1 > \epsilon \geq 0$

$$\text{time}_\epsilon^{pROM}(G, C, m, \text{A}, \sigma_{\text{init}}, \text{h}) \geq \text{time}_\epsilon(G, C, m, \lceil k \cdot \gamma_n \rceil)$$

In other words, if the initial state is roughly $k \cdot w$ bits large (i.e., it's sufficient to store k labels), then the pROM time complexity is as large as the pebbling time complexity of $\mathsf{pebble}(G, C, m)$ for any initial pebbling of size $k \cdot \gamma_n$. Note that the above theorem is basically tight up to the factor γ_n: consider an experiment $\mathsf{time}(G, C, m, \mathsf{P}, P_{\mathsf{init}})$, then we can come up with a state σ_{init} of size $k \cdot w$, namely $\sigma_{\mathsf{init}} = \{\ell_i, i \in P_{\mathsf{init}}\}$, and define A to mimic P, which then implies

$$\mathsf{time}_\epsilon^{\mathsf{pROM}}(G, C, m, \mathsf{A}, \sigma_{\mathsf{init}}, \mathsf{h}) = \mathsf{time}_\epsilon(G, C, m, \mathsf{P}, P_{\mathsf{init}}) \quad \text{with} \quad |\sigma_{\mathsf{init}}| = k \cdot w$$

in particular, if we let $\mathsf{P}, P_{\mathsf{init}}$ be the strategy and initial pebbling of size k minimising time complexity we get

$$\mathsf{time}_\epsilon^{\mathsf{pROM}}(G, C, m, \mathsf{A}, \sigma_{\mathsf{init}}, \mathsf{h}) \geq \mathsf{time}_\epsilon(G, C, m, k) \quad \text{with} \quad |\sigma_{\mathsf{init}}| = k \cdot w$$

Wlog. we will assume that A is deterministic (if A is probabilistic we can always fix some "optimal" coins). Below we prove two claims which imply Theorem 14.

Claim. With probability $1 - 2^{-\Delta}$ over the choice of $\mathsf{h} \leftarrow \mathcal{H}$; If the transcript $T(G, C, m, \mathsf{A}, \sigma_{\mathsf{init}}, \mathsf{h})$ is k-extractable, then

$$|\sigma_{\mathsf{init}}| \geq k \cdot (w - m\log(n) - \log(q)) - \Delta \tag{21}$$

where q is an upper bound on the total number of h queries made by A.

Proof. Let L be an upper bound on the length of queries made by A, so we can assume that the input domain of h is finite, i.e., $\mathsf{h} : \{0,1\}^{\leq L} \rightarrow \{0,1\}^w$. Let $|\mathsf{h}| = 2^L \cdot w$ denote the size of h's function table.

Let $\ell_{i_1}, \ldots, \ell_{i_k}$ be the indices of the k labels (these must not be unique) that can be "extracted", and let h^- denote the function table of h, but where the rows are in a different order (to be defined), and the rows corresponding to the queries that output the labels to be extracted are missing, so $|\mathsf{h}| - |\mathsf{h}^-| = k \cdot w$.

Given the state σ_{init}, the function table of h^- and some extra information α discussed below, we can reconstruct the entire function table of h. As this table is uniform, and a uniform string of length s cannot be compressed below $s - \Delta$ bits except with probability $2^{-\Delta}$, we get that with probability $1 - 2^{-\Delta}$ Eq. (21) must hold, i.e.,

$$|\sigma_{\mathsf{init}}| + |\mathsf{h}^-| + |\alpha| \geq |\mathsf{h}| - \Delta$$

as $|\mathsf{h}| - |\mathsf{h}^-| = k \cdot w$ we get

$$|\sigma_{\mathsf{init}}| \geq k \cdot w - |\alpha| - \Delta$$

It remains to define α and the order in which the values in h^- are stored. For every label to be extracted, we specify on what challenge sequence to run the adversary A, and where exactly in this execution the label we want to extract appears (as part of a query made by A). This requires up to $m\log(n) + \log(q)$ bits for every label to be extracted, so

$$|\alpha| \leq k \cdot (m \cdot \log(n) + \log(q))$$

The first part of h^- now contains the outputs of h in the order in which they are requested by the extraction procedure just outlined (if a query is made twice, then we have to remember it and not simply use the next entry in h^-). Let us stress thaw we only store the w bit long outputs, not the inputs, this is not a problem as we learn the corresponding inputs during the extraction procedure. The entries of h which are not used in this process and are not extracted labels, make up the 2nd part of the h^- table. As we know for which inputs we're still missing the outputs, also here we just have to store the w bit long outputs such that the inputs are the still missing inputs in lexicographic order.

Let us mention that if A behaved nice in the sense that all its queries are on inputs which are actually required to compute the corresponding labels, then we would only need $\log(n)$ bits extra information per label, namely the indices $i_1, \ldots, i_k$. But as A can behave arbitrarily, we can't tell when A actually uses real labels as inputs or some junk, and thus must exactly specify where the real labels to be extracted show up.

Claim. If the transcript $T = T(G, C, m, A, \sigma_{\text{init}}, h)$ is k-extractable (i.e., $ex(T) = k$), then

$$\text{time}^{\text{pROM}}(G, C, m, A, \sigma_{\text{init}}, h) \geq \text{time}(G, C, m, \lceil k \cdot \gamma_n \rceil) \tag{22}$$

and for any $1 > \epsilon \geq 0$

$$\text{time}_\epsilon^{\text{pROM}}(G, C, m, A, \sigma_{\text{init}}, h) \geq \text{time}_\epsilon(G, C, m, \lceil k \cdot \gamma_n \rceil) \tag{23}$$

Proof. We will only prove the first statement Eq. (22). As T is k-extractable, there exist $(P, P^{\updownarrow})$ where $P^{\updownarrow}$ is of weight $\leq \lceil k \cdot \gamma_n \rceil$ such that

$$\text{time}^{\updownarrow}(G, C, m, P, P^{\updownarrow}) = \text{time}^{\text{pROM}}(G, C, m, A, \sigma_{\text{init}}, h)$$

The claim now follows as

$$\text{time}^{\updownarrow}(G, C, m, P, P^{\updownarrow}) \geq \text{time}^{\updownarrow}(G, C, m, \lceil k \cdot \gamma_n \rceil) = \text{time}(G, C, m, \lceil k \cdot \gamma_n \rceil)$$

where the first inequality follows by definition (recall that $|P^{\updownarrow}|_{\updownarrow} \leq \lceil k \cdot \gamma_n \rceil$) and the second by Lemma 3 which states that for time complexity, entangled pebblings are not better than normal ones.

Theorem 14 follow directly from the two claims above.

4.4   The CMC of the Line Graph

Throughout this section $L_n = (V, E), V = [n], E = \{(i, i+1) : i \in [n-1]\}$ denotes the path of length n, and the set of challenge nodes $C = [n]$ contains all verticies. In Sect. 3 we showed that – with overwhelming probability over the choice of a function $h : \{0,1\}^* \to \{0,1\}^w$ – the cumulative parallel entangled pebbling complexity for pebbling n challenges on a path of length n is

$$\text{cc}^{\updownarrow}(L_n, C = [n], n, n) = \Omega\left(n^2/\log^2(n)\right)$$

this then implies a lower bound on the cumulative memory complexity in the pROM against the class $\mathcal{A}^{\downarrow}$ of adversaries which are only allowed to store "encoding" of labels.

$$\mathsf{cmc}_{\mathcal{A}^{\downarrow}}^{\mathrm{pROM}}(L_n, C = [n], n, n) = \Omega\left(w \cdot n^2/\log^2(n)\right)$$

This strengthens previous lower bounds which only proved lower bounds for CC complexity, which then implied security against pROM adversaries that could only store plain labels. In the full version, we show that $\mathsf{cc}^{\downarrow}$ can be strictly lower than cc, thus, at least for some graphs, the ability to store encodings, not just plain labels, can decrease the complexity.

In this section we show a lower bound on $\mathsf{cmc}^{\mathrm{pROM}}(\mathsf{G}, C, m)$, i.e., without making any restrictions on the algorithm. Our bound will again depend on the parameter γ_n from Conjecture 13. We only sketch the proof as it basically follows the proof of Theorem 4.

Theorem 15. *For any* $n \in \mathbb{N}$, *let* $L_n = (V = [n], E = \{(i, i+1) : i \in [n-1]\})$ *be the line of length* n *and* γ_n *be as in Conjecture 13, and the label length* $w = \Omega(n \log n)$, *then*

$$\mathsf{cmc}^{\mathrm{pROM}}(L_n, C = [n], n, \sigma_{\mathrm{init}}) = \Omega\left(w \cdot n^2/\log^2(n) \cdot \gamma_n\right)$$

and for every $\epsilon > 0$

$$\mathsf{cmc}_{\epsilon}^{\mathrm{pROM}}(L_n, C = [n], n, \sigma_{\mathrm{init}}) = \Omega_{\epsilon}\left(w \cdot n^2/\log^2(n) \cdot \gamma_n\right)$$

Proof (sketch). We consider the experiment $\mathsf{computeLabel}(L_n, C, n, \mathsf{A}, \sigma_{\mathrm{init}}, \mathsf{h})$ for the A achieving the minimal $\mathsf{cmc}^{\mathrm{pROM}}$ complexity if h is chosen at random (we can assume A is deterministic). Let $(\mathsf{P}, P_{\mathrm{init}})$ be such that $\mathsf{pebble}^{\downarrow}(L_n, C, n, \mathsf{P}, P_{\mathrm{init}})$ mimics (as defined above) this experiment. By Theorem 9, $\mathsf{cc}^{\downarrow}(L_n, C = [n], n, n) = \Omega\left(n^2/\log^2(n)\right)$, unfortunately – unlike for time complexity – we don't see how this would directly imply a lower bound on $\mathsf{cmc}^{\mathrm{pROM}}$.

Fortunately, although Theorems 4 and 9 are about CC complexity, the proof is based on time complexity: At any timepoint the "potential" of the current state lower bounds the time required to pebble a random challenge, and if the potential is small, then the state has to be large (cf. eq.(16)).

For any $0 \leq i \leq n$ and $\mathbf{c} \in C^i$ let $\sigma_{\mathbf{c}}$ denote the state in the experiment $\mathsf{computeLabel}(L_n, C, n, \mathsf{A}, \sigma_{\mathrm{init}} = \emptyset, \mathsf{h})$ right after the i'th label has been computed by A and conditioned on the first i challenges being $\mathbf{c}$ (as A is deterministic and we fixed the first i challenges, $\sigma_{\mathbf{c}}$ is well defined).

At this point, the remaining experiment is $\mathsf{computeLabel}(L_n, C, n-i, \mathsf{A}, \sigma_{\mathbf{c}}, \mathsf{h})$. Similarly, we let $P_{\mathbf{c}}$ denote the pebbling in the "mimicking" $\mathsf{pebble}^{\downarrow}(L_n, C, n - i, \mathsf{P}, P_{\mathbf{c}})$ experiment after P has pebbled the challenge nodes $\mathbf{c}$. Let $P_{\mathbf{c}}'$ be the entangled pebbling of the smallest possible weight such that there exists a P' such that $\mathsf{pebble}^{\downarrow}(L_n, C, n - i, \mathsf{P}, P_{\mathbf{c}})$ and $\mathsf{pebble}^{\downarrow}(L_n, C, n - i, \mathsf{P}', P_{\mathbf{c}}')$ make the same queries on all possible challenges.

The expected *time* complexity to pebble the $i + 1$'th challenge in $\mathsf{pebble}^{\downarrow}(L_n, C, n-i, \mathsf{P}', P_{\mathbf{c}}')$ – and thus also in $\mathsf{computeLabel}(L_n, C, n-i, \mathsf{A}, \sigma_{\mathbf{c}}, \mathsf{h})$

- is at least $n/6|P'_c|_{\updownarrow}$ by Eq. (16). And by Theorem 14, we can lower bound the size of the state σ_c as (assuming w is sufficiently large)

$$|\sigma_c| \geq \Omega(w \cdot |P'_c|_{\updownarrow}/\gamma_n)$$

The CC cost of computing the next $(i+1)$th label in computeLabel($L_n, C, n - i, A, \sigma_c, h$) – *if we assume that the state remains roughly around its initial size* $|\sigma_c|$ *until the challenge is pebbled* – is roughly (cf. the intuition for the expectation game given in Sect. 3)

$$\frac{n}{2 \cdot |P'_c|_{\updownarrow}} \cdot |\sigma_c| = \Omega\left(\frac{n}{|P'_c|_{\updownarrow}} \cdot \frac{w \cdot |P'_c|_{\updownarrow}}{\gamma_n}\right) = \Omega\left(\frac{n \cdot w}{\gamma_n}\right)$$

As there are n challenges, this would give an $\Omega(w \cdot n^2/\gamma_n)$ bound on the overall CC complexity. Of course the above assumption that the state size never decreases is not true in general, an adversary case always chose to drop most of the pebbles once the challenge is known.

Note that in the above argument we don't actually use the size $|\sigma_c|$ of the current state, but only argue using the potential of the lightest pebbling P'_c necessary to mimic the remaining experiment. Following the same argument as in Theorem 4 (in particular, using Lemma 8) one can show that for a $1/\log(n)$ fraction of the challenges, the potential says within a $\log(n)$ factor of its initial sizes. This argument will lose us a $1/\log^2(n)$ factor in the CC complexity, giving the claimed $\Omega\left(w \cdot n^2/\log^2(n) \cdot \gamma_n\right)$ bound.

Acknowledgments. Joël Alwen, Chethan Kamath, and Krzysztof Pietrzak's research is partially supported by an ERC starting grant (259668-PSPC). Vladimir Kolmogorov is partially supported by an ERC consolidator grant (616160-DOICV). Binyi Chen was partially supported by NSF grants CNS-1423566 and CNS-1514526, and a gift from the Gareatis Foundation. Stefano Tessaro was partially supported by NSF grants CNS-1423566, CNS-1528178, a Hellman Fellowship, and the Glen and Susanne Culler Chair.

This work was done in part while the authors were visiting the Simons Institute for the Theory of Computing, supported by the Simons Foundation and by the DIMACS/Simons Collaboration in Cryptography through NSF grant CNS-1523467.

References

1. Crypto-Currency Market Capitalizations. http://coinmarketcap.com/. Accessed 10 July 2015
2. Almeida, L.C., Andrade, E.R., Barreto, P.S.L. M., Simplicio Jr., M.A.: Lyra: Password-based key derivation with tunable memory and processing costs. Cryptology ePrint Archive, report 2014/030 (2014). http://eprint.iacr.org/2014/030
3. Alwen, J., Serbinenko, V.: High parallel complexity graphs and memory-hard functions. In: Servedio, R.A., Rubinfeld, R. (eds) 47th ACM STOC, pp. 595–603. ACM Press, June 2015
4. Biryukov, A., Dinu, D., Khovratovich, D.: Fast and tradeoff-resilient memory-hard functions for cryptocurrencies and password hashing. Cryptology ePrint Archive, report 2015/430 (2015). http://eprint.iacr.org/2015/430

5. Banik, S., Bogdanov, A., Isobe, T., Shibutani, K., Hiwatari, H., Akishita, T., Regazzoni, F.: Midori: a block cipher for low energy. In: Iwata, T., et al. (eds.) ASIACRYPT 2015. LNCS, vol. 9453, pp. 411–436. Springer, Heidelberg (2015). doi:10.1007/978-3-662-48800-3_17

6. Chang, J., Mishra, S., Kumar Sanadhya, S.: Time memory tradeoff analysis of graphs in password hashing constructions. Preproc. PASSWORDS 14, 256–266 (2014)

7. Dwork, C., Naor, M.: Pricing via processing or combatting junk mail. In: Brickell, E.F. (ed.) CRYPTO 1992. LNCS, vol. 740, pp. 139–147. Springer, Heidelberg (1993)

8. Dwork, C., Naor, M., Wee, H.M.: Pebbling and proofs of work. In: Shoup, V. (ed.) CRYPTO 2005. LNCS, vol. 3621, pp. 37–54. Springer, Heidelberg (2005)

9. Dziembowski, S., Faust, S., Kolmogorov, V., Pietrzak, K.: Proofs of space. In: Gennaro, R., Robshaw, M. (eds.) CRYPTO 2015. LNCS, vol. 9216, pp. 585–605. Springer, Heidelberg (2015)

10. Dziembowski, S., Kazana, T., Wichs, D.: One-time computable self-erasing functions. In: Ishai, Y. (ed.) TCC 2011. LNCS, vol. 6597, pp. 125–143. Springer, Heidelberg (2011)

11. Forler, C., Lucks, S., Wenzel, J.: Catena: a memory-consuming password scrambler. Cryptology ePrint Archive, report 2013/525 (2013). http://eprint.iacr.org/2013/525

12. Jakobsson, M., Juels, A.: Proofs of work, bread pudding protocols. In: Preneel, B., (ed.) Secure Information Networks: Observation of strains. Infect Dis. Ther. 3(1), 35–43.: Communications and Multimedia Security, IFIP TC6/TC11 Joint Working Conference on Communications and Multimedia Security (CMS 1999), 20–21 September 1999, Leuven, Belgium, vol. 152 of IFIP Conference Proceedings, pp. 258–272. Kluwer, 1999 (2011)

13. Kaliski, B.: PKCS #5: Password-based cryptography specification version 2.0 (2000)

14. Lee, C.: Litecoin (2011). https://litecoin.org/

15. Park, S., Pietrzak, K., Kwon, A., Alwen, J., Fuchsbauer, G., Gaži, P.: Spacemint: A cryptocurrency based on proofs of space. Cryptology ePrint Archive, report 2015/528 (2015). http://eprint.iacr.org/2015/528

16. Percival, C. :Stronger key derivation via sequential memory-hard functions (2009). http://www.tarsnap.com/scrypt/scrypt.pdf

17. Corrigan-Gibbs, H., Boneh, D., Schechter, S.: Balloon Hashing: Provably Space-Hard Hash Functions with Data-Independent Access Patterns. Cryptology ePrint Archive, Report 2016/027 (2016). http://eprint.iacr.org/

Anonymous Traitor Tracing: How to Embed Arbitrary Information in a Key

Ryo Nishimaki[1]($\boxtimes$), Daniel Wichs[2], and Mark Zhandry[3]

[1] NTT Secure Platform Laboratories, Tokyo, Japan
nishimaki.ryo@lab.ntt.co.jp
[2] Northeastern University, Boston, USA
wichs@ccs.neu.edu
[3] MIT/Princeton University, Cambridge, USA
mzhandry@princeton.edu

Abstract. In a traitor tracing scheme, each user is given a different decryption key. A content distributor can encrypt digital content using a public encryption key and each user in the system can decrypt it using her decryption key. Even if a coalition of users combines their decryption keys and constructs some "pirate decoder" that is capable of decrypting the content, there is a public tracing algorithm that is guaranteed to recover the identity of at least one of the users in the coalition given black-box access to such decoder.

In prior solutions, the users are indexed by numbers $1, \ldots, N$ and the tracing algorithm recovers the index i of a user in a coalition. Such solutions implicitly require the content distributor to keep a record that associates each index i with the actual identifying information for the corresponding user (e.g., name, address, etc.) in order to ensure accountability. In this work, we construct traitor tracing schemes where all of the identifying information about the user can be embedded directly into the user's key and recovered by the tracing algorithm. In particular, the content distributor does not need to separately store any records about the users of the system, and honest users can even remain anonymous to the content distributor.

The main technical difficulty comes in designing tracing algorithms that can handle an exponentially large universe of possible identities, rather than just a polynomial set of indices $i \in [N]$. We solve this by abstracting out an interesting algorithmic problem that has surprising connections with seemingly unrelated areas in cryptography. We also extend our solution to a full "broadcast-trace-and-revoke" scheme in which the traced users can subsequently be revoked from the system. Depending on parameters, some of our schemes can be based only on the existence of public-key encryption while others rely on indistinguishability obfuscation.

R. Nishimaki—This work was done while the author was visiting Northeastern University.

D. Wichs—Research supported by NSF grants CNS-1347350, CNS-1314722, CNS-1413964. This work was done in part while the author was visiting the Simons Institute for the Theory of Computing, supported by the Simons Foundation and by the DIMACS/Simons Collaboration in Cryptography through NSF grant CNS-1523467.

M. Fischlin and J.-S. Coron (Eds.): EUROCRYPT 2016, Part II, LNCS 9666, pp. 388–419, 2016.
DOI: 10.1007/978-3-662-49896-5_14

1   Introduction

The Traitor-Tracing Problem. Traitor-tracing systems, introduced by Chor et al. [12], are designed to help content distributors identify the origin of pirate decryption boxes (such as pirate cable-TV set-top decoders) or pirate decryption software posted on the Internet.

In the traditional problem description, there is a set of legitimate users with numeric identities $[N] = \{1, \ldots, N\}$ for some (large) polynomial N. Each user $i \in [N]$ is given a different decryption key sk_i. A content distributor can encrypt content under the public key pk of the system and each legitimate user i can decrypt the content with her decryption key sk_i. For example this could model a cable-TV network broadcasting encrypted digital content, where each legitimate customer i is given a set-top decoder with the corresponding decryption key sk_i embedded within it.

One of the main worries in this scenario is that a user might make copies of her key to re-sell or even post in a public forum, therefore allowing illegitimate parties to decrypt the digital content. While this cannot be prevented, it can be deterred by ensuring that such "traitors" are held accountable if caught. To evade accountability, a traitor might modify her secret key before releasing it in the hope that the modified key cannot be linked to her. More generally, a coalition of several traitors might come together and pool the knowledge of all of their secret keys to come up with some "pirate decoder" program capable of decrypting the digital content. Such a program could be made arbitrarily complex and possibly even obfuscated in the hopes that it will be difficult to link it to any individual traitor. A traitor-tracing scheme ensures that no such strategy can succeed – there is an efficient *tracing algorithm* which is given black-box access to any such pirate decoder and is guaranteed to output the numeric identity $i \in [N]$ of at least one of the traitors in the coalition that created the program.

Who Keeps Track of User Info? The traditional problem definition for traitor tracing makes an implicit assumption that there is an external mechanism to keep track of the users in the system and their identifying information in order to ensure accountability. In particular, either the content distributor or some third party would need to keep a record that associates the numeric identities $i \in [N]$ of the users with the actual identifying information (e.g., name, address, etc.). This way, if the tracing algorithm identifies a user with numeric identity i as a traitor, we can link this to an actual person.

Goal: Embedding Information in Keys. The main goal of our work is to create a traitor tracing system where all information about each user is embedded directly into their secret key and there is no need to keep any external record about the honest users of the system. More concretely, this goal translates to having a traitor tracing scheme with a *flexible*, exponential-size universe of identities $\mathcal{ID}^1$. A user's identity id $\in \mathcal{ID}$ can then be a string containing all rele-vant identifying information about the user. The content distributor has a *master secret key* msk, and for any user with identity id $\in \mathcal{ID}$ the content provider can use msk to create a user secret key sk_{id} with this information embedded inside it. The content provider does not need to keep any records about the user after the secret key is given out. If a coalition of traitors gets together and constructs a pirate decoder, the tracing algorithm should recover the entire identity id of a traitor involved in the coalition, which contains all of the information necessary to hold the traitor accountable.

Moreover, if we have such a traitor tracing scheme with an exponentially large universe of identities as described above, it is also possible to construct a fully *anonymous* traitor tracing system where the content provider never learns who the honest users are. Instead of a user requesting a secret key for identity id $\in \mathcal{ID}$ by sending id to the content provider directly, the user and the content provider run a *multiparty computation* (MPC) where the user's input consists of the string id containing all of her identifying information (signed by some external identity verification authority), the content provider's input is msk, and the computation gives the user sk_{id} as an output (provided that the signature verifies) and the content provider learns nothing. This can even be combined with an anonymous payment system such as bit-coin to allow users to anonymously pay for digital content. Surprisingly, this shows that anonymity and traitor tracing are not contradictory goals; we can guarantee anonymity for honest users who keep their decryption keys secret while still maintaining the ability to trace the identities of traitors.

Unfortunately, it turns out that prior approaches to the traitor tracing prob-lem cannot handle large identities and crucially rely on the fact that, in the traditional problem definition, the set of identities $[N]$ is polynomial in size. We first survey the prior work on traitor tracing and then present our new results and techniques that allow us to achieve the above goals.

1.1 Prior Work

Traitor Tracing Overview. Traitor tracing was introduced by Chor et al. [12]. There are many variants of the problem depending on whether the encryption and/or the tracing algorithm are public key or secret key procedures, whether the tracing algorithm is black-box, and whether the schemes are "fully collusion resistant" (no bound on the number of colluding traitors), or whether they are "bounded collusion resistant". See e.g., the works of [6–9,11,13,17,19,29,31–34,37,38] and references within for a detailed overview of prior work.

In this work, we will focus on schemes with a public-key encryption and a public-key and black-box tracing algorithm, and will consider both fully and

---

[1] While schemes with exponential identity spaces are normally referred to as "identity-based", identity-based traitor tracing already has a defined meaning [1]. In particu-lar, the space of identities that are traced in an identity-based traitor tracing scheme is still polynomial. We use the term "flexible" traitor tracing to refer to schemes where the space of identities that can be traced is exponential.

bounded collusion resistance. In all prior systems, the set of legitimate users was fixed to $[N] = \{1, \ldots, N\}$ for some large polynomial N, and the main differences between the prior schemes depends on how various parameters (public key size, secret key size, ciphertext size) scale with the number of users N.

Traitor Tracing via Private Broadcast Encryption (PLBE). Boneh et al. [7] build the first fully collusion resistant traitor tracing scheme where the ciphertext size is $O(\sqrt{N})$, private key size is $O(1)$, public key size is $O(\sqrt{N})$ (we ignore factors that are polynomial in the security parameter but independent of N). The scheme is based on bilinear groups. This work also presents a general approach for building traitor tracing schemes, using an intermediate primitive called *private linear broadcast encryption* (PLBE). We follow the same approach in this work and therefore we elaborate on it now.

A PLBE scheme can be used to create a ciphertext that can only be decrypted by users $i \in [N]$ with $i \leq T$ for some threshold value $T \in \{0, \ldots, N\}$ specified during encryption. Furthermore, the only way to distinguish between a ciphertext created with the threshold value T vs. T' for some $T < T'$ is to have a secret key sk_i with $i \in \{T, \ldots T' - 1\}$ that can decrypt in one case but not the other.

A PLBE scheme can immediately be used as a traitor-tracing scheme. The encryption algorithm of the tracing scheme creates a ciphertext with the threshold $T = N$, meaning that all users can decrypt it correctly. The tracing algorithm gets black-box access to a pirate decoder and does the following: it tries all thresholds $T = 1, \ldots, N$ and tests the decoder on ciphertext created with threshold T until it finds the first such threshold for which there is a "big jump" in the decryption success probability between T and $T - 1$. It outputs the index T as the identity of the traced traitor. The correctness of the above approach can be analyzed as follows. We know that the decoder's success probability on $T = 0$ is negligible (since such ciphertexts cannot be decrypted even given all the keys) and on $T = N$ it is large (by the correctness of the pirate decoder program). Therefore, there must be some threshold T on which there is a big jump in the success probability, but by the privacy property of the PLBE, a big jump can only occur if the secret key sk_T was used in the construction of the pirate decoder. Note that the run-time of this tracing algorithm is $O(N)$.

State of the Art Traitor Tracing via Obfuscation. Recently, Garg et al. [21] and Boneh and Zhandry [9] construct new fully collusion resistant traitor tracing scheme with essentially optimal parameters where key/ciphertext sizes only depend logarithmically on N. The schemes are constructed using the same PLBE framework as in [7] and the main contributions are the construction of a new PLBE scheme with the above parameters. These constructions both rely on indistinguishability obfuscation. More recently, Garg et al. [22] construct a PLBE with polylogarithmic parameters based on simple assumptions on multilinear maps. We note that in all three schemes, the PLBE can be extended to handle flexible (exponential) identity spaces by setting $N = 2^n$ for polynomial n. In this case, encryption and key generation, as well as ciphertext and secret key sizes, will grow polynomially in n. However, a flexible PLBE scheme does not directly yield to a flexible traitor tracing scheme. In particular, the tracing algorithm of [7] cannot be applied in this setting because it will run in exponential time, namely $O(2^n)$.

Broadcast Encryption, Trace and Revoke. We also mention work on a related problem called broadcast encryption. Similar to traitor tracing, such schemes have a collection of users $[N]$. A sender can create a ciphertext that can be decrypted by all of the users of the system *except* for specified set of "revoked users" (which may be colluding). See e.g., [16–18, 20, 24, 26, 32, 34, 39] and references within.

A trace and revoke system is a combination of broadcast encryption and traitor tracing [32, 34]. In other words, once traitors are identified by the tracing algorithm they can also be revoked from decrypting future ciphertexts. Boneh and Waters [8] proposed a fully collusion resistant trace and revoke scheme where the private/public keys and ciphertexts are all of size $O(\sqrt{N})$. It was previously unknown how to obtain fully collusion resistant trace and revoke schemes with logarithmic parameter sizes. Separately, though, it is known how to build both broadcast encryption and traitor tracing with such parameters using obfuscation [9, 21, 41], and one could reasonably expect that it is possible to combine the techniques to obtain a broadcast, trace, and revoke system.

Watermarking. Lastly, we mention related work on watermarking cryptographic functions [14, 15, 35]. These works show how to embed arbitrary data into the secret key of a cryptographic function (e.g., a PRF) in such a way that it is impossible to create any program that evaluates the function (even approximately) but in which the mark is removed. This is conceptually related to our goal of embedding arbitrary data into the secret keys of users in a traitor-tracing scheme. Indeed, one could think of constructing a traitor tracing scheme where we take a standard public-key encryption scheme and give each user a water-marked version of the decryption key containing the user's identity embedded. Unfortunately, this solution does not work with current definitions of watermarking security, where we assume that each key can only be marked once with one piece of embedded data. In the traitor tracing scenario, we would want mark the same key many times with different data for each user. Conversely, solutions to the traitor tracing problem do not yield watermarking schemes since they only require us to embed data in carefully selected secret keys chosen by the scheme designer rather than in arbitrary secret keys chosen by the user.

1.2 Our Results

Our main result is to give new constructions of traitor-tracing schemes that supports a flexibly large space of identities $\mathcal{ID} = [2^n]$ where the parameter n is an arbitrary polynomial corresponding to the bit-length of the string id $\in \mathcal{ID}$ which should be sufficiently large encode all relevant identifying information about the user. The user's secret key $\mathsf{sk}_{\mathsf{id}}$ contains the identity id embedded within it, so there is no need to keep any external record of users. The tracing algorithm recovers all of the identifying information id about a traitor directly from the pirate decoder. We construct such a scheme where the secret key $\mathsf{sk}_{\mathsf{id}}$ is of length $\mathsf{poly}(n)$, which is essentially optimal since it must contain the data id embedded within it. The first scheme we construct also has ciphertexts of size $\mathsf{poly}(n)$ but we then show how to improve this to ciphertexts of constant size independent of n (though still dependent on the security parameter). In the latter scheme, the identity length n need not be specified ahead of time: different users can potentially have different amounts of identifying information included in their key, and there is no restriction on the amount of information that can be included. The schemes are secure against an unbounded number of collusions.

Our schemes are secure assuming the existence of certain types of *private broadcast encryption*, which themselves are special cases of functional encryption (FE). Our work mainly focuses on building traitor tracing from these private broadcast schemes. We then instantiate the private broadcast schemes using recent constructions of FE, which in turn are built from indistinguishability obfuscation (iO) and one-way functions (OWF). An interesting direction for future work is to build private broadcast encryption from milder assumptions such as LWE.

We also construct schemes which are only secure against collusions of size at most q, where the ciphertext size is either of length $O(n)\mathsf{poly}(q)$ assuming only public-key encryption, or of only length $\mathsf{poly}(q)$ independent of n assuming sub-exponential LWE.[2] We also extend the above construction to a full trace and revoke scheme, allowing the content distributor to specify a set of revoked users during encryption. Assuming iO, we get such a scheme where neither the ciphertexts nor the secret keys grow with the set of revoked users.

1.3 Our Techniques

Our high level approach follows that of Boneh et al. [7], using PLBE as an intermediate primitive to construct traitor tracing. There are two main challenges: the first is to construct a PLBE scheme that supports an exponentially large identity space $\mathcal{ID} = [2^n]$ for some arbitrary polynomial n. The second, more interesting challenge, and the main focus of this work, is to construct a tracing algorithm which runs in time polynomial in n rather than $N = 2^n$.

PLBE with Large Identity Space. The work of Boneh and Zhandry [9] already constructs a PLBE scheme where the key/ciphertext size is polynomial in n. Unfortunately, the proof of security relies on a reduction that runs in time polynomial in $N = 2^n$ which is exponential in the security parameter. Thus going through their construction we would need to assume the sub-exponential hardness of iO (and OWFs) to get a secure PLBE. We instead take a different approach, suggested by [21], and construct PLBE directly from (indistinguishability based) functional encryption (FE). For technical reasons detailed below, we

---

[2] The above parameters ignore fixed polynomial factors in the security parameters.

actually need an adaptively secure PLBE scheme, and thus an adaptively secure FE scheme. In the unbounded collusion setting, these can be constructed from iO [2,40] or from simple assumptions on multilinear maps [22]. Alternatively, we get a PLBE scheme which is (adaptively) secure against a *bounded* number of collusions by relying on bounded-collusion FE which can be constructed from any public-key encryption [25] or from sub-exponential LWE if we want succinct ciphertexts [23].

A New Tracing Algorithm and the Oracle Jump-Finding Problem. The more interesting difficulty comes in making the tracing algorithm run in time polynomial in n rather than $N = 2^n$. We can think of the pirate decoder as an oracle that can be tested on PLBE ciphertexts created with various thresholds $T \in \{0, \ldots, N\}$ and for any such threshold T it manages to decrypt correctly with probability p_T. For simplicity, let us think of this as an oracle that on input T outputs the probability p_T directly (since we approximate this value by testing the decoder on many ciphertexts). We know that p_0 is close to 0 and that p_N is the probability that a pirate decoder decrypts correctly, which is large – let's say $p_N = 1$ for simplicity. Moreover, we know that for any T, T' with $T < T'$ the values p_T and $p_{T'}$ are negligibly close *unless* there is a traitor with identity $i \in \{T, \ldots T' - 1\}$, since encryptions with thresholds T and T' are indistinguishable. In particular this means that for any point T at which there is a "jump" so that $|p_T - p_{T-1}|$ is noticeable, corresponds to a traitor. Since we know that the number of traitors in the coalition is bounded by some polynomial, denoted by q, we know that there are at most q jumps in total and that there must be at least one "large jump" with a gap of at least $1/q$. The goal is to find at least one jump. We call this the "oracle jump-finding problem".

An Algorithm for the Oracle Jump-Finding Problem. The tracing algorithm of [7] essentially corresponds to a linear search and tests the oracle on every point $T \in [N]$ and thus takes at least $O(N)$ steps in the worst case to find a jump. When using flexibly large identity universes (that is, taking N to be exponential), the tracing algorithm will therefore run in exponential time. This is true *even if the underlying PLBE is efficient for such identity spaces*, including the PLBEs discussed above. Our goal is to design a better algorithm that takes at most $\mathsf{poly}(n, q)$ steps.

It is tempting to simply substitute binary search in place of linear search. We would first call the oracle on the point $T/2$ and learn $p_{T/2}$. Depending on whether the answer is closer to 0 or 1 we recursively search either the left interval or the right interval. The good news is in each step the size of the interval decreases by half and therefore there would be at most n steps. The bad news is that the gap in probabilities between the left and right end points now also decreases by a half and therefore after i steps we would only be guaranteed that the interval contains a jump with a gap of $2^{-i}/q$ which quickly becomes negligible.

Interestingly, we notice that the same oracle jump-finding problem implicitly appeared in a completely unrelated context in a work of Boyle et al. [10] showing the equivalence of indistinguishability obfuscation and a special case of differing-inputs obfuscation. Using the clever approach developed in the context of that work, we show how to get a $\mathsf{poly}(n, q)$ algorithm for the oracle jump finding problem and therefore an efficient tracing algorithm.

The main idea is to follow the same approach as binary search, but each time that the probability at the mid-point is noticeably far from both end-points we recurse on both the left and the right interval. This guarantees that there is always a large jump with a gap of at least $1/q$ within the intervals being searched. Furthermore, since the number of jumps is at most q we can bound the number of recursive steps in which both intervals need to be searched by q, and therefore guarantee that the algorithm runs in $\mathsf{poly}(n, q)$ steps.

Interestingly, due to our tracing algorithm choosing which T to test based on the results of previous tests, we need our PLBE scheme to be *adaptively* secure, and hence also the underlying FE scheme must be adaptively secure. This was not an issue in [7] for two reasons: (1) their tracing algorithm visits *all* $T \in [N]$, and (2) for polynomial N statically secure and adaptive secure PLBE are equivalent. Fortunately, as explained above, we know how to construct PLBE that is adaptively secure against unbounded collusions from iO or simple multilinear map assumptions. For the bounded collusion setting, we can obtain adaptively secure PLBE from public key encryption following [25].

We note that in an independent work, Kiayias and Tang [28] give another method of tracing in large identity spaces; however their analysis applies only to *random* user identities, and requires a means to verify that the identity out-putted by the tracing algorithm actually corresponds to a one of the generated decryption keys. Our tracing algorithm does not have these limitations.

Tracing More General Decoders. In [7], a pirate decoder is considered "useful" if it decrypts the encryption of a random message with non-negligible probability, and their tracing algorithm is shown to work for such decoders. However, restricting to decoders that work for random messages is unsatisfying, as we would like to trace, say, decoders that work for very particular messages such as cable-TV broadcasts. The analysis of [7] appears insufficient for this setting. Kiayias and Yung [30] consider more general decoders, but their definition inherently places a lower bound on the min-entropy of the plaintext distribution. In our analysis, we show that even if a decoder can distinguish between two particular messages (of the adversary's choice) with non-negligible advantage, then it can be traced. To our knowledge, ours is the first traitor tracing system that can trace such general decoders.

Short Ciphertexts. In the above approach we construct traitor-tracing via a PLBE scheme where the ciphertext is encrypted with respect to some threshold $T \in \{0, \ldots, N\}$. The ciphertext must encode the entire information about T and is therefore of size at least $n = \log N$, which corresponds to the bit-length of the user's identifying information id. In some cases, if the size of id is truly large (e.g., the identifying information might contain a JPEG image of the user) we would want the ciphertext size to be much smaller than n. One trivial option is to first hash the user's identifying information, and use our tracing scheme above on the hashes. However, the tracer would then only learn the hash of the identifying information, and would need to keep track of the information and hashes to actually accuse a user. This prevents the scheme from being used in the anonymous setting.

Instead, we show how to have the tracer learn identifying information in its entirety by generalizing the PLBE approach in a way that lets us divide the user's identity into small blocks. Very roughly, we then trace the value contained in each block one at a time. The ciphertext now only needs to encode the block number that is currently being traced, and a single threshold for that block. This lets us reduce the ciphertext to size to only be proportional to $\log n$ rather than n. To do so we need to generalize the notion of PLBE which also leads to a generalization of the oracle-jump-finding problem and the algorithm that solves it. We note that since we can assume $n < 2^\lambda$, factors logarithmic in n can be absorbed into terms involving the security parameter. Thus our ciphertext size can actually be taken to be independent of the bit length of identities.

We implement our PLBE generalization using FE. As above, we need adaptive security, which corresponds to an adaptively secure FE scheme. We now also need the FE to have *compact* ciphertexts, whose size is independent of the functions being evaluated. In the unbounded collusion setting, a recent construction of Ananth and Sahai [4] shows how to build such an FE from iO. Moreover, in their FE scheme, the function size need not be specified a priori nor known during encryption time, and different secret keys can correspond to functions of different sizes. In our traitor tracing scheme, this translates to there being no a priori bound on the length of identities, and different users can have different amounts of identifying information embedded in their secret keys.

In the bounded collusion setting, we can obtain such an FE from LWE using [23], though the scheme is only statically secure; we then use complexity leveraging to obtain an adaptively secure scheme from sub-exponential LWE.

Trace and Revoke. Finally, we extend our traitor tracing scheme to a trace and revoke system where users can be revoked. It turns out that this problem reduces to the problem of constructing "revocable functional encryption" where the encryption algorithm can specify some revoked users which will be unable to decrypt. The ciphertext size is independent of the size of the revoke list, but we assume that the revoke list is known to all parties. We construct such a scheme from indistinguishability obfuscation using the technique of somewhere statistically binding (SSB) hashing [27]. However, we omit the details about the trace and revoke system due to the limited space. See the full version of this paper [36].

1.4   Outline

In Sect. 2, we give some definitions and notations that we will use in our work. In Sect. 3, we define the oracle jump-finding problem, and show how to efficiently solve it. In Sects. 4 and 5, we use the solution of the jump-finding problem to give our new traitor tracing schemes.

2   Preliminaries

Throughout this work, we will use the notation $[N]$ to mean the positive integers from 1 to N: $[N] = \{1, \ldots, N\}$. We will also use the notation $[M, N]$ to denote the integers form M to N, inclusive. We will use $(M, N]$ as shorthand for $[M+1, N]$. We will use $[M, N]_{\mathbb{R}}$ to denote the *real* numbers between M and N, inclusive.

Next, we will define several of the cryptographic primitives we will be discussing throughout this work. We start with the definition of traitor tracing that we will be achieving. Then, we will define the primitives we will use to construct traitor tracing. In all of our definitions, there is an implicit security parameter λ, and "polynomial time" and "negligible" are with respect to this security parameter.

2.1   Traitor Tracing with Flexible Identities

Here we define traitor tracing. Our definition is similar to that of Boneh, Sahai, and Waters [7], though ours is at least as strong, and perhaps stronger. In particular, our definition allows for tracing pirate decoders that can distinguish between encryptions of any two messages, whereas [7] only allows for tracing pirate decoders that can decrypt random messages. In Sect. 4, we discuss why the analysis in [7] appears insufficient for our more general setting, but nevertheless show that tracing is still possible.

Definition 1. *Let $\mathcal{ID}$ be some collection of identities, and $\mathcal{M}$ a message space. A flexible traitor tracing scheme for $\mathcal{M}, \mathcal{ID}$ is a tuple of polynomial time algorithms* (Setup, KeyGen, Enc, Dec, Trace) *where:*

- Setup() *is a randomized procedure with no input (except the security parameter) that outputs a master secret key* msk *and a master public key* mpk.
- KeyGen(msk, id) *takes as input the master secret* msk *and an identity* id $\in \mathcal{ID}$, *and outputs a secret key* sk$_{id}$ *for* id.
- Enc(mpk, m) *takes as input the master public key* mpk *and a message* m $\in \mathcal{M}$, *and outputs a ciphertext* c.
- Dec(sk$_{id}$, c) *takes as input the secret key* sk$_{id}$ *for an identity* id *and a ciphertext* c, *and outputs a message* m.
- Trace$^{\mathcal{D}}$(mpk, m_0, m_1, q, ϵ) *takes as input the master public key* mpk, *two messages* m_0, m_1, *and parameters* q, ϵ, *and has oracle access to a decoder algorithm* $\mathcal{D}$. *It produces a (possibly empty) list of identities* $\mathcal{L}$.
- **Correctness.** *For any message* m $\in \mathcal{M}$ *and identity* id $\in \mathcal{ID}$, *we have that*

$$\Pr\left[\mathsf{Dec}(\mathsf{sk_{id}}, c) = m : \begin{array}{l} (\mathsf{msk}, \mathsf{mpk}) \leftarrow \mathsf{Setup}(), \mathsf{sk_{id}} \leftarrow \mathsf{KeyGen}(\mathsf{msk}, \mathsf{id}), \\ c \leftarrow \mathsf{Enc}(\mathsf{mpk}, m) \end{array}\right] = 1$$

- **Semantic security.** *Informally, we ask that an adversary that does not hold any secret keys cannot learn the plaintext m. This is formalized by the following experiment between an adversary $\mathcal{A}$ and challenger:*
  - *The challenger runs* $(\mathsf{msk}, \mathsf{mpk}) \leftarrow \mathsf{Setup}()$, *and gives* mpk *to* $\mathcal{A}$.
  - *$\mathcal{A}$ makes a challenge query where it submits two messages m_0^*, m_1^*. The challenger chooses a random bit b, and responds with the encryption of m_b^*: $c^* \leftarrow \mathsf{Enc}(\mathsf{mpk}, m_b^*)$.*
  - *$\mathcal{A}$ produces a guess b' for b. The challenger outputs 1 if $b' = b$ and 0 otherwise.*

  We define the semantic security advantage of $\mathcal{A}$ as the absolute difference between $1/2$ and the probability the challenger outputs 1. The public key encryption scheme is semantically secure if, for all PPT adversaries $\mathcal{A}$, the advantage of $\mathcal{A}$ is negligible.

- **Traceability.** *Consider a subset of colluding users that pool their secret keys and produce a "pirate decoder" that can decrypt ciphertexts. Call a pirate decoder $\mathcal{D}$ "useful" for messages m_0, m_1 if $\mathcal{D}$ can distinguish encryptions of m_0 from m_1 with noticeable advantage. Then we require that such a decoder can be traced using Trace to one of the identities in the collusion. This is formalized using the following game between an adversary $\mathcal{A}$ and challenger, parameterized by a non-negligible function ϵ:*
  - *The challenger runs* $(\mathsf{msk}, \mathsf{mpk}) \leftarrow \mathsf{Setup}()$ *and gives* mpk *to* $\mathcal{A}$.
  - *$\mathcal{A}$ is allowed to make arbitrary keygen queries, where it sends an identity $\mathsf{id} \in \mathcal{ID}$ to the challenger, and the challenger responds with $\mathsf{sk_{id}} \leftarrow \mathsf{KeyGen}(\mathsf{msk}, \mathsf{id})$. The challenger also records the identities queries in a list $\mathcal{L}$.*
  - *$\mathcal{A}$ then produces a pirate decoder $\mathcal{D}$, two messages m_0^*, m_1^*, and a non-negligible value ϵ. Let q be the number of keygen queries made (that is, $q = |\mathcal{L}|$). The challenger computes $\mathcal{T} \leftarrow \mathsf{Trace}^{\mathcal{D}}(\mathsf{mpk}, m_0^*, m_1^*, q, \epsilon)$ as the set of accused users. The challenger says that the adversary "wins" one of the following holds:*
    - *$\mathcal{T}$ contains any identity outside of $\mathcal{L}$. That is, $\mathcal{T} \setminus \mathcal{L} \neq \emptyset$ or*
    - *Both of the following hold:*
      - *$\mathcal{D}$ is ϵ-useful, meaning $\Pr[\mathcal{D}(c) = m_b^* : b \leftarrow \{0,1\}, c \leftarrow \mathsf{Enc}(\mathsf{mpk}, m_b^*)] \geq \frac{1}{2} + \epsilon^3$.*
      - *$\mathcal{T}$ does not contain at least one user inside $\mathcal{L}$. That is, $\mathcal{T} \cap \mathcal{L} = \emptyset$. The challenger then outputs 1 if the adversary wins, and zero otherwise.*

---

[3] Checking the "winning" condition requires computing the probabilities a procedure outputs a particular value, which is in general an inefficient procedure. Thus our challenger as described is not an efficient challenger. However, it is possible to efficiently estimate these probabilities by running the procedure many times, and reporting the fraction of the time the particular value is produced. We could have instead defined our challenger to estimate probabilities instead of determine them exactly, in which case the challenger would be efficient. The resulting security definition would be equivalent.

We define the tracing advantage of $\mathcal{A}$ as the probability the challenger out-puts 1. We say the public key encryption scheme is traceable if, for all PPT adversaries $\mathcal{A}$ and all non-negligible ϵ, the advantage of $\mathcal{A}$ is negligible.

2.2   Private Broadcast Encryption

In our traitor tracing constructions, it will be convenient for us to use a primitive we call *private broadcast encryption*, which is a generalization of the private *linear* broadcast encryption of Boneh et al. [7]. A private broadcast scheme is a broadcast scheme where the recipient set is hidden. Usually, the collection of possible recipient subsets is restricted: for example, in private linear broadcast encryption, the possible recipient sets are simply intervals. It will be useful for us to consider more general classes of recipient sets, especially for our short-ciphertext traitor tracing construction in Sect. 5.

Definition 2. *Let $\mathcal{ID}$ be the set of identities. Let S be a collection of subsets of $\mathcal{ID}$. Let $\mathcal{M}$ be a message space. A Private Broadcast Encryption (PBE) scheme is a tuple of algorithms* (Setup, KeyGen, Enc, Dec) *where:*

- Setup() *is a randomized procedure with no input (except the security parame-ter) that outputs a master secret key* msk *and a master public key* mpk.
- KeyGen(msk, id) *takes as input the master secret* msk *and a user identity* id $\in \mathcal{ID}$. *It outputs a secret key* $\mathsf{sk}_{\mathsf{id}}$ *for* id.
- Enc(mpk, S, m) *takes as input the master public key* mpk, *a secret set* $S \in \mathcal{S}$, *and a message* $m \in \mathcal{M}$. *It outputs a ciphertext* c.
- Dec($\mathsf{sk}_{\mathsf{id}}$, c) *takes as input the secret key* $\mathsf{sk}_{\mathsf{id}}$ *for a user* id, *and a ciphertext* c. *It outputs a message* $m \in \mathcal{M}$ *or a special symbol* $\bot$.
- **Correctness.** *For a secret set* $S \in \mathcal{S}$, *any identity* id $\in S$, *any identity* id$' \notin S$, *any message* $m \in \mathcal{M}$, *we have that*

$$
\Pr\left[\mathsf{Dec}(\mathsf{sk}_{\mathsf{id}}, c) = m : \begin{array}{l} (\mathsf{msk}, \mathsf{mpk}) \leftarrow \mathsf{Setup}(), \mathsf{sk}_{\mathsf{id}} \leftarrow \mathsf{KeyGen}(\mathsf{msk}, \mathsf{id}), \\ c \leftarrow \mathsf{Enc}(\mathsf{mpk}, S, m) \end{array}\right] = 1
$$

$$
\Pr\left[\mathsf{Dec}(\mathsf{sk}_{\mathsf{id}'}, c) = \bot : \begin{array}{l} (\mathsf{msk}, \mathsf{mpk}) \leftarrow \mathsf{Setup}(), \mathsf{sk}_{\mathsf{id}'} \leftarrow \mathsf{KeyGen}(\mathsf{msk}, \mathsf{id}'), \\ c \leftarrow \mathsf{Enc}(\mathsf{mpk}, S, m) \end{array}\right] = 1
$$

In other words, a user id *is "allowed" to decrypt if* id *is in the secret set* S. *We also require that if* id *is not "allowed" (that is, if* id $\notin S$), *then* Dec *outputs* $\bot$.

- **Message and Set Hiding.** *Intuitively, we ask that for* id *that are not explic-itly allowed to decrypt a ciphertext* c, *that the message is hidden. We also ask that nothing is learned about the secret set* S, *except for what can be learned by attempting decryption with various* $\mathsf{sk}_{\mathsf{id}}$ *available to the adversary. These two requirements are formalized by the following experiment between an adversary $\mathcal{A}$ and challenger:*
  - *The challenger runs* (msk, mpk) $\leftarrow$ Setup(), *and gives* mpk *to* $\mathcal{A}$.

- $\mathcal{A}$ *is allowed to make arbitrary keygen queries, where it sends an identity* id $\in \mathcal{ID}$ *to the challenger, and the challenger responds with* $\mathsf{sk}_{\mathsf{id}} \leftarrow$ $\mathsf{KeyGen}(\mathsf{msk}, \mathsf{id})$. *The challenger also records* id *in a list* $\mathcal{L}$.
- *At some point,* $\mathcal{A}$ *makes a single challenge query, where it submits two secret sets* $S_0^*, S_1^* \in \mathcal{S}$, *and two messages* m_0^*, m_1^*. *The challenger flips a random bit* $b \in \{0, 1\}$, *and computes the encryption of* m_b^* *relative to the secret set* S_b^*: $c^* \leftarrow \mathsf{Enc}(\mathsf{mpk}, S_b^*, m_b^*)$. *Then, the challenger makes the following checks, which ensure that the adversary cannot trivially determine* b *from* c^*:
  * *If* $m_0^* \neq m_1^*$, *then successful decryption of the challenge ciphertext would allow determining* b. *Therefore, the challenger requires that none of the identities the adversary has the secret key for can decrypt the ciphertext. In other words, for any* id $\in \mathcal{L}$, id $\notin S_0^*$ *and* id $\notin S_1^*$. *In other words, the sets* $\mathcal{L} \cap S_0^*$ *and* $\mathcal{L} \cap S_1^*$ *must be empty.*
  * *If* $S_0^* \neq S_1^*$, *then successful decryption for* S_b^* *but not for* S_{1-b}^* *would allow for determining* b *(even if* $m_0^* = m_1^*$*). Therefore, the challenger requires that all of the identities the adversary has secret keys for can either decrypt in both cases, or can decrypt in neither. In other words, for any* id $\in L$, id $\notin S_0^* \Delta S_1^*$, *where* Δ *denotes the symmetric difference operator. Notice that this check is redundant if* $m_0^* \neq m_1^*$.

  If either check fails, the challenger outputs a random bit and aborts the game. Otherwise, the challenger sends c^* *to* $\mathcal{A}$.
- $\mathcal{A}$ *is allowed to make additional keygen queries for arbitrary identities* id*, *subject to the constraint that* id *must satisfy the same checks as above: if* $m_0^* \neq m_1^*$, *then* id $\notin S_0^*$ *and* id $\notin S_1^*$, *and if* $S_0^* \neq S_1^*$, *then* id $\notin S_0^* \Delta S_1^*$. *If the adversary tries to query in an* id *that fails the check, the challenger outputs a random bit and aborts the game.*
- $\mathcal{A}$ *outputs a guess* b' *for* b. *The challenger outputs 1 if* $b' = b$ *and 0 otherwise.*

We define the advantage of $\mathcal{A}$ *as the absolute difference between* $1/2$ *and the probability the challenger outputs 1. We say the private broadcast system is secure if, for all PPT adversaries* $\mathcal{A}$, *the advantage of* $\mathcal{A}$ *is negligible.*

For a private broadcast scheme, we call the collection $\mathcal{S}$ of secret sets the *secret class*. We are interested in several metrics for a private broadcast scheme:

- **Ciphertext size.** Notice that the ciphertext, while hiding the secret set S, information-theoretically contains enough information to reveal S: given the secret key for every identity, S can be determined by attempting decryption with every secret key. It must also contain enough information to entirely reconstruct the message m. Thus, we must have $|c| \geq \log |\mathcal{S}| + \log |\mathcal{M}|$. We will say the ciphertext size is *optimal* if $|c| \leq \mathsf{poly}(\lambda, \log |\mathcal{S}|) + \log |\mathcal{M}|$.
- **Secret key size.** Assuming the public and secret classes $\mathcal{P}, \mathcal{S}$ are expressive enough, from the secret key $\mathsf{sk}_{\mathsf{id}}$ for identity id, it is possible to reconstruct the entire identity id by attempting to decrypt ciphertexts meant for various subsets. Therefore, $|\mathsf{sk}_{\mathsf{id}}| \geq \log |\mathcal{ID}|$. We will say the user secret key size is *optimal* if $|\mathsf{sk}_{\mathsf{id}}| \leq \mathsf{poly}(\lambda, \log |\mathcal{ID}|)$.

– **Master key size.** The master public and secret keys do not necessarily encode any information, and therefore could be as short as $O(\lambda)$. We will say the master key sizes are *optimal* if $|\mathsf{msk}|, |\mathsf{mpk}| \leq \mathsf{poly}(\lambda)$.

Notice that in the case where $\mathcal{S} = \{\mathcal{ID}\}$, our notion of private broadcast reduces to the standard notion of (identity-based) broadcast encryption, and the notions of optimal ciphertext, user secret key, and master key sizes coincide with the standard notions for broadcast encryption.

2.3   Functional Encryption

Definition 3. *Let $\mathcal{M}$ be some message space, $\mathcal{Y}$ some other space, and $\mathcal{F}$ be a class of functions $f : \mathcal{M} \to \mathcal{Y}$. A Functional Encryption (FE) scheme for $\mathcal{M}, \mathcal{Y}, \mathcal{F}$ is a tuple of algorithms* (Setup, KeyGen, Enc, Dec) *where:*

– Setup() *is a randomized procedure with no input (except the security parameter) that outputs a master secret key* msk *and a master public key* mpk.
– KeyGen(msk, f) *takes as input the master secret* msk *and a function $f \in \mathcal{F}$. It outputs a secret key* sk_f *for f.*
– Enc(mpk, m) *takes as input the master public key* mpk *and a message $m \in \mathcal{M}$, and outputs a ciphertext c.*
– Dec(sk_f, c) *takes as input the secret key sk_f for a function $f \in \mathcal{F}$ and a ciphertext c, and outputs some $y \in \mathcal{Y}$, or $\perp$.*
– **Correctness.** *For any message $m \in \mathcal{M}$ and function $f \in \mathcal{F}$, we have that*

$$\Pr\left[\mathsf{Dec}(\mathsf{sk}_f, c) = f(m) : \begin{array}{l} (\mathsf{msk}, \mathsf{mpk}) \leftarrow \mathsf{Setup}(), \mathsf{sk}_f \leftarrow \mathsf{KeyGen}(\mathsf{msk}, f), \\ c \leftarrow \mathsf{Enc}(\mathsf{mpk}, m) \end{array} \right] = 1$$

– **Security.** *Intuitively, we ask that the adversary, given secret keys $f_1, \ldots, f_n$, learns $f_i(m)$ for each i, but nothing else about m. This is formalized by the following experiment between an adversary $\mathcal{A}$ and challenger:*
  • *The challenger runs* (msk, mpk) $\leftarrow$ Setup(), *and gives* mpk *to $\mathcal{A}$.*
  • *$\mathcal{A}$ is allowed to make arbitrary keygen queries, where it sends a function $f \in \mathcal{F}$ to the challenger, and the challenger responds with $\mathsf{sk}_f \leftarrow$ KeyGen(msk, f). The challenger also records f in a list $\mathcal{L}$.*
  • *At some point, $\mathcal{A}$ makes a single challenge query, where it submits two messages m_0^*, m_1^*. The challenger checks that $f(m_0^*) = f(m_1^*)$ for all $f \in L$. If the check fails (that is, there is some $f \in L$ such that $f(m_0^*) \neq f(m_1^*)$), then the challenger outputs a random bit and aborts. Otherwise, the challenger flips a random bit $b \in \{0, 1\}$, and responds with the ciphertext $c^* \leftarrow$ Enc(mpk, m_b^*).*
  • *$\mathcal{A}$ is allowed to make additional keygen queries for functions $f \in \mathcal{F}$, subject to the constraint that $f(m_0^*) = f(m_1^*)$.*
  • *$\mathcal{A}$ outputs a guess b' for b. The challenger outputs 1 if $b' = b$ and 0 otherwise.*
  We define the advantage of $\mathcal{A}$ as the absolute difference between $1/2$ and the probability the challenger outputs 1. We say the functional encryption scheme is secure if, for all PPT adversaries $\mathcal{A}$, the advantage of $\mathcal{A}$ is negligible.

For a functional encryption scheme, we will be interested in the size of the various parameters (in addition to the security of the system itself):

- **Ciphertext size.** At a minimum, the ciphertext must information-theoretically encode the entire message (assuming the class $\mathcal{F}$ is expressive enough). Therefore $|c| \geq \log |\mathcal{M}|$. We will consider a scheme to have *optimal* ciphertext size if $|c| \leq \mathrm{poly}(\lambda, \log |\mathcal{M}|)^4$.
- **Secret key size.** The secret key must information-theoretically encode the entire function f, so $|\mathsf{sk}_f| \geq \log |\mathcal{F}|$. However, because we are interested in efficient algorithms, we cannot necessarily represent functions f using $\log |\mathcal{F}|$ bits, and may therefore need larger keys. Generally, f will be a circuit of a certain size, say s. We will say a scheme has *optimal* secret key size if $|\mathsf{sk}_f| \leq \mathrm{poly}(\lambda, s)$.
- **Master key size.** The master public and secret keys do not necessarily encode any information, and therefore could be as short as $O(\lambda)$. We will say the master key sizes are *optimal* if $|\mathsf{msk}|, |\mathsf{mpk}| \leq \mathrm{poly}(\lambda)$.

Construction. A construction of FE that has above properties is proposed by Ananth and Sahai [4]. The construction is based on indistinguishability obfuscation for circuits and one-way function.

3   An Oracle Problem

Here we define the oracle jump finding problem, which abstracts the algorithmic problem underlying both the iO/diO (differing-inputs obfuscation) conversion of [10] as well as the tracing algorithm in this work.

Definition 4. *The (N, q, δ, ϵ) jump finding problem is the following. An adversary chooses a set $C \subseteq [1, N]$ of q unknown points. Then, the adversary provides an oracle $P : [0, N] \to [0, 1]_{\mathbb{R}}$ such that:*

- *$|P(N) - P(0)| > \epsilon$. That is, over the entire domain, P varies significantly.*
- *For any $x, y \in [0, N], x < y$ in interval $(x, y]$ that does not contain any points in C (that is, $(x, y] \cap C = \emptyset$), it must be $|P(x) - P(y)| < \delta$. That is, outside the points in C, P varies very little.*

Our goal is to interact with the oracle P and output some element in C.

A pictorial representation of the jump finding problem is given in Fig. 1.

Notice that if $\epsilon < q\delta$, it is possible to have all adjacent values $P(x-1), P(x)$ be at less than δ apart, even for $x \in C$. Thus it becomes information-theoretically *impossible* to determine an $x \in C$. In contrast, for $\epsilon \geq q\delta$, if we query the oracle on all points there must exist some point x such that $|P(x) - P(x-1)| > \delta$, and this point must therefore belong to C. Therefore, this problem is inefficiently solvable $\epsilon \geq q\delta$. The following shows that for ϵ somewhat larger that $q\delta$, the problem can even be solved efficiently:

---

[4] This property has been referred to as "compactness" [3,5].

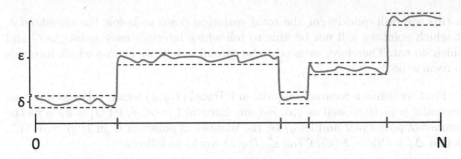

Fig. 1. Example of an oracle P when C contains 4 points. The purple curve represents the outputs of the oracle P on inputs in the interval $[0, N]$. The red hatch marks on the number line indicate the positions of the elements in C. The horizontal dashed lines show that, between the points in C, P is never changes more than δ. At the points in C, P can make arbitrary jumps in either direction.

Theorem 1. *There is a deterministic algorithm* $\mathsf{PTrace}^P(N, q, \delta)$ *that runs in time* $t = \mathsf{poly}(\log N, q)$ *(and in particular makes at most t queries to P) that will output at least one element in C, provided* $\epsilon \geq \delta(2 + (\lceil \log N \rceil - 1)q)$. *Furthermore, the algorithm never outputs an element outside C, regardless of the relationship between ϵ and δ.*

Proof. We assume that $P(N) - P(0) > \epsilon$. The general case can be solved by running our algorithm once, and then running it a second time with the oracle $P'(x) = 1 - P(x)$, and outputting the union of the elements produced. We will also assume $N = 2^n$ is a power of 2, the generalization to arbitrary N being straightforward.

The starting point is the observation that if C contains only a single element x, then this problem is easily solved using binary search. Indeed, we can query P on $0, N/2, N$. If $x \in (0, N/2]$, then there are no points in C that are in $(N/2, N]$, and therefore $P(N) - P(N/2) < \delta$. This implies $P(N/2) - P(0) > \epsilon - \delta > \delta$. Similarly, if $x \in (N/2, N]$, then $P(N/2) - P(0) < \delta < \epsilon - \delta < P(N) - P(N/2)$. Therefore, it is easy to determine which half of $(0, N]$ x lies in. Moreover, on the half that x lies in, P still varies by $\epsilon' = \epsilon - \delta$. Therefore, we can recursively search for x on that half. Each time, we split the interval in which x lies in half, and decrease the total variation on that interval by only an additive δ. Since we perform at most $\log N$ steps in this binary search, the total variation will decrease by at most $\delta \log N$, and our choice of ϵ guarantees that the variation stays greater than δ. Therefore, we can proceed all the way down until we've isolated the point x, which we then output.

The problem arises when C contains more than just a single point. In this case, there may be points in both halves of the interval. If we recurse on both halves, the resulting algorithm will run in time that grows with N as opposed to $\log N$. The other option is to pick a single half-interval arbitrarily, and recurse only on that half. However, if there are points in C among both half-intervals, the variation in each half-interval may decrease by a factor of two. Recursing in this way will quickly cut the total variation down to below the threshold δ, at which point we will not be able to tell which intervals have points in C and which do not. Therefore, we need to be careful in how we choose which intervals to recurse on.

First we define a recursive algorithm $\mathsf{PTrace}_0^P(I, q, \delta)$ which takes as input an interval $I = (a, b]$, as well as q, δ. For any interval $I = (a, b]$, let $|I| = b - a$ be the number of points in I and let q_I be the number of points of C in I: $q_I = |I \cap C|$. Define $\Delta_I = P(b) - P(a)$. $\mathsf{PTrace}_0^P(I, q, \delta)$ works as follows:

- Let $I = (a, b]$. Query P on a, b to obtain $P(a), P(b)$. Compute $\Delta_I = P(b) - P(a)$
- If $\Delta_I \le \delta$, abort and output the empty list $\mathcal{T} = \{\}$
- Otherwise, if $|I| = 1$, output $\mathcal{T} = \{b\}$
- Otherwise, partition I into two equal disjoint intervals I_L, I_R so that $I_L \cap I_R = \emptyset$, $I_L \cup I_R = I$, and $|I_L|, |I_R| = |I|/2$. Run $\mathcal{T}_L = \mathsf{PTrace}_0^P(I_L, q, \delta)$ and $\mathcal{T}_R = \mathsf{PTrace}_0^P(I_R, q, \delta)$. Output $\mathcal{T} = \mathcal{T}_L \cup \mathcal{T}_R$.

We then define PTrace to run PTrace_0 on the entire domain $(0, N]$: $\mathsf{PTrace}^P(N, q, \delta) = \mathsf{PTrace}_0^P((0, N], q, \delta)$. We now make several claims about PTrace_0. The first follows trivially from the definition of PTrace_0:

Claim. Any element outputted by PTrace_0 on interval I must be in $C \cap I$. In particular, any element outputted by PTrace is in C. Moreover, we have that any element s outputted must have $P(s) - P(s-1) > \delta$

Claim. The running time of PTrace is a polynomial in q and in $n = \log N$.

Proof. The running time of PTrace is dominated by the number of calls made to PTrace_0. We observe that the intervals I on which PTrace_0 is potentially called form a binary tree: the root is the entire interval $(0, N]$, the leaves are the singleton intervals $(x - 1, x]$, and each non-leaf node corresponding to interval I has two children corresponding to intervals I_L and I_R that are the left and right halves of I. This tree has $1 + \log N$ levels, where the intervals in level i have size 2^i. Based on the definition of PTrace_0, PTrace_0 is only called on an interval I if I's parent contains at least one point in C, or equivalently that I or its sibling contain at least one point in C. Since there are only q points in C, PTrace is called on at most $2q$ intervals in each level. Thus the total number of calls, and hence the overall running time, is $O(q \log N)$.

Claim. Define $\alpha(I) \equiv \delta(\log |I| + (n - 1)q_I - (n - 2))$ where $n = \log N$. Any call to PTrace_0 with $q_I \ge 1$ and $\Delta_I > \alpha(I)$ will output *some* element.

Proof. If $|I| = 1$ and $q_I = 1$, then $\alpha(I) = \delta((n - 1) - (n - 2)) = \delta$. We already know that if $\Delta_I > \delta = \alpha(I)$, PTrace will output an element. Therefore, the claim holds in the case where $|I| = 1$.

Now assume the claim holds if $|I| \le r$. We prove the case $|I| = r + 1$. Assume $q_I \ge 1$, and running PTrace_0 on I does not give any elements in C. Then running

PTrace$_0$ on I_L and I_R does not give any elements. For now, suppose q_{I_L} and q_{I_R} both positive. By induction this means that $\Delta_{I_L} \leq \alpha(I_L) = \delta(\log |I_L| + (n-1)q_{I_L} - (n-2))$ and $\Delta_{I_R} \leq \alpha(I_R) = \delta(\log |I_R| + (n-1)q_{I_R} - (n-2))$. Recall that $\log |I_R| = \log |I_L| = \log |I| - 1$. Together this means that $\Delta_I \leq \alpha(I_L) + \alpha(I_R) \leq \delta(\log |I| + (n-1)q_I - (n-2) - (n - \log |I|)) = \alpha(I) - (n - \log |I|)$. Since $\log |I| \leq n$, we have that $\Delta_I \leq \alpha(I)$.

Now suppose $q_{I_L} = 0$, which implies $q_{I_R} = q_I > 0$. The case $q_{I_R} = 0$ is handled similarly. Then $\Delta_{I_L} \leq \delta$, and by induction $\Delta_{I_R} \leq \alpha(I_R) = \delta(\log |I| + (n-1)q_I - (n-1))$. Thus $\Delta_I \leq \delta(\log |I| + (n-1)q_I - (n-1) + 1) = \alpha(I)$, as desired. This completes the proof of the claim.   □

Notice that $\alpha((0, N]) = \delta(2 + (n-1)q) \leq \epsilon$. Also notice that by definition $\Delta_{(0,N]} > \epsilon$. Therefore, the initial call to PTrace$_0$ by PTrace outputs *some* element, and that element is necessarily in C.   □

Now we define a related oracle problem, that takes the jump finding problem above, hides the oracle P inside a noisy oracle Q, and only provides us with the noisy oracle Q.

Definition 5. *The (N, q, δ, ϵ) noisy jump finding problem is as follows. An adversary chooses a set $C \subseteq [1, N]$ of q unknown points. The adversary then builds an oracle $P : [0, N] \to [0, 1]_{\mathbb{R}}$ as above, but does not provide it directly. As before, P must satisfy:*

- *$|P(N) - P(0)| > \epsilon$*
- *For any $x, y \in [0, N], x < y$ in interval $(x, y]$ that does not contain any points in C (that is, $(x, y] \cap C = \emptyset$), it must be $|P(x) - P(y)| < \delta$.*

Instead of interacting with P, we interact with a randomized oracle $Q : [0, N] \to \{0, 1\}$ defined as follows: $Q(x)$ chooses and outputs a random bit that is 1 with probability $P(x)$, and 0 otherwise. A fresh sample is chosen for repeated calls to $Q(x)$, and is independent of all other samples outputted by Q. Our goal is to interact with the oracle Q and output some element in C.

Theorem 2. *There is a probabilistic algorithm QTrace$^Q(N, q, \delta, \lambda)$ that runs in time $t = \mathsf{poly}(\log N, q, 1/\delta, \lambda)$ (and in particular makes at most t queries to O) that will output at least one element in C with probability $1 - \mathsf{negl}(\lambda)$, provided $\epsilon > \delta(5 + 2(\lceil \log N \rceil - 1)q)$. Furthermore, the algorithm never outputs an element outside C, regardless of the relationship between ϵ and δ.*

The idea is to, given Q, approximate the underlying oracle P, and run PTrace on the approximated oracle. Similar to the setting above, QTrace works even for "cheating" oracles P, as long as $|P(x) - P(y)| < \delta$ for all queried pairs x, y such that $(x, y]$ contains no points in C. We still need Q to be honestly constructed given P.

Proof. Our basic idea is to use O to simulate an approximation $\hat{P}$ to the oracle P, and then run PTrace using the oracle $\hat{P}$.

QTrace$^Q(N, q, \delta, \epsilon, \lambda)$ works as follows. It simulates PTrace(N, q, δ). Whenever PTrace queries P on input x, QTrace does the following:

- For $i = 1, \ldots, O(\lambda/\delta^2)$, sample $z_i \leftarrow O(x)$
- Output $\hat{p}_x$ as the mean of the z_i.

Then QTrace outputs the output of PTrace.

As PTrace makes $O(q \log N)$ oracle calls to P, QTrace will make $O(\lambda q \log N/\delta^2)$ oracle calls. Moreover, the running time is bounded by this quantity as well. Therefore QTrace has the desired running time.

With probability at least $1 - 2^{-\lambda}$, we have that $|p_x - \hat{p}_x| < \delta/2$ for each x that are queried. This means that, with overwhelming probability, for all intervals $(x, y]$ that do not contain any elements of x, we have that $|p_y - p_x| < \delta$, so $|\hat{p}_y - \hat{p}_x| < 2\delta$ with overwhelming probability. Moreover, $|p_N - p_0| > \epsilon$, so $|\hat{p}_N - \hat{p}_0| > \epsilon - \delta$. Thus with overwhelming probability the oracle $\hat{P}$ seen by PTrace is an instance of the $(N, q, \delta' = 2\delta, \epsilon' = \epsilon - \delta)$ noiseless jump finding problem. Notice that

$$\epsilon' = \epsilon - \delta > \delta(5 + 2(n-1)q) - \delta = (2\delta)(2 + (n-1)q) = \delta'(2 + (n-1)q)$$

Therefore, $\hat{P}$ satisfies the conditions of Theorem 1, and PTrace outputs at least one element in C. QTrace outputs the same element, completing the proof.

Remark 1. We note that PTrace^P and QTrace^Q work even for "cheating" P that do not satisfy $|P(x) - P(y)| < \delta$ for *all* $(x, y]$ which do not intersect C, as long as the property holds for all pairs x, y that where queried by PTrace or QTrace^Q. This will be crucial for traitor tracing.

3.1   The Generalized Jump Finding Problem

Here we define a more general version of the jump finding problem that will be useful for obtaining short-ciphertext traitor tracing. In this version, the domain of the oracle P is an $r \times 2N$ grid that is short but wide (that is, $r \ll N$). The elements in C correspond to non-crossing curves between grid points from the top of the grid to the bottom, which divide the grid into $|C| + 1$ contiguous regions. The probabilities outputted by P are restricted to vary negligibly across each continuous region, but are allowed to vary arbitrary between different regions. The goal is to recover the complete description of *some* curve in C. To help make the problem tractable, we require that each curve is confined to oscillate about an *odd* column of the grid. Such curves can be represented by an integer $s \in [N]$ giving the position $2s - 1$ of the column, and a bit string $b = (b_1, \ldots, b_r) \in \{0, 1\}^r$ specifying which side of the column the curve is on at each row. A pictorial representation of the generalized jump finding problem is given in Fig. 2, and a precise definition is given below.

Definition 6. *The $(N, r, q, \delta, \epsilon)$ generalized jump finding problem is the following. The adversary chooses a set C of q unknown tuples $(s, b_1, \ldots, b_r) \in [N] \times \{0, 1\}^r$ such that the s are distinct. Each tuple $(s, b_1, \ldots, b_r)$ describes a curve between grid points from the top to bottom of the grid $[1, r] \times [0, 2N]$,*

which oscillates about the column at position $2s - 1$, with $b = (b_1, \ldots, b_r)$ speci-
fying which side of the column the curve is on at each row. These curves divide
the grid into $|C| + 1$ contiguous regions. For each pair $(i, x) \in [1, r] \times [0, 2N]$ the
adversary chooses a probability $p_{i,x} \in [0, 1]_{\mathbb{R}}$ such that $p_{i,x}$ varies "minimally"
within each contiguous region. We also require that overall from left to right,
there is "significant" variation of the $p_{i,x}$. Formally, this means:

- *For any pair of pairs of the form $(i, 2x), (j, 2x) \in [1, r] \times [0, 2N]$, $|p_{i,2x} -*
  p_{j,2x}| < \delta$. In other words, since curves in C are restricted to oscillate around
  odd columns, no curve crosses between points on the same even column, so
  each even column lies entirely in a single contiguous region. We therefore
  require that the probabilities associated with any two points on the same even
  column are close.
- *Let C_i be the set of values $2s - b_i$ for tuples in C. C_i is then the set of grid*
  points in the ith row that are immediately to the right of curves in C. For any
  two pairs $(i, x), (i, y) \in [1, r] \times [0, 2N]$ in the same row such that the interval
  $(x, y]$ does not contain any points in C_i then $|p_{i,x} - p_{i,y}| < \delta$. In other words,
  if no curves cross between points in the same row, those points must be in the
  same contiguous region and therefore have close probabilities.
- *We also make the requirement that the probabilities in the 0th column are*
  identical, and the probabilities in the $2N$th column are identical. That is,
  $p_{i,0} = p_{i',0}$ for all $i, i' \in [r]$ and $p_{i,2N} = p_{i',2N}$ for all $i, i' \in [r]$. Define
  $p_0 = p_{i,0}$ and $p_{2N} = p_{i,2N}$.
- *Finally, $|p_{2N} - p_0| > \epsilon$. That is, the 0th and $2N$th columns have very different*
  probabilities.

We are now presented with one of two oracles, depending on the version of the
problem:

- *In the noiseless version, we are given an oracle for the $p_{i,x}$: we are given oracle*
  access to the function $P : [1, r] \times [0, 2N] \to [0, 1]_{\mathbb{R}}$ such that $P(i, x) = p_{i,x}$.
- *In the noisy version, we are given a randomized oracle Q with domain $[1, r] \times*
  [0, 2N]$ that, on input (i, x), outputs 1 with probability $p_{i,x}$. Repeated calls to
  Q on the same x yield a fresh bit sampled independently.

Our goal is to output some element in C.

Theorem 3. *There are algorithms $\mathsf{PTrace}'^P(N, r, q, \delta)$ and $\mathsf{QTrace}'^Q(N, r, q, \delta, \lambda)$*
for the noiseless and noisy versions of the $(N, r, q, \delta, \epsilon)$ generalized jump finding
problem that run in time $\mathsf{poly}(\log N, r, q, 1/\delta)$ and $\mathsf{poly}(\log N, r, q, 1/\delta, \lambda)$, respec-
tively, and output an element in C with overwhelming probability, provided $\epsilon >
\delta(4 + 2(\lceil \log N \rceil - 1)q)$ (for the noiseless case), or $\epsilon > \delta(9 + 4(\lceil \log N \rceil - 1)q)$ (for
the noisy case).

This theorem is proved analogously to Theorems 1 and 2, and appears in
below. Again, PTrace', QTrace' work even if the oracle P is "cheating", as long
as the requirements on P hold for all points queried by PTrace' or QTrace'.

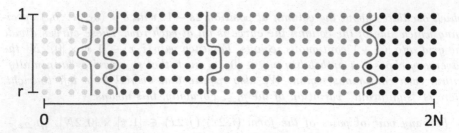

Fig. 2. Example probabilities $p_{i,x}$ when C contains 4 items, $r = 7$, and $N = 15$. The dots represent the various probabilities $p_{i,x}$, where rows are indexed by $i \in [r]$ and columns are indexed by $x \in [0, 2N]$. The shade of the dot at position (i, x) indicates the value of $p_{i,x}$, with darker shade indicating higher $p_{i,x}$. The elements in C describe curves from the top of the grid to the bottom, which are indicated in red in the figure. Notice (1) that the curves in C oscillate around *odd* columns of dots, and (2) that they never intersect, and (3) that the values of the $p_{i,x}$ only vary minimally between the curves in C, and can only have large changes when crossing the curves.

Proof. We prove the noiseless version, extending to the noisy version is a simple extension of Theorem 2. $\mathsf{PTrace}'^P (N, r, q, \delta)$ works as follows:

- First, we determine some of the s for elements in C. Let $P' : [0, N] \to [0, 1]_{\mathbb{R}}$ where $P'(x) = P(1, 2x)$. Notice that $|P'(N) - P'(0)| = |p_{2N} - p_0| > \epsilon$. Moreover, for intervals $(x, y]$ that do not contain any of the s, $|P'(y) - P'(x)| < \delta \leq 2\delta$. Therefore, P' is an instance of the $(N, q, 2\delta, \epsilon)$ problem for $\epsilon > 2\delta(2 + (n-1)q)$. Therefore, we run $\mathsf{PTrace}^{P'} (N, q, \delta')$ to obtain a list $\mathcal{T}$ of s values, with the property that $|P(1, 2x) - P(1, 2x - 2)| = |P'(s) - P'(s - 1)| \geq 2\delta$ for each $s \in \mathcal{T}$.
- For each $s \in \mathcal{T}$, and for each $i \in [r]$, let $b_{s,i} = 1$ if $|P(i, 2s - 2) - P(i, 2s - 1)| > |P(i, 2s - 1) - P(i, 2s)|$, and $b_{s,i} = 0$ otherwise. Let $(s, b_1, \dots, b_r) \in C$ be the tuple corresponding to s. Then the set C_i contains $2s - b_i$, but does not contain $2s - 1 + b_i$, since there is no collision between the s values. Therefore, $|P(2s - 1 + b_i) - P(2s - 2 + b_i)| < \delta$, which means that $|P(2s - b_i) - P(2s - 1 - b_i)| > \delta$. Therefore $b_{s,i} = b_i$
- Output the tuples $(s, b_{s,1}, \dots, b_{s,r})$.

By the analysis above, since PTrace never outputs a value outside of C, PTrace' will never output a tuple corresponding to an identity outside of C. Moreover, if $\epsilon > \delta(4 + 2(n-1)q)$, then PTrace' will output at least one tuple in C. Finally, PTrace' runs in time only slightly worse than PTrace, and is therefore still polynomial time.

4   Tracing with Flexible Identities

Let $(\mathsf{Setup}, \mathsf{KeyGen}, \mathsf{Enc}, \mathsf{Dec})$ be a secure private *linear* broadcast scheme for identity space $\mathcal{ID} = [2^n]$. We now show that such a private broadcast scheme is sufficient for flexible traitor tracing. The Setup, KeyGen, Enc, and Dec algorithms are as follows:

- Setup, KeyGen are inherited from the private broadcast scheme.
- To encrypt a message m, run Enc(mpk, $S = \mathcal{ID}, m$). Call this algorithm Enc$_{TT}$.
- To decrypt a ciphertext c, run Dec(sk$_{id}, c$). Call this algorithm Dec$_{TT}$

Theorem 4. *Let* (Setup, KeyGen, Enc, Dec) *be a secure private broadcast scheme for identity space* $[2^n]$ *and private class* $\mathcal{S} = \{[u]\}_{u \in [0,2^n]}$. *Then there is a polynomial time algorithm* Trace *such that* (Setup, KeyGen, Enc$_{TT}$, Dec$_{TT}$, Trace) *as defined above is a flexible traitor tracing algorithm.*

Proof. Boneh et al. [7] prove this theorem for the case of logarithmic n and for the weaker notion of tracing where the pirate decoder is required to decrypt a random message, as opposed to distinguish between two specific messages of the adversary's choice. Their tracing algorithm gets black-box access to a pirate decoder and does the following: it runs the decoder on encryptions to all sets $[u]$ for $u = 0, \ldots, 2^n$ and determines the success probability of the decoder for each u. It outputs an index u such that there is a "large" gap between the probabilities for $[u - 1]$ and $[u]$ as the identity of the traced traitor. In the analysis, [7] shows that, provided the adversary does not control the identity u, the pirate succeeds with similar probabilities for $[u - 1]$ and $[u]$. To prove this, they run the adversary, answering its secret key queries by making secret key queries to the PLBE challenger. When the adversary outputs a pirate decoder D, they make a PLBE challenge on a random message m and sets $[u]$ and $[u-1]$. Then they run the pirate decoder on the resulting ciphertext, and test whether it decrypts successfully: if yes, then they guess that the ciphertext was encrypted to $[u]$, and guess $[u - 1]$ otherwise. The advantage of this PLBE adversary is exactly the difference in probabilities for decrypting $[u-1]$ and $[u]$. The security of the PLBE scheme shows that this difference must be negligible.

Now, a useful pirate decoder will succeed with high probability on $[2^n]$, and with negligible probability on $[0]$, so there must be some "gap" in probabilities. The above analysis shows that (1) the tracer will find a gap, and (2) that the gap must occur at an identity under the adversary's control.

There are two problems with generalizing to our setting:

- The running time of the tracing algorithm in [7] grows with 2^n as opposed to n, resulting in an exponential-time tracing algorithm when using flexibly-large identities. This is because their tracing algorithm checks the pirate decoder an all identities. We therefore need a tracing algorithm that tests the decoder on a polynomial number of identities. To accomplish this, show that tracing amounts to solving the jump-finding problem in Sect. 3, and we can therefore use our efficient algorithm for the jump-finding problem to trace.
- Since we only ask that the pirate decoder can distinguish two messages, we need to reason about the decoder's "advantage" (decryption probability

minus $1/2$) instead of its decryption probability. In the analysis above, since probabilities are always positive, any "useful" decoder will contribute positively to the PLBE advantage, whereas a "useless" decoder will not detract. However, this crucially relies on the fact that probabilities are positive. In our setting, the advantage is signed and can be both positive and negative, and the contribution of decoders to the PLBE adversary's advantage can cancel out if they have different sign. Thus there is no guarantee that the obtained PLBE adversary has any advantage. To get around this issue, we essentially have our reduction estimate the signed advantage of the pirate decoder, and reject all decoders with negative advantage. The result is that the advantage of all non-rejected decoders is non-negative, and so all decoders contribute positively to the PLBE adversary's advantage.

We now give our proof. Let $\mathcal{A}$ be a potential adversary, let C be the set of colluding parties for which $\mathcal{A}$ obtained secret keys, and $q = |C|$. $\mathcal{A}$ produces a pirate decoder $\mathcal{D}$ and messages m_0, m_1 such that $\mathcal{D}$ can distinguish encryptions of m_0 from encryptions of m_1. Define the quantities

$$p_{\mathsf{id}} = \Pr[\mathcal{D}(c) = b : b \leftarrow \{0,1\}, c \leftarrow \mathsf{Enc}(\mathsf{mpk}, \mathsf{id}, m_b)]$$

for $\mathsf{id} \in \mathcal{S}$, where Enc is the PLBE encryption algorithm. We first will prove two lemmas:

Lemma 1. *Suppose* $(\mathsf{Setup}, \mathsf{KeyGen}, \mathsf{Enc}, \mathsf{Dec})$ *is secure. Fix a non-negligible value* δ. *Suppose an interval* $(\mathsf{id}_L, \mathsf{id}_R]$ *is chosen adversarially after seeing the set* C, *the adversary's secret keys, the pirate decoder* $\mathcal{D}$, *and even the internal state of* $\mathcal{A}$, *and suppose that* $C \cap (\mathsf{id}_L, \mathsf{id}_R] = \emptyset$ *(that is, there are no colluding users in* $(\mathsf{id}_L, \mathsf{id}_R]$*). Then, except with negligible probability,* $|p_{\mathsf{id}_R} - p_{\mathsf{id}_L}| < \delta$.

Proof. We will prove that $p_{\mathsf{id}_R} - p_{\mathsf{id}_L} < \delta$ with overwhelming probability, as proving $p_{\mathsf{id}_L} - p_{\mathsf{id}_R} < \delta$ is almost identical. Suppose towards contradiction that, with non-negligible probability ϵ, $p_{\mathsf{id}_R} - p_{\mathsf{id}_L} \geq \delta$. We then describe an adversary for $(\mathsf{Setup}, \mathsf{KeyGen}, \mathsf{Enc}, \mathsf{Dec})$ that works as follows:

- Run $\mathcal{A}$ on input mpk. Whenever $\mathcal{A}$ makes a keygen query on identity id, make the same keygen query. $\mathcal{A}$ outputs a pirate decoder $\mathcal{D}$.
- Compute estimates $\hat{p}_{\mathsf{id}_R}, \hat{p}_{\mathsf{id}_L}$ for the probabilities p_{id_L} and p_{id_R}, respectively. To compute $\hat{p}_{\mathsf{id}}$, do the following. Take $O(\lambda/\delta^2)$ samples of $\mathcal{D}(c) \oplus b$ where $b \leftarrow \{0,1\}$ and $c \leftarrow \mathsf{Enc}(\mathsf{mpk}, \mathsf{id}, m_b)$, and then output the fraction of those samples that result in 0. Notice that with probability $1 - 2^{-\lambda}$, $|\hat{p}_{\mathsf{id}} - p_{\mathsf{id}}| \leq \delta/4$.
- If $\hat{p}_{\mathsf{id}_R} - \hat{p}_{\mathsf{id}_L} < \frac{1}{2}\delta$, output a random bit and abort. Notice that, with overwhelming probability, $\left|(\hat{p}_{\mathsf{id}_R} - \hat{p}_{\mathsf{id}_L}) - (p_{\mathsf{id}_R} - p_{\mathsf{id}_L})\right| < \delta/2$. Therefore, with overwhelming probability, if we do not abort, $p_{\mathsf{id}_R} - p_{\mathsf{id}_L} > 0$. Moreover, if $p_{\mathsf{id}_R} - p_{\mathsf{id}_L} > \delta$, then $\hat{p}_{\mathsf{id}_R} - \hat{p}_{\mathsf{id}_L} \geq \frac{1}{2}\delta$ holds and we do not abort with overwhelming probability.
- Now choose a random bit b, and make a challenge query on $S_0^* = [\mathsf{id}_L]$, $S_1^* = [\mathsf{id}_R]$, and messages $m_0^* = m_1^* = m_b$.

- Upon receiving the challenge ciphertext c^*, compute $b' = \mathcal{D}(c^*)$. Output 1 if $b' = b$ and 0 otherwise.

Conditioned on no aborts, in the case the challenge ciphertext is encrypted to id_L (resp. id_R), our adversary will output 1 with probability p_{id_L} (resp. p_{id_R}), so our adversary will "win" with probability $\frac{1}{2} + (p_{\mathsf{id}_R} - p_{\mathsf{id}_L})/2$ in this case. Otherwise, during an abort, our adversary wins with probability $1/2$. Moreover, with overwhelming probability, if we do not abort $p_{\mathsf{id}_R} - p_{\mathsf{id}_L} > 0$, and with probability at least $\epsilon - \mathsf{negl}$, we have $p_{\mathsf{id}_R} - p_{\mathsf{id}_L} > \delta/2$. Therefore, a simple computation shows that the adversary "wins" with probability at least $\frac{1}{2} + (\epsilon - \mathsf{negl})(\delta/4 - \mathsf{negl})$, which gives a non-negligible advantage.

Lemma 2. *Suppose* $(\mathsf{Setup}, \mathsf{KeyGen}, \mathsf{Enc}, \mathsf{Dec})$ *is secure. Fix a non-negligible value* δ. *Then, except with negligible probability,* $|p_0 - \frac{1}{2}| < \delta$.

Proof. The proof is similar to the proof of Lemma 1. We will prove that $p_0 - \frac{1}{2} < \delta$ with overwhelming probability, the case $p_0 - \frac{1}{2} > -\delta$ is almost identical. Suppose towards contradiction that, with non-negligible probability ϵ, $p_0 - \frac{1}{2} \geq \delta$. An adversary for $(\mathsf{Setup}, \mathsf{KeyGen}, \mathsf{Enc}, \mathsf{Dec})$ works as follows:

- Run $\mathcal{A}$ on input mpk. Whenever $\mathcal{A}$ makes a keygen query on identity id, make the same keygen query. $\mathcal{A}$ outputs a pirate decoder $\mathcal{D}$.
- Compute estimate $\hat{p}_0$ for p_0 using the algorithm from Lemma 1, so that except with probability $2^{-\lambda}$, $|\hat{p}_0 - p_0| < \delta/2$.
- If $\hat{p}_0 - \frac{1}{2} < \frac{1}{2}\delta$, output a random bit and abort. Notice that, with overwhelming probability, $|(\hat{p}_0 - \frac{1}{2}) - (p_0 - \frac{1}{2})| < \delta/2$. Therefore, with overwhelming probability, if we do not abort, $p_0 - \frac{1}{2} > 0$. Moreover, if $p_0 - \frac{1}{2} > \delta$, with overwhelming probability we do not abort.
- Now make a challenge query on $S_0^* = S_1^* = [0] = \{\}$, and messages $m_0^* = m_0, m_1^* = m_1$.
- Upon receiving the challenge ciphertext c^*, compute $b = \mathcal{D}(c^*)$. Output b

Conditioned on no aborts, our adversary will "win" with probability p_0 in this case. Otherwise, during an abort, our adversary wins with probability $1/2$. Moreover, with overwhelming probability, if we abort $p_0 - \frac{1}{2} > 0$, and with probability at least $\epsilon - \mathsf{negl}$, we have $p_0 - \frac{1}{2} > \delta/2$. Therefore, a simple computation shows that the adversary has non-negligible advantage $(\epsilon - \mathsf{negl})(\delta/2 - \mathsf{negl})$.

Now we define our tracing algorithm $\mathsf{Trace}^{\mathcal{D}}(\mathsf{mpk}, m_0, m_1, q, \epsilon)$. Trace sets $\delta = \epsilon/2(5 + 4(n-2)q)$, and then runs $\mathsf{QTrace}^Q(2^n, q, \delta, \lambda)$ where QTrace is the algorithm from Theorem 2. Whenever QTrace makes a query to Q on identity id, Trace chooses a random bit b, computes the encryption $c \leftarrow \mathsf{Enc}(\mathsf{mpk}, \mathsf{id}, m_b)$ of m_b to the set [id], runs $b' \leftarrow \mathcal{D}(c)$, and responds with 1 if any only if $b = b'$. Define p_{id} to be the probability that $Q(\mathsf{id})$ outputs 1. We now would like to show that Q is an instance of the (N, q, δ, ϵ) noisy jump finding problem, where the set of jumps is the set C. For this it suffices to show that $P(\mathsf{id}) = p_{\mathsf{id}}$ is an instance of the (N, q, δ, ϵ) *noiseless* jump finding problem. By Lemma 2, we have that with overwhelming probability useful $\mathcal{D}$ have $|p_{2^n} - p_0| \geq |\epsilon - \delta| > \epsilon/2$. Moreover, we have that $(\epsilon/2) = \delta(5 + 4(n-2)q)$.

Now we would hope that for any $(\mathsf{id}_L, \mathsf{id}_R]$ that do not contain one of the adversary's points, $|p_{\mathsf{id}_R} - p_{\mathsf{id}_L}| < \delta$. This would seem to follow from Lemma 1. However, we only have this property for $\mathsf{id}_L, \mathsf{id}_R$ that can be efficiently computed. Therefore, $P(\mathsf{id})$ is potentially a cheating oracle. However, since our tracing algorithm is efficient, any query it makes can be efficiently computed, and therefore $|p_{\mathsf{id}_R} - p_{\mathsf{id}_L}| < \delta$ holds (with overwhelming probability) for all *queried* points such that $(\mathsf{id}_L, \mathsf{id}_R]$ does not contain any of the identities in C. Therefore, following Remark 1, we can still invoke Theorem 2, which shows that the following hold:

- QTrace, and hence Trace, runs in polynomial time.
- QTrace, and hence Trace, will with overwhelming probability *not* output an identity outside S.
- If $\mathcal{D}$ is ϵ-useful, then QTrace, and hence Trace, will output *some* element in S (w.h.p.).

Construction. As observed by Garg et al. [21], FE immediately gives a PLBE scheme. Let $\mathcal{F}$ be the set of functions $f_{\mathsf{id}} : S \times \mathcal{M} \to (\mathcal{M} \cup \{\bot\})$ where $f_{\mathsf{id}}(S, m)$ outputs m if $m \in S$ and $\bot$ if $m \notin S$. Let $(\mathsf{Setup}_{FE}, \mathsf{KeyGen}_{FE}, \mathsf{Enc}_{FE}, \mathsf{Dec}_{FE})$ be a FE scheme for this class of functions. The plaintext space $S \times \mathcal{M}$ has size $2^\lambda \times |\mathcal{M}|$, and the function space admits circuits of size $O(\lambda)$. We then immediately obtain a PLBE scheme: to encrypt a message to a set S, simply encrypt the pair (S, m). The secret key for identity id is the secret key for function f_{id}. We use an adaptively secure scheme [2, 21, 40].

Parameter Sizes. In the above conversion, the PLBE scheme inherits the parameter sizes of the functional encryption scheme. Using functional encryption for general circuits, the secret size is $\mathsf{poly}(n)$ and the ciphertext size will similarly grow as $\mathsf{poly}(n, |m|)$. We can make the ciphertext size $|m| + \mathsf{poly}(n)$ by turning the PLBE into a key encapsulation protocol where we use the PLBE to encrypt the key for a symmetric cipher, and then encrypt m using the symmetric cipher. We note that it is inherent that the secret keys and ciphertexts of a PLBE scheme grow with the identity bit length n, as both terms must encode a complete identity. Therefore we obtain a PLBE scheme with essentially optimal parameters:

Corollary 1. *Assuming the existence of iO and OWF, then there exists an adaptively secure traitor tracing scheme whose master key is size is $O(1)$, secret key size is $\mathsf{poly}(n)$, and ciphertext size is $|m| + \mathsf{poly}(n)$.*

Note, however, that the obtained traitor tracing scheme is *not* optimal, as there is no reason ciphertexts in a traitor tracing scheme need to grow with the identity bit-length. The large ciphertexts are inherent to the PLBE approach to traitor tracing, so obtaining smaller ciphertexts necessarily requires a different strategy. In Sect. 5, we give an alternate route to obtaining traitor tracing that does not suffer this limitation, and we are therefore able to obtain an optimal traitor tracing system.

On Bounded Collusions. If we relax the security to bounded-collusion security, then the assumption can be relaxed to PKE using the q-bounded collusion FE scheme of [25].

Corollary 2. *Assume the existence of secure PKE, then there exists a q-bounded collusion-resistant adaptively secure traitor tracing scheme whose master key and secret key sizes are $O(n)\mathsf{poly}(q)$ and ciphertext size is $|m| + O(n)\mathsf{poly}(q)$.*

5   Flexible Traitor Tracing with Short Ciphertexts

We now discuss how to achieve traitor tracing with small ciphertexts that do not grow with the identity size. As noted above, the approach using private *linear* broadcast is insufficient due to having ciphertexts that inherently grow with the identity bit-length. We note that for traitor tracing, secret keys must encode the identities anyway, so they will always be as long as the identities. Therefore the focus here is just on obtaining short ciphertexts. To that end, we introduce a generalization of private linear broadcast that does not suffer from the limitations of the private linear broadcast approach; in particular, the information contained in the ciphertext is much shorter than the identities.

Let $\mathcal{ID}_0 = [2^{t+1}]$ be the set of identity "blocks", and the total identity space $\mathcal{ID} = (\mathcal{ID}_0)^n$ be the set of n-block tuples. Let (Setup, KeyGen, Enc, Dec) be a secure private broadcast scheme for $\mathcal{ID}$, and the secret class $\mathcal{S}$ defined as follows: each set $S_{i,u} \in \mathcal{S}$ is labeled by an index $i \in [n]$ and "identity block" $u \in \mathcal{ID}_0 \cup \{0\}$. $S_{i,u}$ is the set of tuples $\mathsf{id} = (\mathsf{id}_1, \dots, \mathsf{id}_n)$ where $\mathsf{id}_i \leq u$. We call such a private broadcast scheme a *private block linear broadcast encryption* (PBLBE) scheme.

Ideally, we would like to simply add a tracing algorithm on top of (Setup, KeyGen, Enc, Dec) as we did in the previous section. The tracing algorithm would run the tracing algorithm from Sect. 4 on each identity block. For each $i \in [n]$, this gives a list of, say, T_i identity blocks $\mathsf{id}_{j,i} \in \mathcal{ID}_0$ for $j \in [T_i]$, where each of the $\mathsf{id}_{j,i}$ is the ith block of *some* identity owned by the adversary. Repeating this for every i gives a collection of identity blocks for every block number. However, it is not clear how to use these blocks to construct a complete identity in $\mathcal{ID}$. There are two problems:

- How do we argue that the blocks obtained for each index i come from the same set of identities? It may be that, for example when $n = 2$, that the adversary has identities $(\mathsf{id}_{1,1}, \mathsf{id}_{1,2})$ and $(\mathsf{id}_{2,1}, \mathsf{id}_{2,2})$, but tracing for $i = 1$ yields $\mathsf{id}_{1,1}$ whereas tracing $i = 2$ yields $\mathsf{id}_{2,2}$. While we have obtained two of the adversary's blocks, there may not even be a complete identity among the blocks.
- Even if we resolve the issue above, and show that tracing each block number yields blocks from the same set of identities, there is another issue. How to we match up the partial identity blocks? For example, in the case $n = 2$, we may obtain blocks $\mathsf{id}_{1,1}, \mathsf{id}_{2,1}, \mathsf{id}_{1,2}, \mathsf{id}_{2,2}$. However, we have no way of telling if the adversary's identities were $(\mathsf{id}_{1,1}, \mathsf{id}_{1,2})$ and $(\mathsf{id}_{2,1}, \mathsf{id}_{2,2})$, or if they were

$(\mathsf{id}_{1,1}, \mathsf{id}_{2,2})$ and $(\mathsf{id}_{2,1}, \mathsf{id}_{1,2})$. Therefore, while we can obtain the adversary's blocks for the adversary's identities, we cannot actually reconstruct the adversary's identities themselves.

We will now explain a slightly modified scheme and tracing algorithm to rectify the issues above. First, by including a fixed tag τ inside every block of id, we can now identify which blocks belong together simply by matching tags. This resolves the second point above, but still leaves the first. For this, we give a modified tracing algorithm that we can prove always outputs a complete collection of blocks.

We now give the scheme derived from any PBLBE. There will be two identity spaces. Let $\mathcal{ID}' = \{0,1\}^n$ be the identity space for the actual traitor tracing scheme; that is, $\mathcal{ID}'$ is the set of identities that we actually want to recover by tracing. We wish to grow n arbitrarily large without affecting the ciphertext size. The second space will be the space $\mathcal{ID}$ of the underlying PBLBE, which consists of n blocks of $t+1$ bits. In particular, the bit length of the traitor tracing identity space $\mathcal{ID}'$ will be equal to the number of blocks in the PBLBE space. Set $t = \lambda$, so that the bit-length of each block in the PBLBE grows with the security parameter, but crucially not in n. Define $N = 2^t = 2^\lambda$.

- Setup is again inherited from the private broadcast scheme.
- To generate the secret key for an identity $\mathsf{id}' \in \mathcal{ID}'$, write $\mathsf{id}' = (\mathsf{id}'_1, \ldots, \mathsf{id}'_n)$ where $\mathsf{id}'_i \in \{0,1\}$. Choose a random $s \in [N]$, and define the identity $\mathsf{id} = (\mathsf{id}_1, \ldots, \mathsf{id}_n) \in \mathcal{ID}$ where $\mathsf{id}_i = 2s - \mathsf{id}'_i \in \mathcal{ID}_0$. Run the private broadcast keygen algorithm on id, and output the resulting secret key. Call this algorithm KeyGen_{TT}
- $\mathsf{Enc}, \mathsf{Dec}$ are identical to the basic tracing scheme, except that Dec now uses the derived user secret key as defined above. Call these algorithms $\mathsf{Enc}_{TT}, \mathsf{Dec}_{TT}$.

Theorem 5. *Let* $(\mathsf{Setup}, \mathsf{KeyGen}, \mathsf{Enc}, \mathsf{Dec})$ *be a secure private broadcast scheme for identity space* $\mathcal{ID}$ *and private class* $\mathcal{S}$, *where* $\mathcal{ID}, \mathcal{S}$ *are defined as above. Then there is an efficient algorithm* Trace *such that* $(\mathsf{Setup}, \mathsf{KeyGen}, \mathsf{Enc}_{TT}, \mathsf{Dec}_{TT}, \mathsf{Trace})$ *as defined above is a flexible traitor tracing algorithm.*

We prove Theorem 5 using similar techniques as in the proof of Theorem 4, except that the jump finding problem in Sect. 3 does not quite capture the functionality we need. Instead, in Sect. 3.1, we define a *generalized* jump finding problem, and show how to solve it. We then use the solution for the generalized jump finding problem to trace our scheme above.

Proof. We will take an approach very similar to the proof of Theorem 4. We will use a pirate decoder $\mathcal{D}$ to create an oracle Q as in the generalized jump finding problem. Then we run the tracing algorithm QTrace' on this Q, which will output the identities of some the colluders.

Define $Q(i, u)$ to be the randomized procedure that does the following: sample a random bit b, computes the encryption $c \leftarrow \mathsf{Enc}(\mathsf{mpk}, (i, u), m_b)$ of m_b to the set $S_{i,u}$ indexed by $(i, u) \in [n] \times [0, 2N]$, runs $b' \leftarrow \mathcal{D}(c)$, and outputs 1 if and only if $b = b'$. Define $p_{i,u}$ to be the probability that $Q(i, u)$ outputs 1. We now need to show that if $\mathcal{D}$ is useful, then Q satisfies the conditions of Theorem 3.

First, notice that $p_{i,0} = p_{i',0}$ for all $i, i' \in [n]$, since the set indexed by $(i, 0)$ is just the empty set, independent of i. Define $p_0 = p_{i,0}$. Similarly, $p_{i,2N} = p_{2N}$, independent of i, as the set indexed by $(i, 2N)$ is the complete set.

Next, notice that if $\mathcal{D}$ is useful, we have $|p_{2N} - p_0| > \epsilon/2$, similar to Theorem 4. Now set $\delta = \epsilon/(9 + 4(t - 1)q)$ (recall that $N = 2^t$). We have the following:

Lemma 3. *Suppose* (Setup, KeyGen, Enc, Dec) *is secure. Fix a non-negligible value δ. Suppose two pairs $(i, 2x), (j, 2x) \in [n] \times [0, 2N]$ are chosen adversarially after seeing the set C, the adversary's secret keys, the pirate decoder $\mathcal{D}$, and even the internal state of $\mathcal{A}$. Then, except with negligible probability $|p_{i,2x} - p_{j,2x}| < \delta$*

Proof. Let id' be an identity the adversary queries on, with associated tag s. Let $\mathsf{id} = (\mathsf{id}_1, \ldots, \mathsf{id}_n) \in \mathcal{ID}$ where $\mathsf{id}_i = 2s - \mathsf{id}'_i \in \mathcal{ID}_0$ as above. It suffices to show that the set $\mathsf{id} \in S_{i,2x}$ if and only if $\mathsf{id} \in S_{j,2x}$. This is equivalent to the requirement that $2s - \mathsf{id}'_i \leq 2x$ if and only if $2s - \mathsf{id}'_j \leq 2x$. Since $\mathsf{id}'_i, \mathsf{id}'_j$ are binary, this is true. The lemma then follows from the security of the private block linear broadcast scheme.

Next, define C_i to be the set of values $2s - \mathsf{id}'_i$ for identities id' queried by the adversary. Equivalently, C_i is the set of ith blocks of the corresponding identities id. The following also easily follows from the security of private block linear broadcast:

Lemma 4. *Suppose* (Setup, KeyGen, Enc, Dec) *is secure. Fix a non-negligible value δ. Suppose two pairs $(i, x), (i, y) \in [n] \times [0, 2N]$ are chosen adversarially after seeing the set C, the adversary's secret keys, the pirate decoder $\mathcal{D}$, and even the internal state of $\mathcal{A}$, such that the interval $(x, y]$ does not contain any points in C_i. Then $|p_{i,x} - p_{i,y}| < \delta$.*

We now see that the oracle Q corresponds to the $(N, r = n, q, \delta, \epsilon)$-generalized jump finding problem. Here, the hidden set C contains tuples $(s, \mathsf{id}_1, \ldots, \mathsf{id}_n) = (s, \mathsf{id})$ where where $\mathsf{id} \in \mathcal{ID}'$ is one of the adversary's identities, and s is the corresponding tag that was used to generate the secret key for id. Similar to the basic tracing algorithm, the pirate decoder may *cheat*, and the lemmas above may not hold for all possible points. However, they hold for efficiently computable points, and in particular must hold for the points queried by the efficient QTrace' of Theorem 3. Thus, following Remark 1, we can invoke Theorem 3, so QTrace' will produce a non-empty list $\mathcal{L}$ of tuples (s, id) from C. This completes the theorem.

Construction and Parameter sizes. Similar to the case of PLBE, it is straightforward to construct private block linear broadcast encryption from functional encryption, and the PBLBE scheme will inherit the parameter sizes from the FE scheme. We will use $r = \lambda$-bit blocks and n-bit identities. The circuit size needed for the functional encryption scheme is therefore $\mathsf{poly}(n)$, and the plaintext size is $|m| + \mathsf{poly}(\log n)$ (ignoring the security parameter).

Some functional encryption schemes are non-compact, meaning the ciphertext size grows with both the plaintext size and the function size, in which case our ciphertexts will be $|m| + \mathsf{poly}(n)$, no better than the basic tracing system. Instead, we require compact functional encryption, where the ciphertext size is independent of the function size. The original functional encryption scheme of Garg et al. [21] has this property. However, they only obtain static security, and adaptive security is only obtained through complexity leveraging. In a very recent work, Ananth and Sahai [4] show how to obtain adaptively secure functional encryption for Turing machines, and in particular obtain adaptively secure functional encryption that meets our requirements for optimal ciphertext and secret key sizes.

Corollary 3. *Assuming the existence of iO and OWF, there exists an adaptively secure traitor tracing scheme whose master key size is $\mathsf{poly}(\log n)$, secret key size is $\mathsf{poly}(n)$, and ciphertext size is $|m| + \mathsf{poly}(\log n)$.*

On Bounded Collusions. If we relax security to bounded-collusion security, then the underlying assumption can be relaxed to the (sub-exponential) LWE assumption using the succinct FE scheme of [23], which can be made adaptively secure through complexity leveraging.

Corollary 4. *Assume the sub-exponential hardness of the LWE problem with a sub-exponential factor, then there exists a q-bounded collusion-resistant adaptively secure traitor tracing scheme whose master key size is $\mathsf{poly}(\log n, q)$ and secret key size is $\mathsf{poly}(n, q)$ and ciphertext size is $|m| + \mathsf{poly}(\log n, q)$.*

References

1. Abdalla, M., Dent, A.W., Malone-Lee, J., Neven, G., Phan, D.H., Smart, N.P.: Identity-based traitor tracing. In: Okamoto, T., Wang, X. (eds.) PKC 2007. LNCS, vol. 4450, pp. 361–376. Springer, Heidelberg (2007)
2. Ananth, P., Brakerski, Z., Segev, G., Vaikuntanathan, V.: From selective to adaptive security in functional encryption. In: Gennaro, R., Robshaw, M. (eds.) CRYPTO 2015. LNCS, vol. 9216, pp. 657–677. Springer, Heidelberg (2015)
3. Ananth, P., Jain, A.: Indistinguishability obfuscation from compact functional encryption. In: Gennaro, R., Robshaw, M. (eds.) CRYPTO 2015, Part I. LNCS, vol. 9215, pp. 308–326. Springer, Heidelberg (2015)
4. Ananth, P., Sahai, A.: Functional encryption for turing machines. In: Kushilevitz, E., et al. (eds.) TCC 2016-A. LNCS, vol. 9562, pp. 125–153. Springer, Heidelberg (2016). doi:10.1007/978-3-662-49096-9_6
5. Bitansky, N., Vaikuntanathan, V.: Indistinguishability obfuscation from functional encryption. In: Guruswami, V. (ed.) 56th Annual Symposium on Foundations of Computer Science, pp. 171–190. IEEE Computer Society Press, Berkeley, CA, USA, 17–20 October 2015

6. Boneh, D., Franklin, M.K.: An efficient public key traitor scheme (extended abstract). In: Wiener, M. (ed.) CRYPTO 1999. LNCS, vol. 1666, p. 338. Springer, Heidelberg (1999)
7. Boneh, D., Sahai, A., Waters, B.: Fully collusion resistant traitor tracing with short ciphertexts and private keys. In: Vaudenay, S. (ed.) EUROCRYPT 2006. LNCS, vol. 4004, pp. 573–592. Springer, Heidelberg (2006)
8. Boneh, D., Waters, B.: A fully collusion resistant broadcast, trace, and revoke system. In: Juels, A., Wright, R.N., Vimercati, S. (eds.) ACM CCS 2006: 13th Conference on Computer and Communications Security, pp. 211–220. ACM Press, Alexandria, Virginia, USA, 30 October - 3 November 2006
9. Boneh, D., Zhandry, M.: Multiparty key exchange, efficient traitor tracing, and more from indistinguishability obfuscation. In: Garay, J.A., Gennaro, R. (eds.) CRYPTO 2014, Part I. LNCS, vol. 8616, pp. 480–499. Springer, Heidelberg (2014)
10. Boyle, E., Chung, K.-M., Pass, R.: On extractability obfuscation. In: Lindell, Y. (ed.) TCC 2014. LNCS, vol. 8349, pp. 52–73. Springer, Heidelberg (2014)
11. Chabanne, H., Phan, D.H., Pointcheval, D.: Public traceability in traitor tracing schemes. In: Cramer, R. (ed.) EUROCRYPT 2005. LNCS, vol. 3494, pp. 542–558. Springer, Heidelberg (2005)
12. Chor, B., Fiat, A., Naor, M.: Tracing traitors. In: Desmedt, Y.G. (ed.) CRYPTO 1994. LNCS, vol. 839, pp. 257–270. Springer, Heidelberg (1994)
13. Chor, B., Fiat, A., Naor, M., Pinkas, B.: Tracing traitors. IEEE Trans. Inf. Theor. 46(3), 893–910 (2000)
14. Cohen, A., Holmgren, J., Nishimaki, R., Vaikuntanathan, V., Wichs, D.: Watermarking cryptographic capabilities. Cryptology ePrint Archive, Report 2015/1096 (2015). http://eprint.iacr.org/2015/1096
15. Cohen, A., Holmgren, J., Vaikuntanathan, V.: Publicly verifiable software watermarking. Cryptology ePrint Archive, Report 2015/373 (2015). http://eprint.iacr.org/2015/373
16. Dodis, Y., Fazio, N.: Public key broadcast encryption for stateless receivers. In: Feigenbaum, J. (ed.) DRM 2002. LNCS, vol. 2696, pp. 61–80. Springer, Heidelberg (2003)
17. Dodis, Y., Fazio, N.: Public key trace and revoke scheme secure against adaptive chosen ciphertext attack. In: Desmedt, Y. (ed.) PKC 2003. LNCS, vol. 2567, pp. 100–115. Springer, Heidelberg (2003)
18. Dodis, Y., Fazio, N., Kiayias, A., Yung, M.: Scalable public-key tracing and revoking. Distrib. Comput. 17(4), 323–347 (2005)
19. Fiat, A., Tassa, T.: Dynamic traitor tracing. In: Wiener, M. (ed.) CRYPTO 1999. LNCS, vol. 1666, p. 354. Springer, Heidelberg (1999)
20. Gafni, E., Staddon, J., Yin, Y.L.: Efficient methods for integrating traceability and broadcast encryption. In: Wiener, M. (ed.) CRYPTO 1999. LNCS, vol. 1666, pp. 372–387. Springer, Heidelberg (1999)
21. Garg, S., Gentry, C., Halevi, S., Raykova, M., Sahai, A., Waters, B.: Candidate indistinguishability obfuscation and functional encryption for all circuits. In: 54th Annual Symposium on Foundations of Computer Science, pp. 40–49. IEEE Computer Society Press, Berkeley, CA, USA, 26–29 October 2013
22. Garg, S., Gentry, C., Halevi, S., Zhandry, M.: Fully secure functional encryption without obfuscation. In: Kushilevitz, E., et al. (eds.) TCC 2016-A. LNCS, vol. 9563, pp. 480–511. Springer, Heidelberg (2016). doi:10.1007/978-3-662-49099-0_18

23. Goldwasser, S., Kalai, Y.T., Popa, R.A., Vaikuntanathan, V., Zeldovich, N.: Reusable garbled circuits and succinct functional encryption. In: Boneh, D., Roughgarden, T., Feigenbaum, J. (eds.) 45th Annual ACM Symposium on Theory of Computing, pp. 555–564. ACM Press, Palo Alto, CA, USA, 1–4 June 2013

24. Goodrich, M.T., Sun, J.Z., Tamassia, R.: Efficient tree-based revocation in groups of low-state devices. In: Franklin, M. (ed.) CRYPTO 2004. LNCS, vol. 3152, pp. 511–527. Springer, Heidelberg (2004)

25. Gorbunov, S., Vaikuntanathan, V., Wee, H.: Functional encryption with bounded collusions via multi-party computation. In: Safavi-Naini, R., Canetti, R. (eds.) CRYPTO 2012. LNCS, vol. 7417, pp. 162–179. Springer, Heidelberg (2012)

26. Halevy, D., Shamir, A.: The LSD broadcast encryption scheme. In: Yung, M. (ed.) CRYPTO 2002. LNCS, vol. 2442, pp. 47–60. Springer, Heidelberg (2002)

27. Hubacek, P., Wichs, D.: On the communication complexity of secure function evaluation with long output. In: Roughgarden, T. (ed.) ITCS 2015: 6th Innovations in Theoretical Computer Science, pp. 163–172. Association for Computing Machinery, Rehovot, Israel, 11–13 January 2015

28. Kiayias, A., Tang, Q.: Traitor deterring schemes: using bitcoin as collateral for digital content. In: Ray, I., Li, N., Kruegel, C. (eds.) ACM CCS 2015: 22nd Conference on Computer and Communications Security, pp. 231–242. ACM Press, Denver, CO, USA, 12–16 October 2015

29. Kiayias, A., Yung, M.: Traitor tracing with constant transmission rate. In: Knudsen, L.R. (ed.) EUROCRYPT 2002. LNCS, vol. 2332, pp. 450–465. Springer, Heidelberg (2002)

30. Kiayias, A., Yung, M.: Copyrighting public-key functions and applications to black-box traitor tracing. Cryptology ePrint Archive, Report 2006/458 (2006). http://eprint.iacr.org/2006/458

31. Kurosawa, K., Desmedt, Y.G.: Optimum traitor tracing and asymmetric schemes. In: Nyberg, K. (ed.) EUROCRYPT 1998. LNCS, vol. 1403, pp. 145–157. Springer, Heidelberg (1998)

32. Naor, D., Naor, M., Lotspiech, J.: Revocation and tracing schemes for stateless receivers. In: Kilian, J. (ed.) CRYPTO 2001. LNCS, vol. 2139, pp. 41–62. Springer, Heidelberg (2001)

33. Naor, M., Pinkas, B.: Threshold traitor tracing. In: Krawczyk, H. (ed.) CRYPTO 1998. LNCS, vol. 1462, pp. 502–517. Springer, Heidelberg (1998)

34. Naor, M., Pinkas, B.: Efficient trace and revoke schemes. In: Frankel, Y. (ed.) FC 2000. LNCS, vol. 1962, pp. 1–20. Springer, Heidelberg (2001)

35. Nishimaki, R., Wichs, D.: Watermarking cryptographic programs against arbitrary removal strategies. Cryptology ePrint Archive, Report 2015/344 (2015). http://eprint.iacr.org/2015/344

36. Nishimaki, R., Wichs, D., Zhandry, M.: Anonymous traitor tracing: how to embed arbitrary information in a key. Cryptology ePrint Archive, Report 2015/750 (2015). http://eprint.iacr.org/2015/750

37. Safavi-Naini, R., Wang, Y.: Sequential traitor tracing. In: Bellare, M. (ed.) CRYPTO 2000. LNCS, vol. 1880, pp. 316–332. Springer, Heidelberg (2000)

38. Silverberg, A., Staddon, J., Walker, J.L.: Efficient traitor tracing algorithms using list decoding. In: Boyd, C. (ed.) ASIACRYPT 2001. LNCS, vol. 2248, pp. 175–192. Springer, Heidelberg (2001)

39. Tzeng, W.G., Tzeng, Z.J.: A public-key traitor tracing scheme with revocation using dynamic. In: Kim, K. (ed.) PKC 2001. LNCS, vol. 1992, pp. 207–224. Springer, Heidelberg (2001)

40. Waters, B.: A punctured programming approach to adaptively secure functional encryption. In: Gennaro, R., Robshaw, M.J.B. (eds.) CRYPTO 2015. LNCS, vol. 9216, pp. 678–697. Springer, Heidelberg (2015)
41. Zhandry, M.: Adaptively secure broadcast encryption with small system parameters. Cryptology ePrint Archive, Report 2014/757 (2014). http://eprint.iacr.org/2014/757

Unconditionally Secure Computation with Reduced Interaction

Ivan Damgård[1]($^{(\boxtimes)}$), Jesper Buus Nielsen[1], Rafail Ostrovsky[2], and Adi Rosén[3]

[1] Department of Computer Science, Aarhus University, Aarhus, Denmark
ivan@cs.au.dk
[2] UCLA, Los Angeles, USA
[3] CNRS and Université Paris Diderot, Paris, France

Abstract. We study the question of how much interaction is needed for unconditionally secure multiparty computation. We first consider the number of messages that need to be sent to compute a Boolean function with semi-honest security, where all n parties learn the result. We consider two classes of functions called t-difficult and t-very difficult functions, where t refers to the number of corrupted players. For instance, the AND of an input bit from each player is t-very difficult while the XOR is t-difficult but not t-very difficult. We show lower bounds on the message complexity of both types of functions, considering two notions of message complexity called conservative and liberal, where conservative is the more standard one. In all cases the bounds are $\Omega(nt)$. We also show (almost) matching upper bounds for $t = 1$ and functions in a rich class PSM_{eff} including non-deterministic log-space, as well as a stronger upper bound for the XOR function. In particular, we find that the conservative message complexity of 1-very difficult functions in PSM_{eff} is $2n$, while the conservative message complexity for XOR (and $t = 1$) is $2n - 1$. Next, we consider round complexity. It is a long-standing open problem to determine whether all efficiently computable functions can also be efficiently computed in constant-round with *unconditional* security. Motivated by this, we consider the question of whether we can compute any function securely, while minimizing the interaction of *some of* the players? And if so, how many players can this apply to? Note that we still want the standard security guarantees (correctness, privacy, termination) and we consider the standard communication model with secure point-to-point channels. We answer the questions as follows: for passive security, with $n = 2t + 1$ players and t corruptions, up to t players can have minimal interaction, i.e., they send 1 message in the first round to each of the $t + 1$ remaining players and receive one message from each of them in the last round. Using our result on message complexity, we show that this is (unconditionally) optimal. For malicious security with $n = 3t + 1$ players and t corruptions, up to t players can have minimal interaction, and we show that this is also optimal.

1 Introduction

In Multiparty Computation n players want to compute an agreed-upon function on privately held inputs, such that the desired result is correctly computed and

© International Association for Cryptologic Research 2016
M. Fischlin and J.-S. Coron (Eds.): EUROCRYPT 2016, Part II, LNCS 9666, pp. 420–447, 2016.
DOI: 10.1007/978-3-662-49896-5_15

is the only new information released. This should hold even if t players have been actively or passively corrupted by an adversary.

If point-to-point secure channels between players are assumed, any function can be computed with unconditional (perfect) security, against a passive adversary if $n \geq 2t+1$ and against an active adversary if $n \geq 3t+1$. [BGW88, CCD87] If we assume a broadcast channel and accept a small error probability, $n \geq 2t+1$ is sufficient to get active security [RB89].

The protocols behind these results require a number of communication rounds that is proportional to the depth of an (arithmetic) circuit computing the function. One would of course like to compute any function with unconditional security, in constant rounds, and efficiently in terms of the circuit size of the function. This is however a long-standing open problem (note that this is indeed possible if one makes computational assumptions).

This is not only a theoretical question: the methods we typically use in information theoretically secure protocols tend to be computationally much more efficient than the cryptographic machinery we need for computational security. So unconditionally secure protocols are very attractive from a practical point of view, except for the fact that they seem to require a lot of interaction.

It is therefore very natural to ask whether this state of affairs is inherent. How much interaction do we actually need for unconditional security, and can we reduce the interaction needed compared to existing protocols? This type of question was studied in [FKN94, DPP14] in a specific 3-party model where 2 parties have input and a third gets the output. We further detail below some previous work on secure addition, but in general very little is known on this question.

In this paper, we make some progress with respect to two related but different measures of interaction: message complexity and round complexity, in the context of synchronous networks.

Message complexity seems like a very simple measure at first sight: simply count how many messages are sent in the protocol. However, a moment's thought will show that things are a bit more tricky. For instance, what if the protocol varies its communication pattern, so that P_i sometimes (but not always) sends a message to P_j in a certain round? One way to handle this is to declare that the absence of a message is also a signal. This leads to what we call *conservative message complexity*, i.e., we say that if P_i sometimes sends a message to P_j in a certain round, then we consider it to be the case that P_i always sends a message to P_j in this round. This way, we force protocols to have a fixed communication pattern.

However, considering only this measure is not completely satisfying. After all, it could be that one could design protocols with a smaller number of messages by using tricks such as waiting for a certain time before a message is sent, and using the amount of elapsed time as an implicit signal. In real life such an approach could be interesting, as there may be some cost involved in physically moving a message, that is not incurred if one stays silent. Therefore, we also define *liberal message complexity*, where the protocol is only charged for messages that are

	Liberal	Conservative
t-very difficult	$\lceil\frac{n(t+1)-1}{2}\rceil + \frac{n}{2}$	$\lceil\frac{n(t+1)-1}{2}\rceil + n$
t-difficult	$\lceil\frac{n(t+1)-1}{2}\rceil + \frac{n-1}{2}$	$\lceil\frac{n(t+1)-1}{2}\rceil + n - 1$

Fig. 1. Lower bounds.

explicitly sent, and where we consider the *expected* number of messages as well the maximum. We discuss these measures in more detail later, when we define them formally.

Our results are as follows: We consider n players and t semi-honest and static corruptions. We look at statistically secure computation of Boolean functions, where all parties learn the output. We assume secure point to point channels that leak the length of the message sent to the adversary (as any implementation using crypto would do). The ideal functionality for computing the function leaks the output to the adversary only if some party is corrupted, so essentially we ask that the adversary cannot learn anything by doing only traffic analysis.

We consider two classes of functions, called t-very difficult and a larger class called t-difficult. The AND of an input bit from each player, and more generally threshold functions are t-very difficult, whereas the XOR is t-difficult but not t-very difficult.

We show lower bounds for all 4 cases that arise naturally. In all cases the bounds are $\Omega(nt)$. Results are summarized in Fig. 1.

For the case of $t = 1$ we also show upper bounds using perfectly secure protocols, for all functions in a class we call PSM_{eff} which includes non-deterministic log-space and more (see Definition 1 below), as well as a stronger upper bound for the XOR function. Figures 2 and 3 show the lower bounds for $t = 1$ and the upper bounds. We see that we have obtained the exact conservative message complexity for all 1-very difficult functions in PSM_{eff}. This includes, for instance, the AND and thresholds functions in general. We have also obtained the exact conservative and liberal message complexity for XOR (when $t = 1$). Finally we have characterised the liberal message complexity of 1-very difficult functions in PSM_{eff} up to 1/2 message, the exact characterization is left as an open problem.

Some remarks on alternative models are in order: we insist that the number of parties is considered to be constant, even if the security parameter grows. This rules out tricks like secret sharing one's input among a small subset of parties, hoping they are not all corrupt [BGT13, GIPR] (which works for static

	Liberal	Conservative
1-very difficult	$3n/2$	$2n$
1-difficult	$3n/2 - 1/2$	$2n - 1$

Fig. 2. Lower bounds for $t = 1$.

	Liberal	Conservative
PSM_{eff}	$3n/2 + 1/2$	$2n$
XOR	$3n/2 - 1/2$	$2n - 1$

Fig. 3. Upper bounds for $t = 1$.

corruptions, but not for adaptive corruptions). If one is happy with statistical, static, semi-honest security for a large number of parties, then this type of trick can be used to compute simple operations with a poly-log (in n) number of messages. If the communication pattern is fixed, than a quadratic number of messages is required for addition protocols [CK93]. Note that our bounds hold regardless of the number of parties if adaptive security or perfect security is required (and our upper bounds yield perfect security). Therefore the only way to circumvent our lower bounds is to settle for static and statistical security and let the number of parties grow with the security parameter (for adaptive adversary with setup assumptions, see further discussion in [CCG+15]).

Next, we consider round complexity: As mentioned, computing any function with unconditional security, in constant rounds and efficiently in the circuit size of the function is an open problem[1], and providing a positive answer seems to require completely new ideas for protocol design. Motivated by this, we consider the question of whether we can minimize the interaction of *some of* the players? And if so, how many players can this apply to? Note that we still want the standard security guarantees (correctness, privacy, termination). We answer this question as follows: for passive security, with $n = 2t+1$ players and t corruptions, up to t players can have minimal interaction, i.e., they send 1 message in the first round to each of the $t + 1$ remaining players and receive one message from each of them in the last round. Using our result on message complexity, we show that this is (unconditionally) optimal. For malicious security with $n = 3t + 1$ players and t corruptions, up to t players can have minimal interaction, and we show that this is also optimal.

For the purpose of proving the positive result for malicious security, we show a result of independent interest: For the case $n = 3t+1$ and t malicious corruptions, we design a broadcast protocol of the following special form: we can select any subset of t players, who only need to send one message to the other $n-t$ players. After this point, we can do broadcast among the remaining $n - t$ players. Note that we are not guaranteed that we have at most a third corruptions among the $n - t$ players, so we cannot do broadcast from scratch in this set. We find it slightly surprising that we need so little involvement from the t selected players. In particular, they might all be corrupt and hence send completely inconsistent setup values – then, of course, we are saved by the fact that the remaining players are all honest (but they do not know this yet).

---

[1] Using randomizing polynomials [IK00] one can get unconditional security and constant round efficiently in the branching program size of the function, but this does not seem to help beyond NC^1.

2 Preliminaries

We use $\mathbb{N}$ to denote the non-negative integers. For $n \in \mathbb{N}$ we let $[n] = \{1, \ldots, n\}$.

We prove security in the model from [Can00] with unconditional security and a static adversary. We consider a synchronous model with point-to-point perfectly secure channels between each pair of parties, where the length of each message sent is leaked to the adversary. In one round, all parties may send messages to each other. We consider function evaluation between n parties $\mathsf{P}_1, \ldots, \mathsf{P}_n$ with inputs $x_1, \ldots, x_n$ and common output $y = f(x_1, \ldots, x_n)$ for a poly-time n-party function f. In the ideal model, we assume that nothing is leaked to the adversary in case no one is corrupted. We refer to [Can00] for the details of the model.

We say that a protocol has perfect correctness if it always computes the correct result when all parties follow the protocol. We say that a protocol has perfect privacy against t semi-honest corruptions if the ideal world and the real world models have the same distributions even when t parties are passively corrupted, i.e., they follow the protocol but might pool their views of the protocol to learn more than they should. We say that a protocol has statistical privacy against t semi-honest corruptions if the view of the corrupted parties in the ideal world and the real world models have distributions that are statistically close in some security parameter s even if t parties are passively corrupted. We say that a protocol has perfect privacy against t malicious corruptions if the view of the corrupted parties in the ideal world and the real world models have the same distributions even when t parties might deviated from the protocol in a coordinated manner. If the distributions are only statistically close we talk about statistical security against t malicious corruptions.

As is well known, it is possible to implement secure function evaluation of any poly-time n-party function with perfect correctness and perfect privacy against t semi-honest corruptions when $n \geq 2t + 1$. It is possible to implement secure function evaluation of any poly-time n-party function with perfect correctness and perfect privacy against t malicious corruptions when $n \geq 3t+1$, see [BGW88, CCD87].

We will use secure function evaluation protocols for the so-called preprocessing model as tools. In these protocols an incorruptible trusted third party will sample a distribution D to get an n-tuple $(d_1, \ldots, d_n) \leftarrow D$. Then it privately gives d_i to P_i. After the setup phase, the n parties engage in a protocol where they communicate over secure channels. In such pre-processing models there exist appropriate distributions D which will allow to get perfect correctness and perfect privacy against t passive corruptions out of $n = t + 1$ parties. See, e.g., [DZ13] and the references therein.

We also use protocols for the private simultaneous message (PSM) model. For this model an n-party protocol for an n-party function f is given by

$$(R, M_1, \ldots, M_n, g)$$

where R is a distribution with finite support, each M_i is a function, called the message function of party i, and g is function called the reconstruction function.

By perfect correctness of a PSM protocol for an n-party function f we mean that for all r in the support of R and all inputs $(x_1, \ldots, x_n)$ for f it holds that $f(x_1, \ldots, x_n) = g(M_1(x_1, r), \ldots, M_n(x_n, r))$.

By ϵ-privacy of a PSM we mean that there exists a poly-time simulator S such that for all inputs $(x_1, \ldots, x_n)$ for f, $y = f(x_1, \ldots, x_n)$ and a random sample $r \leftarrow R$ it holds that $(M_1(x_1, r), \ldots, M_n(x_n, r))$ and $S(y)$ have statistical distance at most ϵ. If $\epsilon = 0$, then we talk about perfect privacy. If ϵ is negligible we talk about statistical security. Privacy ensures that a party seeing $(M_1(x_1, r), \ldots, M_n(x_n, r))$ learns nothing extra to $y = g(M_1(x_1, r), \ldots, M_n(x_n, r))$.

The PSM model is a generalization of [FKN94] and is defined in [IK97], where they also gave perfectly secure and efficient (poly-time) PSM protocols for a large class of functions including non-deterministic log-space, mod_p L and $\sharp$L. In [IK97] privacy is not formulated via poly-time simulation: the notion only asks that $(M_1(x_1, d_1), \ldots, M_n(x_n, d_n))$ depends only on $f(x_1, \ldots, x_n)$. We need the simulation based notion here, as we prove security in [Can00], which is phrased via efficient simulation. We note that if for a given function f it is possible to compute in poly-time, from an output $y = f(x_1, \ldots, x_n)$, an input $(x'_1, \ldots, x'_n)$ such that $y = f(x'_1, \ldots, x'_n)$ then the notions are equivalent for f. The simulator will simply compute $(x'_1, \ldots, x'_n)$, sample $r \leftarrow R$ and output $(M_1(x_1, r), \ldots, M_n(x_n, r))$. Of course, if inputs are single bits and the number of parties is considered to be constant, such inversion can be done in constant time by trying all possibilities.

In the following, when using PSM protocols, we will consider such efficiently invertible functions f that also have an efficient PSM protocol:

Definition 1. *We will use PSM$_{\mathrm{eff}}$ to denote the class of functions that are efficiently invertible as described above and can be computed by a polynomial time PSM protocol.*

We also use additive secret sharing of bits strings $x \in \{0, 1\}^m$. An additive secret sharings of x between $P_1, \ldots, P_n$ consists of sampling shares $s_1, \ldots, s_n \in (\{0, 1\}^m)^n$ uniformly at random under the only restriction that $x = \oplus_{i=1}^n s_i$, where $\oplus$ denote bit-wise exclusive or. It is easy to show that the distribution of any $n-1$ of the shares is the uniform one on $(\{0, 1\}^m)^{n-1}$ and hence independent of x.

3 Message Complexity

Defining the message complexity of a protocol for the synchronous model with secure channels appropriately is slightly more tricky than one might expect at first, so we address this issue in its own section.

We will first of all need to allow parties to *not* send a message to some party in a given round. Since all parties send messages to all parties in all rounds in [Can00], we need to hack the model a bit for this. We will say that if a party sends the empty string then this counts as not having sent a message. Think of receiving the empty string from P_i as meaning "no message was received from P_i in this round".

This builds up to a subtler point that we demonstrate by an example. Consider the problem where a dealer D is to deal an additive secret sharing of a bit d between n parties $P_1, \ldots, P_n$. What is the average message complexity of this problem? It turns out that if we ignore security for a second, then it is at most $n/2$ if one is not careful. The dealer samples a secret sharing $d = d_1 \oplus \cdots \oplus d_n$. Then for $i = 1, \ldots, n$, if $d_i = 0$ he does not send a message to P_i. If $d_i = 1$, then he sends 1 to P_i. Since d_i is uniformly random it follows from linearity of expectation that he sends an expected $n/2$ messages.

If we consider security, the bound changes. It is the case in [Can00] that the adversary can see the length of a message sent securely. This in particular means that in our setting here, the adversary can see if a message was sent or not between any two parties—it can see the communication pattern. This is a reasonable model, as hiding the presence of a communication is not practical, in particular when we actually do not want to transmit anything when there is no message to be sent.

Of course seeing the communication pattern of the above protocol renders it insecure, but this kind of contrived example shows that in some cases, if we want a very precise measure of message complexity we need to consider protocols with fixed communication patterns, i.e., if P_1 sometimes sends a message to P_2 in round 1, then we consider it the case that P_1 always sends a message to P_2 in round 1, as the absence of the message is a signal.

On the other hand, considering only this measure seems to be not entirely satisfying. We should be intrigued whether or not using tricks as above will allow more efficient protocols, so it makes sense to also consider a notion where we only count messages that are *explicitly* sent.

This will mean that the number of messages may not be the same in all runs of the protocol. When we prove lower bounds it will therefore not be meaningful to consider conservative message complexity. For example, if we can prove that all protocols must with some probability 2^{-s}, where s is the security parameter, send $2^{40}n$ message but that they in all other cases might have to send only $2n$ messages, then we would not consider $2^{40}n$ a very meaningful *lower bound* for the number of messages. When we prove lower bounds we would like to consider expected message complexity, which would turn the lower bound in the just given example into $2n$, as $2^{-s}2^{40}n$ is vanishing in s. We call this liberal communication complexity. Another way to relax the conservative notion is to still only count messages explicitly sent but look at the worst case number over the randomness of the parties. We call this worst case communication complexity. It is obviously in between the conservative and liberal notions and we will at some point only be able to prove an upper bound for the worst case notion (as opposed to the conservative one).

We therefore define three measures of message complexity, a conservative one, a liberal one and a worst case one:

Definition 2 (Conservative Message Complexity). *Let π be an n-party protocol for a synchronous network. Let R be random tapes of all players. By $\mathsf{Msg}_{\mathrm{con}}(\pi)$ we denote the conservative message complexity of π. For all $r \in \mathbb{N}$ and all $i \in [n]$ and all $j \in [n] \setminus \{i\}$ we define $c_{r,i,j}$ to be 1 if there exists an input x for π and randomness R such that when π is run with that input and that randomness, P_i will send a message to P_j in round r. We let $c_{r,i,j} = 0$ otherwise. We let*

$$\mathsf{Msg}_{\mathrm{con}}(\pi) = \sum_{r,i,j} c_{r,i,j}.$$

Note that in the conservative message complexity, even if some player flips a fair coin and sends a message that is independent of it's input, say "hello" to player one if the coin is zero and "hello" to player two, three and four if the coin is one, the conservative message complexity counts this as four messages. A more liberal way to count messages in any specific protocol run and then take expectation or worst case over the random tapes of the parties. We call this liberal message complexity respectively worst case complexity. In the above example, the liberal message complexity of the "hello" messages is two messages and the worst case complexity is three message.

Definition 3 (Liberal Average/Worst-Case Message Complexity). *Let π be an n-party protocol for a synchronous network. For a given run of π on input x and some fixed random tapes R of the parties we define $c_{r,i,j}$ to be 1 if P_i sent a message to P_j in round r. We let $c_{r,i,j} = 0$ otherwise. We let*

$$\mathsf{Msg}(\pi, x, R) = \sum_{r,i,j} c_{r,i,j}$$

and

$$\mathsf{Msg}_{\mathrm{lib}}(\pi) = \max_x \mathrm{E}_R[\mathsf{Msg}(\pi, x, R)].$$

$$\mathsf{Msg}_{\mathrm{wor}}(\pi) = \max_{x,R}[\mathsf{Msg}(\pi, x, R)].$$

It is easy to see that it is always the case that $\mathsf{Msg}_{\mathrm{lib}}(\pi) \leq \mathsf{Msg}_{\mathrm{wor}}(\pi) \leq \mathsf{Msg}_{\mathrm{con}}(\pi)$.

We extend the above notions to the statistical setting by defining them as above for each fixed value of σ and then taking lim sup when this limit is defined. If this limit is not defined, we define the message complexity to be ∞.

4 Lower Bounds

We now proceed to present and prove our lower bounds. We first prove a lower bound on the message complexity of secure function evaluation in the face of semi-honest corruptions. Then we give a lower bound on the individual round complexity in the face of t semi-honest corruptions and then a lower bound on the individual round complexity in the face of t malicious corruptions.

4.1 Message Complexity

We first prove a lower bound on the message complexity of secure function evaluation secure against t semi-honest corruptions. We will prove the bound for a large class of function that we will call t-difficult, and a slightly larger bound for a smaller class called t-very difficult.

First some clarifications: even though we have defined two different ways to count messages, where an empty message counts in one notion and not in the other, in the following, when we say that a message is sent or received, or messages are exchanged, we always refer to non-empty messages.

Very roughly, the intuition we will formalize is as follows: A player whose input matters to the result must somehow communicate his input to the rest of players, in order to enable correct computation of the result by all players. The input cannot be encoded in the communication pattern which is public, so it must follow from the content of messages this player exchanges with other players. On the other hand, a player whose inputs matters has to exchange messages with at least $t + 1$ parties before his input becomes determined. Otherwise he may have talked to only corrupted parties and the protocol would not be private. This already indicates a lower bound of $n(t + 1)/2$ messages (we need to divide by 2 since a message counts as communication for both sender and receiver). But we can do more: we show that *after* the inputs have been fixed, all players must receive information allowing them to determine the result of the computation. Under the liberal message complexity notion, this does not necessarily mean that all players must receive another message, but we can show that in expectation most players must receive a message half the time. So this indicates a lower bound of $n(t + 2)/2$ messages, which is (approximately) what we obtain.

We start with some notation: For an input vector $x = (x_1, \ldots, x_n)$ and a subset $D \subseteq \{1, \ldots, n\}$ and inputs $x_D = \{(j, x'_j)\}_{j \in D}$ for the parties in D we use $x[x_D]$ to denote the vector x with x_j replaced by x'_j for $j \in D$.

Definition 4. *We say that a function f is t-difficult for P_i if the following holds:*

Influence. *There exists two inputs $x^{i,0}$ and $x^{i,1}$ such that $x_j^{i,0} = x_j^{i,1}$ for all $\mathsf{P}_j \neq \mathsf{P}_i$ and such that $f(x^{i,0}) \neq f(x^{i,1})$.*

Uncertainty. *There exists an input $x_i^?$ such that for all subsets $C \subset \{\mathsf{P}_1, \ldots, \mathsf{P}_n\} \setminus \{\mathsf{P}_i\}$ with $|C| = t$ and $D = \{\mathsf{P}_1, \ldots, \mathsf{P}_n\} \setminus (\{\mathsf{P}_i\} \cup C)$ and all inputs x for f there exists $x_D = \{x'_j\}_{j \in D}$ such that $f(x[(i, x_i^?)]) = f(x[x_D])$.*

We say that f is t-difficult if f is t-difficult for all P_i.

Intuitively, if a party has influence, then the function – at least sometimes – depends on the input of that party. If a party P_i has uncertainty, it means that for some input, called $x_i^?$, of P_i, if subset C is corrupt, they will not be able to figure out which input P_i has, no matter what the other inputs were: we can switch P_i's input to anything else and compensate for this by changing the inputs of the other honest parties such that the output is the same. One may think, for instance of the AND function: if P_i has input 0, the output is 0, but the adversary cannot know if this is because P_i or another honest party has a 0.

As examples of t-difficult functions consider the functions where each party has as input a bit and where the output is the AND or the XOR of these n bits. Other examples are general threshold functions, which output 1 iff at least some $0 < t' < n$ parties have input 1.

For a run of a protocol π and a given party P_i and a given point in the protocol we keep track of a set N_i which can be thought of as the parties that P_i has exchanged messages with, but it is defined with a slight twist. From the beginning we set all $N_i = \emptyset$. Whenever P_i sends a message, we update N_i to be the set of parties P_i has sent a message to or received a message from so far in the protocol. The definition is important so let us elaborate:

1. The set N_i is not updated at the time a message is received.
2. The set N_i *is* updated at the time a message is sent.
3. When N_i is updated we add all the messages that were received since the last time is was updated and we also add the outgoing message that triggered the update.

We say that a protocol has t-*floating input* for P_i if at each point in the protocol where $|N_i| \leq t$ it holds that P_i still did not read its input x_i. More formally, if we model P_i as an interactive Turing machine, it means that P_i did not access its input tape. We say that π has t-floating input if it has t-floating input for all parties.

For any run of a protocol we define a *revelation message* to be the message (if it exists) where before the message is sent it holds for at least one P_i that $|N_i| \leq t$ and after the message is received it holds for all P_i that $|N_i| \geq t + 1$. Notice that this implies that it is the size of the set N_i of the *sender* of the revelation message that crosses the threshold t, as N_i is not updated in response to receiving a message.

The *communication pattern* of an execution $\pi(\boldsymbol{x}; R)$ with input vector $\boldsymbol{x}$ and random tape vector R is the transcript seen by the adversary when no parties are corrupted, i.e., who sent a message to whom at which time and the length of those messages, but no contents of the messages and no input or output of any party. We assume that a communication pattern is encoded as a bit string. Let $Q : \{0,1\}^* \to \{0,1\}^*$ be a function on communication patterns. We use $Q(\pi(\boldsymbol{x}))$ to denote the random variable obtained by running π on the input distribution $\boldsymbol{x}$ and uniformly random R and applying Q to the resulting communication pattern and then outputting the output of Q.

Our proof strategy can be summarized as follows: we will first show that a protocol with floating inputs must have a revelation message, and that furthermore, $n - 1$ players must receive a message after the revelation message was sent, with probability at least $1/2$. This is quite straightforward and implies that floating input protocols must satisfy our lower bound. The second step is to show that any secure protocol for a difficult function f can be converted to a floating input protocol with the same message complexity. This is the most complicated part and uses in an essential way that the function is difficult and the assumption in our model that the number of parties does not grow with the security parameter.

Lemma 1 (Input-Independent Communication Pattern). *If π securely implements f with statistical security for t semi-honest corruptions for some $t \geq 0$, then it holds for any two input distributions $\boldsymbol{x}_0$ and $\boldsymbol{x}_1$ and all functions Q on communication patterns that $Q(\pi(\boldsymbol{x}_0))$ and $Q(\pi(\boldsymbol{x}_1))$ are statistically indistinguishable.*

Proof. This follows from the fact that when no parties are corrupted, the adversary still sees the communication pattern of $\pi(\boldsymbol{x}_0)$ and $\pi(\boldsymbol{x}_1)$ and hence can compute and output $Q(\pi(\boldsymbol{x}_0))$ respectively $Q(\pi(\boldsymbol{x}_1))$. However, when no parties are corrupted the simulator has the same view when $\boldsymbol{x}_0$ or $\boldsymbol{x}_1$ is used. The claim then follows from security against 0 semi-honest corruptions.   □

Corollary 1 (Input-Independent Communication Complexity). *If π securely implements f with statistical security for t semi-honest corruptions for some $t \geq 0$, then it holds for any two input distributions $\boldsymbol{x}_0$ and $\boldsymbol{x}_1$ that $\mathsf{Msg}(\boldsymbol{x}_0)$ and $\mathsf{Msg}(\boldsymbol{x}_1)$ are statistically indistinguishable. Here, $\mathsf{Msg}(\boldsymbol{x})$ is the random variable that selects an input according to $\boldsymbol{x}$, runs the protocol and outputs the number of non-empty messages sent.*

Proof. Consider the function on communication patterns outputting the number of non-empty messages sent and then apply Lemma 1.   □

Lemma 2 (Revelation Message). *If π has t-floating input and securely implements f with statistical security for t semi-honest corruptions and f is t-difficult, then it holds for all input distributions $\boldsymbol{x}$ that π has a t-revelation message except with negligible probability.*

Proof. If π does not have a t-revelation message for input distribution $\boldsymbol{x}$, then there exist a party P_i such that with non-negligible probability P_i exchanges messages with at most t parties in $\pi(\boldsymbol{x})$. From Lemma 1 it then follows that it holds for the input distributions $\boldsymbol{x}^{i,0}$ and $\boldsymbol{x}^{i,1}$ from the definition of f being t-difficult that with non-negligible probability P_i exchanges messages with at most t parties in $\pi(\boldsymbol{x}^{i,0})$ and also in $\pi(\boldsymbol{x}^{i,1})$. But since π has t-floating inputs, this implies that with non-negligible probability $\pi(\boldsymbol{x}^{i,0}) = \pi(\boldsymbol{x}^{i,1})$ as the output cannot depend on the input of P_i when P_i did not read its input, and all other parties have the same inputs in $\boldsymbol{x}^{i,0}$ and $\boldsymbol{x}^{i,1}$. However, by assumption $f(\boldsymbol{x}^{i,0}) \neq f(\boldsymbol{x}^{i,1})$ and we have a contradiction with correctness of π.   □

Lemma 3 (Another Message After Revelation Message). *If π has t-floating input and securely implements f with statistical security for t semi-honest corruptions and f is t-difficult, then it holds for all input distributions and all pairs of distinct parties P_j and P_k that in a random run of $\pi(\boldsymbol{x})$ it holds except with negligible probability that when P_k is the sender of the revelation message, then the probability that P_j receives another message after P_k sent the revelation message is at least $\frac{1}{2}$.*

Proof. Assume for the sake of contradiction that there exist $\boldsymbol{x}$ and P_k and $\mathsf{P}_j \neq \mathsf{P}_k$ such that it happens with non-negligible probability that P_k is the sender of the revelation message and that when this happens P_j will receive another message after the revelation message is sent (but not received) with probability at most $\frac{1}{2} - c$, where c is non-negligible. It is a predicate of the communication pattern whether P_k sends the revelation message. It is also a predicate of the communication pattern whether P_j receives another message after the revelation message. Therefore it follows from Lemma 1 that it holds for any input distribution $\boldsymbol{x}$ with non-negligible probability that P_k is the sender of the revelation message and that when this happens then P_j will receive another message after the revelation message is sent (but not received) with probability at most $\frac{1}{2} - c'$, where $c' = c - negl$ is non-negligible as c is non-negligible.

Consider now the particular input distribution which is $\boldsymbol{x}^{k,b}$ for a uniformly random bit b, where $\boldsymbol{x}^{k,0}, \boldsymbol{x}^{k,1}$ are the input vectors guaranteed by the definition of f being t-difficult (P_k has influence). In this case the output of all parties allow to determine the bit b, except with negligible probability. Assume without loss of generality that $f(\boldsymbol{x}^{i,b}) = b$. Let y be the distribution of the output of P_j in a random run on $\boldsymbol{x}^{i,b}$ conditioned on P_j not receiving another message after the revelation message. Notice that y can be sampled by P_j at the time right before the revelation message is sent, by simply assuming that no more messages will be received by P_j. However, at the point before the revelation message is sent P_k did not read its input x_k yet in the protocol, so y is perfectly independent of b. From this it follows that $\Pr[y = 0 \mid b = 0] = \Pr[y = 0 \mid b = 1] = 1 - \Pr[y = 1 \mid b = 1]$, so either $\Pr[y = 0 \mid b = 0] \leq \frac{1}{2}$ or $\Pr[y = 1 \mid b = 1] \leq \frac{1}{2}$. Assume that $\Pr[y = 0 \mid b = 0] \leq \frac{1}{2}$. Since $b = 0$ with probability $\frac{1}{2}$ and P_j receives another message with probability $\frac{1}{2} - c$ it happens with non-negligible probability that $b = 0$ and at the same time P_j does not receive another message and hence outputs according to distribution y, which implies that it happens with non-negligible probability that P_j does not output b, contradicting the correctness of the protocol. If we assume that $\Pr[y = 1 \mid b = 1] \leq \frac{1}{2}$, then a violation of correctness is reached using a symmetric argument. This concludes the proof. $\square$

Lemma 4 (Floating Input). *Let f be a t-difficult n-party function and assume that π is an n-party protocol securely implementing f with statistical correctness and statistical privacy against t semi-honest corruptions. Then there exists a protocol π' with t-floating input which has the same security and is such that for any input distribution, the resulting communication patterns of π and π' are identically distributed.*

Proof. We prove the lemma by constructing π' from π. We prove the lemma for the weaker case where we construct π' where only P_1 has t-floating input. We can then obtain the general case by symmetry and hybrid arguments.

All parties in π' run as in π except P_1 who runs as follows. Initially, run as in π but with input $\beta_1 = x_1^?$ and a uniformly random tape ρ_1. Here, $x_1^?$ is the input value that a exists since f is t-difficult (the uncertainty condition for P_1). If about to send a message which would result in $|N_1| \geq t + 1$, then first apply the following *input patching* procedure: Read the input x_1 and replace ρ_1 with a new random tape r_1 consistent with input x_1 and the communication so far.

Specifically, sample r_1 using rejection sample as follows. Sample r_1 uniformly at random. Let T be the list of messages sent and received by P_1 so far, including who the message was exchanged with and in which round. Run the code of P_1 from π with input x_1 and random tape r_1 and feed P_1 the incoming message from T in the round in which they occurred. If this makes P_1 send the same messages as in T to the same parties and in the same rounds, then accept r_1, otherwise try again. Use $r_1 = \perp$ to denote that no acceptable r_1 exists. We now prove that if π is secure, then π' is secure.

We will actually prove something stronger, which implies that the correctness and the distribution of the communication pattern is also maintained. Namely we will prove that for all input distributions x it holds that the following distributions D_0 and D_1 are statistically indistinguishable: D_0 is obtained by sampling a random run of π on a random input sampled from x and then outputting $((x_1, r_1), (x_2, r_2), \ldots, (x_n, r_n))$, where x_i is the input of P_i and r_i is the random tape used by P_i. D_1 is obtained by sampling a random run of π' on a random input sampled from x and then outputting $((x_1, r_1), (x_2, r_2), \ldots, (x_n, r_n))$, where for $i = 2, \ldots, n$ the value x_i is the input of P_i and r_i is the random tape used by P_i and where x_1 is the input of P_1 and r_1 is the random tape sampled in the input patching procedure. From this it clearly follows that if π is correct, then π' is correct and it follows for all $t' \leq n$ that if π is secure against t' corruptions then π' is also secure against t' corruptions. Notice that to prove the claim for all distributions on x it is sufficient to prove that it holds for all fixed input vectors x, so in the following we assume that x is a fixed value.

Let $x = (x_1, x_2, \ldots, x_n)$. If $x_1 = x_1^?$, then the input patching procedure simply resamples r_1 with the same distribution as β_1 and hence D_0 and D_1 are identical. So, assume that $x_1 \neq x_1^?$ and that D_0 and D_1 are not statistically close. We show how to use this to break the t-security of π. Let $x_0 = (x_1^?, x_2, \ldots, x_n)$ and $x_1 = (x_1, x_2, \ldots, x_n) = x$. We break the analysis into two cases. In case I we assume that $f(x_1^?, x_2, \ldots, x_n) = f(x_1, x_2, \ldots, x_n)$. To avoid confusion, note that the proof of case II in fact implies the result for case I. However, it is instructive to first see the proof of case I as a mental warm-up.

In case I we will run π on x_0 or x_1 and show how to distinguish with non-negligible advantage by corrupting just t parties which do not include P_1. This clearly demonstrates that π is not t-secure, as these t parties have the same inputs and outputs in $f(x_0)$ and $f(x_1)$ as only the input of P_1 differs and because $f(x_0) = f(x_1)$. So, assume that we attack a run of x_b for uniformly random b. The adversary will observe the communication pattern of the protocol. Consider the point where P_1 sends a message that would make $|N_1| > t$ for the first time, and note that P_1 has communicated with at most t parties up to now, call this set of parties C. At this point the adversary corrupts the players in C.[2] Note that all messages sent by P_1 so far was sent to one of these parties. Use D to denote the set of parties which is not in $\{P_1\} \cup C$. Now use rejection sampling to sample a random tape r_1 consistent with the communication between P_1 and

---

[2] when we get to the actual proof in case II, we will construct a static adversary that always corrupts the same set.

the parties in C and input x_i to P_1. Note that if $b = 1$, this samples a string having the same distribution as the random tape used by P_1 in the protocol $\pi(\boldsymbol{x})$. If $b = 0$, then it samples a string having the same distribution as the random tape r_1 sampled by the input fixing procedure in $\pi'(\boldsymbol{x})$. Note that the parties in D have not communicated with P_1, so all the communication leaving the group D is with C. This means that the adversary knows all message going in or out of the group D. It can therefore use rejection sampling to sample a set of uniformly random tapes $\{(j, r_j)\}_{j \in D}$ for the parties in D consistent with the communication between C sand D and P_j for $j \in D$ having input x_j (where x_j is taken from $\boldsymbol{x}$). This perfectly reconstructs the distribution of the state of the parties in D. Then output $((x_1, r_1), (x_2, r_2), \ldots, (x_n, r_n))$. If $b = 0$, this is exactly D_0 and if $b = 1$ it is exactly D_1. But we assumed that D_0 and D_1 are not statistically close so we arrive at a contradiction with t-security of π: since the output is the same in the two cases, a simulator would see no difference between $b = 0$ and $b = 1$.

That brings us to case II. In this case, we can prove as above that if D_0 and D_1 can be distinguished, then we can also distinguish between $\pi(\boldsymbol{x}_0)$ and $\pi(\boldsymbol{x}_1)$ by just corrupting t parties at a point where $|N_1| \le t$. The challenge is that $f(\boldsymbol{x}_0) \ne f(\boldsymbol{x}_1)$, so it does not follow easily from the definition of security that an adversary should *not* be able to distinguish with just this information. We now argue this in a more indirect way.

Consider the following experiment, which is parameterized by an (infinitely powerful) adversary A that outputs one bit:

1. Sample b uniformly at random.
2. Run $\pi(\boldsymbol{x}_b)$ until the point where it is about to happen that $|N_1| > t$. If this point does not occur, then perform the following at the end of the execution of the protocol.
3. Let C be the set of at most t parties defined as in case I above. A corrupts the parties in C. Let V be the joint view of these parties (their inputs, random tapes and messages received). Output $A(C, V)$ (here we abuse notation slightly by using A to denote both the adversary and the (arbitrary) function it calculates on the views).

We now claim that for any A, $\Pr[A(C, V) = b] - \frac{1}{2}$ is negligible. This will imply what we want: Note that one possible choice of A is as follows: use rejection sampling to produce a sample of D_b, exactly as we described in case I above. Then output the best guess at whether the sample came from D_0 or from D_1. Since the claim holds for this particular A, D_0 and D_1 are statistically close.

So assume for the sake of contradiction that there exists A such that $\Pr[A(C, V) = b] - \frac{1}{2}$ is non-negligible.

Note that C may not be the same set in all runs of the protocol. Considering C as a random variable, we have that

$$Pr[A(C,V) = b] - \frac{1}{2}$$

$$= \sum_{C'} Pr[C = C']Pr[A(C,V) = b|C = C'] - \left(\sum_{C'} Pr[C = C']\right)\frac{1}{2}$$

$$= \sum_{C'} Pr[C = C']\left(Pr[A(C,V) = b|C = C'] - \frac{1}{2}\right).$$

Since the number of subsets of the parties is constant as a function of the security parameter, it now follows that we can find a fixed set C' of size at most t such that $Pr[C = C']$ is non-negligible and such that $Pr[A(C',V) = b | C = C'] - \frac{1}{2}$ is non-negligible.

We can then construct a new adversary A' which always corrupts C' and still guesses b with non-negligible advantage: If the set C actually occurring in the protocol equals C', it outputs $A(C',V)$, otherwise it outputs a uniformly random bit. Note that A' makes its guess at a point in time where $N_1 \subseteq C$. A''s advantage is non-negligible because $Pr[A'(C',V) = b | C \neq C'] - \frac{1}{2}$ is negligible — we can only claim negligible here and not 0 as there might be a difference between $Pr[C \neq C' | b = 0]$ and $Pr[C \neq C' | b = 1]$. This difference, however, is negligible by Lemma 1.

We now want to show that such A' does not exist. To avoid ugly notation in the following, we will now use C to denote the set that A' always corrupts.

We start with some notation. Let D be the set of parties not in $C \cup \{P_1\}$. Let $x_1^0 = x_1^?$ and $x_1^1 = x_1$ let x_D^1 be the inputs of the parties in D in x. Let x_D^0 be the inputs x_D for the parties in D given by the definition of P_1 having uncertainty. We therefore have by definition that $f(x_1^0, x_C, x_D^1) = f(x_1^1, x_C, x_D^0)$.

In this notation we have that $x_0 = (x_1^0, x_C, x_D^1)$ and $x_1 = x = (x_1^1, x_C, x_D^1)$. Therefore our job is to prove that A' cannot distinguish $\pi(x_1^1, x_C, x_D^1)$ from $\pi(x_1^0, x_C, x_D^1)$. In the following, for a subset S of the parties, we use $[b, d]_S$ to denote the view of the parties S in an execution of $\pi(x_1^b, x_C, x_D^d)$. To complete the proof we have to show that at any point in the protocol where $N_1 \subseteq C$ it holds that $[0, 1]_C \approx [1, 1]_C$.

For a subset S of the parties, let $[b, d]_S^c$ denote the distribution of their views, conditioned on the parties in S having received at most c messages.

Obviously $[0, 1]_C^0 \approx [1, 1]_C^0$, since before C communicated with any party the view of players in C is just their own inputs and random tapes. We now prove by induction that $[0, 1]_C^c \approx [1, 1]_C^c$ for all constants c, as long as $N_1 \subseteq C$. The latter condition is extremely important because it implies that in all cases we consider, there is no communication between P_1 and D.

We assume that $[0, 1]_C^c \approx [1, 1]_C^c$ and prove that $[0, 1]_C^{c+1} \approx [1, 1]_C^{c+1}$. From the communication pattern being known by the adversary and being indistinguishable in $[0, 1]_C^c$ and $[1, 1]_C^c$ by Lemma 1 we can assume that we know which party P_j sends a message to C in round $c + 1$.

Assume first that $P_j \neq P_1$. Let R_D be the procedure which gets input $[b, 1]_C^c$, and then from the view of the communication between C and D in $[b, 1]_C$ samples a joint state of all parties in D consistent with inputs x_D^1 and that communication and appends this state to $[b, 1]_C$. We have that $R_D([b, 1]_C^c) = [b, 1]_{C,D}^c$ by construction and it follows from the induction hypothesis $[0, 1]_C^c \approx [1, 1]_C^c$ that $R_D([0, 1]_C^c) \approx R_D([1, 1]_C^c)$. So we conclude that in this case (where $\mathsf{P}_j \in D$) $[0, 1]_{C,D}^c \approx [1, 1]_{C,D}^c$. Put another way, given the state of C one can perfectly simulate the state of the parties in D since one knows their inputs and all communication going in and out of D. From the state of the parties in D in $[b, 1]_{C,D}^c$ one can then sample a random run consistent with P_j being the next party to send a message to a party in C. This gives a sample from $[b, 1]_{C,D}^{c+1}$. Since computation (in this case of the next message function) maintains statistical indistinguishability it follows from $[0, 1]_{C,D}^c \approx [1, 1]_{C,D}^c$ that $[0, 1]_{C,D}^{c+1} \approx [1, 1]_{C,D}^{c+1}$. It clearly follows from $[0, 1]_{C,D}^{c+1} \approx [1, 1]_{C,D}^{c+1}$ that $[0, 1]_C^{c+1} \approx [1, 1]_C^{c+1}$.

Assume then that $\mathsf{P}_j = \mathsf{P}_1$. Again, by induction hypothesis we have $[0, 1]_C^c \approx [1, 1]_C^c$. It follows from the security of the protocol that $[0, 1]_C \approx [1, 0]_C$ as the inputs and outputs of the parties in C are the same in the two executions considered and $|C| \leq t$. So in particular we have $[0, 1]_C^c \approx [1, 0]_C^c$. So we conclude by transitivity that $[1, 0]_C^c \approx [1, 1]_C^c$.

Since the next message comes from P_1 we can argue $[1, 0]_C^{c+1} \approx [1, 1]_C^{c+1}$ as we did for the above case, by sampling the state of P_1 from its known input and communication. As we noticed above we have $[0, 1]_C \approx [1, 0]_C$ and therefore in particular $[0, 1]_C^{c+1} \approx [1, 0]_C^{c+1}$. Combining these two we get $[0, 1]_C^{c+1} \approx [1, 1]_C^{c+1}$ as desired. □

Theorem 1. *Let π be the n-party function which securely implements a function f which is t-difficult, with statistical correctness and statistical privacy against t semi-honest corruptions. Then*

$$\mathsf{Msg}_{\mathtt{lib}}(\pi) \geq \lceil (n(t + 1) - 1)/2 \rceil + n/2 - \frac{1}{2}$$

and

$$\mathsf{Msg}_{\mathtt{con}}(\pi) \geq \lceil (n(t + 1) - 1)/2 \rceil + n - 1.$$

Proof. We start by proving the bound for liberal communication complexity. By Lemma 4 we can assume that π has t-floating inputs. From Lemma 2 we then get that π has a revelation message for all input distributions, except with negligible probability. We now want to count the number of send and receive operations that have been executed just before the revelation message is sent. Since $|N_j| \geq t + 1$ for all P_j *after* the revelation message is sent, it follows that after it is sent

$$\sum_{i=1}^{n} |N_i| \geq n(t + 1).$$

Notice that in this sum the revelation message is counted only once, but all other messages might be counted twice. Hence at least $(n(t + 1) - 1)/2 + 1$ messages were sent after the revelation message was sent. Therefore at least $(n(t+1)-1)/2$ messages were sent before the revelation message was sent. Since the number of messages sent is an integer, it follows that at least $\lceil (n(t + 1) - 1)/2 \rceil$ messages were sent. By Lemma 3, after the point where the revelation message is sent by some P_k each other party receives at least one more message with probability at least $\frac{1}{2}$ − negl. By linearity of expectation, this gives at least an expected $(n-1)(\frac{1}{2} - \text{negl}(s))$ more messages. Since n is a constant in s we have that $n\,\text{negl}(s) = \text{negl}(s)$, so $\lim_{s\to\infty}(n-1)(\frac{1}{2} - \text{negl}(s)) = (n-1)\frac{1}{2} = n/2 - \frac{1}{2}$. It is easy to see that for conservative message complexity we get to add $n-1$ instead of $(n-1)/2$: when we consider conservative message complexity, receiving a message with probability $\frac{1}{2}$ counts as 1 towards the message complexity.   □

We say that a function f is *t-very difficult* if it is *t*-difficult and in addition for P_i there exists P_j such that P_i and P_j has an embedded AND in the following sense: There exists an input vector $\boldsymbol{x}$ and inputs x_i^1 and x_i^0 for P_i and inputs x_j^1 and x_j^0 for P_j such that if we set $y_{b,c} = f(\boldsymbol{x}[(i, x_i^b), (j, x_j^c)]$ for $b, c \in \{0, 1\}$, then $y_{0,0} \neq y_{1,1}$ and $y_{0,0} = y_{0,1} = y_{1,0}$. We note that the notion of an embedded AND (or, equivalently, an embedded OR) has been extensively studied in other settings, see [KKMO00] and references therein.) If f is *t*-very difficult we can improve the lower bound by $\frac{1}{2}$ message.

Theorem 2. *Let π be the n-party function which securely implements a function f which is t-very difficult, with statistical correctness and statistical privacy against t semi-honest corruptions. Then*

$$\text{Msg}_{\text{lib}}(\pi) \geq \lceil (n(t+1) - 1)/2 \rceil + n/2$$

and

$$\text{Msg}_{\text{con}}(\pi) \geq \lceil (n(t+1) - 1)/2 \rceil + n.$$

Proof (Sketch). We start by proving the bound for liberal communication complexity. The proof follows the lines of the proof of Theorem 1, so we will only give a sketch. The extra $\frac{1}{2}$ message comes from the fact that we can now argue that even the sender of the revelation message must receive another bit of information after sending the revelation message and therefore must receive another message with probability at least $\frac{1}{2}$. To see this, note that if this was not the case, then it holds for all input distributions, by Lemma 1. Let P_k be the sender of the revelation message and let P_j be the party with which P_k has an embedded AND. Denote an execution of $\pi(\boldsymbol{x}[(j, x_j^b), (j, x_k^c)])$ by $[b, c]$. Assume that P_k receives a message after sending the revelation message with probability less than $\frac{1}{2}$.

In $[b, 0]$ it holds that the view of P_k is independent of b even at the end of the execution as the output and input of P_k are the same in the two executions. That implies that until P_k sends the revelation message it also holds in $[b, 1]$ that the view of P_k is independent of b, as $[b, 0]$ and $[b, 1]$ are perfectly indistinguishable to P_k until P_k actually reads its input. From this it follows that it also holds in $[b, 1]$ that the view of P_k is independent of b *after sending the revelation message*, as reading the input $x_j^c = 1$ cannot change that dependence on b as 1 is a constant and in particular independent of b and the view of P_k so far. But in $[b, 1]$ the output of P_k must be b by the correctness of π. Going from a situation where the view of P_k is independent of b to learning b requires that P_k

receives a message with probability at least $\frac{1}{2}$. When we consider conservative message complexity, receiving a message with probability $\frac{1}{2}$ counts as 1 towards the message complexity.   □

Lower Bounds for Perfect Security and Adaptive Corruption. Our model assume that the number of parties is constant as a function of the security parameter. The only place in our lower bound proofs where we used this assumption is in the proof of Lemma 4. If we consider perfect security, the proof simplifies greatly, and we can easily prove the lemma for any number of parties. Alternatively, if we consider adaptive security, note that the proof first constructs an adaptive adversary that breaks the protocol if our result is false and then converts it to an static adversary using the assumption on a constant number of parties. Therefore it is immediate that the lemma also holds for any number of parties and adaptive security. We conclude that all our lower bounds for this section hold for any number of parties, if we consider perfect or adaptive security.

4.2   Individual Round Complexity

Consider now an n-player protocol π that is executed on a synchronous network. We can define a (possibly empty) set M_π of players with *minimal interaction*, consisting of players whose only communication is to each send a message to a subset of the parties not in M_π and then later, after all parties in M_π have sent all their messages, each receive a message from a subset of the parties not in M_π.

Theorem 3. *Assume $n = 2t + 1$ parties, where each party P_i holds input bit b_i. A protocol π that computes $b_1 \wedge \cdots \wedge b_n$ with perfect correctness and statistical privacy against t semi-honest corruptions must have $|M_\pi| \leq t$.*

Proof. Assume for contradiction that M_π has size $t + 1$. Then we can construct from π a 3-party protocol for players A, B and C, where player A emulates the t players not in M_π, B emulates t of the players in M_π, and C emulates the last player in M_π. Each party will have a single bit as input and will use that bit as input to each of the parties it is emulating. If π is secure, then clearly the 3-party protocol securely computes the AND of the inputs from the 3 players, provided at most 1 is passively corrupt, as corrupting any of A, B and C will corrupt at most t emulated parties. Moreover, the 3 party protocol will have only 4 messages. Namely, the one party from M_π emulated by C will send one message to A and later receive exactly one message from A, as A emulated exactly the parties not in M_π. The same is true for all the emulated players in B, they will all send exactly one message to a player in A and receive back one message from a player in A. Furthermore, since they all send their messages to the players in A before they received any messages from A, we can let B send all the messages as one message. In the same way we can let A return all the messages as one message. Since there is no communication between parties in M_π, there is no communication between B and C. Hence all other communication takes place inside A. However, communicating just 4 message is in contradiction to Theorem 2, which says that 6 messages are required.   □

Theorem 4. *Assume* $n = 3t + 1$ *parties, where each party* P_i *holds input bit* b_i. *A protocol* π *that computes* $b_1 \wedge \cdots \wedge b_n$ *with statistical correctness and statistical privacy against* t *malicious corruptions must have* $|M_\pi| \leq t$.

Proof. If we assume a contradiction we can as above reduce it to the case with $n = 4$ and $t = 1$. We let A simulate t parties with optimal communication complexity. We let B simulate the last party with optimal communication complexity. We let C and D each simulate t of the remaining parties. We set the input of D to be 1 and we denote the inputs of A, B and C by a, b and c. The communication pattern is as follows. First A sends two messages to C and D. Denote the message sent to C by g. At the same time B sends two messages to C and D. Denote the message sent to C by h. By privacy against a semi-honest corruption of C we know that g is independent of a. Clearly the message h is independent of a. Furthermore, since g and h were computed by two different parties which did not communicate before sending these messages, and the parties do not have a source of correlated randomness, g and h are independent. It follows that (g, h) is independent of a. However, by security of one malicious corruption the protocol should still terminate with the correct result if at this point D stops participating in which case C receives no further information. Clearly C cannot always compute the correct result with good probability when its view is independent of a.   □

5   Upper Bounds

In this section we give four constructive upper bounds, one for individual round complexity of secure function evaluation in the face of semi-honest corruptions, then one for individual round complexity of broadcast in the face of malicious corruptions, one for individual round complexity of secure functional evaluation in the face of malicious corruptions, and finally one for message complexity in the face of semi-honest corruptions.

5.1   Individual Round Complexity, Semi-honest Security

We first give a construction with minimal individual round complexity for a group of $t < n/2$ parties in the face of semi-honest corruption.

Theorem 5. *For every poly-time* n-*party function* f, *there exists a poly-time function evaluation protocol computing* f *between* $n = 2t + 1$ *parties with perfect correctness and perfect privacy against* t *semi-honest corruptions, where* t *parties have round complexity two. Specifically, these* t *parties first in parallel each send one message to the* $n - t$ *other parties and then later each receives one message from the same* $n - t$ *parties.*

Proof. We design a protocol where it is the parties $I = \{P_{n-t+1}, \ldots, P_n\}$ which have round complexity two. We denote each of the t parties in I generically by P_i and we denote the parties in $J = \{P_1, \ldots, P_{n-t}\}$ generically by P_j.

Use D to denote the pre-processing distribution of a secure function evaluation protocol for the pre-processing model with $n' = t + 1$ parties and up to t semi-honest corruptions. Let $(D, \pi_{\text{pre-pro}})$ be a protocol for this model with perfect correctness and perfect privacy for t semi-honest corruptions.

Let $\pi_{\text{hon-maj}}$ be a secure function evaluation protocol for the function f for a model with $n = 2t + 1$ parties and assume that it has perfect correctness and perfect privacy against t semi-honest corruptions. Assume that $\pi_{\text{hon-maj}}$ has round complexity ℓ. We can assume that $\pi_{\text{hon-maj}}$ runs as follows in round r: first each party sends one message to each other party which adds this message to its state. Then it applies a round function $R^{i,r}$ which computes the new state of party P_i. The initial state of a party is just its input x_i.

Our protocol π proceeds as follows. First each P_i will additively secret share its input x_i among the parties P_j, i.e., it samples uniformly random shares $x_{i,j}$ for which $x_i = x_{i,1} \oplus \cdots \oplus x_{i,n-t}$ and securely sends $x_{i,j}$ to P_j. At the same time it will for $r = 1, \ldots, \ell$ sample $(d_1^{i,r}, \ldots, d_{n-t}^{i,r}) \leftarrow D$ and send $d_j^{i,r}$ to P_j. Notice that at this point the initial state of each P_i is secret shared among the parties in J. We will keep the invariant that at each round in the protocol $\pi_{\text{hon-maj}}$ the state of P_i in $\pi_{\text{hon-maj}}$ is secret shared among the parties in J. Each round in $\pi_{\text{hon-maj}}$ is emulated as follows.

1. If $P_j \in J$ is to send a message m to $P_k \in J$, then it sends m over the secure channel to P_k.
2. If $P_j \in J$ is to send a message m to $P_i \in I$, then it additively secret shares m among the parties J and this secret sharing is added to the secret shared state of P_i.
3. If $P_i \in I$ is to send a message m to $P_k \in I$, then m is by the invariant already additively secret shared among the parties J. The parties in J can therefore just add this secret sharing to the secret shared state of P_k.
4. If $P_i \in I$ is to send a message m to $P_j \in J$, then m is additively secret shared among the parties J as part of the secret shared state of P_i. The parties in J can therefore reconstruct this message towards P_j.
5. If $P_j \in J$ is to apply the round function $R^{j,r}$, then it simply applies it to its state.
6. If $P_i \in I$ is to apply the round function $R^{i,r}$, then the parties in J uses the preprocessed values $(d_1^{i,r}, \ldots, d_{n-t}^{i,r})$ to do secure function evaluation of the augmented round function $\bar{R}^{i,r}$ which reconstructs the state of P_i from the secret sharing of the state held by the parties in J, then applies $R^{i,r}$ and outputs an additive secret sharing of the new state.

After all ℓ rounds of $\pi_{\text{hon-maj}}$ have been emulated, the secret-shared state of P_i contains its output y_i. The parties in J reconstructs this y_i towards P_i. At this point all n parties received their outputs.

It should be clear that this protocol has perfect correctness, as $\pi_{\text{pre-pro}}$ and $\pi_{\text{hon-maj}}$ both have perfect correctness.

As for perfect privacy, note that if at most t parties are corrupted, then the additive secret sharings among the t parties in J leaks no information, and can indeed be efficiently simulated by just giving all corrupted parties uniformly random shares.

Furthermore, if $P_i \in I$ is honest, then the emulation of P_i in $\pi_{\text{hon-maj}}$ is perfectly private, as P_i is perfectly acting as the trusted third party of the preprocessing model. We can in particular replace the emulation of P_i by an ideal function evaluation of the augmented round function.

Since the additive secret sharing of the inputs and outputs of the augmented round function can be efficiently simulated towards the t corrupted parties without knowing the inputs or outputs, we can replace the ideal evaluation of the augmented round function by an ideal evaluation of the actual round function on the actual state of P_i and then just simulate the secret sharing of the inputs and outputs using uniformly random shares. But having an ideal evaluation of the round function of an honest P_i is exactly the same as just having P_i participate in the protocol. So at this point we have arrived at the protocol $\pi_{\text{hon-maj}}$. Since there are at most t corrupted parties we can then appeal to the security of $\pi_{\text{hon-maj}}$.

Constructing an explicit simulator of π from the simulators of $\pi_{\text{pre-pro}}$ and $\pi_{\text{hon-maj}}$ along the lines of the above sketch is straight forward and we skip the technical details.   □

5.2   Individual Round Complexity, Broadcast

We now turn our attention to the individual round complexity of secure broadcast. Secure broadcast from P_i to the parties $P_1, \ldots, P_n$ is defined to be the secure function evaluation of the function $x_i = f(x_1, \ldots, x_n)$ in the face of malicious corruptions, i.e., P_i communicates x_i to all parties and it is guaranteed that all parties receive the same x_i even if P_i and/or some of the other parties are malicious. By secure broadcast we mean a protocol which allows any of the n parties to broadcast to all the other parties.

It is possible to implement broadcast securely against $t < n/3$ maliciously corrupted parties in a synchronous network with authenticated channels (note that secure channels are not needed for broadcast). It is furthermore possible to do so using a protocol where the honest parties are deterministic. See for instance [BDGK91].

The above protocol is for the setting with $t < n/3$ maliciously corrupted parties. We later need to do broadcast in a setting with $t < n/2$ maliciously corrupted parties. It is actually known that broadcast is impossible in such a setting. We can, however, implement broadcast if we assume $t < n/3$ for just the first round. To show this we need the following lemma.

Lemma 5. *Consider any protocol π for n parties which is perfectly correct and has statistical privacy against t maliciously corrupted parties computing a function f. Assume that $P_{n-t+1}, \ldots, P_n$ have no inputs, i.e., $f(x_1, \ldots, x_n) = g(x_1, \ldots, x_{n-t})$. Assume also that these parties are not to receive outputs.*

Assume furthermore that the protocol remains secure even if all messages sent and received by $P_{n-t+1}, \ldots, P_n$ *are given to the adversary and assume that these parties are deterministic. Then there also exists a protocol* π' *which is statistically correct and has statistical privacy against* t *maliciously corrupted parties computing the function* f *in which* $P_{n-t+1}, \ldots, P_n$ *each sends a message to each of the parties* $P_1, \ldots, P_{n-t}$ *in the first round and then sends or receives no further messages.*

Proof. The parties $I = \{P_1, \ldots, P_{n-t}\}$ will simply emulate the parties $J = \{P_{n-t+1}, \ldots, P_n\}$. Each $P_i \in I$ will run a copy of each $P_j \in J$. Since P_j has no input, the parties P_i will agree on the initial states of all P_j. Whenever P_j wants to send a message, all P_i will know this message and the appropriate receiver will just take that message as if having been sent by P_j. If the receiver is a party $P_j \in J$ all $P_i \in I$ will input the message to their local copy of P_j. In each round all parties $P_i \in I$ apply the deterministic round function of each P_j to their own local copy. This maintains agreement on the state of all the emulated P_j.

The only problematic case is when some $P_i \in I$ wants to send a message m to some $P_j \in J$. In that case P_i must send m to all parties in I such that they can input m to P_j. We have to ensure that P_i sends the same m to all parties in I, or they might end up with inconsistent versions of P_j. We ensure this by letting P_i broadcast the message m. The only problem is that we do not have a broadcast channel. We will therefore let P_j create one using pre-processing. This will be done using the one round of messages that P_j sends in the first round, as detailed now.

It is shown in [PW92] that there exists a protocol (P, π) for the pre-processing model which implements broadcast between n' parties secure against t malicious corruptions for any $t < n'$. We can therefore let each $P_j \in J$ sample $(p_{j,1}, \ldots, p_{j,n'}) \leftarrow P$ and send $p_{j,i}$ securely to P_i. Whenever $P_i \in I$ is to send m to all parties in I, the parties run π on the pre-processed values $(p_{j,1}, \ldots, p_{j,n'})$ and with P_i having input m. Note that each $P_j \in J$ preprocessed his own broadcast channel. This is the broadcast channel that is to be used when message are sent *to* P_j in the emulated protocol. If P_j is honest, the pre-processing is computed as it should, and thus the broadcast protocol will indeed ensure that m is delivered consistently, and hence the emulated P_j will be run correctly and consistently by all honest parties in I. If P_j is corrupted, it might deliver incorrect pre-processed values. In that case the broadcast might not work correctly. In that case the parties in I might get inconsistent views of P_j and might therefore later see inconsistent values of what P_j is sending. This, however, is no worse than the emulated P_j being corrupted and this case only happens when the actual P_j is maliciously corrupted, so the emulated protocol can tolerate this. $\square$

If we plug the protocol from [BDGK91] into the above lemma we get this corollary.

Corollary 2. *There exists a protocol* π_{broad} *for* n *parties which is statistically correct and which allows any party* P_i *(with* $i \leq n-t$*) to broadcast to the parties* $P_1, \ldots, P_{n-t}$. *It is secure against* t *malicious corruptions for* $t < n/3$. *The parties*

$\mathsf{P}_{n-t+1}, \ldots, \mathsf{P}_n$ each sends one message to each of the parties $\mathsf{P}_1, \ldots, \mathsf{P}_{n-t}$ in the first round and otherwise has no communication.

5.3   Individual Round Complexity, Secure Function Evaluation

We now turn our attention to secure function evaluation in the face of malicious corruptions.

Theorem 6. *For every poly-time n-party function f, there exists a poly-time function evaluation protocol computing f between $n = 3t + 1$ parties with statistical correctness and statistical privacy against t maliciously corrupted parties, where t parties have round complexity two. Specifically, these t parties first each sends one message to the $n - t$ other parties in parallel and then later each receives one message from the same $n - t$ parties.*

Proof. As usual, $I = \{\mathsf{P}_1, \ldots, \mathsf{P}_{n-t}\}$ and $J = \{\mathsf{P}_{n-t+1}, \ldots, \mathsf{P}_n\}$. In [RB89] a statistically correct and statistically private protocol for secure function evaluation of any function g is given for the setting with n' parties of which at most $t < n'/2$ parties are maliciously corrupted. The protocol is for the setting with secure point-to-point channels plus a broadcast channel allowing any party to broadcast to the other n' parties. Denote this protocol by π_{RB}. Set $n' = n-t$. We are going to let the parties I run π_{RB} to compute a particular function g derived from f. In doing that they will implement the broadcast channel using π_{broad} from Corollary 2 with the parties in J providing the pre-processing.

We will use a robust secret sharing scheme (sha, rec) for n' parties and $t < n/2$ corruptions to let the parties in J provide inputs. Such a scheme is trivial to derive from, e.g., the verifiable secret sharing scheme constructed in [RB89], and has the following properties:

Privacy. The joined distribution of any t positions from a random sample $(v_1, \ldots, v_{n'}) \leftarrow \mathsf{sha}(v)$ does not depend on the value v.

Robustness. Sample $(v_1, \ldots, v_{n'}) \leftarrow \mathsf{sha}(v)$ for a value v chosen by the adversary. Now give t of the positions v_i to the adversary and let it replace them by v_i'. The positions are chosen by the adversary. For the remaining $n' - t$ positions, let $v_i' = v_i$. Then $\mathsf{rec}(v_1', \ldots, v_{n'}') = v$, except with probability 2^{-s}, where s is the statistical security parameter.

The function g takes $n-t$ inputs, $g(X_1, \ldots, X_{n-t})$, where each X_i is of the form $(x_i, x_{n-t+1,i}, \ldots, x_{n,i})$. It outputs

$$f(x_1, \ldots, x_{n-t}, \mathsf{rec}(x_{n-t+1,1}, \ldots, x_{n-t+1,n-t}), \ldots, \mathsf{rec}(x_{n,1}, \ldots, x_{n,n-t})).$$

The overall protocol then runs as follows.

1. Each $\mathsf{P}_j \in J$ sends the pre-processing needed for π_{broad} to the parties in I and at the same time samples $(x_{j,1}, \ldots, x_{j,n-t}) \leftarrow \mathsf{sha}(x_j)$ and sends $x_{j,i}$ to $\mathsf{P}_i \in I$.
2. Each $\mathsf{P}_i \in I$ computes $X_i = (x_i, x_{n-t+1,i}, \ldots, x_{n,i})$.

3. The parties in I use the pre-processing provided in Step 1 to run π_{broad} and use the emulated broadcast channel to run $\pi_{\text{RB}}(X_1, \ldots, X_{n-t})$.
4. When $\mathsf{P}_i \in I$ learns the output $y = \pi_{\text{RB}}(X_1, \ldots, X_{n-t})$ it sends y to all parties in J.
5. Each party $\mathsf{P}_j \in J$ receives an output y_i from each $\mathsf{P}_i \in I$ and outputs the value y which occurs most often in the list $(y_1, \ldots, y_{n-t})$.

It follows directly from the security of $(\mathsf{sha}, \mathsf{rec})$, π_{broad} and π_{RB} that the protocol is private and that the honest parties in I learn the correct output y, except with negligible probability. Since there are $n' \geq 2t + 1$ parties in I and at most t corrupted parties in I, it follows that there is a majority of honest parties in I. Hence, the honest parties in J will also learn the correct output y. □

5.4   Message Complexity, Semi-honest Security

We now turn our attention to the message complexity of secure function evaluation in the presence of semi-honest corruptions. We consider protocols with n parties which are perfectly secure against t semi-honest corruptions. We present an optimal construction for $t = 1$ for computing functions in PSM_{eff} as defined in Definition 1.

Theorem 7. *For every poly-time n-party function f in PSM_{eff}, there exists a poly-time function evaluation protocol π computing f between n parties with perfect correctness and perfect privacy against $t = 1$ semi-honest corruptions, for which $\mathsf{Msg}_{\text{lib}}(\pi) = (3n + 1)/2$, $\mathsf{Msg}_{\text{wor}}(\pi) \leq \lceil (3n + 1)/2 \rceil$, and $\mathsf{Msg}_{\text{con}}(\pi) = 2n$.*

Proof. We first look at the restricted setting where P_n has no input and is the only player to learn the output, i.e., we look at secure function evaluation of $(\epsilon, \ldots, \epsilon, y) = f(x_1, \ldots, x_n)$, where ϵ is the empty string and $y = h(x_1, \ldots, x_{n-1})$ for an $(n - 1)$-party function h.

Let $(R, M_1, \ldots, M_{n-1})$ be a PSM protocol for h and consider the following protocol π_1.

1. P_1 samples $r \leftarrow R$.
2. P_1 sends r to P_i for $i = 2, \ldots, n - 1$.
3. For $i = 1, \ldots, n - 1$, party P_i sends $m_i = M_i(x_i, r)$ to P_n.
4. P_n outputs $y = g(m_1, \ldots, m_{n-1})$.

Assume that P_n is corrupted. The view of P_n in the real world is

$$(M_1(x_1, r), \ldots, M_{n-1}(x_{n-1}, r))$$

for a random sample $r \leftarrow R$. The view of P_n in the ideal model is

$$y = f(x_1, \ldots, x_n) = h(x_1, \ldots, x_{n-1}) = g(M_1(x_1, r), \ldots, M_{n-1}(x_{n-1}, r)).$$

Privacy then follows from the security of the PSM protocol.

Assume that $P_i \neq P_n$ is corrupted. The view of P_i in the real world is (x_i, r). The view of P_i in the ideal model is x_i. We can simulate the real world view from the ideal view simply by sampling $r \leftarrow R$ and then outputting (x_i, r).

We now extend the above protocol to a protocol π_2 which allows P_n to have an input and where all parties get the output, i.e., we look at secure function evaluation of $y = f(x_1, \ldots, x_n)$. We first present and analyze a simple solution and then later modify it slightly to reduce the number of messages sent. The simple solution is to let P_n additively secret share x_n as $x_n = s_1 \oplus s_2$ and send s_1 to P_1 and send s_2 to P_2. Then apply protocol π_1 to the function

$$h'((x_1, s_1), (x_2, s_2), x_3, \ldots, x_{n-1}) = f(x_1, \ldots, x_{n-1}, s_1 \oplus s_2)$$

and let P_n send the output to all the other parties. We can do this as h' clearly is in non-deterministic log-space if f is in non-deterministic log-space. Note that this simple protocol adds $n+1$ more message. Sending the output y to all parties is obviously secure as this value is also in the view of all parties in the ideal model. Only P_1, P_2 and P_n have any further extra values in the view. The extra values of P_n are s_1 and s_2 such that $x_n = s_1 \oplus s_2$. These are easy to simulate from the view of P_n in the ideal model which includes x_n: simply sample an additive secret sharing of x_n. The extra value of P_1 is s_1. This value is uniformly random and independent of x_n, so it can be simulated by just sampling it uniformly at random. Similarly for P_2.

Since s_1 is uniformly random and independent of x_n, we can save one message in the protocol by letting P_1 pick s_1 uniformly at random and send it to P_n along with the message that it already sends to P_n. The view of all parties will be the same in the modified protocol. The only difference is that the direction of one message was flipped. This gives the following secure protocol.

Let $(R, M_1, \ldots, M_{n-1})$ be a PSM protocol for the function h' described above.

1. P_1 samples $r \leftarrow R$.
2. P_1 sends $m_1 = M_1(x_1, r)$ to P_n along with a uniformly random share s_1.
3. P_n sends $s_2 = x_n \oplus s_1$ to P_2.
4. P_1 sends r to P_i for $i = 2, \ldots, n - 1$.
5. For $i = 2, \ldots, n - 1$, party P_i sends $m_i = M_i(x_i, r)$ to P_n.
6. P_n sends $y = g(m_1, \ldots, m_{n-1})$ to $P_1, \ldots, P_{n-1}$.

To further reduce the message complexity, we will now apply two additional message-reduction tricks. Using the first one we reduce the $2(n-2)$ messages in Steps 4 and 5 to just $n - 1$ messages: Instead of having all parties send to P_n, we will let P_1 send his "PSM-contribution" to P_2, who appends his contribution and sends a message to P_3, etc. until P_n receives everything. In order to make sure that only P_n learns all the contributions, P_1 will send $n - 1$ one-time pads to P_n and also pass them on to the other players who can use them to one-time pad encrypt their contributions.

With the second trick we reduce the number of messages in Step 6 from $n - 1$ to $\lceil (n - 1)/2 \rceil$. We let P_1 choose a random bit w which will be sent to all other players appended to the "PSM-contributions", thus not requiring additional message(s). Now, P_n can communicate the result, y, to the other players in the following way: if $y \oplus w = 0$ then P_n sends a bit 0 to players $P_1, \ldots, P_{\lceil (n-1)/2 \rceil}$ and does not send any message to the players $P_{\lceil (n-1)/2 \rceil + 1}, \ldots P_{n-1}$; otherwise (if $y \oplus w = 1$) then P_n sends a message 0 to the players $P_{\lceil (n-1)/2 \rceil + 1}, \ldots P_{n-1}$ and does not send any message to the players $P_1, \ldots, P_{\lceil (n-1)/2 \rceil}$. Observe that all players can retrieve the computed value y, and that the number of messages sent during that stage is at most $\lceil (n-1)/2 \rceil$. Both tricks can be implemented as follows. We replace steps 4, 5 and 6 by the following procedure:

1. P_1 samples uniformly random bit strings $p_2, \ldots, p_{n-1}$ where p_i has the same length as m_i. He also samples a uniformly distributed bit w. Then P_1 sends $(p_2, \ldots, p_{n-1}), w$ to P_n. This can be done in Step 2 above and therefore does not add another message.
2. P_1 sends $(r, p_2, \ldots, p_{n-1}), w$ to P_2.
3. Then for $i = 2, \ldots, n-1$ party P_i receives $(r, c_2, \ldots, c_{i-1}, p_i, p_{i+1}, \ldots, p_{n-1}), w$ from P_{i-1} and then sends $(r, c_2, \ldots, c_{i-1}, c_i, p_{i+1}, \ldots, p_{n-1}), w$ to P_{i+1}, where $c_i = M_i(x_i, r) \oplus p_i$, except that P_{n-1} does not send r to P_n.
4. Then P_n receives $(c_2, \ldots, c_{n-1})$ from P_{r-1} and for $i = 2, \ldots, n-1$ computes $m_i = c_i \oplus p_i$.
5. P_n computes the result y using the PSM protocol. Now, if $y \oplus w = 0$ then it sends 0 to all of players $P_1, \ldots, P_{\lceil (n-1)/2 \rceil}$ (and no message to the other players). Otherwise (if $f \oplus w = 1$) it sends 0 to all of players $P_{\lceil (n-1)/2 \rceil + 1}, \ldots P_{n-1}$ (and no message to the other players). Each P_i will observes if a message was received from P_n, and, using its index and w, computes y.

It is easy to see that this is perfectly correct. As for perfect security against one semi-honest corruption, consider the values c_i seen by P_j for $i < j < n$. Since P_j does not know p_i, c_i is a one-time pad encryption of m_i. All other values seen by a single party clearly leak no information on the input other than what is implied by y. For a given input, the average number of messages sent by P_n in Stage 5 is $(1/2)(\lceil (n-1)/2 \rceil + \lfloor (n-1)/2 \rfloor) = n/2 - 1/2$. (Whatever the value of w is, at most $\lceil (n-1)/2 \rceil \leq n/2$ messages are sent by P_n at Step 5). The average number of messages sent by the protocol is therefore $n+1+n/2-1/2 = 3n/2+1/2$ (and in the worst case the number of message sent is $n+1+\lceil (n-1)/2 \rceil \leq 3n/2+1$, if n is even.) However, since all parties except P_n may potentially receive a message in the last step, the conservative message complexity is $2n$. $\qquad \square$

If we set $t = 1$ in our previous lower bound for liberal message complexity, we get $3n/2$, matching the upper bound of Theorem 7 except for $1/2$ a message. The conservative message complexity of the protocol in Theorem 7 is clearly $2n$ which matches the lower bound for conservative message complexity of 1-very difficult functions. So we have matching upper and lower bounds for the conservative message complexities of 1-very difficult functions in non-deterministic log space. We leave it as an open problem to find matching bounds for any $t > 1$.

Finally, we consider computing the XOR of one input bit from each player. This is the primary example of a function that is t-difficult but not t-very difficult. We can construct a protocol for this function, secure for $t = 1$ from the proof of Theorem 7: We observe that there is no need for P_n to secret share his input, instead we use the PSM protocol to let P_n learn $b_1 \oplus \cdots \oplus b_{n-1}$. This is secure because this value would anyway follow from the output and P_n's own input. P_n computes the output $b_1 \oplus \cdots \oplus b_n$ and sends it to the other players in the randomised fashion described in the protocol. The liberal and conservative complexities of this protocol are $3n/2 - 1/2$ and $2n - 1$, matching the lower bounds we showed for 1-difficult functions.

Acknowledgements. Work done in part while some of the authors visited Simons Institute. First and second author acknowledge support from the Danish National Research Foundation and The National Science Foundation of China (under the grant 61061130540) for the Sino-Danish Center for the Theory of Interactive Computation, within which part of this work was performed; and also from the CFEM research center (supported by the Danish Strategic Research Council) within which part of this work was performed. The second author was partially supported by the European Research Council Starting Grant 279447, the second partially supported by the European Research Council Advanced Grant MPCPRO. The third author acknowledges partial support by NSF grants 09165174, 1065276, 1118126 and 1136174, US-Israel BSF grant 2008411, OKAWA Foundation Research Award, IBM Faculty Research Award, Xerox Faculty Research Award, B. John Garrick Foundation Award, Teradata Research Award, and Lockheed-Martin Corporation Research Award. This material is also based upon work supported in part by DARPA Safeware program. The views expressed are those of the author and do not reflect the official policy or position of the Department of Defense or the U.S. Government. Research by the fourth author partially supported by ANR project RDAM.

References

[BDGK91] Bar-Noy, A., Deng, X., Garay, J.A., Kameda, T.: Optimal amortized distributed consensus (extended abstract). In: Toueg, S., Spirakis, P.G., Kirousis, L.M. (eds.) WDAG 1991. LNCS, vol. 579, pp. 95–107. Springer, Heidelberg (1991)

[BGT13] Boyle, E., Goldwasser, S., Tessaro, S.: Communication locality in secure multi-party computation. In: Sahai, A. (ed.) TCC 2013. LNCS, vol. 7785, pp. 356–376. Springer, Heidelberg (2013)

[BGW88] Ben-Or, M., Goldwasser, S., Wigderson, A.: Completeness theorems for non-cryptographic fault-tolerant distributed computation (extended abstract). In: Simon, J. (ed.) Proceedings of the 20th Annual ACM Symposium on Theory of Computing, Chicago, Illinois, USA, 2–4 May 1988, pp. 1–10. ACM (1988)

[Can00] Canetti, R.: Security and composition of multiparty cryptographic protocols. J. Cryptology **13**(1), 143–202 (2000)

[CCD87] Chaum, D., Crépeau, C., Damgård, I.B.: Multiparty unconditionally secure protocols (abstract). In: Pomerance, C. (ed.) CRYPTO 1987. LNCS, vol. 293, p. 462. Springer, Heidelberg (1988)

[CCG+15] Chandran, N., Chongchitmate, W., Garay, J.A., Goldwasser, S., Ostrovsky, R., Zikas, V.: The hidden graph model: communication locality and optimal resiliency with adaptive faults. In: Proceedings of the Conference on Innovations in Theoretical Computer Science, ITCS, Rehovot, Israel, 11–13 January 2015, pp. 153–162 (2015)

[CK93] Chor, B., Kushilevitz, E.: A communication-privacy tradeoff for modular addition. Inf. Process. Lett. 45(1), 205–210 (1993)

[DPP14] Data, D., Prabhakaran, M.M., Prabhakaran, V.M.: On the communication complexity of secure computation. In: Garay, J.A., Gennaro, R. (eds.) CRYPTO 2014, Part II. LNCS, vol. 8617, pp. 199–216. Springer, Heidelberg (2014)

[DZ13] Damgård, I., Zakarias, S.: Constant-overhead secure computation of Boolean circuits using preprocessing. In: Sahai, A. (ed.) TCC 2013. LNCS, vol. 7785, pp. 621–641. Springer, Heidelberg (2013)

[FKN94] Feige, U., Kilian, J., Naor, M.: A minimal model for secure computation (extended abstract). In: Proceedings of the Twenty-Sixth Annual ACM Symposium on Theory of Computing, Montréal, Québec, Canada, 23–25 May 1994, pp. 554–563 (1994)

[GIPR] Gonen, M., Ishai, Y., Prabhabkahan, M., Rosulek, M.: Private communication (unpublished work)

[IK97] Ishai,Y., Kushilevitz, E.: Private simultaneous messages protocols with applications. In: ISTCS, pp. 174–184 (1997)

[IK00] Ishai, Y., Kushilevitz, E.: Randomizing polynomials: a new representation with applications to round-efficient secure computation. In: Proceedings of the 41st Annual Symposium on Foundations of Computer Science, pp. 294–304. IEEE (2000)

[KKMO00] Kilian, J., Kushilevitz, E., Micali, S., Ostrovsky, R.: Reducibility and completeness in private computations. SIAM J. Comput. 29(4), 1189–1208 (2000)

[PW92] Pfitzmann, B., Waidner, M.: Unconditional byzantine agreement for any number of faulty processors. In: Finkel, A., Jantzen, M. (eds.) STACS 1992. LNCS, vol. 577, pp. 339–350. Springer, Heidelberg (1992)

[RB89] Rabin, T., Ben-Or, M.: Verifiable secret sharing and multiparty protocols with honest majority (extended abstract). In: Johnson, D.S. (ed.) Proceedings of the 21st Annual ACM Symposium on Theory of Computing, Seattle, Washigton, USA, 14–17 May 1989, pp. 73–85. ACM (1989)

The Exact Round Complexity
of Secure Computation

Sanjam Garg[1]($\boxtimes$), Pratyay Mukherjee[1], Omkant Pandey[2],
and Antigoni Polychroniadou[3]

[1] University of California, Berkeley, USA
{sanjamg,pratyay85}@berkeley.edu
[2] Drexel University, Philadelphia, USA
omkant@drexel.edu
[3] Aarhus University, Aarhus, Denmark
antigoni@cs.au.dk

Abstract. We revisit the *exact* round complexity of secure computation in the multi-party and two-party settings. For the special case of two-parties *without* a simultaneous message exchange channel, this question has been extensively studied and resolved. In particular, Katz and Ostrovsky (CRYPTO '04) proved that 5 rounds are necessary and sufficient for securely realizing every two-party functionality where both parties receive the output. However, the exact round complexity of general multi-party computation, as well as two-party computation *with* a simultaneous message exchange channel, is not very well understood.

These questions are intimately connected to the round complexity of non-malleable commitments. Indeed, the *exact* relationship between the round complexities of non-malleable commitments and secure multi-party computation has also not been explored.

In this work, we revisit these questions and obtain several new results. First, we establish the following main results. Suppose that there exists a k-round non-malleable commitment scheme, and let $k' = \max(4, k + 1)$; then,

- **(Two-party setting with simultaneous message transmission):** there exists a k'-round protocol for securely realizing *every* two-party functionality;
- **(Multi-party setting):** there exists a k'-round protocol for securely realizing the *multi-party coin-flipping* functionality.

As a corollary of the above results, by instantiating them with existing non-malleable commitment protocols (from the literature), we establish

Research supported in part from a DARPA/ARL SAFEWARE award, AFOSR Award FA9550-15-1-0274, and NSF CRII Award 1464397. The views expressed are those of the author and do not reflect the official policy or position of the Department of Defense, the National Science Foundation, or the U.S. Government. Also, Antigoni Polychroniadou received funding from CTIC under the grant 61061130540 and from CFEM supported by the Danish Strategic Research Council. This work was done in part while the authors were visiting the Simons Institute for the Theory of Computing, supported by the Simons Foundation and by the DIMACS/Simons Collaboration in Cryptography through NSF grant #CNS-1523467.

M. Fischlin and J.-S. Coron (Eds.): EUROCRYPT 2016, Part II, LNCS 9666, pp. 448–476, 2016.
DOI: 10.1007/978-3-662-49896-5_16

that **four** rounds are both necessary and sufficient for both the results above. Furthermore, we establish that, for *every multi-party functionality five* rounds are sufficient.

We actually obtain a variety of results offering trade-offs between rounds and the cryptographic assumptions used, depending upon the particular instantiations of underlying protocols.

1   Introduction

The round complexity of secure computation is a fundamental question in the area of secure computation [20,39,40]. In the past few years, we have seen tremendous progress on this question, culminating into constant round protocols for securely computing any multi-party functionality [5,9,10,21,22,27,28,36,38]. These works essentially settle the question of *asymptotic* round complexity of this problem.

The *exact* round complexity of secure computation, however, is still not very well understood[1]. For the special case of two-party computation, Katz and Ostrovsky [26] proved that 5 rounds are necessary and sufficient. In particular, they proved that two-party coin-flipping cannot be achieved in 4 rounds, and presented a 5-round protocol for computing every functionality. To the best of our knowledge, the exact round complexity of multi-party computation has never been addressed before.

The standard model for multi-party computation assumes that parties are connected via authenticated point-to-point channels as well as *simultaneous message exchange* channels where everyone can send messages at the same time. Therefore, in each round, all parties can simultaneously exchange messages.

This is in sharp contrast to the "standard" model for two-party computation where, usually, a simultaneous message exchange framework is not considered. Due to this difference in the communication model, the negative result of Katz-Ostrovsky [26] for 4 rounds, does not apply to the multi-party setting. In particular, a 4 round multi-party coin-flipping protocol might still exist!

In other words, the results of Katz-Ostrovsky only hold for the special case of two parties *without* a simultaneous message exchange channel. The setting of two-party computation *with a simultaneous message exchange channel* has not been addressed before. Therefore, in this work we address the following two questions:

What is the exact *round complexity of secure multi-party computation?*
In the presence of a simultaneous message exchange *channel, what is the* exact *round complexity of secure* two-party *computation?*

These questions are intimately connected to the round complexity of *non-malleable commitments* [12]. Indeed, new results for non-malleable commitments

---

[1] Our rough estimate for the exact round complexity of aforementioned multi-party results in the computational setting is 20–30 rounds depending upon the underlying components and assumptions.

have almost immediately translated to new results for secure computation. For example, the round complexity of coin-flipping was improved by Barak [3], and of every multi-party functionality by Katz et al. [27] based on techniques from non-malleable commitments. Likewise, black-box constructions for constant-round non-malleable commitments resulted in constant-round black-box constructions for secure computation [21,38]. However, all of these results only focus on *asymptotic* improvements and do not try to resolve the exact round complexity, thereby leaving the following fundamental question unresolved:

> *What is the relationship between the* exact *round complexities of non-malleable commitments and secure computation?*

This question is at the heart of understanding the exact round complexity of secure computation in both multi-party, and two-party with simultaneous message transmission.

1.1   Our Contributions

In this work we try to resolve the questions mentioned above. We start by focusing on the simpler case of two-party computation with a simultaneous message exchange channel, since it is a direct special case of the multi-party setting. We then translate our results to the multi-party setting.

Lower bounds for Coin-Flipping. We start by focusing on the following question.

> *How many simultaneous message exchange rounds are necessary for secure two-party computation?*

We show that four *simultaneous message exchange rounds* are necessary. More specifically, we show that:

Theorem (Informal): *Let κ be the security parameter. Even in the simultaneous message model, there does not exist a three-round protocol for the two-party coin-flipping functionality for $\omega(\log \kappa)$ coins which can be proven secure via black-box simulation.*

In fact, as a corollary all of the rounds must be "strictly simultaneous message transmissions", that is, both parties must *simultaneously* send messages in each of the 4 rounds. This is because in the simultaneous message exchange setting, the security is proven against the so called "rushing adversaries" who, in each round, can decide their message after seeing the messages of all honest parties in that round. Consequently, if only one party sends a message for example in the fourth round, this message can be "absorbed" within the third message of this party[2], resulting in a three round protocol.

---

[2] Note that, such absorption is only possible when it maintains the mutual dependency among the messages, in particular does not affect the next-message functions.

Results in the Two-Party Setting with a *Simultaneous Message Exchange* Channel. Next, we consider the task of constructing a protocol for coin-flipping (or any general functionality) in four simultaneous message exchange rounds and obtain a positive result. In fact, we obtain our results by directly exploring the *exact relationship* between the round complexities of non-malleable commitments and secure computation. Specifically, we first prove the following result:

Theorem (Informal): *If there exists a k-round protocol for (parallel) non-malleable commitment,[3] then there exists a k'-round protocol for securely computing every two-party functionality with black-box simulation in the presence of a malicious adversary in the simultaneous message model, where $k' = \max(4, k + 1)$.*

Instantiating this protocol with non-malleable commitments from [36], we get a **four round** protocol for every two-party functionality in the presence of a simultaneous message exchange channel, albeit under a non-standard assumption (adaptive one-way function). However, a recent result by Goyal et al. [23] constructs a non-malleable commitment protocol in *three* rounds from injective one-way functions, although their protocol does not immediately extend to the parallel setting. Instantiating our protocol with such a three-round parallel non-malleable commitment would yield a *four round* protocol under standard assumptions.

Results in the Multi-party Setting. Next, we focus on the case of the multi-party *coin flipping* functionality. We show that a simpler version of our two-party protocol gives a result for multi-party *coin-flipping*:

Theorem (Informal): *If there exists a k-round protocol for (parallel) non-malleable commitments, then there exists a k'-round protocol for securely computing the multi-party coin-flipping functionality with black-box simulation in the presence of a malicious adversary for polynomially many coins where $k' = \max(4, k + 1)$.*

Combining this result with the two-round multi-party protocol of Mukherjee and Wichs [34] (based on the LWE [37]), we obtain a $k' + 2$ round protocol for computing *every multi-party* functionality. Instantiating these protocols with non-malleable commitments from [36], we obtain a **four round** protocol for *coin-flipping* and a **six round** protocol for *every* functionality.

Finally, we show that the coin-flipping protocol for the multi-party setting can be extended to compute what we call the "coin-flipping with committed inputs" functionality. Using this protocol with the two-round protocol of [16] based on indistinguishability obfuscation [17], we obtain a **five round** MPC protocol.

---

[3] Parallel simply means that the man-in-the-middle receives κ non-malleable commitments in *parallel* from the left interaction and makes κ commitments on the right. Almost all known non-malleable commitment protocols satisfy this property.

1.2   Related Work

The round complexity of secure computation has a rich and long history. We only mention the results that are most relevant to this work in the computational setting. Note that, unconditionally secure protocols such as [6,8] are inherently non-constant round. More specifically, the impossibility result of [11] implies that a fundamental new approach must be found in order to construct protocols, that are efficient in the circuit size of the evaluated function, with reduced communication complexity that beat the complexities of BGW, CCD, GMW etc.

For the computational setting and the special case of two party computation, the semi-honest secure protocol of Yao [33,39,40] consists of only three rounds (see Sect. 2). For malicious security[4], a constant round protocol based on GMW was presented by Lindell [31]. Ishai et al. [25] presented a different approach which also results in a constant round protocol.

The problem of exact round complexity of two party computation was studied in the beautiful work of Katz and Ostrovsky [26] who provided a 5 round protocol for computing any two-party functionality. They also ruled out the possibility of a four round protocol for coin-flipping, thus completely resolving the case of two party (albeit without simultaneous message exchange, as discussed earlier). Recently Ostrovsky et al. [35] constructed a different 5-round protocol for the general two-party computation by only relying on *black-box* usage of the underlying trapdoor one-way permutation.

As discussed earlier, the standard setting for two-party computation does not consider simultaneous message exchange channels, and hence the negative results for the two-party setting do not apply to the multi-party setting where simultaneous message exchange channels are standard. To the best of our knowledge, prior to our work, the case of the two-party setting in the presence of a simultaneous message exchange channel was not explored in the context of the exact round complexity of secure computation.

For the multi-party setting, the exact round complexity has remained open for a long time. The work of [5] gave the first constant-round non black-box protocol for honest majority (improved by the black-box protocols of [9,10]). Katz et al. [27], adapted techniques from [3,5,7,12] to construct the first asymptotically round-optimal protocols for any multi-party functionality for the dishonest majority case. The constant-round protocol of [27] relied on non-black-box use of the adversary's algorithm [2]. Constant-round protocols making black-box use of the adversary were constructed by [21,28,36], and making black-box use of one-way functions by Wee in $\omega(1)$ rounds [38] and by Goyal in constant rounds [21]. Furthermore, based on the non-malleable commitment scheme of [21,22] construct a constant-round multi-party coin-tossing protocol. Lin et al. [30] presented a unified approach to construct UC-secure protocols from non-malleable commitments. However, as mentioned earlier, none of the aforementioned works

---

[4] From here on, unless specified otherwise, we are always in the malicious setting by default.

focused on the *exact* round complexity of secure computation based on the round-complexity of non-malleable commitments. For a detailed survey of round complexity of secure computation in the preprocessing model or in the CRS model we refer to [1].

1.3   An Overview of Our Approach

We now provide an overview of our approach. As discussed earlier, we first focus on the two-party setting with a simultaneous message exchange channel.

The starting point of our construction is the Katz-Ostrovsky (KO) protocol [26] which is a *four round* protocol for *one-sided* functionalities, i.e., in that only one party gets the output. Recall that, this protocol does not assume the presence of a simultaneous message exchange channel. At the cost of an extra round, the KO two-party protocol can be converted to a *complete* (i.e. both-sided) protocol where both parties get their corresponding outputs via a standard trick [18] as follows: parties compute a modified functionality in which the first party P_1 learns its output as well as the output of the second party P_2 in an "encrypted and authenticated"[5] form. It then sends the encrypted value to P_2 who can decrypt and verify its output.

A natural first attempt is to adapt this simple and elegant approach to the setting of simultaneous message exchange channel, so that the "encrypted/authenticated output" can somehow be communicated to P_2 simultaneously at the same time when P_2 sends its last message, thereby removing the additional round.

It is not hard to see that any such approach would not work. Indeed, in the presence of malicious adversaries while dealing with a simultaneous message exchange channel, the protocol must be proven secure against "rushing adversaries" who can send their messages after looking at the messages sent by the other party. This implies that, if P_1 could indeed send the "encrypted/authenticated output" message simultaneously with last message from P_2, it could have sent it earlier as well. Now, applying this argument repeatedly, one can conclude that any protocol which does not use the simultaneous message exchange channel necessarily in all of the four rounds, is bound to fail (see Sect. 3). In particular, any such protocol can be transformed, by simple rescheduling, into a 3-round protocol contradicting our lower bound.[6]

This means that we must think of an approach which must use the simultaneous message exchange channel in each round. In light of this, a natural second attempt is to run two executions of a 4-round protocol (in which only one party learns the output) in "opposite" directions. This would allow both parties to learn the output. Unfortunately, such approaches do not work in general since there is no guarantee that an adversarial party would use the same input in both

---

[5] In particular, the encryption prevents P_1 to know P_2's output ensuring output privacy whereas the authentication does not allow P_1 to send P_2 a wrong output.

[6] Recall that we show that (see Theorem 2 for a formal statement) 4 rounds are necessary even with simultaneous message exchange channels.

protocol executions. Furthermore, another problem with this approach is that of "non-malleability" where a cheating party can make its input dependent on the honest party's input: for example, it can simply "replay" back the messages it receives. A natural approach to prevent such attacks is to deploy non-malleable commitments, as we discuss below.

Simultaneous Executions + Non-malleable Commitments. Following the approach discussed above we observe that:

1. A natural direction is to use two simultaneous executions of the KO protocol (or any other similar 4-round protocol) over the simultaneous message exchange channel in opposite directions. Since we have only 4 rounds, a different protocol (such as some form of 2-round semi-honest protocol based on Yao) is not a choice.
2. We must use non-malleable commitments to prevent replay/mauling attacks.

We remark that, the fact that non-malleable commitments come up as a natural tool is not a coincidence. As noted earlier, the multi-party case is well known to be *inherently connected* to non-malleable commitments. Even though our current focus is solely on the two-party case, this setting is essentially (a special case of) the multi-party setting due to the use of the simultaneous message exchange channel. Prior to our work, non-malleable commitments have been used extensively to design multi-party protocols [21,22,29,33]. However, all of these works result in rather poor round complexity because of their focus on asymptotic, as opposed to exact, number of rounds.

To obtain our protocol, we put the above two ideas together, modifying several components of KO[7] to use non-malleable commitments. These components are then put together in a way such that, even though there are essentially two simultaneous executions of the protocol in opposite directions, messages of one protocol cannot be maliciously used to affect the other messages. In the following, we highlight the main ideas of our construction:

1. The first change we make is to the proof systems used by KO. Recall that KO uses the Fiege-Shamir (FS) protocol as a mechanism to "force the output" in the simulation. Our first crucial modification is to consider a variant of the FS protocol in which the verifier gives two non-malleable commitments (nmcom) to two strings σ_1, σ_2 and gives a *witness indistinguishable proof-of-knowledge* (WIPOK) that it knows one of them. These are essentially the simulation trapdoors, but implemented through nmcom instead of a one-way function. This change is actually crucial, and as such, brings in an effect similar to "simulation sound" zero-knowledge.

---

[7] The KO protocol uses a clever combination of garble circuits, semi-honest oblivious transfer, coin-tossing, and WIPOK to ensure that the protocol is executed with a fixed input (allowing at the same time simulation extractability of the input), and relies on the zero-knowledge property of a modified Fiege-Shamir proof to achieve output simulation.

2. The oblivious transfer protocol based on trapdoor permutations and coin-tossing now performs coin-tossing with the help of nmcom instead of simple commitments. This is a crucial change since this allows us to slowly get rid of the honest party's input in the simulation and still argue that the distribution of the adversary's input does not change as a result of this.

   We note that there are many parallel executions on nmcom that take place at this stage, and therefore, we require that nmcom should be non-malleable under many parallel executions. This is indeed true for most nmcom.

3. Finally, we introduce a mechanism to ensure that the two parties use the exact same input in both executions. Roughly speaking, this is done by requiring the parties to prove consistency of messages "across" protocols.

4. To keep the number of rounds to $k + 1$ (or 4 if $k < 3$), many of the messages discussed above are "absorbed" with other rounds by running in parallel.

Multi-party Setting. The above protocol does not directly extend to the multi-party settings. Nevertheless, for the special case of *coin flipping*, we show that a (simplified) version of the above protocol works for the multi-party case. This is because the coin-tossing functionality does not really require any computation, and therefore, we can get rid of components such as oblivious transfer. In fact, this can be extended "slightly more" to also realize the "coin-flipping with committed inputs" since committing the input does not depend on inputs of other parties.

Next, to obtain our result for general functionalities, we simply invoke known results: using [34] with coin-flipping gives us a six round protocol, and using [26] gives a five round result.

2 Preliminaries

Notation. We denote the security parameter by κ. We say that a function $\mu : \mathbb{N} \to \mathbb{N}$ is *negligible* if for every positive polynomial $p(\cdot)$ and all sufficiently large κ's it holds that $\mu(\kappa) < \frac{1}{p(\kappa)}$. We use the abbreviation PPT to denote probabilistic polynomial-time. We often use $[n]$ to denote the set $\{1, ..., n\}$. Moreover, we use $d \leftarrow \mathcal{D}$ to denote the process of sampling d from the distribution $\mathcal{D}$ or, if $\mathcal{D}$ is a set, a uniform choice from it. If $\mathcal{D}_1$ and $\mathcal{D}_2$ are two distributions, then we denote that they are statistically close by $\mathcal{D}_1 \approx_s \mathcal{D}_2$; we denote that they are computationally indistinguishable by $\mathcal{D}_1 \approx_c \mathcal{D}_2$; and we denote that they are identical by $\mathcal{D}_1 \equiv \mathcal{D}_2$. Let V be a random variable corresponding to the distribution $\mathcal{D}$. Sometimes we abuse notation by using V to denote the corresponding distribution $\mathcal{D}$.

We assume familiarity with several standard cryptographic primitives. For notational purposes, we recall here the basic working definitions for some of them. We skip the well-known formal definitions for secure two-party and multi-party computations (see full version for a formal description). It will be sufficient to have notation for the two-party setting. We denote a two party functionality by $F : \{0,1\}^* \times \{0,1\}^* \rightarrow \{0,1\}^* \times \{0,1\}^*$ where $F = (F_1, F_2)$. For every pair of inputs (x, y), the output-pair is a random variable $(F_1(x, y), F_2(x, y))$ ranging over pairs of strings. The first party (with input x) should obtain $F_1(x, y)$ and the second party (with input y) should obtain $F_2(x, y)$. Without loss of generality, we assume that F is deterministic. The security is defined through the ideal/real world paradigm where for adversary $\mathcal{A}$ participating in the real world protocol, there exists an ideal world simulator $\mathcal{S}$ such that for every (x, y), the output of $\mathcal{S}$ is indistinguishable from that of $\mathcal{A}$. See the full version for an extended discussion.

We now recall the definitions for non-malleable commitments as well as some components from the work of Katz-Ostrovsky [26].

2.1   Tag Based Non-malleable Commitments

Let nmcom $= \langle C, R \rangle$ be a k round commitment protocol where C and R represent (randomized) committer and receiver algorithms, respectively. Denote the messages exchanged by $(\mathsf{nm}_1, \ldots, \mathsf{nm}_k)$ where nm_i denotes the message in the i-th round.

For some string $u \in \{0,1\}^\kappa$, tag id $\in \{0,1\}^t$, non-uniform PPT algorithm M with "advice" string $z \in \{0,1\}^*$, and security parameter κ, define (v, view) to be the output of the following experiment: M on input $(1^\kappa, z)$, interacts with C who commits to u with tag id; simultaneously, M interacts with $R(1^\kappa, \widetilde{\mathsf{id}})$ where $\widetilde{\mathsf{id}}$ is arbitrarily chosen by M (M's interaction with C is called the left interaction, and its interaction with R is called the right interaction); M controls the scheduling of messages; the output of the experiment is (v, view) where v denotes the value M commits to R in the right execution unless $\widetilde{\mathsf{id}} = \mathsf{id}$ in which case $v = \bot$, and view denotes the view of M in both interactions.

Definition 1 (Tag based non-malleable commitments). *A commitment scheme* nmcom $= \langle C, R \rangle$ *is said to be non-malleable with respect to commitments if for every non-uniform PPT algorithm M (man-in-the-middle), for every pair of strings $(u_0, u_1) \in \{0,1\}^\kappa \times \{0,1\}^\kappa$, every tag-string id $\in \{0,1\}^t$, every (advice) string $z \in \{0,1\}^*$, the following two distributions are computationally indistinguishable,*

$$(v_0, \mathsf{view}^0) \overset{c}{\approx} (v_1, \mathsf{view}^1).$$

Parallel Non-malleable Commitments. We consider a strengthening of nmcom in which M can receive commitments to m strings on the "left", say $(u_1, \ldots, u_m)$, with tags $(\mathsf{id}_1, \ldots, \mathsf{id}_m)$ and makes m commitments on the "right" with tags $(\widetilde{\mathsf{id}}_1, \ldots, \widetilde{\mathsf{id}}_m)$. We assume that m is a fixed, possibly a-priori bounded, polynomial in the security parameter κ. In the following let $i \in [m], b \in \{0,1\}$: We say that a nmcom is m-bounded parallel non-malleable commitment if for every pair of sequences $\{u_i^b\}$ the random variables $(\{v_i^0\}, \mathsf{view}^0)$ and $(\{v_i^1\}, \mathsf{view}^1)$ are computationally indistinguishable where $\{v_i^b\}$ denote the values committed by M in m sessions on right with tags $\{\widetilde{\mathsf{id}}_i\}$ while receiving parallel commitments to $\{u_i^b\}$ on left with tags $\{\mathsf{id}_i\}$, and view^b denotes M's view.

First Message Binding Property. It will be convenient in the notation to assume that the first message nm_1 of the non-malleable commitment scheme nmcom statistically determines the message being committed. This can be relaxed to only require that the message is fixed before the last round if $k \geq 3$.

2.2   Components of Our Protocol

In this section, we recall some components from the KO protocol [26]. These are mostly standard and recalled here for a better exposition. The only (minor but crucial) change needed in our protocol is to the FLS proof system [13–15] where a *non-malleable* commitment protocol is used by the verifier. For concreteness, let us discuss how to fix these proof systems first.

Modified Feige-Shamir Proof Systems. We use two proof systems: Π_{WIPOK} and Π_{FS}. Protocol Π_{WIPOK} is the 3-round, public-coin, witness-indistinguishable proof-of-knowledge based on the work of Feige et al. [14] for proving graph Hamiltonicity. This proof system proves statements of the form $st_1 \wedge st_2$ where st_1 is fixed at the first round of the protocol, but st_2 is determined only in the last round of the protocol.[8] For concreteness, this proof system is given in the full version.

Protocol Π_{FS} is the 4-round zero-knowledge argument-of-knowledge protocol of Feige and Shamir [15], which allows the prover to prove statement thm, with the modification that the protocol from verifier's side is implemented using nmcom. More specifically,

- Recall that the Feige-Shamir protocol consists of two executions of Π_{WIPOK} in reverse directions. In the first execution, the verifier selects a one-way function f and sets $x_1 = f(w_1)$, $x_2 = f(w_2)$ and proves the knowledge of a witness for $x_1 \vee x_2$. In the second execution, prover proves the knowledge of a witness to the statement thm $\vee$ $(x_1 \vee x_2)$ where thm is the statement to be proven. The rounds of these systems can be somewhat parallelized to obtain a 4-round protocol.
- Our modified system, simply replaces the function f and x_1, x_2 with two executions of nmcom. For convenience, suppose that nmcom has only 3 rounds. Then, our protocol creates the first message of two independent executions of nmcom to strings σ_1, σ_2, denoted by nm_1^1, nm_1^2 respectively, and sets $x_1 = nm_1^1, x_2 = nm_1^2$. The second and third messages of nmcom are sent with the second and third messages of the original FS protocol.

  If nmcom has more than 3 rounds, simply complete the first $k - 3$ rounds of the two executions before the 4 messages of the proof system above are exchanged.
- As before, although Π_{FS} proves statement thm, as noted in [26], it actually proves statements of the form thm $\wedge$ thm' where thm can be fixed in the second round, and thm' in the fourth round. Usually thm is empty and not

---

[8] Typically, st_1 is a empty statement and not usually mentioned; but KO [26] uses a specific, non-empty, statement and so does this work.

mentioned. Indeed, this is compatible with the second Π_{WIPOK} which proves statement of the form $\mathsf{st}_1 \wedge \mathsf{st}_2$, just set $\mathsf{st}_1 = \mathsf{thm}, \mathsf{st}_2 = \mathsf{thm}'$.

For completeness, we describe the full Π_{FS} protocol in the full version.

Components of Katz-Ostrovsky Protocol

The remainder of this section is largely taken from [26] where we provide basic notations and ideas for semi-honest secure two-party computation based on Yao's garbled circuits and semi-honest oblivious transfer (based on trapdoor one-way permutations). Readers familiar with [26] can skip this part without loss in readability.

Semi-honest Secure Two-Party Computation. We view Yao's garbled circuit scheme [32,39] as a tuple of PPT algorithms (GenGC, EvalGC), where GenGC is the "generation procedure" which generates a garbled circuit for a circuit C along with "labels," and EvalGC is the "evaluation procedure" which evaluates the circuit on the "correct" labels. Each individual wire i of the circuit is assigned two labels, namely $Z_{i,0}, Z_{i,1}$. More specifically, the two algorithms have the following format (here $i \in [\kappa], b \in \{0,1\}$):

- $(\{Z_{i,b}\}, \mathsf{GC}_y) \leftarrow \mathsf{GenGC}(1^\kappa, F, y)$: GenGC takes as input a security parameter κ, a circuit F and a string $y \in \{0,1\}^\kappa$. It outputs a *garbled circuit* GC_y along with the set of all *input-wire labels* $\{Z_{i,b}\}$. The *garbled circuit* may be viewed as representing the function $F(\cdot, y)$.
- $v = \mathsf{EvalGC}(\mathsf{GC}_y, \{Z_{i,x_i}\})$: Given a garbled circuit GC_y and a set of input-wire labels $\{Z_{i,x_i}\}$ where $x \in \{0,1\}^\kappa$, EvalGC outputs either an invalid symbol $\perp$, or a value $v = F(x, y)$.

The following properties are required:

Correctness. $\Pr\left[F(x,y) = \mathsf{EvalGC}(\mathsf{GC}_y, \{Z_{i,x_i}\})\right] = 1$ for all F, x, y, taken over the correct generation of $\mathsf{GC}_y, \{Z_{i,b}\}$ by GenGC.
Security. There exists a PPT simulator SimGC such that for any (F, x) and uniformly random labels $\{Z_{i,b}\}$, we have that:

$$(\mathsf{GC}_y, \{Z_{i,x_i}\}) \overset{c}{\approx} \mathsf{SimGC}\left(1^\kappa, F, v\right)$$

where $(\{Z_{i,b}\}, \mathsf{GC}_y) \leftarrow \mathsf{GenGC}(1^\kappa, F, y)$ and $v = F(x, y)$.

In the semi-honest setting, two parties can compute a function F of their inputs, in which only *one* party, say P_1, learns the output, as follows. Let x, y be the inputs of P_1, P_2, respectively. First, P_2 computes $(\{Z_{i,b}\}, \mathsf{GC}_y) \leftarrow \mathsf{GenGC}(1^\kappa, F, y)$ and sends GC_y to P_1. Then, the two parties engage in κ parallel instances of OT. In particular, in the i-th instance, P_1 inputs x_i, P_2 inputs $(Z_{i,0}, Z_{i,1})$ to the OT protocol, and P_1 learns the "output" Z_{i,x_i}. Then, P_1 computes $v = \mathsf{EvalGC}(\mathsf{GC}_y, \{Z_{i,x_i}\})$ and outputs $v = F(x, y)$.
A 3-round, semi-honest, OT protocol can be constructed from enhanced trapdoor permutations (TDP). For notational purposes, define TDP as follows:

Definition 2 (Trapdoor permutations). *Let $\mathcal{F}$ be a triple of PPT algorithms* (Gen, Eval, Invert) *such that if* $\mathrm{Gen}(1^\kappa)$ *outputs a pair* (f, td)*, then* $\mathrm{Eval}(f, \cdot)$ *is a permutation over* $\{0, 1\}^\kappa$ *and* $\mathrm{Invert}(f, \mathrm{td}, \cdot)$ *is its inverse.* $\mathcal{F}$ *is a trapdoor permutation such that for all PPT adversaries A:*

$$\Pr[(f, \mathrm{td}) \leftarrow \mathrm{Gen}(1^\kappa); y \leftarrow \{0, 1\}^\kappa; x \leftarrow A(f, y) : \mathrm{Eval}(f, x) = y] \leq \mu(\kappa).$$

For convenience, we drop (f, td) from the notation, and write $f(\cdot), f^{-1}(\cdot)$ to denote algorithms $\mathrm{Eval}(f, \cdot), \mathrm{Invert}(f, \mathrm{td}, \cdot)$ respectively, when f, td are clear from the context. We assume that $\mathcal{F}$ satisfies (a weak variant of) "certifiability": namely, given some f it is possible to decide in polynomial time whether $\mathrm{Eval}(f, \cdot)$ is a permutation over $\{0, 1\}^\kappa$.

Let H be the *hardcore bit function for κ bits* for the family $\mathcal{F}$; κ hardcore bits are obtained from a single-bit hardcore function h and $f \in \mathcal{F}$ as follows: $\mathsf{H}(z) = h(z) \| h(f(z)) \| \ldots \| h(f^{\kappa-1}(z))$. Informally, $\mathsf{H}(z)$ looks pseudorandom given $f^\kappa(z)$.

The semi-honest OT protocol based on TDP is constructed as follows. Let P_2 hold two strings $Z_0, Z_1 \in \{0, 1\}^\kappa$ and P_1 hold a bit b. In the first round, P_2 chooses trapdoor permutation $(f, f^{-1}) \leftarrow \mathrm{Gen}(1^\kappa)$ and sends f to P_1. Then P_1 chooses two random string $z'_0, z'_1 \leftarrow \{0, 1\}^\kappa$, computes $z_b = f^\kappa(z'_b)$ and $z_{1-b} = z'_{1-b}$ and sends (z_0, z_1) to P_2. In the last round P_2 computes $W_a = Z_a \oplus \mathsf{H}(f^{-\kappa}(z_a))$ where $a \in \{0, 1\}$, H is the hardcore bit function and sends (W_0, W_1) to P_1. Finally, P_2 can recover Z_b by computing $Z_b = W_b \oplus \mathsf{H}(z_b)$.

Putting it altogether, we obtain the following 3-round, semi-honest secure two-party protocol for the single-output functionality F (here only P_1 receives the output):

Protocol $\varPi_{\mathsf{SH}}$. P_1 holds input $x \in \{0, 1\}^\kappa$ and P_2 holds inputs $y \in \{0, 1\}^\kappa$. Let $\mathcal{F}$ be a family of trapdoor permutations and let H be a hardcore bit function. For all $i \in [\kappa]$ and $b \in \{0, 1\}$ the following steps are executed:

Round-1: P_2 computes $(\{Z_{i,b}\}, \mathsf{GC}_y) \leftarrow \mathsf{GenGC}(1^\kappa, F, y)$ and chooses trapdoor permutation $(f_{i,b}, f_{i,b}^{-1}) \leftarrow \mathrm{Gen}(1^\kappa)$ and sends $(\mathsf{GC}_y, \{f_{i,b}\})$ to P_2.

Round-2: P_1 chooses random strings $\{z'_{i,b}\}$, computes $z_{i,b} = f^\kappa(z'_{i,b})$ and $z_{i,1-b} = z'_{i,1-b}$ and sends $\{z_{i,b}\}$ to P_2.

Round-3: P_2 computes $W_{i,b} = Z_{i,b} \oplus \mathsf{H}(f_{i,b}^{-\kappa}(z_{i,b}))$ and sends $\{W_{i,b}\}$ to P_2.

Output: P_1 recovers the labels $Z_{i,x_i} = W_{i,x_i} \oplus \mathsf{H}(z_{i,x_i})$ and computes $v = \mathsf{EvalGC}(\mathsf{GC}_y, \{Z_{i,x_i}\})$ where $v = F(x, y)$

Equivocal Commitment Scheme Eqcom. We assume familiarity with equivocal commitments, and use the following equivocal commitment scheme Eqcom based on any (standard) non-interactive, perfectly binding, commitment scheme com: to commit to a bit x, the sender chooses coins ζ_1, ζ_2 and computes $\mathsf{Eqcom}(x; \zeta_1, \zeta_2) \stackrel{\text{def}}{=} \mathsf{com}(x; \zeta_1) \| \mathsf{com}(x; \zeta_2)$. It sends $\mathsf{C}_x = \mathsf{Eqcom}(x; \zeta_1, \zeta_2)$ to the receiver along with a zero-knowledge proof that C_x was constructed correctly (i.e., that there exist x, ζ_1, ζ_2 such that $\mathsf{C}_x = \mathsf{Eqcom}(x; \zeta_1, \zeta_2)$).

To decommit, the sender chooses a bit b at random and reveals x, ζ_b. Note that a simulator can "equivocate" the commitment by setting $C = \mathsf{com}(x; \zeta_1) \| \mathsf{com}(\overline{x}; \zeta_2)$ for a random bit x, simulating the zero-knowledge proof and then revealing ζ_1 or ζ_2 depending on x and the bit to be revealed. This extends to strings by committing bitwise.

Sketch of the Two-Party KO Protocol. The main component of the two-party KO protocol is Yao's 3-round protocol Π_{SH}, described above, secure against semi-honest adversaries. In order to achieve security against a malicious adversary their protocol proceeds as follows. Both parties commit to their inputs; run (modified) coin-tossing protocols to guarantee that each party obtains random coins which are committed to the other party (note that coin flipping for the side of the garbler P_2 is not needed since a malicious garbler P_2 gains nothing by using non-uniform coins. To force P_1 to use random coins the authors use a 3-round sub-protocol which is based on the work of [4]); and run the Π_{SH} protocol together with ZK arguments to avoid adversarial inconsistencies in each round. Then, simulation extractability is guaranteed by the use of WI proof of knowledge and output simulation by the Feige-Shamir ZK argument of knowledge.

However, since even a ZK argument for the first round of the protocol alone will already require 4 rounds, the authors use specific proof systems to achieve in total a 4-round protocol. In particular, the KO protocol uses a specific WI proof of knowledge system with the property that the statement to be proven need not be known until the last round of the protocol, yet soundness, completeness, and witness-indistinguishability still hold. Also, this proof system has the property that the first message from the prover is computed independently of the statement being proved. Note that their 4-round ZK argument of knowledge enjoys the same properties. Furthermore, their protocol uses an equivocal commitment scheme to commit to the garble circuit for the following reason. Party P_1 may send his round-two message before the proof of correctness for round one given by P_2 is complete. Therefore, the protocol has to be constructed in a way that the proof of correctness for round one completes in round three and that party P_2 reveals the garbled circuit in the third round. But since the proof of security requires P_2 to commit to a garble circuit at the end of the first round, P_2 does so using an equivocal commitment scheme.

3 The Exact Round Complexity of Coin Tossing

In this section we first show that it is impossible to construct two-party (simulatable) coin-flipping for a super-logarithmic number of coins in 3 simultaneous message exchange rounds. We first recall the definition of a simulatable coin flipping protocol using the real/ideal paradigm from [27].

Definition 3 ([27]). *An n-party protocol Π is a simulatable coin-flipping protocol if it is an $(n-1)$-secure protocol realizing the coin-flipping functionality. That is, for every PPT adversary $\mathcal{A}$ corrupting at most $n-1$ parties there exists an expected PPT simulator $\mathcal{S}$ such that the (output of the) following experiments are indistinguishable. Here we parse the result of running protocol Π with adversary $\mathcal{A}$ (denoted by $\mathrm{REAL}_{\Pi,\mathcal{A}}(1^\kappa, 1^\lambda)$) as a pair $(c, \mathsf{view}_\mathcal{A})$ where $c \in \{0,1\}^\lambda \cup \{\perp\}$ is the outcome and $\mathsf{view}_\mathcal{A}$ is the view of the adversary $\mathcal{A}$.*

$\mathrm{REAL}(1^\kappa, 1^\lambda)$	$\mathrm{IDEAL}(1^\kappa, 1^\lambda)$
$c, \mathsf{view}_\mathcal{A} \leftarrow \mathrm{REAL}_{\Pi,\mathcal{A}}(1^\kappa, 1^\lambda)$	$c' \leftarrow \{0,1\}^\lambda$
	$\widetilde{c}, \mathsf{view}_\mathcal{S} \leftarrow \mathcal{S}^\mathcal{A}(c', 1^\kappa, 1^\lambda)$
Output $(c, \mathsf{view}_\mathcal{A})$	If $\widetilde{c} = \{c', \perp\}$ then Output $(\widetilde{c}, \mathsf{view}_\mathcal{S})$
	Else output `fail`

We restrict ourselves to the case of two parties ($n = 2$), which can be extended to any $n > 2$. Below we denote messages in protocol Π which are sent by party P_i to party P_j in the ρ-th round by $\mathsf{m}_{i,j}^{\Pi[\rho]}$.

As mentioned earlier, Katz and Ostrovsky [26] showed that simulatable coin-flipping protocol is impossible in 4 rounds without simultaneous message exchange. Since we will use the result for our proofs in this section, we state their result below without giving their proof.

Lemma 1. *[26, Theorem 1] Let $p(\kappa) = \omega(\log \kappa)$, where κ is the security parameter. Then there does not exist a 4-round protocol without simultaneous message transmission for tossing $p(\kappa)$ coins which can be proven secure via black-box simulation.*

In the following, we state our impossibility result for coin-fliping in 3 rounds of simultaneous message exchange.

Lemma 2. *Let $p(\kappa) = \omega(\log \kappa)$, where κ is the security parameter. Then there does not exist a 3-round protocol with simultaneous message transmission for tossing $p(\kappa)$ coins which can be proven secure via black-box simulation.*

Proof: We prove the above statement by showing that a 3-round simultaneous message exchange protocol can be "rescheduled" to a 4-round *non-simultaneous* protocol which contradicts the impossibility of [26]. Here by rescheduling we mean rearrangement of the messages without violating mutual dependencies among them, in particular without altering the next-message functions.

For the sake of contradiction, assume that there exists a protocol $\Pi_{\text{flip}}^{\leftrightarrow}$ which realizes simulatable coin-flipping in 3 simultaneous message exchange rounds, then we can reschedule it in order to construct a protocol $\Pi_{\text{flip}}^{\rightleftarrows}$ which realizes simulatable coin-flipping in 4 rounds[9] without simultaneous message exchange as follows:

Protocol $\Pi_{\text{flip}}^{\rightleftarrows}$

Round-1: P_1 sends the first message $\text{m}_{1,2}^{\Pi_{\text{flip}}^{\rightleftarrows}[1]} := \text{m}_{1,2}^{\Pi_{\text{flip}}^{\leftrightarrow}[1]}$ to P_2.

Round-2: Party P_2 sends to P_1 the second message
$$\text{m}_{2,1}^{\Pi_{\text{flip}}^{\rightleftarrows}[2]} := (\text{m}_{2,1}^{\Pi_{\text{flip}}^{\leftrightarrow}[1]}, \text{m}_{2,1}^{\Pi_{\text{flip}}^{\leftrightarrow}[2]}).$$

Round-3: Party P_1 sends to P_2 the third message
$$\text{m}_{1,2}^{\Pi_{\text{flip}}^{\rightleftarrows}[3]} := (\text{m}_{1,2}^{\Pi_{\text{flip}}^{\leftrightarrow}[2]}, \text{m}_{1,2}^{\Pi_{\text{flip}}^{\leftrightarrow}[3]}).$$

Round-4: Finally P_2 sends to P_1 the last message
$$\text{m}_{2,1}^{\Pi_{\text{flip}}^{\rightleftarrows}[4]} := \text{m}_{2,1}^{\Pi_{\text{flip}}^{\leftrightarrow}[3]}.$$

We provide a pictorial presentation of the above rescheduling in Fig. 1 for better illustration.

Now, without loss of generality assume that P_1 is corrupted. Then we need to build an expected PPT simulator $\mathcal{S}_{P_1}$ (or simply $\mathcal{S}$) meeting the adequate requirements (according to Definition 1). First note that, since by assumption the protocol $\Pi_{\text{flip}}^{\leftrightarrow}$ is secure (i.e. achieves Definition 1) the following holds: for any corrupt $P_1^{\leftrightarrow}$ executing the simultaneous message exchange protocol $\Pi_{\text{flip}}^{\leftrightarrow}$ there exists an expected PPT simulator $\mathcal{S}^{\leftrightarrow}$ (let us call it the "inner" simulator and $\mathcal{S}$ the "outer" simulator) in the ideal world. So, $\mathcal{S}$ can be constructed using $\mathcal{S}^{\leftrightarrow}$ for a corrupted party $P_1^{\leftrightarrow}$ which can be *emulated* by $\mathcal{S}$ based on P_1. Finally, $\mathcal{S}$ just outputs whatever $\mathcal{S}^{\leftrightarrow}$ returns. $\mathcal{S}$ emulates the interaction between $\mathcal{S}^{\leftrightarrow}$ and $P_1^{\leftrightarrow}$ as follows:

1. On receiving a value $c' \in \{0,1\}^\lambda$ from the ideal functionality, $\mathcal{S}$ runs the inner simulator $\mathcal{S}^{\leftrightarrow}(c', 1^\kappa, 1^\lambda)$ to get the first message $\text{m}_{2,1}^{\Pi_{\text{flip}}^{\leftrightarrow}[1]}$. Notice that in protocol $\Pi_{\text{flip}}^{\leftrightarrow}$ the first message from (honest) party $P_2^{\leftrightarrow}$ does not depend on the first message of the corrupted party $P_1^{\leftrightarrow}$. So, the inner simulator must be able to produce the first message even before seeing the first message of party P_1 (or the emulated party $P_1^{\leftrightarrow}$)[10]. Then it runs P_1 to receive the first message $\text{m}_{1,2}^{\Pi_{\text{flip}}^{\leftrightarrow}[1]}$.

---

[9] The superscript $\leftrightarrow$ stands for the simultaneous message exchange setting and $\rightleftarrows$ for the setting without simultaneous message exchange.

[10] In particular, for so-called "rushing" adversaries, who can wait until receiving the first message and then send its own, the inner simulator must simulate the first message to get the first message from the adversary.

2. Then $\mathcal{S}$ forwards $m_{1,2}^{\Pi_{\text{flip}}^{\leftrightarrow}[1]}$ to the inner simulator which then returns the second simulated message $m_{2,1}^{\Pi_{\text{flip}}^{\leftrightarrow}[2]}$. Now $\mathcal{S}$ can construct the simulated message $m_{2,1}^{\Pi_{\text{flip}}^{\leftarrow}[2]}$ by combining $m_{2,1}^{\Pi_{\text{flip}}^{\leftrightarrow}[2]}$ and $m_{2,1}^{\Pi_{\text{flip}}^{\leftrightarrow}[1]}$ received earlier (see above) which $\mathcal{S}$ then forwards to P_1.

3. In the next step, $\mathcal{S}$ gets back messages $m_{1,2}^{\Pi_{\text{flip}}^{\leftarrow}[3]} = (m_{1,2}^{\Pi_{\text{flip}}^{\leftrightarrow}[2]}, m_{1,2}^{\Pi_{\text{flip}}^{\leftrightarrow}[3]})$ from P_1. It then forwards the second message $m_{1,2}^{\Pi_{\text{flip}}^{\leftrightarrow}[2]}$ to $\mathcal{S}^{\leftrightarrow}$, which then returns the third simulated message $m_{2,1}^{\Pi_{\text{flip}}^{\leftrightarrow}[3]}$. Finally it forwards the third message $m_{1,2}^{\Pi_{\text{flip}}^{\leftrightarrow}[3]}$ to $\mathcal{S}^{\leftrightarrow}$.

4. $\mathcal{S}$ outputs whatever transcript $\mathcal{S}^{\leftrightarrow}$ outputs in the end.

5. Note that, whenever the inner simulator $\mathcal{S}^{\leftrightarrow}$ asks to rewind the emulated $P_1^{\leftrightarrow}$, $\mathcal{S}$ rewinds P_1.

It is not hard to see that the simulator $\mathcal{S}$ emulates correctly the party $P_1^{\leftrightarrow}$ and hence by the security of $\Pi_{\text{flip}}^{\leftrightarrow}$, the inner simulator $\mathcal{S}^{\leftrightarrow}$ returns an indistinguishable (with the real world) view. The key-point is that the re-scheduling of the messages from protocol $\Pi_{\text{flip}}^{\leftrightarrow}$ does not affect the dependency (hence the corresponding next message functions) and hence the correctness and security remains intact in $\Pi_{\text{flip}}^{\leftarrow}$.

We stress that the proof for the case where P_2 is corrupted is straightforward given the above. However, in that case, since P_2's first message depends on the first message of honest P_1, it is mandatory for the inner simulator $\mathcal{S}^{\leftrightarrow}$ to output the first message before seeing anything even in order to run the corrupted P_2 which is not necessary in the above case. As we stated earlier this is possible as the inner simulator $\mathcal{S}^{\leftrightarrow}$ should be able to handle rushing adversaries.

Hence we prove that if the underlying protocol $\Pi_{\text{flip}}^{\leftrightarrow}$ securely realizes simulatable coin-flipping in 3 simultaneous rounds then $\Pi_{\text{flip}}^{\leftarrow}$ securely realizes coin-flipping in 4 non-simultaneous rounds which contradicts the KO lower bound (Lemma 1). This concludes the proof.   □

Going a step further we show that any four-round simultaneous message exchange protocol realizing simulatable coin-flipping must satisfy a necessary property, that is each round must be a strictly simultaneous message exchange round, in other words, both parties must send some "non-redundant" message in each round. By "non-redundant" we mean that the next message from the other party must depend on the current message. Below we show the above, otherwise the messages can be again subject to a "rescheduling" mechanism similar to the one in Lemma 2, to yield a four-round non-simultaneous protocol; thus contradicting Lemma 1. More specifically,

Lemma 3. *Let $p(\kappa) = \omega(\log\kappa)$, where κ is the security parameter. Then there does not exist a 4-round protocol with at least one unidirectional round (i.e. a round without simultaneous message exchange) for tossing $p(\kappa)$ coins which can be proven secure via black-box simulation.*

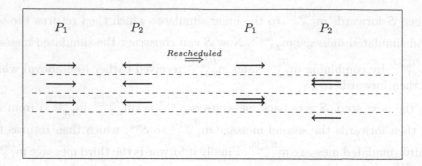

Fig. 1. A 3-round simultaneous protocol rescheduled to a 4-round non-simultaneous protocol.

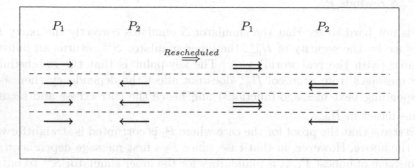

Fig. 2. Rescheduling when P_2 does not send the first message.

Proof: [Proof (Sketch)] We provide a sketch for any protocol with exactly one unidirectional round where only one party, say P_1 sends a message to P_2. Clearly, there can be four such cases where P_2's message is omitted in one of the four rounds. In Fig. 2 we show the case where P_2 does not send the message in the first round, and any such protocol can be re-scheduled (similar to the proof of Lemma 2) to a non-simultaneous 4-round protocol without altering any possible message dependency. This observation can be formalized in a straightforward manner following the proof of Lemma 2 and hence we omit the details. Therefore, again combining with the impossibility from Lemma 1 by [26] such simultaneous protocol can not realize simulatable coin-flipping. The other cases can be easily observed by similar rescheduling trick and therefore we omit the details for those cases.   □

4   Two-Party Computation in the Simultaneous Message Exchange Model

In this section, we present our two party protocol for computing any functionality in the presence of a static, malicious and rushing adversary. As discussed earlier, we are in the simultaneous message exchange channel setting where both parties can simultaneously exchange messages in each round. The structure of this protocol will provide a basis for our later protocols as well.

An overview of the protocol appears in the introduction (Sect. 1). In a high level, the protocol consists of two simultaneous executions of a one-sided (single-output) protocol to guarantee that both parties learn the output. The overall skeleton of the one-sided protocol resembles the KO protocol [26] which uses a clever combination of OT, coin-tossing, and Π_{WIPOK} to ensure that the protocol is executed with a fixed input (allowing at the same time simulation extractability of the input), and relies on the zero-knowledge property of Π_{FS} to "force the output". A sketch of the KO protocol is given in Sect. 2.2. In order to ensure "independence of inputs" our protocol relies heavily on non-malleable commitments. To this end, we change the one-sided protocol to further incorporate non-malleable commitments so that similar guarantees can be obtained even in the presence of the "opposite side" protocol, and we further rely on zero-knowledge proofs to ensure that parties use the same input in both executions.

4.1   Our Protocol

To formally define our protocol, let:

- (GenGC, EvalGC) be the garbled-circuit mechanism with simulator SimGC; $\mathcal{F} = $ (Gen, Eval, Invert) be a family of TDPs with domain $\{0,1\}^\kappa$; H be the hardcore bit function for κ bits; com be a perfectly binding non-interactive commitment scheme; Eqcom be the equivocal scheme based on com, as described in Sect. 2;
- nmcom be a tag based, *parallel*[11] non-malleable commitment scheme for strings, supporting tags/identities of length κ;
- Π_{WIPOK} be the witness-indistinguishable proof-of-knowledge for NP as described in Sect. 2;
- Π_{FS} be the proof system for NP, based on nmcom and Π_{WIPOK}, as described in Sect. 2;
- **Simplifying assumption:** *for notational convenience only,* we assume for now that nmcom consists of exactly three rounds, denoted by $(\mathsf{nm}_1, \mathsf{nm}_2, \mathsf{nm}_3)$. This assumption is removed later (see Remark 1).

We also assume that the first round, nm_1, is from the committer and statistically determines the message to be committed. We use the notation $\mathsf{nm}_1 = \mathsf{nmcom}_1(\mathsf{id}, r; \omega)$ to denote the committer's first message when executing nmcom with identity id to commit to string r with randomness ω.

---

[11] We actually need security against an a-priori bounded number of polynomial executions. Almost all known protocols for nmcom have this additional property.

We are now ready to describe our protocol.

Protocol Π_{2PC}. We denote the two parties by P_1 and P_2; P_1 holds input $x \in \{0,1\}^\kappa$ and P_2 holds input $y \in \{0,1\}^\kappa$. Furthermore, the identities of P_1, P_2 are $\mathsf{id}_1, \mathsf{id}_2$ respectively where $\mathsf{id}_1 \neq \mathsf{id}_2$. Let $F := (F_1, F_2) : \{0,1\}^\kappa \times \{0,1\}^\kappa \to \{0,1\}^\kappa \times \{0,1\}^\kappa$ be the functions to be computed.

The protocol consists of four (strictly) simultaneous message exchange rounds, i.e., both parties send messages in each round. The protocol essentially consists of two simultaneous executions of a protocol in which only one party learns the output. In the first protocol, P_1 learns the output and the messages of this protocol are denoted by (m_1, m_2, m_3, m_4) where (m_1, m_3) are sent by P_1 and (m_2, m_4) are sent by P_2. Likewise, in the second protocol P_2 learns the output and the messages of this protocol are denoted by $(\widetilde{m}_1, \widetilde{m}_2, \widetilde{m}_3, \widetilde{m}_4)$ where $(\widetilde{m}_1, \widetilde{m}_3)$ are sent by P_2 and $(\widetilde{m}_2, \widetilde{m}_4)$ are sent by P_1. Therefore, messages $(m_j, \widetilde{m}_j)$ are exchanged simultaneously in the j-th round, $j \in \{1, \ldots, 4\}$ (see Fig. 3).

We now describe how these messages are constructed in each round below. In the following i always ranges from 1 to κ and b from 0 to 1.

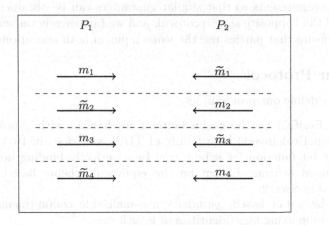

Fig. 3. 2-PC in the simultaneous message exchange model.

Round 1. In this round P_1 sends a message m_1 and P_2 sends a symmetrically constructed message $\widetilde{m}_1$. We first describe how P_1 constructs m_1.

Actions of P_1 :

1. P_1 starts by committing to 2κ random strings $\{(r_{1,0}, r_{1,1}), \ldots, (r_{\kappa,0}, r_{\kappa,1})\}$ using 2κ parallel and independent executions of nmcom with identity id_1. I.e., it uniformly chooses strings $r_{i,b}$, randomness $\omega_{i,b}$, and generates $\mathsf{nm}_1^{i,b}$ which is the first message corresponding to the execution of $\mathsf{nmcom}(\mathsf{id}_1, r_{i,b}; \omega_{i,b})$.

2. P_1 prepares the first message p_1 of Π_{WIPOK}, as well as the first message fs_1 of Π_{FS}.

   For later reference, define st_1 to be the following: $\exists \{(r_i, \omega_i)\}_{i \in [\kappa]}$ s.t.:

$$\forall i : \left(\mathsf{nm}_1^{i,0} = \mathsf{nmcom}_1(\mathsf{id}_1, r_i; \omega_i) \vee \mathsf{nm}_1^{i,1} = \mathsf{nmcom}_1(\mathsf{id}_1, r_i; \omega_i) \right)$$

   Informally, st_1 represents that P_1 "knows" one of the decommitment values for every i.

3. Message m_1 is defined to be the tuple $\left(\{\mathsf{nm}_1^{i,b}\}, \mathsf{p}_1, \mathsf{fs}_1 \right)$.

Actions of P_2 :

   Performs the same actions as P_1 to sample the values $\left\{ (\widetilde{r}_{i,b}, \widetilde{\omega}_{i,b}) \right\}$ and constructs $\widetilde{m}_1 := \left(\{\widetilde{\mathsf{nm}}_1^{i,b}\}, \widetilde{\mathsf{p}}_1, \widetilde{\mathsf{fs}}_1 \right)$ where all $\widetilde{\mathsf{nm}}_1^{i,b}$ are generated with id_2.

   Define the statement $\widetilde{\mathsf{st}}_1$ analogously for these values.

Round 2. In this round P_2 sends a message m_2 and P_1 sends a symmetrically constructed message $\widetilde{m}_2$. We first describe how P_2 constructs m_2.

Actions of P_2 :

1. P_2 generates the second messages $\{\mathsf{nm}_2^{i,b}\}$ corresponding to all executions of nmcom initiated by P_1 (with id_1).

2. P_2 prepares the second message p_2 of the Π_{WIPOK} protocol initiated by P_1.

3. P_2 samples random strings $\{r'_{i,b}\}$ and $(f_{i,b}, f_{i,b}^{-1}) \leftarrow \mathsf{Gen}(1^\kappa)$ for the oblivious transfer executions.

4. P_2 obtains the garbled labels and the circuit for F_1: $(\{Z_{i,b}\}, \ \mathsf{GC}_y) = \mathsf{GenGC}(1^\kappa, F_1, y \ ; \ \Omega)$.

5. P_2 generates standard commitments to the labels, and an equivocal commitment to the garbled circuit: i.e., $\mathsf{C}_{\mathsf{lab}}^{i,b} \leftarrow \mathsf{com}(Z_{i,b}; \omega'_{i,b})$ and $\mathsf{C}_{\mathsf{gc}} \leftarrow \mathsf{Eqcom}(\mathsf{GC}_y; \zeta)$.

6. P_2 prepares the second message fs_2 of the Π_{FS} protocol initiated by P_1. For later reference, define st_2 to be the following: $\exists \left(y, \Omega, \mathsf{GC}_y, \{Z_{i,b}, \omega'_{i,b}\}, \zeta \right)$ s.t.:

   (a) $(\{Z_{i,b}\}, \ \mathsf{GC}_y) = \mathsf{GenGC}(1^\kappa, F_1, y \ ; \ \Omega)$

   (b) $\forall (i, b) : \mathsf{C}_{\mathsf{lab}}^{i,b} = \mathsf{com}(Z_{i,b}; \omega'_{i,b})$

   (c) $\mathsf{C}_{\mathsf{gc}} = \mathsf{Eqcom}(\mathsf{GC}_y; \zeta)$

   (Informally, st_2 is the statement that P_2 performed this step correctly.)

7. Define message $m_2 := \left(\{\mathsf{nm}_2^{i,b}, r'_{i,b}, f_{i,b}, \mathsf{C}_{\mathsf{lab}}^{i,b}\}, \mathsf{C}_{\mathsf{gc}}, \mathsf{p}_2, \mathsf{fs}_2 \right)$.

Actions of P_1 :

   Performs the same actions as P_2 in the previous step to construct the message $\widetilde{m}_2 := \left(\{\widetilde{\mathsf{nm}}_2^{i,b}, \widetilde{r}'_{i,b}, \widetilde{f}_{i,b}, \widetilde{\mathsf{C}}_{\mathsf{lab}}^{i,b}\}, \widetilde{\mathsf{C}}_{\mathsf{gc}}, \widetilde{\mathsf{p}}_2, \widetilde{\mathsf{fs}}_2 \right)$ w.r.t. identity id_2, function F_2, and input x. Define the (remaining) values $\widetilde{f}_{i,b}^{\,-1}, \widetilde{Z}_{i,b}, \widetilde{\omega}'_{i,b}, \mathsf{GC}_x, \widetilde{\Omega}, \widetilde{\zeta}$ and statement $\widetilde{\mathsf{st}}_2$ analogously.

Round 3. In this round P_1 sends a message m_3 and P_2 sends a symmetrically constructed message $\widetilde{m}_3$. We first describe how P_1 constructs m_3.

Actions of P_1 :

1. P_1 prepares the third message $\{\mathsf{nm}_3^{i,b}\}$ of nmcom (with id_1).
2. If any of $\{f_{i,b}\}$ are invalid, P_1 aborts. Otherwise, it invokes κ parallel executions of oblivious transfer to obtain the input-wire labels corresponding to its input x. More specifically, P_1 proceeds as follows:
   - If $x_i = 0$, sample $z'_{i,0} \leftarrow \{0,1\}^\kappa$, set $z_{i,0} = f_{i,0}^\kappa(z'_{i,0})$, and $z_{i,1} = r_{i,1} \oplus r'_{i,1}$.
   - If $x_i = 1$, sample $z'_{i,1} \leftarrow \{0,1\}^\kappa$, set $z_{i,1} = f_{i,1}^\kappa(z'_{i,1})$, and $z_{i,0} = r_{i,0} \oplus r'_{i,0}$.
3. Define st_3 to be the following: $\exists \{(r_i, \omega_i)\}_{i \in [\kappa]}$ s.t. $\forall i$:
   (a) $(\mathsf{nm}_1^{i,0} = \mathsf{nmcom}_1(\mathsf{id}_1, r_i; \omega_i) \land z_{i,0} = r_i \oplus r'_{i,0})$, **or**
   (b) $(\mathsf{nm}_1^{i,1} = \mathsf{nmcom}_1(\mathsf{id}_1, r_i; \omega_i) \land z_{i,1} = r_i \oplus r'_{i,1})$
   Informally, st_3 says that P_1 correctly constructed $\{z_{i,b}\}$.
4. P_1 prepares the final message p_3 of Π_{WIPOK} proving the statement: $\mathsf{st}_1 \land \mathsf{st}_3$.[12] P_1 also prepares the third message fs_3 of Π_{FS}.
5. Define $m_3 := \left(\{\mathsf{nm}_3^{i,b}, z_{i,b}\}, \mathsf{p}_3, \mathsf{fs}_3 \right)$ to P_2.

<u>Actions of P_2:</u>
Performs the same actions as P_1 in the previous step to construct the message $\widetilde{m}_3 := \left(\{\widetilde{\mathsf{nm}}_3^{i,b}, \widetilde{z}_{i,b}\}, \widetilde{\mathsf{p}}_3, \widetilde{\mathsf{fs}}_3 \right)$ w.r.t. identity id_2 and input y. The (remaining) values $\{\widetilde{z}_{i,b}, \widetilde{z}'_{i,b}\}$ and statement $\widetilde{\mathsf{st}}_3$ are defined analogously.

Round 4. In this round P_2 sends a message m_4 and P_1 sends a symmetrically constructed message $\widetilde{m}_4$. We first describe how P_2 constructs m_4.

<u>Actions of P_2:</u>
1. If $\mathsf{p}_3, \mathsf{fs}_3$ are not accepting, P_2 aborts. Otherwise, P_2 completes the execution of the oblivious transfers for every (i, b). I.e., it computes $W_{i,b} = Z_{i,b} \oplus \mathsf{H}(f^{-\kappa}(z_{i,b}))$.
2. Define st_4 to be the following: $\exists\ (y, \Omega, \mathsf{GC}_y, \{Z_{i,b}\}, \omega'_{i,b}, z'_{i,b}, \widetilde{z}'_i)_{i \in [\kappa], b \in \{0,1\}}$ s.t.
   (a) $\forall (i, b)$: $\left(\mathsf{C}_{\mathsf{lab}}^{i,b} = \mathsf{com}(Z_{i,b}; \omega'_{i,b}) \right) \land \left(f_{i,b}^\kappa(z'_{i,b}) = z_{i,b} \right) \land \left(W_{i,b} = Z_{i,b} \oplus \mathsf{H}((z'_{i,b})) \right)$
   (b) $\left((\{Z_{i,b}\}, \mathsf{GC}_y) = \mathsf{GenGC}(1^\kappa, F_1, y\ ;\ \Omega) \right) \land (\mathsf{C}_{\mathsf{gc}} = \mathsf{Eqcom}(\mathsf{GC}_y; \zeta))$
   (c) $\forall i$: $\widetilde{z}_{i, y_i} = \widetilde{f}_{i, y_i}^\kappa(\widetilde{z}'_i)$
   Informally, this means that P_2 performed *both* oblivious transfers correctly.
3. P_2 prepares the final message fs_4 of Π_{FS} proving the statement $\mathsf{st}_2 \land \mathsf{st}_4$.[13]
4. Define $m_4 := \left(\{W_{i,b}\}, \mathsf{fs}_4, \mathsf{GC}_y, \zeta \right)$.

<u>Actions of P_1:</u>
Performs the same actions as P_2 in the previous step to construct the message $\widetilde{m}_4 := \left(\{\widetilde{W}_{i,b}\}, \widetilde{\mathsf{fs}}_4, \mathsf{GC}_x, \widetilde{\zeta} \right)$ and analogously defined statement $\widetilde{\mathsf{st}}_4$.

---

[12] Honest P_1 knows multiple witnesses for st_1. For concreteness, we have to use one of them randomly in the proof.

[13] Recall that Π_{FS} is a modified version of FS protocol: it uses two executions of nmcom to construct its first message, namely, the first message consists of $(\mathsf{nm}_1^1, \mathsf{nm}_1^2)$ corresponding to two executions of nmcom committing to strings σ_1, σ_2 (see Sect. 2).

Output Computation.

P_1's **output:** If any of $(\mathsf{fs}_4, \mathsf{GC}_y, \zeta)$ or the openings of $\{W_{i,b}\}$ are invalid, P_1 aborts. Otherwise, P_1 recovers the garbled labels $\{Z_i := Z_{i,x_i}\}$ from the completion of the oblivious transfer, and computes $F_1(x, y) = \mathsf{EvalGC}(\mathsf{GC}_y, \{Z_i\})$.

P_2's **output:** If any of $(\widetilde{\mathsf{fs}}_4, \mathsf{GC}_x, \widetilde{\zeta})$ or the openings of $\{\widetilde{W}_{i,b}\}$ are invalid, P_2 aborts. Otherwise, P_2 recovers the garbled labels $\{\widetilde{Z}_i := \widetilde{Z}_{i,y_i}\}$ from the completion of the oblivious transfer, and computes $F_2(x, y) = \mathsf{EvalGC}(\mathsf{GC}_x, \{\widetilde{Z}_i\})$.

Remark 1: If nmcom has $k > 3$ rounds, the first $k - 3$ rounds can be performed before the 4 rounds of Π_{2PC} start; this results in a protocol with $k + 1$ rounds. If $k < 3$, then the protocol has only 4 rounds. Also, for large k, it suffices if the first $k - 2$ rounds of nmcom statistically determine the message to be committed; the notation is adjusted to simply use the transcript up to $k - 2$ rounds to define the statements for the proof systems.

Finally, the construction is described for a deterministic F. Known transformations (see [19, Sect. 7.3]) yield a protocol for randomized functionalities, without increasing the rounds.

4.2   Proof of Security

We prove the security of our protocol according to the ideal/real paradigm. We design a sequence of hybrids where we start with the real world execution and gradually modify it until the input of the honest party is not needed. The resulting final hybrid represents the simulator for the ideal world.

Theorem 1. *Assuming the existence of a trapdoor permutation family and a k-round parallel non-malleable commitment schemes, protocol Π_{2PC} securely computes every two-party functionality $F = (F_1, F_2)$ with black-box simulation in the presence of a malicious adversary. The round complexity of Π_{2PC} is $k' = \max(4, k + 1)$.*

Proof: Due to the symmetric nature of our protocol, it is *sufficient* to prove security against the malicious behavior of any party, say P_1. We show that for every adversary $\mathcal{A}$ who participates as P_1 in the "real" world execution of Π_{2PC}, there exists an "ideal" world adversary (simulator) $\mathcal{S}$ such that for all inputs x, y of equal length and security parameter $\kappa \in \mathbb{N}$:

$$\{\mathrm{IDEAL}_{F,\mathcal{S}}(\kappa, x, y)\}_{\kappa, x, y} \overset{c}{\approx} \{\mathrm{REAL}_{\Pi,\mathcal{A}}(\kappa, x, y)\}_{\kappa, x, y}$$

We prove this claim by considering hybrid experiments $H_0, H_1, \dots$ as described below. We start with H_0 which has access to both inputs x and y, and gradually get rid of the honest party's input y to reach the final hybrid.

H_0: Identical to the real execution. More specifically, H_0 starts the execution of $\mathcal{A}$ providing it fresh randomness and input x, and interacts with it honestly by performing all actions of P_2 with uniform randomness and input y. The output consists of $\mathcal{A}$'s view.

By construction, H_0 and the output of $\mathcal{A}$ in the real execution are identically distributed.

H_1: Identical to H_0 except that this hybrid also performs extraction of $\mathcal{A}$'s implicit input x^* from Π_{WIPOK}; in addition, it also extracts the "simulation trapdoor" σ from the first three rounds $(\mathsf{fs}_1, \mathsf{fs}_2, \mathsf{fs}_3)$ of Π_{FS}.[14] More specifically, H_1 proceeds as follows:

1. It completes the first three broadcast rounds exactly as in H_0, and waits until $\mathcal{A}$ either aborts or successfully completes the third round.

2. At this point, H_1 proceeds to extract the witness corresponding to each proof-of-knowledge completed in the first three rounds.

   Specifically, H_1 defines a cheating prover P^* which acts identically to H_0, simulating all messages for $\mathcal{A}$, except those corresponding to (each execution of) Π_{WIPOK} which are forwarded outside. It then applies the extractor of Π_{WIPOK} to obtain the "witnesses" which consists of the following: values $\{(r_i, \omega_i)\}_{i \in [\kappa]}$ which is the witness for $\mathsf{st}_1 \wedge \mathsf{st}_3$, and a value (σ, ω_σ) which is the simulation trapdoor for Π_{FS}.

   If extraction fails, H_1 outputs fail. Otherwise, let $b_i \in \{0,1\}$ be such that $\mathsf{nm}_1^{i,b_i} = \mathsf{nmcom}_1(\mathsf{id}_1, r_i; \omega_i)$. H_1 defines a string $x^* = (x_1^*, \ldots, x_\kappa^*)$ as follows:

$$\text{If } z_{i,b_i} = r_i \oplus r'_{i,b_i} \text{ then } x_i^* = 1 - b_i; \text{ otherwise } x_i^* = b_i$$

3. H_1 completes the final round and prepares the output exactly as H_0.

Claim 1. H_1 *is expected polynomial time, and* H_0, H_1 *are statistically close.*

Proof sketch: This is a (completely) standard proof which we sketch here. Let p be the probability with which $\mathcal{A}$ completes Π_{WIPOK} in the third round, and let trans be the transcript. The extractor for Π_{WIPOK} takes expected time $\mathsf{poly}(\kappa)/p$ and succeeds with probability $1 - \mu(\kappa)$. It follows that the expected running time of H_1 is $\mathsf{poly}(\kappa) + p \cdot \frac{\mathsf{poly}(\kappa)}{p} = \mathsf{poly}(\kappa)$, and its output is statistically close to that of H_0.[15]   ◇

H_2: Identical to H_1 except that this hybrid uses the simulation trapdoor (σ, ω_σ) as the witness to compute fs_4 in the last round. (Recall that fs_4 is the last round of an execution of Π_{WIPOK}).

It is easy to see that H_2 and H_3 are computationally indistinguishable due the WI property of Π_{WIPOK}.

H_3: In this hybrid, we get rid of P_2's input y that is implicitly present in values $\{\tilde{z}_{i,b}\}$ and $\{r_{i,b}\}$ in nmcom (but keep it everywhere else for the time being).

---

[14] Recall that $(\mathsf{fs}_1, \mathsf{fs}_2, \mathsf{fs}_3)$ contains two non-malleable commitments (to values σ_1, σ_2) along with proof-of-knowledge of one of the committed values using Π_{WIPOK}; this execution of Π_{WIPOK} runs in parallel and therefore, it is possible to extract from it at the same time as x^*.

[15] See "witness extended emulation" in [31] for full exposition.

Formally, H_3 is identical to H_2 except that in round 3 it sets $\widetilde{z}_{i,b} = \widetilde{r}_{i,b} \oplus \widetilde{r}'_{i,b}$ for all (i, b).

Claim 2. The *outputs of H_2 and H_3 are computationally indistinguishable.*

Proof. We rely on the non-malleability of nmcom to prove this claim. Let D be a distinguisher for H_2 and H_3.

The high level idea is as follows: first we define two string sequences $\{u^1_{i,b}\}$ and $\{u^2_{i,b}\}$ and a man-in-the-middle M (which incorporates $\mathcal{A}$) and receives non-malleable commitments to one of these sequences in parallel. Then we define a distinguisher D_{nm} which incorporates both M and D, takes as input the value committed by M and its view, and can distinguish which sequence was committed to M. This violates non-malleability of nmcom.

Formally, define a man-in-middle M who receives 2κ nmcom commitments on left and makes 2κ commitments on right as follows:

1. M incorporates $\mathcal{A}$ internally, and proceeds exactly as H_1 by sampling all messages internally except for the messages of nmcom corresponding to P_2. These messages are received from an outside committer as follows. M samples uniformly random values $\{\widetilde{z}_{i,b}\}$ and $\{\widetilde{r}'_{i,b}\}$ and defines $\{u^0_{i,b}\}$ and $\{u^1_{i,b}\}$ as:

$$u^0_{i,y_i} = \widetilde{z}_{i,y_i} \oplus \widetilde{r}'_{i,y_i}, \quad u^0_{i,\overline{y_i}} \leftarrow \{0,1\}^\kappa, \quad u^1_{i,b} = \widetilde{z}_{i,b} \oplus \widetilde{r}'_{i,b} \forall (i, b)$$

   It forwards $\{u^0_{i,b}\}$ and $\{u^1_{i,b}\}$ to the outside committer who commits to one of these sequences in parallel. M forwards these messages to $\mathcal{A}$, and forwards the message given by $\mathcal{A}$ corresponding to nmcom to the outside receiver.

2. After the first three rounds are finished, M halts by outputting its view. In particular, M does not continue further like H_1, it does not extract any values, and does not complete the fourth round. (In fact, M cannot complete the fourth round, since it does not have the witness).

Let $\{v^0_{i,b}\}$ (resp., $\{v^1_{i,b}\}$) be the sequence of values committed by M with id_2 when it receives a commitment to $\{u^0_{i,b}\}$ (resp., $\{u^1_{i,b}\}$) with id_1.

Define the distinguisher D_{nm} as follows: D_{nm} incorporates both M and D. It receives as input a pair $(\{v_{i,b}\}, \mathsf{view})$ and proceeds as follows:

1. D_{nm} parses $v_{i,b}$ to obtain a string σ corresponding to the "trapdoor witness."[16]

2. D_{nm} starts M and feeds him the view view and continues the execution just like H_1. It, however, does not rewind $\mathcal{A}$ (internal to M), instead it uses σ (which is part of its input) and values in view to complete the last round of the protocol.

3. When $\mathcal{A}$ halts, D_{nm} feeds the view of $\mathcal{A}$ to D and outputs whatever D outputs.

It is straightforward to verify that if M receives commitments corresponding to $\{u^0_{i,b}\}$ (resp., $\{u^1_{i,b}\}$) then the output of D_{nm} is identical to that of H_2 (resp., H_3). The claim follows.   ◇

---

[16] Note that, by construction, such a value is guaranteed in both sequences and w.l.o.g. can be the value in the first nmcom.

H_4: Identical to H_3 except that H_4 changes the "inputs of the oblivious transfer" from $(Z_{i,0}, Z_{i,1})$ to $(Z_{i,x_i^*}, Z_{i,x_i^*})$. Formally, in the last round, H_4 sets $W_{i,b} = Z_{i,x_i^*} \oplus H((z'_{i,b}))$ for every (i, b), but does everything else as H_3.

H_3 and H_4 are computationally indistinguishable due to the (indistinguishable) security of oblivious transfer w.r.t. a malicious receiver. This part is identical to the proof in [26], and relies on the fact that one of the two strings for oblivious transfer are obtained by "coin tossing;" and therefore its inverse is hidden, which implies that the hardcore bits look pseudorandom.

H_5: Identical to H_4 except that now we simulate the garbled circuit and its labels for values x^* and $F_1(x^*, y)$. Formally, H_5 starts by proceeding exactly as H_4 up to round 3 except that instead of committing to correct garbled circuit and labels in round 2, it simply commits to random values. After completing round 3, H_5 extracts x^* exactly as in H_4. If extraction succeeds, it sends x^* to the trusted party, receives back $v_1 = F_1(x^*, y)$, and computes $(\{Z_{i,b}\}, GC_*) \leftarrow SimGC(1^\kappa, F_1, x^*, v_1)$. It uses labels $\{Z_{i,x_i^*}\}$ to define the values $\{W_{i,b}\}$ as in H_3, and equivocates C_{gc} to obtain openings corresponding to the simulated circuit GC^*. It then computes fs_4 as before (by using the trapdoor witness (σ, ω_σ)), and constructs $m_4 := (\{W_{i,b}\}, fs_4, GC^*, \zeta)$. It feeds m_4 to $\mathcal{A}$ and finally outputs $\mathcal{A}$'s view and halts.

We claim that H_4 and H_5 are computationally indistinguishable. First observe that the joint distribution of values $(\{C_{lab}^{i,b}\}, C_{gc})$ and GC_y (along with real openings) in H_4 is indistinguishable from the joint distribution of the values $(\{C_{lab}^{i,b}\}, C_{gc})$ and GC^* (along with equivocal openings) in H_5. The two hybrids are identical except for sampling of these values, and can be simulated perfectly given these values from outside. The claim follows.[17]

Observe that H_5 is now independent of the input y. Our simulator $\mathcal{S}$ is H_5. This completes the proof.   □

5   Multi-party Coin Flipping Protocol

In this section, we show a protocol for the multi-party coin-flipping functionality. Since we need neither OT nor garbled circuits for coin-flipping, this protocol is simpler than the the two-party protocol.

At a high level, the multi-party coin flipping protocol Π_{MCF} simply consists of each party "committing" to a random string r, which is opened in the last round along with a simulatable proof of correct opening given to *all* parties independently. The output consists of the $\oplus$ of all strings. This actually does not work directly as stated, but with a few more components, such as equivocal commitment to r for the proof to go through. In particular, we prove the following theorem.

---

[17] Let us note that changing the commitment in second round (from correct garbled labels/circuit to random strings) is performed from the beginning—i.e., in the "main thread" of simulation—therefore the running time stays expected polynomial time as in claim 1.

Theorem 2. *Assuming the existence of a trapdoor permutation family and a k-round protocol for (parallel) non-malleable commitments, then the multi-party protocol Π_{MCF} securely computing the* multi-party coin-flipping *functionality with black-box simulation in the presence of a malicious adversary for polynomially many coins. The round complexity of Π_{MCF} is $k' = \max(4, k+1)$.*

The multi-party coin flipping protocol Π_{MCF} and its security proof can be found in the full version.

5.1   Coin Flipping with Committed Inputs

We now discuss an extension of the coin-flipping functionality which will be useful in the next section. The extension considers a functionality which, in addition to providing a random string to the parties, also "attests" to a commitment to their input.

More specifically, we consider the following setting. Each party P_i has an input string x_i and randomness ρ_i. Let com be a non-interactive perfectly-binding commitment scheme. The Coin Flipping with Committed Inputs functionality $\mathcal{F}_{\mathsf{CF-CI}}$ acts as follows:

1. Each party sends (x_i, ρ_i, c_i) to the functionality where $c_i = \mathsf{com}(x_i; \rho_i)$.
2. Functionality samples a random string r.
3. Functionality tests that for every i, $c_i = \mathsf{com}(x_i; \rho_i)$. If the test succeeds, it sets $y_i = (r, c_i, \mathtt{true})$; otherwise, $y_i = (r, c_i, \mathtt{false})$.
4. Functionality sends $(y_1, \ldots, y_n)$ to all parties.

We claim that a minor modification of our coin-flipping protocol can actually implement $\mathcal{F}_{\mathsf{CF-CI}}$. More details on the new protocol $\Pi_{\mathsf{CF-CI}}$ which implements $\mathcal{F}_{\mathsf{CF-CI}}$ can be found in the full version.

5.2   Results for General Multi-party Functionalities

We now discuss how to obtain protocols for general, as opposed to coin-flipping, functionalities in the multiparty case.

Mukherjee and Wichs [34] construct a 2-round protocol for general multiparty functionalities under the Learning With Errors (LWE) assumption in the CRS model. Combining their protocol with Π_{MCF} (to obtain the CRS), we obtain a protocol for general functionalities with $k'+2$ rounds under the LWE assumption.

Likewise, Garg et al. [16] also construct a 2-round protocol for the same task in the CRS model, under the assumption that general purpose indistinguishability obfuscation exists. Their protocol actually has a special structure: it can be computed in just one round given access to the $\mathcal{F}_{\mathsf{CF-CI}}$ functionality that we have defined above. Consequently, using their protocol with protocol $\Pi_{\mathsf{CF-CI}}$ actually gives a $k'+1$ round protocol.

We thus get the following theorem.

Theorem 3. *Assuming the existence of a trapdoor permutation family and k-round parallel non-malleable commitment schemes, there exists a protocol for securely computing every multiparty functionality in the presence of a malicious adversary such that: (a) the protocol has $k' + 1$ rounds assuming general purpose indistinguishability obfuscation, and (b) $k' + 2$ rounds assuming the LWE assumption where $k' = \max(4, k + 1)$.*

As a corollary of the above theorem, an instantiation of the above protocols with the nmcom scheme in [36] gives a **five** round protocol (assuming indistinguishability obfuscation), and a **six** round protocol (assuming LWE) for general multiparty functionalities.

We note that we can also use the four round protocol of [24] for nmcom; this will result in one extra round and gives seven rounds under LWE, and six under indistinguishability obfuscation.

References

1. Asharov, G., Jain, A., López-Alt, A., Tromer, E., Vaikuntanathan, V., Wichs, D.: Multiparty computation with low communication, computation and interaction via threshold FHE. In: Pointcheval, D., Johansson, T. (eds.) EUROCRYPT 2012. LNCS, vol. 7237, pp. 483–501. Springer, Heidelberg (2012)
2. Barak, B.: How to go beyond the black-box simulation barrier. In: 42nd Annual Symposium on Foundations of Computer Science, pp. 106–115. IEEE Computer Society Press, October 2001
3. Barak, B.: Constant-round coin-tossing with a man in the middle or realizing the shared random string model. In: 43rd Annual Symposium on Foundations of Computer Science, pp. 345–355. IEEE Computer Society Press, November 2002
4. Barak, B., Lindell, Y.: Strict polynomial-time in simulation and extraction. In: 34th Annual ACM Symposium on Theory of Computing, pp. 484–493. ACM Press, May 2002
5. Beaver, D., Micali, S., Rogaway, P.: The round complexity of secure protocols (extended abstract). In: 22nd Annual ACM Symposium on Theory of Computing, pp. 503–513. ACM Press, May 1990
6. Or Ben, M., Goldwasser, S., Wigderson, A.: Completeness theorems for non-cryptographic fault-tolerant distributed computation (extended abstract). In: 20th Annual ACM Symposium on Theory of Computing, pp. 1–10. ACM Press, May 1988
7. Canetti, R., Lindell, Y., Ostrovsky, R., Sahai, A.: Universally composable two-party and multi-party secure computation. In: 34th Annual ACM Symposium on Theory of Computing, pp. 494–503. ACM Press, May 2002
8. Chaum, D., Crépeau, C., Damgård, I.: Multiparty unconditionally secure protocols (extended abstract). In: 20th Annual ACM Symposium on Theory of Computing, pp. 11–19. ACM Press, May 1988
9. Damgård, I.B., Ishai, Y.: Constant-round multiparty computation using a black-box pseudorandom generator. In: Shoup, V. (ed.) CRYPTO 2005. LNCS, vol. 3621, pp. 378–394. Springer, Heidelberg (2005)
10. Damgård, I.B., Ishai, Y.: Scalable secure multiparty computation. In: Dwork, C. (ed.) CRYPTO 2006. LNCS, vol. 4117, pp. 501–520. Springer, Heidelberg (2006)

11. Damgård, I., Nielsen, J.B., Polychroniadou, A.: On the communication required for unconditionally secure multiplication (2015)
12. Dolev, D., Dwork, C., Naor, M.: Non-malleable cryptography (extended abstract). In: 23rd Annual ACM Symposium on Theory of Computing, pp. 542–552. ACM Press, May 1991
13. Feige, U.: Alternative models for zero knowledge interactive proofs. Ph.D thesis (1990)
14. Feige, U., Lapidot, D., Shamir, A.: Multiple noninteractive zero knowledge proofs under general assumptions. SIAM J. Comput. 29(1), 1–28 (1999)
15. Feige, U., Shamir, A.: Witness indistinguishable and witness hiding protocols. In: 22nd Annual ACM Symposium on Theory of Computing, pp. 416–426. ACM Press, May 1990
16. Garg, S., Gentry, C., Halevi, S., Raykova, M.: Two-round secure MPC from indistinguishability obfuscation. In: Lindell, Y. (ed.) TCC 2014. LNCS, vol. 8349, pp. 74–94. Springer, Heidelberg (2014)
17. Garg, S., Gentry, C., Halevi, S., Raykova, M., Sahai, A., Waters, B.: Candidate indistinguishability obfuscation and functional encryption for all circuits. In: 54th Annual Symposium on Foundations of Computer Science, pp. 40–49. IEEE Computer Society Press, October 2013
18. Goldreich, O.: Draft of a chapter on cryptographic protocols. http://www.wisdom.weizmann.ac.il/oded/foc-vol2.html. Accessed June 2003
19. Goldreich, O.: Foundations of Cryptography: Basic Applications, vol. 2. Cambridge University Press, Cambridge (2004)
20. Goldreich, O., Micali, S., Wigderson, A.: How to play any mental game or a completeness theorem for protocols with honest majority. In: STOC, pp. 218–229 (1987)
21. Goyal, V.: Constant round non-malleable protocols using one way functions. In: Fortnow, L., Vadhan, S.P. (eds) 43rd Annual ACM Symposium on Theory of Computing, pp. 695–704. ACM Press, June 2011
22. Goyal, V., Lee, C.-K., Ostrovsky, R., Visconti, I.: Constructing non-malleable commitments: a black-box approach. In: 53rd Annual Symposium on Foundations of Computer Science, pp. 51–60. IEEE Computer Society Press, October 2012
23. Goyal, V., Pandey, O., Richelson, S.: Textbook non-malleable commitments. Manuscript, November 2015
24. Goyal, V., Richelson, S., Rosen, A., Vald, M.: An algebraic approach to non-malleability. In: 55th Annual Symposium on Foundations of Computer Science, pp. 41–50. IEEE Computer Society Press, October 2014
25. Ishai, Y., Prabhakaran, M., Sahai, A.: Founding cryptography on oblivious transfer – efficiently. In: Wagner, D. (ed.) CRYPTO 2008. LNCS, vol. 5157, pp. 572–591. Springer, Heidelberg (2008)
26. Katz, J., Ostrovsky, R.: Round-optimal secure two-party computation. In: Franklin, M. (ed.) CRYPTO 2004. LNCS, vol. 3152, pp. 335–354. Springer, Heidelberg (2004)
27. Katz, J., Ostrovsky, R., Smith, A.: Round efficiency of multi-party computation with a dishonest majority. In: Advances in Cryptology - EUROCRYPT, International Conference on the Theory and Applications of Cryptographic Techniques, Warsaw, Poland, Proceedings, pp. 578–595, 4–8 May 2003
28. Lin, H., Pass, R.: Constant-round non-malleable commitments from any one-way function. In: Fortnow, L., Vadhan, S.P. (eds.) 43rd Annual ACM Symposium on Theory of Computing, pp. 705–714. ACM Press, June 2011

29. Lin, H., Pass, R., Tseng, W.-L.D., Venkitasubramaniam, M.: Concurrent non-malleable zero knowledge proofs. In: Rabin, T. (ed.) CRYPTO 2010. LNCS, vol. 6223, pp. 429–446. Springer, Heidelberg (2010)

30. Lin, H., Pass, R., Venkitasubramaniam, M.: A. unified framework for concurrent security: universal composability from stand-alone non-malleability. In: Mitzenmacher, M. (ed.) 41st Annual ACM Symposium on Theory of Computing, pp. 179–188. ACM Press, May/June 2009

31. Lindell, Y.: Parallel coin-tossing and constant-round secure two-party computation. In: Kilian, J. (ed.) CRYPTO 2001. LNCS, vol. 2139, pp. 171–189. Springer, Heidelberg (2001)

32. Lindell, Y., Pinkas, B.: A proof of security of Yao's protocol for two-party computation. J. Cryptol. 22(2), 161–188 (2009)

33. Lindell, Y., Pinkas, B.: Secure two-party computation via cut-and-choose oblivious transfer. In: Ishai, Y. (ed.) TCC 2011. LNCS, vol. 6597, pp. 329–346. Springer, Heidelberg (2011)

34. Mukherjee, P., Wichs, D.: Two round MPC from LWE via multi-key FHE. IACR Cryptology ePrint Archive 2015:345 (2015)

35. Ostrovsky, R., Richelson, S., Scafuro, A.: Round-optimal black-box two-party computation. In: Gennaro, R., Robshaw, M. (eds.) CRYPTO 2015. LNCS, vol. 9216, pp. 339–358. Springer, Heidelberg (2015)

36. Pandey, O., Pass, R., Vaikuntanathan, V.: Adaptive one-way functions and applications. In: Wagner, D. (ed.) CRYPTO 2008. LNCS, vol. 5157, pp. 57–74. Springer, Heidelberg (2008)

37. Regev, O.: On lattices, learning with errors, random linear codes, and cryptography. In: Gabow, H.N., Fagin, R. (eds.) 37th Annual ACM Symposium on Theory of Computing, pp. 84–93. ACM Press, May 2005

38. Wee, H.: Black-box, round-efficient secure computation via non-malleability amplification. In: 51st Annual Symposium on Foundations of Computer Science, pp. 531–540. IEEE Computer Society Press, October 2010

39. Yao, A.C.-C.: Protocols for secure computations (extended abstract). In: 23rd Annual Symposium on Foundations of Computer Science, pp. 160–164. IEEE Computer Society Press, November 1982

40. Yao, A.C.-C.: How to generate and exchange secrets (extended abstract). In: FOCS, pp. 162–167 (1986)

On the Composition of Two-Prover Commitments, and Applications to Multi-round Relativistic Commitments

Serge Fehr$^{(\boxtimes)}$ and Max Fillinger$^{(\boxtimes)}$

Centrum Wiskunde and Informatica (CWI), Amsterdam, The Netherlands
{serge.fehr,max.fillinger}@cwi.nl

Abstract. We consider the related notions of *two-prover* and of *relativistic* commitment schemes. In recent work, Lunghi *et al.* proposed a new relativistic commitment scheme with a *multi-round sustain phase* that keeps the binding property alive as long as the sustain phase is running. They prove security of their scheme against classical attacks; however, the proven bound on the error parameter is very weak: it blows up *double exponentially* in the number of rounds.

In this work, we give a new analysis of the multi-round scheme of Lunghi *et al.*, and we show a *linear* growth of the error parameter instead (also considering classical attacks only). Our analysis is based on a new *composition theorem* for two-prover commitment schemes. The proof of our composition theorem is based on a better understanding of the binding property of two-prover commitments that we provide in the form of new definitions and relations among them. As an additional consequence of these new insights, our analysis is actually with respect to a strictly *stronger* notion of security than considered by Lunghi *et al.*

1 Introduction

TWO-PROVER COMMITMENT SCHEMES. We consider the notion of *2-prover commitment schemes*, as originally introduced by Ben-Or, Goldwasser, Kilian and Wigderson in their seminal paper [2]. In a 2-prover commitment scheme, the prover (i.e., the entity that is responsible for preparing and opening the commitment) consists of two agents, P and Q, and it is assumed that these two agents cannot communicate with each other during the execution of the protocol. With this approach, the classical and quantum impossibility results [9,11] for unconditionally secure commitment schemes can be circumvented.

A simple 2-prover bit commitment scheme is the scheme proposed by Crépeau *et al.* [5], which works as follows. The verifier V chooses a uniformly random

---

M. Fillinger—Supported by the *NWO Free Competition* grant 617.001.203.

© International Association for Cryptologic Research 2016
M. Fischlin and J.-S. Coron (Eds.): EUROCRYPT 2016, Part II, LNCS 9666, pp. 477–496, 2016.
DOI: 10.1007/978-3-662-49896-5_17

$a \in \{0,1\}^n$ and sends it to P, who replies with $x := y + a \cdot b \in \{0,1\}^n$, where b is the bit to commit to, and $y \in \{0,1\}^n$ is a uniformly random string known (only) to P and Q. Furthermore, "+" is bit-wise XOR, and "·" is scalar multiplication (of the scalar b with the vector a). To open the commitment (to b), Q sends y to V, and V checks if $x + y = a \cdot b$. This scheme is clearly hiding: the commitment $x = y + a \cdot b$ is uniformly random and independent of a no matter what b is. On the other hand, the binding property follows from the observation that in order to open the commitment to $b = 0$, Q needs to announce $y = x$, and in order to open to $b = 1$, he needs to announce $y = x + a$. Thus, in order to open to *both*, he must know x *and* $x + a$, and thus a, which is a contradiction to the no-communication assumption, because a was sent to P only.

RELATIVISTIC COMMITMENT SCHEMES. The idea of *relativistic commitment schemes*, as introduced by Kent [7], is to take a 2-prover commitment scheme as above and enforce the no-communication assumption by means of relativistic effects: place P and Q spatially far apart, and execute the scheme fast enough, so that there is not enough time for them to communicate. The obvious downside of such a relativistic commitment scheme is that the binding property stays alive only for a very short time: the opening has to take place almost immediately after the committing, before the provers have the chance to exchange information. This limitation can be circumvented by considering *multi-round* schemes, where after the actual commit phase there is a *sustain phase*, during which the provers and the verifier keep exchanging messages, and as long as this sustain phase is running, the commitment stays binding (and hiding), until the commitment is finally opened. Such schemes were proposed in [7,8], but they are rather inefficient, and the security analyses are informal (e.g., with no formal security definitions) and of asymptotic nature.

More recently, Lunghi *et al.* [10] proposed a new and simple multi-round relativistic commitment scheme, and provided a rigorous security analysis. Their scheme works as follows (see also Fig. 1). The actual commit protocol is the commit protocol from the Crépeau *et al.* scheme: V sends a uniformly random string $a_0 \in \{0,1\}^n$ to P, who returns $x_0 := y_0 + a_0 \cdot b$. Then, to sustain the commitment, before P has the chance to tell a_0 to Q, V sends a new uniformly random string $a_1 \in \{0,1\}^n$ to Q who replies with $x_1 := y_1 + a_1 \cdot y_0$, where $y_1 \in \{0,1\}^n$ is another random string shared between P and Q, and the multiplication $a_1 \cdot y_0$ is in a suitable finite field. Then, to further sustain the commitment, V sends a new uniformly random string $a_2 \in \{0,1\}^n$ to P who replies with $x_2 := y_2 + a_2 \cdot y_1$, etc. Finally, after the last sustain round where $x_m := y_m + a_m \cdot y_{m-1}$ has been sent to V, in order to finally open the commitment, y_m is sent to V by the other prover. In order to verify the opening, V computes $y_{m-1}, y_{m-2}, \ldots, y_0$ inductively in the obvious way, and checks if $x_0 + y_0 = a_0 \cdot b$.

What is crucial is that in round i (say for odd i), when preparing x_i, the prover Q must not know a_{i-1}, but he is allowed to know $a_1, \ldots, a_{i-2}$. Thus, execution must be timed in such a way that between subsequent rounds there is not enough time for the provers to communicate, but they may communicate over multiple rounds.

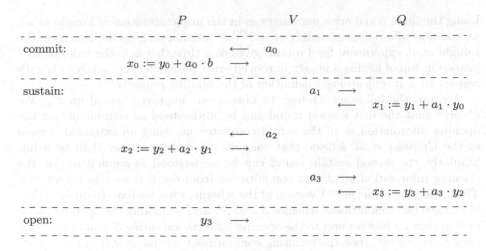

Fig. 1. The Lunghi *et al.* multi-round scheme (for $m = 3$).

As for the security of this scheme, it is obvious that the hiding property stays satisfied up to the open phase: every single message V receives is one-time-pad encrypted. As for the binding property, Lunghi *et al.* prove that the scheme with a m-round sustain phase is ε_m-binding against classical attacks, where ε_m satisfies $\varepsilon_0 = 2^{-n}$ (this is just the standard Crépeau *et al.* scheme) and $\varepsilon_m \leq 2^{-n-1} + \sqrt{\varepsilon_{m-1}}$ for $m \geq 1$. Thus, even when reading this recursive formula liberally by ignoring the 2^{-n-1} term, we obtain

$$\varepsilon_m \lesssim \sqrt[2^m]{\varepsilon_0} = 2^{-\frac{n}{2^m}},$$

i.e., the error parameter blows up *double exponentially* in m.[1] In other words, in order to have a non-trivial ε_m we need that n, the size of the strings that are communicated, is *exponential* in m. This means that Lunghi *et al.* can only afford a very small number of rounds. For instance, in their implementation where they can manage $n = 512$ (beyond that, the local computation takes too long), asking for an error parameter ε_m of approximately 2^{-32}, they can do $m = 4$ rounds.[2] This allows them to keep a commitment alive for 2 ms.

OUR RESULTS. Our main goal is to improve the bound on the binding parameter of the above multi-round scheme. Indeed, our results show that the binding parameter blows up only *linearly* in m, rather than double exponentially. Explicitly, our results show that (for classical attacks)

$$\varepsilon_m \leq (m + 1) \cdot 2^{-\frac{n}{2}+2}.$$

---

[1] Lunghi *et al.* also provide a more complicated recursive formula for ε_m that is slightly better, but the resulting blow-up is still double exponential.

[2] Note that [10] mentions $\varepsilon_m \approx 10^{-5} \approx 2^{-16}$, but this is an error, as communicated to us by the authors, and as can easily be verified. Also, [10] mentions $m = 5$ rounds, but this is because they include the commit round in their counting, and we do not.

Using the same n and error parameter as in the implementation of Lunghi *et al.*, we can now afford approximately $m = 2^{222}$ rounds. Scaling up the 2ms from the Lunghi *et al.* experiment for 4 rounds gives us a time that is in the order of 10^{56} years. On top of having a hugely improved error parameter, our analysis is with respect to a *strictly stronger* definition of the binding property.

We use the following strategy to obtain our improved bound on ε_m. We observe that the first sustain round can be understood as committing on the opening information y_0 of the actual commitment, using an extended version of the Crépeau *et al.* scheme that commits to a *string* rather than to a bit. Similarly, the second sustain round can be understood as committing on the opening information y_1 of that commitment from the first sustain round, etc. Thus, thinking of the $m = 1$ version of the scheme, what we have to prove is that if we have two commitment schemes $\mathcal{S}$ and $\mathcal{S}'$, and we modify the opening phase of $\mathcal{S}$ in that we first commit to the opening information (using $\mathcal{S}'$) and then open that commitment, then the resulting commitment scheme is still binding; note that, intuitively, this is what one would indeed expect. Given such a composition theorem, we can then apply it inductively and conclude security (i.e. the binding property) of the Lunghi *et al.* multi-round scheme.

Our main result is such a general composition theorem, which shows that if $\mathcal{S}$ and $\mathcal{S}'$ are respectively ε- and δ-binding (against classical attacks) then the composed scheme is $(\varepsilon + \delta)$-binding (against classical attacks), under some mild assumptions on $\mathcal{S}$ and $\mathcal{S}'$. Hence, the error parameters simply add up; this is what gives us the linear growth. The proof of our composition theorem crucially relies on a new definition of the binding property of 2-prover commitment schemes, which seems to be handier to work with than the $p_0 + p_1 \leq 1 + \varepsilon$ definition as for instance used by Lunghi *et al.* Our definition formalizes the intuitive requirement that after the commit phase, no matter how the provers behaved, there should exist a bit $\hat{b}$ (or a *string* in case of a string commitment scheme) such that opening the commitment to $b \neq \hat{b}$ fails (with high probability). This new definition is *strictly stronger* than the $p_0 + p_1$ definition, and thus we improve the Lunghi *et al.* result also in that direction.

One subtle issue is that the extended version of the Crépeau *et al.* scheme to strings, as it is used in the sustain phase, is not a fully secure string commitment scheme. The reason is that for *any* y that may be announced in the opening phase, there exists a string s such that $x + y = a \cdot s$; as such, the provers can commit to some fixed string, and then can still decide to either open the commitment to that string (by running the opening phase honestly), or to open it to a random string that is out of their control (by announcing a random y). We deal with this by also introducing a *relaxed* version of the binding property (which we call *fairly-binding*), which captures this limited freedom for the provers, and we show that it is satisfied by the (extended version of the) Crépeau *et al.* scheme and that our composition theorem holds for this relaxed version; finally, we observe that the composed fairly-binding string commitment scheme is a binding *bit* commitment scheme when restricting the domain to a bit.

As such, we feel that our techniques and insights not only give rise to an improved analysis of the Lunghi *et al.* multi-round scheme, but they significantly improve our understanding of the security of 2-prover commitment schemes, and as such are likely to find further applications.

OPEN PROBLEMS. Our work gives rise to a list of interesting and challenging open problems. For instance, our composition theorem only applies to pairs $\mathcal{S}, \mathcal{S}'$ of commitment schemes of a certain restricted form, e.g., only one prover should be involved in the commit phase (as it is the case in the Crépeau *et al.* scheme). Our proof crucially relies on this, but there seems to be no fundamental reason for such a restriction. Thus, we wonder if it is possible to generalize our composition theorem to a larger class of pairs of schemes, or, ultimately, to *all* pairs of schemes (that "fit together").

Also, generalizing our composition theorem to the quantum setting is an interesting open problem. This seems particularly non-trivial because our definition for the binding property does not generalize (immediately) to the quantum setting. Furthermore, in order to obtain security of the Lunghi *et al.* multi-round scheme against quantum attacks, beyond a quantum version of the composition theorem, one also needs to prove security (of the string-commitment version) of the Crépeau *et al.* scheme with respect to a suitable definition of the binding property against quantum attacks.

CONCURRENT WORK. In independent and concurrent work, Chakraborty et al. [3] showed (almost) the same linear bound for the Lunghi *et al.* scheme, but with respect to the original — and thus weaker — notion of security. Their approach is more direct and tailored to the specific scheme; our approach is more abstract and provides more insight, and our result applies much more generally.

2   Preliminaries

2.1   Basic Notation

PROBABILITY DISTRIBUTIONS. For the purpose of this work, a *(probability) distribution* is a function $p : \mathcal{X} \to [0,1]$, $x \mapsto p(x)$, where $\mathcal{X}$ is a finite non-empty set, with the property that $\sum_{x \in \mathcal{X}} p(x) = 1$. For specific choices $x_\circ \in \mathcal{X}$, we tend to write $p(x = x_\circ)$ instead of $p(x_\circ)$. For any subset $\Lambda \subset \mathcal{X}$, called an *event*, the probability $p(\Lambda)$ is naturally defined as $p(\Lambda) = \sum_{x \in \Lambda} p(x)$, and it holds that

$$p(\Lambda) + p(\Gamma) = p(\Lambda \cup \Gamma) + p(\Lambda \cap \Gamma) \le 1 + p(\Lambda \cap \Gamma) \tag{1}$$

for all $\Lambda, \Gamma \subset \mathcal{X}$. For a distribution $p : \mathcal{X} \times \mathcal{Y} \to \mathbb{R}$ on two (or more) variables, probabilities like $p(x = y)$, $p(x = f(y))$, $p(x \ne y)$ etc. are naturally understood as

$$p(x = y) = p(\{(x,y) \in \mathcal{X} \times \mathcal{Y} \mid x = y\}) = \sum_{\substack{x \in \mathcal{X}, y \in \mathcal{Y} \\ \text{s.t. } x = y}} p(x,y)$$

etc., and the *marginals* $p(x)$ and $p(y)$ are given by $p(x) = \sum_y p(x,y)$ and by $p(y) = \sum_x p(x,y)$, respectively. Finally, given that $p(y) > 0$, we write $p(x|y)$ for the *conditional distribution* $p(x|y) := p(x,y)/p(y)$.

PROTOCOLS. In this work, we will consider 3-party (interactive) *protocols*, where the parties are named P, Q and V (the two "provers" and the "verifier"). Such a protocol prot_{PQV} consists of a triple $(\mathsf{prot}_P, \mathsf{prot}_Q, \mathsf{prot}_V)$ of L-round *interactive algorithms* for some $L \in \mathbb{N}$. Each interactive algorithm takes an input, and for every round $\ell \leq L$ computes the messages to be sent to the other algorithms/parties in that round as deterministic functions of its input, the messages received in the previous rounds, and the local randomness. In the same way, the algorithms produce their respective outputs after the last round. We write

$$(out_P \| out_Q \| out_V) \leftarrow (\mathsf{prot}_P(in_P) \| \mathsf{prot}_Q(in_Q) \| \mathsf{prot}_V(in_V))$$

to denote the execution of the protocol prot_{PQV} on the respective inputs in_P, in_Q and in_V, and that the respective outputs out_P, out_Q and out_V are produced. Clearly, for any protocol prot_{PQV} and any input in_P, in_Q, in_V, the probability distribution $p(out_P, out_Q, out_V)$ of the output is naturally well defined.

If we want to make the local randomness explicit, we write $\mathsf{prot}_P[\xi_P](in_P)$ etc., and understand that ξ_P is correctly sampled. We write $\mathsf{prot}_P[\xi_{PQ}](in_P)$ and $\mathsf{prot}_Q[\xi_{PQ}](in_Q)$ to express that prot_P and prot_Q use *the same* randomness, in which case we speak of *joint randomness*.

We can *compose* two interactive algorithms prot_P and prot_P' in the obvious way, by applying prot_P' to the output of prot_P. The resulting interactive algorithm is denoted as $\mathsf{prot}_P' \circ \mathsf{prot}_P$. Composing the respective algorithms of two protocols $\mathsf{prot}_{PQV} = (\mathsf{prot}_P, \mathsf{prot}_Q, \mathsf{prot}_V)$ and $\mathsf{prot}_{PQV}' = (\mathsf{prot}_P', \mathsf{prot}_Q', \mathsf{prot}_V')$ results in the composed protocol $\mathsf{prot}_{PQV}' \circ \mathsf{prot}_{PQV}$.

2.2   2-Prover Commitment Schemes

We formally introduce the notion of 2-prover commitment schemes and discuss the security properties. Defining the binding property is non-trivial; this will be further discussed in Sect. 3.

Definition 2.1. *A 2-prover (string) commitment scheme $\mathcal{S}$ consists of two interactive protocols, the commit protocol $\mathsf{com}_{PQV} = (\mathsf{com}_P, \mathsf{com}_Q, \mathsf{com}_V)$ and the opening protocol $\mathsf{open}_{PQV} = (\mathsf{open}_P, \mathsf{open}_Q, \mathsf{open}_V)$ between the two provers P and Q and the verifier V, with the following syntactics. The commit protocol com_{PQV} uses joint randomness ξ_{PQ} for P and Q and takes a string $s \in \{0,1\}^n$ as input for P and Q (and independent randomness and no input for V), and it outputs a commitment c to V and some state information to P and Q:*

$$(state_P \| state_Q \| c) \leftarrow (\mathsf{com}_P[\xi_{PQ}](s) \| \mathsf{com}_Q[\xi_{PQ}](s) \| \mathsf{com}_V).$$

The opening protocol open_{PQV} uses joint randomness η_{PQ} for P and Q, and outputs a string or a rejection symbol to V, and nothing to P and Q:

$$(\emptyset \| \emptyset \| s) \leftarrow (\mathsf{open}_P[\eta_{PQ}](state_P) \| \mathsf{open}_Q[\eta_{PQ}](state_Q) \| \mathsf{open}_V(c))$$

with $s \in \{0,1\}^n \cup \{\bot\}$. The set $\{0,1\}^n$ is called the domain of $\mathcal{S}$; if $n = 1$ then we refer to $\mathcal{S}$ as a bit commitment scheme instead, and we tend to use b rather than s to denote the committed bit.

Remark 2.2. By convention, we assume throughout the paper that the commitment c output by V equals the *communication* that takes place between V and the provers during the commit phase. This is without loss of generality since, in general, c is computed as a (possibly randomized) function of the communication, which V just as well can apply in the opening phase.

Remark 2.3. Note that we specify that P and Q use *fresh* joint randomness η_{PQ} in the opening phase, and, if necessary, the randomness ξ_{PQ} from the commit phase can be "handed over" to the opening phase via $state_P$ and $state_Q$; this will be convenient later on. Alternatively, one could declare that P and Q *re-use* the joint randomness from the commit phase.

Whenever we refer to such a 2-prover commitment scheme, we take it as understood that the scheme is complete and hiding, as defined below, for "small" values of η and δ. Since our focus will be on the binding property, we typically do not make the parameters η and δ explicit.

Definition 2.4. *A 2-prover commitment scheme is η-complete if in an honest execution V's output s of* open_{PQV} *equals P and Q's input s to* com_{PQV} *except with probability η, for any choice of P and Q's input $s \in \{0,1\}^n$.*

The standard definition for the hiding property is as follows:

Definition 2.5. *A 2-prover commitment scheme is δ-hiding if for any commit strategy $\overline{\mathrm{com}}_V$ and any two strings s_0 and s_1, the respective distributions of the commitments c_0 and c_1, produced as*

$$(state_P \| state_Q \| c_b) \leftarrow (\mathrm{com}_P[\xi_{PQ}](s_b) \| \mathrm{com}_Q[\xi_{PQ}c](s_b) \| \overline{\mathrm{com}}_V)$$

for $b \in \{0,1\}$, have statistical distance at most δ. A 0-hiding scheme is also called perfectly hiding.

Defining the binding property is more subtle. First, note that an *attack* against the binding property consists of an "allowed" commit strategy $\overline{\mathrm{com}}_{PQ} = (\overline{\mathrm{com}}_P, \overline{\mathrm{com}}_Q)$ and an "allowed" opening strategy $\overline{\mathrm{open}}_{PQ} = (\overline{\mathrm{open}}_P, \overline{\mathrm{open}}_Q)$ for P and Q. Any such attack fixes $p(s)$, the distribution of $s \in \{0,1\}^n \cup \{\bot\}$ that is output by V after the opening phase, in the obvious way.

What exactly "allowed" means may depend on the scheme and needs to be specified. Typically, in the 2-prover setting, we only allow strategies $\overline{\mathrm{com}}_{PQ}$ and $\overline{\mathrm{open}}_{PQ}$ with *no communication* at all between the two provers during the course of the scheme, but we may also be more liberal and allow some *well-controlled* communication, as in the Lunghi et al. multi-round scheme. Furthermore, in this work, we focus on *classical* attacks, where $\overline{\mathrm{com}}_P, \overline{\mathrm{com}}_Q, \overline{\mathrm{open}}_P$ and $\overline{\mathrm{open}}_Q$ are classical interactive algorithms as specified in the previous section, with access to joint randomness, but one could also consider *quantum* attacks, where the provers can perform measurements on an entangled quantum state.

A somewhat accepted definition for the binding property of a 2-prover *bit* commitment scheme, as it is for instance used in [5,6] or [10] (up to the factor 2 in the error parameter), is as follows. Here, we assume it has been specified which attacks are *allowed*, e.g., those where P and Q do not communicate during the course of the scheme.

Definition 2.6. *A 2-prover bit commitment scheme is ε-binding in the sense of $p_0 + p_1 \leq 1 + 2\varepsilon$ if for every allowed commit strategy $\overline{\mathrm{com}}_{PQ}$, and for every pair of allowed opening strategies $\overline{\mathrm{open}}^0_{PQ}$ and $\overline{\mathrm{open}}^1_{PQ}$, which fix distributions $p(b_0)$ and $p(b_1)$ for V's respective outputs, it holds that*

$$p(b_0 = 0) + p(b_1 = 1) \leq 1 + 2\varepsilon.$$

In the literature (see e.g. [5] or [10]), the two probabilities $p(b_0 = 0)$ and $p(b_1 = 1)$ above are usually referred to as p_0 and p_1, respectively.

2.3 The $\mathcal{CHSH}^n$ Scheme

Our main example is the bit commitment scheme by Crépeau *et al.* [5] we mentioned in the introduction, and which works as follows. The commit phase com_{PQV} instructs V to sample and send to P a uniformly random $a \in \{0,1\}^n$, and it instructs P to return $x := r + a \cdot b$ to V, where r is the joint randomness, uniformly distributed in $\{0,1\}^n$, b is the bit to commit to, and the opening phase open_{PQV} instructs Q to send $y := r$ to V, and V outputs the (smaller) bit b that satisfies $x + y = a \cdot b$, or $b := \bot$ in case no such bit exists.

It is easy to see that this scheme is 2^{-n}-complete and perfectly hiding (completeness fails in case $a = 0$). For *classical* provers that do not communicate during the course of the scheme, the scheme is 2^{-n-1}-binding in the sense of $p_0 + p_1 \leq 1 + 2^{-n}$, i.e. according to Definition 2.6. As for *quantum* provers, Crépeau *et al.* showed that the scheme is $2^{-n/2}$-binding; this was recently minorly improved to $2^{-(n+1)/2}$ by Sikora et al. [12].

We also want to consider an extended version of the scheme, where the bit b is replaced by a string $s \in \{0,1\}^n$ in the obvious way (where the multiplication $a \cdot s$ is then understood in a suitable finite field), and we want to appreciate this version as a 2-prover *string* commitment scheme. However, it is a priori not clear what is a suitable definition for the binding property, especially because for this particular scheme, the dishonest provers can always honestly commit to a string s, and can then decide to correctly open the commitment to s by announcing $y := r$, or open to a *random* string by announcing a randomly chosen y — any y satisfies $x + y = a \cdot s$ for *some* s (unless $a = 0$, which almost never happens).[3]

Due to its close relation to the CHSH game [4], in particular to the arbitrary-finite-field version considered in [1], we will refer to this *string* commitment scheme as $\mathcal{CHSH}^n$.

---

[3] This could easily be prevented by asking Q to also announce s (rather than letting V compute it), but we want the information announced during the opening phase to fit into the domain of the commitment scheme.

3   On the Binding Property of 2-Prover Commitments

We introduce a new definition for the binding property of 2-prover commitment schemes. In the case of *bit* commitment schemes, it implies Definition 2.6, as we will show. Our new definition is not only stronger, but we also feel that it is closer to the intuition of what is expected from a commitment scheme, and as such it is easier to work with. Indeed, the proof of our composition result is heavily based on our new definition. Also, our new notion is more flexible in terms of tweaking it; for instance, we modify it to obtain a *relaxed* notion for the binding property, which captures the binding property that is satisfied by the string commitment scheme $\mathcal{CHSH}^n$.

Throughout this section, when quantifying over attacks against (the binding property of) a scheme, it is always understood that there is a notion of *allowed* attacks for that scheme (e.g., all attacks for which P and Q do not communicate), and that the quantification is over all such allowed attacks.

3.1   Defining the Binding Property

Intuitively, we say that a scheme is binding if after the commit phase there exists a string $\hat{s}$ so that no matter what the provers do in the opening phase, the verifier will output either $s = \hat{s}$ or $s = \bot$ (except with small probability). Formally, we require that for every possible commit strategy, such a string $\hat{s}$ is uniquely determined by the commitment c and the provers' joint randomness.

Definition 3.1 (Binding property). *A 2-prover commitment scheme S is ε-binding if for every commit strategy $\overline{\mathrm{com}}_{PQ}[\bar{\xi}_{PQ}]$ there exists a function $\hat{s}(\bar{\xi}_{PQ}, c)$ of the joint randomness $\bar{\xi}_{PQ}$ and the commitment c such that for every opening strategy $\overline{\mathrm{open}}_{PQ}$ it holds that $p(s \neq \hat{s}(\bar{\xi}_{PQ}, c) \land s \neq \bot) \leq \varepsilon$. In short:*

$$\forall \overline{\mathrm{com}}_{PQ} \ \exists \hat{s}(\bar{\xi}_{PQ}, c) \ \forall \overline{\mathrm{open}}_{PQ} : p(s \neq \hat{s} \land s \neq \bot) \leq \varepsilon. \tag{2}$$

The string commitment scheme $\mathcal{CHSH}^n$ does *not* satisfy this definition (the bit commitment version does, as we will show): after the commit phase, the provers can still decide to open the commitment to a *fixed* string, chosen before the commit phase, or to a *random* string that is out of their control. We capture this by the following relaxed version of the binding property. In this relaxed version, we allow V's output s to be different from $\hat{s}$ and $\bot$, but in this case the provers should have little control over s: for any *target string* s_o (computed as a function of the provers' randomness), it should be unlikely that $s = s_o$. Formally, this is captured as follows; we will show in Sect. 3.3 that $\mathcal{CHSH}^n$ is fairly-binding in this sense.

Definition 3.2 (Fairly binding property). *A 2-prover commitment scheme S is ε-fairly-binding if for every commit strategy $\overline{\mathrm{com}}_{PQ}[\bar{\xi}_{PQ}]$ there exists a function $\hat{s}(\bar{\xi}_{PQ}, c)$ such that for every opening strategy $\overline{\mathrm{open}}_{PQ}[\bar{\eta}_{PQ}]$ and all functions $s_o(\bar{\xi}_{PQ}, \bar{\eta}_{PQ})$ it holds that $p(s \neq \hat{s}(\bar{\xi}_{PQ}, c) \land s = s_o(\bar{\xi}_{PQ}, \bar{\eta}_{PQ})) \leq \varepsilon$. In short:*

$$\forall \overline{\mathrm{com}}_{PQ} \ \exists \hat{s}(\bar{\xi}_{PQ}, c) \ \forall \overline{\mathrm{open}}_{PQ} \ \forall s_o(\bar{\xi}_{PQ}, \bar{\eta}_{PQ}) : p(s \neq \hat{s} \land s = s_o) \leq \varepsilon. \tag{3}$$

Remark 3.3. By means of standard techniques, one can easily show that it is sufficient for the (fairly) binding property to consider *deterministic* provers. In this case, $\hat{s}$ is a function of c only, and, in the case of fairly-binding, s_o runs over all *fixed* strings.

Remark 3.4. Clearly, the ordinary binding property (i.e., as in Definition 3.1) implies the fairly-binding property. Also, in the case of *bit* commitment schemes it obviously holds that $p(b \neq \hat{b} \wedge b \neq \bot) = p(b \neq \hat{b} \wedge b = 0) + p(b \neq \hat{b} \wedge b = 1)$, and thus the fairly-binding property implies the ordinary one, up to a factor-2 loss. Furthermore, every fairly-binding *string* commitment scheme gives rise to an ordinary-binding *bit* commitment scheme in a natural way, as shown by the following proposition.

Proposition 3.5. *Let S be an ε-fairly-binding string commitment scheme. Fix any two distinct strings $s_0, s_1 \in \{0,1\}^n$ and consider the bit-commitment scheme S' obtained as follows. To commit to $b \in \{0,1\}$, the provers commit to s_b using S, and in the opening phase V checks if $s = s_b$ for some $b \in \{0,1\}$ and outputs this bit if it exists and else outputs $b = \bot$. Then, S' is 2ε-binding.*

Proof. Fix some commit strategy $\overline{com}_{PQ}$ for S' and note that it can also be used to attack S. Thus, there exists a function $\hat{s}(\bar{\xi}_{PQ}, c)$ as in Definition 3.2. We define

$$\hat{b}(\bar{\xi}_{PQ}, c) = \begin{cases} 0 & \text{if } \hat{s}(\bar{\xi}_{PQ}, c) = s_0 \\ 1 & \text{otherwise} \end{cases}$$

Now fix an opening strategy $\overline{open}_{PQ}$ for S', which again is also a strategy against S. Thus, we have $p(\hat{s} \neq s = s_o) \leq \varepsilon$ for any s_o (and in particular $s_o = s_0$ or s_1). This gives us

$$
\begin{aligned}
p(\hat{b} \neq b \neq \bot) &= p(\hat{b} = 1 \wedge b = 0) + p(\hat{b} = 0 \wedge b = 1) \\
&= p(\hat{s} \neq s_0 \wedge s = s_0) + p(\hat{s} = s_0 \wedge s = s_1) \\
&\leq p(\hat{s} \neq s_0 \wedge s = s_0) + p(\hat{s} \neq s_1 \wedge s = s_1), \\
&\leq 2\varepsilon
\end{aligned}
$$

and thus S' is a 2ε-binding bit-commitment scheme. □

Remark 3.6. The proof of Proposition 3.5 generalizes in a straightforward way to k-bit string commitment schemes: given an ε-fairly-binding n-bit string commitment scheme S, for $k < n$, we define a k-bit string commitment scheme S_k as follows: to commit to a k-bit string, the provers pad the string with $n - k$ zeros and then commit to the padded string using S. In the opening phase, the verifier outputs the first k bits of s if the remaining bits in s are all zeros, and $\bot$ otherwise. Then, S' is $2^k \varepsilon$-binding.

3.2   Relation to the Standard Definition

For bit commitment schemes, our binding property implies the $(p_0 + p_1)$-definition.

Theorem 3.7. *A 2-prover bit-commitment scheme that is ε-binding (in the sense of Definition 3.1) is ε-binding in the sense of $p_0 + p_1 \leq 1 + 2\varepsilon$.*

Proof. Consider a scheme that is ε-binding. Fix $\overline{\text{com}}_{PQ}$ and let $\hat{b}(\bar{\xi}_{PQ}, c)$ be a function as promised by Definition 3.1, i.e., such that for every opening strategy $\overline{\text{open}}_{PQ}$ we have $p(b \neq \hat{b} \wedge b \neq \bot) \leq \varepsilon$. Now, fix two opening strategies $\overline{\text{open}}^0_{PQ}$ and $\overline{\text{open}}^1_{PQ}$, and consider the two respective output bits b_0 and b_1. It holds that $p(\hat{b} \neq b_i \neq \bot) \leq \varepsilon$ for $i \in \{0, 1\}$, and thus

$$
\begin{aligned}
p(b_0 = 0) + p(b_1 = 1) &= p(b_0 = 0 \wedge \hat{b} = 0) + p(b_0 = 0 \wedge \hat{b} = 1) \\
&\quad + p(b_1 = 1 \wedge \hat{b} = 0) + p(b_1 = 1 \wedge \hat{b} = 1) \\
&\leq p(\hat{b} = 0) + p(\hat{b} \neq b_0 \neq \bot) + p(\hat{b} \neq b_1 \neq \bot) + p(\hat{b} = 1) \\
&\leq 1 + 2\varepsilon
\end{aligned}
$$

which proves our claim.   □

On the other hand, our Definition 3.1 is *strictly* stronger than the $p_0 + p_1$ based Definition 2.6. Consider the following (artificial and very non-complete) scheme: in the commit phase, V chooses a uniformly random bit and sends it to the provers, and then accepts everything or rejects everything during the opening phase, depending on that bit. Then, $p_0 + p_1 = 1$, yet a commitment can be opened to $1 - \hat{b}$ (no matter how $\hat{b}$ is defined) with probability $\frac{1}{2}$.

Since a non-complete separation example may not be fully satisfying, we note that it can be converted into a complete (but even more artificial) scheme. Fix a "good" (i.e., complete, hiding and binding with low parameters) scheme and call our example scheme above the "bad" scheme. We define a *combined* scheme as follows: at the start, the first prover can request either the "good" or "bad" scheme to be used. The honest prover is instructed to choose the former, guaranteeing completeness. The dishonest prover may choose the latter, so the combined scheme inherits the binding properties of the "bad" scheme: it is binding according to the $(p_0 + p_1)$-definition, but not according to Definition 3.1.

3.3   Security of $\mathcal{CHSH}^n$

In this section, we show that $\mathcal{CHSH}^n$ is a fairly-binding string commitment scheme.[4] To this end, we introduce yet another version of the binding property and show that $\mathcal{CHSH}^n$ satisfies this property. Then we show that this version of the binding property implies the fairly-binding property (up to some loss in the parameter, and under some mild restriction on the scheme).

This new binding property is based on the intuition that it should not be possible to open a commitment to two different values *simultaneously* (except with small probability). For this, we observe that, when considering a commit strategy $\overline{\text{com}}_{PQ}$, as well as *two* opening strategies $\overline{\text{open}}_{PQ}$ and $\overline{\text{open}}'_{PQ}$, we can

---

[4] It is understood that the allowed attacks against $\mathcal{CHSH}^n$ are those where the provers do not communicate during the course of the scheme.

run both opening strategies *simultaneously* on the produced commitment with two (independent) copies of open_V, by applying $\overline{\text{open}}_{PQ}$ and $\overline{\text{open}}'_{PQ}$ to two copies of the respective internal states of P and Q. This gives rise to a *joint* distribution $p(s, s')$ of the respective outputs s and s' of the two copies of open_V.

Definition 3.8 (Simultaneous opening). *A 2-prover commitment scheme S is ε-fairly-binding in the sense of simultaneous opening[5] if for all $\overline{\text{com}}_{PQ}$, all pairs of opening strategies $\overline{\text{open}}_{PQ}$ and $\overline{\text{open}}'_{PQ}$, and all pairs $s_\circ, s'_\circ$ of distinct strings, we have $p(s = s_\circ \wedge s' = s'_\circ) \leq \varepsilon$.*

Remark 3.9. Also for this notion of fairly-binding, it is sufficient to consider *deterministic* strategies, as can easily be seen.

Proposition 3.10. *The commitment scheme $CHSH^n$ is 2^{-n}-fairly-binding in the sense of simultaneous opening.*

Proof. By Remark 3.9, it suffices to consider deterministic attack strategies. Fix a deterministic strategy $\overline{\text{com}}_{PQ}$ and two deterministic opening strategies $\overline{\text{open}}_{PQ}$ and $\overline{\text{open}}'_{PQ}$. The strategy $\overline{\text{com}}_{PQ}$ specifies P's output x as a function $f(a)$ of the verifier's message a. The opening strategies are described by constants y and y'. By definition of $CHSH^n$, $s = s_\circ$ implies $f(a) + y = a \cdot s_\circ$ and likewise, $s' = s'_\circ$ implies $f(a) + y' = a \cdot s'_\circ$. Therefore, $s = s_\circ \wedge s' = s'_\circ$ implies $a = (y - y')/(s_\circ - s'_\circ)$. It thus holds that $p(s = s_\circ \wedge s' = s'_\circ) \leq p(a = (y - y')/(s_\circ - s'_\circ)) \leq \frac{1}{2^n}$, which proves our claim. $\square$

Remark 3.11. It follows directly from (1) that every *bit* commitment scheme that is ε-fairly-binding in the sense of simultaneous opening is ε-binding in the sense of $p_0 + p_1 \leq 1 + 2\varepsilon$. The converse is not true though: the schemes described at the end of Sect. 3.2 again serve as counterexamples.

Theorem 3.12. *Let $S = (\text{com}_{PQV}, \text{open}_{PQV})$ be a 2-prover commitment scheme. If S is ε-fairly-binding in the sense of simultaneous opening and open_V is deterministic, then S is $2\sqrt{\varepsilon}$-fairly-binding.*

Proof. By Remark 3.3, it suffices to consider deterministic strategies for the provers. We fix some deterministic commit strategy $\overline{\text{com}}_{PQ}$ and an enumeration $\{\overline{\text{open}}^i_{PQ}\}_{i=1}^N$ of all deterministic opening strategies. Since we assume that open_V is deterministic, for any fixed opening strategy for the provers, the verifier's output s is a *function* of the commitment c. Thus, for each opening strategy $\overline{\text{open}}^i_{PQ}$ there is a function f_i such that the verifier's output is $s = f_i(c)$. We will now define the function $\hat{s}(c)$ that satisfies the properties required by the fairly-binding property. Our definition depends on a parameter $\alpha > 0$ which we fix later. In order to define $\hat{s}$, we partition the set C of all possible commitments

---

[5] We use "fairly" here to distinguish the notion from a possible "non-fairly" version with $p(\bot \neq s \neq s' \neq \bot) \leq \varepsilon$; however, we do not consider this latter version any further here.

into *disjoint* sets $C = R \cup \bigcup_{s,i} C_{s,i}$ that satisfy the following three properties for every i and every s:

$$C_{s,i} \subseteq f_i^{-1}(\{s\}), \quad p(c \in C_{s,i}) \geq \alpha \text{ or } C_{s,i} = \emptyset, \quad \text{and} \quad p(c \in R \wedge f_i(c) = s) < \alpha.$$

The second property implies that there are at most α^{-1} non-empty sets $C_{s,i}$. It is easy to see that such a partitioning exists: start with $R = C$ and while there exist s and i with $p(c \in R \wedge f_i(c) = s) \geq \alpha$, let $C_{s,i} = \{c \in R \mid f_i(c) = s\}$ and remove the elements of $C_{s,i}$ from R. For any $c \in C$, we now define $\hat{s}(c)$ as follows. We set $\hat{s}(c) = s$ for $c \in C_{s,i}$ and $\hat{s}(c) = 0$ for $c \in R$.

Now fix some opening strategy $\overline{\mathsf{open}}^i_{PQ}$ and a string $s_\circ$, and write s_i for the verifier's output. Using $C_{\neq s_\circ}$ as a shorthand for $\bigcup_{s \neq s_\circ} \bigcup_j C_{s,j}$, we note that if $\hat{s}(c) \neq s_\circ$ then $c \in R \cup C_{\neq s_\circ}$. Thus, it follows that

$$p(s_i \neq \hat{s}(c) \wedge s_i = s_\circ) = p(\hat{s}(c) \neq s_\circ \wedge s_i = s_\circ)$$

$$\leq p\big(c \in (R \cup C_{\neq s_\circ}) \wedge f_i(c) = s_\circ\big)$$

$$= p(c \in R \wedge f_i(c) = s_\circ) + \sum_{s \neq s_\circ, j} p(c \in C_{s,j} \wedge f_i(c) = s_\circ)$$

$$\leq p(c \in R \wedge f_i(c) = s_\circ) + \sum_{\substack{s \neq s_\circ, j \\ \text{s.t. } C_{s,j} \neq \emptyset}} p(f_j(c) = s \wedge f_i(c) = s_\circ)$$

$$< \alpha + \alpha^{-1} \cdot \varepsilon$$

where the final inequality holds because $p(c \in R \wedge f_i(c) = s_\circ) < \alpha$ by the choice of R, because $p(f_j(c) = s \wedge f_i(c) = s_\circ) \leq \varepsilon$ by the assumed binding property, and because the number of non-empty $C_{s,j}$'s is at most $1/\alpha$. It is easy to see that the upper bound $\alpha + \alpha^{-1} \cdot \varepsilon$ is minimized by setting $\alpha = \sqrt{\varepsilon}$. We conclude that $p(s_i \neq \hat{s}(c) \wedge s_i = s_\circ) < 2\sqrt{\varepsilon}$. $\qquad\square$

By combining Theorem 3.7 with Theorem 3.12, we obtain the following statement for the (fairly-)binding property of $\mathcal{CHSH}^n$.

Corollary 3.13. $\mathcal{CHSH}^n$ *is* $2^{-\frac{n}{2}+1}$*-fairly-binding.*

4 Composing Commitment Schemes

4.1 The Composition Operation

We consider two 2-prover commitment schemes $\mathcal{S}$ and $\mathcal{S}'$ of a restricted form, and we compose them to a new 2-prover commitment scheme $\mathcal{S}'' = \mathcal{S} \star \mathcal{S}'$ in a well-defined way; our composition theorem then shows that $\mathcal{S}''$ is secure if $\mathcal{S}$ and $\mathcal{S}'$ are. We start by specifying the restriction to $\mathcal{S}$ and $\mathcal{S}'$ that we impose.

Definition 4.1. *Let $\mathcal{S}$ and $\mathcal{S}'$ be two 2-prover string commitment schemes. We call the pair $(\mathcal{S}, \mathcal{S}')$ eligible if the following three properties hold, or they hold with the roles of P and Q exchanged.*

1. *The commit phase of S is a protocol $\text{com}_{PV} = (\text{com}_P, \text{com}_V)$ between P and V only, and the opening phase of S is a protocol $\text{open}_{QV} = (\text{open}_Q, \text{open}_V)$ between Q and V only. In other words, com_Q and open_P are both trivial and do nothing.[6] Similarly, the commit phase of S' is a protocol com'_{QV} between Q and V only (but both provers may be active in the opening phase).*
2. *The opening phase open_{QV} of S is of the following simple form: Q sends a bit string $y \in \{0,1\}^m$ to V, and V computes s deterministically as $s = \text{Extr}(y, c)$, where c is the commitment.[7]*
3. *The domain of S' contains (or equals) $\{0,1\}^m$.*

Furthermore, we specify that the allowed attacks on S are so that P and Q do not communicate during the course of the entire scheme, and the allowed attacks on S' are so that P and Q do not communicate during the course of the commit phase but there may be limited communication during the opening phase.

An example of an eligible pair of 2-prover commitments is ($\mathcal{CHSH}^n$, $\mathcal{XCHSH}^n$), where $\mathcal{XCHSH}^n$ coincides with scheme $\mathcal{CHSH}^n$ except that the roles of P and Q are exchanged.

Remark 4.2. For an eligible pair (S, S'), it will be convenient to understand open_Q and open_V as *non-interactive* algorithms, where open_Q produces y as its *output*, and open_V takes y as additional *input* (rather than viewing the pair as a protocol with a single one-way communication round).

We now define the composition operation. Informally, committing is done by means of committing using S, and to open the commitment, Q uses open_Q to locally compute the opening information y and he commits to y with respect to the scheme S', and then this commitment is opened (to y), and V computes and outputs $s = \text{Extr}(y, c)$. Formally, this is captured as follows (see also Fig. 2).

Definition 4.3. *Let $S = (\text{com}_{PV}, \text{open}_{QV})$ and $S' = (\text{com}'_{QV}, \text{open}'_{PQV})$ be an eligible pair of 2-prover commitment schemes. Then, their composition $S \star S'$ is defined as the scheme consisting of $\text{com}_{PV} = (\text{com}_P, \text{com}_V)$ and*

$$\text{open}''_{PQV} = (\text{open}'_P, \; \text{open}'_Q \circ \text{com}'_Q \circ \text{open}_Q, \; \text{open}_V \circ \text{open}'_V \circ \text{com}'_V).$$

If in this composition the output in open'_V is $y = \bot$, we define the output of open_V to be $s = \bot$ as well.

When considering attacks against the binding property of the composed scheme $S \star S'$, we declare that the allowed deterministic attacks[8] are those of the form $(\overline{\text{com}}_P, \overline{\text{open}}'_{PQ} \circ \text{ptoq}_{PQ} \circ \overline{\text{com}}'_Q)$, where $\overline{\text{com}}_P$ is an allowed deterministic commit strategy for S, $\overline{\text{com}}'_Q$ and $\overline{\text{open}}'_{PQ}$ are allowed deterministic commit

---

[6] Except that com_Q may output the shared randomness in order to hand it over to the opening protocol open_Q.

[7] Our composition theorem also works for a randomized Extr, but for simplicity, we restrict to the deterministic case.

[8] The allowed *randomized* attacks are then naturally given as those that pick one of the deterministic attacks according to some distribution.

and opening strategies for S', and ptoq_{PQ} is the one-way communication protocol that communicates P's input to Q (see also Fig. 3).[9]

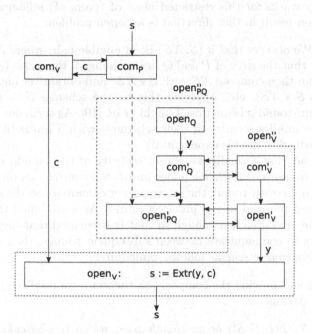

Fig. 2. The composition of $S = (\mathsf{com}_{PV}, \mathsf{open}_{QV})$ and $S' = (\mathsf{com}'_{QV}, \mathsf{open}'_{PQV})$. The dotted arrows indicate communication allowed to the dishonest provers.

Remark 4.4. It is immediate that $S \star S'$ is a commitment scheme in the sense of Definition 2.1, and that it is complete if S and S' are, with the error parameters adding up. Also, the hiding property is obviously inherited from S; however, the point of the composition is to keep the hiding property alive for longer, namely up to before the last round of the opening phase — recall that, using the terminology used in context of relativistic commitments, these rounds of the opening phase up to before the last would then be referred to as the *sustain phase*. We show in Appendix A that $S \star S'$ is hiding up to before the last round, with the error parameters adding up.

It is intuitively clear that $S \star S'$ *should be* binding if S and S' are: committing to the opening information y and then opening the commitment allows the provers to *delay* the announcement of y (which is the whole point of the exercise), but it does not allow them to *change* y, by the binding property of S'; thus, $S \star S'$ should be (almost) as binding as S. This intuition is confirmed by our composition theorem below.

---

[9] This one-way communication models that in the relativistic setting, sufficient time has passed at this point for P to inform Q about what happened during com_P.

Remark 4.5. We point out that the composition $S \star S'$ can be naturally defined for a *larger* class of pairs of schemes (e.g. where *both* provers are active in the commit phase of both schemes), and the above intuition still holds. However, our proof only works for this restricted class of (pairs of) schemes. Extending the composition result in that direction is an open problem.

Remark 4.6. We observe that if $(S, \mathcal{X}S)$ is an eligible pair, where $\mathcal{X}S$ coincides with S except that the roles of P and Q are exchanged, then so is $(\mathcal{X}S, S \star \mathcal{X}S)$. As such, we can then compose $\mathcal{X}S$ with $S \star \mathcal{X}S$, and obtain yet another eligible pair $(S, \mathcal{X}S \star S \star \mathcal{X}S)$, etc. Applying this to the schemes $S = \mathcal{CHSH}^n$, we obtain the multi-round scheme from Lunghi *et al.* [10]. As such, our composition theorem below implies security of their scheme — with a *linear* blow-up of the error term (instead of double exponential).

We point out that formally we obtain security of the Lunghi *et al.* scheme as a *2-prover commitment scheme* under an *abstract restriction* on the provers' communication: in every round, the active prover cannot access the message that the other prover received in the previous round. As such, when the rounds of the protocol are executed fast enough so that it is ensured that there is no time for the provers to communicate between subsequent rounds, then security as a *relativistic commitment scheme* follows immediately.

Before stating and proving the composition theorem, we need to single out one more relevant parameter.

Definition 4.7. *Let (S, S') be an eligible pair, which in particular means that V's action in the opening phase of S is determined by a function* Extr. *We define* $k(S) := \max_{c,s} |\{y \mid \mathrm{Extr}(y, c) = s\}|.$

i.e., $k(S)$ counts the number of ys that are consistent with a given string s (in the worst case). Note that $k(\mathcal{CHSH}^n) = 1$: for every $a, x, s \in \{0,1\}^n$ there is exactly one $y \in \{0,1\}^n$ such that $x + y = a \cdot s$.

4.2 The Composition Theorem

In the following composition theorem, we take it as understood that the assumed respective binding properties of S and S' hold with respect to a well-defined respective classes of allowed attacks.

Theorem 4.8. *Let (S, S') be an eligible pair of 2-prover commitment schemes, and assume that S and S' are respectively ε-fairly-binding and δ-fairly-binding. Then, their composition $S'' = S \star S'$ is $(\varepsilon + k(S) \cdot \delta)$-fairly-binding.*

Proof. We first consider the case $k(S) = 1$. We fix an attack $(\overline{\mathrm{com}}_P, \overline{\mathrm{open}}''_{PQ})$ against S''. Without loss of generality, the attack is deterministic, so $\overline{\mathrm{open}}''_{PQ}$ is of the form $\overline{\mathrm{open}}''_{PQ} = \overline{\mathrm{open}}'_{PQ} \circ \mathrm{ptoq}_{PQ} \circ \overline{\mathrm{com}}'_Q$.

Note that $\overline{\mathrm{com}}_P$ is also a commit strategy for S. As such, by the fairly-binding property of S, there exists a function $\hat{s}(c)$, only depending on $\overline{\mathrm{com}}_P$, so that the property specified in Definition 3.2 is satisfied for every opening strategy $\overline{\text{open}}_Q$ for $\mathcal{S}$. We will show that it is also satisfied for the (arbitrary) opening strategy $\overline{\text{open}}''_{PQ}$ for $\mathcal{S}''$, except for a small increase in ε: we will show that $p(\hat{s}(c) \neq s \wedge s = s_0) \leq \varepsilon + \delta$ for every fixed target string s_0. This then proves the claim.

In order to show this property on $\hat{s}(c)$, we "decompose and reassemble" the attack strategy $(\overline{\text{com}}_P, \overline{\text{open}}'_{PQ} \circ \text{ptoq}_{PQ} \circ \overline{\text{com}}'_Q)$ for $\mathcal{S}''$ into an attack strategy $(\overline{\text{com}}'_Q, \overline{\text{newopen}}'_{PQ})$ for $\mathcal{S}'$ with $\overline{\text{newopen}}'_{PQ}$ formally defined as

$$\overline{\text{newopen}}'_{PQ}[c](\overline{state}'_Q) := \overline{\text{open}}'_{PQ}\big(\overline{state}_P(c) \| (\overline{state}_P(c), \overline{state}'_Q)\big)$$

where

$$(\overline{state}_P(c) \| c) \leftarrow (\overline{\text{com}}_P \| \text{com}_V).$$

Informally, this means that ahead of time, P and Q *simulate* an execution of $(\overline{\text{com}}_P \| \text{com}_V)$ and take the resulting communication/commitment[10] c as shared randomness, and then $\overline{\text{newopen}}'_{PQ}$ computes $\overline{state}_P$ from c as in $\overline{\text{com}}_P$, and runs $\overline{\text{open}}'_{PQ}$ (see Fig. 3).[11] It follows from the fairly-binding property that there is a function $\hat{y}(c')$ of the commitment c' so that $p(\hat{y}(c') \neq y \wedge y = y_0(c)) \leq \delta$ for every function $y_0(c)$.

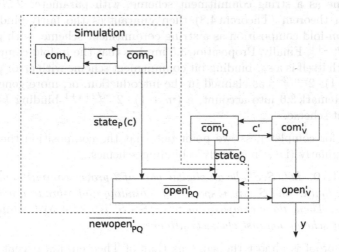

Fig. 3. Constructing the opening strategy $\overline{\text{newopen}}'_{PQ}$ against $\mathcal{S}'$.

The existence of $\hat{y}$ now gives rise to an opening strategy $\overline{\text{open}}_Q$ for $\mathcal{S}$; namely, simulate the commit phase of $\mathcal{S}'$ to obtain the commitment c', and output $\hat{y}(c')$. By Definition 3.2, for $\tilde{s} := \text{Extr}(\hat{y}(c'), c)$ and every s_0, $p(\hat{s}(c) \neq \tilde{s} \wedge \tilde{s} = s_0) \leq \varepsilon$.

---

[10] Recall that by convention (Remark 2.2), the commitment c equals the communication between V and, here, P.

[11] We are using here that Q is inactive during $\overline{\text{com}}_{PQ}$ and P during $\overline{\text{com}}'_{PQ}$, and thus the two "commute".

We are now ready to put things together. Fix an arbitrary target string s_0. For any c we let $y_0(c)$ be the unique string such that $\text{Extr}(y_0(c), c) = s_0$ (and some default string if no such string exists); recall, we assume for the moment that $k(\mathcal{S}) = 1$. Omitting the arguments in $\hat{s}(c), \hat{y}(c')$ and $y_0(c)$, it follows that

$$
\begin{aligned}
p(\hat{s} \neq s \wedge s = s_0) &\leq p(\hat{s} \neq s \wedge s = s_0 \wedge s = \tilde{s}) + p(s = s_0 \wedge s \neq \tilde{s}) \\
&\leq p(\hat{s} \neq \tilde{s} \wedge \tilde{s} = s_0) + p\big(\text{Extr}(y, c) \neq \text{Extr}(\hat{y}, c) \wedge \text{Extr}(y, c) = s_0\big) \\
&\leq p(\hat{s} \neq \tilde{s} \wedge \tilde{s} = s_0) + p(y \neq \hat{y} \wedge y = y_0) \\
&\leq \varepsilon + \delta.
\end{aligned}
$$

Thus, $\hat{s}$ is as required.

For the general case where $k(\mathcal{S}) > 1$, we can reason similarly, except that we then list the $k \leq k(\mathcal{S})$ possibilities $y_0^1(c), \ldots, y_0^k(c)$ for $y_0(c)$, and conclude that $p(s \neq \tilde{s} \wedge s = s_0) \leq \sum_i p(y \neq \hat{y} \wedge y = y_0^i) \leq k(\mathcal{S}) \cdot \delta$, which then results in the claimed bound. $\qquad\square$

Remark 4.9. Putting things together, we can now conclude the security (i.e., the binding property) of the Lunghi *et al.* multi-round commitment scheme. Corollary 3.13 ensures the fairly-binding property of $\mathcal{CHSH}^n$, i.e., the Crépeau *et al.* scheme as a string commitment scheme, with parameter $2^{-n/2+1}$. The composition theorem (Theorem 4.8) then guarantees the fairly-binding property of the m-fold composition as a string commitment scheme, with parameter $(m+1) \cdot 2^{-n/2+1}$. Finally, Proposition 3.5 implies that the m-fold composition of $\mathcal{CHSH}^n$ with itself is a ε_m-binding bit commitment scheme with error parameter $\varepsilon_m = (m+1) \cdot 2^{-n/2+2}$ as claimed in the introduction, or, more generally, and by taking Remark 3.6 into account, a $(m+1) \cdot 2^{-n/2+k+1}$-binding k-bit-string commitment scheme.

Finally, for completeness, we point out that the composition theorem also holds for regularly (i.e., "non-fairly") binding schemes.

Theorem 4.10. *Let $(\mathcal{S}, \mathcal{S}')$ be an eligible pair of 2-prover commitment schemes, and assume that $\mathcal{S}$ and $\mathcal{S}'$ are respectively ε-binding and δ-binding against classical attacks. Then, their composition $\mathcal{S}'' = \mathcal{S} \star \mathcal{S}'$ is a $(\varepsilon + \delta)$-binding 2-prover commitment scheme against classical attacks.*

Proof. The proof is almost the same as that of Theorem 4.8, except that now there are no s_0 and y_0, and in the end we simply conclude that

$$
\begin{aligned}
p(s \neq \hat{s} \wedge s \neq \bot) &\leq p(s \neq \hat{s} \wedge s \neq \bot \wedge s = \tilde{s}) + p(s \neq \tilde{s} \wedge s \neq \bot) \\
&\leq p(\tilde{s} \neq \hat{s} \wedge \tilde{s} \neq \bot) + p(y \neq \hat{y} \wedge y \neq \bot) \\
&\leq \varepsilon + \delta,
\end{aligned}
$$

where the second inequality holds since $y = \bot$ implies $s = \bot$. $\qquad\square$

Acknowledgments. We would like to thank Jędrzej Kaniewski for helpful discussions regarding [10], and for commenting on an earlier version of our work.

A   The Hiding Property of Composed Schemes

We already mentioned that the standard hiding property is not good enough for multi-round relativistic bit commitment schemes, where we want the hiding property to hold until the last round of communication. In this appendix, we define a variation of the hiding property that captures this requirement, and we prove that a composed scheme $S'' = S \star S'$ is hiding "up to the last round" if both S and S' are (with the error parameters adding up).

Definition A.1. *Let* $S = (\text{com}_{PQV}, \text{open}_{PQV})$ *be a 2-prover commitment scheme. We say that* S *is* ε*-hiding until the last round if for any dishonest verifier* V *and any two inputs* s_0 *and* s_1 *to the honest provers, we have* $d(p(v|s_0), p(v|s_1)) \le \varepsilon$, *where* v *is the verifier's view immediately before the last round of communication in* $(\text{open}_{PQ} \| \overline{\text{open}}_V) \circ (\text{com}_{PQ} \| \overline{\text{com}}_V)(s_b \| s_b \| \emptyset)$.

Theorem A.2. *Let* S *be an* ε*-hiding commitment scheme and* S' *a scheme that is* δ*-hiding until the last round. If* (S, S') *is eligible, then the composed scheme* $S'' = S \star S'$ *is* $(\varepsilon + \delta)$*-hiding until the last round.*

Proof. Fix some strategy against the hiding-until-the-last-round property of S''. We consider the distribution $p(v, y, v'|s)$ where s is the string that the provers commit to, v the verifier's view after com_{PQV} has been executed, y the opening information to which Q commits using the scheme S', and v' the verifier's view immediately before the last round of communication. We need to show that $d(p(v'|s_0), p(v'|s_1)) \le \varepsilon + \delta$ for any s_0 and s_1.

First, note that $p(v'|v, y, s_b) = p(v'|v, y)$ since v' is produced by P, Q and V acting on y and v only. From any strategy against S'', we can obtain a strategy against S' by fixing v. Thus, by the hiding property of S', for any y_0 and y_1, we have $d(p(v'|v, y = y_0), p(v'|v, y = y_1)) \le \delta$ and it follows by the convexity of the statistical distance in both arguments that

$$p(v'|v, s_0) = \sum_y p(y|v, s_0)p(v'|v, y) \approx_\delta \sum_y p(y|v, s_1)p(v'|v, y) = p(v'|v, s_1)$$

where we use $\approx_\delta$ to indicate that the two distributions have statistical distance at most δ. Since we have $d(p(v|s_0), p(v|s_1)) \le \varepsilon$ by the hiding property of S, it follows that

$$p(v'|s_0) = p(v, v'|s_0) = p(v|s_0)p(v'|v, s_0) \approx_\delta p(v|s_0)p(v'|v, s_1)$$
$$\approx_\varepsilon p(v|s_1)p(v'|v, s_1) = p(v, v'|s_1) = p(v'|s_1)$$

where the first and last equalities hold because v' contains v since v' is the view of V at a later point in time.   □

References

1. Bavarian, M., Shor, P.W.: Information Causality, Szemerédi-Trotter and Algebraic Variants of CHSH. In: Roughgarden, T. (ed.) ITCS 2015, pp. 123–132. ACM (2015)
2. Ben-Or, M., Goldwasser, S., Kilian, J., Wigderson, A.: Multi-Prover Interactive Proofs: How to Remove Intractability Assumptions. In: Simon, J. (ed.) STOC 1988, pp. 113–131. ACM (1988)
3. Chakraborty, K., Chailloux, A., Leverrier, A : Arbitrarily Long Relativistic Bit Commitment. ArXiv e-prints (2015). http://arxiv.org/abs/1507.00239
4. Clauser, J.F., Horne, M.A., Shimony, A., Holt, R.A.: Proposed Experiment to Test Local Hidden-Variable Theories. Phys. Rev. Lett. **23**, 880–884 (1969)
5. Crépeau, C., Salvail, L., Simard, J.-R., Tapp, A.: Two Provers in Isolation. In: Lee, D.H., Wang, X. (eds.) ASIACRYPT 2011. LNCS, vol. 7073, pp. 407–430. Springer, Heidelberg (2011)
6. Fehr, S., Fillinger, M.: Multi-Prover Commitments Against Non-Signaling Attacks. In: Gennaro, R., Robshaw, M. (eds.) CRYPTO 2015. LNCS, vol. 9216, pp. 403–421. Springer, Heidelberg (2015)
7. Kent, A.: Unconditionally Secure Bit Commitment. Phys. Rev. Lett. **83**(7), 1447–1450 (1999)
8. Kent, A.: Secure Classical Bit Commitment Using Fixed Capacity Communication Channels. J. Cryptology **18**(4), 313–335 (2005)
9. Lo, H.-K., Chau, H.F.: Is quantum bit commitment really possible? Phys. Rev. Lett. **78**, 3410–3413 (1997)
10. Lunghi, T., Kaniewski, J., Bussières, F., Houlmann, R., Tomamichel, M., Wehner, S., Zbinden, H.: Practical Relativistic Bit Commitment. Phys. Rev. Lett. **115**, 30502–30506 (2015)
11. Mayers, D.: Unconditionally Secure Quantum Bit Commitment is Impossible. Phys. Rev. Lett. **18**, 3414–3417 (1997)
12. Sikora, J., Chailloux, A., Kerenidis, I.: Strong Connections Between Quantum Encodings, Non-Locality and Quantum Cryptography. Phys. Rev. A **89**, 22334–22341 (2014)

Computationally Binding Quantum Commitments

Dominique Unruh[✉]

University of Tartu, Tartu, Estonia
unruh@ut.ee

Abstract. We present a new definition of computationally binding commitment schemes in the quantum setting, which we call "collapse-binding". The definition applies to string commitments, composes in parallel, and works well with rewinding-based proofs. We give simple constructions of collapse-binding commitments in the random oracle model, giving evidence that they can be realized from hash functions like SHA-3. We evidence the usefulness of our definition by constructing three-round statistical zero-knowledge quantum arguments of knowledge for all NP languages.

1 Introduction

We study the definition and construction of computationally binding string commitment schemes in the quantum setting. A commitment scheme is a two-party protocol consisting of two phases, the commit and the open phase. The goal of the commitment is to allow the sender to transmit information related to a message m during the commit phase in such a way that the recipient learns nothing about the message (hiding property). But at the same time, the sender cannot change his mind later about the message (binding property). Later, in the open phase, the sender reveals the message m and proves that this was indeed the message that he had in mind earlier. We will focus on non-interactive classical commitments, that is, the commit and open phase consists of a single classical message. However, the adversary who tries to break the binding or hiding property will be a quantum-polynomial-time algorithm. At the first glance, it seems that the definition of the binding property in this setting is straightforward; we just take the classical definition but consider quantum adversaries instead of classical ones:

Definition 1 (Classical-Style Binding – Informal). *No quantum-polynomial-time algorithm A can output, except with negligible probability, a commitment c (i.e., the message sent during the commit phase) as well as two openings u, u' that open c to two different messages m, m'.*

(Formal definition in Sect. 2). Unfortunately, this definition turns out to be inadequate in the quantum setting. Ambainis et al. [1] show the existence of a commitment scheme (relative to a special oracle) such that: The commitment is

© International Association for Cryptologic Research 2016
M. Fischlin and J.-S. Coron (Eds.): EUROCRYPT 2016, Part II, LNCS 9666, pp. 497–527, 2016.
DOI: 10.1007/978-3-662-49896-5_18

classical-style binding. Yet there exists a quantum-polynomial-time adversary A that outputs a commitment c, then expects a message m as input, and then provides valid opening information for c and m. That is, the adversary can open the commitment c to any message of his choosing, even if he learns that message only after committing. This is in clear contradiction to the intuition of the binding property. How is this possible, as Definition 1 says that the adversary cannot produce two different openings for the same commitment? In the construction from [1], the adversary has a quantum state $|\Psi\rangle$ that allows him to compute one opening for a message of his choosing, however, this computation will destroy the state $|\Psi\rangle$. Thus, the adversary cannot compute two openings simultaneously, hence the commitment is classically-binding. But he can open the commitment to an arbitrary message once, which shows that the commitment scheme is basically useless despite being classically-binding.[1]

1.1 Prior Definitions

We now discuss various definitions that appeared in the literature and that circumvent the above limitation of the classical-binding property. (We do not discuss the hiding property here, because that one does not have any comparable problems. See Definition 10 below for the definition of hiding.) In each case, we discuss some limitations of the definitions to motivate the need for a new definition for computationally binding commitments. The reader only interested in our results can safely skip this section.

Sum-Binding. The most obvious solution is to simply require that the adversary cannot open successfully to each of two messages: That is:

Definition 2 (Sum-Binding – Informal). *Consider a bit commitment scheme. (I.e., one can only commit to $m = 0$ or $m = 1$.)*

Given an adversary A, let p_b be the probability that the recipient accepts in the following execution: A commits, then A is given b, and then A provides opening information for message b. A commitment is sum-binding *iff for any quantum-polynomial-time adversary A, $p_0 + p_1 \leq 1 + $ negligible.*

Note that even with an ideal commitment, $p_0 + p_1 = 1$ is possible (the adversary just picks $b := 0$ in the commit phase with probability p_0, and $b := 1$ else). So $p_0 + p_1 \leq 1 + $ *negligible* is the best we can expect if we allow for a negligible probability of an attack. The sum-binding definition has occurred implicitly and explicitly in different variants in [4,6,8,13,15]. We use the name sum-binding here to distinguish it from the other definitions of binding discussed here since it does not have established name.

Although it avoids the attack described above, the sum-binding definition has a number of disadvantages:

---

[1] Note that for classical adversaries, the classical-binding property gives useful guarantees: If an adversary can produce an opening for any message m using some classical algorithm, he can also produce two openings for different messages m, m' by running that algorithm twice.

- It is specific to the bit commitment case. There is no straightforward generalization to the string commitment case (i.e., where the message m does not
  have to be a single bit). See [6] for discussion why obvious approaches fail.
- It is unclear how the definition behaves when we use the commitment several
  times. (I.e., it is not clear how it behaves under composition.) For example,
  given bits $m_1, \ldots, m_n$, what are the security guarantees if we commit to each
  of the m_i? (Be it in parallel, or sequentially.) Basically, we would expect
  that all commitments together form a binding commitment on the string
  $m = m_1 \ldots m_n$, but this is something we cannot even express using the sum-
  binding definition.
- It is not clear how useful sum-binding commitments are as subprotocols in
  larger protocols. That is, is the sum-binding property strong enough to allow
  to prove the security of complex protocols using commitments? While there
  are constructions of sum-binding in the literature (e.g., [13]), we are not
  aware of research where (computational) sum-binding commitments are used
  as subprotocols.

CDMS-Binding. Crépeau et al. [6] suggest a generalization of the sum-binding
property to string commitments. The basic idea is: Instead of bounding $p_0 + p_1 \leq$
$1 + negligible$ where p_m is the probability that the adversary open his commitment
as $m \in \{0,1\}$, we could bound $\sum_m p_m \leq 1 + negligible$ where m ranges over all
bitstrings. However, as discussed in [6], this would be too strong a requirement.
(Basically, this is because the sum $\sum_m p_m$ has exponentially many summands,
so even negligible attack probabilities can add up to large probabilities.) Instead,
they proposed the following definition:

Definition 3 (CDMS-Binding – Informal). *Let F be a family of functions.
Fix a string commitment scheme. For $f \in F$, let $\tilde{p}_y^f$ be the probability that the
recipient accepts in the following execution: A commits. A gets y. A tries to open
the commitment to some m with $f(m) = y$.*

*We call the commitment scheme F-CDMS-binding iff for all adversaries A
and all $f \in F$, we have $\sum_y \tilde{p}_y^f \leq 1 + negligible$.*

Now if all $f \in F$ have a polynomial-size range, the sum $\sum_y \tilde{p}_y^f$ will have polynomially many summands. The intuition behind this definition is that every
function $f \in F$ represents some property of the committed message m (e.g.,
$f(m)$ is the parity of m). Then, if a commitment scheme is F-CDMS-binding,
this intuitively means that the although the adversary might be able to change
his mind about the message m, he cannot change his mind about $f(m)$. (E.g., if
the parity function is in F, this means that the adversary will be committed to
the parity of the message m). [6] successfully used this definition (for a specific
class F) to show that using quantum communication and a commitment, we
can construct an oblivious transfer protocol. (Note however that their protocol
is different and more complex than the original OT protocol from [2]).

Although the CDMS-binding definition generalizes the sum-binding definition to the case of string commitments, it comes with its own challenges:

- The definition is parametrized by a specific family F of functions that specifies in which way the commitment should be binding. This function family has to be chosen dependent on the particular use case. This makes the definition less universal and canonical.
- To the best of our knowledge, no construction of CDMS-binding commitments is known. Crépeau et al. [6] conjecture that the protocol from [7] can be extended to a CDMS-binding one for functions F with small range, but no proof or construction is given.
- It is not known whether the definition is composable. If we commit to messages $m_1, \ldots, m_n$ individually using F-CDMS-binding commitments, does this constitute an F'-CDMS-binding commitment on $m := m_1 \| \ldots \| m_n$? If so, for which F'?
- While CDMS-binding commitments have successfully been used in a larger protocol (namely, the OT protocol from [6]), we believe that in many contexts, the definition is still not very easy to use. At least in classical cryptography, one often uses the fact that it is possible to extract the committed message by rewinding (basically, one runs the open phase, saves the opened message, and rewinds to before the opening phase). It is not clear how to do that with CDMS-binding commitments. For example, it is not clear how one could use CDMS-binding commitments in the construction of sigma-protocols that are quantum arguments of knowledge (as done in Sect. 7 below using our definition of binding commitments).

Perfectly-Binding Commitments. One possibility to solve all the problems mentioned so far is simply to use perfectly-binding commitments.

Definition 4 (Perfectly-Binding – Informal). *A commitment scheme is perfectly-binding if there exists no tuple (c, m, u, m', u') with $m \neq m'$ such that u is a valid opening for c with message m, and u' is a valid opening for c with message m'.*

However, if we restrict ourselves to perfectly-binding commitments, we get the following disadvantages:

- A perfectly-binding commitment cannot be statistically hiding [15]. That is, the hiding property cannot hold against computationally unlimited adversaries. That means that we give up on information-theoretical security for one party just because we do not have a suitable definition for the computationally-binding property. For example, the constructions in [19] are only computational zero-knowledge (not statistical zero-knowledge) because perfectly-binding commitments are used.
- Perfectly-binding commitments cannot be short. That is, the length of the commitment must be as long as the length of the committed message. So by using only perfectly-binding commitments, we may lose efficiency.

UC Commitments. One further possibility is to use commitments that are UC-secure [18]. Since the security of a protocol using a UC-secure commitment can be reduced to the security of the same protocol using an ideal (in particular perfectly-binding) commitment, UC-secure commitments are easy to use. Yet, this solution again comes with disadvantages:

- UC-commitments do not exist without the use of additional setup such as, e.g., a common reference strings (CRS). It is possible to chose the CRS in a pre-computation phase using a coin-toss protocol [12]. But that increases the round complexity of the resulting protocol (and, incidentally, loses the UC security and possibly even the concurrent composability of the resulting protocol).
- In the construction of UC-secure commitment schemes, trapdoors are used that allow the simulator to extract the committed message. This implies that constructions of UC-secure commitment are usually more complex, less efficient, and use stronger computational assumptions.
- At least when using a CRS, UC commitments cannot be short.

Damgård et al. [9] use so-called dual-mode commitments, these are somewhat weaker than UC commitments. Yet, they also use extraction using a trapdoor in the CRS. Hence the disadvantages of UC commitments apply to dual-mode commitments as well.

Q-Binding. Damgård et al. [11] give another definition for computationally binding string commitments. Intuitively, the definition says that an adversary who uses the commitment has negligible advantage in a "betting game" over an adversary that has to use perfect commitments. Here, a betting game is represented as an arbitrary predicate on the opened values in the commitments, and on some random input that the adversary learns only after committing. (E.g., a bet could be: the sum of all opened values equals the random value u that the adversary learns just before opening.) Somewhat more formally:

Definition 5 (Q-Binding – Informal). *For an adversary A and an predicate Q, consider the following game: A outputs commitments $C_1, \ldots, C_N$. Then A gets a random bitstring u. Then A opens a subset $\mathbf{A}$ of the commitments, let $(s_i)_{i \in \mathbf{A}}$ be the contents. A wins if $Q(\mathbf{A}, (s_i)_{i \in \mathbf{A}}, u) = 1$.*

A commitment scheme is Q-binding iff for any quantum-polynomial-time A and any predicate Q, the adversary A wins with probability at most $p_{\mathsf{IDEAL}} + negl$, where p_{IDEAL} is the maximum winning probability when using a perfectly binding commitment.

The definition overcomes some of the problems of the CDMS-binding definition. In particular, there is no need to parametrize the definition with a class F of functions, specifically chosen to fit the use case at hand. Also, the Q-binding definition composes in parallel: if a commitment scheme is Q-binding, then the commitment scheme resulting from committing to each of $m_1, \ldots, m_n$ individually is Q-binding, too. (This should come as no surprise, since the Q-binding definition itself explicitly refers to a polynomial number of parallel copies of the commitment scheme). The definition seems particularly well-suited for commit-and-choose constructions (i.e., where one party commits to a set of values, and the other party selects which of them should be opened), since security when opening a specific subset is built into the definition. [11] give a generic construction for unconditionally hiding Q-binding equivocal trapdoor commitments from a certain class of sigma-protocols. They show that using such commitments, sigma-protocols can be converted into statistical quantum zero-knowledge arguments in the CRS model.

However, their definition also comes with a number of challenges:

- The only construction of unconditionally hiding Q-binding commitments known is actually an equivocal trapdoor commitment. Trapdoor commitments usually need stronger assumptions. Note also that no protocols using non-equivocal Q-binding commitments are known (the zero-knowledge protocols in [11] need the trapdoor because they are constructed following the "no quantum rewinding paradigm"). And, due to the absence of rewinding, the zero-knowledge protocols only work in the CRS model.
- The possibility for parallel composition might be limited: It follows directly from the definition that Q-binding commitments on $m_1, \ldots, m_n$ are a Q-binding commitment on $m = m_1 \ldots m_n$. However, it is not clear what happens if we commit to $m_1, \ldots, m_n$ using *different* Q-binding commitments. (Or the same Q-binding commitment, but using different public keys.)
- The definition is specialized for the commit-and-choose paradigm. It is unclear how it can be used in rewinding-based proofs. (On the other hand, in commit-and-choose situations, Q-binding commitments might be more suitable than those we propose; whether this is the case constitutes future work.)

Summarizing, Q-binding commitments seem to be well suited for commit-and-choose constructions, but for proofs involving rewinding, we need another definition.

DFRSS-Binding. Damgård et al. [10] presented a definition for the unconditional binding property, targeted mainly for the bounded quantum storage model; the following is a direct adaptation of their definition to the computational setting:

Definition 6 (DFRSS-Binding – Adapted). *In a commitment, let V denote the recipient's classical state, and Z the sender's classical state.*

A bit commitment is DFRSS-binding iff for any quantum-polynomial-time sender $\tilde{C}$, there exists a randomized function B' such that the following holds: Let $\tilde{C}$ and the honest recipient execute the commit phase. Compute $b' := B'(V, Z)$. Let $\tilde{C}(b')$ and the honest recipient execute the open phase. Let b denote the opened bit (or $\perp$ if the recipient does not accept). Then $\Pr[b' \neq b]$ is negligible.

In other words, given the classical part of the state of the recipient *and* the sender, it is possible to extract what bit the sender will open to. (The extraction does not have to be efficiently feasible.) The definition can be extended to string commitments by letting B' range over bitstrings.

We have changed the original definition from [10] to refer to quantum-polynomial-time adversaries. (We also reformulated it for easier readability, changing a number of technical details in the process. However, the current definition is in the spirit of the original. And our discussion also applies to the original formulation.)

The definition was originally intended for protocols in the bounded quantum storage model. What happens if we use it in the standard model, i.e., with no limit on the quantum memory of the sender? In this case, it is always possible for the malicious sender to perform all his operations in superposition, and only the recipient will perform measurements. Then, in Definition 6, the register Z will be empty. Hence the definition requires that the committed bit b' can be computed from the recipient's state V alone. This immediately implies that the scheme cannot be statistically hiding, and that the commitments cannot be shorter than the message.

Hence the DFRSS-binding definition shares the drawbacks of the perfectly binding definition, unless we are in the bounded quantum storage model. (We stress that [10] never claimed that the definition should be used outside the bounded quantum storage model.)

1.2 Our Contribution

We give a new definition for the computational-binding property for commitment schemes, called "collapse-binding" (Sect. 2). This definition is composable (several collapse-binding commitments are also collapse-binding together), works well with quantum rewinding (see below), does not conflict with statistical hiding (as perfectly-binding commitments would), allows for short commitments (i.e., the commitment can be shorter than the committed message, in contrast to perfectly-binding commitments, and to extractable commitments in the CRS model). Basically, collapse-binding commitments seem to be in the quantum setting what computationally-binding commitments are in the classical setting.

We show that collision-resistant hash functions are not sufficient for getting collapse-binding or even just sum-binding commitments (Sect. 3), at least when using standard constructions, and relative to an oracle. We present a strengthening of collision-resistant hash functions, "collapsing hash functions" that can serve as a drop-in replacement for collision-resistant hash functions (Sect. 4). Using collapsing hash functions, we show several standard constructions of commitments to be collapse-binding (Sect. 5).

We conjecture that standard cryptographic hash functions such as SHA-3 [17] are collapsing (and thus lead to collapse-binding commitments). We give evidence for this conjecture by proving that the random oracle is a collapsing hash function.

We show that the definition of collapse-binding commitments is usable by extending the construction of quantum proofs of knowledge from [19] (Sect. 7). Their construction uses perfectly-binding commitments (actually, strict-binding, which is slightly stronger) to get proofs of knowledge. We show that when replacing the perfectly-binding commitments with collapse-binding ones, we get statistical zero-knowledge quantum arguments of knowledge. In particular, this shows that collapse-binding commitments work well together with rewinding.

1.3   Our Techniques

Collapse-Binding Commitments. To explain the definition of collapse-binding commitments, first consider a perfectly-binding commitment. That is, when an adversary A outputs a commitment c, there is only one possible message m_c that A can open c to. Hence, if the adversary A outputs a superposition of messages that he can open c to, that superposition will necessarily be in the state $|m_c\rangle$. Hence, we can characterize perfectly-binding commitments by requiring: when an adversary outputs a superposition of messages that he can open the commitment c to, that superposition will necessarily be a single computational basis vector (i.e., no non-trivial superposition).

To express this more formally, consider the circuit in Fig. 1(a). Here the adversary A outputs a commitment c (classical message). Furthermore, he outputs three quantum registers S, U, M. S contains his state. M is supposed to contain a superposition of messages, U a superposition of corresponding opening informations. Then we apply the measurement V_c. This measurement measures whether U, M contain matching opening information/message. More formally, V_c measures whether U, M is a superposition of states $|u, m\rangle$ such that u is valid opening information for message m and commitment

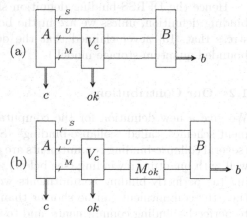

Fig. 1. Games from the definition of collapse-binding commitments.

c. Let $ok = 1$ if the measurement succeeds. Then we feed the registers S, U, M back to the second part B of the adversary. B outputs a classical bit b. As discussed before, a commitment is perfectly-binding iff for all adversaries A, the state of M after measuring $ok = 1$ is a computational basis vector.

The state of a register is a computational basis vector (or, synonymously: is in a collapsed state) iff measuring that register in the computational basis does not change that state. Consider the circuit in Fig. 1(b). Here we added a measurement M_{ok} on M after V_c. M_{ok} is a complete measurement in the computational basis, but is executed only if $ok = 1$. Since M_{ok} disturbs the state of M iff that state is not a computational basis vector, we can rephrase the definition of perfectly-binding commitments:

A commitment is perfectly-binding iff, for all computationally unlimited adversaries A, B, $\Pr[b = 1]$ is equal in Fig. 1(a) and (b) where b is the output (i.e., guess) of B.[2]

Now we are ready to weaken this characterization to get a computational binding property. Basically, we require that the same holds for quantum-polynomial-time adversaries:

Definition 7 (Collapse-Binding – Informal). *A commitment is collapse-binding iff, for all quantum-polynomial-time adversaries A, B, $\Pr[b = 1]$ in Fig. 1(a) is negligibly close to $\Pr[b = 1]$ in Fig. 1(b).*

In other words, with a perfectly-binding commitment, the adversary cannot produce a superposition of different messages that are contained in the commitment. But with a collapse-binding commitment, the adversary is forced to produce a state *that looks like it is not a superposition* of different messages. For the purpose of computational security, this will often be as good.

We quickly explain why collapse-binding commitments work well with quantum rewinding. In the case of quantum rewinding (e.g., in the analysis of proofs of knowledge [19]), one problem is that we might need to run an adversary until he opens a commitment c, then to measure the opened message, and then to go back to an earlier state by applying the inverse of the adversary. The problem is that measuring the opened message will disturb the state of the adversary, and thus make rewinding impossible. Except: if the opened message cannot be distinguished from being already in a collapsed state (as guaranteed by collapse-binding), then measuring the opened message does not disturb the state in a noticeable way and we can rewind. (See the discussion on arguments of knowledge below.)

Constructing Collapse-Binding Commitments. Collapse-binding commitments are useful only if they exist. Perfectly-binding commitments are easily seen to be collapse-binding, but then we cannot have statistically hiding or short commitments. In the classical setting, we get practical computationally-binding commitments from a collision-resistant hash function H. The most obvious construction is to send $c := H(m\|u)$ for uniformly random u of suitable length. We call this the "canonical commitment". The canonical commitment is easily seen to be classical-style binding if H is collision-resistant, and it is statistically hiding if H is a random oracle. To get rid of the random-oracle requirement, we can use a somewhat more complex constructions by Halevi and Micali [14] instead. Unfortunately, both the canonical commitment and the Halevi-Micali commitments are not collapse-binding if H is merely collision-resistant. In fact, relative to a specific oracle and using a specific collision-resistant hash function, there is a total break where the adversary can unveil the commitment to any message of his chosing. To show this, we tweak the technique from [1] to construct a hash function H such that the adversary can sample an image c of H

---

[2] Our exposition above was not very rigorous, but it is easy to see that this is indeed an "if and only if".

together with a quantum state $|\Psi\rangle$ such that: Given the state $|\Psi\rangle$, for any m, the adversary can find a random u with $H(m\|u) = c$. But this process destroys $|\Psi\rangle$, so the adversary cannot find two preimages of c; the hash function is collision-resistant. But the canonical commitment, based on this H, is trivially broken. Similar constructions break the Halevi-Micali commitments.

Since collision-resistance seems too weak a property in the quantum setting (at least for our purposes), we give a strengthening of collision-resistance: collapsing hash functions:

Definition 8 (Collapsing Hash Function – Informal). *An adversary is valid if he outputs a classical value c, and a register M containing a superposition of messages m with $H(m) = c$. We call H collapsing iff no quantum-polynomial-time adversary can distinguish whether we measure M in the computational basis or not, before giving the register M back to the adversary. (This is formalized with games similar to those in Fig. 1.)*

We can show that collapsing hash functions are collision-resistant, and they share a number of structural properties with collision-resistant functions. E.g., injective functions are collapsing, and the composition $H \circ H'$ of collapsing functions is collapsing.

Due to the similarity between the definition of collapsing hash functions and collapse-binding commitments, we can show that the canonical commitment and the Halevi-Micali commitments are collapse-binding if H is collapsing.

However, this leaves the question: do collapsing functions exist in the first place? We conjecture that common industrial hash function like SHA3 [17] are actually collapsing (not only collision-resistant). In fact, we argue that the collapsing property should be a requirement for the design of future hash functions (in the sense that a hash function where the collapsing property is in doubt should not be selected for industry standards), since collision-resistance is not sufficient if we wish to achieve post-quantum secure cryptography. We support our conjecture that sufficiently unstructured functions are collapsing by proving that the random oracle is collapsing:

Random Oracles Are Collapsing. We now sketch on a high level our proof that random oracles are collapsing, or, equivalently, that a random function is collapsing with high probability. In our analysis, we assume that the adversary can query the random oracle on the superposition of different inputs; this is necessary for having a realistic modeling of hash functions [3]. As a first step, we identify a new property, "half-collision resistance":

Definition 9 (Half-Collision Resistance – Informal). *A half-collision of H is a string x such that there exists an $x' \neq x$ with $H(x') = H(x)$. A hash function H is half-collision resistant if no adversary does the following: He outputs a half-collision with non-negligible probability. And he never outputs a non-half-collision. (The adversary may output $\perp$ though.)*

That is, half-collision resistance says that the adversary cannot find non-injective inputs to H without sometimes accidentally outputting injective inputs. We show: if H is half-collision resistant, it is collapsing.

The proof idea is: if H is not collapsing, the adversary can produce a superposition M of messages m with $H(m) = c$ and notice whether M is being measured. The latter implies that M must be a superposition of at least two messages m with $H(m) = c$. Hence by measuring M, the adversary gets a half-collision. Much additional work is needed to make sure that the adversary does not accidentally measure the register M when it is not a nontrivial superposition.

(The half-collision resistance property might be useful independent of the proof that the random oracle is collapsing. When trying to construct collapsing hash functions based on other assumptions, half-collision resistance might be easier to verify since its definition consists of purely classical games.)

Next we construct a random function $H^* : X \to Y$ with $|Y| = \frac{2}{3}|X|$. That is, H^* is slightly compressing. The domain of H^* is partitioned into two sets X_1, X_2 with $|X_1| = 2|X_2|$. H^* is injective on X_2, and 2-to-1 on X_1. Besides those constraints, H^* is uniformly random. We can then show that H^* is half-collision resistant. (Basically, this means that the adversary cannot identify the subset X_1.) Furthermore, we can show that H^* is indistinguishable from a random function $H : X \to Y$. Since H^* is half-collision resistant, it is collapsing. And since H is indistinguishable from H^*, H is collapsing.

We now know that random functions $H : X \to Y$ are collapsing if $|Y| = \frac{2}{3}|X|$ (i.e., if they are slightly compressing). However, we want that H is collapsing for arbitrary X and Y, as long as Y has superpolynomial size. For $|X| \leq |Y|$, H is indistinguishable from a random injection, which in turn is collapsing. The interesting case is $|X| > |Y|$ (namely, when H is compressing). In this case, we show (following an idea from [24]) that H can be written as $H = f_n \circ \cdots \circ f_1$ where all f_i are slightly compressing. Since all f_i are collapsing, so is H. This shows that a random function H is collapsing, in other words, that the random oracle is collapsing (if its range has superpolynomial size).

Quantum Arguments of Knowledge. We illustrate the use of collapse-binding commitments by revisiting the construction of proofs of knowledge from Unruh [19]. Unruh showed that a sigma-protocol (i.e., a particular kind of three round proof system) is a quantum proof of knowledge if it has two properties: *special soundness* (from two interactions with the same first and different second messages one can efficiently compute a witness) and *strict soundness* (the first and second message of a valid interaction determine the third). In the classical setting, only special soundness is needed. In the quantum setting, strict soundness is additionally required to allow for quantum rewinding: In the proof from [19], we run the malicious prover to get his response (the third message). Then we measure the response. Then we rewind the prover (by applying the inverse of the unitary transformation representing the prover). Then we run the prover again to get a second answer. Special soundness then implies that from the two responses, we get a witness. However, we need to make sure that measuring the prover's response before rewinding does not disturb the state (too much). In [19], this follows from strict soundness: strict soundness guarantees that the response is uniquely determined, and thus measuring the response does not disturb the state. To achieve strict soundness, [19] lets the prover commit to all possible responses in the first message using perfectly-binding commitments.[3] The drawback of this solution is that the commitments cannot be statistically hiding, so we cannot get statistical zero-knowledge proofs using the method from [19].

What happens if we replace the perfectly-binding commitments by collapse-binding commitments containing the response? In that case, the response will not necessarily be information-theoretically determined by the first two messages. However, the definition of collapse-binding commitments guarantees that measuring that response will be indistinguishable from not measuring it. Thus, if we measure the response, the state might be disturbed, but it will be computationally indistinguishable from not being disturbed. This is enough for the proof technique from [19] to go through, assuming the prover is computationally limited. The resulting protocol will not be a quantum proof of knowledge, but a quantum argument of knowledge (i.e., secure only against computationally limited provers). But in contrast to [19], the proof system will be statistical zero-knowledge.

To summarize: from collapse-binding commitments (or from collapsing hash functions), we get three-round statistical zero-knowledge quantum arguments of knowledge for all languages in NP (with inverse polynomial knowledge error). To the best of our knowledge, not even three-round statistical zero-knowledge quantum *arguments* were known before.

1.4   Related Work

Commitments. Brassard et al. [4] presented an information-theoretically hiding and binding commitment scheme using quantum communication. However, the protocol was flawed, Mayers [15] showed that information-theoretically hiding and binding commitments are impossible. (This is no contradiction to our results, because our commitments are not information-theoretically binding.) Dumais et al. [13] and Crépeau et al. [7] constructed statistically hiding commitments from quantum one-way permutations/functions, respectively. Their protocols use quantum communication, and are sum-binding. Crépeau et al. [6] generalized the sum-binding definition to string commitments and constructed an OT protocol based on that definition. (However, it is not known whether the protocol composes even sequentially.) Damgård et al. [9] and Unruh [18] showed a much simpler OT protocol to be secure, assuming much stronger commitment definitions in the CRS model, but achieving stronger security notions (sequential composability/UC). Ambainis et al. [1] show that classical-style binding commitments are not necessarily even sum-binding.

---

[3] Actually, "strict-binding commitments" but this distinction is not relevant for this exposition.

Quantum Random Oracles. Random oracles were first explicitly considered in a quantum cryptographic context by Boneh et al. [3] who stressed that the adversary should have superposition access to the random oracle. Zhandry [24] showed that the random oracle is collision-resistant. In contrast, we show (based on his result) that the random oracle is collapsing (a stronger property).

Quantum Rewinding and Proof Systems. Watrous [23] showed how quantum rewinding can be used to prove the security of quantum zero-knowledge protocols. Unruh [19] showed how a different flavor of quantum rewinding can be used for proving the security of quantum proofs of knowledge; we extend their technique to quantum arguments of knowledge.

2 Definitions and Basic Properties

Preliminaries. For the necessary background in quantum computing, see, e.g., [16]. By $|i\rangle$ with $i \in I$ we denote the vectors of the computational basis of the Hilbert space with dimension $|I|$. We also use the symbol $|\cdot\rangle$ to refer to other (non-basis) vectors (e.g., $|\Psi\rangle$). And $\langle\Psi|$ is the conjugate transpose of $|\Psi\rangle$. $\|x\|$ refers to the Euclidean or ℓ^2-norm. We only consider finite dimensional Hilbert spaces. We denote $|+\rangle := \frac{1}{\sqrt{2}}|0\rangle + \frac{1}{\sqrt{2}}|1\rangle$ and $|-\rangle := \frac{1}{\sqrt{2}}|0\rangle - \frac{1}{\sqrt{2}}|1\rangle$. For a linear operator A on a Hilbert space, we denote by $A^\dagger$ its conjugate transpose. We denote by I the identity. We call an operator A on a Hilbert space a projector iff it is an orthogonal projector, i.e., a linear map with $P^2 = P$ and $P = P^\dagger$. By $\mathrm{TD}(\rho, \rho')$ we denote the trace distance between ρ and ρ', and by $F(\rho, \rho')$ the fidelity.

Given an algorithm A, let $x \leftarrow A(y)$ denote the result of running A with inputs y, and assigning the output to x. Let $x \xleftarrow{\$} M$ denote assigning a uniformly random element of M to x. We will use η to denote the security parameter, that is a positive integer that will be passed to all algorithms and adversaries and that indicates the required security level. By $a\|b$ we denote the concatenation of bitstrings a and b.

We call an algorithm quantum-polynomial-time if it is a quantum algorithm and its runtime is bounded by a polynomial in its input length with probability 1. We call an algorithm classical-polynomial-time if it performs only classical operations and its runtime is bounded by a polynomial in its input length with probability 1. We write 1^η for a bitstring (of 1's) of length η. (The latter is useful for making algorithms run in polynomial-time in the length of the security parameter, e.g., $A(1^\eta)$ will run polynomial-time in η.)

Commitments. A commitment scheme $(com, verify)$ consists of a quantum-polynomial-time algorithm com and a deterministic quantum-polynomial-time algorithm $verify$.[4] $(c, u) \leftarrow com(1^\eta, m)$ returns a commitment c and the opening

---

[4] To be practical, those algorithms should of course be classical. We allow quantum-polynomial-time algorithms here to state our results in greater generality.

information u for the message m and security parameter η. c alone is supposed not to reveal anything about m (hiding). To open, we send (m, u) to the recipient who checks whether $verify(1^\eta, c, m, u) = 1$. Both com and $verify$ have classical input and output. com has a well-defined message space MSP_η that also depends on the security parameter η (e.g., $\{0,1\}^\eta$). Furthermore, for technical reasons, we assume that it is possible to find triples (c, m, u) with $verify(1^\eta, c, m, u) = 1$ with probability 1 in quantum-polynomial-time in η.

We first state some standard properties of commitments.

Definition 10. *Let $(com, verify)$ be a commitment scheme. We define:*

- **Perfect completeness:** *$(com, verify)$ has perfect completeness iff for all $m \in \mathsf{MSP}_\eta$, $\Pr[verify(1^\eta, c, m, u) = 1 : (c, u) \leftarrow com(1^\eta, m)] = 1$.*
- **Computational hiding:** *$(com, verify)$ is computationally hiding iff for any quantum-polynomial-time A and any polynomial ℓ, there is a negligible μ such that for any η, any $m_0, m_1 \in \mathsf{MSP}_\eta$ with $|m_0|, |m_1| \le \ell(\eta)$, and any $|\Psi\rangle$,[5] $\left| P_0 - P_1 \right| \le \mu(\eta)$ where $P_i := \Pr[b = 1 : (c, u) \leftarrow com(1^\eta, m_i), b \leftarrow A(1^\eta, |\Psi\rangle, c)]$.*
- **Statistical hiding:** *Like computational hiding, except that we quantify over all A (not just quantum-polynomial-time A).*

Definition 11 (Classical-Style Binding). *A commitment scheme is classical-style binding iff for any quantum-polynomial-time algorithm A, the following is negligible in η: $\Pr[verify(1^\eta, c, m, u) = 1 \wedge verify(1^\eta, c, m', u') = 1 \wedge m \ne m' : (c, m, u, m', u') \leftarrow A(1^\eta)]$.*

Definition 12 (Collapse-Binding). *For algorithms A, B, consider the following games:*

$\mathsf{Game}_1 : \quad (S, M, U, c) \leftarrow A(1^\eta), \ \ ok \leftarrow V_c(M, U), \ \ m \leftarrow M_{ok}(M), \ \ b \leftarrow B(1^\eta, S, M, U)$

$\mathsf{Game}_2 : \quad (S, M, U, c) \leftarrow A(1^\eta), \ \ ok \leftarrow V_c(M, U), \ \ \qquad \qquad \ \ b \leftarrow B(1^\eta, S, M, U)$

Here S, M, U are quantum registers. V_c is a measurement whether M, U contains a valid opening, formally V_c is defined through the projector $\sum_{\substack{m, u \\ verify(1^\eta, c, m, u)=1}} |m\rangle\langle m| \otimes |u\rangle\langle u|$. M_{ok} is a measurement of M in the computational basis if $ok = 1$, and does nothing if $ok = 0$ (i.e., it sets $m := \bot$ and does not touch the register M).

A commitment scheme is collapse-binding *iff for any quantum-polynomial-time algorithms A, B, the difference $\left| \Pr[b = 1 : \mathsf{Game}_1] - \Pr[b = 1 : \mathsf{Game}_2] \right|$ is negligible.*

Instead of measuring using V_c whether the adversary outputs a correct opening information, we can quantify only over adversaries that always output correct opening information. This leads to the following equivalent definition of collapse-binding commitments. This definition is often easier to handle when proving that a given scheme is collapse-binding.

---

[5] $|\Psi\rangle$ is the auxiliary input of A that represents knowledge of A acquired, e.g., in prior protocol runs. One could use a mixed state instead, this would lead to an equivalent definition.

Definition 13 (Collapse-Binding – Variant). *For algorithms A, B, consider the following games:*

$$\text{Game}_1: \quad (S, M, U, c) \leftarrow A(1^\eta), \quad m \leftarrow M_{comp}(M), \quad b \leftarrow B(1^\eta, S, M, U)$$
$$\text{Game}_2: \quad (S, M, U, c) \leftarrow A(1^\eta), \qquad\qquad\qquad\qquad b \leftarrow B(1^\eta, S, M, U)$$

Here S, M, U are quantum registers. $M_{comp}(M)$ is a measurement of M in the computational basis.

We call an adversary (A, B) valid if $\Pr[verify(c, m, u) = 1] = 1$ when running $(S, M, U, c) \leftarrow A(1^\eta)$ and measuring M, U in the computational basis to obtain m, u.

A commitment scheme is collapse-binding *iff for any quantum-polynomial-time valid adversary (A, B), the difference $|\Pr[b = 1 : \text{Game}_1] - \Pr[b = 1 : \text{Game}_2]|$ is negligible.*

In [20], we show Definitions 12 and 13 equivalent, and that the collapse-binding property is preserved under parallel composition of commitments.

3 Commitments from Collision-Resistant Hash Functions

In the following, we will often refer to hash functions. We will always assume that a hash function depends implicitly on the security parameter (in particular, the size of the range can depend on the security parameter). We also assume that the hash function is quantum-polynomial-time computable (in η and the input length).[6] Besides that, we do not assume any further properties such as collision-resistance unless explicitly mentioned.

Definition 14 (Canonical Commitment Scheme). *Given a hash function H and a parameter $\ell_u = \ell_u(\eta)$, the* canonical commitment scheme *for H is:*

- *Message space $\text{MSP}_\eta := \{0, 1\}^*$.*
- *$com_c an(m)$: Pick $u \xleftarrow{\$} \{0, 1\}^{\ell_u}$. Compute $c := H(m\|u)$. Return (c, u).*
- *$verify_{can}(c, m, u)$: Return 1 iff $H(m\|u) = c$.*

It is immediate to see that this scheme is classical-style binding if H is collision-resistant. However, in general it will not be hiding; for example, $H(m\|u)$ could leak the first bit of m. However, it is hiding if H is a random oracle:

Lemma 15. *Fix $\ell_u \geq 0$ and assume that $|Y| \leq 2^{\ell_u/8}$. For a random oracle $H : X \to Y$, the canonical commitment is statistically hiding.*

When using a hash function in the standard model, we can use the following commitment scheme instead:

---

[6] When working in the random oracle model: Quantum-polynomial-time computable given access to the random oracle.

Definition 16 (Bounded-Length Halevi-Micali Commitment [14]). *Fix integers* $\ell = \ell(\eta)$, $n = n(\eta)$. *Let* $L := 4\ell + 2n + 4$. *Let* $H : \{0,1\}^L \to \{0,1\}^\ell$ *be a hash function. Let* $F = F(\eta)$ *be a family of universal hash functions* $f : \{0,1\}^L \to \{0,1\}^n$. *We define the* bounded-length Halevi-Micali commitment $(com_{HMb}, code = verify_{HMb})$ *with* $\mathsf{MSP}_\eta = \{0,1\}^n$ *as:*

- $com_{HMb}(m)$: *Pick* $f \in F$ *and* $u \in \{0,1\}^L$ *uniformly at random, conditioned on* $f(u) = m$. *Compute* $h := H(u)$. *Let* $c := (h, f)$. *Return* (c, u).
- $verify_{HMu}(c, m, u)$ *with* $c = (h, f)$: *Check whether* $f(u) = m$ *and* $h = H(u)$. *If so, return 1.*

Definition 17 (Unbounded Halevi-Micali Commitment [14]). *Fix an integer* $\ell = \ell(\eta)$. *Let* $H : \{0,1\}^* \to \{0,1\}^\ell$ *be a hash function. Let* $L := 6\ell + 4$. *Let* F *be a family of universal hash functions* $f : \{0,1\}^L \to \{0,1\}^\ell$. *We define the* unbounded Halevi-Micali commitment $(com_{HMu}, verify_{HMu})$ *as:*

- $com_{HMu}(m)$: *Pick* $f \in F$ *and* $u \in \{0,1\}^L$ *uniformly at random, conditioned on* $f(u) = H(m)$. *Compute* $h := H(u)$. *Let* $c := (h, f)$. *Return* (c, u).
- $verify_{HMu}(c, m, u)$ *with* $c = (h, f)$: *Check whether* $f(u) = H(m)$ *and* $h = H(u)$. *If so, return 1.*

Theorem 18 (Security of Halevi-Micali [14]). *If* ℓ *is superlogarithmic, then the Halevi-Micali commitment and the bounded-length Halevi-Micali commitment are statistically hiding. If* H *is collision-resistant, then the Halevi-Micali commitment and the bounded-length Halevi-Micali commitment are classical-style binding.*

Note that [14] did not prove the classical-style binding property against *quantum* adversaries. But the (very simple) proof of binding carries over unchanged to the quantum setting (if H is collision-resistant against quantum adversaries). The statistical hiding property holds against unlimited adversaries anyway, thus also against quantum adversaries.

The following theorem shows that collision-resistance does not seem to be enough to make the above constructions secure in the quantum setting, i.e., classical-style binding is all we get.

Theorem 19. *There is an oracle* $\mathcal{O}$ *relative to which there exists a collision-resistant[7] hash function* H *such that the canonical commitment scheme and both Halevi-Micali commitment schemes using* H *admit the following attack:*

There is a quantum-polynomial-time adversary $A^{\mathcal{O}}$ *that outputs a commitment* c, *then expects a bit* b, *and then outputs with overwhelming probability a pair* (m, u) *such that* $verify(c, m, u) = 1$ *and the first bit of* m *is* b.

Clearly, a commitment with that property should not be considered secure. This shows that collision-resistance is too weak a property for constructing commitments in the quantum setting, at least when using standard constructions.

---

[7] H is collision-resistant iff for any quantum-polynomial-time A, $\Pr[x \neq x' \wedge H(x) = H(x') : (x, x') \leftarrow A(1^\eta)]$ is negligible.

The proof [20] uses the oracles constructed in [1]. In a nutshell, those oracles give the adversary access to sets S_y, such that the adversary can perform one single search in S_y for an element with a specific property, but cannot get two elements from the same S_y. Using a suitably constructed hash function H, finding m, u that open c corresponds to a search in S_y. Thus the adversary can use that search to break the binding property. But finding a collision in H corresponds to finding two elements from the same S_y, hence H is collision-resistant.

4   Collapsing Hash Functions

As seen in the previous section, for many protocols collision-resistance is not a sufficiently strong property in the quantum setting. In the following, we propose a strengthening of the collision-resistance property that seems more useful in the quantum setting, namely "collapsing" hash functions. We believe that collapsing hash functions are a natural assumption for real-life hash functions such as SHA-3 etc. This belief is supported by the fact that the random oracle is collapsing (see Sect. 6).

The definition of collapsing hash functions is similar to that of collapsing commitments (Definition 13).

Definition 20 (Collapsing). *For a function H and algorithms A, B, consider the following games:*

$$\text{Game}_1 : \quad (S, M, c) \leftarrow A(1^\eta), \quad m \leftarrow M_{comp}(M), \quad b \leftarrow B(1^\eta, S, M)$$
$$\text{Game}_2 : \quad (S, M, c) \leftarrow A(1^\eta), \quad \qquad \qquad \qquad b \leftarrow B(1^\eta, S, M)$$

Here S, M are quantum registers. $M_{comp}(M)$ is a measurement of M in the computational basis.

We call an adversary (A, B) valid if $\Pr[H(m) = c] = 1$ when we run $(S, M, c) \leftarrow A(1^\eta)$ and measure M in the computational basis as m.

A function H is collapsing *iff for any quantum-polynomial-time valid adversary (A, B), the difference $adv := \big| \Pr[b = 1 : \text{Game}_1] - \Pr[b = 1 : \text{Game}_2] \big|$ is negligible. (We call adv the* advantage*).*

Notice that the definition of collapsing hash functions is inherently quantum, even though the object we consider (the hash function H) is classical. We know of no classical analogue to collapsing hash functions. However, a collapsing hash function will necessarily be collision-resistant, see Lemma 22 below.

We proceed to give a number of useful properties of collapsing hash functions.

Lemma 21. *An injective function H is collapsing with advantage 0.*

Lemma 22. *A collapsing hash function is collision resistant.*

Theorem 23. *If f and g are collapsing, so is $g \circ f$.*

5   Commitments from Collapsing Hash Functions

In Sect. 3 we saw that collision-resistant hash functions are not sufficient for several standard constructions of commitment schemes. We will now show that those same constructions are secure in the quantum setting when using collapsing hash functions instead.

The following theorem allows us to extend the message space of a collapsing commitment by hashing the message with a collapsing hash function. Besides being useful in its own right, we need it in the analysis of the unbounded Halevi-Micali commitment.

Theorem 24. *Let f be a collapsing function. Let $(com, verify)$ be a collapse binding commitment scheme. Let $com_f(1^\eta, m) := com(1^\eta, f(m))$ and $verify_f(1^\eta, c, m, u) = verify(1^\eta, c, f(m), u)$. Then $(com_f, verify_f)$ is a collapse-binding commitment scheme.*

Lemma 25. *If H is collapsing, then the canonical commitment scheme $(com_{can}, verify_{can})$, and the bounded-length Halevi-Micali commitment $(com_{HMb}, code = verify_{HMb})$, and the unbounded Halevi-Micali commitment $(com_{HMu}, verify_{HMu})$ are collapse-binding. (For any choice of the parameters ℓ_u, ℓ, n.)*

We give the proof idea, the full proof is given in [20]. To show that the canonical commitment com_{can} is collapse-binding, we use the characterization of collapse-binding from Definition 13. We need to show that the adversary cannot distinguish between a measurement on register M and no measurement on register M, assuming the adversary outputs M, U containing a superposition of m, u with $verify_{can}(c, m, u) = 1$. The condition $verify_{can}(c, m, u) = 1$ is equivalent to $H(m\|u) = c$. Hence the adversary outputs in M, U a superposition of preimages of c under H. Since H is collapsing, this implies that the adversary cannot distinguish between a measurement on M, U and no measurement on M, U. This also implies (using some additional work) that the adversary cannot distinguish between a measurement on M and no measurement on M. Hence com_{can} is collapse-binding. The Halevi-Micali commitments are handled similarly.

6   Random Oracles Are Collapsing

In Sect. 5 we saw that collapsing hash functions imply collapse-binding commitments. In this section, we explore the existence of collapsing hash functions. Specifically, we show that the random oracle is collapsing. This implies that there are simple collapse-binding commitments in the random oracle model. Furthermore, it supports the assumption that real-life hash functions such as SHA-3 etc. could be collapse-binding. Alternatively, we could also directly start with the assumption that SHA-3 is collapsing, in that setting the constructions from Sect. 5 would not need the random oracle. (In fact, we advocate that a hash function that is not collapsing should not be considered a secure practical hash function, and not recommended for future use.)

For the remainder of this section, X and Y are sets, and $H : X \to Y$ is a random oracle. Furthermore Y is finite, and $X \subseteq \{0,1\}^*$ (finite or infinite). And $q \geq 1$ always refers to an upper bound on the number of oracle queries performed by the adversary. The full proofs are given in [20].

We start by defining a seemingly unrelated property (half-collision resistance) that will turn out to imply the collapsing property. We will need half-collision resistance in our proof that the random oracle is collapsing. However, the concept of half-collision resistance might be of use for constructions in the standard model, too: since half-collision resistance is defined by a classical game, it might be easier to construct hash functions that are half-collision resistant.[8]

Definition 26. *A* half-collision *of a hash function* $f : X \to Y$ *is a value* x *such that* $\exists x' \neq x.f(x) = f(x')$.

An adversary A *has advantage* ε *against* half-collision resistance *iff*

- *with probability* 1, *the output of* A *is a half-collision or* $\perp$, *and*
- *with probability at* ε, A *outputs a half-collision.*

Lemma 27. *If* (A, B) *is valid and has advantage* μ *against the collapsing property of a hash function* f, *then there is an adversary* D *with advantage* $\geq \mu^2/4$ *against the half-collision resistance of* f. *The time-complexity of* D *is linear in that of* (A, B). *(If* f *is given as an oracle,* D *makes* $4q + 4$ *queries to* f *when* (A, B) *makes* q *queries.)*

Proof Sketch: By definition, a valid adversary A will always output in register M a superposition of messages m with $H(m) = c$ (all with the same c). So we have two cases: M contains a superposition of a single message m, or M contains a superposition of several messages that have the same image c, i.e., a superposition of half-collisions. Thus, in the second case, we can find a half-collisions by measuring M. But, an adversary against half-collision resistance must never output a non-half-collision (no false positives). Thus, we need a possibility to test whether M contains only a single message. (In this case, we abort.)

Note that when M contains only a single message, then the adversary B cannot distinguish between a measurement on M and no measurement on M. To exploit this, we run an execution where M is measured and an execution where M is not measured in superposition (roughly speaking), and we make it depend on a control qubit in state $|+\rangle$ which execution is used. Then, in the

---

[8] However, half-collision resistance is strictly stronger than collapsing, at least relative to an oracle, as we show next. Consider an oracle $\mathcal{O}$ picked according to the following distribution: Let $P_0, P_1 : \{0,1\}^n \to \{0,1\}^n$ be random permutations. Let $\mathcal{O}(b\|x) := P_b(x)$ for $b \in \{0,1\}, x \in \{0,1\}^n$. Then every input to $\mathcal{O}$ is a half-collision, thus $\mathcal{O}$ cannot be half-collision resistant. However P_0 and P_1 are indistinguishable from a random function [24], hence $\mathcal{O}$ is indistinguishable from $\mathcal{O}'(b\|x) := H_b(x)$ for random functions H_0, H_1. Note that $\mathcal{O}'$ is a random function, hence $\mathcal{O}'$ is collapsing by Theorem 31. Since $\mathcal{O}$ and $\mathcal{O}'$ are indistinguishable, $\mathcal{O}$ is collapsing as well.

case where M contains only a single message, the control qubit stays unentangled with the rest of the circuit. By measuring whether the qubit is still in state $|+\rangle$, the half-collision resistance adversary can detect whether M contains one or several messages. (It may err and incorrectly assume that M contains only one message, but an error in that direction is permitted.) Thus we have constructed an adversary against half-collision resistance.

Lemma 28. *Assume* $|X| \leq |Y|$. *Then* H *is collapsing with advantage* $O(q^3/|Y|)$.

Proof Sketch. Zhandry [24] *shows that for* $|X| \leq |Y|$, H *can be distinguished from a random injection with probability at most* $O(q^3/|Y|)$. *An injection is collapsing with advantage* 0 *(Lemma 21).*

For the next lemma, we fix some notation first: $[N] := \{1, \ldots, N\}$. For functions $f : [M] \to [N]$ and $g : [M'] \to [N]$, let $f + g : [M + M'] \to [N]$ be defined via $(f + g)(x) := f(x)$ for $x = 1, \ldots, M$ and $(f + g)(x) = g(x - M)$ for $x = M + 1, \ldots, M + M'$. For functions $f : [M] \to [N]$ and $g : [M'] \to [N']$, let $f|g : [M + M'] \to [N + N']$ be defined via $(f|g)(x) := f(x)$ for $x = 1, \ldots, M$ and $(f|g)(x) := g(x - M) + N$ for $x = M + 1, \ldots, M + M'$.

Lemma 29. *Assume that* $M \geq N$. *Let* $\hat{f}, \hat{g} : [N] \to [N]$ *and* $\hat{h} : [M] \to [M]$ *and* $\hat{\varphi} : [N + M] \to [N + M]$ *be uniformly distributed permutations (all independent), and let* $H : [2N + M] \to [N + M]$ *be a uniformly distributed function.*
Then for any q-*query adversary* A,

$$\left| \Pr[A^H = 1] - \Pr[A^{\hat{\varphi} \circ ((\hat{f} + \hat{g})|\hat{h})} = 1] \right| \in O(q^3/N).$$

Proof Sketch: We show this by rewriting $\hat{\varphi} \circ ((\hat{f} + \hat{g})|\hat{h})$ step by step, till it becomes H. In each step, the adversary distinguishes with probability $O(q^3/N)$ (denoted $\approx$ below) or 0 (denoted $\equiv$ below). For this we introduce additional functions $\varphi, v, w, \hat{v}, \hat{a}, \hat{b}, \hat{c}$ of suitable domains/ranges, all independent and uniformly random. The functions with a hat are injections. We compute:

$$\hat{\varphi} \circ ((\hat{f} + \hat{g})|\hat{h}) \approx \varphi \circ ((\hat{f} + \hat{g})|\hat{h}) \equiv (v \circ (\hat{f} + \hat{g})) + (w \circ \hat{h}) \equiv (v \circ (\hat{f} + \hat{g})) + w$$

$$\approx (\hat{v} \circ (\hat{f} + \hat{g})) + w \equiv (\hat{c} \circ \hat{a} \circ (\hat{f} + \hat{g})) + w \approx (\hat{c} \circ \hat{b}) + w$$

$$\equiv \hat{c} + w \approx c + w \equiv H.$$

Most of these equivalences either have elementary proofs, or are reduced to the fact that a random function and a random injection are indistinguishable. We get $H \approx \hat{\varphi} \circ ((\hat{f} + \hat{g})|\hat{h})$ which is the claim of the lemma.

Lemma 30. *Assume that* $|Y| = \lceil \frac{2}{3}|X| \rceil$. *Then* H *is collapsing with advantage* $O(\sqrt{q^3/|X|})$.

Proof Sketch: For simplicity, we consider the case $|Y| = 2N$, $|X| = 3N$. Then, by Lemma 29 with $M := N$, H is indistinguishable from $H^* := \hat{\varphi} \circ ((\hat{f} + \hat{g})|\hat{h})$.

Furthermore, for a random permutation π, H and $H \circ \pi$ are identically distributed, and $H \circ \pi$ is indistinguishable from $H^* \circ \pi$. Thus it is sufficient to show that $H^* \circ \pi$ is collapsing. In turn, by Lemma 27, it is sufficient to show that $H^* \circ \pi$ is half-collision resistant. To show that, observe that the half-collisions of H^* are the inputs $1, \ldots, 2N$, but not $2N + 1, \ldots, 3N$. Thus the half-collisions of $H^* \circ \pi$ are $P := \pi^{-1}(\{1, \ldots, 2N\})$. So, the half-collision resistance adversary has to find elements of P, without false positives, while given oracle access to $H^* \circ \pi$. But $H^* \circ \pi$ is indistinguishable from $H \circ \pi$, so the adversary would also be able to find elements in P given $H \circ \pi$. Since $H \circ \pi$ is a random function, independent of P, the adversary cannot do that without getting false positives. Hence $H^* \circ \pi$ is half-collision resistant and thus collapsing. Hence H is collapsing.

Theorem 31. *Let Y be finite, and $X \subseteq \{0, 1\}^*$ (finite or infinite). Then $H : X \to Y$ is collapsing with advantage $O(\sqrt{q^3/|Y|})$.*

Proof Sketch: H is indistinguishable from a composition $f_n \circ \cdots \circ f_1$ of random functions $f_n : X_n \to Y_n$ with $|X_{n+1}| = |Y_n| = \frac{2}{3}|X_n|$. By Lemma 30, each f_n is collapsing. Thus, by Theorem 23, $f_n \circ \cdots \circ f_1$ is collapsing and hence H is collapsing.

7 Zero-Knowledge Arguments of Knowledge

In this section, we study the security of sigma-protocols. A sigma-protocol is a specific kind three-round proof system in which the verifier's message consists only of random bits. Sigma-protocols play an important role in classical constructions of zero-knowledge proof systems for two reasons: For a number of simple but important languages, sigma-protocols exist. And given sigma-protocols for simple languages, there are efficient constructions for more complex languages. (There are constructions for conjunctions and disjunctions of sigma-protocols, as well as more complex threshold constructions [5].)

In the classical setting, it is relatively simple to give conditions under which sigma-protocols are zero-knowledge proofs of knowledge. In the quantum setting, however, analyzing the security of sigma-protocols turns out to be much harder. Watrous [23] presented a rewinding technique for proving the zero-knowledge property of sigma-protocols (see also Theorem 34 below). Unruh [19] showed that sigma-protocols are quantum proofs of knowledge under a specific additional condition called "strict soundness". This condition requires that the third message ("response") in a valid interaction is uniquely determined by the first two. However, strict soundness is a strong additional assumption. [19] showed how to achieve strict soundness by committing to the response already in the first message. However, the commitment scheme used for this needed to be perfectly-binding (actually, it needed to satisfy a somewhat stronger property, called "strict binding"). In particular, this implies that the commitment scheme cannot be information-theoretically hiding (hence the resulting protocol cannot be statistical zero-knowledge), and we cannot have short commitments (a perfectly-binding commitment will always be at least as long as the message inside).

Furthermore, Ambainis et al. [1] showed that the condition of strict soundness is necessary, at least relative to an oracle. They also showed that even if we assume that strict soundness holds, but only against computationally limited adversaries,[9] the resulting sigma-protocol will, in general, not be a quantum argument of knowledge.[10] Even more, it might not even be a quantum argument. That is, a computationally limited adversary can successfully prove a wrong statement.

In this section we show how we can use collapse-binding commitments as a drop-in replacement for the perfectly-binding commitments in the construction from [19]. One particular consequence is that given collapse-binding hash functions we can construct three-round statistical zero-knowledge quantum arguments of knowledge from sigma-protocols (without using a common-reference string). This assumes the sigma-protocol is statistical honest-verifier zero-knowledge and has special soundness. And that the challenge space (the set from which the verifier picks his random message) is polynomially-bounded. These properties, however, are also needed in the classical setting.

7.1 Interactive Proof Systems

An interactive proof system (P, V) for some relation R consists of two interactive quantum machines P and V that get classical inputs $(x, w) \in R$ and x, respectively. Afterwards, V outputs a bit. For formal definitions see [19]. (In general, P and V can exchange quantum messages, but our concrete constructions below will be classical.)

We consider two important properties of interactive proof systems: First, we want them to be arguments of knowledge. Informally, they should convince the verifier that the prover knows a witness w for the statement x (i.e., $(x, w) \in R$). Second, we want them to be zero-knowledge. Informally, the proof should not leaks anything about the witness besides its existence.

Quantum Arguments of Knowledge. The following definition of quantum arguments of knowledge follows the definition from [22], with one difference: we have formulated security against uniform malicious provers. That is, while in [22] the statement x and the auxiliary input $|\Psi\rangle$ are all-quantified, in our setting they are chosen by an quantum-polynomial-time algorithm Z. The reason we consider only uniform malicious provers here is: A non-uniform adversary can break any non-interactive commitment (with classical messages) that is not already perfectly-binding. (Namely, the auxiliary input can simply contain one

---

[9] I.e., it is hard to find two different valid interactions where the first two messages are equal but the response is different.

[10] Argument and argument of knowledge are the variants of proof and proof of knowledge that consider a computationally limited malicious prover.

commitment and two different openings.) Thus, since we consider only non-interactive commitments in this paper, we need a uniform definition of quantum arguments of knowledge. For a motivation of the remaining definitional choices, see [22].

Definition 32 (Quantum Arguments of Knowledge). *We call an interactive proof system* (P, V) *for a relation* R *(uniformly) quantum-computationally extractable with knowledge error* κ *if there exists a constant* $d > 0$, *a polynomially-bounded function* $p > 0$, *and a quantum-polynomial-time oracle algorithm* K *such that for any unitary quantum-polynomial-time algorithm* P^*, *for any polynomial* ℓ, *and for any quantum-polynomial-time algorithm* Z *(input generator), there exists a negligible* μ *such that for any security parameter* $\eta \in \mathbb{N}$, *we have that*

$$\Pr[\langle \mathsf{P}^*(1^\eta, x, Z), \mathsf{V}(1^\eta, x) \rangle = 1 : (x, Z) \leftarrow \mathsf{Z}(1^\eta)] \geq \kappa(\eta) \implies$$

$$\Pr[(x, w) \in R : (x, Z) \leftarrow \mathsf{Z}(1^\eta), w \leftarrow \mathsf{K}^{\mathsf{P}^*(1^\eta, x, Z)}(1^\eta, x)]$$

$$\geq \tfrac{1}{p(\eta)} \Big(\Pr[\langle \mathsf{P}^*(1^\eta, x, Z), \mathsf{V}(1^\eta, x) \rangle = 1 : (x, Z) \leftarrow \mathsf{Z}(1^\eta)] - \kappa(\eta) \Big)^d - \mu(\eta).$$

Here $\langle \mathsf{P}^*(1^\eta, x, Z), \mathsf{V}(1^\eta, x) \rangle$ *is the output of* V *after an interaction between* P^* *and* V *on the respective inputs* x *and* Z. Z *is a quantum register,* x *is classical, both initialized using the algorithm* Z. *And* $\mathsf{K}^{\mathsf{P}^*(1^\eta, x, Z)}$ *refers to an execution of* K *with black-box access to* $\mathsf{P}^*(1^\eta, x, Z)$. *That is,* K *can apply the unitary* U_x *describing the prover* P^* *and its inverse* $U_x^\dagger$. *(See [19] for a more detailed description of that black-box execution model.)*

Quantum Zero-Knowledge. Roughly speaking, (P, V) is *quantum-computationally zero-knowledge* iff for any quantum-polynomial-time malicious verifier V^*, there exists a quantum-polynomial-time simulator S such that for any $(x, w) \in R$, the output state of S is quantum computationally indistinguishable from the from the output state of V^* in an interaction with $\mathsf{P}(1^\eta, x, w)$.

Similarly, *quantum statistical zero-knowledge* is defined in the same way, except that V^* is not required to be quantum-polynomial-time.

We will not use the definition of quantum zero-knowledge directly, only the imported Theorem 34 from [22] will refer to it. We therefore omit the formal definition and refer to [22].

7.2 Sigma-Protocols

We now introduce sigma-protocols (following [21] with modifications as mentioned in the footnotes). The notions are like the standard classical definitions, all that was done to adopt them to the quantum setting was to make the adversary quantum-polynomial-time.

A *sigma-protocol* for a relation R is a three-message proof system. It is described by its challenge space N_z (where $|N_z| \geq 2$), a classical-polynomial-time prover (P_1, P_2) and a deterministic classical-polynomial-time verifier V. The first message from the prover is $a \leftarrow P_1(1^\eta, x, w)$ and is called the *commitment*, the uniformly random reply from the verifier is $z \xleftarrow{\$} N_z$ (called *challenge*), and the prover answers with $r \leftarrow P_2(1^\eta, x, w, z)$ (the *response*). We assume P_1, P_2 to share state. Finally $V(1^\eta, x, a, z, r)$ outputs whether the verifier accepts.

Definition 33 (Computational Special Soundness). *There is a quantum-polynomial-time algorithm E_Σ (the extractor)[11] such that for any quantum-polynomial-time A, we have that*

$$\Pr[(x, w) \notin R \land z \neq z' \land ok = ok' = 1 : (x, a, z, r, z', r') \leftarrow A(1^\eta),$$
$$ok \leftarrow V(1^\eta, x, a, z, r), \ ok' \leftarrow V(1^\eta, x, a, z', r'), \ w \leftarrow E_\Sigma(1^\eta, x, a, z, r, z', r')]$$

is negligible.

Note that the above is a standard condition expected from sigma-protocols in the classical setting. In contrast, for a sigma-protocol to be a *quantum* proof of knowledge, a much more restrictive condition is required, strict soundness [1,19]. We show below how to circumvent this necessity by adding collapse-binding commitments to the sigma-protocol (at least when we only need a quantum *argument* of knowledge).

We also use the standard properties of honest verifier zero-knowledge (HVZK) and statistical honest-verifier zero-knowledge (SHVZK). They are of secondary importance for the proofs shown in this section, we defer them to [20].

Remark 1. Any sigma-protocol (N_z, P_1, P_2, V) can be seen as an interactive proof (P, V) in a natural way: P sends the output a of P_1 to V. V picks $z \xleftarrow{\$} N_z$ and sends it to P. P sends the resulting output r of P_2 to V. V checks the triple (a, z, r) using V.

The following theorem is shown in [22]:

Theorem 34 (HVZK Implies Zero-Knowledge [22]). *Let $\Sigma = (N_z, P_1, P_2, V)$ be a sigma-protocol. We consider Σ as an interactive proof (P, V), see Remark 1.*

If $|N_z|$ is polynomially-bounded and is SHVZK, then Σ is quantum statistical zero-knowledge. If $|N_z|$ is polynomially-bounded and Σ is HVZK, then Σ is quantum computational zero-knowledge.

Due to this theorem, it will be sufficient to verify that the sigma-protocols we construct are HVZK/SHVZK. We will hence not need to use the definition of quantum zero-knowledge explicitly in the following.

---

[11] [21] requires a classical E_Σ here. By allowing E_Σ to be quantum here, we weaken the notion of computational special soundness slightly, and thus strengthen our results below.

7.3   Constructing Zero-Knowledge Arguments of Knowledge

In [19], the following idea was used to construct quantum proofs of knowledge: We assume a sigma-protocol with special soundness and with polynomial-size $|N_z|$. We convert it into a sigma-protocol with strict soundness as follows: When the prover sends his commitment $a \leftarrow P_1(x, w)$, he additionally sends $com(r_z)$ for all $z \in N_z$ where r_z is the response to the challenge z. When the prover receives the challenge z, he opens $com(r_z)$ instead of sending r_z. If the commitment has the "strict binding" property, the resulting sigma-protocol has strict soundness (without losing the special soundness or HVZK property).[12] Strict binding is a strengthening of perfect binding, it means that not only the message in the commitment is information-theoretically determined, but also the opening information.

Given a sigma-protocol with strict and special soundness, we can show that it is a proof of knowledge. Basically, [19] runs the protocol twice (using the inverse of the unitary malicious prover to rewind) to get two responses r, r' for different challenges $z \neq z'$. The difficulty here is that measuring r can disturb the state of the malicious prover, leading to a corrupt value r'. The trick here is that due to the strict soundness, the value r is essentially uniquely determined, and therefore the measurement does not introduce too much disturbance.[13]

Unfortunately, that technique needs commitments with the strict binding property. First, it is easy to see that strict binding commitments must be longer than the messages they contain. Short strict binding commitments are not possible. Furthermore, the only known construction of strict binding commitments [19] uses quantum 1-1 one-way functions. No candidates for those are known.

We show below that the same technique of committing to the responses works with collapse-binding commitments. The crucial point in the analysis from [19] was that measuring the committed response does not change the state. The collapse-binding property guarantees something slightly weaker: when measuring the committed response, the state may change, but this cannot be noticed by a computationally limited adversary. So with collapse-binding commitments, an analog reasoning as in [19] can be used, except that we get security only against quantum-polynomial-time adversaries. I.e., we get a quantum argument of knowledge. We will now describe this in more detail.

First, we formalize the sigma-protocol in which we commit to the responses:

Definition 35 (Sigma-Protocol with Committed Responses). *Let* (N_z, P_1, P_2, V) *be a sigma-protocol with polynomially-bounded* $|N_z|$. *Let* $(com, verify)$ *be a commitment scheme (with the responses of* (N_z, P_1, P_2, V) *as message space). We construct a sigma-protocol* (N_z, P_1', P_2', V') *as follows:*

---

[12] This part was done only implicitly in [19], in the analysis of the Hamiltonian cycle proof system.

[13] There is some disturbance due to the fact that it is not determined whether r is a valid response or an invalid one.

- $P_1'(1^\eta, x, w)$ runs: $a \leftarrow P_1(1^\eta, x, w)$. For each $z \in N_z$: $r_z \leftarrow P_2(1^\eta, x, w, z)$ [14] and $(c_z, u_z) \leftarrow com(1^\eta, r_z)$. Let $a' := (a, (c_z)_{z \in N_z})$ and return a'.
- $P_2'(1^\eta, x, w, z)$ returns $r' := (r_z, u_z)$.
- $V'(1^\eta, x, a', z, r')$ with $a' = (a, (c_z)_{z \in N_z})$ and $r' = (r, u)$: Check whether $verify(1^\eta, c_z, r, u) = 1$ and $V(1^\eta, a, z, r) = 1$. Iff so, return 1.

We show that the above construction is a quantum argument of knowledge:

Theorem 36 (Quantum Argument of Knowledge). *If (N_z, P_1, P_2, V) is a sigma-protocol with computational special soundness for a relation R, and $(com, verify)$ is collapse-binding, then (N_z, P_1', P_2', V') from Definition 35 is computationally quantum extractable for R with knowledge error $1/\sqrt{|N_z|}$.*

The proof of this theorem will rely on the following lemma from [19]. (That lemma is the core lemma of the rewinding technique from [19]).

Lemma 37 (Extraction via Quantum Rewinding [19]). *Let C be a set with $|C| = c$. Let $(P_i)_{i \in C}$ be projectors. Let $|\Phi\rangle$ be a unit vector. Let $V := \sum_{i \in C} \frac{1}{c} \||P_i|\Phi\rangle\|^2$ and $E := \sum_{i,j \in C, i \neq j} \frac{1}{c^2} \||P_i P_j|\Phi\rangle\|^2$. Then, if $V \geq \frac{1}{\sqrt{c}}$, $E \geq V(V^2 - \frac{1}{c})$.*

Proof of Theorem 36. Recall that any sigma-protocol can be seen as an interactive proof system by Remark 1. Let (P, V) denote the interactive proof system resulting from the sigma-protocol (N_z, P_1', P_2', V'). (In particular, the verifier V sends a random $z \in N_z$, and in the end checks whether $verify(1^\eta, c_z, r, u) = 1$ and $V(1^\eta, a, z, r) = 1$.)

Let P^* denote a malicious prover, i.e., a unitary quantum-polynomial-time algorithm. Since P^* attacks a sigma-protocol, it sends two messages. We can thus assume that P^* is of the following form:

- It operates on quantum registers Z, C, R, U. Here Z contains the internal state of P^* (initialized by algorithm Z). C is the register that will contain the first message $a' = (a, (c_z)_z)$ sent by P^*. R, U contains the second message $r' = (r, u)$ sent by P^*. And C, R, U are initialized with $|0\rangle$.
- The unitary U_x describes the unitary operation of P^* on Z, C during the first invocation of P^*. U_x is parametrized by the classical input x of P^*. The message $a' = (a, (c_z)_z)$ is obtained by measuring C in the computational basis.
- The unitary U_z describes the unitary operation of P^* on Z, R, U during the second invocation of P^*. U_z is parametrized by the challenge z that P^* receives. The message $r' = (r, u)$ is obtained by measuring R and U in the computational basis.

We fix some additional notation for this proof:

- V_z: The projector on R, U onto the span of all $|r, u\rangle$ with $verify(1^\eta, c_z, r, u) = 1$. (That is, V_z measures whether measuring R, U would yield a valid opening of c_z.)

---

[14] We can run P_2 several times using the final state of P_1 because P_1 is classical.

- W_z: The projector on R onto the span of all $|r\rangle$ with $V(1^\eta, a, z, r) = 1$. (That is, W_z measures whether measuring R yields a valid response r for challenge z.)
- $P_z := U_z^\dagger W_z V_z U_z$. Since V_z and W_z are projectors and diagonal in the computational basis, they commute and their product is a projector. And since U_z is a unitary, P_z is a projector (acting on registers Z, R, U).
- $x \leftarrow \mathbf{M}(X)$ denotes that x is assigned the result of measuring the register X in the computational basis.
- $ok \leftarrow P(X)$ means that ok is assigned 1 iff measuring the register X with projector P succeeds. (With P being, e.g., one of V_z, W_z, P_z.)
- We write $U(X)$ or $U(X)$ to mean that the unitary U is applied to the register X. (With U being, e.g., one of U_x, U_z).

With that notation, we can rewrite the success probability of the malicious prover as follows:

$$\begin{aligned}
\Pr_V &:= \Pr[\langle \mathsf{P}^*(1^\eta, x, Z), \mathsf{V}(1^\eta, x)\rangle = 1 : (x, Z) \leftarrow \mathsf{Z}(1^\eta)] \\
&= \Pr[ok_c = ok_v = 1 : (x, Z) \leftarrow \mathsf{Z}(1^\eta), \ U_x(ZC), \ (a, (c_z)_z) \leftarrow \mathbf{M}(C), \\
&\qquad z \overset{\$}{\leftarrow} N_z, \ U_z(ZRU), \ r \leftarrow \mathbf{M}(R), \ u \leftarrow \mathbf{M}(U), \\
&\qquad ok_c = verify(1^\eta, c_z, r, u), \ ok_v = V(1^\eta, a, z, r)] \\
&= \Pr[ok = 1 : (x, Z) \leftarrow \mathsf{Z}(1^\eta), \ U_x(ZC), \ (a, (c_z)_z) \leftarrow \mathbf{M}(C), \ z \overset{\$}{\leftarrow} N_z, \\
&\qquad ok \leftarrow P_z(ZRU)].
\end{aligned}$$

We now construct the extractor $\mathsf{K}^{\mathsf{P}^*(1^\eta, x, Z)}(1^\eta, x)$ required by Definition 32. It operates on quantum registers S, C, R, U as follows:

$$\begin{aligned}
&(x, Z) \leftarrow \mathsf{Z}(1^\eta), \ U_x(ZC), \ (a, (c_z)_z) \leftarrow \mathbf{M}(C), \ z, z' \overset{\$}{\leftarrow} N_z, \ U_z(ZRU), \\
&ok_c \leftarrow V_z(RU), r \leftarrow \mathbf{M}(R), \ U_z^\dagger(ZRU), \ U_{z'}(ZRU), \ r' \leftarrow \mathbf{M}(R), \\
&\qquad w \leftarrow E_\Sigma(1^\eta, x, a, z, r, z', r'), \ \text{return } w.
\end{aligned}$$

Here E_Σ is the extractor of the sigma-protocol (N_z, P_1, P_2, V). This extractor exists because the sigma-protocol has computational special soundness (see Definition 33). Note that K only uses black-box access to P (via the unitaries $U_x, U_z, U_{z'}$ and their inverses).

We will now bound the success probability of the extractor

$$\begin{aligned}
\Pr_E &:= \Pr[(x, w) \in R : w \leftarrow \mathsf{K}^{\mathsf{P}^*(1^\eta, x, Z)}(1^\eta, x)] \\
&= \Pr[(x, w) \in R : (x, Z) \leftarrow \mathsf{Z}(1^\eta), \ U_x(ZC), \ (a, (c_z)_z) \leftarrow \mathbf{M}(C), \ z, z' \overset{\$}{\leftarrow} N_z, \\
&\qquad U_z(ZRU), \ ok_c \leftarrow V_z(RU), \ r \leftarrow \mathbf{M}(R), \ U_z^\dagger(ZRU), \ U_{z'}(ZRU), \\
&\qquad r' \leftarrow \mathbf{M}(R), \ w \leftarrow E_\Sigma(1^\eta, x, a, z, r, z', r')] \\
&= \Pr[(x, w) \in R : (x, Z) \leftarrow \mathsf{Z}(1^\eta), \ U_x(ZC), \ (a, (c_z)_z) \leftarrow \mathbf{M}(C), \ z, z' \overset{\$}{\leftarrow} N_z, \\
&\qquad U_z(ZRU), \ ok_c \leftarrow V_z(RU), \ r \leftarrow \mathbf{M}(R), \ ok_v \leftarrow V(1^\eta, x, a, z, r), \\
&\qquad U_z^\dagger(ZRU), \ U_{z'}(ZRU), \ r' \leftarrow \mathbf{M}(R), \ ok'_v \leftarrow V(1^\eta, x, a, z', r'), \\
&\qquad w \leftarrow E_\Sigma(1^\eta, x, a, z, r, z', r')].
\end{aligned}$$

Due to the computational special soundness of (N_z, P_1, P_2, V), in the previous game, with overwhelming probability, $z \neq z'$ and $ok_v = 1$ and $ok_{v'} = 1$ implies $(x, w) \in R$. Thus there exists a negligible μ_1 such that

$$\Pr_E \geq \Pr[z \neq z' \wedge ok_v = ok'_v = 1 : (x, Z) \leftarrow Z(1^\eta),\ U_x(ZC),\ (a, (c_z)_z) \leftarrow M(C),$$
$$z, z' \overset{\$}{\leftarrow} N_z,\ U_z(ZRU),\ ok_c \leftarrow V_z(RU),\ r \leftarrow M(R),$$
$$ok_v \leftarrow V(1^\eta, x, a, z, r),\ U_z^\dagger(ZRU),\ U_{z'}(ZRU),\ r' \leftarrow M(R),$$
$$ok'_v \leftarrow V(1^\eta, x, a, z', r')] - \mu_1 =: \Pr'_E - \mu_1.$$

Instead of computing $ok_v \leftarrow V(1^\eta, x, a, z, r)$ using the just measured r, we can instead measure whether the register R contains a value r that would make $V(1^\eta, x, a, z, r) = 1$ true. I.e., we can replace $ok_v \leftarrow V(1^\eta, x, a, z, r)$ by a measurement using the projector W_z. Since at that point, R was just measured in the computational basis, the measurement using W_z does not disturb the state of the system. Similarly, we can replace $ok'_v \leftarrow V(1^\eta, x, a, z', r')$ by a measurement using $W_{z'}$. We get:

$$\Pr'_E = \Pr[z \neq z' \wedge ok_v = ok'_v = 1 : (x, Z) \leftarrow Z(1^\eta),\ U_x(ZC),\ (a, (c_z)_z) \leftarrow M(C),$$
$$z, z' \overset{\$}{\leftarrow} N_z,\ U_z(ZRU),\ ok_c \leftarrow V_z(RU),\ r \leftarrow M(R),\ ok_v \leftarrow W_z(R),$$
$$U_z^\dagger(ZRU),\ U_{z'}(ZRU),\ r' \leftarrow M(R),\ ok'_v \leftarrow W_{z'}(R)]$$
$$= \Pr[z \neq z' \wedge ok_v = ok'_v = 1 : (x, Z) \leftarrow Z(1^\eta),\ U_x(ZC),\ (a, (c_z)_z) \leftarrow M(C),$$
$$z, z' \overset{\$}{\leftarrow} N_z,\ U_z(ZRU),\ ok_c \leftarrow V_z(RU),\ r \leftarrow M_{ok_c}(R),\ ok_v \leftarrow W_z(R),$$
$$U_z^\dagger(ZRU),\ U_{z'}(ZRU),\ r' \leftarrow M(R),\ ok'_v \leftarrow W_{z'}(R)].$$

In the last probability, $r \leftarrow M_{ok_c}(R)$ refers to a measurement on R that is only executed if $ok_c = 1$. (And $r := \perp$ otherwise.) The last two probabilities are equal because $M(R)$ and $M_{ok_c}(R)$ only differ if $ok_c = 0$, in which case "$z \neq z' \wedge ok_v = ok'_v = 1$" is false anyway.

Since V_z measures whether R, U contains $|r, u\rangle$ with $verify(1^\eta, c_z, r, u) = 1$, and since $(com, verify)$ is collapse-binding, and since the outcome r is never used, we have that no quantum-polynomial-time adversary can distinguish between "$ok_c \leftarrow V_z(RU),\ r \leftarrow M(R)$" and "$ok_c \leftarrow V_z(RU)$", except with negligible probability. (Cf. Definition 12.) Thus there is a negligible μ_2 such that

$$\Pr'_E \geq \Pr[z \neq z' \wedge ok_v = ok'_v = 1 : (x, Z) \leftarrow Z(1^\eta),\ U_x(ZC),\ (a, (c_z)_z) \leftarrow M(C),$$
$$z, z' \overset{\$}{\leftarrow} N_z,\ U_z(ZRU),\ ok_c \leftarrow V_z(RU),\ ok_v \leftarrow W_z(R),\ U_z^\dagger(ZRU),$$
$$U_{z'}(ZRU),\ r' \leftarrow M(R),\ ok'_v \leftarrow W_{z'}(R)] - \mu_2 =: \Pr''_E - \mu_2.$$

Since $M(R)$ and $W_{z'}(R)$ and $V_{z'}(RU)$ commute, and since adding additional/removing operations after all values z, z', ok_v, ok'_v are fixed does not change the distribution of those values, we have that "$r' \leftarrow M(R), ok'_v \leftarrow W_{z'}(R)$" and "$ok'_c \leftarrow V_{z'}(RU), ok'_v \leftarrow W_z(R), U_{z'}^\dagger(ZRU)$" lead to the same distribution of z, z', ok_v, ok'_v. This justifies (*) in the following calculation:

$$\Pr''_E \overset{(*)}{=} \Pr[z \neq z' \wedge ok_v = ok'_v = 1 : (x, Z) \leftarrow Z(1^\eta),\ U_x(ZC),\ (a, (c_z)_z) \leftarrow M(C),$$
$$z, z' \overset{\$}{\leftarrow} N_z,\ U_z(ZRU),\ ok_c \leftarrow V_z(RU),\ ok_v \leftarrow W_z(R),\ U_z^\dagger(ZRU),$$
$$U_{z'}(ZRU),\ ok'_c \leftarrow V_{z'}(RU),\ ok'_v \leftarrow W_{z'}(R),\ U_{z'}^\dagger(ZRU)]$$

$$\geq \Pr[z \neq z' \wedge ok_c = ok_v = 1 \wedge ok'_c = ok'_v = 1 : (x, Z) \leftarrow \mathsf{Z}(1^\eta),\ U_x(ZC),$$

$$(a, (c_z)_z) \leftarrow \mathsf{M}(C),\ z, z' \xleftarrow{\$} N_z,\ U_z(ZRU),\ ok_c \leftarrow V_z(RU),\ ok_v \leftarrow W_z(R),$$

$$U_z^\dagger(ZRU),\ U_{z'}(ZRU),\ ok'_c \leftarrow V_{z'}(RU),\ ok'_v \leftarrow W_{z'}(R),\ U_{z'}^\dagger(ZRU)]$$

$$= \Pr[z \neq z' \wedge ok = 1 \wedge ok = 1 : (x, Z) \leftarrow \mathsf{Z}(1^\eta),\ U_x(ZC),\ (a, (c_z)_z) \leftarrow \mathsf{M}(C),$$

$$z, z' \xleftarrow{\$} N_z,\ ok \leftarrow P_z(ZRU),\ ok \leftarrow P_{z'}(ZRU)].$$

Let $\alpha_{a'} := \Pr[a' = (a, (c_z)_z)]$ in the previous game, and let $|\psi_{a'}\rangle$ denote the post-measurement-state of registers Z, R, U after the measurement $(a, (c_z)_z) \leftarrow \mathsf{M}(C)$. Then

$$\Pr''_E = \sum_{a'} \alpha_{a'} \underbrace{\sum_{\substack{z, z' \\ z \neq z'}} \frac{1}{|N_z|^2} \left\| P_{z'} P_z |\psi_{a'}\rangle \right\|^2}_{=:E_{a'}}.$$

Furthermore, note that

$$\Pr_V = \sum_{a'} \alpha_{a'} \underbrace{\sum_{z} \frac{1}{|N_z|} \left\| P_z |\psi_{a'}\rangle \right\|^2}_{=:V_{a'}}.$$

Lemma 37 implies that if $V_{a'} \geq 1/\sqrt{|N_z|}$, then $E_{a'} \geq V_{a'}(V_{a'}^2 - 1/|N_z|)$. Or stated differently: $E_{a'} \geq \varphi(V_{a'})$ where $\varphi(x) := 0$ for $x < 1/\sqrt{|N_z|}$ and $\varphi(x) := x(x^2 - 1/|N_z|)$ for $x \geq 1/\sqrt{|N_z|}$. Since φ is convex on $[0, 1]$, by Jensen's inequality we get $\Pr''_E \geq \varphi(\Pr_V)$. In other words $\Pr''_E \geq \Pr_V(\Pr_V^2 - 1/|N_z|)$ whenever $\Pr_V \geq 1/\sqrt{|N_z|}$. Furthermore, the inequalities derived above give $\Pr_E \geq \Pr''_E - \mu$ for $\mu := \mu_1 + \mu_2$. And μ is negligible. It follows that:

$$\Pr_V \geq \frac{1}{\sqrt{N_z}} \quad \Longrightarrow \quad \Pr_E \geq \Pr_V\left(\Pr_V^2 - \frac{1}{|N_z|}\right) - \mu \geq \left(\Pr_V - \frac{1}{\sqrt{|N_z|}}\right)^3 - \mu.$$

Thus (P, V) is quantum-computationally extractable for R with knowledge error $\kappa := 1/\sqrt{|N_z|}$. $\square$

In [20], we additionally show that the resulting protocol is also zero-knowledge. (This only uses the hiding property, and is hence independent of our new definitions).

Theorem 38 (Zero-Knowledge). *If $|N_z|$ is polynomially-bounded, and (N_z, P_1, P_2, V) is HVZK and $(com, verify)$ is computationally hiding, and com is a polynomial-time algorithm, then (N_z, P'_1, P'_2, V') is computational zero-knowledge.*

If $|N_z|$ is polynomially-bounded, and (N_z, P_1, P_2, V) is SHVZK and $(com, verify)$ is statistically hiding, and com is a polynomial-time algorithm, then (N_z, P'_1, P'_2, V') is statistical zero-knowledge.

8   Open Problems

We list some questions for future research:

- We have constructed quantum arguments of knowledge from sigma-protocols by using collapse-binding commitments. However, our construction requires the challenge space N_z of the sigma-protocol to be of polynomially-bounded size. As a consequence, the resulting argument of knowledge will have a noticeable knowledge error; for a negligible knowledge error we need to use sequential repetition, resulting in a proof system with non-constant round complexity. Are there general constructions of arguments of knowledge from sigma-protocols that do not require the challenge space to be polynomially-bounded?
- Can we use collapse-binding commitments to construct a quantum OT protocol? For example, using the construction from [2] or a variation thereof?
- How are the various definitions of computationally binding commitments related? That is, which implications and separations exist between sumbinding, CDMS-binding, collapse-binding, and UC-secure commitments?

Acknowledgements. We thank Ansis Rosmanis for discussions on insecure commitments based on collision-resistant hash functions, and Serge Fehr for discussions on the DFRSS-binding definition. This research by the European Social Fund's Doctoral Studies and Internationalisation Programme DoRa, by the European Regional Development Fund through the Estonian Center of Excellence in Computer Science, EXCS, by European Social Fund through the Estonian Doctoral School in Information and Communication Technology, and by the Estonian ICT program 2011–2015 (3.2.1201.13-0022).

References

1. Ambainis, A., Rosmanis, A., Unruh, D.: Quantum attacks on classical proof systems (the hardness of quantum rewinding). In: FOCS 2014, pp. 474–483. IEEE (2014)
2. Bennett, C.H., Brassard, G., Crépeau, C., Skubiszewska, M.-H.: Practical quantum oblivious transfer. In: Feigenbaum, J. (ed.) CRYPTO 1991. LNCS, vol. 576, pp. 351–366. Springer, Heidelberg (1992)
3. Boneh, D., Dagdelen, Ö., Fischlin, M., Lehmann, A., Schaffner, C., Zhandry, M.: Random oracles in a quantum world. In: Lee, D.H., Wang, X. (eds.) ASIACRYPT 2011. LNCS, vol. 7073, pp. 41–69. Springer, Heidelberg (2011)
4. Brassard, G., Crépeau, C., Jozsa, R., Langlois, D.: A quantum bit commitment scheme provably unbreakable by both parties. In: FOCS 1993, pp. 362–371. IEEE (1993)
5. Cramer, R., Damgård, I.B., Schoenmakers, B.: Proof of partial knowledge and simplified design of witness hiding protocols. In: Desmedt, Y.G. (ed.) CRYPTO 1994. LNCS, vol. 839, pp. 174–187. Springer, Heidelberg (1994)
6. Crépeau, C., Dumais, P., Mayers, D., Salvail, L.: Computational collapse of quantum state with application to oblivious transfer. In: Naor, M. (ed.) TCC 2004. LNCS, vol. 2951, pp. 374–393. Springer, Heidelberg (2004)

7. Crépeau, C., Légaré, F., Salvail, L.: How to convert the flavor of a quantum bit commitment. In: Pfitzmann, B. (ed.) EUROCRYPT 2001. LNCS, vol. 2045, pp. 60–77. Springer, Heidelberg (2001)

8. Crépeau, C., Salvail, L., Simard, J.-R., Tapp, A.: Two provers in isolation. In: Lee, D.H., Wang, X. (eds.) ASIACRYPT 2011. LNCS, vol. 7073, pp. 407–430. Springer, Heidelberg (2011)

9. Damgård, I., Fehr, S., Lunemann, C., Salvail, L., Schaffner, C.: Improving the security of quantum protocols via commit-and-open. In: Halevi, S. (ed.) CRYPTO 2009. LNCS, vol. 5677, pp. 408–427. Springer, Heidelberg (2009)

10. Damgård, I.B., Fehr, S., Renner, R.S., Salvail, L., Schaffner, C.: A tight high-order entropic quantum uncertainty relation with applications. In: Menezes, A. (ed.) CRYPTO 2007. LNCS, vol. 4622, pp. 360–378. Springer, Heidelberg (2007)

11. Damgård, I.B., Fehr, S., Salvail, L.: Zero-knowledge proofs and string commitments withstanding quantum attacks. In: Franklin, M. (ed.) CRYPTO 2004. LNCS, vol. 3152, pp. 254–272. Springer, Heidelberg (2004)

12. Damgård, I., Lunemann, C.: Quantum-secure coin-flipping and applications. In: Matsui, M. (ed.) ASIACRYPT 2009. LNCS, vol. 5912, pp. 52–69. Springer, Heidelberg (2009)

13. Dumais, P., Mayers, D., Salvail, L.: Perfectly concealing quantum bit commitment from any quantum one-way permutation. In: Preneel, B. (ed.) EUROCRYPT 2000. LNCS, vol. 1807, pp. 300–315. Springer, Heidelberg (2000)

14. Halevi, S., Micali, S.: Practical and provably-secure commitment schemes from collision-free hashing. In: Koblitz, N. (ed.) CRYPTO 1996. LNCS, vol. 1109, pp. 201–215. Springer, Heidelberg (1996)

15. Mayers, D.: Unconditionally secure quantum bit commitment is impossible. PRL 78(17), 3414–3417 (1997)

16. Nielsen, M., Chuang, I.: Quantum Computation and Quantum Information, 10th Anniv. edn. Cambridge University Press, Cambridge (2010)

17. NIST: SHA-3 standard: Permutation-based hash and extendable-output functions. Draft FIpPS 202 (2014)

18. Unruh, D.: Universally composable quantum multi-party computation. In: Gilbert, H. (ed.) EUROCRYPT 2010. LNCS, vol. 6110, pp. 486–505. Springer, Heidelberg (2010)

19. Unruh, D.: Quantum proofs of knowledge. In: Pointcheval, D., Johansson, T. (eds.) EUROCRYPT 2012. LNCS, vol. 7237, pp. 135–152. Springer, Heidelberg (2012)

20. Unruh, D.: Computationally binding quantum commitments. IACR ePrint 2015/361 (2015). (full version of this paper)

21. Unruh, D.: Non-interactive zero-knowledge proofs in the quantum random oracle model. In: Oswald, E., Fischlin, M. (eds.) EUROCRYPT 2015. LNCS, vol. 9057, pp. 755–784. Springer, Heidelberg (2015)

22. Unruh, D.: Quantum proofs of knowledge. IACR ePrint 2010/212/20150211:174234 (2015). updated full version of [19]

23. Watrous, J.: Zero-knowledge against quantum attacks. SIAM J. Comput. 39(1), 25–58 (2009)

24. Zhandry, M.: A note on the quantum collision and set equality problems. Quantum Information & Computation 15(7&8), 557–567 (2015)

Structural Lattice Reduction: Generalized Worst-Case to Average-Case Reductions and Homomorphic Cryptosystems

Nicolas Gama[1,2]($\boxtimes$), Malika Izabachène[3], Phong Q. Nguyen[4,5], and Xiang Xie[6]

[1] Laboratoire de Mathématiques de Versailles, UVSQ, CNRS,
Université Paris-Saclay, 78035 Versailles, France
nicolas.gama@uvsq.fr
[2] Inpher, Lausanne, Switzerland
[3] CEA, LIST, 91191 Gif-sur-Yvette Cedex, France
[4] Inria, Paris, France
Phong.Nguyen@inria.fr
[5] CNRS/JFLI and the University of Tokyo, Tokyo, Japan
[6] Huawei Technologies, Shenzhen, China

Abstract. In lattice cryptography, worst-case to average-case reductions rely on two problems: Ajtai's SIS and Regev's LWE, which both refer to a very small class of random lattices related to the group $G = \mathbb{Z}_q^n$. We generalize worst-case to average-case reductions to all integer lattices of sufficiently large determinant, by allowing G to be any (sufficiently large) finite abelian group. Our main tool is a novel generalization of lattice reduction, which we call structural lattice reduction: given a finite abelian group G and a lattice L, it finds a short basis of some lattice $\bar{L}$ such that $L \subseteq \bar{L}$ and $\bar{L}/L \simeq G$. Our group generalizations of SIS and LWE allow us to abstract lattice cryptography, yet preserve worst-case assumptions: as an illustration, we provide a somewhat conceptually simpler generalization of the Alperin-Sheriff-Peikert variant of the Gentry-Sahai-Waters homomorphic scheme. We introduce homomorphic mux gates, which allows us to homomorphically evaluate any boolean function with a noise overhead proportional to the square root of its number of variables, and bootstrap the full scheme using only a linear noise overhead.

1 Introduction

A lattice is a discrete subgroup of $\mathbb{R}^m$. Nearly two decades after its introduction, lattice-based cryptography has emerged as a credible alternative to classical public-key cryptography based on factoring or discrete logarithm. It offers new properties (such as security based on worst-case assumptions) and new functionalities, such as noisy multilinear maps and fully-homomorphic encryption. The worst-case guarantees of lattice-based cryptography come from two problems: Ajtai's *short integer solution* (SIS) [1] and Regev's *learning with errors* (LWE) [37]. These average-case problems are provably as hard as solving certain

© International Association for Cryptologic Research 2016
M. Fischlin and J.-S. Coron (Eds.): EUROCRYPT 2016, Part II, LNCS 9666, pp. 528–558, 2016.
DOI: 10.1007/978-3-662-49896-5_19

lattice problems in the worst case, such as GapSVP (the decision version of the shortest vector problem) and SIVP (finding short lattice vectors).

As noted by Micciancio [25], the SIS problem can be defined as finding short vectors in a random lattice from a class $\mathcal{A}_{n,m,q}$ of m-dimensional integer lattices related to the finite abelian group $G = \mathbb{Z}_q^n$, where n is the dimension of the worst-case lattice problem and q needs to be sufficiently large: any $\mathbf{g} = (g_1, \ldots, g_m) \in G^m$ chosen uniformly at random defines a lattice $\mathcal{L}_{\mathbf{g}} \in \mathcal{A}_{n,m,q}$ formed by all $\mathbf{x} = (x_1, \ldots, x_m) \in \mathbb{Z}^m$ s.t. $\sum_{i=1}^m x_i g_i = 0$ in G; and SIS asks, given $\mathbf{g}$, to find a short (nonzero) $\mathbf{x} \in \mathcal{L}_{\mathbf{g}}$. The class $\mathcal{A}_{n,m,q}$ has an algebraic meaning: for suitable parameters, the distribution of $\mathcal{L}_{\mathbf{g}}$ is statistically close to the uniform distribution over the finite set $\mathcal{L}_{G,m}$ of all full-rank lattices $L \subseteq \mathbb{Z}^m$ such that $\mathbb{Z}^m/L \simeq G$. This suggests that Ajtai's lattices are very rare among all integer lattices: in fact, Nguyen and Shparlinski [31] recently showed that the set $\cup_{G \text{ cyclic}} \mathcal{L}_{G,m}$ of all full-rank integer lattices $L \subseteq \mathbb{Z}^m$ such that $\mathbb{Z}^m/L$ is cyclic (unlike $\mathbb{Z}_q^n$) has natural density $1/[\zeta(6) \prod_{k=4}^m \zeta(k)] \approx 85\%$ (for large m), which implies that Ajtai's classes $\mathcal{A}_{n,m,q}$ form a minority among all integer lattices.

This motivates the natural question of whether other classes of random lattices enjoy similar worst-case to average-case reductions: if we call GSIS the SIS generalization (introduced by Micciancio [25, Definition 5.2]) to any finite abelian group G, does GSIS have similar properties as SIS for other groups than $G = \mathbb{Z}_q^n$? This would imply that the random lattices of $\mathcal{L}_{G,m}$ are also hard. Ajtai (in the proceedings version of [1]) and later Regev [36] noticed that the choice $G = \prod_{i=1}^n \mathbb{Z}_{q_i}$ where the q_i's are distinct prime numbers of similar bit-length also worked. Micciancio [25] gave another choice of G, to obtain a better worst-case to average-case connection (at that time): his G is actually constructed by an algorithm [25, Lemma 2.11] given as input a very special lattice (for which solving the closest vector problem is easy); if the input lattice is $\mathbb{Z}^n$, then $G = (\mathbb{Z}_q)^n$. However, all these choices of G are very special, and it was unknown if the hardness properties held outside a small family of finite abelian groups.

A similar question can be asked for LWE, which is known as a dual problem of SIS, and has been used extensively in lattice-based encryption. However, in order to define GLWE by analogy with GSIS, we need to change the usual definition of LWE based on linear algebra. Any finite abelian group G is isomorphic to its dual group $\hat{G}$ formed by its characters, *i.e.* homomorphisms from G to the torus $\mathbb{T} = \mathbb{R}/\mathbb{Z}$. We define search-GLWE as the problem of learning a character $\hat{s} \in \hat{G}$ chosen uniformly at random, given noisy evaluations of $\hat{s}$ at (public) random points $g_1, \ldots, g_m \in G$, namely one is given g_i and a "Gaussian" perturbation of $\hat{s}(g_i)$ for all $1 \leq i \leq m$. Decisional-GLWE is defined as the problem of distinguishing the previous "Gaussian" perturbations of $\hat{s}(g_i)$ from random elements in $\mathbb{T}$. If $G = (\mathbb{Z}_q)^n$, it can be checked that GLWE is LWE. If $G = \mathbb{Z}_p$ for some large prime p, search-GLWE is a randomized version of Boneh-Venkatesan's *Hidden Number Problem* (HNP) [8] (introduced to study the bit-security of Diffie-Hellman key exchange, but also used in side-channel attacks on discrete-log based signatures [30]), which asks to recover a secret number $s \in \mathbb{Z}_p$, given random $t_1, \ldots, t_m$ chosen uniformly from $\mathbb{Z}_p$ and approximations of each $st_i \bmod p$. Here, randomized means that the approximations given are "Gaussian" perturbations of $st_i \bmod p$. Thus, GLWE captures LWE and the HNP as a single problem, instantiated with different groups. Alternatively, GLWE can be viewed as a lattice problem: solving a randomized version of bounded distance decoding (with "Gaussian" errors) for the dual lattice of $\mathcal{L}_{\mathbf{g}}$.

OUR RESULTS. We show that the worst-case to average-case reductions for SIS and LWE (search and decisional) can be generalized to GSIS and GLWE, provided that G is any sufficiently large finite abelian group, $e.g.$ of order $n^{\Omega(\max(n,\mathrm{rank}(G)))}$ if n is the dimension of the worst-case lattice problem and rank (G) denotes the minimal size of a generating set for G: note that the order of G is the determinant of the average-case lattice. For GSIS and search-GLWE, our reductions are direct from worst-case problems. We transfer decisional-LWE hardness results to decisional-GLWE by generalizing the modulus-dimension switching technique of Brakerski et al. [11].

We believe that our results offer a cleaner high-level picture of worst-case to average-case reductions: previous work tend to focus on quantitative aspects (such as decreasing the worst-case approximation factor, or the parameter q, $etc.$), including work on the ring setting, where one introduces a trade-off between security and efficiency. The ring setting offers more efficient primitives but requires (much) stronger worst-case assumptions: in the ring variants of SIS and LWE, the worst-case lattices are restricted to classes of very special lattices known as ideal lattices.

Our reductions are based on a new tool, which we call structural lattice reduction, and which is of independent interest: Becker et al. [5] recently used it to design new exponential-space algorithms for lattice problems. In lattice reduction, one is given a full-rank lattice $L \subseteq \mathbb{Z}^n$ and wants to find a short basis of L. In our structural lattice reduction, one is further given a finite abelian group G of rank $\leq n$, and wants to find a short basis of some overlattice $\bar{L}$ of L such that $\bar{L}/L \simeq G$ effectively, $i.e.$ there exists an efficiently computable surjective map φ from $\bar{L}$ to G with $\ker \varphi = L$. Our key point is that previous worst-case to average-case reductions ($e.g.$ [11,20]) implicitly used a trivial case[1] of structural lattice reduction: if B is a short basis of a full-rank lattice $L \subseteq \mathbb{Z}^n$ and q is an integer, then $q^{-1}B$ is a short basis of the lattice $\bar{L} = q^{-1}L$ such that $\bar{L}/L \simeq \mathbb{Z}_q^n$, which summarizes the importance of $\mathbb{Z}_q^n$ in SIS and LWE.

Our GSIS reduction shows that in some sense all integer lattices are hard. Indeed, the set of full-rank lattices $L \subseteq \mathbb{Z}^m$ (of sufficiently large co-volume $\geq n^{\Omega(m)}$) can be partitioned based on the finite abelian group $\mathbb{Z}^m/L$, and the reduction implies that each partition cell $\mathcal{L}_{G,m}$ has this worst-case to average-case property: finding short vectors in a lattice chosen uniformly at random from $\mathcal{L}_{G,m}$ is as hard as finding short vectors in any integer lattice of dimension n.

Consider the special case $G = \mathbb{Z}_p$ for a large prime p. Then our GSIS reduction provides the first hardness results for the random lattices in $\mathcal{L}_{\mathbb{Z}_p,m}$ used in many experiments [14,18] to benchmark lattice reduction algorithms, as well as

---

[1] There is a more technical reduction implicitly proposed in [25], but unfortunately too restrictive on the choice of G.

in Darmstadt's SVP internet challenges. And our GLWE reduction provides a general hardness result for the HNP: previously, [11, Corollary 3.4] established the hardness for HNP when the large prime p is replaced by q^n where q is smooth.

Finally, our generalizations of SIS and LWE allow us to abstract (the many) lattice-based schemes based on SIS and/or LWE, where the role of $G = (\mathbb{Z}_q)^n$ was not very explicit in most descriptions (typically based on linear algebra). We believe such an abstraction can have several benefits. First, it can clarify analyses and designs: the El Gamal cryptosystem is arguably better described with an arbitrary group G, rather than by focusing on the historical choice $G = \mathbb{Z}_p^*$; comparisons and analogies with "traditional" public-key cryptography based on factoring or discrete logarithm will be easier. We illustrate this point by providing a somewhat conceptually simpler GLWE-based generalization of the Alperin-Sheriff-Peikert variant [2] of the Gentry-Sahai-Waters homomorphic scheme [21]: this generalization becomes essentially as simple as trapdoor-based fully-homormophic encryption proposals such as [38]. It is based on a GLWE variant of El Gamal encryption, which naturally generalizes Regev's LWE encryption [37]. We also provide a new decryption circuit based on Mux gates, which can bootstrap the system with a polynomial noise overhead, and is arguably simpler than [2]. Second, it opens up the possibility of obtaining more efficient schemes using different choices of G than $G = (\mathbb{Z}_q)^n$. We do not claim that there are better choices than $(\mathbb{Z}_q)^n$, but such a topic is worth investigating, which we leave to future work. Many factors influence efficiency: trapdoor generation, hashing, efficiency of the security reduction, *etc.* For instance, hashing onto $\mathbb{Z}_p$ can sometimes be more efficient than onto $(\mathbb{Z}_q)^n$ for large n, which could be useful in certain settings, like digital signatures.

Furthermore, our abstraction may also be helpful to better understand attacks on GSIS and GLWE. For instance, there are similarities between Bleichenbacher's algorithm [6] for HNP and the BKW algorithm [7] for LWE: by viewing LWE and HNP as two different instances of the same problem GLWE, one can focus on the main ideas. And we note that among several classes of random lattices having a worst-case to average-case reduction, it could be that some are weaker than others, when it comes to the best attack known.

RELATED WORK. Baumslag *et al.* also introduced in [4] group generalizations of LWE for non-commutative groups, but did not obtain hardness result. [16] showed a self-reducibility property for some special non-commutative groups.

OPEN PROBLEMS. Similarly to [11], our strongest hardness result for decisional-GLWE bypasses search-GLWE: a direct search-to-decision equivalence for all sufficiently large G is open. Adapting structural lattice reduction to the ring setting is open: current ring results only address the average-case hardness of very few classes of lattices, and it would be interesting to tackle more classes. Our reductions require the order of G to be large compared to the worst-case lattice dimension, and we would like to minimize this constraint: the GLWE case $G = \mathbb{Z}_2^n$ is essentially LPN, whose hardness is open; here, the order 2^n does not grow quickly enough with respect to the rank n for our reduction. On the other hand, Micciancio and Peikert [27] recently decreased q for SIS.

ROADMAP. Section 2 gives background. Section 3 presents our group generalizations of SIS and LWE. Section 4 presents structural lattice reduction. Sections 5 and 6 show hardness of GSIS and decisional-GLWE. In Sect. 7, we give an example of abstracting lattice cryptography: El Gamal-like encryption and fully-homomorphic encryption from GLWE. Detailed missing proofs can be found in the full version of the paper [17]. In particular, we compare structural reduction with previous work of Ajtai [1] and Micciancio [25]: and show that all previous SIS reductions can be captured by our overlattice framework.

2 Background and Notation

$\mathbb{Z}_q$ denotes $\mathbb{Z}/q\mathbb{Z}$. We use row notation for vectors and matrices. I_n is the $n \times n$ id. matrix. A function $\mathrm{negl}(n)$ is *negligible* if it vanishes faster than any inverse polynomial. $\|B\| = \max_{1 \le i \le n} \|\mathbf{b}_i\|$ is the maximal row norm of a matrix B.

Lattices. A *lattice* L is of the form $L(B) = \{\sum_{i=1}^{n} \alpha_i \mathbf{b}_i,\ \alpha_i \in \mathbb{Z}\}$ for some *basis* $B = (\mathbf{b}_1, \ldots, \mathbf{b}_n)$ of linearly independent vectors in $\mathbb{R}^m$. If $L \subseteq \mathbb{Z}^m$, L is an *integer lattice*. The dimension n of $\mathrm{span}(L)$ is the *dimension* $\dim(L)$ of L. The *(co)-volume* $\mathrm{vol}(L)$ is $\sqrt{\det(BB^t)}$ for any basis B of L. For $1 \le i \le \dim(L)$, $\lambda_i(L)$ is the i-th minimum of L, (smallest radius of the 0-ball containing at least i linearly indep. lattice vectors). The *dual lattice* $L^\times$ is the set of all $\mathbf{u} \in \mathrm{span}(L)$ s.t. $\langle \mathbf{u}, \mathbf{v} \rangle \in \mathbb{Z}$ for all $\mathbf{v} \in L$. If B is a basis of L, its *dual basis* $B^\times = (BB^t)^{-1}B$ is a basis of $L^\times$. For a factor $\gamma = \gamma(n) \ge 1$, GapSVP_γ asks, given $d \ge 0$ and a basis B of an n-dim lattice L, to decide if $\lambda_1(L) \le d$ or $\lambda_1(L) > \gamma d$. $\mathrm{ApproxSIVP}_\gamma$ asks a full-rank family of lattice vectors of norm $\le \gamma \lambda_n(L)$.

Gram-Schmidt Orthogonalization (GSO). The GSO of a lattice basis $B = (\mathbf{b}_1, \ldots, \mathbf{b}_n)$ is the unique decomposition $B = \mu \cdot D \cdot Q$, where μ is a lower triangular matrix with unit diagonal, D is a positive diagonal matrix, and Q has orthonormal rows. We let $B^* = DQ$ whose i-th row $\mathbf{b}_i^*$ is $\pi_i(\mathbf{b}_i)$, where π_i denotes the orthogonal projection of $\mathbf{b}_i$ over $\mathrm{span}\{\mathbf{b}_1, \ldots, \mathbf{b}_{i-1}\}^\perp$. We use the notation $B_{[i,j]}$ for the block $[\pi_i(\mathbf{b}_i), \ldots, \pi_i(\mathbf{b}_j)]$. If $B^\times$ is the dual basis of B and $(B^\times)^*$ denotes its GSO matrix, then $\|(\mathbf{b}_i^\times)^*\| \cdot \|\mathbf{b}_{n-i+1}^*\| = 1$ for $1 \le i \le n$.

(Explicit) Finite Abelian Groups. Any finite abelian group G is isomorphic to a product $\prod_{i=1}^{k} \mathbb{Z}_{q_i}$ of cyclic groups. We call *rank* of G the minimal number of cyclic groups in such decompositions: this should not be confused with the rank of an abelian group. We say that G is *explicit* if one knows $q_1, \ldots, q_k \in \mathbb{N}$ and an isomorphism $\prod_{i=1}^{k} \mathbb{Z}_{q_i} \to G$ computable in poly-time: wlog k is the rank and $q_{i+1} | q_i$. The isomorphism induces k generators $e_1, \ldots, e_k \in G$ s.t. $G = \langle e_1 \rangle \oplus \cdots \oplus \langle e_k \rangle$ and each e_i has order q_i. If the inverse of the isomorphism is also computable in polynomial time, we say that G is *fully-explicit*.

Overlattices. When a lattice $\bar{L}$ contains a sublattice L of the same dimension n, $\bar{L}$ is an *overlattice* of L. Then $\bar{L}/L$ is a finite abelian group of rank $\le n$ and order $\mathrm{vol}(L)/\mathrm{vol}(\bar{L})$. Then we note $\bar{L}/L \overset{\varphi}{\simeq} G$ for some φ, i.e. $\varphi : \bar{L} \to G$ is a surjective morphism s.t $\ker \varphi = L$.

Lattice Reduction. Cai [13] introduced the *basis length* of a lattice L as $\mathrm{bl}(L) = \min_{\text{basis } B} \|B^*\|$. Then: $\lambda_n(L) \geq \mathrm{bl}(L) \geq \lambda_n(L)/\sqrt{n}$, $\mathrm{bl}(L) \geq \lambda_1(L)$, and $\mathrm{bl}(L) \geq \mathrm{vol}(L)^{1/n}$. Lattice reduction can find bases B with small $\|B^*\|$. A basis B is LLL-reduced [23] with factor $\varepsilon_{\mathrm{LLL}} \geq 0$ if its GSO satisfies $|\mu_{i,j}| \leq \frac{1}{2}$ for all $1 \leq j < i$ and $\|\mathbf{b}_i^*\|^2 \leq (1+\varepsilon_{\mathrm{LLL}})(\|\mathbf{b}_{i+1}^*\|^2 + \mu_{i+1,i}\|\mathbf{b}_i^*\|^2)$. Then it is folklore that: $\|B^*\| \leq \left((1+\varepsilon_{\mathrm{LLL}})\sqrt{4/3}\right)^{(n-1)/2} \mathrm{bl}(L)$. Given $\varepsilon_{\mathrm{LLL}} > 0$ and a basis B of a lattice $L \subseteq \mathbb{Z}^n$, LLL [23] outputs an LLL-reduced basis of factor $\varepsilon_{\mathrm{LLL}}$ in time polynomial in $1/\varepsilon_{\mathrm{LLL}}$ and $\mathrm{size}(B)$. Usually, $(1+\varepsilon_{\mathrm{LLL}})\sqrt{4/3} = \sqrt{2}$ or $\varepsilon_{\mathrm{LLL}} = 1/\mathrm{poly}(n)$.

2.1   Gaussian Measures

The statistical distance between two distributions $\mathcal{P}$ and $\mathcal{Q}$ over a domain X is $\Delta(\mathcal{P}, \mathcal{Q}) = \frac{1}{2}\int_{a \in X} |\mathcal{P}(a) - \mathcal{Q}(a)| da$ or $\frac{1}{2}\sum_{a \in X} |\mathcal{P}(a) - \mathcal{Q}(a)|$ when X is discrete. $\mathcal{P}$ and $\mathcal{Q}$ are (statistically) ε-indistinguishable if $\Delta(\mathcal{P}, \mathcal{Q}) < \varepsilon$. We write $\mathbf{y} \leftarrow \mathcal{P}$ (resp. $\leftarrow_\varepsilon \mathcal{P}$) for a sample $\mathbf{y}$ from the distrib. $\mathcal{P}$ (resp. a distribution ε-indistinguishable from $\mathcal{P}$). And $\leftarrow_\approx$ means $\leftarrow_\varepsilon$ for some negligible function ε.

Gaussian Distributions. The *Gaussian Distribution* (over $\mathbb{R}^n$) $\mathcal{D}_{\mathbb{R}^n,\sigma,\mathbf{c}}$ centered at $\mathbf{c} \in \mathbb{R}^n$ of parameter $\sigma \in \mathbb{R}_{\geq 0}$ has a density function proportional to $\rho_{\mathbb{R}^n,\sigma,\mathbf{c}}(\mathbf{x}) = \exp\left(-\pi\|\mathbf{x} - \mathbf{c}\|^2/\sigma^2\right)$. If $\mathbf{c}$ is omitted, then $\mathbf{c} = 0$. For any countable subset $C \subseteq \mathbb{R}^n$ (a lattice L or a coset $\mathbf{x}+L$), $\rho_{\mathbb{R}^n,\sigma,\mathbf{c}}(C)$ is $\sum_{\mathbf{u} \in C} \rho_{\mathbb{R}^n,\sigma,\mathbf{c}}(\mathbf{u})$. The *discrete Gaussian distribution* $\mathcal{D}_{C,\sigma,\mathbf{c}}$ over a lattice or coset $C \subset \mathbb{R}^n$ is $\mathcal{D}_{C,\sigma,\mathbf{c}}(\mathbf{x}) = \rho_{\mathbb{R}^n,\sigma,\mathbf{c}}(\mathbf{x})/\rho_{\mathbb{R}^n,\sigma,\mathbf{c}}(C)$ where $\mathbf{x} \in C$. One can sample efficiently the discrete Gaussian distribution within negligible distance [20,35] or exactly [11]:

Lemma 1. *There is a poly-time algorithm which, given $\mathbf{c} \in \mathbb{Q}^n$, a basis B of a lattice $L \subseteq \mathbb{Q}^n$ and $\sigma \geq \|B^*\| \cdot \sqrt{\ln(2n+4)/\pi}$, samples the dist. $\mathcal{D}_{L,\sigma,\mathbf{c}}$.*

Reciprocally, a short lattice basis is derived from short discrete Gaussian samples:

Proposition 1. *(Corollary of [36, Lemma 14]) Let $\varepsilon > 0$ and $L(B)$ be an n-dim lattice. Given $m = O(n)$ indep. samples $\mathbf{y}_i \leftarrow_\varepsilon \mathcal{D}_{L,s_i}$ s.t. $\sqrt{2}\eta_\varepsilon(L) \leq s_i \leq \sigma$, $1 \leq i \leq m$, one can compute in poly-time a basis C of L s.t. $\|C^*\| \leq \sqrt{n/2\pi} \cdot \max_i s_i$.*

Modular Distributions and Smoothing Parameter. The distributions $\mathcal{D}_{\mathbb{R}^n,\sigma,\mathbf{c}}$ and $\mathcal{D}_{\bar{L},\sigma,\mathbf{c}}$ over an overlattice $\bar{L} \supseteq L$ can be projected modulo L: $\mathcal{D}_{\mathbb{R}^n/L,\sigma,\mathbf{c}}$ (resp. $\mathcal{D}_{\bar{L}/L,\sigma,\mathbf{c}}$) has density $\mathcal{D}_{\mathbb{R}^n,\sigma,\mathbf{c}}(\mathbf{x}+L)$ for $\mathbf{x} \in \mathbb{R}^n/L$ (resp. $\bar{L}/L$). Both $\mathcal{D}_{\mathbb{R}^n/L,\sigma}$ and $\mathcal{D}_{\bar{L}/L,\sigma}$ converge (uniformly) to the uniform distribution when σ increases. This is quantified by the *smoothing parameter* $\eta_\varepsilon(L)$ [28], i.e. the minimal $\sigma > 0$ for $\varepsilon > 0$ s.t. $\rho_{\mathbb{R}^n,\frac{1}{\sigma}}(L^\times \setminus \{0\}) \leq \varepsilon$, i.e. $\left\|\mathcal{D}_{\mathbb{R}^n/L,\sigma}(\mathbf{x}+L) - \frac{1}{\mathrm{vol}(L)}\right\|_\infty \leq \frac{\varepsilon}{\mathrm{vol}(L)}$:

Lemma 2. (see Corollary 2.8 of [20]). *If $\bar{L}$ is an overlattice of L, $\varepsilon \in (0,1/2)$, $\sigma \geq \eta_\varepsilon(L)$ and $\mathbf{c} \in \mathbb{R}^n$, then $\mathcal{D}_{\bar{L}/L,\sigma,\mathbf{c}+L}$ is within stat. distance $\leq 2\varepsilon$ from the uniform distribution over $\bar{L}/L$.*

For any n-dim basis B, $\eta_\varepsilon(L(B)) \leq \eta_\varepsilon(L(B^*)) \leq \eta_\varepsilon(\mathbb{Z}^n) \cdot \|B^*\|$ where $\eta_\varepsilon(\mathbb{Z}^n) \leq \sqrt{\log\left(2n \cdot (1 + \frac{1}{\varepsilon})\right)/\pi}$. In particular, $\eta_\varepsilon(L) \leq \eta_\varepsilon(\mathbb{Z}^n) \cdot \mathrm{bl}(L)$. Finally, we give a technical lemma (proved in App. A.2 of the full version [17]), analogous to [35,37].

Lemma 3. *Let $\mathbb{K} = \mathbb{R}$ or $\mathbb{T}$. Let $c \in \mathbb{R}$, $\mathbf{u} \in \mathbb{R}^n$, $\alpha, \sigma \in \mathbb{R}_{\geq 0}$, $\varepsilon \in (0, 1/2)$ and $\mathbf{z} + L$ be a coset of an n-dim lattice $L \subseteq \mathbb{R}^n$. Assume that $\left(\frac{1}{\sigma^2} + \frac{\|\mathbf{u}\|^2}{\alpha^2}\right)^{-1/2} \geq \eta_\varepsilon(L)$. Then $\mathcal{D}_{\mathbb{K}, \alpha, c + \langle \mathbf{u}, \mathbf{v} \rangle}$ where $\mathbf{v} \leftarrow \mathcal{D}_{\mathbf{z}+L, \sigma}$ is within statistical distance $\leq 4\varepsilon$ from $\mathcal{D}_{\mathbb{K}, \sqrt{\alpha^2 + \sigma^2 \|\mathbf{u}\|^2}, c}$. This still holds when $\mathbb{K} = \frac{1}{N}\mathbb{Z}$ or $\frac{1}{N}\mathbb{Z}/\mathbb{Z}$ if $\alpha \geq \eta_\varepsilon(\frac{1}{N}\mathbb{Z})$.*

3   Lattice Factor Groups and Generalizations of SIS/LWE

3.1   Lattice Factor Groups

If L is a full-rank lattice $\subseteq \mathbb{Z}^m$, its factor group $\mathbb{Z}^m/L$ is a finite abelian group of order $\mathrm{vol}(L)$. For any finite abelian group G, denote by $\mathcal{L}_{G,m}$ the (finite) set of full-rank lattices $L \subseteq \mathbb{Z}^m$ such that $\mathbb{Z}^m/L \simeq G$. The following elementary characterization of $\mathcal{L}_{G,m}$ is a consequence of [33]:

Theorem 1. *Let G be a finite abelian group and L be a full-rank lattice in $\mathbb{Z}^m$. Then $L \in \mathcal{L}_{G,m}$ if and only if G has rank $\leq m$ and there exists $\mathbf{g} = (g_1, \ldots, g_m) \in G^m$ s.t. the g_i's generate G and $L = \mathcal{L}_{\mathbf{g}}$ where $\mathcal{L}_{\mathbf{g}} = \{(x_1, \ldots, x_m) \in \mathbb{Z}^m$ s.t. $\sum_{i=1}^m x_i g_i = 0$ in $G\}$.*

Given G, Algorithm 5 (Appendix A) samples efficiently lattices from the uniform distribution over $\mathcal{L}_{G,m}$, and its correctness follows from Lemma 4. Previously, efficient sampling was only known for $G = \mathbb{Z}_p$ for large prime p [22].

Lemma 4. *Let G be a finite abelian group. Let $\mathbf{g} = (g_1, \ldots, g_m) \in G^m$ be such that the g_i's generate G. Let $\mathbf{h} = (h_1, \ldots, h_m) \in G^m$. Then $\mathcal{L}_{\mathbf{g}} = \mathcal{L}_{\mathbf{h}}$ if and only if there is an automorphism ψ of G such that $h_i = \psi(g_i)$ for all $1 \leq i \leq m$. In such a case, ψ is uniquely determined.*

We note that several implementations of lattice-based cryptography (such as [19]) implicitly used lattices in $\mathcal{L}_{G,m}$ for some large cyclic group G. Recently, Nguyen and Shparlinski [31] showed that such lattices are dominant: the set $\cup_{G \text{ cyclic}} \mathcal{L}_{G,m}$ of all full-rank integer lattices $L \subseteq \mathbb{Z}^m$ such that $\mathbb{Z}^m/L$ is cyclic has natural density $1/[\zeta(6) \prod_{k=4}^m \zeta(k)] \approx 85\%$ (for large m).

3.2   The Group-SIS Problem (GSIS)

Micciancio [25] introduced the Homogeneous SIS problem which is a natural generalization of SIS to an arbitrary finite abelian group G. In this paper, we call it *Group-SIS* problem (GSIS). The parameters are $m \geq 1$ and a bound $\beta > 0$. One picks $\mathbf{g} = (g_1, \ldots, g_m) \in G^m$ uniformly at random. GSIS(G, m, β)

asks to find a non-zero vector $\mathbf{x} \in \mathbb{Z}^m$ s.t. $\sum_{i=1}^m x_i g_i = 0$ and $\|\mathbf{x}\| \leq \beta$. In other words, GSIS asks to find short vectors in random relation lattices $\mathcal{L}_{\mathbf{g}} = \{\mathbf{x} \in \mathbb{Z}^m \text{s.t.} \sum_{i=1}^m x_i g_i = 0\}$. For instance, GSIS$(\mathbb{Z}_q^n, m, \beta)$ is SIS, and GSIS$(\mathbb{Z}_q, m, \beta)$ is finding short vectors in random m-dimensional co-cyclic lattices of volume q. If $\#G$ denotes the order of G, the existence of a GSIS-solution is guaranteed if $\beta \geq \sqrt{m}(\#G)^{1/m}$.

GSIS is connected to $\mathcal{L}_{G,m}$ as follows. As soon as $m \geq n + 2 \log \log \#G + 5$ (resp. $m > 2 \log \#G + 2$), $g_1, \ldots, g_m$ generate G with probability $\geq 1/e$ [24,32] (resp. $\geq 1 - 1/\#G$), in which case $\mathbb{Z}^m/\mathcal{L}_{\mathbf{g}} \simeq G$. In particular, if $m > 2 \log \#G + 2$, the distribution of GSIS lattices $\mathcal{L}_{\mathbf{g}}$ is statistically close to the distribution of Algorithm 5, and therefore the uniform distribution over $\mathcal{L}_{G,m}$, in which case GSIS is equivalent to finding short vectors in random lattices from $\mathcal{L}_{G,m}$.

Finally, we note that to establish hardness of GSIS, it suffices to focus on low-rank groups G. Indeed, if $G' = G \times H$ for some groups G, H, then GSIS over G can trivially be reduced to GSIS over G', by "projecting" G' to G.

3.3   The Group-LWE Problem (GLWE)

We introduce the *Group-LWE* problem (GLWE), using the torus $\mathbb{T} = \mathbb{R}/\mathbb{Z}$ and a finite abelian group G. Let $\hat{G}$ be the dual group of homomorphisms $G \to \mathbb{T}$: it is isomorphic to G but not canonically. If G is explicit, $G = \oplus_{i=1}^k \langle e_i \rangle$ where e_i has order q_i, and $\hat{G}$ is generated by $\hat{e}_1, \ldots, \hat{e}_k$ defined as $\hat{e}_i(\sum_{j=1}^k \alpha_j e_j) = \alpha_i/q_i$ mod 1 where $0 \leq \alpha_j < q_j$.

Let $\mathcal{S}$ be a known distribution over $\hat{G}$. Search-GLWE is the problem of learning a character $\hat{s} \in \hat{G}$ picked from $\mathcal{S}$, given noisy evaluations of $\hat{s}$ at (public) random points $a_1, \ldots, a_m \in G$, namely one is given (for all i's) a_i and a "Gaussian" perturbation of $\hat{s}(a_i)$. Like LWE, several noise distributions are possible. As in [37], we focus on the continuous distribution where $\hat{s}(a)$ is shifted by an error $e \leftarrow \mathcal{D}_{\mathbb{R},\alpha}$. These distributions need to be discretized in order to have a finite representation. In App. B.4 of the full version, we present discrete versions and show that they are at least as hard as the continuous version for suitable parameters, which explains why we only consider the continuous GLWE problem in the rest:

Definition 1. $G = \oplus_{i=1}^k \langle e_i \rangle$ *is an expl. finite abelian group*, $\alpha > 0$ *and* $\hat{s} \in \hat{G}$.

- $A_{G,\alpha}(\hat{s})$ *is the distribution over* $G \times \mathbb{T}$ *defined by choosing* $a \in G$ *uniformly at random, setting* $b \leftarrow \mathcal{D}_{\mathbb{T},\alpha,\hat{s}(a)}$, *and outputting* $(a, b) \in G \times \mathbb{T}$.
- *Search-GLWE*$_{G,\alpha}(\mathcal{S})$ *asks to find* $\hat{s}$ *from* $A_{G,\alpha}(\hat{s})$ *for a fixed* $\hat{s} \leftarrow \mathcal{S}$ *given arbitrarily many independent samples. By finding* $\hat{s}$, *we mean finding* $s_i \in \mathbb{Z}$ *s.t.* $\hat{s} = \sum_{i=1}^k s_i \hat{e}_i$.
- *Decisional-GLWE*$_{G,\alpha}(\mathcal{S})$ *asks to distinguish* $A_{G,\alpha}(\hat{s})$ *from the uniform distribution over* $G \times \mathbb{T}$ *for a fixed* $\hat{s}$ *sampled from* $\mathcal{S}$ *given arbitrarily many independent samples.*

– For $0 < \alpha < 1$, (Search) Decisional-GLWE$_{G,\leq\alpha}(\mathcal{S})$ is the problem of solving (Search) Decisional-GLWE$_{G,\beta}(\mathcal{S})$ for any $\beta \leq \alpha$ respectively, i.e. when the noise parameter is unknown yet $\leq \alpha$, by analogy with LWE.

Search-GLWE$_{G,m,\alpha}(\mathcal{S})$ and Decisional-GLWE$_{G,m,\alpha}(\mathcal{S})$ denote the variants where the algorithms have a bounded number m of samples. If $\mathcal{S}$ is omitted, it is the uniform distribution over $\hat{G}$.

If $G = \mathbb{Z}_q^n$, the canonical representation of G and $\hat{G}$ shows that GLWE is equivalent to the fractional version of Regev's original LWE. If $G = \mathbb{Z}_p$ for some prime p, then $\hat{G}$ can be defined by multiplications: $\hat{s}$ is the homomorphism mapping any $t \in \mathbb{Z}_p$ to $ts/p \mod 1$. Thus, GLWE can be viewed as a randomized version of Boneh-Venkatesan's *Hidden Number Problem* [8]: recover a secret number s mod p, given approximations of st_i mod p for many random integers t_i's. By analogy with LWE (see [11,37]), there is a folklore reduction from (Search) Decisional-GLWE$_{G,\leq\alpha}(\mathcal{S})$ to (Search) Decisional-GLWE$_{G,\alpha}(\mathcal{S})$, respectively.

Lemma 5. *(Adapted from [11, Lemma 2.13]) Let $\mathcal{A}$ be an algorithm for Decisional-GLWE$_{G,m,\alpha}(\mathcal{S})$ (resp. Search) with advantage at least $\varepsilon > 0$. Then there exists an algorithm $\mathcal{B}$ for Decisional-GLWE$_{G,m',\leq\alpha}(\mathcal{S})$ (resp. Search) using oracle access to $\mathcal{A}$ and with advantage $\geq 1/3$, where both m' and its running time are poly$(m, 1/\varepsilon, \log \#G)$.*

Proof. (Sketch, see Appendix B.3 of the full version [17] for a detailed proof). Like in LWE, the basic idea is to add noises in small increments to the distribution obtained from the challenger, and feed it to the oracle solving the Decisional-GLWE$_{G,\alpha}(\mathcal{S})$ (resp. Search) and estimate the behavior of the oracle.   □

4   Structural Lattice Reduction

4.1   Overview

A basic result (following from structure theorems of finitely-generated modules over principal ideal domains) states that for any full-rank sublattice L of a full-rank lattice $\bar{L} \subseteq \mathbb{R}^n$, there is a basis $\bar{B} = (\bar{\mathbf{b}}_1, \ldots, \bar{\mathbf{b}}_n)$ of $\bar{L}$ and integers $q_1 \geq q_2 \geq \cdots \geq q_n \geq 1$ s.t. $B = (q_1\bar{\mathbf{b}}_1, \ldots, q_n\bar{\mathbf{b}}_n)$ is a basis of L. The q_i's can be made unique by selecting powers of prime numbers, or by requiring each q_{i+1} to divide q_i, in which case $q_1, \ldots, q_n$ are the *elementary divisors* of the pair $(\bar{L}, L)$.

In this section, we introduce a lattice reduction converse, which we call *structural lattice reduction*. Lattice reduction asks to find a short basis of a given full-rank lattice $L \subseteq \mathbb{Z}^n$. In structural lattice reduction, one is further given a finite abelian group G of rank $\leq n$, and wants to find a *short* basis of some overlattice $\bar{L}$ of L such that $\bar{L}/L \simeq G$ effectively. More precisely, given a basis B of a full-rank lattice $L \subseteq \mathbb{Z}^n$, a suitable bound $\sigma > 0$ and integers $q_1 \geq \cdots \geq q_k$ defining $G = \mathbb{Z}_{q_1} \times \cdots \times \mathbb{Z}_{q_k}$, one asks to compute a basis $\bar{B}$ of an overlattice $\bar{L} \supseteq L$ such that $\|\bar{B}^*\| \leq \sigma$ and $B = (q_1\bar{\mathbf{b}}_1, \ldots, q_k\bar{\mathbf{b}}_k, \bar{\mathbf{b}}_{k+1}, \ldots, \bar{\mathbf{b}}_n)$ is a basis of L. Interestingly, we do not require the input basis B to have integer or rational coefficients, as long as its Gram-Schmidt coefficients are known with enough precision. Indeed, our structural reduction algorithm can simply focus on finding the rational transformation matrix between $\bar{B}$ and B.

Previous worst-case to average-case reductions implicitly used the group $G = \mathbb{Z}_q^n$, thus $\bar{L} = L/q$. Here, finding a basis $\bar{B}$ of $\bar{L}$ with small $\|\bar{B}^*\|$ is the same as finding a basis $B = q\bar{B}$ of L with small $\|B^*\|$, which is just lattice reduction. However, we obtain new problems and applications by considering different choices of G. In the trivial case $G = \mathbb{Z}_q^n$, $\bar{B} = q^{-1}B$ implies that $\|\bar{B}^*\| = \|B^*\|/q$ where the factor q is exactly $\#G^{1/n}$: this suggests that in general, we might hope to reduce $\|\bar{B}^*\|$ by a factor close to $\#G^{1/n}$, compared to $\|B^*\|$.

Another trivial case of structural lattice reduction is $G = \mathbb{Z}_{q_1} \times \cdots \times \mathbb{Z}_{q_n}$ where the q_i's are distinct positive integers of similar bit-length. If $B = (\mathbf{b}_1, \ldots, \mathbf{b}_n)$ is a basis of $L \subseteq \mathbb{Z}^n$, then $\bar{B} = (q_1^{-1}\mathbf{b}_1, \ldots, q_n^{-1}\mathbf{b}_n)$ generates an overlattice $\bar{L}$ such that $\bar{B}^* = (q_1^{-1}\mathbf{b}_1^*, \ldots, q_n^{-1}\mathbf{b}_n^*)$, and therefore $\|\bar{B}^*\| \leq \|B^*\|/\min_{i=1}^n q_i$. The factor $\min_{i=1}^n q_i$ is close to $\#G^{1/n}$ if the q_i's have similar bit-length. But if the q_i's are unbalanced, such as when $\min_{i=1}^n q_i = 1$, then the bound is much weaker. In particular, the case $G = \mathbb{Z}_p$ for some large prime p looks challenging, as the trivial choice $\bar{B} = (p^{-1}\mathbf{b}_1, \mathbf{b}_2, \ldots, \mathbf{b}_n)$ looks useless: $\bar{L}/L \simeq G$ but $\|\bar{B}^*\|$ is likely to be essentially as big as $\|B^*\|$, because for a typical reduced basis, the first $\|\mathbf{b}_i^*\|$'s have the same size.

4.2 Co-cyclic Lattice Reduction

As a warm-up, we solve structural lattice reduction when the target group G is cyclic of order q, which we call *co-cyclic lattice reduction*. Let $\bar{B}$ be a solution of structural reduction on $(L(B), G, \sigma)$: $C = (q\bar{b}_1, \bar{b}_2, \ldots, \bar{b}_n)$ is a basis of L s.t. $\|\mathbf{c}_1\| \leq q\sigma$ and $\|\mathbf{c}_i^*\| \leq \sigma$ for all $i \geq 2$.

Algorithm 1. Unbalanced Reduction

Input: an $n \times m$ basis B of an integer lattice $L \subseteq \mathbb{Z}^m$ and a target length $\sigma \in \mathbb{Q}^+$. More generally, B can be any n-dimensional projected block $B = B'_{[i,i+n-1]}$ of some basis B' of $L \subseteq \mathbb{Z}^m$.
Output: an $n \times n$ unimodular matrix U such that $C = UB$ satisfies $\|\mathbf{c}_i^*\| \leq \sigma$ for $i \geq 2$ and $\|\mathbf{c}_1\| \leq n\sigma\delta_\sigma(B)$.
1: $C \leftarrow B$, $U \leftarrow I_n$ and compute the Gram-Schmidt matrices μ and C^*
2: If $\|\mathbf{c}_i^*\| \leq \sigma$ for all i, **return** U
3: **for** $i = k - 1$ downto 1 where k is the largest index such that $\|\mathbf{c}_k^*\| > \sigma$ **do**
4:     **if** $\|\mathbf{c}_i^*\| \leq \sigma$ **then**
5:         $\alpha \leftarrow \lfloor -\mu_{i+1,i} \rceil$
6:     **else**
7:         $\alpha \leftarrow \left\lceil -\mu_{i+1,i} + \frac{\|\mathbf{c}_{i+1}^*\|}{\|\mathbf{c}_i^*\|}\sqrt{(\|\mathbf{c}_i^*\|/\sigma)^2 - 1} \right\rceil$
8:     **end if**
9:     $(\mathbf{c}_i, \mathbf{c}_{i+1}) \leftarrow (\mathbf{c}_{i+1} + \alpha \cdot \mathbf{c}_i, \mathbf{c}_i)$, $(\mathbf{u}_i, \mathbf{u}_{i+1}) \leftarrow (\mathbf{u}_{i+1} + \alpha \cdot \mathbf{u}_i, \mathbf{u}_i)$ and update the GS matrices μ and C^*.
10: **end for**
11: **return** U

To find such a basis $\bar{B}$, we first show how to transform B to ensure $\|\mathbf{b}_i^*\| \leq \sigma$ for all $i \geq 2$, using a poly-time algorithm which we call *unbalanced reduction* (see Algorithm 1). This algorithm can be explained as follows: in dimension two, it is easy to make $\mathbf{b}_2^*$ arbitrarily short by lengthening $\mathbf{b}_1$ (adding a suitable multiple of $\mathbf{b}_2$), since $\|\mathbf{b}_1\| \times \|\mathbf{b}_2^*\| = \mathrm{vol}(L)$ is invariant. Unbalanced reduction works by iterating this process on two-dimensional projected lattices, similarly to the classical size-reduction process. However, one would like to make sure that the resulting first basis vector $\mathbf{c}_1$ does not become too large, as follows:

Theorem 2 (Unbalanced Reduction). *Given an n-dim projected block $B = B'_{[i,i+n-1]}$ of a lattice $L \subseteq \mathbb{Z}^m$ and a target $\sigma \in \mathbb{Q}^+$, Algorithm 6 outputs in polynomial time an $n \times n$ unimodular matrix U such that $C = UB$ satisfies $\|\mathbf{c}_1\| \leq n\sigma\delta_\sigma(B)$ and $\|\mathbf{c}_i^*\| \leq \sigma$ for $i \geq 2$, and:*

$$\delta_\nu(B) \leq \delta_\nu(C) \leq \frac{\|\mathbf{c}_1\|}{\sigma\delta_\sigma(B)} \times \delta_\nu(B) \text{ for all } \nu \leq \sigma \tag{1}$$

$$\text{where } \delta_\sigma(B) \underset{\mathrm{def}}{=} \prod_{j=1}^{n} \max\left(1, \|\mathbf{b}_j^*\|/\sigma\right). \tag{2}$$

We call $\delta_\sigma(B)$ the *cubicity-defect* of B relatively to σ: it basically measures by which amount the hypercube of side σ should be scaled up to cover the parallelepiped spanned by $\mathbf{b}_1^*, \ldots, \mathbf{b}_n^*$. The proofs of Theorem 2 and Algorithm 1 can be found in Appendix C.2 of the full version of the paper [17]. Theorem 2 shows that Algorithm 1 solves co-cyclic lattice reduction for $q \geq n\delta_\sigma(B)$. However, this may not be suitable for our applications, since this lower bound depends on B and might be unbounded. To address this issue, we now show that LLL can bound $\delta_\sigma(B)$ depending only on n for appropriate σ:

Theorem 3 (LLL's Cubicity-Defect). *Let L be a full-rank lattice in $\mathbb{R}^n$ and $\sigma \geq ((1 + \varepsilon_{LLL})\sqrt{4/3})^r \cdot \mathrm{bl}(L)$ for some $r \geq 0$. If B is an LLL-reduced basis of L with factor ε_{LLL}, then $\delta_\sigma(B) \leq ((1 + \varepsilon_{LLL})\sqrt{4/3})^{\frac{(n-2r)^2}{8} + \frac{(n-2r)}{4}}$.*

By combining Theorems 2 and 3, we obtain:

Theorem 4 (Co-cyclic Reduction). *Given an $n \times m$ basis of a lattice $L \subseteq \mathbb{Z}^m$, $\varepsilon > 0$ and a rational $\sigma \geq ((1 + \varepsilon_{LLL})\sqrt{4/3})^r \cdot \mathrm{bl}(L)$ for some $r \geq 0$, and an integer $q \geq n((1 + \varepsilon_{LLL})\sqrt{4/3})^{\frac{(n-2r)^2}{8} + \frac{(n-2r)}{4}}$, Algorithm 2 computes a basis $\bar{B}$ of an overlattice $\bar{L} \supseteq L$ in time polynomial in the basis size, σ and $1/\varepsilon$, such that $\|\bar{B}^*\| \leq \sigma$ and $(q\bar{\mathbf{b}}_1, \bar{\mathbf{b}}_2, \ldots, \bar{\mathbf{b}}_n)$ is a basis of L. In particular, $\bar{L}/L \simeq \mathbb{Z}_q$.*

For instance, Theorem 4 with $r = n$ implies that given a lattice L and any cyclic group G of sufficiently large order $2^{\Omega(n^2)}$, one can efficiently obtain a basis $\bar{B}$ of some overlattice $\bar{L}$ of L such that $\bar{L}/L \simeq G$ and $\|\bar{B}^*\| \leq \mathrm{bl}(L)$: by comparison, an LLL-reduced basis only approximates $\mathrm{bl}(L)$ to some exponential factor.

4.3 Arbitrary Groups

Using unbalanced reduction, we prove that for an arbitrary sufficiently large finite abelian group G of rank $\leq n$, given any basis B of the lattice $L \subseteq \mathbb{Z}^n$,

---

Algorithm 2. Co-cyclic Reduction

Input: a basis of a full-rank integer lattice $L \subseteq \mathbb{Z}^n$, a factor $\varepsilon > 0$, and a rational $\sigma \geq ((1 + \varepsilon_{\mathrm{LLL}})\sqrt{4/3})^r \cdot \mathrm{bl}(L)$

for some $r \geq 0$, and an integer $q \geq n((1 + \varepsilon_{\mathrm{LLL}})\sqrt{4/3})^{\frac{(n-2r)^2}{8} + \frac{(n-2r)}{4}}$

Output: a basis $\bar{B}$ of an overlattice $\bar{L}$ such that $\|\bar{B}^*\| \leq \sigma$ and $\bar{L}/L \simeq \mathbb{Z}_q$.

1: Apply Alg. 6 on an LLL-reduced basis with factor $\varepsilon_{\mathrm{LLL}}$ output by the LLL algorithm.

2: **return** $\bar{B} = (\frac{\mathbf{c}_1}{q}, \mathbf{c}_2, \dots, \mathbf{c}_n)$ where C is the basis of L returned by Alg. 6.

---

one can compute a basis $\bar{B}$ of some overlattice $\bar{L}$ of L s.t. $\bar{L}/L \simeq G$ effectively and $\|\bar{B}^*\|$ is essentially lower than $\|B^*\|/\#G^{1/n}$. In particular, $\mathrm{bl}(\bar{L})$ is essentially $\#G^{1/n}$ smaller than $\mathrm{bl}(L)$. Although this is slightly weaker than the result we obtained (in the previous subsection) for cyclic groups G, it is sufficient for our worst-case to average-case reductions.

Algorithm 3. Structural Lattice Reduction

Input: σ, an $n \times m$ basis B of an integer lattice L, and $(q_1, \dots, q_k)$ s.t. $G = \prod_{i=1}^{k} \mathbb{Z}_{q_i}$ satisfies the conditions of Th. 5

Output: an $n \times m$ basis $\bar{B}$ of an overlattice $\bar{L}$ of L such that $\|\bar{B}^*\| \leq \sigma$ and $\bar{L}/L \simeq G$.

1: $C \leftarrow B$

2: **for** $i = 1$ to k **do**

3:     **if** $\|C^*_{[i,n]}\| \leq \sigma$ **return** $\bar{B} = (\frac{\mathbf{c}_1}{q_1}, \dots, \frac{\mathbf{c}_k}{q_k}, \mathbf{c}_{k+1}, \dots, \mathbf{c}_n)$

4:     Compute the smallest $\ell \geq \sigma$ such that $\ell \cdot \delta_\ell(C_{[i,n]}) = q_i \sigma/(n - i + 1)$.

5:     $V \leftarrow \mathrm{UnbalancedReduction}(C_{[i,n]}, \sigma)$ using Alg. 6.

6:     Apply V on $(\mathbf{c}_i, \dots, \mathbf{c}_n)$

7: **end for**

8: **return** $\bar{B} = (\frac{\mathbf{c}_1}{q_1}, \dots, \frac{\mathbf{c}_k}{q_k}, \mathbf{c}_{k+1}, \dots, \mathbf{c}_n)$

Theorem 5 (Structural Lattice Reduction). *Given an $n \times m$ basis B of a lattice $L \subseteq \mathbb{Z}^n$, and $k \leq n$ integers $q_1 \geq \dots \geq q_k$ defining the group $G = \prod_{i=1}^{k} \mathbb{Z}_{q_i}$ s.t. $n^k(\|B^*\|/\sigma)^n \leq \#G$ or:*

$$\#G \geq \frac{n!}{(n-k)!} \delta_\sigma(B) \text{ and for all } i \leq k, \|B^*\|/\sigma \leq q_i/(n + 1 - i)$$

Algorithm 3 outputs in polynomial time in $n, m, \|B\|, \log(q_i)$, a basis $\bar{B}$ of an overlattice $\bar{L} \supseteq L$ such that $\|\bar{B}^*\| \leq \sigma$ and $(q_1 \bar{\mathbf{b}}_1, \dots, q_n \bar{\mathbf{b}}_n)$ is a basis of L where $q_i = 1$ for $i > k$. In particular, $\bar{L}/L \simeq G$.

For instance, the condition $n^k(\|B^*\|/\sigma)^n \leq \#G$ in Theorem 5 means that σ (and therefore $\|\bar{B}^*\|$) can be chosen as low as $n^{k/n}\|B^*\|/(\#G)^{1/n}$. The proof of Theorem 5 can be found in Appendix C.3 of the full version [17]. Intuitively, Algorithm 3 simply applies unbalanced reduction iteratively, cycle by cycle of G.

4.4   Application

Structural reduction finds a short overlattice basis, which can be used to sample short (overlattice) vectors, and provides effective isomorphisms:

Proposition 2. *Let L and $\bar{L}$ be two full-rank lattices such that $\bar{L} \supseteq L$ and $\bar{L}/L \simeq G$ where G is an explicit finite abelian group. Given bases B and $\bar{B}$ of resp. L and $\bar{L}$, one can compute in polynomial time a surjective morphism φ from $\bar{L}$ to G s.t. $\ker \varphi = L$ (i.e. $\bar{L}/L \overset{\varphi}{\simeq} G$), and a "dual" morphism $\varphi^{\times} : L^{\times} \to \hat{G}$ s.t.*

$$[\varphi^{\times}(\mathbf{u})](\varphi(\mathbf{v})) = \langle \mathbf{u}, \mathbf{v} \rangle \quad \mathrm{mod}\ 1 \ \textit{for all } \mathbf{u} \in L^{\times} \textit{ and all } \mathbf{v} \in \bar{L} \tag{3}$$

Furthermore, preimages of $\varphi^{\times}$ can be computed in polynomial time.

5   Hardness of Group-SIS

Our result requires that the finite abelian group G is *explicit* (see Sect. 2).

5.1   Overview

The main idea behind the SIS reduction can be traced back to Mordell's arrithmetical proof [29] of Minkowski's theorem. To prove the existence of short vectors in a full-rank lattice $L \subseteq \mathbb{R}^n$, Mordell implicitly presented an algorithm to find short vectors from (exponentially many) long vectors, as follows. Let $q \geq 1$ be an integer and $\mathbf{w}_1, \ldots, \mathbf{w}_m \in L$ be distinct of norm $\leq R$ where $m > q^n$: for large R, m can be as large as the volume of the R-radius ball divided by the volume of L. Let $\mathbf{v}_i = q^{-1}\mathbf{w}_i \in q^{-1}L$. Since $m > q^n = [(q^{-1}L) : L]$, there are $i \neq j$ such that $\mathbf{v}_i \equiv \mathbf{v}_j \mod L$, i.e. $\mathbf{v}_i - \mathbf{v}_j = q^{-1}(\mathbf{w}_i - \mathbf{w}_j) \in L$ whose (nonzero) norm is $\leq 2R/q$, which is short for appropriate choices of q and R.

This algorithm is not efficient since m is exponential in q, but it can be made polynomial by reducing m to $\mathrm{poly}(n)$, using a $\mathrm{SIS}(m, n, q)$ oracle. Indeed, let L be a full-rank integer lattice in $\mathbb{Z}^n$. The lattice $\bar{L} = q^{-1}L$ is an overgroup of L such that $\bar{L}/L \simeq \mathbb{Z}_q^n = G$ explicitly: there is an efficiently computable surjective morphism $\varphi : \bar{L} \to G$ s.t. $L = \ker \varphi$, e.g. for any basis $(\bar{\mathbf{b}}_1, \ldots, \bar{\mathbf{b}}_n)$ of $\bar{L}$, let $\varphi(\sum_{i=1}^n x_i \bar{\mathbf{b}}_i) = (x_1 \mod q, \ldots, x_n \mod q) \in G$.

Furthermore, if $\bar{B}$ is short enough compared to the minima of L, it is possible to sample short vectors $\mathbf{v}_1, \ldots, \mathbf{v}_m \in \bar{L}$ with Gaussian distribution of parameter as small as $\eta_{\varepsilon}(L)$. Fourier analysis guarantees that for such Gaussian distributions, each projection $g_i = \varphi(\mathbf{v}_i)$ is uniformly distributed over G. This allows us to call an SIS oracle on $(g_1, \ldots, g_m)$, which outputs a short $\mathbf{x} \in \mathbb{Z}^m$ such that $\sum_{i=1}^m x_i g_i = 0$, i.e. $\sum_{i=1}^m x_i \varphi(\mathbf{v}_i) = 0$ which implies that $\mathbf{v} = \sum_{i=1}^m x_i \mathbf{v}_i \in L$. This $\mathbf{v}$ is provably non-zero with overwhelming probability, and is short because the $\mathbf{v}_i$'s and $\mathbf{x}$ are, which concludes the reduction from worst-case SIVP to SIS.

With this formalization, we can replace the SIS oracle by a GSIS oracle if we are able to sample short vectors $\mathbf{v}_1, \ldots, \mathbf{v}_m \in \bar{L}$ with Gaussian distribution, where $\bar{L}/L \simeq G$. And this is exactly what structural lattice reduction ensures. Previous SIS reductions used special choices of $\bar{L}$ and sampled differently short vectors in the overlattice: see Appendix H. of the full version [17] for a comparision with previous works.

5.2   Reducing Worst-Case ApproxSIVP to GSIS

Our main result formalizes the previous sketch and states that for appropriate choices of (G, m, β), if one can solve $\mathrm{GSIS}(G, m, \beta)$ on average, then one can approximate SIVP in the worst case, *i.e.* one can efficiently find short vectors in every n-dimensional lattice:

Algorithm 4. Reducing ApproxSIVP to GSIS

Input: a basis B of a full-rank integer lattice $L \in \mathbb{Z}^n$, a parameter $\sigma \geq \sqrt{2}\,\mathrm{bl}(L)$, a negl. $\varepsilon > 0$, an explicit finite abelian group G satisfying the condition of Th. 6, and an oracle $\mathcal{O}$ solving $\mathrm{GSIS}(G, m, \beta)$ with probability $\geq 1/\mathrm{poly}(n)$.

Output: A set S of n linearly independent vectors of L of norm $\leq \sigma\eta_\varepsilon(\mathbb{Z}^n)\sqrt{n/2\pi}\beta$.

1:   $S \leftarrow \emptyset$.
2:   Call structural reduction (Alg. 3) on (B, G, σ) to get $\bar{B}$ s.t. $\|\bar{B}^*\| \leq \sigma$ and $\varphi : \bar{L} \to G$ (Prop. 2) where $\bar{L} = L(\bar{B})$.
3:   **repeat**
4:     Sample $\mathbf{v}_1, \cdots, \mathbf{v}_m \in \bar{L}$ with distribution $D_{L,\sigma\eta_\varepsilon(\mathbb{Z}^n),0}$ using $\bar{B}$.
5:     $g_i = \varphi(\mathbf{v}_i)$ for $1 \leq i \leq m$, forming a sequence $\mathbf{g} = (g_1, \dots, g_m) \in G^m$.
6:     Call the GSIS-oracle $\mathcal{O}$ on $\mathbf{g}$, which returns $\mathbf{x} = (x_1, \dots, x_m) \in \mathbb{Z}^m$ s.t. $\sum_{i=1}^m x_i g_i = 0$.
7:     $\mathbf{v} \leftarrow \sum_{i=1}^m x_i \mathbf{v}_i \in L$
8:     **if** $\|\mathbf{v}\| \leq \sigma\eta_\varepsilon(\mathbb{Z}^n)\sqrt{n\pi}\beta$ and $\mathbf{v} \notin \mathrm{span}(S)$ **then** $S \leftarrow S \cup \{\mathbf{v}\}$
9:   **until** $\dim(S) = n$
10: Return S

Theorem 6. *Let $n \in \mathbb{N}$ and $\varepsilon = \mathrm{negl}(n)$. Given as input a basis B of a full-rank integer lattice $L \subseteq \mathbb{Z}^n$ and $\sigma \geq \sqrt{2}\,\mathrm{bl}(L)$, and an explicit finite abelian group G of rank $k \leq n$ such that $\#G \geq n^k(\|B^*\|/\sigma)^n$, Algorithm 4 outputs (in random poly-time) n linearly independent vectors of L with norm $\leq \sigma\eta_\varepsilon(\mathbb{Z}^n)\sqrt{n\pi}\beta$, using polynomially many calls to an oracle solving $\mathrm{GSIS}(G, m, \beta)$ with prob. $\geq 1/\mathrm{poly}(n)$.*

In particular, letting $\sigma = \dfrac{\|B\|^*}{2\eta_\varepsilon(\mathbb{Z}^n)\sqrt{n/\pi}\beta}$ gives an incremental version of the reduction, where the output basis is twice as short as the input. This generalizes [28, Theorem 5.9] and [20, Theorem 9.2] with a GSIS oracle instead of SIS. Iterating Theorem 6 until $\sigma = \sqrt{2}\,\mathrm{bl}(L)$ connects GSIS to worst-case ApproxSIVP.

Corollary 1. *Let $n \in \mathbb{N}$ and $\varepsilon = \mathrm{negl}(n)$. Let $(G_n)_{n\in\mathbb{N}}$ be a sequence of explicit finite abelian groups of rank k_n s.t. $\#G_n \leq (\beta_n/\sqrt{m_n})^{m_n}$ for $m_n \in \mathbb{N}$. If $\#G_n \geq n^{k_n}\left(\eta_\varepsilon(\mathbb{Z}^n)\sqrt{2n/\pi}\beta_n\right)^{\max(n,k_n)}$, then using polynomially many calls to an oracle solving $\mathrm{GSIS}(G_n, m_n, \beta_n)$ with prob. $\geq 1/\mathrm{poly}(n)$, one can solve worst-case n-dimensional $\mathrm{ApproxSIVP}_{\eta_\varepsilon(\mathbb{Z}^n)\sqrt{n/\pi}\beta_n}$ in (randomized) poly-time.*

Consider the set of all full-rank integer lattices $\subseteq \mathbb{Z}^m$ of volume $\geq \omega_n = n^m\left(\eta_\varepsilon(\mathbb{Z}^n)\sqrt{2n/\pi}\beta_n\right)^m$. This set can be partitioned as $\cup_G \mathcal{L}_{G,m}$ where G runs over all finite abelian groups of order $\geq \omega_n$ and rank $\leq m$. Each such G satisfies the conditions of Corollary 1, and therefore GSIS over G is as hard as worst-case lattice problems: for any partition cell $\mathcal{L}_{G,m}$, finding short vectors in a random lattice from this cell is as hard as finding short vectors in any n-dim lattice.

6 Hardness of Decisional-Group-LWE

We transfer the following Decisional-LWE hardness results to Decisional-GLWE:

Theorem 7 [34,37]. *Let $n \in \mathbb{N}$, $q_n \geq 1$ be a sequence of integers, and $\alpha_n \in (0,1)$ be a real sequence s.t. $\alpha_n q_n \geq 2\sqrt{n}$. There exists a quantum reduction from worst-case n-dimensional $\text{GapSVP}_{\tilde{O}(n/\alpha_n)}$ to Decisional-GLWE$_{\mathbb{Z}_{q_n}^n, \alpha_n}$. If $q_n \geq 2^{n/2}$ is smooth then there is a classical reduction between them.*

Theorem 8 [11]. *Let $n \in \mathbb{N}$ and $q_n \geq 1$ be a sequence of integers, and let $\alpha_n \in (0,1)$ be a real sequence such that $\alpha_n \geq 2n^{1/4}/2^{\sqrt{n}/2}$. There exists a classical reduction from worst-case $\sqrt{n}$-dimensional $\text{GapSVP}_{\tilde{O}(\sqrt{n}/\alpha_n)}$ to Decisional-GLWE$_{\mathbb{Z}_{q_n}^n, \beta_n}$, where $\beta_n^2 = 10n\alpha_n^2 + \frac{n}{q_n^2} \cdot \omega(\log n)$*

To do so, we reduce Decisional-LWE to Decisional-GLWE using a technique we call group switching. This technique transforms GLWE samples over a group G to another group G', generalizing the modulus-dimension switching technique in [11], which is the special case $G = \mathbb{Z}_q^n$ and $G' = \mathbb{Z}_{q'}^{n'}$. We believe that the group switching technique proposed below is useful to better understand the core idea of the modulus-dimension switching technique.

Before presenting group switching, we note that the modulus-dimension switching technique from [11] implicitly uses a special case of structural lattice reduction. More precisely, Brakerski *et al.* [11] defined a special lattice Λ (see Theorem 3.1 of [11]) to transform LWE samples over $G = \mathbb{Z}_q^n$ to LWE samples over $G' = \mathbb{Z}_{q'}^{n'}$, but the meaning of Λ may look a bit mysterious. The lattice Λ is defined as $\Lambda = \frac{1}{q'}\mathbb{Z}^{n'} \cdot H + \mathbb{Z}^n$ where H is some $n' \times n$ integer matrix: this matrix is actually denoted by G in [11], but this would collide with our notation G for finite abelian groups. And [11] provided a good basis of Λ in special cases. We note that the exact definition of Λ is not important: the quotient $\Lambda/\mathbb{Z}^n$ turns out to be isomorphic to the group $G' = \mathbb{Z}_{q'}^{n'}$, as shown by the transformation mapping $\frac{1}{q'}\mathbf{x} \cdot H + \mathbf{y} \in \Lambda$ to $\mathbf{x} \mod q' \in G'$. Thus, finding a good basis of Λ is actually a special case of structural lattice reduction for the lattice $\mathbb{Z}^n$ and the group G'. Therefore, it is natural to use structural lattice reduction directly (instead of an ad-hoc process) to obtain a more general statement than the modulus-dimension switching technique of [11].

Since we have two groups G and G' and two overlattices $\bar{L}$ and $\bar{L}'$ of $\mathbb{Z}^n$, we have two morphisms $\varphi : \bar{L} \to G$ and $\varphi' : \bar{L}' \to G'$ with $\ker(\varphi) = \ker(\varphi') = \mathbb{Z}^n$. Both morphisms are associated to their dual morphism as in Proposition 2, *i.e.* $\varphi^\times : \mathbb{Z}^n \to \hat{G}$ and $\varphi'^\times : \mathbb{Z}^n \to \hat{G}'$, satisfying $[\varphi'^\times(\mathbf{u})](\varphi'(\mathbf{v})) = \langle \mathbf{u}, \mathbf{v} \rangle \mod 1$ for all $\mathbf{u} \in \mathbb{Z}^n$ and all $\mathbf{v} \in \bar{L}'$ (resp. without primes).

We say that a distribution S over $\mathbb{Z}^n$ is K-bounded if $\Pr_{\mathbf{s} \leftarrow S}[\|\mathbf{s}\| > K] \leq \text{negl}(n)$. By extension, given a (public) morphism $f : \mathbb{Z}^n \to \hat{G}$, we say that a distribution $\mathcal{S}$ over $\hat{G}$ is K-bounded (for f) if it is the image of a K-bounded distribution[2] by f. In the following, we choose $\varphi^\times = f$ and φ its dual morphism

---

[2] Ideally, f should be collision resistant among samples from S. In the classical LWE $(G = \mathbb{Z}_q^n)$, f maps $\mathbf{s} \in \mathbb{Z}^n$ to the secret character $\hat{s} : \mathbf{y} \to 1/q\langle \mathbf{s}, \mathbf{y} \rangle \mod 1$ in $\hat{G}$.

accordingly. Thus, any secret $\hat{s} \leftarrow S$ has with overwhelming probability a preimage $\mathbf{s} \in \mathbb{Z}^n$ of norm $\leq K$. Note that the small $\mathbf{s} \in \mathbb{Z}^n$ may be hard to compute from $\hat{s}$, however what matters is its existence. During group switching, the new secret in $\hat{G}'$ will be $\varphi'^\times(\mathbf{s})$, and the new K-bounded distribution $S' = \varphi'^\times(S)$.

Lemma 6 (Group Switching). *Let G and G' be two finite abelian groups of rank $\leq n$ s.t. G is fully-explicit and G' is explicit. Let $\bar{L}$ be an overlattice of $\mathbb{Z}^n$ such that $\bar{L}/\mathbb{Z}^n \simeq G$. Let $\bar{B}'$ be a basis of an overlattice $\bar{L}'$ of $\mathbb{Z}^n$ such that $\bar{L}'/\mathbb{Z}^n \simeq G'$. Let φ, φ' and $\varphi'^\times$ be defined as in Proposition 2. Let $r \geq \max\left(\sqrt{2}\eta_\varepsilon(\bar{L}), \|\bar{B}'^*\| \cdot \eta_\varepsilon(\mathbb{Z}^n)\right)$, where ε is some negligible function. Then, there is an efficient randomized algorithm which, given as input a sample from $G \times \mathbb{T}$, outputs a sample from $G' \times \mathbb{T}$, with the following properties:*

- *If the input sample has uniform distribution in $G \times \mathbb{T}$, then the output sample has uniform distribution in $G' \times \mathbb{T}$ (except with negligible distance).*
- *If the input is distributed according to $A_{G,\alpha}(\hat{s})$ for some $\hat{s} = \varphi^\times(\mathbf{s})$ s.t. $\mathbf{s} \in \mathbb{Z}^n$ and $\|\mathbf{s}\| \leq K$, then the output distribution is statistically close to $A_{G',\beta}(\hat{s}')$, where $\hat{s}' = \varphi'^\times(\mathbf{s}) \in \hat{G}'$ and $\beta^2 = \alpha^2 + r^2(\|\mathbf{s}\|^2 + K^2) \leq \alpha^2 + 2(rK)^2$.*

By combining Group Switching (Lemma 6) with structural reduction (Theorem 5), one derives a reduction between Decisional-GLWE of two groups G and G':

Corollary 2 (GLWE to GLWE). *Let $n \in \mathbb{N}$ and $0 < \sigma_n < 1$ be a real sequence. Let $(G_n)_{n\in\mathbb{N}}$ and $(G'_n)_{n\in\mathbb{N}}$ be two sequences of finite abelian groups with respective rank $k_n \leq n$ and $k'_n \leq n$ s.t. $\#G_n \geq n^{k_n}(\sqrt{2}/\sigma_n)^n$ (or if $G_n = \mathbb{Z}^n_{q_n}$ where $q_n \geq \sqrt{2}/\sigma_n$) and $\#G'_n \geq n^{k'_n}(1/\sigma_n)^n$. Assume that G_n is fully-explicit and G'_n is explicit. Let S be an arbitrary K_n-bounded distribution over $\mathbb{Z}^n$ and $S = \varphi^\times(S)$ its image by some morphism $\varphi^\times : \mathbb{Z}^n \to \hat{G}_n$, $\alpha_n, \beta_n > 0$ be two real sequences and $\varepsilon = \text{negl}(n)$ satisfying $\beta_n^2 \geq \alpha_n^2 + 2(\sigma_n K_n \cdot \eta_\varepsilon(\mathbb{Z}^n))^2$. Then there is an efficient reduction from Decisional-GLWE$_{G_n, \leq\alpha_n}(S)$ to Decisional-GLWE$_{G'_n, \leq\beta_n}(S')$, where $S' = \varphi'^\times(S)$ for some morphism $\varphi'^\times : \mathbb{Z}^n \to \hat{G}'_n$*

Proof. Given the canonical basis of $\mathbb{Z}^n$ and G_n, structural reduction finds an overlattice $\bar{L}$ together with a basis $\bar{C}$ s.t. $\|\bar{C}^*\| \leq \sigma_n/\sqrt{2}$. Therefore $\sqrt{2}\eta_\varepsilon(\bar{L}) \leq \sigma_n\eta_\varepsilon(\mathbb{Z}^n)$. And structural reduction on G'_n and σ_n gives a short basis $\bar{B}'$ of length $\leq \sigma_n$ and defines $\bar{L}'$. The rest follows immediately from Lemma 6. □

Using the normal form [3] of LWE, namely, if S is the image of the $\alpha_n q_n \sqrt{n}$-bounded distribution $\mathcal{D}_{\mathbb{Z}^n,\alpha_n q_n}$, through the canonical embedding which maps $\mathbf{s} \in \mathbb{Z}^n$ to the character $\hat{s} = \mathbf{y} \to 1/q_n\langle\mathbf{s},\mathbf{y}\rangle \mod 1$, we obtain the quantum/classical hardness of Decisional-GLWE problem for any sufficiently large finite abelian group, together with Theorems 7 and 8:

Corollary 3 (Quantum Hardness of GLWE). *Let $n \in \mathbb{N}$ and $q_n \geq 1$ be a sequence of integers and $(G'_n)_{n\in\mathbb{N}}$ be a sequence of any finite abelian explicit groups such that $\#G'_n \geq n^{k_n}(q_n/\sqrt{2})^n$ where $k_n = \text{rank}(G'_n) \leq n$. Let $\alpha_n, \beta_n \in (0,1)$ be two real sequences such that $\alpha_n q_n \geq 2\sqrt{n}$ and $\beta_n = \alpha_n\sqrt{n} \cdot \omega(\sqrt{\log n})$.*

Then there exists a quantum reduction from worst-case n-dimensional GapSVP$_{\tilde{O}(n/\alpha_n)}$ *to Decisional-GLWE$_{G'_n, \beta_n}$.*

The lower bound on $\#G'_n$ is better than the lower bound on $\#G_n$ in Corollary 1 and for solving Approx-SIVP using a Search-GLWE oracle (see Appendix E.2 of the full version [17]), because group switching relies on structural reduction over $\mathbb{Z}^n$ rather than an arbitrary lattice: the canonical basis of $\mathbb{Z}^n$ is orthonormal, which simplifies the bound of Sect. 4.

Corollary 4 (Classical Hardness of GLWE). *Let $n \in \mathbb{N}$ and $q_n \geq 1$ be a sequence of integers and $(G'_n)_{n \in \mathbb{N}}$ be a sequence of any finite abelian explicit groups such that $\#G'_n \geq n^{k_n}(q_n/\sqrt{2})^n$ where $k_n = \mathrm{rank}\,(G'_n) \leq n$. Let $\alpha_n, \beta_n \in (0,1)$ be two real sequences such that $\alpha_n \geq 2n^{1/4}/2^{\sqrt{n}/2}$ and $\beta_n^2 = n^2 \alpha_n^2 \cdot \omega(\log n) + \frac{n^2}{q_n^2} \cdot \omega(\log^2 n)$. There exists a classical reduction from worst-case $\sqrt{n}$-dimensional* GapSVP$_{\tilde{O}(\sqrt{n}/\alpha_n)}$ *to Decisional-GLWE$_{G'_n, \beta_n}$.*

7 Abstracting Lattice Cryptography: Fully-Homomorphic Encryption from GLWE

We showed that GSIS/GLWE are hard under the same worst-case assumptions as SIS/LWE. This suggests to abstract lattice schemes based on SIS/LWE using an arbitrary finite abelian group G, and check that the security proof carries through. This may lead to a better understanding of the scheme and a clearer presentation: lattice schemes are typically described using matrices and vectors, which our abstraction avoids.

We illustrate this approach with fully-homomorphic encryption. First, we introduce a GLWE-based El Gamal-like encryption scheme, which generalizes Regev's LWE-based encryption [37] and its dual version [20]. Next, we extend this GLWE generalization of Regev's encryption into a somewhat-homomorphic encryption, by carefully abstracting the Alperin-Sheriff-Peikert variant [2] of the Gentry-Sahai-Waters homomorphic scheme [21]. In particular, we show how to evaluate any boolean function with a noise overhead proportional to the square root of its number of variables, how to recognize any regular language with a noise overhead proportional to the length of the tested word, and how to bootstrap the whole system with only a linear noise overhead instead of quadratic in [2].

7.1 A GLWE Variant of El Gamal Encryption

El Gamal encryption combines the one-time pad with Diffie-Hellman. By analogy, we first present a GLWE variant of DH. We consider a (sufficiently large) finite abelian group G and $\mathbf{g} = (g_1, ..., g_m) \in G^m$ chosen uniformly at random. This defines two one-way functions:

- Let $f_{\mathbf{g}} : \mathbb{Z}^m \to G$ be the morphism defined by $f_{\mathbf{g}}(\mathbf{x}) = \sum_{i=1}^{m} x_i.g_i$, where $x_i.g_i$ is defined by the $\mathbb{Z}$-module structure of G. For suitable input distributions $\mathcal{D}$, such as the uniform distribution over $\{0,1\}^m$ or some well-chosen discrete Gaussian distribution, the distribution of $f_{\mathbf{g}}(\mathbf{x})$ becomes statistically close to uniform (e.g. see the left-over-hash lemma), and $f_{\mathbf{g}}$ becomes one-way under GSIS.
- Let $f_{\mathbf{g}}^{\times} : \hat{G} \times \mathbb{T}^m \to \mathbb{T}^m$ defined by $f_{\mathbf{g}}^{\times}(\hat{s}, \mathbf{e}) = (\hat{s}(g_1) + e_1, \ldots, \hat{s}(g_m) + e_m)$: if $\hat{s} \in_R \hat{G}$ and $\mathbf{e}$ is sampled from a suitable distribution such as $\mathcal{D}_{\alpha}^m$, then inverting $f_{\mathbf{g}}^{\times}(\hat{s}, \mathbf{e})$ is search-GLWE, and distinguishing $f_{\mathbf{g}}^{\times}(\hat{s}, \mathbf{e})$ from random is decisional-GLWE.

Consider the bilinear map $\theta : \hat{G} \times \mathbb{Z}^m \to \mathbb{T}$ defined by $\theta(\hat{s}, \mathbf{x}) = \hat{s}(f_{\mathbf{g}}(\mathbf{x}))$. Then $\theta(\hat{s}, \mathbf{x})$ can be efficiently computed from $(\hat{s}, \mathbf{x})$. But it can be computed knowing only $(\hat{s}, f_{\mathbf{g}}(\mathbf{x}))$, or approximately knowing only $(f_{\mathbf{g}}^{\times}(\hat{s}, \mathbf{e}), \mathbf{x})$ by $\sum_{i=1}^{m} c_i x_i$ (where $\mathbf{c} = f_{\mathbf{g}}^{\times}(\hat{s}, \mathbf{e})$), provided that $\mathbf{e}$ and $\mathbf{x}$ are sampled from suitable distributions. This motivates a GLWE noisy key exchange where Alice and Bob compute their own approximation of $\theta(\hat{s}, \mathbf{x})$: Alice picks $\mathbf{x} \in \mathbb{Z}^m$ from some suitable distribution $\mathcal{D}$, and discloses $y = f_{\mathbf{g}}(\mathbf{x})$; Bob picks $\hat{s} \in_R \hat{G}$ and $\mathbf{e}$ from the distribution $\mathcal{D}_{\alpha}^m$, and discloses $\mathbf{c} = f_{\mathbf{g}}^{\times}(\hat{s}, \mathbf{e})$. Alice computes her key as $\sum_{i=1}^{m} c_i x_i$, and Bob computes his key as $\hat{s}(y) + e$ where e is sampled from $\mathcal{D}_{\alpha}$. Both keys are close to $\theta(\hat{s}, \mathbf{x})$. But, as opposed to Diffie-Hellman, Alice and Bob do not have symmetric roles, which leads to two El Gamal cryptosystems by swapping Alice and Bob roles: this is why Regev encryption has a so-called dual variant [20]. We now give a detailed description of the main cryptosystem, which generalizes Regev's [37], and which we use in our fully-homomorphic encryption.

Define the group $H = G \times \mathbb{T}_k$ where $k \in \mathbb{N}^+$ and $\mathbb{T}_k = \frac{1}{2^k}\mathbb{Z}/\mathbb{Z} \subseteq \mathbb{T}$ is a discretized torus.

GLWE.Gen(1^n): Takes as input a security parameter n, it chooses a Gaussian parameter $0 < \alpha < 1$, a (sufficiently large) finite abelian group G and $m \in \mathbb{N}$. Choose $\mathbf{g} = (g_1, \ldots, g_m) \in_R G^m$, $\hat{s} \in_R \hat{G}$ and m Gaussian samples $e_1, \ldots, e_m \leftarrow \mathcal{D}_{\alpha}$. Set the public key $pk = (\mathbf{g}, \mathbf{y}) \in G^m \times \mathbb{T}_k^m$, where $y_i = \hat{s}(g_i) + e_i \in \mathbb{T}$, and the secret key $sk = \hat{s}$, i.e. $\mathbf{y} = f_{\mathbf{g}}^{\times}(\hat{s}, \mathbf{e})$.

GLWE.Enc(pk, μ): Takes as input the public key $pk = (\mathbf{g}, \mathbf{y}) \in G^m \times \mathbb{T}_k^m$ and a message $\mu \in \{0, 1\}$. It selects $\mathbf{x} = (x_1, \ldots, x_m) \in_R \{0, 1\}^m$, and returns $(d, c) \in H$, where $d = f_{\mathbf{g}}(\mathbf{x}) = \sum_{i=1}^{m} x_i g_i \in G$ and $c = \sum_{i=1}^{m} x_i y_i + \mu/2 \in \mathbb{T}_k$. Here, $\sum_{i=1}^{m} x_i y_i$ is Alice's key in the GLWE key exchange. Both d and c use the $\mathbb{Z}$-module structure of G and $\mathbb{T}_k$.

GLWE.Dec($sk, (d, c)$): Returns $\mu = \lfloor 2 \cdot (c - \hat{s}(d)) \rceil \mod 2$ where $sk = \hat{s}$ and $(d, c) \in H$ is the ciphertext.

One obtains a dual scheme by swapping the two one-way functions $f_{\mathbf{g}}$ and $f_{\mathbf{g}}^{\times}$.

Lemma 7 (Correctness). *If $0 < \alpha < 1/(4 \cdot \sqrt{m} \cdot \omega(\sqrt{\log n}))$, the main GLWE public-key encryption scheme decrypts correctly with probability $1 - negl(n)$.*

Proof. We have: $c - \hat{s}(d) = \sum_{i=1}^{m} x_i(\hat{s}(g_i) + e_i) + \mu/2 - \hat{s}(\sum_{i=1}^{m} x_i g_i) = \mu/2 + \sum_{i=1}^{m} x_i e_i$. It is sufficient to show $|\sum_{i=1}^{m} x_i e_i| < 1/4$. Let $w \leq m$ be the Hamming weight of $\mathbf{x}$, we know that $\sum_{i=1}^{m} x_i e_i$ is distributed as $\mathcal{D}_{\sqrt{w}\alpha}$. Therefore, it implies that $|\sum_{i=1}^{m} x_i e_i| < \sqrt{w}\alpha \cdot \omega(\sqrt{\log n})$ with probability $1 - \exp(-\pi \cdot \omega(\log n)) = 1 - \mathrm{negl}(n)$. We obtain that $|\sum_{i=1}^{m} x_i.e_i| < \sqrt{w}\alpha \cdot \omega(\sqrt{\log n}) \leq 1/4$ with probability $1 - \mathrm{negl}(n)$, as desired. $\qquad\square$

Lemma 8 (Security). *If $m \geq 2(\log \#G + k) + \omega(\log n)$ and the $\mathrm{GLWE}_{G,m,\alpha}$ assumption holds, then the main GLWE public-key encryption scheme is IND-CPA secure.*

Proof. $\mathbf{g} \in G^m$ is uniformly distributed. By the $\mathrm{GLWE}_{G,m,\alpha}$ assumption, $\mathbf{y} \in \mathbb{T}_k^m$ is computationally indistinguishable from uniform, hence $(\mathbf{g}, \mathbf{y})$ too. Since $m \geq 2 \cdot \log \#H + \omega(\log n)$ and $\mathbf{x} \in_R \{0,1\}^m$, the left-over-hash lemma ensures that $\sum_{i=1}^{m} x_i(g_i, y_i)$ is computationally indistinguishable from uniform over H, and hence (d, c) too. This proves IND-CPA security. $\qquad\square$

7.2   A GLWE Variant of GSW Homomorphic Encryption

We now show how to generalize the AP variant [2] of GSW [21] Homomorphic encryption. Let $\mathrm{GLWE}(G, \alpha)$ be a black-box instance of GLWE El Gamal encryption over the GLWE group G. All noises are discretized in the torus $\mathbb{T}_k = \frac{1}{2^k}\mathbb{Z}/\mathbb{Z} \subseteq \mathbb{T}$ where $2^k\alpha \approx \eta_\varepsilon(\mathbb{Z})$. The group $H = G \times \mathbb{T}_k$ is of special interest.

First, recall that El Gamal encryption is homomorphic with respect to the group operation. Because $\mathrm{GLWE}(G, \alpha)$ is a noisy variant of El Gamal encryption, it is also homomorphic for a bounded number of XOR. More precisely, any GLWE ciphertext of a message $\mu \in \{0,1\}$ can be written as $c_1 + \mu h_1 \in H$, where $c_1 = \sum_{i=1}^{m} x_i(g_i, y_i) \in H$ is a random ciphertext of 0, and $h_1 = (0, 1/2) \in H$. Here, we use the $\mathbb{Z}$-module structure of H. The GLWE secret key $\hat{s}$ induces a homomorphism $\mathrm{Phase} : H \to \mathbb{T}$ defined as $\mathrm{Phase}((a,b)) = b - \hat{s}(a)$. By definition of GLWE, we have $\mathrm{Phase}((g_i, y_i)) \approx 0$ for all $1 \leq i \leq m$, but $\mathrm{Phase}(h_1) = 1/2$. It follows that the phase of a GLWE ciphertext of a message μ is $\approx \mu/2$, which explains the GLWE decryption procedure: a ciphertext of 0 is close to the kernel of the phase, while a ciphertext of 1 is far away. Because Phase is a homomorphism and h_1 has order 2 in H, if n messages $\mu_1, \ldots, \mu_n \in \{0,1\}$ are GLWE-encrypted, then the sum of these n ciphertexts will de decrypted as $\mu_1 \oplus \cdots \oplus \mu_n$, provided that n is not too large.

To achieve more homomorphic operations, one exploits a special property of lattice problems which is not shared by discrete logarithm problems: with special choices of generators, the SIS one-way function can be inverted. To do so, one first extends h_1 into a generating set of the $\mathbb{Z}$-module H: let $h_2, \ldots, h_\ell \in H$ be such that $\mathbf{h} = (h_1, \ldots, h_\ell)$ is a generating set of H. Recall that the GSIS function $f_{\mathbf{g}}$ from Sect. 7.1 can be defined over any group: here, we use H, so $f_{\mathbf{h}}(\mathbf{x}) = \sum_{i=1}^{\ell} x_i h_i \in H$ for $(x_1, \ldots, x_\ell) \in \mathbb{Z}^\ell$. Since $\mathbf{h}$ generates H, $f_{\mathbf{h}}$ is surjective, and thus, admits a pseudo-inverse $f_{\mathbf{h}}^{-1}$ from H to $\mathbb{Z}^\ell$, such that $f_{\mathbf{h}}(f_{\mathbf{h}}^{-1}(b)) = b$ for any $b \in H$. We also define $F_{\mathbf{h}} : \mathbb{Z}^{\ell \times \ell} \to H^\ell$ by $F_{\mathbf{h}}(\mathbf{X}) = (f_{\mathbf{h}}(\mathbf{x}_1), ..., f_{\mathbf{h}}(\mathbf{x}_\ell))$, where $\mathbf{x}_i$ is the i-th row of $\mathbf{X}$. Accordingly, we define $F_{\mathbf{h}}^{-1} : H^\ell \to \mathbb{Z}^{\ell \times \ell}$.

Given a target in H, finding a short $f_{\mathbf{h}}()$-preimage corresponds to the GSIS problem, which is in general hard, but it becomes easy for special choices of $\mathbf{h}$, like super-increasing knapsacks: following [26], we call *gadget* such a $\mathbf{h}$. We say that $f_{\mathbf{h}}^{-1}()$ is β-bounded for $\mathbf{h}$, if $\|f_{\mathbf{h}}^{-1}(b)\|_\infty \le \beta \in \mathbb{R}^+$ for any $b \in H$. For instance, if the group G is $\mathbb{Z}_N$ where $2^p < N < 2^{p+1}$, a suitable gadget is $\mathbf{h} = ((0, \frac{1}{2}), (0, \frac{1}{4}), ..., (0, \frac{1}{2^k}), (1, 0), (2, 0), ..., (2^p, 0))$, $f_{\mathbf{h}}^{-1}() \in \{0, 1\}^\ell$ can be computed by binary decomposition and is 1-bounded for $\mathbf{h}$. This construction can easily be generalized to any fully-explicit G, using component-wise binary decomposition: if $G = \mathbb{Z}_q^n$, this corresponds to the Flatten/BitDecomp algorithms proposed in [2,21]. However, other algorithms are possible, such as ternary decompositions with preimages in $\{0, \pm 1\}^\ell$.

Given the GLWE encryption scheme (GLWE.Gen, GLWE.Enc, GLWE.Dec) described in Sect. 7.1 as a "black box", we build homomorphic encryption using a gadget $\mathbf{h} \in H^\ell$ whose first element is $(0, \frac{1}{2})$:

GSW.Gen(1^n): Takes as input a security parameter n, it runs the key generation algorithm $(pk, sk) \leftarrow \text{Gen}(1^n)$, where $pk = (\mathbf{g}, \mathbf{y}) \in G^m \times \mathbb{T}_k^m$ and $sk = \hat{s} \in \hat{G}$.

GSW.Enc(pk, μ): Takes as input the public key $pk \in G^m \times \mathbb{T}_k^m$ and a message $\mu \in \{0, 1\}$, it first generates ℓ ciphertexts $c_1 = \text{GLWE.Enc}(pk, 0), ..., c_\ell = \text{GLWE.Enc}(pk, 0)$ of zero, and returns $\mathbf{c} = (c_1, ..., c_\ell) + \mu \cdot \mathbf{h} \in H^\ell$.

This is reminiscent of the GLWE scheme, where a GLWE-ciphertext of a message μ is of the form $c_1 + \mu h_1 \in H$ where c_1 is a random GLWE-ciphertext of 0. Because the first entry of $\mathbf{h}$ is $(0, \frac{1}{2})$, the first entry of $\mathbf{c}$ is a GLWE encryption of μ.

GSW.Dec($sk, \mathbf{c}$): Returns GLWE.Dec($\hat{s}, c_1$) where $sk = \hat{s}$ and $c_1 \in H$ is the first entry of $\mathbf{c}$.

The security of the scheme and the correctness of decryption follow from that of the GLWE cryptosystem:

Lemma 9. *Suppose* (Gen, Enc, Dec) *uses samples from* $GLWE_{G,m,\alpha}$. *If* $m \ge 2(\log \#G + k) + \omega(\log n)$ *and* $0 < \alpha < 1/(4 \cdot \sqrt{m} \cdot \omega(\sqrt{\log n}))$, (GSW.Gen, GSW.Enc, GSW.Dec) *is IND-CPA secure under the* $GLWE_{G,m,\alpha}$ *assumption, and* GSW.Dec *decrypts correctly with probability* $1 - \text{negl}(\lambda)$.

Proof. The proof of IND-CPA security is similar to Lemma 8. Since the first entry of $\mathbf{c}$ is a ciphertext of μ under $\hat{s}$ of the scheme (Gen, Enc, Dec), correctness follows from Lemma 7. $\square$

We now describe our homomorphic operations on ciphertexts, namely how to encode Not, And, and Mux gates. First, we note that the GSW-GLWE scheme inherits the $\oplus$-homomorphic properties of the GLWE scheme. Any circuit can be built using only Not and And elementary gates. We chose to add the Mux ternary gate, which encodes the conditional operator $\text{Mux}(a, b, c) = a?b:c$, because resulting circuits are smaller than NAND-only circuits, all binary gates can be encoded by a single Mux (and a few Not), and it is trivial to batch-convert any truth-table to its corresponding Mux-based binary decision diagram.

Definition 2 (Homomorphic Operations). *For all ciphertexts* $c_1, c_2, c_3 \in H^\ell$, *we define:*

$$GSW.Not(c_1) = h - c_1, \quad GSW.And(c_1, c_2) = F_{c_1}\left(F_h^{-1}(c_2)\right),$$

$$GSW.Mux(c_1, c_2, c_3) = F_{c_1}\left(F_h^{-1}(c_2)\right) + F_{h-c_1}\left(F_h^{-1}(c_3)\right)$$

We express $Xor(a, b)$ as $Mux(a, Not(b), b)$. We naturally extend the Phase homomorphism to H^ℓ as $Phase : H^\ell \to \mathbb{T}^\ell$ defined as $Phase(z) = (b_1 - \hat{s}(a_1), \ldots, b_\ell - \hat{s}(a_\ell)) \in \mathbb{T}^\ell$ where $z = ((a_1, b_1), \ldots, (a_\ell, b_\ell)) \in H^\ell$. Note that a valid ciphertext of a bit μ is of the form $c = z + \mu h$ where its *homogeneous* part z has a small phase. This small $Phase(z) = Phase(c - GSW.Dec(c).h) \in \mathbb{T}^\ell$ will be denoted by $Noise(c)$.

By definition, the decryption function will successfully decrypt any ciphertext $c \in H^\ell$ such that $\|Noise(c)\|_\infty < \frac{1}{4}$, where the max-norm in $\mathbb{T}^\ell$ is taken over all coordinates centered in the interval $(-\frac{1}{2}, \frac{1}{2}]$. This is of course the case of fresh GSW.GLWE ciphertexts, whose Gaussian noise has small parameter α.

We now show that the GSW.Not, GSW.And and GSW.Mux gates amplify the noise only by a small factor if $f_h^{-1}()$ is β-bounded.

Lemma 10 (Worst-Case Noise of Primitive Gates). *Suppose* $f_h^{-1}()$ *is* β-*bounded for some* $\beta \in \mathbb{R}^+$. *Let* $c_1, c_2, c_3 \in H^\ell$ *be three ciphertexts such that* $c_1 = z_1 + \mu_1 \cdot h$, $c_2 = z_2 + \mu_2 \cdot h$ *and* $c_3 = z_3 + \mu_3 \cdot h$, *where* $\|Phase(z_1)\|_\infty \leq B$ *and* $\|Phase(z_2)\|_\infty, \|Phase(z_3)\|_\infty < B'$ *for some* $B, B' \in \mathbb{R}^+$. *Then:*

$$GSW.Not(c_1) = z + NOT(\mu_1) \cdot h \text{ with } \|Phase(z)\|_\infty = B \tag{4}$$

$$GSW.And(c_1, c_2) = z' + (\mu_1 \text{ AND } \mu_2) \cdot h \text{ with } \|Phase(z')\|_\infty \leq \ell\beta B + B' \tag{5}$$

$$GSW.Mux(c_1, c_2, c_3) = z'' + (\mu_1?\mu_2:\mu_3) \cdot h \text{ with } \|Phase(z'')\|_\infty \leq 2\ell\beta B + B' \tag{6}$$

Proof. By definition of GSW, we have $GSW.Not(c_1) = -z_1 + NOT(\mu_1)$, so $z = -z_1$, which proves (4). Then,

$$GSW.And(c_1, c_2) = F_{z_1 + \mu_1 \cdot h}\left(F_h^{-1}(c_2)\right) = F_{z_1}(F_h^{-1}(c_2)) + \mu_1 F_h(F_h^{-1}(c_2))$$

$$= F_{z_1}(F_h^{-1}(c_2)) + \mu_1 \cdot c_2 = \underbrace{F_{z_1}(F_h^{-1}(c_2)) + \mu_1 z_2}_{z'} + \mu_1 \mu_2 \cdot h$$

Letting $z' = F_{z_1}(F_h^{-1}(c_2)) + \mu_1 z_2$, we have $Phase(z') = Phase(z_1) \cdot (F_h^{-1}(c_2))^t + \mu_1 Phase(z_2)$, and therefore $\|Phase(z')\|_\infty \leq \ell \|F_h^{-1}(c_2)\|_\infty \|Phase(z_1)\|_\infty + \|Phase(z_2)\|_\infty \leq \ell\beta B + B'$, which proves (5). Finally, $GSW.Mux(c_1, c_2, c_3)$ is expressed as $GSW.And(c_1, c_2)$ plus $GSW.And(GSW.Not(c_1), c_3)$. By expanding, the expression takes the form $z'' + (\mu_2\mu_1 + \mu_3(1 - \mu_1)) \cdot h$ where $z'' = F_{z_1}(F_h^{-1}(c_2)) + F_{z_1}(F_h^{-1}(c_3)) + \mu_1 z_2 + (1 - \mu_1)z_3$. Thus, $Phase(z'') = Phase(z_1) \cdot (F_h^{-1}(c_2) + F_h^{-1}(c_3)) + \mu_1 Phase(z_2) + (1 - \mu_1)Phase(z_3)$. The norm of the first term is bounded by $2\ell\beta B$ and among the last two terms, only one is non-zero, and its norm is bounded by B'. Finally, the encoded message $\mu_2\mu_1 + \mu_3(1 - \mu_1)$ is precisely $\mu_1?\mu_2:\mu_3$. $\square$

As in [2, Lemma 3.5], we can ensure that the noise of all the entries of a ciphertext have independent Gaussian or Sub-Gaussian distributions. Namely, we say that $f_{\mathbf{h}}^{-1}$ is β-subgaussian if for each $y \in H$, $f_{h}^{-1}(y)$ returns a short Sub-Gaussian vector of parameter $\beta \geq \eta_\varepsilon(\mathcal{L}_{\mathbf{h}})$. The noise propagation analysis of [2, Lemma 3.5] can be extended as follows:

Lemma 11 (All Noises are Sub-gaussian). *Assume that $f_{\mathbf{h}}^{-1}$ is β-subgaussian for $\beta \geq \eta_\varepsilon(\mathcal{L}_{\mathbf{h}})$. In a circuit containing solely* GSW.Not, GSW.And *and* GSW.Mux *gates, and whose inputs are either fresh GLWE ciphertexts or the noiseless ciphertexts 0 and $\mathbf{h}$, the output ciphertext of each individual gate has the form $\mathbf{z} + \mu\mathbf{h}$ where μ is the encoded bit and the ℓ-coordinates of* Phase$(\mathbf{z})$ *are statistically indistinguishable from independent Gaussian samples of $\mathbb{T}_k$. We define the noise parameter $\sigma(\text{Phase}(z))$ as the maximum of these ℓ Gaussian parameters.*

Thus, we may work directly with the square subgaussian parameter of the noise, which follows pythagorean summation.

Lemma 12 (Average Noise of Primitive Gates). *Assume that $f_{\mathbf{h}}^{-1}()$ is $\sqrt{\beta}$-subgaussian for some $\beta > 0$. Let $\mathbf{c}_1 = \mathbf{z}_1 + \mu_1 \cdot \mathbf{h}$, $\mathbf{c}_2 = \mathbf{z}_2 + \mu_2 \cdot \mathbf{h}$, $\mathbf{c}_3 = \mathbf{z}_3 + \mu_3 \cdot \mathbf{h} \in H^\ell$ be three ciphertexts of a circuit satisfying the constraints of Lemma 11, and whose Gaussian parameters satisfy $\sigma(\text{Phase}(\mathbf{z}_1))^2 \leq B$ and $\sigma(\text{Phase}(\mathbf{z}_2))^2, \sigma(\text{Phase}(\mathbf{z}_3))^2 < B'$ for some $B, B' \in \mathbb{R}^+$. Then:*

$$\text{GSW.Not}(\mathbf{c}_1) = \mathbf{z} + \text{NOT}(\mu_1) \cdot \mathbf{h} \text{ with } \sigma(\text{Phase}(\mathbf{z}))^2 = B \tag{7}$$

$$\text{GSW.And}(\mathbf{c}_1, \mathbf{c}_2) = \mathbf{z}' + (\mu_1 \text{ AND } \mu_2) \cdot \mathbf{h} \text{ with } \sigma(\text{Phase}(\mathbf{z}'))^2 \leq \ell\beta B + B' \tag{8}$$

$$\text{GSW.Mux}(\mathbf{c}_1, \mathbf{c}_2, \mathbf{c}_3) = \mathbf{z}'' + (\mu_1?\mu_2{:}\mu_3) \cdot \mathbf{h} \text{ with } \sigma(\text{Phase}(\mathbf{z}''))^2 \leq 2\ell\beta B + B' \tag{9}$$

Since (4), (5) and (6) define the same recurrence as (7), (8) and (9), we will express the end of the paper only in terms of Lemma 10, but all the bounds we obtain on the $\|\text{Noise}\|$ also apply to the $\sigma(\text{Noise})^2$ under Lemma 12.

7.3 Homomorphically Evaluating Arbitrary Functions

The result of the following corollary was already obtained in [2]; it states that in a long chain of And gates where one of the bits is a fresh GLWE-GSW ciphertext, the noise increases in fact linearly instead of exponentially. Here, we invert the operands of the And gates, so that the overall noise in the resulting ciphertext is smaller if one associates long conjunctions on the right.

Corollary 5 (Noise of Conjunctions). *Suppose $f_{\mathbf{h}}^{-1}()$ is β-bounded for some $\beta \in \mathbb{R}^+$. Let $\mathbf{c}_1, \ldots, \mathbf{c}_k \in H^\ell$ be k ciphertexts such that each $\mathbf{c}_i = \mathbf{z}_i + \mu_i \cdot \mathbf{h}$ where $\|\text{Phase}(\mathbf{z}_i)\|_\infty < B$ for some $B \in \mathbb{R}^+$. Then:*

$$\text{GSW.And}(\mathbf{c}_1, \text{GSW.And}(\mathbf{c}_2, \ldots \text{GSW.And}(\mathbf{c}_{k-1}, \mathbf{c}_k))) = \mathbf{z} + (\mu_1\mu_2 \ldots \mu_k) \cdot \mathbf{h}$$

where $\|\text{Phase}(\mathbf{z}\|_\infty \leq k\ell\beta B$.

Proof. Apply (5) by induction on k.   □

Note that any boolean function with k inputs can always be put into disjunctive normal form, *i.e.* a disjoint union of conjunctive terms. One way to homomorphically evaluate the result is to add the ciphertexts of all the terms, which indeed preserves the $\{0, 1\}$ message space. However, with this method, the resulting noise will be proportional to the number of terms in the disjunctive normal form, which may still be exponential in the number of inputs.

By using Mux-gates, we obtain the following corollary, which says that any function can be homomorphically evaluated in a trivial way, where the noise grows proportionally to only the square root of the number of inputs. We recall that the truth table of a boolean function ϕ with k variables is a vector T of length 2^k such that each $T_j = \phi(e_0, \ldots, e_{k-1})$ where $j = \sum e_i 2^{k-1-i}$. The full binary decision diagram (BDD) of ϕ is a circuit representing a binary tree of Mux-gates, of depth k. The bottom level k consists in 2^k leaves $X_{k,j}$, each one is set to T_j. At each intermediate level i, we have 2^i nodes $X_{k,j} = \text{Mux}(\mu_i, X_{i+1,2j+1}, X_{i+1,2j})$. By definition, the root $X_{0,0}$ thus contains $\phi(\mu_0, \ldots, \mu_{k-1})$. See Fig. 1 for an example of truth table and its associated BDD circuit.

Corollary 6 (Evaluating Arbitrary Functions). *Assume that $f_{\mathbf{h}}^{-1}()$ is β-bounded for some $\beta \in \mathbb{R}^+$. Let ϕ be any boolean function with k inputs, and let $\mathbf{c}_1, \ldots, \mathbf{c}_k \in H^\ell$ be k ciphertexts such that each $\mathbf{c}_i = \mathbf{z}_i + \mu_i \cdot \mathbf{h}$ where $\sigma(\mathbf{z}_i)^2 < B$ for some $B \in \mathbb{R}^+$. Then, the Mux-based Binary Decision Diagram of ϕ computes a ciphertext $\mathbf{c} = \mathbf{z} + \phi(\mu_1, \ldots, \mu_k) \cdot \mathbf{h}$ where $\|\mathbf{z}\|_\infty \leq 2k\ell\beta B$.*

Proof. To evaluate the full BDD of ϕ homomorphically, we just replace each leaf $X_{k,j}$ by noiseless ciphertexts $T_j \cdot \mathbf{h}$, each bit μ_i by their encryption $\mathbf{c}_i$, and each Mux gate by GSW.Mux. Apply (6) by induction on the depth, then all nodes $X_{i,j}$ at depth i have a noise bounded by $2(k-i)\beta B$.   □

In the previous corollary, the full BDD tree of the function ϕ contains a number of nodes which is exponential in the number of inputs. If the output noise is indeed really small, the time complexity to evaluate all the gates remains large when the simulated function has many variables. For some useful functions, like the bootstrapping function in the next section, many of the subtrees turn out to be equal. By merging them, the complexity to evaluate the circuit can be significantly reduced.

Corollary 7 (Faster Evaluation of Arbitrary Functions). *Assume that $f_{\mathbf{h}}^{-1}()$ is β-bounded for some $\beta \in \mathbb{R}^+$. Let ϕ be any boolean function with k inputs, and let $\mathbf{c}_1, \ldots, \mathbf{c}_k \in H^\ell$ be k ciphertexts such that each $\mathbf{c}_i = \mathbf{z}_i + \mu_i \cdot \mathbf{h}$ where $\|\text{Phase}(\mathbf{z}_i)\|_\infty < B$ for some $B \in \mathbb{R}^+$. We call $\mathcal{N}(\phi)$ the number of disctinct subtrees in the full Binary Decision Diagram of ϕ. Then we can compute a ciphertext $\mathbf{c} = \mathbf{z} + \phi(\mu_1, \ldots, \mu_k) \cdot \mathbf{h}$ where $\|\text{Phase}(\mathbf{z})\|_\infty \leq 2k\ell\beta B$ by evaluating $\mathcal{N}(\phi)$ homomorphic GSW.Mux-gates.*

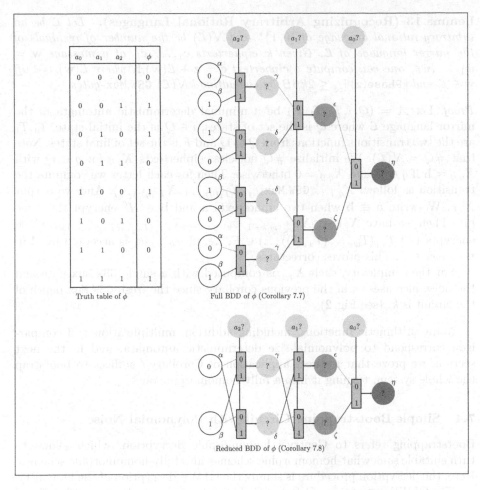

a_0	a_1	a_2	ϕ
0	0	0	0
0	0	1	1
0	1	0	1
0	1	1	0
1	0	0	1
1	0	1	0
1	1	0	0
1	1	1	1

Truth table of ϕ

Full BDD of ϕ (Corollary 7.7)

Reduced BDD of ϕ (Corollary 7.8)

Fig. 1. Homomorphic evaluation of an arbitrary boolean function

Proof. It suffices to evaluate the ciphertext value in the root of the $\mathcal{N}(\phi)$ subtrees by increasing depth. There are at most two different leaves, whose ciphertext values 0 and **h** are given. Whenever we need to evaluate a subtree of non zero depth i, the left and right subtrees have by definition already been fully evaluated, since their depth $i - 1$ is strictly smaller. The root of the current tree is the GSW.Mux of c_i and the two subtrees roots. The last ciphertext to be evaluated is the root of the full tree, which contains the encrypted result. □

In the above corollary, Nerode's partitioning algorithm for reducing deterministic automata can efficiently list the $\mathcal{N}(\phi)$ identical subtrees. Indeed, a binary decision diagram is just the mirror graph of a deterministic accessible automata. More generally, the GSW.Mux gate allows to homomorphically evaluate the transitions of a deterministic automata, which leads to the following lemma.

Lemma 13 (Recognizing Arbitrary Rational Langages). *Let $\mathcal{L}$ be an arbitrary rational language of $\{0,1\}^*$ and $\mathcal{N}(\tilde{\mathcal{L}})$ be the number of residuals of the mirror language of $\mathcal{L}$. Given k ciphertexts $c_1, \ldots, c_k$ of a message $\mathbf{w} = w_1, \ldots, w_k$, one can compute a ciphertext $\mathbf{c} = \mathbf{z} + \mathcal{L}(\mathbf{w}).\mathbf{h}$ where $\mathcal{L}(\mathbf{w}) = 1$ iff $\mathbf{w} \in \mathcal{L}$ and $\|\mathrm{Phase}(\mathbf{z})\|_\infty \leq 2k\beta B$ by evaluating $k\mathcal{N}(\tilde{\mathcal{L}})$ GSW.Mux-gates.*

Proof. Let $\mathcal{A} = (Q, i, T_0, T_1, F)$ be a minimal deterministic automata of the mirror language $\tilde{\mathcal{L}}$ where Q is the set of states, $i \in Q$ is the initial state, T_0, T_1 are the two transitions functions from Q to Q and F is the set of final states. Note that $\#Q = \mathcal{N}(\tilde{\mathcal{L}})$. We initialize $\#Q$ noiseless ciphertexts $X_{q,0}$ for $q \in Q$ with $X_{q,0} = \mathbf{h}$ if $q \in F$ and $X_{q,0} = 0$ otherwise. Then for each letter we compute the transition as follow: $X_{q,j} = \mathrm{GSW.Mux}(c_j, X_{T_1(q),j-1}, X_{T_0(q),j-1})$. And we output $X_{i,k}$. We write $\mathbf{a} \equiv \mathbf{b}$ when two ciphertexts $\mathbf{a}$ and $\mathbf{b} \in H^\ell$ encrypt the same bit. Then we have $X_{i,k} \equiv X_{T_{w_k}(i),k-1} \equiv \cdots \equiv X_{T_{w_1}(T_{w_2}\ldots(T_{w_k}(i))\ldots),0}$, which encrypts 1 iff $T_{w_1}(T_{w_2}\ldots(T_{w_k}(i))\ldots) \in F$, *i.e.* iff $w_k \ldots w_1$ is accepted by $\mathcal{A}$ iff $w_1 \ldots w_k \in \mathcal{L}$. This proves correctness.

For the complexity, each $X_{q,j}$ is computed with a single GSW.Mux gate and the noise increases as in the previous corollary since the fresh-GSW.Mux depth of the circuit is k. (see Fig. 2) $\square$

Many arithmetic functions, including addition, multiplication and comparison correspond to polynomial-size deterministic automata, and in the next section, we prove that a direct application of Corollary 7 suffices to bootstrap the whole system, turning it into a fully homomorphic one.

7.4 Simple Bootstrapping Circuit with Polynomial Noise

Bootstrapping refers to Gentry's homomorphic decryption, which allows to turn suitable somewhat-homomorphic schemes into fully-homomorphic schemes. Here, the decryption procedure is simply the GLWE decryption of the first entry.

The GLWE decryption of $(d, c) \in G \times \mathbb{T}$ consists in computing $c - \hat{s}(d) \in \mathbb{T}$ and deciding whether it is closer to $\frac{1}{2}$ or 0. If the secret $\hat{s}$ has $n-1$ bits $(s_1, \ldots, s_{n-1})$, this sum can be linearized as $c - \sum_{i=1}^{n-1} s_i d_i$ where $c, d_1, \ldots, d_{n-1} \in \mathbb{T}$ are publicly computable. Necessarily n is always $\leq \ell$. Furthermore, if the noise of (d, c) is $\frac{1}{8}$-bounded, these n values can be rounded to their nearest multiple of $\frac{1}{4n}$ without affecting the result of the decryption. Thus, bootstrapping a ciphertext $(d, c) \in H$ is equivalent to homomorphically evaluate on $(s_1, \ldots, s_{n-1})$, its *bootstrapping boolean function* $\phi(x_1, \ldots, x_{n-1})$ which returns the most significant bit of $c' - \sum_{i=1}^{n-1} s_i d_i'$ where $c' = \lfloor 4n(c + \frac{1}{4}) \rfloor, d_i' = \lfloor d_i \rfloor$ are known integers modulo $4n$.

Lemma 14 (Simple Bootstrapping). *Given a GSW ciphertext $\mathbf{c} = \mathbf{z} + \mu\mathbf{h} \in H^\ell$, s.t. $\|\mathrm{Phase}(\mathbf{z})\|_\infty < \frac{1}{8}$, whose first entry is $(d, c) \in H$, its bootstrapping function ϕ satisfies $\mathcal{N}(\phi) \leq 4n^2$. Therefore, if $f_\mathbf{h}^{-1}$ is 1-bounded, given the bootstrapping key $(BK_i)_{i \in [1,n-1]}$ where BK_i encrypts the i-th bit of $\hat{s}$ with $\|\mathrm{Noise}(BK_i)\|_\infty \leq B$, one can compute a ciphertext $\mathbf{c}' = \mathbf{z}' + \mu\mathbf{h}$ of the same message where $\|\mathrm{Phase}(\mathbf{z}')\|_\infty < 2n\ell B$ by evaluating at most $4n^2$ GSW.Mux gates.*

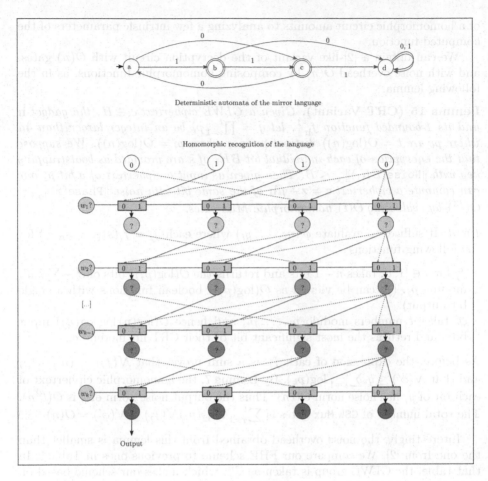

Deterministic automata of the mirror language

Homomorphic recognition of the language

Output

Fig. 2. A Mux-based circuit recognizing a regular language $\mathcal{L}$ and the corresponding deterministic automata of the mirror $\tilde{\mathcal{L}}$.

Proof. The expression of ϕ as a sum proves that for all $(x_1, \ldots, x_k)$ and $(y_1, \ldots, y_k)$ and $(z_{k+1}, \ldots, z_n)$ such that $\sum_{i=1}^{k} x_i d'i = \sum_{i=1}^{k} y_i d'i \mod 4n$, then $\phi(x_1, \ldots, x_k, z_{k+1}, \ldots, z_n) = \phi(y_1, \ldots, y_k, z_{k+1}, \ldots, z_n)$. This proves that for each index $k \in [0, n-1]$, there are at most $4n$ distinct partial functions of ϕ by fixing the first k coordinates. And thus, $\mathcal{N}(\phi) \leq 4n^2$. The rest follows from Corollary 7. □

Recall that under the hypothesis of Lemma 11, the max-norm of the noise can be replaced by its square Gaussian parameter. It follows that the GLWE-GSW scheme is fully homomorphic according to Gentry's blueprint by design, as soon as the initial GLWE Gaussian parameter is $1/\tilde{O}(\ell^{1.5})$, which represents a time *vs* noise trade-off compared to the [2] proposal, and shows that the construction of a homomorphic circuit amounts to analyzing a few intrinsic parameters of the computed function.

We can obtain a [2]-like variant of the decryption circuit with $\tilde{O}(n)$ gates, and with noise overhead $\tilde{O}(n)$ by composing homomorphic functions, as in the following lemma.

Lemma 15 (CRT Variant). *Given a GLWE ciphertext $c \in H$, the gadget $\mathbf{h}$ and its 1-bounded function $f_{\mathbf{h}}^{-1}$, let $q = \prod_{i=1}^{t} p_i$ be an integer larger than $4n$ where p_i are $t = O(\log(n))$ distinct primes where $p_i = O(\log(n))$. We suppose that the encryption of each individual bit BK_i of $\hat{s}$ are provided as bootstrapping key with $\|\mathtt{Noise}(BK_i)\|_{\infty} \leq B$. Then given as input a ciphertext of a bit μ, one can compute a ciphertext $\mathbf{c} = \mathbf{z} + \mu\mathbf{h}$ of the same bit with noise $\|\mathtt{Phase}(\mathbf{z})\|_{\infty} = \tilde{O}(\ell^3)$ by evaluating $\tilde{O}(\ell)$ homomorphic Mux-gates.*

Proof. It suffices to evaluate $\phi'(y_1, \ldots, y_t)$ where each $y_j = f_j(s_1, \ldots, s_{n-1})$ for the following functions:

- f_j for $j \in [1, t]$, takes $n - 1$ bits and returns the $O(\log(p_j))$ bits of $c' - \sum s_i d'_i$ modulo p_j. (f_j can be viewed as $O(\log(p_j))$ boolean functions with a single bit output).
- ϕ' takes t numbers modulo $p_1, \ldots, p_t$, and hence $O(\log(n) \log\log(n))$ input bits, and returns the most significant bit of their CRT lift modulo q.

As before, the expression of each f_j as a sum proves that $\mathcal{N}(f_j) \leq (n-1)p_j$ and that $\mathcal{N}(\phi') \leq q . \sum_{k=1}^{t} \log(p_k)$. By Lemma 7, the homomorphic ciphertext of each bit of y_j has noise norm $\tilde{O}(\ell n)$. Thus the output noise norm of y is $\tilde{O}(\ell^2 n)$. The total number of $\mathtt{GSW.Mux}$ gates is $\sum_{j=1}^{t} \log_2(p_j)\mathcal{N}(f_j) + \mathcal{N}(\phi') = \tilde{O}(n)$ $\square$

Interestingly, the noise overhead obtained from this lemma is smaller than the one from [2]. We compare our FHE scheme to previous ones in Table 1. In that table, the GLWE group is taken as $\mathbb{Z}_q^n$, which makes our scheme based on the standard LWE assumption. In this case, we could take $\ell = O(n \log q)$.

Table 1. Comparisons of LWE-based FHE schemes

Schemes	Primitive gates	#Gates in boots.	Boots. noise overhead
BGV12 [10]	And, Xor, Const.	$\tilde{O}(n^2)$	$n^{O(\log n)}$
Bra12 [9]	And, Xor, Const.	$\tilde{O}(n^2)$	$n^{O(\log n)}$
GSW13 [21]	And, Xor, Nand, Const.	$\tilde{O}(n^2)$	$n^{O(\log n)}$
BV14 [12]	And, Xor, Const.	$\tilde{O}(n^{6/\varepsilon})$	$\tilde{O}(n^{\varepsilon})$
AP14 [2]	And, Not, Const.	$\tilde{O}(n)$	$\tilde{O}(n^2)$
DM15 [15]	Nand, Const.	$\tilde{O}(n)$	$\tilde{O}(n^{1.5})$
Ours	Mux, Not, Const.	$\tilde{O}(n^2)$	$\tilde{O}(n)$
Ours (with CRT)	Mux, Not, Const.	$\tilde{O}(n)$	$\tilde{O}(n^{1.5})$

Acknowledgements. Part of this work has been supported by Fonds Unique Interministériel (FUI) through the CRYPTOCOMP project and the EIT Digital project HC@WORKS, China's 973 Program (Grant 2013CB834205), and NSFC's Key Project (Grant 61133013).

A   Missing Algorithms

All missing algorithms were sketched in the main body. Here, we provide explicit descriptions of these algorithms.

A.1   Algorithm 5: Sampling Lattices of Given Factor Group

Algorithm 5. Sampling lattices of given factor group

Input: Integer $m \geq 1$ and a finite abelian group $G = \mathbb{Z}_{q_1} \times \cdots \times \mathbb{Z}_{q_k}$ such that $1 \leq k \leq m$.
Output: A random lattice from the uniform distribution over $\mathcal{L}_{G,m}$.
1: Generate elements $g_1, \ldots, g_m$ uniformly at random from G until the g_i's generate G.
2: Return the lattice $\mathcal{L}_{\mathbf{g}}$ where $\mathbf{g} = (g_1, \ldots, g_m) \in G^m$.

A.2   Algorithm 1: Unbalanced Reduction

Algorithm 6. Unbalanced Reduction

Input: an $n \times m$ basis B of an integer lattice $L \subseteq \mathbb{Z}^m$ and a target length $\sigma \in \mathbb{Q}^+$. More generally, B can be any n-dimensional projected block $B = B'_{[i,i+n-1]}$ of some basis B' of $L \subseteq \mathbb{Z}^m$.
Output: an $n \times n$ unimodular matrix U such that $C = UB$ satisfies $\|\mathbf{c}_i^*\| \leq \sigma$ for $i \geq 2$ and $\|\mathbf{c}_1\| \leq n\sigma\delta_\sigma(B)$.
1: $C \leftarrow B$, $U \leftarrow I_n$ and compute the Gram-Schmidt matrices μ and C^*
2: If $\|\mathbf{c}_i^*\| \leq \sigma$ for all i, **return** U
3: **for** $i = k - 1$ downto 1 where k is the largest index such that $\|\mathbf{c}_k^*\| > \sigma$ **do**
4:     **if** $\|\mathbf{c}_i^*\| \leq \sigma$ **then**
5:         $\alpha \leftarrow \lfloor -\mu_{i+1,i} \rceil$
6:     **else**
7:         $\alpha \leftarrow \left\lceil -\mu_{i+1,i} + \frac{\|\mathbf{c}_{i+1}^*\|}{\|\mathbf{c}_i^*\|} \sqrt{(\|\mathbf{c}_i^*\|/\sigma)^2 - 1} \right\rceil$
8:     **end if**
9:     $(\mathbf{c}_i, \mathbf{c}_{i+1}) \leftarrow (\mathbf{c}_{i+1} + \alpha \cdot \mathbf{c}_i, \mathbf{c}_i)$, $(\mathbf{u}_i, \mathbf{u}_{i+1}) \leftarrow (\mathbf{u}_{i+1} + \alpha \cdot \mathbf{u}_i, \mathbf{u}_i)$ and update the GS matrices μ and C^*.
10: **end for**
11: **return** U

A.3   Algorithm 7: Bootstrapping

Algorithm 7. Bootstrapping algorithm

Input: A GLWE ciphertext $c \in H$, the gadget $\mathbf{h}$ and its functions $f_{\mathbf{h}}^{-1}$, and the bootstrapping key $(BK_{i,j})_{i \in [1,\ell], j \in [1,n]}$ where $BK_{i,1}, \ldots, BK_{i,n}$ are encryptions of the $n = \log_2(\ell) + 3$ most significant bits of $\mathrm{Phase}(h_i)$.

Output: A GLWE-GSW ciphertext $\mathbf{c}' \in H^{\ell}$ encoding the same bit as c with polynomial noise.

1: $\mathbf{x} \leftarrow f_{\mathbf{h}}^{-1}(c) \in \{0,1\}^{\ell}$
2: $p \leftarrow 0$
3: Set the initial state $(X_{0,0}, \ldots, X_{0,8\ell-1})$ where $X_{i,j} = 1$ iff $j \in [2\ell, 6\ell]$
4: **for each** $i \in [1, \ell]$ s.t. $x_i = 1$ **do**
5:     **for** $j = 1$ to n **do**   ▷ This loop adds $\mathrm{Phase}(h_i)$ to the state
6:         $p \leftarrow p + 1$
7:         **for** $k = 0$ to $8\ell - 1$ **do**   ▷ This loop adds 2^{n-j} to the state iff $BK_{i,j} = 1$
8:             $X_{p,k} \leftarrow \mathrm{GSW.Mux}(BK_{i,j}, X_{p-1,k-2^{n-j} \bmod 8\ell}, X_{p-1,k})$
9:         **end for**
10:     **end for**
11: **end for**
12: **return** $\mathbf{c}' = X_{p,0}$ ▷ This is the final rounding.

References

1. Ajtai, M.: Generating hard instances of lattice problems. In: STOC, pp. 99–108 (1996)
2. Alperin-Sheriff, J., Peikert, C.: Faster bootstrapping with polynomial error. In: Garay, J.A., Gennaro, R. (eds.) CRYPTO 2014, Part I. LNCS, vol. 8616, pp. 297–314. Springer, Heidelberg (2014)
3. Applebaum, B., Cash, D., Peikert, C., Sahai, A.: Fast cryptographic primitives and circular-secure encryption based on hard learning problems. In: Halevi, S. (ed.) CRYPTO 2009. LNCS, vol. 5677, pp. 595–618. Springer, Heidelberg (2009)
4. Baumslag, G., Fazio, N., Nicolosi, A.R., Shpilrain, V., Skeith III, W.E.: Generalized learning problems and applications to non-commutative cryptography. In: Boyen, X., Chen, X. (eds.) ProvSec 2011. LNCS, vol. 6980, pp. 324–339. Springer, Heidelberg (2011)
5. Becker, A., Gama, N., Joux, A.: A sieve algorithm based on overlattices. LMS J. Comput. Math. **17**(A), 49–70 (2014). Cryptology ePrint Archive, report 2013/685
6. Bleichenbacher, D.: On the generation of DSA one-time keys. Draft of 13 September 2004. Short Presentation at the Rump Session of CRYPTO 2005 (2005)
7. Blum, A., Kalai, A., Wasserman, H.: Noise-tolerant learning, the parity problem, and the statistical query model. J. ACM **50**(4), 506–519 (2003)
8. Boneh, D., Venkatesan, R.: Hardness of computing the most significant bits of secret keys in Diffie-Hellman and related schemes. In: Koblitz, N. (ed.) CRYPTO 1996. LNCS, vol. 1109, pp. 129–142. Springer, Heidelberg (1996)
9. Brakerski, Z.: Fully homomorphic encryption without modulus switching from classical GapSVP. In: Canetti, R., Safavi-Naini, R. (eds.) CRYPTO 2012. LNCS, vol. 7417, pp. 868–886. Springer, Heidelberg (2012)

10. Brakerski, Z., Gentry, C., Vaikuntanathan, V.: (Leveled) fully homomorphic encryption without bootstrapping. In: ITCS, pp. 309–325 (2012)
11. Brakerski, Z., Langlois, A., Peikert, C., Regev, O., Stehlé, D.: Classical hardness of learning with errors. In: Proceedings of 45th STOC, pp. 575–584. ACM (2013)
12. Brakerski, Z., Vaikuntanathan, V.: Lattice-based FHE as secure as PKE. In: ITCS, pp. 1–12 (2014)
13. Cai, J.-Y., Theory, A.N.: The complexity of some lattice problems. In: Bosma, W. (ed.) ANTS-IV. LNCS, vol. 1838, pp. 1–32. Springer, Heidelberg (2000)
14. Chen, Y., Nguyen, P.Q.: BKZ 2.0: better lattice security estimates. In: Wang, X., Lee, D.H. (eds.) ASIACRYPT 2011. LNCS, vol. 7073, pp. 1–20. Springer, Heidelberg (2011)
15. Ducas, L., Micciancio, D.: FHEW: bootstrapping homomorphic encryption in less than a second. In: Oswald, E., Fischlin, M. (eds.) EUROCRYPT 2015. LNCS, vol. 9056, pp. 617–640. Springer, Heidelberg (2015)
16. Fazio, N., Iga, K., Nicolosi, A.R., Perret, L., Skeith, W.E.: Hardness of learning problems over burnside groups of exponent 3. Des. Codes Crypt. **75**(1), 59–70 (2015)
17. Gama, N., Izabachène, M., Nguyen, P.Q., Xie, X.: Structural lattice reduction: generalized worst-case to average-case reductions and homomorphic cryptosystems. To appear soon on IACR Cryptology ePrint Archive (2016)
18. Gama, N., Nguyen, P.Q.: Predicting lattice reduction. In: Smart, N.P. (ed.) EUROCRYPT 2008. LNCS, vol. 4965, pp. 31–51. Springer, Heidelberg (2008)
19. Gentry, C., Halevi, S.: Implementing gentry's fully-homomorphic encryption scheme. In: Paterson, K.G. (ed.) EUROCRYPT 2011. LNCS, vol. 6632, pp. 129–148. Springer, Heidelberg (2011)
20. Gentry, C., Peikert, C., Vaikuntanathan, V.: Trapdoors for hard lattices and new cryptographic constructions. In: STOC (2008)
21. Gentry, C., Sahai, A., Waters, B.: Homomorphic encryption from learning with errors: conceptually-simpler, asymptotically-faster, attribute-based. In: Canetti, R., Garay, J.A. (eds.) CRYPTO 2013, Part I. LNCS, vol. 8042, pp. 75–92. Springer, Heidelberg (2013)
22. Goldstein, D., Mayer, A.: On the equidistribution of Hecke points. Forum Math. **15**(2), 165–189 (2003)
23. Lenstra, A.K., Lenstra Jr., H.W., Lovász, L.: Factoring polynomials with rational coefficients. Math. Ann. **261**, 513–534 (1982)
24. Lubotzky, A.: The expected number of random elements to generate a finite group. J. Algebra **257**(2), 452–459 (2002)
25. Micciancio, D.: Almost perfect lattices, the covering radius problem, and applications to Ajtai's connection factor. SIAM J. Comput. **34**(1), 118–169 (2004)
26. Micciancio, D., Peikert, C.: Trapdoors for lattices: simpler, tighter, faster, smaller. In: Pointcheval, D., Johansson, T. (eds.) EUROCRYPT 2012. LNCS, vol. 7237, pp. 700–718. Springer, Heidelberg (2012)
27. Micciancio, D., Peikert, C.: Hardness of SIS and LWE with small parameters. In: Canetti, R., Garay, J.A. (eds.) CRYPTO 2013, Part I. LNCS, vol. 8042, pp. 21–39. Springer, Heidelberg (2013)
28. Micciancio, D., Regev, O.: Worst-case to average-case reductions based on Gaussian measures. SIAM J. Comput. **37**(1), 267–302 (2007)
29. Mordell, L.J.: On some arithmetical results in the geometry of numbers. Compos. Math. **1**, 248–253 (1935)
30. Nguyen, P.Q., Shparlinski, I.E.: The insecurity of the digital signature algorithm with partially known nonces. J. Cryptology **15**(3), 151–176 (2002)

31. Nguyen, P.Q., Shparlinski, I.E.: Counting co-cyclic lattices. CoRR, abs/1505.06429 (2015, preprint)
32. Pak, I.: On probability of generating a finite group (1999, preprint)
33. Paz, A., Schnorr, C.-P.: Approximating integer lattices by lattices with cyclic factor groups. In: Ottmann, T. (ed.) ICALP 1987. LNCS, vol. 267, pp. 386–393. Springer, Heidelberg (1987)
34. Peikert, C.: Public-key cryptosystems from the worst-case shortest vector problem. In: STOC, pp. 333–342. ACM (2009)
35. Peikert, C.: An efficient and parallel gaussian sampler for lattices. In: Rabin, T. (ed.) CRYPTO 2010. LNCS, vol. 6223, pp. 80–97. Springer, Heidelberg (2010)
36. Regev, O.: Lattices in computer science #12: average-case hardness. Regev's Webpage (2004)
37. Regev, O.: On lattices, learning with errors, random linear codes, and cryptography. In: STOC, pp. 84–93 (2005)
38. van Dijk, M., Gentry, C., Halevi, S., Vaikuntanathan, V.: Fully homomorphic encryption over the integers. In: Gilbert, H. (ed.) EUROCRYPT 2010. LNCS, vol. 6110, pp. 24–43. Springer, Heidelberg (2010)

Recovering Short Generators of Principal Ideals in Cyclotomic Rings

Ronald Cramer[1,2]($\boxtimes$), Léo Ducas[1], Chris Peikert[3], and Oded Regev[4]

[1] Cryptology Group, CWI, Amsterdam, The Netherlands
{Ronald.Cramer,Leo.Ducas}@cwi.nl
[2] Mathematical Institute, Leiden University, Leiden, The Netherlands
[3] Department of Computer Science and Engineering, University of Michigan, Michigan, USA
[4] Courant Institute of Mathematical Sciences, New York University, New York, USA

Abstract. A handful of recent cryptographic proposals rely on the conjectured hardness of the following problem in the ring of integers of a cyclotomic number field: given a basis of a principal ideal that is guaranteed to have a "rather short" generator, find such a generator. Recently, Bernstein and Campbell-Groves-Shepherd sketched potential attacks against this problem; most notably, the latter authors claimed a *polynomial-time quantum* algorithm. (Alternatively, replacing the quantum component with an algorithm of Biasse and Fieker would yield a *classical subexponential-time* algorithm.) A key claim of Campbell *et al.* is that one step of their algorithm—namely, decoding the *log-unit* lattice of the ring to recover a short generator from an arbitrary one—is classically efficient (whereas the standard approach on general lattices takes exponential time). However, very few convincing details were provided to substantiate this claim.

In this work, we clarify the situation by giving a rigorous proof that the log-unit lattice is indeed efficiently decodable, for any cyclotomic of prime-power index. Combining this with the quantum algorithm from a recent work of Biasse and Song confirms the main claim of Campbell *et al.* Our proof consists of two main technical contributions: the first is a geometrical analysis, using tools from analytic number theory, of the standard generators of the group of cyclotomic units. The second shows

---

L. Ducas—Supported by an NWO Free Competition Grant.

C. Peikert—Much of this work was done while the author was at the Georgia Institute of Technology. This material is based upon work supported by the National Science Foundation under CAREER Award CCF-1054495, by DARPA under agreement number FA8750-11-C-0096, and by the Alfred P. Sloan Foundation. Any opinions, findings, and conclusions or recommendations expressed in this material are those of the authors and do not necessarily reflect the views of the National Science Foundation, DARPA or the U.S. Government, or the Sloan Foundation. The U.S. Government is authorized to reproduce and distribute reprints for Governmental purposes notwithstanding any copyright notation thereon.

O. Regev—Supported by the Simons Collaboration on Algorithms and Geometry and by the National Science Foundation (NSF) under Grant No. CCF-1320188.

M. Fischlin and J.-S. Coron (Eds.): EUROCRYPT 2016, Part II, LNCS 9666, pp. 559–585, 2016.
DOI: 10.1007/978-3-662-49896-5_20

that for a wide class of typical distributions of the short generator, a standard lattice-decoding algorithm can recover it, given any generator.

By extending our geometrical analysis, as a second main contribution we obtain an efficient algorithm that, given any generator of a principal ideal (in a prime-power cyclotomic), finds a $2^{\tilde{O}(\sqrt{n})}$-approximate shortest vector in the ideal. Combining this with the result of Biasse and Song yields a quantum polynomial-time algorithm for the $2^{\tilde{O}(\sqrt{n})}$-approximate Shortest Vector Problem on principal ideal lattices.

1   Introduction

Over the past several years, *lattices* have emerged as an attractive foundation for cryptography. The most efficient (and potentially practical) lattice-based cryptosystems are related to *ideal lattices*, which correspond to ideals in certain families of rings, e.g., $\mathbb{Z}[X]/(X^{2^k} + 1)$. Representative works include [HPS98, Mic02, LMPR08, Gen09, LPR10].

More recently, a handful of cryptographic constructions have relied directly on *principal* ideals that have *"relatively short"* generators, which serve as secret keys.[1] These include a simplified variant of Gentry's original fully homomorphic encryption scheme [Gen09] due to Smart and Vercauteren [SV10], the closely related Soliloquy encryption scheme [CGS14], and candidate cryptographic multilinear maps [GGH13, LSS14]. Breaking these systems is no harder than solving the following problem, which we call the *Short Generator of a Principal Ideal Problem* (SG-PIP): given some $\mathbb{Z}$-basis of an ideal that is guaranteed to have a "short" generator g, find a sufficiently short generator (not necessarily g itself).

Potential attacks on SG-PIP in certain rings were sketched by Bernstein [Ber14b] and Campbell et al. [CGS14]. The basic structure of the attacks, which appears to be folklore in computational number theory, consists of two main parts:

- First, given a $\mathbb{Z}$-basis of the principal ideal, find some arbitrary (not necessarily short) generator of the ideal. For this task, which is known as the *Principal Ideal Problem* (PIP), the state of the art is an algorithm of Biasse and Fieker [BF14, Bia14], whose running time has only a subexponential $2^{n^{2/3+\epsilon}}$ dependence on n, the degree of the ring (over $\mathbb{Z}$). In addition, building on the recent work of Eisenträger *et al.* [EHKS14], polynomial-time *quantum* algorithms for PIP have recently been described in two independent works [CGS14, BS15], the latter of which provides a fully rigorous treatment.
- Second, transform the generator found in the previous phase into a *short* generator, thereby recovering the secret key, or its functional equivalent. The standard approach casts this task as a *closest vector problem* (CVP) on the Dirichlet "log-unit" lattice.

---

[1] A principal ideal in a commutative ring R is of the form $gR = \{g \cdot r : r \in R\}$ for some $g \in R$, called a *generator* of the ideal.

In this work, we focus entirely on the second phase, i.e., on recovering a short generator from any generator. At first, one might suspect that this is a hard problem: in general, the fastest known algorithms for CVP (even allowing quantum) run in exponential $2^{\Omega(n)}$ time [MV10, ADS15], or in less time but with much weaker guarantees on the solution quality (e.g., [LLL82, Bab85, Sch87]). In addition, Bernstein [Ber14b] suggested an algebraic approach that may yield slightly subexponential running times in number fields having many subfields, but it remains to be seen if this proposal can be carried through. Regardless of the method used, it is not obvious *a priori* whether solving CVP on the log-unit lattice yields a sufficiently short generator; much depends on the geometry of the lattice (in the relevant norm) and the quality of the solution.

A promising observation made by several researchers [CGS14, Ber14a] is that the CVP instances arising in the second phase have some implicit structure: the existence of a "rather short" generator (by choice of the secret key) implies that the target point is "somewhat close" to the log-unit lattice; CVP with such a distance guarantee is more commonly known as *bounded-distance decoding* (BDD) and is sometimes easier than the general case of CVP. Indeed, Garg et al. [GGH13] gave an improved variant of the Gentry-Szydlo algorithm [GS02] which shows that in cyclotomic rings having power-of-two index, BDD on the log-unit lattice is efficiently solvable to within sub-polynomial $n^{-\log\log n}$ distance. However, this threshold is much too small to handle the BDD instances arising in cryptosystems.

Campbell et al. [CGS14] were the first to claim an efficient solution to the second phase above. In more detail, they asserted that in cyclotomic rings having power-of-two index, the second phase can be accomplished simply by decoding the log-unit lattice using a standard algorithm such as LLL [LLL82]. However, this claim was not accompanied by a proof.[2] Nevertheless, experiments in cryptographically relevant choices of dimension have shown that decoding is indeed practically efficient [She14, Sch15], giving strong evidence that the approach of [CGS14] does indeed work.

Contributions. Our first main contribution is a rigorous proof showing that the second phase above can be solved in polynomial time, in any cyclotomic of prime-power index. Our proof is based on classical ideas and results from analytical number theory, along with some techniques from probability theory, and consists of two main technical contributions. First, in Sect. 3 we use standard tools from analytical number theory, such as bounds on *Dirichlet L-series*, to elucidate the geometry of a standard set of generators for the group of *cyclotomic units*. (The cyclotomic units correspond either to the log-unit lattice itself, or to a sublattice whose index is conjectured to be quite small.) Using this geometry, in Sects. 4 and 5 we show that for a wide class of typical distributions

---

[2] The explanation given in [CGS14] is that the secret generator corresponds to a vector that is short relative to the determinant of the log-unit lattice. As far as we can tell, this by itself is not enough to substantiate the claim, as it ignores the geometry of the log-unit lattice and the quality of the output produced by the LLL algorithm.

of the secret generator—e.g., Gaussian-like distributions—the naïve "round-off" lattice-decoding algorithm [Len82, Bab85] (using the standard generators of the cyclotomic units) can be used to efficiently recover the secret short generator, given any generator of the ideal.[3] To complement these results, in Appendix B we give concrete numerical data demonstrating that the second phase succeeds for all practical choices of dimension.

Our second main contribution concerns the questions: in an *arbitrary* principal ideal (of a prime-power cyclotomic), how long can a shortest generator be? And how short of a generator can we find efficiently? In Sect. 6, we show that for an overwhelming majority of principal ideals, the shortest generator is a $2^{\tilde{\Theta}(\sqrt{n})}$ factor longer than the shortest nonzero vector in the ideal. Moreover, one can efficiently find a generator satisfying this bound, given an arbitrary generator. The first of these facts means that the principal ideals used in the aforementioned cryptographic applications are highly atypical, because their shortest generators are also nearly shortest vectors. The second fact implies that the $2^{\tilde{O}(\sqrt{n})}$-approximate Shortest Vector Problem (SVP) on arbitrary principal ideals reduces to the Principal Ideal Problem.

Implications and Discussion. Combining our main contributions with known algorithms for PIP [BF14, Bia14, CGS14, BS15] (which are the computational bottleneck) yields the following two main implications:

- First, there is a quantum polynomial-time, or classical $2^{n^{2/3+\epsilon}}$-time, algorithm for SG-PIP, implying a key-recovery attack for the cryptographic constructions of [SV10, GGH13, LSS14, CGS14].
- Second, there is a quantum polynomial-time algorithm for $2^{\tilde{O}(\sqrt{n})}$-approximate SVP on *principal* ideals in any prime-power cyclotomic. (Note that we do not obtain any improvement over *classical* SVP algorithms, because $2^{n^{2/3}}$ time is sufficient to solve $2^{\tilde{O}(n^{1/3})}$-approximate SVP on arbitrary lattices [Sch87].)

In light of these, an important open problem is to obtain faster classical PIP algorithms, perhaps also using the guarantee that a short generator exists.

A natural question is what effect, if any, these attacks have on other ring-based problems, such as NTRU [HPS98] and ring-LWE [LPR10], which are the heart of many cryptosystems. Specifically, the theoretical foundation of the ring-LWE problem is the conjectured quantum hardness of approximate-SVP on *arbitrary* ideals, usually in a cyclotomic ring and for (near-)polynomial approximation factors. As far as we can tell, the above-described algorithms do not appear to affect this foundation: the first crucially relies on the existence of an "unusually short" generator, the second is inherently limited to relatively large SVP approximation factors, and both apply only to *principal* ideals. An important question is whether these barriers can be overcome, and if so, whether this leads to attacks on ring-LWE or NTRU themselves.

---

[3] Strictly speaking, the polynomial running time of this algorithm depends on a number-theoretic conjecture regarding the class numbers $h^+(m)$; see Sect. 2.4 for details.

In a complementary direction, another interesting question is whether the above attacks can be extended to other families of *non-cyclotomic* rings, such as those suggested in [Ber14b]. For this it may suffice to find (by analysis, computation, or both) a suitably good basis of the log-unit lattice, or of a sublattice of not too large index.

2   Preliminaries

We denote column vectors by lower-case bold letters (e.g., $\mathbf{x}$) and matrices by upper-case bold letters (e.g., $\mathbf{X}$). We often adopt the nonstandard, but very useful, convention of indexing rows and columns by particular finite sets (not necessarily $\{1, \ldots, n\}$), and identify a matrix with its indexed set of column vectors. The canonical scalar product over $\mathbb{R}^n$ and over $\mathbb{C}^n$ is denoted $\langle \cdot, \cdot \rangle$, and $\|\cdot\|$ denotes the Euclidean norm. For a complex number $z \in \mathbb{C}$, $\bar{z}$ denotes its complex conjugate, and $|z| = \sqrt{z \cdot \bar{z}}$ denotes its magnitude.

2.1   Lattices and BDD

A *lattice* $\mathcal{L}$ is a discrete additive subgroup of $\mathbb{R}^n$ for some positive integer n. The *minimum distance* of $\mathcal{L}$ is $\lambda_1(\mathcal{L}) := \min_{\mathbf{v} \in \mathcal{L} \setminus \{0\}} \|\mathbf{v}\|$, the length of a shortest nonzero lattice vector. Every lattice is generated as the integer linear combinations of some (non-unique) $\mathbb{R}$-linearly independent *basis* vectors $\mathbf{B} = \{\mathbf{b}_1, \ldots, \mathbf{b}_k\}$, as $\mathcal{L} = \mathcal{L}(\mathbf{B}) := \{\sum_{j=1}^{k} \mathbb{Z} \cdot \mathbf{b}_j\}$, where $k \leq n$ is called the *rank* of the lattice.

Letting span denote the $\mathbb{R}$-linear span of a set, the *dual* basis $\mathbf{B}^\vee = \{\mathbf{b}_1^\vee, \ldots, \mathbf{b}_k^\vee\} \subset \mathrm{span}(\mathbf{B})$ and dual lattice $\mathcal{L} = \mathcal{L}(\mathbf{B}^\vee)$ are defined to satisfy $\langle \mathbf{b}_j^\vee, \mathbf{b}_{j'} \rangle = \delta_{j,j'}$ for all j, j', where the Kronecker delta $\delta_{j,j'} = 1$ if $j = j'$, and is 0 otherwise. In other words, $\mathbf{B}^t \cdot \mathbf{B}^\vee = (\mathbf{B}^\vee)^t \cdot \mathbf{B}$ is the identity matrix.

In this work we deal with a computational problem on lattices called *bounded-distance decoding* (BDD): given a lattice basis $\mathbf{B} \subset \mathbb{R}^n$ of $\mathcal{L} = \mathcal{L}(\mathbf{B})$ and a target point $\mathbf{t} \in \mathrm{span}(\mathcal{L})$ with the guarantee that $\min_{\mathbf{v} \in \mathcal{L}} \|\mathbf{v} - \mathbf{t}\| \leq r$ for some known $r < \lambda_1(\mathcal{L})/2$, find the unique $\mathbf{v} \in \mathcal{L}$ closest to $\mathbf{t}$ (i.e., such that $\|\mathbf{v} - \mathbf{t}\| \leq r$). In fact, in our context $\mathbf{B}$ and r will be fixed in advance, and $\mathbf{t}$ is the only input that may vary.

A standard approach to solve BDD (and related problems) is the "round-off" algorithm of [Bab85], which simply returns $\mathbf{B} \cdot \lfloor (\mathbf{B}^\vee)^t \cdot \mathbf{t} \rceil$, where the rounding function $\lfloor c \rceil := \lfloor c + \frac{1}{2} \rfloor \in \mathbb{Z}$ is applied to each coordinate independently. (Notice that $(\mathbf{B}^\vee)^t \cdot \mathbf{t}$ is the coefficient vector of $\mathbf{t}$ with respect to basis $\mathbf{B}$.) We recall the following standard fact about this algorithm, and include a brief proof for completeness.

Claim. Let $\mathcal{L} \subset \mathbb{R}^n$ be a lattice with basis $\mathbf{B}$, and let $\mathbf{t} = \mathbf{v} + \mathbf{e} \in \mathbb{R}^n$ for some $\mathbf{v} \in \mathcal{L}$, $\mathbf{e} \in \mathbb{R}^n$. If $\langle \mathbf{b}_j^\vee, \mathbf{e} \rangle \in [-\frac{1}{2}, \frac{1}{2})$ for all j, then on input $\mathbf{t}$ and basis $\mathbf{B}$, the round-off algorithm outputs $\mathbf{v}$.

Proof. Because $\mathbf{v} = \mathbf{B}\mathbf{z}$ for some integer vector $\mathbf{z}$, we have $(\mathbf{B}^\vee)^t \cdot \mathbf{t} = \mathbf{z} + (\mathbf{B}^\vee)^t \cdot \mathbf{e}$, so by hypothesis on the $\langle \mathbf{b}_j, \mathbf{e} \rangle$, we have $\lfloor (\mathbf{B}^\vee)^t \cdot \mathbf{t} \rceil = \mathbf{z}$. The claim follows.

2.2 Circulant Matrices

We recall some standard facts about *circulant* matrices for a finite abelian group $(G, \cdot)$, and their relationship with the *characters* of the group. See e.g., [Lan02] for further details and proofs.

Definition 1 (Circulant Matrix). *For a vector* $\mathbf{a} = (a_g)_{g \in G}$ *indexed by* G, *the* G-*circulant matrix associated with* $\mathbf{a}$ *is the* G-*by-*G *matrix whose* (i, j)th *entry is* $a_{ij^{-1}}$.

Note that the transpose of any G-circulant matrix (associated with $(a_g)_{g \in G}$) is also a G-circulant matrix (associated with $(a_{g^{-1}})_{g \in G}$).

Definition 2 (Character Group). *A character is a group morphism* $\chi \colon G \to \{u \in \mathbb{C} : |u| = 1\}$, *i.e.,* $\chi(g \cdot h) = \chi(g) \cdot \chi(h)$ *for all* $g, h \in G$. *The character group* $(\hat{G}, \cdot)$ *is the set of characters of* G, *with the group operation being the usual multiplication of functions, i.e.,* $(\chi \cdot \psi)(g) = \chi(g) \cdot \psi(g)$.

A basic fact is that $|\hat{G}| = |G|$. Notice that for a character $\chi \in \hat{G}$, we have $\overline{\chi(g)} = \chi(g)^{-1} = \chi(g^{-1})$. We identify χ with the vector $(\chi(g))_{g \in G}$. Then all characters χ have Euclidean norm $\|\chi\| = \sqrt{|G|}$, because

$$\langle \chi, \chi \rangle = \sum_{g \in G} \chi(g) \cdot \overline{\chi(g)} = \sum_{g \in G} 1 = |G|.$$

Moreover, distinct characters χ, ψ are orthogonal:

$$\langle \chi, \psi \rangle = \sum_{g \in G} \chi(g) \cdot \overline{\psi(g)} = \sum_{g \in G} (\chi \cdot \psi^{-1})(g) = 0.$$

Therefore, the complex G-by-$\hat{G}$ matrix

$$\mathbf{P}_G := |G|^{-1/2} \cdot (\chi(g))_{g \in G, \chi \in \hat{G}}$$

is unitary, i.e., $\mathbf{P}_G^{-1} = \mathbf{P}_G^*$, the conjugate transpose of $\mathbf{P}_G$.

Lemma 1. *A complex matrix* $\mathbf{A}$ *is* G-*circulant if and only if the* $\hat{G}$-*by-*$\hat{G}$ *matrix* $\mathbf{P}_G^{-1} \cdot \mathbf{A} \cdot \mathbf{P}_G$ *is diagonal; equivalently, the columns of* $\mathbf{P}_G$ *are the eigenvectors of* $\mathbf{A}$. *If* $\mathbf{A}$ *is the* G-*circulant matrix associated with* $\mathbf{a} = (a_g)_{g \in G}$, *its eigenvalue corresponding to* $\chi \in \hat{G}$ *is* $\lambda_\chi = \langle \mathbf{a}, \chi \rangle = \sum_{g \in G} a_g \cdot \overline{\chi(g)}$.

It follows that every row and column of $\mathbf{A}$ has squared Euclidean norm

$$\|\mathbf{a}\|^2 = \|\mathbf{P}_G^* \cdot \mathbf{a}\|^2 = |G|^{-1} \cdot \sum_{\chi \in \hat{G}} |\lambda_\chi|^2.$$

It also follows that $\mathbf{A}^{-1}$ (when defined) is G-circulant, with eigenvalue λ_χ^{-1} for eigenvector χ.

Proof. Suppose that $\mathbf{A}$ is G-circulant, and let $\chi \in \hat{G}$ be a character of G. Then

$$(\mathbf{A} \cdot \chi)_g = \sum_{h \in G} a_{gh^{-1}} \cdot \chi(h) = \left(\sum_{k \in G} a_k \cdot \overline{\chi(k)} \right) \cdot \chi(g),$$

where in the final equality we have substituted $k = gh^{-1}$ and used $\chi(h) = \overline{\chi(k)} \cdot \chi(g)$. So $\mathbf{A} \cdot \chi = \lambda_\chi \cdot \chi$.

For the other direction, it suffices by linearity to show that $\mathbf{A}_\chi = \mathbf{P}_G \cdot \mathbf{D}_\chi \cdot \mathbf{P}_G^{-1}$ is G-circulant for every $\chi \in \hat{G}$, where $\mathbf{D}_\chi$ is the diagonal $\hat{G}$-by-$\hat{G}$ matrix with 1 in its (χ, χ)th entry and zeros elsewhere. Indeed, by definition of $\mathbf{P}_G$ and because $\mathbf{P}_G^{-1} = \mathbf{P}_G^*$, the (i, j)th entry of $\mathbf{A}_\chi$ is simply $|G|^{-1} \cdot \chi(i) \cdot \overline{\chi(j)} = |G|^{-1} \cdot \chi(ij^{-1})$, which depends only on ij^{-1} as required.

2.3   Dirichlet Characters and L-Series

A *Dirichlet character* χ is a character of $\mathbb{Z}_k^*$ for some positive integer k. Note that if $k|\ell$ then χ induces a character of $\mathbb{Z}_\ell^*$ via the natural morphism $\mathbb{Z}_\ell^* \to \mathbb{Z}_k^*$, so we can equivalently view χ as being defined modulo either k or ℓ. The *conductor* f_χ of χ is the smallest positive f such that χ is induced by a Dirichlet character modulo f. The character is said to be *even* if $\chi(-1) = 1$; note that the even Dirichlet characters correspond with the characters of $\mathbb{Z}_k^*/\{\pm 1\}$. The character is said to be *quadratic* if all its values are real (i.e., ± 1), and it is not the constant 1 character (which is known as the principal character). Following the convention used in [Was97], we often implicitly extend χ to a completely multiplicative function from $\mathbb{Z}$ to $\mathbb{C}$, by considering it as modulo its conductor k (i.e., as a primitive character) and letting $\chi(a) = 0$ if $\gcd(a, k) > 1$.

Definition 3 (Dirichlet L-Series). *For a Dirichlet character χ, the Dirichlet L-function $L(\cdot, \chi)$ is defined as the formal series*

$$L(s, \chi) = \sum_{k \geq 1} \frac{\chi(k)}{k^s}.$$

For any Dirichlet character χ, the series $L(s, \chi)$ is absolutely convergent for all $s \in \mathbb{C}$ with $\Re(s) > 1$. It is also known that $L(1, \chi)$ converges and is nonzero for any non-principal Dirichlet character (i.e., $\chi \neq 1$). We have the following asymptotic bounds on its value; we will only use the lower bounds.

Theorem 1. *There exists a $C > 0$ such that, for any non-quadratic character χ of conductor $f > 1$,*

$$\frac{1}{\ell(f)} \leq |L(1, \chi)| \leq \ell(f) \quad where \, \ell(f) = C \ln f. \tag{1}$$

Moreover, for any quadratic character χ,

$$|L(1, \chi)| \geq \frac{1}{C\sqrt{f}}. \tag{2}$$

Equation (1) can be traced back to Landau [Lan27], and improving the constant C is an active field of research [Lou15]. Equation (2) is also classical and follows from Dirichlet's class number formula (see, e.g., [MV06, Sect. 4.4]). We note that under the Generalized Riemann Hypothesis, the bound in Eq. (1) can be improved to $\ell(f) = C \ln \ln f$, and holds for both quadratic and non-quadratic characters (see, e.g., [LLS15]).

2.4   Cyclotomic Number Fields and the Log-Unit Lattice

Cyclotomic Number Fields. Let L be a field. An element $\zeta \in L$ is a root of unity if $\zeta^m = 1$ for some positive integer m. The order of a root of unity $\zeta \in L$ is the order of the finite multiplicative subgroup of L^* generated by ζ. A primitive mth root of unity in L is a root of unity $\zeta \in L$ of order m. Note that if $\zeta \in L$ is a primitive mth root of unity, then the polynomial $X^m - 1 \in L[X]$ factors as $\prod_{i=0}^{m-1}(X - \zeta^i)$ over $L[X]$. Also note that the complete set of primitive mth roots in L consists of the powers ζ^j for $j \in \mathbb{Z}_m^*$.

An algebraic number field K is an extension field of the rationals $\mathbb{Q}$ such that its dimension $[K : \mathbb{Q}]$ as a $\mathbb{Q}$-vector space (i.e., its degree) is finite. If $\Omega \supset K$ is an extension field such that Ω is algebraically closed over $\mathbb{Q}$, then there are exactly $[K : \mathbb{Q}]$ field embeddings of K into Ω.[4] An algebraic number field is Galois if the order of its automorphism group equals its degree.[5] A number field K is cyclotomic if $K = \mathbb{Q}(\zeta)$ for some root of unity $\zeta \in K$. Its degree is $\varphi(m)$, where $\varphi(\cdot)$ is the Euler totient function and m is the order of ζ, and its ring of integers R is monogenic, i.e., $R = \mathbb{Z}[\zeta]$. We let U denote the cyclic (multiplicative) subgroup of mth roots of unity, which is generated by ζ.

A cyclotomic number field is Galois. If $K = \mathbb{Q}(\zeta)$ is a cyclotomic number field with $\zeta \in K$ an mth primitive root of unity then each automorphism is characterized by the assignment $\zeta \mapsto \zeta^j$ for some $j \in \mathbb{Z}_m^*$. As a consequence, if L is an extension field of a cyclotomic field K, then K is situated uniquely in L. For concreteness, we situate cyclotomic number fields in the complex numbers $\mathbb{C}$. Let m be a positive integer and define $\omega = \omega_m = \exp(2\pi\imath/m) \in \mathbb{C}$. Then ω is a primitive mth root of unity and $K = \mathbb{Q}(\omega)$ is the mth cyclotomic number field. The embeddings of K into the complex numbers (i.e., the automorphisms of K) are denoted σ_j for $j \in \mathbb{Z}_m^*$, where σ_j sends ω to ω^j. The concatenation $\sigma(a) = (\sigma_j(a))_{j \in \mathbb{Z}_m^*}$ of these embeddings is known as the *canonical embedding*, and is used to endow K with a geometry, e.g., $\|a\| := \|\sigma(a)\|$ for any $a \in K$.

Logarithmic Embedding. The embeddings σ_i of K, being complex, come in conjugate pairs, i.e., $\sigma_j(x) = \overline{\sigma_{-j}(x)}$. We will mainly be concerned with their *magnitudes*, so we identify the pairs by indexing over the multiplicative quotient

---

[4] These embeddings are merely ring morphisms $\psi : K \to \Omega$. Each such ψ is automatically injective because K is a field. Also note that any such ψ fixes $\mathbb{Q}$ pointwise.

[5] An automorphism of a field L is a ring isomorphism $\psi \colon L \to L$. The automorphisms of L form a group with functional composition as the group operation.

group $G := \mathbb{Z}_m^*/\{\pm 1\}$. We then have the *logarithmic embedding*, defined as

$$\text{Log}\colon K \to \mathbb{R}^{\varphi(m)/2}$$
$$a \mapsto (\log |\sigma_i(a)|)_{i \in G}.$$

The logarithmic embedding defines a group morphism, mapping the multiplicative group K^* to an additive subgroup of $\mathbb{R}^{\varphi(m)/2}$. The kernel of Log restricted to R^* is $\{\pm 1\} \cdot U$. The Dirichlet Unit Theorem (see [Sam70, Chap. 4.4, Theorem 1]) implies that $\Lambda = \text{Log}(R^*)$, the image of the multiplicative unit group of R under the logarithmic embedding, is a full-rank lattice in the linear subspace of $\mathbb{R}^{\varphi(m)/2}$ orthogonal to the all-1s vector $\mathbf{1}$. We refer to Λ as the *log-unit lattice*.

Cyclotomic Units. Let A be the multiplicative subgroup of K^* generated by $\pm\zeta$ and

$$z_j := \zeta^j - 1, \quad j \in \mathbb{Z}_m \setminus \{0\}.$$

Notice that $z_j = -\zeta^j \cdot z_{-j}$, so z_j and z_{-j} are equivalent modulo $\pm U$; in particular, $\text{Log}(z_j) = \text{Log}(z_{-j})$. The group of *cyclotomic units*, denoted C, is defined by

$$C = A \cap R^*.$$

The z_j given above are not necessarily units in R, and thus do not generate C. However, a closely related generating set, which we call the *canonical generators*, is given by the following lemma. Recall that $G = \mathbb{Z}_m^*/\{\pm 1\}$, and identify it with some canonical set of representatives in $\mathbb{Z}_m^*$.

Lemma 2 (Lemma 8.1 of [Was97]). *Let m be a prime power, and define $b_j := z_j/z_1 = (\zeta^j - 1)/(\zeta - 1)$. The group C of cyclotomic units is generated by $\pm\zeta$ and b_j for $j \in G \setminus \{1\}$.*

Notice that $\text{Log}\,C$ is a sublattice of Λ. As shown below, the index of Λ over $\text{Log}\,C$ is finite. In fact, it is $h^+(m)$, the *class number* of the real subfield $K^+ = \mathbb{Q}(\zeta + \bar\zeta)$, defined as the index of the subgroup of principal fractional ideals in the multiplicative group of all fractional ideals (in K^+). The proof of this theorem is left as Exercise 8.5 in [Was97]. For completeness, we sketch the solution in Appendix A.

Theorem 2. *For a prime power $m > 2$, the index of the log-unit lattice Λ over $\text{Log}\,C$ is*

$$[\Lambda : \text{Log}\,C] = h^+(m).$$

Some Facts and Conjectures Concerning h^+. For our purposes, we need $h^+(m)$ not to be very big. For all power-of-two m up to $m = 256$, and also for $m = 512$ under GRH, it is known that $h^+(m) = 1$ (see [Mil14]). Whether $h^+(m) = 1$ for all power-of-two m is known as Weber's class number problem, and is presented in the literature as a reasonable conjecture.

In the case of odd primes, it also appears that h^+ is quite small. Computations of Schoof [Sch03] and Miller [Mil15] show that $h^+(p) \leq 11$ for all primes $p \leq 241$.

For powers of odd primes it has been conjectured (with support of the Cohen-Lenstra heuristic) that, for all but finitely many pairs (p, ℓ) where p is a prime, $h^+(p^{\ell+1}) = h^+(p^\ell)$ [BPR04]. A direct consequence is that $h^+(p^\ell)$ is bounded for a fixed p and increasing ℓ.

3   Geometry of the Canonical Generators

Throughout this section, let the cyclotomic index m be a prime power. Our goal here is to show that the canonical generators of the cyclotomic units, under the logarithmic embedding, are geometrically well-suited for bounded-distance decoding.

Recalling that $G = \mathbb{Z}_m^*/\{\pm 1\}$ is identified with some set of canonical representatives in $\mathbb{Z}_m^*$ and that $\mathrm{Log}(b_j) = \mathrm{Log}(b_{-j})$, define

$$\mathbf{b}_j = \mathrm{Log}(b_j), \quad j \in G \setminus \{1\},$$

to be the log-embeddings of the canonical generators $b_j = (\zeta^j - 1)/(\zeta - 1)$ defined in Lemma 2. By Lemma 2, these $\mathbf{b}_j$ form a basis of the sublattice $\mathrm{Log}\, C$, which by Theorem 2 has index $h^+(m)$ in Λ.

In order to apply the round-off algorithm and Claim 2.1 with this basis, we bound the norms $\|\mathbf{b}_j^\vee\|$ of the dual basis vectors. The remainder of this section is dedicated to proving the following theorem.

Theorem 3. *Let $m = p^k$ for a prime p, and let $\{\mathbf{b}_j^\vee\}_{j \in G \setminus \{1\}}$ denote the basis dual to $\{\mathbf{b}_j\}_{j \in G \setminus \{1\}}$. Then all $\|\mathbf{b}_j^\vee\|$ are equal, and*

$$\|\mathbf{b}_j^\vee\|^2 \leq 2k|G|^{-1} \cdot (\ell(m)^2 + O(1)) = O(m^{-1} \cdot \log^3 m).$$

To prove the theorem we start by relating the basis vectors $\mathbf{b}_j$ to a certain G-circulant matrix. Recalling that $z_j = \zeta^j - 1$ is the numerator of b_j, define

$$\mathbf{z}_j := \mathrm{Log}(z_j) = \mathbf{b}_j + \mathbf{z}_1. \tag{3}$$

Collect these vectors into a square G-by-G matrix $\mathbf{Z}$ whose jth column is $\mathbf{z}_{j-1}$, and notice that its (i,j)th entry $\log|\omega^{i \cdot j^{-1}} - 1|$ is determined by $ij^{-1} \in G$ alone, so $\mathbf{Z}$ is the G-circulant matrix associated with $\mathbf{z}_1$. For each eigenvector $\chi \in \hat{G}$ of $\mathbf{Z}$, let $\lambda_\chi := \langle \mathbf{z}_1, \chi \rangle$ denote the corresponding eigenvalue.

Lemma 3. *For all $j \in G \setminus \{1\}$ we have*

$$\|\mathbf{b}_j^\vee\|^2 = |G|^{-1} \cdot \sum_{\chi \in \hat{G} \setminus \{1\}} |\lambda_\chi|^{-2}. \tag{4}$$

Proof. Let $\mathbf{z}_j^\vee$ denote the vectors dual to the $\mathbf{z}_j$, i.e., the columns of $\mathbf{Z}^{-t}$. (As shown below in the proof of Theorem 3, $\mathbf{Z}^{-1}$ is indeed well defined because all eigenvalues λ_χ of $\mathbf{Z}$ are nonzero.)

We first claim that $\mathbf{b}_j^\vee$ is simply the projection of $\mathbf{z}_j^\vee$ orthogonal to $\mathbf{1}$, i.e., $\mathbf{b}_j^\vee = \mathbf{z}_j^\vee - |G|^{-1} \cdot \langle \mathbf{z}_j^\vee, \mathbf{1} \rangle \cdot \mathbf{1}$. Indeed, these vectors are all in $\mathrm{span}(\mathbf{b}_{j'})_{j'}$, the space orthogonal to $\mathbf{1}$, and moreover, for all $j, j' \in G \setminus \{1\}$ they satisfy

$$\langle \mathbf{z}_j^\vee - |G|^{-1} \cdot \langle \mathbf{z}_j^\vee, \mathbf{1} \rangle \cdot \mathbf{1}, \mathbf{b}_{j'} \rangle = \langle \mathbf{z}_j^\vee, \mathbf{b}_{j'} \rangle = \langle \mathbf{z}_j^\vee, \mathbf{z}_{j'} - \mathbf{z}_1 \rangle = \delta_{j,j'} - 0.$$

Now,

$$\|\mathbf{b}_j^\vee\|^2 = \|\mathbf{z}_j^\vee\|^2 - |G|^{-1} \cdot \langle \mathbf{z}_j^\vee, \mathbf{1} \rangle^2.$$

Recall by Lemma 1 that $\mathbf{Z}^{-t}$ is the G-circulant matrix associated with $\mathbf{z}_1^\vee$, which has eigenvalue $\lambda_\chi^{-1} = \langle \mathbf{z}_1^\vee, \chi \rangle$ for eigenvector $\chi \in \hat{G}$. By the remarks following Lemma 1, $\|\mathbf{z}_j^\vee\|^2 = |G|^{-1} \cdot \sum_{\chi \in \hat{G}} |\lambda_\chi|^{-2}$. The lemma follows by noting that $\langle \mathbf{z}_j^\vee, \mathbf{1} \rangle = \langle \mathbf{z}_1^\vee, \mathbf{1} \rangle = \lambda_1^{-1}$.

We now provide an upper bound on the right-hand side of Eq. (4). Our proof is similar to the proof that the cyclotomic units have finite index in the full group of units [Was97, Theorem 8.2].

Theorem 4 [Was97, Lemma 4.8 and Theorem 4.9]. *Let χ be an even Dirichlet character of conductor $f > 1$, and let $\omega_f = \exp(2\pi \imath / f) \in \mathbb{C}$. Then*

$$\left| \sum_{a \in \mathbb{Z}_f^*} \overline{\chi(a)} \cdot \log|1 - \omega_f^a| \right| = \sqrt{f} \cdot |L(1, \chi)|.$$

For completeness, we briefly explain how the finite sum on the left hand side gives rise to an L-series, and refer to [Was97] for the details. Using the Taylor expansion

$$\log|1 - x| = -\sum_{k \geq 1} x^k / k,$$

one gets a sum over finitely many a and infinitely many k of terms $\overline{\chi(a)} \cdot \omega_f^{ak}/k$. For a fixed k, the sum over a can easily be rewritten as $\tau(\chi) \cdot \chi(k)/k$, where $\tau(\chi)$ is a Gauss sum (see [Was97, Lemma 4.7]), which makes the Dirichlet L-function apparent.

Corollary 1. *Suppose $f > 1$ divides a prime power m. For any even Dirichlet character χ of conductor f,*

$$\left| \sum_{a \in \mathbb{Z}_m^*} \overline{\chi(a)} \cdot \log|1 - \omega_m^a| \right| = \sqrt{f} \cdot |L(1, \chi)|.$$

Proof. Let $\phi\colon \mathbb{Z}_m^* \to \mathbb{Z}_f^*$ be the map given by reduction modulo f. We have

$$\sum_{a\in\mathbb{Z}_m^*} \overline{\chi(a)} \cdot \log|1-\omega_m^a| = \sum_{a\in\mathbb{Z}_f^*} \overline{\chi(a)} \sum_{\substack{b\in\mathbb{Z}_m^* \\ \phi(b)=a}} \log|1-\omega_m^b|$$

$$= \sum_{a\in\mathbb{Z}_f^*} \overline{\chi(a)} \cdot \log\left|\prod_{\substack{b\in\mathbb{Z}_m^* \\ \phi(b)=a}} (1-\omega_m^b)\right|$$

$$= \sum_{a\in\mathbb{Z}_f^*} \overline{\chi(a)} \cdot \log\left|1-\omega_f^a\right|,$$

where in the last equality we have used the identity $\prod_{i\in\mathbb{Z}_n}(1-\omega_n^i Y) = 1 - Y^n$ and $\omega_m^n = \omega_f$ with $n = m/f$. The claim follows by applying Theorem 4.

We are now ready to complete the proof of the main theorem.

Proof (Proof of Theorem 3). Recall that the characters $\chi \in \hat{G}$ correspond to the even characters of $\mathbb{Z}_m^*$, because $\chi(\pm 1) = 1$. Also recall that by Lemma 1, the eigenvalues are

$$\lambda_\chi = \langle \mathbf{z}_1, \chi \rangle = \sum_{a\in G} \overline{\chi(a)} \cdot \log|1-\omega_m^a| = \frac{1}{2} \sum_{a\in\mathbb{Z}_m^*} \overline{\chi(\pm a)} \cdot \log|1-\omega_m^a|,$$

where the second equality holds because $|1-\omega_m^{-a}| = |1-\omega_m^a|$. Therefore, using Corollary 1 we have

$$|\lambda_\chi| = \frac{1}{2}\sqrt{f_\chi} \cdot |L(1,\chi)|, \tag{5}$$

and so by Lemma 3,

$$\|\mathbf{b}_j^\vee\|^2 = |G|^{-1} \cdot \sum_{\chi\in\hat{G}\setminus\{1\}} |\lambda_\chi|^{-2} = 4|G|^{-1} \cdot \sum_{\chi\in\hat{G}\setminus\{1\}} f_\chi^{-1} \cdot |L(1,\chi)|^{-2}.$$

We first consider the contribution to the sum coming from quadratic characters. When p is an odd prime, there is exactly one quadratic character (see [MV06, Sect. 9.3]), and it is of conductor p, hence by Eq. (2) in Theorem 1, the contribution to the sum is $O(1)$ (assuming it is even; otherwise it does not participate in the sum). In the case $p = 2$ the contribution is also $O(1)$ since there are at most three quadratic characters (see again [MV06, Sect. 9.3]) and their conductor is bounded from above by an absolute constant. Finally, the contribution coming from non-quadratic characters is at most

$$\ell(m)^2 \sum_{\chi\in\hat{G}\setminus\{1\}} f_\chi^{-1} \le \frac{k}{2} \cdot \ell(m)^2,$$

where we used Eq. (1) in Theorem 1 and Claim 3 below.

Claim. Let $m = p^k$ for a prime p. Then, for $G = \mathbb{Z}_m^*/\{\pm 1\}$,

$$\sum_{\chi \in \hat{G} \setminus \{1\}} f_\chi^{-1} \le \frac{k}{2}.$$

Proof. Notice that there are at most f Dirichlet characters of conductor f, at most half of which are even (when $f > 1$), so

$$\sum_{\chi \in \hat{G} \setminus \{1\}} f_\chi^{-1} \le \sum_{\ell=1}^{k} \frac{p^\ell}{2} \cdot \frac{1}{p^\ell} = \frac{k}{2}.$$

4 Algorithmic Implications

The following is our main result about the decoding algorithm, showing that under mild restrictions on the distribution of the short generator, one can recover it from any generator that differs from it by a unit in C. Roughly speaking, the requirement from the distribution is that the *ratios* between its complex embeddings are not too large. We note that since the $\mathbf{v}_i$ below are assumed to be orthogonal to the all-1 vector, the *scale* of the distribution (or variance in the case of Gaussians) is irrelevant: this should not come as a surprise, since, e.g., one can normalize the input generator g' to have algebraic norm 1.

Theorem 5. *Let D be a distribution over $\mathbb{Q}(\zeta)$ with the property that for any tuple of vectors $\mathbf{v}_1, \ldots, \mathbf{v}_{\varphi(m)/2-1} \in \mathbb{R}^{\varphi(m)/2}$ of Euclidean norm 1 that are orthogonal to the all-1 vector $\mathbf{1}$, the probability that $|\langle \mathrm{Log}(g), \mathbf{v}_i \rangle| < c\sqrt{m} \cdot (\log m)^{-3/2}$ holds for all i is at least some $\alpha > 0$, where g is chosen from D and c is a universal constant. Then there is an efficient algorithm that given $g' = g \cdot u$, where g is chosen from D and $u \in C$ is a cyclotomic unit, outputs an element of the form $\pm \zeta^j g$ with probability at least α.*

Proof. The algorithm applies the round-off algorithm from Claim 2.1 to $\mathrm{Log}(g') = \mathrm{Log}(g) + \mathrm{Log}(u)$, using the vectors $\mathbf{b}_j$ (defined and analyzed in Sect. 3) as the basis. By the assumption on D and Theorem 3, with probability at least α the output is $\mathrm{Log}(u) \in \mathrm{Log}(C)$. We next find integer coefficients a_j such that $\mathrm{Log}(u) = \sum a_j \mathbf{b}_j$, and compute $u' = \prod b_j^{a_j}$. Since $\mathrm{Log}(u') = \mathrm{Log}(u)$ it follows that u' must be of the form $\pm \zeta^j u$ for some sign and some j. Therefore, g'/u' is the desired element.

In the next section we show that the condition on D in the theorem is satisfied by several natural distributions.

One possible concern with the above algorithm is that it expects as input $g \cdot u$ for a *cyclotomic unit* $u \in C$, whereas the first phase of the attack described in the introduction, i.e., a PIP algorithm, is only guaranteed to output $g \cdot u$ for an *arbitrary unit* $u \in R^*$. There are several reasons why this should not be an issue. First, as mentioned in Sect. 2, in some cases, e.g., for power-of-2 cyclotomic, it is conjectured that $C = R^*$. More generally, the index of C

in R^*, which we recall is h^+, the class number of the totally real subfield, is often small. In such a case, if we have a list of coset representatives of C in R^*, we can enumerate over all of them and use the algorithm above to recover g, increasing the running time only by a factor of h^+. In order to obtain such a list of representatives, we can use an algorithm for computing the unit group, either classical [BF14] or quantum [EHKS14]. These algorithms are no slower than the known PIP algorithms and moreover, need only be applied once for a given cyclotomic field (as opposed to once for each public key). Alternatively, by running the PIP algorithm multiple times on a basis of a principal ideal with a *known* short generator chosen using the secret key generation algorithm, we can recover a list of representatives for all the cosets that show up as output of the PIP algorithm with non-negligible probability; we can then enumerate over that list.

In the above statement and proof we glossed over issues of precision and assumed for simplicity, as one often does, that the input g' is given exactly. To be fully rigorous, one needs to verify that the algorithm can deal with inputs that are specified with finite precision, and still runs in time polynomial in its input size. Typically, by finite precision one means that the input is given in fixed-point representation, providing additive approximation to the true numbers. Here, however, it is more natural to assume that the input is given in (the strictly more general) floating-point representation, providing multiplicative approximation to the true numbers. Not only is this more natural, but also the known PIP algorithms [BF14,Bia14,BS15] generate an output in this format, or an output that can be easily converted to this format.[6] Luckily, dealing with floating-point inputs is straightforward. First notice that $\mathrm{Log}(g')$ can be written in standard fixed-point representation, and so can $\mathrm{Log}(u)$. The integer coefficients a_j can be stored exactly since they are at most exponential in the input size. Finally, by using a sufficiently good multiplicative approximation of b_j (with the multiplicative error being much less than $1/a_j$), we can obtain an arbitrarily good multiplicative approximation of u'. As a result we get a multiplicative approximation of the desired output g'/u' that can be made essentially as good as the multiplicative approximation of the input g'.

5   Tail Bounds

In this section we show that the condition on D in Theorem 5 is satisfied by two natural distributions: the continuous Gaussian and a wide enough discrete Gaussian (over any lattice). This section is independent of the other sections in this paper, and we avoid the use of notation from algebraic number theory. Instead, we identify elements of K with vectors in $\mathbb{R}^{\varphi(m)}$ by taking the real and the imaginary part of their $\varphi(m)/2$ complex embeddings, i.e., a is mapped to

---

[6] In general number fields (in fact already in quadratic number fields), the use of floating point is *necessary*, since generators are typically doubly exponentially large and so would require exponential time to write down in fixed-point notation.

$(\Re(\sigma_j(a)), \Im(\sigma_j(a)))_{j \in G}$. As a result, all random variables appearing here are real. The results in this section should be easy to extend to other distributions.

We start with Lemma 4, a tail bound on the sum of subexponential random variables. The proof is based on a standard Bernstein argument, and follows the proof in [Ver12] apart from some minor modifications for convenience.

Definition 4. *For $\alpha, \beta > 0$, a random variable X is (α, β)-subexponential if*

$$\mathbb{E}[\cosh(\alpha X)] \leq \beta,$$

where recall that $\cosh(x) := (e^x + e^{-x})/2$.

Lemma 4 (Tail bound). *Let $X_1, \ldots, X_n$ be independent centered (i.e., expectation zero) (α, β)-subexponential random variables. Then, for any $\mathbf{a} = (a_1, \ldots, a_n) \in \mathbb{R}^n$ and every $t \geq 0$,*

$$\Pr\left[\left|\sum a_i X_i\right| \geq t\right] \leq 2\exp\left(-\min\left(\frac{\alpha^2 t^2}{8\beta\|\mathbf{a}\|_2^2}, \frac{\alpha t}{2\|\mathbf{a}\|_\infty}\right)\right).$$

Proof. By scaling, we can assume without loss of generality that $\alpha = 1$. Next, we use the inequality

$$e^{\delta x} - \delta x - 1 \leq (e^{\delta x} - \delta x - 1) + (e^{-\delta x} + \delta x - 1) = 2(\cosh(\delta x) - 1) \leq 2\delta^2(\cosh(x) - 1)$$

which holds for all $-1 \leq \delta \leq 1$ and all $x \in \mathbb{R}$, where the second inequality follows from the Taylor expansion. By applying this inequality to a $(1, \beta)$-subexponential centered random variable X, and taking expectations we see that for all $-1 \leq \delta \leq 1$,

$$\mathbb{E}[\exp(\delta X)] \leq 1 + 2\delta^2 \mathbb{E}[\cosh(X) - 1]$$
$$\leq 1 + 2\delta^2(\beta - 1) \leq \exp(2\delta^2\beta). \tag{6}$$

Using Markov's inequality, we can bound the upper tail probability for any $\lambda > 0$ as

$$\Pr\left[\sum a_i X_i \geq t\right] = \Pr\left[\exp\left(\lambda\sum a_i X_i\right) \geq \exp(\lambda t)\right]$$
$$\leq \exp(-\lambda t) \cdot \mathbb{E}\left[\exp\left(\lambda\sum a_i X_i\right)\right]$$
$$= \exp(-\lambda t) \cdot \prod \mathbb{E}\left[\exp\left(\lambda a_i X_i\right)\right]$$
$$\leq \exp(-\lambda t + 2\beta\lambda^2\|\mathbf{a}\|_2^2),$$

where in the second inequality we used (6) and assumed that $\lambda\|\mathbf{a}\|_\infty \leq 1$. Taking $\lambda = \min(t/(4\beta\|\mathbf{a}\|_2^2), 1/\|\mathbf{a}\|_\infty)$ this bound becomes at most

$$\exp\left(-\min\left(\frac{t^2}{8\beta\|\mathbf{a}\|_2^2}, \frac{t}{2\|\mathbf{a}\|_\infty}\right)\right).$$

We complete the proof by applying the same argument with $-\mathbf{a}$.

The next claim follows immediately from Definition 4.

Claim. If Y is a non-negative random variable such that both $\mathbb{E}[Y]$ and $\mathbb{E}[Y^{-1}]$ are finite, then $\log Y$ is a $(1, \beta)$-subexponential random variable for some $\beta > 0$.

The following is an immediate corollary of the tail bound. It shows that the condition in Theorem 5 holds with overwhelming probability for a continuous Gaussian distribution of any radius that is spherical in the embedding basis. Notice that the parameter r plays no role in the conclusion of the statement.

Lemma 5. *Let* $X_1, \ldots, X_n, X'_1, \ldots, X'_n$ *be i.i.d.* $N(0, r)$ *variables for some* $r >$ 0*, and let* $\hat{X}_i = (X_i^2 + X_i'^2)^{1/2}$*. Then, for any vectors* $\mathbf{a}^{(1)}, \ldots, \mathbf{a}^{(\ell)} \in \mathbb{R}^n$ *of Euclidean norm 1 that are orthogonal to the all-1 vector, and every* $t \geq C$ *for some universal constant* C*,*

$$\Pr\left[\exists j, \left| \sum_i a_i^{(j)} \log(\hat{X}_i) \right| \geq t\right] \leq 2\ell \exp(-t/2).$$

Proof. By union bound, it suffices to prove the lemma for the case $\ell = 1$, and we let $\mathbf{a} = \mathbf{a}^{(1)}$. Since $\sum a_i = 0$, we can assume without loss of generality that $r = 1$. Notice that $\hat{X}_i$ has a chi distribution with 2 degrees of freedom (also known as a Rayleigh distribution) whose density function is given by $xe^{-x^2/2}$ for $x > 0$ and zero otherwise. In particular, it is easy to see that both $\mathbb{E}[\hat{X}_i]$ and $\mathbb{E}[\hat{X}_i^{-1}]$ are finite (both are $\sqrt{\pi/2}$). Therefore, by Claim 5, $\log \hat{X}_i$ is $(1, \beta)$ subexponential for some constant $\beta > 0$. From this it follows that $\hat{X}_i = \log \hat{X}_i - \mathbb{E}[\log \hat{X}_i]$ are *centered* $(1, \beta')$ subexponential random variables for some constant $\beta' > 0$. The result now follows by applying Lemma 4 to $\hat{X}_1, \ldots, \hat{X}_n$, using the bound $\|\mathbf{a}\|_\infty \leq 1$, and the observation that $\sum_i a_i \mathbb{E}[\log \hat{X}_i] = 0$.

In the next lemma we show that small perturbations of the continuous Gaussian distribution still satisfy the condition in Theorem 5.

Lemma 6. *Let* $X = (X_1, \ldots, X_n, X'_1, \ldots, X'_n)$ *be i.i.d.* $N(0, r)$ *variables for some* $r > 0$*, and let* $Y = (Y_1, \ldots, Y_n, Y'_1, \ldots, Y'_n)$ *be a (not necessarily independent) random vector satisfying* $\|Y\|_2 \leq u$ *with probability 1 for some* $u \leq$ $r/(20\sqrt{n})$*. Let* $Z = X + Y$ *and define* $\hat{X}_i, \hat{Y}_i, \hat{Z}_i$ *as before. Then for any vectors* $\mathbf{a}^{(1)}, \ldots, \mathbf{a}^{(\ell)} \in \mathbb{R}^n$ *of Euclidean norm 1 that are orthogonal to the all-1 vector, it holds with constant probability that for all* j*,*

$$\left| \sum_i a_i^{(j)} \log(\hat{Z}_i) \right| \leq 1 + 10 \log \ell.$$

Proof. By Lemma 5 we have that with some constant probability close to 1,

$$\forall j, \left| \sum_i a_i^{(j)} \log(\hat{X}_i) \right| < 10 \log \ell. \tag{7}$$

Moreover, since $\hat{X}_i < r/(10\sqrt{n})$ implies that both $|X_i|$ and $|X_i'|$ are smaller than $r/(10\sqrt{n})$, we see that by independence of X_i, X_i', the probability of the former event is at most c/n for some small constant c. As a result we have that with constant probability close to 1,

$$\forall i, \hat{X}_i > r/(10\sqrt{n}).$$

In the following we assume that these two conditions hold (which happens with constant probability close to 1 by union bound), and bound the effect of Y. Now let $\mathbf{a}$ be one of the vectors in the statement of the lemma. Then,

$$\left| \sum_i a_i \log(\hat{Z}_i) \right| \le \left| \sum_i a_i \log(\hat{X}_i) \right| + \left| \sum_i a_i \log(\hat{Z}_i/\hat{X}_i) \right|$$
$$\le 10 \log \ell + \left| \sum_i a_i \log(\hat{Z}_i/\hat{X}_i) \right|,$$

where we used Eq. (7). Notice that by the triangle inequality (for two-dimensional Euclidean space),

$$\hat{X}_i - \hat{Y}_i \le \hat{Z}_i \le \hat{X}_i + \hat{Y}_i.$$

Since $\hat{Y}_i \le \|Y\|_2 \le u \le r/(20\sqrt{n}) \le \hat{X}_i/2$, and using the inequality $|\log(1+\delta)| \le 2|\delta|$ valid for all $\delta \in [-1/2, 1/2]$,

$$\left| \sum_i a_i \log(\hat{Z}_i/\hat{X}_i) \right| \le \left(\sum_i (\log(\hat{Z}_i/\hat{X}_i))^2 \right)^{1/2}$$
$$\le \left(\sum_i (2\hat{Y}_i/\hat{X}_i)^2 \right)^{1/2}$$
$$\le 20\sqrt{n}/r \cdot \left(\sum_i \hat{Y}_i^2 \right)^{1/2}$$
$$\le 20\sqrt{n}u/r \le 1,$$

where the first inequality follows from Cauchy-Schwarz.

Finally, we consider the spherical (in the embedding basis) discrete Gaussian distribution over an arbitrary lattice $L \subseteq \mathbb{R}^{2n}$. Such distributions show up often in cryptographic constructions (see, e.g., [LPR13]), and often that lattice is the (embedding of the) ring of integers R. For background on the discrete Gaussian distribution and the smoothing parameter, see, e.g., [MR04]. In order to apply Lemma 6 to this distribution, take X to be the continuous Gaussian D_r for some $r \ge 100n\eta_\varepsilon(L)$, and Y the discrete Gaussian $D_{L-X,s}$ over the coset $L - X$ of parameter $s = \eta_\varepsilon(L)$ for some negligible parameter ε. Using Banaszczyk's result [Ban93] we have that with all but exponentially small probability in n,

$\|Y\|_2 \le \sqrt{2n}\eta_\varepsilon(L) \le r/(60\sqrt{n})$. Moreover, by the lemma below, the distribution of $Z = X + Y$ is within negligible statistical distance of the discrete Gaussian distribution $D_{L,r'}$ for $r' = (r^2 + \eta_\varepsilon(L)^2)^{1/2}$. We therefore see that the condition in Theorem 5 holds for the discrete Gaussian distribution $D_{L,r'}$ for any lattice L and any $r' > 200n\eta_\varepsilon(L)$.

Lemma 7 (Special Case of [Pei10, Theorem 3.1]**).** *Let L be a lattice and $r, s > 0$ be such that $s \ge \eta_\varepsilon(L)$ for some $\varepsilon \le 1/2$. Then if we choose $\mathbf{x}$ from the continuous Gaussian D_r and then choose $\mathbf{y}$ from the discrete Gaussian $D_{L-\mathbf{x},s}$ then $\mathbf{x}+\mathbf{y}$ is within statistical distance 8ε of the discrete Gaussian $D_{L,(r^2+s^2)^{1/2}}$.*

6   Shortest Generators of Principal Ideals and an SVP Algorithm

In a principal ideal $\mathcal{I}$, how long (in the Euclidean norm) can the shortest generator be, relative to its algebraic norm? In this section we provide lower and upper bounds showing that for a cyclotomic ring R of prime-power index m, the answer is $\exp(\tilde{\Theta}(\sqrt{m})) \cdot S(\mathcal{I})$, where $S(\mathcal{I}) = \mathrm{N}(\mathcal{I})^{1/\varphi(m)}$ is the dimension-normalized algebraic norm of $\mathcal{I}$, and $\tilde{\Theta}$ hides polylogarithmic factors. (To be precise, the lower bound is under the mild conjecture that $h^+(m) = 2^{O(m)}$; see the end of Sect. 2.4.) By contrast, it is well known (see, e.g., [PR07, Lemmas 6.1 and 6.2]) that the minimum distance (i.e., the length of a shortest nonzero vector) of any ideal is bounded by $\Omega(\sqrt{m}) \cdot S(\mathcal{I})$ and $O(m) \cdot S(\mathcal{I})$, by the arithmetic-mean/geometric-mean inequality and Minkowski's theorem, respectively. Therefore, any algorithm that always outputs a generator when given a principal ideal (e.g., the algorithm analyzed in the previous sections) obtains no better than a $\exp(\tilde{\Omega}(\sqrt{m}))$ approximation factor for the Shortest Vector Problem, in the worst case.

We first show in Sect. 6.1 that upper and lower bounds on shortest generators follow directly from an analysis of the *covering radius* of the log-unit lattice Λ (and its sublattice $\mathrm{Log}\, C$), in the ℓ_∞ and ℓ_1 norms (respectively). Sections 6.2 and 6.3 then prove upper and lower bounds on these covering radii. In fact, the proofs demonstrate more: the lower bound holds for "almost all" principal ideals, and the upper bound is algorithmic in the following sense: given an arbitrary generator (which can be found using the quantum PIP algorithm of [BS15, BS16]), we can efficiently find a generator satisfying the bound, which in particular is a $\exp(\tilde{O}(\sqrt{m}))$-approximate shortest vector in the ideal.

Throughout this section we let $m > 2$ be a prime power, and let $n := |G| = \varphi(m)/2 = \Theta(m)$. Let H be the subspace of $\mathbb{R}^n$ spanned by $\Lambda = \mathrm{Log}\, R^*$ (and by $\mathrm{Log}\, C$, the log embedding of the cyclotomic units), which is the subspace orthogonal to $\mathbf{1}$, the all-1s vector. Define the covering radius of a lattice $\mathcal{L}$ with respect to the ℓ_p norm as

$$\mu^{(p)}(\mathcal{L}) = \max_{\mathbf{x}\in\mathrm{span}(\mathcal{L})} \min_{\mathbf{v}\in\mathcal{L}} \|\mathbf{x} - \mathbf{v}\|_p = \max_{\mathbf{x}\in\mathrm{span}(\mathcal{L})} \min_{\mathbf{v}\in\mathbf{x}+\mathcal{L}} \|\mathbf{v}\|_p.$$

6.1   Relation to Covering Radius

For any $g \in R$, let $\mathcal{I} = gR$. Also let $\mathbf{g} = \mathrm{Log}(g)$ and write it as $\mathbf{g} = s\mathbf{1} + \mathbf{g}_H$ where $\mathbf{g}_H \in H$. Observe that $s = \log S(\mathcal{I})$, because

$$N(\mathcal{I}) = N(g) = \prod_{i \in \mathbb{Z}_m^*} \sigma_i(g) = \prod_{i \in G} |\sigma_i(g)|^2 = \exp(2\langle \mathbf{g}, \mathbf{1}\rangle) = \exp(s \cdot \varphi(m)).$$

Lemma 8. *Let g, $\mathcal{I}$, s, and $\mathbf{g}_H$ be as above. There exists an efficient algorithm that, given g and any $\mathbf{h}_H \in \mathbf{g}_H + \mathrm{Log}\, C$, outputs a generator h of $\mathcal{I}$ such that*

$$\|h\| \leq \sqrt{\varphi(m)} \cdot \exp(\|\mathbf{h}_H\|_\infty) \cdot S(\mathcal{I}).$$

In particular, there exists a generator of Euclidean norm at most $\sqrt{\varphi(m)} \cdot \exp(\mu^{(\infty)}(\mathrm{Log}\, C)) \cdot S(\mathcal{I})$.

Proof. As in the proof of Theorem 5, for simplicity we ignore issues of precision; see the discussion at the end of Sect. 4. The algorithm lets $\mathbf{u} = \mathbf{h}_H - \mathbf{g}_H \in \mathrm{Log}\, C$, computes the coefficients $a_j \in \mathbb{Z}$ such that $\mathbf{u} = \sum a_j \mathbf{b}_j$, and outputs $h = g \cdot \prod b_j^{a_j}$. Because $\mathbf{h} := \mathrm{Log}(h) = \mathrm{Log}(g) + \mathbf{u} = s\mathbf{1} + \mathbf{h}_H$, we have

$$\|h\|^2 = \sum_{i \in \mathbb{Z}_m^*} |\sigma_i(h)|^2 \leq \varphi(m) \cdot \exp(\|\mathbf{h}\|_\infty)^2 = \varphi(m) \cdot \exp(\|\mathbf{h}_H\|_\infty)^2 \cdot S(\mathcal{I})^2.$$

Lemma 9. *There exists a principal ideal $\mathcal{I} \subseteq R$ for which every generator has Euclidean norm at least $\exp(\Omega(\mu^{(1)}(\Lambda)/m)) \cdot S(\mathcal{I})$.*

In fact, the proof shows that a "random principal ideal," i.e., one whose generators correspond to a uniformly random coset of the log-unit lattice, satisfies the above bound with overwhelming probability. (Formalizing this requires a bit more effort; we omit the details.)

Proof. Let $\mathbf{x} + \Lambda \subset H$ be a "deep hole" coset of Λ in the ℓ_1 norm, i.e., one for which $\|\mathbf{v}\|_1 \geq \mu^{(1)}(\Lambda)$ for every $\mathbf{v} \in \mathbf{x} + \Lambda \subset H$. Because the n coordinates of any such $\mathbf{v}$ sum to zero, the sum of the positive coordinates must be exactly $\|\mathbf{v}\|_1/2$, and therefore there must be a coordinate that is at least $\mu^{(1)}(\Lambda)/(2n) = \Omega(\mu^{(1)}(\Lambda)/m)$.

Next, assume for a moment that there exists $g \in R$ for which $\mathbf{g}_H = \mathbf{x}$, where as before we write $\mathbf{g} = \mathrm{Log}(g) = s\mathbf{1} + \mathbf{g}_H$. Then any generator h of the ideal $\mathcal{I} = gR$ satisfies $\mathrm{Log}(h) \in \mathrm{Log}(g) + \Lambda = s\mathbf{1} + \mathbf{x} + \Lambda$, so by the observation above, it must have the claimed Euclidean norm.

To complete the proof, notice that even if there does not exist a g as above, one can find g so as to make $\mathbf{g}_H$ arbitrarily close to $\mathbf{x}$, which suffices for the above analysis. To see this, consider $x = M \cdot \mathrm{Exp}(\mathbf{x})$, where M is a sufficiently large integer and $\mathrm{Exp}(\mathbf{x}) \in \mathrm{Log}^{-1}(\mathbf{x})$ denotes an arbitrary preimage in $K_\mathbb{R} := K \otimes_\mathbb{Q} \mathbb{R}$ of $\mathbf{x}$ under the log embedding (extended to $K_\mathbb{R}$). Then rounding x to a nearest $g \in R$ yields the claim.

6.2   Covering Radius Upper Bound and an SVP Algorithm

Theorem 6. *There is an efficient randomized algorithm that given any vector* $\mathbf{x} \in H$ *outputs a vector* $\mathbf{v} \in \operatorname{Log} C$ *such that* $\|\mathbf{x} - \mathbf{v}\|_\infty = O(\sqrt{m \log m})$ *with high probability.*

Before giving the proof, we mention some implications of the theorem. First, using the fact that $\operatorname{Log} C$ is a sublattice of Λ, we immediately get the following corollary regarding the covering radii of these lattices.

Corollary 2. *For a prime power* m, *we have* $\mu^{(\infty)}(\Lambda) \leq \mu^{(\infty)}(\operatorname{Log} C) \leq O(\sqrt{m \log m})$.

We remark that this corollary can also be obtained directly from Lemma 11 below and the non-trivial result of Banaszczyk and Szarek [BS97] (see also [Ban98]). We also note that if the Komlós conjecture is true, then the $\sqrt{\log m}$ factor in the corollary can be removed.

It follows immediately from the corollary and Lemma 8 that any principal ideal $\mathcal{I}$ has a generator whose Euclidean norm is at most $\exp(O(\sqrt{m \log m})) \cdot S(\mathcal{I})$. This also leads to an efficient quantum algorithm providing a non-trivial approximation to SVP in principal ideals, as described in the following theorem.

Theorem 7. *There is an efficient quantum algorithm that approximates SVP on principal ideal lattices in cyclotomics of prime-power index* m *to within approximation factor* $2^{O(\sqrt{m \log m})}$.

Proof. Given a principal ideal $\mathcal{I}$, first use the efficient quantum algorithm of Biasse and Song [BS15] to recover a generator g of $\mathcal{I}$, and as above, write $\operatorname{Log}(g) = s\mathbf{1} + \mathbf{g}_H$ for $\mathbf{g}_H \in H$. Next, apply Theorem 6 to $\mathbf{g}_H$ and let $\mathbf{v} \in \operatorname{Log} C$ be the output. Finally, apply the algorithm from Lemma 8 with g and $\mathbf{h}_H := \mathbf{g}_H - \mathbf{v}$ to find a generator h whose Euclidean norm is at most $\exp(O(\sqrt{m \log m})) \cdot S(\mathcal{I})$, and output h. It is sufficiently short since, as mentioned at the start of the section, $\lambda_1(\mathcal{I}) = \Omega(\sqrt{m}) \cdot S(\mathcal{I})$ by the arithmetic mean-geometric mean inequality.

For the proof of Theorem 6, we need a simple probabilistic lemma, as well as a bound on the norm of the $\mathbf{b}_j$. For $\alpha \in [0, 1]$, define $S(\alpha)$ as the unique probability distribution on support $\{\alpha, \alpha - 1\}$ with expectation 0 (i.e., it assigns probability $1 - \alpha$ to α and probability α to $\alpha - 1$).

Lemma 10. *Let* $\mathbf{A}$ *be an* $n \times n$ *matrix all of whose rows have Euclidean norm at most* $T > 0$, *and let* $\alpha_1, \dots, \alpha_n \in [0, 1]$ *be arbitrary. Let* $x_1, \dots, x_n$ *be independent with* x_i *distributed as* $S(\alpha_i)$, *and let* $\mathbf{x} = (x_1, \dots, x_n)$. *Then with probability* $\Omega(1/\sqrt{n})$, *both*

$$\|\mathbf{A}\mathbf{x}\|_\infty \leq O(T\sqrt{\log n}) \qquad and \qquad \left| \sum x_i \right| \leq O(1).$$

Proof. Since $S(\alpha)$ is bounded, it is a subgaussian random variable of constant subgaussian norm. (See [Ver12, Sect. 5.2.3] for the definition and properties of subgaussian random variables.) Because the sum of independent subgaussian random variables is also subgaussian (see [Ver12, Lemma 5.9]), $(\mathbf{Ax})_i$ has subgaussian norm $O(T)$ for every $i = 1, \ldots, n$. Therefore, for a large enough universal constant $C > 0$,

$$\Pr\left[|(\mathbf{Ax})_i| > CT\sqrt{\log n}\right] = O(1/n^2),$$

and by a union bound we get

$$\Pr\left[\|\mathbf{Ax}\|_\infty > CT\sqrt{\log n}\right] = O(1/n). \tag{8}$$

Next, by the Berry-Esseen theorem (see, e.g., [O'D14, Sect. 5.2]), since the x_i have expectation 0 and bounded second and third moments, the probability that $|\sum x_i| = O(1)$ is $\Omega(1/\sqrt{n})$. Together with Eq. (8) and the union bound, this completes the proof.

Lemma 11. *Let m be a prime power. Then for all $j \in G$, $\|\mathbf{z}_j\| = O(\sqrt{m})$, where $\mathbf{z}_j$ are the vectors defined in Eq. (3).*

Proof. Notice that

$$\|\mathbf{z}_j\|^2 = \sum_{i \in G} \log^2 |\omega^{ij} - 1| = \sum_{i \in G} \log^2 |\omega^i - 1|$$

$$= \sum_{i \in G} \log^2 |2\sin(\pi i/m)| \le \sum_{i=1}^{\lfloor m/2 \rfloor} \log^2(2\sin(\pi i/m))$$

$$= \sum_{i=1}^{\lfloor m/2 \rfloor} f(i/m), \tag{9}$$

where $f : [0, 1/2] \to \mathbb{R}$ is given by $f(x) = \log^2(2\sin(\pi x))$. Since $f(x) \le \log 2$ for $1/6 \le x \le 1/2$ (recall that $\sin(\pi/6) = 1/2$), the contribution to the sum in Eq. (9) coming from $i > \lfloor m/6 \rfloor$ is at most $O(m)$. It therefore suffices to consider the contribution coming from $i \in \{1, \ldots, \lfloor m/6 \rfloor\}$. Since $\sin(\pi x) \ge 2x$ for $0 \le x \le 1/2$ (as follows from the concavity of sine on $[0, \pi/2]$), that contribution satisfies

$$\sum_{i=1}^{\lfloor m/6 \rfloor} f(i/m) \le \sum_{i=1}^{\lfloor m/6 \rfloor} \log^2(4i/m) \le m \int_0^{1/6} \log^2(4x)dx = O(m),$$

the last equality following from

$$\int_0^y \log^2(x)dx = y(\log^2 y - 2\log y + 2).$$

Proof (Proof of Theorem 6). Given any $\mathbf{y} \in H$, find real coefficients $(a_j)_{j \in G \setminus \{1\}}$ such that $\mathbf{y} = \sum a_j \mathbf{b}_j$. For $j \in G \setminus \{1\}$, let $\alpha_j = (a_j \bmod 1) \in [0, 1)$ be the fractional part of a_j, and let x_j be independent random variables distributed like $S(\alpha_j)$. The algorithm outputs $\mathbf{u} = \sum(a_j - x_j)\mathbf{b}_j$. Notice that $\mathbf{u} \in \operatorname{Log} C$ as desired. To analyze the distance of $\mathbf{u}$ from $\mathbf{y}$, for convenience let x_1 be an independent random variable distributed like $S(0)$ (so $x_1 = 0$ always). Recalling that $\mathbf{b}_j = \mathbf{z}_j - \mathbf{z}_1$, write

$$\mathbf{y} - \mathbf{u} = \sum_{j \in G} x_j(\mathbf{z}_j - \mathbf{z}_1) = \sum_{j \in G} x_j \mathbf{z}_j - \left(\sum_{j \in G} x_j\right) \mathbf{z}_1,$$

and so by the triangle inequality

$$\|\mathbf{y} - \mathbf{u}\|_\infty \leq \left\|\sum_{j \in G} x_j \mathbf{z}_j\right\|_\infty + \left|\sum_{j \in G} x_j\right| \cdot \|\mathbf{z}_1\|_\infty$$

$$\leq \left\|\sum_{j \in G} x_j \mathbf{z}_j\right\|_\infty + \left|\sum_{j \in G} x_j\right| \cdot O(\sqrt{m}),$$

where we used the trivial bound $\|\mathbf{z}_1\|_\infty \leq \|\mathbf{z}_1\|_2$ and applied Lemma 11.[7] We now apply Lemma 10 to the matrix $\mathbf{Z}$ whose columns are the $\mathbf{z}_j$. Since $\mathbf{Z}$ is G-circulant, the Euclidean norms of all its rows and columns are the same, and by Lemma 11 are $O(\sqrt{m})$. We therefore obtain that with probability $\Omega(1/\sqrt{n})$,

$$\|\mathbf{y} - \mathbf{u}\|_\infty \leq O(\sqrt{m \log n}) + O(\sqrt{m}) = O(\sqrt{m \log m}),$$

as desired. The success probability can be amplified by repetition.

6.3   Covering Radius Lower Bound

Let $h' := (h^+)^{1/(n-1)}$, which we recall is conjectured to be constant. Combined with Lemma 9, the theorem below shows that there exists a principal ideal $\mathcal{I} \subseteq R$ for which every generator has Euclidean norm at least $\exp(\Omega(\sqrt{m}/(h' \log m))) \cdot S(\mathcal{I})$.

Theorem 8. *For a prime power m, the log-unit lattice satisfies*

$$\mu^{(1)}(\Lambda) \geq \Omega(m^{3/2}/(h' \log m)).$$

Proof. Using Lemma 12 below,

$$(\det(\operatorname{Log} C))^{1/(n-1)} = \Omega(m^{1/2}/\log m).$$

---

[7] In fact, $\|\mathbf{z}_1\|_\infty = O(\log m)$, but we do not need this.

Since $\det(\Lambda) = \det(\mathrm{Log}\,C)/h^{+}$,[8]

$$(\det(\Lambda))^{1/(n-1)} = \Omega(m^{1/2}/(h'\log m)).$$

The theorem now follows from the fact that

$$\mathrm{vol}(B_1^n \cap H) \le \sqrt{n} \cdot 2^{n-1}/(n-1)! = O(1/n)^{n-1},$$

where $B_1^n := \{\mathbf{x} \in \mathbb{R}^n : \|\mathbf{x}\|_1 \le 1\}$. To prove this inequality, notice that (1) the volume of B_1^{n-1} is $2^{n-1}/(n-1)!$, (2) the projection of $B_1^n \cap H$ on the first $n-1$ coordinates is contained in B_1^{n-1}, and (3) this projection shrinks volumes by $\sqrt{n}$, as can be seen by computing its Jacobian.

Lemma 12. *The determinant of* $\mathrm{Log}\,C$ *satisfies*

$$\det(\mathrm{Log}\,C)^{1/(n-1)} = \Omega(\sqrt{m}/\log m).$$

Proof. Recall from the proof of Lemma 3 that $\mathbf{b}_j^{\vee}$ is the projection of $\mathbf{z}_j^{\vee}$ orthogonal to $\mathbf{1}$. The $|G|$-dimensional full-rank lattice generated by $\{\mathbf{z}_j^{\vee}\}_{j\in G}$ has determinant

$$|\det(\mathbf{Z}^{-t})| = \prod_{\chi \in \widehat{G}} |\lambda_\chi^{-1}|.$$

Next, notice that the shortest vector in the intersection of this lattice with the span of $\mathbf{1}$ is $\mathbf{Z}^{-t}\mathbf{1} = \lambda_1^{-1}\mathbf{1}$, whose Euclidean norm is $\lambda_1^{-1}\sqrt{|G|}$. Therefore, the dual of $\mathrm{Log}\,C$, which is the projection of this lattice orthogonally to $\mathbf{1}$, has determinant

$$|G|^{-1/2} \prod_{\chi \in \widehat{G}\setminus\{1\}} |\lambda_\chi^{-1}|,$$

and therefore

$$\det(\mathrm{Log}\,C) = |G|^{1/2} \prod_{\chi \in \widehat{G}\setminus\{1\}} |\lambda_\chi|$$

$$= |G|^{1/2} \prod_{\chi \in \widehat{G}\setminus\{1\}} \left| \frac{1}{2}\sqrt{f_\chi} \cdot L(1,\chi) \right|, \tag{10}$$

where we used Eq. (5). Letting $m = p^k$ for a prime p, and using Theorem 1, we get that

$$L := \prod_{\chi \in \widehat{G}\setminus\{1\}} |L(1,\chi)| = \Omega((\log m)^{-(n-1-q)} \cdot p^{-q/2})$$

where q denotes the number of even quadratic characters modulo m, which is at most 3 (see [MV06, Sect. 9.3]). We conclude that

$$L^{1/(n-1)} = \Omega(1/\log m). \tag{11}$$

---

[8] We note that $2^{n-1}\det(\Lambda)/\sqrt{n}$ is known as the *regulator*, and a bound similar to what we obtain here can be derived from the Brauer-Siegel theorem [Was97, p. 43]. This leads to a bound that is both somewhat weaker and ineffective.

Next, consider $F = \prod_{\chi \in \hat{G} \setminus \{1\}} f_\chi$. For each $0 < j \leq k$, there are exactly $\varphi(p^j) - \varphi(p^{j-1})$ characters of conductor $f_\chi = p^j$. Exactly half are these are even when p is odd and $j > 1$, and also when $p = 2$ and $j > 2$. When p is odd and $j = 1$ there are $\varphi(p)/2 - 1$ even characters of conductor p, and when $p = 2$ there are no even characters of conductor 2 or 4. Assuming p is odd (the case $p = 2$ being very similar), this leads to

$$
\begin{aligned}
\log_p F &= \sum_{j=1}^{k} j \cdot \frac{\varphi(p^j) - \varphi(p^{j-1})}{2} - \frac{1}{2} \\
&= \frac{k}{2} \cdot \varphi(p^k) - \frac{1}{2} \sum_{j=0}^{k-1} \varphi(p^j) - \frac{1}{2} \\
&= kn - \frac{p-1}{2} \sum_{j=0}^{k-2} p^j - 1 = kn - \frac{p^{k-1}}{2} - \frac{1}{2},
\end{aligned}
$$

and we conclude that

$$
F = m^{n\left(1 - \frac{1}{2k(p-1)} - \frac{1}{2kn}\right)} = \Omega(m)^n. \tag{12}
$$

Plugging (11) and (12) into (10) completes the proof.

Acknowledgments. We thank Dan Bernstein, Jean-François Biasse, Sean Hallgren, Sorina Ionica, Dimitar Jetchev, Paul Kirchner, Shinya Okumara, René Schoof, Alice Silverberg, and Harold M. Stark for comments and many insightful conversations on topics related to this work. We also especially thank Dan Shepherd [She14] for explaining many additional details about the claims made in [CGS14], and for sharing other helpful observations.

A   Proof of Theorem 2

Proof. First, Corollary 4.13 of [Was97] gives that $\mathbb{Z}[\zeta]^*$ is generated by $\mathbb{Z}[\zeta + \overline{\zeta}]^*$ and ζ, so it follows that

$$
\Lambda = \mathrm{Log}\,\mathbb{Z}[\zeta]^* = \mathrm{Log}\,\mathbb{Z}[\zeta + \overline{\zeta}]^*,
$$

since the kernel of Log is the group $\{\pm 1\} \cdot U$.

Next, recall that the group of *cyclotomic units* is defined as $C = A \cap R^*$. We define the group of *real* cyclotomic units as $C^+ = A \cap \mathbb{Z}[\zeta + \overline{\zeta}]^*$. The analogue of Lemma 2 for the real cyclotomic units, also included in Lemma 8.1 of [Was97], says that the group C^+ of real cyclotomic units is generated by -1 and $\zeta^{(1-j)/2} \cdot b_j$. So as above, we obtain that

$$
\mathrm{Log}\,C = \mathrm{Log}\,C^+.
$$

The theorem then follows from the sequence of equalities

$$
[\Lambda : \mathrm{Log}\,C] = \left[\mathrm{Log}\,\mathbb{Z}[\zeta + \overline{\zeta}]^* : \mathrm{Log}\,C^+\right] = \left[\mathbb{Z}[\zeta + \overline{\zeta}]^* : C^+\right] = h^+,
$$

where the second equality follows from $\ker(\mathrm{Log}) \cap C^+ = \ker(\mathrm{Log}) \cap \mathbb{Z}[\zeta + \overline{\zeta}]^*$ $(= \{\pm 1\})$, and the third equality is Theorem 8.2 of [Was97].

B   Numeric Data

The previous sections established *asymptotic* bounds related to the log-embeddings of the cyclotomic units. Figure 1 gives concrete numeric data for several practical (and even impractical) choices of cyclotomic fields. This data confirms that the method works in practice.

$k \ (m = 2^k)$	6	7	8	9	≥ 10
$\|\mathbf{b}_j^\vee\|^{-2}$	5.04	8.56	14.69	25.71	≥ 45.85

$k \ (m = 3^k)$	4	5	≥ 6
$\|\mathbf{b}_j^\vee\|^{-2}$	5.72	13.65	≥ 34.04

$k \ (m = 5^k)$	3	4	≥ 5
$\|\mathbf{b}_j^\vee\|^{-2}$	10.04	36.43	≥ 143

Fig. 1. Lower bounds on the inverse lengths of the dual vectors $\mathbf{b}_j^\vee$ defined in Sect. 3, for various cyclotomics of prime-power index. Larger values correspond to larger decoding distances for the log-embedding of the cyclotomic units.

References

[ADS15] Aggarwal, D., Dadush, D., Stephens-Davidowitz, N.: Solving the closest vector problem in 2^n time - the discrete Gaussian strikes again! In: FOCS, pp. 563–582 (2015)

[Bab85] Babai, L.: On Lovász' lattice reduction, the nearest lattice point problem. Combinatorica **6**(1), 1–13 (1986). Preliminary version in STACS 1985

[Ban93] Banaszczyk, W.: New bounds in some transference theorems in the geometry of numbers. Math. Ann. **296**(4), 625–635 (1993)

[Ban98] Banaszczyk, W.: Balancing vectors and gaussian measures of n-dimensional convex bodies. Random Struct. Algorithms **12**(4), 351–360 (1998)

[Ber14a] Bernstein, D.: Personal Communication. June 2014

[Ber14b] Bernstein, D.: A subfield-logarithm attack against ideal lattices. http://blog.cr.yp.to/20140213-ideal.html, Febuary 2014

[BF14] Biasse, J.-F., Fieker, C.: Subexponential class group, unit group computation in large degree number fields. LMS J. Comput. Math. **17**(suppl. A), 385–403 (2014)

[Bia14]   Biasse, J.-F.: Subexponential time relations in the class group of large degree number fields. Adv. Math. Commun. **8**(4), 407–425 (2014)

[BPR04]   Buhler, J., Pomerance, C., Robertson, L.: Heuristics for class numbers of prime-power real cyclotomic fields. Fields Inst. Commun **41**, 149–157 (2004)

[BS97]   Banaszczyk, W., Szarek, S.J.: Lattice coverings and Gaussian measures of n-dimensional convex bodies. Discrete Comput. Geom. **17**(3), 283–286 (1997)

[BS15]   Biasse, J.-F., Song, F.: A note on the quantum attacks against schemes relying on the hardness of finding a short generator of an ideal in $\mathbb{Q}(\zeta_{2^n})$. In: Technical report –12, The University of Waterloo, Revision of September 28th 2015

[BS16]   Biasse, J.-F., Song, F.: A polynomial time quantum algorithm for computing class groups and solving the principal ideal problem in arbitrary degree number fields. In: SODA (2016)

[CGS14]   Campbell, P., Groves, M., Shepherd, D.: Soliloquy: a cautionary tale. In: ETSI 2nd Quantum-Safe Crypto Workshop, 2014. Available at http://docbox.etsi.org/Workshop/2014/201410_CRYPTO/S07_Systems_and_Attacks/S07_Groves_Annex.pdf

[EHKS14] Eisenträger, K., Hallgren, S., Kitaev, A., Song, F.: A quantum algorithm for computing the unit group of an arbitrary degree number field. In: STOC, pp. 293–302. ACM (2014)

[Gen09]   Gentry, C.: Fully homomorphic encryption using ideal lattices. In: STOC, pp. 169–178 (2009)

[GGH13]   Garg, S., Gentry, C., Halevi, S.: Candidate multilinear maps from ideal lattices. In: Johansson, T., Nguyen, P.Q. (eds.) EUROCRYPT 2013. LNCS, vol. 7881, pp. 1–17. Springer, Heidelberg (2013)

[GS02]   Gentry, C., Szydlo, M.: Cryptanalysis of the revised NTRU signature scheme. In: Knudsen, L.R. (ed.) EUROCRYPT 2002. LNCS, vol. 2332, p. 299. Springer, Heidelberg (2002)

[HPS98]   Hoffstein, J., Pipher, J., Silverman, J.H.: NTRU: a ring-based public key cryptosystem. In: Buhler, J.P. (ed.) ANTS 1998. LNCS, vol. 1423, pp. 267–288. Springer, Heidelberg (1998)

[Lan27]   Landau, E.: Über Dirichletsche Reihen mit komplexen Charakteren. Journal für die reine und angewandte Mathematik **157**, 26–32 (1927)

[Lan02]   Lang, S.: Algebra. Graduate Texts in Mathematics, vol. 211, 3rd edn. Springer, New York (2002)

[Len82]   Lenstra, A.K.: Lattices and factorization of polynomials over algebraic number fields. In: Calmet, J. (ed.) EUROCAM '1982. LNCS, vol. 144, pp. 32–39. Springer, Heidelberg (1982)

[LLL82]   Lenstra, A.K., Lenstra Jr., H.W., Lovász, L.: Factoring polynomials with rational coefficients. Math. Ann. **261**(4), 515–534 (1982)

[LLS15]   Lamzouri, Y., Li, X., Soundararajan, K.: Conditional bounds for the least quadratic non-residue and related problems. Math. Comp. **84**(295), 2391–2412 (2015)

[LMPR08] Lyubashevsky, V., Micciancio, D., Peikert, C., Rosen, A.: SWIFFT: a modest proposal for FFT hashing. In: Nyberg, K. (ed.) FSE 2008. LNCS, vol. 5086, pp. 54–72. Springer, Heidelberg (2008)

[Lou15]   Louboutin, S.: An explicit lower bound on moduli of Dirichlet L-functions at $s = 1$. J. Ramanujan Math. Soc. **30**(1), 101–113 (2015)

[LPR10]   Lyubashevsky, V., Peikert, C., Regev, O.: On ideal lattices, learning with errors over rings. J. ACM **60**(6), 43:1–43:35 (2013)

[LPR13] Lyubashevsky, V., Peikert, C., Regev, O.: A toolkit for ring-LWE cryptography. In: Johansson, T., Nguyen, P.Q. (eds.) EUROCRYPT 2013. LNCS, vol. 7881, pp. 35–54. Springer, Heidelberg (2013)

[LSS14] Langlois, A., Stehlé, D., Steinfeld, R.: GGHLite: more efficient multilinear maps from ideal lattices. In: Nguyen, P.Q., Oswald, E. (eds.) EUROCRYPT 2014. LNCS, vol. 8441, pp. 239–256. Springer, Heidelberg (2014)

[Mic02] Micciancio, D.: Generalized compact knapsacks, cyclic lattices, efficient one-way functions. Comput. Complex. 16(4), 365–411 (2007). Preliminary version in FOCS 2002

[Mil14] Miller, J.C.: Class numbers of totally real fields and applications to the weber class number problem. Acta Arith. 164(4), 381–398 (2014)

[Mil15] Miller, J.C.: Real cyclotomic fields of prime conductor and their class numbers. Math. Comp. 84(295), 2459–2469 (2015)

[MR04] Micciancio, D., Regev, O.: Worst-case to average-case reductions based on Gaussian measures. SIAM J. Comput. 37(1), 267–302 (2007). Preliminary version in FOCS 2004

[MV06] Montgomery, H.L., Vaughan, R.C.: Multiplicative Number Theory I. Cambridge University Press, Cambridge (2006)

[MV10] Micciancio, D., Voulgaris, P.: A deterministic single exponential time algorithm for most lattice problems based on Voronoi cell computations. In: STOC, pp. 351–358 (2010)

[O'D14] O'Donnell, R.: Analysis of Boolean Functions. Cambridge University Press, Cambridge (2014)

[Pei10] Peikert, C.: An efficient and parallel gaussian sampler for lattices. In: Rabin, T. (ed.) CRYPTO 2010. LNCS, vol. 6223, pp. 80–97. Springer, Heidelberg (2010)

[PR07] Peikert, C., Rosen, A.: Lattices that admit logarithmic worst-case to average-case connection factors. In: STOC, pp. 478–487 (2007)

[Sam70] Samuel, P.: Algebraic Theory of Numbers. Hermann, Paris (1970)

[Sch87] Schnorr, C.-P.: A hierarchy of polynomial time lattice basis reduction algorithms. Theor. Comput. Sci. 53, 201–224 (1987)

[Sch03] Schoof, R.: Class numbers of real cyclotomic fields of prime conductor. Math. Comput. 72(242), 913–937 (2003)

[Sch15] Schank, J.: LogCvp, Pari implementation of CVP in $\text{Log}\mathbb{Z}[\zeta_{2^n}]^*$, March 2015. https://github.com/jschanck-si/logcvp

[She14] Shepherd, D.: Personal communication, December 2014

[SV10] Smart, N.P., Vercauteren, F.: Fully homomorphic encryption with relatively small key and ciphertext sizes. In: Nguyen, P.Q., Pointcheval, D. (eds.) Public Key Cryptography. LNCS, vol. 6056, pp. 420–443. Springer, Heidelberg (2010)

[Ver12] Vershynin, R.: Introduction to the non-asymptotic analysis of random matrices. In: Compressed sensing, pp. 210–268. Cambridge University Press, Cambridge (2012). http://www-personal.umich.edu/~romanv/papers/non-asymptotic-rmt-plain.pdf

[Was97] Washington, L.: Introduction to Cyclotomic Fields. Graduate Texts in Mathematics. Springer, New York (1997)

Circuit Compilers with $O(1/\log(n))$ Leakage Rate

Marcin Andrychowicz[1]([✉]), Stefan Dziembowski[1], and Sebastian Faust[2]

[1] University of Warsaw, Warsaw, Poland
marcin.andrychowicz@gmail.com
[2] Ruhr University Bochum, Bochum, Germany

Abstract. The goal of leakage-resilient cryptography is to construct cryptographic algorithms that are secure even if the devices on which they are implemented leak information to the adversary. One of the main parameters for designing leakage resilient constructions is the leakage *rate*, i.e., a proportion between the amount of leaked information and the complexity of the computation carried out by the construction. We focus on the so-called circuit compilers, which is an important tool for transforming any cryptographic algorithm (represented as a circuit) into one that is secure against the leakage attack. Our model is the "probing attack" where the adversary learns the values on some (chosen by him) wires of the circuit.

Our results can be summarized as follows. First, we construct circuit compilers with perfect security and leakage rate $O(1/\log(n))$, where n denotes the security parameter (previously known constructions achieved rate $O(1/n)$). Moreover, for the circuits that have only affine gates we obtain a construction with a constant leakage rate. In particular, our techniques can be used to obtain constant-rate leakage-resilient schemes for refreshing an encoded secret (previously known schemes could tolerate leakage rates $O(1/n)$).

We also show that our main construction is secure against constant-rate leakage in the random probing leakage model, where the leaking wires are chosen randomly.

1 Introduction

Side-channel attacks are an omnipresent threat for the security of cryptographic implementations. In contrast to traditional cryptanalytical attacks that attempt to break the mathematical properties of the cryptographic algorithm, a side-channel adversary targets the implementation by, e.g., observing the running time of a device [29] or measuring its power consumption [30]. An important

M. Andrychowicz and S. Dziembowski—Supported by the WELCOME/2010-4/2 grant founded within the framework of the EU Innovative Economy (National Cohesion Strategy) Operational Programme.
S. Faust—In part supported by the Emmy Noether Program FA 1320/1-1 of the German Research Foundation (DFG).

M. Fischlin and J.-S. Coron (Eds.): EUROCRYPT 2016, Part II, LNCS 9666, pp. 586–615, 2016.
DOI: 10.1007/978-3-662-49896-5_21

countermeasure against side-channel attacks – in particular against power analysis attacks – is the so-called masking countermeasure. A masking scheme randomizes the intermediate values of the computation in order to conceal sensitive information.

Circuit Compilers and the Probing Model. A formalization of the masking countermeasure has been introduced in the seminal work of Ishai et al. [26] with the concept of *leakage resilient circuit compilers*. At a high-level, a circuit compiler takes as input a description of a circuit Γ and compiles it into a protected circuit $\widehat{\Gamma}$ that has the same functionality as Γ but additionally is secure in a well-defined leakage model. One of the most prominent leakage models is the so-called t-threshold probing model of Ishai et al., where the adversary is allowed to observe up to t intermediate values computed by $\widehat{\Gamma}$. The threshold probing model is widely used in practice to analyze the soundness of a masking scheme against higher order attacks [3–5,8,9,22,28,32].

On the Importance of the Leakage Rate. An important parameter to evaluate the security of a masking scheme in the probing model is the value of t. At first sight it may seem that a higher value for t automatically implies a higher level of security as the adversary obtains more knowledge about the internals. To see why such an approach may not always lead to better security imagine two compilers that on an input circuit Γ output the following: the first one produces a circuit Γ_1 that has 1 thousand gates and tolerates leakage of 10 wires, while the second one produces a circuit Γ_2 of size 1 million gates that tolerates leakage of 100 wires. Which construction provides a higher level of security (even discarding the production costs)? The first one has to be implemented on hardware that leaks at most 1 % of its wires, while the second one requires hardware that leaks at most 0.01 % of the wires! Therefore, the second construction is actually weaker (although it "tolerates more leakage" in absolute terms). The above simple example illustrates that in many case it may be more natural to look at the *leakage rate* of a given construction (i.e.: the "amount of leakage" *divided* by the circuit size), than at the "total amount of leakage" itself.[1]

Despite the practical importance of the threshold probing model it is not well-understood how close we can get to the optimal leakage rate of $O(1)$. Indeed, the best known construction is still the circuit compiler of Ishai et al., which remains secure if the adversary learns a $O(1/n)$-fraction of the wires in the transformed circuit $\widehat{\Gamma}$ (for security parameter n). The main contribution of our work is to significantly improve this rate and build new leakage resilient circuit

---

[1] We note that this model still ignores many aspects of the side channel attacks, for example the fact that some operations (like writing bits to the memory) leak more information than some other ones (like the arithmetic operations). We stress that a certain level of abstraction is inevitable in every formal model. Moreover, the fact that a wire w leaks more information than the wires can be reflected by having several copies of w in Γ (where Γ is an input for the circuit compiler, see Sect. 1.1 for more on the circuit compilers).

compilers that achieve $O(1/\log(n))$ leakage rate, and if the circuit Γ is only affine computation we achieve the optimal leakage rate of $O(1)$.

At this point, we want to briefly comment about what we mean by "optimality" of leakage rate and the efficiency of our constructions. First, it shall be clear that we cannot leak everything and, hence, a leakage rate of $O(1)$ is asymptotically optimal. Of course, concretely there can be big differences depending on the hidden constants. Some of our constructions are asymptotically optimal (for affine circuits), but their constants are far from the optimal 1. We believe it is an important question for future work to optimize these constants. We sometimes also talk about optimality in terms of efficiency. In this case, we mean optimality for circuit compilers that compute with encodings and offer information theoretic security. One may get asymptotically better efficiency by using solutions based on FHE.[2]

1.1 The Work of Ishai, Sahai and Wagner

The Threshold Probing Model. Ishai et al. consider two different types of probing adversaries: the first type of adversary is allowed to probe up to t intermediate values in the *entire* transformed circuit $\widehat{\Gamma}$. Notice that this implies that for a growing number of gates in the circuit $\widehat{\Gamma}$ the number of corrupted wires stays the same, and hence the fraction of wires of the circuit that are corrupted decreases. To improve the leakage rate, [26] also considers a significantly stronger adversary where the transformed circuit is structured into so-called *regions* and the adversary can probe up to t wires in each such region. In the following we call the later model the *t-region probing model*, which will be the focus of our work. Notice that in the following a region will correspond to a sub-circuit of the transformed circuit, and we will describe below how we structure the transformed circuit into such regions.

The Circuit Compiler of Ishai, Sahai and Wagner (ISW). Ishai et al. consider Boolean circuits Γ with internal state m, which can, e.g., store the secret key of an AES. The basic ingredient of the compiler of Ishai et al. is a leakage resilient encoding scheme Enc(.) that encodes the computation in $\widehat{\Gamma}$. For a fixed number of probes t define the security parameter $n = O(t)$.[3] In the transformed circuit each intermediate wire (including the state m) which carries a bit x in the original circuit Γ is represented by n wires $\text{Enc}(x) = (X_1, \dots, X_n)$ that are chosen uniformly at random such that $\sum_i X_i = x$ (the sum here represents Boolean XOR, but the encoding scheme can be defined over an arbitrary finite field). Since the above encoding scheme is a perfect $(n-1)$ out of n secret sharing

---

[2] Notice that even for such solutions our construction offers asymptotic improvements over earlier works since the decryption circuit of the FHE scheme has to be protected with an encoding-based circuit compiler.

[3] In the rest of the work we will mostly give a concrete relation between the number of probes t and the security parameter n, which determine the blow-up of the transformed circuit.

scheme, it is easy to see that an adversary that learns up to $t = n - 1$ values of the codeword X obtains no information about the secret x.

The main difficulty in developing a secure circuit compiler is the transformation of the gates, i.e., the transformation of the AND and XOR gate in the case of ISW. In $\widehat{\Gamma}$ gates are represented by so-called *gadgets*. A gadget, e.g., an AND gadget, takes as input two encodings $\mathrm{Enc}(a)$ and $\mathrm{Enc}(b)$ and outputs $\mathrm{Enc}(c)$ such that $c = ab$. Of course, besides correctness, the gadgets have to be specifically tailored such that they withstand t-region probing attacks, where for the ISW construction each gadget corresponds to a region.

Security against t-region probing attacks is formalized by a simulation-based argument. That is, any information that can be learnt by an adversary with t-region probing access to $\widehat{\Gamma}$, can also be obtained by an adversary that has only black-box access to Γ. More formally, Ishai et al. show that for any t-region probing adversary $\mathcal{A}$ against $\widehat{\Gamma}$ there exists a simulator Sim that can simulate answers to all probes with just black-box access to Γ. Notice that the simulation is assumed to be perfect, i.e., the distribution that an adversary obtains by applying a t-probing attack against $\widehat{\Gamma}$ is identical to the distribution that the simulator produces. The ISW compiler achieves security against a probing rate of at least $\Omega(1/n)$ of the wires. In fact, it is easy to see that this bound is tight for the construction of ISW due to the way in which transformed AND gadgets are computed.[4] Hence, to further improve the leakage rate, we need to develop a new circuit transformation.

1.2 Our Contributions

Protecting Affine Circuits. We first consider the seemingly simpler problem of how to protect only affine operations against t-region probing adversaries. We use the simple encoding function described above, i.e., a secret $x \in \mathbb{F}$ is encoded by a random vector $X := (X_1, \ldots, X_n)$ such that $\sum_i X_i = x$. It is easy to see that the addition of two encoded values and multiplication by a constant can be done efficiently requiring only $O(n)$ operations. Hence, if we consider only a single affine operation, then the adversary may learn up to $t = O(n)$ wires from such an operation without violating security. Unfortunately, the above does not easily scale for larger circuits. If we allow the adversary to probe in *each gadget* t wires then the adversary may eventually reveal secret information.

To avoid this problem, the construction of Ishai et al. (and essentially any leakage resilient circuit compiler) uses a so-called **refresh** algorithm that *refreshes* the encoding by introducing new randomness into X, thereby rendering previous leakage on X useless. The basic idea of the algorithm $Y \leftarrow \mathsf{refresh}(X)$ is to sample $Z \leftarrow \mathrm{Enc}(0)$ and compute $Y = X + Z$. Of course, the main difficulty is to generate Z in a way that is secure against t-region probing adversaries.

---

[4] For readers familiar with the construction of [26] the transformation for the AND gate computes on input $A = (A_1, \ldots, A_n)$ and $B = (B_1, \ldots, B_n)$ the values $A_i \cdot B_j$ for all $i, j \in [n]$. Hence, each share A_i appears at least n times and hence it is impossible to obtain leakage rate better than $O(n^{-1})$.

Ishai et al. propose an implementation of the refresh algorithm that requires $O(n^2)$ operations leading to a leakage rate of $O(n^{-1})$. Surprisingly, it is non-trivial to improve the complexity of refresh to $O(n)$ – in fact, we are not aware of any secure refresh algorithm that has the optimal complexity of $O(n)$. The first contribution of our work is to propose the first refreshing algorithm that is perfectly secure against $O(n)$ adversarial chosen probes and has (asymptotically) optimal complexity and randomness usage of $O(n)$. Inspired by the work of Ajtai [1] who studies security against the weaker model of random probing attacks (we will compare our work with the work of Ajtai below), we build our refreshing scheme from expander graphs with constant degree. We emphasize that while our refreshing scheme is similar to the one used by Ajtai, the security proofs differ significantly as we show security in the much stronger model of adaptive probing attacks.

Using the above expander-based scheme for refreshing and combining it with the fact that transformed gadgets for affine operations have complexity $O(n)$, we obtain the first compiler for affine circuits that asymptotically achieves both the *optimal complexity* of $O(n)$ and remains secure against t-region probing adversaries for $t = O(n)$, where each region is of size $O(n)$.

Protecting Any Circuit Against Probing Attacks. To extend our result to work for arbitrary circuits, we need to develop a secure transformation for the multiplication operation. Notice that the transformed multiplication gadget of ISW can be broken when the adversary can probe $\Omega(n)$ wires.[5] Our construction borrows techniques from perfectly secure multiparty computation [14] and combines them in a novel way with our transformation for affine circuits from the previous paragraph. We give some more details below.

Instead of using the simple additive encoding scheme that can be used to protect affine computation, we use a linear secret sharing scheme that has the multiplicative property [11]. Informally speaking, a linear secret sharing scheme is multiplicative if from two encodings X, Y, we can compute the product xy just by computing a linear combination of all the values $Z_i = X_i \cdot Y_i$ and then taking a linear combination of Z_i to compute xy. An example of a secret sharing scheme that satisfies the above property is Shamir's scheme. In Shamir's scheme a secret x from some field $\mathbb{F}$ is encoded into $n = 2t + 1$ shares such that any $(t+1)$ shares can be used to reconstruct the secret x but any subset of at most t shares reveals no information about x. In the following we denote a multiplicative sharing of a secret x with threshold t by $[x]_t$. Notice that since the above encoding scheme is linear, we can easily implement t-region probing resistant addition and multiplication by a constant. We now turn to the description of the protected multiplication operation.

To simplify exposition, let us first assume that the transformed multiplication has access to a leak-free source of randomness that outputs for a random field

---

[5] One may object that by structuring the computation of the ISW AND transformation into regions of size $O(n)$ one can achieve an improved probing rate. However, it is easy to see by a counting argument that such structuring is impossible.

element $r \in \mathbb{F}$ random encodings $[r]_t$ and $[r]_{2t}$ for Shamir's scheme. In this setting, we can use ideas from [14] to carry out the multiplication. On input two vectors $A = [a]_t$ and $B = [b]_t$, first compute $Z_i = X_i \cdot Y_i$. It is easy to see that Z_i defines shares that lie on a polynomial of degree $2t$ with the shared secret being xy, i.e., we have $Z = [xy]_{2t}$. The good news is that the vector Z already yields an encoding of the desired value xy, however, the threshold for Z has increased by a factor 2. To solve this problem we use the encodings $[r]_t$ and $[r]_{2t}$ output by the leak-free component, which enables us to decrease the degree of the encoding $[xy]_t$. Similar techniques have been used in the context of circuit compilers by [2,23], but it is easy to see that their constructions are not secure when $\omega(1/n)$ of the wires are corrupted.

Assuming that $[r]_t$ and $[r]_{2t}$ are produced by the leak-free gates, we can prove that the above construction remains secure in the presence of an adversary that learns up to t wires where $n = 2t + 1$. Of course, for our final transformation we do not want to assume that the computation of $[r]_t$ and $[r]_{2t}$ is done in a leak-free way. Instead, we seek for a t-region probing resistant implementation of it. The crucial observation to achieve this goal is the fact that the encodings $[r]_t$ and $[r]_{2t}$ can be produced by a circuit solely consisting of affine operations (for Shamir's scheme Lagrange polynomial interpolation – but this can be easily generalized). Hence, the problem of protecting arbitrary computation, can be reduced to protecting the affine computation against t-region probing attacks! As the later can be solved by the expander-based transformation described above, we obtain a circuit transformation that works for arbitrary circuits Γ and produces protected circuits $\widehat{\Gamma}$ that remain perfectly secure even if in each region of size $O(n)$ the adversary can learn $O(n/\log(n))$ wires.

The above description omits several technical challenges – in particular, when combining the protected computation operating with the multiplicative secret sharing with our expander-based transformation for affine computation. One problem that we need to address is how to do a secure "conversion" between different types of encodings. More precisely, when we apply our transformation for affine computation in a naive way, then the resulting circuit outputs "encodings of encodings" (so-called "double-encodings"). That is, each share of $[r]_t$ and $[r]_{2t}$ is again encoded using the simple additive sharing used by our affine compiler. Hence, we need to design a t-probing resistant way to "peal-off" the outer layer of the "double-encoding" without revealing the secret value r. To this end, we propose special sub-circuits – so-called tree decoders – that can do the decoding without breaking security in the presence of an adversary that probes a $O(n/\log(n))$-fraction of the wires.[6]

On the Relation to the Noisy Leakage Model. An important leakage model that has recently been considered in several works is the noisy leakage model [6,17,20,31]. The noisy leakage model matches with the engineering perspective as it is closely related to what happens in real-world side-channel attacks based

---

[6] Notice that the tree-decoding is also the technical reason why we do not achieve the optimal rate of $O(1)$.

on the power consumption [31]. Recently, it was shown by Duc et al. [16] that security in the probing model can be translated into security in the noisy leakage model. In particular, using a Chernoff bound and the reduction of [16] security in the t-region probing model implies security in the noisy leakage model of [31]. As the noise parameter is directly related to the leakage rate, by improving the leakage rate, we also get quantitatively better bounds for the Prouff-Rivain noise parameter. More precisely, by applying [16] we directly achieve security when we set the PR noise parameter to $O(1/\log(n)|\mathbb{F}|)$ (compared to $O(1/n|\mathbb{F}|)$). We also show in Sect. 6 that by a more careful analysis our construction actually achieves security for $O(1/|\mathbb{F}|)$ noise level. This is done by showing that our construction is actually secure in the p-random probing model when p is a constant. Using the reduction in [16] and instantiating the multiplicative secret sharing with codes based on algebraic geometry [7], we obtain circuit compilers that are secure for the optimal noise rate of $O(1)$.

On Perfect Security and Adaptive Probing Attacks. We notice that all our result in the t-region probing model achieve perfect security, i.e., there is no statistical error in the theorem statements. This is important, as from a practical point of view such a statistical error term often matters as for small values of the security parameter the error term can be significant.

Another advantage that perfect security has (over statistical security, say) is that one can show that security against adaptive and non-adaptive probing attacks is equivalent. Indeed, in our security analysis we typically consider an adversary that chooses together with the input to the circuit a set of probes $\mathcal{P}$ that specifies for which wires the adversary will learn the corresponding value when the circuit is evaluated. While the adversary can choose a different set of probes $\mathcal{P}$ before each execution of the circuit, most of our analysis does not explicitly allow the adversary to adaptively choose the position of the probes within one execution (i.e., the adversary cannot observe the value of some wire and then depending on the value on that wire decide which wire to probe next). Since all our constructions achieve perfect security, we can apply results from [10] to get security even against fully adaptive probing adversaries.

On the Efficiency of Our Construction. Our basic construction blows up the size of the circuit by a factor of $O(n^3)$. In contrast the construction of Ishai et al. achieves better efficiency and only increases the size of the circuit by a factor of $O(n^2)$. We note that the efficiency of our construction can most likely be improved by a linear factor by using packed secret sharing as the multiplicative encoding scheme (in a similar way as recently done in [2,23]), hence asymptotically achieving the same efficiency as the construction of Ishai et al. We omit the details in this extended abstract.

1.3   Comparison to Other Related Work

Due to space limitations we only compare with the most relevant literature on circuit compilers. Notice also that our focus is not on protecting against active attacks – so-called fault attacks [12,19,25], and hence, we omit a detailed comparison.

The Work of Ishai et al. [26]. Besides the main construction that was already outlined above, Ishai et al. propose a second transformation that achieves improved leakage rate with statistical security (i.e., with a small error probability a probing adversary will break security). In particular, from Theorem 3 of [26] one gets statistical security against $t := O(n)$ probes with a circuit of size $s \cdot O(n \log(n)) \cdot \mathsf{poly}(k)$, where k is the statistical security parameter and s the size of the initial circuit Γ. The above result can be transformed to the t-region probing model considered in our work. In this case, one obtains gadgets of size $n(\log n)\mathsf{poly}(k)$. Since each region is represented by a gadget this yields asymptotically a leakage rate of $O(1/\log n)$, which is as in our paper. There are, however, several important differences:

1. While [26] achieve statistical security, we obtain perfect security. Perfect security is important as it gives full adaptivity for free.
2. The "constant" in $O(1/\log n)$ depends on the statistical security parameter, i.e., it is $\mathsf{poly}(k)$.
3. The results from [26] do not easily generalize to the noisy leakage model. The reason for this comes from the statistical security loss $\mathsf{poly}(k)$ that is hidden in the $O(.)$ notation.
4. Compared to our main construction that has complexity blow-up $O(n^3)$ per multiplication, [26] obtains asymptotically better efficiency of $O(n \log(n))$. We notice, however, that we can trivially improve efficiency of our construction to $O(n^2 \log(n))$, and further efficiency improvements are probably possible using the packed secret sharing.

The Work of Ajtai [1]. At STOC'11 Ajtai proposed a construction that achieves constant rate in the so-called p-random probing model. Ajtai's construction achieves statistical security for "sufficiently" large n and "sufficiently" small constant p, and hence in total a constant fraction of the wires is corrupted. While similar to Ajtai, we use expander graphs to refresh additive encodings, our construction for the transformed multiplication follows a different path. In particular, it is not clear if Ajtai's involved construction for the multiplication operation can be proven secure in the much stronger t-region probing model. Besides the fact that we prove security in the strictly stronger adversarial model, where the adversary can control which wires he wants to corrupt, our construction also improves the efficiency of Ajtai's construction by a factor $O(n \log(n))$ and our security proof is significantly simpler. Hence, one contribution of our work is to simplify and generalize the important work of Ajtai [1].

The Use of Shamir's Secret Sharing in Context of Leakage-Resilient Compilers. Shamir's secret sharing was used in this context before [9,22], however what is achieved there is the leakage rate of $O(1/n)$. Let us stress that the combination of Shamir secret sharing and the expander-based circuit compiler for affine computation was not known before and can be interesting on its own (before it was not known how to get the $O(n)$ overhead and constant fraction rate even for affine computation).

Circuit Compilers in Other Leakage Models. Various other works [13,18,21,27] build circuit compilers in leakage models that are different from the threshold probing model. We notice that all these works achieve security with leakage rate $O(1/n)$ or worse. The work of [13] also gives a nice overview of compilers for the bounded independent leakage model (which is more general than the probing model).

2   Definitions

For two field elements $a, b \in \mathbb{F}$, addition and multiplication in $\mathbb{F}$ are denoted by $a + b$ and ab. For two vectors $A, B \in \mathbb{F}^n$, $A + B$ is the vector-wise addition in $\mathbb{F}$. For a constant $c \in \mathbb{F}$, we denote by $cA = (cA_1, \ldots, cA_n)$, i.e., component-wise multiplication with the constant c. Let $[n] = \{1, \ldots, n\}$ and $[a, b] = \{a, \ldots, b\}$. If $S \subseteq [n]$ and $X \in \mathbb{F}^n$ then $X_S = \{X_i\}_{i \in S}$. We write $M \in \mathbb{F}^{r \times c}$ for a matrix $\{m_{i,j}\}_{i \in [r]}^{j \in [c]}$ with r rows and c columns. For distinct elements $z_1, \ldots, z_r \in \mathbb{F}$ we use $\mathsf{Van}^{r \times c}(z_1, \ldots, z_r)$ to denote the Vandermonde matrix $\{z_i^j\}_{i \in [r]}^{j \in [c]}$.

2.1   Leakage Resilient Encoding Schemes

An important building block to construct a circuit with resilience to leakage is a leakage resilient encoding scheme [15]. An encoding scheme $\Pi = (\mathsf{Enc}, \mathsf{Dec})$ consists of two algorithms. The probabilistic Enc algorithm takes as input some secret $x \in \mathbb{F}$ for a field $\mathbb{F}$ and produces a codeword $X = (X_1, \ldots, X_n) \in \mathbb{F}^n$, where X_i are called the shares of the encoding. The deterministic decoding function Dec takes as input a codeword and outputs the encoded message. A coding scheme satisfies the *correctness* property if for any $x \in \mathbb{F}$ we have $\Pr[\mathsf{Dec}(\mathsf{Enc}(x))) = x] = 1$. Moreover, we want that the encoding scheme is secure against t-probing attacks. An encoding scheme is t-probing secure if for any $x, x' \in \mathbb{F}$ the adversary cannot distinguish t shares of $\mathsf{Enc}(x)$, from t shares of $\mathsf{Enc}(x')$. In this paper we will be interested in two different probing resilient encoding schemes.

Additive Encoding Schemes. The most simple encoding scheme is to encode a secret element $x \in \mathbb{F}$ by a vector X sampled uniformly at random from $\mathbb{F}^n$ such that $\sum_i X_i = x$. Formally, we define the *additive encoding scheme* $\Pi_{n,\mathbb{F}}^{\mathsf{AE}} = (\mathsf{EncAE}, \mathsf{DecAE})$ as:

- $\mathsf{EncAE} : \mathbb{F} \to \mathbb{F}^n$: On input $x \in \mathbb{F}$ choose $X_1, \ldots, X_{n-1}$ uniformly at random and compute $X_n = x - X_1 - \ldots - X_{n-1}$. Output $X = (X_1, \ldots, X_n)$.
- $\mathsf{DecAE} : \mathbb{F}^n \to \mathbb{F}$ works as follows: On input a vector $X \in \mathbb{F}^n$ output $x = \sum_i X_i$.

It is easy to see that any adversary that learns up to $n-1$ shares of X has no knowledge about the secret x.

Encoding Based on Multiplicative Secret Sharing. Additionally to $\Pi_{n,\mathbb{F}}^{\mathsf{AE}}$ which will mainly be used in Sect. 4 to protect affine computation against t-region probing attacks, we need an additional code for protecting *arbitrary* circuits. In particular, we need a linear secret sharing scheme that additionally satisfies the multiplicative property [11]. Informally speaking, a linear secret sharing scheme is multiplicative if from two encodings X, Y, we can compute the product xy just by computing a linear combination of all the values $Z_i = X_i Y_i$. We formally define the encoding scheme $\Pi_{n,t,\mathbb{F},M}^{\mathsf{MSS}} = (\mathsf{EncMSS}, \mathsf{DecMSS})$ with $n = 2t+1$ and M being the generator matrix of the linear code (representing the secret sharing scheme) as follows:

- $\mathsf{EncMSS} : \mathbb{F} \to \mathbb{F}^n$: On input $x \in \mathbb{F}$ choose uniformly at random $(a_1, \ldots, a_t) \leftarrow \mathbb{F}^t$ and compute $X = (x_1, \ldots, x_n) = M \cdot (x, a_1, \ldots, a_t)$. We will often denote encodings of $x \in \mathbb{F}$ using Π^{MSS} with $[x]_t$.
- $\mathsf{DecMSS} : \mathbb{F}^n \to \mathbb{F}$: On input $X \in \mathbb{F}^n$ compute $X \cdot M^{-1} \in \mathbb{F}^{t+1}$, where the first element represents the recovered secret.

We require that $\Pi_{n,t,\mathbb{F},M}^{\mathsf{MSS}}$ is multiplicative meaning that there exists a n-elements vector $R \in \mathbb{F}^n$ such that $\sum_i R_i X_i Y_i$, where all operations are in $\mathbb{F}$.[7] If two encodings $[x]_t$ and $[y]_t$ are multiplied then we obtain $[xy]_{2t}$, where the decoding now requires a slightly adjusted generator matrix $\tilde{M}$.

To simplify exposition, for most of this paper the reader may think of the code as the standard code representing Shamir's secret sharing and as M of the Vandermonde matrix $\mathsf{Van}^{n \times (t+1)}(z_1, \ldots, z_n)$ for distinct elements z_i. Using alternative codes, e.g., packed secret sharing schemes or codes based on algebraic geometry, we can improve the efficiency and the tolerated leakage rate of our construction in the case of random probing from Boolean circuits (we discuss this briefly in Sect. 7). It is easy to see that the encoding scheme $\Pi_{n,t,\mathbb{F},M}^{\mathsf{MSS}}$ based on Shamir's scheme is secure against any t-probing adversary, when $n = 2t+1$. To simplify notation we omit the parameters $\mathbb{F}$ and M and simply denote this scheme $\Pi_{n,t}^{\mathsf{MSS}}$.

2.2 Circuit Transformations

We recall the formalization of circuit transformation of [20,26]. A circuit transformation TR takes as input a security parameter n, a circuit Γ, and an initial state m_0 and produces a new circuit $\hat{\Gamma}$ and a new initial state $\widehat{M_0}$.

The Original Circuit Γ. We assume that the original circuit Γ carries values from an (arbitrary) finite field $\mathbb{F}$ on its wires and is composed of the following gates (in addition to the memory gates which will be discussed later):

---

[7] The above can be generalized but we stick to this simple requirement for simplicity.

- $+, -$, and $*$, which compute, respectively, the sum, difference, and product in $\mathbb{F}$, of their two inputs; moreover, for every $\alpha \in \mathbb{F}$, the constant gate Const_α, which has no inputs and simply outputs α.
- the "coin flip" gate Rand, which has no inputs and produces a uniformly random independently chosen element of $\mathbb{F}$.

Fan-out in Γ is handled by a special Copy gate that takes as input a single value and outputs two copies. Circuits that only contain the above types of gates are called *stateless*.

Stateful Circuits. In addition to the gates described above, stateful circuits also contain memory gates, each of which has a single incoming and a single outgoing edge. Memory gates maintain state between the consecutive executions of the circuit. At any execution of the circuit (called a *round* or a *cycle*), a memory gate sends its current state down its outgoing edge and updates it according to the value of its incoming edge. Let m_i be the state of all memory gates of the circuit after the i-th round and m_0 be the initial state of the circuit. During the i-th round the circuit is run in the state m_{i-1} on the input x_i and the execution results in the output y_i and the new state m_i. The above execution will be denoted as $(y_i, m_i) \leftarrow \Gamma[m_{i-1}](x_i)$ for the circuit Γ. For instance, the state m_0 of an AES circuit may be its secret key.

The Transformed Circuit $\widehat{\Gamma}$. Our circuit transformation TR is encoding-based, i.e., it uses as a main building block an encoding scheme that is resilient to t-probing adversaries. TR takes as input (C, m_0) and outputs a protected state $\widehat{M_0}$ and the description of the protected circuit $\widehat{\Gamma}$. As in earlier work the transformation of the initial state m_0 is easy: instead of storing m_0 we store an encoding of m_0 using a leakage resilient encoding described in the previous section. We denote the transformed state by $\widehat{M_0}$. The transformation of the gates in Γ works gate-by-gate: each gate in the original circuit Γ is represented by a sub-circuit – a so-called *gadget* – that carries out the same computation as the corresponding gate in Γ in encoded form. Notice that the transformed circuit also uses special sub-circuits to encode the input x_i and decode the output of the circuit. As in previous works [20] we deal with this situation with so-called $\mathsf{Decoder}$ and $\mathsf{Encoder}$ gates. These gadgets are simple and just execute the underlying decoding, respectively, encoding function of the underlying leakage resilient encoding scheme.

2.3 Probing Attacks Against Circuits

As discussed in the introduction, we are interested in security against so-called t-region probing adversaries, i.e., adversaries that learn up to t wires in a region of a transformed circuit $\widehat{\Gamma}$. Typically, a region is a sub-circuit of size $O(n)$ (this is the same in the case of the work of [26]) of the transformed circuit, where in most cases in our transformation a region corresponds naturally to a transformed gadget. We will call a set of probes $\mathcal{P}$ t-region admissible if $\mathcal{P}$ contains at most t probes for each region of the transformed circuit.

Security against a t-region probing adversary is formalized by a simulation-based argument and given in Definition 1. To this end, we first define a real and ideal security game shown in Fig. 1. In the following, we use $\mathcal{W}_{\widehat{\Gamma}}(X|Y)$ to denote the wire assignment of $\widehat{\Gamma}$ when run on inputs $X = (x_i, \widehat{M}_{i-1})$ conditioned that the output is $Y = (y_i, \widehat{M}_i)$. The set $\mathcal{P}_i$ denotes the set of wires that the adversary wants to probe in the i-th clock cycle and $\mathcal{P}_i(\mathcal{W}_{\widehat{\Gamma}}(X|Y))$ the leakage during the i-th clock cycle.

Game $\mathsf{Real}_{\mathsf{TR}}(\mathcal{A}, n, \Gamma, m_0)$	**Game** $\mathsf{Ideal}_{\mathsf{TR}}(\mathsf{Sim}, \mathcal{A}, n, \Gamma, m_0)$
$(\widehat{\Gamma}, \widehat{M}_0) \leftarrow \mathsf{TR}(\Gamma, m_0)$	$(\widehat{\Gamma}, \widehat{M}_0) \leftarrow \mathsf{TR}(\Gamma, m_0)$
$(x_1, \mathcal{P}_1) \leftarrow \mathcal{A}(\widehat{\Gamma}, 1^n)$. Set $i = 1$.	$(x_1, \mathcal{P}_1) \leftarrow \mathcal{A}(\widehat{\Gamma}, 1^n)$. Set $i = 1$.
Repeat until the adversary $\mathcal{A}$ holds:	Repeat until the adversary $\mathcal{A}$ holds:
$\quad (y_i, \widehat{M}_i) \leftarrow \widehat{\Gamma}[\widehat{M}_{i-1}](x_i);$	$\quad (y_i, m_i) \leftarrow \Gamma[m_{i-1}](x_i)$
$\quad$ Set $X = (x_i, \widehat{M}_{i-1})$ and $Y = (y_i, \widehat{M}_i);$	$\quad \mathsf{Leak}_i \leftarrow \mathsf{Sim}(x_i, y_i, \mathcal{P}_i)$
$\quad (x_{i+1}, \mathcal{P}_{i+1}) \leftarrow \mathcal{A}(y_i, \mathcal{P}_i(\mathcal{W}_{\widehat{\Gamma}}(X|Y)))$	$\quad (x_{i+1}, \mathcal{P}_{i+1}) \leftarrow \mathcal{A}(y_i, \mathsf{Leak}_i)$
$\quad i = i + 1$	$\quad i = i + 1$
Output $\{\mathcal{P}_i(\mathcal{W}_{\widehat{\Gamma}}((x_i, \widehat{M}_{i-1})|(y_i, \widehat{M}_i)))\}_i$ and $\{(x_i, y_i)\}_i.$	Output $\{\mathsf{Leak}_i\}_i$ and the set $\{(x_i, y_i)\}_i.$

Fig. 1. The real world with the adversary $\mathcal{A}$ observing the computation of the transformed circuit $\widehat{\Gamma}[\widehat{M}_i]$ is shown on the left side. On the right side we describe the simulation.

Definition 1 (Security of Circuit Transformation). *Recall that n is the security parameter. A circuit transformation* TR *is (perfectly) t-region probing secure if for any t-region probing adversary $\mathcal{A}$ there exists a PPT simulator* Sim *such that for any (stateful) circuit Γ with initial state m_0 the distributions* $\mathsf{Real}_{\mathsf{TR}}(\mathcal{A}, n, \Gamma, m_0)$ *and* $\mathsf{Ideal}_{\mathsf{TR}}(\mathsf{Sim}, \mathcal{A}, n, \Gamma, m_0)$ *are identical, where the probabilities are taken over all the coin tosses.*

Leakage from Stateless Circuits. In spirit of earlier works on leakage resilient circuit compilers [20,26] the main difficulty for proving that a compiler satisfies Definition 1 is to show that leakage from stateless transformed circuits can be simulated with probing access to just its encoded inputs and outputs. In the following we will focus on proving such a simulation property for stateless circuits and only provide a high-level discussion how this property can be extended to prove that the circuit transformation is secure according to Definition 1.

We adapt the notion of *reconstructability* from Faust et al. [20] to the probing setting with perfect security. To this end we define a leakage oracle $\Omega(X^{(1)}, X^{(2)}, \ldots)$ for some sequence of encodings $(X^{(1)}, X^{(2)}, \ldots)$. The oracle can be queried on (i, j), and returns the value $X_j^{(i)}$, i.e., the j-th position of the i-th encoding. We will use the notation $\mathsf{Sim}^{\Omega(X^{(1)}, X^{(2)}, \ldots)}$ to denote the run of the simulator Sim with the access to the oracle $\Omega(X^{(1)}, X^{(2)}, \ldots)$. We call the simulator q-bounded if for each of the input encodings given to the oracle he queries at most for q different elements of the encoding.

Definition 2 ((t, q)-region reconstructible). *Let $\widehat{\Gamma}$ be a (transformed) state-less circuit with ς input encodings and producing τ output encodings. We say that a pair of strings (X, Y) is plausible for $\widehat{\Gamma}$ if $\widehat{\Gamma}$ might output $Y = (Y^{(1)}, \ldots, Y^{(\tau)})$ on input $X = (X^{(1)}, \ldots, X^{(\varsigma)})$, i.e., if $\Pr[\widehat{\Gamma}(X) = Y] > 0$. We say that $\widehat{\Gamma}$ is (t, q)-region reconstructible, if for any t-region admissible set of probes $\mathcal{P}$, there exists q-bounded simulator $\mathsf{Sim}_{\widehat{\Gamma}}$ such that for any plausible (X, Y), the following two distributions are identical: $\mathcal{P}(\mathcal{W}_{\widehat{\Gamma}}(X|Y))$ and $\mathsf{Sim}_{\widehat{\Gamma}}^{\Omega(X,Y)}(\mathcal{P})$.*

To better understand the above definition, consider the transformed multiplication gadget. The multiplication gadget takes as input two encoded inputs A, B and produces an encoding C such that $\mathsf{Dec}(\hat{C}) = \mathsf{Dec}(A) \cdot \mathsf{Dec}(B)$. If the multiplication gadget is (t, q)-region reconstructible, then we need to show that for any t-region admissible set of probes $\mathcal{P}$ and any plausible inputs/outputs $((A, B), C)$ there exists a q-bounded simulator Sim such that the following holds: $\mathcal{P}(\mathcal{W}_{\widehat{\Gamma}}((A, B)|C)) \equiv \mathsf{Sim}_{\widehat{\Gamma}}^{\Omega(A,B,C)}(\mathcal{P})$.

In addition to the region-reconstructible property we need that gadgets are *re-randomizing* [20]. Informally, this means that the output encoding of a gadget is independent from the input encodings, except that it encodes the correct result. Before we describe our new circuit compiler we present in the next section our new refreshing scheme that achieves optimal parameters both in terms of complexity and leakage rate.

3   Leakage Resilient Refreshing from Expander Graphs

A fundamental building block of any leakage resilient circuit compiler is a leakage resilient refreshing scheme. Informally, a refreshing scheme updates the encoding of a secret value such that continuous/repeated leakage from the execution of the refresh procedure does not reveal the encoded secret. More precisely, for a secret $x \in \mathbb{F}$ let $X \leftarrow \mathsf{Enc}(x)$ be an encoding of x. A refreshing scheme refresh is a randomized algorithm that takes as input X and outputs $Y \leftarrow \mathsf{refresh}(X)$ such that Y is a fresh encoding of x. Informally, a refreshing scheme refresh is said to be secure if even given continuous probing leakage from the refreshing procedure the adversary cannot distinguish the leakage from an encoding of any two secrets $x, x' \in \mathbb{F}$.

The refreshing procedure of [26] is described by a circuit of size $\Theta(n^2)$ which uses $\Theta(n^2)$ fresh random values per refresh execution and achieves security against a t-probing adversary when $n = 2t + 1$. While it is easy to construct refreshing schemes that achieve security against a $O(1/n)$ fraction of probes per execution, it appears to be much harder to construct a refreshing scheme that achieves the optimal size of $\Theta(n)$ and requires only $\Theta(n)$ random field elements while tolerating $t = \Omega(n)$ probes. This is quite surprising as various candidate schemes look secure at first sight.

As outlined in the introduction the main ingredient of our refreshing scheme (and essentially of most leakage resilient refreshing schemes) is a method to sample form $\mathsf{EncAE}(0)$. Given a "leakage resilient way" to sample $R \leftarrow \mathsf{EncAE}(0)$

we can implement a refreshing algorithm in a simple way: to refresh $X^{(i-1)}$ we compute $X^{(i)} = X^{(i-1)} + R$, where R is sampled from $\mathsf{EncAE}(0)$. Our construction to sample from $\mathsf{EncAE}(0)$ uses a undirected expander graph $\mathcal{G} = (V, E)$, with $V = \{1, \ldots, n\}$ (see, e.g., [24] for an excellent exposition). Informally speaking expander graphs are sparse graphs with strong connectivity properties. Let $\mathcal{G} = (V, E)$ be an undirected graph with V being the set of vertices and E being a set of edges (hence E is a multiset). Assume $\mathcal{G}$ can have self-loops and parallel edges. We define the *edge expansion* of the graph as: $\min_{S \subset V : |S| \leq |V|/2} \frac{|\partial(S)|}{|S|}$, where $\partial(S)$ denotes the *edge boundary of S in $\mathcal{G}$*, i.e., the set of edges with exactly one endpoint in S. We say that an undirected d-regular graph $\mathcal{G}$ is an (d, h)-*expander* if $d > 0$ and its edge expansion is at least h.

To describe our construction we will write the edges of $\mathcal{G}$ as ordered pairs (i, j) where always $i \leq j$. Given such a $\mathcal{G}$ one can construct an arithmetic circuit $\mathsf{RefSamp}_{\mathcal{G}}(1^n)$ (over some additive field $\mathbb{F}$) that produces random additive encodings $(X_1, \ldots, X_n)$ of zero. This is done as follows. The circuit $\mathsf{RefSamp}_{\mathcal{G}}(1^n)$ consists of $|E|$ coin flip gates Rand — to each $e \in E$ we associate one of them. Let r_e denote the output of each Rand_e. To compute the encoding $(X_1, \ldots, X_n)$ we start with each $X_i := 0$ and for every edge $(i, j) \in E$ we add r_e to it, and for every edge $(j, i) \in E$ we subtract r_e from it. In other words each X_i is defined as follows:

$$X_i := \sum_{(i,j) \in E} r_{(i,j)} - \sum_{(j,i) \in E} r_{(j,i)}. \tag{1}$$

The gate-level implementation of the sum computations in (1) is pretty straightforward: we attach a Copy gate to each Rand_e gate. Let v_e and w_e be the output wires of this gate. Then we sum and subtract the appropriate v_e's and w_e's in order to compute the sums in (1). It is easy to see that every $r_{(i,j)}$ is counted twice in the sum $X_1 + \ldots + X_n$: once with a "plus" sign (for the vertex i), and once with a "minus" sign (for the vertex j). Therefore $(X_1, \ldots, X_n)$ is an additive encoding of zero.

3.1 Reconstructibility of $\mathsf{RefSamp}_{\mathcal{G}}$

In this section we show that the circuit $\mathsf{RefSamp}_{\mathcal{G}}$ is (t, q)-region reconstructible for an appropriate choice of t and q. To this end, we start by giving some useful properties about the connectivity of expander graphs and the circuit $\mathsf{RefSamp}_{\mathcal{G}}$. Recall that a connected component of a graph is a subgraph in which any two vertices are connected to each other by a path, and which is connected to no additional vertices. It will be useful to analyze the properties of expanders and their connected components when some number T of their edges is removed (for some parameter T). Call a set of vertices $S \subset V$ *small* if $|S| \leq T/h$, call it *medium* if $T/h < |S| < n - T/h$, and call it *large* otherwise. We can then show the following simple lemma about the sizes of connected components when T vertices are removed from the expander graph. We can then show the following simple lemma about the sizes of connected components when T vertices are removed from the expander graph.

Lemma 1. *Suppose $T < nh/3$ and $\mathcal{G}$ is an (d, h)-expander. Let $\mathcal{G}'$ be an arbitrary graph that resulted from removing up to T edges from $\mathcal{G}$. Then $\mathcal{G}'$ contains exactly one large connected component.*

Proof. We first prove that $\mathcal{G}'$ contains no medium components. We actually show something slightly stronger, namely, that for every medium subset of vertices S there exists an edge in $\mathcal{G}'$ between S and $V \setminus S$. Take such a medium S and consider two cases. First, assume that $S \leq n/2$. From the definition of edge expansion we get that the number x of edges between S and $V \setminus S$ in the original graph $\mathcal{G}$ is equal at least $h \cdot |S|$. Since we assumed that S is medium, thus $|S| > T/h$, and hence $x > T$. It is also easy to see that if $|S| > n/2$, then we can use a symmetric reasoning, as $|S| < n - T/h$ implies that $|V \setminus S| > T/h$. Hence, also in this case we get that $x \geq h \cdot |V \setminus S| > T$. In other words: that there are more than T edges between S and $V \setminus S$ in $\mathcal{G}$. Thus, even if we remove at most T edges from $\mathcal{G}$ there is still one edge remaining. Hence there must be an edge between S and $V \setminus S$ in $\mathcal{G}'$.

Therefore $\mathcal{G}'$ cannot have medium connected components, and hence each connected component has to be either small or large. Recall that we defined a large subgraph to have more than $n - T/h$ vertices. Since we assumed that $T < nh/3$, which implies that $T/h < n/3$, thus a large connected component must have more than $2n/3$ vertices, which means that there can be at most one such a component (as obviously two connected components cannot overlap). To finish the proof we need to show that there is at least one large component. For the sake of contradiction suppose there is no large connected component. Hence, all the connected components need to be small. Let $V_1, \ldots, V_m \subset V$ be these small components. Obviously $|V_1 \cup \cdots \cup V_m| = n$. Since each V_i is such that $|V_i| \leq T/h < n/3$, thus there has to exists j such that $n/3 < |V_1 \cup \cdots \cup V_j| < 2n/3$. Hence $V_1 \cup \cdots \cup V_j$ is a medium set. Therefore, from what we have shown at the beginning of this proof, there has to exist an edge in $\mathcal{G}'$ connecting this union with a vertex outside of it. Hence at least one of the sets $V_1, \ldots, V_j$ cannot be a connected component. This yields a contradiction. $\square$

We next give a lemma that states that after removing edges from the expander graph, the circuit induced by the remaining connected component results into a random additive encoding of a fixed constant value. More technically, we have:

Lemma 2. *Suppose $\mathcal{G}^* = (V^*, E^*)$ is a connected subgraph of $\mathcal{G}$, where $\mathcal{G}$ is as in Lemma 1. Let $(X_1, \ldots, X_n) \leftarrow \mathsf{RefSamp}_{\mathcal{G}}(1^n)$ and let $v_1 \leq \cdots \leq v_m$ be the elements of V^*. Consider an adversary $\mathcal{A}$ that learns all r_e's corresponding to $\mathcal{G}$'s edges that are not in $\mathcal{G}^*$. Note that in particular he knows X_v for every $v \notin V^*$ and can compute $C = \sum_{v \notin V^*} X_v$. Then, from $\mathcal{A}$'s point of view $(X_{v_1}, \ldots, X_{v_m})$ is distributed uniformly over the set $U_m^{-C} := \{(x_{v_1}, \ldots, x_{v_m}) : x_{v_1} + \cdots + x_{v_m} = -C\}$.*

Before we give a proof of Lemma 2 let us first show that the expander based-construction indeed outputs random encodings of 0. To this end, we need the following auxiliary lemma.

Lemma 3. *Let $\mathcal{G}^* = (V^*, E^*)$ be a graph as above except that the set of vertices is a subset of $\{1, \ldots, n\}$. Let $v_1 \leq \cdots \leq v_m$ be the elements of V^*. Suppose $\mathcal{G}^*$ is connected. Then the variable $(Y_{v_1}, \ldots, Y_{v_m}) \leftarrow \mathsf{RefSamp}_{\mathcal{G}^*}(1^n)$ is distributed uniformly over the set $U_m^0 := \{(y_{v_1}, \ldots, y_{v_m}) \in \mathbb{F}^m : y_{v_1} + \ldots + y_{v_m} = 0\}$.*

This fact will be useful, since if $\mathcal{G}'$ results from removing some edges from an expander, then (by Lemma 1) it is guaranteed to contain a large connected component $\mathcal{G}^*$, and hence the variables $Y_{v_1}, \ldots, Y_{v_m}$ obtained by "summing" the r_e's from $\mathcal{G}^*$ will have a uniform distribution over U_m^0.

Proof (of Lemma 3). Induction over m. Consider the base case $m = 1$ first. In this case $\mathcal{G}^*$ contains one node v and no edges. Then clearly $Y_v = 0$ what is distributed uniformly over the set $U_m^0 = \{0\}$.

Now suppose we know that the lemma holds for some m, and let us prove it for $m + 1$. Let v be an arbitrary leaf in an arbitrary spanning tree of $\mathcal{G}^*$. Notice that the graph $\mathcal{G}^*$ with the vertex v (and all edges adjacent to it) removed is connected. To simplify the notation we will assume that $v = v_{m+1}$. Let $R_1, \ldots, R_b$ be all the values produced by the Rand gates corresponding to the edges in $\mathcal{G}^*$ with one endpoint being v_{m+1}. Clearly $Y_{v_{m+1}} = -\sum_{i=1}^b R_i$, and hence it is uniform. On the other hand, by the induction hypothesis $(Y_{v_1}, \ldots, Y_{v_m})$ is uniformly distributed over U_m^0 if one does not consider the edges going to v_{m+1}, i.e., if one does not count the values $R_1, \ldots, R_b$ in the sums. Therefore, if we consider also these values then $(Y_{v_1}, \ldots, Y_{v_m})$ will be uniformly distributed over the set $\{(y_{v_1}, \ldots, y_{v_m}) : y_{v_1} + \cdots + y_{v_m} = \sum_{i=1}^b R_i\}$. Hence, altogether $(Y_{v_1}, \ldots, Y_{v_{m+1}})$ is uniformly distributed over U_{m+1}^0. This concludes the proof. $\square$

The Lemma 2 is a consequence on Lemma 3. The proof is given below.

Proof (of Lemma 2). Look at the graph $\mathcal{G}^{**} := (V, E \setminus E^*)$. Each X_{v_i} can be expressed as $X_{v_i}^* + X_{v_i}^{**}$, where $X_{v_i}^*$ and $X_{v_i}^{**}$ denote the sum of r_e's from respectively $\mathcal{G}^*$ and $\mathcal{G}^{**}$. Since all r_e's that correspond to the edges of $\mathcal{G}^{**}$ are known to $\mathcal{A}$, thus for each v_i he can compute $X_{v_i}^{**}$. Clearly $X_{v_1}^{**} + \cdots + X_{v_m}^{**} = -C$. Moreover, by Lemma 3 the distribution of $(X_{v_1}^*, \ldots, X_{v_m}^*)$ is uniform over U_m^0. Hence the distribution of $(X_{v_1}, \ldots, X_{v_m})$ is uniform over U_m^{-C}. $\square$

Finally, we need the following simple fact, where we denote by $\mathrm{Pr}_{X|Y}$ the conditional distribution of X conditioned on Y.

Lemma 4. *Consider an execution of $(X_1, \ldots, X_n) \leftarrow \mathsf{RefSamp}_{\mathcal{G}}(1^n)$. Let $\{R_e\}_{e \in E}$ denote the random variables corresponding to the r_e values in the circuit computing $\mathsf{RefSamp}_{\mathcal{G}}(1^n)$. Take some $W \subseteq \{1, \ldots, n\}$. Then there exists an efficient procedure that for every input $\{x_i\}_{i \in W}$ produces as output $\{r'_e\}_{e \in E}$ distributed according to the conditional distribution $\mathrm{Pr}_{\{R_e\}_{e \in E} | \forall_{i \in W} X_i = x_i}$.*

Proof. Clearly every X_i is a linear combination o the r_e's. Hence the condition $\forall_{i \in W} X_i = x_i$ can be understood as a system of linear equations (with r_e's being the unknowns), and the set of its solutions is a linear subspace $\mathcal{L}$ whose base can be efficiently computed. To sample a random value of $\mathrm{Pr}_{\{R_e\}_{e \in E} | \forall_{i \in W} X_i = x_i}$ one can simply output a uniform vector from $\mathcal{L}$. $\square$

We are now ready to prove our first technical theorem.

Theorem 1. *Let $n \in \mathbb{N}$ be the security parameter, $\mathcal{G} = (V, E)$ be a d-regular graph with edge expansion $h > 0$. Then for any $t < \frac{nh}{3d}$ the gadget $\mathsf{RefSamp}_{\mathcal{G}}$ treated as one region is (t, q)-region reconstructible for $q = \lfloor td/h \rfloor$.*

The simulator $\mathsf{SimRefSamp}_{\mathcal{G}}^{\Omega(X)}(\mathcal{P})$

1. Compute the set of *compromised edges* $L \subset E$. This set consists of all edges $e \in E$ for which:
   (a) at least one of the input or output wires of the corresponding Copy gate leaks (i.e. is included in the set $\mathcal{P}$), or
   (b) at least one output wires of a $+$ or $-$ gate corresponding to a node incident to e leaks.
2. Compute the graph consisting of uncompromised edges $\mathcal{G}' = (V, E \setminus L)$.
3. Compute the largest (in terms of a number of nodes) connected component in $\mathcal{G}'$ and denote it $\mathcal{G}^* = (V^*, E^*)$.
4. Obtain the values X_i for each $i \in V \setminus V^*$ by querying the oracle $\Omega(X)$. Denote the leaked value of X_i as x_i.
5. Using the procedure from Lemma 4 draw a sequece $\{r_e\}_{e \in E}$ from a conditional distribution $\mathrm{Pr}_{\{R_e\}_{e \in E} | \forall_{i \in V \setminus V^*} X_i = x_i}$.
6. Simulate the execution of $\mathsf{RefSamp}_{\mathcal{G}}(1^n)$ assuming that Rand gates has outputted the sequence $\{r_e\}_{e \in E}$. Let $\mathcal{W}$ be the obtained wire assignment.
7. Output $\mathcal{P}(\mathcal{W})$.

Fig. 2. The $\mathsf{SimRefSamp}_{\mathcal{G}}$ simulator for $\mathsf{RefSamp}_{\mathcal{G}}(1^n)$.

Proof. Let X be a plausible output of $\mathsf{RefSamp}_{\mathcal{G}}(1^n)$, i.e., $\sum_i X_i = 0$. The simulator $\mathsf{SimRefSamp}_{\mathcal{G}}^{\Omega(X)}$ has to simulate the leakage from a t-admissible set of probes $\mathcal{P}$ from the execution of $X \leftarrow \mathsf{RefSamp}_{\mathcal{G}}(1^n)$ with only q-bounded access to its oracle $\Omega(X)$ where $q = \lfloor td/h \rfloor$. We will sketch it now informally, the full description is presented on Fig. 2. The simulator $\mathsf{SimRefSamp}_{\mathcal{G}}$ computes the set of edges $L \subset E$ s.t. the values of the random gates associated with the edges from L are sufficient to compute the values on all leaking wires. Then, it computes the large connected subgraph $\mathcal{G}^* = (V^*, E^*)$ such that the output variables with indices in V^* are independent of the leakage, he then probes the output variables with X_i for $i \in V \setminus V^*$ from its oracle $\Omega(X)$, and simulates a random execution consistent with the probed values X_i.

We start by proving that $\mathsf{SimRefSamp}_{\mathcal{G}}$ is indeed $\lfloor td/h \rfloor$-bounded. To this end we analyse the possible sizes of connected components in the graph $\mathcal{G}'$. It is easy to see that each wire that is revealed according to the set of probes $\mathcal{P}$ increases the set L by at most d elements, and therefore $|L| \leq td$. Since we assumed that $t < nh/(3d)$, thus $|L| < nh/3$. We can therefore apply Lemma 1 to $\mathcal{G}'$ with $T = |L|$. In this way we obtain that the number of vertices in the largest component $\mathcal{G}^*$ in $\mathcal{G}'$ is at least $n - |L|/h$, which is clearly at least $n - td/h$. Therefore the number of vertices in $V \setminus V^*$ is smaller than td/h. Since these are exactly the indexes probed by the simulator $\mathsf{SimRefSamp}_\mathcal{G}$, thus it is $\lfloor td/h \rfloor$-bounded.

The definition of reconstructability states that for each fixed $X_1, \ldots, X_n$ s.t. $X_1 + \ldots + X_n = 0$ the distribution of the leakage in the execution of the real circuit $\mathsf{RefSamp}_\mathcal{G}(1^n)$ assuming that it outputted the sequence $X_1, \ldots, X_n$ is identical to the distribution produced by the simulator $\mathsf{SimRefSamp}_\mathcal{G}(\mathcal{P})$ that uses q-probing leakage from $X_1, \ldots, X_n$. This is equivalent to saying that the *joint* distribution of the output $(X_1, \ldots, X_n)$ and the leakage is the same in the real and simulated case (this follows from the definition of conditional probability). Let us define the two joint distributions more formally by considering two experiments. In the first one the values $(X_1^{\mathsf{REAL}}, \ldots, X_n^{\mathsf{REAL}})$ are sampled using $X^{\mathsf{REAL}} \leftarrow \mathsf{RefSamp}_\mathcal{G}(1^n)$ and the leakage obtained from this execution is denoted by $\mathcal{P}(\mathcal{W}_{\mathsf{RefSamp}_\mathcal{G}}(X^{\mathsf{REAL}}))$, where $\mathcal{P}$ are a set of probes that is t-region admissible. In the simulated case the values $(X_1^{\mathsf{SIM}}, \ldots, X_n^{\mathsf{SIM}})$ are drawn using $X^{\mathsf{SIM}} \leftarrow \mathsf{EncAE}(0)$ and the leakage is computed by the simulator leaking from X^{SIM} and denoted $\mathsf{SimRefSamp}_\mathcal{G}^{\Omega(X^{\mathsf{SIM}})}(\mathcal{P})$. Hence, we need to show that

$$\left(X^{\mathsf{SIM}}, \mathsf{SimRefSamp}_\mathcal{G}^{\Omega(X^{\mathsf{SIM}})}(\mathcal{P}) \right) \equiv \left(X^{\mathsf{REAL}}, \mathcal{P}(\mathcal{W}_{\mathsf{RefSamp}_\mathcal{G}}(X^{\mathsf{REAL}})) \right).$$

First observe that

$$\left(X_{V \setminus V^*}^{\mathsf{SIM}}, \mathsf{SimRefSamp}_\mathcal{G}^{\Omega(X^{\mathsf{SIM}})}(\mathcal{P}) \right) \tag{2}$$

and

$$\left(X_{V \setminus V^*}^{\mathsf{REAL}}, \mathcal{P}(\mathcal{W}_{\mathsf{RefSamp}_\mathcal{G}}(X^{\mathsf{REAL}})) \right) \tag{3}$$

are identically distributed. This is because $\mathsf{SimRefSamp}_\mathcal{G}^{\Omega(X^{\mathsf{SIM}})}(\mathcal{P})$ is computed based on the perfect simulation given in Lemma 4 using the values $X_{V^* \setminus V}^{\mathsf{SIM}}$, which are leaked and hence distributed appropriately. Let U_m^{-C} be as in Lemma 2. Clearly, given (2) the remaining values $X_{V^*}^{\mathsf{SIM}}$ have a uniform distribution over the set $U_{|V^*|}^{-C}$, where $C = \sum_{i \in V \setminus V^*} X_i^{\mathsf{SIM}}$, because they have not been leaked. By Lemma 2 also $X_{V^*}^{\mathsf{REAL}}$ have a uniform distribution over the set $U_{|V^*|}^{-C}$, where $C = \sum_{i \in V \setminus V^*} X_i^{\mathsf{REAL}}$ given $X_{V \setminus V^*}^{\mathsf{REAL}}$ and all the r_e values corresponding to the edges in the set $E \setminus E^*$. Since these values fully determine $\mathcal{P}(\mathcal{W}_{\mathsf{RefSamp}_\mathcal{G}}(X^{\mathsf{REAL}}))$ thus $X_{V^*}^{\mathsf{REAL}}$ have a uniform distribution over the set $U_{|V^*|}^{-C}$ given (3). This finishes the proof. $\qquad\square$

4   Circuits for Affine Computation

In this section we build a circuit transformation $\mathsf{TR}^{\mathsf{Aff}}$ that allows to transform arbitrary circuits implementing affine computation into protected circuits that are resilient to t-region probing adversaries. In the transformed circuit $\widehat{\Gamma}$ that are produced by $\mathsf{TR}^{\mathsf{Aff}}$ each region is represented by a gadget. Hence, if the original circuit Γ has size s then $\widehat{\Gamma} \leftarrow \mathsf{TR}^{\mathsf{Aff}}(1^n, \Gamma)$ has s regions. Notice that we assume that the input and output encoding of each gadget are part of two consecutive regions, and consequently the adversary may leak twice from them.

4.1   The Transformation $\mathsf{TR}^{\mathsf{Aff}}$

Our transformation $\mathsf{TR}^{\mathsf{Aff}}$ is an encoding-based transformation as described in Sect. 2.2. The transformation uses as building blocks the additive encoding scheme Π^{AE}. The initial state m_0 of the original circuit Γ will be stored in encoded form using the code Π^{AE}, i.e., $\widehat{M_0} \leftarrow \mathsf{EncAE}(1^n, m_0)$. One can view the encoded state as an initial encoded input that is given as input to the transformed circuit, and hence security of stateful circuits is just a special case of security of stateless circuits.

We need to define transformations for the basic operations of affine computation. Let Γ be a circuit that takes ς inputs $x_1, \ldots, x_\varsigma$ and produces τ outputs $y_1, \ldots, y_\tau$. The outputs are computed from the inputs using solely the following types of operations:

1. Addition in $\mathbb{F}$ and multiplication by a (known) constant $x \in \mathbb{F}$.
2. The randomness gate Rand that outputs a random element $r \in \mathbb{F}$.
3. The constant gate Const_x that for a constant $x \in \mathbb{F}$ outputs x.
4. The copy gate Copy that for input x outputs two wires carrying the value x. Notice that the Copy gate in Γ is needed for fan-out.

Our transformation $\mathsf{TR}^{\mathsf{Aff}}$ is very simple. Each gate of the above form is replaced by a gadget from Fig. 3. The wires connecting the gadgets are called *wire bundles* and carry the corresponding encoding of the values using the code Π^{AE}. The gadgets presented in Fig. 3 use as a sub-circuit $X \leftarrow \mathsf{RefSamp}_{\mathcal{G}}(1^n)$ for some expander graph $\mathcal{G}$. In the following, we will omit to mention explicitly the graph $\mathcal{G}$ and assume that $\mathcal{G}$ is d-regular with edge expansion h. We will assume that it is fixed once and for all.

4.2 (t, q)-reconstructability of Gadgets in $\mathsf{TR}^{\mathsf{Aff}}$

In this section, we show that the operations of $\mathsf{TR}^{\mathsf{Aff}}$ from Fig. 3 are (t, q)-region reconstructible and re-randomizing. The proofs are given in the full version.

Lemma 5. *Recall that $n \in \mathbb{N}$ is the security parameter and let d and h be constants defining the underlying expander graph on n vertices. For any $t < \frac{nh}{3d}$ we set $q = \lfloor td/h \rfloor$. The gadget PlusAE is (t, q)-region reconstructible and re-randomizing, where the region is defined by the gadget itself.*

We can also show that the remaining gates are region reconstructible.

Lemma 6. *The gadgets MultAE_x, $\mathsf{ConstAE}_x$, CoinAE and CopyAE are (t, q)-region reconstructible and re-randomizing, where the region is defined by each gadget itself.*

The gadgets of the transformation $\mathsf{TR}^{\mathsf{Aff}}$

1. *Transformation for addition in $\mathbb{F}$, i.e., $a + b = c$:* An addition operation in the circuit $\widehat{\Gamma}$ is handled by the gadget $C \leftarrow \mathsf{PlusAE}(A, B)$. On input encodings A, B it computes $Z = A + B$ and samples $Y \leftarrow \mathsf{RefSamp}(1^n)$. Then, it outputs $C = Z + Y$.

2. *Transformation for multiplication with a constant $x \in \mathbb{F}$, i.e., $xa = c$:* Multiplication with a constant $x \in \mathbb{F}$ is handled by the gadget $C \leftarrow \mathsf{MultAE}_x(A)$. For the fixed constant x and on input encoding A, it computes $Z = xA$ (by component-wise multiplication) and samples $Y \leftarrow \mathsf{RefSamp}(1^n)$. Then, it outputs $C = Z + Y$.

3. *Transformation of Rand gate $x \leftarrow \mathbb{F}$:* The transformation for sampling a random element in $\mathbb{F}$ is denoted by $C \leftarrow \mathsf{CoinAE}(1^n)$ in $\widehat{\Gamma}$. The circuit uses n coin gates $C_i \leftarrow \mathsf{Rand}$ and outputs $C = (C_1, \ldots, C_n)$.

4. *Transformation of Const_x gate for some $x \in \mathbb{F}$:* For some x the gadget $C \leftarrow \mathsf{ConstAE}_x$ computes $X = (\mathsf{Const}_x, \mathsf{Const}_0, \ldots, \mathsf{Const}_0)$. Then, it samples $Y \leftarrow \mathsf{RefSamp}(1^n)$ and outputs $C = X + Y$.

5. *Transformation of Copy gate:* The fan-out in $\widehat{\Gamma}$ is handled using the gadget $(B, C) \leftarrow \mathsf{CopyAE}(A)$. On input encoding A, it samples $Y \leftarrow \mathsf{RefSamp}(1^n)$ and $Z \leftarrow \mathsf{RefSamp}(1^n)$. Then, it outputs $B = A + Y$ and $C = A + Z$.

Fig. 3. The transformation $\mathsf{TR}^{\mathsf{Aff}}$ has gadget transformations for each of the elementary operations. $\mathsf{RefSamp}(1^n)$ samples $\mathsf{EncAE}(0)$ using an expander graph.

4.3 Security of Composed Circuits

In this section we discuss briefly that arbitrary composed circuits build from the transformed gadgets defined in Sect. 4.2 are (t, q)-region reconstructible, where in the composed transformed circuit $\widehat{\Gamma}$ each gadget corresponds to a region. We state the lemma in a slightly more general form (similar to Lemma 13 from [20]). This will allow us to later apply it when we consider circuits that are made out of arbitrary transformed gadgets.

Lemma 7. *Recall that n is the security parameter and q and t are functions in n. Let Γ be a stateless circuit over some finite field $\mathbb{F}$ with ς inputs, τ outputs and s gates. Assume that the gates in Γ all have fan-in and fan-out at most 2 elements in $\mathbb{F}$. Let $\Pi = (\mathsf{Enc}, \mathsf{Dec})$ be a $2q$-probing resilient code. Let $\widehat{\Gamma} \leftarrow \mathsf{TR}(1^n, \Gamma)$ be the transformation of Γ based on $\Pi = (\mathsf{Enc}, \mathsf{Dec})$ and let $\widehat{\Gamma}$ be composed from (t, q)-probing reconstructible and re-randomizing gadgets, then $\widehat{\Gamma}$ is (t, q)-probing reconstructible and re-randomizing.*

The proof uses a hybrid argument and is provided in the full version. The above lemma together with Lemmas 5 and 6 immediately implies that any stateless circuit $\widehat{\Gamma} \leftarrow \mathsf{TR}^{\mathsf{Aff}}(1^n, \Gamma)$ is (t, q)-reconstructible for choices of t and q that are given in the lemma below.

Lemma 8. *Recall that n is the security parameter and let d and h be constants defining the underlying expander graph on n vertices. Let Γ be a stateless circuit over field $\mathbb{F}$ using only affine operations. Then, the transformed circuit $\widehat{\Gamma} \leftarrow \mathsf{TR}^{\mathsf{Aff}}(1^n, \Gamma)$ is re-randomizable and (t, q)-reconstructible for $t < \frac{nh}{3d}$ and $q = \lfloor td/h \rfloor$ and regions that correspond to gadgets in $\widehat{\Gamma}$.*

It is easy to see that all transformed gadgets have size $O(n)$ which together with $t < \frac{nh}{3d}$ for constants h and d asymptotically shows that a *constant fraction* of all wires in $\widehat{\Gamma}$ can be learnt by the adversary.

5   Circuits for Arbitrary Computation

To protect non-affine computation, we also need a transformation for multiplication in the underlying field. Before we present our transformation TR, we first discuss a special protected circuit called $\mathsf{RandSamp}(1^n)$ that is mostly produced by $\mathsf{TR}^{\mathsf{Aff}}$ and will be used in the transformed multiplication operation as an important building block. In the following, for some $\tau \in \mathbb{N}$ we let $n = 2\tau + 1$ be the security parameter and require that $|\mathbb{F}| > n$ such that we can use the coding scheme $\Pi_{n,\tau}^{\mathsf{MSS}} = (\mathsf{EncMSS}_{n,\tau}, \mathsf{DecMSS}_{n,\tau})$ based on Shamir secret sharing as described in Sect. 2.1 (as we mentioned we can use other encoding schemes to improve the asymptotic complexity of our construction).

5.1   The Circuit RandSamp

The goal of the circuit $\mathsf{RandSamp}(1^n)$ is to sample correlated randomness that can be used in the transformed multiplication operation even in the presence of a t-region probing adversary. More precisely, the randomized circuit $\mathsf{RandSamp}(1^n)$ takes no inputs and outputs two random encodings $[r]_\tau \leftarrow \mathsf{EncMSS}_{n,\tau}(r)$ and $[r]_{2\tau} \leftarrow \mathsf{EncMSS}_{n,2\tau}(r)^8$, where $r \leftarrow \mathbb{F}$ is a uniformly and independently chosen field element. The main difficulty is to ensure that the computation of $[r]_\tau$ and $[r]_{2\tau}$ do not reveal anything about r even in the presence of a t-region probing adversary. Hence, the goal is to design a circuit that samples these two encodings in an *oblivious* way. Our main observation that enables us to achieve this goal is the fact that $[r]_\tau$ and $[r]_{2\tau}$ can be computed (in a natural way) by an affine circuit Γ' that can be protected using $\mathsf{TR}^{\mathsf{Aff}}$.

A technical difficulty is that the sub-circuit $\mathsf{RandSamp}'(1^n) \leftarrow \mathsf{TR}^{\mathsf{Aff}}(1^n, \Gamma')$ outputs additive encodings of $([r]_\tau, [r]_{2\tau})$, i.e., $(\mathsf{EncAE}([r]_\tau), \mathsf{EncAE}([r]_{2\tau}))$. The protected multiplication operation, however, requires access to $([r]_\tau, [r]_{2\tau})$. To decode one level of the "double-encoding" and obtain the final circuit $\mathsf{RandSamp}$, we append two MultiDecoder sub-circuits to the output of $\mathsf{RandSamp}'$ to decode $\mathsf{EncAE}([r]_\tau)$ and $\mathsf{EncAE}([r]_{2\tau})$, respectively. A MultiDecoder sub-circuit takes as input a double encoding $\mathsf{EncAE}([r]_\tau)$ and outputs $[r]_\tau$ by "peeling off" one layer of the code. More precisely, we let $(U_1, \ldots, U_n) := [r]_\tau$ and $(X^{(1)}, \ldots, X^{(n)}) := (\mathsf{EncAE}(U_1), \ldots, \mathsf{EncAE}(U_n))$. The deterministic MultiDecoder circuit takes as input $(X^{(1)}, \ldots, X^{(n)})$ and outputs $(U_1, \ldots, U_n)$. To this end, it runs n Decoder sub-circuits (corresponding to the decoding function of the code $\Pi_{n,\mathbb{F}}^{\mathsf{AE}}$), where each such sub-circuit takes as input an encoding $X^{(i)}$ and outputs U_i. For the security of our construction it will be important that each such Decoder circuit

---

[8] We present here the parameters n, t to indicate that the value $[r]_{2\tau}$ comes from the encoding $\Pi_{n,2\tau}^{\mathsf{MSS}}$ (and not $\Pi_{n,\tau}^{\mathsf{MSS}}$).

computes the sum of the shares in a natural way by representing the summation as a binary tree. More precisely, the shares of $X^{(i)}$ represent the leaves of the tree, the internal nodes of the tree correspond to the sum of the values assigned to its children and the root is the corresponding result of the decoding procedure. The high-level structure of the RandSamp circuit is given in Fig. 4.

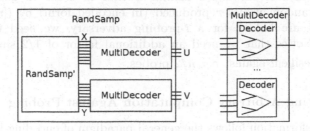

Fig. 4. The architecture of the RandSamp and MultiDecoder circuit. The RandSamp circuit consists of the RandSamp$'$ sub-circuit and two MultiDecoder sub-circuits. Each MultiDecoder circuit consists of n Decoder sub-circuits. Notice that regions in the MultiDecoder circuit does *not* correspond to the Decoder sub-circuits. More precisely, each region in the MultiDecoder circuit consists of n wires — one in each of the Decoder sub-circuits, such that each of them correspond to the same edge in the summing tree. For example, both dotted wires on the figure belong to the same region.

It remains to discuss how RandSamp is structured into regions. First notice that for RandSamp$'$ the structure of the regions is inherited from the compiler TR^{Aff}. Hence, the regions in RandSamp$'$ correspond to a transformed gadget in RandSamp$'$. Next, notice that each of the decoder sub-circuits MultiDecoder has size $\Theta(n^2)$, and we need an appropriate way to structure its computation into regions of size $\Omega(n)$. To illustrate, why for the MultiDecoder we cannot use a natural representation where each region corresponds to a computation of one output value U_i, consider the following example. Let the decoding process of the n encodings be structured into n regions, where each region corresponds to a Decoder gadgets that decodes $X^{(i)}$ into U_i. Unfortunately, however, it is easy to see that already a single probe in each such region allows the adversary to learn the entire output of the MultiDecoder circuit, i.e., the adversary may learn U_i in the i-th region, which allows to recover the secret value r. To prevent this attack, we instead structure the computation of the MultiDecoder in regions of size $O(n)$, where each region corresponds to one node (or one edge) in each of the n Decoder trees.[9] Recall that the MultiDecoder consists of n Decoder trees. The i-th region in MultiDecoder contains the wires associated with the output of the i-th gate in each of the n Decoder trees. Given the above structuring into regions, we can show the following property about the RandSamp circuit.

---

[9] Notice that in reality regions constitute a partition of *wires*, not *gates*. Whenever we say that a particular gate is in a particular region, it simply means that gate's output is in that region.

Lemma 9. *Recall that $n \in \mathbb{N}$ is the security parameter and let d and h be constants defining the underlying expander graph on n vertices. For any $t < \frac{nh}{3d}$ the circuit* RandSamp(1^n) *is (t, q)-region reconstructible for $q = \frac{3}{2}t(\lceil \log(n) \rceil + 1)$, where the regions are defined as described above in the description of* RandSamp$'$ *and the* MultiDecoder *sub-circuit. Moreover,* RandSamp *has circuit size $O(n^3)$.*

A consequence of the above lemma is that in order to guarantee that the encoded random values r produced (in encoded form) by $([r]_\tau, [r]_{2\tau}) \leftarrow$ RandSamp(1^n) are hidden for a t-probing adversary, we need to set: $t < \frac{n}{3(\lceil \log(n) \rceil + 1)}$. Notice that we need an additional factor of $1/2$ since the code Π^{MSS} is only resilient against $\tau < n/2$ probes.

5.2   Protecting Arbitrary Computation Against Probing

Our final transformation follows the general paradigm of encoding-based circuit transformations from Sect. 2, where we use as the underlying code the scheme $\Pi^{\mathsf{MSS}}_{n,\tau}$ with $n = 2\tau + 1$. The initial state m_0 of the circuit is transformed into $\widehat{M_0} \leftarrow$ EncMSS(m_0), and the wires in Γ are represented in $\widehat{\Gamma}$ by wire bundles carrying an encoding of the value carried on the wire in Γ. The transformation for the individual operations is presented in Figs. 5, 6 and 7. In Fig. 5, we present the main ingredient of our new transformation – the transformation for the multiplication operation, which we describe in further detail below.

The $C \leftarrow$ MultSS(A, B) gadget of TR

On input $(A, B) = ([a]_\tau, [b]_\tau)$ proceed as follows:
1. Sample $(U, V) = ([r]_\tau, [r]_{2\tau}) \leftarrow$ RandSamp(1^n) for some random $r \in \mathbb{F}$ as described in Section 5.1. Notice that in a real circuit RandSamp can be implemented by a single sub-circuit that is queried by all MultSS gadgets.
2. For each $i \in [n]$ compute the products $T_i = A_i B_i$ (using n field operations).
3. Compute $W = T + V$ and compute $w = \mathsf{DecMSS}_{2\tau}(W)$ (the decoding uses the constant coefficients of a particular instance of the code Π^{MSS}, cf. Section 2.1).
4. Set $Z := (Z_1, \ldots, Z_n)$ where $Z_i = w$ and compute as output $C = Z - U$.

Fig. 5. The transformation for the multiplication operation in $\mathbb{F}$. RandSamp(1^n) samples $([r]_\tau, [r]_{2\tau})$ as described in Sect. 5.1.

The transformation for the multiplication uses ideas from secure multiparty computation – in particular, the use of $[r]_\tau, [r]_{2\tau}$ that allows to decode $W = T+V$ without revealing sensitive information follows the approach from [14]. There are two important differences to the protocol of [14] – most notably, for our purposes we need to sample $([r]_\tau, [r]_{2\tau})$ in a way that is secure against t-region probing adversaries. Second, in Step 4 of Fig. 5 we use a trivial encoding of the value w with the code Π^{MSS}. In particular, instead of using EncMSS(w) to sample Z, we just use the trivial encoding of w as n-elements vector $Z := (w, \ldots, w)$.

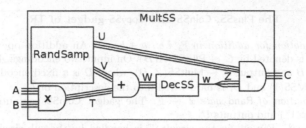

Fig. 6. The architecture of the MultSS circuit. The whole MultSS circuit consists of one region except the RandSamp sub-circuit, which is divided into smaller regions accordingly to the TR^{Aff} compiler.

While clearly this encoding procedure does not offer any security, it guarantees that we can encode w in complexity $O(n)$. This will be relevant when we structure the computation of MultSS into region, which will be explained next.

We structure MultSS into the following regions. The first set of regions corresponds to Step 1 when MultSS queries the external source RandSamp(1^k) for (U, V) and corresponds to the set of regions defined in the previous section. Besides the regions that are naturally inherited from RandSamp, we introduce one additional region that includes all operations of MultSS from Step 2–4. Clearly, this region has size of $O(n)$, which will be important for our security argument.

To complete the description of the transformation it remains to propose constructions for the addition operation, the Rand operation and how to implement fan-out. The transformation is rather straightforward and details are given in Fig. 7. Notice that the transformations from Fig. 7 use the multiplication gadget as a sub-routine to implement a refreshing scheme for the Π^{MSS} encoding scheme. The refreshing algorithm for Π^{MSS} works as follows. We first sample once and for all a fixed encoding $D \leftarrow \text{EncMSS}(1)$, where 1 denotes the multiplicative identity in $\mathbb{F}$. To refresh X, we compute $Z \leftarrow \text{MultSS}(X, D)$.[10]

Finally, notice that each gadget of Fig. 7 represents a single region, where the execution of MultSS$(., D)$ to refresh the output of the gadgets is structured into regions as explained above (and not part of the region of the gadgets itself). This completes the description of the transformation TR and the structuring of computation into regions. We can show the following about the above construction.

Theorem 2. *Let n be the security parameter and d, h be constants defining the underlying expander graph on n vertices. The transformation TR described above is perfectly t-region probing secure for $t < \frac{n}{12(\lceil \log(n) \rceil + 1)}$. Moreover, for a circuit Γ of size s, the transformed circuit $\widehat{\Gamma} \leftarrow \text{TR}(\Gamma, 1^n)$ has size $O(sn^3)$.*

---

[10] We note that the expander-based refreshing from Sect. 3 unfortunately does not easily transfer to a refreshing scheme for the code Π^{MSS}.

The PlusSS, CoinSS and CopySS gadget of TR

1. *Transformation for addition in* $\mathbb{F}$, *i.e.,* $a + b = c$: An addition operation in the circuit $\widehat{\Gamma}$ is denoted by $C \leftarrow \mathsf{PlusSS}(A, B)$. On input two encodings A, B compute $Z = A + B$ and output $C \leftarrow \mathsf{MultSS}(Z, D)$, where D is a fixed encoding of 1, i.e., $D \leftarrow \mathsf{EncMSS}(1)$ and D_i is hard-wired into the description of $\widehat{\Gamma}$.
2. *Transformation of* Rand *gate* $x \leftarrow \mathbb{F}$: The gadget CoinSS computes $(U, V) \leftarrow \mathsf{RandSamp}(1^n)$ and outputs U.
3. *Fan-out in* Γ: Fan-out in the circuit Γ is handled by the sub-circuit $(B, C) \leftarrow \mathsf{CopySS}(A)$ in the transformed circuit $\widehat{\Gamma}$. On input an encoding A, output $B \leftarrow \mathsf{MultSS}(A, D)$ and $C \leftarrow \mathsf{MultSS}(A, D)$.

Fig. 7. The transformation of the remaining operations used by TR. $\mathsf{RandSamp}(1^n)$ samples $([r]_\tau, [r]_{2\tau})$ as described in Sect. 5.1 and MultSS is the transformed multiplication operation from Fig. 5.

We notice that it is straightforward to improve the complexity of the construction to $O(sn^2 \log n)$ using FFT. Moreover, as mentioned in the introduction, further improvements of the efficiency are possible using packed secret sharing.

6   Application to the Noisy Leakage Model

As shown by Duc et al. [16] security in the so-called p-random probing model implies security in the noisy leakage model. In the random probing model the adversary has no control over the choice of the probes and instead corrupts each wire of the circuit independently with a probability p. By applying Chernoff, it is straightforward that security in the threshold probing model with rate r implies security in the random probing model with $p = cr$ for some constant $c < 1$. Hence, applying Theorem 2, we straightforwardly get security in the p-random probing model for $p = O(\log^{-1}(n))$. As argued in the introduction we can further improve p to a constant when we directly prove security in the p-random probing model instead of taking the detour via the much stronger threshold probing model. In particular, we can get the following result.

Theorem 3. *The transformation* TR *described in Sect. 5 is p-random probing secure for a sufficiently small constant $p < 1/12$. For a circuit Γ of size s, the transformed circuit $\widehat{\Gamma} \leftarrow \mathsf{TR}(\Gamma, 1^n)$ has complexity $O(sn^3)$.*

Proof. To distinguish the random probing model from the t-region probing model that we discussed in the last section, we will call the later in the following the t-threshold probing model. To show security against a p-random probing adversary observe that clearly security against a t-region probing adversary for regions of size $O(n)$ and $t = \Omega(n)$ probes implies security in the random probing model for a constant p. This worst-case to average case reduction is a straightforward application of the Chernoff bound. Recall that in our transformation TR from Sect. 5 all parts of the transformed circuit tolerate a constant corruption rate in the threshold probing model[11] except for the MultiDecoder sub-circuits, which are the reason that we only can allow $O(n/\log(n))$ probes (cf. Sect. 5.1). Therefore, to show that our construction achieves security in the random probing model for constant p we only need to show that the MultiDecoder sub-circuits remain secure in the p-random probing model for a constant p. To this end, we need the following fact:

Lemma 10. *Let* MultiDecoder *be a deterministic circuit as described in Sect. 5.1 that takes as input n encodings $X := (X^{(1)}, \ldots, X^{(n)})$ and outputs their decodings $U := (U_1, \ldots, U_n)$. Let $\mathcal{P}$ be a set of probes for* MultiDecoder *drawn by a p-random probing adversary. There exists a simulator* $\mathsf{Sim}_{\mathsf{MultiDecoder}}$ *such that for any plausible inputs $X := (X^{(1)}, \ldots, X^{(n)})$ and corresponding output vector $U := (U_1, \ldots, U_n)$, we have:*

$$\mathcal{P}(\mathcal{W}_{\mathsf{MultiDecoder}}(X|U)) \equiv \mathsf{Sim}_{\mathsf{MultiDecoder}}^{\Omega(X)}(\mathcal{P}).$$

Moreover, for each $i \in [n]$ (independently) we have the following: the probability (over sampling of the set $\mathcal{P}$) that the value $X^{(i)}$ is fully leaked by the $\mathsf{Sim}_{\mathsf{Decoder}}^{\Omega(X)}(\mathcal{P})$ *(i.e., the value $X_j^{(i)}$ is leaked for every $j \in [n]$) is equal at most $\frac{p}{1-p}$.*

The proof is given in the full version.

We now continue the proof of Theorem 3. Note the only requirement we have in Lemma 10 is that for each i (independently) it holds that with probability at least $1 - p/(1 - p)$ the t-th Decoder is not fully covered. Hence, we also need to prove (as in was done in Lemma 9) that not too many of the input encodings to the Decoder are fully leaked by the simulator for the composed circuit RandSamp'. Fix one input encodings $X^{(i)}$ to one of the Decoder sub-circuits. Recall that there are two simulator, which leak from the encoding $X^{(i)}$: $\mathsf{Sim}_{\mathsf{Decoder}}^{\Omega(X)}(\mathcal{P})$ and the simulator for the gadget, which outputs $X^{(i)}$ in RandSamp', which will be denoted Sim'.

Recall that all gadgets except the MultiDecoder sub-circuit are (t, q)-reconstructible for $t = cn$ and $q = c'n$ for an appropriate choice of constants $c, c' < 1/6$. Since all regions are of size $O(n)$ (where the O-notation only hides small constants), there exists a constant $p < p/(1-p) < 1/6$ such that with overwhelming probability a set of probes $\mathcal{P}$ when sampled by a p-random probing adversary is t-region admissible. Therefore, with overwhelming probability (over the choice of $\mathcal{P}$) at most q positions are leaked from $X^{(i)}$ by the simulator Sim' in order to simulate answers to the probes in the part of RandSamp' producing $X^{(i)}$. To simplify the description, we assume that $\mathcal{P}$ produced by the p-random probing adversary is indeed t-admissible, and we do not explicitly mention the bad event when it is not (as this event is negligible anyway).

From Lemma 10 we know that with probability at least $1 - \frac{p}{1-p} \geq \frac{5}{6}$ there exists a random j, s.t. the value $X_j^{(i)}$ is not queried by the simulator

---

[11] This is true for all gadgets of the transformation TR as well as for RandSamp'.

$\mathsf{Sim}_{\mathsf{MultiDecoder}}^{\Omega(X)}(\mathcal{P})$. Notice, that the index j of the share, which is not leaked by the $\mathsf{Sim}_{\mathsf{MultiDecoder}}^{\Omega(X)}(\mathcal{P})$ is uniformly random over $[n]$ due to the symmetry of the MultiDecoder sub-circuit with respect to the input shares indexes[12]. Hence, the probability that the particular value $X_j^{(i)}$ (recall that j was drawn at random) is queried by the Sim′ is equal at most $\frac{q}{n} < c' < \frac{1}{6}$. Therefore, the probability that the encoding $X^{(i)}$ is fully leaked by both simulators is not greater than $\frac{1}{6} + \frac{5}{6} \cdot \frac{1}{6} < \frac{1}{3}$, where the first term in the sum comes from Lemma 10 and the fact that with probability $1/6$ the simulator $\mathsf{Sim}_{\mathsf{MultiDecoder}}$ reveals the entire encoding, and the second term comes from the analysis above (i.e., with probability $5/6$ we have at least one random share $X_j^{(i)}$ that is not queried by $\mathsf{Sim}_{\mathsf{MultiDecoder}}$ and Sim′ only asks for a $1/6$ fraction to its leakage oracle. Given this bound, we can now use again Chernoff to prove that with overwhelming probability (in n) less than $\frac{1}{2}$ of all the input encodings to the MultiDecoder circuit are fully leaked. The rest of the security proof is analogous to the case of the threshold probing adversary. Putting the above together we obtain Theorem 3.   □

We emphasize that the above is mainly a feasibility result and the constant is rather small due to the properties of the expander graph.

7   Extensions

7.1   Security of Boolean Circuits

As outlined in the introduction our transformation TR presented in the last section requires that the computation is carried out over a field $\mathbb{F}$ of size $O(n)$. This implies that the values carried on the wires are from $\mathbb{F}$ and the basic gates used in $\widehat{\varGamma}$ represent the operations from the underlying field $\mathbb{F}$. Notice that the later also means that we require leak-free operations that are of size $O(\log(n) \log\log(n))$, which is required to carry out, e.g., the multiplication in the field $\mathbb{F}$. While we emphasize that this assumption is used by most works that consider leakage resilient circuit transformations, we observe that for our particular construction we can eliminate this assumption by getting slightly weaker parameters (weaker by a constant factor only). The basic idea to achieve this is as follows: instead of using Shamir's secret sharing as underlying code, we can use codes based on algebraic geometry that exhibit the multiplicative property. Such codes are for instance constructed in the work of Chen and Cramer [7]. These codes operate over fields of constant size and hence there basic operations can be implemented by constant size Boolean circuits.

The above is in particular useful for Theorem 3 where we obtain security against constant random probing rate. Using algebraic geometric codes the corruption probability p stays constant even if $\widehat{\varGamma}$ is implemented with Boolean gates – which is optimal.

---

[12] Recall that we assume that n is a power of two and T is then a *full* binary tree. Moreover, the simulator $\mathsf{Sim}_{\mathsf{MultiDecoder}}^{\Omega(X)}(\mathcal{P})$ is also symmetric with respect to the input share indexes. Furthermore, if there is more than one index j, s.t. the value $X_j^{(i)}$ is not leaked by the $\mathsf{Sim}_{\mathsf{MultiDecoder}}^{\Omega(X)}(\mathcal{P})$ we pick one of them uniformly at random.

7.2 From Non-adaptive to Adaptive Security

In our analysis we assumed that for each clock cycle the adversary chooses a set of $\mathcal{P}_i$ that defines what wires leak. This implies that within a clock cycle the adversary is non-adaptive and cannot change the position of his probes, e.g., he cannot learn the first share of an encoding and upon the value of this share decide what wire he wants to probe next. Fortunately, we can easily get fully adaptive security since our construction achieves perfect security against a threshold probing adversary [10]. We stress that the same does not hold for construction that are only statistical secure [10].

References

1. Ajtai, M.: Secure computation with information leaking to an adversary. In: Fortnow, L., Vadhan, S.P. (eds.) Proceedings of the 43rd ACM Symposium on Theory of Computing, STOC 2011, San Jose, CA, USA, 6–8 June 2011, pp. 715–724. ACM (2011)
2. Andrychowicz, M., Damgård, I., Dziembowski, S., Faust, S., Polychroniadou, A.: Efficient leakage resilient circuit compilers. In: Nyberg, K. (ed.) CT-RSA 2015. LNCS, vol. 9048, pp. 311–329. Springer, Heidelberg (2015)
3. Balasch, J., Faust, S., Gierlichs, B.: Inner product masking revisited. In: Oswald, E., Fischlin, M. (eds.) EUROCRYPT 2015. LNCS, vol. 9056, pp. 486–510. Springer, Heidelberg (2015)
4. Barthe, G., Crespo, J.M., Lakhnech, Y., Schmidt, B.: Mind the gap: modular machine-checked proofs of one-round key exchange protocols. In: Oswald, E., Fischlin, M. (eds.) EUROCRYPT 2015, Part II. LNCS, vol. 9057, pp. 689–718. Springer, Heidelberg (2015)
5. Carlet, C., Goubin, L., Prouff, E., Quisquater, M., Rivain, M.: Higher-order masking schemes for S-boxes. In: Canteaut, A. (ed.) FSE 2012. LNCS, vol. 7549, pp. 366–384. Springer, Heidelberg (2012)
6. Chari, S., Jutla, C.S., Rao, J.R., Rohatgi, P.: Towards sound approaches to counteract power-analysis attacks. In: Wiener, M. (ed.) CRYPTO 1999. LNCS, vol. 1666, pp. 398–412. Springer, Heidelberg (1999)
7. Chen, H., Cramer, R.: Algebraic geometric secret sharing schemes and secure multiparty computations over small fields. In: Dwork, C. (ed.) CRYPTO 2006. LNCS, vol. 4117, pp. 521–536. Springer, Heidelberg (2006)
8. Coron, J.-S., Großschädl, J., Vadnala, P.K.: Secure conversion between boolean and arithmetic masking of any order. In: Batina, L., Robshaw, M. (eds.) CHES 2014. LNCS, vol. 8731, pp. 188–205. Springer, Heidelberg (2014)
9. Coron, J.-S., Prouff, E., Roche, T.: On the use of Shamir's secret sharing against side-channel analysis. In: Mangard, S. (ed.) CARDIS 2012. LNCS, vol. 7771, pp. 77–90. Springer, Heidelberg (2013)
10. Cramer, R., Damgård, I., Dziembowski, S., Hirt, M., Rabin, T.: Efficient multiparty computations secure against an adaptive adversary. In: Stern, J. (ed.) EUROCRYPT 1999. LNCS, vol. 1592, pp. 311–326. Springer, Heidelberg (1999)
11. Cramer, R., Damgård, I., Maurer, U.M.: General secure multi-party computation from any linear secret-sharing scheme. In: Preneel, B. (ed.) EUROCRYPT 2000. LNCS, vol. 1807, p. 316. Springer, Heidelberg (2000)

12. Dachman-Soled, D., Kalai, Y.T.: Securing circuits against constant-rate tampering. In: Safavi-Naini, R., Canetti, R. (eds.) CRYPTO 2012. LNCS, vol. 7417, pp. 533–551. Springer, Heidelberg (2012)
13. Dachman-Soled, D., Liu, F.-H., Zhou, H.-S.: Leakage-resilient circuits revisited – optimal number of computing components without leak-free hardware. In: Oswald, E., Fischlin, M. (eds.) EUROCRYPT 2015. LNCS, vol. 9057, pp. 131–158. Springer, Heidelberg (2015)
14. Damgård, I., Ishai, Y., Krøigaard, M.: Perfectly secure multiparty computation and the computational overhead of cryptography. In: Gilbert, H. (ed.) EUROCRYPT 2010. LNCS, vol. 6110, pp. 445–465. Springer, Heidelberg (2010)
15. Davì, F., Dziembowski, S., Venturi, D.: Leakage-resilient storage. In: Garay, J.A., De Prisco, R. (eds.) SCN 2010. LNCS, vol. 6280, pp. 121–137. Springer, Heidelberg (2010)
16. Duc, A., Dziembowski, S., Faust, S.: Unifying leakage models: from probing attacks to noisy leakage. In: Nguyen, P.Q., Oswald, E. (eds.) EUROCRYPT 2014. LNCS, vol. 8441, pp. 423–440. Springer, Heidelberg (2014)
17. Duc, A., Faust, S., Standaert, F.-X.: Making masking security proofs concrete - or how to evaluate the security of any leaking device. In: Oswald, E., Fischlin, M. (eds.) EUROCRYPT 2015. LNCS, vol. 9056, pp. 401–429. Springer, Heidelberg (2015)
18. Dziembowski, S., Faust, S.: Leakage-resilient circuits without computational assumptions. In: Cramer, R. (ed.) TCC 2012. LNCS, vol. 7194, pp. 230–247. Springer, Heidelberg (2012)
19. Faust, S., Pietrzak, K., Venturi, D.: Tamper-proof circuits: how to trade leakage for tamper-resilience. In: Aceto, L., Henzinger, M., Sgall, J. (eds.) ICALP 2011, Part I. LNCS, vol. 6755, pp. 391–402. Springer, Heidelberg (2011)
20. Faust, S., Rabin, T., Reyzin, L., Tromer, E., Vaikuntanathan, V.: Protecting circuits from leakage: the computationally-bounded and noisy cases. In: Gilbert, H. (ed.) EUROCRYPT 2010. LNCS, vol. 6110, pp. 135–156. Springer, Heidelberg (2010)
21. Goldwasser, S., Rothblum, G.N.: How to compute in the presence of leakage. In: 53rd FOCS, pp. 31–40, New Brunswick, NJ, USA, 20–23 October 2012. IEEE Computer Society Press (2012)
22. Goubin, L., Martinelli, A.: Protecting AES with Shamir's secret sharing scheme. In: Preneel, B., Takagi, T. (eds.) CHES 2011. LNCS, vol. 6917, pp. 79–94. Springer, Heidelberg (2011)
23. Grosso, V., Standaert, F.-X., Faust, S.: Masking vs. multiparty computation: how large is the gap for AES? In: Bertoni, G., Coron, J.-S. (eds.) CHES 2013. LNCS, vol. 8086, pp. 400–416. Springer, Heidelberg (2013)
24. Hoory, S., Linial, N., Wigderson, A., Overview, A.: Expander graphs, their applications. Bull. Am. Math. Soc. (N.S) 43, 439–561 (2006)
25. Ishai, Y., Prabhakaran, M., Sahai, A., Wagner, D.: Private circuits II: keeping secrets in tamperable circuits. In: Vaudenay, S. (ed.) EUROCRYPT 2006. LNCS, vol. 4004, pp. 308–327. Springer, Heidelberg (2006)
26. Ishai, Y., Sahai, A., Wagner, D.: Private circuits: securing hardware against probing attacks. In: Boneh, D. (ed.) CRYPTO 2003. LNCS, vol. 2729, pp. 463–481. Springer, Heidelberg (2003)
27. Juma, A., Vahlis, Y.: Protecting cryptographic keys against continual leakage. In: Rabin, T. (ed.) CRYPTO 2010. LNCS, vol. 6223, pp. 41–58. Springer, Heidelberg (2010)

28. Kim, H., Hong, S., Lim, J.: A fast and provably secure higher-order masking of AES S-box. In: Preneel, B., Takagi, T. (eds.) CHES 2011. LNCS, vol. 6917, pp. 95–107. Springer, Heidelberg (2011)
29. Kocher, P.C.: Timing attacks on implementations of Diffie-Hellman, RSA, DSS, and other systems. In: Koblitz, N. (ed.) CRYPTO 1996. LNCS, vol. 1109, pp. 104–113. Springer, Heidelberg (1996)
30. Kocher, P.C., Jaffe, J., Jun, B.: Differential power analysis. In: Wiener, M. (ed.) CRYPTO 1999. LNCS, vol. 1666, pp. 388–397. Springer, Heidelberg (1999)
31. Prouff, E., Rivain, M.: Masking against side-channel attacks: a formal security proof. In: Johansson, T., Nguyen, P.Q. (eds.) EUROCRYPT 2013. LNCS, vol. 7881, pp. 142–159. Springer, Heidelberg (2013)
32. Rivain, M., Prouff, E.: Provably secure higher-order masking of AES. In: Mangard, S., Standaert, F.-X. (eds.) CHES 2010. LNCS, vol. 6225, pp. 413–427. Springer, Heidelberg (2010)

Randomness Complexity of Private Circuits for Multiplication

Sonia Belaïd[1,2]($\boxtimes$), Fabrice Benhamouda[1]($\boxtimes$), Alain Passelègue[1]($\boxtimes$),
Emmanuel Prouff[3,4]($\boxtimes$), Adrian Thillard[1,3]($\boxtimes$), and Damien Vergnaud[1]($\boxtimes$)

[1] ENS, CNRS, INRIA, and PSL, Paris, France
{sonia.belaid,fabrice.benhamouda,alain.passelegue,
adrian.thillard,damien.vergnaud}@ens.fr
[2] Thales Communications and Security, Gennevilliers, France
[3] ANSSI, Paris, France
[4] UPMC, POLSYS, LIP6, Paris, France
emmanuel.prouff@ssi.gouv.fr

Abstract. Many cryptographic algorithms are vulnerable to side channel analysis and several leakage models have been introduced to better understand these flaws. In 2003, Ishai, Sahai and Wagner introduced the d-probing security model, in which an attacker can observe at most d intermediate values during a processing. They also proposed an algorithm that securely performs the multiplication of 2 bits in this model, using only $d(d+1)/2$ random bits to protect the computation. We study the randomness complexity of multiplication algorithms secure in the d-probing model. We propose several contributions: we provide new theoretical characterizations and constructions, new practical constructions and a new efficient algorithmic tool to analyze the security of such schemes.

We start with a theoretical treatment of the subject: we propose an algebraic model for multiplication algorithms and exhibit an algebraic characterization of the security in the d-probing model. Using this characterization, we prove a linear (in d) lower bound and a quasi-linear (non-constructive) upper bound for this randomness cost. Then, we construct a new generic algorithm to perform secure multiplication in the d-probing model that only uses $d + d^2/4$ random bits.

From a practical point of view, we consider the important cases $d \leq 4$ that are actually used in current real-life implementations and we build algorithms with a randomness complexity matching our theoretical lower bound for these small-order cases. Finally, still using our algebraic characterization, we provide a new dedicated verification tool, based on information set decoding, which aims at finding attacks on algorithms for fixed order d at a very low computational cost.

Keywords: Side-channel analysis · Probing model · Randomness complexity · Constructions · Lower bounds · Probabilistic method · Information set decoding · Algorithmic tool

© International Association for Cryptologic Research 2016
M. Fischlin and J.-S. Coron (Eds.): EUROCRYPT 2016, Part II, LNCS 9666, pp. 616–648, 2016.
DOI: 10.1007/978-3-662-49896-5_22

1   Introduction

Most commonly used cryptographic algorithms are now considered secure against classical black-box attacks, when the adversary has only knowledge of their inputs or outputs. Today, it is however well known that their implementations are vulnerable to side-channel attacks, as revealed in the academic community by Kocher in 1996 [16]. These attacks exploit the physical emanations of the underlying device such as the execution time, the device temperature, or the power consumption during the algorithm execution.

To thwart side-channel attacks, many countermeasures have been proposed by the community. Among them, the most widely deployed one is probably *masking* (a.k.a. secret/processing sharing) [8,13], which has strong links with techniques usually applied in secure multi-party computation (see e.g., [5,28]) or private circuits theory [15]. For many kinds of real-life implementations, this countermeasure indeed demonstrated its effectiveness when combined with noise and processing jittering. The idea of the masking approach is to split every single *sensitive* variable/processing, which depends on the secret and on known variables, into several shares. Each share is generated uniformly at random except the last one which ensures that the combination of all the shares is equal to the initial sensitive value. This technique aims at making the physical leakage of one variable independent of the secret and thus useless for the attacker. The tuple of shares still brings information about the shared data but, in practice, the leakages are noisy and the complexity of extracting useful information increases exponentially with the number of shares, the basis of the exponent being related to the amount of noise [8].

In order to formally prove the security of masking schemes, the community has made important efforts to define leakage models that accurately capture the leakage complexity and simultaneously enable to build security arguments. In 2003, Ishai, Sahai, and Wagner introduced the *d-probing model* in which the attacker can observe at most d exact intermediate values [15]. This model is very convenient to make security proofs but does not fit the reality of embedded devices which leak noisy functions of all their intermediate variables. In 2013, Prouff and Rivain extended the noisy leakage model [23], initially introduced by Chari et al. [8], to propose a new one more accurate than [15] but not very convenient for security proofs. The two models [15,23] were later unified by Duc, Dziembowski, and Faust [10] and Duc, Faust, and Standaert [11] who showed that a security proof in the noisy leakage model can be deduced from security proofs in the d-probing model. This sequence of works shows that proving the security of implementations in the d-probing model makes sense both from a theoretical and practical point of view. An implementation secure in the d-probing model is said to satisfy the *d-privacy property* or equivalently to be *d-private* [15] (or secure at order d).

It is worth noting that there is a tight link between sharing techniques, *Multi Party Computation* (MPC) and also *threshold implementations* [6,7,21]. In particular, the study in the classical d-probing security model can be seen as a particular case of MPC with honest players. Furthermore, the threshold implementations manipulate sharing techniques with additional restrictions to thwart further hardware attacks resulting from the leakage of electronic glitches. This problem can itself be similarly seen as a particular case of MPC, with Byzantine players [17].

1.1 Our Problem

Since most symmetric cryptographic algorithms manipulate Boolean values, the most practical way to protect them is generally to implement *Boolean sharing* (a.k.a. *high-order masking*): namely, each sensitive intermediate result x is shared into several pieces, say $d+1$, which are manipulated by the algorithm and whose parity is equal to x. To secure the processing of a function f on a shared data, one must design a so-called *masking scheme* (or formally a *private circuit*) that describes how to build a sharing of $f(x)$ from that of x while maintaining the d-probing security.

In the context of Boolean sharing, we usually separate the protection of linear functions from that of non-linear ones. In particular, at the hardware level, any circuit can be implemented using only two gates: the linear XOR gate and the non-linear AND gate. While the protection of linear operations (e.g., XOR) is straightforward since the initial function f can be applied to each share separately, it becomes more difficult for non-linear operations (e.g., AND). In these cases, the shares cannot be manipulated separately and must generally be processed all together to compute the correct result. These values must then be further protected using additional random bits which results in an important timing overhead.

State-of-the-art solutions to implement Boolean sharing on non-linear functions [9,25] have focused on optimizing the computation complexity. Surprisingly, the amount of necessary random bits has only been in the scope of the seminal paper of Ishai, Sahai and Wagner [15]. In this work, the authors proposed and proved a clever construction (further referred to as ISW multiplication) allowing to compute the multiplication of two shared bits by using $d(d+1)/2$ random bits, that is, half as many random bits as the straightforward solution uses. Their construction has since become a cornerstone of secure implementations [10,12,24,25]. Even if this result is very important, the quantity of randomness remains very expensive to generate in embedded cryptographic implementations. Indeed, such a generation is usually performed using a physical generator followed by a deterministic random bit generator (DRBG). In addition of being a theoretical "chicken-and-egg" problem for this DRBG protection, in practice the physical generator has often a low throughput and the DRBG is also time-consuming. In general, for a DRBG based on a 128-bit block cipher, one call to this block cipher enables to generate 128 pseudorandom bits[1] (see [2]). However, one invocation of the standard AES-128 block cipher with the ISW

---

[1] Actually, the generation of pseudorandom bits roughly corresponds to the execution of a block cipher but we should also consider the regular internal state update.

multiplication requires as much as 30,720 random bits (6 random bytes per multiplication, 4 multiplications per S-box [25]) to protect the multiplications when masked at the low order $d = 3$, which corresponds to 240 preliminary calls to the DRBG.

1.2   Our Contributions

We analyze the quantity of randomness required to define a d-private multiplication algorithm at any order d. Given the sharings $\boldsymbol{a} = (a_i)_{0 \leq i \leq d}$, $\boldsymbol{b} = (b_i)_{0 \leq i \leq d}$ of two bits a and b, the problem we tackle out is to find the minimal number of random bits necessary to securely compute a sharing $(c_i)_{0 \leq i \leq d}$ of the bit $c = ab$ with a d-private algorithm. We limit our scope to the construction of a multiplication based on the sum of shares' products. That is, as in [15], we start with the pairwise products of a's and b's shares and we work on optimizing their sum into $d + 1$ shares with as few random bits as possible. We show that this reduces to studying the randomness complexity of some particular d-private compression algorithm that securely transforms the $(d+1)^2$ shares' products into $d+1$ shares of c. In our study we make extensive use of the following theorem that gives an alternative characterization of the d-privacy:

Theorem 7 *(informal)*. A compression algorithm is d-private if and only if there does not exist a set of ℓ intermediate results $\{p_1, \ldots, p_\ell\}$ such that $\ell \leq d$ and $\sum_{i=1}^{\ell} p_i$ can be written as $\boldsymbol{a}^\mathsf{T} \cdot \boldsymbol{M} \cdot \boldsymbol{b}$ with $\boldsymbol{M}$ being some matrix such that the all-ones vector is in the row space or in the column space of $\boldsymbol{M}$.

From this theorem, we deduce the following lower bound on the randomness complexity:

Theorems 13–14 *(informal)*. If $d \geq 3$ (resp. $d = 2$), then a d-private compression algorithm for multiplication must involve at least $d + 1$ random bits (resp. 2).

This theorem shows that the randomness complexity is in $\Omega(d)$. Following the probabilistic method, we additionally prove the following theorem which claims that there exists a d-private multiplication algorithm with randomness complexity $O(d \cdot \log d)$. This provides a quasi-linear upper bound $O(d \cdot \log d)$ for the randomness complexity, when $d \to \infty$.

Theorem 16 *(informal)*. There exists a d-private multiplication algorithm with randomness complexity $O(d \cdot \log d)$, when $d \to \infty$.

This upper bound is non-constructive: we show that a randomly chosen multiplication algorithm (in some carefully designed family of multiplication algorithms using $O(d \cdot \log d)$ random bits) is d-private with non-zero probability. This means that there exists one algorithm in this family which is d-private.

In order to explicitly construct private algorithms with low randomness, we analyze the ISW multiplication to bring out necessary and sufficient conditions on the use of the random bits. In particular, we identify necessary chainings and we notice that some random bits may be used several times at several locations to protect more shares' products, while in the ISW multiplication, each random bit

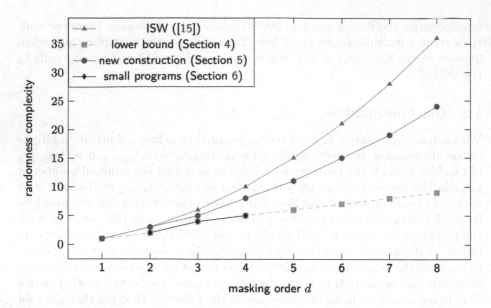

Fig. 1. Randomness complexity of d-private multiplication algorithms

is only used twice. From this analysis, we deduce a new d-private multiplication algorithm requiring $\lfloor d^2/4 \rfloor + d$ random bits instead of $d(d+1)/2$. As a positive side-effect, our new construction also reduces the algorithmic complexity of ISW multiplication (i.e., its number of operations).

Based on this generic construction, we then try to optimize some widely used small order instances. In particular, we bring out new multiplication algorithms, for the orders $d = 2, 3$ and 4, which exactly achieve our proven linear lower bound while maintaining the d-privacy. Namely, we present the optimal multiplication algorithms for orders 2, 3 and 4 when summing the shares' products into $d + 1$ shares. We formally verify their security using the tool provided in [4]. Figure 1 illustrates the randomness complexity of our constructions (for general orders d and small orders) and our lower bound. Note that while the ISW algorithm was initially given for multiplications of bits, it was later extended by Rivain and Prouff in [25] for any multiplication in $\mathbb{F}_{2^n}$. In the following, for the sake of simplicity, we refer to binary multiplications ($n = 1$) for our constructions, but note that all of them can also be adapted to multiplication in $\mathbb{F}_{2^n}$.

Contrary to the ISW algorithm, our new constructions are not directly composable — in the sense of Strong Non-Interferent (SNI) in [3] — at any order. Fortunately, they can still be used in compositions instead of the ISW algorithms at carefully chosen locations. In this paper, we thus recall the different security properties related to compositions and we show that in the AES example, our new constructions can replace half the ISW ones while preserving the d-privacy of the whole algorithm.

Finally, while the tool provided in [4] — which is based on Easycrypt — is able to reveal potential attack paths and formally prove security in the d-probing model with full confidence, it is limited to the verification of small orders ($d = 6$ in our case). Therefore, we propose a new dedicated probabilistic verification tool, which aims at finding attacks in fixed order private circuits (or equivalently masking schemes) at a very low cost. The tool [1] is developed in Sage (Python) [27] and though less generic than [4] it is order of magnitudes faster. It relies on some heuristic assumption (i.e. it cannot be used to actually prove the security) but it usually finds attacks very swiftly for any practical order d. It makes use of information set decoding (a technique from coding theory introduced to the cryptographic community for the security analysis of the McEliece cryptosystem in [20,22]).

2   Preliminaries

This section defines the notations and basic notions that we use in this paper, but also some elementary constructions we refer to. In particular, we introduce the notion of d-private compression algorithm for multiplication and we present its only concrete instance which was proposed by Ishai, Sahai, and Wagner [15].

2.1   Notation

For a set S, we denote by $|S|$ its cardinality, and by $s \xleftarrow{\$} S$ the operation of picking up an element s of S uniformly at random. We denote by $\mathbb{F}_q$ the finite field with q elements. Vectors are denoted by lower case bold font letters, and matrices are denoted by upper case bold font letters. All vectors are column vectors unless otherwise specified. The *kernel* (resp. the *image*) of the linear map associated to a matrix M is denoted by $\ker(M)$ (resp. $\text{im}(M)$). For a vector x, we denote by x_i its i-th coordinate and by $\text{hw}(x)$ its Hamming weight (i.e., the number of its coordinates that are different from 0).

For any fixed $n \geq 1$, let $U_n \in \mathbb{F}_2^{n \times n}$ denote the matrix whose coefficients $u_{i,j}$ equal 1 for all $1 \leq i, j \leq n$. Let $0_{n,\ell} \in \mathbb{F}_2^{n \times \ell}$ denote the matrix whose coefficients are all 0. Let $u_n \in \mathbb{F}_2^n$ denote the vector $(1, \ldots, 1)^\mathsf{T}$ and $0_n \in \mathbb{F}_2^n$ denote the vector $(0, \ldots, 0)^\mathsf{T}$. For vectors $x_1, \ldots, x_t$ in $\mathbb{F}_2^n$ we denote $\langle x_1, \ldots, x_t \rangle$ the vector space generated by the set $\{x_1, \ldots, x_t\}$.

We say that an expression $f(x_1, \ldots, x_n, r)$ functionally depends on the variable r if there exists $a_1, \ldots, a_n$ such that the function $r \mapsto f(a_1, \ldots, a_n, r)$ is not constant.

For an algorithm $\mathcal{A}$, we denote by $y \leftarrow \mathcal{A}(x_1, x_2, \ldots)$ the operation of running $\mathcal{A}$ on inputs $(x_1, x_2, \ldots)$ and letting y denote the output. Moreover, if $\mathcal{A}$ is randomized, we denote by $y \xleftarrow{\$} \mathcal{A}(x_1, x_2, \ldots; r)$ the operation of running $\mathcal{A}$ on inputs $(x_1, x_2, \ldots)$ and with uniform randomness r (or with fresh randomness if r is not specified) and letting y denote the output. The *probability density function* associated to a discrete random variable X defined over S (e.g., $\mathbb{F}_2$) is the function which maps $x \in S$ to $\Pr[X = x]$. It is denoted by $\{X\}$ or by $\{X\}_r$

if there is a need to precise the randomness source r over which the *distribution* is considered.

2.2 Private Circuits

We examine the privacy property in the setting of Boolean circuits and start with the definition of *circuit* and *randomized circuit* given in [15]. A deterministic circuit C is a directed acyclic graph whose vertices are Boolean gates and whose edges are wires. A *randomized circuit* is a circuit augmented with random-bit gates. A random-bit gate is a gate with fan-in 0 that produces a random bit and sends it along its output wire; the bit is selected uniformly and independently of everything else afresh for each invocation of the circuit. From the two previous notions, we may deduce the following definition of a private circuit inspired from [14].

Definition 1 [14]. *A private circuit for $f: \mathbb{F}_2^n \to \mathbb{F}_2^m$ is defined by a triple (I, C, O), where*

- $I: \mathbb{F}_2^n \to \mathbb{F}_2^{n'}$ *is a randomized circuit with uniform randomness ρ and called input encoder;*
- C *is a randomized boolean circuit with input in $\mathbb{F}_2^{n'}$, output in $\mathbb{F}_2^{m'}$, and uniform randomness $r \in \mathbb{F}_2^t$;*
- $O: \mathbb{F}_2^{m'} \to \mathbb{F}_2^m$ *is a circuit, called output decoder.*

We say that C is a d-private implementation of f with encoder I and decoder O if the following requirements hold:

- Correctness: *for any input $w \in \mathbb{F}_2^n$, $\Pr\left[O(C(I(w; \rho); r)) = f(w)\right] = 1$, where the probability is over the randomness ρ and r;*
- Privacy: *for any $w, w' \in \mathbb{F}_2^n$ and any set P of d wires in C, the distributions $\{C_P(I(w; \rho); r)\}_{\rho, r}$ and $\{C_P(I(w'; \rho); r)\}_{\rho, r}$ are identical, where $C_P(I(w; \rho); r)$ denotes the list of the d values on the wires from P.*

Remark 2. It may be noticed that the notions of d-privacy and of security in the d-probing model used, e.g., in [4] are perfectly equivalent.

Unless noted otherwise, we assume I and O to be the following *canonical* encoder and decoder: I encodes each bit-coordinate b of its input w by a block $(b_j)_{0 \leq j \leq d}$ of $d + 1$ random bits with parity b, and O takes the parity of each block of $d + 1$ bits. Each block $(b_j)_{0 \leq j \leq d}$ is called a *sharing* of b and each b_j is called a *share* of b.

From now on, the wires in a set P used to attack an implementation are referred as the *probes* and the corresponding values in $C_P(I(w; \rho); r)$ as the *intermediate results*. To simplify the descriptions, a probe p is sometimes used to directly denote the corresponding result. A set of probes P such that the distributions $\{C_P(I(w; \rho); r)\}_{\rho, r}$ and $\{C_P(I(w'; \rho); r)\}_{\rho, r}$ are *not* identical for some inputs $w, w' \in \mathbb{F}_2^n$ shall be called an *attack*. When the inputs w are clear from the context, the distribution $\{C_P(I(w; \rho); r)\}_{\rho, r}$ is simplified to $\{(p)_{p \in P}\}$.

We now introduce the notions of multiplication algorithm and of d-compression algorithm for multiplication. In this paper, we deeply study d-private multiplication algorithms and d-private compression algorithms for multiplication.

Definition 3. *A multiplication algorithm is a circuit for the multiplication of 2 bits (i.e., with f being the function $f\colon (a,b) \in \mathbb{F}_2^2 \mapsto a \cdot b \in \mathbb{F}_2$), using the canonical encoder and decoder.*

Before moving on to the next notion, let us first introduce a new particular encoder, called *multiplicative*, which has been used in all the previous attempts to build a d-private multiplication algorithm. This encoder takes as input two bits $(a,b) \in \mathbb{F}_2^2$, runs the canonical encoder on these two bits to get $d+1$ random bits $(a_0, \ldots, a_d)$ and $(b_0, \ldots, b_d)$ with parity a and b respectively, and outputs the $(d+1)^2$ bits $(\alpha_{i,j})_{0 \le i,j \le d}$ with $\alpha_{i,j} = a_i \cdot b_j$. Please note that, in particular, we have $a \cdot b = (\sum_{i=0}^{d} a_i) \cdot (\sum_{i=0}^{d} b_i) = \sum_{0 \le i,j \le d} \alpha_{i,j}$.

Definition 4. *A d-compression algorithm for multiplication is a circuit for the multiplication of 2 bits (i.e., with f being the function $f\colon (a,b) \in \mathbb{F}_2^2 \mapsto a \cdot b \in \mathbb{F}_2$), using the canonical decoder and the multiplicative encoder. Moreover, we restrict the circuit C to only perform additions in $\mathbb{F}_2$.*

When clear from the context, we often omit the parameter d and simply say "a compression algorithm for multiplication".

Remark 5. Any d-compression algorithm for multiplication yields a multiplication algorithm, as the algorithm can start by computing $\alpha_{i,j}$ given its inputs $(a_0, \ldots, a_d, b_0, \ldots, b_d)$.

Proposition 6. *A multiplication algorithm $\mathcal{B}$ constructed from a d-compression algorithm for multiplication $\mathcal{A}$ (as in Remark 5) is d-private if and only if the compression algorithm $\mathcal{A}$ is d-private.*

Clearly if $\mathcal{B}$ is d-private, so is $\mathcal{A}$. However, the converse is not straightforward, as an adversary can also probe the input shares a_i and b_i in $\mathcal{B}$, while it cannot in $\mathcal{A}$. The full proof is given in the full version of this paper and is surprisingly hard: we actually use a stronger version of our algebraic characterization (Theorem 7). In the remaining of the paper, we focus on compression algorithms and we do not need to consider probes of the input shares a_i and b_i, which makes notation much simpler.

In the sequel, a d-compression algorithm for multiplication is denoted by $\mathcal{A}(\boldsymbol{a}, \boldsymbol{b}; \boldsymbol{r})$ with $\boldsymbol{r}$ denoting the tuple of uniform random bits used by the algorithm and with $\boldsymbol{a}$ (resp. $\boldsymbol{b}$) denoting the vector of $d+1$ shares of the multiplication operand a (resp. b).

The purpose of the rest of this paper is to investigate how much randomness is needed for such an algorithm to satisfy the d-privacy and to propose efficient or optimal constructions with respect to the consumption of this resource. The number of bits involved in an algorithm $\mathcal{A}(\boldsymbol{a}, \boldsymbol{b}; \boldsymbol{r})$ (i.e., the size of $\boldsymbol{r}$) is called its *randomness complexity* or *randomness cost*.

Algorithm 1. ISW algorithm

Require: sharing $(\alpha_{i,j})_{0 \leq i,j \leq d}$
Ensure: sharing $(c_i)_{0 \leq i \leq d}$
  for $i = 0$ to d **do**
    for $j = i+1$ to d **do**
      $r_{i,j} \xleftarrow{\$} \mathbb{F}_2;$ $t_{i,j} \leftarrow r_{i,j};$ $t_{j,i} \leftarrow r_{i,j} + \alpha_{i,j} + \alpha_{j,i}$
  $c_i \leftarrow \alpha_{i,i}$
  for $i = 0$ to d **do**
    for $j = 0$ to d **do**
      if $i \neq j$ **then**
        $c_i \leftarrow c_i + t_{i,j}$

2.3 ISW Algorithm

The first occurrence of a d-private compression circuit for multiplication in the literature is the ISW algorithm, introduced by Ishai, Sahai, and Wagner in [15]. It is described in Algorithm 1. Its randomness cost is $d(d+1)/2$.

To better understand this algorithm, let us first write it explicitly for $d = 3$:

$$c_0 \leftarrow \alpha_{0,0} + r_{0,1} + r_{0,2} + r_{0,3}$$
$$c_1 \leftarrow \alpha_{1,1} + (r_{0,1} + \alpha_{0,1} + \alpha_{1,0}) + r_{1,2} + r_{1,3}$$
$$c_2 \leftarrow \alpha_{2,2} + (r_{0,2} + \alpha_{0,2} + \alpha_{2,0}) + (r_{1,2} + \alpha_{1,2} + \alpha_{2,1}) + r_{2,3}$$
$$c_3 \leftarrow \alpha_{3,3} + (r_{0,3} + \alpha_{0,3} + \alpha_{3,0}) + (r_{1,3} + \alpha_{1,3} + \alpha_{3,1}) + (r_{2,3} + \alpha_{2,3} + \alpha_{3,2})$$

where, for the security to hold, the terms are added from left to right and where the brackets indicate the order in which the operations must be performed (from d-privacy point of view, the addition is not commutative). In particular, when the brackets gather three terms (e.g., $(r_{0,1} + \alpha_{0,1} + \alpha_{1,0})$), the attacker is allowed to probe two values from left to right (e.g., $r_{0,1} + \alpha_{0,1}$ and $(r_{0,1} + \alpha_{0,1} + \alpha_{1,0})$).

Let us now simplify the description by removing all the $+$ symbols, the assignments $c_i \leftarrow$, and defining $\hat{\alpha}_{i,j}$ as $\alpha_{i,j} + \alpha_{j,i}$ if $i \neq j$ and $\alpha_{i,i}$ if $i = j$. The ISW algorithm for $d = 3$ can then be rewritten as:

$$
\begin{array}{lllllll}
\hat{\alpha}_{0,0} & r_{0,1} & & r_{0,2} & & r_{0,3} & \\
\hat{\alpha}_{1,1} & (r_{0,1} & \hat{\alpha}_{0,1}) & r_{1,2} & & r_{1,3} & \\
\hat{\alpha}_{2,2} & (r_{0,2} & \hat{\alpha}_{0,2}) & (r_{1,2} & \hat{\alpha}_{1,2}) & r_{2,3} & \\
\hat{\alpha}_{3,3} & (r_{0,3} & \hat{\alpha}_{0,3}) & (r_{1,3} & \hat{\alpha}_{1,3}) & (r_{2,3} & \hat{\alpha}_{2,3}).
\end{array}
$$

Please note that the expression of $\hat{\alpha}_{i,j}$ with $i \neq j$ (i.e. $\alpha_{i,j} + \alpha_{j,i}$) is expanded before the actual evaluation, i.e., as in the previous representation, the sum $\alpha_{i,j} + \alpha_{j,i}$ is not evaluated beforehand but evaluated during the processing of $r_{i,j} + \hat{\alpha}_{i,j} = r_{i,j} + \alpha_{i,j} + \alpha_{j,i}$.

3 Algebraic Characterization

In order to reason about the required quantity of randomness in d-private compression algorithms for multiplication, we define an algebraic condition on the security and we prove that an algorithm is d-private if and only if there is no set of probes which satisfies it.

3.1 Matrix Notation

As our condition is algebraic, it is practical to introduce some matrix notation for our probes. We write $\boldsymbol{a} = (a_0, \ldots, a_d)^\mathsf{T}$ and $\boldsymbol{b} = (b_0, \ldots, b_d)^\mathsf{T}$ the vectors corresponding to the shares of the inputs a and b respectively. We also denote by $\boldsymbol{r} = (r_1, \ldots, r_R)^\mathsf{T}$ the vector of the random bits.

We remark that, for any probe p on a compression algorithm for multiplication, p is always an expression that can be written as a sum of $\alpha_{i,j}$'s (with $\alpha_{i,j} = a_i \cdot b_j$) and r_k's, and possibly a constant $c_p \in \mathbb{F}_2$. In other word, we can write p as

$$p = \boldsymbol{a}^\mathsf{T} \cdot \boldsymbol{M_p} \cdot \boldsymbol{b} + \boldsymbol{s_p}^t \cdot \boldsymbol{r} + c_p,$$

with $\boldsymbol{M_p}$ being a matrix in $\mathbb{F}_2^{(d+1) \times (d+1)}$ and $\boldsymbol{s_p}$ being a vector in $\mathbb{F}_2^R$. This matrix $\boldsymbol{M_p}$ and this vector $\boldsymbol{s_p}$ are uniquely defined. In addition, any sum of probes can also be written that way.

Furthermore, if $c_p = 1$, we can always sum the probe with 1 and consider $p + 1$ instead of p. This does not change anything on the probability distribution we consider. Therefore, for the sake of simplicity, we always assume $c_p = 0$ in all the paper.

3.2 Algebraic Condition

We now introduce our algebraic condition:

Condition 1. *A set of probes $P = \{p_1, \ldots, p_\ell\}$ on a d-compression algorithm for multiplication satisfies Condition 1 if and only if the expression $f = \sum_{i=1}^\ell p_i$ can be written as $f = \boldsymbol{a}^\mathsf{T} \cdot \boldsymbol{M} \cdot \boldsymbol{b}$ with $\boldsymbol{M}$ being some matrix such that $\boldsymbol{u}_{d+1}$ is in the row space or the column space of $\boldsymbol{M}$.*

As seen previously, the expression f can always be written as

$$f = \boldsymbol{a}^\mathsf{T} \cdot \boldsymbol{M} \cdot \boldsymbol{b} + \boldsymbol{s}^\mathsf{T} \cdot \boldsymbol{r},$$

for some matrix $\boldsymbol{M}$ and some vector $\boldsymbol{s}$. Therefore, what the condition enforces is that $\boldsymbol{s} = \boldsymbol{0}_R$ (or in other words, f does not functionally depend on any random bit) and the column space or the row space of $\boldsymbol{M}$ contains the vector $\boldsymbol{u}_{d+1}$.

A Weaker Condition. To better understand Condition 1, let us introduce a weaker condition which is often easier to deal with:

Condition 2 (Weak Condition). *A set of probes $P = \{p_1, \ldots, p_\ell\}$ on a d-compression algorithm for multiplication satisfies Condition 2 if and only if the expression $f = \sum_{i=1}^\ell p_i$ does not functionally depend on any r_k and there exists a map $\gamma\colon \{0, \ldots, d\} \to \{0, \ldots, d\}$ such that f does functionally depend on every $(\alpha_{i,\gamma(i)})_{0 \le i \le d}$ or on every $(\alpha_{\gamma(i),i})_{0 \le i \le d}$.*

This condition could be reformulated as $f = \sum_{i=1}^{\ell} p_i$ functionally depends on either all the a_i's or all the b_i's and does not functionally depend on any r_k. It is easy to see that any set P verifying Condition 1 also verifies Condition 2.

3.3 Algebraic Characterization

Theorem 7. *Let $\mathcal{A}$ be a d-compression algorithm for multiplication. Then, $\mathcal{A}$ is d-private if and only if there does not exist a set $P = \{p_1, \ldots, p_\ell\}$ of $\ell \leq d$ probes that satisfies Condition 1. Furthermore any set $P = \{p_1, \ldots, p_\ell\}$ satisfying Condition 1 is an attack.*

Please note that Theorem 7 would not be valid with Condition 2 (instead of Condition 1). A counterexample is given in the full version of this paper.

Proof (Theorem 7).

Direction 1: Left to right. We prove hereafter that if $\mathcal{A}$ is d-private, then there does not exist a set $P = \{p_1, \ldots, p_\ell\}$ of $\ell \leq d$ probes that satisfies Condition 1.

By contrapositive, let us assume that there exists a set $P = \{p_1, \ldots, p_\ell\}$ of at most d probes that satisfies Condition 1. Let M be the matrix such that $f = \sum_{i=1}^{\ell} p_i = a^\mathsf{T} \cdot M \cdot b$ and let us assume, without loss of generality, that u_{d+1} is in the vector subspace generated by the columns of M. We remark that, for any $v \in \mathbb{F}_2^{d+1}$:

$$\Pr\left[a^\mathsf{T} \cdot v = a\right] = \begin{cases} 1 & \text{when } v = u_{d+1} \\ \frac{1}{2} & \text{when } v \neq u_{d+1} \end{cases}$$

by definition of the sharing a of a (probability is taken over a). Thus we have, when $a = 0$ (assuming that b is uniformly random)

$$\Pr\left[f = 0 \mid a = 0\right]$$
$$= \Pr\left[a^\mathsf{T} \cdot M \cdot b = 0 \mid a^\mathsf{T} \cdot u_{d+1} = 0\right]$$
$$= \Pr\left[a^\mathsf{T} \cdot u_{d+1} = 0 \mid a = 0 \text{ and } M \cdot b = u_{d+1}\right] \cdot \Pr\left[M \cdot b = u_{d+1}\right]$$
$$\quad + \sum_{v \in \mathbb{F}_2^{d+1} \setminus \{u_{d+1}\}} \Pr\left[a^\mathsf{T} \cdot v = 0 \mid a = 0 \text{ and } M \cdot b = v\right] \cdot \Pr\left[M \cdot b = v\right]$$
$$= 1 \cdot \Pr\left[M \cdot b = u_{d+1}\right] + \sum_{v \in \mathbb{F}_2^{d+1} \setminus \{u_{d+1}\}} \tfrac{1}{2} \cdot \Pr\left[M \cdot b = v\right]$$
$$= 1 \cdot \Pr\left[M \cdot b = u_{d+1}\right] + \tfrac{1}{2}(1 - \Pr\left[M \cdot b = u_{d+1}\right])$$
$$= \tfrac{1}{2} + \tfrac{1}{2}\Pr\left[M \cdot b = u_{d+1}\right].$$

Similarly, when $a = 1$, we have

$$\Pr\left[f = 0 \mid a = 1\right] = \tfrac{1}{2} - \tfrac{1}{2}\Pr\left[M \cdot b = u_{d+1}\right].$$

As u_{d+1} is in the column space of M, the distribution of $\{f\}$ is not the same when $a = 0$ and when $a = 1$. This implies that the distribution $\{(p_1, \ldots, p_\ell)\}$ is also different when $a = 0$ and $a = 1$. Hence $\mathcal{A}$ is not d-private.

This concludes the proof of the first implication and the fact that any set $P = \{p_1, \ldots, p_\ell\}$ satisfying Condition 1 is an attack.

Direction 2: Right to left. Let us now prove by contradiction that if there does not exist a set $P = \{p_1, \ldots, p_\ell\}$ of $\ell \leq d$ probes that satisfies Condition 1, then $\mathcal{A}$ is d-private.

Let us assume that $\mathcal{A}$ is not d-private. Then there exists an attack using a set of probes $P = \{p_1, \ldots, p_\ell\}$ with $\ell \leq d$. This is equivalent to say that there exists two inputs $(a^{(0)}, b^{(0)}) \neq (a^{(1)}, b^{(1)})$ such that the distribution $\{(p_1, \ldots, p_\ell)\}$ is not the same whether $(a, b) = (a^{(0)}, b^{(0)})$ or $(a, b) = (a^{(1)}, b^{(1)})$.

We first remark that we can consider $0 = a^{(0)} \neq a^{(1)} = 1$, without loss of generality as the $a^{(i)}$'s and the $b^{(i)}$'s play a symmetric role (and $(a^{(0)}, b^{(0)}) \neq (a^{(1)}, b^{(1)})$). Furthermore, we can always choose $b^{(0)} = b^{(1)}$, as if the distribution $\{(p_1, \ldots, p_\ell)\}$ is not the same whether $(a, b) = (0, b^{(0)})$ or $(a, b) = (1, b^{(1)})$, with $b^{(0)} \neq b^{(1)}$, then:

– it is not the same whether $(a, b) = (0, b^{(0)})$ or $(a, b) = (1, b^{(0)})$ (in which case, we could have taken $b^{(1)} = b^{(0)}$), or
– it is not the same whether $(a, b) = (1, b^{(0)})$ or $(a, b) = (1, b^{(1)})$ (in which case, we can just exchange the a's and the b's roles).

To summarize, there exists $b^{(0)}$ such that the distribution $\{(p_1, \ldots, p_\ell)\}$ is not the same whether $(a, b) = (0, b^{(0)})$ or $(a, b) = (1, b^{(0)})$.

In the sequel $b^{(0)}$ is fixed and we call a tuple $(p_1, \ldots, p_\ell)$ satisfying the previous property an *attack tuple*.

We now remark that if $\ell = 1$ or if even the distribution $\{(\sum_{i=1}^{\ell} p_i)\}$ is not the same whether $(a, b) = (0, b^{(0)})$ or $(a, b) = (1, b^{(0)})$ (i.e., $(\sum_{i=1}^{\ell} p_i)$ is an attack tuple), then it follows easily from the probability analysis of the previous proof for the other direction of the theorem, that the set P satisfies Condition 1. The main difficulty is that it is not necessarily the case that $\ell = 1$ or $(\sum_{i=1}^{\ell} p_i)$ is an attack tuple. To overcome it, we use linear algebra.

But first, let us introduce some useful notations and lemmas. We write $\boldsymbol{p}$ the vector $(p_1, \ldots, p_\ell)^\top$ and we say that $\boldsymbol{p}$ is an *attack vector* if and only if $(p_1, \ldots, p_\ell)$ is an attack tuple. Elements of $\boldsymbol{p}$ are polynomials in the a_i's, the b_j's and the r_k's.

Lemma 8. *If $\boldsymbol{p}$ is an attack vector and $\boldsymbol{N}$ is an invertible matrix in $\mathbb{F}_2^{\ell \times \ell}$, then $\boldsymbol{N} \cdot \boldsymbol{p}$ is an attack vector.*

Proof. This is immediate from the fact that $\boldsymbol{N}$ is invertible. Indeed, as a matrix over $\mathbb{F}_2$, $\boldsymbol{N}^{-1}$ is also a matrix over $\mathbb{F}_2$. Hence, multiplying the set of probes $\{\boldsymbol{N} \cdot \boldsymbol{p}\}$ by $\boldsymbol{N}^{-1}$ (which leads to the first set of probes $\{\boldsymbol{p}\}$) can be done by simply computing sums of elements in $\{\boldsymbol{N} \cdot \boldsymbol{p}\}$. Hence, as the distribution of $\{\boldsymbol{p}\}$ differs when $(a, b) = (0, b^{(0)})$ and $(a, b) = (1, b^{(0)})$, the same is true for the distribution $\{\boldsymbol{N} \cdot \boldsymbol{p}\}$. $\square$

We also use the following straightforward lemma.

Lemma 9. *If $(p_1, \ldots, p_\ell)$ is an attack tuple such that the $\ell - t + 1$ random variables $(p_1, \ldots, p_t)$, p_{t+1}, $\ldots$, and p_ℓ are mutually independent, and the distributions of $(p_{t+1}, \ldots, p_\ell)$ is the same for all the values of the inputs (a, b), then $(p_1, \ldots, p_t)$ is an attack tuple.*

Let us consider the matrix $S \in \mathbb{F}_2^{\ell \times R}$ whose coefficients $s_{i,j}$ are defined as $s_{i,j} = 1$ if and only if the expression p_i functionally depends on r_j. In other words, if we write $p_i = a^\mathsf{T} \cdot M_{p_i} \cdot b + s_{p_i}^\mathsf{T} \cdot r$, the i-th row of S is $s_{p_i}^\mathsf{T}$. We can permute the random bits (i.e., the columns of S and the rows of r) such that a row reduction on the matrix S yields a matrix of the form:

$$S' = \begin{pmatrix} 0_{t,t} & 0_{t,\ell-t} \\ I_t & S'' \end{pmatrix}.$$

Let N be the invertible matrix in $\mathbb{F}_2^{\ell \times \ell}$ such that $N \cdot S = S'$. And we write $p' = (p'_1, \ldots, p'_\ell)^\mathsf{T} = N \cdot p$. Then, p' is also an attack vector according to Lemma 8. In addition, for $t < i \le \ell$, p'_i does functionally depend on r_i and no other p'_j does functionally depend on r_j (due to the shape of S'). Therefore, according to Lemma 9, $(p'_1, \ldots, p'_t)$ is an attack tuple.

We remark that $(p'_1, \ldots, p'_t)$ does not functionally depend on any random bit, due to the shape of S'. Therefore, for each $1 \le i \le t$, we can write:

$$p'_i = a^\mathsf{T} \cdot M'_i \cdot b,$$

for some matrix M'_i.

We now need a final lemma to be able to conclude.

Lemma 10. *If $(p'_1, \ldots, p'_t)$ is an attack tuple, then there exists a vector $b^* \in \mathbb{F}_2^{d+1}$ such that u_{d+1} is in the vector space $\langle M'_1 \cdot b^*, \ldots, M'_t \cdot b^* \rangle$.*

Proof. This lemma can be seen as a generalization of the probability analysis in the proof of the first direction of the theorem.

We suppose by contradiction that $(p'_1, \ldots, p'_t)$ is an attack vector but there does not exist a vector $b^* \in \mathbb{F}_2^{d+1}$ such that u_{d+1} is in the vector space $\langle M'_1 \cdot b^*, \ldots, M'_t \cdot b^* \rangle$. Then, for any value $a^{(0)}$, any vector $b^{(0)} \in \mathbb{F}_2^{d+1}$, and any vector $x = (x_1, \ldots, x_t)^\mathsf{T} \in \mathbb{F}_2^t$:

$$\Pr\left[(p'_1, \ldots, p'_t) = (x_1, \ldots, x_t) \mid a = a^{(0)} \text{ and } b = b^{(0)} \right]$$

$$= \Pr\left[(a^\mathsf{T} \cdot M'_1 \cdot b^{(0)}, \ldots, a^\mathsf{T} \cdot M'_t \cdot b^{(0)}) = (x_1, \ldots, x_t) \mid a^\mathsf{T} \cdot u_{d+1} = a^{(0)} \right]$$

$$= \Pr\left[a^\mathsf{T} \cdot B = x^\mathsf{T} \mid a^\mathsf{T} \cdot u_{d+1} = a^{(0)} \right],$$

where B is the matrix whose i-th column is the vector $M'_i \cdot b^{(0)}$. To conclude, we just need to remark that

$$\Pr[a^\mathsf{T} \cdot B = x^\mathsf{T} \mid a^\mathsf{T} \cdot u_{d+1} = 0] = \Pr[a^\mathsf{T} \cdot B = x^\mathsf{T} \mid a^\mathsf{T} \cdot u_{d+1} = 1],$$

which implies that the probability distribution of $(p'_1, \ldots, p'_t)$ is independent of the value of a, which contradicts the fact the $(p'_1, \ldots, p'_t)$ is an attack tuple.

To prove the previous equality, we use the fact that u_{d+1} is not in the column space of B and therefore the value of $a^\mathsf{T} \cdot u_{d+1}$ is uniform and independent of the value of $a^\mathsf{T} \cdot B$ (when a is a uniform vector in $\mathbb{F}_2^{d+1}$). $\qquad\square$

Thanks to Lemma 10, there exists a vector $\boldsymbol{\sigma} = (\sigma_1, \ldots, \sigma_t)^\mathsf{T} \in \mathbb{F}_2^t$ and a vector $\boldsymbol{b}^* \in \mathbb{F}_2^{d+1}$ such that

$$\left(\sum_{i=1}^{t} \sigma_i \cdot \boldsymbol{M}_i' \right) \cdot \boldsymbol{b}^* = \boldsymbol{u}_{d+1}. \tag{1}$$

Let $\boldsymbol{\sigma}'$ be the vector in $\mathbb{F}_2^\ell$ defined by $\boldsymbol{\sigma}'^\mathsf{T} = (\boldsymbol{\sigma}^\mathsf{T} \; \boldsymbol{0}_{\ell-t}^\mathsf{T}) \cdot \boldsymbol{N}$. We have:

$$\boldsymbol{\sigma}'^\mathsf{T} \cdot \boldsymbol{p} = \sum_{i=1}^{t} \sigma_i \cdot p_i' = \sum_{i=1}^{t} \sigma_i \cdot \boldsymbol{a}^\mathsf{T} \cdot \boldsymbol{M}_i' \cdot \boldsymbol{b} = \boldsymbol{a}^\mathsf{T} \cdot \left(\sum_{i=1}^{t} \sigma_i \cdot \boldsymbol{M}_i' \right) \cdot \boldsymbol{b}. \tag{2}$$

Therefore, we can define the set $P' = \{p_i \mid \sigma_i = 1\}$. This set satisfies Condition 1, according to Eqs. (1) and (2).

This concludes the proof.   □

4 Theoretical Lower and Upper Bounds

In this section, we exhibit lower and upper bounds for the randomness complexity of a d-private compression algorithm for multiplication. We first prove an algebraic result and an intermediate lemma that we then use to show that at least $d + 1$ random bits are required to construct a d-private compression algorithm for multiplication, for any $d \geq 3$ (and 2 random bits are required for $d = 2$). Finally, we provide a (non-constructive) proof that for large enough d, there exists a d-private multiplication algorithm with a randomness complexity $O(d \cdot \log d)$.

4.1 A Splitting Lemma

We first prove an algebraic result, stated in the lemma below, that we further use to prove Lemma 12. The latter allows us to easily exhibit attacks in order to prove our lower bounds.

Lemma 11. *Let* $n \geq 1$. *Let* $\boldsymbol{M}_0, \boldsymbol{M}_1 \in \mathbb{F}_2^{n \times n}$ *such that* $\boldsymbol{M}_0 + \boldsymbol{M}_1 = \boldsymbol{U}_n$. *Then, there exists a vector* $\boldsymbol{v} \in \mathbb{F}_2^n$ *such that:*

$$\boldsymbol{M}_0 \cdot \boldsymbol{v} = \boldsymbol{u}_n \quad or \quad \boldsymbol{M}_1 \cdot \boldsymbol{v} = \boldsymbol{u}_n \quad or \quad \boldsymbol{M}_0^\mathsf{T} \cdot \boldsymbol{v} = \boldsymbol{u}_n \quad or \quad \boldsymbol{M}_1^\mathsf{T} \cdot \boldsymbol{v} = \boldsymbol{u}_n.$$

Proof (Lemma 11). We show the above lemma by induction on n.

Base Case: for $n = 1$, $\boldsymbol{M}_0, \boldsymbol{M}_1, \boldsymbol{U} \in \mathbb{F}_2$, so $\boldsymbol{M}_0 + \boldsymbol{M}_1 = 1$, which implies $\boldsymbol{M}_0 = 1$ or $\boldsymbol{M}_1 = 1$ and the claim immediately follows.

Inductive Case: let us assume that the claim holds for a fixed $n \geq 1$. Let us consider two matrices $\boldsymbol{M}_0, \boldsymbol{M}_1 \in \mathbb{F}_2^{(n+1) \times (n+1)}$ such that $\boldsymbol{M}_0 + \boldsymbol{M}_1 = \boldsymbol{U}_{n+1}$.

Clearly, if $\boldsymbol{M}_0$ (or $\boldsymbol{M}_1$) is invertible, then the claim is true (as $\boldsymbol{u}_{n+1}$ is in its range). Then, let us assume that $\boldsymbol{M}_0$ is not invertible. Then, there exists a non-zero vector $\boldsymbol{x} \in \ker(\boldsymbol{M}_0)$. Now, as $\mathrm{im}(\boldsymbol{U}_{n+1}) = \{\boldsymbol{0}_{n+1}, \boldsymbol{u}_{n+1}\}$, if $\boldsymbol{U}_{n+1} \cdot \boldsymbol{x} =$

u_{n+1}, then $M_1 \cdot x = u_{n+1}$ and the claim is true. Hence, clearly, the claim is true if $\ker(M_0) \neq \ker(M_1)$ (with the symmetric remark). The same remarks hold when considering matrices M_0^T and M_1^T.

Hence, the only remaining case to consider is when $\ker(M_0) \neq \{0_{n+1}\}$, $\ker(M_0^\mathsf{T}) \neq \{0_{n+1}\}$ and when $\ker(M_0) = \ker(M_1)$ and $\ker(M_0^\mathsf{T}) = \ker(M_1^\mathsf{T})$. In particular, we have $\ker(M_0) \subseteq \ker(U_{n+1})$ and $\ker(M_0^\mathsf{T}) \subseteq \ker(U_{n+1})$.

Let $x \in \ker(M_0)$ (and then $x \in \ker(M_1)$ as well) be a non-zero vector. Up to some rearrangement of the *columns* of M_0 and M_1 (by permuting some columns), we can assume without loss of generality that $x = (1,\ldots,1,0,\ldots,0)^\mathsf{T}$. Let X denote the matrix $(x, e_2, \ldots, e_{n+1})$ where $e_i = (0,\ldots,0,1,0,\ldots,0)^\mathsf{T}$ is the i-th canonical vector of length $n+1$, so that it has a 1 in the i-th position and 0's everywhere else.

Now, let $y \in \ker(M_0^\mathsf{T})$ (and then $y \in \ker(M_1^\mathsf{T})$ as well) be a non-zero vector, so $y^\mathsf{T} \cdot M_0^\mathsf{T} = 0_{n+1}^\mathsf{T}$. Moreover, up to some rearrangement of the *rows* of M_0 and M_1, we can assume that $y = (1,\ldots,1,0,\ldots,0)^\mathsf{T}$. Let Y denote the matrix $(y, e_2, \ldots, e_{n+1})$.

Please note that rearrangements apply to the columns in the first case and to the rows in the second case, so we can assume without loss of generality that there exists both $x \in \ker(M_0)$ and $y \in \ker(M_0^\mathsf{T})$ with the above form and matrices X and Y are well defined.

We now define the matrices $M_0' = Y^\mathsf{T} \cdot M_0 \cdot X$ and $M_1' = Y^\mathsf{T} \cdot M_1 \cdot X$. We have:

$$M_0' = \begin{pmatrix} y^\mathsf{T} \\ 0_n & I_n \end{pmatrix} \cdot M_0 \cdot \begin{pmatrix} x & 0_n^\mathsf{T} \\ & I_n \end{pmatrix} = \begin{pmatrix} y^\mathsf{T} \\ 0_n & I_n \end{pmatrix} \cdot \begin{pmatrix} 0_{n+1} & M_0^{(1)} \end{pmatrix}$$

where $M_0^{(1)}$ is the matrix extracted from M_0 by removing its first column. Hence:

$$M_0' = \begin{pmatrix} 0 & 0_n^\mathsf{T} \\ 0_n & M_0^{(1,1)} \end{pmatrix}$$

where $M_0^{(1,1)}$ is the matrix extracted from M_0 by removing its first column and its first row. Similar equation holds for M_1' as well. Thus, it is clear that:

$$M_0' + M_1' = \begin{pmatrix} 0 & 0_n^\mathsf{T} \\ 0_n & U_n \end{pmatrix}.$$

Let us consider the matrices M_0'' and M_1'' in $\mathbb{F}_2^{n \times n}$ that are extracted from matrices M_0' and M_1' by removing their first row and their first column (i.e., $M_i'' = M_i'^{(1,1)}$ with the previous notation). Then, it is clear that $M_0'' + M_1'' = U_n$. As matrices in $\mathbb{F}_2^{n \times n}$, by induction hypothesis, there exists $v'' \in \mathbb{F}_2^n$ such that at least one of the 4 propositions from Lemma 11 holds. We can assume without loss of generality that $M_0'' \cdot v'' = u_n$.

Let $v' = \begin{pmatrix} 0 \\ v'' \end{pmatrix} \in \mathbb{F}_2^{n+1}$. Then, we have:

$$M_0' \cdot v' = \begin{pmatrix} 0 & 0_n^\mathsf{T} \\ 0_n & M_0'' \end{pmatrix} \cdot \begin{pmatrix} 0 \\ v'' \end{pmatrix} = \begin{pmatrix} 0_n \cdot v'' \\ M_0'' \cdot v'' \end{pmatrix} = \begin{pmatrix} 0 \\ u_n \end{pmatrix}.$$

Now, let $v = X \cdot v'$ and $w = M_0 \cdot w$, so $Y^\mathsf{T} \cdot w = Y^\mathsf{T} \cdot M_0 X \cdot v' = M_0' \cdot v' = \begin{pmatrix} 0 \\ u_n \end{pmatrix}$. Moreover, as Y is invertible, w is the *unique* vector such that $Y^\mathsf{T} \cdot w = \begin{pmatrix} 0 \\ u_n \end{pmatrix}$. Finally, as the vector u_{n+1} satisfies $Y^\mathsf{T} \cdot u_{n+1} = \begin{pmatrix} 0 \\ u_n \end{pmatrix}$, then $w = u_{n+1}$, and the claim follows for $n+1$, since v satisfies $M_0 \cdot v = w = u_{n+1}$.

Conclusion: The claim follows for any $n \geq 1$, and so does Lemma 11. $\square$

We can now easily prove the following statement that is our main tool for proving our lower bounds, as explained after its proof.

Lemma 12. *Let $\mathcal{A}$ be a d-compression algorithm for multiplication. If there exists two sets S_1 and S_2 of at most d probes such that $s_i = \sum_{p \in S_i} p$ does not functionally depend on any of the random bits, for $i \in \{0, 1\}$, and such that $s_0 + s_1 = a \cdot b$, then $\mathcal{A}$ is not d-private.*

Proof (Lemma 12). Let $\mathcal{A}$, S_0, S_1, s_0 and s_1 defined in the above statement. Then, there exists $M_i \in \mathbb{F}_2^{(d+1) \times (d+1)}$ such that $s_i = a^\mathsf{T} \cdot M_i \cdot b$, for $i \in \{0, 1\}$. Furthermore, as $s_0 + s_1 = a \cdot b = a^\mathsf{T} \cdot U_{d+1} \cdot b$, we have $M_0 + M_1 = U_{d+1}$. Hence, via Lemma 11, there exists $v \in \mathbb{F}_2^{d+1}$ and $i \in \{0, 1\}$ such that $M_i \cdot v = u_{d+1}$ or $M_i^\mathsf{T} \cdot v = u_{d+1}$. This means that u_{d+1} is in the row subspace or in the column subspace of M_i, and therefore, M_i satisfies Condition 1. Therefore, as $|S_i| \leq d$, applying Theorem 7, $\mathcal{A}$ is not d-private. Lemma 12 follows. $\square$

We use the above lemma to prove our lower bounds as follows: for proving that at least $R(d)$ random bits are required in order to achieve d-privacy for a compression algorithm for multiplication, we prove that any algorithm with a lower randomness complexity is not d-private by exhibiting two sets of probes S_0 and S_1 that satisfy the requirements of Lemma 12.

4.2   Simple Linear Lower Bound

As a warm-up, we show that at least d random bits are required, for $d \geq 2$.

Theorem 13. *Let $d \geq 2$. Let us consider a d-compression algorithm for multiplication $\mathcal{A}$. If $\mathcal{A}$ uses only $d-1$ random bits, then $\mathcal{A}$ is not d-private.*

Proof (Theorem 13). Let $r_1, \ldots, r_{d-1}$ denote the random bits used by $\mathcal{A}$. Let $c_0, \ldots, c_d$ denote the outputs of $\mathcal{A}$. Let us define $N \in \mathbb{F}_2^{(d-1) \times d}$ as the matrix whose coefficients $n_{i,j}$ are equal to 1 if and only if c_j functionally depends on r_i, for $1 \leq i \leq d-1$ and $1 \leq j \leq d$. Please note in particular that N does not depend on c_0.

As a matrix over $\mathbb{F}_2$ with d columns and $d-1$ rows, there is necessarily a vector $w \in \mathbb{F}_2^d$ with $w \neq 0_d$ such that $N \cdot w = 0_{d-1}$.

The latter implies that the expression $s_0 = \sum_{i=1}^d w_i \cdot c_i$ does not functionally depend on any of the r_k's. Furthermore, by correctness, we also have that

$s_1 = c_0 + \sum_{i=1}^{d}(1 - w_i) \cdot c_i$ does not functionally depend on any of the r_k's, and $s_0 + s_1 = \sum_{i=0}^{d} c_i = a \cdot b$. Then, the sets of probes $S_0 = \{c_i \mid w_i = 1\}$ and $S_1 = \{c_0\} \cup \{c_i \mid w_i = 0\}$ (whose cardinalities are at most d) satisfy the requirements of Lemma 12, and then, $\mathcal{A}$ is not d-private. Theorem 13 follows. $\square$

4.3 Better Linear Lower Bound

We now show that at least $d + 1$ random bits are actually required if $d \geq 3$.

Theorem 14. *Let $d \geq 3$. Let us consider a d-compression algorithm for multiplication $\mathcal{A}$. If $\mathcal{A}$ uses only d random bits, then $\mathcal{A}$ is not d-private.*

The proof is given in the full version of this paper.

4.4 (Non-constructive) Quasi-Linear Upper Bound

We now construct a d-private compression algorithm for multiplication which requires a quasi-linear number of random bits. More precisely, we show that with non-zero probability, a random algorithm in some family of algorithms (using a quasi-linear number of random bits) is secure, which directly implies the existence of such an algorithm. Note that it is an interesting open problem (though probably difficult) to derandomize this construction.

Concretely, let d be some masking order and R be some number of random bits (used in the algorithm), to be fixed later. For $i = 0, \ldots, d - 1$ and $j = i + 1, \ldots, d$, let us define $\rho(i, j)$ as:

$$\rho(i,j) = \sum_{k=1}^{R} X_{i,j,k} \cdot r_k$$

with $X_{i,j,k} \xleftarrow{\$} \{0,1\}$ for $i = 0, \ldots, d-1$, $j = i+1, \ldots, d$ and $k = 1, \ldots, R$, so that $\rho(i,j)$ is a random sum of all the random bits $r_1, \ldots, r_R$ where each bit appears in $\rho(i,j)$ with probability $1/2$. We also define $X_{d,d,k} = \sum_{i=0}^{d-1} \sum_{j=i+1}^{d} X_{i,j,k}$ and $\rho(d,d)$ as:

$$\rho(d,d) = \sum_{k=1}^{R} X_{d,d,k} \cdot r_k.$$

We generate a (random) algorithm as in Algorithm 2. This algorithm is correct because the sum of all $\rho(i,j)$ is equal to 0.

We point out that we use two kinds of random which should not be confused: the R fresh random bits $r_1, \ldots, r_R$ used in the algorithm to ensure its d-privacy (R is what we really want to be as low as possible), and the random variables $X_{i,j,k}$ used to define a random family of such algorithms (which are "meta"-random bits). In a concrete implementation or algorithm, these latter values are fixed.

Lemma 15. *Algorithm 2 is d-private with probability at least*

$$1 - \binom{(R+3) \cdot d \cdot (d+1)/2}{d} \cdot 2^{-R}$$

over the values of the $X_{i,j,k}$'s.

Algorithm 2. Random algorithm

Require: sharing $(\alpha_{i,j})_{0 \leq i,j \leq d}$
Ensure: sharing $(c_i)_{0 \leq i \leq d}$
 for $i = 1$ to R **do**
  $r_i \xleftarrow{\$} \mathbb{F}_2$
 for $i = 0$ to d **do**
  $c_i \leftarrow \alpha_{i,i}$
  for $j = i + 1$ to d **do**
   $c_i \leftarrow c_i + \rho(i,j) + \alpha_{i,j} + \alpha_{j,i}$ $\triangleright \rho(i,j)$ is not computed first
 $c_d \leftarrow c_d + \rho(d,d)$

Proof (Lemma 15). In order to simplify the proof, we are going to show that, with non-zero probability, there is no set of probes $P = \{p_1, \ldots, p_\ell\}$ with $\ell \leq d$ that satisfies Condition 2. In particular, this implies that, with non-zero probability, there is no set of probes $P = \{p_1, \ldots, p_\ell\}$ with $\ell \leq d$ that satisfies Condition 1, which, via Theorem 7, is equivalent to the algorithm being d-private.

One can only consider sets of exactly d probes as if there is a set of $\ell < d$ probes P' that satisfies Condition 2, one can always complete P' into a set P with exactly d probes by adding $d - \ell$ times the same probe on some input $\alpha_{i,j}$ such that P' initially does not depend on $\alpha_{i,j}$. That is, if M' denotes the matrix such that $\sum_{p' \in P'} p' = a \cdot M' \cdot b$, one could complete P' with any $\alpha_{i,j}$ such that $m'_{i,j} = 0$, so that P, with $\sum_{p \in P} p = a \cdot M \cdot b$ still satisfies Condition 2 if P' initially satisfied the condition.

Thus, let us consider an arbitrary set of d probes $P = \{p_1, \ldots, p_d\}$ and let us bound the probability that P satisfies Condition 2. Let $f = \sum_{i=1}^{d} p_i$. Let us first show that f has to contain at least one $\rho(i,j)$ (meaning that it appears an odd number of times in the sum). Let us assume the contrary, so f does not contain any $\rho(i,j)$. Every $\rho(i,j)$ appears only once in the shares (in the share c_i precisely). Then, one can assume that every probe is made on the same share. Let us assume (without loss of generality) that every probe is made on c_0. If no probe contains any $\rho(0,j)$, then clearly P cannot satisfy Condition 2 as this means that each probe contain at most one $\alpha_{0,j}$, to P cannot contain more than d different $\alpha_{0,j}$. Hence, at least one (so at least two) probe contains at least one $\rho(0,j)$. We note that every probe has one of the following form: either it is exactly a random r_k, a share $\alpha_{0,j}$, a certain $\rho(0,j)$, a certain $\rho(0,j) + \alpha_{0,j}$ or $\rho(0,j) + \alpha_{0,j} + \alpha_{j,0}$, or a subsum (starting from $\alpha_{0,0}$) of c_0. Every form gives at most one $\alpha_{0,j}$ with a new index j except probes on subsums. However, in any subsum, there is always a random $\rho(i,j)$ between $\alpha_{0,j}$ and $\alpha_{0,j+1}$ and one needs to get all the $d+1$ indices to get a set satisfying Condition 2. Then, it is clear that one cannot achieve this unless there is a $\rho(i,j)$ that does not cancel out in the sum, which is exactly what we wanted to show. Now, let $1 \leq k \leq R$ be an integer and let us compute the probability (over the $X_{i,j,k}$'s) that f contains r_k. There exists some set S of pairs (i,j), such that f is the sum of $\sum_{(i,j) \in S} X_{i,j,k} \cdot r_k$ and some other expression not containing any $X_{i,j,k} \cdot r_k$. From the previous point, S is not empty. Furthermore, as there are $d + 1$ outputs $c_0, \ldots, c_d$ and as there are only d probes, S cannot contain all the possible pairs (i, j), and therefore, all the random variables $X_{i,j,k}$ for $(i, j) \in S$ are mutually independent. Therefore, $\sum_{(i,j) \in S} X_{i,j,k}$ is 1 with probability $1/2$ and f functionally depends on the random r_k with probability $1/2$. As there are R possible randoms, f does not functionally depend on any r_k (and then P satisfies Condition 2) with probability $(1/2)^R$.

There are N possibles probes with

$$N \leq \frac{d \cdot (d+1)}{2} + R + (R+2) \cdot \frac{d \cdot (d-1)}{2} \leq (R+3) \cdot \frac{d \cdot (d+1)}{2},$$

as every ρ contains at most R random bits r_k. Also, there are $\binom{N}{d}$ possible sets $P = \{p_1, \ldots, p_d\}$. Therefore, by union bound, the above algorithm is not secure (so there is an attack) with probability at most

$$\binom{N}{d} / 2^R \leq \binom{(R+3) \cdot d \cdot (d+1)/2}{d} \cdot 2^{-R}$$

which concludes the proof of Lemma 15.   □

Theorem 16. *For some $R = O(d \cdot \log d)$, there exists a choice of $\rho(i, j)$ such that Algorithm 2 is a d-private d-compression algorithm for multiplication, when $d \to \infty$.*

We just need to remark that for some $R = O(d \cdot \log d)$, the probability that Algorithm 2 is d-private, according to Lemma 15 is non-zero.

The full proof is given in the full version of this paper.

5   New Construction

The goal of this section is to propose a new d-private multiplication algorithm. Compared to the construction in [15], our construction halves the number of required random bits. It is therefore the most efficient existing construction of a d-private multiplication.

Some rationales behind our new construction may be found in the two following necessary conditions deduced from a careful study of the original work of Ishai, Sahai and Wagner [15].

Lemma 17. *Let $\mathcal{A}(\boldsymbol{a}, \boldsymbol{b}; \boldsymbol{r})$ be a d-compression algorithm for multiplication. Let f be an intermediate result taking the form $f = \boldsymbol{a}^\mathsf{T} \cdot \boldsymbol{M} \cdot \boldsymbol{b} + \boldsymbol{s}^\mathsf{T} \cdot \boldsymbol{r}$. Let t denote the greatest Hamming weight of an element in the vector subspace generated by the rows of $\boldsymbol{M}$ or by the columns of $\boldsymbol{M}$. If $\mathrm{hw}(\boldsymbol{s}) < t - 1$, then $\mathcal{A}(\boldsymbol{a}, \boldsymbol{b}; \boldsymbol{r})$ is not d-private.*

Proof. By definition of $\boldsymbol{s}$, the value $\boldsymbol{a}^\mathsf{T} \cdot \boldsymbol{M} \cdot \boldsymbol{b}$ can be recovered by probing f and then each of the $\mathrm{hw}(\boldsymbol{s}) < t - 1$ random bits on which $\boldsymbol{s}^\mathsf{T} \cdot \boldsymbol{r}$ functionally depends and by summing all these probes. Let $P_1 = \{f, p_1, \ldots, p_j\}$ with $j < t - 1$ denote the set of these at most $t - 1$ probes. Then, we just showed that $f + \sum_{i=1}^{j} p_i = \boldsymbol{a}^{\mathsf{T}} \cdot \boldsymbol{M} \cdot \boldsymbol{b}$.

To conclude the proof, we want to argue that there is a set of at most $d-(t-1)$ probes $P_2 = \{p'_1, \ldots, p'_k\}$ such that $f + \sum_{i=1}^{j} p_i + \sum_{\ell=1}^{k} p'_\ell = \boldsymbol{a}^{\mathsf{T}} \cdot \boldsymbol{M}' \cdot \boldsymbol{b}$, where $\boldsymbol{M}'$ is a matrix such that $\boldsymbol{u}_{d+1}$ is in its row space or in its column space. If such a set P_2 exists, then the set of probes $P_1 \cup P_2$ (whose cardinality is at most d) satisfies Condition 1, and then $\mathcal{A}$ is not d-private, via Theorem 7.

We now use the fact that there is a vector of Hamming weight t in the row space or in the column space of $\boldsymbol{M}$. We can assume (without loss of generality) that there exists a vector $\boldsymbol{w} \in \mathbb{F}_2^{d+1}$ of Hamming weight t in the column subspace of $\boldsymbol{M}$, so that $\boldsymbol{w} = \sum_{j \in J} \boldsymbol{m}_j$, with $J \subseteq \{0, \ldots, d\}$ and $\boldsymbol{m}_j$ the j-th column vector of $\boldsymbol{M}$. Let $i_1, \ldots, i_{d+1-t}$ denote the indices i of $\boldsymbol{w}$ such that $w_i = 0$. Then, let $j \in J$, we claim that $P_2 = \{\alpha_{i_1,j}, \ldots, \alpha_{i_{d+1-t},j}\}$ allows us to conclude the proof. Please note that all these values are probes of intermediate values of $\mathcal{A}$.

Indeed, we have $f + \sum_{i=1}^{j} p_i + \sum_{k=1}^{d+1-t} \alpha_{i_k,j} = \boldsymbol{a}^{\mathsf{T}} \cdot \boldsymbol{M} \boldsymbol{M}' \cdot \boldsymbol{b}$ where all coefficients of $\boldsymbol{M}'$ are the same as coefficients of $\boldsymbol{M}$ except for coefficients in positions $(i_1, j), \ldots, (i_{d+1-t}, j)$ which are the opposite, and now $\sum_{j \in J} \boldsymbol{m}'_j = \boldsymbol{u}_{d+1}$, where $\boldsymbol{m}'_j$ is the j-th column vector of $\boldsymbol{M}'$. Lemma 17 easily follows.   □

In our construction, we satisfy the necessary condition in Lemma 17 by ensuring that any intermediate result that functionally depends on t shares of a (resp. of b) also functionally depends on at least $t - 1$ random bits.

The multiplication algorithm of Ishai, Sahai and Wagner is the starting point of our construction. Before exhibiting it, we hence start by giving the basic ideas thanks to an illustration in the particular case $d = 6$. In Fig. 2 we recall the description of ISW already introduced in Sect. 2.3.

$\hat{\alpha}_{0,0}$	$r_{0,1}$		$r_{0,2}$		$r_{0,3}$		$r_{0,4}$		$r_{0,5}$		$r_{0,6}$	
$\hat{\alpha}_{1,1}$	$(r_{0,1}$	$\hat{\alpha}_{0,1})$	$r_{1,2}$		$r_{1,3}$		$r_{1,4}$		$r_{1,5}$		$r_{1,6}$	
$\hat{\alpha}_{2,2}$	$(r_{0,2}$	$\hat{\alpha}_{0,2})$	$(r_{1,2}$	$\hat{\alpha}_{1,2})$	$r_{2,3}$		$r_{2,4}$		$r_{2,5}$		$r_{2,6}$	
$\hat{\alpha}_{3,3}$	$(r_{0,3}$	$\hat{\alpha}_{0,3})$	$(r_{1,3}$	$\hat{\alpha}_{1,3})$	$(r_{2,3}$	$\hat{\alpha}_{2,3})$	$r_{3,4}$		$r_{3,5}$		$r_{3,6}$	
$\hat{\alpha}_{4,4}$	$(r_{0,4}$	$\hat{\alpha}_{0,4})$	$(r_{1,4}$	$\hat{\alpha}_{1,4})$	$(r_{2,4}$	$\hat{\alpha}_{2,4})$	$(r_{3,4}$	$\hat{\alpha}_{3,4})$	$r_{4,5}$		$r_{4,6}$	
$\hat{\alpha}_{5,5}$	$(r_{0,5}$	$\hat{\alpha}_{0,5})$	$(r_{1,5}$	$\hat{\alpha}_{1,5})$	$(r_{2,5}$	$\hat{\alpha}_{2,5})$	$(r_{3,5}$	$\hat{\alpha}_{3,5})$	$(r_{4,5}$	$\hat{\alpha}_{4,5})$	$r_{5,6}$	
$\hat{\alpha}_{6,6}$	$(r_{0,6}$	$\hat{\alpha}_{0,6})$	$(r_{1,6}$	$\hat{\alpha}_{1,6})$	$(r_{2,6}$	$\hat{\alpha}_{2,6})$	$(r_{3,6}$	$\hat{\alpha}_{3,6})$	$(r_{4,6}$	$\hat{\alpha}_{4,6})$	$(r_{5,6}$	$\hat{\alpha}_{5,6})$

Fig. 2. ISW construction for $d = 6$, with $\hat{\alpha}_{i,j} = \alpha_{i,j} + \alpha_{j,i}$

The first step of our construction is to order the expressions $\hat{\alpha}_{i,j}$ differently. Precisely, to compute the output share c_i (which corresponds, in ISW, to the sum $r_{i,i,} + \sum_{j<i}(r_{j,i} + \hat{\alpha}_{j,i}) + \sum_{j>i} r_{i,j}$ from left to right), we process $r_{i,i,} + \sum_{j<d-i}(r_{i,d-j} + \hat{\alpha}_{i,j}) + \sum_{1 \le j \le i} r_{d-j,i}$ from left to right. Of course, we also put particular care to satisfy the necessary condition highlighted by Lemma 17. This leads to the construction illustrated in Fig. 3.

Then, the core idea is to decrease the randomness cost by reusing some well chosen random bit to protect different steps of the processing. Specifically, for

$$
\begin{aligned}
&\hat{\alpha}_{0,0} &&(r_{0,6} \ \ \hat{\alpha}_{0,6}) &&(r_{0,5} \ \ \hat{\alpha}_{0,5}) &&(r_{0,4} \ \ \hat{\alpha}_{0,4}) &&(r_{0,3} \ \ \hat{\alpha}_{0,3}) &&(r_{0,2} \ \ \hat{\alpha}_{0,2}) &&(r_{0,1} \ \ \hat{\alpha}_{0,1}) \\
&\hat{\alpha}_{1,1} &&(r_{1,6} \ \ \hat{\alpha}_{1,6}) &&(r_{1,5} \ \ \hat{\alpha}_{1,5}) &&(r_{1,4} \ \ \hat{\alpha}_{1,4}) &&(r_{1,3} \ \ \hat{\alpha}_{1,3}) &&(r_{1,2} \ \ \hat{\alpha}_{1,2}) &&r_{0,1} \\
&\hat{\alpha}_{2,2} &&(r_{2,6} \ \ \hat{\alpha}_{2,6}) &&(r_{2,5} \ \ \hat{\alpha}_{2,5}) &&(r_{2,4} \ \ \hat{\alpha}_{2,4}) &&(r_{2,3} \ \ \hat{\alpha}_{2,3}) &&r_{1,2} &&r_{0,2} \\
&\hat{\alpha}_{3,3} &&(r_{3,6} \ \ \hat{\alpha}_{3,6}) &&(r_{3,5} \ \ \hat{\alpha}_{3,5}) &&(r_{3,4} \ \ \hat{\alpha}_{3,4}) &&r_{2,3} &&r_{1,3} &&r_{0,3} \\
&\hat{\alpha}_{4,4} &&(r_{4,6} \ \ \hat{\alpha}_{4,6}) &&(r_{4,5} \ \ \hat{\alpha}_{4,5}) &&r_{3,4} &&r_{2,4} &&r_{1,4} &&r_{0,4} \\
&\hat{\alpha}_{5,5} &&(r_{5,6} \ \ \hat{\alpha}_{5,6}) &&r_{4,5} &&r_{3,5} &&r_{2,5} &&r_{1,5} &&r_{0,5} \\
&\hat{\alpha}_{6,6} &&r_{5,6} &&r_{4,6} &&r_{3,6} &&r_{2,6} &&r_{1,6} &&r_{0,6}
\end{aligned}
$$

Fig. 3. First step of our new construction for $d = 6$, with $\hat{\alpha}_{i,j} = \alpha_{i,j} + \alpha_{j,i}$

any even positive number k, we show that replacing all the random bits $r_{i,j}$ such that $k = j - i$ with a fixed random bit r_k preserves the d-privacy of ISW algorithm. Note, however, that the computations then have to be performed with a slightly different bracketing in order to protect the intermediate variables which involve the same random bits. The obtained construction is illustrated in Fig. 4.

$$
\begin{aligned}
&\hat{\alpha}_{0,0} &&(r_{0,6} \ \ \hat{\alpha}_{0,6} \ \ r_5 \ \ \hat{\alpha}_{0,5}) &&(r_{0,4} \ \ \hat{\alpha}_{0,4} \ \ r_3 \ \ \hat{\alpha}_{0,3}) &&(r_{0,2} \ \ \hat{\alpha}_{0,2} \ \ r_1 \ \ \hat{\alpha}_{0,1}) \\
&\hat{\alpha}_{1,1} &&(r_{1,6} \ \ \hat{\alpha}_{1,6} \ \ r_5 \ \ \hat{\alpha}_{1,5}) &&(r_{1,4} \ \ \hat{\alpha}_{1,4} \ \ r_3 \ \ \hat{\alpha}_{1,3}) &&(r_{1,2} \ \ \hat{\alpha}_{1,2}) &&r_1 \\
&\hat{\alpha}_{2,2} &&(r_{2,6} \ \ \hat{\alpha}_{2,6} \ \ r_5 \ \ \hat{\alpha}_{2,5}) &&(r_{2,4} \ \ \hat{\alpha}_{2,4} \ \ r_3 \ \ \hat{\alpha}_{2,3}) &&r_{1,2} &&r_{0,2} \\
&\hat{\alpha}_{3,3} &&(r_{3,6} \ \ \hat{\alpha}_{3,6} \ \ r_5 \ \ \hat{\alpha}_{3,5}) &&(r_{3,4} \ \ \hat{\alpha}_{3,4}) &&r_3 &&r_3 &&r_3 \\
&\hat{\alpha}_{4,4} &&(r_{4,6} \ \ \hat{\alpha}_{4,6} \ \ r_5 \ \ \hat{\alpha}_{4,5}) &&r_{3,4} &&r_{2,4} &&r_{1,4} &&r_{0,4} \\
&\hat{\alpha}_{5,5} &&(r_{5,6} \ \ \hat{\alpha}_{5,6}) &&r_5 &&r_5 &&r_5 &&r_5 \\
&\hat{\alpha}_{6,6} &&r_{5,6} &&r_{4,6} &&r_{3,6} &&r_{2,6} &&r_{1,6} &&r_{0,6}
\end{aligned}
$$

Fig. 4. Second step of our new construction for $d = 6$, with $\hat{\alpha}_{i,j} = \alpha_{i,j} + \alpha_{j,i}$

Finally, we suppress from our construction the useless repetitions of random bits that appear at the end of certain computations. Hence, we obtain our new construction, illustrated in Fig. 5.

$$
\begin{aligned}
&\hat{\alpha}_{0,0} &&(r_{0,6} \ \ \hat{\alpha}_{0,6} \ \ r_5 \ \ \hat{\alpha}_{0,5}) &&(r_{0,4} \ \ \hat{\alpha}_{0,4} \ \ r_3 \ \ \hat{\alpha}_{0,3}) &&(r_{0,2} \ \ \hat{\alpha}_{0,2} \ \ r_1 \ \ \hat{\alpha}_{0,1}) \\
&\hat{\alpha}_{1,1} &&(r_{1,6} \ \ \hat{\alpha}_{1,6} \ \ r_5 \ \ \hat{\alpha}_{1,5}) &&(r_{1,4} \ \ \hat{\alpha}_{1,4} \ \ r_3 \ \ \hat{\alpha}_{1,3}) &&(r_{1,2} \ \ \hat{\alpha}_{1,2}) &&r_1 \\
&\hat{\alpha}_{2,2} &&(r_{2,6} \ \ \hat{\alpha}_{2,6} \ \ r_5 \ \ \hat{\alpha}_{2,5}) &&(r_{2,4} \ \ \hat{\alpha}_{2,4} \ \ r_3 \ \ \hat{\alpha}_{2,3}) &&r_{1,2} &&r_{0,2} \\
&\hat{\alpha}_{3,3} &&(r_{3,6} \ \ \hat{\alpha}_{3,6} \ \ r_5 \ \ \hat{\alpha}_{3,5}) &&(r_{3,4} \ \ \hat{\alpha}_{3,4}) &&r_3 \\
&\hat{\alpha}_{4,4} &&(r_{4,6} \ \ \hat{\alpha}_{4,6} \ \ r_5 \ \ \hat{\alpha}_{4,5}) &&r_{3,4} &&r_{2,4} &&r_{1,4} &&r_{0,4} \\
&\hat{\alpha}_{5,5} &&(r_{5,6} \ \ \hat{\alpha}_{5,6}) &&r_5 \\
&\hat{\alpha}_{6,6} &&r_{5,6} &&r_{4,6} &&r_{3,6} &&r_{2,6} &&r_{1,6} &&r_{0,6}
\end{aligned}
$$

Fig. 5. Application of our new construction for $d = 6$, with $\hat{\alpha}_{i,j} = \alpha_{i,j} + \alpha_{j,i}$

Before proving that this scheme is indeed d-private, we propose a formal description in Algorithm 3. As can be seen, this new scheme involves $3d^2/2 + d(d + 2)/4 + 2d$ sums if d is even and $3(d^2 - 1)/2 + (d + 1)^2/4 + 3(d + 1)/2$

Algorithm 3. New construction for d-secure multiplication

Require: sharing $(\alpha_{i,j})_{0\leq i,j\leq d}$
Ensure: sharing $(c_i)_{0\leq i\leq d}$

1: **for** $i = 0$ to d **do**   ▷ Random Bits Generation
2:      **for** $j = 0$ to $d - i - 1$ by 2 **do**
3:          $r_{i,d-j} \xleftarrow{\$} \mathbb{F}_2$
4: **for** $j = d - 1$ downto 1 by 2 **do**
5:      $r_j \xleftarrow{\$} \mathbb{F}_2$
6: **for** $i = 0$ to d **do**   ▷ Multiplication
7:      $c_i \leftarrow \alpha_{i,i}$
8:      **for** $j = d$ downto $i + 2$ by 2 **do**
9:          $t_{i,j} \leftarrow r_{i,j} + \alpha_{i,j} + \alpha_{j,i} + r_{j-1} + \alpha_{i,j-1} + \alpha_{j-1,i}; \quad c_i \leftarrow c_i + t_{i,j}$
10:     **if** $i \not\equiv d \pmod 2$ **then**
11:         $t_{i,i+1} \leftarrow r_{i,i+1} + \alpha_{i,i+1} + \alpha_{i+1,i}; \quad c_i \leftarrow c_i + t_{i,i+1}$
12:         **if** $i \equiv 1 \pmod 2$ **then**   ▷ Correction r_i
13:             $c_i \leftarrow c_i + r_i$
14:     **else**
15:         **for** $j = i - 1$ downto 0 **do**   ▷ Correction $r_{i,j}$
16:             $c_i \leftarrow c_i + r_{j,i}$

Algorithm 4. Second-Order Compression Algorithm

Require: sharing $(\alpha_{i,j})_{0\leq i,j\leq 2}$
Ensure: sharing $(c_i)_{0\leq i\leq 2}$

$r_0 \xleftarrow{\$} \mathbb{F}_2; \quad r_1 \leftarrow \mathbb{F}_2$
$c_0 \leftarrow \alpha_{0,0} + r_0 + \alpha_{0,2} + \alpha_{2,0}$
$c_1 \leftarrow \alpha_{1,1} + r_1 + \alpha_{0,1} + \alpha_{1,0}$
$c_2 \leftarrow \alpha_{2,2} + r_0 + r_1 + \alpha_{1,2} + \alpha_{2,1}$

Algorithm 5. Third-Order Compression Algorithm

Require: sharing $(\alpha_{i,j})_{0\leq i,j\leq 3}$
Ensure: sharing $(c_i)_{0\leq i\leq 3}$

$r_0 \xleftarrow{\$} \mathbb{F}_2; \quad r_1 \xleftarrow{\$} \mathbb{F}_2; \quad r_2 \xleftarrow{\$} \mathbb{F}_2; \quad r_3 \xleftarrow{\$} \mathbb{F}_2$
$c_0 \leftarrow \alpha_{0,0} + r_0 + \alpha_{0,3} + \alpha_{3,0} + r_1 + \alpha_{0,2} + \alpha_{2,0}$
$c_1 \leftarrow \alpha_{1,1} + r_2 + \alpha_{1,3} + \alpha_{3,1} + r_1 + \alpha_{1,2} + \alpha_{2,1}$
$c_2 \leftarrow \alpha_{2,2} + r_3 + \alpha_{2,3} + \alpha_{3,2}$
$c_3 \leftarrow \alpha_{3,3} + r_3 + r_2 + r_0 + \alpha_{0,1} + \alpha_{1,0}$

if d is odd. In every case, it also involves $(d + 1)^2$ multiplications and requires the generation of $d^2/4 + d$ random values in $\mathbb{F}_2$ if d is even and $(d^2 - 1)/4 + d$ otherwise (see Table 1 for values at several orders and comparison with ISW).

Proposition 18. *Algorithm 3 is d-private.*

Algorithm 3 was proven to be d-private with the verifier built by Barthe et al. [4] up to order $d = 6$. Furthermore, a pen-and-paper proof for any order d is given in the full version of this paper.

6 Optimal Small Cases

We propose three secure compression algorithms using less random bits than the generic solution given by ISW and than our new solution for the specific small

Table 1. Complexities of ISW, our new d-private compression algorithm for multiplication and our specific algorithms at several orders

Complexities	Algorithm ISW	Algorithm 3	Algorithms 4, 5 and 6
Second-Order Masking			
Sums	12	12	10
Products	9	9	9
Random bits	3	3	2
Third-Order Masking			
Sums	24	22	20
Products	16	16	16
Random bits	6	5	4
Fourth-Order Masking			
Sums	40	38	30
Products	25	25	25
Random bits	10	8	5
d^{th}-Order Masking			
Sums	$2d(d+1)$	$\begin{cases} d(7d+10)/4 & (d \text{ even}) \\ (7d+1)(d+1)/4 & (d \text{ odd}) \end{cases}$	-
Products	$(d+1)^2$	$(d+1)^2$	-
Random bits	$d(d+1)/2$	$\begin{cases} d^2/4+d & (d \text{ even}) \\ (d^2-1)/4+d & (d \text{ odd}) \end{cases}$	-

orders $d = 2, 3$ and 4. These algorithms actually use only the optimal numbers of random bits for these small quantity of probes, as proven in Sect. 4. Furthermore, since they all are dedicated to a specific order d (among 2, 3, and 4), we got use of the verifier proposed by Barthe et al. in [4] to formally prove their correctness and their d-privacy.

Proposition 19. *Algorithms 4, 5, and 6 are correct and respectively 2, 3 and 4-private.*

Table 1 (Sect. 5) compares the amount of randomness used by the new construction proposed in Sect. 5 and by our optimal small algorithms. We recall that each of them attains the lower bound proved in Sect. 4.

7   Composition

Our new algorithms are all d-private, when applied on the outputs of a multiplicative encoder parameterized at order d. We now aim to show how they can be involved in the design of larger functions (e.g., block ciphers) to achieve a global d-privacy. In [3], Barthe et al. introduce and formally prove a method to

---

Algorithm 6. Fourth-Order Compression Algorithm

Require: sharing $(\alpha_{i,j})_{0 \leq i,j \leq 4}$
Ensure: sharing $(c_i)_{0 \leq i \leq 4}$

$$r_0 \xleftarrow{\$} \mathbb{F}_2; \quad r_1 \xleftarrow{\$} \mathbb{F}_2; \quad r_2 \xleftarrow{\$} \mathbb{F}_2; \quad r_3 \xleftarrow{\$} \mathbb{F}_2; \quad r_4 \xleftarrow{\$} \mathbb{F}_2$$

$c_0 \leftarrow \alpha_{0,0} + r_0 + \alpha_{0,1} + \alpha_{1,0} + r_1 + \alpha_{0,2} + \alpha_{2,0}$

$c_1 \leftarrow \alpha_{1,1} + r_1 + \alpha_{1,2} + \alpha_{2,1} + r_2 + \alpha_{1,3} + \alpha_{3,1}$

$c_2 \leftarrow \alpha_{2,2} + r_2 + \alpha_{2,3} + \alpha_{3,2} + r_3 + \alpha_{2,4} + \alpha_{4,2}$

$c_3 \leftarrow \alpha_{3,3} + r_3 + \alpha_{3,4} + \alpha_{4,3} + r_4 + \alpha_{3,0} + \alpha_{0,3}$

$c_4 \leftarrow \alpha_{4,4} + r_4 + \alpha_{4,0} + \alpha_{0,4} + r_0 + \alpha_{4,1} + \alpha_{1,4}$

---

compose small d-private algorithms (a.k.a., *gadgets*) into d-private larger functions. The idea is to carefully refresh the sharings when necessary, according to the security properties of the gadgets. Before going further into the details of this composition, we recall some security properties used in [3].

7.1 Compositional Security Notions

Before stating the new security definitions, we first need to introduce the notion of simulatability. For the sake of simplicity, we only state this notion for multiplication algorithm, but this can easily be extended to more general algorithms.

Definition 20. *A set $P = \{p_1, \ldots, p_\ell\}$ of ℓ probes of a multiplication algorithm can be* simulated *with at most t shares of each input, if there exists two sets $I = \{i_1, \ldots, i_t\}$ and $J = \{j_1, \ldots, j_t\}$ of t indices from $\{0, \ldots, d\}$ and a random function f taking as input $2t$ bits and outputting ℓ bits such that for any fixed bits $(a_i)_{0 \leq i \leq d}$ and $(b_j)_{0 \leq j \leq d}$, the distributions $\{p_1, \ldots, p_\ell\}$ (which implicitly depends on $(a_i)_{0 \leq i \leq d}$, $(b_j)_{0 \leq j \leq d}$, and the random coins used in the multiplication algorithm) and $\{f(a_{i_1}, \ldots, a_{i_t}, b_{j_1}, \ldots, b_{j_t})\}$ are identical.*

We write $f(a_{i_1}, \ldots, a_{i_t}, b_{j_1}, \ldots, b_{j_t}) = f(a_I, b_J)$.

Definition 21. *An algorithm is d-non-interferent (or d-NI) if and only if every set of at most d probes can be simulated with at most d shares of each input.*

While this notion might be stronger than the notion of security we used, all our concrete constructions in Sects. 5 and 6 satisfy it. The proof of Algorithm 3 is indeed a proof by simulation, while the small cases in Sect. 6 are proven using the verifier by Barthe et al. in [4], which directly proves NI.

Definition 22. *An algorithm is d-tight non-interferent (or d-TNI) if and only if every set of $t \leq d$ probes can be simulated with at most t shares of each input.*

While this notion of d-tight non-interference was assumed to be stronger than the notion of d-non-interference in [3], we show hereafter that these two security notions are actually equivalent. In particular, this means that all our concrete constructions are also TNI.

Proposition 23 (*d-NI* ⇔ *d-TNI*). *An algorithm is d-non-interferent if and only if it is d-tight non-interferent.*

Proof. The right-to-left implication is straightforward from the definitions. Let us thus consider the left-to-right direction.

For that purpose, we first need to introduce a technical lemma. Again, for the sake of simplicity, we only consider multiplication algorithm, with only two inputs, but the proof can easily be generalized to any algorithm.   □

Lemma 24. *Let* $P = \{p_1, \ldots, p_\ell\}$ *be a set of* ℓ *probes which can be simulated by the sets* (I, J) *and also by the sets* (I', J'). *Then it can also be simulated by* $(I \cap I', J \cap J')$.

Proof. Let f the function corresponding to I, J and f' the function corresponding to I', J'. We have that for any bits $(a_i)_{0 \leq i \leq d}$ and $(b_j)_{0 \leq j \leq d}$, the distributions $\{p_1, \ldots, p_\ell\}$, $\{f(a_I, b_J)\}$, and $\{f'(a_{I'}, b_{J'})\}$ are identical. Therefore, f does not depend on a_i nor b_j for $i \in I \setminus I'$ and $j \in J \setminus J'$, since f' does not depend on them. Thus, P can be simulated by only shares from $I \cap I', J \cap J'$ (using the function f where the inputs corresponding to a_i and b_j for $i \in I \setminus I'$ and $j \in J \setminus J'$ are just set to zero, for example).   □

We now assume that an algorithm $\mathcal{A}$ is d-NI, that is, every set of at most d probes can be simulated with at most d shares of each input. Now, by contradiction, let us consider a set P with minimal cardinality $t < d$ of probes on $\mathcal{A}$, such that it cannot be simulated by at most t shares of each input. Let us consider the sets I, J corresponding to the intersection of all sets I', J' (respectively) such that the set P can be simulated by I', J'. The sets I, J also simulate P thanks to Lemma 24. Furthermore, by hypothesis, $t < |I| \leq d$ or $t < |J| \leq d$. Without loss of generality, let us suppose that $|I| > t$.

Let i^* be an arbitrary element of $\{0, \ldots, d\} \setminus I$ (which is not an empty set as $|I| \leq d$). Let us now consider the set of probes $P' = P \cup \{a_{i^*}\}$. By hypothesis, P' can be simulated by at most $|P'| = t + 1$ shares of each input. Let I', J' two sets of size at most $t + 1$ simulating P'. These two sets also simulate $P \subseteq P'$, therefore, $I \cap I', J \cap J'$ also simulate P. Furthermore, $i^* \in I$, as all the shares a_i are independent. Since $i^* \notin I$, $|I \cap I'| \leq t$ and $I \cap I' \subsetneq I$, which contradicts the fact that I and J were the intersection of all sets I'', J'' simulating P.   □

Definition 25. *An algorithm $\mathcal{A}$ is d-strong non-interferent (or d-SNI) if and only if for every set $\mathcal{I}$ of t_1 probes on intermediate variables (i.e., no output wires or shares) and every set $\mathcal{O}$ of t_2 probes on output shares such that $t_1 + t_2 \leq d$, the set $\mathcal{I} \cup \mathcal{O}$ of probes can be simulated by only t_1 shares of each input.*

The composition of two d-SNI algorithms is itself d-SNI, while that of d-TNI algorithms is not necessarily d-TNI. This implies that d-SNI gadgets can be directly composed while maintaining the d-privacy property, whereas a so-called *refreshing* gadget must sometimes be involved before the composition of d-TNI algorithms. Since the latter refreshing gadgets consume the same quantity of random values as ISW, limiting their use is crucial if the goal is to reduce the global amount of randomness.

7.2   Building Compositions with Our New Algorithms

In [3], the authors show that the ISW multiplication is d-SNI and use it to build secure compositions. Unfortunately, our new multiplication algorithms are d-TNI but not d-SNI. Therefore, as discussed in the previous section, they can replace only some of the ISW multiplications in secure compositions. Let us take the example of the AES inversion that is depicted in [3]. We can prove that replacing the first ($\mathcal{A}^7$) and the third ($\mathcal{A}^2$) ISW multiplications by d-TNI multiplications (e.g., our new constructions) and moving the refreshing algorithm R in different locations preserves the strong non-interference of the inversion, while benefiting from our reduction of the randomness consumption.

The tweaked inversion is given in Fig. 6. $\otimes$ denotes the d-SNI ISW multiplication, $\cdot^\alpha$ denotes the exponentiation to the power α, $\mathcal{A}^i$ refers to the i-th algorithm or gadget (indexed from left to right), R denotes the d-SNI refreshing gadget, $\mathcal{I}^i$ denotes the set of internal probes in the i-th algorithm, $\mathcal{S}_j^i$ denotes the set of shares from the j inputs of algorithm $\mathcal{A}^i$ used to simulate all further probes. Finally, x denotes the inversion input and $\mathcal{O}$ denotes the set of probes at the output of the inversion. The global constraint for the inversion to be d-SNI (and thus itself composable) is that: $|\mathcal{S}^8 \cup \mathcal{S}^9| \leq \sum_{1 \leq i \leq 9} |\mathcal{I}^i|$, i.e., all the internal probes can be perfectly simulated with at most $\sum_{1 \leq i \leq 9} |\mathcal{I}^i|$ shares of x.

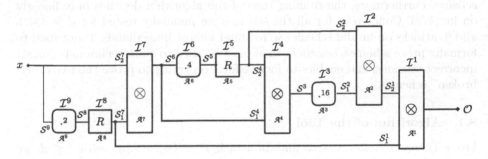

Fig. 6. AES $\cdot^{254}$

Proposition 26. *The AES inversion given in Fig. 6 with $\mathcal{A}^1$ and $\mathcal{A}^4$ being d-SNI multiplications and $\mathcal{A}^2$ and $\mathcal{A}^7$ being d-TNI multiplications is d-SNI.*

Proof. From the d-probing model, we assume that the total number of probes used to attack the inversion is limited to d, that is $\sum_{1 \leq i \leq 9} |\mathcal{I}^i| + |\mathcal{O}| \leq d$. As in [3], we build the proof from right to left by simulating each algorithm. Algorithm $\mathcal{A}^1$ is d-SNI, thus $|\mathcal{S}_1^1|, |\mathcal{S}_2^1| \leq |\mathcal{I}^1|$. Algorithm $\mathcal{A}^2$ is d-TNI, thus $|\mathcal{S}_1^2|, |\mathcal{S}_2^2| \leq |\mathcal{I}^1 + \mathcal{I}^2|$. As explained in [3], since Algorithm $\mathcal{A}^3$ is affine, then $|\mathcal{S}^3| \leq |\mathcal{S}_1^2 + \mathcal{I}^3| \leq |\mathcal{I}^1 + \mathcal{I}^2 + \mathcal{I}^3|$. Algorithm $\mathcal{A}^4$ is d-SNI, thus $|\mathcal{S}_1^4|, |\mathcal{S}_2^4| \leq |\mathcal{I}^4|$. Algorithm $\mathcal{A}^5$ is d-SNI, thus $|\mathcal{S}^5| \leq |\mathcal{I}^5|$. Algorithm $\mathcal{A}^6$ is affine, thus $|\mathcal{S}^6| \leq |\mathcal{S}^5 + \mathcal{I}^6| \leq |\mathcal{I}^5 + \mathcal{I}^6|$. Algorithm $\mathcal{A}^7$ is d-TNI, thus $|\mathcal{S}_1^7|, |\mathcal{S}_2^7| \leq |\mathcal{S}^6 + \mathcal{S}_1^4 + \mathcal{I}^7| \leq$

$|\mathcal{I}^4 + \mathcal{I}^5 + \mathcal{I}^6 + \mathcal{I}^7|$. Algorithm $\mathcal{A}^8$ is d-SNI, thus $|\mathcal{S}^8| \leq |\mathcal{I}^8|$. Algorithm $\mathcal{A}^9$ is affine, thus $|\mathcal{S}^9| \leq |\mathcal{I}^9 + \mathcal{S}^8| \leq |\mathcal{I}^8 + \mathcal{I}^9|$. Finally, all the probes of this inversion can be perfectly simulated from $|\mathcal{S}^9 \cup \mathcal{S}_1^7| \leq |\mathcal{I}^4 + \mathcal{I}^5 + \mathcal{I}^6 + \mathcal{I}^7 + \mathcal{I}^8 + \mathcal{I}^9|$ shares of x, which proves that the inversion is still d-SNI.   □

From Proposition 26, our new constructions can be used to build d-SNI algorithms. In the case of the AES block cipher, half of the d-SNI ISW multiplications can be replaced by ours while preserving the whole d-SNI security.

8   New Automatic Tool for Finding Attacks

In this section, we describe a new automatic tool for finding attacks on compression algorithms for multiplication which is developed in Sage (Python) [27]. Compared to the verifier developed by Barthe *et al.* [4] and based on Easycrypt, to find attacks in practice, our tool is not as generic as it focuses on compression algorithms for multiplication and its soundness is not perfect (and relies on some heuristic assumption). Nevertheless, it is order of magnitudes faster.

A non-perfect soundness means that the algorithm may not find an attack and can only guarantee that there does not exist an attack except with probability ε. We believe that, in practice, this limitation is not a big issue as if ε is small enough (e.g., 2^{-20}), a software bug is much more likely than an attack on the scheme. Furthermore, the running time of the algorithm depends only linearly on $\log(1/\varepsilon)$. Concretely, for all the schemes we manually tested for $d = 3, 4, 5$ and 6, attacks on invalid schemes were found almost immediately. If not used to formally prove schemes, our tool can at least be used to quickly eliminate (most) incorrect schemes, and enables to focus efforts on trying to prove "non-trivially-broken" schemes.

8.1   Algorithm of the Tool

From Theorem 7, in order to find an attack $P = \{p_1, \ldots, p_\ell\}$ with $\ell \leq d$, we just need to find a set $P = \{p_1, \ldots, p_\ell\}$ satisfying Condition 1. If no such set P exists, the compression algorithm for multiplication is d-private.

A naive way to check the existence of such a set P is to enumerate all the sets of d probes. However, there are $\binom{N}{d}$ such sets, with N being the number of intermediate variables of the algorithm. For instance, to achieve 4-privacy, our construction (see Sect. 6) uses $N = 81$ intermediate variables, which makes more than 2^{20} sets of four variables to test. In [4], the authors proposed a faster way of enumerating these sets by considering larger sets which are still independent from the secret. However, their method falls short for the compression algorithms in our paper as soon as $d > 6$, as shown in Sect. 8.4. Furthermore even for $d = 3, 4, 5$, their tool takes several minutes to prove security (around 5 min to check security of Algorithm 3 with $d = 5$) or to find an attack for incorrect schemes, which prevent people from quickly checking the validity of a newly designed scheme.

To counteract this issue, we design a new tool which is completely different and which borrows ideas from coding theory to enumerate the sets of d or less intermediate variables. Let $\gamma_1, \ldots, \gamma_\nu$ be all the intermediate results whose expression functionally depends on at least one random and $\gamma'_1, \ldots, \gamma'_{\nu'}$ be the other intermediate results that we refer to as deterministic intermediate results ($\nu + \nu' = N$). We remark that all the $\alpha_{i,j} = a_i b_j$ are intermediate results and that no intermediate result can functionally depend on more than one shares' product $\alpha_{i,j} = a_i b_j$ without also depending on a random bit. Otherwise, the compression algorithm would not be d-private, according to Lemma 17. As this condition can be easily tested, we now assume that the only deterministic intermediate results are the $\alpha_{i,j} = a_i b_j$ that we refer to as γ'_k in the following. As an example, intermediate results of Algorithm 4 are depicted in Table 2.

Table 2. Intermediate results of Algorithm 4

Non-deterministic ($\nu = 12$)		Deterministic ($\nu' = 9$)	
$\gamma_1 = a_0 b_0 + r_0$	$\gamma_7 = c_1$	$\gamma'_1 = a_0 b_0$	$\gamma'_6 = a_1 b_0$
$\gamma_2 = a_0 b_0 + r_0 + a_0 b_2$	$\gamma_8 = r_1$	$\gamma'_2 = a_0 b_2$	$\gamma'_7 = a_2 b_2$
$\gamma_3 = c_0$	$\gamma_9 = a_2 b_2 + r_1$	$\gamma'_3 = a_2 b_0$	$\gamma'_8 = a_1 b_2$
$\gamma_4 = r_0$	$\gamma_{10} = a_2 b_2 + r_1 + r_0$	$\gamma'_4 = a_1 b_1$	$\gamma'_9 = a_2 b_1$
$\gamma_5 = a_1 b_1 + r_1$	$\gamma_{11} = a_2 b_2 + r_1 + r_0 + a_1 b_2$	$\gamma'_5 = a_0 b_1$	
$\gamma_6 = a_1 b_1 + r_1 + a_0 b_1$	$\gamma_{12} = c_2$		

An attack set $P = \{p_1, \ldots, p_\ell\}$ can then be separated into two sets $Q = \{\gamma_{i_1}, \ldots, \gamma_{i_\delta}\}$ and $Q' = \{\gamma_{i'_1}, \ldots, \gamma_{i'_{\delta'}}\}$, with $\ell = \delta + \delta' \leq d$. We remark that necessarily $\sum_{p \in Q} p$ does not functionally depend on any random value. Actually, we even have the following lemma:

Lemma 27. *Let $\mathcal{A}(\boldsymbol{a}, \boldsymbol{b}; \boldsymbol{r})$ be a compression algorithm for multiplication. Then $\mathcal{A}$ is d-private if and only if there does not exist a set of non-deterministic probes $Q = \{\gamma_{i_1}, \ldots, \gamma_{i_\delta}\}$ with $\delta \leq d$ such that $\sum_{p \in Q} p = \boldsymbol{a}^{\mathsf{T}} \cdot \boldsymbol{M} \cdot \boldsymbol{b}$ where the column space or the row space of $\boldsymbol{M}$ contains a vector of Hamming weight at least $\delta + 1$.*

Furthermore, if such a set Q exists, there exists a set $\{\gamma_{i'_1}, \ldots, \gamma_{i'_{\delta'}}\}$, with $\delta + \delta' \leq d$, such that $P = Q \cup Q'$ is an attack.

Moreover, the lemma is still true when we restrict ourselves to sets Q such that there exists no proper subset $\hat{Q} \subsetneq Q$ such that $\sum_{p \in \hat{Q}} p$ does not functionally depend on any random.

Proof. The two first paragraphs of the lemma can be proven similarly to Lemma 17. Thus, we only need to prove its last part.

By contradiction, let us suppose that there exists a set Q of non-deterministic probes $Q = \{\gamma_{i_1}, \ldots, \gamma_{i_\delta}\}$ such that $\sum_{p \in Q} p = \boldsymbol{a}^{\mathsf{T}} \cdot \boldsymbol{M} \cdot \boldsymbol{b}$ and the column space (without loss of generality, by symmetry of the a_i's and b_i's) of $\boldsymbol{M}$ contains a vector of Hamming weight at least $\delta + 1$, but such that any subset $\hat{Q} \subsetneq Q$ where $\sum_{p \in \hat{Q}} p$ that does not functionally depend on any random. Consequently, the sum $\sum_{p \in \hat{Q}} p = a^{\mathsf{T}} \cdot \hat{M} \cdot b$, is such that the column space (still without loss of generality) of $\hat{M}$ does not contain any vector of Hamming weight at least $|\hat{Q}| + 1$.

First, let us set $\bar{M} = \hat{M} + M$ (over $\mathbb{F}_2$), so $\sum_{p \in Q \setminus \hat{Q}} p = a^{\mathsf{T}} \cdot \bar{M} \cdot b$, as $\sum_{p \in \hat{Q}} p + \sum_{p \in Q \setminus \hat{Q}} p = \sum_{p \in Q} p = a^{\mathsf{T}} \cdot M \cdot b$ and let $\hat{\delta} = |\hat{Q}|$ and $\bar{\delta} = |Q \setminus \hat{Q}| = \delta - \hat{\delta}$. Let also ω, $\hat{\omega}$, and $\bar{\omega}$ be the maximum Hamming weights of the vectors in the column space of M, $\hat{M}$, and $\bar{M}$, respectively. Since $M = \hat{M} + \bar{M}$, then $\omega \leq \hat{\omega} + \bar{\omega}$ and since $\omega > \delta + 1$, and $\delta = \hat{\delta} + \bar{\delta}$, then $\hat{\omega} > \hat{\delta}$ or $\bar{\omega} > \bar{\delta}$. We set $\tilde{Q} = \hat{Q}$ if $\hat{\omega} > \hat{\delta}$, and $\tilde{Q} = Q \setminus \hat{Q}$ otherwise. According to the definitions of $\hat{\delta}$ and $\bar{\omega}$, we have that $\tilde{Q} \subsetneq Q$ is such that $\sum_{p \in Q} p = a^{\mathsf{T}} \cdot \tilde{M} \cdot b$ where the column space of $\tilde{M}$ contains a vector of Hamming weight at least $|\tilde{Q}| + 1$. This contradicts the definition of Q and concludes the proof of the lemma. $\square$

To quickly enumerate all the possible attacks, we first enumerate the sets $Q = \{\gamma_{i_1}, \ldots, \gamma_{i_\delta}\}$ of size $\delta \leq d$ such that $\sum_{p \in Q} p$ does not functionally depend on any random bit (and no proper subset of $\hat{Q} \subsetneq Q$ is such that $\sum_{p \in \hat{Q}} p$ does not functionally depend on any random bit), using *information set decoding*, recalled in the next section. Then, for each possible set Q, we check if the column space or the row space of M (as defined in the previous lemma) contains a vector of Hamming weight at least $\delta + 1$. A naive approach would consist in enumerating all the vectors in the row space and the column space of M. Our tool however uses the two following facts to perform this test very quickly in most cases:

- when M contains at most δ non-zero rows and at most δ non-zero columns, Q does not yield an attack;
- when M contains exactly $\delta + 1$ non-zero rows (resp. columns), that we assume to be the first $\delta + 1$ (without loss of generality), Q yields an attack if and only if the vector $(u_{\delta+1}^{\mathsf{T}}, 0_{d-\delta}^{\mathsf{T}})$ is in the row space (resp. $(u_{\delta+1}, 0_{d-\delta})$ is in the column space) of M (this condition can be checked in polynomial time in d).

8.2   Information Set Decoding and Error Probability

We now explain how to perform the enumeration step of our algorithm using information set decoding. Information set decoding was introduced in the original security analysis of the McEliece cryptosystem in [20,22] as a way to break the McEliece cryptosystem by finding small code words in a random linear code. It was further explored by Lee and Brickell in [18]. We should point out that since then, many improvements were proposed, e.g., in [19,26]. However, for the sake of simplicity and because it already gives very good results, we use the original information set decoding algorithm. Furthermore, it is not clear that the aforementioned improvements also apply in our case, as the codes we consider are far from the Singleton bound.

We assume that random bits are denoted $r_1, \ldots, r_R$. For each intermediate γ_k containing some random bit, we associate the vector $\tau \in \mathbb{Z}_2^R$, where $\tau_i = 1$ if and only if γ_k functionally depends on the random bit r_i. We then consider the matrix $\Gamma \in \mathbb{Z}_2^{R \times \nu}$ whose k-th column is $\boldsymbol{\tau}$. For instance, for Algorithm 4, we have:

$$
\Gamma = \begin{matrix} & \gamma_1 & \gamma_2 & \gamma_3 & \gamma_4 & \gamma_5 & \gamma_6 & \gamma_7 & \gamma_8 & \gamma_9 & \gamma_{10} & \gamma_{11} & \gamma_{12} & \\ & \begin{pmatrix} 1 & 1 & 1 & 1 & 0 & 0 & 0 & 0 & 1 & 1 & 1 & 1 \\ 0 & 0 & 0 & 0 & 1 & 1 & 1 & 1 & 0 & 1 & 1 & 1 \end{pmatrix} & \begin{matrix} r_0 \\ r_1 \end{matrix} \end{matrix}.
$$

For every $\delta \leq d$, enumerating the sets $Q = \{\gamma_{i_1}, \ldots, \gamma_{i_\delta}\}$, such that $\sum_{p \in Q} p$ does not functionally depend on any random, consists in enumerating the vectors $\boldsymbol{x}$ of Hamming weight δ such that $\Gamma \cdot \boldsymbol{x} = \boldsymbol{0}$ (specifically, $\{i_1, \ldots, i_\delta\}$ are the coordinates of the non-zero components of $\boldsymbol{x}$). Furthermore, we can restrict ourselves to vector $\boldsymbol{x}$ such that no vector $\hat{\boldsymbol{x}} < \boldsymbol{x}$ satisfies $\Gamma \cdot \hat{\boldsymbol{x}} = \boldsymbol{0}$ (where $\hat{\boldsymbol{x}} < \boldsymbol{x}$ means that $\hat{\boldsymbol{x}} \neq \boldsymbol{x}$ and for any $1 \leq i \leq \nu$, if $x_i = 0$ then $\hat{x}_i = 0$), since we can restrict ourselves to sets Q such that no proper subset $\hat{Q} \subsetneq Q$ is such that $\sum_{p \in \hat{Q}} p$ does not functionally depend on any random bit. This is close to the problem of finding code words $\boldsymbol{x}$ of small Hamming weight for the linear code of parity matrix Γ and we show this can be solved using information set decoding.

The basic idea is the following one. We first apply a row-reduction to Γ. Let us call the resulting matrix Γ'. We remark that, for any vector $\boldsymbol{x}$, $\Gamma \cdot \boldsymbol{x} = \boldsymbol{0}$ if and only if $\Gamma' \cdot \boldsymbol{x} = \boldsymbol{0}$ and thus we can use Γ' instead of Γ in our problem. We assume in a first time that the first R columns of Γ are linearly independent (recall that the number ν of columns of Γ is much larger than its number R of rows), so that the R first columns of Γ' forms an identity matrix. Then, for any $k^* > R$, if the k^*-th column of Γ' has Hamming weight at most $d - 1$, we can consider the vector $\boldsymbol{x}$ defined as $x_{k^*} = 1$, $x_k = 1$ when $\Gamma'_{k,k^*} = 1$, and $x_k = 0$ otherwise; and this vector satisfies the conditions we were looking for: its Hamming weight is at most d and $\Gamma' \cdot \boldsymbol{x} = \boldsymbol{0}$. That way, we have quickly enumerated all the vectors $\boldsymbol{x}$ of Hamming weight at most d such that $\Gamma' \cdot \boldsymbol{x} = \boldsymbol{0}$ and with the additional property that $x_k = 0$ for all $k > R$ except for at most[2] one index k^*. Without the condition $\Gamma' \cdot \boldsymbol{x} = \boldsymbol{0}$, there are $(\nu - R + 1) \cdot \sum_{i=0}^{d-1} \binom{R}{i} + \binom{R}{d}$ such vectors, as there are $\sum_{i=0}^{d} \binom{R}{i}$ vectors $\boldsymbol{x}$ such that $\mathrm{HW}(\boldsymbol{x}) \leq d$ and $x_k = 0$ for every $k > R$, and there are $(\nu - R) \cdot \sum_{i=0}^{d-1} \binom{R}{i}$ vectors $\boldsymbol{x}$ such that $\mathrm{HW}(\boldsymbol{x}) \leq d$ and $x_k = 1$, for a single $k > R$. In other words, using row-reduction, we have been able to check $(\nu - R + 1) \cdot \sum_{i=0}^{d-1} \binom{R}{i} + \binom{R}{d}$ possible vectors $\boldsymbol{x}$ among at most $\sum_{i=1}^{d} \binom{\nu}{i}$ vectors which could be used to mount an attack, by testing at most $\nu - R$ vectors.[3]

Then, we can randomly permute the columns of Γ and repeat this algorithm. Each iteration would find an attack (if there was one attack) with probability at least $\left((\nu - R + 1) \cdot \sum_{i=0}^{d-1} \binom{R}{i} + \binom{R}{d} \right) \sum_{i=1}^{d} \binom{\nu}{i}$. Therefore, after K iterations,

---

[2] We have seen that for one index k^*, but it is easy to see that, as the first R columns of Γ' form an identity matrix, there does not exist such vector $\boldsymbol{x}$ so that $x_k = 0$ for all $k > R$ anyway.

[3] There are exactly $\sum_{i=1}^{d} \binom{\nu}{i}$ vectors of Hamming weight at most d, but here we recall that we only consider vectors $\boldsymbol{x}$ satisfying the following additional condition: there is no vector $\hat{\boldsymbol{x}} < \boldsymbol{x}$ such that $\Gamma \cdot \hat{\boldsymbol{x}} = \boldsymbol{0}$. We also remark that the vectors $\boldsymbol{x}$ generated by the described algorithm all satisfy this additional condition.

the error probability is only

$$\varepsilon \le \left(1 - \frac{(\nu - R + 1) \cdot \sum_{i=0}^{d-1} \binom{R}{i} + \binom{R}{d}}{\sum_{i=1}^{d} \binom{\nu}{i}}\right)^K,$$

and the required number of iterations is linear with $\log(1/\varepsilon)$, which is what we wanted.

Now, we just need to handle the case when the first R columns of Γ are not linearly independent, for some permuted matrix Γ at some iteration. We can simply redraw the permutation or taking the pivots in the row-reduction instead of taking the first R columns of Γ. In both cases, this may slightly bias the probability. We make the *heuristic assumption* that the bias is negligible. To support this heuristic assumption, we remark that if we iterate the algorithm for all the permutations for which the first R columns of Γ are not linearly independent, then we would enumerate all the vectors x we are interested in, thanks to the additional condition that there is no vector $\hat{x} < x$ such that $\Gamma \cdot \hat{x} = 0$.

8.3 The Tool

The tool takes as input a description of a compression algorithm for multiplication similar to the ones we used in this paper (see Fig. 2 for instance) and the maximum error probability ε we allow, and tries to find an attack. If no attack is found, then the scheme is secure with probability $1 - \varepsilon$. The tool can also output a description of the scheme which can be fed off into the tool in [4].

The source code of the tool and its documentation are provided in [1].

8.4 Complexity Comparison

It is difficult to compare the complexity of our new tool to the complexity of the tool proposed in [4] since it strongly depends on the tested algorithm. Nevertheless, we try to give some values for the verification time of both tools when we intentionally modify our constructions to yield an attack. From order 2 to 4,

Table 3. Complexities of exhibiting an attack at several orders

Time to find an attack			
Order	Target algorithm	Verifier [4]	New tool
$d = 2$	Tweaked Algorithm 4	less than 1 ms	less than 10 ms
$d = 3$	Tweaked Algorithm 5	36 ms	less than 10 ms
$d = 4$	Tweaked Algorithm 6	108 ms	less than 10 ms
$d = 5$	Tweaked Algorithm 3	6.264 s	less than 100 ms
$d = 6$	Tweaked Algorithm 3	26 min	less than 300 ms

we start with our optimal constructions and we just invert two random bits in an output share c_i. Similarly, for orders 5 and 6, we use our generic construction and apply the same small modification. The computations were performed on a Intel(R) Core(TM) i5-2467M CPU @ 1.60 GHz and the results are given in Table 3. We can see that in all the considered cases, our new tool reveals the attack in less than 300 ms while the generic verifier of Barthe et al. needs up to 26 min for order $d = 6$.

Acknowledgments. The authors thank the anonymous reviewers for their constructive comments. This work was supported in part by the French ANR Project ANR-12-JS02-0004 ROMAnTIC, the *Direction Générale de l'Armement* (DGA), the CFM Foundation.

References

1. https://github.com/fabrice102/private_multiplication
2. Barker, E.B., Kelsey, J.M.: Sp 800–90a. recommendation for random number generation using deterministic random bit generators. Technical report, Gaithersburg, MD, USA (2012)
3. Barthe, G., Belaïd, S., Dupressoir, F., Fouque, P.A., Grégoire, B.: Compositional verification of higher-order masking: Application to a verifying masking compiler. Cryptology ePrint Archive, Report 2015/506 (2015). http://eprint.iacr.org/2015/506
4. Barthe, G., Belaïd, S., Dupressoir, F., Fouque, P.-A., Grégoire, B., Strub, P.-Y.: Verified proofs of higher-order masking. In: Oswald, E., Fischlin, M. (eds.) EUROCRYPT 2015. LNCS, vol. 9056, pp. 457–485. Springer, Heidelberg (2015)
5. Ben-Or, M., Goldwasser, S., Kilian, J., Wigderson, A.: Multi-prover interactive proofs: How to remove intractability assumptions. In: 20th ACM STOC, pp. 113–131. ACM Press, May 1988
6. Bilgin, B., Gierlichs, B., Nikova, S., Nikov, V., Rijmen, V.: Higher-order threshold implementations. In: Sarkar, P., Iwata, T. (eds.) ASIACRYPT 2014, Part II. LNCS, vol. 8874, pp. 326–343. Springer, Heidelberg (2014)
7. Bilgin, B., Gierlichs, B., Nikova, S., Nikov, V., Rijmen, V.: A more efficient AES threshold implementation. In: Pointcheval, D., Vergnaud, D. (eds.) AFRICACRYPT 2014. LNCS, vol. 8469, pp. 267–284. Springer, Heidelberg (2014)
8. Chari, S., Jutla, C.S., Rao, J.R., Rohatgi, P.: Towards sound approaches to counteract power-analysis attacks. In: Wiener, M. (ed.) CRYPTO 1999. LNCS, vol. 1666, pp. 398–412. Springer, Heidelberg (1999)
9. Coron, J.-S., Prouff, E., Rivain, M., Roche, T.: Higher-order side channel security and mask refreshing. In: Moriai, S. (ed.) FSE 2013. LNCS, vol. 8424, pp. 410–424. Springer, Heidelberg (2014)
10. Duc, A., Dziembowski, S., Faust, S.: Unifying leakage models: from probing attacks to noisy leakage. In: Nguyen, P.Q., Oswald, E. (eds.) EUROCRYPT 2014. LNCS, vol. 8441, pp. 423–440. Springer, Heidelberg (2014)
11. Duc, A., Faust, S., Standaert, F.-X.: Making masking security proofs concrete. In: Oswald, E., Fischlin, M. (eds.) EUROCRYPT 2015. LNCS, vol. 9056, pp. 401–429. Springer, Heidelberg (2015)

12. Dziembowski, S., Faust, S., Skorski, M.: Noisy leakage revisited. In: Oswald, E., Fischlin, M. (eds.) EUROCRYPT 2015, Part II. LNCS, vol. 9057, pp. 159–188. Springer, Heidelberg (2015)
13. Goubin, L., Patarin, J.: DES and differential power analysis the "duplication" method. In: Koç, Ç.K., Paar, C. (eds.) CHES 1999. LNCS, vol. 1717, pp. 158–172. Springer, Heidelberg (1999)
14. Ishai, Y., Kushilevitz, E., Li, X., Ostrovsky, R., Prabhakaran, M., Sahai, A., Zuckerman, D.: Robust pseudorandom generators. In: Fomin, F.V., Freivalds, R., Kwiatkowska, M., Peleg, D. (eds.) ICALP 2013, Part I. LNCS, vol. 7965, pp. 576–588. Springer, Heidelberg (2013)
15. Ishai, Y., Sahai, A., Wagner, D.: Private circuits: securing hardware against probing attacks. In: Boneh, D. (ed.) CRYPTO 2003. LNCS, vol. 2729, pp. 463–481. Springer, Heidelberg (2003)
16. Kocher, P.C.: Timing attacks on implementations of Diffie-Hellman, RSA, DSS, and other systems. In: Koblitz, N. (ed.) CRYPTO 1996. LNCS, vol. 1109, pp. 104–113. Springer, Heidelberg (1996)
17. Lamport, L., Shostak, R.E., Pease, M.C.: The byzantine generals problem. ACM Trans. Program. Lang. Syst. 4(3), 382–401 (1982)
18. Lee, P.J., Brickell, E.F.: An observation on the security of McEliece's public-key cryptosystem. In: Günther, C.G. (ed.) EUROCRYPT 1988. LNCS, vol. 330, pp. 275–280. Springer, Heidelberg (1988)
19. Leon, J.S.: A probabilistic algorithm for computing minimum weights of large error-correcting codes. IEEE Trans. Inf. Theor. 34(5), 1354–1359 (1988)
20. McEliece, R.J.: A public-key cryptosystem based on algebraic coding theory. DSN Prog. Rep. 42(44), 114–116 (1978)
21. Nikova, S., Rijmen, V., Schläffer, M.: Secure hardware implementation of nonlinear functions in the presence of glitches. J. Cryptology 24(2), 292–321 (2011)
22. Prange, E.: The use of information sets in decoding cyclic codes. IRE Trans. Inf. Theor. 8(5), 5–9 (1962)
23. Prouff, E., Rivain, M.: Masking against side-channel attacks: a formal security proof. In: Johansson, T., Nguyen, P.Q. (eds.) EUROCRYPT 2013. LNCS, vol. 7881, pp. 142–159. Springer, Heidelberg (2013)
24. Reparaz, O., Bilgin, B., Nikova, S., Gierlichs, B., Verbauwhede, I.: Consolidating masking schemes. In: Gennaro, R., Robshaw, M.J.B. (eds.) CRYPTO 2015, Part I. LNCS, vol. 9215, pp. 764–783. Springer, Heidelberg (2015)
25. Rivain, M., Prouff, E.: Provably secure higher-order masking of AES. In: Mangard, S., Standaert, F.-X. (eds.) CHES 2010. LNCS, vol. 6225, pp. 413–427. Springer, Heidelberg (2010)
26. Stern, J.: A method for finding codewords of small weight. In: Cohen, G.D., Wolfmann, J. (eds.) Coding Theory and Applications. LNCS, vol. 388, pp. 106–113. Springer, Heidelberg (1988)
27. The Sage Developers: Sage Mathematics Software (Version 6.8) (2015). http://www.sagemath.org
28. Yao, A.C.C.: Protocols for secure computations (extended abstract). In: 23rd FOCS, pp. 160–164. IEEE Computer Society Press, November 1982

10-Round Feistel is Indifferentiable from an Ideal Cipher

Dana Dachman-Soled, Jonathan Katz, and Aishwarya Thiruvengadam$^{(\boxtimes)}$

University of Maryland, College Park, USA
danadach@ece.umd.edu, {jkatz,aish}@cs.umd.edu

Abstract. We revisit the question of constructing an ideal cipher from a random oracle. Coron et al. (Journal of Cryptology, 2014) proved that a 14-round Feistel network using random, independent, keyed round functions is indifferentiable from an ideal cipher, thus demonstrating the feasibility of such a transformation. Left unresolved is the number of rounds of a Feistel network that are needed in order for indifferentiability to hold. We improve upon the result of Coron et al. and show that 10 rounds suffice.

1 Introduction

The security of practical block ciphers—i.e., pseudorandom permutations—is not currently known to reduce to well-studied, easily formulated, computational problems. Nevertheless, modern block-cipher constructions are far from ad-hoc, and a strong theory for their construction has been developed. An important area of research is to understand the provable security guarantees offered by these classical paradigms.

One of the well-known approaches for building practical block ciphers is to use a *Feistel network* [9], an iterated structure in which key-dependent, "random-looking" *round functions* on $\{0,1\}^n$ are applied in a sequence of rounds to yield a permutation on $\{0,1\}^{2n}$. In analyzing the security that Feistel networks provide, it is useful to consider an information-theoretic setting in which the round functions are instantiated by truly random and independent (keyed) functions. The purpose of such an analysis is to validate the *structural* robustness of the approach. Luby and Rackoff [12] proved that when independent, random round functions are used, a three-round Feistel network is indistinguishable from a random permutation under chosen-plaintext attacks, and a four-round Feistel network is indistinguishable from a random permutation under chosen plaintext/ciphertext attacks.

This work was performed under financial assistance award 70NANB15H328 from the U.S. Department of Commerce, National Institute of Standards and Technology.

D. Dachman-Soled—Work supported in part by NSF CAREER award #1453045. This work was done in part while the author was visiting the Simons Institute for the Theory of Computing, supported by the Simons Foundation and by the DIMACS/Simons Collaboration in Cryptography through NSF grant #1523467.

J. Katz and A. Thiruvengadam—Work supported in part by NSF award #1223623.

© International Association for Cryptologic Research 2016
M. Fischlin and J.-S. Coron (Eds.): EUROCRYPT 2016, Part II, LNCS 9666, pp. 649–678, 2016.
DOI: 10.1007/978-3-662-49896-5_23

In the Luby-Rackoff result, the round functions are secretly keyed and the adversary does not have direct access to them; the security notion considered—namely, *indistinguishability*—is one in which the key of the overall Feistel network is also unknown to the adversary. A stronger notion of security, called *indifferentiability* [14], applies even when the round functions are *public*, and aims to show that a block cipher behaves like an *ideal cipher*, i.e., an oracle for which each key defines an independent, random permutation. Proving indifferentiability is more complex than proving indistinguishability: to prove indifferentiability of a block-cipher construction $\mathbf{BC}$ (that relies on an ideal primitive $\mathcal{O}$) from an ideal cipher $\mathbf{IC}$, one must exhibit a *simulator* $\mathbf{S}$ such that the view of any distinguisher interacting with $(\mathbf{BC}^{\mathcal{O}}, \mathcal{O})$ is indistinguishable from its view when interacting with $(\mathbf{IC}, \mathbf{S}^{\mathbf{IC}})$. For Feistel networks, it is known (see [1,11]) that one can simplify the problem, and focus on indifferentiability of the Feistel network when using random and independent *unkeyed* round functions from a *public random permutation*; an ideal cipher is then obtained by keying the round functions.

In a recent result building on [2,11,16], Coron et al. [1] proved that when using independent, random round functions, a 14-round Feistel network is indifferentiable from a public random permutation. The main question left open by the work of Coron et al. is: precisely how many rounds of a Feistel network are needed for indifferentiability to hold? It is known from prior work [1] that 5 rounds are not sufficient, while (as we have just noted) 14 rounds are. In this work, we narrow the gap and show that 10 rounds suffice.[1]

We provide an overview of our proof, and the differences from that of Coron et al., in Sect. 2.

Concurrent Work. In concurrent and independent work, Dai and Steinberger [4] have also shown indifferentiability of a 10-round Feistel network from an ideal cipher. We provide a brief comparison between our work and theirs in Sect. 2.3.

Subsequent Work. Dai and Steinberger [5] have more recently improved their analysis and shown that an 8-round Feistel network is indifferentiable from an ideal cipher. The true number of rounds needed remains open.

1.1   Other Related Work

Coron et al. [2] claimed that a 6-round Feistel network is indifferentiable from an ideal cipher. Their proof of indifferentiability introduced the *partial chain detection technique* that we also rely on here. Seurin [16] gave a simpler proof of indifferentiability for a 10-round Feistel network, and introduced a clever technique for bounding the simulator complexity. Holenstein et al. [11] later showed that there was a distinguishing attack against the simulator of Coron et al. [2], and a gap in the proof of the 10-round simulator by Seurin [16]; however,

---

[1] Seurin previously claimed that a 10-round Feistel network is indifferentiable from a random permutation [16], but this claim was later retracted by the author [17].

they prove that a 14-round Feistel network is indifferentiable from an ideal cipher by building on prior work as well as incorporating several new techniques.

Ramzan and Reyzin [15] proved that a 4-round Feistel network remains indistinguishable from a random permutation even if the adversary is given access to the middle two round functions. Gentry and Ramzan [10] showed that a 4-round Feistel network can be used to instantiate the random permutation in the Even-Mansour cipher [8], and proved that such a construction is a pseudorandom permutation even if the round functions of the Feistel network are publicly accessible. Dodis and Puniya [7] studied security of the Feistel network in a scenario where the adversary learns intermediate values when the Feistel network is evaluated, and/or when the round functions are unpredictable but not (pseudo)random.

Various relaxations of indifferentiability, such as *public indifferentiability* [7,18] or *honest-but-curious indifferentiability* [6], have also been considered. Dodis and Puniya [6] proved that a Feistel network with super-logarithmic number of rounds is indifferentiable from an ideal cipher in the honest-but-curious setting. Mandal et al. [13] proved that the 6-round Feistel network is publicly indifferentiable from an ideal cipher.

1.2 Organization of the Paper

In Sect. 2 we provide a high-level overview of our proof, and how it differs from the proof of indifferentiability of the 14-round Feistel network [1,11]. After some brief background in Sect. 3, we jump into the technical details, describing our simulator in Sect. 4 and giving the proof of indifferentiability in Sect. 5. Additional discussion and proofs that have been omitted here are available in the full version of this work [3].

2 Overview of Our Proof

We first describe the proof structure used for the proof of indifferentiability of the 14-round Feistel network from an ideal cipher [1,11], and then describe how our proof differs.

2.1 Techniques for the 14-Round Simulator

Consider a naive simulator for an r-round Feistel construction, which responds to distinguisher queries to each of the round functions $\mathbf{F}_1, \ldots, \mathbf{F}_r$, by always returning a uniform value. Unfortunately, there is a simple distinguisher who can distinguish oracle access to $(\mathsf{Feistel}_r^{\mathbf{F}}, \mathbf{F})$ from oracle access to $(\mathbf{P}, \mathbf{S}^{\mathbf{P}})$: The distinguisher queries (x_0, x_1) to the first oracle, receiving (x_r, x_{r+1}) in return, and uses oracle access to the second oracle to evaluate the r-round Feistel and compute (x'_r, x'_{r+1}) on its own, creating a *chain* of queries $(x_1, \ldots, x'_r)$. Note that in the first case $(x_r, x_{r+1}) = (x'_r, x'_{r+1})$ with probability 1, while in the second case the probability that $(x_r, x_{r+1}) = (x'_r, x'_{r+1})$ is negligible.

An approach to addressing the above attack, which essentially gives the high-level intuition for how a successful simulator works, is as follows: If the simulator learns the value of $\mathbf{P}(x_0, x_1) = (x_r, x_{r+1})$ before the distinguisher queries the entire chain, then the simulator assigns values for the remaining queries $\mathbf{F}_i(x_i)$, conditioned on the restriction $\mathsf{Feistel}_r^{\mathbf{F}}(x_0, x_1) = (x_r, x_{r+1})$. More specifically, if there are two consecutive rounds $(i, i+1)$, where $i \in \{1, \ldots, r-1\}$, which have not yet been queried, the simulator adapts its assignments to $\mathbf{F}_i(x_i)$, $\mathbf{F}_{i+1}(x_{i+1})$ to be consistent with $\mathbf{P}(x_0, x_1) = (x_r, x_{r+1})$. When the simulator adapts the assignment of $\mathbf{F}_i(x_i)$ to be consistent with a constraint $\mathbf{P}(x_0, x_1) = (x_r, x_{r+1})$, we say that this value of $\mathbf{F}_i(x_i)$ has been assigned via a *ForceVal* assignment. Further details of the 14-round simulator are discussed below.

Partial Chain Detection and Preemptive Completion. To allow the simulator to preemptively discover $\mathbf{P}(x_0, x_1) = (x_r, x_{r+1})$, the authors fix two "detect zones" which are sets of consecutive rounds $\{1, 2, 13, 14\}$, $\{7, 8\}$. Each time the simulator assigns a value to $\mathbf{F}_i(x_i)$, it also checks whether there exists a tuple of the form $(x_1, x_2, x_{13}, x_{14})$ such that (1) $\mathbf{F}_1(x_1)$, $\mathbf{F}_2(x_2)$, $\mathbf{F}_{13}(x_{13})$, and $\mathbf{F}_{14}(x_{14})$ have all been assigned and (2) $\mathbf{P}(\mathbf{F}_1(x_1) \oplus x_2, x_1) = (x_{14}, \mathbf{F}_{13}(x_{13}) \oplus x_{14})$; or whether there exists a tuple of the form (x_7, x_8) such that $\mathbf{F}_7(x_7)$ and $\mathbf{F}_8(x_8)$ have both been assigned. A pair of consecutive round values (x_k, x_{k+1}) is referred to as a "partial chain," and when a new partial chain is detected in the detect zones described above, it is "enqueued for completion" and will later be dequeued and preemptively completed. When a partial chain is detected due to a detect zone that includes both x_1 and x_r, we say it is a "wraparound" chain. Note that preemptive completion of a chain can cause new chains to be detected and these will then be enqueued for completion. This means that in order to prove indifferentiability, it is necessary to argue that for x_i that fall on multiple completed chains, all restrictions on the assignment of $\mathbf{F}_i(x_i)$ can be *simultaneously* satisfied. In particular the "bad case" will be when some assignment $\mathbf{F}_i(x_i)$ must be adapted via a ForceVal assignment, but an assignment to $\mathbf{F}_i(x_i)$ has previously been made. If such a case occurs, we say the value at an adapt position has been "overwritten." It turns out that to prove indifferentiability, it is sufficient to prove that this occurs with negligible probability.

4-Round Buffer Zone. In order to ensure that overwrites do not occur, the notion of a 4-round buffer zone is introduced in [1,11]. Their simulator has two 4-round buffer zones, corresponding to rounds $\{3, 4, 5, 6\}$ or $\{9, 10, 11, 12\}$. Within the buffer zones, positions $\{3, 6\}$ (respectively, $\{9, 12\}$) are known as the *set uniform positions*, and positions $\{4, 5\}$ (respectively, $\{10, 11\}$) are known as the *adapt positions*. They prove the following property (which we call henceforth the *strong set uniform property*): At the moment a chain is about to be completed, the set uniform positions of the buffer zone are always unassigned. This means that the simulator will always assign uniform values to $\mathbf{F}_3(x_3)$ and $\mathbf{F}_6(x_6)$ (respectively, $\mathbf{F}_9(x_9)$ and $\mathbf{F}_{12}(x_{12})$) immediately before assigning values to $\mathbf{F}_4(x_4)$ and $\mathbf{F}_5(x_5)$ (respectively, $\mathbf{F}_{10}(x_{10})$ and $\mathbf{F}_{11}(x_{11})$) using ForceVal. This ensures that ForceVal overwrites with negligible probability, because $x_4 = x_2 \oplus \mathbf{F}_3(x_3)$ is only determined at the moment $\mathbf{F}_3(x_3)$ is assigned and so the probability that $\mathbf{F}_4(x_4)$ has already been assigned is negligible (a similar argument holds for the other adapt positions).

Rigid Structure. The rigid structure of [1,11] helps their proof in two ways: First, since all assignments across all completed chains are uniform except in the fixed adapt positions $\{4, 5\}$ and $\{10, 11\}$, it is easier to argue about "bad events" occurring. In particular, since the 4-round buffer of one chain ($\{3, 4, 5, 6\}$ or $\{9, 10, 11, 12\}$) cannot overlap with the detect zone of another chain ($\{1, 2, 13, 14\}$ or $\{7, 8\}$), they are able to argue that if a "bad event" occurs while detecting a chain C, then either an equivalent chain was already enqueued or it must have been caused by a uniform setting of $\mathbf{F}_i(x_i)$.

Bounding the Simulator's Runtime. The approach of [1,11] (originally introduced in [2]) is to bound the total number of partial chains that get completed by the simulator. In order to create a partial chain of the form $(x_1, x_2, x_{13}, x_{14})$, it must be the case that $\mathbf{P}(\mathbf{F}_1(x_1) \oplus x_2, x_1) = (x_{14}, \mathbf{F}_{13}(x_{13}) \oplus x_{14})$ and so, intuitively, the distinguisher had to query either $\mathbf{P}$ or $\mathbf{P}^{-1}$ in order to achieve this. Thus, the number of partial chains of the form $(x_1, x_2, x_{13}, x_{14})$ (i.e. wraparound chains) that get detected and completed by the simulator is at most the total number of queries made by the distinguisher. Since there is only a single middle detect zone $\{7, 8\}$, once we have a bound on the number of wraparound chains that are completed, we can also bound the number of completed partial chains of the form (x_7, x_8).

2.2 Our Techniques

We next briefly discuss how our techniques differ from those of the 14-round simulator [1,11], focusing on the four areas discussed above.

Separating Detection from Completion for Wrap-Around Chains. When the distinguisher makes a query $\mathbf{F}_i(x_i)$ to the simulator, our simulator proceeds in two phases: In the first phase, the simulator does not make any queries, but enqueues for completion all partial chains which it *predicts* will require completion. In the second phase, the simulator actually completes the chains and detects and enqueues only on the *middle* detect zone (which in our construction corresponds to rounds $\{5, 6\}$). This simplifies our proof since it means that after the set of chains has been detected in the first phase, the simulator can complete the chains in a manner that minimizes "bad interactions" between partial chains. In particular, in the second phase, the simulator first completes chains C with the property that one of the set uniform positions is "known" and hence could already have been assigned (in the completion of another chain D) before the chain C gets dequeued for completion. (Although this violates the strong set uniform property of [1,11], in our proof we are able to avoid this requirement. See the discussion of the *weak set uniform property* below for further details). The simulator then proceeds to complete (and detect and enqueue) other chains. This allows us to reduce the complexity of our analysis.

Relaxed Properties for the 4-Round Buffer Zone. When a partial chain C is about to be completed, we allow one of the set uniform positions, say $x_{\ell-1}$, to already be assigned, as long as the adapt position x_ℓ adjacent to this set uniform position has not yet been assigned. Chains that exhibit the property where one of the set uniform positions is already assigned before the completion of the chain are said to exhibit the *weak set uniform property*. In Lemma 36, we prove that for chains exhibiting the weak set uniform property, the adapt position is not assigned till the chain is dequeued for completion.

Relaxed Structure. Requiring only the weak set uniform property allows us to consider a more relaxed structure for detect zones and 4-round buffer zones. Instead of requiring that for every chain that gets completed the 4 round buffer positions (i.e., $\{3,4,5,6\}$ or $\{9,10,11,12\}$ in $[1,11]$ are always unassigned, we allow more flexibility in the position of the 4-round buffer. For example, depending on whether the detected chain is of the form (x_1, x_2, x_{10}), (x_1, x_9, x_{10}), or (x_5, x_6), our 4-round buffer will be one of: $\{3,4,5,6\}$ or $\{6,7,8,9\}$, $\{2,3,4,5\}$ or $\{5,6,7,8\}$, $\{1,2,3,4\}$ or $\{7,8,9,10\}$, respectively. This flexibility allows us to reduce the number of rounds. Now, however, the adapt zone of one chain may coincide with the detect zone of another chain. Since there are no dedicated roles for fixed positions, and since partial chains in the middle detect zone are detected during the completion of other chains, we define additional bad events BadlyHitFV and BadlyCollideFV and argue that they occur with low probability. Intuitively, BadlyHitFV captures the event where a FORCEVAL assignment occurs at x_ℓ such that it forms a valid Feistel sub-sequence $x_{\ell-1}$, x_ℓ and $x_{\ell+1}$ where $x_{\ell-1}$ and $x_{\ell+1}$ refer to adjacent positions to x_ℓ that they have already been assigned. This is analogous to the bad event BadlyHit defined in $[1,11]$ with the difference being that BadlyHit refers to a uniform assignment and BadlyHitFV refers to a FORCEVAL assignment. Similarly, BadlyCollideFV captures the event where a FORCEVAL assignment occurs at x_ℓ such that it causes two chains to "collide" at some position. This is analogous to the bad event BadlyCollide defined in $[1,11]$ with the difference being that BadlyCollide refers to a uniform assignment and BadlyCollideFV refers to a FORCEVAL assignment. Furthermore, in order to prove that a new wraparound chain does not get created during the completion of other chains we introduce and bound the probability of a new bad event BadlyCollideP. Intuitively, BadlyCollideP captures the event where a query to the random permutation returns a value (x_0, x_1) such that two chains "collide" on x_1 or returns a value (x_{10}, x_{11}) such that two chains collide on x_{10}.

Balancing Detection with the Simulator's Runtime. There is a clear trade-off between the achieved security bound and the running time of the simulator. If the simulator is too "aggressive" and detects too many chains too early, then we may perhaps achieve better security at the cost of extremely high simulator complexity. In comparison to the construction of $[1,11]$, our construction has more detect zones and, moreover, for wraparound chains, we detect on partial chains consisting of three consecutive queries instead of four consecutive queries. Nevertheless, at a high-level, our proof that the simulator runtime is polynomial follows very similarly to the proof in $[1,11]$. As there, we first bound the number of completed partial chains of the form (x_1, x_2, x_{10}) and (x_1, x_9, x_{10}) (such chains are wraparound chains since they contain both x_1 and x_{10}). Once we have done this, we again have only a single non-wraparound detect zone and so we can follow the argument of [1,11] to bound the number of completed partial chains of the form (x_5, x_6). Once we have a bound on the number of completed partial chains, it is fairly straightforward to bound the simulator complexity.

2.3 Comparison with Concurrent Work

As noted previously, Dai and Steinberger [4] have independently announced the same result we claim here. The starting point of their work is the 10-round simulator proposed by Seurin [16]. They use only two adapt zones (namely, $\{3, 4\}$ and $\{7, 8\}$) and allow the distinguisher to learn the values at both positions surrounding the adapt zones. In contrast, our simulator allows the distinguisher to learn the value at only one of the two positions[2] surrounding the adapt zones; due to our flexible 4-round buffer zone, our adapt zones can be any pair of consecutive rounds except $\{1, 2\}$, $\{5, 6\}$, and $\{9, 10\}$. Additionally, our proof follows the same high-level structure as in [1], whereas Dai and Steinberger present a new proof inspired by changes made to Seurin's simulator [16]. (Their subsequent improvement [5] showing indifferentiability of an 8-round Feistel network from an ideal cipher relies on the observation that detection on wrap-around chains can span only three rounds, rather than four.)

With regard to concrete security, our results are incomparable. Say q is the number of queries made by the distinguisher, and let n be the input/output length of the round functions. Dai and Steinberger [4] show indifferentiability $\epsilon = O(q^8/2^n)$ using a simulator running in time $T = O(q^{10})$; we show indifferentiability $\epsilon = O(q^{12}/2^n)$ using a simulator that runs in time $T = O(q^6)$. It is interesting to observe that both works achieve the same tradeoff for the product $\epsilon \cdot T$.

3   Background

We use the definition of indifferentiability used by the work on 14-round Feistel network [1,11], based on the definition of Maurer, Renner, and Holenstein [14].

Definition 1. *Let $\mathbf{C}$ be a construction that, for any n, accesses functions $\mathbf{F} = (\mathbf{F}_1, \dots, \mathbf{F}_r)$ over $\{0,1\}^n$ and implements an invertible permutation over $\{0,1\}^{2n}$. (We stress that $\mathbf{C}$ allows evaluation of both the forward and inverse directions of the permutation.) We say that $\mathbf{C}$ is* indifferentiable *from a random permutation if there exists a simulator $\mathbf{S}$ and a polynomial t such that for all distinguishers $\mathbf{D}$ making at most $q = \mathsf{poly}(n)$ queries, $\mathbf{S}$ runs in time $t(q)$ and*

$$|\Pr[\mathbf{D}^{\mathbf{C}^{\mathbf{F}}, \mathbf{F}}(1^n) = 1] - \Pr[\mathbf{D}^{\mathbf{P}, \mathbf{S}^{\mathbf{P}}}(1^n) = 1]|$$

---

[2] We refer to that position as the "bad" set uniform position.

is negligible, where $\mathbf{F}$ *are random, independent functions over* $\{0,1\}^n$ *and* $\mathbf{P}$ *is a random permutation over* $\{0,1\}^{2n}$. *(We stress that* $\mathbf{P}$ *can be evaluated in both the forward and inverse directions).*

The r-round Feistel construction, given access to $\mathbf{F} = (\mathbf{F}_1, \ldots, \mathbf{F}_r)$, is defined as follows. Let (L_{i-1}, R_{i-1}) be the input to the i-th round, with (L_0, R_0) denoting the initial input. Then, the output (L_i, R_i) of the i-th round of the construction is given by $L_i := R_{i-1}$ and $R_i := L_{i-1} \oplus \mathbf{F}_i(R_{i-1})$. So, for a r-round Feistel, if the $2n$-bit input is (L_0, R_0), then the output is given by (L_r, R_r).

4   Our Simulator

4.1   Informal Description of the Simulator

The queries to $\mathbf{F}_1, \ldots, \mathbf{F}_{10}$ are answered by the simulator through the public procedure $\mathbf{S}.\mathrm{F}(i,x)$ for $i = 1, \ldots, 10$. When the distinguisher asks a query $\mathrm{F}(i,x)$, the simulator checks to see if the query has already been set. The queries that are already set are held in tables $G_1, \ldots, G_{10}$ as pairs (x,y) such that if $\mathrm{F}(i,x)$ is queried, and if $x \in G_i$, then y is returned as the answer to query $\mathrm{F}(i,x)$. If the query has not already been set, then the simulator adds x to the set A_i^j where j indicates the jth query of the distinguisher. The simulator then checks if $i \in \{1,2,5,6,9,10\}$ (where these positions mark the endpoints of the detect zones) and, if so, checks to see if any new partial chains of the form $(x_9, x_{10}, 9)$, $(x_1, x_2, 1)$, or $(x_5, x_6, 5)$ need to be enqueued. If no new partial chains are detected, the simulator just sets the value of $G_i(x)$ uniformly and returns that value. If new partial chains are detected and enqueued in Q_{enq}, then the simulator evaluates these partial chains "forward" and "backward" as much as possible (without setting any new values of $G_m(\cdot)$) for all $m \in \{1, \ldots, 10\}$. Say the evaluation stopped with $x_m \notin G_m$. Then, the simulator adds x_m to A_m^j and checks if $m \in \{1,2,5,6,9,10\}$ and if so, detects any additional partial chains that form with (x_m, m) and enqueues them for completion if necessary and repeats the process again until no more partial chains are detected.

The chains enqueued for completion during this process are enqueued in queues Q_1, Q_5, Q_6, Q_{10} and Q_{all}. Any chain that has been enqueued in Q_{enq} is also enqueued in Q_{all}. Chains enqueued in Q_b for $b \in \{1,5,6,10\}$ are those that may exhibit the *weak set uniform property*. Specifically, say $C = (x_k, x_{k+1}, k, \ell, g, b)$ is a chain that is enqueued to be adapted at position ℓ i.e. the "adapt" positions for C are at $\ell, \ell + 1$ and the "set uniform" positions are at $\ell - 1, \ell + 2$ with the "set uniform" position that is adjacent to the query that caused C to be enqueued being at "good" set uniform position g and the other "set uniform" position at b. If, at the time of enqueueing, the chain C can be evaluated up to the "bad" set uniform position b and the value of chain C at b, say x_b, is such that $x_b \notin G_b$, then C is enqueued in Q_b. (Note that there are chains that exhibit this property but are not enqueued for completion. These are the chains that belong to the set SimPChains. This is only to simplify the analysis for the bound of the complexity of the simulator. We will later show that ignoring these chains does not affect the simulation and in fact, these chains belong to CompChains at the end of the simulator's run while answering **D**'s j^{th} query).

The completion of enqueued chains starts with the completion of the chains enqueued in Q_b for $b \in \{1,5,6,10\}$. A chain C is dequeued from Q_b and if $C \notin$ CompChains, the simulator "completes" the chain. This process proceeds similarly to the completion process in [1]. The simulator evaluates the chain forward/backward upto the 4-round buffer setting $G_i(x_i)$ values uniformly for any $x_i \notin G_i$ that comes up while evaluating forward/backward. In the 4-round buffer consisting of the "set uniform" positions and the "adapt" positions, the simulator sets the values of C at the set uniform positions uniformly (if they have not already been set) and forces the values at the adapt positions such that evaluation of the Feistel is consistent with the random permutation. (Note that this could possibly lead to a value in $G_i(\cdot)$ getting overwritten. A major technical part of the proof is to show that this happens with negligible probability.) After this process, the simulator places C in the set CompChains along with "equivalent" chains obtained by evaluating C on the detect zone positions i.e. chains of the form (x_k, x_{k+1}, k) for $k = 1,5,9$.

Once the simulator completes the chains enqueued in Q_b for all $b \in \{1,5,6,10\}$, the simulator completes the remaining chains enqueued in Q_{all}. The completion process for the remaining chains enqueued in Q_{all} is the same as the completion process described above except that the simulator detects additional partial chains of the form $(x_5, x_6, 5)$ during the completion and enqueues them in the queue Q_{mid} i.e. during the completion of a chain C in Q_{all}, if an assignment occurs such that $x_k \in G_k$ for some $k \in \{5,6\}$ due to the assignment and $x_k \notin G_k$ before the assignment, then the simulator enqueues the partial chain $(x_5, x_6, 5)$ in Q_{mid} for all $x_{k'} \in G_{k'}$ such that $k' \in \{5,6\}$ and $k \neq k'$. (Note that the assignment could be a FORCEVAL assignment as well.) Finally, the simulator completes all the chains in Q_{mid} that are not already in CompChains. The completion process again is the same as the process described for chains enqueued in Q_b. The simulator then returns the answer $G_i(x)$ to the query $\mathrm{F}(i,x)$.

4.2   Formal Description of the Simulator

The simulator **S** internally uses hashtables $G_1, \ldots, G_{10}$ to store the function values. Additionally, it uses sets $A_1^j, \ldots, A_{10}^j$ for the jth distinguisher query to detect partial chains that need to be completed; these sets store values that would be added to G_i in the future. A queue Q_{enq} to detect partial chains that need to be completed and stores a copy of Q_{enq} in a queue Q_{all} that is used during completion. Queues Q_1, Q_5, Q_6, Q_{10} are used to store the chains in Q_{enq} whose "bad" set uniform position is known at the time of detection. Queue Q_{mid} is used to store new chains of the form $(x_5, x_6, 5)$ that are enqueued during the completion of chains from Q_{all}. A set CompChains is used to remember the chains that have been completed already. Finally, a set SimPChains is used to hold chains of the form $(x_1, x_2, 1)$ and $(x_9, x_{10}, 9)$ that are detected due to $\mathrm{P}/\mathrm{P}^{-1}$ queries made by the simulator. This set is needed only for the purpose of analyzing the complexity of the simulator.

The variables used below are: Queues $Q_{enq}, Q_{all}, Q_1, Q_5, Q_6, Q_{10}, Q_{mid}$; Hashtables $G_1, \ldots, G_{10}$; Sets $A_i^j := \emptyset$ for $i = 1, \ldots, 10$ and $j = 1, \ldots, q$ where q is the maximum number of queries made by the distinguisher, Sets CompChains $:= \emptyset$ and SimPChains $:= \emptyset$. Initialize $j := 0$. The procedure $F(i, x)$ provides the interface to a distinguisher.

```
 1  procedure F(i, x):
 2      j := j + 1
 3      for i ∈ {1, ..., 10} do
 4          A_i^j := ∅
 5      F^ENQ(i, x)
 6      while ¬Q_enq.EMPTY() do
 7          (x_k, x_{k+1}, k, ℓ, g, b) := Q_enq.DEQUEUE()
 8          if (x_k, x_{k+1}, k) ∉ CompChains then
 9              (x_r, x_{r+1}, r) := EvFwdEnq(x_k, x_{k+1}, k, ℓ - 2)
10              if r + 1 = b ∧ x_{r+1} ∉ G_{r+1} then
11                  Q_b.ENQUEUE(x_k, x_{k+1}, k, ℓ, g, b)
12              (x_r, x_{r+1}, r) := EvBwdEnq(x_k, x_{k+1}, k, ℓ + 2)
13              if r = b ∧ x_r ∉ G_r then
14                  Q_b.ENQUEUE(x_k, x_{k+1}, k, ℓ, g, b)
15      for each Q ∈ ⟨Q_1, Q_5, Q_6, Q_{10}, Q_all, Q_mid⟩ do      ▷ processed in that order
16          while ¬Q.EMPTY() do
17              (x_k, x_{k+1}, k, ℓ, g, b) := Q.DEQUEUE()
18              if (x_k, x_{k+1}, k) ∉ CompChains then
19                  (x_{ℓ-2}, x_{ℓ-1}) := EvFwdComp(Q, x_k, x_{k+1}, k, ℓ - 2)
20                  (x_{ℓ+2}, x_{ℓ+3}) := EvBwdComp(Q, x_k, x_{k+1}, k, ℓ + 2)
21                  ADAPT(Q, x_{ℓ-2}, x_{ℓ-1}, x_{ℓ+2}, x_{ℓ+3}, ℓ, g, b)
22                  (x_1, x_2) := EvBwdComp(⊥, x_k, x_{k+1}, k, 1)
23                  (x_5, x_6) := EvFwdComp(⊥, x_1, x_2, 1, 5)
24                  (x_9, x_{10}) := EvFwdComp(⊥, x_1, x_2, 1, 9)
25                  CompChains := CompChains ∪ {(x_1, x_2, 1), (x_5, x_6, 5), (x_9, x_{10}, 9)}
26      F^COMP(⊥, i, x)
27      return G_i(x)


28  procedure EvFwdEnq(x_k, x_{k+1}, k, m):        39          k := k + 1
29      if k = 5 then                              40          flagMid := 0
30          flagMid := 1                           41          return (x_k, x_{k+1}, k)
31      while (k ≠ m) ∧ ((k = 10) ∨
                    F^ENQ(k+1, x_{k+1}) ≠ ⊥)
            do                                     42  procedure EvBwdEnq(x_k, x_{k+1}, k, m):
32          if k = 10 then                         43      if k = 5 then
33              (x_0, x_1) := P^{-1}(x_{10}, x_{11})  44          flagMid := 1
34              k := 0                             45      while (k ≠ m) ∧ ((k = 0) ∨
35          else                                                F^ENQ(k, x_k) ≠ ⊥)
36              if k = 9 ∧ flagMid = 1 then                     do
37                  SimPChains :=                  46          if k = 0 then
                    SimPChains∪ {(x_k, x_{k+1}, k)}  47              (x_{10}, x_{11}) := P(x_0, x_1)
38              x_{k+2} := x_k ⊕ G(k+1, x_{k+1})   48              k := 10
                                                   49          else
```

50   **if** $k = 1 \wedge \mathsf{flagMid} = 1$ **then**
51   SimPChains :=
            SimPChains$\cup \{(x_k, x_{k+1}, k)\}$
52          $x_{k-1} := x_{k+1} \oplus G(k, x_k)$

53        $k := k - 1$
54   flagMid:= 0
55   **return** (x_k, x_{k+1}, k)

56 **procedure** $\mathrm{F}^{\mathrm{ENQ}}(i, x)$:
57   **if** $x \in G_i$ **then**
58   **return** $G_i(x)$
59   **else if** $x \in A_i^j$ **then**
60   **return** $\perp$
61   **else**
62        $A_i^j := \{x\} \cup A_i^j$
63   **if** $i \in \{1, 2, 5, 6, 9, 10\}$ **then**
64            $\mathrm{ENQNEWCHAINS}(i, x)$
65   **return** $\perp$

66 **procedure** $\mathrm{CHKFWD}(x_0, x_1, x_{10})$:
67    $(x'_{10}, x'_{11}) := \mathbf{P}(x_0, x_1)$
68   **return** $x'_{10} \stackrel{?}{=} x_{10}$

69 **procedure** $\mathrm{CHKBWD}(x_{10}, x_{11}, x_1)$:
70    $(x'_0, x'_1) := \mathbf{P}^{-1}(x_{10}, x_{11})$
71   **return** $x'_1 \stackrel{?}{=} x_1$

72 **procedure** $\mathrm{FORCEVAL}(x, y, \ell)$:
73    $G_\ell(x) := y$

74 **procedure** $\mathrm{EVFWDCOMP}($
        $Q, x_k, x_{k+1}, k, m)$:
75   **while** $k \neq m$ **do**
76   **if** $k = 10$ **then**
77            $(x_0, x_1) := \mathbf{P}^{-1}(x_{10}, x_{11})$
78            $k := 0$
79   **else**
80            $x_{k+2} := x_k \oplus$
                $\mathrm{F}^{\mathrm{COMP}}(Q, k+1, x_{k+1})$
81            $k := k + 1$
82   **return** (x_m, x_{m+1})

83 **procedure** $\mathrm{EVBWDCOMP}($
        $Q, x_k, x_{k+1}, k, m)$:
84   **while** $k \neq m$ **do**
85   **if** $k = 0$ **then**
86            $(x_{10}, x_{11}) := \mathbf{P}(x_0, x_1)$
87            $k := 10$
88   **else**
89            $x_{k-1} := x_{k+1} \oplus$
                $\mathrm{F}^{\mathrm{COMP}}(Q, k, x_k)$
90            $k := k - 1$
91   **return** (x_m, x_{m+1})

92 **procedure** $\mathrm{F}^{\mathrm{COMP}}(Q, i, x)$:
93   **if** $x \notin G_i$ **then**
94        $G_i(x) \leftarrow \{0, 1\}^n$
95   **if** $Q \neq \perp \wedge Q = Q_{\mathrm{all}} \wedge i \in \{5, 6\}$
            then
96            $\mathrm{ENQNEWMIDCHAINS}(i, x)$
97   **return** $G_i(x)$

98 **procedure** $\mathrm{ENQNEWMIDCHAINS}(i, x)$:
99   **if** $i = 5$ **then**
100   **for all** $(x_5, x_6) \in \{x\} \times G_6$ **do**
101            $Q_{\mathrm{mid}}.\mathrm{ENQUEUE}(x_5, x_6, 5, 2, 4, 1)$
102   **if** $i = 6$ **then**
103   **for all** $(x_5, x_6) \in G_5 \times \{x\}$ **do**
104            $Q_{\mathrm{mid}}.\mathrm{ENQUEUE}(x_5, x_6, 5, 8, 7, 10)$

105 **procedure** $\mathrm{ENQNEWCHAINS}(i, x)$:
106   **if** $i = 1$ **then**
107   **for all** $(x_9, x_{10}, x_1) \in (G_9 \cup A_9^j) \times G_{10} \times \{x\}$ **do**
108   **if** $\mathrm{CHKBWD}(x_{10}, G_{10}(x_{10}) \oplus x_9, x_1)$ **then**
109   **if** $(x_9, x_{10}, 9) \notin \mathsf{SimPChains}$ **then**
110                    $Q_{\mathrm{enq}}.\mathrm{ENQUEUE}(x_9, x_{10}, 9, 3, 2, 5)$
111                    $Q_{\mathrm{all}}.\mathrm{ENQUEUE}(x_9, x_{10}, 9, 3, 2, 5)$
112   **if** $i = 2$ **then**
113   **for all** $(x_{10}, x_1, x_2) \in (G_{10} \cup A_{10}^j) \times G_1 \times \{x\}$ **do**
114   **if** $\mathrm{CHKFWD}(x_2 \oplus G_1(x_1), x_1, x_{10})$ **then**

```
115              if (x₁, x₂, 1) ∉ SimPChains then
116                  Q_enq.ENQUEUE(x₁, x₂, 1, 4, 3, 6)
117                  Q_all.ENQUEUE(x₁, x₂, 1, 4, 3, 6)
118          if i = 5 then
119              for all (x₅, x₆) ∈ {x} × (G₆ ∪ A₆ʲ) do
120                  Q_enq.ENQUEUE(x₅, x₆, 5, 2, 4, 1)
121                  Q_all.ENQUEUE(x₅, x₆, 5, 2, 4, 1)
122          if i = 6 then
123              for all (x₅, x₆) ∈ (G₅ ∪ A₅ʲ) × {x} do
124                  Q_enq.ENQUEUE(x₅, x₆, 5, 8, 7, 10)
125                  Q_all.ENQUEUE(x₅, x₆, 5, 8, 7, 10)
126          if i = 9 then
127              for all (x₉, x₁₀, x₁) ∈ {x} × G₁₀ × (G₁ ∪ A₁ʲ) do
128                  if CHKBWD(x₁₀, G₁₀(x₁₀) ⊕ x₉, x₁) then
129                      if (x₉, x₁₀, 9) ∉ SimPChains then
130                          Q_enq.ENQUEUE(x₉, x₁₀, 9, 6, 8, 5)
131                          Q_all.ENQUEUE(x₉, x₁₀, 9, 6, 8, 5)
132          if i = 10 then
133              for all (x₁₀, x₁, x₂) ∈ {x} × G₁ × (G₂ ∪ A₂ʲ) do
134                  if CHKFWD(x₂ ⊕ G₁(x₁), x₁, x₁₀) then
135                      if (x₁, x₂, 1) ∉ SimPChains then
136                          Q_enq.ENQUEUE(x₁, x₂, 1, 7, 9, 6)
137                          Q_all.ENQUEUE(x₁, x₂, 1, 7, 9, 6)

138  procedure ADAPT(Q, x_{ℓ-2}, x_{ℓ-1}, x_{ℓ+2}, x_{ℓ+3}, ℓ, g, b):
139      flagMidAdapt0 := 0
140      flagMidAdapt1 := 0
141      F^COMP(Q, ℓ - 1, x_{ℓ-1})
142      x_ℓ := x_{ℓ-2} ⊕ G_{ℓ-1}(x_{ℓ-1})
143      if (Q = Q_all) ∧ (ℓ = 5 ∨ ℓ = 6) ∧ (x_ℓ ∉ G_ℓ) then
144          flagMidAdapt0 := 1
145      F^COMP(Q, ℓ + 2, x_{ℓ+2})
146      x_{ℓ+1} := x_{ℓ+3} ⊕ G_{ℓ+2}(x_{ℓ+2})
147      if (Q = Q_all) ∧ (ℓ + 1 = 5 ∨ ℓ + 1 = 6) ∧ (x_{ℓ+1} ∉ G_{ℓ+1}) then
148          flagMidAdapt1 := 1
149      FORCEVAL(x_ℓ, x_{ℓ+1} ⊕ x_{ℓ-1}, ℓ)
150      if flagMidAdapt0 = 1 then
151          ENQNEWMIDCHAINS(ℓ, x_ℓ)
152      FORCEVAL(x_{ℓ+1}, x_ℓ ⊕ x_{ℓ+2}, ℓ + 1)
153      if flagMidAdapt1 = 1 then
154          ENQNEWMIDCHAINS(ℓ + 1, x_{ℓ+1})
```

5   Proof of Indifferentiability

Let Feistel denote the 10-round Feistel construction, let $\mathbf{F}$ be 10 independent random functions with domain and range $\{0,1\}^n$, and let $\mathbf{P}$ denote a random permutation on $\{0,1\}^{2n}$. Let $\mathbf{S}$ denote the simulator from the previous section. We prove:

Theorem 2. *The probability that a distinguisher* **D** *making at most* q *queries outputs 1 in an interaction with* $(\mathbf{P}, \mathbf{S^P})$ *and the probability that it outputs 1 in an interaction with* $(\mathsf{Feistel^F}, \mathbf{F})$ *differ by at most* $O(q^{12}/2^n)$. *Moreover,* **S** *runs in time* $O(q^6)$ *except with probability* $O(q^{12}/2^n)$.

For the rest of the paper, fix a distinguisher **D** making at most q queries.

5.1 Proof Overview

Our proof structure utilizes four hybrid experiments $H_1, \ldots, H_4$ as in the proof of indifferentiability of the 14-round Feistel network [1,11]. Hybrid H_1 denotes the scenario in which **D** interacts with $(\mathbf{P}, \mathbf{S^P})$, and H_4 denotes the scenario in which **D** interacts with $(\mathsf{Feistel^F}, \mathbf{F})$. To prove indifferentiability, we show that the difference between the probability **D** outputs 1 in H_1 and the probability **D** outputs 1 in H_4 is at most $\mathrm{poly}(q)/2^n$.

In H_2, the random permutation **P** is replaced with a two-sided random function **R**. Following [1,11], we first bound the simulator complexity in hybrid H_2 and use that to bound the simulator's complexity in H_1.

Next, we define certain "bad events" that can occur in an execution of H_2, and show that these events occur with low probability. We then show that as long as these events do not occur in an execution of H_2, then certain "good" properties hold; in particular, we can prove that for every call to $\textsc{ForceVal}(x, \cdot, j)$ that occurs in the execution, we have $x \notin G_j$ before the call. If this is true, we say that "$\textsc{ForceVal}$ does not overwrite." This is the main technical part of the proof and can be found in Sect. 5.3.2.

In H_3, the two-sided random function **R** is replaced with the 10-round Feistel construction. The distinguisher interacts with $(\mathsf{Feistel}, \hat{\mathbf{S}}^{\mathsf{Feistel^+}})$ where $\mathsf{Feistel^+}$ is the Feistel construction with additional procedures $\textsc{ChkFwd}$ and $\textsc{ChkBwd}$. Given the "good" properties that were proven in Sect. 5.3.2, we prove that H_2 and H_3 are indistinguishable. The proof follows exactly along the lines of the proof in [1,11].

Finally, in H_4, the distinguisher interacts with $(\mathsf{Feistel^F}, \mathbf{F})$ and hence accesses the random functions **F** directly instead of through the simulator. We prove that H_3 and H_4 are indistinguishable similar to the proof of [1,11].

Due to space constraints, we omit some of the proofs in the following sections. The omitted proofs can be found in the full version [3].

5.2 Indistinguishability of the First and Second Experiments

In H_2, we replace the random permutation with the two-sided random function **R**, and **D** interacts with $(\mathbf{R}, \hat{\mathbf{S}}^\mathbf{R})$. The simulator $\hat{\mathbf{S}}$ in H_2 is exactly the same as the simulator **S** described in Sect. 4.2 except that it implements procedures $\hat{\mathbf{S}}.\textsc{ChkFwd}$ and $\hat{\mathbf{S}}.\textsc{ChkBwd}$ by calling the procedures $\mathbf{R}.\textsc{ChkFwd}$ and $\mathbf{R}.\textsc{ChkBwd}$ that are provided by **R** (described below).

The two-sided function **R** maintains a hashtable P containing elements of the form $(\downarrow, x_0, x_1)$ and $(\uparrow, x_{10}, x_{11})$. Whenever $\mathbf{R}.P(x_0, x_1)$ is queried, **R** checks

```
 1  procedure P(x_0, x_1):              12  procedure P^{-1}(x_10, x_11):
 2      if (↓, x_0, x_1) ∉ P then       13      if (↑, x_10, x_11) ∉ P then
 3          (x_10, x_11) ←$ {0,1}^{2n}  14          (x_0, x_1) ←$ {0,1}^{2n}
 4          P(↓, x_0, x_1) := (x_10, x_11)  15          P(↑, x_10, x_11) := (x_0, x_1)
 5          P(↑, x_10, x_11) := (x_0, x_1)  16          P(↓, x_0, x_1) := (x_10, x_11)
 6      return P(↓, x_0, x_1)           17      return P(↑, x_10, x_11)

 7  procedure CHKFWD(x_0, x_1, x_10):   18  procedure CHKBWD(x_10, x_11, x_1):
 8      if (↓, x_0, x_1) ∈ P then       19      if (↑, x_10, x_11) ∈ P then
 9          (x'_10, x'_11) := P(↓, x_0, x_1)  20          (x'_0, x'_1) := P(↑, x_10, x_11)
10          return x'_10 =? x_10        21          return x'_1 =? x_1
11      return false                    22      return false
```

Fig. 1. Random two-sided function **R**.

if $(\downarrow, x_0, x_1) \in P$ and if so, answers accordingly. Otherwise, an independent uniform output (x_{10}, x_{11}) is picked and $(\downarrow, x_0, x_1)$ as well as $(\uparrow, x_{10}, x_{11})$ are added to P, mapping to each other. In addition to P and P^{-1}, **R** contains the procedures $\text{CHKFWD}(x_0, x_1, x_{10})$ and $\text{CHKBWD}(x_{10}, x_{11}, x_1)$.[3] $\text{CHKFWD}(x_0, x_1, x_{10})$ works as follows: If $(\downarrow, x_0, x_1) \in P$, it returns true if $(\downarrow, x_0, x_1)$ maps to (x_{10}, x_{11}) for some value of $x_{11} \in \{0,1\}^n$ and false otherwise. Procedure $\text{CHKBWD}(x_{10}, x_{11}, x_1)$ works as follows: If $(\uparrow, x_{10}, x_{11}) \in P$, it returns true if $(\uparrow, x_{10}, x_{11})$ maps to (x_0, x_1) for some value of $x_0 \in \{0,1\}^n$ and false otherwise. The pseudocode for the two-sided random function **R**, using hashtable P, is as follows:

The proof of indistinguishability of H_1 and H_2 can be found in the full version [3]. In particular, we prove the following statements regarding the the indistinguishability of H_1 and H_2 and the simulator complexity.

Lemma 3. *The probability that* **D** *outputs 1 in* H_1 *differs from the probability that it outputs 1 in* H_2 *by at most* $\frac{2 \cdot 10^{15} q^{12}}{2^n}$.

Lemma 4. *In* H_1, *the simulator runs for at most* $O(q^6)$ *steps and makes at most* $3.2 \times (10q)^6$ *queries except with probability at most* $\frac{10^{15} q^{12}}{2^n}$.

We will prove some properties of H_2 in the following section that will be useful to prove the indistinguishability of the second and third experiments.

5.3 Properties of H_2

We introduce some definitions and establish some properties of executions in H_2. The definitions here follow closely along the lines of the definitions in [1,11].

---

[3] This is similar to the CHECK procedure in [1,11].

A *partial chain* is a triple $(x_k, x_{k+1}, k) \in \{0,1\}^n \times \{0,1\}^n \times \{0,\ldots,10\}$. If $C = (x_k, x_{k+1}, k)$ is a partial chain, we let $C[1] = x_k$, $C[2] = x_{k+1}$, and $C[3] = k$.

Definition 5. *Fix tables $G = \hat{\mathbf{S}}.G$ and $P = \mathbf{R}.P$ in an execution of H_2, and let $C = (x_k, x_{k+1}, k)$ be a partial chain. We define functions* next, prev, val$^+$, val$^-$, *and* val *as follows:*

```
1  procedure next(x_k, x_{k+1}, k):
2    if k < 10 then
3      if x_{k+1} ∉ G_{k+1} then
4        return ⊥
5      x_{k+2} := x_k ⊕ G_{k+1}(x_{k+1})
6      return (x_{k+1}, x_{k+2}, k+1)
7    else if k = 10 then
8      if (↑, x_10, x_11) ∉ P then
9        return ⊥
10     (x_0, x_1) := P(↑, x_10, x_11)
11     return (x_0, x_1, 0)

12 procedure prev(x_k, x_{k+1}, k):
13   if k > 0 then
14     if x_k ∉ G_k then
15       return ⊥
16     x_{k-1} := x_{k+1} ⊕ G_k(x_k)
17     return (x_{k-1}, x_k, k-1)
18   else if k = 0 then
19     if (↓, x_0, x_1) ∉ P then
20       return ⊥
21     (x_10, x_11) := P(↓, x_0, x_1)
22     return (x_10, x_11, 10)
```

```
1  procedure val_i^+(C):
2    while (C ≠⊥) ∧ (C[3] ∉ {i-1, i}) do
3      C := next(C)
4    if C =⊥ then return ⊥
5    if C[3] = i then return C[1]
6    else return C[2]

7  procedure val_i^-(C):
8    while (C ≠⊥) ∧ (C[3] ∉ {i-1, i}) do
9      C := prev(C)
10   if C =⊥ then return ⊥
11   if C[3] = i then return C[1]
12   else return C[2]
```

```
1  procedure val_i(C):
2    if val_i^+(C) ≠⊥ then return val_i^+(C)
3    else return val_i^-(C)
```

We say that $\bot \notin G_i$ for $i \in \{1,\ldots,10\}$. So, if val$_i(C) \notin G_i$, then either val$_i(C) = \bot$ or val$_i(C) \neq \bot$ and val$_i(C) \notin G_i$.

Definition 6. *For a given set of tables G, P, two partial chains C, D are* equivalent *(denoted $C \equiv D$) if they are in the reflexive, transitive closure of the relations given by* next *and* prev.

So, two chains C and D are equivalent if $C = D$, or if D can be obtained by applying next and prev finitely many times to C.

Definition 7. *The set of* table-defined chains *contains all chains C for which* next$(C) \neq \bot$ *and* prev$(C) \neq \bot$.

Definition 8. *A chain $C = (x_k, x_{k+1}, k, \ell, g, b)$ is called an* enqueued chain *if C is enqueued for completion. For such an enqueued chain, we define* next(C)

as the procedure next *applied to the partial chain* (x_k, x_{k+1}, k) *i.e.* $\text{next}(C) := \text{next}(x_k, x_{k+1}, k)$. *The procedures* prev, val$^+$, val$^-$ *and* val *on an enqueued chain* C *are defined in a similar manner.*

Definition 9. *The set* Q_{all}^* *contains chains that are enqueued in* Q_{all} *but not in* Q_1, Q_5, Q_6, Q_{10}.

Definition 10. *We say a* uniform assignment *to* $G_k(x_k)$ *occurs when the simulator sets* $G_k(x_k)$ *through an assignment* $G_k(x_k) \leftarrow \{0,1\}^n$, *i.e., a uniform value is chosen from the set of n-bit strings and* $G_k(x_k)$ *is assigned that value.*

A uniform assignment to $G_k(x_k)$ occurs in line 94 of the simulator's execution. In particular, if $G_k(x_k)$ is set through a FORCEVAL$(x_k, \cdot, k)$ call, then it is not a uniform assignment.

Definition 11. *We say a* uniform assignment *to* P *occurs in a call to* $\mathbf{R}.P(x_0, x_1)$ *if* $(\downarrow, x_0, x_1) \notin P$ *when the call is made and* $P(\downarrow, x_0, x_1)$ *is set through the assignment* $P(\downarrow, x_0, x_1) := (x_{10}, x_{11})$ *where* (x_{10}, x_{11}) *is chosen uniformly from the set of 2n-bit strings.*

Similarly, it occurs in a call to $\mathbf{R}.P^{-1}(x_{10}, x_{11})$ *if* $(\uparrow, x_{10}, x_{11}) \notin P$ *when the call is made and* $P(\uparrow, x_{10}, x_{11})$ *is set through the assignment* $P(\uparrow, x_{10}, x_{11}) := (x_0, x_1)$ *where* (x_0, x_1) *is chosen uniformly from the set of 2n-bit strings.*

A uniform assignment to $P(\downarrow, x_0, x_1)$ occurs in line 4 of $\mathbf{R}$ in Fig. 1 and a uniform assignment to $P(\uparrow, x_{10}, x_{11})$ occurs in line 15 of $\mathbf{R}$ in Fig. 1.

In the following section, we define a set of "bad" events, and show that these occur with negligible probability. Following that, we analyze execution of the experiment assuming that none of these bad events occur.

In the remainder of the section, we let $T = O(q^2)$ be an upper bound on the sizes of G_i and P as well as the upper bound on the number of enqueued chains and hence, the number of calls to the ADAPT procedure in an execution of H_2. The derivation of the bound on T and the proof of the lemmas below can be found in the full version [3].

5.3.1   Bad Executions

Definition 12. *We say that event* BadP *occurs in* H_2 *if either:*

- *Immediately after choosing* (x_{10}, x_{11}) *in a call to* $\mathbf{R}.P(\cdot, \cdot)$, *either* $(\uparrow, x_{10}, x_{11})$ $\in P$ *or* $x_{10} \in G_{10}$.
- *Immediately after choosing* (x_0, x_1) *in a call to* $\mathbf{R}.P^{-1}(\cdot, \cdot)$, *either* $(\downarrow, x_0, x_1) \in$ P *or* $x_1 \in G_1$.

Lemma 13. *The probability of event* BadP *in* H_2 *is at most* $2T^2/2^n$.

A partial chain $C = (x_k, x_{k+1}, k)$ that has been enqueued by our simulator may not get table-defined till it is completed since it is possible that $x_k \in G_k$ while $x_{k+1} \in A_{k+1}^j$ for some j but not in G_{k+1}. Hence, we augment the definitions of BadlyHit and BadlyCollide given in [1,11] to refer to interactions with enqueued chains and refer to the augmented definitions as BadlyHit$^+$ and BadlyCollide$^+$.

Definition 14. *We say that event* BadlyHit$^+$ *occurs in* H_2 *if either:*

- *Immediately after a uniform assignment to* $G_k(x_k)$, *there is a partial chain* (x_k, x_{k+1}, k) *such that* $\mathsf{prev}(\mathsf{prev}(x_k, x_{k+1}, k)) \neq \perp$.
- *Immediately after a uniform assignment to* $G_k(x_k)$, *there is a partial chain* $(x_{k-1}, x_k, k-1)$ *such that* $\mathsf{next}(\mathsf{next}(x_{k-1}, x_k, k-1)) \neq \perp$.

and the relevant partial chain is either table-defined or an enqueued chain in Q_{all}.

Lemma 15. *The probability of event* BadlyHit$^+$ *in* H_2 *is at most* $40\,T^3/2^n$.

Definition 16. *We say that event* BadlyCollide$^+$ *occurs in* H_2 *if a uniform assignment to* $G_i(x_i)$ *is such that there exist two partial chains* C *and* D *such that for some* $\ell \in \{0, \dots, 11\}$ *and* $\sigma, \rho \in \{+, -\}$ *all of the following are true:*

- *Immediately before the assignment,* C *and* D *are not equivalent.*
- *Immediately before the assignment,* $\mathsf{val}_\ell^\sigma(C) = \perp$ *or* $\mathsf{val}_\ell^\rho(D) = \perp$.
- *Immediately after the assignment,* $\mathsf{val}_\ell^\sigma(C) = \mathsf{val}_\ell^\rho(D) \neq \perp$.

and one of the following is true:

- *Immediately after the assignment,* C *and* D *are table-defined.*
- *Immediately after the assignment,* C *is table-defined and* D *is a chain enqueued in* Q_{all}.
- *C and* D *are chains enqueued in* Q_{all}.

Lemma 17. *The probability of event* (BadlyCollide$^+$ $\wedge$ ¬BadlyHit$^+$ $\wedge$ ¬BadP) *in* H_2 *is at most* $21160\,T^5/2^n$.

Definition 18. *We say that event* BadlyCollideP *occurs in* H_2 *if either:*

- *A uniform assignment* $P(\downarrow, x_0, x_1) := (x_{10}, x_{11})$ *is such that there exist partial chains* C, D *such that for some* $\sigma, \rho \in \{+, -\}$ *the following are all true:*
  - *Immediately before the assignment,* C *and* D *are not equivalent.*
  - *Immediately before the assignment,* $\mathsf{val}_{10}^\sigma(C) = \perp$ *or* $\mathsf{val}_{10}^\rho(D) = \perp$.
  - *Immediately after the assignment,* $\mathsf{val}_{10}^\sigma(C) = \mathsf{val}_{10}^\rho(D) = x_{10} \neq \perp$.
  and one of the following conditions hold:
  - *Before the assignment,* C *and* D *are chains in* Q_{all}^*.
  - *Immediately after the assignment,* C *and* D *are table-defined.*
  - *Before the assignment,* C *is a chain enqueued in* Q_{all} *and immediately after the assignment,* D *is table-defined.*
- *A uniform assignment* $P(\uparrow, x_{10}, x_{11}) := (x_0, x_1)$ *is such that there exist partial chains* C, D *such that for some* $\sigma, \rho \in \{+, -\}$ *the following are all true:*
  - *Immediately before the assignment,* C *and* D *are not equivalent.*
  - *Immediately before the assignment,* $\mathsf{val}_1^\sigma(C) = \perp$ *or* $\mathsf{val}_1^\rho(D) = \perp$.
  - *Immediately after the assignment,* $\mathsf{val}_1^\sigma(C) = \mathsf{val}_1^\rho(D) = x_1 \neq \perp$.
  and one of the following conditions hold:
  - *Before the assignment,* C *and* D *are chains in* Q_{all}^*.

- *Immediately after the assignment, C and D are table-defined.*
- *Before the assignment, C is a chain enqueued in Q_{all} and immediately after the assignment, D is table-defined.*

Lemma 19. *The probability of event* BadlyCollideP *in H_2 is at most $314\,T^5/2^n$.*

Proof. Consider the case that after a uniform choice of (x_0, x_1) leading to an assignment $P(\uparrow, x_{10}, x_{11}) := (x_0, x_1)$, event BadlyCollideP occurs. The value $\mathsf{val}_1^-(C)$ for a chain C does not change due to the assignment since it is a $P(\uparrow, x_{10}, x_{11})$ assignment and $\mathsf{val}_1^-(C)$ can change only due to a $P(\downarrow, x_0, x_1)$ assignment by definition of $\mathsf{val}^-(\cdot)$.

Suppose that $\mathsf{val}_1^+(C) = \perp$ and $\mathsf{val}_1^-(D) \neq \perp$ before the assignment and after the assignment $\mathsf{val}_1^+(C) = \mathsf{val}_1^-(D) = x_1$. The value $\mathsf{val}_1^-(D)$ does not change due to the assignment as mentioned above. So, the probability that $\mathsf{val}_1^+(C) = \mathsf{val}_1^-(D) = x_1$ is 2^{-n}.

Suppose that $\mathsf{val}_1^+(C) = \mathsf{val}_1^+(D) = \perp$ before the assignment and after the assignment $\mathsf{val}_1^+(C) = \mathsf{val}_1^+(D) = x_1$. For this to happen, $\mathsf{val}_{10}(C) = \mathsf{val}_{10}(D) = x_{10}$ and $\mathsf{val}_{11}(C) = \mathsf{val}_{11}(D) = x_{11}$ implying that C and D are equivalent chains. So, the probability of this event is 0.

Suppose that $\mathsf{val}_1^+(C) = \perp$ and $\mathsf{val}_1^+(D) \neq \perp$ before the assignment and after the assignment $\mathsf{val}_1^+(C) = \mathsf{val}_1^+(D) = x_1$. Now, the value of $\mathsf{val}_1^+(D)$ stays the same after the assignment (even if BadP occurs). So, the probability that $\mathsf{val}_1^+(C) = \mathsf{val}_1^+(D) = x_1$ is 2^{-n}.

The analysis for the other case follows similarly. There are at most T assignments of the form $P(\uparrow, x_{10}, x_{11})$ or $P(\downarrow, x_0, x_1)$. There are at most $11T^2$ possibilities for a chain to be table-defined before the assignment and T possibilities for a chain to be table-defined after the assignment but not before. There are at most T chains enqueued for completion in Q_{all}. So, the probability of event BadlyCollideP is at most $\left(T \cdot ((11T^2 + T)^2 + T^2 + T \cdot (11T^2 + T)) \cdot 2\right) \cdot 2^{-n}$.

Definition 20. *We say event* BadlyHitFV *occurs in H_2 if a uniform assignment to $G_s(x_s)$ that occurs in a call* $\text{ADAPT}(Q, x_{\ell-2}, x_{\ell-1}, x_{\ell+2}, x_{\ell+3}, \ell, g, b)$, *for some $s \in \{g, b\}$ one of the following happens (where we let $C = (x_{\ell-2}, x_{\ell-1}, \ell - 2)$):*

- $s = \ell + 2$ *and the following holds:*
  - *Immediately before the assignment,* $\mathsf{val}_{\ell+1}^-(C) = \perp$.
  - *Immediately after the assignment,* $\mathsf{val}_{\ell+1}^-(C) \neq \perp$.
  - *Immediately after the assignment,* $y := \mathsf{val}_{\ell-1}^-(C) \oplus \mathsf{val}_{\ell+1}^-(C)$ *is such that $x'_{\ell+1} \oplus x'_{\ell-1} = y$ for some $x'_{\ell+1} \in G_{\ell+1}$ and $x'_{\ell-1} \in G_{\ell-1}$.*
- $s = \ell - 1$ *and the following holds:*
  - *Immediately before the assignment,* $\mathsf{val}_{\ell}^+(C) = \perp$.
  - *Immediately after the assignment,* $\mathsf{val}_{\ell}^+(C) \neq \perp$.
  - *Immediately after the assignment,* $y := \mathsf{val}_{\ell+2}(C) \oplus \mathsf{val}_{\ell}^+(C)$ *is such that $x'_{\ell+2} \oplus x'_{\ell} = y$ for some $x'_{\ell+2} \in G_{\ell+2}$ and $x'_{\ell} \in G_{\ell}$.*

Lemma 21. *The probability of event* BadlyHitFV *in H_2 is at most $2\,T^3/2^n$.*

Proof. Consider the first case where $s = \ell + 2$. Note that for a chain C with $s = \ell + 2$ the "value" at the adapt position $\ell + 1$ is set as $\mathsf{val}_{\ell+1}(C) := \mathsf{val}_{\ell+3}(C) \oplus G_s(\mathsf{val}_s(C))$ where $\mathsf{val}_{\ell+3}(C) \neq \perp$ is one of the arguments to ADAPT. Since the assignment to $G_s(x_s)$ happens inside the ADAPT call, $\mathsf{val}^-_{\ell+1}(C) = \perp$ until the assignment and $\mathsf{val}^-_{\ell+1}(C) \neq \perp$ immediately after the assignment.

Now, $y := \mathsf{val}_{\ell-1}(C) \oplus \mathsf{val}^-_{\ell+1}(C)$. Note that $\mathsf{val}_{\ell-1}(C) \neq \perp$ since $\mathsf{val}_{\ell-1}(C) = x_{\ell-1}$ is one of the arguments of the ADAPT procedure. So, for $y := \mathsf{val}_{\ell-1}(C) \oplus \mathsf{val}_{\ell+3}(C) \oplus G_s(\mathsf{val}_s(C))$ to be such that $y = x'_{\ell-1} \oplus x'_{\ell+1}$ where $x'_{\ell-1} \in G_{\ell-1}$ and $x'_{\ell+1} \in G_{\ell+1}$, y needs to take one of $T^2/2^n$ values. Note that there are at most T such calls to ADAPT by assumption. So, the probability of the first case is at most $T^3/2^n$. The analysis for the second case is analogous.

Definition 22. *We say that event* BadlyCollideFV *occurs in* H_2 *if a uniform assignment to* $G_s(x_s)$ *that occurs in a call to* ADAPT$(Q, x_{\ell-2}, x_{\ell-1}, x_{\ell+2}, x_{\ell+3}, \ell, g, b)$, *for some* $s \in \{g, b\}$ *the following happens (where we let* $C = (x_{\ell-2}, x_{\ell-1}, \ell - 2)$ *and* D *is a chain in* Q^*_{all}*):*

- $s = \ell + 2$, *and for some* $(k, k') \in \{(\ell-1, \ell+1), (\ell+1, \ell-1)\}$ *the following holds:*
  - *Immediately before the assignment,* $\mathsf{val}^-_{\ell+1}(C) = \perp$ *and* $\mathsf{val}_k(D) \neq \perp$.
  - *Immediately after the assignment,* $\mathsf{val}^-_{\ell+1}(C) \neq \perp$.
  - *Immediately after the assignment,* $y := \mathsf{val}_{\ell-1}(C) \oplus \mathsf{val}^-_{\ell+1}(C)$ *is such that* $x \oplus y = \mathsf{val}_k(D)$ *for some* $x \in G_{k'}$.
- $s = \ell - 1$, *and for some* $(k, k') \in \{(\ell, \ell+2), (\ell+2, \ell)\}$ *the following holds:*
  - *Immediately before the assignment,* $\mathsf{val}^+_\ell(C) = \perp$ *and* $\mathsf{val}_k(D) \neq \perp$.
  - *Immediately after the assignment,* $\mathsf{val}^+_\ell(C) \neq \perp$.
  - *Immediately after the assignment,* $y := \mathsf{val}_{\ell+2}(C) \oplus \mathsf{val}^+_\ell(C)$ *is such that* $x \oplus y = \mathsf{val}_k(D)$ *for some* $x \in G_{k'}$.

Lemma 23. *The probability of event* BadlyCollideFV *in* H_2 *is at most* $4T^3/2^n$.

Proof. Consider the first case where $s = \ell + 2$. Note that during the ADAPT call the "value" at the adapt position $\ell + 1$ is set as $\mathsf{val}_{\ell+1}(C) := \mathsf{val}_{\ell+3}(C) \oplus G_s(\mathsf{val}_s(C))$ where $\mathsf{val}_{\ell+3}(C) \neq \perp$ is one of the arguments to ADAPT. Since the assignment to $G_s(x_s)$ happens inside the ADAPT call, $\mathsf{val}^-_{\ell+1}(C) = \perp$ until the assignment and $\mathsf{val}^-_{\ell+1}(C) \neq \perp$ immediately after the assignment.

Now, $y := \mathsf{val}_{\ell-1}(C) \oplus \mathsf{val}^-_{\ell+1}(C)$. Note that $\mathsf{val}_{\ell-1}(C) \neq \perp$ since it is one of the arguments of ADAPT. Also note that if $\mathsf{val}_k(D) \neq \perp$ before the assignment, then $\mathsf{val}_k(D)$ does not change due to the assignment. Say $k = \ell-1$ and $k' = \ell+1$. So, for $y := \mathsf{val}_{\ell-1}(C) \oplus \mathsf{val}_{\ell+3}(C) \oplus G_s(x_s)$ to be such that $y = x \oplus \mathsf{val}_{\ell-1}(D)$ where $x \in G_{\ell+1}$, the value y would have to take one of $T^2/2^n$ values. (This is because T is the upper bound on the number of chains enqueued in Q_{all} by assumption and on the size of $G_{\ell+1}$.) Similarly for the case where $k = \ell+1$ and $k' = \ell - 1$. So, for a single call to ADAPT where $s = \ell + 2$, we have that the probability that the event occurs is $2T^2/2^n$. There are at most T calls to ADAPT by assumption and hence, the probability of the first case is at most $2T^3/2^n$.

The analysis for the second case is analogous.

We say an execution of H_2 is *good* if none of BadP, BadlyHit$^+$, BadlyCollide$^+$, BadlyCollideP, BadlyHitFV, or BadlyCollideFV occur. Lemmas 13–23 imply:

Lemma 24. *The probability that an execution of H_2 is good is $1 - O(T^5)/2^n$.*

5.3.2   Properties of Good Executions

Notation. For a chain $C = (x_k, x_{k+1}, k, \ell, g, b)$ that is enqueued for completion, the "adapt positions" are at ℓ, $\ell+1$. These positions are those where the simulator uses FORCEVAL$(\cdot, \cdot, \ell)$ and FORCEVAL$(\cdot, \cdot, \ell+1)$ to force the values at $G_\ell(\cdot)$ and $G_{\ell+1}(\cdot)$. Also, for the chain C, the "set uniform" positions are at $\ell-1$, $\ell+2$. (These are the buffer zones that surround the adapt positions.) One of these "set uniform" positions is adjacent to the query that caused the chain to be enqueued and this position is denoted by g and referred to as the "good" set uniform position. The other "set uniform" position is referred to as the "bad" set uniform position. Note that $g, b \in \{\ell-1, \ell+2\}$ and $g \neq b$; Let a be the adapt position that is adjacent to "bad" set uniform position. So, if $b = \ell-1$, then $a = \ell$; Else, if $b = \ell + 2$, $a = \ell + 1$. Consider a call ADAPT$(x_{\ell-2}, x_{\ell-1}, x_{\ell+2}, x_{\ell+3}, \ell, g, b)$, if $b = \ell - 1$ define $x_a = x_\ell$ as $x_\ell := x_{\ell-2} \oplus G_{\ell-1}(x_{\ell-1})$ if $x_{\ell-1} \in G_{\ell-1}$, and $x_\ell = \perp$ otherwise. Analogously, if $b = \ell + 2$, define $x_a = x_{\ell+1} := x_{\ell+3} \oplus G_{\ell+2}(x_{\ell+2})$ if $x_{\ell+2} \notin G_{\ell+2}$ and $x_{\ell+1} = \perp$ otherwise.

Also, for a chain C enqueued in Q_b we say *adapting is safe* if just before the call to ADAPT for C, we have $x_g \notin G_g$ and $x_a \notin G_a$. Analogously, for a chain C in Q_{all}^* or Q_{mid} we say *adapting is safe* if just before the call to ADAPT for C, we have $x_{\ell-1} \notin G_{\ell-1}$ and $x_{\ell+2} \notin G_{\ell+2}$. Also, we loosely use the statement $C \in$ CompChains where $C = (x_k, x_{k+1}, k, \ell, g, b)$ to mean that $(x_k, x_{k+1}, k) \in$ CompChains.

High-level Overview. The aim of this section is to prove that during a good execution of H_2, every call to FORCEVAL$(x, \cdot, a)$ is such that $x \notin G_a$, i.e., to prove that a FORCEVAL call does not "overwrite."

To prove that FORCEVAL does not "overwrite," we prove that for every call to ADAPT that occurs during the completion of a chain $C = (x_k, x_{k+1}, k, \ell, g, b)$, we have $\text{val}_g(C) \notin G_g$ before the call and if C is enqueued in Q_b, $\text{val}_a(C) \notin G_a$ before the call; else, $\text{val}_b(C) \notin G_b$ before the call i.e. every call to ADAPT is "safe". In order to prove the above statements, we will prove that at the time a chain C is enqueued in Q_{all}, $\text{val}_g(C) = \perp$ and if C is a chain enqueued in Q_b for some $b \in \{1, 5, 6, 10\}$, then $\text{val}_b(C) \notin G_b$; else, $\text{val}_b(C) = \perp$ when C was enqueued. Similarly, if a chain C is enqueued in Q_{mid}, then just before the assignment that precedes C being enqueued occurs, we will prove that $\text{val}_g(C) = \perp$ and $\text{val}_b(C) = \perp$. We also need to prove properties of equivalent chains in order to prove that if a chain equivalent to C has been completed before C, then $C \in$ CompChains when it is dequeued. All of this put together will help us prove that FORCEVAL does not "overwrite" (Theorem 39). While the structure explained above is similar to the structure of the proof in [1,11], the major difference is in how we prove the properties of chains at the time they are enqueued. This is due to the fact that we separate enqueueing from completion in our simulation.

Due to space constraints, we state some lemmas without proofs, and refer to the full version of our work for details [3].

Properties of Equivalent Chains

Lemma 25. *Consider a good execution of H_2. Suppose that at some point in the execution, two partial chains C and D are equivalent. Then there exists a sequence of partial chains $C_1, \ldots, C_r$ such that*

- *$C = C_1$ and $D = C_r$, or else $D = C_1$ and $C = C_r$,*
- *for $r \geq 2$, $C_i = \mathsf{next}(C_{i-1})$ and $C_{i-1} = \mathsf{prev}(C_i)$ for all $i \in \{2, \ldots, r\}$,*
- *for $r \geq 3$, $C_2, \ldots, C_{r-1}$ is table-defined,*
- *$D = (\mathsf{val}_j^\rho(C), \mathsf{val}_{j+1}^\rho(C), j)$ where $\mathsf{val}_j^\rho(C) \neq \bot$ and $\mathsf{val}_{j+1}^\rho(C) \neq \bot$ for some $\rho \in \{+, -\}$,*
- *$C = (\mathsf{val}_k^\sigma(D), \mathsf{val}_{k+1}^\sigma(D), k)$ where $\mathsf{val}_k^\sigma(D) \neq \bot$ and $\mathsf{val}_{k+1}^\sigma(D) \neq \bot$ for some $\sigma \in \{+, -\}$.*

Lemma 26. *Consider some point in a good execution of H_2 and assume that $x \notin G_j$ before every call to $\mathrm{FORCEVAL}(x, \cdot, j)$ prior to this point in the execution. Then, if the partial chains $C = (x_k, x_{k+1}, k)$ with $k \in \{1, 5, 9\}$ and $D = (x'_m, x'_{m+1}, m)$ with $m \in \{1, 5, 9\}$ are equivalent at this point in the execution, then $C \in \mathsf{CompChains}$ if and only if $D \in \mathsf{CompChains}$.*

Properties of Enqueued Chains

Recall that $\{1, 5, 6, 10\}$ are "bad" set uniform positions.

Lemma 27. *Say a chain $C = (x_k, x_{k+1}, k, \ell, g, b)$ is enqueued to be completed in Q_b. Then at the time C is enqueued, $\mathsf{val}_g(C) = \bot$ and $\mathsf{val}_b(C) \notin G_b$.*

Effects of a Call to ForceVal

For the following lemmas, note that $g, b \in \{\ell - 1, \ell + 2\}$ and $g \neq b$.

Lemma 28. *In a good execution of H_2, let $x_{\ell-1} \notin G_{\ell-1}$ (respectively $x_{\ell+2} \notin G_{\ell+2}$) immediately before a call $\mathrm{ADAPT}(Q, x_{\ell-2}, x_{\ell-1}, x_{\ell+2}, x_{\ell+3}, \ell, g, b)$. Then, before the call to $\mathrm{FORCEVAL}(x_\ell, \cdot, \ell)$ (respectively $\mathrm{FORCEVAL}(x_{\ell+1}, \cdot, \ell+1)$) in that ADAPT call, we have $x_\ell \notin G_\ell$ (respectively $x_{\ell+1} \notin G_{\ell+1}$).*

The lemma above immediately gives us the following corollary.

Corollary 29. *Consider a call $\mathrm{ADAPT}(Q, x_{\ell-2}, x_{\ell-1}, x_{\ell+2}, x_{\ell+3}, \ell, g, b)$ in a good execution of H_2 and assume that adapting was safe for all chains C that were dequeued before this ADAPT call. Then, before the call to $\mathrm{FORCEVAL}(x_\ell, \cdot, \ell)$ and $\mathrm{FORCEVAL}(x_{\ell+1}, \cdot, \ell+1)$ that occurs in $\mathrm{ADAPT}(Q, x_{\ell-2}, x_{\ell-1}, x_{\ell+2}, x_{\ell+3}, \ell, g, b)$, we have $x_\ell \notin G_\ell$ and $x_{\ell+1} \notin G_{\ell+1}$ respectively.*

Lemma 30. *Suppose that $x_{\ell-1} \notin G_{\ell-1}$ (respectively $x_{\ell+2} \notin G_{\ell+2}$) immediately before a call $\text{ADAPT}(Q, x_{\ell-2}, x_{\ell-1}, x_{\ell+2}, x_{\ell+3}, \ell, g, b)$ in a good execution of H_2. Then, if C is a table-defined chain before the call to ADAPT, $\text{val}_i(C)$ for $i \in \{1, \ldots, 10\}$ stays constant during the call to $\text{FORCEVAL}(x_\ell, \cdot, \ell)$ (respectively $\text{FORCEVAL}(x_{\ell+1}, \cdot, \ell+1))$.*

Lemma 31. *Suppose that $x_{\ell-1} \notin G_{\ell-1}$ (respectively $x_{\ell+2} \notin G_{\ell+2}$) immediately before a call $\text{ADAPT}(Q, x_{\ell-2}, x_{\ell-1}, x_{\ell+2}, x_{\ell+3}, \ell, g, b)$ in a good execution of H_2. Then, if C is a chain enqueued in Q_{all}, $\text{val}_i(C)$ for $i \in \{1, \ldots, 10\}$ stays constant during the call to $\text{FORCEVAL}(x_\ell, \cdot, \ell)$ (respectively $\text{FORCEVAL}(x_{\ell+1}, \cdot, \ell+1))$ that occurs in the ADAPT call.*

Lemma 32. *Consider a call to $\text{ADAPT}(Q, x_{\ell-2}, x_{\ell-1}, x_{\ell+2}, x_{\ell+3}, \ell, g, b)$ in a good execution of H_2 for some $Q \in \{Q_1, Q_5, Q_6, Q_{10}\}$. Assume that adapting was safe for all chains C that were dequeued from Q_1, Q_5, Q_6, Q_{10} before this ADAPT call. If $x_a \notin G_a$ and $x_g \notin G_g$ (where a is the adapt position adjacent to the "bad" set uniform position) before the ADAPT call, then if C is a chain enqueued in Q_{all}, $\text{val}_i(C)$ for $i \in \{1, \ldots, 10\}$ stays constant during the call to $\text{FORCEVAL}(x_a, \cdot, a)$ that occurs in the ADAPT call.*

Additional Properties of Enqueued Chains

For the following lemma, if a chain $C = (x_k, x_{k+1}, k, \ell, g, b)$ is enqueued in Q_{mid}, then the assignment $G_i(x_i)$ that precedes C being enqueued happens either in lines 19, 149 or 152 of the simulator's execution.

Lemma 33. *Suppose that a chain $C = (x_k, x_{k+1}, k, \ell, g, b)$ is enqueued in Q_{mid} during a good execution of H_2 such that no chain equivalent to C has been enqueued for completion so far. Suppose also that adapting has been safe for every chain dequeued from Q_1, Q_5, Q_6, Q_{10} or Q_{all}^* so far. Then $\text{val}_g(C) = \bot$ and $\text{val}_b(C) = \bot$ just before the assignment $G_i(x_i)$ that precedes C being enqueued. Also, $\text{val}_9(C) = \text{val}_2(C) = \bot$ just before the assignment $G_i(x_i)$ that precedes C being enqueued.*

Proof. Say a chain $C = (x_5, x_6, 5, 2, 4, 1)$ is enqueued in Q_{mid} with $g = 4$ and $b = 1$. Then, the assignment $G_5(x_5)$ that precedes the enqueueing of C is such that $x_5 \notin G_5$ before the assignment, by construction of the simulator. Otherwise, $\text{ENQNEWMIDCHAINS}(5, x_5)$ is not called. Hence, $\text{val}_4^-(C) = \bot$ just before the assignment $G_5(x_5)$ that precedes C being enqueued. Also, since $\text{val}_4^-(C) = \bot$, we have $\text{val}_1^-(C) = \bot$.

Before we prove $\text{val}_4^+(C) = \bot$ and $\text{val}_1^+(C) = \bot$ (and hence, $\text{val}_4(C) = \bot$ and $\text{val}_1(C) = \bot$), we make the following observation. If a partial chain $(x_5, x_6, 5)$ is enqueued in Q_{mid} such that no equivalent chain has been enqueued previously, by construction of the simulator, either (1) $\text{val}_5(D) = x_5$ for a chain D belonging to Q_{all}^* where $\text{val}_5(D) = \bot$ when D was enqueued or (2) $\text{val}_6(E) = x_6$ for a chain E enqueued in Q_{all}^* where $\text{val}_6(E) = \bot$ when E was enqueued or (3) both. In other words, either $x_5 \notin G_5 \cup A_5^t$ or $x_6 \notin G_6 \cup A_6^t$ or both when $Q_{\text{enq}}.\text{EMPTY}() = \text{true}$ in line 6 of the simulator's execution after $\mathbf{D}$'s t^{th} query.

Consider a chain $C = (x_5, x_6, 5, 2, 4, 1)$ which was enqueued in Q_{mid} such that no chain equivalent to C was enqueued previously. Such a chain C is enqueued in Q_{mid}, when $x_6 \in G_6$, $\text{val}_5(C) = \text{val}_5(D) = x_5$ and $x_5 \in G_5$ right before C was enqueued (and not earlier) where D is a chain belonging to Q^*_{all} and $x_5 \in G_5$ due to the completion of D.

For $\text{val}_1(C) \neq \perp$ at the time of the assignment that precedes the enqueueing of C, we need $\text{val}_1^+(C) \neq \perp$. Then, in particular, we have that $x_7 := \text{val}_7(C) \in G_7$ and $x_8 := \text{val}_8(C) \in G_8$ (otherwise, $\text{val}_9^+(C) = \perp$ implying that $\text{val}_1^+(C) = \perp$).

Consider the partial chains $C = (x_5, x_6, 5)$, $C_1 = (x_6, x_7, 6)$ and $C_2 = (x_7, x_8, 7)$. For $\text{val}_9^+(C) \neq \perp$ just before the assignment that precedes the enqueueing of C, we need (1) $C_1 = \text{next}(C)$, $C_2 = \text{next}(C_1)$ (and hence, $x_6 \in G_6$ and $x_7 \in G_7$) and (2) $x_5 = \text{val}_5(D)$ for a chain D in Q^*_{all} and (3) $x_8 \in G_8$ or $x_8 = \text{val}_8(E)$ of a chain E enqueued in Q_{all}. Note that this condition is not true at the time the simulator finished enqueueing chains in Q_{all} since we have either $x_5 \notin G_5 \cup A_5^t$ or $x_6 \notin G_6 \cup A_6^t$ or both. Hence, the conditions must have been met during the completion of chains in Q_{all}. Consider the last assignment that was made before all the above conditions were met.

Consider the case that when the last assignment (such that all the conditions listed above were met immediately after this assignment) happened, the chain C_1 was already table-defined. Now, if the assignment was a P/P^{-1} assignment, then BadP occurred. It cannot be a FORCEVAL assignment since FORCEVAL does not change the value of a chain enqueued in Q_{all} by Lemmas 31 and 32. If it were a uniform assignment to $G_i(x_i)$, then, BadlyCollide$^+$ occurred.

Consider the case that when the last assignment (such that all the conditions listed above were met immediately after this assignment) happened, the chain C_1 was not table-defined before the assignment but table-defined immediately after. Recall that if $C_1 = (x_6, x_7, 6)$ is table-defined then $x_6 \in G_6$ and $x_7 \in G_7$. So, the assignment was either to $G_6(x_6)$ or $G_7(x_7)$.

Consider the case that it set $G_7(x_7)$. If this were a uniform assignment to $G_7(x_7)$, then BadlyCollide$^+$ occurred since $C_1 (\equiv C)$ and E are not equivalent as no chain equivalent to C has been enqueued previously. If this were a FORCEVAL assignment, then BadlyCollideFV occurred. This is because 7 is an adapt position only for partial chains that are either of the form (a) $X = (x_9, x_{10}, 9)$ such that $(x_9, x_{10}, 9, 6, 8, 5)$ belongs to Q^*_{all}. By assumption for chains in Q^*_{all}, we have $\text{val}_5(X) \notin G_5$ before the ADAPT call for such a chain or, (b) $Y = (x_1, x_2, 1)$ such that $(x_1, x_2, 1, 7, 9, 6)$ is enqueued in Q_6. In this case, the adapt position 7 is adjacent to the "bad" set uniform position 6. By assumption for chains enqueued in Q_6, we have $\text{val}_9(Y) \notin G_9$ before the ADAPT call for such a chain. Hence, BadlyCollideFV occurred due to the assignment $G_5(\text{val}_5(X))$ or $G_9(\text{val}_9(Y))$ that occurs in the ADAPT call. The analysis for the case when $G_6(x_6)$ is set is similar. So, the above conditions are not met for a chain C to be enqueued in Q_{mid}. Hence, for such a chain $C = (x_5, x_6, 5, 2, 4, 1)$, $\text{val}_9^+(C) = \perp$ just before the assignment that caused C to be enqueued. Since $\text{val}_9^+(C) = \perp$ and $\text{val}_4^-(C) = \perp$ before the assignment, we have $\text{val}_4(C) = \perp$, $\text{val}_9(C) = \perp$ and $\text{val}_1(C) = \perp$ just before the assignment that precedes C being enqueued. The analysis for the case where $C = (x_5, x_6, 5, 8, 7, 10)$ is analogous.

Lemma 34. *Consider a good execution of* H_2. *Just before the execution of line 27 during the simulator's execution, if adapting was safe for every chain dequeued from* $Q_1, Q_5, Q_6, Q_{10}, Q_{all}^*$ *or* Q_{mid} *so far, then it holds that:*

   i. if $x_9 \in G_9$, $x_{10} \in G_{10}$, $x_1 \in G_1$ *such that* **R**.CHKBWD$(x_{10}, x_9 \oplus G_{10}(x_{10}), x_1) =$ true, *then* $(x_9, x_{10}, 9) \in$ CompChains.

   ii. if $x_1 \in G_1$, $x_2 \in G_2$, $x_{10} \in G_{10}$ *such that* **R**.CHKFWD$(x_2 \oplus G_1(x_1), x_1, x_{10}) =$ true, *then* $(x_1, x_2, 1) \in$ CompChains.

   iii. if $x_5 \in G_5$, $x_6 \in G_6$, *then* $(x_5, x_6, 5) \in$ CompChains.

Proof. We start by proving (i). For a triple (x_9, x_{10}, x_1), we say that "condition holds" if (x_9, x_{10}, x_1) is such that $x_9 \in G_9$, $x_{10} \in G_{10}$, $x_1 \in G_1$ and **R**.CHKBWD$(x_{10}, x_9 \oplus G_{10}(x_{10}), x_1) =$ true. Also, we refer to the partial chain $(x_9, x_{10}, 9)$ as the partial chain associated with the triple (x_9, x_{10}, x_1). So, our aim is to prove that for every triple (x_9, x_{10}, x_1) such that condition holds, the associated partial chain $(x_9, x_{10}, 9) \in$ CompChains. Assume that the lemma has held right before (and hence immediately after) line 27 of the simulator's execution while answering the distinguisher's $(t-1)^{th}$ query to F$(\cdot, \cdot)$. Let the distinguisher ask its tth query F(k, x). The aim is to prove that at line 27 of the simulator's execution while answering the distinguisher's t^{th} query to F$(\cdot, \cdot)$, if a triple $T^* = (x_9, x_{10}, x_1)$ is such that condition holds, then the partial chain $C^* = (x_9, x_{10}, 9)$ associated with the triple is such that $C^* \in$ CompChains. Note that the distinguisher could have made queries to P/P^{-1} between the $(t-1)^{th}$ and t^{th} queries to F$(\cdot, \cdot)$; but if those queries resulted in condition being true, then BadP occurred.

Suppose that there exists a triple T^* such that condition holds at line 27 of the simulator's execution while answering the distinguisher's t^{th} query. If condition held at the end of simulator's execution while answering the previous distinguisher query, then by assumption that the lemma has held so far, the partial chain C^* associated with the triple T^* is such that $C^* \in$ CompChains. If condition held at the end of the simulator's execution of the current query t (and not at the end of the previous query), we differentiate cases where the associated partial chain C^* was enqueued for completion during the simulator's execution while answering the tth query and when it's not.

Consider the case where a chain equivalent to C^* was enqueued in Q_{all} during the simulator's execution while answering the distinguisher's current query. If $C^* = (x_9, x_{10}, 9)$ was enqueued during the tth query, then $(x_9, x_{10}, 9) \in$ CompChains by construction of the simulator. Note also that chains in SimPChains are not enqueued for completion by the simulator. By definition of the set SimPChains, these chains are such that they are equivalent to a chain of the form $(x_5, x_6, 5)$ that has been enqueued for completion. Since BadP does not occur and FORCEVAL does not overwrite, the equivalence holds when $(x_5, x_6, 5) \in$ CompChains and hence, by Lemma 26, such a chain in SimPChains is placed in CompChains as well. By the same argument, if a chain equivalent to C^* has been enqueued for completion, then too $C^* \in$ CompChains by the end of the simulator's execution of the current query. So, if a chain equivalent to

C^* was enqueued for completion or was in SimPChains during the simulator's execution while answering the current query t, then $C^* \in$ CompChains.

Consider the case where no chain equivalent to C^* was enqueued in Q_{all} and $C^* \notin$ SimPChains during the simulator's execution while answering the distinguisher's current query. We differentiate between the cases where (1) $C =$ next$(C^*) \neq \perp$, next$(C) \neq \perp$ when Q_{enq}.EMPTY$() = $ true in line 6 of the simulator's execution when answering the distinguisher's t^{th} query and (2) when it's not.

Consider the case when $C = $ next$(C^*) \neq \perp$ and next$(C) \neq \perp$ at the time the simulator stops enqueueing chains in Q_{all} i.e. when Q_{enq}.EMPTY$() = $ true in line 6 of the simulator's execution when answering the distinguisher's t^{th} query. This implies that $x_{10} \in G_{10}$ and $(\uparrow, x_{10}, x_{11}) \in P$ where $x_{11} := x_9 \oplus G_{10}(x_{10})$ and hence, $C = (x_{10}, x_{11}, 10)$ is table-defined at the time the simulator stops enqueueing chains in Q_{all}. Since the triple T^* is such that the associated partial chain $C^* = (x_9, x_{10}, 9)$ was not enqueued for completion and not in SimPChains, we have that either (a) $x_9 \notin G_9 \cup A_9^t$ or (b) $x_1 \notin G_1 \cup A_1^t$ when Q_{enq}.EMPTY$() = $ true in line 6. For the condition to be true, we need $x_1 \in G_1$ and $x_9 \in G_9$ and hence, we have that condition does not hold for the triple T^* when Q_{enq}.EMPTY$() = $ true in line 6. Consider the case where $x_1 \notin G_1 \cup A_1^t$. For $x_1 \in G_1$ to be true by the end of the simulator's execution while answering the distinguisher's t^{th} query, it must be the case that val$_1(D) = $ val$_1(C) = x_1$ at some point for a chain D that has been enqueued in Q_{all} or Q_{mid}. Before analyzing the case that val$_1(D) = $ val$_1(C) = x_1$ occurs, we make the following observations. Firstly, C and D are not equivalent as $C \equiv C^*$ and no chain equivalent to C^* (including itself) has been enqueued. Secondly, for all chains D that have been enqueued in Q_{all}, val$_1(D) \neq x_1$ when enqueued since $x_1 \notin G_1 \cup A_1^t$. Now, if val$_1(D) \neq x_1$ and val$_1(D) \neq \perp$, it cannot be that val$_1(D) = x_1$ at a later point since FORCEVAL does not overwrite and BadP does not occur. Hence, if val$_1(D) = x_1$ at a later point, then val$_1(D) = \perp$ when enqueued. Similarly, for all chains D that have been enqueued in Q_{mid} val$_1(D) = \perp$ just before the assignment that precedes the enqueueing of D by Lemma 33. Since BadlyHit$^+$ and BadlyHitFV do not occur, val$_1(D) = \perp$ at the time D is enqueued. Now, if val$_1(D) = $ val$_1(C) = x_1$, then this is during the completion of some chain E during the simulator's execution while answering the distinguisher's t^{th} query. Consider the last assignment before val$_1(D) = $ val$_1(C) = x_1$ was true. This cannot be a uniform assignment to $G_i(x_i)$ since then BadlyCollide$^+$ occurred. This cannot be due to a uniform assignment to P since then BadP or BadlyCollideP occurred. This cannot be a FORCEVAL assignment since that would contradict Lemmas 30, 31 or 32. The analysis for the case where $x_9 \notin G_9 \cup A_9^t$ when the simulator stops enqueueing chains in Q_{all} is analogous. So, if C was table-defined when the simulator stops enqueueing chains in Q_{all}, then condition does not hold for the triple T^* at the end of the simulator's execution of the current query.

Consider the case when either next$(C^*) = \perp$ or $C = $ next$(C^*) \neq \perp$ and next$(C) = \perp$ at the time the simulator stops enqueueing chains in Q_{all} i.e. when Q_{enq}.EMPTY$() = $ true in line 6 of the simulator's execution when answering the distinguisher's t^{th} query. Now if the triple $T^* = (x_9, x_{10}, x_1)$ is such that condition holds by the end of the simulator's execution of the current query, then it must be the case that $\text{next}(C^*) \neq \perp$ and $\text{next}(\text{next}(C^*)) \neq \perp$ by the end of the simulator's execution. In particular, it means that the partial chain $\text{next}(C^*) = C = (x_{10}, x_{11}, 10)$ where $x_{11} := x_9 \oplus G_{10}(x_{10})$ is table-defined (with $\text{val}_1(C) = x_1$) by the end of the simulator's execution. Note that at the moment that C becomes table-defined either $x_1 \notin G_1$ or $x_9 \notin G_9$ as otherwise either BadP or BadlyHit$^+$ occurred. Furthermore, immediately before the assignment that causes C to be table-defined we have either $\text{val}_1(C) = \perp$ or $\text{val}_9(C) = \perp$ and immediately after the assignment, we have $\text{val}_9(C) \neq \perp$ and $\text{val}_1(C) \neq \perp$ by definition. Say $\text{val}_1(C) = \perp$ immediately before the assignment that caused C to be table-defined and $\text{val}_1(C)(= x_1) \neq \perp$ immediately after. For $x_1 \in G_1$ to be true by the end of the simulator's execution while answering the distinguisher's t^{th} query, it must be the case that $\text{val}_1(D) = \text{val}_1(C) = x_1$ at some point for a chain D that has been enqueued in Q_{all} or Q_{mid}. Consider the last assignment before $\text{val}_1(D) = \text{val}_1(C) = x_1$ was true. The rest of the analysis proceeds similarly to the analysis above. The case when $\text{val}_9(C) = \perp$ immediately before the assignment that caused C to be table-defined and $\text{val}_9(C)(= x_9) \neq \perp$ immediately after follows in a similar fashion. So, if $\text{next}(C^*) = \perp$ or if $\text{next}(C^*) \neq \perp$ and $\text{next}(\text{next}(C^*)) = \perp$ when the simulator stops enqueueing chains in Q_{all}, then too the condition does not hold for the triple T^* at the end of the simulator's execution of the current query. Summarizing, if a chain equivalent to C^* was not enqueued in Q_{all} and $C^* \notin \text{SimPChains}$ during the simulator's execution while answering the distinguisher's current query, then condition does not hold for the triple T^* at the end of the simulator's execution of the current query.

The proof of (ii) follows exactly along the lines of the proof of (i) given above.

The proof of (iii) is as follows. Let $\mathbf{D}$ ask its tth query $F(k, x)$. Just before the simulator returns $G_k(x)$ in line 27, let the lemma be false and let this be the first time that the lemma does not hold implying that there exists $x_5 \in G_5$, $x_6 \in G_6$ such that $(x_5, x_6, 5) \notin \text{CompChains}$.

If the lemma has held so far, in particular it has held right before (and immediately after) line 27 of the simulator's execution while answering $\mathbf{D}$'s $(t-1)^{th}$ query to $F(\cdot, \cdot)$. Note that the distinguisher could have made queries to P/P^{-1} between the $(t-1)^{th}$ and t^{th} queries to $F(\cdot, \cdot)$; but those queries cannot result in $x_5 \in G_5$ or $x_6 \in G_6$.

So, $x_5 \in G_5$, $x_6 \in G_6$ such that $(x_5, x_6, 5) \notin \text{CompChains}$ happened during the simulator's execution while answering $\mathbf{D}$'s t^{th} query. Now, if $(x_5, x_6, 5)$ were enqueued for completion during the tth query then $(x_5, x_6, 5) \in \text{CompChains}$. If a chain equivalent to $(x_5, x_6, 5)$ were enqueued for completion during the tth query, then $(x_5, x_6, 5) \in \text{CompChains}$. This is because equivalent chains are placed in CompChains simultaneously since BadP does not occur and FORCEVAL does not overwrite. So, for $x_5 \in G_5$, $x_6 \in G_6$ such that $(x_5, x_6, 5) \notin \text{CompChains}$ to be true, the simulator did not enqueue this partial chain. (Note that chains of the type $(x_5, x_6, 5)$ are not added to SimPChains).

Let $x_6 \in G_6$, and say an assignment occurs such that before the assignment $x_5 \notin G_5$, but after the assignment $x_5 \in G_5$ leading to the creation of a partial chain of the form $(x_5, x_6, 5)$ with $x_5 \in G_5, x_6 \in G_6$. (The analysis for the other case is analogous.) Such an assignment can happen only by completion of a chain in Q_1, Q_5, Q_6, Q_{10} or completion of a chain in Q^*_{all}. We analyze these next.

Case 1: An assignment happens to $G_5(x_5)$ during the completion of a chain C enqueued in Q_b where $b \in \{1, 5, 6, 10\}$ and $x_6 \in G_6$ before this assignment. Now, if $x_6 \in G_6$ before assignment causing $x_5 \in G_5$, then either $x_6 \in G_6$ before $\mathbf{D}$'s t-th query or $x_6 \in G_6$ due to the completion of a chain D enqueued in Q_1, Q_5, Q_6, Q_{10} and dequeued before C. Again, by construction of the simulator, chains C that are enqueued in Q_b are such that either $\text{val}_5(C) \in A_5^t$ or $\text{val}_5(C) \in G_5$ at the time C was enqueued and similarly, chains D that are enqueued in Q_b are such that either $\text{val}_6(D) \in A_6^t$ or $\text{val}_6(D) \in G_6$ at the time D was enqueued. Since BadP does not occur and FORCEVAL does not overwrite, $\text{val}_5(C) = x_5 \in A_5^t$ (since $x_5 \notin G_5$ before this assignment) and $\text{val}_6(D) = x_6 \in G_6 \cup A_6^t$. And so, $(x_5, x_6, 5)$ is enqueued for completion by construction of simulator.

Case 2: An assignment happens to $G_5(x_5)$ during the completion of a chain C in Q^*_{all} and $x_6 \in G_6$ before this assignment. If $x_6 \in G_6 \cup A_6^t$ and $x_5 \in A_5^t$ when the simulator enqueues chains in Q_{all}, then $(x_5, x_6, 5)$ is enqueued for completion in Q_{all}. Else, $(x_5, x_6, 5)$ is enqueued for completion in Q_{mid}.

This completes the proof.

Lemma 35. *Consider a good execution of H_2. If a chain $C = (x_k, x_{k+1}, k, \ell, g, b)$ belongs to Q^*_{all} such that at the time C is enqueued, adapting was safe for every chain dequeued from Q_1, Q_5, Q_6, Q_{10}, Q^*_{all} or Q_{mid} so far, then $\text{val}_b(C) = \perp$ and $\text{val}_g(C) = \perp$ at the time C is enqueued.*

Proof. Say $C = (x_9, x_{10}, 9, 3, 2, 5)$ is enqueued where the query preceding the chain's enqueueing is $G_1(x_1)$ where $\text{val}_1(C) = x_1$. Then, by definition of simulator, $x_1 \notin G_1$ as otherwise, ENQNEWCHAINS$(1, x_1)$ is not called. So, $\text{val}_2^+(C) = \perp$. Now, we claim that $\text{val}_5^-(C) \notin G_5$. This is because if $\text{val}_5^-(C) \in G_5$, then $\text{val}_6^-(C) \in G_6$ since otherwise, $\text{val}_5^-(C) = \perp$. This implies that the partial chain $(x_5, x_6, 5)$ where $x_5 = \text{val}_5^-(C)$ and $x_6 = \text{val}_6^-(C)$ is such that $x_5 \in G_5$ and $x_6 \in G_6$. Hence, by Lemma 34, we have that $(x_5, x_6, 5) \in \text{CompChains}$ since no new G_i assignments have been issued between the moment the simulator returned the answer (line 27 of its execution) and the moment when a chain C is enqueued in Q_{all}. However, since BadP does not occur, this means that $x_1 \in G_1$ contradicting the first statement. Thus, we have that $\text{val}_5^-(C) \notin G_5$. Now, $\text{val}_5^+(C) = \perp$ since $\text{val}_2^+(C) = \perp$. So, $\text{val}_5(C) \notin G_5$.

Since C is not enqueued in Q_1, Q_5, Q_6, Q_{10}, we have $\text{val}_5(C) = \perp$ when C is enqueued. So $\text{val}_2(C) = \perp$ and $\text{val}_5(C) = \perp$, where $g = 2$ and $b = 5$. The other cases are analogous.

ForceVal$(x, \cdot, j)$ does not Overwrite $G_j(x)$

Lemma 36. *Let $C = (x_k, x_{k+1}, k, \ell, g, b)$ be a partial chain enqueued in Q_1, Q_5, Q_6 or Q_{10} during a good execution of H_2. At the moment $C = (x_k, x_{k+1}, k, \ell, g, b)$ is dequeued, assume that adapting was safe for every chain C' in Q^*_{all} or Q_{mid} dequeued so far. Then,*

– *At the moment $C = (x_k, x_{k+1}, k, \ell, g, b)$ is dequeued, $C \in$ CompChains, or*
– *Just before the call to ADAPT for C, $\mathsf{val}_g(C) \notin G_g$ and $\mathsf{val}_a(C) \notin G_a$ (where a is the adapt position adjacent to the "bad" set uniform position b).*

Proof. Assume that the lemma has held until the moment that a chain $C = (x_k, x_{k+1}, k, \ell, g, b)$ is dequeued. Note that if the lemma has held until now we have that for every call to FORCEVAL$(x, \cdot, j)$ so far, $x \notin G_j$ by Corollary 29.

Consider the case that at the moment C was dequeued there is a chain D equivalent to C that was dequeued before C. Now, if D was dequeued before C, then $D \in$ CompChains by construction of the simulator. If C and D are equivalent chains such that $D \in$ CompChains, then $C \in$ CompChains by Lemma 26.

Let us consider the case where no chain equivalent to C was dequeued before C was dequeued. Say $C \notin$ CompChains when dequeued. Note that if we prove $\mathsf{val}_g(C) \notin G_g$ and $\mathsf{val}_a(C) \notin G_a$ at the time C was dequeued, we have that $\mathsf{val}_g(C) \notin G_g$ and $\mathsf{val}_a(C) \notin G_a$ just before the call to ADAPT for C since otherwise BadP or BadlyHit$^+$ occurred.

By Lemma 27, we have that $\mathsf{val}_g(C) = \perp$ at the time C was enqueued. If $\mathsf{val}_g(C) \in G_g$ at the time C was dequeued, then this was due to the completion of a chain D which was enqueued in $Q_{b'}$ where $b' \in \{1, 5, 6, 10\}$ due to the same distinguisher query as C and dequeued(and completed) before C such that $\mathsf{val}_g(C) = \mathsf{val}_g(D) \neq \perp$.

Consider the last assignment that was made before $\mathsf{val}_g(C) = \mathsf{val}_g(D) \neq \perp$ was true. This cannot have been a uniform assignment to $G_i(x_i)$ since that implies that BadlyCollide$^+$ occurred. This is because C and D are not equivalent(by assumption) and C and D are both enqueued for completion in Q_{all} and either $\mathsf{val}_g(C) = \perp$ or $\mathsf{val}_g(D) = \perp$ before the assignment(otherwise this is not the last assignment before $\mathsf{val}_g(C) = \mathsf{val}_g(D) \neq \perp$) and $\mathsf{val}_g(C) = \mathsf{val}_g(D) \neq \perp$ after the assignment.

The assignment cannot have been of the form $P(\downarrow, x_0, x_1) = (x_{10}, x_{11})$ or $P(\uparrow, x_{10}, x_{11}) = (x_0, x_1)$ since then BadP occurred. The assignment cannot have been a FORCEVAL query. This is because from Lemmas 32 and 31 we have that FORCEVAL does not change $\mathsf{val}_i(C)$ for a chain C enqueued in Q_{all} (including those enqueued in Q_1, Q_5, Q_6, Q_{10}) during completion of chains in Q_1, Q_5, Q_6, Q_{10}.

Now, consider the argument for $\mathsf{val}_a(C) \notin G_a$ when C is dequeued. By Lemma 27, we have that $\mathsf{val}_b(C) \notin G_b$ and $\mathsf{val}_g(C) = \perp$ at the time C was enqueued, implying that $\mathsf{val}_a(C) = \perp$ when C was enqueued (where a is the adapt position adjacent to "bad" set uniform position). The argument for this case follows similar to the one above for $\mathsf{val}_g(C)$.

Lemma 37. *Consider a good execution of H_2. Let $C = (x_k, x_{k+1}, k, \ell, g, b)$ be a partial chain in Q^*_{all}. At the moment $C = (x_k, x_{k+1}, k, \ell, g, b)$ is dequeued, assume that adapting was safe for every chain C' in Q_{mid} dequeued so far. Then,*

– *At the moment C is dequeued, $C \in$ CompChains or,*
– *Just before the call to ADAPT for C, $\mathsf{val}_{\ell-1}(C) \notin G_{\ell-1}$ and $\mathsf{val}_{\ell+2}(C) \notin G_{\ell+2}$.*

Lemma 38. *Consider a good execution of H_2. Let $C = (x_k, x_{k+1}, k, \ell, g, b)$ be a partial chain enqueued in Q_{mid}. Then,*

- *At the moment C is dequeued, $C \in$ CompChains, or*
- *Just before the call to* ADAPT *for C,* $\mathsf{val}_{\ell-1}(C) \notin G_{\ell-1}$ *and* $\mathsf{val}_{\ell+2}(C) \notin G_{\ell+2}$.

Theorem 39 (No Overwrites). *In a good execution of H_2, for any call to* FORCEVAL$(x, \cdot, j)$ *we have $x \notin G_j$ before the call.*

Proof. Combining the result of Lemmas 36, 37 and 38 with Corollary 29, we have that for every call to FORCEVAL$(x, \cdot, j)$, $x \notin G_j$ before the call.

5.4 Indistinguishability of H_2 and H_4

Relying on the properties of good executions of H_2 from the previous section, we prove that H_2 and H_4 are indistinguishable.

Lemma 40. *The probability that a distinguisher $\mathbf{D}$ outputs 1 in H_2 differs at most by $O(q^{10})/2^n$ from the probability that it outputs 1 in H_3.*

Lemma 41. *The probability that a distinguisher outputs 1 in H_3 differs by at most by $O(q^{10})/2^n$ from the probability that it outputs 1 in H_4.*

This concludes the proof.

Acknowledgments. We thank Vanishree Rao for collaboration during the early stages of this work.

References

1. Coron, J.S., Holenstein, T., Künzler, R., Patarin, J., Seurin, Y., Tessaro, S.: How to build an ideal cipher: the indifferentiability of the feistel construction. J. Cryptology **29**(1), 61–114 (2014)
2. Coron, J.-S., Patarin, J., Seurin, Y.: The random oracle model and the ideal cipher model are equivalent. In: Wagner, D. (ed.) CRYPTO 2008. LNCS, vol. 5157, pp. 1–20. Springer, Heidelberg (2008)
3. Dachman-Soled, D., Katz, J., Thiruvengadam, A.: 10-round Feistel is indifferentiable from an ideal cipher (2015). http://eprint.iacr.org/2015/876
4. Dai, Y., Steinberger, J.P.: Indifferentiability of 10-round Feistel networks (2015). http://eprint.iacr.org/2015/874
5. Dai, Y., Steinberger, J.P.: Indifferentiability of 8-round Feistel networks (2015). http://eprint.iacr.org/2015/1069
6. Dodis, Y., Puniya, P.: On the relation between the ideal cipher and the random oracle models. In: Halevi, S., Rabin, T. (eds.) TCC 2006. LNCS, vol. 3876, pp. 184–206. Springer, Heidelberg (2006)
7. Dodis, Y., Puniya, P.: Feistel networks made public, and applications. In: Naor, M. (ed.) EUROCRYPT 2007. LNCS, vol. 4515, pp. 534–554. Springer, Heidelberg (2007)

8. Even, S., Mansour, Y.: A construction of a cipher from a single pseudorandom permutation. In: Imai, H., Rivest, R.L., Matsumoto, T. (eds.) ASIACRYPT 1991. LNCS, pp. 210–224. Springer, Heidelberg (1993)

9. Feistel, H.: Cryptography and computer privacy. Sci. Am. **228**(5), 15–23 (1973)

10. Gentry, C., Ramzan, Z.: Eliminating random permutation oracles in the even-mansour cipher. In: Lee, P.J. (ed.) ASIACRYPT 2004. LNCS, vol. 3329, pp. 32–47. Springer, Heidelberg (2004)

11. Holenstein, T., Künzler, R., Tessaro, S.: The equivalence of the random oracle model and the ideal cipher model, revisited. In: Fortnow, L., Vadhan, S.P. (eds.) 43rd ACM STOC. pp. 89–98. ACM Press, June 2011

12. Luby, M., Rackoff, C.: How to construct pseudorandom permutations from pseudo-random functions. SIAM J. Comput. **17**(2), 373–386 (1988)

13. Mandal, A., Patarin, J., Seurin, Y.: On the public indifferentiability and correlation intractability of the 6-round feistel construction. In: Cramer, R. (ed.) TCC 2012. LNCS, vol. 7194, pp. 285–302. Springer, Heidelberg (2012)

14. Maurer, U.M., Renner, R.S., Holenstein, C.: Indifferentiability, impossibility results on reductions, and applications to the random oracle methodology. In: Naor, M. (ed.) TCC 2004. LNCS, vol. 2951, pp. 21–39. Springer, Heidelberg (2004)

15. Ramzan, Z., Reyzin, L.: On the round security of symmetric-key cryptographic primitives. In: Bellare, M. (ed.) CRYPTO 2000. LNCS, vol. 1880, pp. 376–393. Springer, Heidelberg (2000)

16. Seurin, Y.: Primitives et Protocoles Cryptographiques à Sécurité Prouvée. PH.D. thesis, Versailles University (2009)

17. Seurin, Y.: A note on the indifferentiability of the 10-round Feistel construction (2011). http://eprint.iacr.org/2015/903

18. Yoneyama, K., Miyagawa, S., Ohta, K.: Leaky random oracle (extended abstract). In: Baek, J., Bao, F., Chen, K., Lai, X. (eds.) ProvSec 2008. LNCS, vol. 5324, pp. 226–240. Springer, Heidelberg (2008)

Indifferentiability of Confusion-Diffusion Networks

Yevgeniy Dodis[1](✉), Martijn Stam[2], John Steinberger[3], and Tianren Liu[4]

[1] Courant Institute, New York University, New York, USA
dodis@cs.nyu.edu
[2] Department of Computer Science, University of Bristol, Bristol, UK
csxms@bristol.ac.uk
[3] Institute for Interdisciplinary Information Sciences,
Tsinghua University, Beijing, China
jpsteinb@gmail.com
[4] MIT, Cambridge, USA
liutianren@gmail.com

Abstract. We show the first positive results for the indifferentiability security of the confusion-diffusion networks (which are extensively used in the design of block ciphers and hash functions). In particular, our result shows that a constant number of confusion-diffusion rounds is sufficient to extend the domain of a public random permutation.

1 Introduction

In this work we simultaneously address the following two questions:

- **Question 1:** secure domain extension of a *public* random permutation.
- **Question 2:** theoretical soundness of Shannon's (or Feistel's) confusion-diffusion paradigm.

DOMAIN EXTENSION OF RPs. The question of domain extension of various cryptographic primitives, such as encryption, signatures, message authentication codes, pseudorandom functions (PRFs), pseudorandom permutations (PRPs), etc., is one of the fundamental questions in cryptography.

In this paper we address a similar question for a *public* random permutation. Namely, given one (or a constant number of) n-bit random permutation(s) $P : \{0,1\}^n \to \{0,1\}^n$, and a number $w \geq 2$, build a wn-bit random permutation $\mathcal{Z} : \{0,1\}^{wn} \to \{0,1\}^{wn}$. This question is clearly natural and interesting it is own right, but also seems extremely relevant in practice. Indeed, the random

Y. Dodis—Partially supported by gifts from VMware Labs and Google, and NSF grants 1319051, 1314568, 1065288, 1017471.

J. Steinberger—Supported by National Basic Research Program of China Grant 2011CBA00300, 2011CBA00301, the National Natural Science Foundation of China Grant 61033001, 61361136003, and by the China Ministry of Education grant number 20121088050.

M. Fischlin and J.-S. Coron (Eds.): EUROCRYPT 2016, Part II, LNCS 9666, pp. 679–704, 2016.
DOI: 10.1007/978-3-662-49896-5_24

permutation model (RPM) has recently received a lot of attention [2,13,26,29], starting to "compete with" and perhaps even "overtake" the more well known random oracle model (ROM) and the ideal cipher model (ICM). Aside from elegance, one of the reasons for this renewed attention comes from the fact that one can abstract the design of both the block-cipher standard AES and the new SHA-3 standard Keccak as being in the RPM. Namely, AES can be viewed as a 10-round *key-alternating* cipher applied to a concrete ("random-looking") permutation, while SHA-3 can be viewed as applying a "sponge" mode of operation [2] to a similarly "random-looking" permutation. In fact, in his invited talk at Eurocrypt'13, the designer of both AES and SHA-3 Joan Daemen claimed that the RPM is much closer to the existing practice of designing hash functions and block ciphers than either the ROM or ICM, challenging the cryptographic community to switch to the RPM.

Of course, one must now build those "random looking permutations" $\mathcal{Z}$ on relatively large domains (perhaps from 128 bits, like AES-128, to 1600 bits, like Keccak, or even longer). In practice, we have two well-known methods for accomplishing such a goal. The first method is based on applying several rounds of the Feistel network to some (not necessarily invertible) round functions. In our (public) setting, this method was theoretically analyzed only recently by Holenstein et al. [18] (building on an earlier work of [11]), who showed that a 14-round Feistel network is indeed sufficient for building a random permutation (RP), provided the round functions are modeled as (easily made) independent random oracles (ROs). Although very important in theory (i.e., showing the equivalence between ROM and RPM), this method does not seem to be used in practice, as it appears almost as hard,—if not *harder*,—to design "random-looking" non-invertible round functions on large domains as it is to design the desired random-looking permutation $\mathcal{Z}$.

CONFUSION-DIFFUSION PARADIGM. Instead, practitioners use the second method,—the *confusion-diffusion* (CD) paradigm,[1]—which directly connects our motivating Questions 1 and 2. The idea of CD goes back to the seminal paper of Feistel [15] and even[2] back to Shannon [28]. Abstractly, one splits the input x to $\mathcal{Z}$ into several shorter blocks $x_1 \ldots x_w$, and then alternates the following two steps for several rounds: (a) *Confusion*, which consists of applying some

---

[1] This is closely related to the substitution-permutation network (SPN) paradigm. Historically, though, the term SPN usually refers to the design of block ciphers as opposed to a single permutation, where one also XORs some key material in between successive CD rounds. To avoid confusion, we will stick with the term CD and not use the term SPN.

[2] Shannon [28] introduces "confusion" and "diffusion" into the cryptographic lexicon while Feistel [15] articulates the modern notion of a confusion-diffusion network, crediting Shannon with inspiration. There are some notable gaps between Shannon and the modern viewpoint. In particular Shannon does not seem to view confusion as a local operation, nor does he advocate repeatedly alternating steps of "confusion" and "diffusion". Instead, Shannon seems to view confusion and diffusion as globally desirable attributes of a cryptographic mixing operation.

fixed short permutations $P_1 \ldots P_w$ (called *S-boxes*) to $x_1, \ldots x_w$; and (b) *Diffusion*, which consists of applying some "mixing" non-cryptographic permutation $\pi(y_1 \ldots y_w)$ (typically, carefully chosen linear function, sometimes also called *D-box*) to the results $y_1 \ldots y_w$ of step (a).

Despite its extensive use in practice, the CD paradigm received extremely little attention from the theoretical cryptographic community.[3] A notable exception is a beautiful work of Miles and Viola [23], who only looked at the secret-key setting—where the permutations $P_1, \ldots, P_w$ are secret— and also primarily considered the "weaker-than-indistinguishability" properties which can be proven about CD (and, more generally, SPN networks). In contrast, we are interested in the public setting, where the permutations P_i are modeled as RPs, and seek to examine the *indifferentiability* properties [9,21] of the CD paradigm. This leads us to the following more precise reformulation of our motivating Questions 1 and 2:

- **Main Question:** *Analyze indifferentiability of the confusion-diffusion paradigm as a way to extend the domain of a (constant number of) random permutation(s).* More precisely, for how many rounds r, and under what conditions on the D-boxes $\pi_1 \ldots \pi_r$, is the r-round CD paradigm indifferentiable from an nw-bit random permutation $\mathcal{Z}$?

Before presenting our results, we make a few remarks. First, we will model the "small permutations" P_i as both random and independent. The independence assumption is crucially used in our current proofs, but does not appear necessary. Unfortunately, the proofs we have are already extremely involved, so we feel this initial simplification is justified. We notice similar abstractions are made by most other papers in the area (including the seminal Luby-Rackoff paper [20] and the indifferentiability results of [11,18]), though one hopes it might be lifted in future work.

As for modeling P_i as random, it seems inherent if we want to build a random permutation $\mathcal{Z}$; e.g., we cannot build it from "nothing" (as this implies $P \neq NP$ and more), and it seems unlikely that any weaker assumption on the P_i will work. However, it does come with an important caveat: the best security bound ε we can naturally get with this approach will certainly be $\varepsilon \gg 2^{-n}$, where n is the domain of the S-boxes P_i. In practice, however, the S-boxes use a very small value of n (e.g., $n = 8$ for the AES), partly so that S-boxes can be easily and efficiently implemented as lookup tables. With such a small value of n, however, our bounds appear "practically meaningless", irrespective of the number of queries q made by the attacker. This means that none of our results would be directly applicable to any of the "practical" permutations $\mathcal{Z}$ used in the existing hash functions and block ciphers. Still, we believe establishing "structural soundness" of the CD paradigm is an important conceptual contribution—and an overdue sanity check—even with this serious (and inherent) limitation.

---

[3] Of course, there is a lot of cryptanalytic work in the area whose survey is beyond the scope of this work.

1.1   Overview of Our Results

We give a sequence of results establishing the soundness of the CD paradigm as a method for domain-extension of random permutations. Our indifferentiability results include CD networks of 5, 6, 7, 9, 10 and 11 rounds. These networks achieve different security levels, different query complexities, and place different combinatorial requirements on the D-boxes as well, even within the same network. Figure 1 summarizes the main bounds achieved for each network length, up to lower-order (e.g., logarithmic) factors, and subject to various caveats to be shortly explained.

rounds	(flags)	D-boxes	ε ($w = 2$)	q_S ($w = 2$)	ε ($w > 2$)	q_S ($w > 2$)
5	(000)	arbitrary	$q^4/2^n$	q^4	$q^{2w}/2^n$	q^{w^2}
6	(100)	arbitrary	$q^2/2^n$	q^2	$q^2/2^n$	q^w
7	(110)	arbitrary	$q^2/2^n$	q	$q^2/2^n$	q
5	(000)	GF(2^n)-linear	$q^4/2^n$	q^4	1	$-$
6	(100)	GF(2^n)-linear	$[q^{10/3}/2^n]$	$[q^{10/3}]$	1	$-$
7	(110)	GF(2^n)-linear	$[q^5/2^n]$	$[q^{25/9}]$	1	$-$
9	(001)	GF(2^n)-linear	$q^4/2^n$	q^4	$q^{2w}/2^n$	q^{w^2}
10	(101)	GF(2^n)-linear	$[q^{10/3}/2^n]$	$[q^{10/3}]$	$[q^4/2^n]$	$[q^{2w-1/2}]$
11	(111)	GF(2^n)-linear	$[q^5/2^n]$	$[q^{25/9}]$	$[q^6/2^n]$	$[q^4]$

Fig. 1. Summary of security ε and simulator query complexity q_S (as functions of block length n, width w, and the number of distinguisher queries q) across our six main simulators with arbitrary or GF(2^n)-linear permutations. Entries in square brackets are not known, with the value inside the brackets being conjectured based on current best-known bounds. Constants and logarithmic factors are elided for simplicity. The meaning of the bit sequence next to each round number is explained in Sect. 5.

To read Fig. 1, recall that n is the block length of the S-boxes and that w is the width of the network. Moreover, q is the number of distinguisher queries, ε is the simulator's security (i.e., the indistinguishability of the real and simulated worlds), and lastly q_S is the simulator's query complexity (see Sect. 2 for definitions). The meaning of the 3-bit sequence next to each round number in the left column will be explained in Sect. 5.

The first three rows of Fig. 1 present our best bounds (i.e., assuming a "smart" choice of the D-boxes) when the D-boxes are not restricted to be GF(2^n)-linear. One can observe that in this part of the table the bounds for $w = 2$ are simply obtained by plugging in $w = 2$ to the general bounds. The best simulator here, at 7 rounds, achieves "birthday" security of $q^2/2^n$ and an essentially optimal query complexity of q.

Concerning the first three rows of Fig. 1 there is only one caveat: for some networks (specifically those of 6 and 7 rounds, and for both $w = 2$ and $w > 2$) the bounds presume that certain D-boxes have low "conductance"—a new critical property that we introduce and elaborate on below. Such permutations are known to exist on probabilistic grounds, but so far we do not know any explicit constructions, and building explicit permutations with low conductance is indeed one of the more interesting open problems raised by our work.

The last six rows of Fig. 1 concern the case when all D-boxes in the network are required to be $GF(2^n)$-linear, which is a case of interest because it aligns with most practical constructions. In this case, the simulators at width $w > 2$ and at 5, 6 and 7 rounds are not secure at all: $\varepsilon = 1$. In this regime, indeed, our simulator places certain combinatorial requirements on some of the D-boxes that are not satisfied by any $GF(2^n)$-linear permutation. Fortunately, these requirements can be relaxed by using four more (i.e., 9, 10 or 11) rounds, so that $GF(2^n)$-linear D-boxes become possible again for those round numbers at $w > 2$.

Unfortunately, except for the bounds at 5 rounds for $w = 2$ and the (rather poor) bounds at 9 rounds for $w > 2$, remaining bounds in the $GF(2^n)$-linear section of the table are speculative, this being again related to "conductance"— specifically, the issue is that the lowest possible conductance of a $GF(2^n)$-linear permutations is not currently known. We show some nontrivial lower bounds on the conductance of *generic* $GF(2^n)$-linear permutations in the full version [8], but we have no similar nontrivial upper bounds! The conjectured bounds contained in Fig. 1 are obtained by using our lower bounds as a guess for the actual lowest possible conductance, and by rounding up some inconvenient exponents.[4] The "true" values in the lower part of the table may well turn out to be *lower* than our conjectured values (if *non-generic* $GF(2^n)$-linear permutations with low conductance turn out to exist) or *higher* (if even better lower bounds on the conductance of all $GF(2^n)$-linear permutations are proved).

Still in the same part of the table, one can also note that going from 6 to 7 rounds or from 10 to 11 rounds entails a *decrease* in security. In our simulator, indeed, the actual purpose of adding the extra round (from 6 to 7 or from 10 to 11) is to improve the query complexity!

COMBINATORIAL PROPERTIES OF D-BOXES. We note that our general theorem statement (found in Sect. 4) makes no distinction between linear and nonlinear cases; it simply expresses the security, query complexity and runtime of the simulator as a function of various combinatorial metrics (such as conductance mentioned above) of the diffusion permutations that are present in the network. Different metrics matter for different D-boxes, depending on their position in the networks, leading to a subtle (but also modular and fine-grained) result. In our opinion, the identification of precise (and distinct!) combinatorial requirements for each D-box of the network is one of the interesting contributions of this work,

---

[4] If trivial upper bounds on conductance are applied instead, the rows for the 6- and 7-round networks with $w = 2$ in the second part of the table become the same as the row for the 5-round network in that half, while the rows for the 10- and 11-round networks become the same as the row for the 9-round network.

as it potentially allows future constructions to optimize the design of D-boxes layer by layer, with respect to the specific metrics targeted by each layer.

Moreover, some of our metrics entirely disappear when more rounds are added (specifically, by going from 5, 6, 7 rounds to 9, 10, 11 rounds respectively), leading to a relaxation of the conditions on the D-boxes at a larger number of rounds. For example, as already mentioned, security cannot be achieved at 5, 6 or 7 rounds with $w > 2$ and with $GF(2^n)$-linear D-boxes, but can (unconditionally) be achieved by adding 4 more rounds to each of these networks, because a certain metric that no $GF(2^n)$-linear permutation satisfies at $w > 2$ is no longer needed after the addition of the extra four rounds. In fact, although our table for general D-boxes (first 3 rows) in Fig. 1 does not include networks of 9, 10 and 11 rounds (since these networks would have the same security and simulator efficiency as the included networks for 5, 6 and 7 rounds, respectively) such networks are nonetheless considered by our main result, and might indeed be interesting from the point of view of increased efficiency. Namely, the weakened requirements on the D-boxes could potentially make these networks more cheap/fast/space efficient than their shorter, more combinatorially demanding counterparts! (Further such options, at even more rounds than considered by our main result, are explored in the paper's full version [8]).

In all we identify four combinatorial metrics on D-boxes that are useful for our purposes, these being (and for lack of better terminology) "entry-wise randomized preimage resistance" (RPR), "entry-wise randomized collision resistance" (RCR), "conductance" and its cousin "all-but-one conductance". The full definitions for these metrics can be found in Sect. 4. The first two metrics—RPR and RCR—are relatively unsurprising for our type of proof. (Briefly, they concern experiments in which all but one of the w input wires are fixed, and the final D-box input wire is drawn at random; the probability of a certain event occuring on the output wires should be low). Moreover, there is not much "mystery" in RPR and RCR, since it happens that one can construct explicit permutations that achieve essentially optimal bounds for these metrics.[5]

Indeed, we consider conductance to be a more novel and interesting metric. (All-but-one conductance is, conceptually at least, very closely related to conductance). We expand on this key metric now.

CONDUCTANCE. Briefly, conductance is a function of the number of queries q; the conductance of a permutation $\pi : \{0,1\}^{wn} \to \{0,1\}^{wn}$ at q queries is the maximum over all possible pairs of cartesian products $(U_1 \times \cdots \times U_w, V_1 \times \cdots \times V_w)$, where $U_i, V_i \subseteq \{0,1\}^n$ and $|U_i| = |V_i| = q$ for each $1 \le i \le w$, of the numbers of pairs $(\mathbf{x}, \mathbf{y}) \in \{0,1\}^{wn} \times \{0,1\}^{wn}$ such that

$$\pi(\mathbf{x}) = \mathbf{y} \quad \text{and} \quad (\mathbf{x}, \mathbf{y}) \in (U_1 \times \cdots \times U_w, V_1 \times \cdots \times V_w)$$

---

[5] On the other hand, an interesting research direction might be to find explicit constructions of RPR and RCR permutations that achieve higher speeds than our own naïve constructions!.

In other words, one can choose q different values on each input and on each output wire, and one counts the number of intput-output pairs $(\mathbf{x}, \mathbf{y})$ that "entirely fit" inside the induced cartesian products.

Now if one imagines the D-box π to be sandwiched between two rounds of S-boxes—as it will be in the network—the relevance of conductance to our setting can easily be guessed: U_i corresponds to the set of values that are queried outputs of the i-th S-box in the round before π, V_i corresponds to the set of values that are queried inputs of the i-th S-box in the round after π, and the conductance is an upper bound on the number of "all consistent input-output pairs that can be assembled" from these S-box queries under π's mapping. Intuitively, if the number of such pairs is low, the job of the indifferentiability simulator is easier, as it has to worry about fewer "consistent chains" that the distinguisher is trying to assemble.

It is easy to see from the definition that conductance for any permutation $\pi : \{0,1\}^{wn} \to \{0,1\}^{wn}$ lies between q and q^w. In [8] we show that a random permutation has conductance close to $wqn \approx q$ and that the conductance of a generic $\mathrm{GF}(2^n)$-linear permutation is at least $q^{2-1/(2w-1)}$, which is always super-linear in q. An already-mentioned corollary is that, and at least with respect to our current simulator, having linear D-boxes seems to cause strictly worse security than what is achievable by arbitrary (albeit currently non-constructive!) D-boxes. Since it may well turn out that low conductance is instantiable by D-boxes that are no slower than the current $\mathrm{GF}(2^n)$-linear D-boxes, our work raises, among others, the question of whether $\mathrm{GF}(2^n)$-linearity is really the right choice for D-boxes in the CD paradigm.

SUMMARY. Overall, we show the first positive results for the indifferentiability security of the confusion-diffusion paradigm which is extensively used in the design of block ciphers and hash functions. Our result shows that a constant number of confusion-diffusion rounds is sufficient to extend the domain of a public random permutation. In the process, we reduced the indifferentiability properties of the CD network (for a variety of rounds between 5 and 11) to natural and novel combinatorial properties of the D-boxes (such as conductance), which we hope will lead to a better understanding of the confusion-diffusion networks, and will be useful and the future design and analysis of block ciphers and hash functions.

1.2 Other Related Work

The question of domain extension ideal primitives was considered by [9, 22] for the setting of public random functions (ROM), and by [10] for the setting of block ciphers (ICM). While none if these domain extensions directly apply to the RPM (e.g., the result of [10] crucially relies of the existence of a key for the "small ideal cipher"), they can be composed with the series of results showing the equivalence between RPM, ICM and ROM [1, 9, 11–14, 18] to achieve various theoretical domain extension methods in the RPM. For example, one can get a "small RO" from "small RP" [12, 13], extend domain of RO [9], and apply the

14-round Feistel construction to get a large-domain RP [18] (many other combinations of prior results will suffice as well). However, all such combinations of prior work will be *much* less efficient (and elegant) than our natural construction, and, more importantly, such results will not correspond to the way random permutations are built in real life.

The domain extension of *secret-key* random permutations is well studied: examples include PEP [4], XCB [16], HCTR [32], HCH [5] and TET [17] (and even the original Feistel constructions [20,24] could be viewed as domain doubling techniques in this setting). However, it is easy to see that none of those constructions provide the indifferentiability property in the public permutation setting.

Finally, the design of public permutations is related in spirit to the area of *white-box cryptography* [6,7], with the idea to "obfuscate" key-dependent parts of the cipher and publish them as lookup tables, making the entire construction public. We refer to [3] for an excellent discussion of this big area of research, as well as a survey of many cryptanalytic efforts attacking various SPN designs with linear diffusion layers.

2   Definitions

BASIC NOTATIONS. We write $[w]$ for the set of integers $\{1, \ldots, w\}$. Elements of $\{0,1\}^{wn}$ are written with bold letters such as $\mathbf{x}$, $\mathbf{y}$; the i-th n-bit block of $\mathbf{x} \in \{0,1\}^{wn}$, $1 \le i \le w$, is written $\mathbf{x}[i]$.

CONFUSION-DIFFUSION NETWORKS. Fix integers $w, n, r \in \mathbb{N}$. Let

$$\mathcal{P} = \{P_{i,j} : (i,j) \in [r] \times [w]\}$$

be an array of rw permutations from $\{0,1\}^n$ to $\{0,1\}^n$, i.e., $P_{i,j}$ is a permutation from $\{0,1\}^n$ to $\{0,1\}^n$ for each $i \in [r]$ and each $j \in [w]$. Also let

$$\overline{\pi} = (\pi_1, \ldots, \pi_{r-1})$$

be an arbitrary sequence of $r - 1$ permutations, each from $\{0,1\}^{wn}$ to $\{0,1\}^{wn}$.

Given $\mathcal{P}$ and $\mathbf{x} \in \{0,1\}^{wn}$ we let

$$P_i(\mathbf{x})$$

denote the value in $\{0,1\}^{wn}$ obtained by applying the permutations $P_{i,1}, \ldots, P_{i,w}$ blockwise to $\mathbf{x}$. In other words, $P_i : \{0,1\}^{wn} \to \{0,1\}^{wn}$ is defined by setting

$$P_i(\mathbf{x})[j] = P_{i,j}(\mathbf{x}[j])$$

for all $j \in [w]$. It is obvious that P_i is a permutation of $\{0,1\}^{wn}$.

Given $\mathcal{P}$ and $\overline{\pi}$, we define the permutation $P = P[\mathcal{P}, \overline{\pi}]$ from $\{0,1\}^{wn}$ to $\{0,1\}^{wn}$ as the composition

$$P[\mathcal{P}, \overline{\pi}] = P_r \circ \pi_{r-1} \circ \ldots \circ P_2 \circ \pi_1 \circ P_1.$$

I.e.,

$$P[\mathcal{P}, \overline{\pi}](\mathbf{x}) = P_r(\pi_{r-1}(\ldots P_2(\pi_1(P_1(\mathbf{x})))\ldots))$$

for $\mathbf{x} \in \{0,1\}^{wn}$. We call $P[\mathcal{P}, \overline{\pi}]$ the *confusion-diffusion network* built from $\mathcal{P}$ and $\overline{\pi}$. The permutations in $\mathcal{P}$ are variously called the *confusion permutations* or *S-boxes*. The permutations in $\overline{\pi}$ are variously called the *diffusion permutations* or *D-boxes*.

The values n, w and r will be called the *wire length*, the *width* and the *number of rounds* respectively.

In practice, the S-boxes are implemented by "convoluted" or "random-like" permutations while the D-boxes are implemented by "easy" (typically linear) permutations that are cryptographically weak. In our indifferentiability model, described next, the S-boxes are modeled as random permutations while the D-boxes are publically fixed parameters of the network.

INDIFFERENTIABILITY. Let C be a construction making calls to an ideal set of primitives $\mathcal{P}$, which we notate as $C^{\mathcal{P}}$. Let $\mathcal{Z}$ be an ideal primitive with the same interface as $C^{\mathcal{P}}$ (e.g., $\mathcal{Z}$ is a random permutation if $C^{\mathcal{P}}$ implements a permutation). Indifferentiability is meant to capture the intuitive notion that the construction $C^{\mathcal{P}}$ is "just as good" as $\mathcal{Z}$, in some precise sense. The definition involves a simulator:

Definition 1. *An (oracle) circuit C with access to a set of ideal primitives $\mathcal{P}$ is (t_S, q_S, ε)-indifferentiable from an ideal primitive $\mathcal{Z}$ if there exists a simulator S such that*

$$\Pr\left[D^{C^{\mathcal{P}}, \mathcal{P}} = 1\right] - \Pr\left[D^{\mathcal{Z}, S^{\mathcal{Z}}}\right] \leq \varepsilon$$

for every distinguisher D making at most q_0 queries to its oracles, and such that S runs in total time t_S and makes at most q_S queries to $\mathcal{Z}$. Here t_S, q_S and ε are functions of q_0.

We note that in the "real world" D has oracle access to the construction $C^{\mathcal{P}}$ as well as to the primitives $\mathcal{P}$; in the "ideal world" $C^{\mathcal{P}}$ is replaced by the ideal primitive $\mathcal{Z}$ and the ideal primitives $\mathcal{P}$ are replaced by the simulator S. Thus, S's job is to make $\mathcal{Z}$ look like $C^{\mathcal{P}}$ by inventing "answers that fit" for D's queries to the primitives in $\mathcal{P}$. For this, S requires query access to $\mathcal{Z}$ (notated as $S^{\mathcal{Z}}$); on the other hand, S does not get to see which queries D is making to $\mathcal{Z}$.

Informally, $C^{\mathcal{P}}$ is *indifferentiable* from $\mathcal{Z}$ if it is (t_S, q_S, ε)-indifferentiable for "reasonable" values of (t_S, q_S, ε). An essential composition theorem [9,21] states that any cryptosystem that is secure when implemented with $\mathcal{Z}$ remains secure if $\mathcal{Z}$ is replaced with $C^{\mathcal{P}}$, if $C^{\mathcal{P}}$ is indifferentiable from $\mathcal{Z}$. However, the class of adversaries with respect to which the cryptosystem's security is defined must be a class that is large enough to accomodate the simulator S from Definition 1. See, e.g., [25] for a dramatic example in which indifferentiability fails completely.

In our setting "$\mathcal{P}$ will be $\mathcal{P}$" (i.e., the set of ideal primitives $\mathcal{P}$ will be the set of wr independent random permutations discussed in the previous subsection), while $C^{\mathcal{P}}$ will be $P[\mathcal{P}, \overline{\pi}]$. (As explained, the diffusion permutations $\overline{\pi}$ are a fixed, publically known parameter of the construction). Consequently, $\mathcal{Z}$ (matching

$C^{\mathcal{P}}$'s syntax) will be a random permutation from $\{0,1\}^{wn}$ to $\{0,1\}^{wn}$. Like all permutation oracles, $\mathcal{Z}$ can be queried in both forward and backward directions.

3   Attack on Two-Round Confusion-Diffusion Networks

In this section we outline a simple distinguishing attack that shows confusion-diffusion networks of two rounds or less cannot be indifferentiable from a random permutation. Unfortunately we could not find a similarly general attack for networks with three rounds, which leaves open the possibility that 3- or 4-round confusion-diffusion network might already be indifferentiable.

The attack on 2-round networks requires $w \geq 2$, which is indeed a trivial requirement since if $w = 1$ then a 1-round network is already indifferentiable from a random permutation.

For concreteness we sketch the attack with $w = 2$. The confusion-diffusion network then has four S-boxes labeled $P_{i,j}$ for $(i,j) \in [2] \times [2]$ and one diffusion permutation $\pi : \{0,1\}^{2n} \to \{0,1\}^{2n}$. The S-boxes in the first round are $P_{1,j}$, $j \in [2]$, the S-boxes in the second round are $P_{2,j}$, $j \in [2]$.

We will say the distinguisher "rejects" if it believes that it is in the simulated world; "accepts" if it believes it is in the real world.

The distinguishing attack is as follows:

1. The distinguisher randomly chooses $\mathbf{x} \in \{0,1\}^{2n}$ and queries $\mathcal{Z}(\mathbf{x})$, where $\mathcal{Z} : \{0,1\}^{2n} \to \{0,1\}^{2n}$ is the random permutation, obtaining $\mathbf{y} \in \{0,1\}^{2n}$ as answer.
2. The distinguisher make the two S-box queries $P_{1,1}(\mathbf{x}[1])$ and $P_{2,1}^{-1}(\mathbf{y}[1])$ receiving answers $a \in \{0,1\}^n$ and $b \in \{0,1\}^n$ respectively.
3. If there exists no pair of values (c,d) such that $\pi(a\|c) = (b\|d)$, the distinguisher rejects.
4. If there exists a pair of values (c,d) such that $\pi(a\|c) = (b\|d)$, the distinguisher chooses any such pair, queries $P_{1,2}^{-1}(c)$ obtaining answer t, and accepts if and only if $\mathcal{Z}(\mathbf{x}[1]\|t)[1] = \mathbf{y}[1]$.

It is clear that the distinguisher always accepts in the real world. We now argue that the simulator has negligible chance of making the distinguisher accept.

It is helpful to picture the simulator as knowing the distinguisher's attack. Moreover, we can be generous to the simulator and give both $\mathbf{x}[1]$ and $\mathbf{y}[1]$ to the simulator before requesting the answers a and b from the simulator.

By choosing a and b, the simulator knows which of options 3 and 4 the distinguisher will execute, so the simulator is essentially choosing between these options when it chooses a and b.

Obviously, case 3 is no good for the simulator; moreover, the simulator has no further information on $\mathbf{x}$ and $\mathbf{y}$ besides $\mathbf{x}[1]$ and $\mathbf{y}[1]$, from which it is computationally infeasible, if $\mathcal{Z}$ is a random permutation, to locate a value t such that $\mathcal{Z}(\mathbf{x}[1]\|t)[1] = \mathbf{y}[1]$, and which rules out case 4. The simulator is therefore doomed.

4   Combinatorial Definitions

In this section (re-)define the four combinatorial metrics on diffusion permutations mentioned in the introduction, these being *entry-wise randomized preimage resistance* (RPR), *entry-wise randomized collision resistance* (RCR), *conductance* and *all-but-one-conductance*.

Properties are defined unidirectionally: π might satisfy a property while π^{-1} does not.

Given $\pi : \{0,1\}^{wn} \rightarrow \{0,1\}^{wn}$, a vector $\mathbf{x} \in \{0,1\}^{wn}$ and indices $j, j' \in [w]$, we let

$$\pi^{\mathbf{x}}_{j,j'} : \{0,1\}^n \rightarrow \{0,1\}^n$$

be the function from $\{0,1\}^n$ to $\{0,1\}^n$ obtained by restricting the i-th block of input of π, $i \neq j$, to $\mathbf{x}[i]$, by replacing $\mathbf{x}[j]$ with the input $x \in \{0,1\}^n$, and by considering only the j'-th block of output. (The value $\mathbf{x}[j]$ being, thus, immaterial to $\pi^{\mathbf{x}}_{j,j'}$, since it is replaced by the input).

ENTRY-WISE RANDOMIZED PREIMAGE RESISTANCE AND ENTRY-WISE RANDOMIZED COLLISION RESISTANCE. The *entry-wise randomized preimage resistance* (RPR) of π is denoted $\mathsf{MaxPreim}(\pi)$, and defined as

$$\mathsf{MaxPreim}(\pi) = \max_{\mathbf{x},j,h,y} |\{x \in \{0,1\}^n : \pi^{\mathbf{x}}_{j,h}(x) = y\}|$$

while the *entry-wise randomized collision resistance* (RCR) of π is denoted $\mathsf{MaxColl}(\pi)$, and defined as

$$\mathsf{MaxColl}(\pi) = \max_{\mathbf{x},\mathbf{x}',j,h} |\{x \in \{0,1\}^n : \pi^{\mathbf{x}}_{j,h}(x) = \pi^{\mathbf{x}'}_{j,h}(x)\}|$$

where the latter maximum is taken over all tuples $\mathbf{x}, \mathbf{x}', j, h$ such that $\mathbf{x}[j'] \neq \mathbf{x}'[j']$ for some $j' \neq j$. Then by definition

$$\Pr_{x}[\pi^{\mathbf{x}}_{j,h}(x) = y] \leq \frac{\mathsf{MaxPreim}(\pi)}{2^n}$$

for all $\mathbf{x} \in \{0,1\}^{wn}$, $y \in \{0,1\}^n$, and $j, h \in [w]$, where the probability is computed over a uniform choice of $x \in \{0,1\}^n$, and

$$\Pr_{x}[\pi^{\mathbf{x}}_{j,h}(x) = \pi^{\mathbf{x}'}_{j,h}(x)] \leq \frac{\mathsf{MaxColl}(\pi)}{2^n}$$

for all $\mathbf{x}, \mathbf{x}' \in \{0,1\}^{wn}$, $j, h \in [w]$, such that $\mathbf{x}[j'] \neq \mathbf{x}'[j']$ for at least one $j' \neq j$.

Small values of $\mathsf{MaxPreim}(\pi)$ and of $\mathsf{MaxColl}(\pi)$ are better. It is easy to construct permutations with $\mathsf{MaxPreim}(\pi) = 1$ (which is optimal): simply use a linear permutation $\pi : \mathrm{GF}(2^n)^w \rightarrow \mathrm{GF}(2^n)^w$ whose associated matrix (a $w \times w$ matrix with entries in $\mathrm{GF}(2^n)$) has all nonzero entries. Constructing permutations with small values of $\mathsf{MaxColl}(\pi)$ is a bit more involved; this is done in the full version [8].

An interesting research direction would be to devise faster-than-linear RPR permutations and faster-than-polynomial RCR permutations. (Our construction of RCR permutations [8] uses finite field operations).

CONDUCTANCE AND ALL-BUT-ONE CONDUCTANCE. The *conductance* $\mathrm{Cond}_\pi(q)$ of a permutation π was defined in Sect. 1.1. The notion all-but-one conductance is essentially the same as conductance, except that one coordinate position in either the input or output is ignored. The all-but-one conductance of a permutation π at q queries is denoted $\mathrm{aboCond}_\pi(q)$.

Formally, given a permutation $\pi : (\{0,1\}^n)^w \to (\{0,1\}^n)^w$, we define the *all-but-one conductance* of π by

$$\mathrm{Cond}_\pi^{h,+}(q) = \max_{\substack{U_1,\ldots,U_w,V_1,\ldots,V_w \subseteq \{0,1\}^n \\ |U_1| = \cdots = |V_w| = q}} |\{(\mathbf{x},\mathbf{y}) : \mathbf{y} = \pi(\mathbf{x}), \mathbf{x}[j] \in U_j \; \forall j \in [w], \mathbf{y}[j] \in V_j \; \forall j \in [w] \setminus h\}|,$$

$$\mathrm{Cond}_\pi^{h,-}(q) = \max_{\substack{U_1,\ldots,U_w,V_1,\ldots,V_w \subseteq \{0,1\}^n \\ |U_1| = \cdots = |V_w| = q}} |\{(\mathbf{x},\mathbf{y}) : \mathbf{y} = \pi(\mathbf{x}), \mathbf{x}[j] \in U_j \; \forall j \in [w] \setminus h, \mathbf{y}[j] \in V_j \; \forall j \in [w]\}|$$

$$\mathrm{aboCond}_\pi(q) = \max\left(\max_{h \in [w]}(\mathrm{Cond}^{h,+}(\pi,q)), \max_{h \in [w]}(\mathrm{Cond}^{h,-}(\pi,q))\right)$$

(Here the first two definitions are for all $h \in [w]$, and we use $\forall$ in postfix notation). Thus the set V_h is immaterial in the definition of $\mathrm{Cond}^{h,+}$, while the set U_h is immaterial in the definition of $\mathrm{Cond}^{h,-}$.

In [8] we show that the conductance and all-but-one conductance of a random permutation π are both roughly qwn, which is essentially $q\log(q)$ since q is exponential in n. Constructing explicit permutations with low conductance and low all-but-one conductance remains a nice open problem.

5   Network Nomenclature and Main Result

In this section we start with a syntax-oriented description of the confusion-diffusion networks for which our results are obtained. We follow with the formal statement of our main theorem.

While we mentioned six different networks in the introduction (cf. Fig. 1), our full results actually encompass two extra networks, of 6 and 10 rounds (but structurally different from the "other" 6 and 10 round networks); these two extra networks are not mentioned in Fig. 1 because they offer no structural, security or query complexity advantages over other networks. However we include them what follows since they come anyway "for free", so that our result will subsume eight different networks. (Moreover the proof is easier to write this way, since the eight different networks happen to arise as the result of setting three independent boolean flags).

NETWORK NOMENCLATURE. A *round* of a confusion-diffusion network refers to a round of S-boxes. More precisely, all S-box permutations $P_{i,j}$ with the same value of i lie in the same *round* of the network.

Since, say, the middle round of our 5-round confusion-diffusion network plays the same structural role (with respect to our simulator) as the middle round in our 9-round network, it makes more sense to designate rounds according to their structural role instead of by their round number (as the latter will keep changing from network to network, even while the structural purpose of the round stays the same).

For this purpose, we replace the array $\mathcal{P} = \{P_{i,j}\}$ of $r \times w$ random permutations with an array $\mathcal{Q}$ of $12 \times w$ random permutations where each "round" (value of the index i) is designated by a different alphabet letter. Specifically, we let

$$\mathcal{Q} = \{F_j, G_j, I_j, D_j, J_j, B_j, A_j, C_j, K_j, E_j, L_j, H_j : j \in [w]\} \qquad (1)$$

be a set of $12w$ random permutations, where each permutation is thus indexed by an alphabet letter from the set $\{A, \ldots, L\}$ as well as by an index $j \in [w]$.

Having traded the set of indices $\{i : i \in [r]\}$ (the possible round numbers) for the set of letters $\{A, \ldots, L\}$, a "round" will henceforth mean a member of the latter set, i.e., a "round" means one of the letters $A, \ldots, L$.

Not all rounds will be used for all confusion-diffusion networks, and we describe which rounds appear in which networks below. (See also Fig. 2 below as an aid). However two rounds always appear in the same order as they appear listed in $\mathcal{Q}$ above (cf. (1)), when they both appear in a network.

In more detail, our eight different confusion-diffusion networks correspond to the eight different possible settings of three independent boolean flags called XtraMiddleRnd, XtraOuterRnd and XtraUntglRnds. The rounds that appear in each network, as a function of these boolean flags, are as follows:

$$\begin{cases} A & \text{if XtraMiddleRnd is off} \\ B, C & \text{if XtraMiddleRnd is on} \end{cases}$$

$$\begin{cases} G, H & \text{if XtraOuterRnd is off} \\ F, G, H & \text{if XtraOuterRnd is on} \end{cases}$$

$$\begin{cases} D, E & \text{if XtraUntglRnds is off} \\ I, D, J, K, E, L & \text{if XtraUntglRnds is on} \end{cases}$$

As can be seen, toggling either of XtraMiddleRnd or XtraOuterRnd "costs" one extra round, whereas toggling XtraUntglRnds costs four extra rounds. Hence, and because the network has 5 rounds when all flags are off, the number of rounds in the network will be

$$5 + \text{XtraMiddleRnd} + \text{XtraOuterRnd} + 4 \cdot \text{XtraUntglRnds}$$

which spans the integers 5, 6, 6, 7, 9, 10, 10, 11. In Fig. 1 the flag bits appear in the order XtraMiddleRnd, XtraOuterRnd, XtraUntglRnds, hence the two "missing" combinations are those where XtraMiddleRnd/XtraOuterRnd are off/on. In addition, the top half of the table is missing all rows with XtraUntglRnds = **true** since toggling this flag does not (significantly) improve security or query complexity when the D-boxes can be arbitrary and, in particular, when the D-boxes can be RCR.

For example, our 11-round network consists of the rounds

$$F, G, I, D, J, B, C, K, E, L, H$$

in this order. (We refer to Fig. 2 further down). The 10-round network with XtraMiddleRnd = **false** consists of the rounds

$$F, G, I, D, J, A, K, E, L, H$$

in this order as well. All other networks can be obtained by removing rounds from one of these two sequences. In more detail, round F is removed to un-toggle XtraOuterRnd and rounds I, J, K, L are removed to un-toggle XtraUntglRnds.

We will also rename the generic diffusion permutations $\overline{\pi} = (\pi_1, \ldots, \pi_r)$ according to their structural roles in the diffusion network. Specifically, we let

$$\overline{\pi} = (\nu, \pi_G, \pi_I, \pi_J, \pi_B, \tau, \pi_C, \pi_K, \pi_L, \pi_H)$$

where each element in the sequence $\overline{\pi}$ is a permutation from $\{0, 1\}^{wn}$ to $\{0, 1\}^{wn}$. (Thus, we redefine $\overline{\pi}$ to a specific sequence of diffusion permutations). In the 11-round confusion-diffusion network, diffusion permutations appear interleaved with the S-box rounds in the order

$$F-\nu-G-\pi_G-I-\pi_I-D-\pi_J-J-\pi_B-B-\tau-C-\pi_C-K-\pi_K-E-\pi_L-L-\pi_H-H$$

(i.e., the S-box round consisting of the parallel application of the permutations F_j is followed by the diffusion permutation ν, and so on), whereas in the 10-round network with XtraMiddleRnd= **false** the diffusion permutations appear in the order

$$F-\nu-G-\pi_G-I-\pi_I-D-\pi_J-J-\pi_B-A-\pi_C-K-\pi_K-E-\pi_L-L-\pi_H-H$$

with τ dropped. From either of these configurations one can un-toggle XtraOuter-Rnd by dropping $F-\nu-$ and one can un-toggle XtraUntglRnds by dropping $I-\pi_I-$, $-\pi_J-J$, $K-\pi_K-$ and $-\pi_L-L$. For example, our 9-round confusion-diffusion network has the order

$$G-\pi_G-I-\pi_I-D-\pi_J-J-\pi_B-A-\pi_C-K-\pi_K-E-\pi_L-L-\pi_H-H$$

whereas the 5-round and 6-round network with XtraMiddleRnd toggled respectively have order

$$G-\pi_G-D-\pi_B-A-\pi_C-E-\pi_H-H$$
$$G-\pi_G-D-\pi_B-B-\tau-C-\pi_C-E-\pi_H-H$$

and so on.

In summary, the confusion-diffusion network under consideration is a function of the confusion permutations $\mathcal{Q}$, of the diffusion permutation vector $\overline{\pi}$ and of

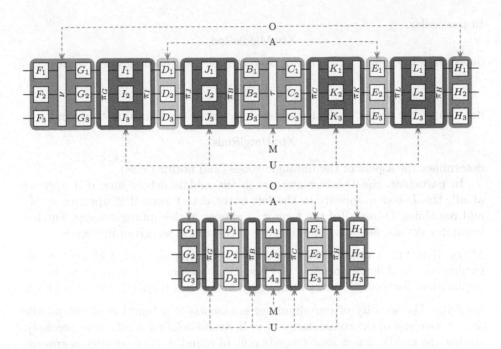

Fig. 2. Emplacement of the outer detect (O), adapt (A), middle detect (M) and untangle (U) zones for the 11- and 5-round simulators. The adapt zones always consist of rounds D and E.

the three boolean flags XtraMiddleRnd, XtraOuterRnd and XtraUntglRnds. For brevity we write this network as

$$P[\mathcal{Q}, \overline{\pi}]$$

keeping the three boolean flags implicit. Depending on the value of the flags some permutations in $\mathcal{Q}$ and/or $\overline{\pi}$ are of course unused. In particular, we assume that unused permutations in $\mathcal{Q}$ are simply ignored for the purpose of the indifferentiability experiment (i.e., these unused permutations are not accessible as oracles).

Semantically, moreover, our simulator divides the confusion rounds and diffusion permutations into nine "zones" of four different types, to wit, one *middle detect* zone (M), left and right *outer detect* zones (O), four *untangle* zones (U) and two *adapt* zones (A). Each zone consists of one or more contiguous rounds and/or diffusion permutations, with every round and every diffusion permutation belonging to exactly one zone. Figure 2 shows how the zones appear in the 11- and 5-round networks, while the zoning of other networks can be interpolated from these, given that each of our zones either appears the same as in the 11-round network or the same as in the 5-round network. (For example, if XtraMiddleRnd is off, then round A exists and rounds B and C do not, so the middle detect zone consists of round A only, as in the 5-round simulator; etc).

In particular,

XtraMiddleRnd

determines the aspect of the middle detect zone (and nothing else), while

XtraOuterRnd

determines the aspect of the left outer detect zone (and nothing else), and

XtraUntglRnds

determines the aspect of the untangle zones (and nothing else).

In particular, the D-box τ appears in the middle detect zone if it appears at all, the D-box ν appears in the left outer detect zone if it appears at all, and remaining D-boxes (of the form $\pi_{...}$) appear in the untangle zones. Further simulator details, including the purpose of the zones, are given in Sect. 6.

MAIN RESULT: TAKEAWAY POINTS. Before giving our full technical result (which can be difficult to parse the first time around), we summarize its key implications for security, query complexity, and D-box properties at a high level:

Security. The security of our simulator is essentially a function of the middle detect zone and of the conductance of τ. If XtraMiddleRnd is off, more precisely, so that the middle detect zone consists only of round A, then security is approximately q^{2w}/N where $N = 2^n$, whereas if XtraMiddleRnd is on, the security improves (essentially) to q^2/N, assuming that τ has conductance $\sim q$.

Query Complexity. The simulator's query complexity is determined by the left outer detect zone and by the middle detect zone. The query complexity is approximately q^{w^2} if neither of XtraMiddleRnd or XtraOuterRnd is toggled, is approximately q^w if exactly one of these two flags is toggled, and is approximately q if both flags are toggled, where the bounds quoted in the last two cases assume that τ and/or ν both have conductance $\sim q$.

Combinatorial Requirements on the Diffusion Permutations. The diffusion permutations τ and ν must have low conductance and low all-but-one-conductance. (Near q in order to have bounds approaching those of Fig. 1). The combinatorial requirements on the permutations $\pi_{...}$ are different depending on whether XtraUntglRnds is toggled or not:

- if XtraUntglRnds is off the permutations in the untangle zones must be both RPR- and RCR-resistant (or else the above-quoted security bounds are not achieved); in particular, if $w > 2$ then these permutations *cannot* be $GF(2^n)$-linear, because, as mentioned above, $GF(2^n)$-linear permutations are not RCR-resistant for $w > 2$;
- if XtraUntglRnds is on then the permutations in the untangle zones need only be RPR-resistant; in particular, these permutations can now be $GF(2^n)$-linear even for $w > 2$.

One can also summarize the situation orthogonally, according to which boolean flags have which effect. Namely:

- toggling XtraMiddleRnd impacts security and query complexity
- toggling XtraOuterRnd impacts query complexity
- toggling XtraUntglRnds reduces the cryptographic requirements on the permutations $\pi_{...}$

It is natural to wonder whether even further extension of the middle, outer left, or untangle zones can be beneficial. The short answer to this question, that we consider in detail in [8], is yes; namely, by adding more rounds to these zones one can further reduce the combinatorial requirements on the diffusion permutations, leading, e.g., to networks in which the permutations in the untangle zones are not even RPR-resistant.

As a final high-level comment, a potentially striking feature of our simulator is the left-right asymmetry that arises when XtraOuterRnd is toggled. As explained in Sect. 6, adding an extra round to the right-hand outer detect zone (assuming XtraOuterRnd is already set) does not further decrease the simulator's query complexity, nor improve security. Nonetheless, such an extra zone can be used to reduce the simulator's *space complexity* while maintaining essentially the same security and query complexity. (See the discussion after Lemma 52 in the full version). Hence, while the benefit of adding an additional round to the right outer detect zone is somewhat technical, such an extra round does have a potential justification in terms of simulator design.

MAIN RESULT. We recall that $\overline{\pi} = (\nu, \pi_G, \pi_I, \pi_J, \pi_B, \tau, \pi_C, \pi_K, \pi_L, \pi_H)$ is the vector of diffusion permutations, not necessarily all of which are used. In order to more succinctly state the main result, we define

$$\mathsf{MaxColl}(\overline{\pi}) = \max(\mathsf{MaxColl}(\pi_G), \mathsf{MaxColl}(\pi_B^{-1}), \mathsf{MaxColl}(\pi_C), \mathsf{MaxColl}(\pi_H^{-1}))$$
$$\mathsf{MaxPreim}(\overline{\pi}) = \max(\mathsf{MaxPreim}(\pi_G), \mathsf{MaxPreim}(\pi_B^{-1}), \mathsf{MaxPreim}(\pi_C), \mathsf{MaxPreim}(\pi_H^{-1}),$$
$$\mathsf{MaxPreim}(\pi_I), \mathsf{MaxPreim}(\pi_J^{-1}), \mathsf{MaxPreim}(\pi_K), \mathsf{MaxPreim}(\pi_L^{-1}))$$
$$\mathsf{MaxCoPr}(\overline{\pi}) = \max(\mathsf{MaxColl}(\overline{\pi}), \mathsf{MaxPreim}(\overline{\pi}))$$

where π^{-1} denotes the inverse of π.

Moreover we define

$$\alpha(q) = \begin{cases} (2q)^w & \text{if XtraMiddleRnd is off,} \\ \mathsf{Cond}_\tau(2q) & \text{if XtraMiddleRnd is on,} \end{cases}$$

$$\beta(q) = \begin{cases} (q + \alpha(q))^w & \text{if XtraOuterRnd is off,} \\ \mathsf{Cond}_\nu(q + \alpha(q)) & \text{if XtraOuterRnd is on.} \end{cases}$$

The definitions of $\alpha(q)$ and $\beta(q)$ might seem annoyingly technical right now. In Sect. 6 we provide more digestible semantic explanations for $\alpha(q)$ and $\beta(q)$.

Theorem 1. *Let* $N = 2^n$. *The confusion-diffusion network* $P[\mathcal{Q}, \overline{\pi}]$ *achieves* (t_S, q_S, ε)-*indifferentiability from a random permutation* $\mathcal{Z} : \{0,1\}^{wn} \rightarrow \{0,1\}^{wn}$

for ε equal to

$$\frac{\beta(q)(q + \alpha(q))^w}{N^w - q - \alpha(q)} + \frac{1}{N^w} + \frac{4w(q + \alpha(q))^2}{N - q - \alpha(q)}$$

$$+ \frac{4wq \operatorname{aboCond}_\tau(2q)}{N - 2q} \qquad\qquad \textit{if } \mathsf{XtraMiddleRnd} \textit{ is on}$$

$$+ \frac{2w(q + \alpha(q)) \operatorname{aboCond}_\nu(q + \alpha(q))}{N - q - \alpha(q)} \qquad \textit{if } \mathsf{XtraOuterRnd} \textit{ is on}$$

$$+ \frac{4w\alpha(q)(q + \alpha(q))\mathsf{MaxCoPr}(\overline{\pi})}{N - q - \alpha(q)} \qquad \textit{if } \mathsf{XtraUntglRnds} \textit{ is off}$$

$$+ \frac{6w(q + \alpha(q))^2 \mathsf{MaxPreim}(\overline{\pi})}{N - q - \alpha(q)} \qquad \textit{if } \mathsf{XtraUntglRnds} \textit{ is on}$$

and for $q_S = \beta(q)$, $t_S = O(w(q + \alpha(q))^w)$. Here $q = q_0(1 + rw)$ where q_0 is the number of distinguisher queries and $r \in \{5, 6, 7, 9, 10, 11\}$ is the number of rounds in the confusion-diffusion network.

The proof of this result is in the full version [8].

Parsing the Security Bound. In order to get a rough feel for the security bound of Theorem 1 it is helpful to make the order-of-magnitude approximations

$$\mathsf{MaxPreim}(\overline{\pi}) = \mathsf{MaxColl}(\overline{\pi}) \approx O(1)$$
$$\mathsf{Cond}_\tau(2q) = \operatorname{aboCond}_\tau(2q) \approx q$$
$$\mathsf{Cond}_\nu(q + \alpha(q)) = \operatorname{aboCond}_\nu(q + \alpha(q)) \approx \alpha(q).$$

With these approximations in place, and given $q \ll N$ (in fact we can assume $q \leq N^{1/2}$, since the security bound is void otherwise) it easy to verify that the largest terms in Theorem 1 are of the order

$$\frac{\alpha(q)^2}{N}$$

and which is, therefore, a first approximation to the security ε that we achieve. Since

$$\alpha(q) = \begin{cases} (2q)^w & \text{if } \mathsf{XtraMiddleRnd} \text{ is off,} \\ q & \text{if } \mathsf{XtraMiddleRnd} \text{ is on,} \end{cases}$$

under the order-of-magnitude approximations given above, we find

$$\varepsilon \approx \begin{cases} (2q)^{2w}/N & \text{if } \mathsf{XtraMiddleRnd} \text{ is off,} \\ q^2/N & \text{if } \mathsf{XtraMiddleRnd} \text{ is on} \end{cases}$$

for the security, to a first approximation. On the other hand we find

$$q_S = \beta(q) \approx \begin{cases} (2q)^{w^2} & \text{if } \mathsf{XtraMiddleRnd}/\mathsf{XtraOuterRnd} \text{ are off/off} \\ (2q)^w & \text{if } \mathsf{XtraMiddleRnd}/\mathsf{XtraOuterRnd} \text{ are off/on} \\ (2q)^w & \text{if } \mathsf{XtraMiddleRnd}/\mathsf{XtraOuterRnd} \text{ are on/off} \\ q & \text{if } \mathsf{XtraMiddleRnd}/\mathsf{XtraOuterRnd} \text{ are on/on} \end{cases}$$

for the query complexity, again to a first approximation.

On the other hand, it is relatively easy to see that $\mathsf{MaxColl}(\pi) = 2^n = N$ for any linear permutation $\pi : \mathrm{GF}(2^n)^w \to \mathrm{GF}(2^n)^w$ as long as $w > 2$. In this case $\mathsf{MaxCoPr}(\pi) = N$, and Theorem 1 becomes void if XtraUntglRnds is off. Thus one of the main reasons for toggling XtraUntglRnds might be to enable the use of linear diffusion permutations, or any other (fast) family of permutations that have small entry-wise randomized preimage resistance (but potentially large entry-wise randomized collision resistance).

6   Simulator Overview

CONTEXT. We start with some very high-level description and reminder-of-purpose of our simulator. For this discussion it will be more convenient if we momentarily revert to indexing the S-boxes by coordinate pairs (i, j) where $i \in [r]$ the round number and $j \in [w]$ the layer number, with r being the number of rounds and w being the width. The diffusion permutation between the i-th and $(i + 1)$-th rounds will again be denoted π_i as well.

The simulator is responsible for answering queries to the S-boxes, and has access to a random permutation oracle $\mathcal{Z} : \{0, 1\}^{wn} \to \{0, 1\}^{wn}$ that is being independently accessed by the distinguisher. The simulator's job is to keep the S-box answers compatible with $\mathcal{Z}$ in the sense that it looks to the distinguisher as if $\mathcal{Z}$ is implemented by the confusion-diffusion network.

For each pair $(i, j) \in [r] \times [w]$ the simulator maintains a pair of tables $P_{i,j}$ and $P_{i,j}^{-1}$, each containing 2^n entries of n bits each, in which the simulator keeps a record of "what it has already decided" about the (i, j)-th S-box. Initially the tables are blank, meaning that $P_{i,j}(x) = P_{i,j}^{-1}(y) = \perp$ for all $x, y \in \{0, 1\}^n$. The simulator sets $P_{i,j}(x) = y$, $P_{i,j}^{-1}(y) = x$ to indicate that the (i, j)-th S-box maps x to y. The simulator never overwrites values in $P_{i,j}$ or in $P_{i,j}^{-1}$ and always keeps these two tables consistent. Hence $P_{i,j}$ encodes a partial matching (or "partial permutation") from $\{0, 1\}^n$ to $\{0, 1\}^n$ from which edges are never subtracted. We also note that the edges in $P_{i,j}$ are a *superset* of those queries that the distinguisher has made to the (i, j)-th S-box or to its inverse (i.e., $P_{i,j}$ contains the answers to those queries, and possibly more).

By analogy with the notation of Sect. 2 we write

$$P_i(\mathbf{x}) = \mathbf{y} \tag{2}$$

if $\mathbf{x}, \mathbf{y} \in \{0, 1\}^{wn}$ are vectors such that $P_{i,j}(\mathbf{x}[j]) = \mathbf{y}[j]$ for all $j \in [w]$. Note that (2) is a time-dependent statement, in the sense that the tables $P_{i,j}$ keep accruing entries as the distinguishing experiment proceeds. For example, (2) is initially false for all i and all vectors $\mathbf{x}$, $\mathbf{y}$. Moreover P_i is not an actual table maintained by the simulator—i.e., (2) is "just notation".

A sequence of vectors $(\mathbf{x}^1, \mathbf{y}^1, \ldots, \mathbf{x}^r, \mathbf{y}^r)$ is called a *completed path*[6] if $P_i(\mathbf{x}^i) = \mathbf{y}^i$ for $i = 1, \ldots, r$ and if $\pi_i(\mathbf{y}^i) = \mathbf{x}^{i+1}$ for $i = 1, \ldots, r-1$. The set of completed paths is also time-dependent. The vectors $\mathbf{x}^1$ and $\mathbf{y}^r$ are called the *endpoints* of the path.

We informally say that the distinguisher *completes a path* if it makes queries to the simulator that form a completed path. (There are many different possible ways to order such a set of queries, obviously). One can picture the distinguisher as trying to complete paths in various devious ways (typically, reusing the same queries as part of different paths), and checking that the path endpoints are each time compatible with $\mathcal{Z}$.

The simulator's job, in response, is to run ahead of the distinguisher and preemptively complete paths that it thinks the distinguisher is interested in, such as to make these paths are compatible with $\mathcal{Z}$. The simulator's dilemma is that it must choose under which conditions to complete a path; if it waits too long, or completes paths in only highly specialized cases, it may find itself trapped in a contradiction (typically, while trying to complete several paths at once); but if it is too trigger-happy, having a very large number of conditions under which it will choose to complete a path, the simulator runs the risk creating[7] an out-of-control chain reaction of path completions.

Essentially the simulator must be safe, but in a smart enough way that it avoids (out-of-control) chain reactions. We will informally refer to the problem of showing that no out-of-control chain reactions occur as the problem of *simulator termination*.

SIMULATOR ZONES. As already mentioned our simulator divides the confusion rounds and diffusion permutations into nine zones of four different types, shown in Fig. 2 for the 11- and 5-round networks.

Specifically, the "middle zone" is

$$\begin{cases} A & \text{if XtraMiddleRnd is off} \\ B\text{--}\tau\text{--}C & \text{if XtraMiddleRnd is on} \end{cases}$$

depending only on XtraMiddleRnd; the left and right outer detect zones are

$$\begin{cases} G, H & \text{if XtraOuterRnd is off} \\ F\text{--}\nu\text{--}G, H & \text{if XtraOuterRnd is on} \end{cases}$$

---

[6] This definition, made for the sake of expository convenience, is superceded further down, where we redefine "completed path" by adding the requirement that the endpoints be compatible with $\mathcal{Z}$, i.e., that $\mathcal{Z}(\mathbf{x}^1) = \mathbf{y}^r$.

[7] Indeed, the simulator makes no distinction between those entries in its tables $P_{i,j}$ that are the direct result of an distinguisher query, and those which it created on its own while pre-emptively completing paths. It seems very hard to leverage such a distinction. Note for example that the distinguisher may know values in $P_{i,j}$ without having made the relevant queries, simply by virtue of knowing how the simulator works.

depending only on XtraOuterRnd; the four "untangle zones" are

$$\begin{cases} \pi_G, \pi_B, \pi_C, \pi_H & \text{if XtraUntglRnds is off} \\ \pi_G\text{-}I\text{-}\pi_I, \pi_J\text{-}J\text{-}\pi_B, \pi_C\text{-}K\text{-}\pi_K, \pi_L\text{-}L\text{-}\pi_H & \text{if XtraUntglRnds is on} \end{cases}$$

depending only on XtraUntglRnds; the two "adapt" zones, finally, are rounds D and E in all networks.

We observe that zones (specifically, untangle zones with XtraUntglRnds = **false**) might consist solely of diffusion permutations. This contrasts with semantically similar zoning systems for Feistel network simulators and key-alternating simulators [18,19,27] for which the zones always contain at least one ideal component. We also observe that in the minimalist 5-round simulator, each round and each diffusion permutation corresponds to an individual zone.

TABLE NOTATION, COLUMNS AND MATCHING QUERY SETS. We revert to identifying rounds with letters in $\{A, \ldots, L\}$. Under this notation tables $P_{i,j}$, $P_{i,j}^{-1}$ described above become, e.g., tables A_j and A_j^{-1}. Thus for each round $T \in \{A, \ldots, L\}$ that is "operational" (as will depend on the flag settings) the simulator maintains tables T_j, T_j^{-1} for $1 \leq j \leq w$, as already described above under the notation $P_{i,j}, P_{i,j}^{-1}$.

We write $T(\mathbf{x}) = \mathbf{y}$ if $T_j(\mathbf{x}[j]) = \mathbf{y}[j]$ for each $j \in [w]$, and for all $T \in \{A, \ldots, L\}$. We also write $T(\mathbf{x}) = \perp$ if $T_j(\mathbf{x}[j]) = \perp$ for at least one value of j. The notation $T^{-1}(\mathbf{y})$ is analogous, with T^{-1} being the inverse of T. As emphasized above this notation is really "notation only" in the sense that the simulator does not maintain such things as tables T or T^{-1}.

A non-null entry in table T_j will be called an $(S\text{-box})$ *query*. More formally, an S-box query is a quadruple (T, j, x, y) where $T \in \{A, \ldots, L\}$, $j \in [w]$, $x, y \in \{0,1\}^n$, such that $T_j(x) = y$.

A set of w queries with the same value of T but with different values of j will be called a *column* or a *T-column* when we wish to emphasize the round. The (unique) vectors $\mathbf{x}, \mathbf{y}$ such that $(T, j, \mathbf{x}[j], \mathbf{y}[j])$ is a query in the column for each $j \in [w]$ are called the *input* and *output* of the column respectively. We note that a column is uniquely determined by either its input or output.

Two columns are *adjacent* if their rounds are adjacent. (E.g., a B-column and a C-column are adjacent). Two adjacent columns are *matching* if $\pi(\mathbf{y}) = \mathbf{x}$, where $\mathbf{y}$ is the output of the first column, where $\mathbf{x}$ the input of the second column, and where π is the diffusion permutation between the two rounds.

A pair of columns from the first and last round of the confusion-diffusion network are likewise *matching* if $\mathcal{Z}(\mathbf{x}) = \mathbf{y}$, where $\mathbf{x}$ is the input to the first-round column (either an F- or G-column) and $\mathbf{y}$ is the output of the last-round column (the H-column).

The notion of matching columns naturally extends to sequences of columns from consecutive rounds of length greater than two. (The first and last round of the network are always considered adjacent). If a set of matching columns encompasses all rounds we call the set a *completed path*. Thus, since the first and last column of a network are considered adjacent, completed paths are compatible with $\mathcal{Z}$ by definition.

The set of queries in a matching sequence of columns of any length is called a *matching set* of queries. We will also refer to the queries of a single column as a matching set, considering that column as a matching sequence of columns of length 1.

One can also observe that two different completed paths cannot contain the same column. Indeed, each D-box is a permutation and each round implements a partial permutation as well.

SIMULATOR OPERATION AND TERMINATION ARGUMENT. Our simulator applies a design paradigm pioneered by Seurin [27] that has also been used in other simulators for Feistel networks or key-alternating ciphers [18,19]. As for all such simulators, our simulator completes two types of paths, where one type of path is triggered by a "middle detect zone" and the other is triggered by an "outer detect zone".

In more detail, our simulator pre-emptively completes a path[8] for every matching set of queries in the middle detect zone (such a matching set will consist of either one or two columns, depending on whether XtraMiddleRnd is toggled) and also for every matching set of queries in the two outer detect zones, considered as a single consecutive set of columns (the latter kind of matching set will thus consist of either two or three columns, depending on whether XtraOuterRnd is toggled).

Crucially, one can show that, unless some bad event of negligible probability occurs, each outer-triggered path completion is associated to a unique pre-existing query made by the distinguisher to $\mathcal{Z}$. Since the distinguisher makes only q queries in total, this means that at most q outer-triggered path completions occur, with high probability. In fact our simulator aborts if it counts[9] more than q outer-triggered path completions, so in our case at most q outer-triggered path completions occur with probability 1.

Moreover, a middle-triggered path completion does not add any new queries to the middle detect zone. This means that all queries in the middle detect zone can be chalked up to one of two causes: (1) queries directly made by the distinguisher, (2) queries made by the simulator during outer-triggered path completions.

Cause (1) accounts for at most q queries to each S-box and, as just seen, cause (2) accounts for at most q queries as well. Hence a middle detect zone S-box is never queried at more then $2q$ points, i.e., the table T_j for $T \in \{A, B, C\}$ never has more than $2q$ non-$\perp$ entries.

---

[8] The phrase "completes a path" is informal at this point, as there are generally many different ways to complete a path. (E.g., where to "adapt" a path to make it compatible with $\mathcal{Z}$, etc). More details follow below.

[9] This means the simulator knows the value of q beforehand, which introduces a small amount of non-uniformity into the simulator. One could remove this non-uniformity—i.e., not tell the simulator the value of q beforehand—at the cost of a more complicated theorem statement and proof. But the fact that essentially all security games allow adversaries that know the value of q for which they want to carry out an attack makes the issue a bit moot.

In particular, if XtraMiddleRnd is off, the latter implies that no more than $(2q)^w$ middle-triggered path completions occur, or one for every possible combination of an entry from each of the tables $A_1, \ldots, A_w$. If XtraMiddleRnd is on, on the other hand, than at most $\mathrm{Cond}_\tau(2q)$ middle-triggered path completions occur, as is easy to see per the definition[10] of conductance. In other words, $\alpha(q)$ (cf. Sect. 5) is an upper bound on the number of middle-triggered path completions. In fact, $\alpha(q)$ is an upper bound for the *total* number of path completions performed (including also outer path completions), since each completed path is also associated to a unique set of matching queries from the middle detect zone.

As for all S-boxes outside the middle detect zone, their queries can also be chalked up to one of two sources, namely direct distinguisher queries and path completions. There are at most q direct distinguisher queries, and each completed path contributes at most 1 query to each S-box, so each S-box outside the middle detect zone ends up with at most $q + \alpha(q)$ queries.

The simulator, moreover, only queries $\mathcal{Z}$ in order to either complete paths or else in order to detect outer-triggered path completions. This implies the number of distinct simulator queries to $\mathcal{Z}$ is upper bounded by the number of matching sets of queries in the left-hand outer detect zone. Indeed each completed path is obviously associated to a matching set of queries in the left-hand outer detect zone; and for the purpose of outer-triggered path detection, it is easy to see that the simulator only needs to query $\mathcal{Z}$ at most once for each such matching set as well by maintaining a table[11] of queries already made to $\mathcal{Z}$. If XtraOuterRnd is off, thus, the number of simulator queries to $\mathcal{Z}$ will be at most $(q + \alpha(q))^w$; whereas if XtraOuterRnd is on, the same number will be at most $\mathrm{Cond}_\nu(q + \alpha(q))$. The simulator query complexity is thus at most $\beta(q)$ (cf. Sect. 5).

7   Extensions

In [8] we discuss the extension of our main result to even longer networks. In particular, we show that each of our four combinatorial properties are specialized (extreme) cases of more general adversarial games involving entire segments of CD networks. We give a more general theorem (albeit without proof) in terms of these properties defined on network segments. The use of more rounds, then, enables the D-boxes to become even weaker (provably so in the case of the untangle rounds, conjecturally so for the detect zones), and thus even faster. This leaves many exciting possibilities/questions for future work.

---

[10] More precisely, let $U_j = \{y \in \{0,1\}^n : B_j^{-1}(y) \neq \bot\}$, let $V_j = \{x \in \{0,1\}^n : C_j(x) \neq \bot\}$ at the end of the distinguishing experiment. Then $|U_j|, |V_j| \leq 2q$ for each j, and the number of middle-triggered path completions that have occurred is at most

$$\{(\mathbf{x}, \mathbf{y}) : \tau(\mathbf{x}) = \mathbf{y}, \mathbf{x} \in \prod_j U_j, \mathbf{y} \in \prod_j V_j\} \leq \mathrm{Cond}_\tau(2q).$$

[11] Thus, in particular, adding an extra round to the right-hand detect zone would not further reduce the query complexity, since the query complexity is only proportional to the number of matching query sets for the left-hand outer detect zone.

Another possible improvement is to reduce the number of independent S-boxes. In a private note (unpublished), we have proved an analogous result where the same S-box is used at each round. To make this variant indifferentiable from a random permutation over $\{0,1\}^{wn}$, the D-boxes need to satisfy stronger combinatorial properties. For example, the current construction needs a D-box with entry-wise randomized preimage resistance (RPR): roughly, if one of the input blocks is uniform random and the other input blocks are fixed, then each output block is fairly random. In the variant construction with identical S-boxes, the D-boxes must satisfy a generalization of RPR: roughly, if some of the input blocks contain the same uniform random value and the other input blocks are fixed, then each output block is fairly random. We also found that a minor modification to our D-boxes would satisfy these generalized properties. Additionally, in this variant construction the security (distinguisher's advantage) ε and simulator query complexity q_S are worse than the current construction. The query complexity is increased by a factor of w^{w^2}, w^w or w (depending on the flags), while the security deteriorates by a factor of w.

References

1. Andreeva, E., Bogdanov, A., Dodis, Y., Mennink, B., Steinberger, J.P.: On the indifferentiability of key-alternating ciphers. In: Canetti, R., Garay, J.A. (eds.) CRYPTO 2013, Part I. LNCS, vol. 8042, pp. 531–550. Springer, Heidelberg (2013)
2. Bertoni, G., Daemen, J., Peeters, M., Van Assche, G.: On the indifferentiability of the sponge construction. In: Smart [31] pp. 181–197
3. Biryukov, A., Bouillaguet, C., Khovratovich, D.: Cryptographic schemes based on the ASASA structure: black-box, white-box, and public-key (Extended Abstract). In: Sarkar, P., Iwata, T. (eds.) ASIACRYPT 2014. LNCS, vol. 8873, pp. 63–84. Springer, Heidelberg (2014)
4. Chakraborty, D., Sarkar, P.: A new mode of encryption providing a tweakable strong pseudo-random permutation. In: Robshaw, M. (ed.) FSE 2006. LNCS, vol. 4047, pp. 293–309. Springer, Heidelberg (2006)
5. Chakraborty, D., Sarkar, P.: HCH: a new tweakable enciphering scheme using the hash-encrypt-hash approach. In: Barua, R., Lange, T. (eds.) INDOCRYPT 2006. LNCS, vol. 4329, pp. 287–302. Springer, Heidelberg (2006)
6. Chow, S., Eisen, P.A., Johnson, H., van Oorschot, P.C.: White-box cryptography and an AES implementation. In: Nyberg, K., Heys, H.M. (eds.) SAC 2002. LNCS, vol. 2595, pp. 250–270. Springer, Heidelberg (2003)
7. Chow, S., Eisen, P., Johnson, H., van Oorschot, P.C.: A white-box DES implementation for DRM applications. In: Feigenbaum, J. (ed.) DRM 2002. LNCS, vol. 2696, pp. 1–15. Springer, Heidelberg (2003)
8. Dodis, Y., Tianren, L., Stam, M., Steinberger, J.: Indifferentiability of Confusion-Diffusion Networks, IACR eprint archive 2015/680. (Full version of this paper.)
9. Coron, J.-S., Dodis, Y., Malinaud, C., Puniya, P.: Merkle-damgård revisited: how to construct a hash function. In: Shoup, V. (ed.) CRYPTO 2005. LNCS, vol. 3621, pp. 430–448. Springer, Heidelberg (2005)
10. Coron, J.-S., Dodis, Y., Mandal, A., Seurin, Y.: A domain extender for the ideal cipher. In: Micciancio, D. (ed.) TCC 2010. LNCS, vol. 5978, pp. 273–289. Springer, Heidelberg (2010)

11. Coron, J.-S., Patarin, J., Seurin, Y.: The random oracle model and the ideal cipher model are equivalent. In: Wagner [32], pp. 1–20
12. Dodis, Y., Pietrzak, K., Puniya, P.: A new mode of operation for block ciphers and length-preserving macs. In: Smart [31], pp. 198–219
13. Dodis, Y., Reyzin, L., Rivest, R.L., Shen, E.: Indifferentiability of permutation-based compression functions and tree-based modes of operation, with applications to MD6. In: Dunkelman, O. (ed.) FSE 2009. LNCS, vol. 5665, pp. 104–121. Springer, Heidelberg (2009)
14. Dodis, Y., Ristenpart, T., Shrimpton, T.: Salvaging merkle-damgård for practical applications. In: Joux, A. (ed.) EUROCRYPT 2009. LNCS, vol. 5479, pp. 371–388. Springer, Heidelberg (2009)
15. Feistel, H.: Cryptographic coding for data-bank privacy. IBM Technical report RC-2827, 18 March 1970
16. Fluhrer, S.R., McGrew, D.A.: The extended codebook (XCB) mode of operation. Technical report 2004/078, IACR eprint archive (2004)
17. Halevi, S.: Invertible universal hashing and the TET encryption mode. In: Menezes, A. (ed.) CRYPTO 2007. LNCS, vol. 4622, pp. 412–429. Springer, Heidelberg (2007)
18. Holenstein, T., Künzler, R., Tessaro, S.: The equivalence of the random oracle model and the ideal cipher model, revisited. In: Fortnow, L., Vadhan, S.P. (eds.), Proceedings of the 43rd ACM Symposium on Theory of Computing, STOC 2011, San Jose, CA, USA, pp. 89–98. ACM, 6–8 June 2011
19. Lampe, R., Seurin, Y.: How to construct an ideal cipher from a small set of public permutations. In: Sako, K., Sarkar, P. (eds.) ASIACRYPT 2013, Part I. LNCS, vol. 8269, pp. 444–463. Springer, Heidelberg (2013)
20. Luby, M., Rackoff, C.: How to construct pseudorandom permutations and pseudorandom functions. SIAM J. Comput. 17(2), 373–386 (1988)
21. Maurer, U.M., Renner, R.S., Holenstein, C.: Indifferentiability, impossibility results on reductions, and applications to the random oracle methodology. In: Naor, M. (ed.) TCC 2004. LNCS, vol. 2951, pp. 21–39. Springer, Heidelberg (2004)
22. Maurer, U.M., Tessaro, S.: Domain extension of public random functions: beyond the birthday barrier. In: Menezes, A. (ed.) CRYPTO 2007. LNCS, vol. 4622, pp. 187–204. Springer, Heidelberg (2007)
23. Miles, E., Viola, E.: Substitution-permutation networks, pseudorandom functions, and natural proofs. In: Safavi-Naini, R., Canetti, R. (eds.) CRYPTO 2012. LNCS, vol. 7417, pp. 68–85. Springer, Heidelberg (2012)
24. Naor, M., Reingold, O.: On the construction of pseudorandom permutations: Luby-Rackoff revisited. J. Cryptology 12(1), 29–66 (1999). Preliminary Version: STOC
25. Ristenpart, T., Shacham, H., Shrimpton, T.: Careful with composition: limitations of the indifferentiability framework. In: Paterson, K.G. (ed.) EUROCRYPT 2011. LNCS, vol. 6632, pp. 487–506. Springer, Heidelberg (2011)
26. Rogaway, P., Steinberger, J.P.: Constructing cryptographic hash functions from fixed-key blockciphers. In: Wagner [32], pp. 433–450
27. Seurin, Y.: Primitives et protocoles cryptographiques à sécurité prouvée. Ph.D. thesis, Université de Versailles Saint-Quentin-en-Yvelines, France (2009)
28. Shannon, C.E.: Communication theory of secrecy systems. Bell Syst. Technical J. 28(4), 656–715 (1949). www.cs.ucla.edu/jkong/research/security/shannon.html, www3.edgenet.net/dcowley/docs.html
29. Shrimpton, T., Stam, M.: Building a collision-resistant compression function from non-compressing primitives. In: Aceto, L., Damgård, I., Goldberg, L.A., Halldórsson, M.M., Ingólfsdóttir, A., Walukiewicz, I. (eds.) ICALP 2008, Part II. LNCS, vol. 5126, pp. 643–654. Springer, Heidelberg (2008)

30. Smart, N.P. (ed.): EUROCRYPT 2008. LNCS, vol. 4965. Springer, Heidelberg (2008)
31. Boneh, D., Halevi, S., Hamburg, M., Ostrovsky, R.: Circular-secure encryption from decision diffie-hellman. In: Wagner, D. (ed.) CRYPTO 2008. LNCS, vol. 5157, pp. 108–125. Springer, Heidelberg (2008)
32. Wang, P., Feng, D., Wu, W.: HCTR: a variable-input-length enciphering mode. In: Feng, D., Lin, D., Yung, M. (eds.) CISC 2005. LNCS, vol. 3822, pp. 175–188. Springer, Heidelberg (2005)

Fair and Robust Multi-party Computation Using a Global Transaction Ledger

Aggelos Kiayias[1], Hong-Sheng Zhou[2(✉)], and Vassilis Zikas[3]

[1] National and Kapodistrian University of Athens, Athens, Greece
aggelos@di.uoa.gr
[2] Virginia Commonwealth University, Richmond, USA
hszhou@vcu.edu
[3] Rensselaer Polytechnic Institute, Troy, USA
vzikas@cs.rpi.edu

Abstract. Classical results on secure multi-party computation (MPC) imply that fully secure computation, including fairness (either all parties get output or none) and robustness (output delivery is guaranteed), is impossible unless a majority of the parties is honest. Recently, cryptocurrencies like Bitcoin where utilized to leverage the fairness loss in MPC against a dishonest majority. The idea is that when the protocol aborts in an unfair manner (i.e., after the adversary receives output) then honest parties get compensated by the adversarially controlled parties.

Our contribution is three-fold. First, we put forth a new formal model of secure MPC with compensation and show how the introduction of suitable ledger and synchronization functionalities makes it possible to describe such protocols using standard interactive Turing machines (ITM) circumventing the need for the use of extra features that are outside the standard model as in previous works. Second, our model, is expressed in the universal composition setting with global setup and is equipped with a composition theorem that enables the design of protocols that compose safely with each other and within larger environments where other protocols with compensation take place; a composition theorem for MPC protocols with compensation was not known before. Third, we introduce the first robust MPC protocol with compensation, i.e., an MPC protocol where not only fairness is guaranteed (via compensation) but additionally the protocol is guaranteed to deliver output to the parties that get engaged and therefore the adversary, after an initial round of deposits, is not even able to mount a denial of service attack without having to suffer a monetary penalty. Importantly, our robust MPC protocol requires only a *constant* number of (coin-transfer and communication) rounds.

1 Introduction

Secure multiparty computation (MPC) enables a set of parties to evaluate the output of a known function $f(\cdot)$ on inputs they privately contribute to the protocol execution. The design of secure MPC protocols, initiated with the seminal

© International Association for Cryptologic Research 2016
M. Fischlin and J.-S. Coron (Eds.): EUROCRYPT 2016, Part II, LNCS 9666, pp. 705–734, 2016.
DOI: 10.1007/978-3-662-49896-5_25

works of Yao [31] and Goldreich et al. [21] has evolved to a major effort in computer security engineering. Beyond privacy, a secure MPC protocol is highly desirable to be *fair* (either all parties learn the output or none) and *robust* (the delivery of the output is guaranteed and the adversary cannot mount a "denial of service" against the protocol). Achieving fairness and robustness in a setting where there is an arbitrary number of corruptions, as desirable as it may appear, is prohibited by strong impossibility results stemming from the work of Cleve [16] who showed that coin-flipping is infeasible in any setting where there is no honest majority among parties that execute the protocol. These impossibility results, combined with the importance of the properties that they prevent, strongly motivate the exploration of alternate – yet still realistic – models that would enable fair and robust MPC protocols.

With the advent of Bitcoin [28] and other decentralized cryptocurrencies, the works of [1,2,7,27] showed a new direction for circumvention of the impossibility results regarding the fairness property: enforcing fairness could be achieved through imposing monetary penalties. In this setting a breach of fairness by the adversary is still possible but it results in the honest parties collecting a compensation in a way that is determined by the protocol execution. At the same time, in case fairness is not breached, it is guaranteed that no party loses any money (despite the fact that currency transfers may have taken place between the parties). The rationale here is that a suitable monetary penalty suffices in most practical scenarios to force the adversary to operate in the protocol fairly.

While the main idea of fairness with penalties sounds simple enough, its implementation proves to be quite challenging. The main reason is that the way a crypto-currency operates does not readily provide a trusted party that will collect money from all participants and then either return it or redistribute it according to the pre-agreed penalty structure. This is because crypto-currencies are decentralized and hence no single party is ever in control of a money transfer beyond the owner of a set of coins. The mechanism used in [1,2,7,27] to circumvent the above problem is the capability[1] of the Bitcoin network to issue transactions that are "time-locked", i.e., become valid only after a specific time and prior to that time may be superseded by other transactions that are posted in the public ledger. Superseded time-locked transactions become invalid and remain in the ledger without ever being redeemed.

While the above works are an important step for the design of MPC protocols with properties that circumvent the classical impossibility results, several critical open questions remain to be tackled; those we address herein are as follows.

Our Results. Our contribution is three-fold. First, we put forth a new formal model of secure MPC with compensation and we show how the introduction of suitable ledger and synchronization functionalities makes it possible to express completely such protocols using standard interactive Turing machines (ITM) circumventing the need for the use of extra features that are outside the standard model (in comparison, the only previous model [7] resorted to specialized ITM's

---

[1] Note that this feature is currently not fully supported.

that utilize resources outside the computational model[2]). Second, our model is equipped with a composition theorem that enables the design of protocols that compose safely with each other and within larger environments where other protocols with compensation take place; a composition theorem for this class of protocols was not known before and requires a new framework for synchronization in the global UC setting that can be of independent interest. Third, we introduce the first robust MPC protocol with compensation, i.e., an MPC protocol where not only fairness is guaranteed (via compensation) but additionally the protocol is guaranteed to deliver output to the parties that get engaged and therefore the adversary is not even able to mount a denial of service attack without having to suffer a monetary penalty. In more details we have the following.

- We put forth a new model that utilizes two ideal functionalities and expresses the ledger of transactions and a clock in the sense of [24] that is connected to the ledger and enables parties to synchronize their protocol interactions. Our ledger functionality enable us to abstract all the necessary features of the underlying cryptocurrency. Contrary to the only previous formalization approach [7,27], our modeling allows the entities that participate in an MPC execution to be regular interactive Turing machines (ITM) and there is no need to equip them with additional physical features such as "safes" and "locks." Furthermore the explicit inclusion of the clock functionality (which is only alluded to in [7,27]) and a synchronous framework for protocol design given such clock reveal the exact dependencies between the ledger and the clock functionality that are necessary in order for MPC with compensation protocols to be properly described. We express our model within a general framework that we call Q-fairness and robustness and may be of independent interest as it can express meaningful relaxations of fairness and robustness in the presence of a global ideal functionality.
- We prove a composition theorem that establishes that protocols in our framework are secure in a universally composable fashion. Our composition proof treats the clock and ledger functionalities as global setups in the sense of [11,13]. We emphasize that this is a critical design choice: the fact that the ledger is a global functionality ensures that any penalties that are incurred to the adversary that result to credits towards the honest parties will be globally recognized. This should be contrasted to an approach that utilizes regular ideal functionalities which may be only accessible within the scope of a single protocol instance and hence any penalty bookkeeping they account may vanish with the completion of the protocol. Providing a composition theorem for MPC protocols with compensation was left as an open question in [7].
- We finally present a new protocol for fair and robust secure MPC with compensation. Our robustness property guarantees that once the protocol passes

---

[2] An ITM with the special features of "wallet" and "safe" was introduced in [7] to express the ability of ITM's to store and transfer "coins." Such coins were treated as physical quantities that were moved between players but also locked in safes in a way that parties were then prevented to use them in certain ways (in other words such safes were not local but were affected from external events).

an initial round of deposits, parties are guaranteed to obtain output or be compensated. This is in contrast to fair MPC with compensation [1,2,7,27] where the guarantee is that compensation takes place only in case the adversary obtains output while an honest party does not. To put it differently, it is feasible for the adversary to lead the protocol to a deadlock where no party receives output however the honest parties have wasted resources by introducing transactions in the ledger. We remark that it is in principle possible to upgrade the protocols of [1,2,7,27] to the robust MPC setting by having them perform an MPC with identifiable abort, cf. [21,23], (in such protocol the party that causes the abort can be identified and excluded from future executions). However even using such protocol the resulting robust MPC with compensation will need in the worst case a *linear* number of deposit/communication rounds in the number of malicious parties. Contrary to that, our robust protocol can be instantiated so that it requires a constant number of deposit/communication rounds independently of the number of parties that are running the protocol. Our construction uses time-locked transactions in a novel way to ensure that parties do progress in the MPC protocol or otherwise transactions are suitably revertible to a compensation for the remaining parties. The structure of our transactions is quite more complex than what can be presently supported by bitcoin; we describe in high-level how our protocol can be implemented via Ethereum[3] contracts.

Related Work. In addition to the previous works [1,2,7,27] in fair MPC with compensation, very recently, Ruffing et al. [30] address equivocation issues via penalty mechanism, and design decentralized "non-equivocation" contracts.

There are a number of other works that attempted to circumvent the impossibility results for fairness in the setting of dishonest majority by considering alternate models. Contrary to the approach based on cryptocurrencies these works give an advantage to the protocol designer with respect to the adversarial strategy for corruption. For instance, in [18] a rational adversary is proposed and the protocol designer is privy to the utility function of the adversary. In [3] a reputation system is used and the protocol designer has the availability of the reputation information of the parties that will be engaged in the protocol. Finally in [17] a two tiered model is proposed where the protocol designer is capable of distinguishing two distinct sets of servers at the onset of the computation that differ in terms of their corruptibility.

Global setups were first put forth in [11] motivated by notion of deniability in cryptographic protocols. In our work we utilize global functionalities for universal composition (without the deniability aspect) as in [13] where a similar approach was taken for the case of the use of the random oracle as a global setup functionality for MPC.

Fairness was considered from the resource perspective, cf. [8,19,29], where it is guaranteed due to the investment of proportional resources between the parties running the protocol, and the optimistic perspective, cf. [4,5,9], where

---

[3] http://www.ethereum.org.

a trusted mediator can be invoked in the case of an abort. We finally note that without any additional assumptions, due to the impossibility results mentioned above, one can provide fairness only with certain high probability that will be affecting the complexity of the resulting protocol, see, e.g., [22] and references therein.

In concurrent and independent work, Kosba et al. [26] propose a framework for composable protocols based on a ledger. and explore a notion of fairness with compensation. Our work goes beyond fairness and provides a treatment of robustness. Furthermore we provide a synchronous framework with a global clock (of independent interest) that uses the ledger as a *global* setup to achieve fairness and robustness and we prove a composition theorem for our framework.

Organization. We start with preliminaries in Sect. 2. Then in Sects. 3 and 4, we lay down a formal framework for designing composable fair protocols in the presence of globally available trusted resources. In Sect. 3, we introduce two shared functionalities $\bar{\mathcal{G}}_{\text{CLOCK}}$ and $\bar{\mathcal{G}}_{\text{LEDGER}}$ respectively to formulate the trust resources that are provided by Bitcoin-like systems. Subsequently, in Sect. 4, we put forth a new formal framework for secure MPC with compensation: we introduce the notions of Q-fairness, and Q-robustness via wrapper functionalities; we then consider the realization of such wrapper functionalities, and further provide a composition theorem. In Sect. 5, we present a protocol in our new framework to achieve our new notions of fairness and robustness. We refer the reader to the full version of our work [25] for a discussion about implementing our protocol within Ethereum, supplementary material for Sects. 2 and 3, and for the formal proofs of our theorems.

2   Preliminaries

Throughout the paper we assume an (often implicit) security parameter denoted as κ. For a number $n \in \mathbb{N}$ we denote by $[n]$ the set $[n] = \{1, \ldots, n\}$ and denote by 0^n (resp. 1^n) the all-zero (resp. all-ones) string of length n. For a randomized algorithm Alg we denote by $\text{Alg}(x; r)$ the output of Alg on input x and random coins r. To avoid always explicitly writing the coins r, we shall denote by $y \xleftarrow{\$} \text{Alg}(x)$ the operation of running Alg on input x (and uniformly random coins) and storing the output on variable y. We write $f : X \xrightarrow{\$} Y$ to denote a probabilistic function with domain X and range Y. We use the standard definition of *negligible* and *overwhelming* (e.g., see [20]).

For a multiparty function $f : (\{0,1\}^* \cup \{\lambda\})^n \to (\{0,1\}^* \cup \{\bot\})^n$ for parties in $\mathcal{P} = \{p_1, \ldots, p_n\}$ and for a set $\mathcal{P} \subseteq \mathcal{P}$, we denote by $f|_{|\mathcal{P}'|}$ the restriction of f to the parties in $\mathcal{P}'$, namely, if each $p_i \in \mathcal{P}'$ has input x_i, then the output of $f|_{|\mathcal{P}'|}$ is the output of f evaluated on inputs x_i for each $p_i \in \mathcal{P}'$ and $x_j = \lambda$ for each $p_j \in \mathcal{P} \backslash \mathcal{P}'$.

We describe our results in the extension of Canetti's UC framework [10] to allow for global setups, known as GUC [11]. As argued above, this is the natural model to consider execution in the present of a globally synchronized clock and a ledger/bulletin board. Consistently with the (G)UC notation, we denote local (UC) functionalities by calligraphic letters, as in $\mathcal{F}$, and add a bar to denote global functionalities, as in $\bar{\mathcal{G}}$. Furthermore, we denote by ϕ, the dummy protocol. Note that in GUC ϕ might receive inputs for its (UC) hybrids and/or for the global setup, where an implicit mechanism is assumed to allow the environment to define the intended recipient of each submitted input to ϕ. For a protocol π, a (local) UC functionality $\mathcal{F}$ and a global setup $\bar{\mathcal{G}}$ we denote by $\text{EXEC}_{\pi,\mathcal{A},\mathcal{Z}}^{\bar{\mathcal{G}},\mathcal{F}}$ the output of the environment $\mathcal{Z}$ in an execution of π having hybrid access to $\bar{\mathcal{G}}$ and $\mathcal{F}$ in the presence of adversary $\mathcal{A}$. We assume some familiarity with the UC and/or the GUC framework.

Correlated Randomness as a Sampling Functionality. Our protocols are in the *correlated randomness* model, i.e., they assume that the parties initially, before receiving their inputs, receive appropriately correlated random strings. In particular, the parties jointly hold a vector $\boldsymbol{R} = (R_1, \ldots, R_n) \in (\{0,1\}^*)^n$, where P_i holds R_i, drawn from a given efficiently samplable distribution $\mathcal{D}$. This is, as usual, captured by giving the parties initial access to an ideal functionality $\mathcal{F}_{\text{CORR}}^{\mathcal{D}}$, known as a *sampling functionality*, which, upon receiving a default input from any party, samples $\boldsymbol{R}$ from $\mathcal{D}$ and distributes it to the parties (see [25] for details). Hence, a protocol in the correlated randomness model is formally an $\mathcal{F}_{\text{CORR}}^{\mathcal{D}}$-hybrid protocol. Formally, a sampling functionality $\mathcal{F}_{\text{CORR}}^{\mathcal{D}}$ is parameterized by an efficiently computable sampling distribution $\mathcal{D}$ and the (ID's of the parties in) the player set $\mathcal{P}$.

3 Model

In this section and next section, we lay down a formal framework for designing composable fair protocols in the presence of globally available trusted resources. We introduce in the current section, shared (in the sense of the GUC model [11]) functionalities $\bar{\mathcal{G}}_{\text{CLOCK}}$ and $\bar{\mathcal{G}}_{\text{LEDGER}}$ respectively to formulate the trust resources that are provided by Bitcoin-like systems. We stress that these two functionalities can be thought of as a single global functionality and in our description are allowed to communicate. Nonetheless, we choose to describe then as two separate functionalities, because as we argue, the clock $\bar{\mathcal{G}}_{\text{CLOCK}}$ can also be used alone (without $\bar{\mathcal{G}}_{\text{LEDGER}}$) to naturally model synchronous computation with a global notion of time.

3.1   Global Clock Functionality and Synchronous Protocol Executions

In this section we describe how to model execution of synchronous protocols that can access a global-clock setup. This is an adaptation of the original idea by Katz et al. [24], where a clock was modelled as UC functionality that is local to the calling protocol, and is of independent interest as a model for the design of synchronous protocols. In addition to being a more realistic model for capturing time in UC, the notion of the global clock allows for synchronous execution of any protocols that choose to use it.

Before defining our clock, we recall the reader the clock and model of synchronous execution from [24] and then highlight the main differences. The clock in [24] is a UC functionality that keeps an indicator bit b originally set to 0. The parties can send to the clock special "update" messages, and once the clock sees that all honest parties agree to update the state it sets $b := b \oplus 1$. The clock then continues to receive "update" messages, and again, once it sees that all honest parties have requested to update after the last switch of the bit b it switches it again. To make sure that the adversary is given enough activations, whenever the clock receives an "update" message from the honest party it notifies the adversary. In addition to "update" messages, the parties can send the clock a "read" message which the clock replies with the current value of b.

The use of such a clock to keep a round structure is as follows: Whenever a party observes a switch of the bit b, it interprets it as a round advance. Thus, a synchronous protocol with access to such a clock is executed as follows. In each round, every party performs all its protocol instructions for the current round, and at the end sends an "update" message to the clock; from the point where the party updates (its round has finished) it queries ("reads") the clock with each following activation to detect when all parties have also finished their rounds (i.e., when the value of b switches). Once this happens, the party starts its next protocol round.

An issue with the above clock is that in order to execute two protocols using the same clock we need to make use of the joint-state UC theorem [15]. Instead, in this work we take an alternative modelling approach and define a shared clock functionality $\bar{\mathcal{G}}_{\text{CLOCK}}$. This functionality can be viewed as a shared version of the clock functionality which was defined by Katz et al. [24]. The main intuition behind our clock functionality is that all honest parties can use it to ensure that they proceed with their rounds at the same pace. On a high level, the clock operates as follows: any party that wishes to be synchronised with the global clock can send (REGISTER, sid) to the clock and subsequently it can send it (CLOCK-UPDATE, sid) commands, where sid is $\bar{\mathcal{G}}_{\text{CLOCK}}$'s identifier. The clock stores a global-time counter τ (initially set to 0), and as soon it is instructed by all currently honest parties and by associated shared functionalities[4] to advance the time (i.e., receives (CLOCK-UPDATE, sid) it increases its state-counter τ by 1.

The main difference between our formulation and that by Katz et al. [24] is that in [24] the clock is a UC functionality which is local to a single protocol and waits for an "update" message by every honest party to advance its state; however, here we intend to have the clock to be accessed *globally* and used by arbitrary protocols. Therefore we give the power to the environment to define the clock's speed. Indeed, if there are no associated shared functionalities, the environment can instruct dummy parties to send inputs (CLOCK-UPDATE, sid) to $\bar{\mathcal{G}}_{\text{CLOCK}}$ and advance the clock whenever it wishes. An additional difference is

---

[4] Certain global functionalities, such as the ledger defined in the following section, might depend on time and, therefore, need to be synchronized with the clock.

that in [24], the clock state is binary while here, in our formulation, the state τ is a positive integer which indicates the time that has passed from point zero (i.e., from the beginning of time).

Next, we elaborate and explain how to use the global clock to design synchronous protocols. We remark that the model of synchronous protocol execution of [24] cannot be used in our setting as the environment can make the clock advance before honest parties have time to take actions in any round. Indeed, in the ideal setting the environment can keep sending (CLOCK-UPDATE, sid) to the dummy parties, which will forward it to the clock making its state to advance; to make sure that the protocol is indistinguishable, honest parties would have to do the same, thereby giving away the activations that they need for executing their protocol instructions such as send and receive operations.[5] This might, at first, seem like a bug but it is in fact a feature. It captures the fact that since time is a quantity that should be in the control of the environment, if the environment chooses to advance time too fast then some protocol might not have enough time to perform their operations for each round, and might therefore need to give up.

To make sure that the environment cannot exploit such fast-forwarding of the clock we use the following idea: We allow the clock to receive from honest parties or (non-shared) ideal functionalities a special (CLOCK-FAST) message, which makes it set an internal indicator from 0 to 1. This indicator will be added onto the response of the clock to CLOCK-READ queries, and will make any synchronous protocol or corresponding functionality that reads the clock and observes this indicator being set to one to immediately terminate with a default value. This way we ensure that an environment that tries such a fast-forward distinguishing attack will be forced to make any synchronous protocol behave in a default way, a behavior which, as we see, is easily imitated in the ideal world. The detailed description of the clock functionality can be found in Fig. 1.

We stress that having a global $\bar{\mathcal{G}}_{\text{CLOCK}}$-hybrid model makes the mode of execution of synchronous protocols more intuitive compared to [24]. Here is how synchronous protocols are executed in this setting. First, as is the case in real-life synchronous protocols, we assume that the protocol participants have agreed on the starting time τ_0 of their protocol and also on the duration of each round.[6] We abstract this knowledge by assuming the parties know a function Round2Time : $\mathbb{Z} \to \mathbb{Z}$ which maps protocol rounds to time (according to the global clock) in which the round should be completed. For $\rho \in \mathbb{Z}$, Round2Time(ρ) is the time in which the ρth round of the protocol should be completed. To make sure that no party proceeds to round $\rho + 1$ of the protocol before all honest parties have completed round ρ, we require that any two protocol rounds are at least two clock-ticks apart (see [24] for a discussion); formally, for all $\rho \geq 0$, it holds that Round2Time$(\rho + 1) \geq$ Round2Time$(\rho) + 2$.

---

[5] The communication channels we are using are fetch-type bounded delivery channels as in [24]. In such channels, the receiver needs to issue "fetch"-requests which are answered only if a message is ready for delivery. We refer to [24] for details.

[6] Different protocols might proceed at a different pace.

Functionality $\bar{\mathcal{G}}_{\text{CLOCK}}$

Shared functionality $\bar{\mathcal{G}}_{\text{CLOCK}}$ is globally available to all participants. The shared functionality is parameterized with variables τ, a bit $d_{\bar{\mathcal{G}}_{\text{LEDGER}}}$ a set $\mathcal{P}'$ and a bit **fast** and is associated with a ledger shared functionality $\bar{\mathcal{G}}_{\text{LEDGER}}$.

Initially, $\tau := 0$, $d_{\bar{\mathcal{G}}_{\text{LEDGER}}} := 0$, **fast** $:= 0$ and $\mathcal{P}' := \emptyset$.

- Upon receiving (REGISTER, sid) from some party P, set $\mathcal{P}' := \mathcal{P}' \cup \{P\}$ and if P was not registered before, set $d_P := 0$; subsequently, forward (REGISTER, sid, P) to $\mathcal{A}$.
- Upon receiving (CLOCK-UPDATE, sid) from $\bar{\mathcal{G}}_{\text{LEDGER}}$ set $d_{\bar{\mathcal{G}}_{\text{LEDGER}}} := 1$ and forward (CLOCK-UPDATE, sid, $\bar{\mathcal{G}}_{\text{LEDGER}}$) to $\mathcal{A}$
- Upon receiving (CLOCK-UPDATE, sid) from some honest party $P \in \mathcal{P}'$ set $d_i := 1$; then if $d_{\bar{\mathcal{G}}_{\text{LEDGER}}} := 1$ and $d_P = 1$ for all honest parties in $\mathcal{P}'$, then set $\tau := \tau + 1$ and reset $d_{\bar{\mathcal{G}}_{\text{LEDGER}}} := 0$ and $d_P := 0$ for all parties in $\mathcal{P}'$. Forward (CLOCK-UPDATE, sid, P) to $\mathcal{A}$.
- Upon receiving (CLOCK-READ, sid) from any participant (including the environment, the adversary, or any ideal—shared or local—functionality) return (CLOCK-READ, sid, τ, **fast**) to the requestor.
- Upon receiving (CLOCK-FAST) from any honest party or ideal functionality, set **fast** $:= 1$.

Fig. 1. The clock functionality

A synchronous protocol in the above setting proceeds as follows where the parties keep locally track of the current round ρ in the protocol they are in:

- Upon receiving a (CLOCK-UPDATE, sid) input (from its environment) where sid is the ID of $\bar{\mathcal{G}}_{\text{CLOCK}}$, party P_i forwards it to $\bar{\mathcal{G}}_{\text{CLOCK}}$.
- Upon receiving a (CLOCK-READ, sid) input (from its environment), party P_i forwards it to $\bar{\mathcal{G}}_{\text{CLOCK}}$ and outputs the response to the environment.
- Upon receiving a (CLOCK-FAST) input (from its environment), party P_i forwards it to $\bar{\mathcal{G}}_{\text{CLOCK}}$.
- Upon receiving any message (INPUT, sid') where sid' is the session ID of a protocol P_i is involved in, do the following: Send (CLOCK-READ, sid) to $\bar{\mathcal{G}}_{\text{CLOCK}}$ and denote the response by (CLOCK-READ, sid, τ, **fast**); if **fast** $= 1$ then output CLOCK-FAST to the environment. Otherwise do:
  - if $\tau \leq \text{Round2Time}(\rho - 1)$ halt;
  - else, if $\text{Round2Time}(\rho-1) < \tau \leq \text{Round2Time}(\rho)$ execute the next pending round$-\rho$ instruction (if all the instructions for round ρ are finished halt.)
  - else, if $\tau > \text{Round2Time}(\rho)$ and there are still pending instructions for the current round, send (CLOCK-FAST) to $\bar{\mathcal{G}}_{\text{CLOCK}}$.
  - else, i.e., if $\tau > \text{Round2Time}(\rho)$ and P_i has completed all round-ρ instruction, then set $\rho := \rho + 1$ and halt.

It is easy to verify that the above mode of operation will guarantee that the parties are never out-of-sync, since as soon as the first party issues a CLOCK-FAST message for the clock, all synchronous protocols will enter the mode of outputting CLOCK-FAST for every input that the environment hands them (that is not intended for the clock). However, there is one more thing that needs to be taken care of. Since in the real-world the parties go to a default mode (where they output CLOCK-FAST to every query) when the environment does not give them sufficient time, this should also be the case in the ideal world. To achieve this we use another idea inspired by the guaranteed termination functionality from [24]: Let π be a synchronous protocol with round-to-time function Round2Time : $\mathbb{Z} \to \mathbb{Z}$, where in each round, each party needs exactly m activations to perform its instructions[7]. We introduce a wrapper $\tilde{\mathcal{W}}$ which, at a high level, forwards messages to and from its wrapped functionality but stores a round-index and checks, as the protocol would, that every party issues to the wrapped functionality, at least m activations for each round ρ in the intended interval. If this is not the case the wrapper sends (CLOCK-FAST) to $\bar{\mathcal{G}}_{\mathrm{CLOCK}}$ and responds with CLOCK-FAST form that point on. The detailed description can be found in the full version [25].

3.2 Global Ledger Functionality

Functionality $\bar{\mathcal{G}}_{\mathrm{LEDGER}}$ provides the abstraction of a public ledger in Bitcoin-like systems (e.g., Bitcoin, Litecoin, Namecoin, Ethereum, etc.). Intuitively, the public ledger could be accessed globally by protocol parties or other entities including the environment $\mathcal{Z}$. Protocol parties or the environment can generate transactions; and these valid transactions will be gathered by a set of ledger maintainers (e.g., miners in Bitcoin-like systems) in a certain order as the state of the ledger. More concretely, whenever the ledger maintainers receive a vector of transactions $\mathtt{tx}$, they first add the transactions in a buffer, assuming they are valid with respect to the existing transactions and the state of the ledger; thus, in this way a vector of transactions is formed in the buffer. After a certain amount of time, denoted by $\mathtt{T}$, which will be also referred to as a *ledger round*, all transactions in the buffer will be "glued" into the ledger state in the form of a block. The adversary is allowed to permute the buffer prior to its addition to the ledger. In Bitcoin, $\mathtt{T}$ is 10 min (approximately); thus in about every 10 min, a new block of transactions will be included into the ledger, and the ledger state will be updated correspondingly.

To enable the ledger to be aware of time, the ledger maintainers are allowed to "read" the state of another publicly available functionality $\bar{\mathcal{G}}_{\mathrm{CLOCK}}$ defined above. Furthermore, to ensure that the ledger is activated at least once in each time-tick[8] (i.e., each advance of the $\bar{\mathcal{G}}_{\mathrm{CLOCK}}$ state) we have the ledger, with every message it gets from a party other than the adversary, send a (CLOCK-UPDATE, sid) message to $\bar{\mathcal{G}}_{\mathrm{CLOCK}}$. (Recall that, as defined, $\bar{\mathcal{G}}_{\mathrm{CLOCK}}$ always waits for at least one such message from the ledger before advancing its time counter.)

---

[7] One can make any synchronous protocol have this form by introducing dummy instructions.

[8] This is essential to ensure that updates are done in a time-consistent manner.

Functionality $\bar{\mathcal{G}}_{\text{LEDGER}}$

Shared functionality $\bar{\mathcal{G}}_{\text{LEDGER}}$ is globally available to all participants. The shared functionality is parameterized with a predicate Validate, a constant T, and variables state, buffer and counter.

Initially, state $:= \varepsilon$, buffer $:= \varepsilon$, and counter $:= 0$.

- Upon receiving (SUBMIT, sid, tx) from some participant, If Validate(state, (buffer, tx)) $= 1$, then set buffer $:=$ buffer$||$tx. Go to *State Extend*.
- Upon receiving (READ, sid) from a party P or $\mathcal{A}$, if P is honest set $b =$ state else set $b =$ (state, buffer).
  1. Execute *State Extend*.
  2. Return (READ, sid, b) to the requestor.
- Upon receiving (PERMUTE, sid, π) from $\mathcal{A}$ apply permutation π on the elements of buffer.

State Extend: Send (CLOCK-READ, sid) to $\bar{\mathcal{G}}_{\text{CLOCK}}$ and receive (CLOCK-READ, sid, τ) from $\bar{\mathcal{G}}_{\text{CLOCK}}$. If $|\tau - \text{T} \cdot \text{counter}| > \text{T}$, then set state $:=$ state$||$Blockify(τ, buffer) and buffer $:= \varepsilon$ and counter $:=$ counter $+ 1$. Subsequently, send (CLOCK-UPDATE, sid) to $\bar{\mathcal{G}}_{\text{CLOCK}}$ where sid is the ID of $\bar{\mathcal{G}}_{\text{CLOCK}}$.

Fig. 2. The public ledger functionality.

We remark that all gathered transactions should be "valid" which is defined by a predicate Validate. In different systems, predicate Validate will take different forms. For example, in the Bitcoin system, the predicate Validate should make sure that for each newly received transaction that transfers v coins from the original wallet address address$_o$ to the destination wallet address address$_d$, the original wallet address address$_o$ should have v or more than v coins, and the transaction should be generated by the original wallet holder (as shown by the issuance of a digital signature). Furthermore, prior to each vector of transactions becoming block, the vector is passed through a function Blockify($\cdot$) that homogenizes the sequence of transactions in the form of a block. Moreover, in some systems like Bitcoin, it may add a special transaction called a "coinbase" transaction that implements a reward mechanism for the ledger maintainers.

In Fig. 2 we provide the details of the ledger functionality.

4   Q-Fairness and Q-Robustness

In this section, we provide a formal framework for secure computation with fair and robust compensation. In the spirit of [19], our main tool is a wrapper functionality. Our wrapper functionality is equipped with a predicate $Q_{\bar{\mathcal{G}}}$ which is used to make sure that the outcome of the protocol execution is consistent with appropriate conditions on the state of the global setup $\bar{\mathcal{G}}$. Intuitively, the predicate $Q_{\bar{\mathcal{G}}}$ works as a filter, such that if certain "bad" event occurs (e.g., an abort), then the wrapped functionality will restrict the simulators influence. More concretely, the predicate $Q_{\bar{\mathcal{G}}}$ has three modes $Q_{\bar{\mathcal{G}}}^{\text{Init}}$, $Q_{\bar{\mathcal{G}}}^{\text{Dlv}}$ and, $Q_{\bar{\mathcal{G}}}^{\text{Abt}}$, where $Q_{\bar{\mathcal{G}}}^{\text{Init}}$ specifies under which condition (on the global setup's state) the protocol should start executing; $Q_{\bar{\mathcal{G}}}^{\text{Dlv}}$ specifies under which condition parties should receive their output; and $Q_{\bar{\mathcal{G}}}^{\text{Abt}}$ specifies under which condition the simulator is allowed to force parties to abort. With foresight $Q_{\bar{\mathcal{G}}}^{\text{Init}}$ will ensure that the protocol is executed only if all honest participants have enough coins; $Q_{\bar{\mathcal{G}}}^{\text{Dlv}}$ will ensure that honest parties do not lose coins if they execute the protocol; and $Q_{\bar{\mathcal{G}}}^{\text{Abt}}$ will ensure that honest parties might be forced to an "unfair" abort (i.e., where the adversary has received his output) only if they are compensated by earning coins (from the corrupted parties). We will call an implementation of a wrapped version of $\mathcal{F}$ a Q-fair implementation of $\mathcal{F}$.[9]

Our definition of $Q_{\bar{\mathcal{G}}}$-fairness can be instantiated with respect to any global setup that upon receiving a READ symbol (from any protocol participant or functionality) it returns its public state **trans**. Concretely, let $\bar{\mathcal{G}}$ be global ideal functionality and let $Q_{\bar{\mathcal{G}}}$ a predicate, as above, with respect to such $\bar{\mathcal{G}}$. Let also $\mathcal{F}$ be a non-reactive functionality[10] which allows for fair evaluation of a given function (SFE) in the sense of [19], i.e., it has two modes of delivering output: (i) delayed delivery: (DELIVER, sid, m, P) signifying delayed output delivery[11] of m to party P, (ii) fair delivery: (FAIR-DELIVER, sid, $(m, P_{i_1}), \ldots, (m, P_{i_k}), (m_{\mathcal{S}}, \mathcal{S})$) that results in simultaneous delivery of outputs $m_{i_1}, \ldots m_{i_k}$ to parties $P_{i_1}, \ldots, P_{i_k}$ and output $m_{\mathcal{S}}$ to $\mathcal{S}$. We note that (G)UC does not have an explicit mechanism for simultaneous delivery of outputs. Thus, when we refer to simultaneous delivery of a vector $(m_{i_1}, \ldots, m_{i_k})$ to parties $P_{i_1}, \ldots, P_{i_k}$, respectively, we imply that the functionality prepares all the output to be delivered in a "fetch mode" as defined in [24]; that is:

- The functionality registers the pairs $(m_{i_1}, P_{i_1}), \ldots, (m, P_{i_k})$ as "ready to fetch" and sends the set $\{(m_{i_j}, P_{i_j}) | P_{i_j}$ is corrupted$\}$ to $\mathcal{S}$.
- Upon receiving an input (FETCH-OUTPUT, P_i) from party P_i, if a message (m_i, P_i) has been registered as "ready to fetch" then remove it from the "ready to fetch" set and output it to P_i (if more than one such messages are registered, deliver and remove from the "ready to fetch" set the first, chronologically, registered such pair); otherwise send (FETCH-OUTPUT, P_i) to $\mathcal{S}$.

4.1 $Q_{\bar{\mathcal{G}}}$-Fairness

The wrapper functionality $\mathcal{W}$ that will be used in the definition of Q-fair (secure) computation is given in Fig. 3. The intuition is as follows: Prior to handing

---

[9] We note that whenever it is clear from the context we may drop the subscript $\bar{\mathcal{G}}$ in $Q_{\bar{\mathcal{G}}}, Q_{\bar{\mathcal{G}}}^{\text{Init}}, Q_{\bar{\mathcal{G}}}^{\text{Dlv}}, Q_{\bar{\mathcal{G}}}^{\text{Abt}}$.

[10] A non-reactive functionality does not accept any input from honest parties after generating output.

[11] Delayed output delivery is a standard (G)UC mechanism where the adversary is allowed to schedule the output at a time of its choosing.

inputs to the (wrapped) functionality $\mathcal{F}$, the parties can request the wrapper to generate on their behalf a *resource-setup* (by executing an associated resource-setup generation algorithm Gen) which allows them to update the global setup $\bar{\mathcal{G}}$; this resource setup consists of a public component $RS^{\text{pub}}_{P,\text{sid}}$ and a private component $RS^{\text{priv}}_{P,sid}$.[12] Both these values are given to the simulator, and the public component is handed to the party.

From the point when parties receive their inputs the Q predicate is used as a filter to specify the wrapper's behavior and add the fairness guarantees. More concretely, upon receiving an input from a party, the wrapper checks on the global setup to ensure that Q^{Init} is true, and if it is not true it aborts (i.e., sets all honest parties' outputs to $\bot$ and blocks any communication between $\mathcal{F}$ and the adversary). This means that if the environment has not set up the experiment properly,[13] then the experiment will not be executed and the wrapped functionality will become useless. This formally resolves the question "What happens if some party does not have sufficient coins to play the protocol?" which leads to some ambiguity in existing bitcoin-based definitions of computation with fair compensation [7].

The predicates Q^{Dlv} and Q^{Abt} are used to filter out attempts of the simulator to deliver outputs or abort when Q^{Dlv} and Q^{Abt} are violated.[14] Concretely, any such attempt will be ignored if the corresponding predicate is not satisfied.

Intuitively, by requiring the protocol to implement such a wrapped version of a functionality, we will ensure that the parties might only abort if Q^{Abt} is true, and might output a valid (non-$\bot$) value if Q^{Dlv}. As we shall see in Sect. 4.2, by a trivial modification of the fairness wrapper, we can capture a stronger property which we will call Q-robustness; the latter, roughly, guarantees that honest parties which start the protocol will either receive their output (and Q^{Dlv} being true) or will abort and increase their revenue. (I.e., there is no way for the adversary to make the protocol abort after the first honest party has sent its first input-dependent message).

Definition 1. *We say protocol π realizes functionality $\mathcal{F}$ with $Q_{\bar{\mathcal{G}}}$-fairness with respect to global functionality $\bar{\mathcal{G}}$, provided the following statement is true. For all adversaries $\mathcal{A}$, there is a simulator $\mathcal{S}$ so that for all environments $\mathcal{Z}$ it holds:*

$$\text{EXEC}^{\bar{\mathcal{G}}}_{\pi,\mathcal{A},\mathcal{Z}} \approx \text{EXEC}^{\bar{\mathcal{G}},\mathcal{W}_{Q,\bar{\mathcal{G}}}(\mathcal{F})}_{\mathcal{S},\mathcal{Z}}$$

More generally, the protocol σ realizes $\mathcal{H}$ with $Q'_{\bar{\mathcal{G}}}$ fairness using a functionality $\mathcal{F}$ with fairness $Q_{\bar{\mathcal{G}}}$ provided that for all adversaries $\mathcal{A}$, there is a simulator

---

[12] In the case of bitcoin-like ledgers these will correspond to a wallet (public-key) and a corresponding secret key.

[13] In the case of a bitcoin-ledger this corresponds to the environment not transferring to some protocol-related wallet sufficient funds to execute the protocol.

[14] As we will see, in bitcoin-like instantiations, Q^{Dlv} will be satisfied when no honest party has a negative balance, and Q^{Abt} will be satisfied when every honest party has a (strictly) positive balance.

Wrapper Functionality $\mathcal{W}_{Q,\bar{\mathcal{G}}}(\mathcal{F})$

The wrapper $\mathcal{W}_{Q,\bar{\mathcal{G}}}(\mathcal{F})$ interacts with a set of parties $\mathcal{P} = \{P_1, \ldots, P_n\}$, the adversary $\mathcal{S}$ and the environment $\mathcal{Z}$, as well as shared functionality $\bar{\mathcal{G}}$. It is parameterized with a predicate $Q = (Q^{\text{Init}}, Q^{\text{Dlv}}, Q^{\text{Abt}})$ and a resource-setup generating algorithm $\text{Gen} : 1^* \xrightarrow{\$} (\{0,1\}^*)^2$ and wraps any given non-reactive n-party functionality $\mathcal{F}$ with the two output-delivery modes (delayed and fair) described in Section 4.1. The functionality also keeps an indicator bit b, initially set to 0, indicating whether or not $\mathcal{S}$ is blocked from sending messages to $\mathcal{F}$.

– *Allocating Resources.* Upon receiving (ALOCATE, sid) from a party P, if a message (ALOCATE, sid) has already been received for P then ignore it; else send (COINS, sid, P) to $\mathcal{S}$ and upon receiving (COINS, sid, P, r) from $\mathcal{S}$ compute $(RS^{\text{pub}}_{P,\text{sid}}, RS^{\text{priv}}_{P,\text{sid}}) \leftarrow \text{Gen}(1^\kappa; r)$ and sends a delayed output (DELIVER, sid, $RS^{\text{pub}}_{P,\text{sid}}, P$) to P.
– Upon receiving any message M from $\mathcal{F}$ to be delivered to its simulator, if $b = 0$ forward M to $\mathcal{S}$.
– Upon receiving a message (FORWARD, M) from $\mathcal{S}$, if $b = 0$ then forward M to $\mathcal{F}$ as a message coming from its simulator.
– *Receiving input for $\mathcal{F}$.* Upon receiving (INPUT, sid, x) from a party P, send READ to $\bar{\mathcal{G}}$, denote the response by trans and if $\neg Q^{\text{Init}}(RS^{\text{pub}}_{P,\text{sid}}, \text{trans})$ then set $b := 1$ and issue a message (FAIR-DELIVER, sid, $(\bot, P_1), \ldots, (\bot, P_n), (\bot, \mathcal{S})$) (i.e., simultaneously deliver $\bot$ to all parties and ignore all future messages except (FETCH-OUTPUT, ·) messages. Otherwise, forward (INPUT, sid, x) to $\mathcal{F}$ as input for P.
– *Generating delayed output.* Upon receiving a message from $\mathcal{F}$ marked (DELIVER, sid, m, P) forwards m to party P via delayed output.
– *Registering fair output.* Upon receiving a message from $\mathcal{F}$ that is marked for fair delivery (FAIR-DELIVER, sid, mid, $(m_1, P_{i_1}), \ldots, (m_k, P_{i_k}), (m_{\mathcal{S}}, \mathcal{S})$), it forwards (mid, $P_{i_1}, \ldots, P_{i_k}, m_{\mathcal{S}}$) to $\mathcal{S}$.
– Q-*fair delivery.* Upon receiving (Q-DELIVER, sid, mid) from $\mathcal{S}$ then provided that a message (mid, ...) has been delivered to $\mathcal{S}$ operate as follows. For each pair of the form (m, P) associated with mid: Let $L = \{(m, P) | P$ is uncorrupted$\}$. Send $\{(m, P) | P$ is corrupted$\}$ to $\mathcal{S}$. (If some currently honest P becomes corrupted later on, remove (m, P) from sending and send (m, P) to $\mathcal{S}$.) Subsequently perform the following.
  • On input a message (DELIVER, sid, mid, P) from $\mathcal{S}$, provided that the record mid contains the pair $(m, P) \in L$, send READ to $\bar{\mathcal{G}}$, denote the response by trans and if $\neg Q^{\text{Dlv}}(\text{sid}, P, RS^{\text{pub}}_{P,\text{sid}}, \text{trans})$ then ignore the message. Else, remove (m, P) from L and register (m, P) as "ready to fetch".
  • On input a message (ABORT, sid, mid, P) from $\mathcal{S}$, provided that the record mid contains the pair $(m, P) \in L$, send READ to $\bar{\mathcal{G}}$, denote the response by trans and if $\neg Q^{\text{Abt}}(\text{sid}, P, RS^{\text{pub}}_{P,\text{sid}}, \text{trans})$ then ignore the message. Else, remove (m, P) from L and register $(\bot, P)$ as "ready to fetch".
– Upon receiving an input (FETCH-OUTPUT, P) from party P, if a message (m, P) has been registered as "ready to fetch" then remove it from the "ready to fetch" set and output it to P_i (if more than one such messages are registered, deliver and remove from the "ready to fetch" set the first, chronologically, registered such pair); otherwise send (FETCH-OUTPUT, P_i) to $\mathcal{S}$.

Fig. 3. The Q-Fairness wrapper functionality.

$\mathcal{S}$ so that for all environments $\mathcal{Z}$, it holds:

$$\text{EXEC}_{\pi,\mathcal{A},\mathcal{Z}}^{\bar{\mathcal{G}},\mathcal{W}_{\mathsf{Q},\bar{\mathcal{G}}}(\mathcal{F})} \approx \text{EXEC}_{\mathcal{S},\mathcal{Z}}^{\bar{\mathcal{G}},\mathcal{W}_{\mathsf{Q}',\bar{\mathcal{G}}}(\mathcal{H})}$$

We note that, both protocol π and the functionality $(\mathcal{W}_{\mathsf{Q},\bar{\mathcal{G}}}(\mathcal{F}),\bar{\mathcal{G}})$ are with respect to the global functionality[15] $\bar{\mathcal{G}}$. By following the very similar proof idea in [11], we can prove the following lemma and theorem:

Lemma 1. *Let* $\mathsf{Q}_{\bar{\mathcal{G}}}$ *be a predicate with respect to global functionality* $\bar{\mathcal{G}}$. *Let* π *be a protocol that realizes the functionality* $\mathcal{F}$ *with* $\mathsf{Q}_{\bar{\mathcal{G}}}$-*fairness. Let* σ *be a protocol in* $(\mathcal{W}_{\mathsf{Q},\bar{\mathcal{G}}}(\mathcal{F}),\bar{\mathcal{G}})$-*hybrid world. Then for all adversaries* $\mathcal{A}$, *there is a simulator* $\mathcal{S}$ *so that for all environments* $\mathcal{Z}$, *it holds*

$$\text{EXEC}_{\sigma^\pi,\mathcal{A},\mathcal{Z}}^{\bar{\mathcal{G}}} \approx \text{EXEC}_{\sigma,\mathcal{S},\mathcal{Z}}^{\bar{\mathcal{G}},\mathcal{W}_{\mathsf{Q},\bar{\mathcal{G}}}(\mathcal{F})}$$

Theorem 1. *Let* $\mathsf{Q}_{\bar{\mathcal{G}}}$ *and* $\mathsf{Q}'_{\bar{\mathcal{G}}}$ *be predicates with respect to global functionality* $\bar{\mathcal{G}}$. *Let* π *be a protocol that realizes the functionality* $\mathcal{F}$ *with* $\mathsf{Q}_{\bar{\mathcal{G}}}$-*fairness. Let* σ *be a protocol in* $(\mathcal{W}_{\mathsf{Q},\bar{\mathcal{G}}}(\mathcal{F}),\bar{\mathcal{G}})$-*hybrid world that realizes the functionality* $\mathcal{H}$ *with* $\mathsf{Q}'_{\bar{\mathcal{G}}}$-*fairness. Then for all adversaries* $\mathcal{A}$, *there is a simulator* $\mathcal{S}$ *so that for all environments* $\mathcal{Z}$ *it holds:*

$$\text{EXEC}_{\sigma^\pi,\mathcal{A},\mathcal{Z}}^{\bar{\mathcal{G}}} \approx \text{EXEC}_{\mathcal{S},\mathcal{Z}}^{\bar{\mathcal{G}},\mathcal{W}_{\mathsf{Q}',\bar{\mathcal{G}}}(\mathcal{H})}$$

Is the Ledger Functionality Sufficient for Q Fairness? We will construct secure computation protocols based on the ledger functionality $\bar{\mathcal{G}}_{\text{LEDGER}}$ together with other trusted setups. We may wonder if we can construct secure computation protocol from $\bar{\mathcal{G}}_{\text{LEDGER}}$ only. The answer if negative. Indeed, we prove the following statement.

Theorem 2. *Let* $\mathsf{Q}_{\bar{\mathcal{G}}}$ *be a predicate with respect to global functionality* $\bar{\mathcal{G}} = \bar{\mathcal{G}}_{\text{LEDGER}}$. *There exists no protocol in the* $\bar{\mathcal{G}}_{\text{LEDGER}}$ *hybrid world which realizes the commitment functionality* $\mathcal{F}_{\text{COM}}$ *with* $\mathsf{Q}_{\bar{\mathcal{G}}}$ *fairness.*

The proof idea is very similar to the well-known Canetti-Fischlin [12] impossibility proof and can be found in [25].

4.2 $\mathsf{Q}_{\bar{\mathcal{G}}}$-Robustness

The above wrapper $\mathcal{W}$ allows the simulator to delay delivery of messages arbitrarily. Thus, although the predicates do guarantee the promised notion of fairness, the resulting functionality lacks the other relevant property that we discussed in the introduction, namely robustness. In the following we define Q-robustness which will ensure that if any party starts executing the protocol on its input (i.e., the protocol does not abort due to lack of resources for some party), then every honest party is guaranteed to either receive its output without loosing

---

[15] In GUC framework [11], this is also called, $\bar{\mathcal{G}}$-subroutine respecting.

revenue, or receive bottom and a compensation. This property can be obtained by modifying the wrapper $\mathcal{W}$ using an idea from [24] so that in addition to the global-setup-related guarantees induced by predicate Q, it also preserves the guaranteed termination property of the wrapped functionality.[16]

More concretely, in [24], a functionality was augmented to have guaranteed termination, by ensuring that given appropriately many activations (i.e., dummy inputs), from its honest interface, it computes its output.[17] In the same spirit, a wrapper which ensures Q-robustness is derived from $\mathcal{W}$ via the following modification: As soon as a fair-output is registered (i.e., upon the wrapper receiving (FAIR-DELIVER, sid, mid, $(m_1, P_{i_1}), \ldots, (m_k, P_{i_k}), (m_{\mathcal{S}}, \mathcal{S})$) from its inner functionality) it initiates a counter $\lambda = 0$ and an indicator variable $\lambda_{i_j} := 0$ for each $P_{i_j} \in \{P_{i_1}, \ldots, P_{i_k}\}$; whenever a message is received from some $P_{i_j} \in \{P_{i_1}, \ldots, P_{i_k}\}$, the wrapper sets $\lambda_{i_j} := 1$ and does the following check: if $\lambda_{i_j} = 1$ for all $P_{i_j} \in \{P_{i_1}, \ldots, P_{i_k}\}$ then increase $\lambda := \lambda + 1$ and reset $\lambda_{i_j} = 0$ for all $P_{i_j} \in \{P_{i_1}, \ldots, P_{i_k}\}$. As soon as λ reaches a set threshold T, the wrapper simultaneously delivers each $((m_1, P_{i_k}), \ldots, (m_k, P_{i_k})$ (i.e., prepares them to be fetched) without waiting for the simulator and does not accept any inputs other than (FETCH-OUTPUT, $\cdot$) from that point on. When this happens, we will say that the wrapper *reached its termination limit*. We denote by $\hat{\mathcal{W}}^T$ the wrapper from Fig. 3 modified as described above. Note that the wrapper is parameterized by the termination threshold T.

The intuition why this modification ensures guaranteed termination is the same as in [24]: if the environment wishes the experiment to terminate, the it can make it terminate irrespective of the simulator's strategy. Thus a protocol which realizes such a wrapper should also have such a guaranteed termination (the adversary cannot stall the computation indefinitely.)

Definition 2. *We say protocol π realizes functionality $\mathcal{F}$ with $Q_{\bar{\mathcal{G}}}$-robustness with respect to global functionality $\bar{\mathcal{G}}$, provided the following statement is true. There exists a threshold T such that for all adversaries $\mathcal{A}$, there is a simulator $\mathcal{S}$ so that for all environments $\mathcal{Z}$ it holds:*

$$\mathrm{EXEC}^{\bar{\mathcal{G}}}_{\pi, \mathcal{A}, \mathcal{Z}} \approx \mathrm{EXEC}^{\bar{\mathcal{G}}, \hat{\mathcal{W}}^T_{Q, \bar{\mathcal{G}}}(\mathcal{F})}_{\mathcal{S}, \mathcal{Z}}.$$

Moreover, whenever the wrapper reaches its termination limit, then for the state trans *of the global setup $\bar{\mathcal{G}}$ upon termination it holds that $Q^{Dlv}_{\bar{\mathcal{G}}}(\mathsf{sid}, P, RS^{\mathrm{pub}}_{P, \mathsf{sid}}, \mathsf{trans})$ for every party $P \in \mathcal{P}$.*

The composition theorems for Q-fairness from Sect. 4.1 can be adapted in a straight-forward manner to Q-robustness. The statements and proofs are as in the previous section and are omitted. We note in passing that since the wrapper $\hat{\mathcal{W}}$ is in fact a wrapper which restricts the behavior of $\mathcal{S}$ on top of the restrictions

---

[16] That is, we want to ensure that if the functionality $\mathcal{F}$ has guaranteed termination then the wrapped functionality will also have guaranteed termination.

[17] Of course, the simulator needs to be given sufficiently many activation so that he can provide its own inputs and perform the simulation (please see [24] for details).

which are applied by the Q-fairness wrapper $\mathcal{W}$, a protocol which is Q-robustness is also Q-fair with respect to the same predicate Q.

4.3 Computation with Fair/Robust Compensation

We are now ready to instantiate the notion of Q-fairness with a compensation mechanism. For the case when $\bar{\mathcal{G}}$ corresponds to a Bitcoin-like ledger, e.g., $\bar{\mathcal{G}} = \bar{\mathcal{G}}_{\text{LEDGER}}$, and $Q_{\bar{\mathcal{G}}}$ provides compensation of c coins, where $c > 0$, in the case of an abort, the resource-setup generation algorithm Gen a pair of (address, sk) where address is a bitcoin address and sk is the corresponding secret-key and the predicate $Q_{\bar{\mathcal{G}}}^{\text{coin}} = (Q_{\bar{\mathcal{G}}}^{\text{C-Init}}, Q_{\bar{\mathcal{G}}}^{\text{C-Dlv}}, Q_{\bar{\mathcal{G}}}^{\text{C-Abt}})$ operates as follows. On input a session ID sid, a party id P, a wallet address $RS_{P,\text{sid}}^{\text{pub}}$, and a string trans which is parsed as a bitcoin ledger that contains transactions:[18]

- $Q_{\bar{\mathcal{G}}}^{\text{C-Init}}$ outputs true if and only if the balance of all transactions (both incoming and outgoing) that concern $RS_{P,\text{sid}}^{\text{pub}}$ in trans and carry the meta-data sid is higher than a fixed pre-agreed initialization amount.[19]
- $Q_{\bar{\mathcal{G}}}^{\text{C-Dlv}}$ outputs true if and only if the balance of all transactions (both incoming and outgoing) that concern $RS_{P,\text{sid}}^{\text{pub}}$ in trans and carry the meta-data sid is greater or equal to 0.
- $Q_{\bar{\mathcal{G}}}^{\text{C-Abt}}$ outputs true if and only if the balance of all transactions (both incoming and outgoing) that concern $RS_{P,\text{sid}}^{\text{pub}}$ in trans and carry the meta-data sid is greater or equal to a fixed pre-agreed compensation amount.

If a protocol π realizes a functionality $\mathcal{F}$ with $Q_{\bar{\mathcal{G}}}^{\text{coin}}$-fairness (resp. $Q_{\bar{\mathcal{G}}}^{\text{coin}}$-robustness), i.e., with respect to the global functionality $\bar{\mathcal{G}}_{\text{LEDGER}}$, we say that π realizes $\mathcal{F}$ with fair compensation (resp. with robust compensation). Because our results are proved for $Q_{\bar{\mathcal{G}}}^{\text{coin}}$, to keep the notation simple in the remainder of the paper we might drop the superscript from $Q_{\bar{\mathcal{G}}}^{\text{coin}}$, i.e., we write Q or $Q_{\bar{\mathcal{G}}}$ instead of $Q_{\bar{\mathcal{G}}}^{\text{coin}}$.

5 Our $Q_{\bar{\mathcal{G}}}^{\text{Coin}}$-Robust Protocol Compiler

In this section we present our fair and robust protocol compiler. Our compiler compiles a synchronous protocol π_{SH} which is secure (i.e., private) against a corrupted majority in the semi-honest correlated randomness model (e.g., an OT-hybrid protocol where the OT's have been pre-computed) into a protocol π which is secure with fair-compensation in the malicious correlated randomness model. The high-level idea is the following: We first compile π_{SH} into a protocol in

---

[18] Transactions in trans can also be marked with metadata.

[19] In our construction $Q_{\bar{\mathcal{G}}}^{\text{C-Init}}$ will check additional properties for the initial set of transactions that concern $RS_{P,\text{sid}}^{\text{pub}}$; specifically, not only that a fixed amount μ is present but also that it is distributed in a special way.

the malicious correlated randomness model, which is executed over a broadcast channel and is secure with publicly identifiable abort. (Roughly, this means that someone observing the protocol execution can decide, upon abort, which party is not executing its code.) This protocol is then transformed into a protocol with fair compensation as follows: Every party (after receiving his correlated randomness setup) posts to the ledger transactions that the other parties can claim only if they, later, post transactions that prove that they follow their protocol. Transactions that are not claimed this way are returned to the source address; thus, if some party does not post such a proof it will not be able to claim the corresponding transaction, and will therefore leave the honest parties with a positive balance as their transactions will be refunded. Observe that these are not standard Bitcoin transactions, but they have a special format which is described in the following.

Importantly, the protocol we describe is guaranteed to either produce output in as many (Bitcoin) rounds as the rounds of the original malicious protocol, or to compensate all honest parties. This *robustness* property is achieved by a novel technique which ensures that once the honest parties make their initial transaction, the adversary has no way of preventing them from either computing their output or being compensated. Informally, our technique consists of splitting the parties into "islands" depending on the transactions they post (so that all honest parties are on the same island) and then allowing them to either compute the function within their island, or if they abort to get compensated. (The adversary has the option of being included or not in the honest parties' island.)

5.1 MPC with Publicly Identifiable Abort

As a first step in our compiler we invoke the semi-honest to malicious with identifiable abort compiler of Ishai, Ostrovsky, and Zikas [23] (hereafter referred to as the *IOZ compiler*). This compiler takes a semi-honest protocol π_{SH} in the correlated randomness model and transforms it to a protocol in the malicious correlated randomness model (for an appropriate setup) which is secure with identifiable abort, i.e., when it aborts, every party learns the identity of a corrupted party. The compiler in [23] follows the so called GMW paradigm [21], which in a nutshell has every party commit to its input and randomness for executing the semi-honest protocol π_{SH} and then has every party run π_{SH} over a broadcast channel, where in each round ρ every party broadcasts his round ρ messages and proves in zero-knowledge that the broadcasted message is correct, i.e., that he knows the input and randomness that are consistent with the initial commitments and the (public) view of the protocol so far. The main difference of the IOZ compiler and the GMW compiler is that the parties are not only committed to their randomness, but they are also committed to their entire setup string, i.e., their private component of the correlated randomness. In the following, for the sake of completeness, we enumerate some key properties of the resulting maliciously secure protocol π_{Mal} (which is based on the compiler in [23]) that will be important for our construction:

- Every party is committed to his setup, i.e., the part of the correlated randomness it holds. That is, every party P_i receives from the setup his randomness (which we refer to as P_i's *private component* of the setup) along with one-to-many commitments[20] on the private components of all parties. Without loss of generality, we also assume that a common-reference string (CRS) and a public-key infrastructure (PKI) are included in every party's setup. We refer to the distribution of this correlated randomness as $\mathcal{D}_{\text{Mal}}$.
- The protocol π_{Mal} uses *only* the broadcast channel for communication.
- Given the correlated randomness setup, the protocol π_{Mal} is completely deterministic. This is achieved in [23] by ensuring that all the randomness used in the protocol, even the one needed for the zero-knowledge proofs, is part of the private components that are distributed by the sampling functionality.[21]
- π_{Mal} starts off by having every party broadcast a one-time pad encryption of its input with its (committed) randomness and a NIZK that it knows the input and randomness corresponding to the broadcasted message.
- By convention, the next-message function of π_{Mal} is such that if in any round the transcript seen by a party is an aborting transcript, i.e., is not consistent with an accepting run of the semi-honest protocol, then the party outputs $\perp$. Recall that the identifiable abort property ensures that in this case every party will also output the identity of a malicious party (the same for all parties).
- There is a (known) upper bound on the number ρ_c of rounds of π_{Mal}.

We remark that, given appropriate setup, the IOZ-compiler achieves information-theoretic security, and needs therefore to build information-theoretic commitments and zero-knowledge proofs. As in this work we are only after computational security, we modify the IOZ compiler so that we use (computationally) UC secure one-to-many commitments [14] and computationally UC secure non-interactive zero-knowledge proofs (NIZKs) instead if their information-theoretic instantiation suggested in [23]. Both the UC commitment and the NIZKs can be built in the CRS model. Moreover, the use of UC secure instantiations of zero-knowledge and commitments ensures that the resulting protocol will be (computationally) secure.

Using the Setup Within a Subset of Parties. A standard property of many protocols in the correlated-randomness model is that once the parties in $\mathcal{P}$ have received the setup, any subset $\mathcal{P}' \subset \mathcal{P}$ is able to use it to perform a computation of a $|\mathcal{P}'|$-party function amongst them while ignoring parties in $\mathcal{P} \setminus \mathcal{P}'$. More concretely, assume the parties in $\mathcal{P}$ have been handed a setup allowing them to execute some protocol π for computing any $|\mathcal{P}|$-party function f; then for any $\mathcal{P}' \subseteq \mathcal{P}$, the parties in $\mathcal{P}'$ can use their setup within a protocol $\pi|_{\mathcal{P}'}$ to compute any $|\mathcal{P}'|$-party function $f|_{|\mathcal{P}'|}$. This property which will prove very useful for obtaining computation with robustness or compensation, is also satisfied

---

[20] These are commitments that can be opened so that every party agrees on whether or not the opening succeeded.

[21] As an example, the challenge for the zero-knowledge proofs is generated by the parties opening appropriate parts of their committed random strings.

by the IOZ protocol, as the parties in $\mathcal{P}'$ can simply ignore the commitments (public setup component) corresponding to parties in $\mathcal{P}\backslash\mathcal{P}'$. It should be noted that this is not an inherent property of the correlated randomness model: e.g., protocols based on threshold encryption do not immediately satisfy this property (as players would have to readjust the threshold).

Making Identifiability Public. The general idea of our protocol is to have every party issue transactions by which he commits to transferring a certain amount of coins per party for each protocol round. All these transactions are issued at the beginning of the protocol execution. Every party can claim the "committed" coins transferred to him associated to some protocol round ρ only under the following conditions: (1) the claim is posted in the time-interval corresponding to round ρ; (2) the party has claimed all his transferred coins associated to the previous rounds; and (3) the party has posted a transaction which includes his valid protocol message for round ρ.

In order to ensure that a party cannot claim his coins unless he follows the protocol, the ledger (more concretely the validation predicate) should be able to check that the party is indeed posting its valid next message. In other words, in each round ρ, P_i's round-ρ message acts as a witness for P_i claiming all the coins committed to him associated with this round ρ. To this direction we make the following modification to the protocol: Let $f(x_1,\ldots,x_n) = (y_1,\ldots,y_n)$ denote the n-party function we wish to compute, and let f^{+1} be the $(n+1)$-party function which takes input x_i from each P_i, $i \in [n]$, and no input from P_{n+1} and outputs y_i to each P_i and a special symbol (e.g., 0) to P_{n+1}. Clearly, if π_{SH} is a semi-honest n-party protocol for computing f over broadcast, then the $n+1$ protocol π_{SH}^{+1} (in which every P_i with $i \in [n]$ executes π_{SH} and P_{n+1} simply listens to the broadcast channel and outputs 0) is a semi-honest secure protocol for f^{+1}.

Now if π_{Mal}^{+1} denotes the $(n+1)$-party malicious protocol which results by applying the above modified IOZ compiler on the $(n+1)$-party semi-honest protocol π_{SH}^{+1} for computing the function f^{+1}, then, by construction this protocol computes function f^{+1} with identifiable abort and has the following additional properties:

- Party P_{n+1} does not make any use of his private randomness whatsoever; this is true because he broadcasts no messages and simply verifies the broadcasted NIZKs.
- If some party P_i, $i \in [n]$ deviates from running π_{SH} with the correlated (committed) randomness as distributed from the sampling functionality, then this is detected by all parties, including P_{n+1} (and protocol π_{Mal}^{+1} aborts identifying P_i as the offender). This follows by the soundness of the NIZK which P_i needs to provide proving that he is executing π_{SH} in every round.

Due to P_{n+1}'s role as an observer who gets to decide if the protocol is successful (P_{n+1} outputs 0) or some party deviated (P_{n+1} observes that the corresponding NIZK verification failed) in the following we will refer to P_{n+1} in the above protocol as the *judge*. The code of the judge can be used by anyone who has the public setup and wants to follow the protocol execution and decide whether it should abort or not given the parties' messages. Looking ahead, the judge's code in the protocol will be used by the ledger to decide whether or not a transaction that claims some committed coins is valid.

5.2 Special Transactions Supported by Our Ledger

In this section we specify the Validate and the Blockify predicates that are used for achieving our protocol's properties. More specifically, our protocol uses the following type of transactions which transfer v coins from wallet address_i to wallet address_j conditioned on a statement Σ:

$$\mathbb{B}_{v,\text{address}_i,\text{address}_j,\Sigma,\text{aux},\sigma_i,\tau} \tag{1}$$

where σ_i is a signature of the transaction, which can be verified under wallet address_i; τ is the time-stamp, i.e., the current value of the clock when this transaction is added to the state by the ledger—note that this timestamp is added by the ledger and not by the users,—$\text{aux} \in \{0,1\}^*$ is an arbitrary string[22]; and the statement Σ consists of three arguments, i.e., $\Sigma = (\text{arg1}, \text{arg2}, \text{arg3})$, which are processed by the Validate predicate in order to decide if the transaction is valid (i.e., if it will be included in the ledger's next block).

The Validate *Predicate.* The validation happens by processing the arguments of Σ in a sequential order, where if while processing of some argument the validation rejects, algorithm Validate stops processing at that point and this transaction is dropped. The arguments are defined/processed as follows:

TIME-RESTRICTIONS: The first argument is a pair $\text{arg1} = (\tau_-, \tau_+) \in \mathbb{Z} \times (\mathbb{Z}^+ \cup \{\infty\})$ of points in time. If $\tau_- > \tau_+$ then the transaction is invalid (i.e., it will be dropped by the ledger). Otherwise, before time τ_- the coins in the transaction "remain" blocked, i.e., no party can spend them; from time τ_- until time τ_+, the money can be spent by the owner of wallet address_j provided that the spending statement satisfies also the rest of the requirements/arguments in statement Σ (listed below). After time τ_+ the money can be spent by the owner of wallet address_i without any additional restrictions (i.e., the rest of the arguments in Σ are not parsed). As a special case, if $\tau_+ = \infty$ then the transferred coins can be spent from address_j at any point (provided the spending statement is satisfied); we say then that the transaction is *time-unrestricted,*[23] otherwise we say that the transaction is *time restricted.*

SPENDING LINK: Provided that the processing of the first argument, as above, was not rejecting, the Validate predicate proceeds to the second argument, which is a unique "anchor", $\text{arg2} = \alpha \in \{0,1\}^*$. Informally, this serves as a unique identifier for linked transactions[24]; that is, when $\alpha \neq \perp$, then the

---

[22] This string will be included to the Ledger's state as soon as the transaction is posted and can be, therefore, referred to by other spending statements.

[23] This is the case with standard Bitcoin transactions.

[24] Looking ahead arg2 will be used to point to specific transactions of a protocol instance. The mechanism may be simulated by generating multiple addresses however it is more convenient for the protocol description and for this reason we adopt it.

Validate algorithm of the ledger looks in the ledger's state and buffer to confirm that the balance of transactions to/from the wallet address address_i with this anchor arg2 is at least $v' \geq v$ coins. That is, the sum of coins in the state or in the buffer with receiver address address_i and anchor arg2 minus the sum of coins in the state or in the buffer with sender address address_i and anchor arg2 is greater equal to v. If this is not the case then the transaction is rendered invalid; otherwise the validation of this argument succeeds and the algorithm proceeds to the next argument.

STATE-DEPENDENT CONDITION: The last argument to be validated is arg3, which is a relation $\mathcal{R} : \mathcal{S} \times \mathcal{B} \times \mathcal{T} \rightarrow \{0,1\}$, where $\mathcal{S}$, $\mathcal{B}$, and $\mathcal{T}$ are the domains of possible ledger-states, ledger-buffers, and transactions, respectively (in a given encoding). This argument defines which type of transactions can spend the coins transferred in the current transaction. That is, in order to spend the coins, the receiver needs to submit a transaction $\text{tx} \in \mathcal{T}$ such that $\mathcal{R}(\text{state}, \text{buffer}, \text{tx}) = 1$ at the moment when tx is to be validated and inserted in the buffer. In our construction this is the part of the transaction that we will take advantage to detect cheating (and thus $\mathcal{R}$ will encode a NIZK verifier etc.).

We point out that as with standard Bitcoin transactions, the validation predicate will always also check validity of the signature σ_i with respect to the wallet address_i. Moreover, the standard Bitcoin-like transactions can be trivially casted as transactions of the above type by setting $\alpha = \perp$ and $\Sigma = ((0, \infty), \perp, \mathcal{R}_\emptyset)$, where $\mathcal{R}_\emptyset$ denotes the relation which is always true.

To simplify the structure of our special transactions and ease their implementation, we impose the following additional constraints: whenever a time-restriction is given, i.e., $\text{arg}_1 = (\tau_-, \tau_+)$ then it must be that $\alpha \neq \perp$. Furthermore, if a time-restricted transaction is present with anchor α from address_1 to address_2, the only transactions that are permitted with anchor α in the ledger would be time-unrestricted transactions originating from either address_2 within the specified time-window, or address_1 after the specified time window.

The Blockify *Algorithm.* This algorithm simply groups transactions in the current buffer and adds a timestamp from the current round. We choose to ignore any additional functionality (e.g., such as a reward mechanism for mining that is present in typical cryptocurrencies — however such mechanism can be easily added independently of our results).

5.3   The Protocol

Let π_{Mal}^{+1} denote the protocol described in Sect. 5.1. Let $\text{Round2Time}(1)$ denote the time in which the parties have agreed to start the protocol execution. Without loss of generality we assume that $\text{Round2Time}(1) > \text{T} + 1$ where T is the number clock ticks for each block generation cf. Fig. 2.[25] Furthermore, for simplicity, we

---

[25] That is we assume that at least one ledger rounds plus one extra clock-ticks have passed from the beginning of the time.

assume that each party P_i receives its input x_i with its first activation from the environment at time Round2Time(1) (if some honest party does not have an input by that time it will execute the protocol with a default input, e.g., 0).

Informally, the protocol proceeds as follows: In a pre-processing step, before the parties receive input, the parties invoke the sampling functionality for π_{Mal}^{+1} to receive their correlated randomness.[26] The public component of this randomness includes their protocol-associated wallet address$_i$ which they output (to the environment). This corresponds to the resources allocation step in the Q-robustness wrapper $\hat{\mathcal{W}}$. The environment is then expected to submit ρ_c special (as above) transactions for each pair of parties $P_i \in \mathcal{P}$ and $P_j \in \mathcal{P}$; the source wallet-address for each such transaction is P_i's, i.e., address$_i$ and the target wallet-address for is P_j's, i.e., address$_j$, and the corresponding anchors are as follows: $\alpha_{i,j,\rho} = (\text{pid}, i, j, \rho)$, for $(i,j,\rho) \in [n]^2 \times [\rho_c]$, where[27] pid is the (G)UC protocol ID for π_{Mal}^{+1}. Since by assumption, Round2Time(1) $> \text{T} + 1$, the environment has sufficient time to submit these transaction so that by the time the protocol starts they have been posted on the ledger.

At time Round2Time(1) the parties receive their inputs and initiate the protocol execution by first checking that sufficient funds are allocated to their wallets linked to the protocol executions by appropriate anchors, as above. If some party does not have sufficient funds then it broadcasts an aborting message and all parties abort.[28] This aborting in case of insufficient funds is consistent with the behavior of the wrapper $\hat{\mathcal{W}}$ when $\text{Q}_{\hat{\mathcal{G}}}^{\text{C-Init}}$ is false. Otherwise, parties make the special transactions that commit them (see below) into executing the protocol, and then proceed into claiming them one-by-one by executing their protocol in a round-by-round fashion.

Note that each protocol round lasts one ledger round so that the parties have enough time to claim their transactions. This means that Round2Time($i + 1$) $-$ Round2Time(i) $\geq \text{T}$, which guarantees that any transaction submitted for round ρ, $\rho = 1, \ldots, \rho_c - 1$, of the protocol, has been posted on the ledger by the beginning of round $\rho + 1$. Observe that by using a constant round protocol π_{Mal}^{+1} (e.g., the modified compiled protocol from [23] instantiated with a constant round semi-honest protocol) we can ensure that our protocol will terminate in a constant number of ledger rounds and every honest party will either receive its input, or will have a positive balance in its wallet.

Remark 1 (On availability of funds). Unlike existing works, we choose to explicitly treat the issue of how funds become available to the protocol by making the off-line transfers external to the protocol itself (i.e., the environment takes care of them). However, the fact that the environment is in charge of "pouring" money into the wallets that are used for the protocol does not exclude that the

---

[26] In an actual application, the parties will use an unfair protocol for computing the correlated randomness. As this protocol has no inputs, an abort will not be unfair (i.e., the simulator can always simulate the view of the adversary in an aborting execution.).

[27] Recall that we assume $|\mathcal{P}| = n$.

[28] Note that this is a fair abort, i.e., no party has spent time into making transactions.

parties might be actually the ones having done so. Indeed, the environment's goal is to capture everything that is done on the side of, before, or after the protocol, including other protocols that the parties might have participated in. By giving the environment enough time to ensure these transactions are posted we ensure that some honest party not having enough funds corresponds to an environment that makes the computation abort (in a fair way and only in the pre-processing phase, before the parties have invested time into posting protocol transactions).

Here is how we exploit the power of our special transactions in order to arrange that the balance of honest parties is positive in case of an abort. We require that the auxiliary string of a transaction of a party P_j which claims a committed transaction for some round ρ includes his ρ-round protocol message. We then have the relation of this transaction be such that it evaluates to 1 if only if this is indeed P_j's next message. Thus, effectively the validate predicate implements the judge in π_{Mal}^{+1} and can, therefore, decide if some party aborted: if some party broadcasts a message that would make the judge abort, then the validate predicate drops the corresponding transaction and all claims for committed transactions corresponding to future rounds, thus, all other parties are allowed to reclaim their committed coins starting from the next round.

Before we give the protocol description there is a last question: how is the ledger able to know which parties should participate in the protocol? Here is the problem: The adversary might post in the first round (as part of the committing transaction for the first round) a fake, maliciously generated setup. Since the ledger is not part of the correlated randomness sampling, it would be impossible to decide which is the good setup. We solve this issue by the following technique that is inspired by [6]: The ledger[29] groups together parties that post the same setup; these parties form "islands", i.e., subsets of $\mathcal{P}$. For each such subset $\mathcal{P}' \subseteq \mathcal{P} \cup \{P_{n+1}\}$ which includes the judge P_{n+1}, the ledger acts as if the parties in $\mathcal{P}'$ are executing the protocol $\pi_{\text{Mal}}^{+1}|_{\mathcal{P}'}$ (which, recall, is the restriction of π_{Mal}^{+1} to the parties in $\mathcal{P}'$) for computing the $|\mathcal{P}'|$-party function $f^{+1}|_{\mathcal{P}'}(\boldsymbol{x})$ defined as follows: let the function to be computed be $f(\boldsymbol{x})$, where $\boldsymbol{x} = (x_1, \ldots, x_n)$, and f^{+1} be as above, then $f^{+1}|_{\mathcal{P}'}(\boldsymbol{x}) = f^{+1}(\boldsymbol{x}_{\mathcal{P}'})$ where $\boldsymbol{x}_{\mathcal{P}'} = (x'_1, \ldots, x'_n)$ with $x'_i = x_i$ for $P_i \in \mathcal{P}'$ and x'_i being a default value for every $P_i \notin \mathcal{P}'$. This solves the problem as all honest parties will be in the same island $\mathcal{P}' \subset \mathcal{P}$ (as they will all post the same value for public randomness); thus if the adversary chooses not to post this value on behalf of some corrupted party, he is effectively setting this party's input to a default value, a strategy which is easily simulatable. (Of course, the above solution will allow the adversary to also have "islands" of only corrupted parties that might execute the protocol, but this is also a fully simulatable strategy and has no effect on fair-compensation whatsoever—corrupted parties are not required to have a positive balance upon abort).

---

[29] Throughout the following description, we say that the ledger does some check to refer to the process of checking a corresponding relation, as part of validating a special transaction.

The final protocol $\pi_{\mathtt{Mal}}^{\mathbb{B}}$ is detailed in the following. The protocol ID is sid. The function to be computed is $f(x_1, \ldots, x_n)$. The protocol parties are $\mathcal{P} = \{P_1, \ldots, P_n\}$. We assume all parties have registered with the clock functionality in advance and are therefore synchronized once the following steps start.

Phase 1: Setup Generation
Time $\tau_{-2} = \mathtt{Round2Time}(1) - \mathtt{T} - 2$:

The parties invoke the sampling functionality for $\mathcal{D}_{\mathtt{Mal}}$, i.e., every party $P_i \in \mathcal{P}$ starts off by sending the sampling functionality a message $(\mathtt{REQUEST}, \mathtt{sid})$; the sampling functionality returns $(R_i^{\mathtt{priv}}, R^{\mathtt{pub}})$ to P_i where $R_i^{\mathtt{priv}}$ is P_i's private component (including all random coins he needs to run the protocol, along with his signing key sk_i) of the setup and $R^{\mathtt{pub}}$ is the public component (the same for every party P_j) which includes the vector of UC commitments $(\mathsf{Com}_1, \ldots, \mathsf{Com}_n)$, where for $j \in [n]$, Com_j is a commitment to $R_j^{\mathtt{priv}}$, along with a vector of public (verification) keys $(\mathsf{vk}_1, \ldots, \mathsf{vk}_n)$ corresponding to the signing keys $(\mathsf{sk}_1, \ldots, \mathsf{sk}_n)$ and a common reference string CRS. Every party outputs its own public key, as its wallet address for the protocol, i.e., $\mathsf{address}_i = \mathsf{vk}_i$.

Phase 2: Inputs and Protocol Execution
Time $\tau_{-1} = \mathtt{Round2Time}(1) - 1$:

Every party $P_i \in \mathcal{P}$ receives its input x_i ($x_i = 0$ if no input is received in the first activation of P_i for time $\mathtt{Round2Time}(1)$) and does the following to check that it has sufficient fund available: P_i reads the current state from the ledger. If the state does not include for each $(i, j, \rho) \in [n]^2 \times [\rho_c]$ a transaction $\mathbb{B}_{c, \mathsf{address}, \mathsf{address}_i, \Sigma_{i,j,\rho}^0, \mathsf{aux}_{i,j,\rho}^0, \sigma, \tau}$, for some arbitrary $\mathsf{address}$ and where $\Sigma_{i,j,\rho}^0 = ((0, \infty), (\mathsf{sid}, i, j, \rho), \mathcal{R}_{\emptyset})$ then P_i broadcasts $\bot$ and every party aborts the protocol execution with output $\bot$ (i.e., no party does anything from that point on. Recall that ρ_c is the upper bound on the number of rounds of $\pi_{\mathtt{Mal}}^{+1}$, cf. Sect. 5.1.

Time $\tau_0 = \mathtt{Round2Time}(1)$:

Every P_i submits to the ledger the following "commitment" transactions:[30]

1. For each $P_j \in \mathcal{P}$: $\mathbb{B}_{c, \mathsf{address}_i, \mathsf{address}_j, \Sigma_{i,j,1}, \mathsf{aux}_{i,j,1}, \sigma, \tau}$, where $\mathsf{aux}_{i,j,1} = R^{\mathtt{pub}}$ and $\Sigma_{i,j,1} = (\mathsf{arg1}_{i,j,1}, \mathsf{arg2}_{i,j,1}, \mathsf{arg3}_{i,j,1})$ with
   - $\mathsf{arg1}_{i,j,1} = (\mathtt{Round2Time}(1) + \mathtt{T}, \mathtt{Round2Time}(1) + 2\mathtt{T} - 1)$
   - $\mathsf{arg2}_{i,j,1} = (\mathsf{sid}, i, j, 1)$
   - $\mathsf{arg3}_{i,j,1} = \mathcal{R}_{i,j,1}$ defined as follows: Let $\mathcal{P}^{+1} = \mathcal{P} \cup \{P_{n+1}\}$, where P_{n+1} denotes the judge, be the player set implicit in $R^{\mathtt{pub}}$,[31] and let $\mathcal{P}_i^{+1} \subseteq \mathcal{P}^{+1}$ denote the island of party i including the judge, i.e., the set of parties (wallets), such that in the first block posted after time

---

[30] Recall that, by definition of the clock, every party has as much time as it needs to complete all the steps below before the clock advances time.

[31] Recall that $R^{\mathtt{pub}}$ includes commitments to all parties' private randomness (including the judge's P_d) used for running the protocol, which is an implicit representation of the player set.

Round2Time(1) all parties $P_k \in \mathcal{P}_i^{+1}$ have exactly one transaction for every $P_j \in \mathcal{P}$ with $\text{arg1}_{k,j,1} = (\text{Round2Time}(1) + \text{T}, \text{Round2Time}(1) + 2\text{T} - 1)$, $\text{arg2}_{k,j,1} = (\text{sid}, k, j, 1)$, and $\text{aux}_{k,j,1}^1 = R^{\text{pub}}$. Furthermore, let $\pi_{\text{Mal}}^{+1}|_{\mathcal{P}_i^{+1}}$ be the protocol with public identifiability for computing $f^{+1}|_{\mathcal{P}_i^{+1}}$, described above and denote by $R^{\text{pub}}|_{\mathcal{P}_i^{+1}}$ the restriction of the public setup to the parties in $\mathcal{P}_i^{+1}$. Then $\mathcal{R}_{i,j,1}(\text{state}, \text{buffer}, \text{tx}) = 1$ if and only if the protocol of the judge with public setup $R^{\text{pub}}|_{\mathcal{P}_i^{+1}}$ accepts the auxiliary string aux_{tx} in tx as P_i's first message in $\pi_{\text{Mal}}^{+1}|_{\mathcal{P}_i^{+1}}$ (i.e., it does not abort in the first round).

2. For each protocol round $\rho = 2, \ldots, \rho_c$ and each $P_j \in \mathcal{P}$: each party posts the transaction: $\mathbb{B}_{c, \text{address}_i, \text{address}_j, \Sigma_{i,j,\rho}, \text{aux}_{i,j,\rho}^1, \sigma, \tau}$, where $\text{aux}_{i,j,\rho}^1 = R^{\text{pub}}$ and $\Sigma_{i,j,\rho} = (\text{arg1}, \text{arg2}, \text{arg3})$ with
   - $\text{arg1} = (\text{Round2Time}(\rho) + \text{T}, \text{Round2Time}(\rho + 1) + 2\text{T} - 1)$
   - $\text{arg2} = (\text{sid}, i, j, \rho)$.
   - $\text{arg3} = \mathcal{R}_{i,j,\rho}$ defined as follows: Let $\mathcal{P}_i^{+1}$, $\pi_{\text{Mal}}^{+1}|_{\mathcal{P}_i^{+1}}$ be defined as above (and assume $\mathcal{P}_i^{+1} = \{P_{i_1}, \ldots, P_{i_m}\}$). Then $\mathcal{R}_{i,j,\rho}(\text{state}, \text{buffer}, \text{tx}) = 1$ if and only if, for each $r = 1, \ldots, \rho - 1$ and each party $P_{i_k} \in \mathcal{P}_i^{+1}$, the state state includes transactions in which the auxiliary input is $\text{aux}_{i_k, r}$ and the protocol of the judge with public setup $R^{\text{pub}}|_{\mathcal{P}_i^{+1}}$, and transcript $(\text{aux}_{i_1,1}, \ldots, \text{aux}_{i_m,1}), \ldots, (\text{aux}_{i_1,\rho-1}, \ldots, \text{aux}_{i_m,\rho-1})$, accepts the auxiliary string aux in tx as P_i's next (ρ-round) message in $\pi_{\text{Mal}}^{+1}|_{\mathcal{P}_i^{+1}}$ (i.e., it does not abort in the ρ-th round).

Phase 3: Claiming Committed Transactions/Executing the Protocol
Time $\tau \geq \text{Round2Time}(1)$:
For each $\rho = 1, \ldots, \rho_c + 1$, every P_i does the following at time $\text{Round2Time}(\rho)$:
1. If $\tau = \text{Round2Time}(\rho_c + 1)$ then go to Step 4; otherwise do the following:
2. Read the ledger's state, and compute $\mathcal{P}_i^{+1}$, $\pi_{\text{Mal}}^{+1}|_{\mathcal{P}_i^{+1}}$ as above.
3. If the state state is not aborting for $\mathcal{P}_i^{+1} = \{P_{i_1}, \ldots, P_{i_m}\}$, i.e., it includes for each $r = 1, \ldots, \rho - 1$ and each party $P_{i_k} \in \mathcal{P}_i^{+1}$ a transaction in which the auxiliary input is $\text{aux}_{i_k, r}$ such that P_i executing $\pi_{\text{Mal}}^{+1}|_{\mathcal{P}_i^{+1}}$ with public setup $R^{\text{pub}}|_{\mathcal{P}_i^{+1}}$, private setup R_i^{priv}, and transcript $(\text{aux}_{i_1,1}, \ldots, \text{aux}_{i_m,1}), \ldots, (\text{aux}_{i_1,\rho-1}, \ldots, \text{aux}_{i_m,\rho-1})$ for the first $r - 1$ rounds does not abort, then compute P_i's message for round ρ, denoted as msg_ρ, and submit to the ledger for each $P_k \in \mathcal{P}_i^{+1}$ a transaction $\mathbb{B}_{c, \text{address}_i, \text{address}, \Sigma'_{k,i,\rho}, \text{aux}_{k,i,\rho}^\rho, \sigma, \tau}$, where $\text{aux}_{k,i,\rho}^\rho = \text{msg}_\rho$, address is the address that was the input of the first transaction with link (sid, i, k, ρ) and $\Sigma'_{k,i,\rho} = (\text{arg1}, \text{arg2}, \text{arg3})$ instantiated as follows: $\text{arg1} = (0, \infty)$; $\text{arg2} = (\text{sid}, k, i, \rho)$; $\text{arg3} = \mathcal{R}_\emptyset$. For each such transaction posted enter (sid, k, i, ρ) in a set of "claimed" transactions CLAIM_i.
4. Otherwise, i.e., if the state state is aborting, then prepare for each round $r = 1, \ldots, \rho - 1$, and each $P_k \in \mathcal{P}$ a transaction by which the committed transaction towards P_k corresponding to round r is claimed back to address_i, i.e., $\mathbb{B}_{c, \text{address}_k, \text{address}_i, \Sigma, \text{aux}, \sigma, \tau}$, where $\text{aux} = \bot$ and

$\Sigma = (\texttt{arg1}, \texttt{arg2}, \texttt{arg3})$ instantiated as follows: $\texttt{arg1} = (0, \infty)$; $\texttt{arg2} = (\text{sid}, i, k, r)$; $\texttt{arg3} = \mathcal{R}_{\emptyset}$. The above transaction is posted as long as it is not claimed already, i.e., $(\text{sid}, i, k, r) \in \text{CLAIM}_i$ in a previous step.

This completes the description of the protocol. The protocol terminates in $O(\rho_c)$ ledger rounds. A depiction of the transactions that are associated with a protocol round is given in Fig. 4.

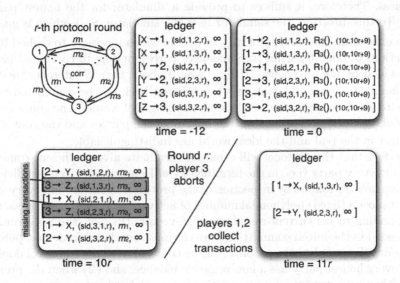

Fig. 4. The transactions associated with the first round r of our protocol compiler. $R_i(\cdot)$ is a relation which is true given the r-th round message of P_i (for the given correlated randomness and previous messages); m_i is the message of player P_i for round r. Player 3 aborts in the r-th round of the protocol and players 1,2 collect their reward.

Observe that by using a constant-round protocol $\pi_{\texttt{Mal}}$ [23], we obtain a protocol with constantly many ledger rounds. Furthermore, as soon as an honest party posts a protocol-related transaction, he is guaranteed to either receive his output or have a positive balance (of at least c coins) after $O(\rho_c)$ ledger rounds. The following theorem states the achieved security. We assume the protocol is executed in the synchronous model of Sect. 3.1.

Theorem 3. *Let* $\bar{\mathcal{G}} = (\bar{\mathcal{G}}_{\text{LEDGER}}, \bar{\mathcal{G}}_{\text{CLOCK}})$, *The above protocol in the* $(\bar{\mathcal{G}}, \mathcal{F}_{\text{CORR}}^{\mathcal{D}_{\texttt{Mal}}})$-*hybrid world realizes* $\tilde{\mathcal{W}}(\mathcal{F})$ *with robust compensation.*

Proof (sketch). We first prove that the above protocol is simulatable, by sketching the corresponding simulator $\mathcal{S}$. If the protocol aborts already before the parties make their transactions, then the simulator can trivially simulate such an abort, as he needs to just receive the state of the ledger and see if all wallets corresponding to honest parties have sufficient funds to play the protocol. In the following we show that the rest of the protocol (including the ledger's contents) can be simulated so that if there is an abort, honest parties' wallets have a positive balance as required by Q fairness. First we observe that the simulator S can easily decide the islands in which the parties are split, as he internally simulates the sampling functionality. Any island other than the one of honest parties (all honest parties will be in the same island because they will post transactions including the same public setup-component) is trivially simulatable as it only consists of adversarial parties and no guarantee is given about their wallets by Q-fairness. Therefore, it suffices to provide a simulator for the honest parties' island. To this direction, the simulator uses the simulator $S_{\pi_{\mathsf{Mal}}^{+1}}$ which is guaranteed to exist from the security of π_{Mal}^{+1} to decide which messages to embed in the transactions of honest parties (the messages corresponding to corrupted parties are provided by the adversary). If $S_{\pi_{\mathsf{Mal}}^{+1}}$ would abort, then S interacts the ideal functionality to abort and continues by claiming back all the committed transactions to the honest parties' wallets, as the protocol would. The soundness of the simulation of $S_{\pi_{\mathsf{Mal}}^{+1}}$ ensure that the output of the parties and the contents of the ledger in the real and the ideal world are indistinguishable.

The fact that the protocol will eventual terminate given sufficient rounds of activating every party (i.e., in the terminology of Definition 2, given a sufficiently high threshold T) follows by inspection of the protocol: in each round every party needs at most a (fixed) polynomial number of activations to post the transactions corresponding to his current-round message-vector. (In fact, the polynomial is only needed in the initial committing-transactions round and from that point on it is linear). To complete the proof, we argue that **(1)** when the protocol does not abort, every honest party has a non-negative balance, and **(2)** when the protocol aborts, then honest parties have a positive balance of at least c coins as required by predicate Q for the simulator to be able to complete its simulation and deliver the (possibly aborting) outputs. These properties are argued as follows:

Property (1): The parties that are not in the honest parties' islands cannot claim any transaction that honest parties make towards them as the ledger will see they as not in the island and reject them. Thus by the last round every honest party will have re-claimed all transactions towards parties not in his island. As far as parties in the honest island are concerned, if no abort occurs then every party will claim all the transactions from parties in his island, and therefore his balance will be 0.

Property (2): Assume that the protocol aborts because some (corrupted) P_i broadcasts an inconsistent message in some round ρ. By inspection of the protocol one can verify that honest parties will be able to claim all transaction-commitments done to them up to round ρ (as they honestly execute their protocol) plus all committed transactions that they made for rounds $\rho + 1 \ldots, \rho_c$. Additionally, because P_i broadcasts an inconsistent message in round ρ, he will be unable to claim transactions of honest parties done from round ρ and on; these bitcoins will be reclaimed by the honest parties, thus giving their wallets a positive balance of at least c coins.

Acknowledgements. The first author was supported by ERC project CODAMODA # 259152, and the third author was supported partly by the Swiss National Science Foundation (SNF) Ambizione grant PZ00P2-142549. This work was done (in part) while the authors were visiting the Simons Institute for the Theory of Computing, supported by the Simons Foundation and by the DIMACS/Simons Collaboration in Cryptography through NSF grant #CNS-1523467 and (in part) when visiting the National Kapodistrian University of Athens. The authors thank Andrew Miller for helpful discussions.

References

1. Andrychowicz, M., Dziembowski, S., Malinowski, D., Mazurek, L.: Fair two-party computations via the bitcoin deposits. In: 1st Workshop on Bitcoin Research 2014. Assocation with Financial Crypto (2014). http://eprint.iacr.org/2013/837
2. Andrychowicz, M., Dziembowski, S., Malinowski, D., Mazurek, L.: Secure multi-party computations on bitcoin. In: 2014 IEEE Symposium on Security and Privacy, pp. 443–458. IEEE Computer Society Press, May 2014
3. Asharov, G., Lindell, Y., Zarosim, H.: Fair and efficient secure multiparty computation with reputation systems. In: Sako, K., Sarkar, P. (eds.) ASIACRYPT 2013, Part II. LNCS, vol. 8270, pp. 201–220. Springer, Heidelberg (2013)
4. Asokan, N., Schunter, M., Waidner, M.: Optimistic protocols for fair exchange. In: ACM CCS 1997, pp. 7–17. ACM Press, April 1997
5. Asokan, N., Shoup, V., Waidner, M.: Optimistic fair exchange of digital signatures. In: Nyberg, K. (ed.) EUROCRYPT 1998. LNCS, vol. 1403, pp. 591–606. Springer, Heidelberg (1998)
6. Barak, B., Canetti, R., Lindell, Y., Pass, R., Rabin, T.: Secure computation without authentication. In: Shoup, V. (ed.) CRYPTO 2005. LNCS, vol. 3621, pp. 361–377. Springer, Heidelberg (2005)
7. Bentov, I., Kumaresan, R.: How to use bitcoin to design fair protocols. In: Garay, J.A., Gennaro, R. (eds.) CRYPTO 2014, Part II. LNCS, vol. 8617, pp. 421–439. Springer, Heidelberg (2014)
8. Boneh, D., Naor, M.: Timed commitments. In: Bellare, M. (ed.) CRYPTO 2000. LNCS, vol. 1880, pp. 236–254. Springer, Heidelberg (2000)
9. Cachin, C., Camenisch, J.L.: Optimistic fair secure computation. In: Bellare, M. (ed.) CRYPTO 2000. LNCS, vol. 1880, pp. 93–111. Springer, Heidelberg (2000)
10. Canetti, R.: Universally composable security: a new paradigm for cryptographic protocols. In: 42nd FOCS, pp. 136–145. IEEE Computer Society Press, October 2001
11. Canetti, R., Dodis, Y., Pass, R., Walfish, S.: Universally composable security with global setup. In: Vadhan, S.P. (ed.) TCC 2007. LNCS, vol. 4392, pp. 61–85. Springer, Heidelberg (2007)
12. Canetti, R., Fischlin, M.: Universally composable commitments. In: Kilian, J. (ed.) CRYPTO 2001. LNCS, vol. 2139, pp. 19–40. Springer, Heidelberg (2001)
13. Canetti, R., Jain, A., Scafuro, A.: Practical UC security with a global random oracle. In: Ahn, G.-J., Yung, M., Li, N. (eds.) ACM CCS 2014, pp. 597–608. ACM Press, November 2014
14. Canetti, R., Lindell, Y., Ostrovsky, R., Sahai, A.: Universally composable two-party and multi-party secure computation. In: 34th ACM STOC, pp. 494–503. ACM Press, May 2002

15. Canetti, R., Rabin, T.: Universal composition with joint state. In: Boneh, D. (ed.) CRYPTO 2003. LNCS, vol. 2729, pp. 265–281. Springer, Heidelberg (2003)
16. Cleve, R.: Limits on the security of coin flips when half the processors are faulty (extended abstract). In: Hartmanis, J. (ed.) STOC, pp. 364–369. ACM (1986)
17. Garay, J.A., Gelles, R., Johnson, D.S., Kiayias, A., Yung, M.: A little honesty goes a long way. In: Dodis, Y., Nielsen, J.B. (eds.) TCC 2015, Part I. LNCS, vol. 9014, pp. 134–158. Springer, Heidelberg (2015)
18. Garay, J.A., Katz, J., Maurer, U., Tackmann, B., Zikas, V.: Rational protocol design: cryptography against incentive-driven adversaries. In: 54th FOCS, pp. 648–657. IEEE Computer Society Press, October 2013
19. Garay, J.A., MacKenzie, P.D., Prabhakaran, M., Yang, K.: Resource fairness and composability of cryptographic protocols. In: Halevi, S., Rabin, T. (eds.) TCC 2006. LNCS, vol. 3876, pp. 404–428. Springer, Heidelberg (2006)
20. Goldreich, O.: Foundations of Cryptography: Basic Tools, vol. 1. Cambridge University Press, Cambridge (2001)
21. Goldreich, O., Micali, S., Wigderson, A.: How to play any mental game or a completeness theorem for protocols with honest majority. In: Aho, A. (ed.) 19th ACM STOC, pp. 218–229. ACM Press, May 1987
22. Gordon, S.D., Katz, J.: Complete fairness in multi-party computation without an honest majority. In: Reingold, O. (ed.) TCC 2009. LNCS, vol. 5444, pp. 19–35. Springer, Heidelberg (2009)
23. Ishai, Y., Ostrovsky, R., Zikas, V.: Secure multi-party computation with identifiable abort. In: Garay, J.A., Gennaro, R. (eds.) CRYPTO 2014, Part II. LNCS, vol. 8617, pp. 369–386. Springer, Heidelberg (2014)
24. Katz, J., Maurer, U., Tackmann, B., Zikas, V.: Universally composable synchronous computation. In: Sahai, A. (ed.) TCC 2013. LNCS, vol. 7785, pp. 477–498. Springer, Heidelberg (2013)
25. Kiayias, A., Zhou, H.-S., Zikas, V.: Fair and robust multi-party computation using a global transaction ledger. Cryptology ePrint Archive, Report 2015/574 (2015). http://eprint.iacr.org/2015/574
26. Kosba, A., Miller, A., Shi, E., Wen, Z., Papamanthou, C.: Hawk: The blockchain model of cryptography and privacy-preserving smart contracts. Cryptology ePrint Archive, Report 2015/675, (2015). http://eprint.iacr.org/2015/675
27. Kumaresan, R., Bentov, I.: How to use bitcoin to incentivize correct computations. In: Ahn, G.-J., Yung, M., Li, N. (eds.) ACM CCS 2014, pp. 30–41. ACM Press, November 2014
28. Nakamoto, S.: Bitcoin: A peer-to-peer electronic cash system (2008). http://bitcoin.org/bitcoin.pdf
29. Pinkas, B.: Fair secure two-party computation. In: Biham, E. (ed.) EUROCRYPT 2003. LNCS, vol. 2656, pp. 87–105. Springer, Heidelberg (2003)
30. Ruffing, T., Kate, A., Schröder, D.: Liar, liar, coins on fire!: penalizing equivocation by loss of bitcoins. In: Ray, I., Li, N., Kruegel, C. (eds.) ACM CCS 2015, pp. 219–230. ACM Press, October 2015
31. Yao, A.C.-C.: Protocols for secure computations (extended abstract). In: 23rd FOCS, pp. 160–164. IEEE Computer Society Press, November 1982

Two Round Multiparty Computation via Multi-key FHE

Pratyay Mukherjee[1](✉) and Daniel Wichs[2]

[1] University of California, Berkeley, USA
pratyay85@berkeley.edu
[2] Northeastern University, Boston, USA
wichs@ccs.neu.edu

Abstract. We construct a general multiparty computation (MPC) protocol with only two rounds of interaction in the common random string model, which is known to be optimal. In the honest-but-curious setting we only rely on the learning with errors (LWE) assumption, and in the fully malicious setting we additionally assume the existence of non-interactive zero knowledge arguments (NIZKs). Previously, Asharov et al. (EUROCRYPT '12) showed how to achieve three rounds based on LWE and NIZKs, while Garg et al. (TCC '14) showed how to achieve the optimal two rounds based on indistinguishability obfuscation, but it was unknown if two rounds were possible under standard assumptions without obfuscation.

Our approach relies on *multi-key fully homomorphic encryption (MFHE)*, introduced by Lopez-Alt et al. (STOC '12), which enables homomorphic computation over data encrypted under different keys. We present a construction of MFHE based on LWE that significantly simplifies a recent scheme of Clear and McGoldrick (CRYPTO '15). We then extend this construction to allow for a one-round distributed decryption of a multi-key ciphertext. Our entire MPC protocol consists of the following two rounds:

1. Each party individually encrypts its input under its own key and broadcasts the ciphertext. All parties can then homomorphically compute a multi-key encryption of the output.
2. Each party broadcasts a partial decryption of the output using its secret key. The partial decryptions can be combined to recover the output in plaintext.

P. Mukherjee—Research supported in part from DARPA Safeware Award W911NF15C0210, AFOSR Award FA9550-15-1-0274 and NSF CRII Award 1464397. The views expressed are those of the author and do not reflect the official policy or position of the Department of Defense, the National Science Foundation, or the U.S. Government. Part of the work was done when this author was a PhD Fellow at Aarhus University supported by ERC starting grant 279447.

D. Wichs—Research supported by NSF grants CNS-1347350, CNS-1314722, CNS-1413964.

M. Fischlin and J.-S. Coron (Eds.): EUROCRYPT 2016, Part II, LNCS 9666, pp. 735–763, 2016.
DOI: 10.1007/978-3-662-49896-5_26

1   Introduction

Multiparty Computation. Secure multiparty computation (MPC) allows multiple parties to evaluate an arbitrary function over their inputs *privately*, without revealing anything about their inputs to each other beyond the function's output. This problem was initially studied by Yao [34,35], in the case of *two honest-but-curious parties* (who follow the protocol honestly but hope to learn information from its execution) and later by Goldreich, Micali and Wigderson [19] in the case of an arbitrary number of *fully malicious* parties (who can deviate arbitrarily from the specified protocol execution). By now, MPC is a fundamental part of cryptography and a subject of intense study.

One of the main challenges is to optimize the efficiency of MPC protocols. In this work, our main focus will be on constructing MPC protocols with the optimal round complexity.[1]

Round Complexity of MPC. We refer the reader to [3] for a comprehensive overview of prior work on round complexity of MPC. In the honest-but-curious setting, it was known how to achieve a constant number of rounds assuming the existence of oblivious transfer [2,5,22,25,27]. However, the concrete constants were not explicitly stated and they seem to require at least 4 rounds. These protocols can also be compiled into secure constructions in the fully malicious setting with only a constant number of additional rounds by using coin-flipping and concurrent zero-knowledge proofs [20,25–27]. In the plain model and the fully malicious setting, there is a known lower bounds of 5 rounds for two party computation[2], albeit in non-simultaneous message model where no broadcast channel is available [24]. A very recent work [15] shows a similar lower bound of 4 rounds assuming broadcast channel. However, in the honest-but-curious setting or even in the fully malicious setting with a common random string (CRS) the above lower bound does not hold and there is only a simple lower bound of 2 rounds [21]. In this work, we will assume the CRS model.

Recently, a result of Asharov et al. [3] showed how to achieve a 3 round MPC protocol in the CRS model, by relying on techniques from fully homomorphic encryption (FHE). Their construction achieves semi-honest security under the *learning with errors* (LWE) assumption, and fully malicious security (in the universal composability (UC) framework) by further assuming the existence of non-interactive zero knowledge arguments (NIZKs). The construction also yields a 2 round protocol in the public-key infrastructure (PKI) model, but it was left as an open problem to achieve 2 rounds in the CRS model.

---

[1] We assume a broadcast communication channel and in each round of a protocol all parties broadcast a message to all other parties. Each honest party must broadcast their round i message right away prior to receiving the round i messages of other parties. On the other hand, we assume a "rushing" adversary that can wait to collect the round i messages of all honest parties prior to selecting the round i messages of the corrupted parties.

[2] In their non-simultaneous model no broadcast channel is present, that is in a round only one party sends message to another.

Even more recently, the results of Garg et al. [14,16] achieve a 2 round MPC protocol in the CRS model by relying on *indistinguishability obfuscation* (iO) and statistically sound NIZKs. On a high level, the main idea of that work is to have each party obfuscates its "next-message" function, after an initial round where the parties commit to their input. Making this work under the iO assumption is non-trivial and requires much care. However, this approach appears to crucially rely on obfuscation and does not easily lend itself to instantiations under simpler assumptions.

The main open question left by these works is whether 2 round MPC is achievable under more "standard" cryptographic assumptions, without relying on obfuscation.

Our Result. In this work, we construct a 2 round MPC protocol in the CRS model. We achieve honest-but-curious security under only the LWE assumption, and fully malicious security (in the UC framework) by additionally assuming the existence of NIZKs. As our main technical result, which may be of independent interest, we show how to construct a multi-key fully homomorphic encryption scheme with a one-round threshold decryption protocol.

2 Overview of Our Techniques

We now give an overview of our techniques by first describing how to construct MPC from multi-key FHE with threshold decryption, and then how to construct the latter from LWE.

2.1 MPC via Threshold (Multi-key) FHE

MPC via Threshold FHE. We begin with the approach of Asharov et al. [3] (variants of which were used in many preceding works [6,7,11–13,23,31]) for constructing MPC based on fully homomorphic encryption (FHE). At a high level, this approach is based on the following simple template:

1. The parties first run a secure distributed protocol for the "threshold key-generation" of an FHE scheme to agree on a common public key pk and a secret sharing of the corresponding secret key sk so that each party holds one share, and all shares are needed to recover sk.
2. Each party i then broadcasts an encryption of its input x_i under the common public key pk. Note that no individual party or incomplete set of parties can decrypt this ciphertext and so the privacy of the input is maintained. At the end of this round, each party can homomorphically compute the desired function f on the received ciphertexts and derive a common output ciphertext which encrypts $y = f(x_1, \ldots, x_N)$.

3. The parties run a secure distributed protocol for "threshold decryption" using their shares of the secret key sk to decrypt the output ciphertext and recover the output y in plaintext.[3]

Secure protocols for threshold key-generation and decryption can be implemented generically for any FHE scheme by using general MPC techniques, but this would require many rounds. Instead [3] show that specific FHE schemes by Brakerski, Gentry and Vaikuntanathan [8,9] based on the LWE assumption have a "key homomorphic" property which can be leveraged to get distributed key-generation and decryption protocols consisting of one round each. Therefore, when instantiated with these schemes, the above template results in a 3 round MPC protocol.[4]

MPC via Threshold Multi-key FHE. One could hope to shave off an additional round from the above template by using *multi-key fully homomorphic encryption* (MFHE), recently introduced by Lopez-Alt, Tromer and Vaikuntanathan [28]. An MFHE schemes allows parties to independently encrypt their data under different individually chosen keys, while still allowing homomorphic computations over such ciphertexts. The output of such homomorphic computation is a "multi-key ciphertext" which cannot be decrypted by any single party individually (as this would violate semantic security of the other parties) but can be decrypted by the parties jointly using the combination of all their secret keys. The work of [28] constructed such an MFHE scheme based on (a variant of) the NTRU assumption.

Using MFHE, we naturally get the following simplified template for MPC:

1. Each party individually chooses its own MFHE key pair (pk_i, sk_i), encrypts its input x_i under pk_i, and broadcasts the resulting ciphertext. At the end of this round, each party can homomorphically compute the desired function f on the received ciphertexts and derive a common multi-key ciphertext which encrypts the output $y = f(x_1, \ldots, x_N)$.
2. The parties run a secure distributed protocol for "threshold decryption" using their secret keys sk_i to decrypt the multi-key ciphertext and recover the output y in plaintext.

As before, a distributed threshold decryption can be implemented generically using general MPC techniques, but this would require many rounds. Unfortunately, the MFHE scheme of [28] does not appear to admit any simpler threshold

---

[3] Throughout this work, we use the term "threshold" to denote distributed schemes where *all* parties are needed to perform an operation and security is maintained from any incomplete subset of parties.

[4] We note that one the main challenges in the work of [3] is to implement the threshold generation of the FHE "evaluation key" which has a complex structure in the FHE schemes of [8,9]. This could be vastly simplified using a more recent FHE scheme of Gentry-Sahai-Waters [18] which does not require an evaluation key. However, this would still not improve the final round complexity of the MPC construction below 3 rounds.

decryption protocol and therefore it is not known how to use this scheme to get a 2 round MPC.

A recent work of Clear and McGoldrick [10] gives an alternate construction of MFHE based on the LWE assumption, by cleverly adapting an FHE scheme of Gentry, Sahai and Waters [18]. We first present a significantly simplified construction of MFHE from LWE and give a stand-alone presentation of this scheme. We then show that this scheme admits a simple 1-round threshold decryption protocol. This threshold decryption protocol only satisfies a weak notion of security which doesn't allow us to directly plug it into the above template for MPC. However, we show that we can make this approach work with only minor additional modifications.

As in [3], we show that our basic scheme (based on LWE) achieves security in *semi-malicious* setting, which is a strengthening of the honest-but-curious setting, where parties follow the protocol specification but can choose their random coins adversarially. By using NIZKs, we can then compile such a scheme into one which is secure in the *fully malicious setting* (and even universally composable) without additional rounds.

2.2 Constructing Threshold Multi-key FHE

We now give a high-level description of the MFHE construction and the threshold decryption protocol. We begin by describing a recent FHE construction by Gentry, Sahai and Waters (GSW) [18] using the notation and exposition of [1]. Then describe how to convert it into a MFHE scheme. Finally, we discuss how to perform threshold decryption.

Public Short Preimage Matrix. Before we describe the GSW encryption, we state a useful fact from [29] which we heavily rely on in the construction.

Lemma 1 ([29]). *For any $m \geq n\lceil \log q \rceil$ there exists a fixed efficiently computable matrix $\mathbf{G} \in \mathbb{Z}_q^{n \times m}$ and an efficiently computable deterministic "short preimage" function $\mathbf{G}^{-1}(\cdot)$ satisfying the following. On input a matrix $\mathbf{M} \in \mathbb{Z}_q^{n \times m'}$ for any m', the function $\mathbf{G}^{-1}(\mathbf{M})$ outputs a bit-matrix $\mathbf{G}^{-1}(\mathbf{M}) \in \{0,1\}^{m \times m'}$ such that $\mathbf{GG}^{-1}(\mathbf{M}) = \mathbf{M}$.*

We can think of $\mathbf{G}$ as a special matrix with a "public trapdoor" that allows us to solve the *short integer solution (SIS)* problem. For those familiar with GSW encryption, multiplication by $\mathbf{G}$ is the $\mathsf{BitDecomp}^{-1}$ operation and the function $\mathbf{G}^{-1}(\cdot)$ is called $\mathsf{BitDecomp}$, but we can ignore the low-level detail of how this is implemented. Note that $\mathbf{G}^{-1}(\cdot)$ is not itself a matrix but rather an efficiently computable function.

Gentry-Sahai-Waters (GSW) FHE. Firstly, choose a random public matrix $\mathbf{B} \in \mathbb{Z}_q^{(n-1) \times m}$ where $m = O(n \log q)$. We can think of this as a common public parameter used by all parties. A public/secret key pair is chosen by selecting a random vector $\mathbf{s} \in \mathbb{Z}_q^{n-1}$ and setting $\mathbf{b} = \mathbf{sB} + \mathbf{e}$ where $\mathbf{e}$ is some short "error vector". We set the secret key to $\mathbf{t} = (-\mathbf{s}, 1) \in \mathbb{Z}_q^n$ and the public key to the matrix

$$\mathbf{A} := \begin{bmatrix} \mathbf{B} \\ \mathbf{b} \end{bmatrix} \in \mathbb{Z}_q^{n \times m}$$

which ensures that $\mathbf{tA} = \mathbf{e} \approx \mathbf{0}$ (throughout the introduction, we use $\approx$ to hide "short" values).

A valid GSW ciphertext of a bit $\mu \in \{0, 1\}$ with respect to a secret key $\mathbf{t}$ is a matrix $\mathbf{C} \in \mathbb{Z}_q^{n \times m}$ such that $\mathbf{tC} \approx \mu \mathbf{tG}$. To encrypt a bit μ using the public key $\mathbf{A}$ we set $\mathbf{C} = \mathbf{AR} + \mu \mathbf{G}$ where $\mathbf{R} \in \{0, 1\}^{m \times m}$ is chosen as a random bit-matrix. This ensures that the result is a valid encryption of μ under the secret key $\mathbf{t}$ since $\mathbf{tC} = \mathbf{tAR} + \mu \mathbf{tG} \approx \mu \mathbf{tG}$.

Given two valid GSW ciphertexts $\mathbf{C}_1, \mathbf{C}_2$ encrypting the bits μ_1, μ_2 with respect to a secret key $\mathbf{t}$ we can perform homomorphic addition by setting $\mathbf{C}^+ = \mathbf{C}_1 + \mathbf{C}_2$ and multiplication by setting $\mathbf{C}^\times = \mathbf{C}_1 \mathbf{G}^{-1}(\mathbf{C}_2)$. It is a simple exercise to check that $\mathbf{tC}^+ \approx (\mu_1 + \mu_2)\mathbf{tG}$ and $\mathbf{tC}^\times \approx (\mu_1 \mu_2)\mathbf{tG}$. This allows us to homomorphically evaluate any circuit, subject to the error not getting "too large".

Finally, to decrypt a ciphertext $\mathbf{C}$ we set $\mathbf{w} := (0, \ldots, 0, \lceil q/2 \rceil)$ and compute $v = \mathbf{tCG}^{-1}(\mathbf{w}^T)$. If $\mathbf{C}$ is a valid encryption of μ under $\mathbf{t}$ then $v \approx \mu \lceil q/2 \rceil$. We recover μ by checking whether v is closer to 0 or to $q/2$.

Multi-key Variant of GSW. We now describe how to convert the above GSW FHE into a multi-key FHE. For simplicity, let's assume that we only have $N = 2$ parties, but everything extends naturally to any polynomial number of parties N. We assume that the matrix $\mathbf{B}$ of the GSW encryption scheme is a common public parameter which is used by all parties.

The two parties choose independent GSW secret keys $\mathbf{t}_1 = (-\mathbf{s}_1, 1), \mathbf{t}_2 = (-\mathbf{s}_2, 1)$ and compute the corresponding public key components $\mathbf{b}_1 = \mathbf{s}_1 \mathbf{B} + \mathbf{e}_1$ and $\mathbf{b}_2 = \mathbf{s}_2 \mathbf{B} + \mathbf{e}_2$ using the common (and random) $\mathbf{B}$. We let

$$\mathbf{A}_1 := \begin{bmatrix} \mathbf{B} \\ \mathbf{b}_1 \end{bmatrix}, \qquad \mathbf{A}_2 := \begin{bmatrix} \mathbf{B} \\ \mathbf{b}_2 \end{bmatrix}$$

be the two GSW public keys for parties 1 and 2 respectively.

Now assume that the two parties independently encrypt some data under their respective keys. Unfortunately, we will not get anything meaningful by naively attempting to perform the GSW homomorphic operations on these ciphertexts under different keys. Instead, our goal will be to first convert both ciphertexts into a "common format" that will allow us to perform homomorphic operations over them.

In particular, we define a "combined secret key" $\widehat{\mathbf{t}} = (\mathbf{t}_1, \mathbf{t}_2) \in \mathbb{Z}_q^{2n}$ as the concatenation of the two individual secret keys. Our goal will be to take a ciphertext $\mathbf{C} \in \mathbb{Z}_q^{n \times m}$ which encrypts a bit μ with respect to the secret key of a single party (along with some helper information specified later) and expand it into multi-key ciphertext $\widehat{\mathbf{C}} \in \mathbb{Z}_q^{2n \times 2m}$ which encrypts μ with respect to the combined secret key $\widehat{\mathbf{t}}$. In particular, a multi-key encryption of a bit μ satisfies $\widehat{\mathbf{t}}\widehat{\mathbf{C}} \approx \mu\widehat{\mathbf{t}}\widehat{\mathbf{G}}$ where $\widehat{\mathbf{G}} = \begin{bmatrix} \mathbf{G} & \mathbf{0} \\ \mathbf{0} & \mathbf{G} \end{bmatrix} \in \mathbb{Z}_q^{2n \times 2m}$ is an expanded public matrix with a corresponding short preimage function $\widehat{\mathbf{G}}^{-1}(\cdot)$. Once we do this, we can expand all ciphertexts under individual keys into multi-key ciphertexts under the key $\widehat{\mathbf{t}}$ and then perform homomorphic operations on the multi-key ciphertexts just like in basic GSW scheme (just with larger parameters $n' = 2n, m' = 2m$). Therefore, the only challenge is how to perform the above "ciphertext expansion step".

Ciphertext Expansion. To perform ciphertext expansion, we use a new primitive called "masking scheme" introduced by Clear and McGoldrick in [10]. Let $\mathbf{C}$ be a GSW encryption of some bit μ. A masking scheme allows party 1 to create some additional helper information $\mathcal{U}$ about the ciphertext $\mathbf{C}$ at encryption time and release the tuple $(\mathcal{U}, \mathbf{C})$ while keeping the semantic security of the message intact. This information is completely independent of party 2 whose identity is unknown at encryption time. Later, if we are given the public key $\mathbf{A}_2$ for party 2, we can use the information $\mathcal{U}$ to create a matrix $\mathbf{X}$ such that $\mathbf{t}_1\mathbf{X} + \mathbf{t}_2\mathbf{C} \approx \mu\mathbf{t}_2\mathbf{G}$ where $\mathbf{t}_2$ is the secret key of party 2. This allows us to perform ciphertext expansion by creating the expanded ciphertext:

$$\widehat{\mathbf{C}} = \begin{bmatrix} \mathbf{C} & \mathbf{X} \\ \mathbf{0} & \mathbf{C} \end{bmatrix}$$

so that,

$$\widehat{\mathbf{t}}\widehat{\mathbf{C}} = [\, \mathbf{t}_1\mathbf{C} \,,\, \mathbf{t}_1\mathbf{X} + \mathbf{t}_2\mathbf{C} \,] \approx [\, \mu\mathbf{t}_1\mathbf{G}, \mu\mathbf{t}_2\mathbf{G} \,] = \mu\widehat{\mathbf{t}}\widehat{\mathbf{G}}.$$

We can similarly expand the individually created ciphertexts of party 2 and then perform GSW style homomorphic operations on the expanded ciphertexts.[5] Therefore, the only thing left to do is to construct such a "masking scheme" which we briefly describe below.

A Masking Scheme for GSW. The masking scheme consists of party 1 creating tuple $(\mathcal{U}, \mathbf{C})$ where $\mathbf{C}$ is a GSW encryption of the message μ under its own public key $pk_1 = \mathbf{A}_1$ so that

$$\mathbf{C} := \mathbf{A}_1\mathbf{R} + \mu\mathbf{G} = \begin{bmatrix} \mathbf{B}\mathbf{R} \\ \mathbf{b}_1\mathbf{R} \end{bmatrix} + \mu\mathbf{G}$$

for some random matrix $\mathbf{R} \in \{0,1\}^{m \times m}$. The additional helper information $\mathcal{U}$ consists of m^2 GSW encryptions of each of the scalars $\{\mathbf{R}[a, b]\}_{a \in [m], b \in [m]}$ under

---

[5] In the actual scheme involving N parties we first expand the single-key ciphertext of each party into a multi-key ciphertext (under the concatenated keys of all the parties) and subsequently perform homomorphic operations on the expanded ciphertexts.

the public key pk_1. It is easy to show that the pair $(\mathcal{U}, \mathbf{C})$ computationally hides μ by relying on semantic security of the GSW scheme.

Later, assume we are given the public key $\mathbf{A}_2 := \begin{bmatrix} \mathbf{B} \\ \mathbf{b}_2 \end{bmatrix}$ for party 2, corresponding to a secret key $\mathbf{t}_2 = (-\mathbf{s}_2, 1)$. Then

$$\mathbf{t}_2 \mathbf{C} = -\mathbf{s}_2 \mathbf{B} \mathbf{R} + \mathbf{b}_1 \mathbf{R} + \mu \mathbf{t}_2 \mathbf{G} \approx (\mathbf{b}_1 - \mathbf{b}_2) \mathbf{R} + \mu \mathbf{t}_2 \mathbf{G}$$

since $\mathbf{b}_2 \approx \mathbf{s}_2 \mathbf{B}$. The value $\mathbf{t}_2 \mathbf{C}$ corresponds to decrypting the GSW ciphertext $\mathbf{C}$ with the "incorrect" secret key $\mathbf{t}_2$ and it yields the correct value $\mu \mathbf{t}_2 \mathbf{G}$ *except* that it is "masked" by the additional term $(\mathbf{b}_1 - \mathbf{b}_2) \mathbf{R}$.

Our goal is to come up with a matrix $\mathbf{X}$ for which $\mathbf{t}_1 \mathbf{X} \approx (\mathbf{b}_2 - \mathbf{b}_1) \mathbf{R}$ and therefore adding $\mathbf{t}_1 \mathbf{X} + \mathbf{t}_2 \mathbf{C} \approx \mu \mathbf{t}_2 \mathbf{G}$ as desired. One can do this by homomorphically combining the m^2 ciphertexts contained in $\mathcal{U}$, which encrypt each of the scalars $\mathbf{R}[a, b]$ of the matrix $\mathbf{R}$ under $\mathbf{t}_1$, to get a "pseudo ciphertext" $\mathbf{X}$ which acts like an encryption of the vector $(\mathbf{b}_2 - \mathbf{b}_1) \mathbf{R}$ in the sense that $\mathbf{t}_1 \mathbf{X} \approx (\mathbf{b}_2 - \mathbf{b}_1) \mathbf{R}$. This is not a standard homomorphic operation yielding a standard ciphertext – for example, the output is a vector rather than a scalar – but the idea for how to do this is very similar to the way we do standard GSW homomorphic operations. We skip the details of this step in the introduction, and refer the reader to Sect. 5.1 for details.

Threshold Decryption of Multi-key GSW. A multi-key GSW ciphertext encrypting a bit μ with respect to the expanded secret key $\widehat{\mathbf{t}} = (\mathbf{t}_1, \ldots, \mathbf{t}_N)$ corresponding to N parties, is a matrix $\widehat{\mathbf{C}} \in \mathbb{Z}_q^{nN \times mN}$ such that $\widehat{\mathbf{t}} \widehat{\mathbf{C}} \approx \mu \widehat{\mathbf{t}} \widehat{\mathbf{G}}$.

If we were given all of the secret keys $\widehat{\mathbf{t}} = (\mathbf{t}_1, \ldots, \mathbf{t}_N)$ simultaneously, we could decrypt this ciphertext using the GSW decryption procedure, scaled up to the larger dimension: let $\widehat{\mathbf{w}} = (0, \ldots, 0, \lceil q/2 \rceil) \in \mathbb{Z}_q^{nN}$ and compute $v = \widehat{\mathbf{t}} \widehat{\mathbf{C}} \widehat{\mathbf{G}}^{-1}(\widehat{\mathbf{w}}^T) \approx \mu \lceil q/2 \rceil$.

However, our goal is to design a distributed decryption protocol, where the parties collaboratively decrypt μ without revealing their secret keys to each other. We do this as follows. Let's think of $\widehat{\mathbf{C}}$ as consisting of N matrices $i \in \mathbb{Z}_q^{n \times mN}$ stacked on top of each other. Then each party i uses its secret key $\mathbf{t}_i$ to output a "partial decryption" $p_i = \mathbf{t}_i \widehat{\mathbf{C}}^{(i)} \widehat{\mathbf{G}}^{-1}(\widehat{\mathbf{w}}^T) + e_i$ where e_i is some "medium-sized smudging error". This error is needed to smudge out any information about the error contained in the ciphertext $\widehat{\mathbf{C}}$, which might contain sensitive information beyond just the plaintext bit. These partial decryptions can be combined to compute $\sum_i p_i \approx v \approx \mu \lceil q/2 \rceil$ and therefore recover the plaintext bit μ.

The above process satisfies the following security notion: given the ciphertext $\widehat{\mathbf{C}}$, the bit μ that it encrypts, and the secret keys $\{\mathbf{t}_i : i \neq j\}$ of all-but-one of the parties, we can simulate the partial decryption p_j of party j without knowing its secret key $\mathbf{t}_j$. Intuitively, this property says that the partial decryption p_j cannot reveal too much information about $\mathbf{t}_j$.

The above security property of partial decryption is tricky to use since it allows us to simulate the partial decryption of only one party at a time. Nevertheless, we show that this security property of threshold decryption is sufficient in the context of implementing MPC.

2.3 Road-Map Through the Paper

We begin by giving a definition of multi-key FHE (MFHE) first and then MFHE with threshold decryption in Sect. 4. Then in Sect. 5 we construct such a scheme from the LWE and in Sect. 6 we show how to construct MPC from such a scheme. These two sections are independent of each other and can be read in any order.

3 Preliminaries

Throughout, we let λ denote the *security parameter* and $\mathsf{negl}(\lambda)$ denote a negligible function. We represent elements in $\mathbb{Z}_q$ as integers in the range $(-q/2, q/2]$. Let $\mathbf{x} = (x_1, \ldots, x_n) \in \mathbb{Z}^n$ be a vector. We use the notation $\mathbf{x}[i]$ to denote the ith component scalar. Similarly for a matrix $\mathbf{M} \in \mathbb{Z}^{n \times m}$ we use $\mathbf{M}[i,j]$ to denote the scalar element located in the i-th row and the j-th column. In general, vectors are represented as single row matrices. The infinity norm (often called simply *norm*) of a vector $\mathbf{x}$ is defined as $\|\mathbf{x}\|_\infty = \max_i(|\mathbf{x}[i]|)$. The norm of matrices is defined similarly. An n-dimensional all-zero vector is usually denoted by 0^n and similarly $0^{n \times m}$ denotes an all-zero matrix.

For two distributions X, Y, over a finite domain Ω, the *statistical distance* between X and Y is defined by $\Delta(X,Y) \overset{\text{def}}{=} \frac{1}{2} \sum_{\omega \in \Omega} |X(\omega) - Y(\omega)|$. If X, Y are distribution ensembles parameterized by the security parameter, we write $X \overset{\text{stat}}{\approx} Y$ if the quantity $\Delta(X,Y)$ is negligible. Similarly, we write $X \overset{\text{comp}}{\approx} Y$ if they are computationally indistinguishable. We write $\omega \leftarrow X$ to denote that ω is sampled at random according to distribution X. We write $\omega \leftarrow \Omega$ to denote that it is sampled *uniformly at random* from the set Ω. For a distribution ensemble $\chi = \chi(\lambda)$ over the integers, and integers bounds $B = B(\lambda)$, we say that χ is B-*bounded* if $\Pr_{x \leftarrow \chi(\lambda)}[|x| \leq B(\lambda)] = 1$.

We rely on the following lemma, which says that adding large noise "smudges out" any small values (see e.g., [4] for proof).

Lemma 2 (Smudging Lemma). *Let $B_1 = B_1(\lambda)$, and $B_2 = B_2(\lambda)$ be positive integers and let $e_1 \in [-B_1, B_1]$ be a fixed integer. Let $e_2 \leftarrow [-B_2, B_2]$ be chosen uniformly at random. Then the distribution of e_2 is statistically indistinguishable from that of $e_2 + e_1$ as long as $B_1/B_2 = \mathsf{negl}(\lambda)$.*

Learning with Errors. The *decisional learning with errors (LWE)* problem, introduced by Regev [33], is defined as follows.

Definition 1 (LWE [33]). *Let λ be the security parameter, $n = n(\lambda), q = q(\lambda)$ be integers and let $\chi = \chi(\lambda)$, be distributions over $\mathbb{Z}$. The $\mathsf{LWE}_{n,q,\chi}$ assumption says that for any polynomial $m = m(\lambda)$ we have*

$$(\mathbf{A}, \mathbf{s}\mathbf{A} + \mathbf{e}) \stackrel{comp}{\approx} (\mathbf{A}, \mathbf{z})$$

where $\mathbf{A} \leftarrow \mathbb{Z}_q^{n \times m}$, $\mathbf{s} \leftarrow \mathbb{Z}_q^n$, $\mathbf{e} \leftarrow \chi^m$ and $\mathbf{z} \leftarrow \mathbb{Z}_q^m$.

The works of [32, 33] show that the LWE problem is as hard as approximating the shortest vector problem in lattices (for appropriate parameters). The version of the LWE assumption that we need here is that for any polynomial $p = p(\lambda)$ there is a polynomial $n = n(\lambda)$, a modulus $q = q(\lambda)$ of singly-exponential size, and a distribution $\chi = \chi(\lambda)$ such that χ is B_χ-bounded and $q \geq 2^p B_\chi$ such that $\mathsf{LWE}_{n,q,\chi}$ holds. This is as hard as approximating the shortest vector with sub-exponential approximation factors.

4 Defining Threshold Multi-key FHE

4.1 Multi-key FHE (MFHE)

We start with our definition of (leveled) multi-key FHE which is adapted from the definition given by Lopez-Alt, Tromer and Vaikuntanathan [28] with some minor differences which reflect differences in the properties achieved by the schemes of [28] and [10]. On the positive side, in the scheme of [10] the number of parties N need not be known ahead of time during key generation or encryption. On the negative side, the scheme of [10] requires some common public parameters that are available to the parties during key generation.

Below we call any ciphertext which is associated with multiple keys an "expanded" ciphertext. Also, the ciphertexts that are generated by the encryption procedure (and thus corresponds to a single key) are called "fresh" ciphertexts, and the expanded ciphertexts that are output by the homomorphic evaluations are called "evaluated" ciphertexts.

Definition 2 (Multi-key (Leveled) FHE). *A multi-key (leveled) FHE is a tuple of algorithms $\mathsf{MFHE} = (\mathsf{Setup}, \mathsf{Keygen}, \mathsf{Encrypt}, \mathsf{Expand}, \mathsf{Eval}, \mathsf{Decrypt})$ described as follows:*

- $\mathsf{params} \leftarrow \mathsf{Setup}(1^\lambda, 1^d)$: *Setup takes as input the security parameter λ and the circuit depth d and outputs the system parameters params. We assume that all the other algorithms take params as an input implicitly.*
- $(sk, pk) \leftarrow \mathsf{Keygen}(\mathsf{params})$: *Output secret key sk and public key pk.*
- $c \leftarrow \mathsf{Encrypt}(pk, \mu)$: *On input pk and some message μ output a ciphertext c.*
- $\widehat{c} \leftarrow \mathsf{Expand}((pk_1, \dots, pk_N), i, c)$: *Given a sequence of N public-keys and a fresh ciphertext c under the i-th key pk_i, it outputs an "expanded" ciphertext $\widehat{c}$.*
- $\widehat{c} := \mathsf{Eval}(\mathsf{params}, \mathcal{C}, (\widehat{c}_1, \dots, \widehat{c}_\ell))$: *Given a (description of) boolean circuit $\mathcal{C}$ of depth $\leq d$ along with ℓ expanded ciphertexts $(\widehat{c}_1, \dots, \widehat{c}_\ell)$, outputs an evaluated ciphertext $\widehat{c}$.*

- $\mu := \mathsf{Decrypt}(\mathsf{params}, (sk_1, \ldots, sk_N), c)$: *On input some ciphertext $\widehat{c}$ and a sequence of N secret keys output a message μ.*

We require the following properties:

Semantic security of encryption: *For any polynomial $d = d(\lambda)$ and any two messages μ_0, μ_1 the following distributions are computationally indistinguishable:*

$$(\mathsf{params}, pk, \mathsf{Encrypt}(pk, \mu_0)) \overset{\text{comp}}{\approx} (\mathsf{params}, pk, \mathsf{Encrypt}(pk, \mu_1))$$

where $\mathsf{params} \leftarrow \mathsf{Setup}(1^\lambda, 1^d), (sk, pk) \leftarrow \mathsf{Keygen}(\mathsf{params})$.

Correctness and compactness: *Let* $\mathsf{params} \leftarrow \mathsf{Setup}(1^\lambda, 1^d)$. *Consider any sequences of N correctly generated key pairs $\{(pk_i, sk_i) \leftarrow \mathsf{Keygen}(\mathsf{params})\}_{i \in [N]}$ and any ℓ-tuple of messages $(\mu_1, \ldots, \mu_\ell)$. For any sequence of indices $(I_1, \ldots, I_\ell)$ where each $I_i \in [N]$ let $\{c_i \leftarrow \mathsf{Encrypt}(pk_{I_i}, \mu_i)\}_{i \in [\ell]}$ be encryptions of the messages μ_i under the I_i-th public key and let $\widehat{c}_i \leftarrow \mathsf{Expand}((pk_1, \ldots, pk_N), I_i, c_i)\}_{i \in [\ell]}$ be the corresponding expanded ciphertexts. Let $\mathcal{C}$ be any (boolean) circuit of depth $\leq d$ and let $\widehat{c} := \mathsf{Eval}(\mathcal{C}, (\widehat{c}_1, \ldots, \widehat{c}_\ell)$ be the evaluated ciphertext. Then the following holds:*

CORRECTNESS OF EXPANSION: $\forall\, i \in [\ell]$, $\mathsf{Decrypt}((sk_1, \ldots, sk_N), \widehat{c}_i) = \mu_i$.
CORRECTNESS OF EVALUATION: $\mathsf{Decrypt}((sk_1, \ldots, sk_N), \widehat{c}) = \mathcal{C}(\mu_1, \ldots, \mu_\ell)$.
COMPACTNESS: *There exists a polynomial $p(\cdots)$ such that $|\widehat{c}| \leq p(\lambda, d, N)$. In other words the size of $\widehat{c}$ should be independent of $\mathcal{C}$ and ℓ, but can depend on λ, d and N.*

Public-Coin Parameter Generation. By default, we will consider schemes where the Setup algorithm is "public-coin" meaning that its randomness is included in its output. For such algorithms, we can derive params from a common random string.

4.2 Threshold Decryption for MFHE

We now define a multi-key FHE which supports a one-round threshold distributed decryption protocol. Such a protocol consists of two components: (1) given an expanded ciphertext (possibly evaluated) $\widehat{c}$ each party can compute a partial decryption using its secret key sk_i, (2) there is a way to combine the partial decryptions computed by each party to recover the plaintext.

Definition 3. *A Threshold multi-key FHE scheme (TMFHE) is a multi-key FHE scheme with two additional algorithms* $\mathsf{MFHE.PartDec}, \mathsf{MFHE.FinDec}$ *described as follows:*

- $p_i \leftarrow \mathsf{MFHE.PartDec}(\widehat{c}, (pk_1, \ldots, pk_N), i, sk_i)$: *On input an expanded ciphertext under a sequence of N keys and the i-th secret key output a partial decryption p_i.*

- $\mu \leftarrow$ MFHE.FinDec($p_1, \ldots, p_N$): *On input N partial decryption output the plaintext μ.*

Along with the properties of multi-key FHE we require the scheme to satisfy the following properties.

Correctness and Simulation: *Let* params $\leftarrow$ Setup($1^\lambda, 1^d$). *Consider any sequences of N correctly generated key pairs $\{(pk_i, sk_i) \leftarrow$ Keygen(params)$\}_{i \in [N]}$ and any ℓ-tuple of messages $(\mu_1, \ldots, \mu_\ell)$. For any sequence of indices $(I_1, \ldots, I_\ell)$ where each $I_i \in [N]$ let $\{c_i \leftarrow$ Encrypt(pk_{I_i}, μ_i)$\}_{i \in [\ell]}$ be encryptions of the messages μ_i under the I_i-th public key and let $\widehat{c}_i \leftarrow$ Expand($(pk_1, \ldots, pk_N), I_i, c_i$)$\}_{i \in [\ell]}$ be the corresponding expanded ciphertexts. Let $\mathcal{C}$ be any (boolean) circuit of depth $\leq d$ and let $\widehat{c} :=$ Eval($\mathcal{C}, (\widehat{c}_1, \ldots, \widehat{c}_\ell)$) be the evaluated ciphertext.*

CORRECTNESS OF DECRYPTION: *The following holds with probability 1:*

$$\text{MFHE.FinDec}(\widehat{c}, (p_1, \ldots, p_N)) = \mathcal{C}(\mu_1, \ldots, \mu_\ell)$$

where $\{p_i \leftarrow$ MFHE.PartDec($\widehat{c}, (pk_1, \ldots, pk_N), i, sk_i$)$\}_{i \in [N]}$ are the partial decryptions.

SIMULATABILITY OF PARTIAL DECRYPTION: *There exists a PPT simulator $\mathcal{S}^{thr}$ which, on input and index $i \in [N]$ and all but the i-th keys $\{sk_j\}_{j \in [N] \setminus \{i\}}$ the evaluated ciphertext $\widehat{c}$ and the output message $\mu := \mathcal{C}(\mu_1, \ldots, \mu_\ell)$ produces a simulated partial decryption $p'_i \leftarrow \mathcal{S}^{thr}(\mu, \widehat{c}, i, \{sk_j\}_{j \in [N] \setminus \{i\}})$ such that*

$$p_i \overset{stat}{\approx} p'_i$$

where $p_i \leftarrow$ MFHE.PartDec($\widehat{c}, (pk_1, \ldots, pk_N), i, sk_i$). Note that the randomness is only over the random coins of the simulator and the MFHE.PartDec procedure and all other values are assumed to be fixed (and known).

The simulatability of partial decryptions property says that we can simulate the partial decryption p_i produced by a single party i given the plaintext value μ and the secret keys of all other parties. Ideally, we would have a stronger definition that allows us to simulate the partial decryptions $\{p_i\}_{i \in S}$ of any subset of the parties S given the secret keys of all other parties (rather than just a single values), but unfortunately we do not know how to achieve this type of security. It turns out that, with a little additional work, the given definition suffices in our MPC construction.

5 Constructing Threshold Multi-key FHE from LWE

We now show how to construct threshold multi-key FHE from LWE. The construction proceeds in four parts. First, we present the GSW encryption scheme along with a non-standard but useful homomorphic property that it satisfies. Secondly, we define the notion of a masking scheme for GSW and show how to construct it. Thirdly, we use GSW and the masking scheme to construct multi-key FHE. Finally, we show to perform threshold decryption for this scheme.

5.1 GSW Fully Homomorphic Encryption

We now describe the GSW fully homomorphic encryption scheme.

- params $\leftarrow$ GSW.SetUp($1^\lambda, 1^d$): Choose a lattice dimension parameters $n = n(\lambda, d)$ and B_χ-bounded error distribution $\chi = \chi(\lambda, d)$ and a modulus q of size $q = B_\chi 2^{\omega(d\lambda \log \lambda)}$ such that $\mathsf{LWE}_{n-1,q,\chi,B_\chi}$ holds.[6] Choose $m = n \log(q) + \omega(\log \lambda)$. Finally choose a random matrix $\mathbf{B} \in \mathbb{Z}_q^{n-1 \times m}$. Output params := $(q, n, m, \chi, B_\chi, \mathbf{B})$. We stress that all the other algorithms implicitly get params as input even if we usually do not write this explicitly.
- GSW.Keygen(params): We separately describe two sub-algorithms to generate secret-key and pubic-key respectively:
  - GSW.SKGen(params): Sample $\mathbf{s} \xleftarrow{\$} \mathbb{Z}_q^{n-1}$. Output $sk = \mathbf{t} = (-\mathbf{s}, 1) \in \mathbb{Z}_q^n$.
  - GSW.PKGen(params, sk): Sample $\mathbf{e} \leftarrow \chi^m$. Set $\mathbf{b} := \mathbf{sB} + \mathbf{e} \in \mathbb{Z}_q^m$. Output $pk = \mathbf{A}$ where, $\mathbf{A} \in \mathbb{Z}_q^{n \times m}$ is defined as $\mathbf{A} := \begin{bmatrix} \mathbf{B} \\ \mathbf{b} \end{bmatrix}$
- GSW.Encrypt(pk, μ): Choose a short random matrix as the randomness $\mathbf{R} \xleftarrow{\$} \{0,1\}^{m \times m}$. Then output the encryption of message $\mu \in \{0,1\}$ as $\mathbf{C} \in \mathbb{Z}_q^{n \times m}$ where,

$$\mathbf{C} := \mathbf{AR} + \mu\mathbf{G}$$

- GSW.Decrypt($sk, \mathbf{C}$): Let $\mathbf{t} := sk$. Define a vector $\mathbf{w} \in \mathbb{Z}_q^n$ as follows:

$$\mathbf{w} = [0, \ldots, 0, \lceil q/2 \rceil]$$

Then compute $v = \mathbf{tCG}^{-1}(\mathbf{w}^T) \in \mathbb{Z}_q^m$. Finally output $\mu' = \left| \left| \frac{v}{q/2} \right| \right|$ as the decrypted message.

- On input two ciphertexts $\mathbf{C}_1, \mathbf{C}_2 \in \mathbb{Z}_q^{n \times m}$ we can define homomorphic addition, multiplication:
  - GSW.Add($\mathbf{C}_1, \mathbf{C}_2$): Output $\mathbf{C}_1 + \mathbf{C}_2 \in \mathbb{Z}_q^{n \times m}$.
  - GSW.Mult($\mathbf{C}_1, \mathbf{C}_2$): Output the matrix product $\mathbf{C}_1\mathbf{G}^{-1}(\mathbf{C}_2) \in \mathbb{Z}_q^{n \times m}$.
  This also allows us to compute a homomorphic NAND gate by outputting $\mathbf{G} - \mathbf{C}_1\mathbf{G}^{-1}(\mathbf{C}_2)$.

We sketch the proof of the following theorem for completeness.

Theorem 1. ([18]). *The scheme described above is a secure FHE under the* $\mathsf{LWE}_{n-1,q,\chi,B_\chi}$ *assumption.*

---

[6] The size of q here is bigger than needed for GSW encryption alone in order to support our extensions.

Security. The proof of semantic security consists of two steps. First, we can use the LWE assumption to replace the public key $pk = \mathbf{A}$ with a uniformly random matrix in $\mathbb{Z}_q^{n \times m}$. Then we can use the leftover hash lemma to replace the ciphertext $\mathbf{C} := \mathbf{AR} + \mu\mathbf{G}$ with a uniformly random value $\mathbf{C}'$. We refer the reader to [18] for details.

Correctness. To analyze correctness, it is helpful to define the following notion of a "noisy ciphertext".

Definition 4 *(β-noisy ciphertext). A β-noisy ciphertext of some message μ under secret-key $sk = \mathbf{t} \in \mathbb{Z}_q^n$ is a matrix $\mathbf{C} \in \mathbb{Z}_q^{n \times m}$ such that: $\mathbf{tC} = \mu\mathbf{tG} + \mathbf{e}$ for some $\mathbf{e}$ with $\|\mathbf{e}\|_\infty \leq \beta$.*

Encryption: Consider a fresh ciphertext $\mathbf{C} = \mathbf{AR} + \mu\mathbf{G}$ which is generated by encrypting some message μ with some public key $\mathbf{A}$ with corresponding secret key $\mathbf{t}$. First recall that $\mathbf{tA} = \mathbf{e}$ such that $\|\mathbf{e}\|_\infty \leq B_\chi$. Therefore $\mathbf{tC} = \mathbf{e}' + \mu\mathbf{tG}$ where $\mathbf{e}' = \mathbf{eR}$ which implies $\|\mathbf{e}'\|_\infty \leq mB_\chi$. Hence $\mathbf{C}$ is mB_χ-noisy encryption of μ under $\mathbf{t}$. Let us call this value initial noise or $\beta_{init} = mB_\chi$.

Evaluation: Let $\mathbf{C}_1$ and $\mathbf{C}_2$ be two ciphertexts which are β_1 and β_2 noisy encryption of $\mu_1, \mu_2 \in \{0, 1\}$ under the key $\mathbf{t}$ respectively, so that: $\mathbf{tC}_1 = \mathbf{e}_1 + \mu_1\mathbf{tG}$ and $\mathbf{tC}_2 = \mathbf{e}_2 + \mu_2\mathbf{tG}$ with $\|\mathbf{e}_1\|_\infty \leq \beta_1, \|\mathbf{e}_2\|_\infty \leq \beta_2$.

- *Addition:* Then their addition will result in a ciphertext $\mathbf{C}^{(+)} = \mathbf{C}_1 + \mathbf{C}_2$ such that, $\mathbf{tC}^{(+)} = \mathbf{e}' + (\mu_1 + \mu_2)\mathbf{tG}$ where $\mathbf{e}' = \mathbf{e}_1 + \mathbf{e}_2$. Clearly this is $\beta_1 + \beta_2$-noisy.

- *Multiplication:* On the other hand the multiplication would produce a ciphertext $\mathbf{C}^{(\times)} = \mathbf{C}_1\mathbf{G}^{-1}(\mathbf{C}_2)$ such that $\mathbf{tC}^{(\times)} = \mathbf{e}'' + \mu_1\mu_2\mathbf{G}$ where $\mathbf{e}'' = \mathbf{e}\mathbf{G}^{-1}(\mathbf{C}_2) + \mu_1\mathbf{e}_2$. Clearly $\|\mathbf{e}''\|_\infty \leq (m\beta_1 + \beta_2)$ and the ciphertext $\mathbf{C}^{(\times)}$ is $(m\beta_1 + \beta_2)$-noisy. The same calculation holds for NAND gates.

Decryption: Let $\mathbf{C}$ be a β-noisy encryption of μ so that: $\mathbf{tC} = \mathbf{e} + \mu\mathbf{tG}$ where $\|\mathbf{e}\|_\infty = \beta$. Then $v = \mathbf{tCG}^{-1}(\mathbf{w}^T) = e' + \mu(q/2)$ such that $e' = \langle \mathbf{e}, \mathbf{G}^{-1}(\mathbf{w}^T)\rangle$. Clearly $\|e'\|_\infty \leq m\beta$. Now one can observe that decryption works correctly as long as $\|e'\|_\infty < q/4$. Therefore correctness holds as long as $\beta < q/(4m)$. We call this value $\beta_{max} := q/(4m)$.

Consider evaluating a (boolean) circuit of depth d consisting of NAND gates. It takes input fresh ciphertexts (β_{init}-noisy) and each level multiplies the noise by a factor of at most $(m + 1)$. Therefore, the final output is β_{final}-noisy ciphertexts where $\beta_{final} = (m + 1)^d\beta_{init}$. To ensure correctness of decryption we need $\beta_{final} \leq \beta_{max}$ meaning $B_\chi 4m^2(m + 1)^d < q$ which is satisfied by our choice of parameters. This concludes the proof.

Homomorphic Linear Combinations and Pseudo Encryption. We now define an additional homomorphic operation. This operation takes as input GSW ciphertexts $\mathbf{C}_{i,j}$ encrypting the individual entries $\mathbf{M}[i,j]$ of some matrix $\mathbf{M} \in \mathbb{Z}_q^{m \times m}$ under a secret key $\mathbf{t}$. It also takes a plaintext vector $\mathbf{v} \in \mathbb{Z}_q^m$ which specifies the homomorphic function to be computed. The operations outputs a "pseudo ciphertext" $\mathbf{C}_{lc}$ which we can think of as a pseudo encryption of the vector $\mathbf{v}\mathbf{M}$, meaning that $\mathbf{t}\mathbf{C}_{lc} \approx \mathbf{v}\mathbf{M}$. Note that the "pseudo ciphertext" $\mathbf{C}_{lc}$ cannot be correctly decrypted (we can only recover something close to $\mathbf{v}\mathbf{M}$ but not the exact value) nor can we further perform any of the standard GSW homomorphic operations on it.

Property 1 (**Linear combination**). Let $\mathbf{M} \in \{0,1\}^{m \times m}$ be a matrix and for $i \in [m], j \in [m]$ let $\mathbf{C}_{i,j} \in \mathbb{Z}_q^{n \times m}$ be a β-noisy GSW encryption of $\mathbf{M}[i,j]$ under a secret key $\mathbf{t} \in \mathbb{Z}_q^n$. Let $\mathbf{v} \in \mathbb{Z}_q^m$ be some vector (not necessarily short). Then there is a polynomial-time deterministic algorithm

$$\mathbf{C}_{lc} = \mathsf{GSW.LComb}((\mathbf{C}_{1,1}, \ldots, \mathbf{C}_{m,m}), \mathbf{v})$$

which outputs $\mathbf{C}_{lc} \in \mathbb{Z}_q^{n \times m}$ such that $\mathbf{t}\mathbf{C}_{lc} = \mathbf{v}\mathbf{M} + \mathbf{e}$ where $\|\mathbf{e}\|_\infty \leq m^3 \beta$.

Implementation. The algorithm $\mathsf{GSW.LComb}((\mathbf{C}_{1,1}, \ldots, \mathbf{C}_{m,m}), \mathbf{v})$ is implemented as follows:

1. For each $i \in [m], j \in [m]$ define a matrix $\mathbf{Z}_{i,j} \in \mathbb{Z}_q^{n \times m}$ as follows:

$$\mathbf{Z}_{i,j}[a,b] := \begin{cases} \mathbf{v}[i] & \text{when } a = n \text{ and } b = j \\ 0 & \text{otherwise} \end{cases}$$

In other words $\mathbf{Z}_{i,j}$ will have 0 everywhere except the n-th (final) row and j-th column where it has the value $\mathbf{v}[i]$.

2. Now output $\mathbf{C}_{lc} \in \mathbb{Z}_q^{n \times m}$ where: $\mathbf{C}_{lc} = \displaystyle\sum_{i=1, j=1}^{m,m} \mathbf{C}_{i,j} \mathbf{G}^{-1}(\mathbf{Z}_{i,j})$

Correctness. Correctness follows because,

$$\mathbf{t}\mathbf{C}_{lc} = \mathbf{t} \sum_{i,j} \mathbf{C}_{i,j} \mathbf{G}^{-1}(\mathbf{Z}_{i,j})$$

$$= \sum_{i,j} (\mathbf{M}[i,j]\mathbf{t}\mathbf{G} + \mathbf{e}_{i,j}) \mathbf{G}^{-1}(\mathbf{Z}_{i,j})$$

$$= \sum_{i,j} (\mathbf{M}[i,j]\mathbf{t}\mathbf{Z}_{i,j} + \mathbf{e}'_{i,j})$$

$$= \mathbf{t} \sum_{i,j} \mathbf{M}[i,j]\mathbf{Z}_{i,j} + \sum_{i,j} \mathbf{e}'_{i,j}$$

$$= (-\mathbf{s}, 1) \begin{bmatrix} 0^{n-1} \\ \mathbf{v}\mathbf{M} \end{bmatrix} + \mathbf{e}'' = \mathbf{v}\mathbf{M} + \mathbf{e}''$$

where $\mathbf{e}_{i,j}$ is the noise contained in $\mathbf{C}_{i,j}$ which is of magnitude $\|\mathbf{e}_{i,j}\|_\infty \leq \beta$, $\mathbf{e}'_{i,j} = \mathbf{e}_{i,j}\mathbf{G}^{-1}(\mathbf{Z}_{i,j})$ has magnitude $\|\mathbf{e}_{i,j}\|_\infty \leq m\beta$, and finally $\mathbf{e}'' = \displaystyle\sum_{i,j} \mathbf{e}'_{i,j}$ has magnitude $\|\mathbf{e}''\|_\infty \leq m^3 \beta$.

5.2 A Masking Scheme for GSW

We now define and show how to construct a "masking scheme" for GSW, which serves as the main component of the multi-key FHE scheme. Intuitively, a masking scheme allows us to take a GSW public key $pk = \mathbf{A}$ (having a corresponding secret key $\mathbf{t}$) and a bit μ and output a pair of values $(\mathcal{U}, \mathbf{C})$ such that $\mathbf{C}$ is a GSW encryption of μ with pk and $\mathcal{U}$ is an auxiliary value such that (1) the pair $(\mathcal{U}, \mathbf{C})$ computationally hide μ (just like $\mathbf{C}$ alone) and (2) later, given another GSW public key $pk = \mathbf{A}'$ (having a corresponding secret key $\mathbf{t}'$) we can compute a matrix $\mathbf{X}$ such that $\mathbf{t}\mathbf{X} + \mathbf{t}'\mathbf{C} = \mu\mathbf{t}'\mathbf{G}$.

Property 2 **(GSW Masking Scheme).** There exists a pair of algorithms (UniEnc, Extend):

- UniEnc(μ, pk): On input a message $\mu \in \{0, 1\}$ and a GSW public key pk it generates a pair $(\mathcal{U}, \mathbf{C})$ where $\mathbf{C} \in \mathbb{Z}_q^{n \times m}$ and $\mathcal{U} \in \{0, 1\}^*$.
- Extend($\mathcal{U}, pk, pk'$): On input $\mathcal{U}$ and GSW public keys pk, pk' it outputs $\mathbf{X} \in \mathbb{Z}_q^{n \times m}$.

for which the following properties holds:

- SEMANTIC SECURITY: For any polynomial $d = d(\lambda)$ security of GSW encryption implies that:

$$(\mathsf{params}, pk, \mathsf{UniEnc}(0, pk)) \stackrel{\mathrm{comp}}{\approx} (\mathsf{params}, pk, \mathsf{UniEnc}(1, pk))$$

  where $\mathsf{params} \leftarrow \mathsf{GSW.SetUp}(1^\lambda, 1^d)$, $(sk, pk) \leftarrow \mathsf{GSW.Keygen}(\mathsf{params})$.
- CORRECTNESS: Let $\mathsf{params} \leftarrow \mathsf{GSW.SetUp}(1^\lambda, 1^d)$ and let $(sk = \mathbf{t}, pk), (sk' = \mathbf{t}', pk')$ be two independent key pairs generated with $\mathsf{GSW.Keygen}(\mathsf{params})$. For any $\mu \in \{0, 1\}$ let $(\mathcal{U}, \mathbf{C}) \leftarrow \mathsf{UniEnc}(\mu, pk)$ and $\mathbf{X} \leftarrow \mathsf{Extend}(\mathcal{U}, pk, pk')$. Then

$$\mu := \mathsf{GSW.Decrypt}(sk, \mathbf{C}) \quad \text{and} \quad \mathbf{t}\mathbf{X} + \mathbf{t}'\mathbf{C} = \mu\mathbf{t}'\mathbf{G} + \mathbf{e}$$

  where $\|\mathbf{e}\|_\infty \leq \beta_{mask}$ for $\beta_{mask} := (m^4 + m)B_\chi$.

Instantiation. We now show how to implement such masking scheme.

- UniEnc(pk, μ): On input a message μ and a public key pk the algorithm outputs $\mathcal{U}$, which is a m^2-tuple of matrices in $\mathbb{Z}_q^{n \times m}$, and $\mathbf{C} \in \mathbb{Z}_q^{n \times m}$ as follows.
  1. Let $\mathbf{A} = pk$. Set $\mathbf{C} \leftarrow \mathsf{GSW.Encrypt}(pk, \mu) \in \mathbb{Z}_q^{n \times m}$ so that $\mathbf{C} = \mathbf{AR} + \mu\mathbf{G}$ where $\mathbf{R} \in \{0, 1\}^{m \times m}$ is the encryption randomness.
  2. Encrypt each element of the random matrix $\mathbf{R}$ (chosen in Step 1) to get m^2 ciphertexts: $\mathbf{V}^{(a,b)} \leftarrow \mathsf{GSW.Encrypt}(pk, \mathbf{R}[a, b])$. Set $\mathcal{U} := (\mathbf{V}^{(1,1)}, \ldots, \mathbf{V}^{(m,m)}) \in (\mathbb{Z}_q^{n \times m})^{(m^2)}$.
- Extend($\mathcal{U}, pk, pk'$): On input a $\mathcal{U} \in (\mathbb{Z}_q^{n \times m})^{(m^2)}$ and public keys pk, pk' the algorithm computes $\mathbf{X} \in \mathbb{Z}_q^{n \times m}$ as follows:
  1. Parse $pk = \mathbf{A} = \begin{bmatrix} \mathbf{B} \\ \mathbf{b} \end{bmatrix}$, $pk' = \mathbf{A}' = \begin{bmatrix} \mathbf{B} \\ \mathbf{b}' \end{bmatrix}$ and, $\mathcal{U} = (\{\mathbf{V}^{(a,b)}\}_{a,b \in [m]})$.

2. Set $\mathbf{X} = \mathsf{GSW.LComb}((\mathbf{V}^{(1,1)}, \ldots, \mathbf{V}^{(m,m)}), \mathbf{b}' - \mathbf{b})$.

Semantic Security. The view of the attacker is the following distribution:

$$\left(\mathsf{params}, \mathbf{A}, \mathbf{C}, \mathcal{U} = \left(\ \mathbf{V}^{(11)}, \ldots, \mathbf{V}^{(m,m)} \right) \right)$$

generated via params $\leftarrow$ $\mathsf{GSW.SetUp}(1^\lambda, 1^d)$, $(sk, pk = \mathbf{A})$ $\leftarrow$ $\mathsf{GSW.Keygen}(\mathsf{params})$ and $(\mathbf{C}, \mathcal{U}) \leftarrow \mathsf{UniEnc}(pk, \mu)$, where either $\mu = 0$ or $\mu = 1$. We prove semantic security of the masking scheme by relying on the semantic security of the underlying GSW scheme. The proof consists of the following hybrids:

- Firstly, we modify each of the ciphertexts $\mathbf{V}^{(a,b)}$ so that instead of being GSW encryptions of $\mathbf{R}[a, b]$, we just choose them as GSW encryptions of 0. This just relies on semantic security of GSW encryption.
- Secondly, we also choose $\mathbf{C}$ as a GSW encryption of 0. This also just follows from the semantic security of GSW encryption, since after the first step no information about the randomness $\mathbf{R}$ is given out.

Finally, this distribution is completely independent of the bit μ which concludes the proof of semantic security.

Correctness. Let $((sk = \mathbf{t}, pk = \mathbf{A}), (sk' = \mathbf{t}', pk' = \mathbf{A}'))$ be two correctly generated GSW key-pairs. Now recall that, $sk = \mathbf{t} = (-\mathbf{s}, 1) \in \mathbb{Z}_q^n$, and $sk' = \mathbf{t}' = (-\mathbf{s}', 1) \in \mathbb{Z}_q^n$; $pk = \mathbf{A} = \begin{bmatrix} \mathbf{B} \\ \mathbf{b} \end{bmatrix} \in \mathbb{Z}_q^{n \times m}$, $pk' = \mathbf{A}' = \begin{bmatrix} \mathbf{B} \\ \mathbf{b}' \end{bmatrix} \in \mathbb{Z}_q^{n \times m}$ where $\mathbf{b} = \mathbf{s}\mathbf{B} + \mathbf{e}$, $\mathbf{b}' = \mathbf{s}'\mathbf{B} + \mathbf{e}'$ with $\|\mathbf{e}\|_\infty, \|\mathbf{e}'\|_\infty \leq \beta_\chi$.

Furthermore, for any message μ let $(\mathcal{U}, \mathbf{C}) \leftarrow \mathsf{UniEnc}(pk, \mu)$ and $\mathbf{X} \leftarrow \mathsf{Extend}(\mathcal{U}, pk, pk')$ where $\mathcal{U} = (\mathbf{V}^{(1,1)}, \ldots, \mathbf{V}^{(nm)})$. Then it is easy to see that $\mu := \mathsf{GSW.Decrypt}(sk, \mathbf{C})$ which implies that $\mathbf{C} = \mathbf{A}\mathbf{R} + \mu\mathbf{G} = \begin{bmatrix} \mathbf{B} \\ \mathbf{b} \end{bmatrix} \mathbf{R} + \mu\mathbf{G}$ for some $\mathbf{R} \in \{0, 1\}^{m \times m}$ and hence

$$\begin{aligned}
\mathbf{t}'\mathbf{C} &= (-\mathbf{s}', 1) \begin{bmatrix} \mathbf{B} \\ \mathbf{b} \end{bmatrix} \mathbf{R} + \mu\mathbf{t}'\mathbf{G} \\
&= -\mathbf{s}'\mathbf{B}\mathbf{R} + \mathbf{b}\mathbf{R} + \mu\mathbf{t}'\mathbf{G} \\
&= -(\mathbf{b}' - \mathbf{e}')\mathbf{R} + \mathbf{b}\mathbf{R} + \mu\mathbf{t}'\mathbf{G} \\
&= (\mathbf{b} - \mathbf{b}')\mathbf{R} + \mu\mathbf{t}'\mathbf{G} + \mathbf{e}_C
\end{aligned}$$

where $\mathbf{e}_C = \mathbf{e}'\mathbf{R}$ has norm $\|\mathbf{e}_C\|_\infty = m\beta_\chi$.

On the other hand, by the correctness of linear combinations, we have:

$$\mathbf{t}\mathbf{X} = (\mathbf{b}' - \mathbf{b})\mathbf{R} + \mathbf{e}_X$$

where $\|\mathbf{e}_X\|_\infty = m^4\beta_\chi$.

Combining these equations, we get $\mathbf{t}\mathbf{X} + \mathbf{t}'\mathbf{C} = \mu\mathbf{t}'\mathbf{G} + \mathbf{e}^*$ where $\|\mathbf{e}^*\|_\infty \leq (m^4 + m)\beta_\chi$ as claimed.

5.3 Construction of Multi-key FHE

First recall the fixed matrix $\mathbf{G} \in \mathbb{Z}_q^{n \times m}$ that played an important role for the earlier construction and analysis. In this section we define an "expanded matrix" $\widehat{\mathbf{G}}_N \in \mathbb{Z}_q^{nN \times mN}$ as:

$$\widehat{\mathbf{G}}_N = \begin{bmatrix} \mathbf{G} & \cdots\cdots & \mathbf{0} \\ \mathbf{0} & \mathbf{G} & \cdots & \vdots \\ \vdots & \cdots & \mathbf{G} & \mathbf{0} \\ \mathbf{0} & \cdots\cdots & & \mathbf{G} \end{bmatrix}$$

We note that there exists a corresponding efficiently computable function $\widehat{\mathbf{G}}_N^{-1}(\cdot)$ such that for any $m' \in \mathbb{N}$ any matrix $\mathbf{M} \in \mathbb{Z}_q^{nN \times m'}$, $\widehat{\mathbf{G}}_N^{-1}(\mathbf{M}') \in \{0,1\}^{mN \times mN}$ is "short" and $\widehat{\mathbf{G}}_N \widehat{\mathbf{G}}_N^{-1}(\mathbf{M}) = \mathbf{M}$. Such $\widehat{\mathbf{G}}_N^{-1}(\cdot)$ can be computed using $\mathbf{G}^{-1}(\cdot)$ in the natural way.

Construction. Now we describe our multi-key FHE construction.

- MFHE.SetUp$(1^\lambda, 1^d)$: Run the set-up algorithm of GSW to generate the parameters:
  $$\mathsf{params} := (q, n, m, \chi, B_\chi, \mathbf{B}) \leftarrow \mathsf{GSW.SetUp}(1^\lambda, 1^d).$$

- MFHE.Keygen(params): Run the key-generation algorithm of GSW to generate:
  $$sk := \mathbf{t} \leftarrow \mathsf{GSW.SKGen}(\mathsf{params}) \qquad pk := \mathbf{A} \leftarrow \mathsf{GSW.PKGen}(\mathsf{params}, sk)$$

- MFHE.Encrypt(pk, μ): Execute the following steps:
  - Just use the masking scheme: $(\mathcal{U}, \mathbf{C}) \leftarrow \mathsf{UniEnc}(\mu, pk)$.
  - Output the pair $c := (\mathcal{U}, \mathbf{C})$ as the ciphertext for μ.

- MFHE.Expand$((pk_1, \ldots, pk_N), i, c)$: On receiving a sequence of public-keys $(pk_1, \ldots, pk_N)$ and a fresh ciphertext $c = (\mathcal{U}, \mathbf{C})$ under the public key pk_i run the Extend algorithm for all pk_j where $i \neq j$.
  - For $j \in \{pk_1, \ldots, pk_N\} \setminus \{i\}$, compute $\mathbf{X}_j \leftarrow \mathsf{Extend}(\mathcal{U}, pk_i, pk_j)$.
  - Then define a matrix $\widehat{\mathbf{C}} \in \mathbb{Z}_q^{nN \times mN}$ as a concatenation of N^2 sub-matrices where each sub-matrix $\mathbf{C}_{a,b} \in \mathbb{Z}_q^{n \times m}$ for $a, b \in [N]$ is defined as:

$$\mathbf{C}_{a,b} := \begin{cases} \mathbf{C} & \text{when } a = b \\ \mathbf{X}_j & \text{when } a = i \neq j \text{ and } b = j \\ \mathbf{0}^{n \times m} & \text{otherwise} \end{cases}$$

  For reader's convenience we provide a pictorial representation of $\widehat{\mathbf{C}}$ in Fig. 1:
  Finally output $\widehat{c} := \widehat{\mathbf{C}}$ as the expanded ciphertext.

- MFHE.Eval$(\mathsf{params}, \mathcal{C}, (\widehat{c}_1, \ldots, \widehat{c}_\ell))$ On input ℓ expanded ciphertexts simply use the GSW homomorphic evaluation algorithms namely GSW.Add and GSW.Mult, albeit with expanded dimensions $n' = nN$ and $m' = mN$ and the expanded $\widehat{\mathbf{G}}_N, \widehat{\mathbf{G}}_N^{-1}$ (in place of n, m and $\mathbf{G}, \mathbf{G}^{-1}$).

$$\text{Row } i \longrightarrow \begin{bmatrix} \mathbf{C} & 0 & \cdots & 0 & 0 \\ 0 & \ddots & \cdots & \cdots & \vdots \\ \vdots & 0 & 0 & \cdots & 0 \\ \mathbf{X}_1 & \cdots & \mathbf{C} & \cdots & \mathbf{X}_N \\ 0 & \cdots & 0 & 0 & \vdots \\ \vdots & \cdots & \cdots & \ddots & 0 \\ 0 & 0 & \cdots & 0 & \mathbf{C} \end{bmatrix}$$

$$\underset{\text{Column } i}{\uparrow}$$

Fig. 1. Structure of the expanded ciphertext $\widehat{\mathbf{C}}$

- MFHE.Decrypt(params, $(sk_1, \ldots, sk_N), c$): On input a ciphertext $c = \widehat{\mathbf{C}}$ and the sequence of secret keys $(sk_1, \ldots, sk_N)$ parse each $\mathbf{t}_i := sk_i$ and then construct the joint secret key by horizontally appending all the secret-keys in sequence $\widehat{\mathbf{t}} = [\widehat{\mathbf{t}}_1 \ \widehat{\mathbf{t}}_2 \ \cdots \ \widehat{\mathbf{t}}_N] \in \mathbb{Z}_q^{nN}$. Then run the GSW decryption algorithm albeit with expanded dimensions $n' = nN$ and $m' = mN$ and the expanded $\widehat{\mathbf{G}}_N, \widehat{\mathbf{G}}_N^{-1}$ (in place of n, m and $\mathbf{G}, \mathbf{G}^{-1}$).

Correctness and Security of MFHE Construction

Theorem 2. *The scheme described above is a secure MFHE under the* $\mathsf{LWE}_{n-1,q,\chi,B_\chi}$ *assumption (with the same parameters as we defined for GSW encryption).*

Semantic Security. The semantic security of the above multi-key FHE follows directly from that of the GSW masking scheme.

Correctness of Expansion. Consider a sequences of N key pairs $((sk_1 = \mathbf{t}_1, pk_1), \ldots, (sk_N = \mathbf{t}_N, pk_N))$ correctly generated by running the key-generation as $\{(pk_i, sk_i) \leftarrow \mathsf{MFHE.Keygen}(\mathsf{params})\}_{i \in [N]}$. Now suppose for any message μ and any $i \in [N]$ we have a ciphertext $c \leftarrow \mathsf{MFHE.Encrypt}(pk_i, \mu)$ under the i-th key and the corresponding expanded ciphertext $\widehat{\mathbf{C}} \leftarrow \mathsf{MFHE.Expand}((pk_1, \ldots, pk_N), i, c)$ as shown in Fig. 1. Let $\widehat{\mathbf{t}} = [\mathbf{t}_1, \ldots, \mathbf{t}_N]$. Then

$$\widehat{\mathbf{t}}\widehat{\mathbf{C}} = [\mathbf{t}_i\mathbf{X}_1 + \mathbf{t}_1\mathbf{C}, \ldots, \mathbf{t}_i\mathbf{C}, \ldots, \mathbf{t}_i\mathbf{X}_N + \mathbf{t}_N\mathbf{C}]$$
$$= [\mu\mathbf{t}_1\mathbf{G} + \mathbf{e}_1, \ldots, \mu\mathbf{t}_i\mathbf{G} + \mathbf{e}_i, \ldots, \mu\mathbf{t}_N\mathbf{G} + \mathbf{e}_N]$$
$$= \mu\widehat{\mathbf{t}}\widehat{\mathbf{G}} + [\mathbf{e}_1, \ldots, \mathbf{e}_N]$$

where $\|\mathbf{e}_i\|_\infty \leq mB_\chi$ by the correctness of GSW encryption and for $j \neq i$, $\|\mathbf{e}_j\|_\infty \leq (m^4 + m)B_\chi$ by the correctness of the GSW masking scheme. Therefore,

$\widehat{\mathbf{t}}\widehat{\mathbf{C}} = \mu\widehat{\mathbf{t}}\widehat{\mathbf{G}} + \mathbf{e}$ where $\|\mathbf{e}\|_\infty \leq (m^4 + m)B_\chi$. Let's call this value $\beta'_{init} = (m^4 + m)B_\chi = 2^{O(\log \lambda)}B_\chi$. The correctness of GSW encryption is guaranteed as long as $\beta'_{init} \leq q/(4m')$ which holds with the choice of q we defined.

Correctness of Evaluation. Let $\widehat{\mathbf{C}}_1, \ldots, \widehat{\mathbf{C}}_\ell$ be expanded ciphertexts corresponding to bit $\mu_1, \ldots, \mu_\ell$ so that, by the above correctness property, $\widehat{\mathbf{t}}\widehat{\mathbf{C}}_i = \mu_i\widehat{\mathbf{t}}\widehat{\mathbf{G}} + \mathbf{e}_i$ where $\|\mathbf{e}_i\|_\infty \leq \beta'_{init}$. If $\widehat{\mathbf{C}}$ is the output of a homomorphic evaluation of a circuit $\mathcal{C}$ of depth d over the above ciphertexts such that $\mu = \mathcal{C}(\mu_1, \ldots, \mu_\ell)$ then by the correctness of GSW homomorphic evaluation with scaled up parameters $n' = nN, m' = mN$ we have $\widehat{\mathbf{t}}\widehat{\mathbf{C}} = \mu\widehat{\mathbf{t}}\widehat{\mathbf{G}} + \mathbf{e}$ where $\|\mathbf{e}\|_\infty \leq \beta'_{init}(m' + 1)^d = (m^4 + m)B_\chi(mN + 1)^d$. Let's call this value $\beta'_{final} = B_\chi(m^4 + m)(mN + 1)^d = 2^{O(d \log \lambda)}B_\chi$. The correctness of GSW encryption is guaranteed as long as $\beta'_{final} \leq q/(4m')$ which holds with the choice of q we defined.

5.4   Threshold Decryption for Multi-key FHE

We now show how to implement threshold decryption for the MFHE construction outlined in the previous section.

MFHE.PartDec$(\widehat{c}, (pk_1, \ldots, pk_N), i, sk_i)$: On input an expanded ciphertext $\widehat{c} = \widehat{\mathbf{C}} \in \mathbb{Z}_q^{nN \times mN}$ under a sequence of keys $(pk_1, \ldots, pk_N)$ and the i-th secret key $sk_i = \mathbf{t}_i \in \mathbb{Z}_q^n$ do the following:

- Parse $\widehat{\mathbf{C}}$ as consisting of N sub-matrices $\widehat{\mathbf{C}}^{(i)} \in \mathbb{Z}_q^{n \times mN}$ such that

$$\widehat{\mathbf{C}} = \begin{bmatrix} \widehat{\mathbf{C}}^{(1)} \\ \vdots \\ \widehat{\mathbf{C}}^{(N)} \end{bmatrix}.$$

- Define $\widehat{\mathbf{w}} \in \mathbb{Z}_q^{nN}$ as $\widehat{\mathbf{w}} = [0, \ldots, 0, \lceil q/2 \rceil]$.
- Then compute $\gamma_i = \mathbf{t}_i\widehat{\mathbf{C}}^{(i)}\widehat{\mathbf{G}}^{-1}(\widehat{\mathbf{w}}^T) \in \mathbb{Z}_q$ and output $p_i = \gamma_i + e_i^{sm} \in \mathbb{Z}_q$ where $e_i^{sm} \xleftarrow{\$} [-B_{smdg}^{dec}, -B_{smdg}^{dec}]$ is some random "smudging noise" where $B_{smdg}^{dec} = 2^{d\lambda \log \lambda}B_\chi$.

MFHE.FinDec$(p_1, \ldots, p_N)$: Given $p_1, \ldots, p_N$, compute the sum $p := \sum_{i=1}^N p_i$. Output $\mu := \left|\left\lceil \frac{p}{q/2} \right\rfloor\right|$.

Correctness and Simulation Security

Theorem 3. *The above threshold decryption procedures for MFHE satisfy correctness and (statistical) simulation security.*

Correctness. Here the entire scheme is same as MFHE except the decryption. So if $\widehat{\mathbf{C}}$ is an evaluated ciphertext encrypting a bit μ and the secret keys are $\widehat{\mathbf{t}} = [\widehat{\mathbf{t}}_1, \ldots, \widehat{\mathbf{t}}_N]$ then, by the analysis used for non-threshold correctness, we have

$$\widehat{\mathbf{t}}\widehat{\mathbf{C}} = \sum_{i \in [N]} \mathbf{t}_i \widehat{\mathbf{C}}^{(i)} = \mu \widehat{\mathbf{t}}\widehat{\mathbf{G}} + \mathbf{e}$$

where $\|\mathbf{e}\|_\infty \leq \beta'_{final} = (m^4 + m)B_\chi(mN + 1)^d$. Therefore if the partial decryptions p_i are computed as specified we have:

$$\sum_{i \in [N]} p_i = \sum_{i \in [N]} \gamma_i + \sum_{i \in [N]} e_i^{sm} = \sum_{i \in [N]} \mathbf{t}_i \widehat{\mathbf{C}}^{(i)} \widehat{\mathbf{G}}^{-1}(\widehat{\mathbf{w}}^T) + e^{sm}$$

$$= (\mu \widehat{\mathbf{t}}\widehat{\mathbf{G}} + \mathbf{e})\widehat{\mathbf{G}}^{-1}(\widehat{\mathbf{w}}^T) + e^{sm}$$

$$= \mu \lceil q/2 \rceil + e' + e^{sm}$$

where $e^{sm} = \sum_{i \in [N]} e_i^{sm}$ has norm $|e^{sm}| \leq N B_{smdg}^{dec} = 2^{O(d\lambda \log \lambda)} B_\chi$ and $e' = \mathbf{e}\widehat{\mathbf{G}}^{-1}(\widehat{\mathbf{w}}^T)$ has norm $|e'| \leq \beta'_{final} mN = 2^{O(d \log \lambda)} B_\chi$. Since $q = 2^{\omega(d\lambda \log \lambda)} B_\chi$ we have $|e' + e^{sm}| < q/4$ and correctness holds.

Simulatability: The simulator $\mathcal{S}^{thr}(\mu, \widehat{\mathbf{C}}, i, \{\mathbf{t}_j\}_{j \in [N] \setminus \{i\}})$, on input the secrets keys $\{\mathbf{t}_j\}_{j \neq i}$ the evaluated ciphertext $\widehat{\mathbf{C}} \in \mathbb{Z}_q^{nN \times mN}$ and the output value $\mu = \mathcal{C}(\mu_1, \ldots, \mu_\ell)$ encrypted in $\widehat{\mathbf{C}}$ outputs the *simulated partial decryption*:

$$p_i' = \mu \lceil q/2 \rceil + e_i^{sm} - \sum_{i \neq j} \gamma_j \tag{1}$$

for $e_i^{sm} \xleftarrow{\$} [-B_{smdg}^{dec}, B_{smdg}^{dec}]$ where $\gamma_j = \mathbf{t}_j \widehat{\mathbf{C}}^{(j)} \widehat{\mathbf{G}}^{-1}(\widehat{\mathbf{w}}^T)$.

To see the indistinguishability note that, by the same calculation as used to argue correctness, we know that $\sum_{j \in [N]} \gamma_j = \mu \lceil q/2 \rceil + e'$ where $|e'| \leq \beta'_{final} mN = 2^{O(d \log \lambda)} B_\chi$. Therefore if $p_i = \gamma_i + e_i^{sm}$ is the *real partial decryption* then

$$p_i = \mu \lceil q/2 \rceil + e' + e_i^{sm} - \sum_{i \neq j} \gamma_j$$

The difference between the real value p_i and the simulated value p_i' is the noise e' of norm $|e'| = 2^{O(d \log \lambda)} B_\chi$. But by the smudging Lemma 2, the distributions of e_i^{sm} and $e_i^{sm} + e'$ are statistically close since $e_i^{sm} \xleftarrow{\$} [-B_{smdg}^{dec}, -B_{smdg}^{dec}]$ where $B_{smdg}^{dec} = 2^{d\lambda \log \lambda} B_\chi$ so that $B_{smdg}^{dec}/|e'| \geq 2^\lambda$. Therefore the simulated partial decryption and the real one are statistically indistinguishable.

5.5 Bootstrapping

Note that the above MFHE scheme is leveled i.e., it depends on the multiplicative depth of the circuit to be computed. However, this dependency can be avoided easily by boot-strapping and assuming circular security. We briefly describe the straightforward procedure and omit the details.

During key generation, each party i chooses a key pair (sk_i, pk_i) and uses the MFHE scheme to encrypt the secret key sk_i under pk_i bit-by-bit.[7] It appends these encryptions to the public key. Later, given a sequence of public keys $\{pk_1, \ldots, pk_N\}$ anyone can create an expanded multi-key encryption of each sk_i using the MFHE expansion procedure. This allows us to use Gentry's boot-strapping technique [17] to "refresh" a highly noisy multi-key ciphertext by homomorphically computing the MFHE decryption procedure. Therefore, to compute a circuit of *arbitrary* depth, we only need to set the parameters of the MFHE scheme so as to be evaluate circuits of some *fixed* depth $d + 1$ where d is the depth of the MFHE decryption procedure.

Note that, by circular security it is assured that an encryption of a secret key under itself is semantically secure which implies that the semantic security of the above modified MFHE scheme remains intact.

6   Secure MPC via Threshold MFHE

Basic Template. We now present a protocol for general MPC, using any threshold multi-key fully homomorphic scheme. The protocol is based on the template discussed in the introduction which we recall below:

1. Each party individually chooses its own MFHE key pair (pk_i, sk_i), encrypts its input x_i under pk_i, and broadcasts the resulting ciphertext. At the end of this round, each party can homomorphically compute the desired function f on the received ciphertexts and derive a common multi-key ciphertext which encrypts the output $y = f(x_1, \ldots, x_N)$.
2. The parties run a distributed protocol for "threshold decryption" using their secret keys sk_i to decrypt the multi-key ciphertext and recover the output y in plaintext. In particular each party first generates partial decryptions p_i from the common (evaluated) ciphertext $\widehat{c}$ and then broadcasts them. Finally each party, on receiving all those partial decryptions can compute the final decryption y.

Our goal is to prove the security of this protocol (as least in the honest-but-curious setting, as a start). The natural attempt to construct a MPC simulator S would be to first use the simulator of threshold decryption, S^{thr} to replace the correct partial decryptions p_i with simulated ones p'_i and then use semantic security of the encryption to replace each ciphertext (broadcast in the first round) by encryptions of 0.

The Problem. Unfortunately, we notice that the simulatability of the threshold decryption does not suffice when there is more than one honest party. Essentially,

---

[7] We ignore the algebraic structure of the secret key here and assume each element in $\mathbb{Z}_q$ can be represented as a $\lceil \log(q) \rceil + 1$ binary string.

our definition of simulation security for threshold decryption only allows us to simulate the partial decryption of a single party at a time while knowing the secret keys of all other parties. We cannot, however, simultaneously simulate the partial decryptions of (even) two honest parties without knowing either of their secret keys.

Solution. Essentially we solve the above problem by two steps. We first show that the "basic" protocol as described above is already secure when there is exactly one honest party. Then, later in Sect. 6.2 we extend the basic protocol to another protocol which can handle any arbitrary number of corruption. The extended protocol additionally requires only pseudorandom functions (PRFs) and thus no new assumptions are used. Combining, we get a protocol which securely realizes any functionality against any arbitrary number of corruptions. Below we provide the basic protocol from any MFHE scheme and prove security against exactly $N-1$ corruptions. Later in Sect. 6.2 we present the extension in detail.

Semi-Malicious Security. Following [3], we will actually prove that the above protocol satisfies something called "semi-malicious" security which is stronger than honest-but-curious. Intuitively, it means that adversarial parties need to follow the protocol specification, but can use arbitrary values for their random coins. In fact, the adversary only needs to decide on the input and the random coins to use for each party in each round at the time that the party sends the first message[8]. We will then rely on a theorem of [3] showing that one can compile any such protocol which is secure in the semi-malicious setting into one that is secure in the fully malicious setting, without adding any rounds, by using non-interactive zero-knowledge proofs (NIZKs).

6.1 Protocol Secure Against Exactly $N-1$ Corruptions

The protocol, given in Fig. 2, realizes general multiparty computation for any polynomial-time deterministic functions f which produces a common output for all parties. It does so with respect to a static *semi-malicious attackers* corrupting exactly $N-1$ parties. Formally we prove the following theorem.

Theorem 4. *Let f be a poly-time computable deterministic function with N inputs and 1 output. Let* MFHE = (Setup, Keygen, Encrypt, Expand, Eval, PartDec, FinDec) *be a multi-key FHE scheme with threshold decryption. Then the protocol π_f described in Fig. 2 UC-realizes the function f against any static semi-malicious adversary corrupting exactly $N-1$ parties.*

Proof: The correctness of the protocol follows in a straightforward way from the correctness of the underlying threshold MFHE scheme.

To prove security basically we need to construct an efficient (PPT) simulator $\mathcal{S}$ for any adversary corrupting exactly $N-1$ parties. Let $\mathcal{A}$ be a static semi-malicious adversary and P_h be the *only* honest party. The simulator simulates the protocol execution on behalf of the honest party P_h as follows.

---

[8] See the full version [30] for a formal definition.

Let $f : (\{0,1\}^{\ell_{in}})^N \to \{0,1\}^{\ell_{out}}$ be the function to compute. Let d be the depth of the circuit for f.

Preprocessing. Run setup $\leftarrow$ MFHE.Setup$(1^\lambda, 1^d)$. All the parties share the common setup.

Input: Each party P_k has input $\mathbf{x}_k \in \{0,1\}^{\ell_{in}}$.

The Protocol:

Round I. Each party P_k executes the following steps.
- Generate a key-pair $(sk_k, pk_k) \leftarrow$ MFHE.Keygen(setup).
- Encrypt the message bit-by-bit:

$$\{c_{k,j} \leftarrow \text{MFHE.Encrypt}(pk_k, \mathbf{x}_k[j])\}_{j \in [\ell_{in}]}.$$

- Broadcast the public-key and the ciphertexts $(pk_k, \{c_{k,j}\}_{j \in [\ell_{in}]})$.

Round II. Each party P_k on receiving values $\{pk_i, c_{i,j}\}_{i \in [N] \setminus \{k\}, j \in [\ell_{in}]}$ executes the following steps:
- First expand each $c_{i,j}$:

$$\{\widehat{c}_{i,j} \leftarrow \text{MFHE.Expand}((pk_1, \ldots, pk_N), i, c_{i,j})\}_{i \in [N], j \in [\ell_{in}]}$$

- Run the evaluation algorithm to generate the evaluated ciphertext:

$$\{\widehat{c}_j \leftarrow \text{MFHE.Eval}(f_j, (\widehat{c}_{1,1}, \ldots, \widehat{c}_{N,\ell_{in}}))\}_{j \in [\ell_{out}]}.$$

where f_j is the boolean function for j-th bit of the output of f.
- Finally all the parties concurrently take part in one-round threshold decryption to obtain the output message bit-by-bit as follows:
  - Each P_k computes the partial decryption for all $j \in [\ell_{out}]$:

$$p_k^{(j)} \leftarrow \text{MFHE.PartDec}(\widehat{c}_j, (pk_1, \ldots, pk_N), k, sk_k)$$

  - P_k broadcasts all the values $\{p_k^{(j)}\}_{j \in [\ell_{out}]}$.

Output. On receiving all the values $\{p_i^{(j)}\}_{i \in [N], j \in [\ell_{out}]}$ run the final decryption to obtain the j-th output bit: $\{y_j \leftarrow \text{MFHE.FinDec}(p_1^{(j)}, \ldots, p_N^{(j)})\}_{j \in [\ell_{out}]}$. Output $y = y_1 \ldots y_{\ell_{out}}$.

Fig. 2. π_f: A basic MPC protocol for f secure against $N-1$ corruptions

The Simulator. In round-I, the simulator encrypts 0s instead of the real input bits of the honest party P_h. After round-I it gets the inputs and the secret keys of the $N-1$ corrupt parties from the "witness tape". It gives these inputs to the ideal functionality and receives the output bits y_j for each $j \in [\ell_{out}]$. At this point it can also compute the evaluated ciphertexts $\widehat{c}_j$. Then it computes the simulated partial decryptions for the honest party $\widetilde{p}_h^{(j)} \leftarrow \mathcal{S}^{thr}(y_j, \widehat{c}_j, h, \{sk_i\}_{i \in [N] \setminus \{h\}})$ and broadcast those in round-II instead of correctly computed partial decryptions $p_h^{(j)}$ generated via MFHE.PartDec$(\cdots)$.

Hybrid Games. We now define a series of *hybrid games* that will be used to prove the indistinguishability of the real and ideal worlds:

$$\text{IDEAL}_{\mathcal{F},\mathcal{S},\mathcal{Z}} \stackrel{comp}{\approx} \text{REAL}_{\pi,\mathcal{A},\mathcal{Z}} \tag{2}$$

The output of each game is always just the output of the environment.

The game $REAL_{\pi,\mathcal{A},\mathcal{Z}}$: This is exactly an execution of the protocol π in the real world with environment $\mathcal{Z}$ and semi-malicious adversary $\mathcal{A}$.

The game $HYB^1_{\pi,\mathcal{A},\mathcal{Z}}$: In this game, we modify the real world experiment as follows. Assume (as a mental experiment) that P_h is given the all the secret keys $\{sk_i\}_{i\in[N]\setminus\{h\}}$ (as written on the "witness tape" of the adversary) after round I. In the second round, instead of broadcasting a correctly generated partial decryptions $p_h^{(j)}$ generated via MFHE.PartDec($\cdots$), it broadcasts simulated ones $\{\tilde{p}_h^{(j)} \leftarrow \mathcal{S}^{thr}(y_j, \hat{c}_j, h_1, \{sk_i\}_{i\neq h})\}_{j\in[\ell_{out}]}$.

The game $IDEAL_{\mathcal{F},\mathcal{S},\mathcal{Z}}$: This is similar to the game $HYB^1_{\pi,\mathcal{A},\mathcal{Z}}$ except instead of encrypting its real input, P_h now broadcasts encryption of 0s in the first round.

Claim 1. $\text{REAL}_{\pi,\mathcal{A},\mathcal{Z}} \stackrel{stat}{\approx} \text{HYB}^1_{\pi,\mathcal{A},\mathcal{Z}}$

Proof: Notice that, the only change between those experiments are that, the partial decryption of party P_h is generated through simulator $\mathcal{S}^{thr}$ instead of correctly using MFHE.PartDec. By simulatability of threshold decryption the partial decryptions are statistically indistinguishable hence so are the experiments. ∎

Claim 2. $\text{HYB}^1_{\pi,\mathcal{A},\mathcal{Z}} \stackrel{comp}{\approx} \text{IDEAL}_{\mathcal{F},\mathcal{S},\mathcal{Z}}$

Proof: The only change between those experiments are in generating encryptions of party P_h. By semantic security of the underlying MFHE the encryptions are computationally indistinguishable. Hence the experiments are also computationally indistinguishable. Note that here it is possible to use the semantic security as the partial decryptions of P_h in both the experiments are simulated and hence independent of the secret key sk_h. ∎

This concludes the proof of the theorem. ∎

6.2 An Extended Protocol for Arbitrary Many Corruptions

In this section we construct an "extended" MPC protocol $\hat{\pi}_f$ which securely computes any function f against any semi-malicious adversary that can corrupt any $t \in [N]$ parties. We do so by relying on the "basic" MPC protocol π_f from the previous section, which is secure against a semi-malicious adversary that corrupts exactly $N-1$ parties. To compute a function f, our extedned protocol simply runs the basic protocol $\pi_{\hat{f}}$ on an extended function $\hat{f}$ defined as follows.

Definition 5 *(Extended function).* *For any polynomial* $\ell_{in}, \ell_{out}, N \in \mathbb{N}$ *let* $f : \{\{0,1\}^{\ell_{in}}\}^N \rightarrow \{0,1\}^{\ell_{out}}$ *be a poly-time computable function and* PRF $:$ $\{0,1\}^\lambda \times [N] \rightarrow \{0,1\}^{\ell_{in}}$ *be a PRF. Then we define an extended function* $\widehat{f} : \{\{0,1\}^{\ell_{in}} \times \{1,2,3\} \times \{0,1\}^\lambda\}^N \rightarrow \{0,1\}^{\ell_{out}}$ *which takes as input* $((\mathbf{x}_1, \mathtt{mode}_1, \mathbf{z}_1), \ldots, (\mathbf{x}_N, \mathtt{mode}_N, \mathbf{z}_N))$ *and does the following:*

- *If* $\forall\ i \in [N]$, $\mathtt{mode}_i = 1$ *then output* $f(\mathbf{x}_1, \ldots, \mathbf{x}_N)$.
- *If* $\exists$ *unique* $i \in [N]$ *such that* $\mathtt{mode}_i = 2$ *then let* $K := \mathbf{z}_i$. *For all* $j \in [N]$:
  - *If* $\mathtt{mode}_j = 3$ *then set* $\mathbf{x}'_j := \mathsf{PRF}(K, j) \oplus \mathbf{x}_j$.
  - *Else set* $\mathbf{x}'_j := \mathbf{x}_j$.
  Output $f(\mathbf{x}'_1, \ldots, \mathbf{x}'_N)$.
- *Otherwise output* $0^{\ell_{out}}$.

Roughly speaking, the extended function does the same thing as the original function if all the inputs have $\mathtt{mode}_i = 1$. However, if there is one special party with $\mathtt{mode}_i = 2$ then the function uses a PRF key $K = \mathbf{z}_i$ provided by that party to "decrypt" the inputs of all the parties with $\mathtt{mode}_j = 3$.

We define an "extended protocol" $\widehat{\pi}_f$ in Fig. 3. It essentially just runs the original basic protocol $\pi_{\widehat{f}}$ with an extended function $\widehat{f}$ and appropriately extended inputs.

Let $f : \{\{0,1\}^{\ell_{in}}\}^N \rightarrow \ell_{out}$ be the function we wish to compute and let and PRF $:$ $\{0,1\}^\lambda \times [N] \rightarrow \{0,1\}^{\ell_{in}}$ be a PRF. Let $\widehat{f} : \{\{0,1\}^{\ell_{in}} \times \{1,2,3\} \times \{0,1\}^\lambda\}^N \rightarrow \{0,1\}^{\ell_{out}}$ be the corresponding extended function (Definition 5). Let $\pi_{\widehat{f}}$ be the protocol from Figure 2 applied to the extended function $\widehat{f}$. The extended protocol $\widehat{\pi}_f$ is defined as follows:

Setup: The setup is the same as the that of the protocol $\pi_{\widehat{f}}$.

Input: Each party P_k has input $\mathbf{x}_k \in \{0,1\}^{\ell_{in}}$. Additionally each party sets $\mathtt{mode}_k :=$ 1, $\mathbf{z}_k := 0$, and defines its extended input as $\widehat{\mathbf{x}}_k := (\mathbf{x}_k, \mathtt{mode}_k, \mathbf{z}_k) \in \{0,1\}^{\widehat{\ell_{in}}}$ where $\widehat{\ell_{in}} = \ell_{in} + \lambda + 2$. [a]

Protocol: The parties run the protocol $\pi_{\widehat{f}}$ using the extended inputs $\{\widehat{\mathbf{x}}_k\}_{k \in [N]}$. They output whatever $\pi_{\widehat{f}}$ outputs.

[a] Here any 0 denotes a string of 0s of appropriate size. We abuse notation for simplification.

Fig. 3. $\widehat{\pi}_f$: Extended protocol secure against any number of semi-malicious corruptions.

The following theorem states that the extended protocol $\widehat{\pi}_f$ is secure against any arbitrary number of semi-malicious corruptions. The proof is deferred to the full version [30].

Theorem 5. *Let* f *be a function with* N *inputs and 1 output. Let* $\mathsf{PRF} : \{0,1\}^\lambda \times$ $[N] \rightarrow \{0,1\}^{\ell_{out}}$ *be a PRF. Then, under the LWE assumption, the protocol* $\widehat{\pi}_f$ *shown in Fig. 3 UC-realizes* f *against a static semi-malicious adversary that can corrupt any number of parties.*

6.3 Extensions and Applications

Generalized Functionalities. Our protocol (Fig. 2) considers deterministic functionalities where all the parties receive the same output. One can extend that to handle randomized functionalities and individual output in a straightforward manner using known standard techniques just like [4]. We refer to [4] for more details.

Fully Malicious Adversary. Our protocol protects only against semi-malicious adversaries. However, since we are in the CRS model such protocol can be generically converted to one secure against fully-malicious adversary using non-interactive zero-knowledge (NIZK) arguments. For more detail on this again we refer to [4].

Communication Complexity. Although our main focus was on round complexity, we mention that our scheme also achieves essentially optimal communication complexity which is only proportional to the total input size, output size and circuit depth. We can get rid of the reliance on circuit depth by using bootstrapping and relying on circular security: each party would simply send a GSW encryption of its secret key under its public key and then we would perform a boostrapping step after each homomorphic operation to reduce the noise in the ciphertext.

Computation on the Web. Our results also relate to the idea of "computation on the web" [21] where parties can't interact with each other but can only interact with some central website without further coordination. Using our scheme (or any 2 round protocol) each party needs to log in twice: once to give its ciphertext to the sever and once to give a partial decryption of the output.

7 Conclusions

We have shown how to implement MPC with only two rounds of interaction by relying on the LWE assumption (and NIZKs for malicious security). Several interesting open problems remain. Firstly, is possible to get a 2 round MPC protocol under general assumptions such as the existence of oblivious transfer? Secondly, is it possible to get a protocol that achieves adaptive security? A recent work of [16] does this using indistinguishability obfuscation (iO) but it remains an open problem to do this using more standard assumptions such as LWE. Lastly, it would be interesting to get a 2 round protocol in the honest-but-curious model without a CRS. One way to achieve this would be to a build a threshold multi-key FHE without any common public parameters.

References

1. Alperin-Sheriff, J., Peikert, C.: Faster bootstrapping with polynomial error. In: Garay, J.A., Gennaro, R. (eds.) CRYPTO 2014, Part I. LNCS, vol. 8616, pp. 297–314. Springer, Heidelberg (2014)

2. Applebaum, B., Ishai, Y., Kushilevitz, E.: Computationally private randomizing polynomials and their applications. In: IEEE Conference on Computational Complexity, pp. 260–274 (2005)
3. Asharov, G., Jain, A., López-Alt, A., Tromer, E., Vaikuntanathan, V., Wichs, D.: Multiparty computation with low communication, computation and interaction via threshold FHE. In: Pointcheval, D., Johansson, T. (eds.) EUROCRYPT 2012. LNCS, pp. 483–501. Springer, Heidelberg (2012)
4. Asharov, G., Jain, A., Wichs, D.: Multiparty computation with low communication, computation and interaction via threshold fhe. Cryptology ePrint Archive, Report 2011/613 2011. http://eprint.iacr.org/
5. Beaver, D., Micali, S., Rogaway, P.: The round complexity of secure protocols (extended abstract). In: STOC, pp. 503–513 (1990)
6. Bendlin, R., Damgård, I.: Threshold decryption and zero-knowledge proofs for lattice-based cryptosystems. In: Micciancio, D. (ed.) TCC 2010. LNCS, vol. 5978, pp. 201–218. Springer, Heidelberg (2010)
7. Bendlin, R., Damgård, I., Orlandi, C., Zakarias, S.: Semi-homomorphic encryption and multiparty computation. In: Paterson, K.G. (ed.) EUROCRYPT 2011. LNCS, vol. 6632, pp. 169–188. Springer, Heidelberg (2011)
8. Brakerski, Z., Gentry, C., Vaikuntanathan, V.: Fully homomorphic encryption without bootstrapping. In: ITCS (2012)
9. Brakerski, Z., Vaikuntanathan, V.: Efficient fully homomorphic encryption from (standard) lwe. In: FOCS (2011)
10. Clear, M., McGoldrick, C.: Multi-identity and multi-key leveled FHE from learning with errors. In: Gennaro, R., Robshaw, M. (eds.) CRYPTO 2015. LNCS, vol. 9216, pp. 630–656. Springer, Heidelberg (2015)
11. Cramer, R., Damgård, I.B., Nielsen, J.B.: Multiparty computation from threshold homomorphic encryption. In: Pfitzmann, B. (ed.) EUROCRYPT 2001. LNCS, vol. 2045, pp. 280–300. Springer, Heidelberg (2001)
12. Damgård, I.B., Nielsen, J.B.: Universally composable efficient multiparty computation from threshold homomorphic encryption. In: Boneh, D. (ed.) CRYPTO 2003. LNCS, vol. 2729, pp. 247–264. Springer, Heidelberg (2003)
13. Matthew, K.: Franklin and stuart haber: joint encryption and message-efficient secure computation. J. Cryptology 9(4), 217–232 (1996)
14. Garg, S., Gentry, C., Halevi, S., Raykova, M.: Two-round secure MPC from indistinguishability obfuscation. In: Lindell, Y. (ed.) TCC 2014. LNCS, vol. 8349, pp. 74–94. Springer, Heidelberg (2014)
15. Garg, S., Mukherjee, P., Pandey, O., Polychroniadou, A.: The exact round complexity of secure computation. Manuscript, October 2015
16. Garg, S., Polychroniadou, A.: Two-round adaptively secure MPC from indistinguishability obfuscation. In: Dodis, Y., Nielsen, J.B. (eds.) TCC 2015, Part II. LNCS, vol. 9015, pp. 614–637. Springer, Heidelberg (2015)
17. Gentry, C.: Fully homomorphic encryption using ideal lattices. In: STOC, pp. 169–178 (2009)
18. Gentry, C., Sahai, A., Waters, B.: Homomorphic encryption from learning with errors: conceptually-simpler, asymptotically-faster, attribute-based. In: Canetti, R., Garay, J.A. (eds.) CRYPTO 2013, Part I. LNCS, vol. 8042, pp. 75–92. Springer, Heidelberg (2013)
19. Goldreich, O., Micali, S., Wigderson, A.: How to play any mental game or a completeness theorem for protocols with honest majority. In: STOC, pp. 218–229 (1987)

20. Goyal, V.: Constant round non-malleable protocols using one way functions. In: STOC, pp. 695–704 (2011)
21. Halevi, S., Lindell, Y., Pinkas, B.: Secure computation on the web: computing without simultaneous interaction. In: Rogaway, P. (ed.) CRYPTO 2011. LNCS, vol. 6841, pp. 132–150. Springer, Heidelberg (2011)
22. Ishai, Y., Kushilevitz, E.: Randomizing polynomials: A new representation with applications to round-efficient secure computation. In: FOCS, pp. 294–304 (2000)
23. Jakobsson, M., Juels, A.: Mix and match: secure function evaluation via ciphertexts. In: Okamoto, T. (ed.) ASIACRYPT 2000. LNCS, vol. 1976, pp. 162–177. Springer, Heidelberg (2000)
24. Katz, J., Ostrovsky, R.: Round-optimal secure two-party computation. In: Franklin, M. (ed.) CRYPTO 2004. LNCS, vol. 3152, pp. 335–354. Springer, Heidelberg (2004)
25. Katz, J., Ostrovsky, R., Smith, A.: Round Efficiency of Multi-party Computation with a Dishonest Majority. In: Biham, E. (ed.) EUROCRYPT 2003. LNCS, pp. 578–595. Springer, Heidelberg (2003)
26. Lin, H., Pass, R.: Constant-round non-malleable commitments from any one-way function. In: STOC, pp. 705–714 (2011)
27. Lindell, Y.: Parallel coin-tossing and constant-round secure two-party computation. In: Kilian, J. (ed.) CRYPTO 2001. LNCS, vol. 2139, pp. 171–189. Springer, Heidelberg (2001)
28. López-Alt, A., Tromer, E., Vaikuntanathan, V.: On-the-fly multiparty computation on the cloud via multikey fully homomorphic encryption. In: Karloff, H.J., Pitassi, T. (eds.) Proceedings of the 44th Symposium on Theory of Computing Conference, STOC, New York, NY, USA, 19–22 May 2012, pp. 1219–1234. ACM (2012)
29. Micciancio, D., Peikert, C.: Trapdoors for lattices: simpler, tighter, faster, smaller. In: Pointcheval, D., Johansson, T. (eds.) EUROCRYPT 2012. LNCS, vol. 7237, pp. 700–718. Springer, Heidelberg (2012)
30. Mukherjee, P., Wichs, D.: Two round multiparty computation via multi-key FHE. Cryptology ePrint Archive, Report 2015/345 (2015). http://eprint.iacr.org/
31. Myers, S., Sergi, M., Shelat, A.: Threshold fully homomorphic encryption and secure computation. In: eprint /454 (2011)
32. Peikert, C.: Public-key cryptosystems from the worst-case shortest vector problem: extended abstract. In: STOC, pp. 333–342 (2009)
33. Regev, O.: On lattices, learning with errors, random linear codes, and cryptography. In: STOC, pp. 84–93 (2005)
34. Andrew Chi-Chih Yao: Protocols for secure computations (extended abstract). In: FOCS, pp. 160–164 (1982)
35. Andrew Chi-Chih Yao: How to generate and exchange secrets (extended abstract). In: FOCS, pp. 162–167 (1986)

Post-zeroing Obfuscation: New Mathematical Tools, and the Case of Evasive Circuits

Saikrishna Badrinarayanan[1]([⊠]), Eric Miles[1], Amit Sahai[1],
and Mark Zhandry[2,3]

[1] Center for Encrypted Functionalities, UCLA, Los Angeles, USA
{saikrishna,enmiles,sahai}@cs.ucla.edu
[2] MIT, Cambridge, USA
[3] Princeton University, Princeton, USA
mzhandry@princeton.edu

Abstract. Recent devastating attacks by Cheon et al. [Eurocrypt'15] and others have highlighted significant gaps in our intuition about security in candidate multilinear map schemes, and in candidate obfuscators that use them. The new attacks, and some that were previously known, are typically called "zeroizing" attacks because they all crucially rely on the ability of the adversary to create encodings of 0.

In this work, we initiate the study of *post-zeroizing obfuscation*, and we obtain a key new mathematical tool to analyze security in a post-zeroizing world. Our new mathematical tool allows for analyzing polynomials constructed by the adversary when given encodings of randomized matrices arising from a general matrix branching program. This technique shows that the types of encodings an adversary can create are much more restricted than was previously known, and is a crucial step toward achieving post-zeroizing security. We also believe the technique is of independent interest, as it yields efficiency improvements for existing schemes – efficiency improvements that have already found application in other settings.

Finally, we show how to apply our new mathematical tool to the special case of evasive functions. We show that our obfuscator survives *all known attacks* on the underlying multilinear maps, by proving that no top-level encodings of 0 can be created by a generic-model adversary. Previous obfuscators (for both evasive and general functions) were either analyzed in a less-conservative "pre-zeroizing" model that *does not* capture recent attacks, or were proved secure relative to assumptions that no longer have any plausible instantiation due to zeroizing attacks.

This paper subsumes a previous work of Sahai and Zhandry [35].
A. Sahai—Supported in part by a DARPA/ONR PROCEED award, NSF grants 1228984, 1136174, 1118096, 1065276,0916574 and 0830803, a Xerox Faculty Research Award, a Google Faculty Research Award, an equipment grant from Intel, and an Okawa Foundation Research Grant. This material is based upon work supported by the Defense Advanced Research Projects Agency through the U.S. Office of Naval Research under Contract N00014-11-1-0389. The views expressed are those of the author and do not reflect the official policy or position of the Department of Defense, the National Science Foundation, or the U.S. Government.

M. Fischlin and J.-S. Coron (Eds.): EUROCRYPT 2016, Part II, LNCS 9666, pp. 764–791, 2016.
DOI: 10.1007/978-3-662-49896-5_27

1 Introduction

Over the past three years, *all* candidate constructions [16–18,22,27] of multilinear maps, also called graded encoding schemes, have been shown to suffer from "zeroizing" attacks [9,11,13–15,18,27,28,32] — and these attacks have in many cases been devastating.

Given this state of affairs, one would expect that the most-studied application of graded encodings schemes – indistinguishability obfuscation [7,20] – would be similarly devastated. However, quite surprisingly, until our work, *none* of the zeroizing attacks placed current obfuscation schemes over prime-order[1] graded encodings in jeopardy. In this paper, we ask: *Why is this the case?* Given the profound level of interest in obfuscation over the past two years [1,4,6,10,12,19,20,24,25,30,33,34,37], and given that so far all proposed obfuscation schemes rely on graded encoding schemes, we believe this question is of paramount importance. And indeed, before our work, no security analysis for obfuscation used a model or assumption that took into account the impact of zeroizing attacks.

Long-Term Vision. This paper seeks to initiate a research program whose aim is to build fully secure obfuscation schemes out of *weakened* graded encodings schemes – graded encoding schemes that are subject to zeroizing attacks. As the research cycle of construct-and-attack over the past 3 years has shown, building fully secure graded encoding schemes is a challenging task. Thus, our approach is to take a pessimistic view and see if, in fact, even weakened forms of graded encoding schemes suffice for constructing fully secure obfuscation. We note that even if future constructions of graded encoding schemes are successful in avoiding zeroizing attacks, the research program initiated by our work would still be valuable, because it will help to identify the minimal security properties actually needed by graded encoding schemes to achieve secure obfuscation. This could lead to greater efficiency.

The central contribution of our work is a new mathematical tool that characterizes when an adversary can set up the most basic requirement for a zeroizing attack, namely a top-level encoding of zero. Furthermore, we present this mathematical tool in a very general form, which even has consequences for efficiency of prime-order obfuscation constructions. We believe that our characterization lemma will prove valuable in the long-term study of both guiding research into new attacks on graded encodings as well as building secure obfuscation from weak graded encodings. We demonstrate this by applying our lemma to the case

---

[1] We note that certain *simplified* versions of obfuscation schemes over composite-order graded encoding schemes [37] have been broken by zeroizing attacks [15]. There are no published methods for converting the composite-order obfuscation schemes of [4,37] to the prime-order setting. Furthermore, zeroizing attacks over prime-order graded encoding schemes discovered prior to our work typically applied when the multiplicity of zero achieved at the top level is greater than zero, and if a prime-order conversion was attempted, care would need to be taken to ensure that such higher multiplicities do not occur.

of evasive circuits, for which we can show security even using only extremely weak graded encodings whose security completely breaks down when a top-level encoding of zero is found.

Background - Obfuscation. Obfuscation is a cryptographic tool that offers a powerful capability: software that can keep a secret. That is, consider a piece of software that makes use of a secret to perform its computation. Obfuscation allows us to transform this software so that it can be run *publicly*: anyone can obtain the full code of the program, run it, and see its outputs, but no one can learn anything about the embedded secret, beyond what can be learned by examining the outputs of the program.

The first candidate construction for a general-purpose obfuscator was given by Garg, Gentry, Halevi, Raykova, Sahai, and Waters [20]. This construction, and all subsequent works constructing candidate obfuscators [1,4,6,12,24,30,33,37], are built on top of another cryptographic primitive called a graded encoding scheme. In a graded encoding scheme, plaintext elements are encoded at various levels, and can be added and multiplied subject to algebraic restrictions relating to these levels. Further, there is a "top" level at which one can test whether an element encodes 0.

Background - Zeroizing Attacks. The many zeroizing attacks differ somewhat in their details, but each attack obeys the algebraic restrictions imposed by the graded encodings schemes, and critically they all share the need to create top-level 0-encodings. Indeed, many such encodings are needed for each attack, and the attacks require that these encodings have further structure.

Several of the works constructing candidate obfuscators prove security in an idealized "generic multilinear model" that seeks to capture the algebraic restrictions imposed by the graded encoding scheme candidates. However, the known zeroizing attacks use extra information that is provided by the zero-testing procedure, which is not captured in the standard generic model. Thus, a proof of security in the generic multilinear model by itself is no longer a persuasive argument of security. In particular, it is now crucial to gain a better understanding of exactly what types of top-level 0-encodings can be constructed in prime-order graded encoding schemes.

Our Contribution. We introduce a new mathematical tool for analyzing how an adversary can create 0-encodings given a set of randomized matrices. This tool both shows that the types of 0-encodings an adversary can create are much more restricted than was previously known, and that the adversary's behavior can be so controlled in a much richer set of circumstances than was previously known. We stress that this new tool, Theorem 4, was not present in any previous work, and allows for a much more fine-grained analysis of the adversary's behavior in prime-order settings than was previously available.

Briefly, we first consider an obfuscator $\mathcal{O}$ that can use much wider class of matrix branching programs than was previously known, most notably this class includes matrix branching programs that involve low-rank matrices. Theorem 4 shows that any polynomial p over (the encodings in) the obfuscation $\mathcal{O}(f)$ can be efficiently mapped to a poly-size set of inputs X such that p evaluates to an encoding of 0 if and only if *every* $x \in X$ satisfies $f(x) = 0$. For context, previous works both could not handle the case of low-rank matrices, and only gave a map that allowed the evaluation of p to be *simulated* given the set $\{f(x) \mid x \in X\}$, but did not show the stronger precise characterization of 0-encodings that we obtain. We stress that we do not know of any simpler way of obtaining such a characterization.

We now elaborate on Theorem 4, how it is proved, and how previous works that did not consider zeroing attacks did not need and did not achieve such a theorem. Following that, we mention two applications of this new theorem, namely improving the efficiency of obfuscation, and obfuscating *evasive* functions in a model that captures all known attacks on graded encoding schemes.

1.1   Our Techniques

As stated above, the main technical challenge in our paper is to show that any polynomial p over the obfuscation $\mathcal{O}(f)$ can be efficiently mapped to a set of inputs X such that p evaluates to an encoding of 0 if and only if every $x \in X$ satisfies $f(x) = 0$.

One ingredient in our paper is the notion of strong straddling sets from [30], as this tool allows us to show that low-level encodings of 0 can be efficiently transformed into top-level encodings of 0. Thus, the only obstacle that remains is to prove Theorem 4 for top-level encodings.

The Technical Barrier – Kilian's Statistical Simulation. Before we proceed to provide intuition about our proof, let us consider the technical roots of how security was shown in previous works. In every paper constructing secure obfuscation for matrix branching programs so far [1,6,12,20,24,30,33] and in every different model that has been considered, one theorem has played a starring role in all security analyses: Kilian's statistical simulation theorem [29]. As relevant here, Kilian's theorem considers the setting where we randomize each matrix in a sequence of matrices as follows:

$$\widehat{\mathbf{B}_i} = \mathbf{R}_{i-1}^{-1} \mathbf{B}_i \mathbf{R}_i$$

where $\mathbf{R}_i$ are random invertible matrices for $i \in [\ell - 1]$, and identity otherwise. Note that this randomization does not affect the iterated product. Then, for any particular input x, if the iterated product is M, Kilian's theorem states that we can statistically simulate the collection of matrices $\{\widehat{\mathbf{B}_i}\}_{i \in [\ell]}$ knowing only M but with no knowledge of the original matrices $\{\mathbf{B}_i\}$.

Kilian's statistical simulation theorem has been a keystone in all previous analyses of obfuscation: in one way or another, all previous security analyses for obfuscation methods have found some way to isolate the adversary's view of the obfuscation to a single input. Once this isolation was accomplished, Kilian's theorem provided the assurance that the adversary's view of the obfuscation, as it related to this single input, only encoded information about the output of the computation within M, and nothing more.

However, Kilian's statistical simulation theorem only allows for simulation. It does not rule out the possibility that an encoding of 0 may result no matter what the function outputs on the input in question. Indeed, it is not hard to construct an obfuscator that is secure in the generic model but allows for encodings of 0 even when the function being obfuscated always outputs 1. Moreover, Kilian's theorem only applies when the branching program matrices are full-rank. Indeed, if the matrices are allowed to be arbitrary rank, then it is *impossible* to simulate each of the matrices just given the product M, as there is no way to determine what the rank of each matrix should be, nor the ranks of various subproducts of matrices. (In the next subsection, we discuss the efficiency benefits of allowing low-rank matrices.) Because of this impossibility, we know of no way to generalize Kilian's theorem or its proof to obtain our theorem.

Our Approach. To obtain our result, we directly analyze what kinds of polynomials an adversary can generate using multilinear operations. We model the multilinear setting as follows. There is a universe set $[\ell]$, and for every subset $S \subseteq [\ell]$, we have a copy of $\mathbb{Z}_q$ that we name G_S. Then, the adversary has access to the following operations:

- ADD: $G_S \times G_S \rightarrow G_S$, for every subset $S \subseteq [\ell]$.
- MULT: $G_S \times G_T \rightarrow G_{S \cup T}$, for every pair $S, T \subseteq [\ell] : S \cap T = \emptyset$.
- ZeroTest: $G_{[\ell]} \rightarrow \{\text{TRUE}, \text{FALSE}\}$.

This is sometimes called the "asymmetric" multilinear setting, is natively supported by known instantiations of prime-order graded encoding schemes [18], and was used in previous works. Observe that in this setting, if the adversary is given a matrix entirely encoded in $G_{\{1\}}$, then for example it is not possible for it to compute the rank of this matrix. This is because no two entries within this matrix can be multiplied together, since they both reside in the same group $G_{\{1\}}$, and multiplication is only possible across elements of groups corresponding to disjoint index sets.

Even though we do not rely on Kilian's simulation theorem, our obfuscator uses a matrix randomization scheme that is essentially[2] identical to the one used when applying Kilian's randomization. Our analysis then proceeds by considering the most general polynomial that the adversary can construct in $G_{[\ell]}$. More precisely, we consider every possible monomial m that can exist over the matrix entries that are given to the adversary, and we associate each such monomial m with a coefficient α_m that the adversary could potentially choose arbitrarily. Thus, the adversary's polynomial is a giant sum $p = \Sigma_m \alpha_m m$ over all these potential monomials.

---

[2] Because we consider rectangular matrices in general, we do need to modify this slightly. Also, for technical simplification, we consider the adjugate matrix rather than the inverse. However, for the purposes of this technical overview, these variations can be ignored.

Observe that the adversary can only extract useful information from this polynomial by passing it to ZeroTest, thereby determining if it is zero or not. However, recall that the randomizing matrices $\{\mathbf{R}_i\}$ are chosen uniformly during obfuscation. Therefore, by the Schwartz-Zippel lemma, we know that unless the adversary's polynomial p is the zero polynomial over the entries of the $\mathbf{R}_i$ matrices, ZeroTest will declare the polynomial to be nonzero with overwhelming probability. So, we restrict ourselves to analyzing adversarial polynomials that are identically zero over the entries of the $\mathbf{R}_i$ matrices.

Our new analysis differs at a fundamental level from Kilian's analysis. At the heart of the analysis is an argument based on the structure of random square matrices R and their inverses R^{-1} that allows us to argue about how terms that arise in R^{-1} can be cancelled using terms from R. In particular, we use the fact that:

$$R_{i,\ell}^{-1} = \frac{1}{\det(R)} \sum_{\sigma:\sigma(i)=\ell} \text{sign}(\sigma) \left(\prod_{t \neq i} R_{\sigma(t),t} \right)$$

Our analysis is obtained by carefully considering different types of permutations σ that arise in the expression above, and how different permutations interfere with each other. (This exemplifies our conceptual departure from the proof of Kilian's theorem.) Our analysis shows that multilinear polynomials that allow for cancellation of R and R^{-1} terms are extremely constrained.

From this analysis, we conclude that any adversarial polynomial that is identically zero over the entries of the $\{\mathbf{R}_i\}$ matrices must in fact be the result of an honest iterated matrix multiplication (or a constant multiple thereof), which corresponds to evaluating $f(x)$ for some input x. In other words, such an adversarial polynomial will result in an encoding of 0 if and only if $f(x) = 0$, as desired. Even though the analysis as presented here is not efficient, we are still able to use it to yield an efficient simulator in our generic model. At a high level, this is done by using the Schwartz-Zippel lemma to "weed out" most adversarial polynomials without needing to examine their structure at all.

1.2   Applications

We now discuss two applications of our new analysis tool.

Efficiency of Obfuscation. Current techniques, while being asymptotically polynomial-time, lead to incredibly inefficient implementations of obfuscation. For example, the recent implementation of Apon et al. [2] for obfuscating (only) 16-bit point functions resulted in a 31 GB obfuscated program, which took over 6 h to generate and about 11 min to run on each input.

A major source of inefficiency is that the direct application of current obfuscators to circuits requires overhead that grows exponentially with the depth. This occurs because the level of multilinearity required grows exponentially with the depth, while current multilinear map candidates have complexity that grows polynomially with the level of multilinearity.

The work of Garg et al. [20] shows that, nevertheless, such a "core" obfuscator can be used to obfuscate general (high depth) circuits with a polynomial overhead through a "bootstrapping" procedure (see also [3,12,26]). However, bootstrapping based on existing core obfuscators entails overheads that are asymptotically polynomial but easily reach above 2^{100}. Such large overheads primarily arise due to the depth of the circuit processed by the core obfuscator (though, asymptotically, this circuit has depth logarithmic in the security parameter). Indeed, similarly large overheads arise when attempting to apply the core obfuscator to other programs represented in circuit form, since few interesting and non-learnable families of circuits have depth below, say, 50.

This suggests that practical implementations of obfuscation will only be able to handle functionalities that require a polynomial level of multilinearity, and not exponential. One such class of functionalities are those computable by small matrix branching programs, where evaluation corresponds to evaluating an iterated matrix product. This class of functionalities includes, among others, finite automata.

Unfortunately, natural representations of finite automata and other simple programs as branching programs require *low-rank* matrices. Though these representations can be made full rank by using much larger matrices, this results in substantial efficiency loss. The reason for this, intuitively, is that branching programs with invertible matrices model *reversible* computation, whereas general computation allows for previous states to be forgotten. While it is possible to convert an irreversible computation into a reversible one, the cost is a significant loss in efficiency. The ability to handle low-rank matrices is thus crucial to obtaining efficient obfuscators even for simple functionalities. As detailed in the preceding subsection, all previous constructions critically relied on full-rank matrices.

Armed with our new tool (Theorem 4), our construction *no longer requires full-rank matrices*, and even non-square matrices are allowed.[3] That is, we show for the first time how to obfuscate matrix branching programs that are represented with low-rank, rectangular matrices. This leads to more efficient obfuscators, even beyond previous works that lack a post-zeroizing proof of security; for details, see the full version of this paper. Our analysis also extends to other settings besides obfuscation: for example, Boneh et al. [8] rely on our analysis to obtain implementable constructions of order-revealing encryption.

Obfuscating Evasive Circuits in a Post-zeroizing Model. We view Theorem 4 as the first step on a path towards achieving obfuscation in a post-zeroizing world. As a "proof of concept" for this goal, we construct an obfuscator for a natural class of functions that, for the first time, is provably secure in a model that captures *all* known attacks on graded encoding schemes.

---

[3] We do require a mild natural technical condition, called *non-shortcutting*, on the branching program. Non-shortcutting can be achieved generically on any branching program with minimal overhead.

In previous works that prove security in a generic model, the graded encoding scheme's zero-test procedure is modeled as a Boolean function (i.e. one that returns a yes/no answer). In candidate constructions however, a successful zero-test actually returns an algebraic element in the ring of encodings, and this fact is crucially exploited in the zeroizing attacks. By contrast, our new model considers *any* encoding of 0 to be a complete break, thereby capturing these attacks.

We show how to obfuscate *evasive* functions [5] in this model, namely functions for which it is hard to find an input that evaluates to 0. (Typically one defines evasive functions as having hidden 1-outputs, but in terms of their functionality this is only a semantic difference.) A natural example of an evasive function is the "password check" function (typically called a point function), which evaluates to 0 on only a single, secret input. Obfuscating general evasive functions has many applications, including most notably obfuscating important classes of software patches that check for rare inputs on which the unpatched software is known to misbehave (see [5] for further discussion).

Prior to our work, except as a special case of general obfuscation, the only work that considered obfuscating general classes of evasive functions is that of [5]. However, the positive results in [5] were based on assumptions over approximate multilinear maps that are now known to be false when instantiated with current multilinear map candidates. Furthermore, the positive results in [5] did not consider completely arbitrary distributions of evasive circuits, as we do here.

Using our new analysis techniques, we prove the following.

Theorem 1 (informal). *There exists a PPT obfuscator $\mathcal{O}$ such that, for any evasive function family $\mathcal{C}$ on n-bit inputs and any efficient generic-model adversary $\mathcal{A}$,*

$$\Pr\left[\mathcal{A}(\mathcal{O}(C)) \text{ constructs an encoding of } 0\right] < \mathsf{negl}(n)$$

where the probability is over the choice of $C \leftarrow \mathcal{C}$ and the coins of $\mathcal{A}$ and $\mathcal{O}$.

Theorem 1 in particular implies the first *witness encryption* scheme [21] with a generic model proof that captures zeroizing attacks[4]. Indeed, in the original witness encryption protocol of [21] the attacker *can* produce top-level encodings of zero, and therefore the protocol is not secure in the post-zeroizing model. Subsequent witness encryption protocols [23,36] also allow top-level encodings of zero to be constructed.

In proving Theorem 1, we show that the "bootstrapping" theorem of [20] extends to the setting of evasive functions. (As mentioned above, this theorem transforms a core obfuscator for a "small" class of functions into an obfuscator for all efficient functions.) We observe that the proof of this theorem only uses the core obfuscator on evasive functions, and we show that it holds only assuming the core obfuscator's security on such functions. In particular, we show that

---

[4] When building witness encryption from obfuscation, witness encryption security only requires the obfuscator to be secure when obfuscating functions that always evaluate to 0, which are in particular evasive.

Theorem 1 applies to all evasive functions and not only those on which the core obfuscator operates. Interestingly, the more recent bootstrapping technique of Applebaum [3] cannot be used for our purposes, because it inherently produces encodings of 0 regardless of the function being obfuscated.

Directions for Future Work. The obvious next step is to consider obfuscating non-evasive functions. To do so, we will need to look precisely at the kinds of post-zero-test information that can be obtained using zeroizing attacks during zero testing for general (non-evasive) functions. We note that our paper answers a critical first question toward this goal: we show that in our scheme, the *only* way that the adversary can create top-level encodings of zero are the prescribed ways of evaluating the function at a particular input. This is a necessary first step in understanding what kinds of information can arise in the general case, and whether this information can lead to more sophisticated attacks.

Subsequent work by a subset of the authors [31] has shown how to attack candidate iO schemes (including the one here), when implemented with the [18] multilinear map candidate, by further analyzing the polynomials that correspond to honest evaluations of the obfuscated function. However, we remark that this attack still crucially relies on encodings of 0 (corresponding to 0-outputs of the function), and as a result it cannot be mounted when the function being obfuscated is evasive.

Organization. In Sect. 2 we give some preliminary definitions and background information. In Sect. 3 we define our obfuscator for matrix branching programs. In Sect. 4 we prove the key technical theorem that analyzes adversarially-constructed polynomials over the obfuscation. The proof of VBB security is outlined in Sect. 5 (due to space limitations, the complete proof is deferred to the full version of this paper). In Sect. 6 we prove that, when obfuscating evasive functions, no encodings of zero can be created.

2   Preliminaries

2.1   Evasive Circuits

We define evasive circuit collections as in Barak et al. [5], except that in our definition it is hard to find a 0-output (typically one says that it is hard to find a 1-output).

Definition 1. *A function family* $\{C_\ell\}_{\ell \in \mathbb{N}}$ *is evasive if for every oracle-aided adversary* $\mathcal{A}^{(\cdot)}$ *that makes at most* $\mathsf{poly}(\ell)$ *queries on input* 1^ℓ, *and every* $\ell \in \mathbb{N}$:

$$\Pr_{C \leftarrow \mathcal{C}_\ell} \left[C\left(\mathcal{A}^C\left(1^\ell\right)\right) = 0 \right] = \mathsf{negl}(\ell).$$

$\{C_\ell\}_{\ell \in \mathbb{N}}$ *is evasive with auxiliary input* Aux *for a (possibly randomized) function* $\mathsf{Aux} : \mathcal{C}_\ell \to \{0,1\}^*$ *if* $\mathcal{A}$ *additionally receives* $\mathsf{Aux}(C)$ *when its oracle is* C.

2.2 Obfuscation

We now give the definition of virtual black-box obfuscation in an idealized model, identical to the model studied in Barak et al. [6] and Ananth et al. [1], with one exception: we also consider giving both the adversary and simulator an auxiliary input determined by the program.

Definition 2 (Virtual Black-Box Obfuscation in an $\mathcal{M}$-idealized model). *For a (possibly randomized) oracle $\mathcal{M}$, a circuit class $\{\mathcal{C}_\ell\}_{\ell \in \mathbb{N}}$, and an efficiently computable deterministic function $\mathsf{Aux}_\ell : \mathcal{C}_\ell \to \{0,1\}^{t_\ell}$, we say that a uniform PPT oracle machine $\mathcal{O}$ is a "Virtual Black-Box" Obfuscator for $\{\mathcal{C}_\ell\}_{\ell \in \mathbb{N}}$ in the $\mathcal{M}$-idealized model with respect to auxiliary information Aux_ℓ, if the following conditions are satisfied:*

- *Functionality: For every $\ell \in \mathbb{N}$, every $C \in \mathcal{C}_\ell$, every input x to C, and for every possible coins for $\mathcal{M}$:*

$$\Pr[(\mathcal{O}^{\mathcal{M}}(C))(x) \neq C(x)] \leq \mathrm{negl}(|C|) \ ,$$

  where the probability is over the coins of C.
- *Polynomial Slowdown: there exist a polynomial p such that for every $\ell \in \mathbb{N}$ and every $C \in \mathcal{C}_\ell$, we have that $|\mathcal{O}^{\mathcal{M}}(C)| \leq p(|C|)$.*
- *Virtual Black-Box: for every PPT adversary $\mathcal{A}$ there exist a PPT simulator Sim, and a negligible function μ such that for all PPT distinguishers D, for every $\ell \in \mathbb{N}$ and every $C \in \mathcal{C}_\ell$:*

$$\left| \Pr\left[D\left(\mathcal{A}^{\mathcal{M}}\left(\mathcal{O}^{\mathcal{M}}(C), \mathsf{Aux}_\ell(C) \right) \right) = 1, \right] - \Pr\left[D\left(\mathsf{Sim}^C\left(1^{|C|}, \mathsf{Aux}_\ell(C) \right) \right) = 1 \right] \right| \leq \mu(|C|) \ ,$$

  where the probabilities are over the coins of D, $\mathcal{A}$, Sim, $\mathcal{O}$ and $\mathcal{M}$.

Note that in this model, both the obfuscator and the evaluator have access to the oracle $\mathcal{M}$ but the function family that is being obfuscated does not have access to $\mathcal{M}$.

We also define the average-case version of VBB obfuscation, which is the correct security notion when obfuscating evasive circuit collections.

Definition 3 (Average-case Virtual Black-Box Obfuscation in an $\mathcal{M}$-idealized model). *Let $\mathcal{M}$, $\{\mathcal{C}_\ell\}_{\ell \in \mathbb{N}}$, and Aux_ℓ be as in Definition 2. We say that a uniform PPT oracle machine $\mathcal{O}$ is an average-case Virtual Black-Box Obfuscator for $\{\mathcal{C}_\ell\}_{\ell \in \mathbb{N}}$ in the $\mathcal{M}$-idealized model with respect to auxiliary information Aux_ℓ, if it satisfies all properties in Definition 2 except that in the Virtual Black-Box property the probabilities are over a uniform choice of $C \leftarrow \mathcal{C}_\ell$ (as opposed to $\forall C \in \mathcal{C}_\ell$).*

Definition 4 (Average-case Indistinguishability Obfuscation in an $\mathcal{M}$-idealized model). *For a (possibly randomized) oracle $\mathcal{M}$, a circuit class $\{\mathcal{C}_\ell\}_{\ell \in \mathbb{N}}$, we say that a uniform PPT oracle machine $\mathcal{O}$ is an Average-case Indistinguishability Obfuscator for $\{\mathcal{C}_\ell\}_{\ell \in \mathbb{N}}$ in the $\mathcal{M}$-idealized model if the following conditions are satisfied:*

– *Functionality: Same as in the definition of VBB.*
– *Polynomial Slowdown: Same as in the definition of VBB.*
– *Indistinguishability: For every PPT Distinguisher D, there exists a negligible function μ such that the following holds: for every $\ell \in \mathbb{N}$, for a uniform choice of circuit $C \in \mathcal{C}_\ell$ and for every pair of circuits $C_0, C_1 \in \mathcal{C}_\ell$ that compute the same function as C, we have:*

$$\left| \Pr\left[D(\mathcal{O}^{\mathcal{M}}(C_0)) = 1 \right] - \Pr\left[D(\mathcal{O}^{\mathcal{M}}(C_1)) = 1 \right] \right| \leq \mu(|C|) ,$$

where the probabilities are over the coins of D, $\mathcal{O}$, $\mathcal{M}$ and the choice of C.

Note that in this model, both the obfuscator and the evaluator have access to the oracle $\mathcal{M}$ but the function family that is being obfuscated does not have access to $\mathcal{M}$.

2.3   Branching Programs

Here we define the main type of branching program we consider. A detailed description of other types of branching programs, and how to build these branching programs from other computational models, can be found in the full version of this paper.

Definition 5. *A dual-input generalized matrix branching program of length ℓ and shape $(d_0, d_1, \ldots, d_\ell) \in (\mathbb{Z}^+)^{\ell+1}$ for n-bit inputs is given by a sequence*

$$BP = \left(\mathsf{inp}_0, \mathsf{inp}_1, \{\mathbf{B}_{i,b_0,b_1}\}_{i \in [\ell], b_0, b_1 \in \{0,1\}} \right)$$

where $\mathbf{B}_{i,b_0,b_1} \in \mathbb{Z}^{d_{i-1} \times d_i}$ are $d_{i-1} \times d_i$ matrices, and $\mathsf{inp} : [\ell] \to [n]$ is the evaluation function of BP. BP defines the following three functions:

– $BP_{arith} : \{0,1\}^n \to \mathbb{Z}^{d_0 \times d_\ell}$ *computed as* $BP_{arith}(x) = \displaystyle\prod_{i=1}^{n} \mathbf{B}_{i, x_{\mathsf{inp}_0(i)}, x_{\mathsf{inp}_1(i)}}$

– $BP_{bool} \quad : \quad \{0,1\}^n \quad \to \quad \{0,1\}^{d_0 \times d_\ell} \quad$ *computed   as*
$BP_{bool}(x)_{j,k} = \begin{cases} 0 & \text{if } BP_{arith}(x)_{j,k} = 0 \\ 1 & \text{if } BP_{arith}(x)_{j,k} \neq 0 \end{cases}$

– $BP_{bool(q)} \quad : \quad \{0,1\}^n \quad \to \quad \{0,1\}^{d_0 \times d_\ell} \quad$ *computed   as* $BP_{bool(q)}(x)_{j,k} \quad =$
$\begin{cases} 0 & \text{if } BP_{arith}(x)_{j,k} = 0 \mod q \\ 1 & \text{if } BP_{arith}(x)_{j,k} \neq 0 \mod q \end{cases}$

A matrix branching program is t-bounded if $|BP_{arith}(x)_{j,k}| \leq t$ for all x, j, k.

Next, we define a notion of non-shortcutting for matrix branching programs, which roughly states that it is not possible to determine any of the output components of $BP_{arith/bool}$ without carrying out the entire matrix product. In the case $d_0 = d_\ell = 1$ (so that the branching program outputs just a single element), this translates to requiring that no strict sub-product $\prod_{i=i_0}^{i_1} \mathbf{B}_{i,x_{\mathsf{inp}_0(i)},x_{\mathsf{inp}_1(i)}}$ for $(i_0, i_1) \neq (1, n)$ of the overall matrix product evaluates to an all-zero matrix. Clearly, if some sub-product evaluates to zero, the entire product would evaluate to zero, and so the evaluation could stop after computing just the sub-product. We call this a short-cut, and non-shortcutting is the requirement that there are no shortcuts for any inputs. In the more general case of arbitrary d_0, d_ℓ, the condition becomes slightly more technical, and is given below:

Definition 6. *A dual-input generalized matrix branching program is non-shortcutting if, for any input x, and any $j \in [d_0]$ and any $k \in [d_\ell]$, the following holds:*

$$\mathbf{e}_j^T \cdot \left(\prod_{i=1}^{\ell-1} \mathbf{B}_{i,x_{\mathsf{inp}_0(i)},x_{\mathsf{inp}_1(i)}}\right) \neq 0^{d_{\ell-1}} \quad and \quad \left(\prod_{i=2}^{\ell} \mathbf{B}_{i,x_{\mathsf{inp}_0(i)},x_{\mathsf{inp}_1(i)}}\right) \cdot \mathbf{e}_k \neq 0^{d_1}$$

where $\mathbf{e}_j$ and $\mathbf{e}_k$ are the jth and kth standard basis vectors of the correct dimension. Equivalently, each row of the product $\prod_{i=1}^{\ell-1} \mathbf{B}_{i,x_{\mathsf{inp}_0(i)},x_{\mathsf{inp}_1(i)}}$ and each column of the product $\prod_{i=2}^{\ell} \mathbf{B}_{i,x_{\mathsf{inp}_0(i)},x_{\mathsf{inp}_1(i)}}$ has at least one non-zero entry.

Matrix Branching Program Samplers. We now define a matrix branching program sampler (MBPS). Roughly, an MBPS is a procedure that takes as input a modulus q, and outputs a matrix branching program BP. However, we will be interested mainly in the function $BP_{bool(q)}$.

Definition 7. *A matrix branching program sampler (MBPS) is a possibly randomized procedure BP^S that takes as input a modulus q satisfying $q > t$ for some bound t. It outputs a matrix branching program.*

Fact 2. *Any matrix branching program BP with bound t can trivially be converted into a matrix branching program sampler BP^S with the same bound t, such that if $BP' \leftarrow BP^S(q)$, then $BP'_{bool(q)}(x) = BP_{bool}(x)$.*

2.4 The Ideal Graded Encoding Model

In this section, we describe the ideal graded encoding model. This section has been taken almost verbatim from [1,6]. All parties have access to an oracle $\mathcal{M}$, implementing an ideal graded encoding. The oracle $\mathcal{M}$ implements an idealized and simplified version of the graded encoding schemes from [18]. The parties are provided with encodings of various elements at different levels. They are allowed to perform arithmetic operations of addition/multiplication and testing equality to zero as long as they respect the constraints of the multilinear setting. We start by defining an algebra over the elements.

Definition 8. *Given a ring R and a universe set $\mathbb{U}$, an element is a pair (α, S) where $\alpha \in R$ is the value of the element and $S \subseteq \mathbb{U}$ is the index of the element. Given an element e we denote by $\alpha(e)$ the value of the element, and we denote by $S(e)$ the index of the element. We also define the following binary operations over elements:*

- *For two elements e_1, e_2 such that $S(e_1) = S(e_2)$, we define $e_1 + e_2$ to be the element $(\alpha(e_1) + \alpha(e_2), S(e_1))$, and $e_1 - e_2$ to be the element $(\alpha(e_1) - \alpha(e_2), S(e_1))$.*
- *For two elements e_1, e_2 such that $S(e_1) \cap S(e_2) = \emptyset$, we define $e_1 \cdot e_2$ to be the element $(\alpha(e_1) \cdot \alpha(e_2), S(e_1) \cup S(e_2))$.*

We will often use the notation $[\alpha]_S$ to denote the element (α, S). Next, we describe the oracle $\mathcal{M}$. $\mathcal{M}$ is a stateful oracle mapping elements to "generic" representations called *handles*. Given handles to elements, $\mathcal{M}$ allows the user to perform operations on the elements. $\mathcal{M}$ will implement the following interfaces:

Initialization. $\mathcal{M}$ will be initialized with a ring R, a universe set $\mathbb{U}$, and a list L of initial elements. For every element $e \in L$, $\mathcal{M}$ generates a handle. We do not specify how the handles are generated, but only require that the value of the handles are independent of the elements being encoded, and that the handles are distinct (even if L contains the same element twice). $\mathcal{M}$ maintains a handle table where it saves the mapping from elements to handles. $\mathcal{M}$ outputs the handles generated for all the elements in L. After $\mathcal{M}$ has been initialized, all subsequent calls to the initialization interface fail.

Algebraic Operations. Given two input handles h_1, h_2 and an operation $\circ \in \{+, -, \cdot\}$, $\mathcal{M}$ first locates the relevant elements e_1, e_2 in the handle table. If any of the input handles does not appear in the handle table (that is, if the handle was not previously generated by $\mathcal{M}$) the call to $\mathcal{M}$ fails. If the expression $e_1 \circ e_2$ is undefined (i.e., $S(e_1) \neq S(e_2)$ for $\circ \in \{+, -\}$, or $S(e_1) \cap S(e_2) \neq \emptyset$ for $\circ \in \{\cdot\}$) the call fails. Otherwise, $\mathcal{M}$ generates a new handle for $e_1 \circ e_2$, saves this element and the new handle in the handle table, and returns the new handle.

Zero Testing. Given an input handle h, $\mathcal{M}$ first locates the relevant element e in the handle table. If h does not appear in the handle table (that is, if h was not previously generated by $\mathcal{M}$) the call to $\mathcal{M}$ fails. If $S(e) \neq \mathbb{U}$, the call fails. Otherwise, $\mathcal{M}$ returns 1 if $\alpha(e) = 0$, and returns 0 if $\alpha(e) \neq 0$.

2.5   Straddling Set Systems

We use the *strong* straddling set system of [30], which modifies the straddling set system of [6] to obtain a denser intersection graph between the subsets. This extra power is used in Sect. 6 when showing that the adversary cannot create low-level encodings of 0.

Definition 9 (Strong straddling set system). *A strong straddling set system with n entries is a collection of sets $\mathbb{S} = \{S_{i,b} : i \in [n], b \in \{0,1\}\}$ over a universe $\mathbb{U}$, such that $\cup_{i \in [n]} S_{i,0} = \mathbb{U} = \cup_{i \in [n]} S_{i,1}$, and the following holds.*

- *(Collision at universe.) If $C, D \subseteq \mathbb{S}$ are distinct non-empty collections of disjoint sets such that $\bigcup_{S \in C} S = \bigcup_{S \in D} S$, then $\exists b \in \{0,1\}$ such that $C = \{S_{i,b}\}_{i \in [n]}$ and $D = \{S_{i,1-b}\}_{i \in [n]}$.*

- *(Strong intersection.)* For every $i, j \in [n]$, $S_{i,0} \cap S_{j,1} \neq \emptyset$.

We will need the following simple lemma.

Lemma 1. *Let* $\mathbb{S} = \{S_{i,b} : i \in [n], b \in \{0,1\}\}$ *be a strong straddling set system over a universe* $\mathbb{U}$. *Then for any* $T \subsetneq \mathbb{U}$ *that can be written as a disjoint union of sets from* $\mathbb{S}$, *there is a unique* $b \in \{0,1\}$ *such that* $T = \bigcup_{i \in I} S_{b,i}$ *for some* $I \subseteq [n]$.

Proof. By the second property of Definition 9, any pairwise disjoint collection of sets from $\mathbb{S}$ must be either all of the form $S_{i,0}$ or all of the form $S_{i,1}$. If there are two sets $I_0, I_1 \subseteq [n]$ such that $\bigcup_{i \in I_0} S_{i,0} = T = \bigcup_{i \in I_1} S_{i,1}$, then by the first property of Definition 9 we must have $T = \mathbb{U}$ which contradicts our assumption.

We use the following construction from [30].

Construction 3 (Strong straddling set system). *Define* $\mathbb{S} = \{S_{i,b} : i \in [n], b \in \{0,1\}\}$ *over a universe* $\mathbb{U} = \{1, 2, ..., n^2\}$ *as follows for all* $1 \leq i \leq n$.

$$S_{i,0} = \{n(i-1)+1, n(i-1)+2, \ldots, ni\} \qquad S_{i,1} = \{i, n+i, 2n+i, \ldots, n(n-1)+i\}$$

3 Obfuscator for Low-Rank Branching Programs

We now describe our obfuscator for generalized matrix branching programs. Our obfuscator is essentially the same as the obfuscator of Ananth et al. [1]. The differences are as follows:

- We view branching programs as including the bookends. While the bookends of previous works did not depend on the input, they can in our obfuscator. However, for [1], this distinction is superficial: the bookends of [1] can be "absorbed" into the branching program by merging them with the left-most and right-most matrices of the branching program. This does not change functionality, since this merging always happens during evaluation, and it does not change security, since the adversary can perform the merging himself.
- We allow our branching program to have singular and rectangular matrices. We do, however, require the branching program to be non-shortcutting. Note that a branching program with square invertible internal matrices and non-zero bookend vectors, such as in [1], necessarily is non-shortcutting.
- We allow branching programs to output multiple bits — that is, the function computed by our obfuscated program will be BP_{bool}, which is a matrix of 0/1 entries. In order to prove security, we will have to perform additional randomization. However, in the case of single-bit outputs, this additional randomization is redundant.

Input. The input to our obfuscator is a dual-input matrix branching program sampler BP^S of length ℓ, shape $(d_0, d_1, \ldots, d_\ell)$, and bound t. The first step is to choose a large prime q for the graded encodings. Then sample $BP \leftarrow BP^S(q)$. Write

$$BP = (\mathsf{inp}_0, \mathsf{inp}_1, \{\mathbf{B}_{i,b_0,b_1}\})$$

We require BP^S to output BP satisfying the following properties:

- BP is non-shortcutting.
- For each i, $\mathsf{inp}_0(i) \neq \mathsf{inp}_1(i)$
- For each pair $(j, k) \in [n]^2$, there exists an $i \in [\ell]$ such that $(\mathsf{inp}_0(i), \mathsf{inp}_1(i)) = (j, k)$ or $(\mathsf{inp}_1(i), \mathsf{inp}_0(i)) = (j, k)$

For ease of notation in our security proof, we will also assume that each input bit is used exactly m times, for some integer m. In other words, for each $i \in [n]$, the sets $\mathsf{ind}(i) = \{j : \mathsf{inp}_b(j) = i$ for some $b \in \{0, 1\}\}$ have the same size. This requirement, however, is not necessary for security.

Step 1: Randomize BP. First, similar to previous works, we use Kilian [29] to randomize BP, obtaining a randomized branching program BP'. This is done as follows.

- Let q be a sufficiently large prime of $\Omega(\lambda)$ bits.
- For each $i \in [\ell - 1]$, choose a random matrix $\mathbf{R}_i \in \mathbb{Z}_q^{d_i \times d_i}$. Set $\mathbf{R}_0, \mathbf{R}_\ell$ to be identity matrices of the appropriate size. Define

$$\widehat{\mathbf{B}_{i,b_0,b_1}} = \mathbf{R}_{i-1}^{adj} \cdot \mathbf{B}_{i,b_0,b_1} \cdot \mathbf{R}_i$$

- For each $s \in [d_0]$, choose a random β_s and set $\mathbf{S}$ to be the $d_0 \times d_0$ diagonal matrix with the β_s along the diagonal. For each $t \in [d_\ell]$, choose a random γ_t and set $\mathbf{T}$ to be the $d_\ell \times d_\ell$ diagonal matrix with γ_t along the diagonal. Set

$$\mathbf{C}_{1,b_0,b_1} = \mathbf{S} \cdot \widehat{\mathbf{B}_{1,b_0,b_1}} \quad \mathbf{C}_{\ell,b_0,b_1} = \widehat{\mathbf{B}_{1,b_0,b_1}} \cdot \mathbf{T} \quad \mathbf{C}_{i,b_0,b_1} = \widehat{\mathbf{B}_{i,b_0,b_1}} \text{ for each } i \in [2, \ell - 1]$$

We note that this additional randomization step is not present in previous works, but is required to handle multi-bit outputs

- For each $i \in [\ell]$, $b_0, b_1 \in \{0, 1\}$, choose a random $\alpha_{i,b_0,b_1} \in \mathbb{Z}_p$, and define

$$\mathbf{D}_{i,b_0,b_1} = \alpha_{i,b_0,b_1} \mathbf{C}_{i,b_0,b_1}$$

Then define $BP' = (\mathsf{inp}_0, \mathsf{inp}_1, \{\mathbf{D}_{i,b_0,b_1}\})$. Observe that $BP'_{bool(q)}(x) = BP_{bool(q)}(x)$ for all x.

Step 2: Create Set Systems. Consider a universe $\mathbb{U}$, and a partition $\mathbb{U}_1, \ldots, \mathbb{U}_\ell$ of $\mathbb{U}$ into equal sized disjoint sets: $|\mathbb{U}_i| = 2m - 1$. Let $\mathbb{S}^j$ be a straddling set system over the elements of $\mathbb{U}_j$. Note that $\mathbb{S}^j$ will have m entries, corresponding to the number of times each input bit is used. We now associate the elements of $\mathbb{S}_j$ to the indicies of BP that depend on x_j:

$$\mathbb{S}^j = \{S_{k,b}^j : k \in \mathsf{ind}(j), b \in \{0, 1\}\}$$

Next, we associate a set to each element output by the randomization step. Recall that in a dual-input relaxed matrix branching program, each step depends on two fixed bits in the input defined by the evaluation functions inp_0 and inp_1. For each step $i \in [n], b_0, b_1 \in \{0,1\}$, we define the set $S(i, b_0, b_1)$ using the straddling sets for input bits $inp_1(i)$ and $inp_2(i)$ as follows:

$$S_{i,b_0,b_1} = S_{i,b_0}^{inp_0(i)} \cup S_{i,b_1}^{inp_1(i)}$$

Step 3: Initialization. $\mathcal{O}$ initializes the oracle $\mathcal{M}$ with the ring $\mathbb{Z}_p$ and the universe $\mathbb{U}$. Then it asks for the encodings of the following elements:

$$\{(\mathbf{D}_{i,b_0,b_1}[j,k], S_{i,b_0,b_1})\}_{i\in[\ell],b_0,b_1\in\{0,1\},j\in[d_{i-1}],k\in[d_i]}$$

$\mathcal{O}$ receives a list of handles back from $\mathcal{M}$. Let $[\beta]_S$ denote the handle for (β, S), and for a matrix M, let $[M]_S$ denote the matrix of handles $([M]_S)[j,k] = [M[j,k]]_S$. Thus, $\mathcal{O}$ receives the handles:

$$\left\{ [\mathbf{D}_{i,b_0,b_1}]_{S_{i,b_0,b_1}} \right\}_{i\in[\ell],b_0,b_1\in\{0,1\}}$$

Output. $\mathcal{O}(BP^S)$ outputs these handles, along with the length ℓ, shape $d_0, \ldots, d_\ell$, and input functions inp_0, inp_1, as the obfuscated program. Denote the resulting obfuscated branching program as $BP^{\mathcal{O}}$.

Evaluation. To evaluate $BP^{\mathcal{O}}$ on input x, use the oracle $\mathcal{M}$ to add and multiply encodings in order to compute the product

$$h = \left[\prod_{i\in[\ell]} \mathbf{D}_{i,x_{inp_0(i)},x_{inp_1(i)}} \right]_{\mathbb{U}} = \prod_{i\in[\ell]} \left[\mathbf{D}_{i,x_{inp_0(i)},x_{inp_1(i)}} \right]_{S_{i,x_{inp_0(i)},x_{inp_1(i)}}}$$

h is a $d_0 \times d_\ell$ matrix of encodings relative to $\mathbb{U}$. Next, use $\mathcal{M}$ to test each of the components of h for zero, obtaining a matrix $h_{bool} \in \{0,1\}^{d_0 \times d_\ell}$. That is, if the zero test on returns a 1 on $h[s,t]$, $h_{bool}[s,t]$ is 0, and if the zero test returns a 0, $h_{bool}[s,t]$ is 1.

Correctness of Evaluation. The following shows that all calls to the oracle $\mathcal{M}$ succeed:

Lemma 2 (Adapted from [1]). *All calls made to the oracle $\mathcal{M}$ during obfuscation and evaluation succeed.*

It remains to show that the obfuscated program computes the correct function. Fix an input x, and define $b_c^i = x_{inp_c(i)}$ for $i \in [\ell], c \in \{0,1\}$. From the description above, $BP^{\mathcal{O}}$ outputs 0 at position $[s,t]$ if and only if

$$0 = \left(\prod_{i\in[\ell]} \mathbf{D}_{i,b_0^i,b_1^i} \right)[s,t] = \beta_s \gamma_t \left(\prod_{i\in[\ell]} \alpha_{i,b_0^i,b_1^i} \mathbf{R}_{i-1}^{adj} \cdot \mathbf{B}_{i,b_0^i,b_1^i} \cdot \mathbf{R}_i \right)[s,t]$$

$$= \beta_s \gamma_t \left(\left(\prod_{i\in[\ell]} \alpha_{i,b_0^i,b_1^i} \right) \left(\prod_{i\in[\ell]} \mathbf{B}_{i,b_0^i,b_1^i} \right) \right)[s,t] = \left(\beta_s \gamma_t \prod_{i\in[\ell]} \alpha_{i,b_0^i,b_1^i} \right) (BP_{arith}(x)[s,t])$$

With high probability $\beta_s, \gamma_t, \alpha_{i,b_0,b_1} \neq 0$, meaning $BP_{arith}(x)[s,t] = 0 \mod q$ if and only if the zero test procedure on position $[s,t]$ gives 0. Therefore, $BP^{\mathcal{O}}(x) = BP_{bool(q)}(x)$ for the branching program BP sampled from BP^S.

4   Polynomials on Kilian-Randomized Matrices

In this section, we prove a theorem about polynomials on the Kilian-randomized matrices from the previous section. Our high level goal is to show polynomials the adversary tries to construct other than the correct matrix products will be useless to the adversary. In this section, we focus on a simpler case where the polynomial is only over matrices corresponding to a single input. In the following section, we use the results of this section to prove the general case.

Previous works showed the single-input case using Kilian simulation [6,12], or a variant of it [1,33]. Namely, these works queried the function oracle to determine what the result of the matrix product $P(x)$ should be. Then, they tested the polynomial on random matrices, subject to the requirement that the product equaled $P(x)$, to see what the result was. Unfortunately, this step of the analysis does indicate what the outputs of the polynomial may be, only that they can be simulated. If the polynomial were to output zero, this would correspond to the adversary obtaining a zero encoding, which would violate security in our post-zeroing model.

Moreover, previous works crucially relied on the fact that the matrices the polynomial is tested on come from the same distribution as the matrices would in the branching program. This requires the branching program to consist of square invertible matrices. However, we need to be able to handle generalized matrix branching programs with rectangular and low-rank matrices.

In light of the two issues above, we need to replace the Kilian randomization theorem with a new theorem suitable in our setting.

Let $d_1, \ldots, d_{n-1}$ be positive integers and $d_0 = d_n = 1$. Let $\widehat{\mathbf{A}}_k$ for $k \in [n]$ be $d_{k-1} \times d_k$ matrices of variables.

Definition 10. Let $d_k, \widehat{\mathbf{A}}_k$ be as above. Consider a multilinear polynomial p on the variables in $\{\widehat{\mathbf{A}}_k\}_{k \in [n]}$. We call p allowable if each monomial in the expansion of p contains at most one variable from each of the $\widehat{\mathbf{A}}_k$.

As an example of an allowable polynomial, consider the *matrix product polynomial* $\widehat{\mathbf{A}}_1 \cdot \widehat{\mathbf{A}}_2 \cdots \widehat{\mathbf{A}}_n$.

Now fix a field $\mathbb{F}$, and let $\mathbf{A}_k \in \mathbb{F}^{d_{k-1} \times d_k}$ for $k = 1, \ldots, n$ be a collection of matrices over $\mathbb{F}$. Let $\mathbf{R}_k$ be $d_k \times d_k$ matrices of variables for $k \in [n]$, and let $\mathbf{R}_k^{adj}$ be the adjugate matrix of $\mathbf{R}_k$. Let $\mathbf{R}_0 = \mathbf{R}_{n+1} = 1$. Now suppose we set

$$\widehat{\mathbf{A}}_k = \mathbf{R}_{k-1}^{adj} \cdot \mathbf{A}_k \cdot \mathbf{R}_k$$

Theorem 4. Let $\mathbb{F}, d_k, \mathbf{A}_k, \mathbf{R}_k, \widehat{\mathbf{A}}_k$ be as above. Consider an allowable polynomial p in the $\widehat{\mathbf{A}}_k$, and suppose p, after making the substitution $\widehat{\mathbf{A}}_k = \mathbf{R}_{k-1}^{adj} \cdot \mathbf{A}_k \cdot \mathbf{R}_k$, is identically 0 as a polynomial over the $\mathbf{R}_k$. Then the following is true:

- If $\mathbf{A}_1 \cdot \mathbf{A}_2 \cdots \cdot \mathbf{A}_n \neq 0$, then p is identically zero as a polynomial over its formal variables, namely the $\widehat{\mathbf{A}}_k$.
- If $\mathbf{A}_1 \cdot \mathbf{A}_2 \cdots \cdot \mathbf{A}_n = 0$ but

$$\mathbf{A}_1 \cdot \mathbf{A}_2 \cdots \cdot \mathbf{A}_{n-1} \neq 0^{1 \times d_n}$$
$$\mathbf{A}_2 \cdots \cdot \mathbf{A}_{n-1} \cdot \mathbf{A}_n \neq 0^{d_2 \times 1}$$

then p, as a polynomial over the $\widehat{\mathbf{A}}_k$, is a constant multiple of the matrix product polynomial $\widehat{\mathbf{A}}_1 \cdot \widehat{\mathbf{A}}_2 \cdots \cdot \widehat{\mathbf{A}}_n$.

Proof. If $n = 1$, there are no $\mathbf{R}_k$ matrices, a single $\mathbf{A}_1$ matrix of dimension 1×1, with entry a. Then $p = p(a) = ca$ for some constant c. As a polynomial over the (non-existent) $\mathbf{R}_i$ matrices, p is just a constant polynomial, so $p = 0$ means $ca = 0$. In the first case above, $a \neq 0$, so $c = 0$, meaning p is identically 0. The second case above is trivially satisfied since the matrix product polynomial is also a constant.

We will assume that $\mathbf{A}_1$ is non-zero in every coordinate. At the end of the proof, we will show this is without loss of generality.

Now we proceed by induction on n. Assume Theorem 4 is proved for $n - 1$. Consider an arbitrary allowable polynomial p. We can write p as

$$p = \sum_{j_1, i_2, j_2, \ldots, j_n, i_{n+1}} \alpha_{j_1, i_2, \ldots, j_{n-1}, i_n} \widehat{A}_{1,1,j_1} \widehat{A}_{2,i_2,j_2} \cdots \widehat{A}_{n-1,i_{n-1},j_{n-1}} \widehat{A}_{n,i_n,1}$$

where $i_{k+1}, j_k \in [d_k]$, and $\widehat{A}_{k,i,j}$ is the (i, j) entry of the matrix $\widehat{\mathbf{A}}_k$. From this point forward, for convenience, we will no longer explicitly refer to the bounds d_k on the i_{k+1}, j_k.

Now we can expand p in terms of the R_1 matrix:

$$p = \sum_{j_1, i_2, j_2, \ldots, j_n, i_{n+1}, m, \ell} \alpha_{j_1, i_2, \ldots, j_{n-1}, i_n} A_{1,1,m} R_{1,m,j_1} R_{1,i_2,\ell}^{adj} (\mathbf{A}_2 \cdot \mathbf{R}_2)_{\ell, j_2} \widehat{A}_{3,i_3,j_3} \cdots \widehat{A}_{n,i_n,1}$$
$$= \sum_{j,i,\ell,m} \alpha'_{j,i,\ell} A_{1,1,m} R_{1,m,j} R_{1,i,\ell}^{adj}$$

where

$$\alpha'_{j,i,\ell} = \sum_{j_2, \ldots, j_n, i_{n+1}} \alpha_{j,i,\ldots,j_{n-1},i_n} (\mathbf{A}_2 \cdot \mathbf{R}_2)_{\ell, j_2} \widehat{A}_{3,i_3,j_3} \cdots \widehat{A}_{n,i_n,1}$$

Recall that

$$R_{1,i,\ell}^{adj} = \sum_{\sigma : \sigma(i) = \ell} \mathsf{sign}(\sigma) \left(\prod_{t \neq i} R_{1,\sigma(t),t} \right)$$

where the sum is over all permutations satisfying $\sigma(i) = \ell$. Thus we can write p as

$$p = \sum_{j,i,\sigma,m} \mathsf{sign}(\sigma) \alpha'_{j,i,\sigma(i)} A_{1,1,m} R_{1,m,j} \left(\prod_{t \neq i} R_{1,\sigma(t),t} \right)$$

Now, since p is identically zero as a polynomial over the $\mathbf{R}_k$ matrices, it must be that for each product $R_{1,m,j}\left(\prod_{t\neq i}R_{1,\sigma(t),t}\right)$, the coefficient of the product (which is a polynomial over the $\mathbf{R}_k : k \geq 2$ matrices) must be identically 0. We now determine the coefficients.

First, we examine the types of products of entries in $\mathbf{R}_1$ that are possible. Products can be thought of as arising from the following process. Choose a permutation σ, which corresponds to selecting d_1 entries of $\mathbf{R}_1$ such that each row and column of $\mathbf{R}_1$ contain exactly one selected entry. Then, for some i, unselect the selected entry from column i and instead select any entry from $\mathbf{R}_1$ (possibly selecting the same entry twice). We observe that the following products are possible:

- $\prod_t R_{1,\sigma(t),t}$ for a permutation σ. This corresponds to re-selecting the unselected entry from column i. The resulting list of entries determines the permutation σ used to select the original entries (since it is identical to the original list), but allows the column i of the un-selected/re-selected entry to vary. Thus in the summation above, this fixes σ, $j = i$ and $m = \sigma(i)$, but allows i to vary over all values, corresponding to the fact that if we remove any entry and replace it with itself, the result is independent of which entry we removed. Call such products *well-formed*. Well-formed products give the following equation:

$$\sum_i \alpha'_{i,i,\sigma(i)} A_{1,1,\sigma(i)} = 0 \text{ for all } \sigma \tag{1}$$

- $R_{1,m,j}\prod_{t\neq i}R_{1,\sigma(t),t}$ where $j \neq i$ and $m \neq \sigma(i)$. This corresponds to, after un-selecting the entry in column i, selecting a another entry that is in both a different row and a different column. Note that, given final list of selected entries, it is possible to determine the newly selected entry as the unique selected entry that shares both a column with another selected entry and a row with another selected entry. It is also possible to determine the unselected entry as the only entry that shares no column nor row with another entry. Therefore, the original entry selection is determined as well. Thus, in the summation above, the selected entries fix σ, i, j, and m. In other words, there is no other selection process that gives the same list of entries from $\mathbf{R}_1$. We call such products *malformed type 1*. Malformed type 1 products have the coefficient

$$\alpha'_{j,i,\sigma(i)} A_{1,1,m}$$

Given any $i, j \neq i, m, \ell \neq m$, pick σ so that $\sigma(i) = \ell$. Since $A_{1,1,m} \neq 0$ for all m, this gives

$$\alpha'_{j,i,\ell} = 0 \text{ for all } i, j \neq i, \ell \tag{2}$$

- $R_{1,m,i}\prod_{t\neq i}R_{1,\sigma(t),t}$ where $m \neq \sigma(i)$. This corresponds to, after un-selecting the entry $R_{1,\sigma(i),i}$, selecting a different entry $R_{1,m,i}$ in the same column. Let i', m', σ' be some other selection process that leads to the same product.

Given the final selection of entries, it is possible to determine $m' = m$ as the only row with two selected entries. It is also possible to determine $\sigma'(i') = \sigma(i)$ as the only row with no selected entries (though i' has not been determined yet). Moreover, i' must be one of the two columns selected in row m, call the other i''. All entries outside of these two rows must have come from the original selection of entries, so this determines $\sigma'(t) = \sigma(t)$ on all inputs outside of i, i''. Notice that if $i = i'$, then σ' agrees with σ on $d_1 - 1$ entries, and since they are both permutations, this sets $\sigma' = \sigma$. In this case, $(i', m', \sigma') = (i, m, \sigma)$.

Otherwise $i' \neq i$, so $i'' = i$, which leaves $\sigma'(i) = \sigma(i') = m$. At this point, σ' is fully determined as $\sigma \circ (i \ i')$ where $(i \ i')$ is the transposition swapping i and i'. Therefore, there are two possibilities leading to this product, one corresponding to i and the other corresponding to i'.

We call these products *malformed type 2*. Notice that σ' and σ only differ by a transposition swapping i and i', and so they have opposite parity, meaning the corresponding terms in p have the opposite sign. Given $i, i' \neq i, m, \ell \neq m$, choose σ so that $\sigma(i) = \ell$. This gives us $(\alpha'_{i,i,\ell} - \alpha'_{i',i',\ell})A_{1,1,m} = 0$. Since $A_{1,1,m} \neq 0$ for all m, we therefore have that $\alpha'_{i,i,\ell} = \alpha'_{i',i',\ell}$ for all i, i'. We can thus choose β_ℓ such that:

$$\alpha'_{i,i,\ell} = \beta_\ell \text{ for all } i, \ell \tag{3}$$

$- R_{1,\sigma(i),j} \prod_{t \neq i} R_{1,\sigma(t),t}$ where $j \neq i$. We call such products *malformed type 3*. The coefficients of these products are linear combinations of the $\alpha'_{i,j,\ell}$ for $i \neq j$, which we already know to be 0. Therefore, these equations are redundant, and we will not need to consider them.

Setting $\sigma(i) = i$ in Eq. 1 and combining with Eq. 3, we have that

$$\sum_\ell \beta_\ell A_{1,1,\ell} = 0 \tag{4}$$

Now we can expand $\alpha'_{j,i,\ell}$ and β_i in Eqs. 2 and 4, obtaining:

$$0 = \alpha'_{i,j,\ell} = \sum_{j_2,i_3,\ldots,j_{n-1},i_n} \alpha_{j,i,j_2,i_3,\ldots,j_{n-1},i_n} (\mathbf{A}_2 \cdot \mathbf{R}_2)_{\ell,j_2} \widehat{A}_{3,i_3,j_3} \cdots \widehat{A}_{n,i_n,1} \text{ for all } \ell, i, j \neq i \tag{5}$$

$$0 = \sum_\ell \beta_\ell A_{1,1,\ell} = \sum_{\ell,j_2,i_3,\ldots,j_{n-1},i_n} \alpha_{i,i,j_2,i_3,\ldots,j_{n-1},i_n} A_{1,1,\ell} (\mathbf{A}_2 \cdot \mathbf{R}_2)_{\ell,j_2} \widehat{A}_{3,i_3,j_3} \cdots \widehat{A}_{n,i_n,1}$$

$$= \sum_{j_2,i_3,\ldots,j_{n-1},i_n} \alpha_{i,i,j_2,i_3,\ldots,j_{n-1},i_n} (\mathbf{A}_1 \cdot \mathbf{A}_2 \cdot \mathbf{R}_2)_{1,j_2} \widehat{A}_{3,i_3,j_3} \cdots \widehat{A}_{n,i_n,1} \text{ for all } i \tag{6}$$

Now we invoke the inductive step multiple times. Let $\mathbf{A}_{2,\ell}$ be the ℓth row of $\mathbf{A}_2$, and let $\widehat{\mathbf{A}_{2,\ell}} = \mathbf{A}_{2,\ell} \cdot \mathbf{R}_2$. Since $\mathbf{A}_2 \cdot \mathbf{A}_3 \ldots \mathbf{A}_n \neq 0$, there is some ℓ such that $\mathbf{A}_{2,\ell} \cdot \mathbf{A}_3 \ldots \mathbf{A}_n \neq 0$. Then the matrices $\mathbf{A}_{2,\ell}, \mathbf{A}_3, \ldots, \mathbf{A}_n$ satisfy the first set of requirements of Theorem 4 for $n - 1$. Moreover, the right side of Eq. 5

gives an allowable polynomial that is identically zero as a polynomial over the $\mathbf{R}_k, k \geq 2$, and therefore, by induction, it is identically 0 as a polynomial over $\widehat{\mathbf{A}_{2,\ell}}, \widehat{\mathbf{A}_3}, \ldots, \widehat{\mathbf{A}_n}$. This shows us that

$$\alpha_{j,i,j_2,i_3,\ldots,j_{n-1},i_n} = 0 \text{ for all } j \neq i \tag{7}$$

Next, Let $\mathbf{A}_2' = \mathbf{A}_1 \cdot \mathbf{A}_2$, and let $\widehat{\mathbf{A}_2'} = \mathbf{A}_2' \cdot \mathbf{R}_2$. There are two cases:

- $\mathbf{A}_1 \cdot \mathbf{A}_2 \cdots \mathbf{A}_n \neq 0$. Then $\mathbf{A}_2' \cdot \mathbf{A}_3 \cdots \mathbf{A}_n \neq 0$. Therefore, $\mathbf{A}_2', \mathbf{A}_3, \ldots, \mathbf{A}_n$ satisfy the first set of requirements in Theorem 4. Moreover, for each i, Eq. 6 gives an allowable polynomial that is identically zero as a polynomial over the $\mathbf{R}_k, k \geq 2$. Therefore, by induction, the polymomial is identically zero as a polynomial over $\widehat{\mathbf{A}_2'}, \widehat{\mathbf{A}_3}, \ldots, \widehat{\mathbf{A}_n}$. This means

$$\alpha_{i,i,j_2,i_3,\ldots,j_{n-1},i_n} = 0 \text{ for all } i$$

Combining with Eq. 7, we have that all the α values are 0. Therefore p is identically zero as a polynomial over the $\widehat{\mathbf{A}_1}, \widehat{\mathbf{A}_2}, \ldots, \widehat{\mathbf{A}_n}$.
- $\mathbf{A}_1 \cdot \mathbf{A}_2 \cdots \mathbf{A}_n = 0$. Then $\mathbf{A}_2' \cdot \mathbf{A}_3 \cdots \mathbf{A}_n = 0$. However, $\mathbf{A}_2' \cdot \mathbf{A}_3 \cdots \mathbf{A}_{n-1} = \mathbf{A}_1 \cdot \mathbf{A}_2 \cdots \mathbf{A}_{n-1} \neq 0$ and $\mathbf{A}_3 \ldots \mathbf{A}_4 \cdots \mathbf{A}_n \neq 0$ (since otherwise $\mathbf{A}_2 \cdots \mathbf{A}_3 \cdots \mathbf{A}_n = 0$, contradicting the assumptions of Theorem 4). Therefore, $\mathbf{A}_2', \mathbf{A}_3, \ldots, \mathbf{A}_n$ satisfy the second set of requirements in Theorem 4. By induction, for each i, the polynomial in Eq. 6 must therefore be a multiple $\gamma_i \widehat{\mathbf{A}_2'} \cdot \widehat{\mathbf{A}_3} \cdots \widehat{\mathbf{A}_n}$ of the matrix product polynomial. This is equivalent to

$$\alpha_{i,i,j_2,i_3,\ldots,j_{n-1},i_n} = 0 \text{ if } j_k \neq i_{k+1} \text{ for any } k$$
$$\alpha_{i,i,i_3,i_3,\ldots,i_n,i_n} = \gamma_i$$

This means we can write

$$\alpha_{j,i,\ell}' = 0 \text{ for all } j \neq i \text{ (by Eq. 7 and the definition of } \alpha_{i,j,\ell}')$$
$$\alpha_{i,i,\ell}' = \gamma_i \sum_{i_3,\ldots,i_n} (\mathbf{A}_2 \cdot \mathbf{R}_2)_{\ell,i_3} \widehat{A}_{3,i_3,i_4} \ldots \widehat{A}_{n,i_n,1} = \gamma_i (\mathbf{A}_2 \cdot \mathbf{A}_3 \cdots \mathbf{A}_n)_{\ell,1}$$

Since $\alpha_{i,i,\ell}' = \beta_\ell$ for all i and the product $\mathbf{A}_2 \cdot \mathbf{A}_3 \cdots \mathbf{A}_n$ is non-zero, we have that $\gamma_i = \gamma$ is the same for all i. Therefore,

$$\alpha_{i,i,i_3,i_3,\ldots,i_n,i_n} = \gamma \text{ for all } i, i_3, \ldots, i_n$$

meaning p is a multiple of the matrix product polynomial, as desired.

It remains to show the case where $\mathbf{A}_1$ has zero entries. Since $\mathbf{A}$ is non-zero (as a consequence of our assumptions), and $\mathbf{A}$ is a single row vector, it is straightforward to build an invertible matrix $\mathbf{B}$ such that $\mathbf{A}_1' = \mathbf{A}_1 \cdot \mathbf{B}$ is non-zero in every coordinate.

Let $\mathbf{A}_2' = \mathbf{B}^{-1}\mathbf{A}_2$. Let $\mathbf{R}_1' = \mathbf{B}^{-1} \cdot \mathbf{R}_1$, $\widehat{\mathbf{A}_1'} = \mathbf{A}_1' \cdot \mathbf{R}_1' = \widehat{\mathbf{A}_1}$, and $\widehat{\mathbf{A}_2'} = (\mathbf{R}_1')^{adj} \cdot \mathbf{A}_2' \cdot \mathbf{R}_2 = \widehat{\mathbf{A}_2}$. Now $\mathbf{A}_1', \mathbf{A}_2', \mathbf{A}_3, \ldots, \mathbf{A}_n$ satisfy the same conditions of Theorem 4 as the original $\mathbf{A}_k$. Moreover, p is still allowable as a polynomial over $\widehat{\mathbf{A}'_1}, \widehat{\mathbf{A}'_2}, \widehat{\mathbf{A}_3}, \ldots \widehat{\mathbf{A}_n}$. Moreover, we can relate p as a polynomial over $\mathbf{R}_k$ to p as a polynomial over $\mathbf{R}'_1, \mathbf{R}_2, \ldots, \mathbf{R}_{n-1}$ by a linear transformation on the $\mathbf{R}_1$ variables. Therefore, p is identically zero as a polynomial over the $\mathbf{R}_k$ if and only if it is identically zero as a polynomial over $\mathbf{R}'_1, \mathbf{R}_2, \ldots, \mathbf{R}_n$. Thus we can invoke Theorem 4 on $\mathbf{A}'_1, \mathbf{A}'_2, \ldots, \mathbf{A}_n$ using the same polynomial p, and arrive at the desired conclusion. This completes the proof.

5   Sketch of VBB Security Proof

We now explain how to use Theorem 4 to prove the VBB security of our obfuscator. Due to space constraints, the complete proof is deferred to the full version of this paper. In this sketch, we pay special attention to the steps in our proof that deviate from previous works [1,6]. We also state a definition and lemma that will be used in Sect. 6 to prove that encodings of zero cannot be created when the function being obfuscated is evasive.

The adversary is given an obfuscation of a branching program BP, which consists of a list of handles corresponding to elements in the graded encoding. The adversary can operate on these handles using the graded encoding interface, which allows performing algebraic operations and zero testing. Our goal is to build a simulator that has oracle access only to the output of BP, and is yet able to simulate all of the handles and interfaces seen by the adversary. Formally, we prove the following theorem.

Theorem 5. *If BP^S outputs non-shortcutting branching programs, then for any PPT adversary $\mathcal{A}$, there is a PPT simulator Sim such that*

$$\left| \Pr[\mathcal{A}^\mathcal{M}(\mathcal{O}^\mathcal{M}(BP^S)) = 1] - \Pr_{BP \leftarrow BP^S}[\mathsf{Sim}^{BP}(\ell, d_0, \ldots, d_\ell, \mathsf{inp}_0, \mathsf{inp}_1) = 1] \right| < \mathsf{negl}.$$

The simulator will choose random handles for all of the encodings in the obfuscation, leaving the actual entries of the $\mathbf{D}_{i,b_0,b_1}$ as formal variables[5]. Simulating the algebraic operations is straightforward; the bulk of the security analysis goes in to answering zero-test queries. Any handle the adversary queries the zero test oracle on corresponds to some polynomial p on the variables $\mathbf{D}_{i,b_0,b_1}$, which the adversary can determine by inspecting the queries made by the adversary so far.

The simulator's goal is to decide if p evaluates to zero, when the formal variables in the $\mathbf{D}_{i,b_0,b_1}$ are set to the values in the randomized matrix branching program BP'. However, the simulator does not know BP', and must instead determine if p gives zero knowing only the outputs of BP.

The analysis of [1,6] first simplifies the problem of determining if p evaluates to zero, using Lemma 3 below.

---

[5] The simulator does not know the branching program, and so it has no way of actually sampling the $\mathbf{D}_{i,b_0,b_1}$.

Definition 11. *A single-input element for an input x is a polynomial p_x whose variables are the $\mathbf{C}_{i,x_{\mathrm{inp}_0(i)},x_{\mathrm{inp}_1(i)}}$ matrices, and p_x is allowable in the sense of Definition 10: each monomial in the expansion of p_x contains exactly one variable from each of the $\mathbf{C}_{i,x_{\mathrm{inp}_0(i)},x_{\mathrm{inp}_1(i)}}$ matrices.*

Lemma 3 (Adapted from [1,6]). *Any polynomial p over the obfuscation $\mathcal{O}^{\mathcal{M}}(BP^S)$ can be efficiently decomposed into a sum $p = \sum_{x \in D} \alpha_x p_x$, where $\alpha_x = \prod_{i \in [\ell]} \alpha_{i,x_{\mathrm{inp}_0(i)},x_{\mathrm{inp}_1(i)}}$, each p_x is a single-input element for input x, and $|D|$ is polynomial in the circuit size of p.*

Due to the independence of the α_x variables, it can be shown that p evaluates to zero iff each of the polynomials p_x do. Thus Lemma 3, along with some extra analysis of our own to handle multi-bit outputs, reduces the general problem to the following simpler problem. There is an unknown sequence of matrices $\mathbf{A}_i \in \mathbb{Z}_q^{d_{i-1} \times d_i}$ for $i \in [\ell]$, where $d_0 = d_\ell = 1$ (the shapes of the $\mathbf{A}_i$ ensure that the product $\prod_{i \in [\ell]} \mathbf{A}_i$ is valid and results in a scalar). We are also given an allowable polynomial p' on matrices of random variables $\widehat{\mathbf{A}_i}$. Our goal is to determine, if the $\widehat{\mathbf{A}_i}$ are set to the Kilian-randomized matrices $\widehat{\mathbf{A}_i} = \mathbf{R}_{i-1} \cdot \mathbf{A} \cdot \mathbf{R}_i^{adj}$, whether or not p' evaluates to zero. We note that by applying the Schwartz-Zippel lemma, it suffices to decide if p' is *identically* zero, when considered a polynomial over the formal variables $\mathbf{R}_i$.

It is not hard to see that this simpler problem is impossible in general: p' could be the polynomial computing the iterated matrix product $\prod_{k \in [\ell]} \widehat{\mathbf{A}_i}$, which is equal to $\prod_{i \in [\ell]} \mathbf{A}_i$. Therefore, to decide if p' is identically zero in this case, we at a minimum need to know if $\prod_{i \in [\ell]} \mathbf{A}_i$ evaluates to 0.

The analysis shows that the $\mathbf{A}_i$ are actually equal to $\mathbf{B}_{i,x_{\mathrm{inp}_0(i)},x_{\mathrm{inp}_1(i)}}$ for some (known) input x, where $\mathbf{B}_{i,b_0,b_1}$ are the matrices in the branching program BP. Therefore, we can determine if $\prod_{i \in [\ell]} \mathbf{A}_i = 0$ by querying the BP oracle on x. In the case where p' is the iterated matrix product, this allows us to determine if p' is identically 0. What about other, more general, polynomials p'?

In previous works, $\mathbf{A}_1$ and $\mathbf{A}_\ell$ are bookend vectors, and the $\mathbf{A}_i$ for $k \in [2, \ell-1]$ are square invertible matrices. In this setting, Kilian's statistical simulation theorem allows us to sample from the distribution of $\widehat{\mathbf{A}_i}$ knowing only the product of the $\mathbf{A}_i$, but not the individual values. Then we can apply p' to the sample, and the Schwartz-Zippel lemma shows that p' will evaluate to zero, with high probability, if and only if it is identically zero. This allows deciding if p' is identically zero.

In our case, we cannot sample from the correct distribution of $\widehat{\mathbf{A}_i}$. Instead, we observe that our branching program is non-shortcutting, which means the $\mathbf{A}_i$ and p' satisfy the requirements of Theorem 4. Theorem 4 implies something remarkably strong: if p' is not (a multiple of) the iterated matrix product, it *cannot possibly* be identically zero as a polynomial over the formal variables $\mathbf{R}_k$. Thus, we first decide if p' is a multiple of the iterated matrix product, which is possible using the Schwartz-Zippel lemma. If p' is a multiple, then we know it is identically zero if and only if the product $\prod_{i \in [\ell]} \mathbf{A}_i$ is zero, and we know whether this product is zero by using our BP oracle.

6 Obfuscating Evasive Functions with No Zero Encodings

In this section we show that when the obfuscator of Sect. 3 is applied to an *evasive* function, any poly-time adversary will have only negligible probability in constructing an encoding of 0.

Definition 12. *We say that an adversary $\mathcal{A}$ constructs an encoding of 0 if it ever receives a handle h from $\mathcal{M}$ such that (a) h maps to an encoding of 0 in $\mathcal{M}$'s table, and (b) the polynomial that produced the encoding is not identically zero as a polynomial over its formal variables.*

Theorem 6. *Let $\mathcal{O}$ be the obfuscator from Sect. 3, and let BP^S sample an evasive function family. Then for any PPT adversary $\mathcal{A}$:*

$$\Pr\left[\mathcal{A}^{\mathcal{M}}(\mathcal{O}^{\mathcal{M}}(BP^S)) \text{ constructs an encoding of } 0\right] < \mathsf{negl}(\ell).$$

One can never prevent an adversary from constructing a trivial encoding of 0 by computing $e - e$ for some encoding e that it has. (More generally, any identically zero polynomial will produce a trivial encoding of 0.) However in all candidate constructions of graded encoding schemes, such an operation always produces *the integer* 0, which contains no information. Indeed, it seems unlikely that a plausible candidate would not have this property.

To prove Theorem 6, we first show that any element that is not at the top level $\mathbb{U}$ can be "completed" to the top level by multiplying with other basic elements output by the obfuscator. This is a consequence of our use of strong straddling sets.

Definition 13. *For $i \in [\ell]$ and $b \in \{0,1\}$, an element encoded at level S_{j,b_0,b_1} implies $x_i = b$ if either $\mathsf{inp}_0(j) = i$ and $b_0 = b$ or $\mathsf{inp}_1(j) = i$ and $b_1 = b$.*

Lemma 4. *Let $R := \{[\mathbf{D}_{i,b_0,b_1}]_{S_{i,b_0,b_1}}\}$ be the basic elements output by the obfuscator $\mathcal{O}$, and let $[r]_S$ be any valid element created by a polynomial p over R.*
   Then there exists a set of elements $R' \subseteq R$ such that $[r]_S \times \prod_{z \in R'} z$ is a valid element at level $\mathbb{U}$, and further R' can be efficiently found.

Proof. We say that p *touches* layer $j \in [n]$ if any leaf of p is a basic element from layer j (cf. [30, Definition 4.2]). S uniquely determines the layers touched by p and vice versa (though not necessarily the specific matrices touched in each layer); in particular, p touches every layer iff $S = \mathbb{U}$. Thus we construct R' to contain one basic element from each layer that is not touched by p. If $S = \mathbb{U}$ then the lemma holds trivially with $R' := \emptyset$, so assume $S \neq \mathbb{U}$ and let $J \subseteq [n]$ be the set of layers not touched by p. Let $I := \{\mathsf{inp}_0(j), \mathsf{inp}_1(j) \mid j \in J\} \subseteq [\ell]$ be the set of all indices that are read in some untouched layer.

We claim that there is a sequence $(b_i)_{i \in I} \in \{0,1\}^{|I|}$ such that for every $i \in I$, p's leaves do not contain any basic element that implies $x_i = 1 - b_i$. Fix any $i \in I$. Recall that $\mathbb{U}_i \subset \mathbb{U}$ is the universe set for index i, and note that we must have $\mathbb{U}_i \not\subseteq S$ because some layer that reads index i is untouched. If $\mathbb{U}_i \cap S = \emptyset$, then p's leaves do not contain a basic element that implies $x_i = 0$ nor one that implies $x_i = 1$; in this case we can take $b_i = 0$. If instead $\mathbb{U}_i \cap S \not\subseteq \{\emptyset, \mathbb{U}_i\}$, then by Lemma 1 there is a unique $b_i \in \{0, 1\}$ for which there exists $J' \subset [n]$ such that

$$\mathbb{U}_i \cap S = \bigcup_{j' \in J'} S^i_{j', b_i}.$$

(Recall that each S^i_{j', b_i} comes from the strong straddling set system over $\mathbb{U}_i$.) Thus p's leaves do not contain any basic element that implies $x_i = 1 - b_i$.

Finally let R' contain, for each $j \in J$, an arbitrary entry from the $(b_{\mathsf{inp}_0(j)}, b_{\mathsf{inp}_1(j)})$th matrix in layer j. Formally, $R' := \{\mathbf{D}_{j, b_{\mathsf{inp}_0(j)}, b_{\mathsf{inp}_1(j)}}[0, 0] \mid j \in J\}$ which can be efficiently computed given e. Then $[r]_S \times \prod_{z \in R'} z$ is valid by construction, and it is at level $\mathbb{U}$ because it touches every layer.

We now prove the main theorem of this section. The proof uses the simulator Sim of Theorem 5 in a non-black-box way, and specifically relies on properties of the decomposition $p = \sum_x \alpha_x p_x$ given by Lemma 3.

Proof *(Proof of Theorem 6).* For any PPT adversary $\mathcal{A}$, denote

$$\mathcal{P}'(\mathcal{A}) := \Pr\left[\mathcal{A}^\mathcal{M}(\mathcal{O}^\mathcal{M}(BP^S)) \text{ constructs a level-}\mathbb{U} \text{ encoding of } 0\right].$$

We first show that if $\mathcal{P}'(\mathcal{A})$ is a noticeable function of ℓ for some PPT $\mathcal{A}$, then BP^S cannot be evasive, in contradiction to our assumption. Next we use Lemma 4 to remove the assumption that $\mathcal{A}$'s encoding of 0 is at level $\mathbb{U}$.

Let $f \leftarrow BP^S$ denote the function being obfuscated. Let $\mathcal{A}$ be any PPT, and let Sim denote the corresponding simulator given by Theorem 5. We construct a new adversary $\mathcal{B}$, with oracle access to f, that finds an input x such that $f(x) = 0$.

$\mathcal{B}^f(1^\ell)$:

1. Run Sim^f, which itself is running $\mathcal{A}$, up until the point where $\mathcal{A}$ constructs a level-$\mathbb{U}$ encoding.
2. Decompose $p = \sum_{x \in D} \alpha_x p_x$ as in Lemma 3. Check if $f(x) = 0$ for any $x \in D$. If so, stop and output x; otherwise, continue running Sim until $\mathcal{A}$'s next level-$\mathbb{U}$ encoding, and repeat.
3. If Sim halts, then output a random $x \in \{0, 1\}^\ell$.

Note that $\mathcal{B}$'s simulation of $\mathcal{A}$'s view is correct up to statistical distance $\mathsf{negl}(\ell)$, because Sim's is. The proof of Theorem 5 establishes that for any level-$\mathbb{U}$ p constructed by $\mathcal{A}$,

$$\Pr[p \text{ is an encoding of 0 but some } p_x \text{ is not}] < \mathsf{negl}(\ell).$$

Further, Theorem 4 establishes that if p_x is not identically zero (and some p_x must not be since p is not), then p_x is a multiple of the honest matrix product polynomial corresponding to input x. Thus p_x is an encoding of 0 iff $f(x) = 0$, and we have established $\forall$ PPT $\mathcal{A}$ $\exists$ PPT $\mathcal{B}$:

$$\Pr\left[f\left(\mathcal{B}^f(1^\ell)\right) = 0\right] \geq \mathcal{P}'(\mathcal{A}) - \mathsf{negl}(\ell). \tag{8}$$

Finally, let

$$\mathcal{P}(\mathcal{A}) := \Pr\left[\mathcal{A}^{\mathcal{M}}(\mathcal{O}^{\mathcal{M}}(BP^S)) \text{ constructs an encoding of } 0\right]$$

be the probability that we want to bound. We claim that $\forall$ PPT $\mathcal{A}$ $\exists$ PPT $\mathcal{A}'$: $\mathcal{P}'(\mathcal{A}') \geq \mathcal{P}(\mathcal{A})$. Namely $\mathcal{A}'$ runs $\mathcal{A}$, and for every encoding $[r]_S$ with $S \neq \mathbb{U}$ created by $\mathcal{A}$, $\mathcal{A}'$ also creates the level-$\mathbb{U}$ encoding $[r']_\mathbb{U} := [r]_S \times \prod_{z \in R'} z$ guaranteed by Lemma 4. Note that if $[r]_S$ encodes 0 then $[r']_\mathbb{U}$ must encode 0 as well, so we have $\mathcal{P}'(\mathcal{A}') \geq \mathcal{P}(\mathcal{A})$. Combining this with (8), we complete the proof: if $\exists$ PPT $\mathcal{A}$ such that $\mathcal{P}(\mathcal{A})$ is a noticeable function of ℓ, then BP^S does not sample an evasive function family.

In the full version of this paper, we show that, via the bootstrapping technique of [12,20], an obfuscator for log-depth evasive circuits implies an obfuscator for all poly-size evasive circuits.

References

1. Ananth, P.V., Gupta, D., Ishai, Y., Sahai, A.: Optimizing obfuscation: avoiding Barrington's theorem. In: Proceedings of the 2014 ACM SIGSAC Conference on Computer and Communications Security, pp. 646–658 (2014)
2. Apon, D., Huang, Y., Katz, J., Malozemoff, A.J.: Implementing cryptographic program obfuscation. Cryptology ePrint Archive, Report 2014/779 (2014). http://eprint.iacr.org/
3. Applebaum, B.: Bootstrapping obfuscators via fast pseudorandom functions. In: Sarkar, P., Iwata, T. (eds.) ASIACRYPT 2014, Part II. LNCS, vol. 8874, pp. 162–172. Springer, Heidelberg (2014)
4. Applebaum, B., Brakerski, Z.: Obfuscating circuits via composite-order graded encoding. In: Dodis, Y., Nielsen, J.B. (eds.) TCC 2015, Part II. LNCS, vol. 9015, pp. 528–556. Springer, Heidelberg (2015)
5. Barak, B., Bitansky, N., Canetti, R., Kalai, Y.T., Paneth, O., Sahai, A.: Obfuscation for evasive functions. In: Lindell, Y. (ed.) TCC 2014. LNCS, vol. 8349, pp. 26–51. Springer, Heidelberg (2014)
6. Barak, B., Garg, S., Kalai, Y.T., Paneth, O., Sahai, A.: Protecting obfuscation against algebraic attacks. In: Nguyen, P.Q., Oswald, E. (eds.) EUROCRYPT 2014. LNCS, vol. 8441, pp. 221–238. Springer, Heidelberg (2014)
7. Barak, B., Goldreich, O., Impagliazzo, R., Rudich, S., Sahai, A., Vadhan, S.P., Yang, K.: On the (im)possibility of obfuscating programs. In: Kilian, J. (ed.) CRYPTO 2001. LNCS, vol. 2139, pp. 1–18. Springer, Heidelberg (2001)
8. Boneh, D., Lewi, K., Raykova, M., Sahai, A., Zhandry, M., Zimmerman, J.: Semantically secure order-revealing encryption: multi-input functional encryption without obfuscation. In: Oswald, E., Fischlin, M. (eds.) EUROCRYPT 2015. LNCS, vol. 9057, pp. 563–594. Springer, Heidelberg (2015)
9. Boneh, D., Wu, D.J., Zimmerman, J.: Immunizing multilinear maps against zeroizing attacks. Cryptology ePrint Archive, Report 2014/930 (2014). http://eprint.iacr.org/
10. Boneh, D., Zhandry, M.: Multiparty key exchange, efficient traitor tracing, and more from indistinguishability obfuscation. In: Garay, J.A., Gennaro, R. (eds.) CRYPTO 2014, Part I. LNCS, vol. 8616, pp. 480–499. Springer, Heidelberg (2014)

11. Brakerski, Z., Gentry, C., Halevi, S., Lepoint, T., Sahai, A., Tibouchi, M.: Cryptanalysis of the quadratic zero-testing of GGH. Cryptology ePrint Archive, Report 2015/845 (2015). http://eprint.iacr.org/
12. Brakerski, Z., Rothblum, G.N.: Virtual black-box obfuscation for all circuits via generic graded encoding. In: Lindell, Y. (ed.) TCC 2014. LNCS, vol. 8349, pp. 1–25. Springer, Heidelberg (2014)
13. Cheon, J.H., Han, K., Lee, C., Ryu, H., Stehlé, D.: Cryptanalysis of the multilinear map over the integers. In: Oswald, E., Fischlin, M. (eds.) EUROCRYPT 2015. LNCS, vol. 9056, pp. 3–12. Springer, Heidelberg (2015)
14. Cheon, J.H., Lee, C., Ryu, H.: Cryptanalysis of the new clt multilinear maps. Cryptology ePrint Archive, Report 2015/934 (2015). http://eprint.iacr.org/
15. Coron, J.-S., et al.: Zeroizing without low-level zeroes: new MMAP attacks and their limitations. In: Gennaro, R., Robshaw, M. (eds.) CRYPTO 2015. LNCS, vol. 9216, pp. 247–266. Springer, Heidelberg (2015)
16. Coron, J.-S., Lepoint, T., Tibouchi, M.: Practical multilinear maps over the integers. In: Canetti, R., Garay, J.A. (eds.) CRYPTO 2013, Part I. LNCS, vol. 8042, pp. 476–493. Springer, Heidelberg (2013)
17. Coron, J.-S., Lepoint, T., Tibouchi, M.: New multilinear maps over the integers. In: Gennaro, R., Robshaw, M. (eds.) CRYPTO 2015. LNCS, vol. 9216, pp. 267–286. Springer, Heidelberg (2015)
18. Garg, S., Gentry, C., Halevi, S.: Candidate multilinear maps from ideal lattices. In: Johansson, T., Nguyen, P.Q. (eds.) EUROCRYPT 2013. LNCS, vol. 7881, pp. 1–17. Springer, Heidelberg (2013)
19. Garg, S., Gentry, C., Halevi, S., Raykova, M.: Two-round secure MPC from indistinguishability obfuscation. In: Lindell, Y. (ed.) TCC 2014. LNCS, vol. 8349, pp. 74–94. Springer, Heidelberg (2014)
20. Garg, S., Gentry, C., Halevi, S., Raykova, M., Sahai, A., Waters, B.: Candidate indistinguishability obfuscation and functional encryption for all circuits. In: FOCS, pp. 40–49 (2013)
21. Garg, S., Gentry, C., Sahai, A., Waters, B.: Witness encryption and its applications. In: STOC (2013)
22. Gentry, C., Gorbunov, S., Halevi, S.: Graph-induced multilinear maps from lattices. In: Dodis, Y., Nielsen, J.B. (eds.) TCC 2015, Part II. LNCS, vol. 9015, pp. 498–527. Springer, Heidelberg (2015)
23. Gentry, C., Lewko, A., Waters, B.: Witness encryption from instance independent assumptions. In: Garay, J.A., Gennaro, R. (eds.) CRYPTO 2014, Part I. LNCS, vol. 8616, pp. 426–443. Springer, Heidelberg (2014)
24. Gentry, C., Lewko, A.B., Sahai, A., Waters, B.: Indistinguishability obfuscation from the multilinear subgroup elimination assumption. In: IEEE Symposium on Foundations of Computer Science FOCS, pp. 151–170 (2015)
25. Goldwasser, S., et al.: Multi-input functional encryption. In: Nguyen, P.Q., Oswald, E. (eds.) EUROCRYPT 2014. LNCS, vol. 8441, pp. 578–602. Springer, Heidelberg (2014)
26. Goyal, V., Ishai, Y., Sahai, A., Venkatesan, R., Wadia, A.: Founding cryptography on tamper-proof hardware tokens. In: Micciancio, D. (ed.) TCC 2010. LNCS, vol. 5978, pp. 308–326. Springer, Heidelberg (2010)
27. Halevi, S.: Graded encoding, variations on a scheme. IACR Cryptology ePrint Archive 2015, 866 (2015)
28. Hu, Y., Jia, H.: Cryptanalysis of GGH map. IACR Cryptology ePrint Archive 2015, 301 (2015)

29. Kilian, J.: Founding cryptography on oblivious transfer. In: STOC, pp. 20–31 (1988)
30. Miles, E., Sahai, A., Weiss, M.: Protecting obfuscation against arithmetic attacks. IACR Cryptology ePrint Archive 2014, 878 (2014)
31. Miles, E., Sahai, A., Zhandry, M.: Annihilation attacks for multilinear maps: Cryptanalysis of indistinguishability obfuscation over ggh13. Cryptology ePrint Archive, Report 2016/147 (2016). http://eprint.iacr.org/
32. Minaud, B., Fouque, P.A.: Cryptanalysis of the new multilinear map over the integers. Cryptology ePrint Archive, Report 2015/941 (2015). http://eprint.iacr.org/
33. Pass, R., Seth, K., Telang, S.: Indistinguishability obfuscation from semantically-secure multilinear encodings. In: Garay, J.A., Gennaro, R. (eds.) CRYPTO 2014, Part I. LNCS, vol. 8616, pp. 500–517. Springer, Heidelberg (2014)
34. Sahai, A., Waters, B.: How to use indistinguishability obfuscation: deniable encryption, and more. In: Symposium on Theory of Computing (STOC), pp. 475–484 (2014)
35. Sahai, A., Zhandry, M.: Obfuscating low-rank matrix branching programs. IACR Cryptology ePrint Archive 2014, 773 (2014). http://eprint.iacr.org/2014/773
36. Zhandry, M.: How to avoid obfuscation using witness PRFs. In: Kushilevitz, E., Malkin, T. (eds.) TCC 2016-A. LNCS, vol. 9563, pp. 421–448. Springer, Heidelberg (2016). doi:10.1007/978-3-662-49099-0_16
37. Zimmerman, J.: How to obfuscate programs directly. In: Oswald, E., Fischlin, M. (eds.) EUROCRYPT 2015. LNCS, vol. 9057, pp. 439–467. Springer, Heidelberg (2015)

New Negative Results on Differing-Inputs Obfuscation

Mihir Bellare[1]([✉]), Igors Stepanovs[1], and Brent Waters[2]

[1] Department of Computer Science and Engineering,
University of California San Diego, San Diego, USA
{mihir,istepano}@eng.ucsd.edu
[2] Department of Computer Science, University of Texas at Austin, Austin, USA
bwaters@cs.utexas.edu
https://cseweb.ucsd.edu/~mihir/
https://cseweb.ucsd.edu/~istepano/
https://www.cs.ucsb.edu/~bwaters/

Abstract. We provide the following negative results for differing-inputs obfuscation (diO): (1) If sub-exponentially secure one-way functions exist then sub-exponentially secure diO for TMs does not exist (2) If in addition sub-exponentially secure iO exists then polynomially secure diO for TMs does not exist.

1 Introduction

Differing-inputs obfuscation (diO) is a natural extension of indistinguishability obfuscation (iO). It has been conjectured that candidate constructions of iO also met diO. Based on this, diO has been exploited in applications. Garg, Gentry, Halevi and Wichs (GGHW) [28] showed that if something they called "special purpose" obfuscation exists, then diO does not. This has put diO in an ambiguous and contentious position, some people arguing that GGHW is evidence diO does not exist, others saying that perhaps it does and it is special-purpose obfuscation that does not exist. This paper uses a new approach to give powerful evidence that the first camp is right, meaning it is indeed diO that does not exist, by showing this to be true under weaker and more standard assumptions than special-purpose obfuscation. We show (1) If sub-exponentially secure one-way functions exist then sub-exponentially secure diO for TMs does not exist (2) If in addition sub-exponentially secure iO exists then polynomially secure diO for TMs does not exist.

1.1 Background

The notion of program obfuscation that is most intuitive and appealing is that an obfuscated program should be no more useful than an oracle for the program itself. Formalized as VBB obfuscation (vbbO), it was shown impossible in the sense that there is no obfuscator that will successfully VBB obfuscate all programs [7,36]. Further negative results about vbbO were given in [17,33].

© International Association for Cryptologic Research 2016
M. Fischlin and J.-S. Coron (Eds.): EUROCRYPT 2016, Part II, LNCS 9666, pp. 792–821, 2016.
DOI: 10.1007/978-3-662-49896-5_28

In the face of this, Barak et al. [7] suggested other, weaker notions of obfuscation that appeared not to succumb to their counter-examples and might therefore be achievable. The most prominent were indistinguishability obfuscation (iO) and its extension, differing-input obfuscation (diO). The first asks that obfuscations of functionally equivalent programs are indistinguishable. The second is a natural computational relaxation: even if the programs are not functionally equivalent, as long as it is hard, given the programs, to find an input on which they differ, then the obfuscations of the programs are indistinguishable. The underlying intuition is that if one can find a differing input for the programs, one can clearly distinguish their obfuscations. In iO this is excluded information theoretically, by saying there does not exist such an input, while in diO it is excluded computationally, by saying such an input might exist but is hard to find. On the surface both might appear equally reasonable, since the vbbO negative results do not apply to either. But this turns out not to be true.

These intriguing notions lay dormant for many years, for two reasons. First, that one could not prove these notions unachievable did not mean they were achievable. Second, they seemed quite weak; even if they were achievable, what could one do with them? An answer to the first question came with candidate constructions of iO [6,27,30,43]. An answer to the second came when Sahai and Waters showed how to use iO towards many ends [45]. Since then, applications of iO and diO have ballooned.

In these applications, a crucial role is played by *auxiliary information*. The modern definitions of iO and diO used in these applications [1,12,20,27,45] consider a program sampler S that spits out a pair P_0, P_1 of programs *together with associated auxiliary information aux*. The sampler is said to produce functionally equivalent programs if P_0 and P_1 agree on all inputs. The sampler is said to be difference-secure if an adversary given P_0, P_1, aux cannot find an input x such that $P_0(x) \neq P_1(x)$ except with small probability. The obfuscation game picks a challenge bit b and gives you (the adversary) an obfuscation $\overline{P}$ of P_b under the obfuscator Obf, *together with aux*. Your task (as the adversary) is to guess b. Obf is called iO-secure if you have small advantage for all samplers producing functionally equivalent programs, and diO-secure if you have small advantage for all difference-secure samplers. Adversaries are always polynomial time, but probabilities referred to as "small" may be sub-exponentially so or negligible. Programs may be TMs or circuits. This leads to a collection of variant notions.

1.2 The GGHW Result

Let Obf be an obfuscator. GGHW [28] provide a program sampler S for which they show, under certain assumptions, that diO-security of Obf fails, which means that (1) the sampler is difference secure under these assumptions, but (2) there is a way to distinguish the obfuscations under Obf of the two programs returned by the sampler given the auxiliary information. Their approach is to have the sampler first generate a signing and verification key pair (sk, vk) for a signature scheme meeting the standard notion of unforgeability [34]. The program P_1 takes a message m and candidate signature σ and accepts iff σ is a valid signature on m under vk. The program P_0 will take in the same inputs, but it will always reject. Clearly the programs P_0 and P_1 differ exactly on the input pairs (m, σ) where σ is a valid signature of m under vk. Next, the sampler creates a third program P_2 that has hardwired the secret signing key sk and takes as input a (smaller) program $\overline{P}$. It hashes $\overline{P}$ using a CRHF to get a message m, and uses sk to get a signature σ on m. It then runs the $\overline{P}$ on (m, σ) and outputs 1 if $\overline{P}$ accepts on these inputs. Finally, S creates auxiliary information aux consisting of an obfuscation P_2^* of P_2. This obfuscation is not under the given obfuscator Obf, but under some other assumed "special purpose" obfuscator Obf* whose role and properties will emerge in the following.

To serve as a counterexample it should both (1) be possible, using the auxiliary information P_2^*, to distinguish between obfuscations under Obf of P_0 and P_1, and (2) be difficult, given P_0, P_1, P_2^*, to find an input on which P_0 and P_1 differ. The first property follows trivially from the design. An adversary given the auxiliary information P_2^* and a challenge program $\overline{P}$ that is either an obfuscation of P_0 or P_1 can distinguish these cases by simply feeding the program $\overline{P}$ as an input to P_2^*. If $\overline{P}$ is an obfuscation of P_1 then, when P_2^* runs $\overline{P}$ on the message and valid signature that P_2^* creates, $\overline{P}$ will accept. But if $\overline{P}$ is an obfuscation of P_0, then P_2^* will reject.

In contrast it is much more difficult to establish the second property, namely that it is hard to find an input on which P_0, P_1 differ *even in the presence of the auxiliary information* P_2^*. The difficulty stems from the latter. In the absence of aux the property follows straightforwardly from the security of the signature scheme, as a differing input is exactly a valid message-signature pair, and would amount to a signature forgery. However, since the obfuscated differentiating program P_2^* has embedded in it the secret signing key sk it is not clear how to prove that it is hard to find signatures in the presence of P_2^*.

Recall that P_2^* was an obfuscation, under some un-specified obfuscator Obf*, of P_2. GGHW [28] simply conjecture that there exists some obfuscator Obf* that will hide the secret key sk sufficiently well that it is hard to find a differing input for P_0, P_1, meaning to find a valid message-signature pair, even given P_2^*. While they were unable to prove this conjecture under any standard obfuscation definitions such as iO or even vbbO, they were able to partially justify their conjecture with a heuristic analysis. Their analysis replaces the adversary's access to the obfuscated program P_2^* with an oracle that performs the same functionality. In this world the adversary no longer has direct access to an object containing sk and GGHW are able to demonstrate differing inputs security of S by a fairly straightforward reduction to the underlying security of the signature scheme.

The GGHW result certainly creates significant questions regarding the use of diO. Arguably, the primary reason for using the diO security definition over vbbO is that no impossibility results like those of [7,17,33,36] are known for diO. However, if the GGHW conjecture holds, then this is no longer true and the perceived benefit of diO versus vbbO is significantly reduced (The benefit is not eliminated, since even if there exist functionalities that cannot be diO obfuscated, it is still possible that there are functionalities that can be diO obfuscated but not VBB obfuscated.). At the same time, the heuristic used to justify the GGHW counterexample is itself much stronger than assuming diO — namely their analysis relies on modeling the differentiating obfuscated program as an oracle.

1.3   Our Approach

We introduce a new approach to proving the impossibility of diO. In contrast to the prior work, we analyze our sampler under *concrete assumptions* that replace the GGHW conjecture. We now explain the intuition behind our approach as well as the obstacles we had to overcome.

Let Obf be an obfuscator that we assume, towards a contradiction, is diO-secure. At the highest level our approach is similar to GGHW. We build a program sampler S that produces programs P_0, P_1 and auxiliary information P_2^* consisting of an obfuscation of a program P_2 under an obfuscator Obf*. As in GGHW, the sampler generates a signing and verification key pair (sk, vk) for an underlying signature scheme DS, and program P_0 always rejects. Likewise, P_1 takes as input a candidate message-signature pair (m, σ) and checks its validity under the signature verification program DS.Ver with key vk. The auxiliary information continues to be the obfuscation P_2^*, under an obfuscator Obf*, of a program P_2, where P_2 hardwires the secret signing key sk. P_2 takes as input a program $\overline{P}$ of a certain maximum length, and uses $m = \overline{P}$ as the message it signs, and runs $\overline{P}$ on m and the signature, accepting if this accepts. The important difference now however is that Obf* is not some new type of obfuscator as in GGHW. Rather Obf* is *assumed to be only an iO-secure obfuscator*.

It continues to be easy, using the auxiliary information P_2^*, to distinguish between obfuscations under Obf of P_0 and P_1. The main issue is to prove that it is difficult, given P_0, P_1, P_2^*, to find an input on which P_0 and P_1 differ. The hurdle here continues to be the same, namely that the auxiliary information program P_2^* embeds the secret signing key sk. This precludes reducing to the security of the signature scheme in an obvious way. To prove security we will show that it is computationally difficult to generate a signature on any message. We do this via a hybrid argument that steps through every possible message one by one. Since our hybrid steps through the entire message space we base our security on assumptions of sub-exponential hardness.

To execute our strategy we will replace the generic signature scheme of GGHW with a special type of puncturable signature scheme that we call a *consistent puncturable* signature scheme. Given a "master" secret key sk, it should be possible to create a punctured version sk_{m^*} of the key, for a given message m^*, that can be used to sign any message $m \neq m^*$ but even given which it is hard to produce a signature on m^*. So far this is a special type of policy-based [9], functional [21] or delegatable [5] signatures, these themselves analogues of the notions of puncturable, constrained and functional PRFs [19,21,40]. The additional consistency requirement is that the signatures of $m \neq m^*$ produced under the master key and the punctured key should be the same. Note that only deterministic puncturable signature schemes can be consistent, but the former is not a sufficient condition. We show in Sect. 3 that such signature schemes can be built from iO and one-way functions. While making a standard signature scheme deterministic is trivial via the use of PRFs, our challenge is making the punctured and master versions of the key produce consistent signatures.

Our hybrid now proceeds as follows. We step through each program (message) P^* and show that it is computationally difficult to produce a signature on P^*. We do this by first replacing the obfuscation of P_2 with an obfuscation of a program P_{2,P^*} that works as follows. On all inputs $P \neq P^*$ the program P_{2,P^*} behaves as P_2 with the exception that it uses a punctured version of the signing key sk_{P^*}. On input P^* its output is hardwired to be whatever the output of $P_2(P^*)$ was. We observe that if indistinguishability obfuscation holds, then no poly-time attacker can distinguish between obfuscations of programs P_2 and P_{2,P^*}. This follows since the two programs share the same output on every input. On every $P \neq P^*$ the master and punctured keys will produce the same signature that they feed into P, and on input P^* program P_{2,P^*} is hardwired to behave the same as P_2. Since it is hard to distinguish between obfuscations of these two programs, it should be no easier to output a signature on message P^* when P_2 is obfuscated to get the auxiliary information aux than it is when P_{2,P^*} is obfuscated. However, in the latter case the security of the puncturable signature scheme guarantees this is hard.

Note that since we assumed a diO-secure obfuscator Obf to start our proof by contradiction, an iO-secure obfuscator, which we use both directly and to build consistent punctured signatures, is provided for free and is not an extra assumption. This means the only assumption we need is a sub-exponentially hard one-way function. More precisely, this is the case for sub-exponential diO, while for polynomial diO the iO assumption will be an extra one.

While the text above outlines our main approach, there are several important factors that still must be taken into account. First, we notice that P_1 should be capable of verifying a signature on a message that is an obfuscation of P_1 and thus longer than P_1 itself. For this reason we need to view P_0 and P_1 as Turning Machines (TMs) that can process inputs longer than their own descriptions.

Next, our complexity leveraging argument requires that the advantage ϵ of any PT attacker on the signature scheme multiplied by the message space be negligible. To satisfy this using sub-exponential hardness assumptions we must use a verification key vk that is larger than the programs P_0, P_1. However, this creates a circularity problem under the obvious strategy of having P_1 actually contain vk to verify the messages! We circumvent this issue by the use of a UOWHF [42], also called a target collision-resistant (TCR) hash function [10], that hashes a separate verification program as follows. We construct a program P_{ver} that takes as input a candidate message-signature pair (m, σ) and uses an embedded verification key vk to either accept or reject it. Now P_1 takes P'_{ver} as an additional input and uses it to check the candidate message-signature pairs, rather than storing vk and performing the verification itself. P_1 hardwires the hash h of P_{ver} under a TCR hash function, and rejects unless the hash h' of P'_{ver} matches its hardwired hash h. This ensures that only P_{ver} can be used to verify the signatures. We analyze security by adding a hybrid step at the beginning using the UOWHF security. We emphasize that the argument using our UOWHF is outside of the complexity leveraging part of our hybrid.

The above is a very high-level description, and the devil is in the details that the body of the paper sorts out. The circularity issues, summarized via Fig. 6, have to be dealt with very carefully. A critical element of dealing with them is that *different primitives are run with different values of the security parameter.* Thus, while the convention is that the security parameter in a proof remains λ throughout, our constructions will feature $n(\lambda)$ as the security parameter in certain places, with n a polynomial that is carefully defined based on other parameters. Another subtlety is that the success of this program depends on the details of how sub-exponential security is defined. Specifically (cf. Sect. 2) we use "uniform" rather than "pointwise" definitions in the language of [8]. The latter showed them equivalent in the usual setting of negligible functions but they are not known to be equivalent in the sub-exponential setting.

1.4 Discussion and Related Work

Sub-exponential Security. Our assumptions and conclusions both involve sub-exponential hardness and one might ask about the validity of such assumptions and the value of such conclusions. Empirical evidence, at least, says that when problems are hard, they are sub-exponentially hard. Natural problems do not appear to be polynomially but not sub-exponentially hard except in rare cases [3]. Indeed sub-exponential hardness is frequently assumed in cryptography, especially recently [24,31,35]. In particular it is unlikely that polynomially-secure diO exists but sub-exponentially secure diO does not, so ruling out the latter is significant in terms of evidence against diO. Similarly it is unlikely that polynomially secure OWFs exist but sub-exponentially secure ones do not, so assuming the latter is reasonable.

Bounded Versus Unbounded Inputs. In this work we provide negative results about the existence of differing-inputs obfuscators for TMs that can take arbitrarily long inputs. Our results do not rule out the possibility of constructing diO for TMs with a-priori bounded input-lengths.

Implications. Note that [1,20] build diO for TMs with unbounded inputs from circuit diO and SNARKs [14,15]. This means that if SNARKs exist then our negative results for TM diO extend to circuit diO. Also [38] build diO for TMs with unbounded inputs from public-coin diO for NC^1, fully homomorphic encryption with decryption in NC^1 and public-coin SNARKs. Our results would imply that (if the assumptions we make hold) one of these three primitives does not exist.

Constructions and Applications of diO. Differing-inputs obfuscation has proven to be a powerful tool using which we have built new primitives. In some cases it has later been possible to reduce the assumption to iO or other diO variants, but sometimes at the cost of weakening the conclusion and usually at the cost of increased complexity and difficulty. All this motivates understanding whether or not diO is achievable.

diO for circuits is used in [1,20] to achieve adaptively-secure FE (Functional Encryption) and extractable witness encryption. It was later shown in [16,23,41] how to build TM iO from circuit iO but the conclusion was weaker. Adaptively-secure FE from iO did emerge but the solutions were more complex than the ones from diO [2,47].

Boyle et al. [20] show that iO implies diO for samplers outputting circuits that differ on only polynomially-many inputs. Our counter-examples and results do not apply to this type of diO. Differing input obfuscation is used as a tool in [12], via the result of [20], to give hardcore functions with polynomially-many output bits from any injective one-way function and iO, and is used as an assumption to extend this result to arbitrary one-way functions. It is used similarly as a tool in [22].

Ishai et al. [38] define public-coin diO, by relaxing the notion of diO to require that only public random coins can be used to build challenge programs and the corresponding auxiliary information. Our negative results do not apply to public-coin diO. Public-coin diO is a valuable notion but it doesnt take away from the interest in proving impossibility of diO because diO precedes public-coin diO and there are works that still rely on it, and there could be interesting new applications from diO but not from public-coin diO. Furthermore, our techniques might help understand the possibility of public-coin diO.

A variant of diO was also used as an assumption in a result in [29].

Consistent Signature Schemes. Some of the prior work focuses on constructing digital signature schemes with properties that are similar to the ones we require from consistent signature schemes. The known primitives include: functional signatures [21], policy-based signatures [9] and operational signatures [4], the latter subsuming the preliminary work on delegatable signatures [5]. However, none of the proposed constructions of these primitives satisfy the consistency requirement which requires that the master and punctured signing keys produce the same signatures for all messages except for the punctured message, and which is crucial for our impossibility result.

We get consistent puncturable signatures from OWFs and iO, which in our context effectively means from OWFs since our proof assumes diO towards a contradiction and thus gets iO for free. Our definition of consistent puncturable signatures is novel, but our construction follows Sahai-Waters signatures [45]. Consistent puncturable signatures are also implied by splittable signatures [41], which are built based on an injective PRG and iO. Injective PRGs are not known to be implied by OWFs so the assumption is stronger than ours. However, [18] build injective OWFs from OWFs and iO, and also say that, due to an observation of Boyle et al. [20], the injective PRG of [41] can be replaced with an injective OWF. By this route one can get consistent puncturable signatures from OWFs and iO. However our construction is direct, substantially simpler and self contained. Consistent puncturable signatures can also be constructed from constrained verifiable PRFs [25, 26]. The latter are achievable from κ-Multilinear DDH assumption. In our context, this would be an additional assumption since it is not known to be implied by diO.

2   Preliminaries

Notation. Let $\mathbb{N} = \{0, 1, 2, \ldots\}$ be the set of non-negative integers. We denote by $\lambda \in \mathbb{N}$ the security parameter and by 1^λ its unary representation. If $x \in \{0, 1\}^*$ is a string then $|x|$ denotes its length. If $x \in \{0, 1\}^*$ is a string and

$\ell \in \mathbb{N}$ such that $|x| \leq \ell$ then $\langle x \rangle_\ell$ denotes the string of length ℓ that is built by padding x with leading zeros. If X is a finite set, we let $x \leftarrow_\$ X$ denote picking an element of X uniformly at random and assigning it to x. Algorithms may be randomized unless otherwise indicated. Running time is worst case. "PT" stands for "polynomial-time," whether for randomized algorithms or deterministic ones. If A is an algorithm, we let $y \leftarrow A(x_1, \ldots; r)$ denote running A with random coins r on inputs $x_1, \ldots$ and assigning the output to y. We let $y \leftarrow_\$ A(x_1, \ldots)$ be the result of picking r at random and letting $y \leftarrow A(x_1, \ldots; r)$. We let $[A(x_1, \ldots)]$ denote the set of all possible outputs of A when invoked with inputs $x_1, \ldots$. We say that $f \colon \mathbb{N} \to \mathbb{R}$ is negligible if for every positive polynomial p, there exists $\lambda_p \in \mathbb{N}$ such that $f(\lambda) < 1/p(\lambda)$ for all $\lambda \geq \lambda_p$. We use the code based game playing framework of [11] (See Fig. 1 for an example.). By $G^{\mathcal{A}}(\lambda)$ we denote the event that the execution of game G with adversary $\mathcal{A}$ and security parameter λ results in the game returning true.

Uniform and Pointwise Security Definitions. There are two common ways to formalize security definitions – by using different order of quantification. Let GAME be a security game, and let $\mathsf{Adv}_{\mathcal{A}}^{\mathsf{game}}(\lambda)$ be the advantage of a PT adversary $\mathcal{A}$ winning in this game with security parameter λ. Consider the following two alternative definitions of sub-exponential security. A *uniform* definition requires that there is a constant $0 < \epsilon < 1$ such that for every PT adversary $\mathcal{A}$ there exists $\lambda_{\mathcal{A}} \in \mathbb{N}$ such that $\mathsf{Adv}_{\mathcal{A}}^{\mathsf{game}}(\lambda) \leq 2^{-\lambda^\epsilon}$ for all $\lambda \geq \lambda_{\mathcal{A}}$. A *pointwise* definition requires that for every PT adversary $\mathcal{A}$ there exist $0 < \epsilon < 1$ and $\lambda_{\mathcal{A}} \in \mathbb{N}$ such that $\mathsf{Adv}_{\mathcal{A}}^{\mathsf{game}}(\lambda) \leq 2^{-\lambda^\epsilon}$ for all $\lambda \geq \lambda_{\mathcal{A}}$. These definitions differ in the order of quantification between ϵ and $\mathcal{A}$. In this work, we use uniform security definitions. For the case of polynomial security, Bellare [8] proved that uniform and pointwise definitions are equivalent. It is not known whether the equivalence also holds for the above definitions of sub-exponential security.

Circuits and Turing Machines. We say that P is a program if it is either a circuit or a Turing Machine (TM), and we denote the size of its binary representation by $|P|$. We assume that any program P takes a single input string x; if P is defined to take multiple inputs $x_1, \ldots$ then running P on an input x is implicitly assumed to parse $(x_1, \ldots) \leftarrow x$ and run $P(x_1, \ldots)$.

Game $\mathrm{OW}_\mathsf{F}^{\mathcal{F}}(\lambda)$	Game $\mathrm{TCR}_\mathsf{H}^{\mathcal{H}}(\lambda)$	Game $\mathrm{PPRF}_\mathsf{G}^{\mathcal{G}}(\lambda)$
$fk \leftarrow_\$ \mathsf{F.Kg}(1^\lambda)$	$(x_0, st) \leftarrow_\$ \mathcal{H}_1(1^\lambda)$	$b \leftarrow_\$ \{0,1\}$; $gk \leftarrow_\$ \mathsf{G.Kg}(1^\lambda)$
$x \leftarrow_\$ \mathsf{F.In}(\lambda)$	$hk \leftarrow_\$ \mathsf{H.Kg}(1^\lambda)$	$b' \leftarrow_\$ \mathcal{G}^{\mathrm{CH}}(1^\lambda)$; Return $(b = b')$
$y \leftarrow \mathsf{F.Ev}(1^\lambda, fk, x)$	$x_1 \leftarrow_\$ \mathcal{H}_2(1^\lambda, st, hk)$	$\mathrm{CH}(x^*)$
$x' \leftarrow_\$ \mathcal{F}(1^\lambda, fk, y)$	$h_0 \leftarrow \mathsf{H.Ev}(1^\lambda, hk, x_0)$	$gk^* \leftarrow_\$ \mathsf{G.PKg}(1^\lambda, gk, x^*)$
$y' \leftarrow \mathsf{F.Ev}(1^\lambda, fk, x')$	$h_1 \leftarrow \mathsf{H.Ev}(1^\lambda, hk, x_1)$	If $b = 1$ then $r^* \leftarrow \mathsf{G.Ev}(1^\lambda, gk, x^*)$
Return $(y = y')$	$\mathsf{win}_0 \leftarrow (x_0 \neq x_1)$	else $r^* \leftarrow_\$ \mathsf{G.Out}(\lambda)$
	$\mathsf{win}_1 \leftarrow (h_0 = h_1)$	Return (gk^*, r^*)
	Return $(\mathsf{win}_0 \wedge \mathsf{win}_1)$	

Fig. 1. Games defining one-wayness of function family F, target collision-resistance of function family H and puncturable-PRF security of function family G.

We say that circuits C_0, C_1 are functionally equivalent, written $C_0 \equiv C_1$, if they have the same number of inputs $\ell \in \mathbb{N}$ and if $C_0(x) = C_1(x)$ holds for all $x \in \{0,1\}^\ell$. We say that TMs M_0, M_1 are functionally equivalent, and denote it by $M_0 \equiv M_1$, if both $M_0(x)$ and $M_1(x)$ halt on all $x \in \{0,1\}^*$ and if $M_0(x) = M_1(x)$ for all $x \in \{0,1\}^*$.

If M is a TM and $t \in \mathbb{N}$ then $y \leftarrow \mathsf{UTM}^t_M(x_1, \ldots)$ denotes running M on inputs $x_1, \ldots$ and assigning the output to y; if $M(x_1, \ldots)$ does not halt within t steps, then $\mathsf{UTM}^t_M(x_1, \ldots)$ returns 0. If M is a TM and $x \in \{0,1\}^*$ is a string such that M halts on input x, we use $\mathsf{time}(M, x)$ to denote the number of steps that are required for it to halt.

Let P be any circuit or any TM that halts on all inputs. For any $s \in \mathbb{N}$ such that $|P| \leq s$ let $\mathsf{Pad}_s(P)$ denote P padded to have size s, meaning that $\mathsf{Pad}_s(P)$ and P are of the same type (i.e. both are circuits or TMs) and $\mathsf{Pad}_s(P) \equiv P$. We assume that P can be padded to any size larger or equal to $|P|$.

<u>Function Families.</u> A family of functions F specifies PT algorithms F.Kg and F.Ev, where F.Ev is deterministic. Associated to F is a collection if input sets F.In and a collection of output sets F.Out, defining all valid inputs and outputs for each of security parameters. Key generation algorithm F.Kg takes 1^λ to return a key fk. Evaluation algorithm F.Ev takes $1^\lambda, fk$ and an input $x \in \mathsf{F.In}(\lambda)$ to return $\mathsf{F.Ev}(1^\lambda, fk, x) \in \mathsf{F.Out}(\lambda)$. We say that F is injective if the function $\mathsf{F.Ev}(1^\lambda, fk, \cdot)\colon \mathsf{F.In}(\lambda) \to \mathsf{F.Out}(\lambda)$ is injective for all $\lambda \in \mathbb{N}$ and $fk \in [\mathsf{F.Kg}(1^\lambda)]$.

<u>Puncturable Function Families.</u> A *puncturable* function family G specifies (beyond the usual algorithms) additional PT algorithms G.PKg and G.PEv, where G.PEv is deterministic. Punctured key generation algorithm G.PKg takes 1^λ, a key $gk \in [\mathsf{G.Kg}(1^\lambda)]$ and a target input $x^* \in \mathsf{G.In}(\lambda)$ to return a "punctured" key gk^*. Punctured evaluation algorithm G.PEv takes $1^\lambda, gk^*$ and an input $x \in \mathsf{G.In}(\lambda)$ to return $\mathsf{G.PEv}(1^\lambda, gk^*, x) \in \mathsf{G.Out}(\lambda)$. The correctness condition requires that $\mathsf{G.PEv}(1^\lambda, gk^*, x) = \mathsf{G.Ev}(1^\lambda, gk, x)$ for all $\lambda \in \mathbb{N}$, $gk \in [\mathsf{G.Kg}(1^\lambda)]$, $x^* \in \mathsf{G.In}(\lambda)$, $gk^* \in [\mathsf{G.PKg}(1^\lambda, gk, x^*)]$ and $x \in \mathsf{G.In}(\lambda) \setminus \{x^*\}$.

<u>One-Way Functions.</u> Consider game OW of Fig. 1 associated to a function family F and an adversary $\mathcal{F}$, where $\mathsf{F.In}(\lambda)$ is required to be finite for all $\lambda \in \mathbb{N}$. For $\lambda \in \mathbb{N}$ let $\mathsf{Adv}^{\mathsf{ow}}_{\mathsf{F}, \mathcal{F}}(\lambda) = \Pr[\mathsf{OW}^{\mathcal{F}}_{\mathsf{F}}(\lambda)]$. Let $\delta\colon \mathbb{N} \to \mathbb{R}$ be any function. We say that F is δ-OW-secure if for every PT adversary $\mathcal{F}$ there exists $\lambda_{\delta, \mathcal{F}} \in \mathbb{N}$ such that $\mathsf{Adv}^{\mathsf{ow}}_{\mathsf{F}, \mathcal{F}}(\lambda) \leq \delta(\lambda)$ for all $\lambda \geq \lambda_{\delta, \mathcal{F}}$. We say that F is sub-exponentially OW-secure if it is $2^{-(\cdot)^\epsilon}$-OW-secure for some $0 < \epsilon < 1$.

<u>Target Collision-Resistant Functions.</u> Consider game TCR of Fig. 1 associated to a function family H and an adversary $\mathcal{H}$. For $\lambda \in \mathbb{N}$ let $\mathsf{Adv}^{\mathsf{tcr}}_{\mathsf{H}, \mathcal{H}}(\lambda) = \Pr[\mathsf{TCR}^{\mathcal{H}}_{\mathsf{H}}(\lambda)]$. Let $\delta\colon \mathbb{N} \to \mathbb{R}$ be any function. We say that H is δ-TCR-secure if for every PT adversary $\mathcal{H}$ there exists $\lambda_{\delta, \mathcal{H}} \in \mathbb{N}$ such that $\mathsf{Adv}^{\mathsf{tcr}}_{\mathsf{H}, \mathcal{H}}(\lambda) \leq \delta(\lambda)$ for all $\lambda \geq \lambda_{\delta, \mathcal{H}}$. We say that H is sub-exponentially TCR-secure if it is $2^{-(\cdot)^\epsilon}$-TCR-secure for some $0 < \epsilon < 1$. Target collision-resistant hash functions were introduced by Naor and Yung [42] under the name of Universal One-Way Hash Functions (UOWHF). [10] redefined the corresponding security notion under the name of *target collision-resistance*.

TCR-secure function families can be built from one-way functions, by combining the following results. First, [37,44] (see also [39]) proposed constructions of TCR-secure *compression function* families with fixed input and output lengths. More formally, they show how to build a function family H' such that $H'.\text{In}(\cdot) = \{0,1\}^{p_{in}(\cdot)}$ and $H'.\text{Out}(\cdot) = \{0,1\}^{p_{out}(\cdot)}$, where p_{in}, p_{out} are some polynomials such that $p_{in}(\lambda) \geq p_{out}(\lambda)$ for all $\lambda \in \mathbb{N}$. Next, [10,46] showed how to use any TCR-secure compression function H' with fixed input length in order to build another TCR-secure function family H for arbitrary, bounded variable-length inputs, meaning that $H.\text{In}(\lambda) = \bigcup_{i \leq p(\lambda)} \{0,1\}^i$ and $H.\text{Out}(\lambda) = H'.\text{Out}(\lambda)$ for some function $p\colon \mathbb{N} \to \mathbb{N}$ all $\lambda \in \mathbb{N}$.

Puncturable PRFs. Consider game PPRF of Fig. 1 associated to a puncturable function family G and an adversary $\mathcal{G}$, where $G.\text{Out}(\lambda)$ is required to be finite for all $\lambda \in \mathbb{N}$ and $\mathcal{G}$ is required to make exactly one oracle query to CH. For $\lambda \in \mathbb{N}$ let $\text{Adv}^{pprf}_{G,\mathcal{G}}(\lambda) = 2\Pr[\text{PPRF}^{\mathcal{G}}_G(\lambda)] - 1$. Let $\delta\colon \mathbb{N} \to \mathbb{R}$ be any function. We say that G is a δ-PPRF-secure if for every PT adversary $\mathcal{G}$ there exists $\lambda_{\delta,\mathcal{G}} \in \mathbb{N}$ such that $\text{Adv}^{pprf}_{G,\mathcal{G}}(\lambda) \leq \delta(\lambda)$ for all $\lambda \geq \lambda_{\delta,\mathcal{G}}$. We say that G is sub-exponentially PPRF-secure if it is $2^{-(\cdot)^\epsilon}$-PPRF-secure for some $0 < \epsilon < 1$. Puncturable PRFs were concurrently and independently introduced in [19,21,40]. They can be built by extending the standard PRF construction of Goldreich, Goldwasser and Micali [32].

Digital Signature Schemes. A digital signature scheme DS defines PT algorithms DS.Kg, DS.Sig, DS.Ver, where DS.Ver is deterministic. Associated to DS is a collection of input sets DS.In and a collection of output sets DS.Out, defining all valid messages and signatures for each of security parameters. Key generation algorithm DS.Kg takes 1^λ to return a signing key sk and a verification key vk. Signing algorithm DS.Sig takes $1^\lambda, sk$ and a message $m \in DS.\text{In}(\lambda)$ to return a signature $\sigma \in DS.\text{Out}(\lambda)$. Verification algorithm DS.Ver takes $1^\lambda, vk, m, \sigma$ to return a decision $d \in \{1,0\}$ regarding whether σ is a valid signature of m under vk, where 1 is returned if σ is a valid and 0 otherwise. The correctness condition requires that $DS.\text{Ver}(1^\lambda, vk, m, \sigma) = 1$ for all $\lambda \in \mathbb{N}$, $(sk, vk) \in [DS.\text{Kg}(1^\lambda)]$, $m \in DS.\text{In}(\lambda)$ and $\sigma \in [DS.\text{Sig}(1^\lambda, sk, m)]$. We say that a digital signature scheme DS is deterministic if its signing algorithm DS.Sig is deterministic.

Obfuscators. An obfuscator is a PT algorithm Obf that on input 1^λ and a program P returns a program $\overline{P}$ of the same type as P such that $\overline{P} \equiv P$. We say that

Game $\text{DIFF}^{\mathcal{D}}_S(\lambda)$	Game $\text{IO}^{\mathcal{O}}_{\text{Obf},S}(\lambda)$
$(P_0, P_1, aux) \leftarrow_\$ S(1^\lambda)$	$b \leftarrow_\$ \{0,1\}\,;\ (P_0, P_1, aux) \leftarrow_\$ S(1^\lambda)$
$x \leftarrow_\$ \mathcal{D}(1^\lambda, P_0, P_1, aux)$	$\overline{P} \leftarrow_\$ \text{Obf}(1^\lambda, P_b)$
Return $(P_0(x) \neq P_1(x))$	$b' \leftarrow_\$ \mathcal{O}(1^\lambda, \overline{P}, aux)\,;$ Return $(b = b')$

Fig. 2. Games defining difference-security of program sampler S and iO-security of program obfuscator Obf relative to program sampler S.

Obf is a circuit obfuscator if it obfuscates circuits, and we say that Obf is a TM obfuscator if it obfuscates TMs. Note that according to our definition of functionally equivalent programs, obfuscation is not defined for TMs that do not halt on some inputs. The polynomial slowdown condition requires that for every TM obfuscator Obf there is a polynomial $p\colon \mathbb{N} \times \mathbb{N} \to \mathbb{N}$ such that for every TM M that halts on all inputs and for every input $x \in \{0,1\}^*$, we have $\mathrm{time}(\overline{\mathrm{M}}, x) \le p(\lambda, \mathrm{time}(\mathrm{M}, x))$ for all $\lambda \in \mathbb{N}$ and $\overline{\mathrm{M}} \in [\mathrm{Obf}(1^\lambda, \mathrm{M})]$. An analogous slowdown condition trivially holds for any PT circuit obfuscator.

In this work, we discuss indistinguishabilty obfuscation (iO) and differing-inputs obfuscation (diO). The study of these obfuscation notions was initiated in [7]. Later [27,45] showed how to build and use the former, whereas [1,20] provided results on the latter. We extend the definitional framework of [12] that uses classes of program samplers to capture different variants of security notions for iO and diO. Specifically, our definitions allow for a unified treatment of polynomial and sub-exponential security of both circuit and TM obfuscation.

Program Samplers. A circuit sampler is a PT algorithm $\mathsf{S}^{\mathsf{circ}}$ that on input 1^λ returns a triple $(\mathrm{C}_0, \mathrm{C}_1, aux)$, where $\mathrm{C}_0, \mathrm{C}_1$ are circuits of the same size, number of inputs and number of outputs, and aux is a string. A TM sampler is a PT algorithm S^{tm} that on input 1^λ returns a triple $(\mathrm{M}_0, \mathrm{M}_1, aux)$, where $\mathrm{M}_0, \mathrm{M}_1$ are TMs of the same size, and aux is a string. We require that $\mathrm{M}_0(x)$ and $\mathrm{M}_1(x)$ halt for all $\lambda \in \mathbb{N}$, $(\mathrm{M}_0, \mathrm{M}_1, aux) \in [\mathsf{S}^{\mathsf{tm}}(1^\lambda)]$ and $x \in \{0,1\}^*$. We say that S is a program sampler if it is either a circuit sampler or a TM sampler.

Classes of Program Samplers. We say that a program sampler S produces functionally equivalent programs if $\Pr\left[\, \mathrm{P}_0 \equiv \mathrm{P}_1 \ : \ (\mathrm{P}_0, \mathrm{P}_1, aux) \leftarrow\!\!{}_\$ \, \mathsf{S}(1^\lambda)\,\right] = 1$ for all $\lambda \in \mathbb{N}$. Let $\boldsymbol{S}^{\mathsf{circ}}_{\mathsf{eq}}$ be the class of all circuit samplers that produce functionally equivalent circuits, and let $\boldsymbol{S}^{\mathsf{tm}}_{\mathsf{eq}}$ be the class of all TM samplers that produce functionally equivalent TMs. Consider game DIFF of Fig. 2 associated to a program sampler S and an adversary $\mathcal{D}$. For $\lambda \in \mathbb{N}$ let $\mathsf{Adv}^{\mathsf{diff}}_{\mathsf{S},\mathcal{D}}(\lambda) = \Pr[\mathrm{DIFF}^{\mathcal{D}}_{\mathsf{S}}(\lambda)]$. Let $\delta\colon \mathbb{N} \to \mathbb{R}$ be any function. We say that S is δ-DIFF-secure if for every PT adversary $\mathcal{D}$ there exists $\lambda_{\delta,\mathcal{D}} \in \mathbb{N}$ such that $\mathsf{Adv}^{\mathsf{diff}}_{\mathsf{S},\mathcal{D}}(\lambda) \le \delta(\lambda)$ for all $\lambda \ge \lambda_{\delta,\mathcal{D}}$. We say that S is sub-exponentially DIFF-secure if it is $2^{-(\cdot)^\epsilon}$-DIFF-secure for some $0 < \epsilon < 1$. Let $\boldsymbol{S}^{\mathsf{circ}}_{\delta\text{-diff}}$ be the class of all δ-DIFF-secure circuit samplers, and let $\boldsymbol{S}^{\mathsf{tm}}_{\delta\text{-diff}}$ be the class of all δ-DIFF-secure TM samplers. Informally, difference-security of a program sampler S means that given its output $(\mathrm{P}_0, \mathrm{P}_1, aux)$, it is hard to find an input on which the programs P_0 and P_1 differ.

Indistinguishability Obfuscation and Differing-Inputs Obfuscation. Consider game IO of Fig. 2 associated to an obfuscator Obf, a program sampler S and an adversary $\mathcal{O}$. For $\lambda \in \mathbb{N}$ let $\mathsf{Adv}^{\mathsf{io}}_{\mathsf{Obf},\mathsf{S},\mathcal{O}}(\lambda) = 2\Pr[\mathrm{IO}^{\mathcal{O}}_{\mathsf{Obf},\mathsf{S}}(\lambda)] - 1$. Let $\delta\colon \mathbb{N} \to \mathbb{R}$ be any function. Let $\boldsymbol{S}$ be a class of program samplers. We say that Obf is δ-$\boldsymbol{S}$-secure if for every program sampler $\mathsf{S} \in \boldsymbol{S}$ and for every PT adversary $\mathcal{O}$ there exists $\lambda_{\delta,\mathsf{S},\mathcal{O}} \in \mathbb{N}$ such that $\mathsf{Adv}^{\mathsf{io}}_{\mathsf{Obf},\mathsf{S},\mathcal{O}}(\lambda) \le \delta(\lambda)$ for all $\lambda \ge \lambda_{\delta,\mathsf{S},\mathcal{O}}$. We say that Obf is sub-exponentially $\boldsymbol{S}$-secure if it is $2^{-(\cdot)^\epsilon}$-$\boldsymbol{S}$-secure for some $0 < \epsilon < 1$.

We say that Obf is a sub-exponentially secure indistinguishability obfuscator for TMs (resp. circuits) if there exists $0 < \epsilon < 1$ such that Obf is $2^{-(\cdot)^\epsilon}$-$\boldsymbol{S}^{\mathsf{tm}}_{\mathsf{eq}}$-

secure (resp. $2^{-(\cdot)^\epsilon}$-S_{eq}^{circ}-secure). We say that Obf is a differing-inputs obfuscator for TMs (resp. circuits) if for every negligible function $\gamma \colon \mathbb{N} \to \mathbb{R}$ there exists a negligible function $\nu \colon \mathbb{N} \to \mathbb{R}$ such that Obf is ν-$S_{\gamma\text{-diff}}^{tm}$-secure (resp. ν-$S_{\gamma\text{-diff}}^{circ}$-secure). Note that ν-$S_{\gamma\text{-diff}}^{tm}$-security may be unachievable if there exists an infinite number of security parameters $\lambda \in \mathbb{N}$ such that $\gamma(\lambda) > \nu(\lambda)$. We say that Obf is a sub-exponentially secure differing-inputs obfuscator for TMs (resp. circuits) if for every $0 < \epsilon_0 < 1$ and $\gamma = 2^{-(\cdot)^{\epsilon_0}}$ there exists $0 < \epsilon_1 < 1$ such that Obf is $2^{-(\cdot)^{\epsilon_1}}$-$S_{\gamma\text{-diff}}^{tm}$-secure (resp. $2^{-(\cdot)^{\epsilon_1}}$-$S_{\gamma\text{-diff}}^{circ}$-secure).

Note that according to our definitions, a sub-exponentially secure differing-inputs obfuscator is not necessarily a polynomially-secure differing-inputs obfuscator. Namely, the former guarantees no security with respect to δ-DIFF-secure program samplers when δ is negligible but not sub-exponentially small. This observation can be used to strengthen our definition of sub-exponentially secure diO. We chose to use the weaker definition, which is simpler to define and which makes our impossibility results stronger.

3 Consistent Puncturable Digital Signature Schemes

We start by defining *consistent puncturable digital signature schemes* that will be used for our impossibility results in Sect. 4. Our construction follows Sahai-Waters signatures [45], and we prove its security assuming OWF and iO.

Informally, a puncturable digital signature scheme allows to 'puncture' its signing key sk at an arbitrary message m^*. The resulting punctured secret key sk^*, punctured at m^*, allows to produce signatures for all messages except for m^*. The puncturability property is similar to the one of puncturable PRFs. We say that a puncturable digital signature scheme is *consistent* if its secret signing key sk and every possible punctured signing key sk^*, that can be derived from sk, deterministically produce the same signatures for all messages except for the punctured message.

We now define a security notion, informally, requiring that no PT adversary should be able to forge a valid signature for the punctured message. The natural formalization of this security notion requires *selective* unforgeability, meaning that an adversary has to choose a message m^* at which the original signing key sk should be punctured. Having received the corresponding pair of punctured signing key sk^* and verification key vk, the goal of the adversary is to produce a valid signature for m^* with respect to the verification key.

Puncturable Digital Signature Schemes. A *puncturable* digital signature scheme DS specifies (beyond the algorithms associated to digital signatures schemes) additional PT algorithms DS.PKg, DS.PSig, where DS.PSig is deterministic. Punctured key generation algorithm DS.PKg takes 1^λ, a signing key $sk \in$ [DS.Kg(1^λ)] and a message $m^* \in$ DS.In(λ) to return a "punctured" signing key sk^*. Punctured signing algorithm DS.PSig takes $1^\lambda, sk^*$ and a message $m \in$ DS.In(λ) to return a signature $\sigma \in$ DS.Out(λ). We say that puncturable digital signature scheme DS is *consistent* if DS.Sig$(1^\lambda, sk, m) =$ DS.PSig$(1^\lambda, sk^*, m)$ for all $\lambda \in \mathbb{N}$, $(sk, vk) \in$ [DS.Kg(1^λ)], $m^* \in$ DS.In(λ), $sk^* \in$ [DS.PKg$(1^\lambda, sk, m^*)$]

$$\boxed{\begin{array}{l} \underline{\text{Game PSUFCMA}_{\mathsf{DS}}^{\mathcal{U}}(\lambda)} \\[4pt] (m^*, st) \leftarrow\!\!\$\, \mathcal{U}_1(1^\lambda) \;;\; (sk, vk) \leftarrow\!\!\$\, \mathsf{DS.Kg}(1^\lambda) \\[4pt] sk^* \leftarrow\!\!\$\, \mathsf{DS.PKg}(1^\lambda, sk, m^*) \;;\; \sigma^* \leftarrow\!\!\$\, \mathcal{U}_2(1^\lambda, st, vk, sk^*) \\[4pt] d \leftarrow \mathsf{DS.Ver}(1^\lambda, vk, m^*, \sigma^*) \;;\; \text{Return } (d = 1) \end{array}}$$

Fig. 3. Game defining selective unforgeability of puncturable digital signature scheme DS under chosen message attack.

and $m \in \mathsf{DS.In}(\lambda) \backslash \{m^*\}$. Note that DS can be consistent only if it is deterministic. More precisely, both DS.Sig and DS.PSig should be deterministic. However, determinism is a necessary but not a sufficient condition.

Punctured Selective Unforgeability Under Chosen Message Attack. Consider game PSUFCMA of Fig. 3 associated to a puncturable digital signature scheme DS and an adversary $\mathcal{U}$. For $\lambda \in \mathbb{N}$ let $\mathsf{Adv}_{\mathsf{DS},\mathcal{U}}^{\mathsf{psufcma}}(\lambda) = \Pr[\mathsf{PSUFCMA}_{\mathsf{DS}}^{\mathcal{U}}(\lambda)]$. Let $\delta \colon \mathbb{N} \to \mathbb{R}$ be any function. We say that DS is δ-PSUFCMA-secure if for every PT adversary $\mathcal{U}$ there exists $\lambda_{\delta,\mathcal{U}} \in \mathbb{N}$ such that $\mathsf{Adv}_{\mathsf{DS},\mathcal{U}}^{\mathsf{psufcma}}(\lambda) \leq \delta(\lambda)$ for all $\lambda \geq \lambda_{\delta,\mathcal{U}}$. We say that DS is sub-exponentially PSUFCMA-secure if it is $2^{-(\cdot)^\epsilon}$-PSUFCMA-secure for some $0 < \epsilon < 1$.

Our Construction. We build a consistent puncturable digital signature scheme DS from a PPRF G, an indistinguishability obfuscator Obf and a OWF F. Our main observation is that a PPRF key gk can be used as a secret key for DS. In order to obtain a punctured key for DS, we puncture gk accordingly. The correctness condition of puncturable PRFs guarantees that DS is consistent. We build a verification key by obfuscating a circuit that embeds the PPRF key gk and a OWF key fk. The circuit takes a message-signature pair (m, σ) and returns 1 if $\mathsf{F.Ev}(1^\lambda, fk, \sigma) = \mathsf{F.Ev}(1^\lambda, fk, \mathsf{G.Ev}(1^\lambda, gk, m))$; it returns 0 otherwise.

Puncturable Digital Signature Scheme PUNC-DS. Let $s \colon \mathbb{N} \to \mathbb{N}$ be a polynomial. Let G be a puncturable function family. Let F be a function family such that $\mathsf{F.In} = \mathsf{G.Out}$. Let Obf be a circuit obfuscator. We build a consistent puncturable digital signature scheme $\mathsf{DS} = \mathsf{PUNC\text{-}DS}[\mathsf{G}, \mathsf{F}, \mathsf{Obf}, s]$ as follows. Let $\mathsf{DS.In}(\lambda) = \mathsf{G.In}(\lambda)$ and $\mathsf{DS.Out}(\lambda) = \mathsf{G.Out}(\lambda)$ for all $\lambda \in \mathbb{N}$, and let Fig. 4 define the puncturable digital signature scheme DS. We say that DS is *well-defined* if $s(\lambda) \geq |C_{1^\lambda, gk, fk}|$ for all $\lambda \in \mathbb{N}$, $gk \in [\mathsf{G.Kg}(1^\lambda)]$ and $fk \in [\mathsf{F.Kg}(1^\lambda)]$.

The following says that a PSUFCMA-secure, consistent punctured digital signature scheme can be built assuming OWF and iO.

Theorem 1. *Let* G *be a sub-exponentially PPRF-secure function family such that* $\mathsf{G.In}(\lambda), \mathsf{G.Out}(\lambda) \subseteq \bigcup_{i \leq p_0(\lambda)} \{0,1\}^i$ *for some polynomial* p_0 *and all* $\lambda \in \mathbb{N}$. *Let* F *be a sub-exponentially OW-secure function family such that* $\mathsf{F.In} = \mathsf{G.Out}$ *and* $\mathsf{F.Out}(\lambda) \subseteq \bigcup_{i \leq p_1(\lambda)} \{0,1\}^i$ *for some polynomial* p_1 *and all* $\lambda \in \mathbb{N}$. *Let* Obf *be a sub-exponentially* $\mathbf{S}_{\mathsf{eq}}^{\mathsf{circ}}$-*secure circuit obfuscator. Then there is a polynomial*

Algorithm DS.Kg(1^λ)	Algorithm DS.PKg($1^\lambda, gk, m^*$)
$gk \leftarrow\!\!\$ \, G.Kg(1^\lambda)$; $fk \leftarrow\!\!\$ \, F.Kg(1^\lambda)$	Return G.PKg($1^\lambda, gk, m^*$)
$\overline{C} \leftarrow\!\!\$ \, Obf(1^\lambda, Pad_{s(\lambda)}(C_{1^\lambda, gk, fk}))$	Algorithm DS.Ver($1^\lambda, \overline{C}, m, \sigma$)
Return $(gk, \overline{C})$	
	Return $\overline{C}(m, \sigma)$
Circuit $C_{1^\lambda, gk, fk}(m, \sigma)$	
	Algorithm DS.Sig($1^\lambda, gk, m$)
$\sigma' \leftarrow G.Ev(1^\lambda, gk, m)$	
$y' \leftarrow F.Ev(1^\lambda, fk, \sigma')$	Return G.Ev($1^\lambda, gk, m$)
If $(y' = F.Ev(1^\lambda, fk, \sigma))$ then return 1	Algorithm DS.PSig($1^\lambda, gk^*, m$)
Else return 0	Return G.PEv($1^\lambda, gk^*, m$)

Fig. 4. Puncturable digital signature scheme $DS = PUNC\text{-}DS[G, F, Obf, s]$.

$s \colon \mathbb{N} \to \mathbb{N}$ *such that the following is true. Let* $DS = PUNC\text{-}DS[G, F, Obf, s]$. *Then (1)* DS *is well-defined, and (2)* DS *is sub-exponentially* PSUFCMA-*secure.*

In order to prove that DS is PSUFCMA-secure, we show that an adversary can not find the value of $G.Ev(1^\lambda, gk, m^*)$ for a challenge message m^*, even given the obfuscated verification-key circuit that contains gk. In the proof, we puncture gk at m^* to get a punctured key gk^*, and construct a functionally equivalent verification-key circuit that embeds gk^* along with $y^* = F.Ev(1^\lambda, fk, G.Ev(1^\lambda, gk, m^*))$. The new verification key accepts σ as a valid signature for m^* if and only if $y^* = F.Ev(1^\lambda, fk, \sigma)$, whereas the verification of signatures for all other messages $m \neq m^*$ remains the same. First, we use the iO-security of Obf to switch the verification circuits. Then we use the PPRF-security of G, followed by the OWF-security of F to show that no adversary can find the value of $G.Ev(1^\lambda, gk, m^*)$ from gk^* and y^*. The proof is given in [13].

4 Impossibility of Differing-Inputs Obfuscation for TMs

In this section we show that differing-inputs obfuscation for Turing Machines is impossible. In order to disprove sub-exponentially secure diO for TMs, we assume only the existence of sub-exponentially secure one-way functions. Furthermore, we show that polynomially secure diO for TMs is also impossible, additionally assuming sub-exponentially secure iO.

We construct a sub-exponentially difference-secure TM sampler, meaning that given a pair of TMs produced by this sampler it is hard to find an input on which these TMs produce different outputs. The proof of difference-security is the core part of our work. It requires to carefully specify how to choose parameters for our sampler in a way that does not introduce any circular dependencies. Besides proving difference-security, we also show that there exists an adversary that can distinguish between obfuscations of TMs that are produced by the sampler regardless of the used obfuscator. Together these claims imply the impossibility of diO for TMs.

The Blueprint for Impossibility Results. The first attack on differing-inputs obfuscation was presented by Garg, Gentry, Halevi and Wichs (GGHW) [28]. They introduced a novel *special-purpose obfuscation* assumption and showed that it contradicts diO. Our impossibility result follows the high-level idea from their work, but we achieve it using concrete assumptions. We now explain the core ideas of our impossibility result, which closely follow those of GGHW.

We construct a TM sampler S^{tm} that returns TMs M^0, M^1 along with an auxiliary information string aux. The sampler generates a key pair (sk, vk) for a digital signature scheme DS, and its output depends on these keys. TM M^0 returns 0 on every input. TM M^1 returns 1 if and only if it gets a valid message-signature pair as input, corresponding to the verification key vk; it returns 0 otherwise. The auxiliary information string aux is an iO-obfuscation of a TM M^{aux}. The latter embeds the signing key sk and takes a TM $\overline{M}$ as input, which for our purpose will normally be a diO-obfuscation of M^0 or M^1. M^{aux} returns the result of running $\overline{M}$ on a message-signature pair that is produced using its embedded signing key sk.

In order to determine whether a TM $\overline{M}$ is an obfuscation of M^0 or M^1, one can run M^{aux} with $\overline{M}$ as input. According to the construction of M^{aux}, it will return $b \in \{0, 1\}$ if and only if $\overline{M}$ is an obfuscation of M^b. To prove difference-security of S^{tm}, we will show that it is hard to find a valid message-signature pair given (M^0, M^1, aux). The main technical challenge of the proof is to show that aux (the obfuscation of M^{aux}) properly hides the embedded signing key sk, which does not naturally follow from the security of indistinguishability obfuscation.

Turing Machine Sampler TM-SAMP. Let $s_0, \ell, n, t_0, t_1, s_1 \colon \mathbb{N} \to \mathbb{N}$ be polynomials. Let $\mathsf{Obf}^{tm}_{eq}, \mathsf{Obf}^{tm}_{diff}$ be TM obfuscators. Let H be a function family such that $\mathsf{H.In}(\lambda) = \{0, 1\}^*$ and $\mathsf{H.Out}(\lambda) \subseteq \bigcup_{i \le p_0(\lambda)} \{0, 1\}^i$ for some polynomial p_0 and all $\lambda \in \mathbb{N}$. Let DS be a deterministic digital signature scheme such that $\mathsf{DS.In}(\lambda) =$

TM Sampler $S^{tm}(1^\lambda)$

$(sk, vk) \leftarrow\!\!{\scriptstyle\$}\ \mathsf{DS.Kg}(1^{n(\lambda)})$
$hk \leftarrow\!\!{\scriptstyle\$}\ \mathsf{H.Kg}(1^\lambda)$
$h \leftarrow \mathsf{H.Ev}(1^\lambda, hk, M^{ver}_{1^\lambda, vk})$
$M_0 \leftarrow \mathsf{Pad}_{s_0(\lambda)}(M^0)$
$M_1 \leftarrow \mathsf{Pad}_{s_0(\lambda)}(M^1_{1^\lambda, hk, h})$
$M_{aux} \leftarrow \mathsf{Pad}_{s_1(\lambda)}(M^{aux}_{1^\lambda, sk, vk})$
$aux \leftarrow\!\!{\scriptstyle\$}\ \mathsf{Obf}^{tm}_{eq}(1^{n(\lambda)}, M_{aux})$
Return (M_0, M_1, aux)

TM $M^{ver}_{1^\lambda, vk}(m, \sigma)$

If $(|m| \ne \ell(\lambda))$ then return 0
Return $\mathsf{DS.Ver}(1^{n(\lambda)}, vk, \langle m \rangle_{\ell(n(\lambda))}, \sigma)$

TM $M^0(M, 1^t, m, \sigma)$

Return 0

TM $M^1_{1^\lambda, hk, h}(M, 1^t, m, \sigma)$

$h' \leftarrow \mathsf{H.Ev}(1^\lambda, hk, M)$
If $(h' \ne h)$ then return 0
Return $\mathsf{UTM}^t_M(m, \sigma)$

TM $M^{aux}_{1^\lambda, sk, vk}(\overline{M})$

If $(|\overline{M}| \ne \ell(\lambda))$ then return 0
$\sigma \leftarrow \mathsf{DS.Sig}(1^{n(\lambda)}, sk, \langle \overline{M} \rangle_{\ell(n(\lambda))})$
$d \leftarrow \mathsf{UTM}^{t_1(\lambda)}_{\overline{M}}(M^{ver}_{1^\lambda, vk}, 1^{t_0(\lambda)}, \overline{M}, \sigma)$
Return d

Fig. 5. TM sampler $S^{tm} = \mathsf{TM\text{-}SAMP}[\mathsf{Obf}^{tm}_{diff}, \mathsf{H}, \mathsf{DS}, \mathsf{Obf}^{tm}_{eq}, s_0, \ell, n, t_0, t_1, s_1]$.

$\{0,1\}^{\ell(\lambda)}$ and $\mathsf{DS.Out}(\lambda) \subseteq \bigcup_{i \leq p_1(\lambda)} \{0,1\}^i$ for some polynomial p_1 and all $\lambda \in \mathbb{N}$. We build a TM sampler $\mathsf{S}^{\mathsf{tm}} = \mathsf{TM\text{-}SAMP}[\mathsf{Obf}_{\mathsf{diff}}^{\mathsf{tm}}, \mathsf{H}, \mathsf{DS}, \mathsf{Obf}_{\mathsf{eq}}^{\mathsf{tm}}, s_0, \ell, n, t_0, t_1, s_1]$ as defined in Fig. 5. We say that S^{tm} is *well-defined* if $s_0(\lambda) \geq |\mathsf{M}^0|$, $s_0(\lambda) \geq |\mathsf{M}^1_{1^\lambda, hk, h}|$, $\ell(n(\lambda)) \geq \ell(\lambda)$, $t_0(\lambda) \geq \mathsf{time}(\mathsf{M}^{\mathsf{ver}}_{1^\lambda, vk}, (m, \sigma))$, $t_1(\lambda) \geq \mathsf{time}(\overline{\mathsf{M}}, (\mathsf{M}^{\mathsf{ver}}_{1^\lambda, vk}, 1^{t_0(\lambda)}, \overline{\mathsf{M}}, \sigma))$ and $s_1(\lambda) \geq |\mathsf{M}^{\mathsf{aux}}_{1^\lambda, sk, vk}|$ for all $\lambda \in \mathbb{N}$, $hk \in [\mathsf{H.Kg}(1^\lambda)]$, $h \in \mathsf{H.Out}(\lambda)$, $\mathsf{M} \in \{\mathsf{M}^0, \mathsf{M}^1_{1^\lambda, hk, h}\}$, $\overline{\mathsf{M}} \in [\mathsf{Obf}_{\mathsf{diff}}^{\mathsf{tm}}(1^\lambda, \mathsf{Pad}_{s_0(\lambda)}(\mathsf{M}))]$, $(sk, vk) \in [\mathsf{DS.Kg}(1^{n(\lambda)})]$, $m \in \{0,1\}^{\ell(\lambda)}$ and $\sigma \in \mathsf{DS.Out}(n(\lambda))$.

Core Design Ideas Behind TM-SAMP. Note that TM $\mathsf{M}^{\mathsf{aux}}_{1^\lambda, sk, vk}$ takes as input an obfuscated TM $\overline{\mathsf{M}}$ and computes the signature σ for message $\langle \overline{\mathsf{M}} \rangle_{\ell(n(\lambda))}$, where the latter denotes $\overline{\mathsf{M}}$ padded to size $\ell(n(\lambda))$. It then uses a Universal Turing Machine UTM to simulate $\overline{\mathsf{M}}$ on input x for the duration of $t_1(\lambda)$ steps, where $x = (\mathsf{M}^{\mathsf{ver}}_{1^\lambda, vk}, 1^{t_0(\lambda)}, \overline{\mathsf{M}}, \sigma)$. The idea of computing a signature on a message that depends on $\overline{\mathsf{M}}$ was already proposed in GGHW [28], with the goal of avoiding a trivial attack against the difference-security of the sampler. Specifically, if a fixed message-signature pair $(m_{\mathsf{ch}}, \sigma_{\mathsf{ch}})$ was used for all inputs of $\mathsf{M}^{\mathsf{aux}}_{1^\lambda, sk, vk}$, then a difference-security adversary could construct a sequence of TMs that each reveals a single bit of $(m_{\mathsf{ch}}, \sigma_{\mathsf{ch}})$ when used as an input $\overline{\mathsf{M}}$ to $\mathsf{M}^{\mathsf{aux}}_{1^\lambda, sk, vk}$. This would allow adversary to recover the message-signature pair bit-by-bit.

Turing Machine $\mathsf{M}^1_{1^\lambda, hk, h}$ takes an input $x = (\mathsf{M}, 1^t, m, \sigma)$, where M is a TM, 1^t is the unary representation of some integer $t \in \mathbb{N}$, and (m, σ) is a message-signature pair. We use a target collision-resistant function family H in order to ensure that $\mathsf{M}^1_{1^\lambda, hk, h}$ can return 1 only if $\mathsf{M} = \mathsf{M}^{\mathsf{ver}}_{1^\lambda, vk}$. This is achieved by embedding a key hk for H and the value $h = \mathsf{H.Ev}(1^\lambda, hk, \mathsf{M}^{\mathsf{ver}}_{1^\lambda, vk})$ into $\mathsf{M}^1_{1^\lambda, hk, h}$, and by returning 0 whenever $h \neq \mathsf{H.Ev}(1^\lambda, hk, \mathsf{M})$. If $\mathsf{M} = \mathsf{M}^{\mathsf{ver}}_{1^\lambda, vk}$ is satisfied, then $\mathsf{M}^1_{1^\lambda, hk, h}$ uses a Universal Turing Machine UTM to simulate M on input (m, σ) for the duration of t steps. TM $\mathsf{M}^{\mathsf{ver}}_{1^\lambda, vk}$ is designed to return 1 if and only if its input $x = (m, \sigma)$ is a valid message-signature pair with respect to a verification key vk for the digital signature scheme DS. Our impossibility results require the choice of DS to depend on the construction of $\mathsf{M}^1_{1^\lambda, hk, h}$, so embedding vk directly into the latter would have introduced a circular dependency between the two. Instead we have to resort to the above approach of embedding vk into a separate TM.

According to our definitions, two TMs can be functionally equivalent only if both of them halt on all inputs. The notion of functional equivalence is further used for the definitions of program samplers and obfuscation. This means that whenever a TM needs to simulate the code of another TM, it is required to use a Universal Turing Machine UTM and specify the number of steps for the simulation. Otherwise, the simulated TMs would not be guaranteed to halt.

Parameters of TM-SAMP. Figure 6 shows the dependencies between all schemes and parameters that will be used to instantiate the construction of TM-SAMP in Theorem 2. Let us introduce the notation that is used in this picture. For any two entities A and B, an arrow from A to B means that the construction, or the choice, of B depends on A. The relations are transitive, meaning that we do not

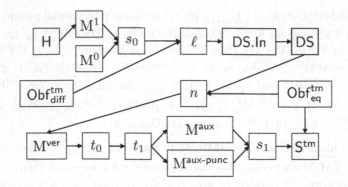

Fig. 6. Parameter dependencies in TM-SAMP for the proof of Theorem 2.

draw a direct arrow from A to B in the case if B is already reachable from A. TM $M^{aux\text{-}punc}$ will be used only for the proof of security and is defined in Fig. 7.

The construction of TM-SAMP is parameterized by polynomials s_0, s_1, t_0, t_1, ℓ and n. Polynomials s_0, s_1 denote the size to which some of our TMs must be padded prior to obfuscating them. This stems from our definition of program samplers that are required to return programs of the same size. Polynomials t_0, t_1 are used to indicate the number of steps that must be done when simulating various TMs using a Universal Turing Machine UTM. Our definition of a well-defined instantiation of TM-SAMP specifies lower bounds for t_0, t_1 that ensure the correctness of the attack that we will design against the sub-exponential (d)iO-security of Obf_{diff}^{tm} with respect to S^{tm}. Polynomial ℓ will be defined to upper-bound the size of any obfuscation $\overline{M}$ of programs M^0 and $M^1_{1^\lambda, hk, h}$, when obfuscator Obf_{diff}^{tm} is used. Note that $M^{aux}_{1^\lambda, sk, vk}$ rejects all inputs $\overline{M}$ of size different than $\ell(\lambda)$; our attack will pad all obfuscations of M^0 and $M^1_{1^\lambda, hk, h}$ to size $\ell(\lambda)$, using the padding operator $Pad_{\ell(\lambda)}(\cdot)$ that is assumed to produce functionally equivalent TMs as per Sect. 2. Polynomial n is used to set security parameters for schemes DS and Obf_{eq}^{tm}. Specifically, if the TM sampler S^{tm} is instantiated with a security parameter $\lambda \in \mathbb{N}$, then its construction uses these two schemes, each with the security parameter $n(\lambda)$.

In order for our proof of difference-security to work, if a $2^{-(\cdot)^\epsilon}$-security is assumed for either of DS or Obf_{eq}^{tm}, then the choice of polynomial n will depend on ϵ. This leads to an inconvenient dependency: DS uses $n(\lambda)$ as its security parameter, but the choice of polynomial n depends on the choice of DS. Ideally, we would have liked to choose a digital signature scheme DS such that $DS.Out(n(\lambda)) = \{0,1\}^{\ell(\lambda)}$, because DS is used to sign messages that are TMs of size $\ell(\lambda)$. However, since we do not know n ahead of choosing DS, we require that for all $\lambda \in \mathbb{N}$ we have $DS.Out(\lambda) = \{0,1\}^{\ell(\lambda)}$ and $\ell(n(\lambda)) \geq \ell(\lambda)$, resulting in $DS.Out(n(\lambda)) = \{0,1\}^{\ell(n(\lambda))}$. We then use an injective string padding to map TMs (i.e. their string representations) of length $\ell(\lambda)$ into strings of length $\ell(n(\lambda))$. The injectivity of padding is necessary for the proof of difference-security of S^{tm}. In order to ensure that the requirement $\ell(n(\lambda)) \geq \ell(\lambda)$ is satisfied, we will choose polynomials ℓ, n such that $\ell(\lambda+1) \geq \ell(\lambda)$ and $n(\lambda) \geq \lambda$ for all $\lambda \in \mathbb{N}$.

Limitations and Extensions. Our definition of TM samplers in Sect. 2 requires them to return TMs that halt on all inputs. One could argue that this definition is still insufficient for the purpose of obfuscation. Namely, a sampler can produce TMs that have significantly different running times, and it might not be reasonable to expect an obfuscator to properly hide the difference in the running times. We note that this does not hinder our results because we can artificially alter our TMs M^0 and $M^1_{1^\lambda,hk,h}$ to have the same running times, by adding void instructions to the definition of M^0.

The construction of TM-SAMP uses a TM obfuscator Obf^{tm}_{eq} that in our theorem statements will be assumed to be sub-exponentially S^{tm}_{eq}-secure. It is used to produce auxiliary information by obfuscating TMs $M^{aux}_{1^\lambda,sk,vk}$ and $M^{aux-punc}_{1^\lambda,sk^*,vk,m',b}$. We use a TM obfuscator for readability, but we note that a sub-exponentially S^{circ}_{eq}-secure *circuit* obfuscator could be used instead. There are no circular dependencies preventing us from redefining these two TMs as circuits.

According to Fig. 6, the size of M^{aux} depends on the maximum size of TMs M^0 and $M^1_{1^\lambda,hk,h}$, and in particular it might be larger than these TMs. This means that our impossibility result might not hold if we restrict our attention to TM samplers whose auxiliary information strings aux are required to be shorter than the size of the corresponding TMs M^0 and M^1. GGHW [28] circumvent this limitation in their impossibility result by using a CRHF to compute and then sign a hash of the TM that is passed inside their auxiliary-information program, rather than signing the TM itself. Our proof techniques do not seem to be compatible with such approach.

Impossibility Results. We now formally state our results. Theorem 2 shows how to choose parameters for TM-SAMP such that the resulting TM sampler is simultaneously well-defined and difference-secure. Theorem 3 shows that any well-defined instantiation of TM-SAMP produces TMs that can not be securely obfuscated.

Theorem 2. *Let* Obf^{tm}_{diff} *be a TM obfuscator. Let* H *be a sub-exponentially TCR-secure function family such that* $\mathsf{H.In}(\lambda) = \{0,1\}^*$ *and* $\mathsf{H.Out}(\lambda) \subseteq \bigcup_{i \leq p_0(\lambda)} \{0,1\}^i$ *for some polynomial* p_0 *and all* $\lambda \in \mathbb{N}$. *Then there are polynomials* $s_0, \ell \colon \mathbb{N} \to \mathbb{N}$ *such that the following is true. Let* DS *be a sub-exponentially PSUFCMA-secure, consistent puncturable digital signature scheme such that* $\mathsf{DS.In}(\lambda) = \{0,1\}^{\ell(\lambda)}$ *and* $\mathsf{DS.Out}(\lambda) \subseteq \bigcup_{i \leq p_1(\lambda)} \{0,1\}^i$ *for some polynomial* p_1 *and all* $\lambda \in \mathbb{N}$. *Let* Obf^{tm}_{eq} *be a sub-exponentially* S^{tm}_{eq}-*secure TM obfuscator. Then there are polynomials* $n, t_0, t_1, s_1 \colon \mathbb{N} \to \mathbb{N}$ *such that the following is true. Let* $S^{tm} = $ TM-SAMP $[\mathsf{Obf}^{tm}_{diff}, \mathsf{H}, \mathsf{DS}, \mathsf{Obf}^{tm}_{eq}, s_0, \ell, n, t_0, t_1, s_1]$. *Then (1)* S^{tm} *is well-defined, and (2)* S^{tm} *is sub-exponentially DIFF-secure.*

We defer the proof of Theorem 2 until after we show how to use this theorem to state and prove our main claims regarding the impossibility of differing-inputs obfuscation for TMs.

Theorem 3. *Let* $s_0, \ell, n, t_0, t_1, s_1 \colon \mathbb{N} \to \mathbb{N}$ *be polynomials. Let* $\mathsf{Obf}_{\mathsf{eq}}^{\mathsf{tm}}, \mathsf{Obf}_{\mathsf{diff}}^{\mathsf{tm}}$ *be TM obfuscators. Let* H *be a function family with* $\mathsf{H.In}(\lambda) = \{0,1\}^*$ *and* $\mathsf{H.Out}(\lambda) \subseteq \bigcup_{i \leq p_0(\lambda)} \{0,1\}^i$ *for some polynomial* p_0 *and all* $\lambda \in \mathbb{N}$. *Let* DS *be a deterministic digital signature scheme such that* $\mathsf{DS.In}(\lambda) = \{0,1\}^{\ell(\lambda)}$ *and* $\mathsf{DS.Out}(\lambda) \subseteq \bigcup_{i \leq p_1(\lambda)} \{0,1\}^i$ *for some polynomial* p_1 *and all* $\lambda \in \mathbb{N}$. *Let* $\mathsf{S}^{\mathsf{tm}} = \mathsf{TM\text{-}SAMP}\ [\mathsf{Obf}_{\mathsf{diff}}^{\mathsf{tm}}, \mathsf{H}, \mathsf{DS}, \mathsf{Obf}_{\mathsf{eq}}^{\mathsf{tm}}, s_0, \ell, n, t_0, t_1, s_1]$. *Assume that* S^{tm} *is well-defined. Then there exists a PT adversary* $\mathcal{O}$ *such that* $\mathsf{Adv}_{\mathsf{Obf}_{\mathsf{diff}}^{\mathsf{tm}}, \mathsf{S}^{\mathsf{tm}}, \mathcal{O}}^{\mathsf{io}}(\lambda) = 1$.

Proof (Theorem 3). We build a PT adversary $\mathcal{O}$ against the (d)iO-security of $\mathsf{Obf}_{\mathsf{diff}}^{\mathsf{tm}}$ relative to S^{tm} as follows:

$$\frac{\text{Adversary } \mathcal{O}(1^\lambda, \overline{\mathsf{M}}, aux)}{\overline{\mathsf{M}}_{\mathsf{aux}} \leftarrow aux;\ b' \leftarrow \overline{\mathsf{M}}_{\mathsf{aux}}(\mathsf{Pad}_{\ell(\lambda)}(\overline{\mathsf{M}}));\ \text{Return } b'}$$

Adversary $\mathcal{O}$ takes $1^\lambda, \overline{\mathsf{M}}, aux$ as input, where $\overline{\mathsf{M}}$ is an obfuscation of either TM M^0 or TM $\mathsf{M}^1_{1^\lambda, hk, h}$ that was produced by the obfuscator $\mathsf{Obf}_{\mathsf{diff}}^{\mathsf{tm}}$ in game $\mathsf{IO}_{\mathsf{Obf}_{\mathsf{diff}}^{\mathsf{tm}}, \mathsf{S}^{\mathsf{tm}}}^{\mathcal{O}}(\lambda)$, and aux is an auxiliary information string. The goal of $\mathcal{O}$ is to guess which of M^0 and $\mathsf{M}^1_{1^\lambda, hk, h}$ was obfuscated. It should return 0 if $\overline{\mathsf{M}}$ is an obfuscation of M^0, and it should return 1 otherwise.

Adversary $\mathcal{O}$ parses auxiliary information string aux into a TM $\overline{\mathsf{M}}_{\mathsf{aux}}$. The latter is an obfuscation of TM $\mathsf{M}^{\mathsf{aux}}_{1^\lambda, sk, vk}$, which was computed in S^{tm} using obfuscator $\mathsf{Obf}_{\mathsf{eq}}^{\mathsf{tm}}$. Next, $\mathcal{O}$ pads $\overline{\mathsf{M}}$ to construct a functionally equivalent TM of size $\ell(\lambda)$ and passes it as input to $\overline{\mathsf{M}}_{\mathsf{aux}}$. According to the construction of $\mathsf{M}^{\mathsf{aux}}_{1^\lambda, sk, vk}$, the latter returns 1 if and only if $\overline{\mathsf{M}}$ is an obfuscation of TM $\mathsf{M}^1_{1^\lambda, hk, h}$. Adversary $\mathcal{O}$ returns the same value to win the game. This concludes the proof of Theorem 3.

Next, Theorem 4 shows the impossibility of a polynomially secure diO, and Theorem 5 shows the impossibility of a sub-exponentially secure diO.

Theorem 4. *Let* Obf *be a Turing Machine obfuscator. Assume the existence of sub-exponentially secure one-way functions and sub-exponentially secure indistinguishability obfuscation for Turing Machines. Then* Obf *is not a differing-inputs obfuscator.*

We now prove Theorem 4. Let $\mathsf{Obf}_{\mathsf{eq}}^{\mathsf{tm}}$ be a sub-exponentially $\mathsf{S}_{\mathsf{eq}}^{\mathsf{tm}}$-secure TM obfuscator. Theorem 1 shows how to build a sub-exponentially PSUFCMA-secure, consistent puncturable digital signature scheme DS assuming only sub-exponentially secure OWF and sub-exponentially secure iO. For a moment, assume that we can build a TCR-secure function family H with $\mathsf{H.In}(\lambda) = \{0,1\}^*$ for all $\lambda \in \mathbb{N}$ just from sub-exponentially secure OWFs (which is not known to be true, and we address this below). Then according to Theorem 2, we can build a TM sampler S^{tm} that is (1) well-defined and (2) sub-exponentially DIFF-secure. But Theorem 3 shows that there exists an efficient adversary that breaks the IO-security of Obf with respect to S^{tm}. Therefore, Obf is not a differing-inputs obfuscator.

In order to build a TCR-secure function family H from a sub-exponentially secure OWF, the statements of Theorems 2 and 3 can be relaxed to require $\mathsf{H.ln}(\lambda) = \{0,1\}^{2^\lambda}$ for all $\lambda \in \mathbb{N}$. This change will still ensure the correctness of S^{tm}, which requires that H can process inputs of length $|\mathsf{M}^{ver}_{1^\lambda,vk}|$. The size of $\mathsf{M}^{ver}_{1^\lambda,vk}$ in our construction is bounded polynomially in the security parameter. But the reason we have to use a hash function that can process inputs of arbitrary, super-polynomially bounded lengths is because the size of $\mathsf{M}^{ver}_{1^\lambda,vk}$ is not known prior to fixing H (as shown in Fig. 6).

As noted in Sect. 2, Shoup [46] shows how to build a TCR-secure function family H for arbitrary, bounded variable-length inputs from any TCR-secure compression function family with fixed input size. The latter is shown to be achievable from OWFs by Rompel [44]. We note that the key size of Shoup's construction grows logarithmically with the maximum input length of the constructed function family, which is still polynomially bounded in the case of H that was proposed above. Furthermore, the super-polynomial bound on the message lengths does not introduce any difficulties for the security reduction of Shoup's construction. This is because the loss of security during the reduction depends on the length of the messages that are chosen by a PT adversary, rather than by the (super-polynomial) bound on the messages supported by the scheme.

This concludes the proof of Theorem 4. Note that we ruled out the existence of polynomially-secure differing-inputs obfuscation even with respect to *sub-exponentially* secure TM samplers, which is a stronger version of difference-security than the one required by our definition of polynomially-secure differing-inputs obfuscation.

Theorem 5. *Let* Obf *be a Turing Machine obfuscator. Assume the existence of sub-exponentially secure one-way functions. Then* Obf *is not a sub-exponentially secure differing-inputs obfuscator.*

To prove Theorem 5, assume for a contradiction that Obf is a sub-exponentially secure differing-inputs obfuscator. According to our definitions, it implies the existence of sub-exponentially secure indistinguishability obfuscation. The rest of the proof is identical to the proof of Theorem 4. It results in constructing a sub-exponentially difference-secure TM sampler S^{tm} that can not be securely obfuscated by Obf. Thus, we get a contradiction.

Finally, we now prove Theorem 2.

Proof (Theorem 2). We start by proving part (1) of the theorem. Specifically, we choose polynomials $s_0, \ell, n, t_0, t_1, s_1 : \mathbb{N} \to \mathbb{N}$ such that S^{tm} is well-defined.

We now specify polynomials $s_0, \ell : \mathbb{N} \to \mathbb{N}$. For any $\lambda \in \mathbb{N}$ let $s_0(\lambda)$ be a polynomial upper bound on $\max(|\mathsf{M}^0|, |\mathsf{M}^1_{1^\lambda,hk,h}|)$ where the maximum is over all $hk \in [\mathsf{H.Kg}(1^\lambda)]$ and $h \in \mathsf{H.Out}(\lambda)$. For any $\lambda \in \mathbb{N}$ let $\ell(\lambda)$ be a polynomial upper bound on $\max(|\overline{\mathsf{M}}|)$ such that $\ell(\lambda) \leq \ell(\lambda+1)$, where the maximum is over all $hk \in [\mathsf{H.Kg}(1^\lambda)]$, $h \in \mathsf{H.Out}(\lambda)$, $\mathsf{M} \in \{\mathsf{M}^0, \mathsf{M}^1_{1^\lambda,hk,h}\}$ and $\overline{\mathsf{M}} \in [\mathsf{Obf}^{tm}_{diff}(1^\lambda, \mathsf{Pad}_{s_0(\lambda)}(\mathsf{M}))]$. Note that the requirement that $\ell(\lambda) \leq \ell(\lambda+1)$ for all $\lambda \in \mathbb{N}$ is trivially achievable by removing all terms with negative coefficients from the polynomial.

We now specify a constant $0 < \epsilon < 1$ for which we will prove that S^{tm} is $2^{-(\cdot)^\epsilon}$-DIFF-secure. Let $0 < \epsilon_{\mathrm{tcr}} < 1$ be a constant such that H is $2^{-(\cdot)^{\epsilon_{\mathrm{tcr}}}}$-TCR-secure. Let $0 < \epsilon_{\mathrm{psuf}} < 1$ be a constant such that DS is $2^{-(\cdot)^{\epsilon_{\mathrm{psuf}}}}$-PSUFCMA-secure. Let $0 < \epsilon_{\mathrm{io}} < 1$ be a constant such that $\mathsf{Obf}^{\mathrm{tm}}_{\mathrm{eq}}$ is $2^{-(\cdot)^{\epsilon_{\mathrm{io}}}}$-$\mathsf{S}^{\mathrm{tm}}_{\mathrm{eq}}$-secure. Let $\epsilon = \min(\frac{1}{2}\epsilon_{\mathrm{tcr}}, \epsilon_{\mathrm{psuf}}, \epsilon_{\mathrm{io}})$.

We now specify polynomial $n \colon \mathbb{N} \to \mathbb{N}$. For any $\lambda \in \mathbb{N}$ let $n(\lambda) = (2\lambda + \ell(\lambda) + 3)^{\lceil 1/\epsilon \rceil}$. Note that for any $\lambda \in \mathbb{N}$ we have $n(\lambda) \geq \lambda$, and earlier we required that $\ell(\lambda + 1) \geq \ell(\lambda)$ for all $\lambda \in \mathbb{N}$. It follows that $\ell(n(\lambda)) \geq \ell(\lambda)$ for all $\lambda \in \mathbb{N}$, as required for S^{tm} to be well-defined. Let Inv_n be a deterministic, PT algorithm that takes $1^{\lambda'}$ to return the smallest $\lambda \in \mathbb{N}$ such that $n(\lambda) \geq \lambda'$. We note that n is injective, implying that $\mathsf{Inv}_n(1^{n(\lambda)}) = \lambda$ for all $\lambda \in \mathbb{N}$.

We now specify polynomials $t_0, t_1, s_1 \colon \mathbb{N} \to \mathbb{N}$. For any $\lambda \in \mathbb{N}$ let $t_0(\lambda)$ be a polynomial upper bound on the maximum running time of $\mathsf{M}^{\mathrm{ver}}_{1^\lambda, vk}(m, \sigma)$ where the maximum is over all $(sk, vk) \in [\mathsf{DS.Kg}(1^{n(\lambda)})]$, $m \in \{0,1\}^{\ell(\lambda)}$ and $\sigma \in \mathsf{DS.Out}(n(\lambda))$. For any $\lambda \in \mathbb{N}$ let $t_1(\lambda)$ be a polynomial upper bound on the maximum running time of $\overline{\mathsf{M}}(\mathsf{M}^{\mathrm{ver}}_{1^\lambda, vk}, 1^{t_0(\lambda)}, \overline{\mathsf{M}}, \sigma)$ where the maximum is over all $hk \in [\mathsf{H.Kg}(1^\lambda)]$, $h \in \mathsf{H.Out}(\lambda)$, $\mathsf{M} \in \{\mathsf{M}^0, \mathsf{M}^1_{1^\lambda, hk, h}\}$, $\overline{\mathsf{M}} \in [\mathsf{Obf}^{\mathrm{tm}}_{\mathrm{diff}}(1^\lambda, \mathsf{Pad}_{s_0(\lambda)}(\mathsf{M}))]$, $(sk, vk) \in [\mathsf{DS.Kg}(1^{n(\lambda)})]$ and $\sigma \in \mathsf{DS.Out}(n(\lambda))$. For any $\lambda \in \mathbb{N}$ let $s_1(\lambda)$ be a polynomial upper bound on $\max(|\mathsf{M}^{\mathrm{aux}}_{1^\lambda, sk, vk}|, |\mathsf{M}^{\mathrm{aux\text{-}punc}}_{1^\lambda, sk^*, vk, m', b}|)$ where the TM $\mathsf{M}^{\mathrm{aux\text{-}punc}}_{1^\lambda, sk^*, vk, m', b}$ is defined in Fig. 7 and where the maximum is over all $(sk, vk) \in [\mathsf{DS.Kg}(1^{n(\lambda)})]$, $m' \in \{0,1\}^{\ell(\lambda)}$, $sk^* \in [\mathsf{DS.PKg}(1^{n(\lambda)}, sk, \langle m' \rangle_{\ell(n(\lambda))})]$ and $b \in \{0,1\}$.

We proceed to prove part (2) of Theorem 2, namely that S^{tm} is $2^{-(\cdot)^\epsilon}$-DIFF-secure. The main challenge of the proof is to show that the signing key sk of DS can not be extracted from an obfuscation of TM $\mathsf{M}^{\mathrm{aux}}_{1^\lambda, sk, vk}$, meaning that the $\mathsf{S}^{\mathrm{tm}}_{\mathrm{eq}}$-secure obfuscator $\mathsf{Obf}^{\mathrm{tm}}_{\mathrm{eq}}$ is sufficient to hide sk. In our proof this is implicit. The core idea of the proof is to consider the exponential number of messages from $\mathsf{DS.In}(n(\lambda))$ and for each of them we argue that a PT adversary is unlikely to produce a signature for this message. This implies that it is hard to find an input on which TMs M^0 and $\mathsf{M}^1_{1^\lambda, hk, h}$ return different outputs.

Let $\mathcal{D}$ be a PT adversary. Consider the games and associated TMs of Fig. 7. Lines not annotated with comments are common to all games. Game $\mathsf{G}_0(\lambda)$ is equivalent to $\mathsf{DIFF}^{\mathcal{D}}_{\mathsf{S}^{\mathrm{tm}}}(\lambda)$, so for all $\lambda \in \mathbb{N}$ we have

$$\mathsf{Adv}^{\mathrm{diff}}_{\mathsf{S}^{\mathrm{tm}}, \mathcal{D}}(\lambda) = \Pr[\mathsf{G}_0(\lambda)]. \tag{1}$$

Let us discuss the transitions between hybrid games that will be used in our proof. Let $\lambda \in \mathbb{N}$. In order to transition from game $\mathsf{G}_0(\lambda)$ to game $\mathsf{G}_{1,0}(\lambda)$ we claim that if adversary $\mathcal{D}$ wins in game $\mathsf{DIFF}^{\mathcal{D}}_{\mathsf{S}^{\mathrm{tm}}}(\lambda)$ then it must return a differing-input $x = (\mathsf{M}, 1^t, m, \sigma)$ such that $\mathsf{M} = \mathsf{M}^{\mathrm{ver}}_{1^\lambda, vk}$. Otherwise, one could use this adversary to break the TCR-security of H. Next, we consider an exponential number of games, going from game $\mathsf{G}_{1,0}(\lambda)$ to game $\mathsf{G}_{1,2^{\ell(\lambda)}}(\lambda)$. Each game corresponds to a unique value of message m that can be taken as input by TM

Fig. 7. Games for proof of Theorem 2.

$M^{ver}_{1^\lambda, vk}$. For any $i \in \{0, 1, \ldots, 2^{\ell(\lambda)}\}$, adversary $\mathcal{D}$ wins in game $G_{1,i}(\lambda)$ if and only if it returns $x = (M, 1^t, m, \sigma)$ such that $M = M^{ver}_{1^\lambda, vk}$, $m \geq i$ and $M_0(x) \neq M_1(x)$. According to this definition, it is impossible to win game $G_{1,2^{\ell(\lambda)}}(\lambda)$ because TM $M^{ver}_{1^\lambda, vk}$ rejects whenever it takes a message m as input such that $|m| \neq \ell(\lambda)$ (whereas the length of m in this game is required to be at least $\ell(\lambda) + 1$). We now need to show that for each $i \in \{0, 1, \ldots, 2^{\ell(\lambda)} - 1\}$ the success probabilities of adversary $\mathcal{D}$ in games $G_{1,i}(\lambda)$ and $G_{1,i+1}(\lambda)$ are sub-exponentially close.

Let $i \in \{0, 1, \ldots, 2^{\ell(\lambda)} - 1\}$. We split the transition from game $G_{1,i}(\lambda)$ to game $G_{1,i+1}(\lambda)$ into three steps. Specifically, we consider a sequence of games $G_{1,i}(\lambda)$, $G_{1,i,A}(\lambda)$, $G_{1,i,B}(\lambda)$ and $G_{1,i+1}(\lambda)$. Games $G_{1,i,A}(\lambda)$ and $G_{1,i,B}(\lambda)$ generate

aux as an obfuscation of TM $M^{\text{aux-punc}}_{1^\lambda,sk^*,vk,m',b}$ instead of an obfuscation of TM $M^{\text{aux}}_{1^\lambda,sk,vk}$, where $m' = i$ and the used obfuscator is $\text{Obf}^{\text{tm}}_{\text{eq}}$. As opposed to TM $M^{\text{aux}}_{1^\lambda,sk,vk}$, note that TM $M^{\text{aux-punc}}_{1^\lambda,sk^*,vk,m',b}$ contains a punctured signing key sk^* for DS that is punctured at message $m^* = \langle m' \rangle_{\ell(n(\lambda))}$. Both TMs are defined to produce the same outputs on all inputs such that $\overline{M} \neq m'$, which is achieved because the punctured digital signature scheme DS is assumed to be consistent (Recall that the latter requires that sk and sk^* return the same signatures for all messages except m^*.). Furthermore, TM $M^{\text{aux-punc}}_{1^\lambda,sk^*,vk,m',b}$ is hardwired to return $b = M^{\text{aux}}_{1^\lambda,sk,vk}(m')$ on input $\overline{M} = m'$, meaning that the TMs are functionally equivalent. We use it to claim that the success probabilities of adversary $\mathcal{D}$ in games $G_{1,i}(\lambda)$ and $G_{1,i,A}(\lambda)$— and in games $G_{1,i,B}(\lambda)$ and $G_{1,i+1}(\lambda)$ —are sub-exponentially close. Namely, if $\mathcal{D}$ can distinguish between any pair of these games with a better than sub-exponentially small probability, then one can use $\mathcal{D}$ to break the iO-security of obfuscator $\text{Obf}^{\text{tm}}_{\text{eq}}$.

It remains to discuss the transition from game $G_{1,i,A}(\lambda)$ to game $G_{1,i,B}(\lambda)$. The difference between these games is that the former requires $m \geq i$ as a part of its winning condition, whereas the later requires $m \geq i + 1$. Both of these games set aux to be an obfuscation of TM $M^{\text{aux-punc}}_{1^\lambda,sk^*,vk,m',b}$, where sk^* is punctured at $m^* = \langle m' \rangle_{\ell(n(\lambda))}$ and $m' = i$. Note that adversary $\mathcal{D}$ can only have a different success probability in both games if it is capable of forging a signature on message m^* given any information it might be able to extract from TM $M^{\text{aux-punc}}_{1^\lambda,sk^*,vk,m',b}$. However, $M^{\text{aux-punc}}_{1^\lambda,sk^*,vk,m',b}$ does not contain any information that could help to forge the signature for message m^* (only bit b depends on the challenge signature, but $\mathcal{D}$ can attempt to guess it). Therefore, we can use the PSUFCMA-security of DS to bound the difference in adversary's success probability when transitioning between games $G_{1,i,A}(\lambda)$ and $G_{1,i,B}(\lambda)$.

Below we will prove the following claims:

<u>Claim 1.</u> There exists a PT adversary $\mathcal{H}$ against the TCR-security of H such that for all $\lambda \in \mathbb{N}$ we have

$$\Pr[G_0(\lambda)] - \Pr[G_{1,0}(\lambda)] \leq \text{Adv}^{\text{tcr}}_{\text{H},\mathcal{H}}(\lambda). \tag{2}$$

<u>Claim 2.</u> There exist TM samplers $S^{\text{tm}}_0, S^{\text{tm}}_1$ and a PT adversary $\mathcal{O}$ against the iO-security of $\text{Obf}^{\text{tm}}_{\text{eq}}$ relative to S^{tm}_0 and S^{tm}_1, such that for all $\lambda \in \mathbb{N}$ we have

$$\sum_{i=0}^{2^{\ell(\lambda)}-1} (\Pr[G_{1,i}(\lambda)] - \Pr[G_{1,i,A}(\lambda)]) \leq 2^{\ell(\lambda)} \cdot \text{Adv}^{\text{io}}_{\text{Obf}^{\text{tm}}_{\text{eq}},S^{\text{tm}}_0,\mathcal{O}}(n(\lambda)), \tag{3}$$

$$\sum_{i=0}^{2^{\ell(\lambda)}-1} (\Pr[G_{1,i,B}(\lambda)] - \Pr[G_{1,i+1}(\lambda)]) \leq 2^{\ell(\lambda)} \cdot \text{Adv}^{\text{io}}_{\text{Obf}^{\text{tm}}_{\text{eq}},S^{\text{tm}}_1,\mathcal{O}}(n(\lambda)). \tag{4}$$

<u>Claim 3.</u> There exists a PT adversary $\mathcal{U}$ against the PSUFCMA-security of DS such that for all $\lambda \in \mathbb{N}$ we have

$$\sum_{i=0}^{2^{\ell(\lambda)}-1} (\Pr\left[\, G_{1,i,A}(\lambda)\,\right] - \Pr\left[\, G_{1,i,B}(\lambda)\,\right]) \leq 2^{\ell(\lambda)+1} \cdot \mathsf{Adv}_{\mathsf{DS},\mathcal{U}}^{\mathsf{psufcma}}(n(\lambda)). \tag{5}$$

Finally, we claim that no adversary can win against $G_{1,2^{\ell(\lambda)}}(\lambda)$. Let $x = (\mathrm{M}, 1^t, m, \sigma)$ be the output of adversary $\mathcal{D}$ in game $G_{1,2^{\ell(\lambda)}}(\lambda)$. Adversary $\mathcal{D}$ wins the game if the following three conditions are simultaneously true: $\mathrm{M}^0(x) \neq \mathrm{M}_{1^\lambda,hk,h}^1(x)$, $\mathrm{M} = \mathrm{M}_{1^\lambda,vk}^{\mathsf{ver}}$ and $|m| > \ell(\lambda)$. The first condition requires $\mathrm{M}_{1^\lambda,hk,h}^1(x)$ to return 1. The second condition means that $\mathrm{M}_{1^\lambda,hk,h}^1(x)$ will return the output of $\mathrm{M}_{1^\lambda,vk}^{\mathsf{ver}}(m,\sigma)$. However, according to the third condition, the latter returns 0. Therefore, for any $\lambda \in \mathbb{N}$ we have

$$\Pr[G_{1,2^{\ell(\lambda)}}(\lambda)] = 0. \tag{6}$$

We now show that there exists $\lambda_{\mathcal{D}} \in \mathbb{N}$ such that for all $\lambda \geq \lambda_{\mathcal{D}}$ we have $\mathsf{Adv}_{\mathsf{S}^{\mathsf{tm}},\mathcal{D}}^{\mathsf{diff}}(\lambda) \leq 2^{-\lambda^\epsilon}$. By definition, this means that S^{tm} is $2^{-(\cdot)^\epsilon}$-DIFF-secure.

$$\mathsf{Adv}_{\mathsf{S}^{\mathsf{tm}},\mathcal{D}}^{\mathsf{diff}}(\lambda) = (\Pr[G_0(\lambda)] - \Pr[G_{1,0}(\lambda)])$$

$$+ \sum_{i=0}^{2^{\ell(\lambda)}-1} (\Pr[G_{1,i}(\lambda)] - \Pr[G_{1,i+1}(\lambda)]) + \Pr[G_{1,2^{\ell(\lambda)}}(\lambda)] \tag{7}$$

$$\leq \mathsf{Adv}_{\mathsf{H},\mathcal{H}}^{\mathsf{tcr}}(\lambda) + 2^{\ell(\lambda)} \cdot \mathsf{Adv}_{\mathsf{Obf}_{\mathsf{eq}}^{\mathsf{ftm}},\mathsf{S}_0^{\mathsf{tm}},\mathcal{O}}^{\mathsf{io}}(n(\lambda))$$

$$+ 2^{\ell(\lambda)+1} \cdot \mathsf{Adv}_{\mathsf{DS},\mathcal{U}}^{\mathsf{psufcma}}(n(\lambda)) + 2^{\ell(\lambda)} \cdot \mathsf{Adv}_{\mathsf{Obf}_{\mathsf{eq}}^{\mathsf{ftm}},\mathsf{S}_1^{\mathsf{tm}},\mathcal{O}}^{\mathsf{io}}(n(\lambda)) \tag{8}$$

$$\leq 2^{-\lambda^{\epsilon_{\mathsf{tcr}}}} + 2^{\ell(\lambda)} \cdot \left(2^{-n(\lambda)^{\epsilon_{\mathsf{io}}}} + 2 \cdot 2^{-n(\lambda)^{\epsilon_{\mathsf{psuf}}}} + 2^{-n(\lambda)^{\epsilon_{\mathsf{io}}}} \right) \tag{9}$$

$$\leq 2^{-\lambda^{\epsilon_{\mathsf{tcr}}}} + 2^{\ell(\lambda)+1} \cdot \left(2^{-n(\lambda)^{\epsilon_{\mathsf{io}}}} + 2^{-n(\lambda)^{\epsilon_{\mathsf{psuf}}}} + 2^{-n(\lambda)^{\epsilon_{\mathsf{io}}}} \right) \tag{10}$$

$$\leq 2^{-\lambda^{2\epsilon}} + 2^{\ell(\lambda)+1} \cdot 3 \cdot 2^{-n(\lambda)^\epsilon} \tag{11}$$

$$= 2^{-\lambda^{2\epsilon}} + 2^{\ell(\lambda)+1+\log_2 3-(2\lambda+\ell(\lambda)+3)^{\lceil 1/\epsilon \rceil \cdot \epsilon}} \tag{12}$$

$$\leq 2^{-\lambda^{2\epsilon}} + 2^{-(2\lambda)^\epsilon} \tag{13}$$

$$\leq 2^{-(2\lambda)^\epsilon} + 2^{-(2\lambda)^\epsilon} = 2^{1-(2\lambda)^\epsilon} \tag{14}$$

$$\leq 2^{-\lambda^\epsilon}. \tag{15}$$

Let $\lambda_{\mathcal{H}} \in \mathbb{N}$ such that $\mathsf{Adv}_{\mathsf{H},\mathcal{H}}^{\mathsf{tcr}}(\lambda) \leq 2^{-\lambda^{\epsilon_{\mathsf{tcr}}}}$ for all $\lambda \geq \lambda_{\mathcal{H}}$. Let $\lambda_{\mathcal{U}} \in \mathbb{N}$ such that $\mathsf{Adv}_{\mathsf{DS},\mathcal{U}}^{\mathsf{psufcma}}(\lambda) \leq 2^{-\lambda^{\epsilon_{\mathsf{psuf}}}}$ for all $\lambda \geq \lambda_{\mathcal{U}}$. For $b \in \{0,1\}$ let $\lambda_{\mathsf{S}_b^{\mathsf{tm}},\mathcal{O}} \in \mathbb{N}$ be such that $\mathsf{Adv}_{\mathsf{Obf}_{\mathsf{eq}}^{\mathsf{ftm}},\mathsf{S}_b^{\mathsf{tm}},\mathcal{O}}^{\mathsf{io}}(\lambda) \leq 2^{-\lambda^{\epsilon_{\mathsf{io}}}}$ for all $\lambda \geq \lambda_{\mathsf{S}_b^{\mathsf{tm}},\mathcal{O}}$.

Equation (7) follows from Eq. (1) for all $\lambda \in \mathbb{N}$. Equation (8) follows from Eqs. (2)–(6) for all $\lambda \in \mathbb{N}$. Equation (9) holds for all $\lambda \in \mathbb{N}$ such that $\lambda \geq \lambda_{\mathcal{H}}$ and

$n(\lambda) \geq \max(\lambda_{S_0^{tm},\mathcal{O}}, \lambda_{\mathcal{U}}, \lambda_{S_1^{tm},\mathcal{O}})$. Equation (10) holds for all $\lambda \in \mathbb{N}$. Equation (11) is obtained by expanding ϵ according to its definition, namely by using the following relations: $2\epsilon \leq \epsilon_{tcr}$, $\epsilon \leq \epsilon_{psuf}$ and $\epsilon \leq \epsilon_{io}$. Equation (12) is obtained by expanding $n(\lambda)$ according to its definition. Equation (13) holds for all $\lambda \in \mathbb{N}$, because for any polynomial $\ell \colon \mathbb{N} \to \mathbb{N}$, any constant $0 < \epsilon < 1$ and all $\lambda \in \mathbb{N}$ we have

$$\ell(\lambda) + 1 + \log_2 3 - (2\lambda + \ell(\lambda) + 3)^{\lceil 1/\epsilon \rceil \cdot \epsilon}$$
$$\leq \ell(\lambda) + 1 + \log_2 3 - (2\lambda + \ell(\lambda) + 3)$$
$$< -2\lambda \leq -(2\lambda)^{\epsilon}.$$

Equation (14) holds for all $\lambda \in \mathbb{N}$ such that $\lambda^{2\epsilon} \geq (2\lambda)^{\epsilon}$, requiring that $\lambda \geq 2$. Equation (15) holds for all $\lambda \in \mathbb{N}$ such that $1 - 2^{\epsilon}\lambda^{\epsilon} \leq -\lambda^{\epsilon}$, requiring that $\lambda \geq \left(\frac{1}{2^{\epsilon}-1}\right)^{1/\epsilon}$. Therefore, it suffices to set

$$\lambda_{\mathcal{D}} = \max\left(\lambda_{\mathcal{H}}, \mathsf{Inv}_n(1^{\lambda_{S_0^{tm},\mathcal{O}}}), \mathsf{Inv}_n(1^{\lambda_{\mathcal{U}}}), \mathsf{Inv}_n(1^{\lambda_{S_1^{tm},\mathcal{O}}}), 2, \left\lceil (2^{\epsilon}-1)^{-1/\epsilon} \right\rceil\right).$$

This completes the proof. We now prove Claims 1–3.

Proof of Claim 1. We build a PT adversary $\mathcal{H}$ against the TCR-security of H such that for all $\lambda \in \mathbb{N}$ we have $\Pr[G_0(\lambda)] - \Pr[G_{1,0}(\lambda)] \leq \mathsf{Adv}_{\mathsf{H},\mathcal{H}}^{tcr}(\lambda)$.

Adversary $\mathcal{H}_1(1^{\lambda})$	Adversary $\mathcal{H}_2(1^{\lambda}, st, hk)$
$(sk, vk) \leftarrow_{\$} \mathsf{DS.Kg}(1^{n(\lambda)})$	$(sk, vk) \leftarrow st$; $h \leftarrow \mathsf{H.Ev}(1^{\lambda}, hk, \mathrm{M}_{1^{\lambda},vk}^{ver})$
$st \leftarrow (sk, vk)$	$\mathrm{M}_0 \leftarrow \mathsf{Pad}_{s_0(\lambda)}(\mathrm{M}^0)$; $\mathrm{M}_1 \leftarrow \mathsf{Pad}_{s_0(\lambda)}(\mathrm{M}_{1^{\lambda},hk,h}^1)$
Return $(\mathrm{M}_{1^{\lambda},vk}^{ver}, st)$	$aux \leftarrow_{\$} \mathsf{Obf}_{eq}^{tm}(1^{n(\lambda)}, \mathsf{Pad}_{s_1(\lambda)}(\mathrm{M}_{1^{\lambda},sk,vk}^{aux}))$
	$(\mathrm{M}, 1^t, m, \sigma) \leftarrow_{\$} \mathcal{D}(1^{\lambda}, \mathrm{M}_0, \mathrm{M}_1, aux)$; Return M

Let $x = (\mathrm{M}, 1^t, m, \sigma)$ be an output of adversary $\mathcal{D}$ in games $G_0(\lambda)$ and $G_{1,0}(\lambda)$ (note that the input distribution of $\mathcal{D}$ is the same in both games). If these games produce different outcomes for the same x, it means that $\mathrm{M}^0(x) \neq \mathrm{M}_{1^{\lambda},hk,h}^1(x)$ and $\mathrm{M} \neq \mathrm{M}_{1^{\lambda},vk}^{ver}$. According to the construction of M^0 and $\mathrm{M}_{1^{\lambda},hk,h}^1$ it follows that $\mathsf{H.Ev}(1^{\lambda}, hk, \mathrm{M}_{1^{\lambda},vk}^{ver}) = \mathsf{H.Ev}(1^{\lambda}, hk, \mathrm{M})$. Whenever this happens, adversary $\mathcal{H}$ wins in game $\mathrm{TCR}_{\mathsf{H}}^{\mathcal{H}}(\lambda)$ by returning $x_0 = \mathrm{M}_{1^{\lambda},vk}^{ver}$ and $x_1 = \mathrm{M}$. This proves the claim.

Proof of Claim 2. We build TM samplers S_0^{tm}, S_1^{tm} and a PT adversary $\mathcal{O}$ against the iO-security of Obf_{eq}^{tm} relative to S_0^{tm} and S_1^{tm}, such that for all $\lambda \in \mathbb{N}$ we have $\sum_{i=0}^{2^{\ell(\lambda)}-1} \left(\Pr[G_{1,i}(\lambda)] - \Pr[G_{1,i,A}(\lambda)]\right) \leq 2^{\ell(\lambda)} \cdot \mathsf{Adv}_{\mathsf{Obf}_{eq}^{tm},S_0^{tm},\mathcal{O}}^{io}(n(\lambda))$ and $\sum_{i=0}^{2^{\ell(\lambda)}-1} \left(\Pr[G_{1,i,B}(\lambda)] - \Pr[G_{1,i+1}(\lambda)]\right) \leq 2^{\ell(\lambda)} \cdot \mathsf{Adv}_{\mathsf{Obf}_{eq}^{tm},S_1^{tm},\mathcal{O}}^{io}(n(\lambda))$.

Below, on the left we (simultaneously) define the TM samplers S_0^{tm} and S_1^{tm} that differ at the commented lines and have the uncommented lines in common. On the right, we define the PT adversary $\mathcal{O}$.

TM Samplers $S_0^{tm}(1^{\lambda'})$, $S_1^{tm}(1^{\lambda'})$	Adversary $\mathcal{O}(1^{\lambda'}, \overline{M}, aux)$
$\lambda \leftarrow Inv_n(1^{\lambda'})$; $i \leftarrow_\$ \{0,1\}^{\ell(\lambda)}$	$\lambda \leftarrow Inv_n(1^{\lambda'})$
$(sk, vk) \leftarrow_\$ DS.Kg(1^{n(\lambda)})$	$a\tilde{u}x \leftarrow \overline{M}$
$hk \leftarrow_\$ H.Kg(1^\lambda)$; $h \leftarrow H.Ev(1^\lambda, hk, M_{1^\lambda,vk}^{ver})$	$(\tilde{M}_0, \tilde{M}_1, vk, z) \leftarrow aux$
$\tilde{M}_0 \leftarrow Pad_{s_0(\lambda)}(M^0)$; $\tilde{M}_1 \leftarrow Pad_{s_0(\lambda)}(M_{1^\lambda,hk,h}^1)$	$x \leftarrow_\$ \mathcal{D}(1^\lambda, \tilde{M}_0, \tilde{M}_1, a\tilde{u}x)$
$m' \leftarrow \langle i \rangle_{\ell(\lambda)}$; $b \leftarrow M_{1^\lambda,sk,vk}^{aux}(m')$	$(M, 1^t, m, \sigma) \leftarrow x$
$m^* \leftarrow \langle m' \rangle_{\ell(n(\lambda))}$; $sk^* \leftarrow_\$ DS.PKg(1^{n(\lambda)}, sk, m^*)$	$d_0 \leftarrow (\tilde{M}_0(x) \neq \tilde{M}_1(x))$
$M_{aux} \leftarrow Pad_{s_1(\lambda)}(M_{1^\lambda,sk,vk}^{aux})$	$d_1 \leftarrow (M = M_{1^\lambda,vk}^{ver})$
$M_{aux\text{-}punc} \leftarrow Pad_{s_1(\lambda)}(M_{1^\lambda,sk^*,vk,m',b}^{aux\text{-}punc})$	If $(d_0 \wedge d_1 \wedge m \geq z)$
$M_1 \leftarrow M_{aux}$; $M_0 \leftarrow M_{aux\text{-}punc}$; $z \leftarrow i$ // S_0^{tm}	Then return 1
$M_0 \leftarrow M_{aux}$; $M_1 \leftarrow M_{aux\text{-}punc}$; $z \leftarrow i+1$ // S_1^{tm}	Else return 0
$aux \leftarrow (\tilde{M}_0, \tilde{M}_1, vk, z)$; return (M_0, M_1, aux)	

We now show that $S_0^{tm}, S_1^{tm} \in \boldsymbol{S}_{eq}^{tm}$, meaning that these samplers produce functionally equivalent TMs. Both samplers return TMs $M_{1^\lambda,sk,vk}^{aux}$ and $M_{1^\lambda,sk^*,vk,m',b}^{aux\text{-}punc}$ that are padded to size $s_1(\lambda)$. First, observe that $M_{1^\lambda,sk,vk}^{aux}$ contains a signing key sk for DS, whereas $M_{1^\lambda,sk^*,vk,m',b}^{aux\text{-}punc}$ contains the corresponding punctured signing key sk^*, punctured at $m^* = \langle m' \rangle_{\ell(n(\lambda))}$, and a bit b that is equal to $M_{1^\lambda,sk,vk}^{aux}(m')$. According to the definition of a consistent puncturable digital signature scheme, keys sk and sk^* produce the same signatures for all $m \in DS.In(n(\lambda))\backslash\{m^*\}$. Note that both $M_{1^\lambda,sk,vk}^{aux}$ and $M_{1^\lambda,sk^*,vk,m',b}^{aux\text{-}punc}$ compute a signature for an $\ell(n(\lambda))$-bit string $\langle \overline{M} \rangle_{\ell(n(\lambda))}$ that is built from the $\ell(\lambda)$-bit input string $\overline{M}$ by padding it with leading zeros, which is an injective padding. Since m^* can only be built by padding m', these TMs are equivalent for all inputs in $\overline{M} \in \{0,1\}^{\ell(\lambda)}\backslash\{m'\}$. Furthermore, notice that $M_{1^\lambda,sk^*,vk,m',b}^{aux\text{-}punc}$ returns $b = M_{1^\lambda,sk,vk}^{aux}(m')$ on input m', so these TMs are equivalent for *all* inputs.

Let $\lambda \in \mathbb{N}$. For any $b \in \{0,1\}$ consider game $IO_{Obf_{eq}^{tm},S_b^{tm}}^{\mathcal{O}}(n(\lambda))$. Let i_b denote the value of i sampled by TM sampler S_b^{tm}. For any $i \in \{0,1,\ldots,2^{\ell(\lambda)}-1\}$ we have $Pr[i_b = i] = 2^{-\ell(\lambda)}$, and hence

$$Adv_{Obf_{eq}^{tm},S_b^{tm},\mathcal{O}}^{io}(n(\lambda)) = 2 \cdot Pr[IO_{Obf_{eq}^{tm},S_b^{tm}}^{\mathcal{O}}(n(\lambda))] - 1$$

$$= 2 \cdot \sum_{i=0}^{2^{\ell(\lambda)}-1} \left(Pr[i_b = i] \cdot Pr[IO_{Obf_{eq}^{tm},S_b^{tm}}^{\mathcal{O}}(n(\lambda)) \mid i_b = i] \right) - 1$$

$$= 2 \cdot 2^{-\ell(\lambda)} \cdot \sum_{i=0}^{2^{\ell(\lambda)}-1} Pr[IO_{Obf_{eq}^{tm},S_b^{tm}}^{\mathcal{O}}(n(\lambda)) \mid i_b = i] - 1. \tag{16}$$

Finally, observe that for any $i \in \{0,1,\ldots,2^{\ell(\lambda)}-1\}$ we have the following by construction:

$$2 \cdot Pr[IO_{Obf_{eq}^{tm},S_0^{tm}}^{\mathcal{O}}(n(\lambda)) \mid i_0 = i] - 1 = Pr[G_{1,i}(\lambda)] - Pr[G_{1,i,A}(\lambda)],$$

$$2 \cdot Pr[IO_{Obf_{eq}^{tm},S_1^{tm}}^{\mathcal{O}}(n(\lambda)) \mid i_1 = i] - 1 = Pr[G_{1,i,B}(\lambda)] - Pr[G_{1,i+1}(\lambda)].$$

Claim 2 follows from (16) together with the two equations above.

<u>Proof of Claim 3.</u> We build a PT adversary $\mathcal{U}$ against the PSUFCMA-security of DS such that for all $\lambda \in \mathbb{N}$ we have $\sum_{i=0}^{2^{\ell(\lambda)}-1} (\Pr[G_{1,i,A}(\lambda)] - \Pr[G_{1,i,B}(\lambda)]) \leq 2^{\ell(\lambda)+1} \cdot \mathsf{Adv}^{\mathsf{psufcma}}_{\mathsf{DS},\mathcal{U}}(n(\lambda))$.

Adversary $\mathcal{U}_1(1^{\lambda'})$	Adversary $\mathcal{U}_2(1^{\lambda'}, st, vk, sk^*)$
$\lambda \leftarrow \mathsf{Inv}_n(1^{\lambda'})$	$\lambda \leftarrow \mathsf{Inv}_n(1^{\lambda'})$; $m' \leftarrow st$; $b \leftarrow\!\!\text{\$}\ \{0,1\}$
$m' \leftarrow\!\!\text{\$}\ \{0,1\}^{\ell(\lambda)}$	$hk \leftarrow\!\!\text{\$}\ \mathsf{H.Kg}(1^\lambda)$; $h \leftarrow \mathsf{H.Ev}(1^\lambda, hk, \mathrm{M}^{\mathsf{ver}}_{1^\lambda, vk})$
$m^* \leftarrow \langle m' \rangle_{\ell(n(\lambda))}$	$\mathrm{M}_0 \leftarrow \mathsf{Pad}_{s_0(\lambda)}(\mathrm{M}^0)$; $\mathrm{M}_1 \leftarrow \mathsf{Pad}_{s_0(\lambda)}(\mathrm{M}^1_{1^\lambda, hk, h})$
$st \leftarrow m'$	$aux \leftarrow\!\!\text{\$}\ \mathsf{Obf}^{\mathsf{tm}}_{\mathsf{eq}}(1^{n(\lambda)}, \mathsf{Pad}_{s_1(\lambda)}(\mathrm{M}^{\mathsf{aux\text{-}punc}}_{1^\lambda, sk^*, vk, m', b}))$
Return (m^*, st)	$x \leftarrow\!\!\text{\$}\ \mathcal{D}(1^\lambda, \mathrm{M}_0, \mathrm{M}_1, aux)$; $(\mathrm{M}, 1^t, m, \sigma) \leftarrow x$
	$d_0 \leftarrow (\mathrm{M}_0(x) \neq \mathrm{M}_1(x))$; $d_1 \leftarrow (\mathrm{M} = \mathrm{M}^{\mathsf{ver}}_{1^\lambda, vk})$
	If $(d_0 \wedge d_1 \wedge m = m')$ then return σ else return $\bot$

Let $\lambda \in \mathbb{N}$. Consider the value of m' sampled by $\mathcal{U}_1$ in game $\mathrm{PSUFCMA}^{\mathcal{U}}_{\mathsf{DS}}(n(\lambda))$. For any $i \in \{0, 1, \ldots, 2^{\ell(\lambda)} - 1\}$ it holds that $\Pr[m' = i] = 2^{-\ell(\lambda)}$. Hence,

$$\mathsf{Adv}^{\mathsf{psufcma}}_{\mathsf{DS},\mathcal{U}}(n(\lambda)) = \sum_{i=0}^{2^{\ell(\lambda)}-1} \left(\Pr[m' = i] \cdot \Pr[\,\mathrm{PSUFCMA}^{\mathcal{U}}_{\mathsf{DS}}(n(\lambda)) \mid m' = i\,]\right)$$

$$= 2^{-\ell(\lambda)} \cdot \sum_{i=0}^{2^{\ell(\lambda)}-1} \Pr[\,\mathrm{PSUFCMA}^{\mathcal{U}}_{\mathsf{DS}}(n(\lambda)) \mid m' = i\,]. \qquad (17)$$

Now observe that for any $i \in \{0, 1, \ldots, 2^{\ell(\lambda)} - 1\}$ we also have

$$\Pr[\,\mathrm{PSUFCMA}^{\mathcal{U}}_{\mathsf{DS}}(n(\lambda)) \mid m' = i\,] \geq \frac{1}{2} \cdot (\Pr[G_{1,i,A}(\lambda)] - \Pr[G_{1,i,B}(\lambda)]). \qquad (18)$$

Let $x = (\mathrm{M}, 1^t, m, \sigma)$ be an output of adversary $\mathcal{D}$ in games $G_{1,i,A}(\lambda)$ and $G_{1,i,B}(\lambda)$ (note that the input distribution of $\mathcal{D}$ is the same in both games). If these games produce different outcomes for the same x, it means that $\mathrm{M}^0(x) \neq \mathrm{M}^1_{1^\lambda, hk, h}(x)$, $\mathrm{M} = \mathrm{M}^{\mathsf{ver}}_{1^\lambda, vk}$ and $m = i$. According to the construction of M^0 and $\mathrm{M}^1_{1^\lambda, hk, h}$ it follows that $(\langle m \rangle_{\ell(n(\lambda))}, \sigma)$ is a valid message-signature pair for the digital signature scheme DS with verification key vk.

Whenever the above happens, adversary $\mathcal{U}$ wins in game $\mathrm{PSUFCMA}^{\mathcal{U}}_{\mathsf{DS}}(n(\lambda))$ by forging a valid signature σ for message m^*, given that the following two conditions are satisfied. First, it is only true if adversary $\mathcal{U}$ sampled $m' = i$. Second, in order to build TM $\mathrm{M}^{\mathsf{aux\text{-}punc}}_{1^\lambda, sk^*, vk, m', b}$, adversary $\mathcal{U}$ has to compute $b = \mathrm{M}^{\mathsf{aux}}_{1^\lambda, sk, vk}(m')$. Since $\mathcal{U}$ does not know sk, instead it has to guess the value of $b \in \{0, 1\}$. Hence, $\mathcal{U}$ can perfectly simulate the games with probability $\frac{1}{2}$.

Claim 3 follows from (17) and (18).

Acknowledgments. Bellare and Stepanovs were supported in part by NSF grants CNS-1526801 and CNS-1228890, ERC Project ERCC FP7/615074 and a gift from

Microsoft. Waters was supported in part by NSF grants CNS-1228599 and CNS-1414082, DARPA SafeWare, a Google Faculty Research award, the Alfred P. Sloan Fellowship, a Microsoft Faculty Fellowship and a Packard Foundation Fellowship. We thank the Eurocrypt 2016 reviewers for their comments.

References

1. Ananth, P., Boneh, D., Garg, S., Sahai, A., Zhandry, M.: Differing-inputs obfuscation and applications. Cryptology ePrint Archive, Report 2013/689 (2013). http://eprint.iacr.org/2013/689
2. Ananth, P., Brakerski, Z., Segev, G., Vaikuntanathan, V.: From selective to adaptive security in functional encryption. In: Gennaro, R., Robshaw, M. (eds.) CRYPTO 2015. LNCS, vol. 9216, pp. 657–677. Springer, Heidelberg (2015)
3. Applebaum, B., Barak, B., Wigderson, A.: Public-key cryptography from different assumptions. In: Schulman, L.J. (ed.) 42nd ACM STOC, pp. 171–180. ACM Press, June 2010
4. Backes, M., Dagdelen, O., Fischlin, M., Gajek, S., Meiser, S., Schröder, D.: Operational signature schemes. Cryptology ePrint Archive, Report 2014/820 (2014). http://eprint.iacr.org/2014/820
5. Backes, M., Meiser, S., Schröder, D.: Delegatable functional signatures. Cryptology ePrint Archive, Report 2013/408 (2013). http://eprint.iacr.org/2013/408
6. Barak, B., Garg, S., Kalai, Y.T., Paneth, O., Sahai, A.: Protecting obfuscation against algebraic attacks. In: Nguyen, P.Q., Oswald, E. (eds.) EUROCRYPT 2014. LNCS, vol. 8441, pp. 221–238. Springer, Heidelberg (2014)
7. Barak, B., Goldreich, O., Impagliazzo, R., Rudich, S., Sahai, A., Vadhan, S.P., Yang, K.: On the (im)possibility of obfuscating programs. In: Kilian, J. (ed.) CRYPTO 2001. LNCS, vol. 2139, pp. 1–18. Springer, Heidelberg (2001)
8. Bellare, M.: A note on negligible functions. J. Cryptol. 15(4), 271–284 (2002)
9. Bellare, M., Fuchsbauer, G.: Policy-based signatures. In: Krawczyk, H. (ed.) PKC 2014. LNCS, vol. 8383, pp. 520–537. Springer, Heidelberg (2014)
10. Bellare, M., Rogaway, P.: Collision-resistant hashing: towards making UOWHFs practical. In: Kaliski Jr., B.S. (ed.) CRYPTO 1997. LNCS, vol. 1294, pp. 470–484. Springer, Heidelberg (1997)
11. Bellare, M., Rogaway, P.: The security of triple encryption and a framework for code-based game-playing proofs. In: Vaudenay, S. (ed.) EUROCRYPT 2006. LNCS, vol. 4004, pp. 409–426. Springer, Heidelberg (2006)
12. Bellare, M., Stepanovs, I., Tessaro, S.: Poly-many hardcore bits for any one-way function and a framework for differing-inputs obfuscation. In: Sarkar, P., Iwata, T. (eds.) ASIACRYPT 2014, Part II. LNCS, vol. 8874, pp. 102–121. Springer, Heidelberg (2014)
13. Bellare, M., Stepanovs, I., Waters, B.: New negative results on differing-inputs obfuscation. Cryptology ePrint Archive, Report 2016/162 (2016). http://eprint.iacr.org/2016/162
14. Bitansky, N., Canetti, R., Chiesa, A., Goldwasser, S., Lin, H., Rubinstein, A., Tromer, E.: The hunting of the SNARK. Cryptology ePrint Archive, Report 2014/580 (2014). http://eprint.iacr.org/2014/580
15. Bitansky, N., Canetti, R., Chiesa, A., Tromer, E.: From extractable collision resistance to succinct non-interactive arguments of knowledge, and back again. In: Goldwasser, S. (ed.) ITCS 2012, pp. 326–349. ACM, January 2012

16. Bitansky, N., Garg, S., Lin, H., Pass, R., Telang, S.: Succinct randomized encodings and their applications. In: Servedio, R.A., Rubinfeld, R. (eds.) 47th ACM STOC, pp. 439–448. ACM Press, June 2015

17. Bitansky, N., Paneth, O.: On the impossibility of approximate obfuscation and applications to resettable cryptography. In: Boneh, D., Roughgarden, T., Feigenbaum, J. (eds.) 45th ACM STOC, pp. 241–250. ACM Press, June 2013

18. Bitansky, N., Paneth, O., Wichs, D.: Perfect structure on the edge of chaos. In: Kushilevitz, E., Malkin, T. (eds.) TCC 2016-A. LNCS, vol. 9562, pp. 474–502. Springer, Heidelberg (2016). doi:10.1007/978-3-662-49096-9_20

19. Boneh, D., Waters, B.: Constrained pseudorandom functions and their applications. In: Sako, K., Sarkar, P. (eds.) ASIACRYPT 2013, Part II. LNCS, vol. 8270, pp. 280–300. Springer, Heidelberg (2013)

20. Boyle, E., Chung, K.-M., Pass, R.: On extractability obfuscation. In: Lindell, Y. (ed.) TCC 2014. LNCS, vol. 8349, pp. 52–73. Springer, Heidelberg (2014)

21. Boyle, E., Goldwasser, S., Ivan, I.: Functional signatures and pseudorandom functions. In: Krawczyk, H. (ed.) PKC 2014. LNCS, vol. 8383, pp. 501–519. Springer, Heidelberg (2014)

22. Brzuska, C., Mittelbach, A.: Using indistinguishability obfuscation via UCEs. In: Sarkar, P., Iwata, T. (eds.) ASIACRYPT 2014, Part II. LNCS, vol. 8874, pp. 122–141. Springer, Heidelberg (2014)

23. Canetti, R., Holmgren, J., Jain, A., Vaikuntanathan, V.: Succinct garbling and indistinguishability obfuscation for RAM programs. In: Servedio, R.A., Rubinfeld, R. (eds.) 47th ACM STOC, pp. 429–437. ACM Press, June 2015

24. Canetti, R., Lin, H., Tessaro, S., Vaikuntanathan, V.: Obfuscation of probabilistic circuits and applications. In: Dodis, Y., Nielsen, J.B. (eds.) TCC 2015, Part II. LNCS, vol. 9015, pp. 468–497. Springer, Heidelberg (2015)

25. Chandran, N., Raghuraman, S., Vinayagamurthy, D.: Constrained pseudorandom functions: Verifiable and delegatable. Cryptology ePrint Archive, Report 2014/522 (2014). http://eprint.iacr.org/2014/522

26. Fuchsbauer, G.: Constrained verifiable random functions. In: Abdalla, M., De Prisco, R. (eds.) SCN 2014. LNCS, vol. 8642, pp. 95–114. Springer, Heidelberg (2014)

27. Garg, S., Gentry, C., Halevi, S., Raykova, M., Sahai, A., Waters, B.: Candidate indistinguishability obfuscation and functional encryption for all circuits. In: 54th FOCS, pp. 40–49. IEEE Computer Society Press, October 2013

28. Garg, S., Gentry, C., Halevi, S., Wichs, D.: On the implausibility of differing-inputs obfuscation and extractable witness encryption with auxiliary input. In: Garay, J.A., Gennaro, R. (eds.) CRYPTO 2014, Part I. LNCS, vol. 8616, pp. 518–535. Springer, Heidelberg (2014)

29. Gentry, C., Halevi, S., Raykova, M., Wichs, D.: Outsourcing private RAM computation. In: 55th FOCS, pp. 404–413. IEEE Computer Society Press, October 2014

30. Gentry, C., Lewko, A.B., Sahai, A., Waters, B.: Indistinguishability obfuscation from the multilinear subgroup elimination assumption. In: Guruswami, V. (ed.) 56th FOCS, pp. 151–170. IEEE Computer Society Press, October 2015

31. Gentry, C., Lewko, A., Waters, B.: Witness encryption from instance independent assumptions. In: Garay, J.A., Gennaro, R. (eds.) CRYPTO 2014, Part I. LNCS, vol. 8616, pp. 426–443. Springer, Heidelberg (2014)

32. Goldreich, O., Goldwasser, S., Micali, S.: How to construct random functions. J. ACM **33**(4), 792–807 (1986)

33. Goldwasser, S., Kalai, Y.T.: On the impossibility of obfuscation with auxiliary input. In: 46th FOCS, pp. 553–562. IEEE Computer Society Press, October 2005
34. Goldwasser, S., Micali, S., Rivest, R.L.: A digital signature scheme secure against adaptive chosen-message attacks. SIAM J. Comput. **17**(2), 281–308 (1988)
35. Gorbunov, S., Vaikuntanathan, V., Wee, H.: Attribute-based encryption for circuits. In: Boneh, D., Roughgarden, T., Feigenbaum, J. (eds.) 45th ACM STOC, pp. 545–554. ACM Press, June 2013
36. Hada, S.: Zero-knowledge and code obfuscation. In: Okamoto, T. (ed.) ASIACRYPT 2000. LNCS, vol. 1976, pp. 443–457. Springer, Heidelberg (2000)
37. Haitner, I., Holenstein, T., Reingold, O., Vadhan, S., Wee, H.: Universal one-way hash functions via inaccessible entropy. In: Gilbert, H. (ed.) EUROCRYPT 2010. LNCS, vol. 6110, pp. 616–637. Springer, Heidelberg (2010)
38. Ishai, Y., Pandey, O., Sahai, A.: Public-coin differing-inputs obfuscation and its applications. In: Dodis, Y., Nielsen, J.B. (eds.) TCC 2015, Part II. LNCS, vol. 9015, pp. 668–697. Springer, Heidelberg (2015)
39. Katz, J., Koo, C.-Y.: On constructing universal one-way hash functions from arbitrary one-way functions. Cryptology ePrint Archive, Report 2005/328 (2005). http://eprint.iacr.org/2005/328
40. Kiayias, A., Papadopoulos, S., Triandopoulos, N., Zacharias, T.: Delegatable pseudorandom functions and applications. In: Sadeghi, A.-R., Gligor, V.D., Yung, M. (eds.) ACM CCS 2013, pp. 669–684. ACM Press, November 2013
41. Koppula, V., Lewko, A.B., Waters, B.: Indistinguishability obfuscation for turing machines with unbounded memory. In: Servedio, R.A., Rubinfeld, R. (eds.) 47th ACM STOC, pp. 419–428. ACM Press, June 2015
42. Naor, M., Yung, M.: Universal one-way hash functions and their cryptographic applications. In: 21st ACM STOC, pp. 33–43. ACM Press, May 1989
43. Pass, R., Seth, K., Telang, S.: Indistinguishability obfuscation from semantically-secure multilinear encodings. In: Garay, J.A., Gennaro, R. (eds.) CRYPTO 2014, Part I. LNCS, vol. 8616, pp. 500–517. Springer, Heidelberg (2014)
44. Rompel, J.: One-way functions are necessary and sufficient for secure signatures. In: 22nd ACM STOC, pp. 387–394. ACM Press, May 1990
45. Sahai, A., Waters, B.: How to use indistinguishability obfuscation: deniable encryption, and more. In: Shmoys, D.B. (ed.) 46th ACM STOC, pp. 475–484. ACM Press, May/June 2014
46. Shoup, V.: A composition theorem for universal one-way hash functions. In: Preneel, B. (ed.) EUROCRYPT 2000. LNCS, vol. 1807, pp. 445–452. Springer, Heidelberg (2000)
47. Waters, B.: A punctured programming approach to adaptively secure functional encryption. In: Gennaro, R., Robshaw, M. (eds.) CRYPTO 2015. LNCS, vol. 9216, pp. 678–697. Springer, Heidelberg (2015)

Automated Unbounded Analysis
of Cryptographic Constructions
in the Generic Group Model

Miguel Ambrona$^{(\boxtimes)}$, Gilles Barthe, and Benedikt Schmidt

IMDEA Software Institute, Madrid, Spain
{miguel.ambrona,gilles.barthe,benedikt.schmidt}@imdea.org

Abstract. We develop a new method to automatically prove security statements in the Generic Group Model as they occur in actual papers. We start by defining (i) a general language to describe security definitions, (ii) a class of logical formulas that characterize how an adversary can win, and (iii) a translation from security definitions to such formulas. We prove a Master Theorem that relates the security of the construction to the existence of a solution for the associated logical formulas. Moreover, we define a constraint solving algorithm that proves the security of a construction by proving the absence of solutions.

We implement our approach in a fully automated tool, the gga$^\infty$ tool, and use it to verify different examples from the literature. The results improve on the tool by Barthe et al. (CRYPTO'14, PKC'15): for many constructions, gga$^\infty$ succeeds in proving standard (unbounded) security, whereas Barthe's tool is only able to prove security for a small number of oracle queries.

1 Introduction

The gold standard in provable security is to demonstrate security in the standard model. However, proofs in the standard model sometimes rely on non-standard hardness assumptions. In such situations, it is essential to prove that the hardness assumptions used in the security proofs meet some minimal requirements, for instance the absence of algebraic attacks. The accepted method for validating new DDH-like assumptions is to show absence of generic attacks, i.e. attacks that solely exploit the underlying algebraic structure, using the Generic Group Model [32,33,35,38] or its bilinear and multilinear variants [11,17]. The Generic Group Model provides an algebraic setting for describing a wide class of DDH-like assumptions, and is supported by so-called Master Theorems that give a purely algebraic condition that ensures the security of an assumption in the Generic Group Model (or its variants). Very roughly, the proof of the Master Theorems uses the Schwartz-Zippel Lemma to prove a security reduction between the Generic Group Model and a Symbolic Generic Group Model, in which the security experiment is purely deterministic. Security in the Symbolic Generic Group Model is trivially equivalent to a purely algebraic condition. For

M. Fischlin and J.-S. Coron (Eds.): EUROCRYPT 2016, Part II, LNCS 9666, pp. 822–851, 2016.
DOI: 10.1007/978-3-662-49896-5_29

instance, the algebraic condition for a decisional assumption requires to prove that the two sets of polynomials extracted from the left and right games have the same linear dependencies. Therefore, and unavoidably, the difficulty of checking the algebraic condition increases as the assumption becomes more complex, as witnessed by unfortunate failures [24,28,39]. For some recent hypotheses, several pages of error-prone calculations are required for proving that the algebraic condition holds, and several authors have used computer algebra systems to carry part of the verifications. These examples suggest the importance of building general tools to assist proofs of security assumptions in the Generic Group Model. One such tool is the Generic Group Analyzer [11], which uses SMT solvers and computer algebra systems to analyze DDH-like assumptions. The tool takes as input a description of an assumption and either returns an algebraic attack or a concrete probability bound if the assumption is secure. The Generic Group Analyzer primarily works for non-interactive assumptions, in which the adversary can only call the oracles which perform the algebraic operations.

The Generic Group Model can also be used for proving the security of cryptographic constructions, such as signature schemes and algebraic MACs, against algebraic attacks. In this context, the adversary has access to oracles for performing signatures, verification, *etc.* The Generic Group Analyzer also provides support for such problems, but is inherently limited to oracles which do not take handles to group elements as inputs. This support can be used for analyzing simple interactive assumptions. Subsequent extensions of the Generic Group Analyzer overcome this limitation by providing support for oracles that take handles as inputs, and by allowing adversaries to make a bounded number of oracle queries [12]. Using this extension, Barthe et al. [12] synthesize (in the Type II-setting) structure-preserving signatures that are secure against adversaries that can make a bounded number of signing queries. Their approach is based on an algebraic characterization of security, using a vector space whose dimension increases by one for each query. Therefore, their approach is limited to a small number of queries, and an alternative approach must be used for proving security notions which do not impose a bound on the number of queries.

The first main contribution of this paper is to extend the Master Theorem to a general setting where adversaries can make arbitrarily many queries to oracles with group inputs, and where the winning conditions can be described using a rich language. As for simpler Master Theorems, our Master Theorem yields a sufficient condition for the security of cryptographic constructions. However, this simpler condition cannot be expressed in finite-dimensional linear algebra: informally, each adversarial query to an oracle taking group elements as inputs increases the dimension of the system to be analyzed, and therefore allowing arbitrarily many queries leads to a system that is not finite-dimensional. As a consequence, the algebraic approach of the Generic Group Analyzer cannot be used to analyze automatically sufficient conditions given by the Master Theorem.

The second main contribution of this paper is an automated method for proving the validity of these conditions, using a combination of methods from constraint solving, computer algebra, and symbolic cryptography. Building on these two contributions, we implement an analyzer which subsumes the Generic Group Analyzer for interactive assumptions and is able to analyze many cryptographic constructions, including signatures and message authentication codes.

Technical Overview. In more detail, our contributions are as follows.

First, we define a language to express security experiments in the Generic Group Model where the adversary can make an unbounded number of queries to oracles; moreover, our model allows oracles to take group values as inputs. In addition, we define a rich language of winning conditions. We then establish a Master Theorem, which states that a generic algorithm is secure with respect to a security goal expressed using our language of winning conditions, if the constraint system extracted from the security experiment, given by the algorithm and the winning condition, has no computable solution. Informally, the notion of computable solution provides an algebraic counterpart to the notion of deducibility used in the symbolic (a.k.a. Dolev-Yao) approach to cryptography; more technically, this notion is based on an inductive definition of the adversary's knowledge throughout execution of the algorithm. From a broader perspective, our Master Theorem provides a novel light on the relationship between different cryptographic models, by showing a general relationship between the Generic Group Model and the symbolic model. Note that, for the sake of simplicity, we focus on group settings with bilinear pairings; however, we believe that our model and Master Theorem can naturally extend to the case of multilinear maps.

Second, we define an automated method for proving the absence of computable solutions of constraint systems. Our language of constraints supports algebraic expressions that are generally not considered by prior work on the symbolic model. Therefore, we cannot use previous constraint-solving methods developed for reasoning about cryptographic protocols in the symbolic model. Rather, we define a specialized method which combines general purpose algebraic computations and specialized steps. The algebraic computations are performed using Gröbner bases, whereas the specialized steps include simplifications related to big operators and case distinctions. The latter can be used to add new equations to constraint systems and thus to trigger new simplifications. Case distinctions are an essential ingredient for the success of our method: they yield compact proofs that follow the structure of pen-and-paper arguments found in the literature. Of course, the use of case distinctions is not new in automated deduction; it is at the core of Staalmarck's method, an empirically successful method for propositional logic. However, its use in our setting appears to be new.

Third, we implement our method and evaluate its effectiveness on a sizable set of case studies. Our tool uses off-the-shelf computer algebra systems to perform Gröbner bases computations. However, it draws its efficiency from a finely tuned heuristics for carrying case distinctions. We evaluate our tool on structure-preserving signatures, in all settings (Type I, Type II and Type III). Our tool is able to prove unbounded security of many structure-preserving signatures from the literature, as well as of the algebraic MACs from Chase, Meiklejohn and

Zaverucha [18], and of the short randomizable signatures from Pointcheval and Sanders [34]. Furthermore, it also proves unbounded security for most of the examples proved 2-time secure in [12] (these examples were generated automatically using synthesis techniques). Moreover, we also adapt the synthesis tool from [12] to generate structure-preserving signatures in the Type III setting and use our tool to prove security for more than a 100 such schemes.

Related Work. The Generic Group Model was introduced by Nechaev [33], Shoup [38] and Maurer [32], following distinct but equivalent approaches [29]. The original approach by Nechaev and Shoup lets the adversary access a randomly selected representation of group elements; in contrast, Maurer's approach requires the adversary to perform all algebraic operations via oracles, and uses handles as symbolic representations of group elements known to the adversary. We opt for the second approach, for its distinctively symbolic flavour. These works establish lower complexity bounds for the generic discrete logarithms and the generic hardness of Diffie-Hellman like assumptions. As for us, they use the Schwartz-Zippel Lemma for transforming their original problem into an algebraic one. This approach was extended by Boneh, Boyen and Goh [17]. First, their Generic Group Model focuses on bilinear groups. Second, they consider a general class of assumptions, and provide the first Master Theorem, which provides a systematic method for extracting algebraic conditions of security from assumptions. Their Master Theorem was subsequently extended in many directions. The most relevant works are those that involve the use of computer tools for verifying algebraic conditions. Notably, Freeman [23] verifies the hardness of two assumptions using Magma.

Shoup [38] and Schnorr and Jakobsson [36,37] were among the first to use the Generic Group Model for proving the security of crytographic constructions. Specifically, Shoup proves (generic) security of an identification scheme, whereas Schnorr and Jakobsson consider signed ElGamal encryption and blind discrete log signatures. More recently, the Generic Group Model has also become an important tool for analyzing the security of pairing-based cryptographic constructions. Chase, Meiklejohn and Zaverucha [18] propose a class of algebraic MACs and prove their generic security. Several authors use the Generic Group Model for proving the generic security of structure-preserving signatures [1]. Groth [26] proposes new fully-structure-preserving signatures [6] and proves their generic security. Similarly, Fuchsbauer, Hanser and Slamanig [25] define a structure-preserving signature on equivalence classes and prove its generic security. Furthermore, the Generic Group Model gives a convenient setting for establishing lower bounds on the complexity of structure-preserving signatures [2,4,5,12]. In a similar spirit, the Generic Group Model has been used for proving the correctness of translations of signature schemes from Type I to Type III [3,5,7].

It is also worth pointing to a recent examination of the efficiency of pairing-based implementations. Based on a practical evaluation of the efficiency of state-of-the-art implementations of pairings, Chatterjee and Menezes [19] argue that

Type III pairings are more efficient than their Type II counterparts, and should be favoured in implementations. Their observation justifies the need to transpose existing results and tools for the Type II setting to the Type III setting, and has motivated the application of our methods to the latter.

Several works have developed or used tools for reasoning about the Generic Group Model. As already mentioned, the Generic Group Analyzer [11] implements an automated method for analyzing assumptions. Moreover, a subsequent extension of the analyzer [12] supports the automated analysis of security of structure-preserving security against adversaries that make a bounded number of queries. In practice, the tool only terminates for small bounds on the number of queries. While these works are the most closely related to ours, there have been previous works that apply computer tools to the Generic Group Model. Barthe, Cederquist and Tarento [9,15] were the first to use formal verification tools for analyzing the security of hardness assumptions and cryptographic constructions in the Generic Group Model. Their work uses the Coq proof assistant, and provides no support for automation. Freeman [23] reports on using computer algebra systems to prove the validity of new hardness assumptions in the Generic Group Model. Beyond the Generic Group Model, there exist several tools for synthesizing constructions, such as encryption schemes, modes of operations, tweakable blockciphers, and structure-preserving signatures in the Type II setting [10,12,27,31], automated transformation of existing constructions, including signature schemes [3,7,8], and verification of security proofs [13,14,16]. In particular, [14] introduce AutoG&P, a highly automated framework for proving the security of pairing-based cryptographic primitives; the focus of [14] is on encryption schemes, but their methods are also applicable to signatures and MACs. AutoG&P and gga$^\infty$ are complementary in two different ways. First, gga$^\infty$ focuses on full automation in the Generic Group Model while AutoGP provides partial automation in the Standard model. Second, and more interestingly, some of our techniques for equational reasoning could be used to achieve more automation in AutoG&P, whereas it could be possible to use techniques from AutoG&P as a fallback solution when full automation fails in gga$^\infty$.

2 Preliminaries

In this section, we give some background on bilinear groups and define the notation used throughout the paper.

2.1 Bilinear Groups

We consider bilinear groups $\mathcal{G} = (\mathbb{G}_1, \mathbb{G}_2, \mathbb{G}_t, e : \mathbb{G}_1 \times \mathbb{G}_2 \to \mathbb{G}_t)$. For Type I, $\mathbb{G}_1 = \mathbb{G}_2$ and for Type II, there is an additional isomorphism $\Psi : \mathbb{G}_2 \to \mathbb{G}_1$. We use additive notation for all three groups and use P_1, P_2, P_t to denote their generators. For $a \in \mathbb{F}_p$, we use $[\![a]\!]_i$ to denote the implicit representation aP_i of a in $\mathbb{G}_i$ following [21].

2.2   Notation

We define $aS = \{as \mid s \in S\}$ and $SS' = \{ss' \mid s \in S \wedge s' \in S'\}$. For a set S, we write S^* to denote vectors of elements in S. We define $[n]$ as the range $\{1, \ldots, n\}$ for an arbitrary $n \in \mathbb{N}$. We use $\boldsymbol{v}$ to denote a vector and $\boldsymbol{v}_{(i)}$ to denote the i-th element. We assume given a set of *uniform variables* UVar, a set of *handle variables* HVar $=$ HVar$_1 \uplus$ HVar$_2 \uplus$ HVar$_t$, a set of *parameter variables* PVar, and a set of *index variables* IVar. We use $ty(h) \in \{1, 2, t\}$ to denote the type of a handle variable, i.e., $ty(h) = i$ iff $h \in$ HVar$_i$.

We use $R[\boldsymbol{X}^{\pm 1}]$ to denote the set of *Laurent polynomials* over the ring R with variables in $\boldsymbol{X}$. We also use the shorthand $R[\boldsymbol{Y}, \boldsymbol{X}^{\pm 1}]$ for $(R[\boldsymbol{Y}])[\boldsymbol{X}^{\pm 1}]$ to denote nested polynomial rings. We use a similar notation $Mon[\boldsymbol{X}^{\pm 1}, \boldsymbol{Y}]$ for *Laurent monomials*. We write $\deg_V(M)$ to denote the *degree* of V in the Laurent monomial M. We write $coeff_M(F)$ to denote the coefficient of the Laurent monomial M in the Laurent polynomial F.

For a term t possibly containing variables, we write $t[x \mapsto t']$ to denote the result of substituting all occurrences of the variable x in t with t'. A context C is a term with a distinguished variable $\square$ which denotes a hole that can be filled in by an arbitrary term. We assume the hole occurs exactly once in a context. We use $C[t]$ to denote the term obtained by plugging t into $C's$ hole.

3   Translating Security Experiments into Constraints

In this section, we first present a language to define security experiments in the Generic Group Model. Next, we define the language of winning constraints. Winning constraints are formulas that characterize if an adversary can win a security experiment. Finally, we present a translation procedure from security experiments to winning constraints.

3.1   Security Experiment Definition

We first present the language that we use to define security experiments. Afterwards, we define the corresponding games in the Generic Group Model and the symbolic group model (see [11]). We will exploit that the generic and symbolic games are indistinguishable and use the symbolic game to perform our analysis.

Definition 1 *(Security experiment). A security experiment is defined by a tuple* $SE = (t, ainp, odef, wcond)$ *where*

- *the* group type *is defined by* $t \in \{\text{I}, \text{II}, \text{III}\}$,
- *the* adversary input *is defined by* $ainp = (\boldsymbol{X}, (\boldsymbol{F_1}, \boldsymbol{F_2}, \boldsymbol{F_t}))$ *for*
  - global uniform variables $\boldsymbol{X} \in \text{UVar}^*$ *and*
  - input polynomials $\boldsymbol{F_i} \in \mathbb{Z}[\boldsymbol{X}^{\pm 1}]^*$,
- *the* oracle *is defined by* $odef = (\boldsymbol{a}, \boldsymbol{h}, \boldsymbol{R}, (\boldsymbol{H_1}, \boldsymbol{H_2}, \boldsymbol{H_t}))$ *for*
  - arguments $\boldsymbol{a} \in \text{PVar}^*$ *and oracle handles* $\boldsymbol{h} \in \text{HVar}^*$,[1]

---

[1] Handle variables are typed, i.e., for all $j \in [|\boldsymbol{h}|]$, it holds that $ty(\boldsymbol{h}_{(j)}) \in \{1, 2, t\}$.

- oracle uniform variables $R \in \mathsf{UVar}^*$, *and*
- oracle polynomials $H_i \in \mathbb{Z}[X^{\pm 1}, R^{\pm 1}, a, h]^*$, *and*
- *the* winning condition *is defined by* $wcond = (\hat{a}, \hat{H}, W^=, W^{\neq})$ *for*
    - winning arguments $\hat{a} \in \mathsf{PVar}^*$ *and* winning handles $\hat{h} \in \mathsf{HVar}^*$, *and*
    - winning (in)equalities $W^=, W^{\neq} \in \mathbb{Z}[X^{\pm 1}, R^{\pm 1}, a, h, \hat{a}, \hat{h}]^*$.

Intuitively, the *adversary input* represents the values given initially to the adversary. This usually includes the public parameters and the public keys. The *oracle* is defined by *arguments* and *oracle handles* that represent the oracle input; *uniform variables* that denote randomness sampled by the oracle; and *oracle polynomials* that denote the oracle response. Finally, the *winning condition* is defined by *winning arguments* that represent the forgery that the adversary must produce; and *winning (in)equalities* that characterize valid forgeries.

We define the corresponding generic group game $\mathsf{G}^{\mathsf{gen}}(SE)$ as follows:

1. Sample the vector $x \in (\mathbb{F}_p^{\times})^{|X|}$, compute the adversary inputs $\llbracket F_i(x) \rrbracket_i \in \mathbb{G}_i^{|F_i|}$ (for $i \in \{1, 2, \mathsf{t}\}$), and call the adversary $\mathcal{A}$ with the corresponding handles.

2. The adversary $\mathcal{A}$ can perform q_g queries to perform group operations (for group type t), an unbounded number of equality queries, and q queries to an oracle that implements *odef*. The oracle for *odef* takes scalars $v \in \mathbb{F}_p^{|a|}$ for a and a vector of handles to group elements U for h. We use u to denote the discrete logarithms of U, i.e., for all $j \in [|h|]$, $\llbracket u_{(j)} \rrbracket_i = U_{(j)}$ where $i = ty(h_{(j)})$. Then it samples $r \in (\mathbb{F}_p^{\times})^{|R|}$ and returns handles to $\llbracket H_i(x, v, u, r) \rrbracket_i \in \mathbb{G}_i^{|H_i|}$. We use $v^{(j)}, u^{(j)}, r^{(j)}$ to denote the corresponding values used in the j-th query.

3. The adversary $\mathcal{A}$ returns scalars $\hat{v} \in \mathbb{F}_p^{|\hat{a}|}$ for $\hat{a}$ and handles to group elements $\hat{U}$ for $\hat{h}$. Again, we denote the discrete logarithms of $\hat{U}$ with $\hat{u}$. The adversary wins if for $\bowtie \in \{=, \neq\}$, $w \in W^{\bowtie}$, and $j \in [q]$, it holds that $w(x, r^{(j)}, v^{(j)}, u^{(j)}, \hat{v}, \hat{u}) \bowtie 0$.

Note that additional care must be taken to ensure that the oracles and winning conditions are efficiently computable using scalar multiplication, addition, application of isomorphisms, and application of bilinear maps. For example, it is possible to specify an oracle that takes a handle to an element $\llbracket v \rrbracket_{\mathsf{t}} \in \mathbb{G}_{\mathsf{t}}$ and returns $\llbracket v \rrbracket_1 \in \mathbb{G}_1$, which cannot be efficiently computed in most bilinear groups of interest.

The symbolic game $\mathsf{G}^{\mathsf{sym}}(SE)$ is defined similarly, but internally uses Laurent polynomials $f(X)$ instead of group elements $\llbracket f(x) \rrbracket_i$. It is completely deterministic since it uses formal variables X to represent the initially sampled values and indexed formal variables $R^{(j)}$ to represent the values sampled in the oracle.

Formally, we define $\mathsf{G}^{\mathsf{sym}}(SE)$ as follows:

1. Store the polynomials $F_i(X) \in \mathbb{Z}[X^{\pm 1}]^{|F_i|}$ in the list for the group $\mathbb{G}_i$ (for $i \in \{1, 2, \mathsf{t}\}$) and call the adversary $\mathcal{A}$ with the corresponding handles.

2. The oracles for group operations and equality checks provide the same interface as in the generic model, but perform all computations in the ring of Laurent polynomials. The oracle for *odef* takes (in the j-th query) scalars $v \in \mathbb{F}_p^{|a|}$ for $\boldsymbol{a}$ and handles to polynomials

$$\boldsymbol{u} \in \mathbb{Z}[\boldsymbol{X}^{\pm 1}, (\boldsymbol{R}^{(1)})^{\pm 1}, \ldots, (\boldsymbol{R}^{(j-1)})^{\pm 1}]^{|h_i|}$$

for $\boldsymbol{h}$. It returns handles to polynomials

$$\boldsymbol{H_i}(\boldsymbol{X}, \boldsymbol{v}, \boldsymbol{u}, \boldsymbol{R}^{(j)}) \in \mathbb{Z}[\boldsymbol{X}^{\pm 1}, (\boldsymbol{R}^{(1)})^{\pm 1}, \ldots, (\boldsymbol{R}^{(j)})^{\pm 1}]^{|H_i|}.$$

3. The adversary $\mathcal{A}$ returns scalars $\hat{\boldsymbol{v}} \in \mathbb{F}_p^{|\hat{a}|}$ for $\hat{\boldsymbol{a}}$ and handles to polynomials

$$\hat{\boldsymbol{u}} \in \mathbb{Z}[\boldsymbol{X}^{\pm 1}, (\boldsymbol{R}^{(1)})^{\pm 1}, \ldots, (\boldsymbol{R}^{(q)})^{\pm 1}]^{|\hat{h}_i|}$$

for $\hat{\boldsymbol{h}}$. He wins if for $\bowtie \in \{=, \neq\}$, $w \in \boldsymbol{W}^{\bowtie}$, and $j \in [q]$, it holds that $w(\boldsymbol{X}, \boldsymbol{R}^{(j)}, \boldsymbol{v}^{(j)}, \boldsymbol{u}^{(j)}, \hat{\boldsymbol{v}}, \hat{\boldsymbol{u}}) \bowtie 0$.

Setup $\mathcal{P}(1^\lambda)$: Return $PP = (p, \mathbb{G}_1, \mathbb{G}_2, \mathbb{G}_t, e) \leftarrow \mathcal{G}(1^\lambda)$ where $\mathcal{G}$ is a polynomial time algorithm that on input 1^λ returns a description of a bilinear map in the Type III setting with groups of order p for a λ-bit prime p.

Key generation $\mathcal{K}(PP)$:
  Choose $v, w \leftarrow \mathbb{F}_p^\times$ and compute $VK = (PP, V, W)$ and $SK = (PP, v, w)$ as

$$V \leftarrow [\![v]\!]_1 \text{ and } W \leftarrow [\![w]\!]_1.$$

Signing $\mathcal{S}_{SK}(M)$:
  For $M = [\![m]\!]_2 \in \mathbb{G}_2$ choose $r \leftarrow \mathbb{F}_p^\times$ and compute the signature (T_1, T_2, S) as

$$T_1 \leftarrow [\![r]\!]_1, \ T_2 = [\![r]\!]_2, \text{ and } S \leftarrow [\![mv + w + r^2]\!]_2.$$

Verification $\mathcal{V}_{VK}(M, S)$:
  Accept if and only if $T_1 \in \mathbb{G}_1$, $M, T_2, S \in \mathbb{G}_2$,

$$e([\![1]\!]_1, S) = e(V, M) + e(W, [\![1]\!]_2) + e(T_1, T_2), \text{ and } e(T_1, [\![1]\!]_2) = e([\![1]\!]_1, T_2).$$

Fig. 1. SPS-scheme from [19] in Type III setting.

Example 1. We can formalize the EUF-CMA security of the scheme in Fig. 1 using the security experiment $SE = (t, ainp, odef, wcond)$ defined as follows:

- the group type is $t = \mathrm{III}$
- the adversary input is $ainp = (\boldsymbol{X}, (\boldsymbol{F_1}, \boldsymbol{F_2}, \boldsymbol{F_t}))$ where
  - $\boldsymbol{X} = (v, w)$ (for $v, w \in \mathsf{UVar}$), $\boldsymbol{F_1} = (1, v, w)$, $\boldsymbol{F_2} = (1)$, $\boldsymbol{F_t} = (1)$
- the *oracle* is $odef = (\boldsymbol{a}, \boldsymbol{h}, \boldsymbol{R}, (\boldsymbol{H_1}, \boldsymbol{H_2}, \boldsymbol{H_t}))$ where

- $a = ()$, $h = (m)$ (for $m \in \mathsf{HVar}_2$),
- $R = (r)$ (for $r \in \mathsf{UVar}$)
- $H_1 = (r)$, $H_2 = (r,\ mv + w + r^2)$, $H_t = ()$
- the winning condition is $wcond = (\hat{a}, \hat{h}, W^=, W^{\neq})$ where
  - $\hat{a} = ()$, $\hat{h} = (\hat{m}, \hat{t}_1, \hat{t}_2, \hat{s})$ and (for $\hat{t}_1 \in \mathsf{HVar}_1$ $\hat{m}, \hat{t}_2, \hat{s} \in \mathsf{HVar}_2$),
  - $W^= = (\hat{s} - \hat{m}v - w - \hat{t}_1\hat{t}_2,\ \hat{t}_1 - \hat{t}_2)$, $W^{\neq} = (\hat{m} - m^{(j)})$ ∎

3.2   Winning Constraints

We first define the language of winning constraints, a class of formulas that can be used to characterize if an adversary can win the symbolic game $\mathsf{G}^{\mathsf{sym}}(SE)$. Then we define the set of solutions of a winning constraint and present a set of simplication rules that preserve the set of solutions.

$$
\begin{array}{lll}
\mathcal{C} ::= \exists i \notin K.\mathcal{C} \mid \mathcal{C} & & \text{constraint} \\[4pt]
\mathcal{C}' ::= \mathcal{C}' \wedge \mathcal{C}' \mid \forall k \notin K.\mathcal{C}' \mid \mathcal{E} = 0 \mid \mathcal{E} \neq 0 & & \text{non-existential constraint} \\[4pt]
\mathcal{E} ::= \mathcal{E} + \mathcal{E} \mid \mathcal{E} * \mathcal{E} \mid -\mathcal{E} \mid \mathsf{Coeff}_{\mathcal{M}}(\mathcal{E}) & & \text{expression} \\[4pt]
\quad \mid \sum_{k \notin K} \mathcal{E} \mid \mathcal{R} \mid \mathcal{R}^{-1} \mid \mathcal{P} \mid \mathcal{V} \mid 1 \mid 0 & & \\[8pt]
\mathcal{M} ::= \mathcal{M} * \mathcal{M} \mid \mathcal{R} \mid \mathcal{R}^{-1} \mid 1 & & \text{monomial over uniform variables} \\[4pt]
\mathcal{R} ::= R_{[k]} \mid R & & \text{(indexed) uniform variable } (R \in \mathsf{UVar}) \\[4pt]
\mathcal{P} ::= \rho_{[k]} \mid \rho & & \text{(indexed) parameter } (\rho \in \mathsf{PVar}) \\[4pt]
\mathcal{V} ::= Y_{[k]} & & \text{indexed handle variable } (Y \in \mathsf{HVar})
\end{array}
$$

Fig. 2. Grammar for winning constraints (for $k \in \mathsf{IVar}, K \subset \mathsf{IVar}$). For every $\mathsf{Coeff}(\mathcal{E})$, $\mathcal{E}$ does not contain the symbol Coeff.

Definition 2 *(Winning constraints). The language of winning constraints is defined by the grammar given in Fig. 2. We distinguish between* bound index variables *and* free index variables *depending on whether they are bound by* $\forall/\Sigma$. *We write* $\mathrm{ivars}(\mathcal{C})$ *to denote the free index variables in the constraint* $\mathcal{C}$.

Intuitively, atomic constraints $\mathcal{E} = 0$ represent polynomial equalities. In the quantifications $\forall k \notin K$ and $\sum_{k \notin K}$, the index variable k ranges over all elements in $[q]$ except for the valuations of the index variables in K. Uniform variables $R/R_{[k]}$ are treated like formal variables, parameters $\rho/\rho_{[k]}$ can be instantiated with integers, handle variables $Y_{[k]}$ can be instantiated with Laurent polynomials over uniform variables, and the arithmetic operations are interpreted in the ring

$$eval_s(c) = \begin{cases} eval_{s_{k,K,1}}(c') \vee \ldots \vee eval_{s_{k,K,q-|K|}}(c') & \text{for } c = \exists k \notin K.c' \\ eval_{s_{k,K,1}}(c') \wedge \ldots \wedge eval_{s_{k,K,q-|K|}}(c') & \text{for } c = \forall k \notin K.c' \\ eval_s(c_1) \otimes eval_s(c_2) & \text{for } c = c_1 \otimes c_2 \\ eval_{s_{k,K,1}}(c') + \ldots + eval_{s_{k,K,q-|K|}}(c') & \text{for } c = \sum_{k \notin K} c' \\ \delta(\rho) & \text{for } c = \rho \\ \chi(\rho, \sigma(i)) & \text{for } c = \rho_{[i]} \\ \xi(Y, \sigma(i)) & \text{for } c = Y_{[i]} \\ R^e & \text{for } c = R^e, e \in \{+1, -1\} \\ R^e_{[\sigma(i)]} & \text{for } c = R^e_{[i]}, e \in \{+1, -1\} \\ coeff_{\sigma(\mathcal{M})}(eval_s(\mathcal{E})) & \text{for } c = \mathsf{Coeff}_{\mathcal{M}}(\mathcal{E}) \\ c & \text{for } c \in \{0, 1\} \end{cases}$$

Fig. 3. Definition of the evaluation function $eval_s$ for $s = (p, q, \sigma, \delta, \chi, \xi)$, where $R \in \mathsf{UVar}$, $\otimes \in \{=, \neq, \wedge, *, +\}$ are interpreted as the corresponding boolean operations/arithmetic operations in the ring of Laurent polynomials over $\mathbb{F}_p$ and $s_{k,K,i}$ defined as follows. Let $\{v_1, \ldots v_{q-|K|}\} = [q] \setminus \sigma(K)$, then $s_{k,K,i} = (p, q, \sigma', \delta, \chi, \xi)$ where $\sigma' = \sigma[k \mapsto v_i]$ for $i \in \{1, \ldots, q - |K|\}$.

of Laurent polynomials over $\mathbb{F}_p$ for a prime p. An expression $\mathsf{Coeff}_{\mathcal{M}}(\mathcal{E})$ represents the coefficient of the monomial $\mathcal{M}$ in the expression $\mathcal{E}$ after the parameters and handle variables in $\mathcal{E}$ are instantiated. The resulting Laurent polynomial after instantiation contains only (indexed) uniform variables. Formally, the set of solutions of a winning constraint is defined as follows.

Definition 3 (*Solutions of winning constraints*). *A structure $s = (p, q, \sigma, \delta, \chi, \xi)$ for a prime number p, a natural number q, a valuation $\sigma : \mathsf{IVar} \to [q]$ for (free) index variables, valuations $\delta : \mathsf{PVar} \to \mathbb{F}_p$ and $\chi : \mathsf{PVar} \times [q] \to \mathbb{F}_p$ for the parameters, and a valuation $\xi : \mathsf{HVar} \times [q] \to \mathbb{F}_p[\mathsf{UVar}^{\pm 1}, \mathsf{UVar}^{\pm 1}_{[1]}, \ldots, \mathsf{UVar}^{\pm 1}_{[q]}]$ for the handle variables is a solution for a winning constraint $\mathcal{C}$ if $eval_s(\mathcal{C}) = true$ for the function $eval$ defined in Fig. 3.*

3.3 Translation from Security Experiments to Winning Constraints

We define the translation function to convert a security experiment definition into winning constraints. The translation is sound and complete with respect to a certain class of solutions. Roughly, this means that there is an efficient attacker[2] on the security experiment in the Generic Group Model with non-negligible winning probability iff there is a solution for the translated winning constraints where handle variables are instantiated with "computable" Laurent polynomials.

To simplify the presentation, we assume that for all security experiments in Type II, it holds that $F_2 \subseteq F_1$ and $H_2 \subseteq H_1$ which allows us to ignore the

---

[2] More precisely, an attacker that performs a polynomial number of queries q_g and q.

isomorphism Ψ. Similarly, we assume for Type I that $F_1 = F_2$ and $H_1 = H_2$ which allows us to ignore that $\mathbb{G}_1 = \mathbb{G}_2$.

First, note that $W^{\bowtie} \subset \mathbb{Z}[X^{\pm 1}, R^{\pm 1}, a, h, \hat{a}, \hat{h}]$ where $X, R \in \mathsf{UVar}^*$, $a, \hat{a} \in \mathsf{PVar}^*$, and $h, \hat{h} \in \mathsf{HVar}^*$. For an index variable $j \in \mathsf{IVar}$, we write $R_{[j]}$ to denote the vector $(R_{(1)[j]}, \ldots, R_{(|R|)[j]})$ of indexed uniform variables. Similarly, we write $a_{[j]}$ and $h_{[j]}$. For our translation, we instantiante each winning handle variable $\hat{h}_{(u)} \in \mathsf{HVar}_1 \cup \mathsf{HVar}_2$ with a linear combination of polynomials in the adversary input and in the oracle output. Formally, we define the vector E of expressions as follows. For $u \in [|\hat{h}|]$ such that $\hat{h}_{(u)} \in \mathsf{HVar}_1$ and $l = |H_1|$, we define

$$E_{(u)} = \rho^{(1,u,1)} F_{1(1)}(X) + \ldots + \rho^{(1,u,|F_1|)} F_{1(|F_1|)}(X) +$$
$$\sum_k \tau_{[k]}^{(1,u,1)} H_{1(1)}(X, R_{[k]}, a_{[k]}, h_{[k]}) + \ldots +$$
$$\sum_k \tau_{[k]}^{(1,u,l)} H_{1(l)}(X, R_{[k]}, a_{[k]}, h_{[k]})$$

where $\rho^{(1,u,n)}$ and $\tau^{(1,u,n)}$ are distinct fresh parameter variables. For $u \in [|\hat{h}|]$ such that $\hat{h}_{(u)} \in \mathsf{HVar}_2$, we define $E_{(u)}$ analogously. For $u \in [|\hat{h}|]$ such that $\hat{h}_{(u)} \in \mathsf{HVar}_{\mathsf{t}}$, we define $E_{(u)}$ analogously additionally taking products of polynomials from $\mathbb{G}_1$ and $\mathbb{G}_2$ into account. We define the winning constraint derived from SE as

$$toConstr(SE) = \bigwedge_{w \in W^{\bowtie}} \forall j. \left(w(X, R_{[j]}, a_{[j]}, h_{[j]}, \hat{a}, E) \bowtie 0 \right).$$

A priori, the notion of solution for winning constraints does not restrict the set of Laurent polynomials that can be used to instantiate the handle variables in $h_{[j]}$. Since we are only interested in solutions where the instantiations of handle variables are computable, we now define the notion of constrained solution.

Definition 4 *(Constrained solutions of winning constraints). A solution is con-strained by sequences of sets* $\{K_j^{(i)}\}_{j \in \mathbb{N}}$ *of Laurent polynomials (for* $i \in \{1, 2, \mathsf{t}\}$*) if for all* $i \in \{1, 2, \mathsf{t}\}$, $Y \in \mathsf{HVar}_i$, *and* $j \in [q]$, *it holds that* $\xi(Y, j) \in K_j^{(i)}$.

Since we are interested in solutions constrained by computable Laurent poly-nomials, we next define the sequences of computable polynomials. We use $\langle S \rangle$ to denote the vector space over $\mathbb{F}_p$ generated by S.

Definition 5 *(Computable polynomials). The sequences of computable polyno-mials for a security experiment*

$$SE = (t, X, (F_1, F_2, F_{\mathsf{t}})), (a, h, R, (H_1, H_2, H_{\mathsf{t}})), wcond)$$

are defined as follows:

$$\mathcal{K}_0^{SE,(i)} = \langle toSet(F_i) \rangle \qquad\qquad for\ i \in \{1, 2\}$$
$$\mathcal{K}_0^{SE,(\mathsf{t})} = \langle toSet(F_{\mathsf{t}}) \cup (\mathcal{K}_0^{SE,(1)} * \mathcal{K}_0^{SE,(2)}) \rangle$$

$$\mathcal{K}_{j+1}^{SE,(i)} = \langle \mathcal{K}_j^{SE,(i)} \cup \qquad\qquad\qquad\qquad\qquad for\ j \geq 0, i \in \{1,2\}$$

$$\{H(\boldsymbol{X}, \boldsymbol{v}, \boldsymbol{E}, \boldsymbol{R}^{(j+1)}) \mid H \in \boldsymbol{H}_i \land$$

$$\boldsymbol{v} \in \mathbb{F}_p^{|a|} \land |\boldsymbol{E}| = |\boldsymbol{h}| \land \boldsymbol{E}_{(u)} \in \mathcal{K}_j^{SE,(ty(h_{(u)}))}\})$$

$$\mathcal{K}_{j+1}^{SE,(t)} = \langle \mathcal{K}_j^{SE,(t)} \cup (\mathcal{K}_{j+1}^{SE,(1)} * \mathcal{K}_{j+1}^{SE,(2)}) \cup \qquad\qquad for\ j \geq 0$$

$$\{H(\boldsymbol{X}, \boldsymbol{v}, \boldsymbol{E}, \boldsymbol{R}^{(j+1)}) \mid H \in \boldsymbol{H}_t \land$$

$$\boldsymbol{v} \in \mathbb{F}_p^{|a|} \land |\boldsymbol{E}| = |\boldsymbol{h}| \land \boldsymbol{E}_{(u)} \in \mathcal{K}_j^{SE,(ty(h_{(u)}))}\})$$

The definition is always valid for Type III. For Types I and II, it is valid under the previously stated assumptions on $\boldsymbol{F}_i$ and $\boldsymbol{H}_i$. We say a solution s is an SE-computable solution if it is constrained by $(\mathcal{K}_j^{SE,(i)})_{j,i}$.

Theorem 1 *(Soundness and Completeness of Translation). Let $p \approx 2^\lambda$ and q_g, q polynomial in λ. Then the winning probability in the generic group game $\mathsf{G}^{gen}(SE)$ with a group of order p is negligible in λ for all adversaries that perform at most q_g (resp. q) queries iff there is no SE-computable solution for $toConstr(SE)$.*

Proof (Sketch). For all concrete values of q_g, q, and SE we can use the master theorem for interactive assumptions from [11] (more precisely, the extended version for handles from [22]) to obtain an algebraic criterion that is equivalent to the security of the construction. By unfolding the definitions of $toConstr$ and $eval$, we can verify that the criterion is true for all bounds on the number of oracle-queries iff there is no SE-computable solution for $toConstr(SE)$. □

Example 2. The translation of the security experiment for the example in Fig. 1 to winning constraints is

$$\hat{S} - \hat{M} * V - W - \hat{T}_1 * \hat{T}_2 = 0 \quad \land \quad \hat{T}_1 - \hat{T}_2 = 0 \quad \land \quad \forall k.\ \hat{M} - M_{[k]} \neq 0$$

where $V, W, R \in \mathsf{UVar}$ and $M \in \mathsf{HVar}_2$, $\mu, \mu', \mu'', \rho, \rho', \rho'', \rho''', \tau, \tau', \tau'', \gamma, \gamma', \gamma'' \in \mathsf{PVar}$, and $\hat{M}, \hat{S}_1, \hat{S}_2, \hat{S}_3$ are defined as

$$\hat{M} = \mu + \sum_k \mu'_{[k]} * R_{[k]} + \sum_k \mu''_{[k]} * (M_{[k]} * V + W + R_{[k]}^2),$$

$$\hat{T}_1 = \rho + \sum_k \rho'_{[k]} * R_{[k]} + \rho'' * V + \rho''' * W,$$

$$\hat{T}_2 = \tau + \sum_k \tau'_{[k]} * R_{[k]} + \sum_k \tau''_{[k]} * (M_{[k]} * V + W + R_{[k]}^2),\ \text{and}$$

$$\hat{S} = \gamma + \sum_k \gamma'_{[k]} * R_{[k]} + \sum_k \gamma''_{[k]} * (M_{[k]} * V + W + R_{[k]}^2).$$

We first outline the sequence of computable monomials for $\mathbb{G}_1$:

$$\mathcal{K}_0^{SE,(2)} = \langle 1, V, W \rangle$$
$$\mathcal{K}_1^{SE,(2)} = \langle \mathcal{K}_0^{SE,(2)} \cup \{R_{[1]}\} \rangle$$
$$\mathcal{K}_2^{SE,(2)} = \langle \mathcal{K}_1^{SE,(2)} \cup \{R_{[2]}\} \rangle$$
$$\cdots$$

For $\mathbb{G}_2$, the sequence looks as follows:

$$\mathcal{K}_0^{SE,(2)} = \langle 1 \rangle$$
$$\mathcal{K}_1^{SE,(2)} = \langle \mathcal{K}_0^{SE,(2)} \cup \{1, R_{[1]}, \overbrace{V + W + R_{[1]}^2}^{:=f_1}\} \rangle$$
$$\mathcal{K}_2^{SE,(2)} = \langle \mathcal{K}_1^{SE,(2)} \cup \{R_{[2]}, R_{[1]} * V + W + R_{[2]}^2, f_1 * V + W + R_{[2]}^2\} \rangle$$
$$\cdots$$

For $\mathbb{G}_t$, only the first line of the definition (computable earlier or product of computable in $\mathbb{G}_1$ and computable in $\mathbb{G}_2$) is non-empty. ∎

4 Constraint Solving

In this section, we define an algorithm that takes a winning constraint and tries to derive a contradiction thereby showing that the winning constraint has no solution. Our algorithm uses constraint solving rules to perform a complete search for solutions using simplification rules and case distinctions. We first give the rules and then describe a strategy to apply the rules in Sect. 5. We begin by describing a set of simplification rules for constraints that exploit logical equivalences to bring a constraint into a simplified form. Next, we describe a set of rules for introducing and simplifying Coeff constraints. Then, we describe our rules for performing case distinctions followed by describing a procedure for equational simplification based on Gröbner Basis techniques. We conclude by giving a worked out example.

4.1 Constraint Solving Rules and Soundness

We use the notation $C \rightsquigarrow_{SE} C_1 \vee \ldots \vee C_k$ to denote the constraint solving rule that "simplifies" the constraint C into the disjunction of constraints $C_1, \ldots, C_k$. The constraint solving rule might depend on the security experiment SE. Our rules are sound in the following sense: If there exists an SE-solution s for C, then there is an $i \in \{1, \ldots, k\}$ such that there exists an SE-solution s' for C_i. The solution s' is usually very similar to s, but might, for example, perform an additional query with trivial parameters. We use $C \rightsquigarrow_{SE} \bot$ to denote that C can be simplified to the empty disjunction, which is equivalent to false.

We say a constraint C is contradictory if there is either a rule $C \rightsquigarrow_{SE} \bot$ or there is a rule $C \rightsquigarrow_{SE} C_1 \vee \ldots \vee C_k$ such that for all $i \in \{1, \ldots, k\}$, the constraint C_i is contradictory. Since all rules are sound, we obtain that if C is contradictory, then C has no solution.

4.2 Simplification Rules

To exploit the equivalence $e = e'$ given in Fig. 5, we define a corresponding constraint solving rule $C[e] \rightsquigarrow_{SE} C[e']$ for each of them. The rules up to and including the equivalences for Coeff can be used to bring every winning constraint into simplified form (see Fig. 4). Additionally, we assume given rules for the axioms of commutative rings with respect to $0, 1, *$ and $+$.

The remaining rules are useful to enable the application of other rules. The first remaining set of rules allows to swap binders, which might be required before applying rules that expect a certain binder to be in outermost position. To preserve the well-formedness of constraints, we adapt the index exception sets K as shown below. The second remaining set of rules allows us to add exceptions to binders. This might also benefit the applicability of other rules.

4.3 Introducing and Simplifying Coeff Constraints

In this section, we describe how to introduce and simplify constraints that involve Coeff expressions. To define our constraint solving rules, we define three functions that filter variables in monomials.

The functions

- $umon : Mon[\mathsf{UVar}^{\pm 1}, \mathsf{HVar}, \mathsf{PVar}] \rightarrow Mon[\mathsf{UVar}^{\pm 1}]$,
- $hmon : Mon[\mathsf{UVar}^{\pm 1}, \mathsf{HVar}, \mathsf{PVar}] \rightarrow Mon[\mathsf{HVar}]$, and
- $pmon : Mon[\mathsf{UVar}^{\pm 1}, \mathsf{HVar}, \mathsf{PVar}] \rightarrow Mon[\mathsf{PVar}]$.

keep the exponents for the desired type of variables and set the exponents of all other variables to zero.

$$
\begin{aligned}
\mathcal{C}_{simp} &::= \exists k \notin K.\mathcal{C}_{simp} \mid \mathcal{C}^{\wedge} & \text{existential quantification} \\
\mathcal{C}^{\wedge} &::= \mathcal{C}^{\forall} \wedge \ldots \wedge \mathcal{C}^{\forall} & \text{conjunction} \\
\mathcal{C}^{\forall} &::= \forall k \notin K.\mathcal{C}^{\forall} \mid \mathcal{C}^{eq} & \text{universal quantification} \\
\mathcal{C}^{eq} &::= \mathcal{E}^{+} = 0 \mid \mathcal{E}^{+} \neq 0 & \text{(in)equality} \\
\mathcal{E}^{+} &::= \mathcal{E}^{\Sigma} + \ldots + \mathcal{E}^{\Sigma} \mid 0 & \text{sum} \\
\mathcal{E}^{\Sigma} &::= \sum_{k \notin K} \mathcal{E}^{\Sigma} \mid -\mathcal{E}^{*} \mid \mathcal{E}^{*} \mid \mathsf{Coeff}_{\mathcal{M}}(\mathcal{E}^{*}) & \text{symbolic sum} \\
\mathcal{M} &::= \mathcal{M} * \mathcal{M} \mid R^{\pm 1} \mid R^{\pm 1}_{[k]} \mid 1 & \text{monomial over uniform variables} \\
\mathcal{E}^{*} &::= \mathcal{E}^{pv} * \ldots * \mathcal{E}^{pv} \mid 1 & \text{monomials} \\
\mathcal{E}^{pv} &::= \rho_{[k]} \mid \rho \mid R^{\pm 1} \mid R^{\pm 1}_{[k]} \mid Y_{[k]} & \text{parameter/variable}
\end{aligned}
$$

Fig. 4. Grammar for simplified winning constraints where $\rho \in \mathsf{PVar}, R \in \mathsf{UVar}, Y \in \mathsf{HVar}, k \in \mathsf{IVar}$. Conjunctions, sums, and products cannot by empty, but they can have a single argument. All bound variables must occur in the body. A monomial never contains a uniform variable and its inverse and never contains 1 unless it is equal to 1.

$$(\forall k \notin K.\mathcal{C}_1 \wedge \mathcal{C}_2) = (\forall k \notin K.\mathcal{C}_1) \wedge (\forall k \notin K.\mathcal{C}_2) \tag{equiv-1}$$

$$(\forall k \notin K.\mathcal{C}_1) = \mathcal{C}_1 \qquad\qquad \text{if } k \notin ivars(\mathcal{C}_1) \tag{equiv-2}$$

$$\mathcal{E}_1 * \mathcal{E}_2 = \mathcal{E}_2 * \mathcal{E}_1 \tag{equiv-3}$$

$$-(\mathcal{E}_1 + \mathcal{E}_2) = (-\mathcal{E}_1) + (-\mathcal{E}_2) \tag{equiv-4}$$

$$\sum_{k \notin K}(\mathcal{E}_1 + \mathcal{E}_2) = (\sum_{k \notin K}\mathcal{E}_1) + (\sum_{k \notin K}\mathcal{E}_2) \tag{equiv-5}$$

$$(\sum_{k \notin K}\mathcal{E}_1) * \mathcal{E}_2 = (\sum_{k \notin K}\mathcal{E}_1 * \mathcal{E}_2) \tag{equiv-6}$$

$$-(\sum_{k \notin K}\mathcal{E}) = (\sum_{k \notin K}-\mathcal{E}) \tag{equiv-7}$$

$$((-\mathcal{E}_1) * \mathcal{E}_2) = -(\mathcal{E}_1 * \mathcal{E}_2) \tag{equiv-8}$$

$$-(-\mathcal{E}) = \mathcal{E} \tag{equiv-9}$$

$$\mathcal{R} * \mathcal{R}^{-1} = 1 \tag{equiv-10}$$

$$\mathsf{Coeff}_\mathcal{M}(\mathcal{E}_1 + \mathcal{E}_2) = \mathsf{Coeff}_\mathcal{M}(\mathcal{E}_1) + \mathsf{Coeff}_\mathcal{M}(\mathcal{E}_2) \tag{equiv-11}$$

$$\mathsf{Coeff}_\mathcal{M}(\sum_{k \notin K}\mathcal{E}) = \sum_{k \notin K}\mathsf{Coeff}_\mathcal{M}(\mathcal{E}) \qquad \text{if } ivars(\mathcal{M}) \subseteq K \tag{equiv-12}$$

$$\mathsf{Coeff}_\mathcal{M}(-\mathcal{E}) = -\mathsf{Coeff}_\mathcal{M}(\mathcal{E}) \tag{equiv-13}$$

$$\exists k_1 \notin K_1. \exists k_2 \notin K_2.\mathcal{C} = \exists k_2 \notin K_2'. \exists k_1 \notin K_1'.\mathcal{C} \tag{swap-1}$$

$$\forall k_1 \notin K_1. \forall k_2 \notin K_2.\mathcal{C} = \forall k_2 \notin K_2'. \forall k_1 \notin K_1'.\mathcal{C} \tag{swap-2}$$

$$\sum_{k_1 \notin K_1}\sum_{k_2 \notin K_2}\mathcal{E} = \sum_{k_2 \notin K_2'}\sum_{k_1 \notin K_1'}\mathcal{E} \tag{swap-3}$$

$$\forall k \notin K.\mathcal{C} = (\forall k \notin K \cup \{k^*\}.\mathcal{C}) \wedge \mathcal{C}[k \mapsto k^*] \qquad \text{if } k^* \notin K \tag{split-1}$$

$$C[\sum_{k \notin K}\mathcal{E}] = C[(\sum_{k \notin K \cup \{k^*\}}\mathcal{E}) + \mathcal{E}[k \mapsto k^*]] \qquad \begin{array}{l}\text{where C} \\ \text{defines } k^* \neq k' \\ \text{forall } k' \in K\end{array} \tag{split-2}$$

Fig. 5. Equivalences for simplifying constraints where K_2' is defined as $K_2 \setminus \{k_1\}$ and K_1' is defined as $K_1 \cup \{k_2\}$ if $k_1 \in K_2$ and K_1 otherwise.

$$C[\mathcal{E} = 0] \leadsto_{SE} C[\mathcal{E} = 0 \wedge (\forall i_1 \notin K_1, \ldots, i_l \notin K_l. \, \mathsf{Coeff}_{\mathcal{M}}(\mathcal{E}) = 0)] \quad \text{(coeff-1)}$$
$$\text{if } \{i_1, \ldots, i_l\} \cap \mathit{ivars}(\mathcal{E}) = \emptyset \text{ and } \mathcal{E} \text{ does not contain } \mathsf{Coeff}$$

$$C[\mathsf{Coeff}_{\mathcal{M}}(\mathcal{E})] \leadsto_{SE} C[\mathit{pmon}(\mathcal{E})] \quad \text{if } \mathit{hmon}(\mathcal{E}) = 1 \text{ and } \mathcal{M} = \mathit{umon}(\mathcal{E}) \quad \text{(coeff-2)}$$

$$C[\mathsf{Coeff}_{\mathcal{M}}(\mathcal{E})] \leadsto_{SE} C[0] \quad \text{if } \mathsf{contMon}_{\mathcal{M}/\mathit{umon}(\mathcal{E})}(\mathit{hmon}(\mathcal{E})) \leadsto_{SE} \bot \quad \text{(coeff-3)}$$
$$\text{and } C \text{ assures } \mathit{ivars}(\mathcal{M}) \cap \mathit{ivars}(\mathcal{E}) = \emptyset$$

Fig. 6. Rules for introducing and simplifying Coeff expressions

The constraint solving rules are given in Fig. 6. The first rule exploits that if a polynomial is equal to zero, then when interpreting the polynomial as a polynomial over uniform variables, the coefficients for all monomials must be zero. The remaining two rules allow to simplify Coeff expressions. The first rule deals with the case where $\mathcal{E}$ does not contain any handle variables and $\mathcal{M}$ is equal to the monomial over uniform variables contained in $\mathcal{E}$. The second rule deals with the case where it is possible to prove that there is no (SE-computable) instantiation of the handle variables in $\mathcal{E}$ such that the resulting Laurent polynomial contains the monomial $\mathcal{M}$. The rule makes uses the contMon constraint. We will present the rules for showing that such a constraint is contradictory in the next section.

Example 3. Consider the constraint Γ such that

$$\Gamma = (\sum_j \rho_{[j]} R_{[j]} = 0) \wedge \Gamma'$$

We can simplify the constraint as follows:

$$\Gamma \leadsto_{SE} \Gamma \wedge \forall i. \, \mathsf{Coeff}_{R_{[i]}} (\sum_j \rho_{[j]} R_{[j]}) = 0 \qquad \text{[coeff-1]}$$

$$\leadsto_{SE} \Gamma \wedge \forall i. \, \mathsf{Coeff}_{R_{[i]}} ((\sum_{j \notin \{i\}} \rho_{[j]} R_{[j]}) + \rho_{[i]} R_{[i]}) = 0 \qquad \text{[split-2]}$$

$$\leadsto_{SE} \Gamma \wedge \forall i. \, \mathsf{Coeff}_{R_{[i]}} (\sum_{j \notin \{i\}} \rho_{[j]} R_{[j]}) + \mathsf{Coeff}_{R_{[i]}} (\rho_{[i]} R_{[i]}) = 0 \qquad \text{[equiv-11]}$$

$$\leadsto_{SE} \Gamma \wedge \forall i. \, \mathsf{Coeff}_{R_{[i]}} (\sum_{j \notin \{i\}} \rho_{[j]} R_{[j]}) + \rho_{[i]} = 0 \qquad \text{[coeff-2]}$$

$$\leadsto_{SE} \Gamma \wedge \forall i. \, (\sum_{j \notin \{i\}} \mathsf{Coeff}_{R_{[i]}} (\rho_{[j]} R_{[j]})) + \rho_{[i]} = 0 \qquad \text{[equiv-12]}$$

$$\leadsto_{SE} \Gamma \wedge \forall i. \, (\sum_{j \notin \{i\}} 0) + \rho_{[i]} = 0 \qquad \text{[coeff-3]}$$

$$\leadsto_{SE} \Gamma \wedge \forall i. \, \rho_{[i]} = 0 \qquad \text{[equiv-ring]}$$

For the step using [coeff-3], we exploit that $\mathsf{contMon}_{R_{[i]}/R_{[j]}}(1) \leadsto_{SE} \bot$ and that $j \notin \{i\}$ ensures that these index variables will never be instantiated with the same value in the given context. We will give the required rules in the next section. Then, our Gröbner-Basis based simplification algorithm will replace $\rho_{[j]}$ by 0 in Γ for arbitrary index variables j. ∎

Proving Coeff to be zero for all SE solutions. In this section, we describe a method to check if $\mathsf{Coeff}_{\mathcal{M}}(\mathcal{E})$ can be simplified to 0, i.e., for all SE-computable solutions $s = (p, q, \sigma, \delta, \chi, \xi)$, it holds that $coeff_{\sigma(\mathcal{M})}(eval_s(\mathcal{E})) = 0$. As in previous sections, we describe our approach for Type III, but stress that it can be adapted to Type I and Type II, e.g., by transforming the security experiment to make the isomorphisms redundant. We assume that the oracle definitions are efficiently computable and only return handles to elements of $\mathbb{G}_1$ and $\mathbb{G}_2$. Furthermore, we assume that the winning condition only uses handles to elements of $\mathbb{G}_1$ and $\mathbb{G}_2$. This covers most cryptographic constructions of interest (including all SPS schemes). In this case, we never have to deal with handle variables from HVar_t and for $i \in \{1, 2\}$, the polynomials $\boldsymbol{H}_i$ defining the oracle return values contain only handle variables from HVar_i. We distinguish three cases for $\mathsf{contMon}_{\mathcal{M}}(\mathcal{E})$: (i) $\deg(\mathcal{E}) = 0$, (ii) $\deg(\mathcal{E}) = 1$, and (iii) $\deg(\mathcal{E}) > 1$.

Case (i): We use the rule

$$\mathsf{contMon}_{\mathcal{M}}(1) \leadsto_{SE} \bot \quad \text{if } \mathcal{M} \neq 1.$$

Here, we require that distinct index variables must be instantiated with distinct values, which is ensured by the side condition of the Coeff-(3) rule.

Case (ii): We have $\mathcal{E} = Y_{[j]}$ for $Y \in \mathsf{HVar}_i$, $j \in \mathsf{IVar}$, and $i \in \{1, 2\}$. We must prove that the monomial $\mathcal{M}$ is not computable in i before query j, i.e., it is impossible (in the symbolic group model) to obtain a handle h for $\mathbb{G}_i$ that points to a polynomial F with $m \in mons(F)$ before the j-th oracle query. We perform a proof by contradiction that covers all cases on how a given monomial $\mathcal{M}$ can be computed. We write $\mathsf{canMult}_{i, \{j_1, \ldots, j_n\}}(m)$ if it is possible to perform the multiplication of a given monomial with m using oracle queries with query-indices distinct from $\{j_1, \ldots, j_n\}$. For example, if we have an oracle that returns a handle to $Y * R_{[j]} + W$ in $\mathbb{G}_1$ (where $Y \in \mathsf{HVar}_1, R, W \in \mathsf{UVar}$), then $\mathsf{canMult}_{1, \{j_1\}}(R_{[j_2]} * R_{[j_3]})$ is true since we can call the oracle for indices j_2 and j_3 to perform a multiplication with $R_{[j_2]}$ and $R_{[j_3]}$. In contrast, $\mathsf{canMult}_{1, \{j_1\}}(R_{[j_1]} * R_{[j_2]} * R_{[j_3]})$ is false because we cannot multiply with $R_{[j_1]}$ if using the oracle for query index j_1 is forbidden. To formalize this reasoning, we define a set of rules to reduce a constraint $\mathsf{contMon}_m(Y_{[j]})$ to a disjunction of constraints $\mathsf{canMult}_{i, J}(m)$ such that $ivars(m) = \emptyset$.

We define the set $\mathcal{SM}_i^{SE}$ of *start monomials for a security experiment SE and group index i* as $mons(\boldsymbol{F}_i) \cup (mons(\boldsymbol{H}_i) \cap Mon[\mathsf{UVar}^{\pm 1}])$ where the $\boldsymbol{H}_i$ are considered as polynomials over handle and uniform variables. We define the set $\mathcal{TM}_i^{SE}$ of *transformation monomials for a security experiment SE and a group index i* as $\{m \mid Y * m \in mons(\boldsymbol{H}_i) \wedge Y \in \mathsf{HVar}_i\} \subseteq Mon[\mathsf{UVar}^{\pm 1}]$. For both sets, we partition the previously defined sets into $\mathcal{SM}_i^{SE} = \mathcal{SM}_{i,glob}^{SE} \uplus \mathcal{SM}_{i,orcl}^{SE}$ and

$\mathcal{TM}_i^{SE} = \mathcal{TM}_{i,glob}^{SE} \uplus \mathcal{TM}_{i,orcl}^{SE}$ where the *glob*-sets contain all monomials that contain only global uniform variables and the *orcl*-sets contain all monomials that contain at least one oracle uniform variable. For monomials m, we write $m[j]$ to denote the monomial where all oracle uniform variables Y are replaced with their indexed versions $Y_{[j]}$. We also use the same notation for sets of monomials.

$\mathsf{contMon}_{\tilde{m}}(Y_{[j]}) \rightsquigarrow_{SE}$ $\qquad\qquad\qquad\qquad\qquad\qquad\qquad\qquad$ [contMon-1]

$\qquad \mathsf{canMult}_{i,\{j\}}(\tilde{m}/m_1) \vee \ldots \vee \mathsf{canMult}_{i,\{j\}}(\tilde{m}/m_l) \vee$

$\qquad \mathsf{canMult}_{i,\{j,j_1\}}(\tilde{m}/\hat{m}_1[j_1]) \vee \ldots \vee \mathsf{canMult}_{i,\{j,j_1\}}(\tilde{m}/\hat{m}_{\hat{l}}[j_1]) \vee$

$\qquad \ldots \vee$

$\qquad \mathsf{canMult}_{i,\{j,j_n\}}(\tilde{m}/\hat{m}_1[j_n]) \vee \ldots \vee \mathsf{canMult}_{i,\{j,j_n\}}(\tilde{m}/\hat{m}_{\hat{l}}[j_n])$

$\qquad\qquad$ if $Y \in \mathsf{HVar}_i$, $\{m_1,\ldots,m_l\} = \mathcal{SM}_{i,glob}^{SE}$,

$\qquad\qquad$ $\{\hat{m}_1,\ldots,\hat{m}_{\hat{l}}\} = \mathcal{SM}_{i,orcl}^{SE}$, and $\{j_1,\ldots,j_n\} = ivars(\tilde{m}) \setminus \{j\}$.

$\mathsf{canMult}_{i,J}(\tilde{m}) \rightsquigarrow_{SE}$ $\qquad\qquad\qquad\qquad\qquad\qquad\qquad\qquad$ [contMon-2]

$\qquad \mathsf{canMult}_{i,J\cup\{j\}}(\tilde{m}/m_1[j]) \vee \ldots \vee \mathsf{canMult}_{i,J\cup\{j\}}(\tilde{m}/m_l[j])$

$\qquad\qquad$ if $\{m_1,\ldots,m_l\} = \mathcal{TM}_{i,orcl}^{SE}$ and $j \in ivars(\tilde{m}) \setminus J$.

$\mathsf{canMult}_{i,J}(\tilde{m}) \rightsquigarrow_{SE} \bot$ $\qquad\qquad\qquad\qquad\qquad\qquad\qquad\quad$ [contMon-3]

$\qquad\qquad$ if $J \cap ivars(\tilde{m}) \neq \emptyset$

Fig. 7. Rules for dealing with $\mathsf{contMon}$. We use m/m' to denote the corresponding reduced Laurent monomial

We can now define the rules given in Fig. 7. The first rule captures that to compute the monomial $\tilde{m}$ in i before query j, the adversary must start with a monomial m' (in $m_1,\ldots,m_l,\hat{m}_1[j_1],\ldots$) and then use oracle queries to achieve an indirect multiplication of m' by $\tilde{m}/m'$. Here, the monomials m_i are either monomials included in the adversary input or monomials included in the oracle return values that do not depend on handles and do not contain oracle uniform variables. The monomials $\hat{m}_i[j_u]$ are monomials included in the oracle return values that do not depend on handles and that contain oracle uniform variables. The set of forbidden query indices for the indirect multiplication takes into account that j can never be used and that j_u cannot be used if a monomial with index j_u is used as the start monomial.

The second rule is applicable whenever $\tilde{m}$ contains an indexed uniform variable $R_{[j]}$ such that $j \notin J$. In this case, the j-th query must be used to perform an indirect multiplication that cancels out $R_{[j]}$ and we perform a case distinction on all monomial multiplications containing oracle uniform variables that can be performed by the oracle. For all cases where this step does not cancel out *all* variables indexed with j, we can use the third rule that formalizes the following fact: If the j-th query is forbidden, there is no way to cancel out a uniform variable with index j.

It is not hard to see that we can reduce all constraints to $\mathsf{canMult}_{i,J}(\tilde{m})$ such that $ivars(\tilde{m}) = \emptyset$: If $ivars(\tilde{m})$ non-empty, then either there is a $j \in ivars(\tilde{m}) \cap J$ and we can conclude with the last rule or we can apply the second rule and add an index $j \in ivars(\tilde{m})$ to J. To check if a constraint $\mathsf{canMult}_{i,J}(\tilde{m})$ with $ivars(\tilde{m}) = \emptyset$ is unsatisfiable, we translate the constraint into a system of linear equations that formalizes the following idea. Let $\{m_1, \ldots, m_l\} = \mathcal{TM}_{i,glob}^{SE}$, then all indirect multiplications that do not introduce indexed uniform variables are of the form

$$m_1^{\delta_1} * \ldots * m_l^{\delta_l}$$

for $\delta_i \in \mathbb{N}$. This corresponds to using the i-th transformation δ_i times to achieve a multiplication with $m_i^{\delta_i}$. To check if there exist $\delta_1, \ldots, \delta_l \in \mathbb{N}$ such that

$$\tilde{m} = m_1^{\delta_1} * \ldots * m_l^{\delta_l}$$

we check if the linear system of equations

$$\deg_{V_1}(\tilde{m}) = \deg_{V_1}(m_1) * \delta_1 + \ldots + \deg_{V_1}(m_l) * \delta_l$$
$$\ldots$$
$$\deg_{V_n}(\tilde{m}) = \deg_{V_n}(m_1) * \delta_1 + \ldots + \deg_{V_n}(m_l) * \delta_l$$

has a solution over $\mathbb{N}$ where $\{V_1, \ldots, V_n\}$ is the set of uniform variables that occur in $\tilde{m}, m_1, \ldots, m_l$.

Case (iii): The last case can be handled by generalizing the previous case. We sketch how to achieve this, the full description will be included in the full version of this paper. We have $\mathcal{E} = (Y_1)_{[j_1]} * \ldots * (Y_n)_{[j_n]}$ for $Y_u \in \mathsf{HVar}_{i_u}$, $j_u \in \mathsf{IVar}$, and $i_u \in \{1, 2\}$. To extend the method from Case (ii), we use adapted set of start monomials and transformation monomials that take cancellations between these values for the different handles into account. For example, the set of transformation monomials is the product of transformation monomial sets for $j_1, \ldots, j_n$ also allowing any set to be replaced by $\{1\}$.

Example 4. We will show that $\mathsf{contMon}_{R_{[i]}/V}(M_{[k]})$ is contradictory for the security experiment SE defined in Example 1. Note that $M_{[k]} \in \mathsf{HVar}_2$ and the monomial sets for this group are:

$$\mathcal{SM}_{2,glob}^{SE} = \{1, W\} \qquad\qquad \mathcal{SM}_{2,orcl}^{SE} = \{R, R^2\}$$
$$\mathcal{TM}_{2,glob}^{SE} = \{V\} \qquad\qquad \mathcal{TM}_{2,orcl}^{SE} = \emptyset$$

By applying the first rule in Fig. 7 we have:

$\mathsf{contMon}_{R_{[i]}/V}(M_{[k]}) \rightsquigarrow_{SE}$

$\qquad \mathsf{canMult}_{2,\{k\}}(R_{[i]}V^{-1}) \vee \mathsf{canMult}_{2,\{k\}}(R_{[i]}V^{-1}W^{-1}) \vee \qquad$ (div. by 1 and W)

$\qquad \mathsf{canMult}_{2,\{k,i\}}(V^{-1}) \vee \mathsf{canMult}_{2,\{k,i\}}(V^{-1}R_{[i]}^{-1}) \qquad\qquad$ (div. by $R_{[i]}$ and $R_{[i]}^2$)

Now, since $\mathcal{TM}_{2,orcl}^{SE} = \emptyset$, the second rule in Fig. 7 gives us:

$$\mathsf{canMult}_{2,\{k\}}(R_{[i]}V^{-1}) \leadsto_{SE} \bot$$
$$\mathsf{canMult}_{2,\{k\}}(R_{[i]}V^{-1}W^{-1}) \leadsto_{SE} \bot$$

Additionally,

$$\mathsf{canMult}_{2,\{k,i\}}(V^{-1}R_{[i]}^{-1}) \leadsto_{SE} \bot$$

because $\{k,i\} \cap ivars(V^{-1}R_{[i]}^{-1}) \neq \emptyset$. Our problem has been reduced to compute

$$\mathsf{canMult}_{2,\{k,i\}}(V^{-1})$$

so we define the system of equations:

$$\deg_V(V^{-1}) = \deg_V(V) * \delta_1$$

where $\delta_1 \in \mathbb{N}$. The equation is $-1 = 1 * \delta_1$ and it reduces to $\bot$. This analysis proves that $\mathsf{contMon}_{R_{[i]}/V}(M_{[k]}) \leadsto_{SE} \bot$, i.e., the handle variable $M_{[k]}$ cannot contain the monomial $R_{[i]}/V$.

4.4 Case Distinctions and Contradictions

The rules for case distinctions and contradictions are given in Fig. 8. The first rule is applicable whenever we can express the left-hand-side of an equality with 0 as a product of the two factors $\mathcal{E}_1$ and $\mathcal{E}_2$. Since we reason about elements of an integral domain, we can conclude that at least one of the factors must be equal to 0. The second rule formalizes that if $\mathcal{C}'$ is true for some i, then it is either true for some $i \neq j$ or it is true for $i = j$. The third rule formalizes that for all expressions $\mathcal{E}$, the expression is either equal to 0 or not. We only apply this rule with an $\mathcal{E}$ that already occurs as a subterm of C. In most cases $\mathcal{E} = \rho$ for $\rho \in \mathsf{PVar}$. The final case distinction rule deals with indexed parameter variables $\rho_{[i]}$. Either $\rho_{[i]}$ is equal to zero for all indices not in K or there is an index j not in K such that $\rho_{[j]}$ is not zero. The rule uses Δ to denote all existential bindings in the constraint.

The two contradiction rules are straightforward. The first rule states that a non-zero constant c is not equal to zero. We keep track of applications of this rule to obtain a lower bound on the prime p for which our proof is valid. The second rule just formalizes that zero is always equal to itself.

4.5 Gröbner Basis Simplification

Before applying the Gröbner Basis simplification, we ensure that all $\forall$-quantifiers use the same binders Δ and that all index exception sets are maximal for Δ. This might require renaming of variables, extending the index exception sets, and introducing unused variables. For the $\sum$-binders $\hat{\Delta}_u$, we assume for all u, v that (i) $\hat{\Delta}_u = \hat{\Delta}_v$, (ii) $\hat{\Delta}_u$ is a prefix of $\hat{\Delta}_v$, or (iii) vice versa.

$$C[\mathcal{E}_1 * \mathcal{E}_2 = 0] \rightsquigarrow_{SE} C[\mathcal{E}_1 = 0] \vee C[\mathcal{E}_2 = 0] \qquad \text{[dist-1]}$$

$$C[\exists i \notin K. \mathcal{C}'] \rightsquigarrow_{SE} \begin{array}{l} C[\exists i \notin K \cup \{j\}. \mathcal{C}'] \\ \vee C[\mathcal{C}'[i \mapsto j]] \end{array} \qquad \text{if } j \notin K \qquad \text{[dist-2]}$$

$$C[\mathcal{C}'] \rightsquigarrow_{SE} \begin{array}{l} C[\mathcal{C}' \wedge \mathcal{E} = 0] \\ \vee C[\mathcal{C}' \wedge \mathcal{E} \neq 0] \end{array} \qquad \text{where } \mathcal{E} \text{ arbitrary} \qquad \text{[dist-3]}$$

$$\exists \Delta. \mathcal{C}' \rightsquigarrow_{SE} \begin{array}{l} \exists \Delta. (\forall i \notin K. \rho_{[i]} = 0) \wedge \mathcal{C}' \\ \vee \exists \Delta, j \notin K. \rho_{[j]} \neq 0 \wedge \mathcal{C}' \end{array} \qquad \begin{array}{l} \text{where } K \text{ arbitrary} \\ \text{and } j \notin ivars(\Delta) \\ \cup ivars(\mathcal{C}') \end{array} \qquad \text{[dist-4]}$$

$$C[c = 0] \rightsquigarrow_{SE} \bot \qquad \text{if } c \in \mathbb{Z} \setminus \{0\} \qquad \text{[false-1]}$$

$$C[0 \neq 0] \rightsquigarrow_{SE} \bot \qquad \text{[false-2]}$$

Fig. 8. Rules for performing case distinctions and contradictions.

The resulting constraint system can be rearranged to have the following form

$$\exists \nabla. (\forall \Delta. \mathcal{E}_1 = 0) \wedge \ldots \wedge (\forall \Delta. \mathcal{E}_l = 0) \wedge$$
$$(\forall \Delta. \hat{\mathcal{E}}_1 \bowtie_1 0) \wedge \ldots \wedge (\forall \Delta. \hat{\mathcal{E}}_{\hat{i}} \bowtie_{\hat{i}} 0)$$

where the $\mathcal{E}_u$ are expressions that do not contain handle variables, uniform variables, or Coeff expressions, which we call parameter equality polynomials. The $\hat{\mathcal{E}}_u$ denote the remaining expressions. We want to move all the $\mathcal{E}_u$ under a single quantifier for simplification. To take renamings of the bound variables into account, we ensure beforehand that for all $\mathcal{E}_u$ and all permutations of the $\forall$-bound variables, the resulting expression is already included. For example, given

$$\forall j_1, j_2 \notin \{j_1\}. \rho_{[j_1]} * \rho'_{[j_2]} = 0 \wedge \forall j_1, j_2 \notin \{j_1\}. \rho_{[j_2]} * \rho'_{[j_1]} - \alpha = 0$$

it is usually useful to add at least the permutation

$$\forall j_1, j_2 \notin \{j_1\}. \rho_{[j_1]} * \rho'_{[j_2]} - \alpha = 0$$

before moving everything under a common quantifier since this yields the shared monomial $\rho_{[j_1]} * \rho'_{[j_2]}$. After moving the parameter equality polynomials under the same quantifier, we get:

$$\exists \nabla. (\forall \Delta. \mathcal{E}_1 = 0 \wedge \ldots \wedge \mathcal{E}_l = 0) \wedge$$
$$(\forall \Delta. \hat{\mathcal{E}}_1 \bowtie_1 0) \wedge \ldots \wedge (\forall \Delta. \hat{\mathcal{E}}_{\hat{i}} \bowtie_{\hat{i}} 0)$$

Now, we move non-indexed parameters in monomials out of the $\sum$-binder and consistently replace non-bound parameters and $\sum$-expressions with variables X_v. We call the corresponding mapping σ and use g_u to denote polynomial resulting from $\mathcal{E}_u$. We can revert this abstraction process by applying σ, i.e.,

$\sigma(g_u) = \mathcal{E}_u$. Next, we compute the Gröbner Basis (over $\mathbb{Z}$) of the ideal $\langle g_1, \ldots, g_l \rangle$ which we denote with $I = \langle g'_1, \ldots, g'_{l'} \rangle$. By the properties of the Gröbner Basis, we know that

$$(g_1 = 0 \wedge \ldots \wedge g_l = 0) \Leftrightarrow (g'_1 = 0 \wedge \ldots \wedge g'_{l'} = 0)$$

and hence

$$(\mathcal{E}_1 = 0 \wedge \ldots \wedge \mathcal{E}_l = 0) \Leftrightarrow (\mathcal{E}'_1 = 0 \wedge \ldots \wedge \mathcal{E}'_{l'} = 0)$$

for $\mathcal{E}'_u = \sigma(g_u)$ which we exploit to simplify the parameter equality polynomials. For computing the Gröbner Basis, we use a monomial order that prefers to eliminate abstracted $\sum$ expressions. Next, we use the Gröbner Basis to simplify the expressions $\forall \Delta. \hat{\mathcal{E}}_u \bowtie_u 0$. If $\hat{\mathcal{E}}_u$ uses all variables in Δ, we use an extension σ' of σ to abstract $\hat{\mathcal{E}}_u$ to the polynomial f and define f' as the result of reducing f modulo the Gröbner Basis I. As before, we define the simplified $\hat{\mathcal{E}}'_u$ as $\sigma'(f')$. Often, it is very useful to also simplify below $\sum$-binders. We use an example to illustrate how this works.

Example 5. Assume $\nabla = j_1$, $\Delta = j_2 \notin \{j_1\}$, $I = \langle X_1 * X_2 \rangle$, $\sigma = \{X_1 \mapsto \rho_{[j_1]}, X_1 \mapsto \rho'_{[j_2]}\}$, and

$$\mathcal{E}_1 = (\sum_{j_3 \notin \{j_1\}} \rho_{[j_1]} * \rho'_{[j_3]} = 0).$$

Then we use $\forall j_2 \notin \{j_1\}. \rho_{[j_1]} * \rho'_{[j_2]} = 0$ to rewrite $\rho_{[j_1]} * \rho'_{[j_3]}$ to 0 below $\sum_{j_3 \notin \{j_1\}}$ by instantiating j_2 with j_3 (both have the same exception j_1). ∎

4.6   Example: Proof of EUF-CMA for SPS

In this section show how our constraint solving rules can be used to prove (unbounded) EUF-CMA security of the signature scheme in Fig. 1. The winning constraints for the associated security experiment SE are already given in Example 2. To prove EUF-CMA security in the Generic Group Model, we must show that the following constraint has no SE-computable solution

$$\gamma + \sum_k \gamma'_{[k]} * R_{[k]} + \sum_k \gamma''_{[k]} * (M_{[k]} * V + W + R^2_{[k]})$$

$$- ((\tau + \sum_k \tau'_{[k]} * R_{[k]} + \sum_k \tau''_{[k]} * (M_{[k]} * V + W + R^2_{[k]}))$$

$$* (\rho + \sum_k \rho'_{[k]} * R_{[k]} + \rho'' * V + \rho''' * W) + \hat{M} * V + W) = 0 \quad (1)$$

$$\wedge \quad \rho + \sum_k \rho'_{[k]} * R_{[k]} + \rho'' * V + \rho''' * W$$

$$- (\tau + \sum_k \tau'_{[k]} * R_{[k]} + \sum_k \tau''_{[k]} * (M_{[k]} * V + W + R^2_{[k]})) = 0 \quad (2)$$

$$\wedge \quad \forall k. \hat{M} - M_{[k]} \neq 0 \quad (3)$$

where $\hat{M}$ is defined as

$$\hat{M} = \mu + \sum_k \mu'_{[k]} * R_{[k]} + \sum_k \mu''_{[k]} * (M_{[k]} * V + W + R^2_{[k]}).$$

Instead of immediately simplifying everything using the equivalences in Fig. 5, we first apply the rule [coeff-1] where $\mathcal{M} = R^2_{[i]}$ and $\mathcal{E}$ is the Eq. (2). After simplifying the resulting Coeff expressions (see Example 4), we get the new equation $\forall i. - \tau''_{[i]} = 0$. Our Gröbner Basis simplification replaces every occurrence of τ''_i by 0. This results in the following new constraint:

$$\gamma + \sum_k \gamma'_{[k]} * R_{[k]} + \sum_k \gamma''_{[k]} * (M_{[k]} * V + W + R^2_{[k]})$$
$$- ((\tau + \sum_k \tau'_{[k]} * R_{[k]}) * (\rho + \sum_k \rho'_{[k]} * R_{[k]} + \rho'' * V + \rho''' * W)$$
$$+ \hat{M} * V + W) = 0 \qquad (1)$$

$$\wedge \quad \rho + \sum_k \rho'_{[k]} * R_{[k]} + \rho'' * V + \rho''' * W - (\tau + \sum_k \tau'_{[k]} * R_{[k]}) = 0 \qquad (2)$$

$$\wedge \quad \forall k. \hat{M} - M_{[k]} \neq 0 \qquad (3)$$

Now, we can apply the rule [coeff-1] where $\mathcal{E}$ is the left hand side of Eq. (2) and for different monomials $\mathcal{M}$, we obtain the following new equations:

$$\rho - \tau = 0 \qquad \qquad \text{for } \mathcal{M} = 1$$
$$\forall k. \rho'_{[k]} - \tau'_{[k]} = 0 \qquad \qquad \text{for } \mathcal{M} = R_{[k]}$$
$$\rho'' = 0 \qquad \qquad \text{for } \mathcal{M} = V$$
$$\rho''' = 0 \qquad \qquad \text{for } \mathcal{M} = W$$

After this, we basically got rid of Eq. (2) and our Gröbner Basis simplification yields:

$$\gamma + \sum_k \gamma'_{[k]} * R_{[k]} + \sum_k \gamma''_{[k]} * (M_{[k]} * V + R^2_{[k]} + W)$$
$$- (\tau^2 + (2 \sum_k \tau * \tau'_{[k]} * R_{[k]}) + \sum_{k,k' \notin \{k\}} \tau'_{[k]} * \tau'_{[k']} * R_{[k]} * R_{[k']}$$
$$+ \sum_k \tau'^2_{[k]} * R^2_{[k]} + \hat{M} * V + W) = 0 \qquad (1)$$

$$\wedge \quad \forall k. \hat{M} - M_{[k]} \neq 0 \qquad (2)$$

We now apply the rule [coeff-1] where $\mathcal{E}$ is expression in Eq. (1) obtaining the following new equations:

$$\wedge \quad \sum_k \gamma''_{[k]} - 1 = 0 \qquad \qquad \text{for } \mathcal{M} = W \qquad (3)$$

$$\wedge \quad \forall k. \gamma''_{[k]} - \tau'^2_{[k]} = 0 \qquad \qquad \text{for } \mathcal{M} = R^2_{[k]} \qquad (4)$$

$$\wedge \quad \forall k. \forall k' \notin \{k\}. 2 * \tau'_{[k]} * \tau'_{[k']} = 0 \qquad \qquad \text{for } \mathcal{M} = R_{[k]} R_{[k']} \qquad (5)$$

Then, we apply the rule [dist-4] with $K = \emptyset$ to perform a case distinction on the parameter τ':

$$\forall k. \tau'_{[k]} = 0 \wedge \Gamma \qquad \text{(case 1)}$$
$$\vee \quad \exists k^*. \tau'_{[k^*]} \neq 0 \wedge \Gamma \qquad \text{(case 2)}$$

Here, Γ represents the conjunction of our previous five equations. In case 1, the Gröbner Basis simplification results in the system

$$\gamma + \sum_k \gamma'_{[k]} * R_{[k]} - \tau^2 - \hat{M} * V - W = 0 \tag{1}$$
$$\wedge \quad \forall k. \hat{M} - M_{[k]} \neq 0 \tag{2}$$
$$\wedge \quad -1 = 0 \tag{3}$$

which simplifies to $\bot$ after applying rule [false-1] to Eq. (3).
In case 2, Gröbner Basis simplification yields:

$$\exists k^*.$$
$$\gamma + \sum_k \gamma'_{[k]} * R_{[k]} + M_{[k^*]} * V - \tau^2 - 2\tau R_{[k^*]} - \hat{M} * V \tag{1}$$
$$\wedge \quad \forall k. \hat{M} - M_{[k]} \neq 0 \tag{2}$$

We apply the rule [coeff-1] where $\mathcal{E}$ is the left hand side of Eq. (1) for different monomials as $\mathcal{M}$, obtaining:

$$\gamma - \tau^2 = 0 \qquad \text{for } \mathcal{M} = 1$$
$$\forall k \notin \{k^*\}. \gamma'_{[k]} = 0 \qquad \text{for } \mathcal{M} = R_{[k]}$$
$$\gamma'_{[k^*]} - 2\tau = 0 \qquad \text{for } \mathcal{M} = R_{[k^*]}$$

After simplifying the system, we obtain:

$$M_{[k^*]} * V - \hat{M} * V = 0 \tag{1}$$
$$\wedge \quad \forall k. \hat{M} - M_{[k]} \neq 0 \tag{2}$$

Applying the rule [dist-1] to Eq. (1) we obtain two cases:

$$\exists k^*. \qquad\qquad\qquad \exists k^*.$$
$$V = 0 \qquad\qquad\qquad M_{[k^*]} - \hat{M} = 0$$
$$\wedge \quad \forall k. \hat{M} - M_{[k]} \neq 0 \qquad \wedge \quad \forall k. \hat{M} - M_{[k]} \neq 0$$

$$\text{(case 2.1)} \qquad\qquad\qquad \text{(case 2.2)}$$

In case 2.1, after applying [coeff-1] for $\mathcal{M} = V$ to the first equation and simplifying, we obtain the equation $1 = 0$ that reduces to $\bot$ according to rule [false-1].

Finally, in case 2.2 we apply the rule [split-2] and we get the system:

$$M_{[k^*]} - \hat{M} = 0$$
$$\wedge \quad \forall k \notin \{k^*\}.\, \hat{M} - M_{[k]} \neq 0$$
$$\wedge \quad \hat{M} - M_{[k^*]} \neq 0$$

Our Gröbner Basis simplification will reduce it to,

$$0 \neq 0 \wedge (\forall k \notin \{k^*\}.\, \hat{M} - M_{[k]} \neq 0)$$

which reduces to $\perp$ according to rule [false-2].

5   Implementation and Case Studies

We have implemented the described algorithm in the gga^∞ tool[3] and have evaluated its effectiveness and performance on cryptographic constructions from the literature (presented in Table 1) and automatically synthesized schemes (presented in Table 2). The source code is written in OCaml and uses the computer algebra system SAGE [40] for Gröbner Basis computations and the SMT solver Z3 [20] for checking the satisfiability of linear equations over the natural numbers. Although the code reproduces the algorithm as it is described in this paper, it also implements some optimizations and additional rules to derive contradictions, that will be further explained in the full version of this paper.

The tool takes an input file such as the one shown in Fig. 9 and performs a proof search using our constraint solving rules guided by a heuristic. If the search is successful, the tool returns a representation of the proof tree. To ensure termination, we establish a timeout of 1000 s.

```
group_setting 3.

sample V,W.
input [V,W] in G1.

oracle o1(M:G2) =
  sample R;
  return [ R ] in G1,
         [ R, M*V + R^2 + W] in G2.

win (wM:G2, wT1:G1, wT2:G2, wS:G2) =
  ( (forall i: wM <> M_i) /\ wT1 = wT2 /\ wS = V*wM + wT1*wT2 + W ).
```

Fig. 9. Input file for the Type III re-randomizable SPS scheme from Fig. 1

---

[3] Source code and case studies at http://generic-group-analyzer.github.io/.

5.1 Case Studies

We analyze the security of cryptographic constructions from the literature and collect the results in Table 1. The first five entries do not require support for oracles that take handles and are therefore also in the scope of the tool presented in [11]. For the first four entries, both the tool from [11] and gga^{∞} prove unbounded security. For the fifth example, gga^{∞} succeeds, whereas the tool from [11] fails to find a proof.

The remaining examples are all outside the scope of the tool from [11]. First, we analyze the Message Authentication Codes proposed in [18]. They propose two MACs (instead of public key signatures) as the basis for their anonymous credential system. One of them is proven secure in the Generic Group Model and the other under the decisional Diffie-Hellman (DDH) assumption. Our tool confirms the first proof and finds a proof in the Generic Group Model for the second construction[4].

We also prove security for a number of structure-preserving signature schemes. First, we analyze the scheme proposed in [2] for bilinear groups of Type III.

Then, we analyze the re-randomizable scheme from [4] for Type II and Type III. Next, we prove sEUF-CMAsecurity of the unified SPS signature scheme proposed in [5], which is secure in all three settings. We also prove EUF-CMA security of its re-randomizable version (randomization tokens are given to the adversary). Later, we analyze the translation of the scheme for Type III proposed in [19]. We also consider the Type II scheme from [12].

Finally, we analyze two instances of fully structure-preserving signature schemes proposed in [26].

Table 1. Case studies (last column denotes time for fully automated proof).

Reference	Scheme	Property	Time
Lysyanskaya et al. '99 [30]	LRSW assumption	Valid	$2s$
Abe et al. '11 [5]	One-time SPS in Type I	OT-EUF-CMA	$1s$
Pointcheval et al. '15 [34]	Assumption 1	Valid	$1s$
"	Assumption 2	Valid	$1s$
"	Multi-message sign. scheme ($r = 3$)	EUF-CMA	$1s$
Chase et al. '13 [18]	MAC_{GGM} (messages length ≤ 3)	UF-CMVA	$1s$
"	MAC_{DDH} (messages length ≤ 3)	UF-CMVA	$3s$
Abe et al. '11 [2]	SPS scheme, messages in $\mathbb{G}_1 \times \mathbb{G}_2$	sEUF-CMA	$22s$
Abe et al. '14 [4]	Re-random. SPS for msg. in $\mathbb{G}_2$	EUF-CMA	$6s$
Abe et al. '14 [5]	Unified SPS scheme	sEUF-CMA	$5s$
"	Unified SPS scheme (with tokens)	EUF-CMA	$11s$
Chatterjee et al. '15 [19]	Type III randomizable SPS	EUF-CMA	$3s$
Barthe et al. '15 [12]	Re-randomizable SPS in Type III	EUF-CMA	$6s$
Groth '15 [26]	Fully comb. $\text{SPS}_{b=0}$ ($m, n = 1$)	EUF-CMA	$8s$
"	Fully comb. $\text{SPS}_{b=1}$ ($m, n = 1$)	sEUF-CMA	$8s$

---

[4] This is of course implied by the pen-and-paper proof under the DDH assumption.

Table 2. Synthesis results for SPS schemes in Type II and Type III with $r, v, w \xleftarrow{\$} \mathbb{Z}_p$, verification keys $V = g_1^v, W = g_1^w \in \mathbb{G}_1$, message $M = g_2^m \in \mathbb{G}_2$ and signatures $S_1 = g_1^{s_1} \in \mathbb{G}_1$, $S_2 = g_2^{s_2}, S_3 = g_2^{s_3} \in \mathbb{G}_2$.

Search Space		Results	
Verification equations	First signature elements	2-secure	∞-secure
II $s_3 = f(r, v, w, m)$	$S_2 = [\![r]\!]_2,$	1	1
$s_3 s_2 = f(r, v, w, m)$	$S_2 = [\![r]\!]_2,$	12	9
$s_3(s_2 - w) = f(r, v, w, m)$	$S_2 = [\![r + w]\!]_2,$	14	8
III $s_1 = s_2 \wedge \quad s_3 = f(r, v, w, m)$	$S_1 = [\![r]\!]_1, S_2 = [\![r]\!]_2$	2	2
$s_1 = s_2 \wedge s_1 s_3 = f(r, v, w, m)$	$S_1 = [\![r]\!]_1, S_2 = [\![r]\!]_2$	117	75
$s_1 s_2 = 1 \wedge s_1 s_3 = f(r, v, w, m)$	$S_1 = [\![r]\!]_1, S_2 = [\![r^{-1}]\!]_2$	39	22
		185	117

To evaluate our tool on a wider range of examples, we also make use of the synthesis tool for structure-preserving signature schemes presented in [12]. We take the existing results for Type II from [12] and use our tool to analyze (unbounded) EUF-CMA-security for all schemes where the tool from [12] succeeds to prove 2-EUF-CMA security. We also extend the synthesis tool to generate new schemes in Type III and apply our tool to those schemes that can be proven 2-EUF-CMA secure with the tool from [12]. The results for both Type II and Type III are summarized in Table 2. We classify the schemes in different groups, depending on the shape of the verification equations (first column). The column 2-*secure* represents the number of schemes of each group that are proven 2-EUF-CMA secure using the tool from [12], while the column ∞-*secure* represents the number of schemes of each group that are proven EUF-CMA secure using our tool (for all bounds that are polynomial in the security parameter).

Acknowledgements. This work is supported in part by ONR grant N00014-12-1-0914, Madrid regional project S2009TIC-1465 PROMETIDOS, and Spanish national projects TIN2009-14599 DESAFIOS 10, and TIN2012-39391-C04-01 Strongsoft. The research of Schmidt has received funds from the European Commissions Seventh Framework Programme Marie Curie Cofund Action AMAROUT II (grant no. 291803).

References

1. Abe, M., Fuchsbauer, G., Groth, J., Haralambiev, K., Ohkubo, M.: Structure-preserving signatures and commitments to group elements. In: Rabin, T. (ed.) CRYPTO 2010. LNCS, vol. 6223, pp. 209–236. Springer, Heidelberg (2010)
2. Abe, M., Groth, J., Haralambiev, K., Ohkubo, M.: Optimal structure-preserving signatures in asymmetric bilinear groups. In: Rogaway, P. (ed.) CRYPTO 2011. LNCS, vol. 6841, pp. 649–666. Springer, Heidelberg (2011)

3. Abe, M., Groth, J., Ohkubo, M., Tango, T.: Converting cryptographic schemes from symmetric to asymmetric bilinear groups. In: Garay, J.A., Gennaro, R. (eds.) CRYPTO 2014, Part I. LNCS, vol. 8616, pp. 241–260. Springer, Heidelberg (2014)

4. Abe, M., Groth, J., Ohkubo, M., Tibouchi, M.: Structure-preserving signatures from type II pairings. In: Garay, J.A., Gennaro, R. (eds.) CRYPTO 2014, Part I. LNCS, vol. 8616, pp. 390–407. Springer, Heidelberg (2014)

5. Abe, M., Groth, J., Ohkubo, M., Tibouchi, M.: Unified, minimal and selectively randomizable structure-preserving signatures. In: Lindell, Y. (ed.) TCC 2014. LNCS, vol. 8349, pp. 688–712. Springer, Heidelberg (2014)

6. Abe, M., Kohlweiss, M., Ohkubo, M., Tibouchi, M.: Fully structure-preserving signatures and shrinking commitments. In: Oswald, E., Fischlin, M. (eds.) EURO-CRYPT 2015. LNCS, vol. 9057, pp. 35–65. Springer, Heidelberg (2015)

7. Akinyele, J.A., Garman, C., Hohenberger, S.: Automating fast and secure translations from type-I to type-III pairing schemes. In: Proceedings of the 22nd ACM SIGSAC Conference on Computer and Communications Security, CCS 2015, pp. 1370–1381. ACM, New York (2015)

8. Akinyele, J.A., Green, M., Hohenberger, S.: Using SMT solvers to automate design tasks for encryption, signature schemes. In: Sadeghi, A.-R., Gligor, V.D., Yung, M. (eds.) 20th Conference on Computer and Communications Security, ACM CCS 2013, 4–8 November 2013, Berlin, Germany, pp. 399–410. ACM Press (2011)

9. Barthe, G., Cederquist, J., Tarento, S.: A machine-checked formalization of the generic model and the random oracle model. In: Basin, D., Rusinowitch, M. (eds.) IJCAR 2004. LNCS (LNAI), vol. 3097, pp. 385–399. Springer, Heidelberg (2004)

10. Barthe, G., Crespo, J.M., Grégoire, B., Kunz, C., Lakhnech, Y., Schmidt, B., Zanella Béguelin, S.: Fully automated analysis of padding-based encryption in the computational model. In: Sadeghi, A.-R., Gligor, V.D., Yung, M. (eds.) 20th Conference on Computer and Communications Security, ACM CCS 2013, 4–8 November 2013, Berlin, Germany, pp. 1247–1260. ACM Press (2011)

11. Barthe, G., Fagerholm, E., Fiore, D., Mitchell, J., Scedrov, A., Schmidt, B.: Automated analysis of cryptographic assumptions in generic group models. In: Garay, J.A., Gennaro, R. (eds.) CRYPTO 2014, Part I. LNCS, vol. 8616, pp. 95–112. Springer, Heidelberg (2014)

12. Barthe, G., Fagerholm, E., Fiore, D., Scedrov, A., Schmidt, B., Tibouchi, M.: Strongly-optimal structure preserving signatures from type II pairings: synthesis and lower bounds. In: Katz, J. (ed.) PKC 2015. LNCS, vol. 9020, pp. 355–376. Springer, Heidelberg (2015)

13. Barthe, G., Grégoire, B., Heraud, S., Béguelin, S.Z.: Computer-aided security proofs for the working cryptographer. In: Rogaway, P. (ed.) CRYPTO 2011. LNCS, vol. 6841, pp. 71–90. Springer, Heidelberg (2011)

14. Barthe, G., Grégoire, B., Schmidt, B.: Automated proofs of pairing-based cryptography. In: Proceedings of the 22nd ACM SIGSAC Conference on Computer and Communications Security, 12–16 October 2015, Denver, CO, USA, pp. 1156–1168 (2015)

15. Barthe, G., Tarento, S.: A machine-checked formalization of the random oracle model. In: Filliâtre, J.-C., Paulin-Mohring, C., Werner, B. (eds.) TYPES 2004. LNCS, vol. 3839, pp. 33–49. Springer, Heidelberg (2006)

16. Blanchet, B.: A computationally sound mechanized prover for security protocols. In: IEEE Symposium on Security and Privacy, 21–24 May 2006, Berkeley, California, USA, pp. 140–154. IEEE Computer Society Press (2006)

17. Boneh, D., Boyen, X., Goh, E.-J.: Hierarchical identity based encryption with constant size ciphertext. In: Cramer, R. (ed.) EUROCRYPT 2005. LNCS, vol. 3494, pp. 440–456. Springer, Heidelberg (2005)

18. Chase, M., Meiklejohn, S., Zaverucha, G.: Algebraic MACs, keyed-verification anonymous credentials. In: Ahn, G.-J., Yung, M., Li, N. (eds.) 21st Conference on Computer and Communications Security, ACM CCS 2014, 3–7 November 2014, Scottsdale, AZ, USA, pp. 1205–1216. ACM Press (2011)

19. Chatterjee, S., Menezes, A.: Type 2 structure-preserving signature schemes revisited. In: Iwata, T., et al. (eds.) ASIACRYPT 2015. LNCS, vol. 9452, pp. 286–310. Springer, Heidelberg (2015). doi:10.1007/978-3-662-48797-6_13

20. de Moura, L., Bjørner, N.S.: Z3: an efficient SMT solver. In: Ramakrishnan, C.R., Rehof, J. (eds.) TACAS 2008. LNCS, vol. 4963, pp. 337–340. Springer, Heidelberg (2008)

21. Escala, A., Herold, G., Kiltz, E., Ràfols, C., Villar, J.: An algebraic framework for Diffie-Hellman assumptions. In: Canetti, R., Garay, J.A. (eds.) CRYPTO 2013, Part II. LNCS, vol. 8043, pp. 129–147. Springer, Heidelberg (2013)

22. Fagerholm, E.: Automated analysis in generic groups. Ph.D. thesis, University of Pennsylvania (2015)

23. Freeman, D.M.: Converting pairing-based cryptosystems from composite-order groups to prime-order groups. In: Gilbert, H. (ed.) EUROCRYPT 2010. LNCS, vol. 6110, pp. 44–61. Springer, Heidelberg (2010)

24. Fuchsbauer, G.: Breaking existential unforgeability of a signature scheme from asiacrypt 2014. Cryptology ePrint Archive, Report 2014/892 (2014). http://eprint.iacr.org/2014/892

25. Fuchsbauer, G., Hanser, C., Slamanig, D.: EUF-CMA-secure structure-preserving signatures on equivalence classes. Cryptology ePrint Archive, Report 2014/944 (2014). http://eprint.iacr.org/2014/944

26. Groth, J.: Efficient fully structure-preserving signatures for large messages. In: Iwata, T., et al. (eds.) ASIACRYPT 2015. LNCS, vol. 9452, pp. 239–259. Springer, Heidelberg (2015). doi:10.1007/978-3-662-48797-6_11

27. Hoang, V.T., Katz, J., Malozemoff, A.J.: Automated analysis and synthesis of authenticated encryption schemes. In: Proceedings of the 22nd ACM SIGSAC Conference on Computer and Communications Security, 12–16 October 2015, Denver, CO, USA, pp. 84–95 (2015)

28. Hwang, J.Y., Lee, D.H., Yung, M.: Universal forgery of the identity-based sequential aggregate signature scheme. In: Li, W., Susilo, W., Tupakula, U.K., Safavi-Naini, R., Varadharajan, V. (eds.) 4th ACM Symposium on Information, Computer and Communications Security, ASIACCS 2009, 10–12 March 2009, Sydney, Australia, pp. 157–160. ACM Press (2011)

29. Jager, T., Schwenk, J.: On the equivalence of generic group models. In: Baek, J., Bao, F., Chen, K., Lai, X. (eds.) ProvSec 2008. LNCS, vol. 5324, pp. 200–209. Springer, Heidelberg (2008)

30. Lysyanskaya, A., Rivest, R.L., Sahai, A., Wolf, S.: Pseudonym systems. In: Heys, H., Adams, C. (eds.) SAC 1999. LNCS, vol. 1758, pp. 184–199. Springer, Heidelberg (2000)

31. Malozemoff, A.J., Katz, J., Green, M.D.: Automated analysis and synthesis of block-cipher modes of operation. In: IEEE 27th Computer Security Foundations Symposium, CSF 2014, 19–22 July 2014, Vienna, Austria, pp. 140–152 (2014)

32. Maurer, U.M.: Abstract models of computation in cryptography. In: Smart, N.P. (ed.) Cryptography and Coding 2005. LNCS, vol. 3796, pp. 1–12. Springer, Heidelberg (2005)

33. Nechaev, V.: Complexity of a determinate algorithm for the discrete logarithm. Math. Notes 55(2), 165–172 (1994)
34. Pointcheval, D., Sanders, O.: Short randomizable signatures. In: Sako, K. (ed.) CT-RSA 2016. LNCS, vol. 9610, pp. 111–126. Springer, Heidelberg (2016)
35. Rupp, A., Leander, G., Bangerter, E., Dent, A.W., Sadeghi, A.-R.: Sufficient conditions for intractability over black-box groups: generic lower bounds for generalized DL and DH problems. In: Pieprzyk, J. (ed.) ASIACRYPT 2008. LNCS, vol. 5350, pp. 489–505. Springer, Heidelberg (2008)
36. Schnorr, C.-P.: Security of blind discrete log signatures against interactive attacks. In: Qing, S., Okamoto, T., Zhou, J. (eds.) ICICS 2001. LNCS, vol. 2229, pp. 1–12. Springer, Heidelberg (2001)
37. Schnorr, C.-P., Jakobsson, M.: Security of signed ElGamal encryption. In: Okamoto, T. (ed.) ASIACRYPT 2000. LNCS, vol. 1976, pp. 73–89. Springer, Heidelberg (2000)
38. Shoup, V.: Lower bounds for discrete logarithms and related problems. In: Fumy, W. (ed.) EUROCRYPT 1997. LNCS, vol. 1233, pp. 256–266. Springer, Heidelberg (1997)
39. Szydlo, M.: A note on chosen-basis decisional Diffie-Hellman assumptions. In: Di Crescenzo, G., Rubin, A. (eds.) FC 2006. LNCS, vol. 4107, pp. 166–170. Springer, Heidelberg (2006)
40. The Sage Developers. Sage Mathematics Software (Version 6.8) (2015). http://www.sagemath.org

Multi-input Functional Encryption in the Private-Key Setting: Stronger Security from Weaker Assumptions

Zvika Brakerski[1], Ilan Komargodski[1(✉)], and Gil Segev[2]

[1] Weizmann Institute of Science, 76100 Rehovot, Israel
{zvika.brakerski,ilan.komargodski}@weizmann.ac.il
[2] Hebrew University of Jerusalem, 91904 Jerusalem, Israel
segev@cs.huji.ac.il

Abstract. We construct a general-purpose *multi-input* functional encryption scheme in the private-key setting. Namely, we construct a scheme where a functional key corresponding to a function f enables a user holding encryptions of $x_1, \ldots, x_t$ to compute $f(x_1, \ldots, x_t)$ but nothing else. This is achieved starting from any general-purpose private-key *single-input* scheme (without any additional assumptions), and is proven to be *adaptively secure* for any constant number of inputs t. Moreover, it can be extended to a super-constant number of inputs assuming that the underlying single-input scheme is sub-exponentially secure.

Instantiating our construction with existing single-input schemes, we obtain multi-input schemes that are based on a variety of assumptions (such as indistinguishability obfuscation, multilinear maps, learning with errors, and even one-way functions), offering various trade-offs between security and efficiency.

Previous and concurrent constructions of multi-input functional encryption schemes either rely on stronger assumptions and provided weaker security guarantees (Goldwasser et al. [EUROCRYPT '14], and Ananth and Jain [CRYPTO '15]), or relied on multilinear maps and could be proven secure only in an idealized generic model (Boneh et al. [EUROCRYPT '15]). In comparison, we present a general transformation that simultaneously relies on weaker assumptions and guarantees stronger security.

Z. Brakerski—Supported by the Israel Science Foundation (Grant No. 468/14) and by the Alon Young Faculty Fellowship.

I. Komargodski—Research supported in part by a grant from the Israel Science Foundation, the I-CORE Program of the Planning and Budgeting Committee, BSF and the Israeli Ministry of Science and Technology.

G. Segev—Supported by the European Union's 7th Framework Program (FP7) via a Marie Curie Career Integration Grant, by the Israel Science Foundation (Grant No. 483/13), by the Israeli Centers of Research Excellence (I-CORE) Program (Center No. 4/11), by the US-Israel Binational Science Foundation (Grant No. 2014632), and by a Google Faculty Research Award.

© International Association for Cryptologic Research 2016
M. Fischlin and J.-S. Coron (Eds.): EUROCRYPT 2016, Part II, LNCS 9666, pp. 852–880, 2016.
DOI: 10.1007/978-3-662-49896-5_30

1 Introduction

The emerging vision of functional encryption [14,31,32] extends the traditional "all-or-nothing" view of encryption schemes. Specifically, functional encryption schemes offer additional flexibility by supporting restricted decryption keys. These keys allow users to learn specific functions of the encrypted data, without learning any additional information. Building upon the early examples of functional encryption schemes for restricted function families (such as identity-based encryption [11,20,34]), extensive research is currently devoted to the construction of functional encryption schemes offering a variety of expressive families of functions (see, for example, [2,4,5,9,10,14,16,19,21,22,25,26,30–32,36]).

Until very recently, research on functional encryption has focused on the case of *single-input* functions. In a single-input functional encryption scheme, a functional key sk_f corresponding to a function f enables a user holding an encryption of a value x to compute $f(x)$, while not revealing any additional information on x. In many scenarios, however, dealing only with single-input functions is insufficient, and a more general framework allowing *multi-input* functions is required.

Goldwasser et al. [24] recently introduced the notion of a *multi-input* functional encryption scheme. In such a scheme, a functional key corresponding to a t-input function f enables a user holding encryptions of $x_1,\ldots,x_t$ to compute $f(x_1,\ldots,x_t)$ without learning any additional information on the x_i's. The work of Goldwasser et al. and their new notion are very well-motivated by a wide range of applications based on mining aggregate information from several different data sources. These include, for example, running SQL queries on encrypted databases, computing over encrypted data streams, non-interactive differentially-private data release, and order-revealing encryption (all of which are relevant in both the public-key setting and the private-key one [12]).

Goldwasser et al. presented a rigorous framework for capturing the security of multi-input schemes in the public-key setting and in the private-key one. In addition, relying on indistinguishability obfuscation and one-way functions [8,21,29], they constructed the first multi schemes. In terms of functionality, their schemes are extremely expressive, supporting all multi-input functions that are computable by bounded-size circuits. In terms of security, however, their private-key scheme satisfies a weak selective notion, which does not allow the adversary to access an encryption oracle (which is quite crippling in the private-key setting), and requires an a-priori bound on the number of challenge ciphertexts (the ciphertext length in their scheme depends on the number of challenge ciphertexts).

Following the work of Goldwasser et al. [24], a private-key multi-input functional encryption scheme that satisfies a more standard notion of security (one that allows access to an encryption oracle) was constructed by Boneh et al. [12]. Their scheme is based on multilinear maps, and is proven secure in the idealized generic multilinear map model. In addition, in an independent and concurrent work, Ananth and Jain [5] constructed a *selectively-secure* multi-input private-key functional encryption scheme based on any general-purpose *public key* functional encryption scheme (as an intermediate step in constructing an indistinguishability obfuscator).

Thus, constructions of multi-input functional encryption schemes in the private-key setting have so far either relied on stronger assumptions and provided weaker security guarantees [5, 24][1], or could be proven secure only in an idealized generic model [12].

1.1   Our Contributions

In this paper we present a construction of private-key multi-input functional encryption from *any* general-purpose *private-key single-input* functional encryption scheme (without introducing any additional assumptions). The resulting scheme supports any set of efficiently-computable functions, and provides adaptive security in the standard model for any constant number of inputs. We prove the following theorem:

Theorem 1.1. *Assuming the existence of any private-key single-input selectively-secure functional encryption scheme, for any constant $t \geq 2$ there exists a private-key t-input adaptively-secure functional encryption scheme.*

Moreover, assuming that the underlying private-key single-input scheme is sub-exponentially secure, our resulting scheme provides adaptive security for a *super-constant* number of inputs (we refer the reader to Sect. 1.3 for more details). Following [1, 19], our scheme provides not only message privacy, but in fact a unified notion that captures both message privacy and function privacy (this notion is known as *full security* – see Sect. 2.3 for more details).

Instantiations. Instantiating our construction with existing private-key single-input schemes, we obtain new multi-input schemes based on a variety of assumptions in the standard model. Specifically, we obtain schemes that are secure for an unbounded number of encryption and key-generation queries based on indistinguishability obfuscation or multilinear maps. In addition, if the number of encryption and key-generation queries is a-priori bounded, we can rely on much milder assumptions such as learning with errors [25] or even the existence of one-way functions or low-depth pseudorandom generators [26]. See Sect. 2.2 for further discussion.

Comparison with Previous and Concurrent Work. Compared to the previous work of Goldwasser et al. [24] and Boneh et al. [12], our work yields stronger security guarantees and at the same time relies solely on a necessary assumption. Specifically, whereas Goldwasser et al. and Boneh et al. rely on indistinguishability obfuscation and multilinear maps, respectively, we rely on the existence of any general-purpose private-key single-input scheme, which is obviously necessary. Moreover, whereas the scheme of Goldwasser et al. provides a selective notion of security which, in addition, does not allow adversaries to access an

---

[1] In terms of assumptions, the recent work of Asharov and Segev [7] shows that indistinguishability obfuscation and public-key functional encryption are significantly stronger primitives than private-key functional encryption. We refer the reader to Sect. 1.1 for a more elaborate discussion.

encryption oracle and requires an a-priori bound on the number of challenge ciphertexts, and the scheme of Boneh et al. is proved secure only in an idealized generic model that does not properly capture real-world adversaries, our scheme provides adaptive security in the standard model for any number of challenge ciphertexts.

Compared to the concurrent work of Ananth and Jain [5], our work again yields stronger security guarantees while relying on a weaker assumption. Specifically, whereas the construction of Ananth and Jain relies on *public-key* functional encryption and guarantees *selective* security (where, in addition, the adversary is not allow to access an encryption oracle), our construction relies on *private-key* functional encryption and guarantees *full* security. From the technical point of view, the scheme of Ananth and Jain is essentially "Step 1" of our approach (see Sect. 1.3), which was sufficient (together with additional techniques and assumptions) for constructing their obfuscator. The vast majority of our efforts in this paper are devoted for providing better security while simultaneously relying on weaker assumptions, as mentioned above.

In terms of assumptions, the recent work of Asharov and Segev [7] shows that private-key functional encryption is much weaker than *any* public-key primitive (in particular, it is much weaker than public-key functional encryption). Specifically, they show that using the currently-known techniques it is impossible to use a private-key functional encryption scheme for constructing even a key-agreement protocol (and therefore, in particular, it is impossible to construct a public-key encryption scheme or a public-key functional encryption scheme).

Finally, we note that in addition to introducing the notion of a multi-input functional encryption scheme, Goldwasser et al. [24] introduced the more general notion of a *multi-client* multi-input functional encryption scheme. In such a scheme, each input coordinate is associated with its own encryption key, and security should be satisfied for all coordinates whose encryption keys are not known to the adversary. In this paper we do not consider this more general notion, and an interesting open problem is to extend our approach to the multi-client setting.

1.2   Additional Related Work

Extensive research has been devoted to the study of functional encryption, and for concreteness we focus here only on those previous efforts that are directly relevant to the techniques used in this paper.

Function-Private Functional Encryption. The security guarantees of functional encryption typically focus on *message privacy*. Intuitively, message privacy asks that a functional key sk_f does not help in distinguishing encryptions of two messages, m_0 and m_1, as long as $f(m_0) = f(m_1)$. In various cases, however, it is also useful to consider *function privacy* [1,13,19,35], asking that a functional key sk_f does not reveal any unnecessary information on the function f. Specifically, in the private-key setting, function privacy asks that an encryption of a message m does not help in distinguishing two functional keys, sk_{f_0} and sk_{f_1}, as long as

$f_0(m) = f_1(m)$. Brakerski and Segev [19] recently showed that any private-key functional encryption scheme can be generically transformed into one that satisfies a unified notion of security, referred to as *full security*, which considers both message privacy and function privacy.

Other than being a useful notion for various applications, function privacy was found useful as a building block in the construction of several functional encryption schemes [4,30]. One of the key insights that we utilize in this work is that function-private functional encryption allows to successfully apply proof techniques "borrowed" from the indistinguishability obfuscation literature (including, for example, a variant of the punctured programming approach of Sahai and Waters [33]).

Key-Encapsulation Techniques in Functional Encryption. Key encapsulation (also known as "hybrid encryption") is an extremely useful approach in the design of encryption schemes, both for improved efficiency and for improved security. Specifically, key encapsulation typically means that instead of encrypting a message m under a fixed key sk, one can instead sample a random key k, encrypt m under k and then encrypt k under sk. Recently, Ananth et al. [4] showed that key encapsulation is useful also in the setting of functional encryption. They showed that it can be used to transform any selectively-secure functional encryption scheme into an adaptively-secure one (in both the public-key setting and the private-key one). Their construction and proof technique hint that key encapsulation techniques may in fact be a general tool that is useful in the design of functional encryption schemes. Our constructions incorporate key encapsulation techniques, and exhibit additional strengths of this technique in the context of functional encryption schemes. Specifically, as discussed in Sect. 1.3, we use key encapsulation techniques to create "sufficient independence" between combinations of different ciphertexts, a crucial ingredient in our constructions (see Sect. 1.3 for a detailed comparison between our technique and that of Ananth et al.).

Multi-input Functional Encryption Schemes and Obfuscation. An important aspect in studying multi-input functional encryption schemes is its tight connection to indistinguishability obfuscation. Goldwasser et al. [24] showed that the following three primitives are equivalent: (1) selectively-secure *private*-key multi-input functional encryption scheme with polynomially many inputs, (2) selectively-secure *public*-key two-input functional encryption scheme, and (3) indistinguishability obfuscation. The works of Ananth and Jain [5] and Ananth, Jain and Sahai [6] show how to construct a selectively-secure private-key multi-input functional encryption scheme with polynomially many inputs (and thereby an indistinguishability obfuscator) from any sub-exponentially-secure *public-key* single-input functional encryption scheme.[2]

---

[2] Bitansky and Vaikuntanathan [10] achieved the same result (an indistinguishability obfuscator) as [5] using a similar construction (at least conceptually) while relying essentially on the same assumptions. However, they construct an indistinguishability obfuscator directly without going through the equivalence to multi-input functional encryption scheme.

1.3   Overview of Our Constructions and Techniques

In this section we provide a high-level overview of our constructions. For concreteness, we focus here mainly on two-input schemes, and then briefly discuss the generalization of our approach to more than two inputs (we refer the reader to Appendix A for the generalization to t-input schemes for $t \geq 2$). In what follows, we start by briefly describing the functionality and security properties of two-input schemes in the private-key setting. Then, we explain the main ideas underlying our constructions. We emphasize that the forthcoming overview is very high-level and ignores many technical details. For the full details we refer to Sects. 3 and 4.

Functionality and Security. In a private-key two-input functional encryption scheme, the master secret key msk of the scheme is used for encrypting any messages x and y (separately) to the first and second coordinates, respectively, and for generating functional keys for two-input functions. A functional key sk_f corresponding to a function f enables to compute $f(x, y)$ given $\mathsf{Enc}(x)$ and $\mathsf{Enc}(y)$. Building upon the previous notions of security for private-key multi-input functional encryption schemes [12,24], we consider a strengthened notion of security that combines both message privacy and function privacy (as in [1,19] for single-input schemes), to which we refer as *full security*.[3] Specifically, we consider *adaptive* adversaries that are given access to "left-or-right" key-generation and encryption oracles. These oracles operate in one out of two modes corresponding to a randomly-chosen bit b. The key-generation oracle receives as input pairs of the form (f_0, f_1) and outputs a functional key for f_b. The encryption oracle receives as input pairs of the form (x_0, x_1) for the first coordinate, or (y_0, y_1) for the second coordinate, and outputs an encryption of x_b or y_b. We require that no efficient adversary can guess the bit b with probability noticeably higher than $1/2$, as long as for each such three queries (f_0, f_1), (x_0, x_1) and (y_0, y_1) it holds that $f_0(x_0, y_0) = f_1(x_1, y_1)$.

Intuition: Input Aggregation. Given a two-input function $f(\cdot, \cdot)$, one can view f as a single-input function, f^*, that takes a tuple (x, y), which we denote by $x \| y$ to avoid confusion, and computes $f^*(x \| y) = f(x, y)$. Using a single-input scheme, we can generate a functional key for the function f^*. We thus remain with the problem of *aggregating the input*. That is, we need to be able to encrypt inputs x and y, such that given $\mathsf{Enc}(x)$ and $\mathsf{Enc}(y)$ it is possible to compute $\mathsf{Enc}(x \| y)$. At a very high-level, this is achieved by having the encryption of x be an "aggregator": To encrypt x, we will generate a functional key for the

---

[3] We consider a unified notion capturing both message privacy and function privacy not only as a useful feature for various applications. In fact, the function privacy of the resulting two-input scheme plays a crucial role when extending our results to more than two inputs.

function $\mathsf{AGG}_x(\cdot)$, that on input y outputs an encryption of $x\|y$.[4] There are many technical difficulties in realizing this intuition, as we explain in the remainder of this section.

Step 1: Functional Keys as Ciphertexts. Given any private-key single-input functional encryption scheme, 1FE, the first step in our transformation is to use both its ciphertexts and its functional keys as ciphertexts for a two-input scheme 2FE: An encryption of a message x to the first coordinate is a functional key sk_x corresponding to a certain functionality that depends on x, and an encryption of a message y to the second coordinate is simply an encryption of y. Intuitively, the hope is that the function privacy of 1FE will hide x, and that the message privacy of 1FE will hide y. More specifically, a first attempt towards realizing this intuition is as follows:

1. The master secret key consists of two keys, $\mathsf{msk}_{\mathsf{in}}$ and $\mathsf{msk}_{\mathsf{out}}$, for the single-input scheme 1FE. The key $\mathsf{msk}_{\mathsf{in}}$ is used for encryption, and the key $\mathsf{msk}_{\mathsf{out}}$ is used to decryption.
2. An encryption of a message x to the first coordinate is a functional key $\mathsf{sk}_{x,\mathsf{msk}_{\mathsf{out}}}$ that is generated using $\mathsf{msk}_{\mathsf{in}}$ and corresponds to the following functionality: Given an input y, it outputs an encryption $\mathsf{Enc}_{\mathsf{msk}_{\mathsf{out}}}(x\|y)$ of x concatenated with y under $\mathsf{msk}_{\mathsf{out}}$. An encryption of a message y to the second coordinate is simply an encryption $\mathsf{Enc}_{\mathsf{msk}_{\mathsf{in}}}(y)$ of y under $\mathsf{msk}_{\mathsf{in}}$.
3. A functional key for a two-input function f is a functional key that is generated using $\mathsf{msk}_{\mathsf{out}}$ for the function f when viewed as a single-input function.
4. Given a functional key for a function f, and two encryptions $\mathsf{sk}_{x,\mathsf{msk}_{\mathsf{out}}}$ and $\mathsf{Enc}_{\mathsf{msk}_{\mathsf{in}}}(y)$, we first apply $\mathsf{sk}_{x,\mathsf{msk}_{\mathsf{out}}}$ on $\mathsf{Enc}_{\mathsf{msk}_{\mathsf{in}}}(y)$ to obtain $\mathsf{Enc}_{\mathsf{msk}_{\mathsf{out}}}(x\|y)$, and then apply the functional key for f on $\mathsf{Enc}_{\mathsf{msk}_{\mathsf{out}}}(x\|y)$.

It is straightforward to verify that the above scheme indeed provides the required functionality of a two-input scheme. Proving its security, however, does not seem to go through: When "attacking" the key $\mathsf{msk}_{\mathsf{out}}$, we clearly cannot embed it in the encryptions $\mathsf{sk}_{x,\mathsf{msk}_{\mathsf{out}}}$ generated to the first coordinate. A typical approach for dealing with such a difficulty (e.g., [4,19,30]) is to embed all possibly-needed encryptions under $\mathsf{msk}_{\mathsf{out}}$ inside the ciphertexts of the two-input scheme (so that the key $\mathsf{msk}_{\mathsf{out}}$ will not be explicitly needed). Note, however, that when an adversary makes T encryption queries there may be roughly T^2 different pairs of the form (x,y), and these T^2 pairs cannot be embedded into T ciphertexts (we note that $T = T(\lambda)$ may be any polynomial and it is not known in advance).

An additional approach is to use a *public-key* functional encryption scheme for the role played by $\mathsf{msk}_{\mathsf{out}}$ (i.e., replacing $\mathsf{sk}_{x,\mathsf{msk}_{\mathsf{out}}}$ with $\mathsf{sk}_{x,\mathsf{pk}_{\mathsf{out}}}$). Although

---

[4] A somewhat related functionality was recently considered by Iovino and Zebrowski [27] who introduced the notion of *mergeable* functional encryption, where one can publicly transform encryptions, $\mathsf{Enc}(x)$ and $\mathsf{Enc}(y)$, of two values into an encryption $\mathsf{Enc}(x\|y)$ of their concatenation. They show how to construct such a scheme for two inputs building on the *specific* construction of [21] and assuming strong notions of obfuscation. In comparison, our approach applies to many inputs (as discussed below), and is based on minimal assumptions.

this solution allows to prove security, we view it as a "warm-up solution" as we would like to avoid relying on a stronger primitive than necessary. Specifically, we would like to rely on private-key functional encryption and not on public-key function encryption (as recently shown by Asharov and Segev [7], private-key functional encryption is significantly weaker than any public-key primitive).

Step 2: Selective Security via "One-Sided" Key Encapsulation. Our approach for resolving the difficulty described uses key-encapsulation techniques in functional encryption. Our main idea here is that when encrypting a message x, we sample a fresh key $\mathsf{msk}^\star$ for the single-input scheme, and output two components: $\mathsf{Enc}_{\mathsf{msk}_{\mathsf{out}}}(\mathsf{msk}^\star)$ and $\mathsf{sk}_{x,\mathsf{msk}^\star}$. Given an encryption $\mathsf{Enc}_{\mathsf{msk}_{\mathsf{in}}}(y)$ of a message y, the component $\mathsf{sk}_{x,\mathsf{msk}^\star}$ enables to compute $\mathsf{Enc}_{\mathsf{msk}^\star}(x||y)$. In addition, a functional key for a function f is now generated using $\mathsf{msk}_{\mathsf{out}}$ for the following functionality: Given an input $\mathsf{msk}^\star$, it outputs a functional key for f (viewed as a single-input function) using $\mathsf{msk}^\star$. This enables to compute $f(x,y)$ given $\mathsf{Enc}_{\mathsf{msk}^\star}(x||y)$ and provides the required functionality.

This "one-sided" key encapsulation enables us to prove a selectively-secure variant of our notion of security.[5] In this variant we require adversaries to specify their encryption queries in advance, and they are then given adaptive access to the left-or-right key-generation oracle. The main idea underlying the proof of security is that our one-sided key encapsulation approach yields sufficient independence and allows attacking the x's one by one, by attacking their corresponding encapsulated keys. Focusing on one message x and its encapsulated key $\mathsf{msk}^\star$, an adversary that make T encryption queries $y_1, \ldots, y_T$ to the second coordinate induces only T pairs $\{(x, y_i)\}_{i \in [T]}$ (instead of T^2 pairs as above). Moreover, given that the encryption queries are chosen in advance, we can embed an encryption of $x||y_i$ under $\mathsf{msk}^\star$ inside the encryption of each y_i. This way the key $\mathsf{msk}^\star$ is not explicitly needed, and thus can be attacked (while not affecting any of the other x's).

As discussed in Sect. 1.2, key-encapsulation techniques have been introduced into the setting of functional encryption by Ananth et al. [4]. Our approach builds upon and significantly extends their initial observations, and enables us to create "sufficient independence" between combinations of different ciphertexts, a crucial ingredient in our constructions.

This enables us to construct a selectively-secure two-input scheme from any selectively-secure single-input one (we refer the reader to Sect. 3 for the scheme and its proof of security). Note, however, that this approach is limited to selective adversaries: embedding an encryption of $x||y_i$ inside the encryption of y_i requires knowing x before the adversary queries for the encryption of y_i.

Step 3: Adaptive Security via "Two-Sided" Key Encapsulation. Next, we present a general transformation from selective security to adaptive security (in fact, to our stronger notion of full security). Specifically, we rely on two building blocks: (1) any private-key *selectively-secure two-input* scheme, and (2) any

---

[5] "One-sided" here refers to the fact that the encapsulated key $\mathsf{msk}^\star$ is generated only from the side of the x's.

private-key *adaptively-secure single-input* scheme (recall that in the single-input setting, selective security implies adaptive security [4]). For this transformation we introduce a new technique which we call "two-sided" key encapsulation, where each pair of messages x and y has its own encapsulated key $\mathsf{msk}^\star$. This, more subtle approach, enables us to "attack" a specific pair of messages each time, since each such pair uses a different encapsulated key: If x is known before y then we embed $x\|y$ inside the encryption of y, and if x is known after y then we embed $x\|y$ inside the encryption of x. This leaves the problem of how to realize this idea of two-sided key encapsulation. Our two-sided key encapsulation works as follows.

1. The master secret key consists of two keys: A master secret key $\mathsf{msk}_\mathsf{out}$ for a selectively-secure two-input scheme, and a master secret key msk_in for an adaptively-secure single-input scheme.
2. An encryption of a message y consists of two components: $\mathsf{Enc}_{\mathsf{msk}_\mathsf{out}}(t)$ and $\mathsf{Enc}_{\mathsf{msk}_\mathsf{in}}(y, t)$, where t is a fresh random tag.
3. An encryption of a message x consists of two components: $\mathsf{Enc}_{\mathsf{msk}_\mathsf{out}}(s)$ and $\mathsf{sk}_{x,s}$, where s is a fresh random tag. The functional key $\mathsf{sk}_{x,s}$ is generated using msk_in and corresponds to the following functionality: Given an input (y, t), derive $\mathsf{msk}^\star = \mathsf{PRF}(s, t)$,[6] and output $\mathsf{Enc}_{\mathsf{msk}^\star}(x\|y)$.
4. A functional key for a function f is generated using $\mathsf{msk}_\mathsf{out}$ for the following functionality: Given *two inputs*, s and t, derive $\mathsf{msk}^\star = \mathsf{PRF}(s, t)$, and output a functional key for f (viewed as a single-input function) using $\mathsf{msk}^\star$.

The crucial observation is that the master secret key $\mathsf{msk}_\mathsf{out}$ of the two-input selectively-secure scheme is used for encrypting random tags, whereas the plaintext itself is always encrypted using the master secret key msk_in of the adaptively-secure single-input scheme. This enables us to prove the full security of the resulting scheme (we refer the reader to Sect. 4 for the scheme and its proof of security).

Comparison to the Selective-to-Adaptive Transformation of Ananth et al. [4]. Our two-sided key encapsulation technique shows that the usability of key-encapsulation in the context of functional encryption, demonstrated by Ananth et al. [4], can be significantly extended. Whereas their generic transformation from selective security to adaptive security for single-input scheme uses a rather direct form of key encapsulation, our approach requires a significantly more structured one in which the encapsulated key is not determined at the time of encryption, but rather generated "freshly" (in a pseudorandom manner) for any two messages x and y as above.

Specifically, Ananth et al. encrypted a message m under a selectively-secure key msk, by sampling a fresh master secret key $\mathsf{msk}^\star$ for a "one-time" adaptively-secure scheme, encrypted m under $\mathsf{msk}^\star$ and then encrypted $\mathsf{msk}^\star$ under msk. This direct encapsulation does not seem to extend to the two-input setting,

---

[6] More accurately, the key $\mathsf{msk}^\star$ is computed by applying the setup algorithm of 1FE with randomness $\mathsf{PRF}(s, t)$.

as applying it independently in each coordinate seems to hurt both the security and the functionality of the scheme. By introducing our two-sided key-encapsulation idea we are able to balance between the need for using key encapsulation in each coordinate and the need for generating sufficient independence between different pairs of messages.

Step 4: Generalization to t-input Schemes. The generalization of our result to t-input schemes, for $t \geq 2$, consists of two components. The first component is a construction that uses any $(t-1)$-input scheme for building a selectively-secure t-input scheme, for any $t \geq 2$. The second component is a construction that uses any selectively-secure t-input scheme and a fully-secure $(t-1)$-input scheme for building a fully-secure t-input scheme. Thus, for obtaining a fully-secure t-input scheme from any single-input scheme, one can iteratively apply both components alternately t times. This is illustrated in Fig. 1 for the case $t = 3$ (and the same illustration generalizes to any $t > 3$ in a straightforward manner).

This iterative application of our components places a restriction on the number of supported inputs. In general, each such application may result in a polynomial blow-up in the parameters of the scheme. Therefore, $t - 1$ applications may result in a blow-up of $\lambda^{2^{O(t)}}$ which must be kept polynomial. Without any additional assumptions, this implies that t can be any fixed constant. Assuming, in addition, that the underlying single-input scheme is sub-exponentially secure, the number of inputs can be made super-constant. Specifically, for any constant $0 < \epsilon < 1$, when instantiating the underlying single-input scheme with security parameter $\tilde{\lambda} = 2^{(\log \lambda)^\epsilon}$, the first component can be iteratively applied to reach $t = \Theta(\log \log \lambda)$ inputs. Obtaining a generic transformation that supports a super-constant number of inputs without assuming sub-exponential security (or an alternative form of "succinctness") is left as an open problem.

1.4 Paper Organization

The remainder of this paper is organized as follows. In Sect. 2 we provide an overview of the notation, definitions, and tools underlying our constructions. In Sect. 3 we present a construction of a selectively-secure two-input functional encryption scheme from any single-input scheme. In Sect. 4 we present a construction of a fully-secure two-input functional encryption scheme from any selectively-secure one. In Appendix A we generalize our approach to t-input schemes for $t \geq 2$. In the full version [18] we provide the formal proofs of our theorems from Sects. 3 and 4, and from Appendix A.

2 Preliminaries

In this section we present the notation and basic definitions that are used in this work. For a distribution X we denote by $x \leftarrow X$ the process of sampling a value x from the distribution X. Similarly, for a set $\mathcal{X}$ we denote by $x \leftarrow \mathcal{X}$ the process of sampling a value x from the uniform distribution over $\mathcal{X}$.

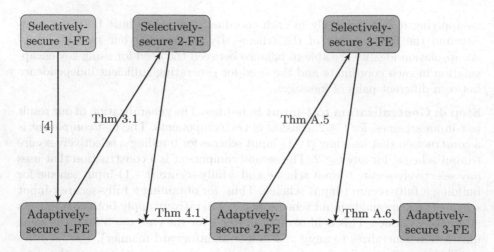

Fig. 1. An illustration of the required iterative applications of our two transformations for obtaining an adaptively-secure three-input scheme based on any selectively-secure single-input scheme.

For a randomized function f and an input $x \in \mathcal{X}$, we denote by $y \leftarrow f(x)$ the process of sampling a value y from the distribution $f(x)$. For an integer $n \in \mathbb{N}$ we denote by $[n]$ the set $\{1, \ldots, n\}$. A function $\mathsf{neg} : \mathbb{N} \rightarrow \mathbb{R}$ is *negligible* if for every constant $c > 0$ there exists an integer N_c such that $\mathsf{neg}(\lambda) < \lambda^{-c}$ for all $\lambda > N_c$. Two sequences of random variables $X = \{X_\lambda\}_{\lambda \in \mathbb{N}}$ and $Y = \{Y_\lambda\}_{\lambda \in \mathbb{N}}$ are *computationally indistinguishable* if for any probabilistic polynomial-time algorithm $\mathcal{A}$ there exists a negligible function $\mathsf{neg}(\cdot)$ such that $\left| \Pr[\mathcal{A}(1^\lambda, X_\lambda) = 1] - \Pr[\mathcal{A}(1^\lambda, Y_\lambda) = 1] \right| \leq \mathsf{neg}(\lambda)$ for all sufficiently large $\lambda \in \mathbb{N}$. Throughout the paper, we denote by λ the security parameter.

2.1 Pseudorandom Functions

Let $\{\mathcal{K}_\lambda, \mathcal{X}_\lambda, \mathcal{Y}_\lambda\}_{\lambda \in \mathbb{N}}$ be a sequence of sets and let $\mathsf{PRF} = (\mathsf{PRF.Gen}, \mathsf{PRF.Eval})$ be a function family with the following syntax:

- $\mathsf{PRF.Gen}$ is a probabilistic polynomial-time algorithm that takes as input the unary representation of the security parameter λ, and outputs a key $K \in \mathcal{K}_\lambda$.
- $\mathsf{PRF.Eval}$ is a deterministic polynomial-time algorithm that takes as input a key $K \in \mathcal{K}_\lambda$ and a value $x \in \mathcal{X}_\lambda$, and outputs a value $y \in \mathcal{Y}_\lambda$.

The sets $\mathcal{K}_\lambda$, $\mathcal{X}_\lambda$, and $\mathcal{Y}_\lambda$ are referred to as the *key space*, *domain*, and *range* of the function family, respectively. For easy of notation we may denote by $\mathsf{PRF.Eval}_K(\cdot)$ or $\mathsf{PRF}_K(\cdot)$ the function $\mathsf{PRF.Eval}(K, \cdot)$ for $K \in \mathcal{K}_\lambda$. The following is the standard definition of a pseudorandom function family.

Definition 2.1 (Pseudorandomness). *A function family* $\mathsf{PRF} = (\mathsf{PRF.Gen}, \mathsf{PRF.Eval})$ *is* pseudorandom *if for every probabilistic polynomial-time algorithm*

$\mathcal{A}$ there exits a negligible function neg($\cdot$) such that

$$\mathsf{Adv}_{\mathsf{PRF},\mathcal{A}}(\lambda) \overset{\mathsf{def}}{=} \left| \Pr_{K \leftarrow \mathsf{PRF.Gen}(1^\lambda)} \left[\mathcal{A}^{\mathsf{PRF.Eval}_K(\cdot)}(1^\lambda) = 1 \right] - \Pr_{f \leftarrow F_\lambda} \left[\mathcal{A}^{f(\cdot)}(1^\lambda) = 1 \right] \right|$$

$$\leq \mathsf{neg}(\lambda),$$

for all sufficiently large $\lambda \in \mathbb{N}$, where F_λ is the set of all functions that map $\mathcal{X}_\lambda$ into $\mathcal{Y}_\lambda$.

In addition to the standard notion of a pseudorandom function family, we rely on the seemingly stronger (yet existentially equivalent) notion of a *puncturable* pseudorandom function family [15,17,28,33]. In terms of syntax, this notion asks for an additional probabilistic polynomial-time algorithm, PRF.Punc, that takes as input a key $K \in \mathcal{K}_\lambda$ and a set $S \subseteq \mathcal{X}_\lambda$ and outputs a "punctured" key K_S. The properties required by such a puncturing algorithm are captured by the following definition.

Definition 2.2 (Puncturable PRF). *A pseudorandom function family* PRF = (PRF.Gen, PRF.Eval, PRF.Punc) *is* puncturable *if the following properties are satisfied:*

1. **Functionality:** *For all sufficiently large* $\lambda \in \mathbb{N}$, *for every set* $S \subseteq \mathcal{X}_\lambda$, *and for every* $x \in \mathcal{X}_\lambda \setminus S$ *it holds that*

$$\Pr_{\substack{K \leftarrow \mathsf{PRF.Gen}(1^\lambda); \\ K_S \leftarrow \mathsf{PRF.Punc}(K,S)}} [\mathsf{PRF.Eval}_K(x) = \mathsf{PRF.Eval}_{K_S}(x)] = 1.$$

2. **Pseudorandomness at punctured points:** *Let* $\mathcal{A} = (\mathcal{A}_1, \mathcal{A}_2)$ *be any probabilistic polynomial-time algorithm such that* $\mathcal{A}_1(1^\lambda)$ *outputs a set* $S \subseteq \mathcal{X}_\lambda$, *a value* $x \in S$, *and state information* state. *Then, for any such* $\mathcal{A}$ *there exists a negligible function* neg($\cdot$) *such that*

$$\mathsf{Adv}_{\mathsf{PRF},\mathcal{A}}(\lambda) \overset{\mathsf{def}}{=} |\Pr[\mathcal{A}_2(K_S, \mathsf{PRF.Eval}_K(x), \mathsf{state}) = 1]$$
$$- \Pr[\mathcal{A}_2(K_S, y, \mathsf{state}) = 1]|$$
$$\leq \mathsf{neg}(\lambda)$$

for all sufficiently large $\lambda \in \mathbb{N}$, *where* $(S, x, \mathsf{state}) \leftarrow \mathcal{A}_1(1^\lambda)$, $K \leftarrow \mathsf{PRF.Gen}(1^\lambda)$, $K_S = \mathsf{PRF.Punc}(K, S)$, *and* $y \leftarrow \mathcal{Y}_\lambda$.

For our constructions we rely on pseudorandom functions that need to be punctured only at one point (i.e., in both parts of Definition 2.2 it holds that $S = \{x\}$ for some $x \in \mathcal{X}_\lambda$). As observed by [15,17,28,33] the GGM construction [23] of PRFs from any one-way function can be easily altered to yield such a puncturable pseudorandom function family.

2.2   Private-Key Single-Input Functional Encryption

A private-key single-input functional encryption scheme over a message space $\mathcal{X} = \{\mathcal{X}_\lambda\}_{\lambda \in \mathbb{N}}$ and a function space $\mathcal{F} = \{\mathcal{F}_\lambda\}_{\lambda \in \mathbb{N}}$ is a quadruple (FE.S, FE.KG, FE.E, FE.D) of probabilistic polynomial-time algorithms. The setup algorithm FE.S takes as input the unary representation 1^λ of the security parameter $\lambda \in \mathbb{N}$ and outputs a master-secret key msk. The key-generation algorithm FE.KG takes as input a master-secret key msk and a single-input function $f \in \mathcal{F}_\lambda$, and outputs a functional key sk_f. The encryption algorithm FE.E takes as input a master-secret key msk and a message $x \in \mathcal{X}_\lambda$, and outputs a ciphertext ct. In terms of correctness we require that for all sufficiently large $\lambda \in \mathbb{N}$, for every function $f \in \mathcal{F}_\lambda$ and message $x \in \mathcal{X}_\lambda$ it holds that FE.D(FE.KG(msk, f), FE.E(msk, x)) = $f(x)$ with all but a negligible probability over the internal randomness of the algorithms FE.S, FE.KG, and FE.E.

In terms of security, we rely on the private-key variant of the existing indistinguishability-based notions for message privacy and function privacy. In fact, following [1,19], our notion of security combines both message privacy and function privacy. When formalizing this notion it would be convenient to use the following standard notion of a *left-or-right oracle*.

Definition 2.3 (Left-or-right oracle). *Let* $\mathcal{O}(\cdot, \cdot)$ *be a probabilistic two-input functionality. For each* $b \in \{0, 1\}$ *we denote by* $\mathcal{O}_b$ *the probabilistic three-input functionality* $\mathcal{O}_b(k, z_0, z_1) \overset{\text{def}}{=} \mathcal{O}(k, z_b)$.

Intuitively, a private-key functional-encryption scheme is secure if encryptions of messages $x_1, \ldots, x_T$ together with functional keys corresponding to functions $f_1, \ldots, f_T$ reveal essentially no information other than the values $\{f_i(x_j)\}_{i,j \in [T]}$. We consider an adaptive notion of security, to which we refer to as *full security*, in which adversaries are given adaptive access to left-or-right encryption and key-generation oracles.

Definition 2.4 (Full security [1,19]). *A private-key single-input functional encryption scheme* FE = (FE.S, FE.KG, FE.E, FE.D) *over a message space* $\mathcal{X} = \{\mathcal{X}_\lambda\}_{\lambda \in \mathbb{N}}$ *and a function space* $\mathcal{F} = \{\mathcal{F}_\lambda\}_{\lambda \in \mathbb{N}}$ *is fully secure if for any probabilistic polynomial-time adversary* $\mathcal{A}$ *there exists a negligible function* $\mathsf{neg}(\cdot)$ *such that*

$$\mathsf{Adv}_{\mathsf{FE},\mathcal{A},\mathcal{F}}^{\mathsf{full1FE}}(\lambda) \overset{\text{def}}{=} \left| \Pr\left[\mathcal{A}^{\mathsf{KG}_0(\mathsf{msk},\cdot,\cdot),\mathsf{Enc}_0(\mathsf{msk},\cdot,\cdot)}(1^\lambda) = 1 \right] \right.$$
$$\left. - \Pr\left[\mathcal{A}^{\mathsf{KG}_1(\mathsf{msk},\cdot,\cdot),\mathsf{Enc}_1(\mathsf{msk},\cdot,\cdot)}(1^\lambda) = 1 \right] \right|$$
$$\leq \mathsf{neg}(\lambda)$$

$(f_0, f_1) \in \mathcal{F}_\lambda \times \mathcal{F}_\lambda$ *and* $(x_0, x_1) \in \mathcal{X}_\lambda \times \mathcal{X}_\lambda$ *with which* $\mathcal{A}$ *queries the left-or-right key-generation and encryption oracles, respectively, it holds that* $f_0(x_0) = f_1(x_1)$. *Moreover, the probability is taken over the choice of* msk $\leftarrow$ FE.S(1^λ) *and the internal randomness of* $\mathcal{A}$.

Known Constructions. Private-key single-input functional encryption schemes that satisfy the above notion of full security and support circuits of any a-priori bounded polynomial size are known to exist based on a variety of assumptions.

Ananth et al. [4] gave a generic transformation from selective-message (or selective-function) security to full security. Moreover, Brakerski and Segev [19] showed how to transform any message-private functional encryption scheme into a functional encryption scheme which is fully secure, and the resulting scheme inherits the security guarantees of the original one. Therefore, based on [4,19], given any selective-message (or selective-function) message-private functional encryption scheme we can generically obtain a fully-secure scheme. This implies that schemes that are fully secure for any number of encryption and key-generation queries can be based on indistinguishability obfuscation [21,36], differing-input obfuscation [3,16], and multilinear maps [22]. In addition, schemes that are fully secure for a bounded number $T = T(\lambda)$ of encryption and key-generation queries can be based on the Learning with Errors (LWE) assumption (where the length of ciphertexts grows with T and with a bound on the depth of allowed functions) [25], based on pseudorandom generators computable by small-depth circuits (where the length of ciphertexts grows with T and with an upper bound on the circuit size of the functions) [26], and even based on one-way functions (for $T = 1$) [26].

2.3 Private-Key Two-Input Functional Encryption

In this section we define the functionality and security of private-key *two-input* functional encryption scheme (we refer the reader to Appendix A for the generalization to t-input schemes for any $t \geq 2$). Let $\mathcal{X} = \{\mathcal{X}_\lambda\}_{\lambda \in \mathbb{N}}$, $\mathcal{Y} = \{\mathcal{Y}_\lambda\}_{\lambda \in \mathbb{N}}$, and $\mathcal{Z} = \{\mathcal{Z}_\lambda\}_{\lambda \in \mathbb{N}}$ be ensembles of finite sets, and let $\mathcal{F} = \{\mathcal{F}_\lambda\}_{\lambda \in \mathbb{N}}$ be an ensemble of finite two-ary function families. For each $\lambda \in \mathbb{N}$, each function $f \in \mathcal{F}_\lambda$ takes as input two strings, $x \in \mathcal{X}_\lambda$ and $y \in \mathcal{Y}_\lambda$, and outputs a value $f(x, y) \in \mathcal{Z}_\lambda$. A private-key two-input functional encryption scheme Π for $\mathcal{F}$ consists of four probabilistic polynomial time algorithm Setup, Enc, KG and Dec, described as follows.

- Setup(1^λ) – The setup algorithm takes as input the security parameter λ, and outputs a master secret key msk.
- Enc(msk, m, i) – The encryption algorithm takes as input a master secret key msk, message input m, and an index i $\in [2]$, where $m \in \mathcal{X}_\lambda$ if i $= 1$ and $m \in \mathcal{Y}_\lambda$ if i $= 2$. It outputs a ciphertext ct$_i$.
- KG(msk, f) – The key-generation algorithm takes as input a master secret key msk and a function $f \in \mathcal{F}_\lambda$, and outputs a functional key sk$_f$.
- Dec(sk$_f$, ct$_1$, ct$_2$) – The (deterministic) decryption algorithm takes as input a functional key sk$_f$ and two ciphertexts ct$_1$ and ct$_2$, and outputs a string $z \in \mathcal{Z}_\lambda \cup \{\bot\}$.

Definition 2.5 (Correctness). *A private-key two-input functional encryption scheme $\Pi = $ (Setup, Enc, KG, Dec) for $\mathcal{F}$ is correct if there exists a negligible* *function* $\mathsf{neg}(\cdot)$ *such that for every* $\lambda \in \mathbb{N}$, *for every* $f \in \mathcal{F}_\lambda$, *and for every* $(x, y) \in \mathcal{X}_\lambda \times \mathcal{Y}_\lambda$, *it holds that*

$$\Pr\left[\mathsf{Dec}(\mathsf{sk}_f, \mathsf{Enc}(\mathsf{msk}, x, 1), \mathsf{Enc}(\mathsf{msk}, y, 2)) = f(x, y)\right] \geq 1 - \mathsf{neg}(\lambda),$$

where $\mathsf{msk} \leftarrow \mathsf{Setup}(1^\lambda)$, $\mathsf{sk}_f \leftarrow \mathsf{KG}(\mathsf{msk}, f)$, *and the probability is taken over the internal randomness of* $\mathsf{Setup}, \mathsf{Enc}$ *and* KG.

Intuitively, we say that a two-input scheme is secure if for any two pairs of messages (x_0, x_1) and (y_0, y_1) that are encrypted with respect to indices $\mathsf{i} = 1$ and $\mathsf{i} = 2$, respectively, and for every pair of functions (f_0, f_1), the triplets $(\mathsf{sk}_{f_0}, \mathsf{Enc}(\mathsf{msk}, x_0, 1), \mathsf{Enc}(\mathsf{msk}, y_0, 2))$ and $(\mathsf{sk}_{f_1}, \mathsf{Enc}(\mathsf{msk}, x_1, 1), \mathsf{Enc}(\mathsf{msk}, y_1, 2))$ are computationally indistinguishable as long as $f_0(x_0, y_0) = f_1(x_1, y_1)$ (note that this considers both message privacy and function privacy). The formal notions of security build upon this intuition and capture the fact that an adversary may in fact hold many functional keys and ciphertexts, and may combine them in an arbitrary manner. As in the case of single-input schemes, we formalize our notions of security using left-or-right key-generation and encryption oracles. Specifically, for each $b \in \{0, 1\}$ and $\mathsf{i} \in \{1, 2\}$ we let $\mathsf{KG}_b(\mathsf{msk}, f_0, f_1) \overset{\mathsf{def}}{=} \mathsf{KG}(\mathsf{msk}, f_b)$ and $\mathsf{Enc}_b(\mathsf{msk}, (m_0, m_1), \mathsf{i}) \overset{\mathsf{def}}{=} \mathsf{Enc}(\mathsf{msk}, m_b, \mathsf{i})$. Before formalizing our notions of security we define the notion of a *valid two-input adversary*.

Definition 2.6 (Valid two-input adversary). *A probabilistic polynomial-time algorithm* $\mathcal{A}$ *is a* valid two-input adversary *if for all private-key two-input functional encryption schemes* $\Pi = (\mathsf{Setup}, \mathsf{KG}, \mathsf{Enc}, \mathsf{Dec})$ *over a message space* $\mathcal{X} \times \mathcal{Y} = \{\mathcal{X}_\lambda\}_{\lambda \in \mathbb{N}} \times \{\mathcal{Y}_\lambda\}_{\lambda \in \mathbb{N}}$ *and a function space* $\mathcal{F} = \{\mathcal{F}_\lambda\}_{\lambda \in \mathbb{N}}$, *for all* $\lambda \in \mathbb{N}$ *and* $b \in \{0, 1\}$, *and for all* $(f_0, f_1) \in \mathcal{F}_\lambda$, $((x_0, x_1), 1) \in \mathcal{X}_\lambda \times \mathcal{X}_\lambda \times \{1\}$ *and* $((y_0, y_1), 1) \in \mathcal{Y}_\lambda \times \mathcal{Y}_\lambda \times \{2\}$ *with which* $\mathcal{A}$ *queries the left-or-right key-generation and encryption oracles, respectively, it holds that* $f_0(x_0, y_0) = f_1(x_1, y_1)$.

We consider two notions of security for two-input schemes, both of which combine message privacy and function privacy. The first notion, *full security*, considers adversaries that have adaptive access to both the encryption oracle and the key-generation oracle. The second notion, *selective-message security*, considers adversaries that must specify all of their encryption queries in advance, but can then have adaptive access to the key-generation oracle. Full security clearly implies selective-message security, and our work shows that the two notions are in fact equivalent for multi-input schemes.

Definition 2.7 (Full security). *A private-key two-input functional encryption scheme* $\Pi = (\mathsf{Setup}, \mathsf{KG}, \mathsf{Enc}, \mathsf{Dec})$ *over a message space* $\mathcal{X} \times \mathcal{Y} = \{\mathcal{X}_\lambda\}_{\lambda \in \mathbb{N}} \times \{\mathcal{Y}_\lambda\}_{\lambda \in \mathbb{N}}$ *and a function space* $\mathcal{F} = \{\mathcal{F}_\lambda\}_{\lambda \in \mathbb{N}}$ *is* fully secure *if for any valid two-input adversary* $\mathcal{A}$ *there exists a negligible function* $\mathsf{neg}(\cdot)$ *such that*

$$\mathsf{Adv}_{\Pi, \mathcal{F}, \mathcal{A}}^{\mathsf{full2FE}} \overset{\mathsf{def}}{=} \left| \Pr\left[\mathsf{Exp}_{\Pi, \mathcal{F}, \mathcal{A}}^{\mathsf{full2FE}}(\lambda) = 1\right] - \frac{1}{2} \right| \leq \mathsf{neg}(\lambda),$$

for all sufficiently large $\lambda \in \mathbb{N}$, *where the random variable* $\mathsf{Exp}_{\Pi, \mathcal{F}, \mathcal{A}}^{\mathsf{full2FE}}(\lambda)$ *is defined via the following experiment:*

1. $\mathsf{msk} \leftarrow \mathsf{Setup}(1^\lambda)$, $b \leftarrow \{0,1\}$.
2. $b' \leftarrow \mathcal{A}^{\mathsf{KG}_b(\mathsf{msk},\cdot,\cdot),\mathsf{Enc}_b(\mathsf{msk},(\cdot,\cdot),\cdot)}(1^\lambda,)$.
3. If $b' = b$ then output 1, and otherwise output 0.

Definition 2.8 (Selective-message security). *A private-key two-input functional encryption scheme* $\Pi = (\mathsf{Setup}, \mathsf{KG}, \mathsf{Enc}, \mathsf{Dec})$ *over a message space* $\mathcal{X} \times \mathcal{Y} = \{\mathcal{X}_\lambda\}_{\lambda \in \mathbb{N}} \times \{\mathcal{Y}_\lambda\}_{\lambda \in \mathbb{N}}$ *and a function space* $\mathcal{F} = \{\mathcal{F}_\lambda\}_{\lambda \in \mathbb{N}}$ *is selective-message secure if for any valid two-input adversary* $\mathcal{A} = (\mathcal{A}_1, \mathcal{A}_2)$ *there exists a negligible function* $\mathsf{neg}(\lambda)$ *such that*

$$\mathsf{Adv}_{\Pi,\mathcal{F},\mathcal{A}}^{\mathsf{sel2FE}} \overset{\mathsf{def}}{=} \left| \Pr\left[\mathsf{Exp}_{\Pi,\mathcal{F},\mathcal{A}}^{\mathsf{sel2FE}}(\lambda) = 1 \right] - \frac{1}{2} \right| \le \mathsf{neg}(\lambda),$$

for all sufficiently large $\lambda \in \mathbb{N}$, *where the random variable* $\mathsf{Exp}_{\Pi,\mathcal{F},\mathcal{A}}^{\mathsf{sel2FE}}(\lambda)$ *is defined via the following experiment:*

1. $(\vec{x}, \vec{y}, \mathsf{state}) \leftarrow \mathcal{A}_1(1^\lambda)$, *where* $\vec{x} = ((x_1^0, x_1^1), \ldots, (x_T^0, x_T^1))$ *and* $\vec{y} = ((y_1^0, y_1^1), \ldots, (y_T^0, y_T^1))$.
2. $\mathsf{msk} \leftarrow \mathsf{Setup}(1^\lambda)$, $b \leftarrow \{0,1\}$.
3. $\mathsf{ct}_{1,i} \leftarrow \mathsf{Enc}(\mathsf{msk}, x_i^b, 1)$ *and* $\mathsf{ct}_{2,i} \leftarrow \mathsf{Enc}(\mathsf{msk}, y_i^b, 2)$ *for* $i \in [T]$.
4. $b' \leftarrow \mathcal{A}_2^{\mathsf{KG}_b(\mathsf{msk},\cdot,\cdot)}(1^\lambda, \mathsf{ct}_{1,1}, \ldots, \mathsf{ct}_{1,T}, \mathsf{ct}_{2,1} \ldots, \mathsf{ct}_{2,T}, \mathsf{state})$.
5. If $b' = b$ then output 1, and otherwise output 0.

Our definitions of a two-input functional encryption scheme is inspired by the definition of [12]. It is a natural generalization of the single-input case and gives rise to an order-revealing encryption. Moreover, as a concrete motivation, a t-input scheme according to the above definition is enough to construct indistinguishability obfuscation for circuits with t input bits [24].[7]

Additional natural ways to define two-input functional encryptions schemes exist. Specifically, Goldwasser et al. [24] considered two such definitions. The first allows to encrypt a message m independently of an index $i \in [2]$. Thus, given a key for a two-input function f and encryptions of two messages x and of y, one can compute both $f(x, y)$ and $f(y, x)$. Hence, this definition requires a stronger "validity requirement" (see Definition 2.6), which means it can support less functionalities. A construction which satisfies our (indexed) definition can be easily transformed into one which satisfies the above (non-indexed) definition by encrypting each message with respect to both indices.

The second, referred to as "multi-client", considers each index as a different "client" and gives each of them his own secret key. In this setting, their security game is quite different, and in particular, an adversary is allowed to obtain the secret keys of a subset of the clients of his choice. The approach underlying our schemes does not seem to directly extend to the multi-client setting, and we leave it as an interesting path for future exploration.

---

[7] Indeed, [5] get a construction of a t-input scheme for any $t \ge 1$ which implies an indistinguishability obfuscator. Our construction falls short from being generalized to such extent (however, it relies on weaker assumptions).

3   A Selectively-Secure Two-Input Scheme from Any Single-Input Scheme

In this section we construct a private-key two-input functional encryption scheme that is selectively secure. Let $\mathcal{F} = \{\mathcal{F}_\lambda\}_{\lambda \in \mathbb{N}}$ be a family of two-ary functionalities, where for every $\lambda \in \mathbb{N}$ the set $\mathcal{F}_\lambda$ consists of functions of the form $f : \mathcal{X}_\lambda \times \mathcal{Y}_\lambda \to \mathcal{Z}_\lambda$. Our construction relies on the following building blocks:

1. A private-key single-input functional encryption scheme 1FE = (1FE.S, 1FE.KG, 1FE.E, 1FE.D).
2. A pseudorandom function family PRF = (PRF.Gen, PRF.Eval).

As discussed in Sect. 1.1, we assume that the scheme 1FE is sufficiently expressive in the sense that 1FE supports the function family $\mathcal{F}$ (when viewed as a family of single-input functions), the evaluation procedure of the pseudorandom function family PRF, the encryption and key-generation procedures of the private-key functional encryption scheme 1FE, and a few additional basic operations. Our scheme $2\mathsf{FE}^{\mathsf{sel}} = (2\mathsf{FE}^{\mathsf{sel}}.\mathsf{S}, 2\mathsf{FE}^{\mathsf{sel}}.\mathsf{KG}, 2\mathsf{FE}^{\mathsf{sel}}.\mathsf{E}, 2\mathsf{FE}^{\mathsf{sel}}.\mathsf{D})$ is defined as follows.

- **The setup algorithm.** On input the security parameter 1^λ the setup algorithm $2\mathsf{FE}^{\mathsf{sel}}.\mathsf{S}$ samples $\mathsf{msk}_{\mathsf{out}}, \mathsf{msk}_{\mathsf{in}} \leftarrow 1\mathsf{FE}.\mathsf{S}(1^\lambda)$ and outputs $\mathsf{msk} = (\mathsf{msk}_{\mathsf{out}}, \mathsf{msk}_{\mathsf{in}})$.
- **The key-generation algorithm.** On input the master secret key msk and a function $f \in \mathcal{F}_\lambda$, the key-generation algorithm $2\mathsf{FE}^{\mathsf{sel}}.\mathsf{KG}$ samples a random string $z \leftarrow \{0,1\}^\lambda$ and outputs $\mathsf{sk}_f \leftarrow 1\mathsf{FE}.\mathsf{KG}(\mathsf{msk}_{\mathsf{out}}, D_{f,\perp,z,\perp})$, where $D_{f,\perp,z,\perp}$ is a single-input function that is defined in Fig. 2.
- **The encryption algorithm.** On input the master secret key msk, a message m and an index $\mathsf{i} \in [2]$, the encryption algorithm $2\mathsf{FE}^{\mathsf{sel}}.\mathsf{E}$ has two cases:
  - If $(m, \mathsf{i}) = (x, 1)$, it samples a master secret key $\mathsf{msk}^\star \leftarrow 1\mathsf{FE}.\mathsf{S}(1^\lambda)$, a PRF key $K \leftarrow \mathsf{PRF}.\mathsf{Gen}(1^\lambda)$, and a random string $s \in \{0,1\}^\lambda$, and then outputs a pair $(\mathsf{ct}_1, \mathsf{sk}_1)$ defined as follows:

    $$\mathsf{ct}_1 \leftarrow 1\mathsf{FE}.\mathsf{E}(\mathsf{msk}_{\mathsf{out}}, (\mathsf{msk}^\star, K, 0))$$
    $$\mathsf{sk}_1 \leftarrow 1\mathsf{FE}.\mathsf{KG}(\mathsf{msk}_{\mathsf{in}}, \mathsf{AGG}_{x,\perp,0,s,\mathsf{msk}^\star,K}),$$

    where $\mathsf{AGG}_{x,\perp,0,s,\mathsf{msk}^\star,K}$ is a single-input function that is defined in Fig. 3.
  - If $(m, \mathsf{i}) = (y, 2)$, it samples a random string $t \in \{0,1\}^\lambda$, and outputs

    $$\mathsf{ct}_2 \leftarrow 1\mathsf{FE}.\mathsf{E}(\mathsf{msk}_{\mathsf{in}}, (y, \perp, t, \perp, \perp)).$$

- **The decryption algorithm.** On input a functional key sk_f and two ciphertexts, $(\mathsf{ct}_1, \mathsf{sk}_1)$ and ct_2, the decryption algorithm $2\mathsf{FE}^{\mathsf{sel}}.\mathsf{D}$ computes $\mathsf{ct}' = 1\mathsf{FE}.\mathsf{D}(\mathsf{sk}_1, \mathsf{ct}_2)$, $\mathsf{sk}' = 1\mathsf{FE}.\mathsf{D}(\mathsf{sk}_f, \mathsf{ct}_1)$ and outputs $1\mathsf{FE}.\mathsf{D}(\mathsf{sk}', \mathsf{ct}')$.

The correctness of the above scheme with respect to any family of two-ary functionalities follows in a straightforward manner from the correctness of the underlying functional encryption scheme 1FE. Specifically, consider any pair of

$D_{f_0,f_1,z,u}((\mathsf{msk}^\star, K, w))$:
1. If $\mathsf{msk}^\star = \perp$, output u and HALT.
2. Compute $r = \mathsf{PRF.Eval}(K, z)$.
3. Output $1\mathsf{FE.KG}(\mathsf{msk}^\star, C_{f_w}; r)$.

$C_f((x, y))$:
1. Output $f(x, y)$.

Fig. 2. The single-input functions $D_{f_0,f_1,z,u}$ and C_f.

$\mathsf{AGG}_{x_0,x_1,a,s,\mathsf{msk}^\star,K}((y_0, y_1, t, s', v))$:
1. If $s' = s$ output v and HALT.
2. Compute $r = \mathsf{PRF.Eval}(K, t)$.
3. Output $1\mathsf{FE.E}(\mathsf{msk}^\star, (x_a, y_a); r)$.

Fig. 3. The single-input function $\mathsf{AGG}_{x_0,x_1,a,s,\mathsf{msk}^\star,K}$.

messages x and y and any function f. The encryption of x with respect to the index i $=1$ and the encryption of y with respect to the index i $= 2$ result in ciphertexts $(\mathsf{ct}_1, \mathsf{sk}_1)$ and ct_2, respectively. Using the correctness of the scheme 1FE, by executing $1\mathsf{FE.D}(\mathsf{sk}_1, \mathsf{ct}_2)$ we obtain an encryption ct' of the message (x, y) under the key $\mathsf{msk}^\star$. In addition, by executing $1\mathsf{FE.D}(\mathsf{sk}_f, \mathsf{ct}_1)$ we obtain a functional key sk' for C_f under the key $\mathsf{msk}^\star$. Therefore, executing $1\mathsf{FE.D}(\mathsf{sk}', \mathsf{ct}')$ outputs the value $C_f((x, y)) = f(x, y)$ as required.

The following theorem captures the security of the scheme, stating that under suitable assumptions on the underlying building blocks, the two-input scheme $2\mathsf{FE}^{\mathsf{sel}}$ is selective-message secure (see Definition 2.8). We refer the reader to the full version [18] for the complete proof.

Theorem 3.1. *Assuming that (1) 1FE is fully secure, and (2) PRF is a pseudorandom function family, then $2\mathsf{FE}^{\mathsf{sel}}$ is selective-message secure.*

We note that for proving that $2\mathsf{FE}^{\mathsf{sel}}$ is selective-message secure it suffices to require selective-message security from 1FE. However, given the generic transformations of Ananth et al. [4] (from selective security to adaptive security) and of Brakerski and Segev [19] (from message security to full security), for simplifying the proof of Theorem 3.1 we assume that 1FE is fully secure. In addition, when assuming that 1FE is fully secure, the scheme $2\mathsf{FE}^{\mathsf{sel}}$ can be shown to satisfy a notion of security that seems in between selective-message security and full security. Specifically, this notion considers adversaries that first have adaptive access to encryptions only for the first coordinate, and then have adaptive access to encryptions only for the second coordinate (while having adaptive access to the key-generation oracle throughout the experiment). However, given our generic transformation from selective-message security to full security for multi-input schemes (see Sect. 4), for simplifying the proof of Theorem 3.1 we focus on proving selective-message security.

In addition, for concreteness we focus on the unbounded case where the underlying scheme supports an unbounded (i.e., not fixed in advance) number of key-generation queries and encryption queries. More generally, the proof of Theorem 3.1 shows that if the scheme corresponding to $\mathsf{msk}_{\mathsf{out}}$ supports T_1 encryption queries and T_2 key-generation queries, the scheme corresponding to $\mathsf{msk}_{\mathsf{in}}$ supports T_3 encryption queries and T_4 key-generation queries, and the scheme corresponding to each msk^* supports T_5 encryption queries and T_6 key-generation queries, then the resulting scheme $\mathsf{2FE}^{\mathsf{sel}}$ supports $\min\{T_1, T_4, T_5\}$ encryption queries with respect to index $i = 1$, $\min\{T_3, T_5\}$ encryption queries with respect to index $i = 2$ and $\min\{T_2, T_6\}$ key-generation queries. When the polynomials $T_1, \ldots, T_6$ are known in advance (i.e., do not depend on the adversary), such schemes are known to exist based on the LWE assumption or even only one-way functions (see Sect. 2.2 for a more elaborated discussion of the existing schemes).

4   From Selective to Adaptive Security for Two-Input Schemes

In this section we show how to transform any private-key selective-message secure two-input functional encryption scheme (see Definition 2.8) into a fully secure one (see Definition 2.7). Our construction relies on the following building blocks:

1. A private-key single-input functional encryption scheme $\mathsf{1FE} = (\mathsf{1FE.S}, \mathsf{1FE.KG}, \mathsf{1FE.E}, \mathsf{1FE.D})$.
2. A private-key two-input functional encryption scheme $\mathsf{2FE}^{\mathsf{sel}} = (\mathsf{2FE}^{\mathsf{sel}}.\mathsf{S}, \mathsf{2FE}^{\mathsf{sel}}.\mathsf{KG}, \mathsf{2FE}^{\mathsf{sel}}.\mathsf{E}, \mathsf{2FE}^{\mathsf{sel}}.\mathsf{D})$.
3. A puncturable pseudorandom function family $\mathsf{PRF} = (\mathsf{PRF.Gen}, \mathsf{PRF.Eval}, \mathsf{PRF.Punc})$.

We assume that the schemes $\mathsf{1FE}$ and $\mathsf{2FE}^{\mathsf{sel}}$ are sufficiently expressive in the sense that they support the function family $\mathcal{F}$ (when viewed as a family of single-input functions), the evaluation procedure of the pseudorandom function family PRF, the setup, encryption and key-generation procedures of the scheme $\mathsf{1FE}$, and a few additional basic operations. The scheme $\mathsf{2FE} = (\mathsf{2FE.S}, \mathsf{2FE.KG}, \mathsf{2FE.E}, \mathsf{2FE.D})$ is defined as follows.

- **The setup algorithm.** On input the security parameter 1^λ the setup algorithm $\mathsf{2FE.S}$ samples $\mathsf{msk}_1 \leftarrow \mathsf{1FE.S}(1^\lambda)$ and $\mathsf{msk}_2 \leftarrow \mathsf{2FE}^{\mathsf{sel}}.\mathsf{S}(1^\lambda)$ and then outputs $\mathsf{msk} = (\mathsf{msk}_1, \mathsf{msk}_2)$.
- **The key-generation algorithm.** On input the master secret key msk and a function $f \in \mathcal{F}_\lambda$, the key-generation algorithm $\mathsf{2FE.KG}$ outputs $\mathsf{sk}_f \leftarrow \mathsf{2FE}^{\mathsf{sel}}.\mathsf{KG}(\mathsf{msk}_2, D_{f,\perp,1,\perp,\perp,\perp})$, where $D_{f,\perp,1,\perp,\perp,\perp}$ is a two-input function that is defined in Fig. 4.
- **The encryption algorithm.** On input the master secret key msk, a message m and an index $i \in [2]$, the encryption algorithm $\mathsf{2FE.E}$ has two cases:

- If $(m, i) = (x, 1)$, it samples $s \leftarrow \{0, 1\}^\lambda$ uniformly at random, three PRF keys $K^{enc}, K^{key}, K^{msk} \leftarrow \mathsf{PRF.Gen}(1^\lambda)$ and outputs a pair $(\mathsf{ct}_1, \mathsf{sk}_1)$ defined as follows:

$$\mathsf{ct}_1 \leftarrow 2\mathsf{FE}^{sel}.\mathsf{E}(\mathsf{msk}_2, (K^{msk}, K^{key}, s, 0), 1)$$
$$\mathsf{sk}_1 \leftarrow 1\mathsf{FE.KG}(\mathsf{msk}_1, \mathsf{AGG}_{x, \perp, 0, s, K^{msk}, K^{enc}, \perp, \perp})$$

where the single-input function $\mathsf{AGG}_{x, \perp, 0, s, K^{msk}, K^{enc}, \perp, \perp}$ is defined in Fig. 5.
- If $(m, i) = (y, 2)$, it samples $t \leftarrow \{0, 1\}^\lambda$ uniformly at random and outputs a pair $(\mathsf{ct}_2, \mathsf{ct}_3)$ defined as follows:

$$\mathsf{ct}_2 \leftarrow 2\mathsf{FE}^{sel}.\mathsf{E}(\mathsf{msk}_2, (1, t), 2)$$
$$\mathsf{ct}_3 \leftarrow 1\mathsf{FE.E}(\mathsf{msk}_1, (y, \perp, 1, t, \perp, \perp)).$$

- **The decryption algorithm.** On input a functional key sk_f and two ciphertexts $(\mathsf{ct}_1, \mathsf{sk}_1)$ and $(\mathsf{ct}_2, \mathsf{ct}_3)$, the decryption algorithm $2\mathsf{FE.D}$ first computes the value $\mathsf{sk}' = 2\mathsf{FE}^{sel}.\mathsf{D}(\mathsf{sk}_f, \mathsf{ct}_1, \mathsf{ct}_2)$, then it computes the value $\mathsf{ct}' = 1\mathsf{FE.D}(\mathsf{sk}_1, \mathsf{ct}_3)$, and finally it outputs $1\mathsf{FE.D}(\mathsf{sk}', \mathsf{ct}')$.

$D_{f_0, f_1, c, s', t', u}((K^{msk}, K^{key}, s, \mathsf{thr}), (c', t))$:
1. If $s' = s$ and $t' = t$, output u and HALT.
2. Compute $r = \mathsf{PRF.Eval}(K^{msk}, t)$.
3. Compute $r' = \mathsf{PRF.Eval}(K^{key}, t)$.
4. Compute $\mathsf{msk}_{s,t} = 1\mathsf{FE.S}(1^\lambda; r)$.
5. If $c \leq \mathsf{thr}$ and $c' \leq \mathsf{thr}$ set $f = f_1$.
6. Else (if $c > \mathsf{thr}$ or $c' > \mathsf{thr}$) set $f = f_0$.
7. Output $1\mathsf{FE.KG}(\mathsf{msk}_{s,t}, C_f; r')$.

$C_f((x, y))$:
1. Output $f(x, y)$.

Fig. 4. The two-input function $D_{f_0, f_1, c, s', t', u}$ and the single-input function C_f.

The correctness of the above scheme with respect to any family of two-ary functionalities follows in a straightforward manner from the correctness of the underlying functional encryption schemes 1FE and $2\mathsf{FE}^{sel}$. Specifically, consider any pair of messages x and y and any function f. The encryption of x with respect to the index i $=1$ and the encryption of y with respect to the index i $= 2$ result in ciphertexts $(\mathsf{ct}_1, \mathsf{sk}_1)$ and $(\mathsf{ct}_2, \mathsf{ct}_3)$, respectively. Using the correctness of the scheme $2\mathsf{FE}^{sel}$, by executing $2\mathsf{FE}^{sel}.\mathsf{D}(\mathsf{sk}_f, \mathsf{ct}_1, \mathsf{ct}_2)$ we obtain a functional key sk' for C_f under the key $\mathsf{msk}_{s,t}$. In addition, by executing $1\mathsf{FE.D}(\mathsf{sk}_1, \mathsf{ct}_3)$ we obtain a an encryption ct' of (x, y) under the key $\mathsf{msk}_{s,t}$. Therefore, executing $1\mathsf{FE.D}(\mathsf{sk}', \mathsf{ct}')$ outputs the value $C_f((x, y)) = f(x, y)$ as required.

$\mathsf{AGG}_{x_0,x_1,\mathsf{thr},s,K^{\mathsf{msk}},K^{\mathsf{enc}},t',v'}((y_0,y_1,\mathsf{c},t,s',u'))$:
1. If $t' = t$ output v' and HALT.
2. If $s' = s$ output u' and HALT.
3. Compute $r = \mathsf{PRF.Eval}(K^{\mathsf{msk}},t)$.
4. Compute $r' = \mathsf{PRF.Eval}(K^{\mathsf{enc}},t)$.
5. Compute $\mathsf{msk}_{s,t} = \mathsf{1FE.S}(1^\lambda;r)$.
6. If $\mathsf{c} \le \mathsf{thr}$ set $x = x_1$ and $y = y_1$.
7. Else (if $\mathsf{c} > \mathsf{thr}$) set $x = x_0$ and $y = y_0$.
8. Output $\mathsf{1FE.E}(\mathsf{msk}_{s,t},(x,y);r')$.

Fig. 5. The single-input function $\mathsf{AGG}_{x_0,x_1,\mathsf{thr},s,K^{\mathsf{msk}},K^{\mathsf{enc}},t',v'}$.

The following theorem captures the security of the scheme. This theorem states that under suitable assumptions on the underlying building blocks, the two-input scheme 2FE is fully secure (see Definition 2.7). We refer the reader to the full version [18] for the complete proof.

Theorem 4.1. *Assuming that (1) 1FE is fully secure, (2) 2FE^{sel} is selective-message secure, and (3) PRF is a puncturable pseudorandom function family, then 2FE is fully secure.*

As in Sect. 3, for concreteness we focus on the unbounded case where the underlying schemes, 1FE and 2FE^{sel}, support an unbounded (i.e., not fixed in advance) number of key-generation queries and encryption queries. More generally, the proof of Theorem 4.1 shows that if the scheme corresponding to msk_1 supports T_1 encryption queries and T_2 key-generation queries, the scheme corresponding to msk_2 supports $T_3^{(1)}$ encryption queries with respect to index $\mathsf{i} = 1$ and $T_3^{(2)}$ encryption queries with respect to index $\mathsf{i} = 2$, and T_4 key-generation queries, and the scheme corresponding to each $\mathsf{msk}_{s,t}$ supports a *single* encryption query and T_5 key-generation queries, then the resulting scheme 2FE supports $\min\{T_2, T_3^{(1)}\}$ encryption queries with respect to index $\mathsf{i} = 1$, $\min\{T_1, T_3^{(2)}\}$ encryption queries with respect to index $\mathsf{i} = 2$ and $\min\{T_4, T_5\}$ key-generation queries. When the polynomials $T_1, T_2, T_3^{(1)}, T_3^{(2)}, T_4$ and T_5 are known in advance (i.e., do not depend on the adversary), such schemes are known to exist based on the LWE assumption or even only one-way functions (see Sect. 2.2 for a more elaborated discussion of the existing schemes).

Acknowledgments. We thank Eylon Yogev for various insightful discussions and the EUROCRYPT '16 reviewers for their useful comments.

A   Generalization to $t \ge 2$ Inputs

In this section we generalize our results to more than two inputs. In Appendix A.1 we generalize the definitions introduced in Sect. 2.3, and in Appendices A.2

and A.3 we generalize the constructions from Sects. 3 and 4, respectively. More precisely, in Appendix A.2 we show how to obtain a *selectively-secure* t-input scheme assuming any fully secure $(t-1)$-input scheme. Then, in Appendix A.3 we show how to obtain a *fully-secure* t-input scheme assuming any fully-secure $(t-1)$-input scheme and a selectively-secure t-input scheme.

A.1   Private-Key t-Input Functional Encryption

In this section we generalize the framework introduced in Sect. 2.3 to the general case of t-input schemes (Sect. 2.3 dealt with the case $t = 2$).

For $i \in [t]$ let $\mathcal{X}_i = \{(\mathcal{X}_i)_\lambda\}_{\lambda \in \mathbb{N}}$ be an ensemble of finite sets, and let $\mathcal{F} = \{\mathcal{F}_\lambda\}_{\lambda \in \mathbb{N}}$ be an ensemble of finite t-ary function families. For each $\lambda \in \mathbb{N}$, each function $f \in \mathcal{F}_\lambda$ takes as input t strings, $x_1 \in (\mathcal{X}_1)_\lambda, \ldots, x_t \in (\mathcal{X}_t)_\lambda$, and outputs a value $f(x_1, \ldots, x_t) \in \mathcal{Z}_\lambda$. A private-key t-input functional encryption scheme Π for $\mathcal{F}$ consists of four probabilistic polynomial time algorithm Setup, Enc, KG and Dec, described as follows. The setup algorithm $\mathsf{Setup}(1^\lambda)$ takes as input the security parameter λ, and outputs a master secret key msk. The encryption algorithm $\mathsf{Enc}(\mathsf{msk}, m, \mathsf{i})$ takes as input a master secret key msk, a message m, and an index $\mathsf{i} \in [t]$, where $m \in (\mathcal{X}_\mathsf{i})_\lambda$, and outputs a ciphertext ct_i. The key-generation algorithm $\mathsf{KG}(\mathsf{msk}, f)$ takes as input a master secret key msk and a function $f \in \mathcal{F}_\lambda$, and outputs a functional key sk_f. The (deterministic) decryption algorithm Dec takes as input a functional key sk_f and t ciphertexts, $\mathsf{ct}_1, \ldots, \mathsf{ct}_t$, and outputs a string $z \in \mathcal{Z}_\lambda \cup \{\perp\}$.

Definition A.1 (Correctness). *A private-key t-input functional encryption scheme $\Pi = (\mathsf{Setup}, \mathsf{Enc}, \mathsf{KG}, \mathsf{Dec})$ for $\mathcal{F}$ is correct if there exists a negligible function $\mathsf{neg}(\cdot)$ such that for every $\lambda \in \mathbb{N}$, for every $f \in \mathcal{F}_\lambda$, and for every $(x_1, \ldots, x_t) \in (\mathcal{X}_1)_\lambda \times \cdots \times (\mathcal{X}_t)_\lambda$, it holds that*

$$\Pr\left[\mathsf{Dec}(\mathsf{sk}_f, \mathsf{Enc}(\mathsf{msk}, x_1, 1), \ldots, \mathsf{Enc}(\mathsf{msk}, x_t, t)) = f(x_1, \ldots, x_t)\right] \geq 1 - \mathsf{neg}(\lambda),$$

where $\mathsf{msk} \leftarrow \mathsf{Setup}(1^\lambda)$, $\mathsf{sk}_f \leftarrow \mathsf{KG}(\mathsf{msk}, f)$, and the probability is taken over the internal randomness of Setup, Enc and KG.

Next, we generalize the security definitions from Sect. 2.3 to the t-input case. As in Sect. 2.3, we start by defining the notion of a *valid t-input adversary*. Then, we define *full security* and *selective-message security*.

Definition A.2 (Valid t-input adversary). *A probabilistic polynomial-time algorithm $\mathcal{A}$ is a valid t-input adversary if for all private-key t-input functional encryption schemes $\Pi = (\mathsf{Setup}, \mathsf{KG}, \mathsf{Enc}, \mathsf{Dec})$ over a message space $\mathcal{X}_1 \times \cdots \times \mathcal{X}_t = \{(\mathcal{X}_1)_\lambda\}_{\lambda \in \mathbb{N}} \times \cdots \times \{(\mathcal{X}_t)_\lambda\}_{\lambda \in \mathbb{N}}$ and a function space $\mathcal{F} = \{\mathcal{F}_\lambda\}_{\lambda \in \mathbb{N}}$, for all $\lambda \in \mathbb{N}$ and $b \in \{0, 1\}$, and for all $(f_0, f_1) \in \mathcal{F}_\lambda$ and $((x_i^0, x_i^1), i) \in \mathcal{X}_i \times \mathcal{X}_i \times \{i\}$ (where $i \in [t]$) with which $\mathcal{A}$ queries the left-or-right key-generation and encryption oracles, respectively, it holds that $f_0(x_1^0, \ldots, x_t^0) = f_1(x_1^1, \ldots, x_t^1)$.*

Definition A.3 (Full security). *A private-key t-input functional encryption scheme* $\Pi = (\mathsf{Setup}, \mathsf{KG}, \mathsf{Enc}, \mathsf{Dec})$ *over a message space* $\mathcal{X}_1 \times \cdots \times \mathcal{X}_t = \{(\mathcal{X}_1)_\lambda\}_{\lambda \in \mathbb{N}} \times \cdots \times \{(\mathcal{X}_t)_\lambda\}_{\lambda \in \mathbb{N}}$ *and a function space* $\mathcal{F} = \{\mathcal{F}_\lambda\}_{\lambda \in \mathbb{N}}$ *is fully secure if for any valid t-input adversary* $\mathcal{A}$ *there exists a negligible function* $\mathsf{neg}(\cdot)$ *such that*

$$\mathsf{Adv}_{\Pi,\mathcal{F},\mathcal{A}}^{\mathsf{fullFE}_t} \overset{\text{def}}{=} \left| \Pr\left[\mathsf{Exp}_{\Pi,\mathcal{F},\mathcal{A}}^{\mathsf{fullFE}_t}(\lambda) = 1 \right] - \frac{1}{2} \right| \leq \mathsf{neg}(\lambda),$$

for all sufficiently large $\lambda \in \mathbb{N}$, *where the random variable* $\mathsf{Exp}_{\Pi,\mathcal{F},\mathcal{A}}^{\mathsf{fullFE}_t}(\lambda)$ *is defined via the following experiment:*

1. $\mathsf{msk} \leftarrow \mathsf{Setup}(1^\lambda),\ b \leftarrow \{0,1\}$.
2. $b' \leftarrow \mathcal{A}^{\mathsf{KG}_b(\mathsf{msk},\cdot,\cdot),\mathsf{Enc}_b(\mathsf{msk},(\cdot,\cdot),\cdot)}(1^\lambda)$.
3. *If* $b' = b$ *then output 1, and otherwise output 0.*

Definition A.4 (Selective-message security). *A private-key t-input functional encryption scheme* $\Pi = (\mathsf{Setup}, \mathsf{KG}, \mathsf{Enc}, \mathsf{Dec})$ *over a message space* $\mathcal{X}_1 \times \cdots \times \mathcal{X}_t = \{(\mathcal{X}_1)_\lambda\}_{\lambda \in \mathbb{N}} \times \cdots \times \{(\mathcal{X}_t)_\lambda\}_{\lambda \in \mathbb{N}}$ *and a function space* $\mathcal{F} = \{\mathcal{F}_\lambda\}_{\lambda \in \mathbb{N}}$ *is selective-message secure if for any valid t-input adversary* $\mathcal{A} = (\mathcal{A}_1, \mathcal{A}_2)$ *there exists a negligible function* $\mathsf{neg}(\lambda)$ *such that*

$$\mathsf{Adv}_{\Pi,\mathcal{F},\mathcal{A}}^{\mathsf{selFE}_t} \overset{\text{def}}{=} \left| \Pr\left[\mathsf{Exp}_{\Pi,\mathcal{F},\mathcal{A}}^{\mathsf{selFE}_t}(\lambda) = 1 \right] - \frac{1}{2} \right| \leq \mathsf{neg}(\lambda),$$

for all sufficiently large $\lambda \in \mathbb{N}$, *where the random variable* $\mathsf{Exp}_{\Pi,\mathcal{F},\mathcal{A}}^{\mathsf{selFE}_t}(\lambda)$ *is defined via the following experiment:*

1. $(\vec{x_1}, \ldots, \vec{x_t}, \mathsf{state}) \leftarrow \mathcal{A}_1(1^\lambda)$, *where* $\vec{x_i} = ((x_{i,1}^0, x_{i,1}^1), \ldots, (x_{i,T}^0, x_{i,T}^1))$ *for* $i \in [t]$.
2. $\mathsf{msk} \leftarrow \mathsf{Setup}(1^\lambda),\ b \leftarrow \{0,1\}$.
3. $\mathsf{ct}_{i,j} \leftarrow \mathsf{Enc}(\mathsf{msk}, x_{i,j}^b, 1)$ *for* $i \in [t]$ *and* $j \in [T]$.
4. $b' \leftarrow \mathcal{A}_2^{\mathsf{KG}_b(\mathsf{msk},\cdot,\cdot)}(1^\lambda, \{\mathsf{ct}_{i,j}\}_{i \in [t], j \in [T]}, \mathsf{state})$.
5. *If* $b' = b$ *then output 1, and otherwise output 0.*

A.2   A Selectively-Secure t-Input Scheme from any (t − 1)-Input Scheme

In this section we generalize the construction from Sect. 3 by presenting a construction of a selectively-secure t-input scheme assuming any fully-secure $(t - 1)$-input scheme. Let $\mathcal{F} = \{\mathcal{F}_\lambda\}_{\lambda \in \mathbb{N}}$ be a family of t-input functionalities, where for every $\lambda \in \mathbb{N}$ the set $\mathcal{F}_\lambda$ consists of functions of the form $f : (\mathcal{X}_1)_\lambda \times \cdots \times (\mathcal{X}_t)_\lambda \to \mathcal{Z}_\lambda$. Our construction relies on the following building blocks:

1. A private-key single-input functional encryption scheme $\mathsf{FE}_1 = (\mathsf{FE}_1.\mathsf{S}, \mathsf{FE}_1.\mathsf{KG}, \mathsf{FE}_1.\mathsf{E}, \mathsf{FE}_1.\mathsf{D})$.

2. A private-key $(t-1)$-input functional encryption scheme $\mathsf{FE}^{\mathsf{sel}}_{t-1} = (\mathsf{FE}^{\mathsf{sel}}_{t-1}.\mathsf{S},$ $\mathsf{FE}^{\mathsf{sel}}_{t-1}.\mathsf{KG}, \mathsf{FE}^{\mathsf{sel}}_{t-1}.\mathsf{E}, \mathsf{FE}^{\mathsf{sel}}_{t-1}.\mathsf{D})$.
3. A pseudorandom function family $\mathsf{PRF} = (\mathsf{PRF.Gen}, \mathsf{PRF.Eval})$.

Our scheme $\mathsf{FE}^{\mathsf{sel}}_t = (\mathsf{FE}^{\mathsf{sel}}_t.\mathsf{S}, \mathsf{FE}^{\mathsf{sel}}_t.\mathsf{KG}, \mathsf{FE}^{\mathsf{sel}}_t.\mathsf{E}, \mathsf{FE}^{\mathsf{sel}}_t.\mathsf{D})$ is defined as follows.

- **The setup algorithm.** On input the security parameter 1^λ the setup algorithm $\mathsf{FE}^{\mathsf{sel}}_t.\mathsf{S}$ samples $\mathsf{msk}_{\mathsf{out}} \leftarrow \mathsf{FE}_1.\mathsf{S}(1^\lambda), \mathsf{msk}_{\mathsf{in}} \leftarrow \mathsf{FE}^{\mathsf{sel}}_{t-1}.\mathsf{S}(1^\lambda)$ and outputs $\mathsf{msk} = (\mathsf{msk}_{\mathsf{out}}, \mathsf{msk}_{\mathsf{in}})$.
- **The key-generation algorithm.** On input the master secret key msk and a function $f \in \mathcal{F}_\lambda$, the key-generation algorithm $\mathsf{FE}^{\mathsf{sel}}_t.\mathsf{KG}$ samples a random string $z \leftarrow \{0,1\}^\lambda$ and outputs $\mathsf{sk}_f \leftarrow \mathsf{FE}_1.\mathsf{KG}(\mathsf{msk}_{\mathsf{out}}, D_{f,\perp,z,\perp})$, where $D_{f,\perp,z,\perp}$ is a single-input function that is defined in Fig. 6.
- **The encryption algorithm.** On input the master secret key msk, a message m and an index $\mathsf{i} \in [t]$, the encryption algorithm $\mathsf{FE}^{\mathsf{sel}}_t.\mathsf{E}$ has two cases:
  - If $(m, \mathsf{i}) = (x_1, 1)$, it samples a master secret key $\mathsf{msk}^\star \leftarrow \mathsf{FE}^{\mathsf{sel}}_{t-1}.\mathsf{S}(1^\lambda)$, a PRF key $K \leftarrow \mathsf{PRF.Gen}(1^\lambda)$, and a random string $s \in \{0,1\}^\lambda$, and then outputs a pair $(\mathsf{ct}_1, \mathsf{sk}_1)$ defined as follows:

$$\mathsf{ct}_1 \leftarrow \mathsf{FE}_1.\mathsf{E}(\mathsf{msk}_{\mathsf{out}}, (\mathsf{msk}^\star, K, 0))$$
$$\mathsf{sk}_1 \leftarrow \mathsf{FE}^{\mathsf{sel}}_{t-1}.\mathsf{KG}(\mathsf{msk}_{\mathsf{in}}, \mathsf{AGG}_{x_1, \perp, 0, s, \mathsf{msk}^\star, K}),$$

  where $\mathsf{AGG}_{x, \perp, 0, \mathsf{msk}^\star, K}$ is a $(t-1)$-input function that is defined in Fig. 7.
  - If $(m, \mathsf{i}) = (x_i, \mathsf{i})$ where $\mathsf{i} \in \{2, \dots, t\}$, it samples a random string $\tau_i \in \{0,1\}^\lambda$, and outputs

$$\mathsf{ct}_i \leftarrow \mathsf{FE}^{\mathsf{sel}}_{t-1}.\mathsf{E}(\mathsf{msk}_{\mathsf{in}}, (x_i, \perp, \tau_i, \perp, \perp), \mathsf{i}-1).$$

- **The decryption algorithm.** On input a functional key sk_f and ciphertexts $(\mathsf{ct}_1, \mathsf{sk}_1), \mathsf{ct}_2, \dots, \mathsf{ct}_t$, the decryption algorithm $\mathsf{FE}^{\mathsf{sel}}_t.\mathsf{D}$ computes $(\mathsf{ct}'_2, \dots, \mathsf{ct}'_t) = \mathsf{FE}^{\mathsf{sel}}_{t-1}.\mathsf{D}(\mathsf{sk}_1, (\mathsf{ct}_2, \dots, \mathsf{ct}_t))$, $\mathsf{sk}' = \mathsf{FE}_1.\mathsf{D}(\mathsf{sk}_f, \mathsf{ct}_1)$ and outputs $\mathsf{FE}^{\mathsf{sel}}_{t-1}.\mathsf{D}(\mathsf{sk}', (\mathsf{ct}'_2, \dots, \mathsf{ct}'_t))$.

$D_{f_0, f_1, z, u}((\mathsf{msk}^\star, K, w))$:
1. If $\mathsf{msk}^\star = \perp$, output u and HALT.
2. Compute $r = \mathsf{PRF.Eval}(K, z)$.
3. Output $\mathsf{FE}^{\mathsf{sel}}_{t-1}.\mathsf{KG}(\mathsf{msk}^\star, C_{f_w}; r)$.

$C_f((x_1, x_2), x_3, \dots, x_t)$:
1. Output $f(x_1, \dots, x_t)$.

Fig. 6. The single-input function $D_{f_0, f_1, z, u}$ and the $(t-1)$-input function C_f.

$\mathsf{AGG}_{x_1^0, x_1^1, a, s, \mathsf{msk}^\star, K}((x_2^0, x_2^1, \tau_2, s_2, v_2), \dots, (x_t^0, x_t^1, \tau_t, s_t, v_t)):$
1. If $s_2 = \cdots = s_t = s$ output $(v_2, \dots, v_t)$ and HALT.
2. Set $x_i = x_i^a$ for all $i \in [t]$.
3. Compute $r_i = \mathsf{PRF.Eval}(K, \tau_i)$ for $2 \leq i \leq t$.
4. Output $\qquad\qquad (\mathsf{FE}_{t-1}^{\mathsf{sel}}.\mathsf{E}(\mathsf{msk}^\star, (x_1, x_2), 1; r_2), \mathsf{FE}_{t-1}^{\mathsf{sel}}.\mathsf{E}(\mathsf{msk}^\star, x_3, 2; r_3), \dots,$
$\mathsf{FE}_{t-1}^{\mathsf{sel}}.\mathsf{E}(\mathsf{msk}^\star, x_t, t-1; r_t)).$

Fig. 7. The $(t-1)$-input function $\mathsf{AGG}_{x_1^0, x_1^1, a, s, \mathsf{msk}^\star, K}$.

Theorem A.5. *Assuming that (1)* FE_1 *is fully secure, (2)* $\mathsf{FE}_{t-1}^{\mathsf{sel}}$ *is selective-message secure, and (3)* PRF *is a pseudorandom function family, then* $\mathsf{FE}_t^{\mathsf{sel}}$ *is selective-message secure.*

As in Theorem 3.1, we note that for proving that $\mathsf{FE}_t^{\mathsf{sel}}$ is selective-message secure it suffices to require selective-message security from FE_1. However, given the generic transformation for single-input schemes [4,19] (from selective security to adaptive security and from message security to full security, respectively), for simplifying the proof of Theorem A.5 we assume that FE_1 is fully secure. We refer the reader to the full version [18] for the complete proof.

A.3 From Selective to Adaptive Security for t-Input Schemes

In this section we generalize the construction from Sect. 4 to get a fully-secure t-input functional encryption scheme assuming any fully-secure $(t-1)$-input functional encryption scheme and any selectively-secure t-input functional encryption scheme. Our construction relies on the following building blocks:

1. A private-key single-input functional encryption scheme $\mathsf{FE}_1 = (\mathsf{FE}_1.\mathsf{S}, \mathsf{FE}_1.\mathsf{KG}, \mathsf{FE}_1.\mathsf{E}, \mathsf{FE}_1.\mathsf{D})$.
2. A private-key $(t-1)$-input functional encryption scheme $\mathsf{FE}_{t-1} = (\mathsf{FE}_{t-1}.\mathsf{S}, \mathsf{FE}_{t-1}.\mathsf{KG}, \mathsf{FE}_{t-1}.\mathsf{E}, \mathsf{FE}_{t-1}.\mathsf{D})$.
3. A private-key t-input functional encryption scheme $\mathsf{FE}_t^{\mathsf{sel}} = (\mathsf{FE}_t^{\mathsf{sel}}.\mathsf{S}, \mathsf{FE}_t^{\mathsf{sel}}.\mathsf{KG}, \mathsf{FE}_t^{\mathsf{sel}}.\mathsf{E}, \mathsf{FE}_t^{\mathsf{sel}}.\mathsf{D})$.
4. A puncturable pseudorandom function family $\mathsf{PRF} = (\mathsf{PRF.Gen}, \mathsf{PRF.Eval}, \mathsf{PRF.Punc})$.

The scheme $\mathsf{FE}_t = (\mathsf{FE}_t.\mathsf{S}, \mathsf{FE}_t.\mathsf{KG}, \mathsf{FE}_t.\mathsf{E}, \mathsf{FE}_t.\mathsf{D})$ is defined as follows.

– **The setup algorithm.** On input the security parameter 1^λ the setup algorithm $\mathsf{FE}_t.\mathsf{S}$ samples $\mathsf{msk}_{t-1} \leftarrow \mathsf{FE}_{t-1}.\mathsf{S}(1^\lambda)$ and $\mathsf{msk}_t \leftarrow \mathsf{FE}_t^{\mathsf{sel}}.\mathsf{S}(1^\lambda)$ and then outputs $\mathsf{msk} = (\mathsf{msk}_{t-1}, \mathsf{msk}_t)$.
– **The key-generation algorithm.** On input the master secret key msk and a function $f \in \mathcal{F}_\lambda$, the key-generation algorithm $\mathsf{FE}_t.\mathsf{KG}$ outputs $\mathsf{sk}_f \leftarrow \mathsf{FE}_t^{\mathsf{sel}}.\mathsf{KG}(\mathsf{msk}_t, D_{f, \perp, 1, \underbrace{\perp, \dots, \perp}_{t \text{ times}}, \perp})$, where $D_{f, \perp, 1, \underbrace{\perp, \dots, \perp}_{t \text{ times}}, \perp}$ is a t-input function that is defined in Fig. 8.

- **The encryption algorithm.** On input the master secret key msk, a message m and an index $i \in [2]$, the encryption algorithm $\mathsf{FE}_{t-1}.\mathsf{E}$ has two cases:
  - If $(m, i) = (x_1, 1)$, it samples $\tau_1 \leftarrow \{0,1\}^\lambda$ uniformly at random, three PRF keys $K^{\mathsf{enc}}, K^{\mathsf{key}}, K^{\mathsf{msk}} \leftarrow \mathsf{PRF}.\mathsf{Gen}(1^\lambda)$ and outputs a pair $(\mathsf{ct}_1, \mathsf{sk}_1)$ defined as follows:

$$\mathsf{ct}_1 \leftarrow \mathsf{FE}_t^{\mathsf{sel}}.\mathsf{E}(\mathsf{msk}_t, (K^{\mathsf{msk}}, K^{\mathsf{key}}, \tau_1, \underbrace{0, \ldots, 0}_{t-1 \text{ times}}), 1)$$

$$\mathsf{sk}_1 \leftarrow \mathsf{FE}_{t-1}.\mathsf{KG}(\mathsf{msk}_{t-1}, \mathsf{AGG}_{x_1, \perp, \underbrace{0, \ldots, 0}_{t-1 \text{ times}}, \tau_1, K^{\mathsf{msk}}, K^{\mathsf{enc}}, \underbrace{\perp, \ldots, \perp}_{t-1 \text{ times}}, \perp})$$

where the single-input function $\mathsf{AGG}_{x_1, \perp, \underbrace{0, \ldots, 0}_{t-1 \text{ times}}, \tau_1, K^{\mathsf{msk}}, K^{\mathsf{enc}}, \underbrace{\perp, \ldots, \perp}_{t-1 \text{ times}}, \perp}$ is defined in Fig. 9.
  - If $(m, i) = (x_i, i)$ and $i > 1$, it samples $\tau_i \leftarrow \{0,1\}^\lambda$ uniformly at random and outputs a pair $(\mathsf{ct}_i, \mathsf{ct}_i')$ defined as follows:

$$\mathsf{ct}_i \leftarrow \mathsf{FE}_t^{\mathsf{sel}}.\mathsf{E}(\mathsf{msk}_t, (1, \tau_i), i)$$
$$\mathsf{ct}_i' \leftarrow \mathsf{FE}_{t-1}.\mathsf{E}(\mathsf{msk}_{t-1}, (x_i, \perp, 1, \tau_i, \underbrace{\perp, \ldots, \perp}_{t-1 \text{ times}}, \perp), i-1).$$

- **The decryption algorithm.** On input a functional key sk_f and t cipher-texts $(\mathsf{ct}_1, \mathsf{sk}_1)$ and $(\mathsf{ct}_2, \mathsf{ct}_2'), \ldots, (\mathsf{ct}_t, \mathsf{ct}_t')$, the decryption algorithm $\mathsf{FE}_t.\mathsf{D}$ first computes the value $\mathsf{sk}' = \mathsf{FE}_t^{\mathsf{sel}}.\mathsf{D}(\mathsf{sk}_f, \mathsf{ct}_1, \ldots, \mathsf{ct}_t)$, then it computes the value $\mathsf{ct}' = \mathsf{FE}_{t-1}.\mathsf{D}(\mathsf{sk}_1, \mathsf{ct}_2', \ldots, \mathsf{ct}_t')$, and finally it outputs $\mathsf{FE}_1.\mathsf{D}(\mathsf{sk}', \mathsf{ct}')$.

$D_{f_0, f_1, c, \tau_1', \ldots, \tau_t', u}((K^{\mathsf{msk}}, K^{\mathsf{key}}, \tau_1, \mathsf{thr}_2, \ldots, \mathsf{thr}_t), (c_2, \tau_2), \ldots, (c_t, \tau_t))$:
1. If $\tau_i' = \tau_i$ for all $i \in [t]$, output u and HALT.
2. Compute $r = \mathsf{PRF}.\mathsf{Eval}(K^{\mathsf{msk}}, \tau_2 \ldots \tau_t)$.
3. Compute $r' = \mathsf{PRF}.\mathsf{Eval}(K^{\mathsf{key}}, \tau_2 \ldots \tau_t)$.
4. Compute $\mathsf{msk}_{\tau_1, \ldots, \tau_t} = \mathsf{FE}_1.\mathsf{S}(1^\lambda; r)$.
5. For $i = 1, \ldots, t$ do:
   (a) If $c_i < \mathsf{thr}_i$ then set $f = f_1$ and exit loop.
   (b) If $c_i > \mathsf{thr}_i$ then set $f = f_0$ and exit loop.
   (c) If $c_i = \mathsf{thr}_i$ and $i < t$ continue to next iteration (with $i = i+1$).
   (d) If $c_i = \mathsf{thr}_i$ and $i = t$ set $f = f_1$.
6. Output $\mathsf{FE}_1.\mathsf{KG}(\mathsf{msk}_{\tau_1, \ldots, \tau_t}, C_f; r')$.

$C_f((x_1, \ldots, x_t))$:
1. Output $f(x_1, \ldots, x_t)$.

Fig. 8. The t-input function $D_{f_0, f_1, c, \tau_1', \ldots, \tau_t', u}$ and the single-input function C_f.

$\mathbf{AGG}_{\varpi_1^0,\varpi_1^1,\mathsf{thr}_2,\ldots,\mathsf{thr}_t,\tau_1,K^{\mathsf{msk}},K^{\mathsf{enc}},\tau_{1,2},\ldots,\tau_{1,t},u_1}$

$((x_2^0, x_2^1, c_2, \tau_2, \tau_{2,1}, \ldots, \tau_{2,t}, u_2), \ldots,$

$(x_t^0, x_t^1, c_t, \tau_t, \tau_{t,1}, \ldots, \tau_{t,t}, u_t))$:

1. If $\exists i \in [t]$ such that $\forall j \in [t] \setminus \{i\}$ it holds that $\tau_{i,j} = \tau_j$, then output u_i and HALT.
2. Compute $r = \mathsf{PRF.Eval}(K^{\mathsf{msk}}, \tau_2 \ldots, \tau_t)$.
3. Compute $r' = \mathsf{PRF.Eval}(K^{\mathsf{enc}}, \tau_2 \ldots, \tau_t)$.
4. Compute $\mathsf{msk}_{\tau_1,\ldots,\tau_t} = \mathsf{FE}_1.\mathsf{S}(1^\lambda; r)$.
5. For $i = 1, \ldots, t$ do:
    (a) If $c_i < \mathsf{thr}_i$ then set $x_i = x_i^1$ for all $i \in [t]$ and exit loop.
    (b) If $c_i > \mathsf{thr}_i$ then set $x_i = x_i^0$ for all $i \in [t]$ and exit loop.
    (c) If $c_i = \mathsf{thr}_i$ and $i < t$ continue to next iteration (with $i = i+1$).
    (d) If $c_i = \mathsf{thr}_i$ and $i = t$ set $x_i = x_i^1$ for all $i \in [t]$.
6. Output $\mathsf{FE}_1.\mathsf{E}(\mathsf{msk}_{\tau_1,\ldots,\tau_t}, (x_1 \ldots, x_t); r')$.

Fig. 9. The t-input function $\mathsf{AGG}_{x_1^0,x_1^1,\mathsf{thr}_2,\ldots,\mathsf{thr}_t,\tau_1,K^{\mathsf{msk}},K^{\mathsf{enc}},\tau_{1,2}',\ldots,\tau_{1,t}',u_1}$.

The following theorem captures the security of the scheme. This theorem states that under suitable assumptions on the underlying building blocks, the t-input scheme FE_t is fully private (see Definition 2.7). We refer the reader to the full version [18] for the complete proof.

Theorem A.6. *Let $t > 1$ be any fixed integer. Assuming that (1) FE_1 is fully secure, (2) FE_{t-1} is fully secure, (3) $\mathsf{FE}_t^{\mathsf{sel}}$ is selective-message secure, and (4) PRF is a puncturable pseudorandom function family, then FE_t is fully secure.*

We note that the proof of Theorem A.6 assumes that t is a fixed constant. The reason for this limitation is that the number of hybrids in the proof of security is $\lambda^{O(t)}$, where λ is the security parameter, which is polynomial for any constant t. If we assume that the underlying building blocks are sub-exponentially secure, then the proof of Theorem A.6 can be used for a super-constant number of inputs.

References

1. Agrawal, S., Agrawal, S., Badrinarayanan, S., Kumarasubramanian, A., Prabhakaran, M., Sahai, A.: Function private functional encryption and property preserving encryption: New definitions and positive results. Cryptology ePrint Archive, Report 2013/744 (2013)
2. Agrawal, S., Gorbunov, S., Vaikuntanathan, V., Wee, H.: Functional encryption: new perspectives and lower bounds. In: Canetti, R., Garay, J.A. (eds.) CRYPTO 2013, Part II. LNCS, vol. 8043, pp. 500–518. Springer, Heidelberg (2013)
3. Ananth, P., Boneh, D., Garg, S., Sahai, A., Zhandry, M.: Differing-inputs obfuscation and applications. Cryptology ePrint Archive, Report 2013/689 (2013)

4. Ananth, P., Brakerski, Z., Segev, G., Vaikuntanathan, V.: From selective to adaptive security in functional encryption. In: Gennaro, R., Robshaw, M. (eds.) CRYPTO 2015. LNCS, vol. 9216, pp. 657–677. Springer, Heidelberg (2015)
5. Ananth, P., Jain, A.: Indistinguishability obfuscation from compact functional encryption. In: Gennaro, R., Robshaw, M. (eds.) CRYPTO 2015. LNCS, vol. 9215, pp. 308–326. Springer, Heidelberg (2015)
6. Ananth, P., Jain, A., Sahai, A.: Achieving compactness generically: Indistinguishability obfuscation from non-compact functional encryption. Cryptology ePrint Archive, Report 2015/730 (2015)
7. Asharov, G., Segev, G.: Limits on the power of indistinguishability obfuscation and functional encryption. In: Proceedings of the 56th Annual IEEE Symposium on Foundations of Computer Science, pp. 191–209 (2015)
8. Barak, B., Goldreich, O., Impagliazzo, R., Rudich, S., Sahai, A., Vadhan, S.P., Yang, K.: On the (im)possibility of obfuscating programs. J. ACM 59(2), 6 (2012)
9. Bellare, M., O'Neill, A.: Semantically-secure functional encryption: possibility results, impossibility results and the quest for a general definition. In: Abdalla, M., Nita-Rotaru, C., Dahab, R. (eds.) CANS 2013. LNCS, vol. 8257, pp. 218–234. Springer, Heidelberg (2013)
10. Bitansky, N., Vaikuntanathan, V.: Indistinguishability obfuscation from functional encryption. In: Proceedings of the 56th Annual IEEE Symposium on Foundations of Computer Science, pp. 171–190 (2015)
11. Boneh, D., Franklin, M.: Identity-based encryption from the weil pairing. In: Kilian, J. (ed.) CRYPTO 2001. LNCS, vol. 2139, pp. 213–229. Springer, Heidelberg (2001)
12. Boneh, D., Lewi, K., Raykova, M., Sahai, A., Zhandry, M., Zimmerman, J.: Semantically secure order-revealing encryption: multi-input functional encryption without obfuscation. In: Oswald, E., Fischlin, M. (eds.) EUROCRYPT 2015. LNCS, vol. 9057, pp. 563–594. Springer, Heidelberg (2015)
13. Boneh, D., Raghunathan, A., Segev, G.: Function-private identity-based encryption: hiding the function in functional encryption. In: Canetti, R., Garay, J.A. (eds.) CRYPTO 2013, Part II. LNCS, vol. 8043, pp. 461–478. Springer, Heidelberg (2013)
14. Boneh, D., Sahai, A., Waters, B.: Functional encryption: definitions and challenges. In: Ishai, Y. (ed.) TCC 2011. LNCS, vol. 6597, pp. 253–273. Springer, Heidelberg (2011)
15. Boneh, D., Waters, B.: Constrained pseudorandom functions and their applications. In: Sako, K., Sarkar, P. (eds.) ASIACRYPT 2013, Part II. LNCS, vol. 8270, pp. 280–300. Springer, Heidelberg (2013)
16. Boyle, E., Chung, K.-M., Pass, R.: On extractability obfuscation. In: Lindell, Y. (ed.) TCC 2014. LNCS, vol. 8349, pp. 52–73. Springer, Heidelberg (2014)
17. Boyle, E., Goldwasser, S., Ivan, I.: Functional signatures and pseudorandom functions. In: Krawczyk, H. (ed.) PKC 2014. LNCS, vol. 8383, pp. 501–519. Springer, Heidelberg (2014)
18. Brakerski, Z., Komargodski, I., Segev, G.: Multi-input functional encryption in the private-key setting: Stronger security from weaker assumptions. Cryptology ePrint Archive, Report 2015/158 (2015)
19. Brakerski, Z., Segev, G.: Function-private functional encryption in the private-key setting. In: Dodis, Y., Nielsen, J.B. (eds.) TCC 2015, Part II. LNCS, vol. 9015, pp. 306–324. Springer, Heidelberg (2015)
20. Cocks, C.: An identity based encryption scheme based on quadratic residues. In: Honary, B. (ed.) Cryptography and Coding 2001. LNCS, vol. 2260, pp. 360–363. Springer, Heidelberg (2001)

21. Garg, S., Gentry, C., Halevi, S., Raykova, M., Sahai, A., Waters, B.: Candidate indistinguishability obfuscation and functional encryption for all circuits. In: Proceedings of the 54th Annual IEEE Symposium on Foundations of Computer Science, pp. 40–49 (2013)
22. Garg, S., Gentry, C., Halevi, S., Zhandry, M.: Functional encryption without obfuscation. In: Kushilevitz, E., Malkin, T. (eds.) TCC 2016-A. LNCS, vol. 9563, pp. 480–511. Springer, Heidelberg (2016). doi:10.1007/978-3-662-49099-0_18
23. Goldreich, O., Goldwasser, S., Micali, S.: How to construct random functions. J. ACM **33**(4), 792–807 (1986)
24. Goldwasser, S., et al.: Multi-input functional encryption. In: Nguyen, P.Q., Oswald, E. (eds.) EUROCRYPT 2014. LNCS, vol. 8441, pp. 578–602. Springer, Heidelberg (2014)
25. Goldwasser, S., Kalai, Y., Popa, R.A., Vaikuntanathan, V., Zeldovich, N.: Reusable garbled circuits and succinct functional encryption. In: Proceedings of the 45th Annual ACM Symposium on Theory of Computing, pp. 555–564 (2013)
26. Gorbunov, S., Vaikuntanathan, V., Wee, H.: Functional encryption with bounded collusions via multi-party computation. In: Safavi-Naini, R., Canetti, R. (eds.) CRYPTO 2012. LNCS, vol. 7417, pp. 162–179. Springer, Heidelberg (2012)
27. Iovino, V., Zebrowski, K.: Mergeable functional encryption. Cryptology ePrint Archive, Report 2015/103 (2015)
28. Kiayias, A., Papadopoulos, S., Triandopoulos, N., Zacharias, T.: Delegatable pseudorandom functions and applications. In: Proceedings of the 20th Annual ACM Conference on Computer and Communications Security, pp. 669–684 (2013)
29. Komargodski, I., Moran, T., Naor, M., Pass, R., Rosen, A., Yogev, E.: One-way functions and (im)perfect obfuscation. In: Proceedings of the 55th Annual IEEE Symposium on Foundations of Computer Science, pp. 374–383 (2014)
30. Komargodski, I., Segev, G., Yogev, E.: Functional encryption for randomized functionalities in the private-key setting from minimal assumptions. In: Dodis, Y., Nielsen, J.B. (eds.) TCC 2015, Part II. LNCS, vol. 9015, pp. 352–377. Springer, Heidelberg (2015)
31. O'Neill, A.: Definitional issues in functional encryption. Cryptology ePrint Archive, Report 2010/556 (2010)
32. Sahai, A., Waters, B.: Slides on functional encryption (2008). http://www.cs.utexas.edu/~bwaters/presentations/files/functional.ppt
33. Sahai, A., Waters, B.: How to use indistinguishability obfuscation: deniable encryption, and more. In: Proceedings of the 46th Annual ACM Symposium on Theory of Computing, pp. 475–484 (2014)
34. Shamir, A.: Identity-based cryptosystems and signature schemes. In: Blakely, G.R., Chaum, D. (eds.) CRYPTO 1984. LNCS, vol. 196, pp. 47–53. Springer, Heidelberg (1985)
35. Shen, E., Shi, E., Waters, B.: Predicate privacy in encryption systems. In: Reingold, O. (ed.) TCC 2009. LNCS, vol. 5444, pp. 457–473. Springer, Heidelberg (2009)
36. Waters, B.: A punctured programming approach to adaptively secure functional encryption. In: Gennaro, R., Robshaw, M. (eds.) CRYPTO 2015. LNCS, vol. 9216, pp. 678–697. Springer, Heidelberg (2015)

Non-malleable Codes for Bounded Depth, Bounded Fan-In Circuits

Marshall Ball[1]([✉]), Dana Dachman-Soled[2], Mukul Kulkarni[2], and Tal Malkin[1]

[1] Columbia University, New York, NY, USA
marshall@cs.columbia.edu
[2] University of Maryland, College Park, MD, USA

Abstract. We show how to construct efficient, unconditionally secure non-malleable codes for bounded output locality. In particular, our scheme is resilient against functions such that any output bit is dependent on at most n^δ bits, where n is the total number of bits in a codeword and $0 \le \delta < 1$ a constant. Notably, this tampering class includes NC^0.

1 Introduction

Non-malleable codes were first introduced by Dziembowski, Pietrzak and Wichs [24] as an extension of error-correcting codes. Whereas error-correcting codes provide the guarantee that (if not too many errors occur) the receiver can recover the original message from a corrupted codeword, non-malleable codes are essentially concerned with security. In other words, correct decoding of corrupted codewords is not guaranteed (nor required), but it is instead guaranteed that adversarial corruptions cannot influence the output of the decoding in a way that depends on the original message: the decoding is either correct or *independent* of the original message.

The main application of non-malleable codes is in the setting of tamper-resilient computation. Indeed, as suggested in the initial work of Dziembowski et al. [24], non-malleable codes can be used to encode a secret state in the memory of a device such that a tampering adversary interacting with the device does not learn anything more than the input-output behavior. Unfortunately, it is impossible to construct non-malleable codes secure against arbitrary tampering, since the adversary can always apply the tampering function that decodes the entire codeword to recover the message m and then re-encodes a related message m'. Thus, non-malleable codes are typically constructed against limited classes of tampering functions $\mathcal{F}$. Indeed, given this perspective, error correcting codes can be viewed as a special case of non-malleable codes, where the class of tampering functions, $\mathcal{F}$, consists of functions which can only modify some fraction of the input symbols. Since non-malleable codes have a weaker guarantee than error correcting codes, there is potential to achieve non-malleable codes against much broader classes of tampering functions $\mathcal{F}$.

Indeed, there has been a large body of work on constructing non-malleable codes against classes of tampering functions which can potentially change every

© International Association for Cryptologic Research 2016
M. Fischlin and J.-S. Coron (Eds.): EUROCRYPT 2016, Part II, LNCS 9666, pp. 881–908, 2016.
DOI: 10.1007/978-3-662-49896-5_31

bit of the codeword, and for which no error correcting is possible. In particular, much attention has been given in the literature to the bit-wise tampering class (cf. [9,17,24]), where each bit of the codeword can be tampered individually, and its generalization, the split state tampering class (cf. [1–3,13,14,23,31]), where the codeword is split into two or more parts, each of which can be tampered individually (and independently of other blocks). One goal in this line of papers is to bring down the number of states, preferably to just two split states. Another goal is to increase the *rate* of the code, defined as the ratio k/n where k is the length of the original message and n is the length of the codeword outputted by the encoding algorithm. A constant-rate code is preferred, with the best possible rate being 1.

However, beyond some non-explicit (randomized) or inefficient constructions for more general classes (cf. [13,24,26]), almost all known results are only for function classes that are split state or "compartmentalized". There are a few exceptions, providing explicit and efficient non-malleable codes against *non-compartmentalized* classes of functions, including Chabanne et al. [10]—who address certain types of linear tampering functions—and Agrawal et al. [5,6]—who address the class of functions that can permute the bits of a codeword, flip individual bits, or set them to 0 or 1.

Other than the above results, achieving (explicit and efficient) non-malleable codes against natural tampering classes that are not split-state is a fundamental open question in this area, and is the focus of our paper.

1.1 Our Results

In this work, we devise explicit, efficient, and unconditionally secure non-malleable codes against a powerful tampering class which includes all bounded-depth circuits with bounded fan-in and unbounded fan-out. Specifically, we consider the class Local^{ℓ_o}, consisting of all functions $f : \{0,1\}^n \to \{0,1\}^n$ that can be computed with output locality $\ell_o(n)$, where each output bit depends on at most $\ell_o(n)$ input bits. Note that this class includes all fan-in-b circuits of depth at most $\log_b \ell_o$.

The class of bounded depth circuits is natural both as a complexity class and in the context of practical tampering attacks, where it seems to realistically capture the capabilities of a tampering adversary who has *limited time to tamper with memory* before the memory gets overwritten and/or refreshed. Moreover, the class of bounded output locality functions is a natural class in its own right, and is in fact much broader, including arbitrarily complex functions (even those outside of P), as long as the output locality constraint is maintained; we do not impose any constraints on the number or type of gates in the circuit. Finally, as we discuss below, our constructions actually hold for an even broader class, that also includes all split state functions, and beyond. We prove the following.

Main Theorem (informal): *For any $\ell_o = o(\frac{n}{\log n})$, there is an explicit, unconditionally secure non-malleable code for* Local^{ℓ_o}*, which encodes a $2k$ bit string into a string of length $n = \Theta(k\ell_o)$ bits. The encoding and decoding run in time polynomial in n, namely* $\mathrm{poly}(k, \ell_o)$.

This construction can be instantiated for any $\ell_o = o(n/\log n)$, and the resulting code has rate $\Theta(1/\ell_o)$. In general, since the output length is $n = \Theta(k\ell_o)$ bits, this may result in super-polynomial length encoding. However, using sublinear locality n^δ yields an efficient code. We highlight this, as well as the special cases of *constant depth* circuits (a subset of $\mathrm{Local}^{O(1)}$), in the following.

Corollaries: *There are efficient, explicit, and unconditionally secure non-malleable codes for the following classes:*

- $\mathrm{Local}^{n^\delta}$ *for any constant $0 \le \delta < 1$, with inverse-polynomial rate.*
- NC^0*with rate $\Theta(1/\ell_o)$ for any $\ell_o = \omega(1)$.*
- NC^0_c *for any constant c, with constant rate.*

The first corollary follows by instantiating the main theorem with $\ell_o = n^\delta$, the second by using any ℓ_o that is super constant (e.g., $\log^*(n)$), and the third by using $\ell_o = 2^c$ (a constant).

While our result for NC^0 correspond to constant depth circuits, the first corollary above implies as a special case that the code is also non-malleable against any $\delta \log n$ depth NC circuit, for any constant $0 \le \delta < 1$. Note that, since separations between P and NC^1 are not known, constructing (unconditional) non-malleable codes against NC^1 is unlikely, since an attacker in P can always decode and re-encode a related message, thus immediately breaking non-malleability.

Intermediate Results for (Input-and-Output) Local Functions. To prove our results, we use the concept of *non-malleable reduction*, introduced by Aggarwal et al. [2]. Informally, a class of functions $\mathcal{F}$ reduces to the class $\mathcal{G}$, if there is an encoding and decoding algorithms satisfying the following: applying the encoding to the input, then applying any $f \in \mathcal{F}$, and then applying the decoding, can be simulated by directly applying some function $g \in \mathcal{G}$ to the input. [2] prove that in this case a non-malleable code for $\mathcal{G}$ can be used to construct one for $\mathcal{F}$, and further prove a composition theorem, providing an elegant and powerful way to construct non-malleable codes.

Following this technique, we start by proving two separate results, and compose them (together with known results for the class of split state functions), to obtain a restricted variant of the main theorem above. We then use the same ideas to show a single construction allowing for a better combined analysis that achieves the full main theorem (via reduction to the class of split state functions). We believe our techniques are more comprehensible presented in this modular fashion, and the intermediate results are of independent interest.

First, we consider the class $\mathrm{Local}^{\ell_o}_{\ell_i}$ of local functions, with output locality ℓ_o as well as input locality ℓ_i (namely each input bit influences at most ℓ_i output bits). This class includes bounded-depth circuits with bounded fan-in and bounded fan-out. Our first intermediate result shows that the class $\text{Local}_{\tilde{O}(\sqrt{n})}^{\tilde{O}(\sqrt{n})}$ (and in fact a larger, leaky version of it) can be non-malleably reduced to the class of split state functions. Plugging in known results for non-malleable split state codes, we obtain a non-malleable code for this class. Our second result shows a non-malleable reduction of the class $\text{Local}^{\tilde{O}(\sqrt{n})}$ to the above class (thus giving a non-malleable code for functions with output locality $\tilde{O}(\sqrt{n})$). Finally, we combine the encoding schemes presented previously to a single encoding scheme (via a reduction to split state class), and improve the analysis to show resilience against $o(n/\log n)$ output locality.

We remark that our first technical result for (input and output) local functions is of independent interest, and although as stated it is strictly weaker than our output-local results, the construction can have advantages in terms of complexity and concrete parameters, and has stronger resilience to tampering functions that are both local and split-state, as we discuss next. We believe that both $\text{Local}_{\ell_i}^{\ell_o}$ and Local^{ℓ_o} are interesting classes, capturing natural types of tampering adversaries.

Extended Classes: Combining with Split State and Beyond. Our results are in fact broader than presented so far. First, every one of our results works not only for the class of functions claimed, but also for any split state function. This is because for all of our schemes, encoding is applied independently on each half of the input, and thus can handle a split-state tampering function trivially.

Furthermore, our intermediate result for (input-output) local functions can handle any function that applies arbitrary changes *within* each part and has bounded input and output locality *between* the two parts (this class is much broader than all functions that are either split state or local). More precisely, we can handle functions where any bit on the left affects at most $\tilde{O}(\sqrt{n})$ bits on the right (and vice-versa), and any bit on the left is affected by at most $\tilde{O}(\sqrt{n})$ bits on the right (and vice-versa).

Finally, our constructions can also handle some leakage to the tampering function, capturing an adversary that first leaks some bits, and can then select a tampering function. For our input-output local tampering result, the leakage can be a constant fraction of the bits, while for our output-local tampering result, the leakage is more limited.

Relation of Our Class to Previous Work. As mentioned above, almost all previous results presenting explicit and efficient non-malleable codes, do so for a split state tampering class (with two or more states). These classes are a special case of ours, as we explained, which is not surprising given that we use results for split state functions as a starting point to prove our result. As for the exceptions that go beyond split state, we note that the class of functions that permute the bits or apply bitwise manipulations, introduced by [5], is also a special case of our class, as it is a subset of Local^1 (in fact, even a subset of Local_1^1). The restricted linear tampering class considered by [10], on the other hand, seems incomparable to our class of output-local functions.

Thus, in terms of the tampering class captured, our results simultaneously encompass (and significantly extend) almost all previously known explicit, efficient constructions of non-malleable codes (we are aware of only one exception). This is not the case in terms of the *rate*, where several previous works focus on optimizing the rate for smaller classes of functions (e.g., [14] achieve rate $1 - o(1)$ non-malleable codes for bit-wise tampering functions), while we only achieve a constant rate for these classes.

We also mention that the original work of Dziembowski et al. [24] already considered the question of constructing non-malleable codes against the class $\mathrm{Local}^{\delta \cdot n}$, where n is the length of the codeword and $\delta < 1$ is a constant. We emphasize, however, that in [24] (and an improvement in [13]), they showed a construction of non-malleable codes against $\mathrm{Local}^{\delta \cdot n}$ in a *non-standard, random oracle model* where the encoding and decoding functions make queries to a random oracle, but the adversarial tampering function *does not* query the random oracle. Our work shows that it is possible to construct non-malleable codes for $\mathrm{Local}^{\delta \cdot n}$ for $\delta = o(1/\log n)$ in the standard model, with *no* random oracle.

On Randomized Decoding. Our constructions require the decoding function of the non-malleable code to be randomized. We note that, unlike the case of error correcting codes and encryption schemes, deterministic decoding for non-malleable codes *does not* seem to be without loss of generality, even in the case where the encoding scheme enjoys perfect correctness. To see why, note that while perfect correctness guarantees that all possible coins of the decoding algorithm produce the same output on a *valid codeword*, correctness provides no guarantees in the case when the codeword is corrupted and so it is not possible to derandomize by simply fixing an arbitrary sequence of coins for the decoder. Moreover, since the decoder holds no secret key in the non-malleable codes setting, it is also not possible to derandomize the decoding process by including the key of a pseudorandom function in the secret key. Since the encoding procedure must be randomized, and since non-malleable codes are only secure in the one-time setting—each time the codeword is decoded it must immediately be refreshed by re-encoding the original message—we believe that allowing randomized decoding is the natural and "correct" definition for non-malleable codes (although the original definition required deterministic decoding).

Interestingly, we can combine our technical building blocks into a construction of non-malleable codes against Local^{ℓ_o} for any $\ell_o \leq n^{1/4}$, using *deterministic* decoding. Unfortunately, when compared to our construction utilizing *randomized* decoding, this construction has a lower rate of $O(1/\ell_o{}^2)$ (instead of $O(1/\ell_o)$), and due to that also lower output locality that can be supported ($\mathrm{Local}^{n^{1/4}}$ instead of $\mathrm{Local}^{n^\delta}$ or $\mathrm{Local}^{o(n/\log n)}$ without efficiency).

We therefore leave as an interesting open question to resolve whether ramdomized decoding is *necessary* for achieving security against certain tampering classes, $\mathcal{F}$, or whether there is a generic way to derandomize decoding algorithms for non-malleable codes.

1.2 Technical Overview

We give a high level technical overview of our constructions. We use as an underlying tool a so called "reconstructable probabilistic encoding scheme", a code that can correct a constant fraction of errors (denoted c^{err}), and enjoys some additional secret-sharing like properties: given a (smaller) constant fraction c^{sec} of the codeword gives no information on the encoded message, and can in fact be completed to a (correctly distributed) encoding of any message. This or similar tools were used in previous works either implicitly or explicitly, e.g., the construction based on Reed Solomon codes and Shamir secret sharing with Berlekamp-Welch correction, as used already in [8] is a RPE (any small enough subset of shares is distributed uniformly at random, and any such collection of shares can be extended to be the sharing of any message of our choice). Other RPE schemes with possibly improved parameters can be constructed from, e.g., [15,16,19,21].

Handling Local Functions. Local functions are functions that have both small input and small output locality (i.e. each input bit affects a small number of output bits and each output bit depends on a small number of input bits). Our goal is to show a *non-malleable reduction* from a class of local functions with appropriate parameters, to the class of split-state functions. Loosely speaking, a non-malleable reduction from a class $\mathcal{F}$ to a class $\mathcal{G}$, is a pair (E, D) of encoding/decoding functions along with a *reduction* that transforms every $f \in \mathcal{F}$ into a distribution G_f over functions $g \in \mathcal{G}$, such that for every x, the distributions $\mathsf{D}(f(\mathsf{E}(x)))$ and $G_f(x)$ are statistically close. In the case of reductions to split-state, we let $x = (\mathsf{L}, \mathsf{R})$ where $\mathsf{L}, \mathsf{R} \in \{0,1\}^k$. We want to construct (E, D) such that, informally, given any local f, the effect of applying f to the encoding $\mathsf{E}(x)$ and then decoding $\mathsf{D}(f(\mathsf{E}(x)))$, can be simulated by applying some split state function $g = (g_L, g_R)$ directly to $x = (\mathsf{L}, \mathsf{R})$.

We will use an encoding that works on each half of the input separately, and outputs $E(\mathsf{L}, \mathsf{R}) = (E^L(\mathsf{L}), E^R(\mathsf{R})) = (s^\mathsf{L}, s^\mathsf{R})$, where $|s^\mathsf{L}| = n_L, |s^\mathsf{R}| = n_R$ (we will refer to these as "left" and "right" sides, though as we will see they will not be of equal lengths, and we will have appropriately designed decoding algorithms for each part separately). Now for any given local f, consider $f(s^\mathsf{L}, s^\mathsf{R}) = (f^L(s^\mathsf{L}, s^\mathsf{R}), f^R(s^\mathsf{L}, s^\mathsf{R}))$. Clearly, if f^L only depended on s^L and f^R only depended on s^R, we would be done (as this would naturally correspond to a distribution of split state functions on the original $x = (\mathsf{L}, \mathsf{R})$). However, this is generally not the case, and we need to take care of "cross-effects" of s^R on f^L and s^L on f^R.

Let's start with f^L, and notice that if its output locality is at most ℓ_o, then at most $n_L \ell_o$ bits from s^R could possibly influence the output of f^L. Thus, we will use E^R that is an RPE with $n_L \ell_o \leq c^{sec} n_R$. This means that we can just fix the relevant $n_L \ell_o$ bits from $s^\mathsf{R} = E^R(\mathsf{R})$ randomly (and independently of R), and now f^L will only depend on s^L, while s^R can still be completed to a correctly distributed encoding of R. Note that this requires making the right side larger than the left side ($n_R \geq \frac{n_L \ell_o}{c^{sec}}$).

Now let's consider f^R. Clearly we cannot directly do the same thing we did for f^L, since that technique required n_R to be much longer than n_L, while applying it here would require the opposite. Instead, we will take advantage of the smaller size on the left, and its limited input locality. Specifically, if the input locality of f^L is ℓ_i, then at most $n_L \ell_i$ bits on the right side can be influenced by s^L. A first (failed) attempt would be to just make sure that the encoding on the right can correct up to $n_L \ell_i$ errors, and hope that we can therefore set s^L arbitrarily when computing f^R and the resulting encoding would still be decoded to the same initial value R. While this argument works if the only changes made to s^R (a valid codeword) are caused by the "crossing bits" from s^L, it fails to take into account that f^R can in fact apply other changes inside s^R, and so it could be that s^R is malformed in such a way that applying f^R will cause it to decode differently in a way that crucially depends on s^L. The issue here seems to be that there is an exact threshold for when the decoding algorithm succeeds or not, and thus the function can be designed so that f^R is just over or under the threshold depending on the left side.

To overcome this problem, we use randomized decoding and a "consistency check" technique introduced in [15], and a forthcoming version by the same authors [16], in a different context. Roughly speaking, we make the right side encoding redundant, so that *any* large enough subset of bits is enough to recover R. An RPE has this property due its error correction capabilities. The decoding algorithm will decode via the first such subset, but will check a *random* subset of bits were consistent with a particular corrected codeword. This will yield similar behavior, regardless of which subset is used to decode. This construction has various subtleties, but they are all inherited from previous work, so we do not explain them here. The main point is that, like in [15,16], while the real decoding algorithm uses the *first* subset large enough, it can be simulated by using *any* other large enough subset.

Now, using the fact that "large enough" is not too large, and that at most $n_L \ell_i$ bits on the right side can be influenced by s^L, we can show that with high probability, there is a large enough "clean" subset of s^R that has *no* influence from s^L. The real decoding algorithm could be simulated by a decoding that uses this clean subset, which in turn means that the output of the decoding on $f^R(s^L, s^R)$ is in fact independent of s^L, as needed.

Putting the above together provides us the first result, namely a non-malleable reduction from local to split state functions. We note that the proof above in fact works for a more general class of functions (a fact we will use in our second construction). In particular, the first part requires a limit on the output locality of f^L, and the second part requires a limit on the output locality of f^R and the input locality of f^L, where all of these only refer to "cross-over" influences (within each part separately f can be arbitrary). Moreover, due to our use of encoding, security is maintained even with leakage, as long as the leakage is a constant fraction of bits on the left and a constant fraction on the right, independently. Similarly, security is maintained even when a constant fraction of bits on the left do not adhere to the input locality bound.

Removing Input Locality. We next present a non-malleable reduction from output local functions (which have no restriction on input locality) to local functions. Now let f be an output local tampering function. Since the input and output to f are the same size, note that the *average* input locality of f can be bounded by its output locality, ℓ_o. Our local construction above requires low input locality for the left side, but also requires the left side to be much shorter than the right side. Unfortunately, what this means is that the input locality of *all* bits on the left side of the local encoding described above can be far higher than average. So, in order to bound the average input locality of the left side, we must increase the length of the left side, but this destroys our analysis from the first construction.

In order to achieve the best of both worlds, our idea is to construct a non-malleable reduction which increases the size of the left side of the underlying local encoding by adding dummy inputs. The "relevant" inputs, which correspond to bits of the left side of the underlying local encoding, are placed randomly throughout the left side of the new encoding. The idea is that since the adversary does not know which bit positions on the left side are "relevant," it cannot succeed in causing too many "relevant" positions to have input locality that is too much above average.

But now, in order to decode properly, the decoding algorithm must be able to recover these "relevant" locations, without sharing a secret state with the encoding algorithm (which is disallowed in the standard non-malleable codes setting). In order to do this, the first idea is to encode the relevant positions on the left side of the new encoding in an additional string, which is then appended to the left side during the new encoding procedure. Unfortunately, it is not clear how to make this work: Since this additional string is long, it can depend on a large number of input bits from both the left and right sides; on the other hand, in order to obtain a reduction from output local to local functions, the reduction must be able to recover this (possibly tampered) additional string so that it "knows" which output bits of $\widetilde{X}^{\mathsf{L}}$ are relevant.

The solution is to use a PRG with a short seed. The seed of the PRG is now the additional string that is appended to the left side and the output of the PRG yields an indicator string which specifies the "relevant" locations for decoding. Note that now since the PRG seed of length r is short, we can, using the leakage resilient properties of the underlying local code, leak all $r \cdot \ell_o \leq c^{\mathsf{sec}} \cdot n_L \leq c^{\mathsf{sec}} \cdot n_R$ number of bits affecting these output locations from both the left and right sides.

Moreover, because the tampering attacker is very limited, in the sense that it must choose the tampering function before learning any information about the output of the PRG, we are able to show that Nisan's PRG (see Definition 12), an *unconditional* PRG is sufficient for our construction. Thus, our construction does not rely on any computational assumption.

Improving the Parameters. Ultimately the technique sketched above and presented in the body of the paper imposes two restrictions on output locality (modulo smaller terms): (1) $n_L \ell_o \leq n_R$ (2) $\ell_o \approx \ell_i \leq n_L$. Together these restrictions imply tolerance against output locality of approximately $\sqrt{n}$. The first restriction follows from the asymmetric encoding to handle bits on the left dependent on the right. The second restriction results from handling bits on the left of affecting the right side's consistency check.

To bypass this $\sqrt{n}$ barrier, we consider the two encoding schemes as a single scheme. Then in analysis, we can use the pseudorandom hiding of the left side encoding to relax the second bound. Namely, with high probability only a small portion of the left side RPE affects the consistency check, even if the consistency check and/or output locality is large with respect to n_L. This simple change in analysis gives resilience against $o(n/\log n)$ output locality.

1.3 Other Related Work

The concept of non-malleability was introduced by Dolev, Dwork and Naor [22] and has been applied widely in cryptography since. Although it was defined in computational setting, most recent work on non-malleability has been in the information-theoretic setting. The study of non-malleable codes was inspired by error-correcting codes and early works on tamper resilience [27–29].

Dziembowski, Pietrzak and Wichs [24] motivated and formalized the notion of non-malleable codes. They showed the existence of such codes for the class of all bit-wise independent functions (which can be viewed as split state with n parts). Later work on split state classes improved this by reducing the number of states, increasing the rate, or adding desirable features to the scheme. For example, [23] presented an information theoretic non-malleable code for 1-bit messages against 2 state split state functions, followed by [3], who gave an information-theoretic construction for k-bit messages using results from additive combinatorics. A constant rate construction for a constant (>2) number of states was provided in [3,12]. This line of research culminated with the result of [2], who used their reduction-based framework to achieve constant rate codes for two state split-state functions (using several intermediate constructions against various classes of functions). [1] improve this to (optimal) rate 1 non-malleable codes for two states, in the computational setting.

Beyond the above and other results constructing explicit efficient codes, there are several inefficient, existential or randomized constructions for much more general classes of functions (sometimes presented as efficient construction in a random-oracle model). In particular, Dziembowski et al. [24] gave an existential proof for the existence non-malleable codes secure against any 'small-enough' tampering family ($<2^{2^n}$). [13,26] give randomized construction of non-malleable codes against bounded poly-size circuits (where the bound on the circuit size is selected prior to the code).

Several other variants and enhanced models were considered. For example, [17], in the context of designing UC secure protocols via tamperable hardware tokens, consider a variant of non-malleable codes which has *deterministic* encryption and decryption. It is interesting to note the contrast between their restriction to deterministic encoding (and decoding) and our relaxation to randomized decoding (and encoding). They provide inefficient general constructions and efficient constructions for bit-wise functions and generalizations. [31], in the context of securing cryptographic protocols against continual split-state leakage and tampering, provide a (computational) non-malleable code for split state functions, in the CRS model. This was one of the first works using the split state model for tampering. [11,20] consider a variant of non-malleable codes that is also locally decodable and updatable. [25] allow continual tampering, and [4] allow for bounded leakage model. As discussed previously, [10] considers a subclass of linear tampering functions. We guide the interested reader to [30,31] for illustrative discussion of various models.

2   Preliminaries

2.1   Notation

Firstly, we present some standard notations that will be used in what follows. For any positive integer n, $[n] := \{1, \ldots, n\}$. If $x = (x_1, \ldots, x_n) \in \Sigma^n$ (for some set Σ), then $x_{i:j} := (x_i, x_{i+1}, \ldots, x_{j-1}, x_j)$ for $i \leq j$. If Σ is a set, then $\Sigma^\Sigma := \{f : \Sigma \to \Sigma\}$, the set of all functions from Σ to Σ. We say two vectors $x, y \in \Sigma^n$ are ε-far if they disagree on at least $\varepsilon \cdot n$ indices, $|\{i : x_i \neq y_i\}| \geq \varepsilon n$. Conversely, we say two vectors $x, y \in \Sigma^n$ are $(1-\varepsilon)$-close if they agree on at least $(1-\varepsilon) \cdot n$ indices, $|\{i : x_i = y_i\}| \geq (1-\varepsilon)n$. Alternatively, for $x, y \in \mathrm{GF}(2)^n$ define their distance to be $d(x, y) := \frac{\|x+y\|_0}{n}$. (I.e. x and y are ε-far if $d(x, y) \geq \varepsilon$.) We take the statistical distance between two distributions, A and B, over a domain X to be $\Delta(A, B) := 1/2 \sum_{x \in X} |A(x) - B(x)|$. We say A and B are statistically indistinguishable, $A \overset{s}{\approx} B$, if $\Delta(A, B)$ is negligible, in some parameter appropriate to the domain.

2.2   Non-malleable Codes and Reductions

Definition 1 (Coding Scheme). *[24] A Coding scheme, (E, D), consists of a randomized encoding function $\mathsf{E} : \{0,1\}^k \mapsto \{0,1\}^n$ and a randomized decoding function $\mathsf{D} : \{0,1\}^n \mapsto \{0,1\}^k \cup \{\bot\}$ such that $\forall x \in \{0,1\}^k, \Pr[\mathsf{D}(\mathsf{E}(x)) = x] = 1$ (over randomness of E and D).*

We note that this definition differs from the original one given in [24], in that we allow the decoding to be randomized, while they required deterministic decoding. While this technically weakens our definition (and a code with deterministic decoding would be preferable), we feel that allowing randomized decoding fits the spirit and motivation of non-malleable codes, and possibly is "the right" definition (which was simply not used before because it was not needed by previous constructions). More importantly, it may allow for a wider classes of functions.

This difference (allowing randomized decoding) also applies to the rest of the section, but all the previous results (in particular, Theorem 1) go through in exactly the same way, as long as we have independent randomness in all encoding and decoding.

Originally, non-malleable codes were defined in the following manner:

Definition 2 (Non-Malleable Code). *[2] Let $\mathcal{F}$ denote a family of tampering functions. Let $\mathsf{E} : B \to A$, $\mathsf{D} : A \to B$ be a coding scheme. For all $f \in \mathcal{F}$ and all $x \in B$ define:*

$$Tamper_x^f := \{c \leftarrow \mathsf{E}(x); \tilde{c} \leftarrow f(c); \tilde{x} \leftarrow \mathsf{D}(\tilde{c}); output : \tilde{x}\}.$$

Then, (E, D) is an ε-non-malleable code with respect to $\mathcal{F}$, if there exists a distribution D_f over $\{0, 1\}^k \cup \{\perp, same\}$ such that $\forall x \in B$, the statistical distance between

$$Sim_x^f := \{\tilde{x} \leftarrow D_f; output : x \text{ if } \tilde{x} = same \ \& \ \tilde{x}, otherwise\},$$

and $Tamper_x^f$ is at most ε.

The above of definition has its origins in [24]. Dziembowski, Pietrzak, and Wichs required the simulator to be efficient. Aggarwal et al. demonstrated that the above relaxation is, in fact, equivalent for deterministic decoding. Allowing decoding to be randomized does not affect their proof. For this reason, we will not concern ourselves with the efficiency of a simulator (or, equivalently, sampling relevant distributions) for the remainder of this paper.

Aggarwal et al. provide a simpler alternative to the above simulation-based definition, which they prove equivalent. [2] Their definition is based on the notion of non-malleable reduction, which we will use.

Definition 3 (Non-Malleable Reduction). *[2] Let $\mathcal{F} \subset A^A$ and $\mathcal{G} \subset B^B$ be some classes of functions. We say $\mathcal{F}$ reduces to $\mathcal{G}$, $(\mathcal{F} \Rightarrow \mathcal{G}, \varepsilon)$, if there exists an efficient (randomized) encoding function $\mathsf{E} : B \to A$, and an efficient (randomized) decoding function $\mathsf{D} : A \to B$, such that*

(a) $\forall x \in B, \Pr[\mathsf{D}(\mathsf{E}(x)) = x] = 1$ (over the randomness of E, D).
(b) $\forall f \in \mathcal{F}, \exists G : \forall x \in B, \Delta(\mathsf{D}(f(\mathsf{E}(x))); G(x)) \leq \varepsilon$, where G is a distribution over $\mathcal{G}$ and $G(x)$ denotes the distribution $g(x)$, where $g \leftarrow G$.

If the above holds, then (E, D) is an $(\mathcal{F}, \mathcal{G}, \varepsilon)$-non-malleable reduction.

Definition 4 (Non-Malleable Code). *[2] Let NM_k denote the set of trivial manipulation functions on k-bit strings, consisting of the identity function $id(x) = x$ and all constant functions $f_c(x) = c$, where $c \in \{0, 1\}^k$.*

A coding scheme (E, D) defines an $(\mathcal{F}, k, \varepsilon)$-non-malleable code, if it defines an $(\mathcal{F}, \mathrm{NM}_k, \varepsilon)$-non-malleable reduction.

Aggarwal et al. also prove the following useful theorem for composing non-malleable reductions.

Theorem 1 (Composition). *[2] If $(\mathcal{F} \Rightarrow \mathcal{G}, \varepsilon_1)$ and $(\mathcal{G} \Rightarrow \mathcal{H}, \varepsilon_2)$, then $(\mathcal{F} \Rightarrow \mathcal{H}, \varepsilon_1 + \varepsilon_2)$.*

We note that the proof given in [2] goes through unchanged with randomized decoding.

2.3   Tampering Families

Definition 5 (Split-State Model). *[24] The* split-state model, SS_k, *denotes the set of all functions:*

$$\{f = (f_1, f_2) : \ f(x) = (f_1(x_{1:k}) \in \{0,1\}^k, f_2(x_{k+1:2k}) \in \{0,1\}^k) \ for \ x \in \{0,1\}^{2k}\}.$$

Theorem 2 (Split-State Non-malleable Codes with Constant Rate). *[2]   There exists an efficient, explicit* $(SS_{O(k)}, k, 2^{-\Omega(k)})$ *non-malleable code,* (E_{SS}, D_{SS}).

We next define a class of local functions, where the number of input bits that can affect any output bit (input locality) and the number of output bits that depend on an input bit (output locality) are restricted. Loosely speaking, an input bit x_i affects the output bit y_j if for any boolean circuit computing f, there is a path in the underlying DAG from x_i to y_j. The formal definitions are below, and our notation follows that of [7].

Definition 6. *We say that a bit x_i affects the boolean function f,*
if $\exists \{x_1, x_2, \cdots x_{i-1}, x_{i+1}, \cdots x_n\} \in \{0,1\}^{n-1}$ such that,
$f(x_1, x_2, \cdots x_{i-1}, 0, x_{i+1}, \cdots x_n) \neq f(x_1, x_2, \cdots x_{i-1}, 1, x_{i+1}, \cdots x_n)$.
   Given a function $f = (f_1, \ldots, f_n)$ (where each f_j is a boolean function), we say that input bit x_i affects output bit y_j, or that output bit y_j depends on input bit x_i, if x_i affects f_j.

Definition 7 (Output Locality). *A function $f : \{0,1\}^n \to \{0,1\}^n$ is said to have* output locality m *if every output bit f_i is dependent on at most m input bits.*

Definition 8 (Input Locality). *A function $f : \{0,1\}^n \to \{0,1\}^n$ is said to have* input locality ℓ *if every input bit f_i is affects at most ℓ output bits.*

Definition 9 (Local Functions). *[7] A function $f : \{0,1\}^n \to \{0,1\}^n$ is said to be (m, ℓ)-local, $f \in Local_\ell^m$, if it has input locality ℓ and output locality m. We denote the class $Local_n^m$ (namely no restriction on the input locality) by $Local^m$.*

The above notions can be generalized to function ensembles $\{f_n : \{0,1\}^n \to \{0,1\}^n\}_{n \in \mathbb{Z}}$ with the following corresponding locality bound generalizations: $\ell(n), m(n)$.

Recall that NC^0 is the class of functions where each output bit can be computed by a boolean circuit with constant depth and fan-in 2 (namely in constant parallel time). It is easy to see that $NC^0 \subseteq Local^{O(1)}$.

2.4   Reconstructable Probabilistic Encoding Scheme

Reconstructable Probabilistic Encoding (RPE) schemes were defined by Choi et al. (in an in-submission journal version of [15], as well as in [16]), extending a definition given by Decatur, Goldreich and Ron [21]. Informally, this is an error correcting code, which has an additional secrecy property and reconstruction property. The secrecy property allows a portion of the output to be revealed without leaking any information about the encoded message. The reconstruction property allows, given a message and a partial codeword for it, to reconstruct a complete consistent codeword. Thus, this is a combination of error correcting code and secret sharing, similar to what has been used in the literature already starting with Ben-Or, Goldwasser, and Wigderson [8].

Definition 10 (Binary Reconstructable Probabilistic Encoding). *[15, 16] We say a triple* (E, D, Rec) *is a* binary reconstructable probabilistic encoding scheme *with parameters* $(k, n, c^{\text{err}}, c^{\text{sec}})$*, where* $k, n \in \mathbb{N}$*,* $0 < c^{\text{err}}, c^{\text{sec}} < 1$*, if it satisfies the following properties:*

1. ***Error correction.*** E : $\{0,1\}^k \to \{0,1\}^n$ *is an efficient probabilistic procedure, which maps a message* $m \in \{0,1\}^k$ *to a distribution over* $\{0,1\}^n$*. If we let* $\mathcal{W}$ *denote the support of* E*, any two strings in* $\mathcal{W}$ *are* $2c^{\text{err}}$*-far. Moreover,* D *is an efficient procedure that given any* $w' \in \{0,1\}^n$ *that is* $(1 - \epsilon)$*-close to some string* w *in* $\mathcal{W}$ *for any* $\epsilon \leq c^{\text{err}}$*, outputs* w *along with a consistent* m*.*

2. ***Secrecy of partial views.*** *For all* $m \in \{0,1\}^k$ *and all sets* $S \subset [n]$ *of size* $\leq \lfloor c^{\text{sec}} \cdot n \rfloor$*, the projection of* E(m) *onto the coordinates in* S*, as denoted by* E(m)$|_S$*, is identically distributed to the uniform distribution over* $\{0,1\}^{\lfloor c^{\text{sec}} n \rfloor}$*.*

3. ***Reconstruction from partial views.*** Rec *is an efficient procedure that given any set* $S \subset [n]$ *of size* $\leq \lfloor c^{\text{sec}} \cdot n \rfloor$*, any* $I \in \{0,1\}^n$*, and any* $m \in \{0,1\}^k$*, samples from the distribution* E(m) *with the constraint* $\forall i \in S, E(m)_i = I_i$*.*

Choi et al. show that a construction of Decatur, Goldreich, and Ron [21] meets the above requirements.

Lemma 1. *[15, 16] For any* $k \in \mathbb{N}$*, there exists constants* $0 < c^{\text{rate}}, c^{\text{err}}, c^{\text{sec}} < 1$ *such that there is a binary RPE scheme with parameters* $(k, c^{\text{rate}}k, c^{\text{err}}, c^{\text{sec}})$*.*

Remark 1. To achieve longer encoding lengths ck, with the same c^{err} and c^{sec} parameters, one can simply pad the message to an appropriate length.

Specifically, Decatur, Goldreich and Ron [21] construct a probabilistic encoding scheme that possesses the first two properties listed above. Moreover, since the construction they present, instantiates E with a linear error-correcting code, we have that property (3) holds. (Any linear error-correcting code has efficient reconstruction.)

These are the parameters we use here, but we believe it may be possible to achieve a better rate if we use parameters based on the recent result of Coretti et al. [18] (see also [14]).

2.5 Boolean Function Restrictions

The following two definitions are special cases of Boolean function restrictions. It will be convenient to have explicit notation for restrictions of Boolean functions f where the input/output of the function f has a particular form.

Definition 11 (Restriction). *For a vector* $v \in \{0, 1, *\}^n$ *and a Boolean function* $f : \{0, 1\}^n \to \{0, 1\}^n$ *the restriction of f to v, $\tilde{f}|_v$ is defined as $\tilde{f}|_v(x) = f(z)$ where,*

$$z_i = \begin{cases} x_i & v_i = * \\ v_i & v_i \in \{0, 1\} \end{cases}$$

Let $f : D \to \{0, 1\}^r$ be a function. Then, we denote by f_i the function which outputs the i-th output bit of f. Let $f : D \to \{0, 1\}^r$ be a function and let $v \in \{0, 1\}^r$ be a vector. Then, we denote by f_v the function which outputs all f_i such that $v_i = 1$.

2.6 Pseudorandom Generators of Space-Bounded Computation

Definition 12. *[32] A generator* $\mathsf{prg} : \{0, 1\}^m \to (\{0, 1\}^n)^k$ *is a pseudorandom generator for space(w) and block size n with parameter ε if for every finite state machine, Q, of size 2^w over alphabet $\{0, 1\}^n$ we have that*

$$|\Pr_y[Q \text{ accepts } y] - \Pr_x[Q \text{ accepts } \mathsf{prg}(x)]| \leq \varepsilon$$

where y is chosen uniformly at random in $(\{0, 1\}^n)^k$ and x in $\{0, 1\}^m$.

Theorem 3. *[32] There exists a fixed constant $c > 0$ such that for any $w, k \leq cn$ there exists an (explicit) pseudorandom generator* $\mathsf{prg} : \{0, 1\}^{O(k)+n} \to \{0, 1\}^{n2^k}$ *for space(w) with parameter 2^{-cn}. Moreover, prg can be computed in polynomial time (in $n, 2^k$).*

3 Non-malleable Codes for $\text{Local}_{\ell_i(n)}^{\ell_o(n)}$

Theorem 4. (E, D) *is a* $(\text{Local}_{\ell_i(k)}^{\ell_o(k)} \Rightarrow \mathsf{SS}_k, \mathsf{negl}(k))$-*non-malleable reduction given the following parameters for* $\text{Local}_{\ell_i(k)}^{\ell_o(k)}$ *(where $c^{\mathrm{rate}}, c^{\mathrm{err}}, c^{\mathrm{sec}}$ are taken from Lemma 1):*

- $\ell_o \leq \frac{c^{\mathrm{rate}} c^{\mathrm{sec}} k}{\log^2(k)}$.
- $\ell_i \leq 12\ell_o / c^{\mathrm{sec}}$.
- $n := c^{\mathrm{rate}} \frac{k^2}{\log^2(k)} + c^{\mathrm{rate}} k = O\left(\frac{k^2}{\log^2(k)}\right)$.

Putting together Theorem 4 with Theorems 1 and 2, we obtain the following.

Corollary 1. $(\mathsf{E} \circ \mathsf{E}_{\mathsf{SS}}, \mathsf{D}_{\mathsf{SS}} \circ \mathsf{D})$ *is a* $(\text{Local}_\ell^\ell, k, \mathsf{negl}(k))$-*non-malleable code with rate $\Theta(1/\ell)$, where $\ell = \tilde{O}(\sqrt{n})$.*

Remark 2. The reduction presented below is, in fact, a $(\text{XLocal}_\ell^\ell \Rightarrow \mathsf{SS}_k, \mathsf{negl}(k))$-non-malleable reduction, where $\ell = \tilde{O}(\sqrt{n})$ and XLocal_ℓ^ℓ is the following class of functions $f : \{0, 1\}^{n_L + n_R} \to \{0, 1\}^{n_L + n_R}$:

- For $i = 1, \ldots, n_L$, there are at most ℓ indices $j \in \{n_L + 1, \ldots, n_L + n_R\}$ such that the i-th input bit affects f_j. And, for $i = n_L + 1, \ldots, n_L + n_R$, there are at most ℓ indices $j \in \{1, \ldots, n_L\}$ such that the i-th input bit affects f_j.
- For $i = 1, \ldots, n_L$, there are at most ℓ indices $j \in \{n_L + 1, \ldots, n_L + n_R\}$ such that the f_i-th is affected by the j-th input bit. And, for $i = n_L+1, \ldots, n_L+n_R$, there are at most ℓ indices $j \in \{1, \ldots, n_L\}$ such that the f_i-th is affected by the j-th input bit.

In other words, the reduction holds for a generalized variant of split state tampering where we only restrict locality with respect to the opposite side, and allow arbitrary locality *within* each side. n_L and n_R are the lengths of the left and right side codewords, respectively.

We construct an encoding scheme (E, D) summarized in Fig. 1 and parametrized below. We then show that the pair (E, D) is an $(\mathrm{Local}_{\ell_i(k)}^{\ell_o(k)}, \mathsf{SS}_k, \mathsf{negl}(k))$-non-malleable reduction. This immediately implies that given a non-malleable encoding scheme $(\mathsf{E}^{\mathsf{ss}}, \mathsf{D}^{\mathsf{ss}})$ for class SS_k (where SS is the class of split-state functions), the encoding scheme $\Pi = (\mathsf{E}^{\mathsf{bd}}, \mathsf{D}^{\mathsf{bd}})$, where $\mathsf{E}^{\mathsf{bd}}(m) := \mathsf{E}(\mathsf{E}^{\mathsf{ss}}(m))$ and $\mathsf{D}^{\mathsf{bd}}(s) := \mathsf{D}^{\mathsf{ss}}(\mathsf{D}(s))$ yields a non-malleable code against $\mathrm{Local}_{\ell_i(k)}^{\ell_o(k)}$.

We parametrize our construction for $\mathrm{Local}_{\ell_i(k)}^{\ell_o(k)} \Rightarrow \mathsf{SS}_k$ with the following:

- $(\mathsf{E}_L, \mathsf{D}_L)$ parametrized by $(k, n_L, c_L^{\mathsf{err}}, c_L^{\mathsf{sec}}) := (k, c^{\mathsf{rate}}k, c^{\mathsf{err}}, c^{\mathsf{sec}})$ where $c^{\mathsf{err}}, c^{\mathsf{sec}}, c^{\mathsf{rate}}$ are taken from Lemma 1.
- $n_{\mathsf{check}} := \log^2(k)$.
- $\ell_{\mathsf{sec}} := \sqrt{\frac{cn_L}{n_{\mathsf{check}}}} = \Theta(\frac{\sqrt{k}}{\log(k)})$.
- $(\mathsf{E}_R, \mathsf{D}_R)$ parametrized by $(k, n_R, c_R^{\mathsf{err}}, c_R^{\mathsf{sec}}) := (k, \frac{\ell_o c^{\mathsf{rate}}k}{c^{\mathsf{sec}}}, c^{\mathsf{err}}, c^{\mathsf{sec}})$.
- $n := \ell_o c^{\mathsf{rate}}k + c^{\mathsf{rate}}k = O(\frac{k^2}{\log^2(k)})$.

Note that this setting of parameters is taken with our forthcoming reduction in mind. (See Corollary 2 and Theorem 5.) One may take any parametrization for which (a) such RPEs exist, (b) $(1 - c^{\mathsf{err}}/4)^{n_{\mathsf{check}}}$ is negligible in k, and (c) Observation 1 (below) is satisfied. For certain applications, parametrization other than ours may be advantageous.

Let $f(s^{\mathsf{L}}, s^{\mathsf{R}}) = (f^{\mathsf{L}}(s^{\mathsf{L}}, s^{\mathsf{R}}), f^{\mathsf{R}}(s^{\mathsf{L}}, s^{\mathsf{R}}))$, where $(s^{\mathsf{L}}, s^{\mathsf{R}}) \in \{0,1\}^{n_L} \times \{0,1\}^{n_R}$ and $f^{\mathsf{L}}(s^{\mathsf{L}}, s^{\mathsf{R}}) \in \{0,1\}^{n_L}$ and $f^{\mathsf{R}}(s^{\mathsf{L}}, s^{\mathsf{R}}) \in \{0,1\}^{n_R}$.

- Let $\mathcal{S}_{\mathsf{R} \to \mathsf{L}}$ denote the set of positions j such that input bit s_j^{R} affects the output of f^{L}.
- Let $\mathcal{S}_{\mathsf{L} \to \mathsf{R}}$ denote the set of positions i such that input bit s_i^{L} affects the output of f^{R}.
- For $J \subset [n_R]$, let $\mathcal{S}_{\mathsf{L} \to \mathsf{R}}^J$ denote the set of positions i such that input bit s_i^{L} affects the output of f_j^{R} for some $j \in J$.
- For a set $R_{\mathsf{check}} \subseteq n_R$ of size n_{check}, let $\mathcal{S}_{\mathsf{check}}$ denote the set of positions i such that input bit s_i^{L} affects the output of f_ℓ^{R} for some $\ell \in R_{\mathsf{check}}$.

The sets defined above are illustrated in Fig. 2. We observe the following immediate facts about their sizes:

Let (E_L, D_L, Rec_L) be a binary reconstructable probabilistic encoding scheme with parameters $(k, n_L, c_L^{\text{err}}, c_L^{\text{sec}})$ and let (E_R, D_R, Rec_R) be a binary reconstructable probabilistic encoding scheme with parameters $(k, n_R, c_R^{\text{err}}, c_R^{\text{sec}})$.

Also let $\ell_{\text{sec}}, n_{\text{check}}$ be parameters.

$E(x := (L, R))$:

1. Compute $(s_1^L, \ldots, s_{n_L}^L) \leftarrow E_L(L)$ and $(s_1^R, \ldots, s_{n_R}^R) \leftarrow E_R(R)$.
2. Output the encoding $(s^L, s^R) := ([s_i^L]_{i \in [n_L]}, [s_i^R]_{i \in [n_R]})$.

$D(\sigma := (\sigma^L, \sigma^R))$:

1. Let $(\sigma^L, \sigma^R) := ([\sigma_i^L]_{i \in [n_L]}, [\sigma_i^R]_{i \in [n_R]})$.
2. Compute $((w_1^L, \ldots, w_{n_L}^L), L) \leftarrow D_L(\sigma_1^L, \ldots, \sigma_{n_L}^L)$. If the decoding fails, set $L := \perp$.
3. **(decoding-check on right)** Let $t := \lceil n_R(1 - c_R^{\text{err}}/4) \rceil$ Define $\sigma'^R := \sigma_1'^R, \ldots, \sigma_{n_R}'^R$ as follows: Set $\sigma_\ell'^R := \sigma_\ell^R$ for $\ell = 1, \ldots, t$. Set $\sigma_\ell'^R := 0$ for $\ell = t+1, \ldots, n_R$. Compute $((w_1^R, \ldots, w_{n_R}^R), R) \leftarrow D_R(\sigma_1'^R, \ldots, \sigma_t'^R)$. If the decoding fails or $(w_1^R, \ldots, w_{n_R}^R)$ is not $c_R^{\text{err}}/4$-close to $(\sigma_1^R, \ldots, \sigma_t^R)$, set $R := \perp$.
4. **(codeword-check on right)** Pick a random subset $R_{\text{check}} \subset [n_R]$ of size $n_{\text{check}} < c_R^{\text{sec}} \cdot n_R$. For all $\ell \in R_{\text{check}}$, check that $\sigma_\ell^R = w_\ell^R$. If the check fails, set $R := \perp$.
5. **(output)** Output $x := (L, R)$.

Fig. 1. THE $(\text{Local}_{\ell_i(k)}^{\ell_o(k)}, \text{SS}, \text{negl}(k))$-NON-MALLEABLE REDUCTION (E, D)

Observation 1. *For $f \in \text{Local}_{\ell_i}^{\ell_o}$, we have the following:*

1. *There is some set $J^* \subset [n_R]$ such that $|J^*| = t$ and $|S_{L \to R}^{J^*}| = 0$ (from now on, J^* denotes the lexicographically first such set). (Since $|S_{L \to R}| \leq \ell_i \cdot n_L \leq n_R - t$.)*
2. *By choice of parameters $n_L, n_{\text{check}}, c_L^{\text{sec}}$, we have that $|S_{\text{check}}| \leq n_L \cdot c_L^{\text{sec}}$. (Since $S_{\text{check}} \leq \ell_o \cdot n_{\text{check}}$.)*
3. *By choice of parameters $n_L, n_R, c_R^{\text{sec}}$, we have that $|S_{R \to L}| \leq \ell_o \cdot n_L \leq n_R \cdot c_R^{\text{sec}}$.*

Now, for every $f \in \text{Local}_{\ell_i}^{\ell_o}$, we define the distribution G_f over SS_k. A draw from G_f is defined as follows:

- Choose a random subset $R_{\text{check}} \subseteq [n_R]$ of size n_{check}.
- Choose vectors $I^L \in \{0,1\}^{n_L} \times \{*\}^{n_L}$, $I^R \in \{*\}^{n_L} \times \{0,1\}^{n_R}$ uniformly at random.
- Let J^* be the subset of $[n_R]$ as described in Observation 1.
- The split-state tampering function $g := (g_L, g_R) \in \text{SS}_k$ has I^L, I^R hardcoded into it and is specified as follows:

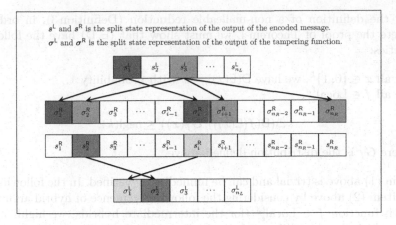

Fig. 2. The adversary chooses tampering function $f = (f^{\mathsf{L}}, f^{\mathsf{R}}) \in \mathrm{Local}_{\ell_i(k)}^{\ell_o(k)}$ which takes inputs $(s^{\mathsf{L}}, s^{\mathsf{R}})$ and produces outputs $(\sigma^{\mathsf{L}}, \sigma^{\mathsf{R}})$. The highlighted bits of s^{L} and s^{R} are the "bad" bits. E.g. note that bits s_2^{R} and s_i^{R} *affect* the output bits σ_2^{L} and σ_1^{L} respectively after f^{L} is applied to $(s^{\mathsf{L}}, s^{\mathsf{R}})$. Thus we add 2 and i to the set $\mathcal{S}_{\mathsf{R} \to \mathsf{L}}$. Similarly, the bits s_1^{L} and s_3^{L} *affect* the bits $\{\sigma_1^{\mathsf{R}}, \sigma_i^{\mathsf{R}}\}$ and the bits $\{\sigma_2^{\mathsf{R}}, \sigma_{i+1}^{\mathsf{R}}, \sigma_{n_R,}^{\mathsf{R}}\}$ respectively after the tampering function f^{R} is applied to $(s^{\mathsf{L}}, s^{\mathsf{R}})$. We therefore add 1 to the sets $\mathcal{S}_{\mathsf{L} \to \mathsf{R}}^1$ and $\mathcal{S}_{\mathsf{L} \to \mathsf{R}}^i$, while we add 3 to the sets $\mathcal{S}_{\mathsf{L} \to \mathsf{R}}^2, \mathcal{S}_{\mathsf{L} \to \mathsf{R}}^{i+1}$ and $\mathcal{S}_{\mathsf{L} \to \mathsf{R}}^{n_R}$. We also add both 1 and 3 to the set $\mathcal{S}_{\mathsf{L} \to \mathsf{R}}$.

$g_L(\mathsf{L})$:

1. (apply tampering and plain decode on left) Let $s^{\mathsf{L}} := \mathrm{Rec}(\mathcal{S}_{\mathrm{check}}, I^{\mathsf{L}}, \mathsf{L})$. Let $(\sigma_1^{\mathsf{L}}, \ldots, \sigma_{n_L}^{\mathsf{L}}) := f^{\mathsf{L}}|_{I^{\mathsf{R}}}(s^{\mathsf{L}})$. Compute $((w_1^{\mathsf{L}}, \ldots, w_{n_L}^{\mathsf{L}}), \widetilde{\mathsf{L}}) \leftarrow D_L(\sigma_1^{\mathsf{L}}, \ldots, \sigma_{n_L}^{\mathsf{L}})$. If the decoding fails, set $\widetilde{\mathsf{L}} :=\perp$.
2. (output) Output $\widetilde{\mathsf{L}}$.

$g_R(\mathsf{R})$:

1. (apply tampering and decoding-check on right) Let $s^{\mathsf{R}} = (s_1^{\mathsf{R}}, \ldots, s_{n_R}^{\mathsf{R}}) := \mathrm{Rec}(\mathcal{S}_{\mathsf{R} \to \mathsf{L}}, I^{\mathsf{R}}, \mathsf{R})$. Let $(\sigma_1^{\mathsf{R}}, \ldots, \sigma_{n_R}^{\mathsf{R}}) := f^{\mathsf{R}}|_{I^{\mathsf{L}}}(s^{\mathsf{R}})$. Define $\sigma'^{\mathsf{R}} := \sigma_1'^{\mathsf{R}}, \ldots, \sigma_{n_R}'^{\mathsf{R}}$ as follows: Set $\sigma_\ell'^{\mathsf{R}} := \sigma_\ell^{\mathsf{R}}$ for $\ell \in [J^*]$. Set $\sigma_\ell'^{\mathsf{R}} := 0$ for $\ell \notin [J^*]$. Compute $((w_1^{\mathsf{R}}, \ldots, w_{n_R}^{\mathsf{R}}), \widetilde{\mathsf{R}}) \leftarrow D_R(\sigma_1'^{\mathsf{R}}, \ldots, \sigma_t'^{\mathsf{R}})$. If the decoding fails or $(w_1^{\mathsf{R}}, \ldots, w_{n_R}^{\mathsf{R}})$ is not $c_R^{\mathrm{err}}/4$-close to $(\sigma_1^{\mathsf{R}}, \ldots, \sigma_{n_R}^{\mathsf{R}})$, then set $\widetilde{\mathsf{R}} :=\perp$.
2. (codeword-check on right) For all $\ell \in R_{\mathrm{check}}$, check that $\sigma_\ell^{\mathsf{R}} = w_\ell^{\mathsf{R}}$. If the check fails, set $\widetilde{\mathsf{R}} :=\perp$.
3. (output) Output $\widetilde{\mathsf{R}}$.

– Output $g = (g_L, g_R)$.

Whenever Rec is run above, we assume that enough positions are set by $\mathcal{S}$ such that there is only a single consistent codeword. If this is not the case, then additional positions are added to $\mathcal{S}$ from $I^{\mathsf{L}}, I^{\mathsf{R}}$, respectively.

By the definition of a non-malleable reduction (Definition 3), in order to complete the proof of Theorem 4, we must show that (E, D) have the following properties:

1. For all $x \in \{0,1\}^k$, we have $\mathsf{D}(\mathsf{E}(x)) = x$ with probability 1.
2. For all $f \in \mathsf{Local}_{\ell_i}^{\ell_o}$,

$$\Delta(\mathsf{D}(f(\mathsf{E}(x))); G_f(x)) \leq \mathsf{negl}(k),$$

where G_f is the distribution defined above.

Item (1) above is trivial and can be immediately verified. In the following, we prove Item (2) above by considering the following sequence of hybrid arguments for each function $f \in \mathsf{Local}_{\ell_i}^{\ell_o}$ (for the intermediate hybrids, we highlight the step in which they are different from the desired end distributions).

Hybrid H_0. This is the original distribution $\mathsf{D}(f(\mathsf{E}(x)))$
Hybrid H_1. H_1 corresponds to the distribution $\mathsf{D}'(f(\mathsf{E}(x)))$, where D' is defined as follows:

$\mathsf{D}(\sigma := (\sigma^\mathsf{L}, \sigma^\mathsf{R}))$:

1. (**plain decode on left**) Let $(\sigma^\mathsf{L}, \sigma^\mathsf{R}) := ([\sigma_i^\mathsf{L}]_{i \in [n_L]}, [\sigma_\ell^\mathsf{R}]_{\ell \in [n_R]})$. Compute $((w_1^\mathsf{L}, \ldots, w_{n_L}^\mathsf{L}), \mathsf{L}) \leftarrow \mathsf{D}_L(\sigma_1^\mathsf{L}, \ldots, \sigma_{n_L}^\mathsf{L})$. If the decoding fails, set $\mathsf{L} := \bot$.

---

2. (**decoding-check on right**) Define $\sigma'^\mathsf{R} := \sigma_1'^\mathsf{R}, \ldots, \sigma_{n_R}'^\mathsf{R}$ as follows: Set $\sigma_\ell'^\mathsf{R} := \sigma_\ell^\mathsf{R}$ for $\ell \in J^*$ and $\sigma_\ell'^\mathsf{R} := 0$ for $\ell \notin J^*$, where $J^* \subseteq [n_R]$ is the lexicographically first set such that $|J^*| = t$ and $|\mathcal{S}_{\mathsf{L} \to \mathsf{R}}^{J^*}| = 0$. Compute $((w_1^\mathsf{R}, \ldots, w_{n_R}^\mathsf{R}), \mathsf{R}) \leftarrow \mathsf{D}_R(\sigma_1'^\mathsf{R}, \ldots, \sigma_{t_R}'^\mathsf{R})$. If the decoding fails or $(w_1^\mathsf{R}, \ldots, w_{n_R}^\mathsf{R})$ is not $c_R^{\mathsf{err}}/4$-close to $(\sigma_1^\mathsf{R}, \ldots, \sigma_{t_R}^\mathsf{R})$, set $\mathsf{R} := \bot$.

---

3. (**codeword-check on right**) For all $\ell \in R_\mathsf{check}$, check that $\sigma_\ell^\mathsf{R} = w_\ell^\mathsf{R}$. If the check fails, set $\mathsf{R} := \bot$.
4. (**output**) Output $x := (\mathsf{L}, \mathsf{R})$.

Note that the only difference between D and D' is that in **decoding-check on right**, σ^R is decoded from J^*, instead of the first n_check positions.

Claim.
$$H_0 \overset{s}{\approx} H_1.$$

Proof. Let $\delta := \frac{c_R^{\mathsf{err}}}{4}$. Additionally, define

$$\rho(n_R, \delta, n_\mathsf{check}) := \frac{\binom{(1-\delta)n_R}{n_\mathsf{check}}}{\binom{n_R}{n_\mathsf{check}}}.$$

Notice that our parametrization of n_{check}, δ yields $\rho(n_R, \delta, n_{\text{check}}) = \text{negl}(k)$.

$$\frac{\binom{(1-\delta)n_R}{n_{\text{check}}}}{\binom{n_R}{n_{\text{check}}}} = \frac{((1-\delta)n_R)!n_{\text{check}}!(n_R - n_{\text{check}})!}{n_{\text{check}}!((1-\delta)n_R - n_{\text{check}})!n_R!}$$

$$= \left(\frac{(1-\delta)n_R}{n_R}\right)\left(\frac{(1-\delta)n_R - 1}{n_R - 1}\right) \cdots \left(\frac{(1-\delta)n_R - n_{\text{check}} + 1}{n_R - n_{\text{check}} + 1}\right)$$

$$\leq (1-\delta)^{n_{\text{check}}},$$

where the last inequality follows due to the fact that for $i \in \{0, \ldots, n_{\text{check}} - 1\}$, $\frac{(1-\delta)n_R - i}{n_R - i} \leq (1-\delta)$. Since $(1-\delta) < 1$ is a constant, we can set $n_{\text{check}} = \omega(\log(k))$.

Note that correctness still holds for D' with probability 1.

We want to show that for every $\boldsymbol{\sigma} = (\boldsymbol{\sigma}^L, \boldsymbol{\sigma}^R) \leftarrow f(E(x))$, $D(\boldsymbol{\sigma}) = D'(\boldsymbol{\sigma})$ with high probability, over the coins of D, D'.

Let $D := (D^L, D^R)$ (respectively, $D' := (D'^L, D'^R)$), where D^R (respectively, D'^R) correspond to the right output of the decoding algorithm. Notice that only decoding on the right changes. So, it suffices to show that for each $(\boldsymbol{\sigma}^L, \boldsymbol{\sigma}^R)$ in the support of the distribution $f(E(x))$,

$$\Pr[D|_{\sigma^L}(\boldsymbol{\sigma}^R) = D'|_{\sigma^L}(\boldsymbol{\sigma}^R)] \geq 1 - \text{negl}(n), \tag{3.1}$$

where the probabilities are taken over the coins of D, D'.

Let $\mathcal{W}$ denote the set of all valid codewords for the given reconstructable probabilistic encoding scheme with parameters $k, n_R, c_R^{\text{err}}, c_R^{\text{sec}}, \text{GF}(2)$). For $x \in \text{GF}(2)^{n_R}$, define its distance from $\mathcal{W}$ to be $d(x, \mathcal{W}) := \min_{w \in \mathcal{W}} d(x, w)$.

To analyze (3.1), we define the following set of instances (which intuitively corresponds to the set of instances on which both $D|_{\sigma^L}$ and $D'|_{\sigma^L}$ are likely to output $\perp$).

$$\Pi_\perp := \{\boldsymbol{\sigma}^R \in \{0, 1\}^{n_R} \quad d(\boldsymbol{\sigma}, \mathcal{W}) \geq \delta\}.$$

So, now consider the two cases:

- Suppose $\boldsymbol{\sigma}^R \in \Pi_\perp$. Then, both $D(\boldsymbol{\sigma}^R)$ and $D'(\boldsymbol{\sigma}^R)$ will fail the codeword-check with probability $\geq 1 - \rho(n_R, \delta, n_{\text{check}})$.
- Suppose $\boldsymbol{\sigma}^R \notin \Pi_\perp$. Then, $\exists w \in \mathcal{W}$ such that $d(\boldsymbol{\sigma}^R, w) \leq \delta$. Moreover, in both D and D' it must be the case that $\boldsymbol{\sigma}'^R$ is $c^{\text{err}}/2$-close to w. (Because $\delta + (n_R - t)/n_R \leq c^{\text{err}}/2$). So both D and D' must decode to the same w. Fix a set of coins for D and D'. Therefore, when D and D' are run with the same coins, all comparisons made during the codeword-check are identical, and thus the probability (over the coins of D, D') that the codeword-check fails in D and D' is identical.

So for any $\boldsymbol{\sigma} = (\boldsymbol{\sigma}^L, \boldsymbol{\sigma}^R)$, $\Delta(\{D(\boldsymbol{\sigma})\}, \{D'(\boldsymbol{\sigma})\}) = \Delta(\{D^R|_{\sigma^L}(\boldsymbol{\sigma}^R)\}, \{D'^R|_{\sigma^L}(\boldsymbol{\sigma}^R)\}) \leq \rho(n_R, \delta, n_{\text{check}})$. Therefore, $\Delta(\{D(f(E(x)))\}, \{D'(f(E(x)))\}) \leq \rho(n_R, \delta, n_{\text{check}})$.

Hybrid H_2. H_2 corresponds to the distribution $G'(x)$, where G'_f is a distribution over functions $g' = (g'_L, g'_R)$ defined as follows:

- Choose a random subset $R_{\text{check}} \subseteq [n_R]$ of size n_{check}.

> - Choose vectors $I^{\mathsf{L}} \in \{0,1\}^{n_L}$, $I^{\mathsf{R}} \in \{0,1\}^{n_R}$ in the following way: $I^{\mathsf{L}} \leftarrow E_L(\mathsf{L})$, $I^{\mathsf{R}} \leftarrow E_R(\mathsf{R})$.

- Let J^* be the subset of $[n_R]$ as described in Observation 1.
- The split-state tampering function $g := (g_L, g_R) \in SS_k$ has $I^{\mathsf{L}}, I^{\mathsf{R}}$ hardcoded into it and is specified as follows:

$g_L(\mathsf{L})$:

1. **(apply tampering and plain decode on left)** Let $s^{\mathsf{L}} := \mathsf{Rec}(\mathcal{S} := \mathcal{S}_{\text{check}}, I^{\mathsf{L}}, \mathsf{L})$. Let $(\sigma_1^{\mathsf{l}}, \ldots, \sigma_{n_L}^{\mathsf{l}}) := f^{\mathsf{L}}|_{I^{\mathsf{R}}}(s^{\mathsf{L}})$. Compute $((w_1^{\mathsf{l}}, \ldots, w_{n_L}^{\mathsf{l}}), \widetilde{\mathsf{L}}) \leftarrow D_L(\sigma_1^{\mathsf{l}}, \ldots, \sigma_{n_L}^{\mathsf{l}})$. If the decoding fails, set $\widetilde{\mathsf{L}} :=\perp$.
2. **(output)** Output $\widetilde{\mathsf{L}}$.

$g_R(\mathsf{R})$:

1. **(apply tampering and decoding-check on right)** Let $s^{\mathsf{R}} = (s_1^{\mathsf{R}}, \ldots, s_{n_R}^{\mathsf{R}}) := \mathsf{Rec}(\mathcal{S}_{\mathsf{R}\to\mathsf{L}}, I^{\mathsf{R}}, \mathsf{R})$. Let $(\sigma_1^{\mathsf{R}}, \ldots, \sigma_{n_R}^{\mathsf{R}}) := f^{\mathsf{R}}|_{I^{\mathsf{L}}}(s^{\mathsf{R}})$. Define $\sigma'^{\mathsf{R}} := \sigma_1'^{\mathsf{R}}, \ldots, \sigma_{n_R}'^{\mathsf{R}}$ as follows: Set $\sigma_\ell'^{\mathsf{R}} := \sigma_\ell^{\mathsf{R}}$ for $\ell \in [J^*]$. Set $\sigma_\ell'^{\mathsf{R}} := 0$ for $\ell \notin [J^*]$. Compute $((w_1^{\mathsf{R}}, \ldots, w_{n_R}^{\mathsf{R}}), \widetilde{\mathsf{R}}) \leftarrow D_R(\sigma_1'^{\mathsf{R}}, \ldots, \sigma_t'^{\mathsf{R}})$. If the decoding fails or $(w_1^{\mathsf{R}}, \ldots, w_{n_R}^{\mathsf{R}})$ is not $c_R^{\text{err}}/4$-close to $(\sigma_1^{\mathsf{R}}, \ldots, \sigma_{n_R}^{\mathsf{R}})$, then set $\widetilde{\mathsf{R}} :=\perp$.
2. **(codeword-check on right)** For all $\ell \in R_{\text{check}}$, check that $\sigma_\ell^{\mathsf{R}} = w_\ell^{\mathsf{R}}$. If the check fails, set $\widetilde{\mathsf{R}} :=\perp$.
3. **(output)** Output $\widetilde{\mathsf{R}}$.

- Output $g = (g_L, g_R)$.

Note that the only difference between G_f and G_f' is that $I^{\mathsf{L}} \leftarrow E_L(\mathsf{L}), I^{\mathsf{R}} \leftarrow E_R(\mathsf{R})$ are chosen honestly, instead of being chosen uniformly at random. Furthermore, note that $g' = (g_L', g_R')$ are not split-state, since g_L' depends on I^{R} and g_R' depends on I^{L}.

Claim.

$$H_1 \equiv H_2.$$

The claim can be verified by inspection.

Hybrid H_3. Hybrid H_3 is simply the distribution $G_f(x)$, defined previously.

Claim.

$$H_2 \equiv H_3.$$

Note that the result of f^{R} only depends on the bits in J^* and R_{check}. Moreover, $f_{\chi_{J^* \cup R_{\text{check}}}}^{\mathsf{R}}$ only depends on s^{R}, $[s_i^{\mathsf{l}}]_{i \in \mathcal{S}_{\text{check}}}$. Moreover, note that f^{L} depends only on s^{L}, $[s_i^{\mathsf{R}}]_{i \in \mathcal{S}_{\mathsf{R}\to\mathsf{L}}}$. Since by Observation 1, we have that $|\mathcal{S}_{\text{check}}| \leq n_L \cdot c_L^{\text{sec}}$ and $|\mathcal{S}_{\mathsf{R}\to\mathsf{L}}| \leq n_R \cdot c_R^{\text{sec}}$, the claim follows from the secrecy property of the reconstructable probabilistic encoding scheme.

3.1   Extending to Leaky Local

The construction from Sect. 3 is actually secure against a slightly larger class of tampering functions beyond $\text{Local}^{\ell_o}_{\ell_i}$ functions, which we call LL, or "Leaky Local." Notice that the parameters given above (as in Observation 1) in fact yield:

1. $|\mathcal{S}^{J^*}_{\mathsf{L}\to\mathsf{R}}| + |\mathcal{S}_{\mathsf{check}}| = |\mathcal{S}_{\mathsf{check}}| \le n_L \cdot \frac{c^{\mathsf{sec}}_L}{3}$.
2. $|\mathcal{S}^+_{\mathsf{R}\to\mathsf{L}}| \le \ell_o \cdot n_L \le n_R \cdot \frac{2c^{\mathsf{sec}}_R}{3}$.

It is not too hard to see that we can leak $1/3$ of the security threshold, on both the left and right, to a tampering adversary. Given this leakage, the adversary can then select a tampering function from the subset of Local^{ℓ_o} where all but a fraction of the first n_L bits have input locality ℓ_i. Note that the input locality restrictions are only needed on the left portions of codewords in the above proof. We formalize this new class of tampering functions as follows.

Definition 13. *Let* $\text{LL} \subseteq \{\{0,1\}^{n_L} \times \{0,1\}^{n_R} \to \{0,1\}^{n_L} \times \{0,1\}^{n_R}\}$, *Leaky Local, be the set of functions* $\{\psi_{f,h_1,h_2}\}$, *parametrized by functions* (f, h_1, h_2), *where* $\psi_{f,h_1,h_2}(s^{\mathsf{L}}, s^{\mathsf{R}}) := C_{univ}(f(h_1(s^{\mathsf{L}}), h_2(s^{\mathsf{R}})), s^{\mathsf{L}}, s^{\mathsf{R}})$, f *outputs a circuit* C *and* C_{univ} *is a universal circuit that computes the output of the circuit* C *on input* $(s^{\mathsf{L}}, s^{\mathsf{R}})$. *Moreover, we require that* f, h_1, h_2 *have the following form:*

- *On input* $s^{\mathsf{L}} \in \{0,1\}^{n_L}$, h_1 *outputs a subset of* $c^{\mathsf{err}}_L/3$ *of its input bits.*
- *On input* $s^{\mathsf{R}} \in \{0,1\}^{n_R}$, h_2 *outputs a subset of* $c^{\mathsf{err}}_R/3$ *of its input bits.*
- *On input* $h_1(s^{\mathsf{L}}), h_2(s^{\mathsf{L}}) \in \{0,1\}^{c^{\mathsf{err}}_L/3} \times \{0,1\}^{c^{\mathsf{err}}_R/3}$, f *outputs a circuit* C : $\{0,1\}^{n_L} \times \{0,1\}^{n_R} \to \{0,1\}^{n_L} \times \{0,1\}^{n_R}$, *where* C *has output-locality* ℓ_o. *Of the first* n_L *input bits, all but at most* $c^{\mathsf{err}}_L/3$-*fraction have input-locality at most* ℓ_i.

The following corollary can be easily verified.

Corollary 2. $(\mathsf{E}\circ\mathsf{E}_{\mathsf{SS}}, \mathsf{D}_{\mathsf{SS}}\circ\mathsf{D})$ *is an* $(\text{LL}, \text{SS}_k, \text{negl}(k))$-*non-malleable reduction.*

4   Extending to $\text{Local}^{m(n)}$

We now state our theorem for $\text{Local}^{m(n)}$ tampering functions, or bounded fan-in bounded-depth circuits.

Theorem 5. $(\mathsf{E}', \mathsf{D}')$ *is a* $(\text{Local}^{\ell_o{}'} \Rightarrow \text{LL}, \text{negl}(n))$-*non-malleable reduction given the following parameters for* $\text{Local}^{\ell_o{}'}$:

- $\ell_o{}' := c^{\mathsf{sec}}/12 \cdot \ell_i$, *where* ℓ_i *is the input locality of* LL,
- $\mathsf{E}' : \{0,1\}^n \to \{0,1\}^N$, *where* $N = n_{in} + 2n - n_L$, *and* $r = \log^4(k)$, *where* n *is the output length of* LL *and* n_L *is the length of the left output of* LL.

We construct an encoding scheme (E', D') summarized in Fig. 3 and parametrized below. In brief, our encoding simply distributes the bits of the left input pseudorandomly in a string comparable in length to the right input. We then append a short description of where the encoding is hiding, a seed to pseudorandom generator.

We then show that the pair (E', D') is an $(\mathsf{Local}^{\ell_o'}, \mathsf{LL}, \mathsf{negl}(n))$-non-malleable reduction. Combined with our previous construction, this immediately implies that given a non-malleable encoding scheme (E^{ss}, D^{ss}) for SS_k, the encoding scheme $\widehat{\Pi} = (\widehat{E^{bd}}, \widehat{D^{bd}})$, where $\widehat{E^{bd}}(m) := E'(E(E^{ss}(m)))$ and $\widehat{D^{bd}}(s) := D^{ss}(D(D'(s)))$ yields the following corollary, a non-malleable code against $\mathsf{Local}^{\ell_o'}$.

Corollary 3. (E', D') *yields, with previous results, a* $(\mathsf{Local}^{\tilde{O}(\sqrt{n})}, k, \mathsf{negl}(k))$-*non-malleable reduction with sublinear rate, where* $n = \Theta(\frac{k^2}{\log^2(k)})$.

Remark 3. As before, the encoding scheme presented below is independent on the left and right. Therefore, our reduction holds for not just for $\mathsf{Local}^{\ell_o'}$ but additionally any split-state function, independent on each side, trivially.

We parametrize our construction for $\mathsf{Local}^{\ell_o'} \Rightarrow \mathsf{LL}$ with the following:

- $r := \log^4(k)$
- $\tau := 2(n - n_L)$, where n is the length of the output of LL and n_L is the length of the left output of LL.

Now, for every $\mu \in \mathsf{Local}^{\ell_o'}$ where $\mu(\zeta, X^L, x^R) := (\mu^\zeta(\zeta, X^L, x^R), \mu^L(\zeta, X^L, x^R), \mu^R(\zeta, X^L, x^R))$ we define the distribution G_μ over LL. A draw from G_μ is defined as follows:

- Choose $\zeta \leftarrow \{0,1\}^r$ uniformly at random. Compute $y := \mathsf{prg}(\zeta)$, where $y = y_1, \ldots, y_\tau$. For $i \in [\tau]$, compute Compute $\rho_i := \phi(y_i)$.
- If ρ has less than n_L number of ones, then set h_1, h_2, f all to the constant function 0.
- Otherwise, choose vector $I^L \in \{0,1\}^{+n_R}$ such that $\forall i$ such that $1 \le i \le$ if $\rho_i = 1$ then $I_i^L = *$ and otherwise, I_i^L is chosen uniformly at random.
- The function h_1 is defined as follows: h_1 outputs the bits in input x^L that affect the output bits of μ^ζ (at most $r \cdot \ell_o' \le c_L^{sec}/3 \cdot n_L$).
- The function h_2 is defined as follows: h_2 outputs the bits in x^R that affect the output bits of μ^ζ (at most $r \cdot \ell_o' \le c_R^{sec}/3 \cdot n_R$).
- The function f is defined as follows:
  - f computes $\widetilde{\zeta}$, given ζ and the output of h_1, h_2.
  - f computes $\widetilde{y} := \mathsf{prg}(\widetilde{\zeta})$, where $\widetilde{y} = \widetilde{y}_1, \ldots, \widetilde{y}_\tau$.
  - For $i \in [\tau]$, f computes $\widetilde{\rho}_i := \phi(\widetilde{y}_i)$.
  - Let $\widetilde{\rho}^* \in \{0,1\}^\tau$ be defined as follows: For $i \in [\mathsf{pos}^*], \widetilde{\rho}^* = \widetilde{\rho}$; for $\mathsf{pos}^* < i \le \tau, \widetilde{\rho}^* = 0$, where pos^* is the index of the n_L-th one in $\widetilde{\rho}$ (and is set to τ if no such index exists).
  - Let $\mu^{L,\zeta}$ (resp. $\mu^{R,\zeta}$) correspond to the function $\mu^L(\zeta, X^L, x^R)$ (resp. $\mu^R(\zeta, X^L, x^R)$)), which has ζ hardcoded in it.

Let prg be a pseudorandom generator for space bounded computations (see Definition 12), with inputs of length r and outputs of length $\log(\tau) \cdot \tau$. Let $G(\zeta)$ be defined as follows:

1. Compute $y := \mathsf{prg}(\zeta)$.
2. Divide pseudorandom tape y into blocks of bit strings $y_1, \ldots, y_\tau$. Let ϕ be the randomized function that chooses a bit $b \in \{0, 1\}$ with bias $p := 3n_L/2\tau$. For $i \in [\tau]$, let $\rho_i = \phi(y_i)$, where y_i is the explicit randomness of ϕ. Let $\rho = \rho_1, \ldots, \rho_\tau$. Let num denote the number of positions of ρ that are set to 1.
3. If $\mathsf{num} < n_L$, set $\rho := 1_L^n 0^{\tau - n_L}$.
4. Otherwise, flip all but the first n_L 1's in ρ to 0.
5. Output ρ.

Let $\mathsf{E}' : \{0,1\}^n \to \{0,1\}^N$ and $\mathsf{D}' : \{0,1\}^N \to \{0,1\}^n$.

$\mathsf{E}'(x^L := x_1^L, \ldots, x_{n_L}^L, x^R)$:

1. Choose $\zeta \leftarrow \{0,1\}^r$ uniformly at random. Choose $\zeta \leftarrow \{0,1\}^r$ uniformly at random. Compute $\rho := G(\zeta)$.
2. For $j \in [\mathsf{num}]$, let pos_j denote the j-th position i such that $\rho_i = 1$.
3. Let $X^L \in \{0,1\}^\tau$ be defined in the following way: For $j \in [n_L]$, $X_{\mathsf{pos}_j}^L := x_j^L$. In all other locations, X_i^L is set uniformly at random.
4. Output the encoding (ζ, X^L, x^R).

$\mathsf{D}'(Z := (\widetilde{\zeta}, \widetilde{X}^L, \widetilde{x}^R))$:

1. (**Recover** $\widetilde{\rho}$) Let $\widetilde{\rho} := G(\widetilde{\zeta})$. Let $\widetilde{\mathsf{num}} \geq n_L$ denote the number of ones in $\widetilde{\rho} := \widetilde{\rho}_1, \ldots, \widetilde{\rho}_\tau$.
2. (**Recover** x) For $j \in [\widetilde{\mathsf{num}}]$, let pos_j denote the j-th position i such that $\widetilde{\rho}_i = 1$.
3. Let $\widetilde{x}_j^L \in \{0,1\}^{n_L}$ be defined in the following way: For $j \in [\min(\widetilde{\mathsf{num}}, n_L)]$, $\widetilde{x}_j^L := \widetilde{X}_{\mathsf{pos}_j}^L$.
4. (**output**) Output $(\widetilde{x}^L, \widetilde{x}^R)$.

Fig. 3. THE $(\mathsf{Local}^{\ell_o'}, \mathsf{LL}, \mathsf{negl}(n))$-NON-MALLEABLE REDUCTION $(\mathsf{E}', \mathsf{D}')$

- Let C be the circuit corresponding to the following restriction: $((\mu^{L,\zeta}|_{I^L})_{\widetilde{\rho}^*}, \mu^{R,\zeta}|_{I^L})$.
- If C is in LL, then f outputs C. Otherwise, f outputs the constant function 0.

By the definition of a non-malleable reduction (Definition 3), in order to complete the proof of Theorem 5, we must show that $(\mathsf{E}', \mathsf{D}')$ has the following properties:

1. For all $x \in \{0,1\}^n$, we have $\mathsf{D}'(\mathsf{E}'(x)) = x$ with probability 1.

2. For all $\mu \in \text{Local}^{\ell_o'}$,

$$\Delta(\mathsf{D}'(\mu(\mathsf{E}'(x))); G_\mu(x)) \leq \mathsf{negl}(n),$$

where G_μ is the distribution defined above.

Item (1) above is trivial and can be immediately verified.

In the following, we prove Item (2), above, by noting that the statistical distance $\Delta(\mathsf{D}'(\mu(\mathsf{E}'(x))); G_\mu(x))$ is upper bounded by the probability that either ρ does not contain at least n_L number of ones or C is not in LL.

We first argue that if ρ is chosen uniformly at random, then the probability that either of these events occurs is negligible and then show that the same must be true when ρ is chosen via a PRG with appropriate security guarantees.

Clearly, by multiplicative Chernoff bounds, if ρ is chosen uniformly at random, then the probability that ρ contains less than n_L ones is negligible. We now show that the probability that $C \notin \mathsf{LL}$ is negligible. If $C \notin \mathsf{LL}$, it means that more than $c_L^{\mathsf{sec}}/3$ number of positions i in X^{L} are such that (1) X_i^{L} has "high input locality" (i.e. input locality greater than $12/c_L^{\mathsf{sec}} \cdot \ell_o' = \ell_i$) (2) $\rho_i = 1$.

Since the adversary first specifies the tampering function μ, all positions in X^{L} with "high input locality" are determined. Note that, by choice of parameters (since $\tau \geq N/2$), there can be at most $c_L^{\mathsf{sec}} \cdot \tau/6$ number of positions in X^{L} with "high input locality". Since $p = 3n_L/2\tau$, we *expect* $c_L^{\mathsf{sec}} \cdot n_L/4$ number of positions i in X^{L} where (1) X_i^{L} has "high input locality" and (2) $\rho_i = 1$. Therefore, by multiplicative Chernoff bounds, the probability that more than $c_L^{\mathsf{sec}} \cdot n_L/3$ number of positions i in X^{L} are such that (1) X_i^{L} has "high input locality" and (2) $\rho_i = 1$ is negligible.

We now argue that these events must also occur with negligible probability when ρ is pseudorandom. Assume the contrary, then the following is a distinguisher T that can distinguish truly random strings y from strings $y := \mathsf{prg}(\zeta)$ with non-negligible probability.

T is a circuit that has a string $w \in \{0,1\}^\tau$ hardwired into it (non-uniform advice). w corresponds to the high input locality positions determined by the tampering function μ that was chosen by the adversary A. Intuitively, w is the string that causes A to succeed in breaking security of the non-malleable code with highest probability.

On input $y = y_1, \ldots, y$ (where either $y := \mathsf{prg}(\zeta)$ or y is chosen uniformly at random), $T(y)$ does the following:

1. Set $count_1 = 0$, $count_2 = 0$.
2. For $i = 1$ to :
   (a) Run $\phi(y_i)$ to obtain ρ_i.
   (b) If $\rho_i = 1$, set $count_2 := count_2 + 1$
   (c) If $\rho_i = 1$ and $w_i = 1$, set $count_1 := count_1 + 1$.
3. If $count_1 > c_L^{\mathsf{sec}} \cdot n_L/3$ or $count_2 < n_L$, output 0. Otherwise, output 1.

T can clearly be implemented by a read-once, Finite State Machine (FSM) with $2^{O(\log^2(\tau))}$ number of states. However, note that by Theorem 3, prg is a pseudorandom generator for space $\log^3(k)$ with parameter $2^{-\log^3(k)}$. Thus, existence of distinguisher T as above, leads to contradiction to the security of the Nisan PRG.

5 Achieving Resilience Against $o(n/\log n)$ Output Locality

Here we sketch how to improve parameters. We refer readers to the full paper for the complete proof.

Theorem 6. *There exists an explicit* $(\mathsf{Local}^{\ell_o} \Rightarrow SS, \mathsf{negl}(n))$-*non-malleable reduction,* $(\mathsf{E} : \{0,1\}^{2k} \to \{0,1\}^n, \mathsf{D} : \{0,1\}^n \to \{0,1\}^{2k})$, *for any* $\ell_o = o(n/\log n)$ *where* $n = O(\ell_o k)$.

Roughly, the reduction (E, D) is simply a composition of the reductions presented previously. Recall that the encoding scheme is independent on the left and right, $\mathsf{E}(\mathsf{L}, \mathsf{R}) = (\mathsf{E}_L(\mathsf{L}), \mathsf{E}_R(\mathsf{R}))$. The left side, $\mathsf{E}_L(\mathsf{L})$, is comprised of a seed for a PRG that describes where to pseudorandomly embedded a (small) RPE of L and that very embedding. The right side, $\mathsf{E}_R(\mathsf{R})$, is simply a (longer) RPE of R. Decoding is the same as before as well. The parameters are slightly different, but we will gloss over that here.

To prove the theorem, we analyze the composed encoding schemes as a single reduction. As mentioned in the introduction, the idea is to use the PRG to "free up" the restrictions relating the size of the left RPE (previously denoted by n_L) and ℓ_o that is an artifact of the piecewise analysis.

Recall that our encoding scheme is comprised of three blocks: (1) the PRG seed, (2) the "hidden" left side encoding, and (3) the right side encoding. First, (as in the previous section) we claim that a number of good things happen if the left side is "hidden" in a large block in a truly random way. Namely, we have that, with respect to the tampering function, only a small fraction of bits in the hidden left-side RPE is either (1) of high input locality, (2) effects bits in the right-side's consistency check or (3) effects the PRG seed used in decoding. (1) Implies that there exists a "safe" subset to simulate decoding from (as before), and (2) and (3) allow us to relax the bounds on locality. Next, we use a hybrid argument to essentially disconnect influence between the 3 blocks of our encoding (that is dependent on the underlying message, (L, R)).

We will present the "good" event described above and sketch the hybrid argument.

Definition 14 (informal). *The event* Good_f *occurs if for tampering function* $f \in \mathsf{Local}^{\ell_o}$ *all of the following hold:*

1. ρ *contains at least* n_L *ones, where* n_L *is the length of the left side RPE.*
2. $|S|$ *is below the security threshold of the left-side RPE, where S is the set of bits in the "hidden" RPE of L that have (1) "high" input locality, (2) effect the consistency check on the right (consider this chosen secretly and randomly at the time of encoding), or (3) effect the PRG seed used in decoding.*

3. *There is some (large enough) set, J^*, of bits that is not effected by any bit in the RPE of* L *which does not have "high" input locality.*
4. *The bits on the right that effect the output of decoding on the left is below the security threshold of the RPE.*

Claim. Suppose ρ is chosen truly at random (ones occurring with bias $p = 3n_L/2\tau$). Then for every $f \in \text{Local}^{\ell_o}$, $\Pr[\text{Good}_f] \geq 1 - \text{negl}(n)$.

The first two items follow from Chernoff bounds. The main difference in the new analysis is that the hidden RPE of L is now very small size. Whereas previously the events (2), (3) described in the second item held simply because the total number of bits on the left affecting the consistency check and PRG seed was below the security threshold of the left RPE, now, since the left RPE is now very small, we must rely on Chernoff bounds and the fact that the relevant bits are hidden to argue that (2) and (3) hold. The second two items are similar to Observation 1, given item (2).

Next as in the previous section, we argue that with high probability, the pseudorandomness of the PRG is sufficient to obtain that the event Good_f holds even when ρ is chosen via the PRG (instead of being truly random). This gives us the bounds on the "bad" bits in the output of the encoding of the left input, L, mentioned previously.

Now we are in essentially a similar situation to the proof of Theorem 4 and we can apply a very similar sequence of hybrids.

First, we use hybrids to effectively sample the bits on the left and the right that effect some other block, or are in the "bad" set S. By our claim above, all of these sets together will be below the security properties of the respective RPEs. So, the distribution over the randomness of the encoding procedure will be identical, for any message.

Second, we use hybrids to effectively simulate decoding on the right from the set J^* that is not effected by the RPE on the left. This completes the proof.

Acknowledgments. We thank Seung Geol Choi and Hoeteck Wee for sharing with us an in-submission journal version of [15], as well as the manuscript [16]. We also thank Yevgeniy Dodis for helpful discussions and clarifications regarding [2] and other previous work. Finally, we thank Eran Tromer for enlightening discussions on practical tampering attacks, which inspired the class of attacks considered in this work.

This work was done in part while all authors were visiting the Simons Institute for the Theory of Computing, supported by the Simons Foundation and by the DIMACS/Simons Collaboration in Cryptography through NSF grant #CNS-1523467. The first and fourth authors are supported in part by the Defense Advanced Research Project Agency (DARPA) and Army Research Office (ARO) under Contract #W911NF-15-C-0236, and NSF grants #CNS-1445424 and #CCF-1423306. The second and third authors are supported by an NSF CAREER award #CNS-1453045 and by a Ralph E. Powe Junior Faculty Enhancement Award. Any opinions, findings and conclusions or recommendations expressed are those of the authors and do not necessarily reflect the views of the Defense Advanced Research Projects Agency, Army Research Office, the National Science Foundation, or the U.S. Government.

References

1. Aggarwal, D., Agrawal, S., Gupta, D., Maji, H.K., Pandey, O., Prabhakaran, M.: Optimal computational split-state non-malleable codes. In: Kushilevitz, E., Malkin, T. (eds.) TCC 2016-A. LNCS, vol. 9563, pp. 393–417. Springer, Heidelberg (2016). doi:10.1007/978-3-662-49099-0_15

2. Aggarwal, D., Dodis, Y., Kazana, T., Obremski, M.: Non-malleable reductions and applications. In: Servedio, R.A., Rubinfeld, R. (eds.) 47th ACM STOC, Portland, OR, USA, 14–17 June 2015, pp. 459–468. ACM Press (2015)

3. Aggarwal, D., Dodis, Y., Lovett, S.: Non-malleable codes from additive combinatorics. In: Shmoys, D.B. (ed.) 46th ACM STOC, NY, USA, May 31–June 3, 2014, pp. 774–783 (2014)

4. Aggarwal, D., Dziembowski, S., Kazana, T., Obremski, M.: Leakage-resilient non-malleable codes. Cryptology ePrint Archive, Report 2014/807 (2014). http://eprint.iacr.org/2014/807

5. Agrawal, S., Gupta, D., Maji, H.K., Pandey, O., Prabhakaran, M.: Explicit non-malleable codes against bit-wise tampering and permutations. In: Gennaro, R., Robshaw, M. (eds.) CRYPTO 2015. LNCS, vol. 9215, pp. 538–557. Springer, Heidelberg (2015). doi:10.1007/978-3-662-47989-6_26

6. Agrawal, S., Gupta, D., Maji, H.K., Pandey, O., Prabhakaran, M.: A rate-optimizing compiler for non-malleable codes against bit-wise tampering and permutations. In: Dodis, Y., Nielsen, J.B. (eds.) TCC 2015, Part I. LNCS, vol. 9014, pp. 375–397. Springer, Heidelberg (2015)

7. Applebaum, B.: Cryptography in Constant Parallel Time. Information Security and Cryptography. Springer, Heidelberg (2014). http://dx.doi.org/10.1007/978-3-642-17367-7

8. Ben-Or, M., Goldwasser, S., Wigderson, A.: Completeness theorems for non-cryptographic fault-tolerant distributed computation (extended abstract). In: 20th ACM STOC, Chicago, Illinois, USA, 2–4 May 1988, pp. 1–10. ACM Press (1998)

9. Chabanne, H., Cohen, G.D., Flori, J., Patey, A.: Non-malleable codes from the wire-tap channel. CoRR abs/1105.3879 (2011). http://arxiv.org/abs/1105.3879

10. Chabanne, H., Cohen, G.D., Patey, A.: Secure network coding and non-malleable codes: protection against linear tampering. In: Proceedings of the 2012 IEEE International Symposium on Information Theory, ISIT 2012, Cambridge, MA, USA, 1–6 July 2012, pp. 2546–2550. IEEE (2012). http://dx.doi.org/10.1109/ISIT.2012.6283976

11. Chandran, N., Kanukurthi, B., Raghuraman, S.: Information-theoretic local non-malleable codes and their applications. In: Kushilevitz, E., Malkin, T. (eds.) TCC 2016-A. LNCS, vol. 9563, pp. 367–392. Springer, Heidelberg (2016). doi:10.1007/978-3-662-49099-0_14

12. Chattopadhyay, E., Zuckerman, D.: Non-malleable codes against constant split-state tampering. In: 55th FOCS, Philadelphia, PA, USA, 18–21 October 2014, pp. 306–315. IEEE Computer Society Press (2014)

13. Cheraghchi, M., Guruswami, V.: Capacity of non-malleable codes. In: Naor, M. (ed.) ITCS, Princeton, NJ, USA, 12–14 January 2014, pp. 155–168. ACM (2014)

14. Cheraghchi, M., Guruswami, V.: Non-malleable coding against bit-wise and split-state tampering. In: Lindell, Y. (ed.) TCC 2014. LNCS, vol. 8349, pp. 440–464. Springer, Heidelberg (2014)

15. Choi, S.G., Dachman-Soled, D., Malkin, T., Wee, H.M.: Black-box construction of a non-malleable encryption scheme from any semantically secure one. In: Canetti, R. (ed.) TCC 2008. LNCS, vol. 4948, pp. 427–444. Springer, Heidelberg (2008)

16. Choi, S.G., Dachman-Soled, D., Malkin, T., Wee, H.: A note on improved, black-box constructions of non-malleable encryption from semantically-secure encryption. Manuscript (2015)
17. Choi, S.G., Kiayias, A., Malkin, T.: BiTR: built-in tamper resilience. In: Lee, D.H., Wang, X. (eds.) ASIACRYPT 2011. LNCS, vol. 7073, pp. 740–758. Springer, Heidelberg (2011)
18. Coretti, S., Dodis, Y., Tackmann, B., Venturi, D.: Non-malleable encryption: Simpler, shorter, stronger. Cryptology ePrint Archive, Report 2015/772 (2015). http://eprint.iacr.org/2015/772
19. Coretti, S., Dodis, Y., Tackmann, B., Venturi, D.: Non-malleable encryption: simpler, shorter, stronger. In: Kushilevitz, E., Malkin, T. (eds.) TCC 2016-A. LNCS, vol. 9562, pp. 306–335. Springer, Heidelberg (2016). doi:10.1007/978-3-662-49096-9_13
20. Dachman-Soled, D., Liu, F.-H., Shi, E., Zhou, H.-S.: Locally decodable and updatable non-malleable codes and their applications. In: Dodis, Y., Nielsen, J.B. (eds.) TCC 2015, Part I. LNCS, vol. 9014, pp. 427–450. Springer, Heidelberg (2015)
21. Decatur, S.E., Goldreich, O., Ron, D.: Computational sample complexity. SIAM J. Comput. **29**(3), 854–879 (2000)
22. Dolev, D., Dwork, C., Naor, M.: Nonmalleable cryptography. SIAM J. Comput. **30**(2), 391–437 (2000)
23. Dziembowski, S., Kazana, T., Obremski, M.: Non-malleable codes from two-source extractors. In: Canetti, R., Garay, J.A. (eds.) CRYPTO 2013, Part II. LNCS, vol. 8043, pp. 239–257. Springer, Heidelberg (2013)
24. Dziembowski, S., Pietrzak, K., Wichs, D.: Non-malleable codes. In: Yao, A.C.C. (ed.) ICS, 5–7 January 2010, pp. 434–452. Tsinghua University Press, Tsinghua University, Beijing, China (2010)
25. Faust, S., Mukherjee, P., Nielsen, J.B., Venturi, D.: Continuous non-malleable codes. In: Lindell, Y. (ed.) TCC 2014. LNCS, vol. 8349, pp. 465–488. Springer, Heidelberg (2014)
26. Faust, S., Mukherjee, P., Venturi, D., Wichs, D.: Efficient non-malleable codes and key-derivation for poly-size tampering circuits. In: Nguyen, P.Q., Oswald, E. (eds.) EUROCRYPT 2014. LNCS, vol. 8441, pp. 111–128. Springer, Heidelberg (2014)
27. Gennaro, R., Lysyanskaya, A., Malkin, T., Micali, S., Rabin, T.: Algorithmic tamper-proof (ATP) security: theoretical foundations for security against hardware tampering. In: Naor, M. (ed.) TCC 2004. LNCS, vol. 2951, pp. 258–277. Springer, Heidelberg (2004)
28. Ishai, Y., Prabhakaran, M., Sahai, A., Wagner, D.: Private circuits II: keeping secrets in tamperable circuits. In: Vaudenay, S. (ed.) EUROCRYPT 2006. LNCS, vol. 4004, pp. 308–327. Springer, Heidelberg (2006)
29. Ishai, Y., Sahai, A., Wagner, D.: Private circuits: securing hardware against probing attacks. In: Boneh, D. (ed.) CRYPTO 2003. LNCS, vol. 2729, pp. 463–481. Springer, Heidelberg (2003)
30. Kalai, Y.T., Kanukurthi, B., Sahai, A.: Cryptography with tamperable and leaky memory. In: Rogaway, P. (ed.) CRYPTO 2011. LNCS, vol. 6841, pp. 373–390. Springer, Heidelberg (2011)
31. Liu, F.-H., Lysyanskaya, A.: Tamper and leakage resilience in the split-state model. In: Safavi-Naini, R., Canetti, R. (eds.) CRYPTO 2012. LNCS, vol. 7417, pp. 517–532. Springer, Heidelberg (2012)
32. Nisan, N.: Pseudorandom generators for space-bounded computation. Combinatorica **12**(4), 449–461 (1992)

Author Index

Printed in the United States
By Bookmasters